Student Friendly
Quantum Field Theory
Second Edition

Basic Principles and Quantum Electrodynamics

Robert D. Klauber

Student Friendly Quantum Field Theory, 2nd Edition (with pedagogic improvements and corrections)
Basic Principles and Quantum Electrodynamics

Copyright © of Robert D. Klauber

Published by
Sandtrove Press
Fairfield, Iowa
sandtrovepress@gmail.com

Cover by Aalto Design

Second Edition December 2013 (third revision with improvements and corrections September 2018)

Library of Congress Control Number: 2013920914
ISBN: Hard cover 978-0-9845139-4-9
         Soft cover   978-0-9845139-5-6

Printed in the United States of America

To

the students

May they find this the easiest, and thus the most efficient,
physics text to learn from that they have ever used.

"Of all the communities available to us there is not one I would want to devote myself to, except for the society of the true searchers, which has very few living members at any time."

Albert Einstein

Table of Contents

Table of Wholeness Charts

Preface to the Second Edition

In the eight months since publication of the first edition, I have received many excellent suggestions from readers for making certain parts clearer and easier to understand. Though a new edition of a text typically comes out years after the prior one, I would be remiss to wait any longer to include those suggestions in a second edition.

The changes made encompass the re-wording of several sections and the addition of a dozen new pages spread throughout the text, all of which should improve the learning experience for many. These modifications are posted on the book website (see URL on pg. xvi opposite pg. 1), for the benefit of those using the first edition.

Things to note about newly added material

To facilitate and simplify communication between users of different editions, all equations and sections in the first edition have the same numbers in the second edition. Where new equations have been inserted into the second edition between two first edition equations, the new ones are numbered with that of the preceding first edition equation augmented with +1, +2, +3, etc. Where new pages have been inserted, they are numbered with the preceding (first edition) page number augmented with letters, a, b, etc. For example,

New pages inserted inside chapter	New page numbers	New equations inserted	New equation numbers
Two pgs between pgs 327 and 328	327a and 327b	after (13-29), before (13-30)	(13-29)+1, (13-29)+2, etc

Where material has been added at the end of a chapter, equations are given new numbers that are simply incremented over the last numbered equation in that chapter of the first edition. For example,

New appendix at end of chapter	New page numbers	New equations inserted	New equation numbers
Two pgs inserted before problems in Chap. 16, on pgs. 430-431	431a and 431b	after last equation in chapter, (16-120)	(16-121) (16-122), etc

Hence, at a glance, a reader can tell what has been inserted since the first edition, and confusion in communication between users of different editions should be minimized.

On page 203, in the middle of Chap. 7, new material has been added, but since it was not enough to justify insertion of additional pages, the remainder of that chapter (from page 203 to 212) has page numbering which is off by half a page or so from the first edition. From Chap. 8 onward, second edition page numbering returns to mirroring that of the first edition. Other than this one area of Chap. 7, a reader of one edition should be able to reference material by simply giving the relevant page number to a reader of the other edition.

In addition to new material being added, a significant number of typographical errors have been corrected. These are also listed on the book web site for the benefit of first edition users.

Robert D. Klauber
November 2013

Addendum, September 2018: Appendix F has been added to Chap. 10. See pgs. 285b to 285d.

Acknowledgments for the Second Edition

I am simply incapable of adequately expressing my gratitude to Tom Bartholet, Luc Longtin, Jimmy Snyder, and Holger Teutsch, each of whom I consider a candidate for the title of world's greatest technical book editor. The four of them have offered exceptional insights into pedagogic improvement of various sections, pointed out typographical errors, and generally worked tirelessly to help make this book better. I am indebted to, and thank, them deeply.

I also thank Sebastian Allende, Sukruti Bansai, Ben Balmforth, Lou Biegeleisen, Juan José Bigeón, John Davidson, Pavel Fadeev, Bill Foster, Thomas Fowler, Vasudev Godbole, Michael Heiss, Kurt Huddleston, Michael Heiss, Michael Hyams, Zhang Juenjie, Dory Kodeih, Jason Koeller, Michael Koren, Steffen Leger, John Lynch, Jeff Magill, Ian Marshall, Doug McKenzie, Pete Morcos, Marius Paraschiv, Ramon Salazar, Andrew Solomon, Jeroen Spandaw, Brian Stephanik, Allan Tameshtit, Harm van der Lek, and Qingzhong Wu for other excellent suggestions and corrections.

Additionally, several of those who reviewed and helped edit the first edition, Chris Locke, Christian Maennel, Mike Worsell, and David Scharf, helped once again, and I thank them sincerely for their continued support.

Preface

"All of physics is either impossible or trivial.
It is impossible until you understand it, and then it becomes trivial."

Ernest Rutherford

This book is

1. an attempt to make learning quantum field theory (QFT) as easy, and thus as efficient, as is humanly possible,
2. intended, first and foremost, for new students of QFT, and
3. an introduction to only the most fundamental and central concepts of the theory, particularly as employed in quantum electrodynamics (QED).

It is not

1. orthodox (pedagogically),
2. an exhaustive treatment of QFT,
3. concise (lacking extensive explanation),
4. written for seasoned practitioners in the field, or
5. a presentation of the latest, most modern approach to it.

Students planning a career in field theory will obviously have to move on to more advanced texts, after they digest the more elementary material presented herein. This book is intended to provide a solid foundation in the most essential elements of the theory, nothing more.

In my own teaching experience, and in the course of researching pedagogy, I have come to see that "learning" has at its basis a fundamental three-in-one structure. The wholeness of learning is composed of

    i)   the knowledge to be learned,
    ii) the learner, and
    iii) the process of learning itself.

It seems unfortunate that physics and physics textbooks have too often been almost solely concerned with the *knowledge* of physics and only rarely concerned with *those who are learning it* or *how they could best go about learning*. However, there are signs that this situation may be changing somewhat, and I hope that this book will be one stepping stone in that direction.

In writing this book, I have repeatedly tried to visualize the learning process as a new learner would. This viewpoint is one we quickly lose when we, as teachers and researchers, gain familiarity with a given subject, and yet it is a perspective we must maintain if we are to be effective educators. To this end, I have solicited guidance and suggestions from professional educators (those who make learning and education, *per se*, their central focus in life), and more importantly, from those studying QFT for the first time. In addition, I have used my own notes, compiled when I was first studying the theory myself, in which I carefully delineated ways the subject could be presented in a more student-friendly manner. In this sense, the text incorporates "peer instruction", a pedagogic tool of recognized, and considerable, merit, wherein students help teach fellow students who are learning the same subject.

It is my sincere hope that the methodologies I have employed herein have helped me to remain sympathetic to, and in touch with, the perspective of a new learner. Of course, different students find different teaching techniques to have varying degrees of transparency, so there are no hard and fast rules. However, I do believe that most students would consider many of the following principles, which I have employed in the text, to be of pedagogic value.

1) Brevity avoided

Conciseness is typically a horror for new students trying to fathom unfamiliar concepts. While it can be advantageous in some arenas, it is almost never so in education. Unfortunately, being succinct, has, in scientific/technical circles, become a goal unto itself, extending even into pedagogy – an area for which it was never suited.

In this book, I have gone to great lengths to avoid conciseness and to present extensive explanations. I often take a paragraph or more for what other authors cover in a single sentence. I do this because I learned a long time ago that the thinnest texts were the hardest. Thicker ones covering the same material actually took less time to get through, and I understood them better, because the authors took time and space to elaborate, rather than leave significant gaps.

Such gaps often contain ambiguities or possibilities for misunderstanding that the author has overlooked and left unresolved. Succinctness may impress peers but can be terribly misleading and frustrating for students.

2) Holistic previews

The entire book, each chapter, and many sections begin with simple, non-mathematical overviews of the material to be covered. These allow the student to gain a qualitative understanding of the "big picture" before he or she plunges into the rigors of the underlying mathematics.

Doing physics is a lot like doing a jig-saw puzzle. We assemble bits and pieces into small wholes and then gradually merge those small wholes into greater ones, until ultimately, we end up with the "big picture." Seeing the picture on the puzzle box before we start has immense value in helping us put the whole thing together. We know the blue goes here, the green there, and the boundary of the two, somewhere in between. Without that picture preview to guide us, the entire job becomes considerably more difficult, more tedious, and less enjoyable. In this book, the holistic previews are much like the pictures on the puzzle boxes. The detail is not there, but the essence of the final goal is. These overviews should eliminate, or at least minimize, the "lost in a maze of equations" syndrome by providing a "birds-eye road map" of where we have come from, and where we are going. By so doing we not only will keep sight of the forest in spite of the trees, but will also have a feeling, from the beginning, for the relevance of each particular topic to the overriding structure of the wholeness of knowledge in which it is embedded.

3) Schematic diagram summaries (wholeness charts)

Enhancing the "birds-eye road map" approach are block diagram summaries, which I call *wholeness charts*, so named because they reveal in chart form the underlying connections that unite various aspects of a given theory into a greater whole. Unlike the chapter previews, these are often mathematical and contain considerable theoretical depth.

Learning a computer program line-by-line is immensely harder than learning it with a block diagram of the program, showing major sections and sub-sections, and how they are all interrelated. There is a structure underlying the program, which is its essence and most important aspect, but which is not obvious by looking directly at the program code itself.

The same is true in physics, where line-by-line delineation of concepts and mathematics corresponds to program code, and in this text, wholeness charts play the role of block diagrams. In my own learning experiences, in which I constructed such charts myself from my books and lecture notes, I found them to be invaluable aids. They coalesced a lot of different information into one central, compact, easy-to-see, easy-to-understand, and easy-to-reference framework.

The specific advantages of wholeness charts are severalfold.

First, in learning any given material we are seeking, most importantly, an understanding of the kernel or conceptual essence, i.e., the main idea(s) underlying all the text. A picture is worth a thousand words, and a wholeness chart is a "snapshot" of those thousand words.

Second, although the charts can summarize in-depth mathematics and concepts, they can be used to advantage even when reading through material for the first time. The holistic overview perspective can be more easily maintained by continual reference to the schematic as one learns the details.

Third, comparison with similar diagrams in related areas can reveal parallel underlying threads running through seemingly diverse phenomena. (See, for example, Summary of Classical Mechanics Wholeness Chart 2-2 and Summary of Quantum Mechanics Wholeness Chart 2-5 in Chap. 2, pgs. 20-21 and 30-31.) This not only aids the learning process but also helps to reveal some of the subtle workings and unified structure inherent in Mother Nature.

Further, review of material for qualifying exams or any other future purpose is greatly facilitated. It is much easier to refresh one's memory, and even deepen understanding, from one or two summary sheets, rather than time consuming ventures through dozens of pages of text. And by copying all of the wholeness charts herein and stapling them together, you will have a pretty good summary of the entire book.

Still further, the charts can be used as quick and easy-to-find references to key relations at future times, even years later.

4) Reviews of background material

In situations where development of a given idea depends on material studied in previous courses (e.g., quantum mechanics) short reviews of the relevant background subject matter are provided, usually in chapter introductory sections or later on, in special boxes separate from the main body of the text.

5) Only basic concepts without peripheral subjects

I believe it is of primary importance in the learning process to focus on the fundamental concepts first, to the exclusion of all else. The time to branch out into related (and usually more complex) areas is *after* the core knowledge is assimilated, *not during* the assimilation period.

All too often, students are presented with a great deal of new material, some fundamental, other more peripheral or advanced. The peripheral/advanced material not only consumes precious study time, but tends to confuse the student with regard to what precisely is essential (what he or she *must* understand), and what is not (what it would be *nice if* he or she also understood at this point in their development).

As one example, for those familiar with other approaches to QFT, this book does not introduce concepts appropriate to weak interactions, such as ϕ^4 theory, before students have first become grounded in the more elementary theory of quantum electrodynamics.

This book, by careful intention, restricts itself to only the most core principles of QFT. Once those principles are well in hand, the student should then be ready to glean maximum value from other, more extensive, texts.

6) Optimal "return on investment" exercises

All too often students get tied up, for what seem interminable periods, working through problems from which minimum actual learning is reaped. Study time is valuable, and spending it engulfed in great quantities of algebra and trigonometry is probably not its best use.

I have tried, as best I could, to design the exercises in this book so that they consume minimum time but yield maximum return. Emphasis has been placed on gleaning an understanding of concepts without getting mired down.

Later on, when students have become practicing researchers and time pressure is not so great, there will be ample opportunities to work through more involved problems down to every minute algebraic detail. If they are firmly in command of the *concepts* and *principles* involved, the calculations, though often lengthy, become trivial. If, however, they never got grounded in the fundamentals because study time was not efficiently used, then research can go slowly indeed.

7) Many small steps, rather than fewer large ones

Professional educators have known for some time now that learning progresses faster and more profoundly when new material is presented in small bites. The longer, more moderately sloped trail can get one to the mountaintop much more readily than the agonizing climb up the nearly vertical face.

Unfortunately, from my personal experience as a student, it often seemed like my textbooks were trying to take me up the steepest grade. I sincerely hope that those using this book do not have this experience. I have made every effort to include each and every relevant step in all derivations and examples.

In so doing, I have sought to avoid the common practice of letting students work out significant amounts of algebra that typically lies "between the lines". The thinking, as I understand it, is that students are perfectly capable of doing that themselves, so "why take up space with it in a text?"

My answer is simply that including those missing steps makes the learning process more efficient. If it takes the author ten minutes to write out two or three more lines of algebra, then it probably takes the student twenty minutes to do so, provided he/she is not befuddled (which is not rare, and in which case, it can take a great deal longer). That ten minutes spent by the author saves hundreds, or even thousands, of student readers twenty minutes, or more, each. Multiply that by the number of times such things occur per chapter and the number of chapters per book, and we are talking enormous amounts of student time saved.

Students learn very little, if anything, doing algebra. They recapture a lot of otherwise wasted time that can be used for actual learning, if the author types out the missing lines.

8) Liberal use of simple concrete examples

Professional educators have also known for quite some time that abstract concepts are best taught by leading into them with simple, physically visualizable examples. Further, understanding is deepened, broadened, and solidified with even more such concrete examples.

Some may argue that a more formal mathematical approach is preferable because it is important to have a profound, not superficial, understanding. While I completely agree that a profound understanding is essential, it is my experience that the mathematically rigorous introduction, more often than not, has quite the opposite result. (Ask any student about this.) Further, to know any field profoundly we must know it from all angles. We must know the underlying mathematics in detail *plus* we must have a grasp on what it all means in the real world, i.e., how the relevant systems behave, how they parallel other types of systems with which we are already familiar, etc. Since we have to cover the whole range of knowledge from abstract to physical anyway, it seems best to start with the end of the spectrum most readily apprehensible (i.e., the visualizable, concrete, and analogous) and move on from there.

This methodology is employed liberally in this book. It is hoped that so doing will ameliorate the "what is going on?" frustration common among students who are introduced to conceptually new ideas almost solely via routes heavily oriented toward abstraction and pure mathematics.

In this context it is relevant that Richard Feynman, in his autobiography, notes,

> "I can't understand anything in general unless I'm carrying along in my mind a specific example and watching it go....(Others think) I'm following the steps mathematically but that's not what I'm doing. I have the specific, physical example of what (is being analyzed) and I know from instinct and experience the properties of the thing."

I know from my own experience that I learn in the same way, and I have a suspicion that almost everyone else does as well. Yet few *teach* that way. This book is an attempt to teach in that way.

9) Margin overview notes

Within a given section of any textbook, one group of paragraphs can refer to one subject, another group to another subject. When reading material for the first time, not knowing exactly where one train of the author's thought ends and a different one begins can oftentimes prove confusing. In this book, each new idea not set off with its own section heading is highlighted, along with its central message, by notations in the margins. In this way, emphasis is once again placed on the overview, the "big picture" of each topic, even on the subordinate levels within sections and subsections.

Additionally, the extra space in the margins can be used by students to make their own notes and comments. In my own experience as a student I found this practice to be invaluable. My own remarks written in a book are, almost invariably, more comprehensible to me when reviewing later for exams or other purposes than are those of the author.

10) Definitions and key equations emphasized

As a student, I often found myself encountering a term that had been introduced earlier in the text, but not being clear on its exact meaning, I had to search back through pages clumsily trying to find the first use of the word. In this book, new terminology is underlined when it is introduced or defined, so that it "jumps out" at the reader later when trying to find it again.

In addition, key equations – the ones students really need to know – have borders around them.

11) Non-use of terms like "obvious", "trivial", etc.

The text avoids use of emotionally debilitating terms such as "obvious", "trivial", "simple", "easy", and the like to describe things that may, after years of familiarity, be easy or obvious to the author, but can be anything but that to the new student. (See "A Nontrivial Manifesto" by Matt Landreman, *Physics Today*, March 2005, 52-53.)

The job I have undertaken here has been a challenging one. I have sought to produce a physics textbook which is relatively lucid and transparent to those studying quantum field theory for the first time. In so doing, I have employed some decidedly non-traditional tactics, and so anticipated resistance from main stream publishers, who typically have motivations for wanting to do things the way they have been done before. Their respective missions do not seem, at least to me, to be focused primarily on optimizing the process of conveying knowledge.

As an example, a good friend of mine submitted a graduate level physics text manuscript, with student friendly notes in the margins, to one of the world's top academic publishers. He was ordered to remove the margin notes before they would publish the book. Not wanting to fight (and lose) this kind of battle over methodologies I employ, and consider essential in making students' work easier, I have chosen a different route.

I also anticipate resistance from some physics professors who may consider the book too verbose and too simple. I only ask them to try it and let their students be the judges. The proof will be in the pudding. If comprehension comes more quickly and more deeply, then the approach taken here will be vindicated.

If you are a student now, appreciate the pedagogic methodologies used in this book, and end up one day writing a text of your own, I hope you will not forget what advantage you once gained from those methodologies. I hope you will use them in your own book. Above all, I hope your presentation will be profuse with elucidation and not terse.

Good luck to the new students of quantum field theory! May their studies be personally rewarding and professionally fruitful.

Robert D. Klauber
February 2013

Prerequisites

Quantum field theory takes off where the following subjects end. Those beginning this book should be reasonably well versed in them, at the levels described below.

Quantum Mechanics

An absolute minimum of two undergraduate quarters, but far more preferably, an additional two graduate level quarters. Some exposure to relativistic quantum mechanics would be advantageous but is not necessary. Optimal level of proficiency: Eugen Merzbacher's *Quantum Mechanics* (John Wiley) or a similar book.

Classical Mechanics

A semester at the graduate level. Topics covered should include the Lagrangian formulation (for particles, and importantly, also for fields), the Legendre transformation, special relativity, and classical scattering. A familiarity with Poisson brackets would be helpful. Optimal level of proficiency: Herbert Goldstein's *Classical Mechanics* (Addison-Wesley) or similar.

Electromagnetism

Two quarters at the undergraduate level plus two graduate quarters. Areas studied should comprise Maxwell's equations, conservation laws, e/m wave propagation, relativistic treatment, Maxwell's equations in terms of the four potential. Optimal level of proficiency: John David Jackson's *Classical Electrodynamics* (John Wiley) or similar.

Math/Relativity

Advantageous but not essential, as it is covered in the appendix of Chap. 2: Exposure to covariant and contravariant coordinates, and metric tensors, for orthogonal 4D systems, at the level found in Jackson's chapters on special relativity.

Acknowledgements

*"You cannot live a perfect day without doing something
for someone who will never be able to repay you."*

John Wooden
Hall of Fame UCLA basketball coach.

The people who reviewed, edited, made suggestions for, and corrected draft portions of this book had many candidate perfect days. There is no way I can repay them.

I am most indebted to three, Chris Locke, Christian Maennel and Mike Worsell, who read every word and made innumerable great contributions. Close behind on my gratitude list are Carlo Marino, David Scharf, Jean-Louis Sicaud, and Jon Tyrrell, each of whom read most of the text and provided a substantial number of valuable suggestions and corrections. David, Jon, and Morgan Orcutt deserve further heartfelt thanks for working most of the problems (and finding errors in several of them).

Others making significant, much appreciated contributions include Martin Bäker, Jim Bogan, Ben Brenneman, Brad Carlile, Bill Cohwig, Trevor Daniels, Saurya Das, Lorenzo Del Re, Tony D'Esopo, Paul Drechsel, Michael Gildner, Esteban Herrera, Phil Jones, Ruth Kastner, Lorek Krzysztof, Claude Liechti, Rattan Mann, Lorenzo Massimi, Enda McGlynn, Gopi Rajagopal, Javier Rubio, Girish Sharma, and Dennis Smoot.

Many years before I started writing this text, I fell in debt to my teachers, Robin Ticciati and John Hagelin, who guided me through my earliest sojourns into the quantum theory of fields, and earned both my respect and deep gratitude. Robin, in particular, was generous well beyond the call of duty, in granting me numerous one-on-one sessions to discuss various aspects of the theory.

Non-technical, but nonetheless vital support came from my wonderful wife Susan. I cannot thank her enough for her patience, understanding, love, and unswerving devotion throughout the days, weeks, months, and years I spent writing ... and re-writing. Last mentioned, yet anything but least, are my amazing and caring parents, without whose support and many, many sacrifices, I would never have gained the education I did, and thus, never have written this book. Thank you, mom and dad.

This book, whatever it is, would be substantially less without these people.

Regardless, any errors or insufficiencies that may still remain are my responsibility, and mine alone.

The website for this book is

www.quantumfieldtheory.info

It contains presentations of advanced topics as well as a list of corrections and improvements to this printing. Please use the site to report any errors you might find and to suggest ways to make material presented in this book easier for students to understand.

Problem solutions booklet

Solutions to the problems in this book are provided in the booklet *Solutions to Problems for Student Friendly Quantum Field Theory*. The website above contains a link to where the booklet can be obtained.

Chapter 1

Bird's Eye View

Well begun is half done.
Old Proverb

1.0 Purpose of the Chapter

Before starting on any journey, thoughtful people study a map of where they will be going. This allows them to maintain their bearings as they progress, and not get lost en route. This chapter is like such a map, a schematic overview of the terrain of quantum field theory (QFT) without the complication of details. You, the student, can get a feeling for the theory, and be somewhat at home with it, even before delving into the "nitty-gritty" mathematics. Hopefully, this will allow you to keep sight of the "big picture", and minimize confusion, as you make your way, step-by-step, through this book.

1.1 This Book's Approach to QFT

There are two main branches to (ways to do) quantum field theory called

- the canonical quantization approach, and

- the path integral (many paths, sum over histories, or functional quantization) approach.

The first of these is considered by many, and certainly by me, as the easiest way to be introduced to the subject, since it treats particles as objects that one can visualize as evolving along a particular path in spacetime, much as we commonly think of them doing. The path integral approach (which goes by several names), on the other hand, treats particles and fields as if they were simultaneously traveling all possible paths, a difficult concept with even more difficult mathematics behind it.

This book is primarily devoted to the canonical quantization approach, though I have provided a simplified, brief introduction to the path integral approach in Chap. 18 near the end. Students wishing to make a career in field theory will eventually need to become well versed in both.

1.2 Why Quantum Field Theory?

The quantum mechanics (QM) courses students take prior to QFT generally treat a single particle such as an electron in a potential (e.g., square well, harmonic oscillator, etc.), and the particle retains its integrity (e.g., an electron remains an electron throughout the interaction.) There is no general way to treat transmutations of particles, such as that of a particle and its antiparticle annihilating one another to yield neutral particles such as photons (e.g., $e^- + e^+ \rightarrow 2\gamma$.) Nor is there any way to describe the decay of an elementary particle such as a muon into other particles (e.g. $\mu^- \rightarrow e^- + \nu + \bar{\nu}$, where the latter two symbols represent neutrino and antineutrino, respectively).

Limitation of original QM: no transmutation of particles

Here is where QFT comes to the rescue. It provides a means whereby particles can be annihilated, created, and transmigrated from one type to another. In so doing, its utility surpasses that provided by ordinary QM.

QFT: transmutation included

There are other reasons why QFT supersedes ordinary QM. For one, it is a relativistic theory, and thus more all encompassing. Further, as we will discuss more fully later on, the straightforward extrapolation of non-relativistic quantum mechanics (NRQM) to relativistic quantum mechanics (RQM) results in states with negative energies, and in the early days of quantum theory, these were quite problematic. We will see in subsequent chapters how QFT resolved this issue quite nicely.

Energies <0 RQM yes QFT no

1.3 How Quantum Field Theory?

As an example of the type of problem QFT handles well, consider the interaction between an electron and a positron that produces a muon and anti-muon, i.e., $e^- + e^+ \rightarrow \mu^- + \mu^+$, as shown in

QFT example:
$e^- + e^+$
$\rightarrow \mu^- + \mu^+$
scattering

Fig. 1-1. At event x_2, the electron and positron annihilate one another to produce a photon. At event x_1, this photon is transmuted into a muon and an anti-muon. Antiparticles like positrons and anti-muons are represented by lines with arrows pointing opposite their direction of travel through time. The seemingly strange, reverse order of numbering here, i.e., $2 \rightarrow 1$, is standard in QFT.

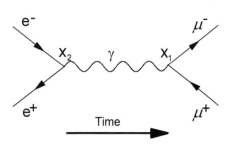

Figure 1-1. $e^- + e^+ \rightarrow \mu^- + \mu^+$
Scattering

Note that we can think of this interaction as an annihilation (destruction) of the electron and the positron at x_2 accompanied with creation of a photon, and that followed by the destruction of the photon accompanied by creation of an muon and anti-muon at x_1. Unlike the incoming and outgoing particles in this example, the photon here is not a "real" particle, but transitory, short-lived, and undetectable, and is called a <u>virtual particle</u> (which mediates the interaction between real particles.)

What we seek and what, as students eventually see, QFT delivers, is a mathematical relationship, called a <u>transition amplitude</u>, describing a transition from an initial set of particles to a final set (i.e., an interaction) of the sort shown pictorially via the <u>Feynman diagram</u> of Fig. 1-1. It turns out that the square of the absolute value of the transition amplitude equals the probability of finding (upon measurement) that the interaction occurred. This is similar to the square of the absolute value of the wave function in NRQM equaling the probability density of finding the particle.

QFT employs creation and destruction operators acting on states (i.e., kets), and these creation/destruction operators are part of the transition amplitude. We illustrate the general idea with the following grossly oversimplified transition amplitude, reflecting the interaction process of Fig. 1-1. Be cautioned that we have omitted, and even distorted a bit, some ultimately essential ingredients in the following, to make it simpler, and easier, to grasp the fundamental concept.

Introductory overview with significant liberties taken from rigorous treatment in order to simplify as much as possible

$$\text{Transition amplitude} \atop \text{of Fig. 1-1} = {}_{final}\left\langle \mu^+ \mu^- \right| \underset{\substack{\text{a constant} \\ \text{from theory}}}{K} \left(\bar{\psi}_c^{\mu^-} A_d \psi_c^{\mu^+} \right)\left(\bar{\psi}_d^{e^+} A_c \psi_d^{e^-} \right) \left| e^+ e^- \right\rangle_{initial} . \quad (1\text{-}1)$$

The ket $|e^+ e^-\rangle$ represents the incoming e^- and e^+; the bra $\langle \mu^+ \mu^- |$, the outgoing muon and anti-muon. K is a constant determined by theory. $\psi_d^{e^-}$ is an operator that destroys the e^- in the ket; $\bar{\psi}_d^{e^+}$, an operator that destroys the e^+. $\psi_c^{\mu^+}$ creates an anti-muon in the ket. $\bar{\psi}_c^{\mu^-}$ creates a muon. A_c is an operator that creates a virtual photon, and A_d is an operator that destroys that virtual photon, with the lines underneath indicating that the photon is virtual and propagates (in Fig. 1-1 from x_2 to x_1). The mathematical procedure and symbolism (lines underneath) representing this virtual particle (photon here) process, as shown in (1-1), is called a <u>contraction</u>. When the virtual particle is represented as a mathematical function, it is known as the <u>Feynman propagator</u> or simply, <u>the propagator</u>, because it represents the propagation of a virtual particle from one event to another.

Feynman propagator

Note what happens to the ket part of the transition amplitude as we proceed, step-by-step, through the interaction process. First, the incoming particles (in the ket) are destroyed by the destruction operators, so at an intermediate point, we have

Destruction operators leave vacuum ket times a numeric factor

$$\text{Fig. 1-1 transition amplitude} = {}_{final}\left\langle \mu^+ \mu^- \right| K \left(\bar{\psi}_c^{\mu^-} A_d \psi_c^{\mu^+} \right)\left(A_c \right) K_2 \left| 0 \right\rangle , \quad (1\text{-}2)$$

where the destruction operators have acted on the original ket to leave the vacuum ket $|0\rangle$ (no particles left) with a purely numeric factor K_2 in front of it. The value of this factor is determined by the formal mathematics of QFT.

In the next step after (1-2), the virtual photon propagator, due to the creation operator A_c, creates a virtual photon (at x_2 in Fig. 1-1) that then propagates (from x_2 to x_1 in the figure) and then, via A_d, is annihilated. This process leaves the vacuum ket still on the right along with an additional numeric factor, which comes out of the formal mathematics, and which we designate below as K_γ.

Propagator action leaves only another numeric factor

$$\text{Fig. 1-1 transition amplitude} = {}_{final}\left\langle \mu^+ \mu^- \right| K \left(\bar{\psi}_c^{\mu^-} \psi_c^{\mu^+} \right) K_\gamma K_2 \left| 0 \right\rangle \quad (1\text{-}3)$$

The remaining creation operators then create a muon and anti-muon out of the vacuum. This leaves us with the newly created ket $|\mu^+\mu^-\rangle$ times a numeric factor K_1 in front. The ket and the bra now represent the same multi-particle state (same particles in the same states), so their inner product (the bracket) is not zero (as it would be if they were different states). Nor are there any operators left, but only numeric quantities, so we can move them outside the bracket without changing anything. Thus,

$$\text{Fig. 1-1 transition amplitude} = {}_{final}\langle \mu^+\mu^- | K K_1 K_\gamma K_2 | \mu^+\mu^-\rangle_{final}$$

$$= {}_{final}\langle \mu^+\mu^- | \underbrace{S_{Fig1-1}}_{\substack{\text{just a number}\\\text{without operators}}} | \mu^+\mu^-\rangle_{final} = S_{Fig1-1}\underbrace{{}_{final}\langle \mu^+\mu^- \| \mu^+\mu^-\rangle_{final}}_{1} = S_{Fig1-1}, (1\text{-}4)$$

where we note the <u>important point</u> that *in QFT the bracket of a multiparticle state* (inner product of multiparticle state with itself) such as that shown in (1-4) is defined so it *always equals unity*. Note, that if we had ended up with a ket different than the bra, the inner product would be zero, because the two (different) states, represented by the bra and ket, would be orthogonal. Examples are

$$\underbrace{\langle \mu^+\mu^- \| \mu^-\mu^-\rangle}_{\substack{\text{two}\\\text{muons}}} = 0, \quad \underbrace{\langle \mu^-\mu^+ | | e^+e^-\rangle}_{\substack{\text{electron \&}\\\text{positron}}} = 0, \quad \text{and} \quad \underbrace{\langle \mu^-\mu^+ \| \gamma \rangle}_{\substack{\text{single}\\\text{photon}}} = 0. \quad (1\text{-}5)$$

Further, the inner product of any bra and ket with the same particle types but different states (e.g., different momentum for at least one particle in the bra from the ket) equals zero.

The whole process of Fig. 1-1 can be pictured as simply an evolution, or progression, of the original state, represented by the ket, to the final state, represented by the bra. At each step along the way, the operators act on the ket to change it into the next part of the progression. When we get to the point where the ket is the same as the bra, the full transition has been made, and the bracket then equals unity. What is left is our transition amplitude for the scattering of Fig. 1-1.

$$\text{Fig 1-1 probability of interaction} = S^\dagger_{Fig1-1}S_{Fig1-1} = \left| S_{Fig1-1} \right|^2 . \quad (1\text{-}6)$$

The quantity $K K_1 K_\gamma K_2 = S_{Fig1-1}$ arising in (1-4) depends on particle momenta, spins, and masses, as well as the inherent strength of the electromagnetic interaction, all of which one would rightly expect to play a role in the probability of an interaction taking place. There are other subtleties, including some integration over x_1 and x_2, that have been suppressed, and even distorted a bit, above in order to convey the essence of the transition amplitude as simply as possible.

From the interaction probability, scattering cross sections can be calculated.

1.4 From Whence Creation and Destruction Operators?

In NRQM, the solutions to the relevant wave equation, the Schrödinger equation, are states (particles or kets.) Surprisingly, the *solutions to the relevant wave equations in QFT are not states* (not particles.) In QFT, it turns out that these solutions are *actually operators that create and destroy states.* (The operators used in relations like (1-1) are actually just such solutions.) Different solutions exist that create or destroy every type of particle and antiparticle. In this unexpected (and, for students, often strange at first) twist lies the power of QFT.

1.5 Overview: The Structure of Physics and QFT's Place Therein

Students are often confused over the difference (and whether or not there is a difference) between relativistic quantum mechanics (RQM) and QFT. The following discussion, summarized below in Wholeness Chart 1-1, should help to distinguish them.

1.5.1 Background: Poisson Brackets and Quantization

Classical particle theories contain rarely used entities called <u>Poisson brackets</u>, which, though it would be nice, are not necessary for you to completely understand at this point. (We will show their precise mathematical form in Chap. 2.) What you should realize now is that Poisson brackets are mathematical manipulations of certain pairs of properties (dynamical variables like position and momentum) that bear a striking resemblance to commutators in quantum theories. For example, the

Poisson bracket for position X (capital letters will designate Cartesian coordinates in this book) and momentum p_X, symbolically expressed herein as $\{X, p_X\}$, is non-zero (and equal to one), but the Poisson bracket for Y and p_X equals zero.

Shortly after NRQM theory had been worked out, theorists, led by Paul Dirac, realized that for each pair of quantum operators that had non-zero (zero) commutators, the corresponding pair of classical dynamical variables also had non-zero (zero) Poisson brackets. They had originally arrived at NRQM by taking classical dynamical variables as operators, and that led, in turn, to the non-zero commutation relations for certain operators (which result in other quantum phenomena such as uncertainty.) But it was soon recognized that one could do the reverse. One could, instead, take the classical Poisson brackets over into quantum commutation relations first, and because of that, the dynamical variables turn into operators. (Take my word for this now, but after reading the next section, do Prob. 6 at the end of this chapter, and you should understand it better.)

The process of extrapolating from classical theory to quantum theory became known as quantization. Apparently, for many, the specific process of starting with Poisson brackets and converting them to commutators was considered the more elegant way to quantize.

Quantization: Poisson brackets become commutators

1.5.2 First vs. Second Quantization

Classical mechanics has both a non-relativistic and a relativistic side, and each contains a theory of particles (localized entities, typically point-like objects) and a theory of fields (entities extended over space). All of these are represented in the first row of Wholeness Chart 1-1. Properties (dynamical variables) of entities in classical particle theories are *total* values, such as object mass, charge, energy, momentum, etc. Properties in classical field theories are *density* values, such as mass and charge density, or field amplitude at a point, etc. that generally vary from point to point. Poisson brackets in field theories are similar to those for particle theories, except they entail *densities* of the respective dynamical variables, instead of total values.

Branches of classical mechanics

With the success of quantization in NRQM, people soon thought of applying it to relativistic particle theory and found they could deduce RQM in the same way. Shortly thereafter they tried applying it to relativistic field theory, the result being QFT. The term first quantization came to be associated with *particle* theories (and is sometimes call particle quantization). The term second quantization became associated with *field* theories (and is sometimes called field quantization).

1st quantization is for particles; 2nd is for fields

In quantizing, we also assume the classical Hamiltonian (total or density value) has the same quantum form. We can summarize all of this as follows.

First Quantization (Particle Theories)
   1) Assume the quantum particle Hamiltonian has the same form as the classical particle Hamiltonian.
   2) Replace the classical Poisson brackets for conjugate properties with commutator brackets (divided by $i\hbar$),e.g.,

$$\{X_i, p_j\} = \delta_{ij} \;\Rightarrow\; [X_i, p_j] = X_i p_j - p_j X_i = i\hbar \delta_{ij} \;. \tag{1-7}$$

In doing (1-7), the classical properties (dynamical variables) of position and its conjugate 3-momentum become quantum non-commuting operators.

Second Quantization (Field Theories)
   1) Assume the quantum field Hamiltonian density has the same form as the classical field Hamiltonian density.
   2) Replace the classical Poisson brackets for conjugate property densities with commutator brackets (divided by $i\hbar$), e.g.

$$\left\{\phi_r\left(\mathbf{x},t\right), \pi_s\left(\mathbf{y},t\right)\right\} = \delta_{rs}\delta\left(\mathbf{x}-\mathbf{y}\right) \;\Rightarrow\; \left[\phi_r\left(\mathbf{x},t\right), \pi_s\left(\mathbf{y},t\right)\right] = i\hbar\delta_{rs}\delta\left(\mathbf{x}-\mathbf{y}\right), \tag{1-8}$$

where π_s is the conjugate momentum density of the field ϕ_s, different values for r and s mean different fields, and $\mathbf{x}$ and $\mathbf{y}$ represent different 3D position vectors. In doing (1-8), the classical field dynamical variables become quantum field non-commuting operators (and this, as we will see, has major ramifications for QFT.)

Note that the specific quantization we are talking about here (both first and second) is called <u>canonical quantization</u>, because, in both the Poisson brackets and the commutators, we are using (in classical mechanics terminology) <u>canonical variables</u>. For example, p_x is called the *canonical momentum* of X. (It is sometimes also called the *conjugate momentum*, as we did above, or the *generalized momentum* of X.)

Our approach here: canonical quantization

This differs from the form of quantization used in the path integral approach (see Sect. 1.1 on page 1) to QFT, which is known as <u>functional quantization</u>, because the path integral approach employs mathematical quantities known as *functionals* (See Chap. 18 near the end of the book for a brief introduction to this alternative method of doing QFT.)

Path integral approach to QFT: functional quantization

1.5.3 The Whole Physics Enchilada

All of the above two sections is summarized in Wholeness Chart 1-1. In using it, the reader should be aware that, depending on context, the term <u>quantum mechanics</u> (<u>QM</u>) can mean i) only non-relativistic ("ordinary") quantum mechanics (NRQM), or ii) the entire realm of quantum theories including NRQM, RQM, and QFT. In the leftmost column of the chart, we employ the second of these.

Note that because quantum field applications usually involve photons or other relativistic particles, non-relativistic quantum field theory (NRQFT) is not widely applicable and thus rarely taught, at least not at elementary levels. However, in some areas where non-relativistic approximations can suffice, such as condensed matter physics, NRQFT can be useful because calculations are simpler. The term "quantum field theory" (QFT) as used in the physics community generally means "relativistic QFT", and our study in this book is restricted to that.

Wholeness Chart 1-1. The Overall Structure of Physics

	Non-relativistic		**Relativistic**	
	Particle	**Field**	**Particle**	**Field**
Classical mechanics (non-quantum)	Newtonian particle theory	Newtonian field theory (continuum mechanics + gravity), e/m (quasi-static)	Relativistic macro particle theory	Relativistic macro field theory (continuum mechanics + e/m + gravity)
Properties (Dynamical variables) $\Downarrow$ Operators	$\Downarrow$ 1st quantization $\Downarrow$	$\Downarrow$ 2nd quantization $\Downarrow$	$\Downarrow$ 1st quantization $\Downarrow$	$\Downarrow$ 2nd quantization $\Downarrow$
Quantum mechanics	NRQM	NRQFT rarely taught.	RQM	QFT (not gravity)

As an aside, quantum theories of gravity, such as superstring theory and loop quantum gravity, are not included in the chart, as QFT in its standard model form cannot accommodate gravity. Thus, the relativity in QFT is special, but not general, relativity.

Conclusions: RQM is similar to NRQM in that both are particle theories. They differ in that RQM is relativistic. RQM and QFT are similar in that both are relativistic theories. They differ in that QFT is a field theory and RQM is a particle theory.

1.6 Comparison of Three Quantum Theories

NRQM employs the (non-relativistic) Schrödinger equation, whereas RQM and QFT must employ relativistic counterparts sometimes called <u>relativistic Schrödinger equations</u>. Students of QFT soon learn that each spin type (spin 0, spin ½, and spin 1) has a different relativistic Schrödinger equation. For a given spin type, that equation is the same in RQM and in QFT, and hence, both theories have the same form for the solutions to those equations.

Different spin types→ different wave equations

The difference between RQM and QFT is in the meaning of those solutions. In RQM, the solutions are interpreted as states (particles, such as an electron), just as in NRQM. In QFT, though it may be initially disorienting to students previously acclimated to NRQM, the solutions turn out not to be states, but rather operators that create and destroy states. Thus, QFT can handle transmutation of particles from one kind into another (e.g., muons into electrons, by destroying the original muon and creating the final electron), whereas NRQM and RQM cannot. Additionally, the problem of negative energy state solutions in RQM does not appear in QFT, because, as we will see, the creation and destruction operator solutions in QFT create and destroy both particles and anti-particles. Both of these have positive energies.

Additionally, while RQM (and NRQM) are amenable primarily to single particle states (with some exceptions), QFT better, and more easily, accommodates multi-particle states.

In spite of the above, there are some contexts in which RQM and QFT may be considered more or less the same theory, in the sense that QFT encompasses RQM. By way of analogy, classical relativistic particle theory is inherent within classical relativistic field theory. For example, one could consider an extended continuum of matter which is very small spatially, integrate the mass density to get total mass, the force/unit volume to get total force, etc., resulting in an analysis of *particle* dynamics. The field theory contains within it, the particle theory.

In a somewhat similar way, QFT deals with relativistic states (kets), which are essentially the same states dealt with in RQM. QFT, however, is a more extensive theory and can be considered to encompass RQM within its structure.

And in both RQM and QFT (as well as NRQM), operators act on states in similar fashion. For example, the expected energy measurement is determined the same way in both theories, i.e.,

$$\bar{E} = \langle \phi | H | \phi \rangle, \tag{1-9}$$

with similar relations for other observables.

These similarities and differences, as well as others, are summarized in Wholeness Chart 1-2. The chart is fairly self explanatory, though we augment it with a few comments. You may wish to follow along with the chart as you read them (below).

The different relativistic Schrödinger equations for each spin type are named after their founders (see names in chart.) We will cover each in depth. At this point, you have to simply accept that in QFT their solutions are operators that create and destroy states (particles). We will soon see how this results from the commutation relation assumption of 2nd quantization (1-8).

With regard to phenomena, I recall wondering, as a student, why some of the fundamental things I studied in NRQM seemed to disappear in QFT. One of these was bound state phenomena, such as the hydrogen atom. None of the introductory QFT texts I looked at even mentioned, let alone treated, it. It turns out that QFT can, indeed, handle bound states, but elementary courses typically don't go there. Neither will we, as time is precious, and other areas of study will turn out to be more fruitful. Those other areas comprise scattering (including inelastic scattering where particles transmute types), deducing particular experimental results, and vacuum energy.

I also once wondered why spherical solutions to the wave equations are not studied, as they play a big role in NRQM, in both scattering and bound state calculations. It turns out that scattering calculations in QFT can be obtained to high accuracy with the simpler plane wave solutions. So, for most applications in QFT, they suffice.

Wave packets, as well, can seem nowhere to be found in QFT. Like the other things mentioned, they too can be incorporated into the theory, but simple sinusoids (of complex numbers) serve us well in almost all applications. So, wave packets, too, are generally ignored in introductory (and most advanced) courses.

Wave function collapse, a much discussed topic in NRQM, is generally not a topic of focus in QFT texts. It does, however, play a key, commonly hidden role, which is discussed herein in Sects. 7.4.3 and 7.4.4, pgs. 196-197.

The next group of blocks in the chart points out the scope of each theory with regard to the four fundamental forces. Nothing there should be too surprising.

The final blocks note the similarities and differences between forces (interactions) in the different theories. As in classical theory, in all three quantum theories, interactions comprise forces that change the momentum and energy of particles. However, in QFT alone, interactions can also involve changes in type of particle, such as shown in Fig. 1-1. At event x_2, the electron and positron are changed into a photon, and in the process energy and momentum is transferred to the photon.

Wholeness Chart 1-2. Comparison of Three Theories

	NRQM	**RQM**	**QFT**
Wave equation	Schrödinger	Klein-Gordon (spin 0) Dirac (spin ½) Proca (spin 1) Special case of Proca: ↙ Maxwell (spin 1 massless)	Same as RQM at left
Solutions to wave equation	States	States	Operators that create and destroy states
Negative energy?	No	Yes	No
Particles per state	Single[*]	Single[*]	Multi-particle
Expectation values	$\bar{\mathcal{O}} = \langle \phi \vert \mathcal{O} \vert \phi \rangle$	As at left, but relativistic.	As at left in RQM.
Phenomena:			
1. bound states	Yes, non-relativistic	Yes, relativistic	Yes (usually not studied in introductory courses)
2. scattering			
a. elastic	a. Yes	a. Yes	a. Yes
b. inelastic (transmutation)	b. No (though some models can estimate)	b. Yes and no. (i.e., cumbersome and only for particle/antiparticle creation & destruction.)	b. Yes
3. decay			
a. composite particles	a. Yes (tunneling)	a. Yes	a. Yes
b. elementary particles	b. No	b. No	b. Yes
4. vacuum energy	No	No	Yes
Coordinates 1. Cartesian (plane waves)	Free, 1D potentials, particles in "boxes"	As at left	Used primarily for free particles, particles in "boxes", and scattering.
2. Spherical (spherical waves)	Bound states and scattering.	As at left.	Not usually used in elementary courses.
Wave Packets	Yes	Yes	Yes, but rarely used. Not taught in intro courses.
Wave function collapse	Yes	Yes	Yes, but not usually noted and behind the scenes. (See Sects. 7.4.3 & 4)

Interaction types			
1. e/m	No, though can pseudo model	As at left	Yes
2. weak	No	No	Yes
3. strong	No*	No*	Yes
4. gravity	No	No	Not as of this edition date.
Interaction nature			
Transfers energy & momentum?	Yes	Yes	Yes
Can change particle type?	No	No	Yes

*Some caveats exist for this chart. For example, NRQM and RQM can handle certain multiparticle states (e.g. hydrogen atom), but QFT generally does it more easily and more extensively. And the strong force can be modeled in NRQM and RQM by assuming a Yukawa potential, though a truly meaningful handling of the interaction can only be achieved via QFT.

1.7 Major Components of QFT

There are four major components of QFT, and this book (after the first two foundational chapters) is divided into four major parts corresponding to them. These are:

The four major parts of QFT

1. Free (non-interacting) fields/particles

    The field equations (relativistic Schrödinger equations) have no interaction terms in them, i.e., no forces are involved. The solutions to the equations are <u>free field solutions</u>.

2. Interacting fields/particles

    In principle, one would simply add the interaction terms to the free field equations and find the solutions. As it turns out, however, doing this is intractable, at best (impossible, at least in closed form, is a more accurate word). A trick employed in interaction theory actually lets us use the free field solutions of 1 above, so those solutions end up being quite essential throughout all of QFT.

3. Renormalization

    If you are reading this text, you have almost certainly already heard of the problem with infinities popping up in the early, naïve QFT calculations. The calculations referred to here are specifically those of the transition amplitude (1-4), where some of the numeric factors, if calculated straightforwardly, turn out to be infinite. Renormalization is the mathematical means by which these infinites are tamed, and made finite.

4. Application to experiment

    The theory of parts 1, 2, and 3 above are put to practical use in determining interaction probabilities and from them, scattering cross sections, decay probabilities (half-lives, etc.), and certain other experimental results. Particle decay is governed by the weak force, so we will not do anything with that in the present volume, which is devoted solely to quantum electrodynamics (QED), involving only the electromagnetic force.

1.8 Points to Keep in Mind

When the word "field" is used classically, it refers to an entity, like fluid wave amplitude, **E**, or **B**, that is spread out in space, i.e., has different values at different places. By that definition, the wave function of ordinary QM, or even the particle state in QFT, is a field. But, it is important to realize that in quantum terminology, the word "field" means an operator field, which creates and destroys particle states. States (= particles = wave functions = kets) are *not* considered fields in that context.

Terminology "field" = operator in QFT

In this text, the symbol e, representing the magnitude of charge on an electron or positron, is always positive. The charge on an electron is $-e$.

Symbol e >0

1.9 Big Picture of Our Goal

The big picture of our goal is this. We want to understand Nature. To do so, we need to be able to predict the outcomes of particle accelerator scattering experiments, certain other experimental results, and elementary particle half-lives. To do these things, we need to be able to calculate probabilities for each to occur. To do that, we need to be able to calculate transition amplitudes for specific elementary particle interactions. And for that, we need first to master a fair amount of theory, based on the postulates of quantization.

Our goal: predict scattering and decay seen in Nature

We will work through the above steps in reverse. Thus, our immediate goal is to learn some theory in Parts 1 and 2. Then, how to formulate transition amplitudes, also in Part 2. Necessary refinements will take up Part 3, with experimental application in Part 4.

Steps to our goal

2nd quantization postulates $\rightarrow$ QFT theory $\rightarrow$ transition amplitude calculation $\rightarrow$ probability $\rightarrow$ scattering, decay, other experimental results $\rightarrow$ confirmation of QFT

Steps to our goal

In this book our goal is a bit limited, as we will examine a part – an essential part – of the big picture. We will i) develop the fundamental principles of QFT, ii) use those principles to derive quantum electrodynamics (QED), the theory of electromagnetic quantum interactions, and iii) apply the theory of QED to electromagnetic scattering and other experiments. We will not examine herein the more advanced theories of weak and strong interactions, which play essential roles in particle decay, most present day high energy particle accelerator experiments, and composite particle (e.g., proton) structure. Weak and strong interaction theories build upon the foundation laid by QED. First things first.

Our goal in this book: basic QFT principles and QED, theory and experiments

1.10 Summary of the Chapter

Throughout this book, we will close each chapter with a summary, emphasizing its most salient aspects. However, the present chapter is actually a summary (in advance) of the entire book and all of QFT. So, you, the reader, can simply look back in this chapter to find appropriate summaries. These should include Sect. 1.1 (This Book's Approach to QFT), the transition amplitude relations of Eqs. (1-1) though (1-6), Sect.1.5.2 (1st and 2nd Quantization), Wholeness Chart 1-1 (The Overall Structure of Physics), Wholeness Chart 1-2 (Comparison of Three Theories), and Sect. 1.9 (Big Picture of Our Goal).

1.11 Suggestions?

If you have suggestions to make the material anywhere in this book easier to learn, or if you find any errors, please let me know via the web site address for this book posted on pg. xvi (opposite pg.1). Thank you.

1.12 Problems

As there is not much in the way of mathematics in this chapter, for most of it, actual problems are not really feasible. However, you may wish to try answering the questions in 1 to 5 below without looking back in the chapter. Doing Prob. 6 can help a lot in understanding first quantization.

1. Draw a Feynman diagram for a muon and anti-muon annihilating one another to produce a virtual photon, which then produces an electron and a positron. Using simplified symbols to represent more complex mathematical quantities (that we haven't studied yet), show how the transition amplitude of this interaction would be calculated.

2. Detail the basic aspects of first quantization. Detail the basic aspects of second quantization, then compare and contrast it to first quantization. In second quantization, what is analogous to position in first quantization? What is analogous to particle 3-momentum?

3.   Construct a chart showing how non-relativistic theories, relativistic theories, particles, fields, classical theory, and quantum theory are interrelated.

4.   For NRQM, RQM, and QFT, construct a chart showing i) which have states as solutions to their wave equations, ii) how to calculate expectation values in each, iii) which can handle bound states, inelastic scattering, elementary particle decay, and vacuum fluctuations, iv) which can treat the following interactions: i) e/m, ii) weak, iii) strong, and iv) gravity.

5.   What are the four major areas of study making up QFT?

6.   Using the corresponding Poisson bracket relation $\{X, p_x\} = 1$, we deduce, from first quantization postulate #2, that, quantum mechanically, $[X, p_x] = i\hbar$. For this commutator acting on a function ψ, i.e., $[X, p_x]\,\psi = i\hbar\psi$, determine, assuming p_x is a single term, what form p_x must have. Is this an operator? Does it look like what you started with in elementary QM, and from which you then derived the commutator relation above? Can we go either way?

Then, take the eigenvalue problem $H\psi = E\psi$, and use the same form of the Hamiltonian H as used in classical mechanics (i.e., $p^2/2m + V$), with the operator form you found for p above. This last step is the other part of first quantization (see page 4).

Did you get the time independent Schrödinger equation? (You should have.) Do you see how, by starting with the Poisson brackets and first quantization, you can derive the basic assumptions of NRQM, i.e., that dynamical variables become operators, the form of those operators, and even the time independent Schrödinger equation, itself? We won't do it here, but from that point, one could deduce the time dependent Schrödinger equation, as well.

Chapter 2

Foundations

Tiger got to hunt. Bird got to fly.
Man got to ask himself "why, why, why?".
Tiger got to rest. Bird got to land.
Man got to tell himself he understand.

The Book of Bonkonon in
Cat's Cradle by Kurt Vonnegut

2.0 Chapter Overview

In this chapter, we will cover the mathematical and physical foundations underlying quantum field theory to be sure you, the reader, are prepared and fit enough to traverse the rest of the book. The first cornerstone of these foundations is a new system of units, called natural units, which is common to QFT, and once learned, simplifies mathematical relations and calculations.

Topics covered after that comprise the notation used in this book, a comparison of classical and quantum waves, variational methods, classical mechanics in a nutshell, different "pictures" in quantum mechanics, and quantum theories in a nutshell. Whereas Chap. 1 was strictly an overview of what you will study, much of this chapter is an overview of what you have already studied, structured to make its role in our work more transparent. The rest is material you will need to know before we leap into the formal development of quantum field theory, beginning in Chap 3.

2.1 Natural Units and Dimensions

The Gaussian system (an extension of cgs devised for use in electromagnetism that takes the vacuum permittivity ε_0 and permeability μ_0 values as unity) has been common in NRQM, although standard international units (SI) [essentially, MKS for electromagnetism] are also used. Another is the Heaviside-Lorentz system, which is similar to the Gaussian system except it is structured to eliminate factors of 4π found in the Gaussian form of Maxwell's equations. (See Chap. 5.)

Natural units are another set of units that arise "naturally" in relativistic elementary particle physics. QFT uses them almost exclusively, they are the units we employ in this book, and we will see how they arise below.

Natural units are "natural" and used in QFT

2.1.1 Deducing a System of Units

Convenient systems of units start with arbitrary definitions for units of certain fundamental quantities and derive the remaining units from laws of nature. To see how this works, assume we know three basic laws of nature and we want to devise a system of units from scratch. We will do this first for the cgs system and then for natural units.

Any system of units: defined units + laws of nature → additional derived units

The three laws are:

1. The distance L traveled by a photon is the speed of light multiplied by its time of travel. $L = ct$.

2. The energy of a massive particle is equal to its mass (at rest) m times the speed of light squared. $E = mc^2$.

3. The energy of a photon is proportional to its frequency f. The constant of proportionality is Planck's constant h. $E = hf$ or re-expressed as $E = \hbar\omega$.

2.1.2 Deducing the cgs System

The cgs system takes its fundamental dimensions to be length, mass, and time. It then defines standard units of each of these dimensions to be the centimeter, the gram, and the second, respectively. With these standards and the laws of nature, dimensions and units are then derived for all other quantities science deals with.

For example, from law number one above, the speed of light in the cgs system is known to have dimensions of length/time and units of centimeters/second. Further, by measuring the time it takes for light to travel a certain distance we can get a numerical value of 3×10^{10} cm/s.

cgs: cm, g, s defined. Other units derived from laws of nature

From law number two, the dimensions of energy are mass-length^2/time^2 and the units are g-cm^2/s^2. We use shorthand by calling this an erg.

From number three, $\hbar$ has dimensions of energy-time and units of g-cm^2/s, or for short, erg-s. It, like the speed of light, can be measured by experiment and is found to be 1.0545×10^{-27} erg-s.

The point is this. We started with three pre-defined quantities (length, mass, and time) and derived the rest using the laws of nature. Of course, other laws could be used to derive other quantities ($\mathbf{F}=m\mathbf{a}$ for force, etc.). We only use three laws here for simplicity and brevity.

2.1.3 Deducing Natural Units

With natural units we do much the same thing as was done for the cgs system. We start with three pre-defined quantities and derive the rest. The trick here is that we choose different quantities and define *both* their dimensions and their units in a way that suits our purposes best.

Instead of starting with length, mass, and time, we start with c, $\hbar$, and energy. We then get even trickier. We take both c and $\hbar$ to have numerical values of one. In other words, just as someone once took an arbitrary distance to call a centimeter and gave it a numerical value of one, or an arbitrary interval of time to call a second and gave it a value of one, we now take whatever amount nature gives us for the speed of light and call it one in our new system. We do the same thing for $\hbar$. (This, in fact, is why the system is called *natural,* i.e., because we use *nature's* amounts for these things to use as our basic units of measure and not some amount arbitrarily chosen by us.)

Natural units: $\hbar = c = 1$ and energy defined. Other units derived from laws of nature.

We then get even trickier still. We take c and $\hbar$ to be dimensionless, as well. Since c (or any velocity) is distance divided by time, we find, in developing our new system, that length and time must therefore have the same units.

Note that the founders of the cgs system could have done the same type of thing if they had wanted to. If they had started with velocity as dimensionless they would have derived length and time as having the same dimensions, and we might now be speaking of time as measured in centimeters rather than seconds. Alternatively, they could have first decided instead that time and space would be measured in the same units and then derived velocity as a dimensionless quantity. The only difference in these two alternative approaches would have been in choice of which units were considered fundamental and which were derived. In any event this was not done, not because it was invalid, but because it was simply not convenient.

In particle physics, however, it does become convenient, and so we define $c=1$ and dimensionless. It is also convenient to define $\hbar =1$ dimensionless for similar reasons.

With energy, our third fundamental quantity, we stay more conventional. We give it a dimension (energy), and we give it units of <u>mega-electron-volts</u>, i.e., <u>MeV</u> = 1 million eV. (We know from other work "how much" an electron-volt is just as the devisors of the metric system knew "how much" one second was.) As with everything else, we do this because it will turn out to be advantageous.

Energy in natural units: electron volts (MeV convenient)

Note now what happens with our three fundamental entities defined in this way. From law of nature number two with $c=1$ dimensionless, mass has the same units as energy and the same numerical value as well. So an electron with 0.511 MeV rest energy also has 0.511 MeV rest mass. Because mass and energy are exactly the same thing in natural units, this dimension has come to be referred to commonly as "mass" (i.e., M) rather than "energy" even though the units remain as MeV.

From law of nature number three with $\hbar =1$ dimensionless, the dimensions for ω are M (instead of s^{-1} as in cgs), and hence time has dimension M^{-1} and units of (MeV)$^{-1}$. Similarly, from law number one, length has inverse mass dimensions and inverse MeV units as well. Units and dimensions for all other quantities can be derived from other laws of nature, just as was done in the cgs system.

So, by starting with different fundamental quantities and dimensions, we derive a different (more convenient for particle physics) system of units. Because we started with only one of our three fundamental entities having a dimension, the entire range of quantities we will deal with will be expressible in terms of that one dimension or various powers thereof.

2.1.4 The Hybrid Units System

When doing theoretical work, natural units are the most streamlined, and thus, usually the quickest and easiest. They are certainly the most common. When carrying out experiments or making calculations that relate to the real world, however, it is often necessary to convert to units which can be measured most readily. In particle physics applications, one typically uses *centimeters, seconds, and MeV*. Note this is a <u>hybrid system</u> and is not quite the same as cgs. (Energy is expressed in ergs in cgs.) It is convenient though, since energy in natural units is MeV, and no conversion is needed for it. Converting other quantities is necessary, however, and there is a little trick for doing it.

Hybrid units used in experiments: cm, s, MeV

2.1.5 Converting from One System to Another

To do the conversion trick alluded to above, we first have to note two things: i) in natural units any quantity can be multiplied or divided by c or $\hbar$ any number of times without changing either its numerical value or its dimensions, and ii) a quantity is the same thing, the same total amount, regardless of what system it is expressed in terms of.

To illustrate, suppose we determine a theoretical value for some time interval in natural units to be 10^{16} (MeV)$^{-1}$. What is its measurable value in seconds? To find out, observe that

$t = 10^{16}$ (MeV)$^{-1} \times \hbar = 10^{16}$ (MeV)$^{-1}$ where $\hbar = 1$, and all quantities are in natural units.

But the above relation can be expressed in terms of the hybrid MeV-cm-s system also. The actual amount of time will stay the same, only the units used to express it, and the numerical value it has in those units, will change. So let's simply change $\hbar$ to its value in the hybrid system, $\hbar = 6.58 \times 10^{-22}$ MeV-s. Then,

$t = 10^{16}$ (MeV)$^{-1} \times \hbar = 10^{16}$ (MeV)$^{-1} \times 6.58 \times 10^{-22}$ MeV-s $= 6.58 \times 10^{-6}$ s.

The same time interval is described as either 10^{16} (MeV)$^{-1}$ or 6.58×10^{-6} seconds depending on our system of units.

The <u>moral here</u> is that we can simply multiply or divide any quantity we like (which is expressed in natural units) by c and/or $\hbar$ (expressed in MeV-cm-s units) as many times as is necessary to get the units we know that quantity should have in the MeV-cm-s system.

Multiply natural units by powers of $\hbar$ and/or c to get hybrid units

2.1.6 Mass and Energy in the Hybrid and Natural Systems

As mentioned, the hybrid system is not the same as the cgs system, even though both use centimeters and seconds. In the cgs system, energy is measured in ergs and mass in grams. In the hybrid system, energy is measured in MeV and mass in unfamiliar, and never used, units. (See Wholeness Chart 2-1 below.) It may be confusing, but when experimentalists talk of mass, energy, length, and time, they like to use the hybrid system, *yet* they commonly refer to mass in MeV. For example, in high energy physics, the mass of the electron is commonly referred to as 0.511 MeV, rather than hybrid (unfamiliar) or cgs (gram) mass units. Hopefully, Wholeness Chart 2-1 will help to keep all of this straight.

Mass is MeV in natural units. Commonly expressed the same way even when other system of units used.

Though we have used MeV (1 million eV) for energy in hybrid and natural units throughout this chapter, energy is also commonly expressed in keV (kilo electron volts), GeV (giga electron volts = 1 billion eV), and TeV (tera electron volts = 1 trillion eV). It is, of course, simple to convert any of these to, and from, MeV.

Wholeness Chart 2-1. Conversions between Natural, Hybrid, and cgs Numeric Quantities

Natural Units		Hybrid Units		cgs Units	
$c = \hbar = 1$		$c = 2.99 \times 10^{10}$ cm/s $\hbar = 6.58 \times 10^{-22}$ MeV-s $\hbar c = 1.973 \times 10^{-11}$ MeV-cm		conversion factor $F = 1.602 \times 10^{-6}$ ergs/MeV	
Quantity, **units of (MeV)^M**	**M**	**Multiply ← value** **by ↓ to get →**	**in** **MeV-cm-s**	**Multiply ← value** **by ↓ to get →**	**in** **cgs**
energy	1	1	MeV	F	ergs
mass, m	1	$1/c^2$	MeV-s^2/cm^2	F	erg-s^2/cm^2 = gs
length	-1	$\hbar c$	cm	1	cm
time	-1	$\hbar$	s	1	s
velocity	0	c	cm/s	1	cm/s
acceleration, **a**	1	$c/\hbar$	cm/s^2	1	cm/s^2
force	2	$m\mathbf{a}$ factors = $1/c\hbar$	MeV/cm	F	ergs/cm = dynes
$\hbar\ (=1)$	0	$\hbar$	MeV-s	F	erg-s
Hamiltonian	1	1	MeV	F	ergs
Hamiltonian density	4	$1/(\hbar c)^3$	MeV/cm^3	F	ergs/cm^3
Lagrangian	1	1	MeV	F	ergs
Lagrangian density	4	$1/(\hbar c)^3$	MeV/cm^3	F	ergs/cm^3
action S	0	$\hbar$	MeV-s	F	erg-s
fine structure constant	0	1	unitless	1	unitless
cross section	-2	$(\hbar c)^2$	cm^2	1	cm^2

2.1.7 Summary of Natural, Hybrid, and cgs Units

To summarize the three systems of units we have discussed.

cgs: cm,s,g fundamental, other quantities derived from laws of nature

hybrid: cm,s,MeV fundamental, other quantities derived from laws of nature

natural: $c,\hbar$,MeV fundamental (c and $\hbar$ unitless and unit magnitude; 1 MeV = an amount we know from other work), other quantities derived from laws of nature

Conversion of algebraic relations

cgs or hybrid to natural: Put $c = \hbar = 1$. e.g., $E = mc^2 \rightarrow m$; $p_x = \hbar k_x \rightarrow k_x$.

natural to cgs or hybrid: Easiest just to remember, or look up, relations. e.g., $E = m \rightarrow mc^2$.
Can instead insert factors of c and $\hbar$ needed on each side to balance units. e.g.,
E(energy units) $= m$(energy-s^2/cm^2 units) $\times$?, where ? must be c^2.

Conversion of numeric quantities

natural to hybrid to cgs: go from left to right in Wholeness Chart 2-1.

cgs to hybrid to natural: go from right to left, dividing rather than multiplying.

Note in the chart, that the Lagrangian and Hamiltonian densities in cgs have energy/(length)³ dimensions. In natural units these become (energy)⁴ or (mass)⁴. The action is the integral of the Lagrangian density over space and time. In cgs this is energy-time; in natural units it is M^0.

2.1.8 QFT Approach to Units

QFT starts with familiar relations for quantities from the cgs system, e.g., $p_x = \hbar k_x$, and then expresses them in terms of natural units, e.g, $p_x = k_x$.. The theory is then derived, and predictions for scattering and decay interactions made, in terms of natural units. Finally, before comparing these predictions to experiment, they are converted to the hybrid system, which is the system experimentalists use for measurement.

In <u>summary</u>:

relations in cgs $\rightarrow$ same relations in natural units $\rightarrow$ develop theory in natural units $\rightarrow$
predict experiment in natural units $\rightarrow$ same predictions in hybrid (MeV-cm-s) units.

How QFT uses different systems of units

The first arrow above is easy. Just set $c = \hbar = 1$. For the last arrow, use Wholeness Chart 2-1. All of the other arrows are what the remainder of this book is all about.

You may wonder if this conversion to natural units is really all that worthwhile, as its primary value seems to be in saving the extra effort of writing out c and $\hbar$ in all our equations (which do occur with monotonous regularity.) You may have a point on that. More importantly, the essential mathematical structure of the resulting equations, and the fundamentals of the underlying physics, is more clearly seen without the clutter of relatively unimportant unit scaling factors.

Regardless, natural units are what everyone working in QFT uses, so you should resign yourself to getting used to them

2.2 Notation

We shall use a notation defining <u>contravariant components</u> x^μ of the 4D position vector as 3D Cartesian coordinates X_i plus ct (see Appendix A if you are not comfortable with this), i.e.,

Contravariant 4D position components for us = 3D Cartesian coordinates plus time

$$x^\mu = \begin{bmatrix} x^0 \\ x^1 \\ x^2 \\ x^3 \end{bmatrix} = \begin{bmatrix} ct \\ X_1 \\ X_2 \\ X_3 \end{bmatrix} = [ct, X_i]^T , \quad \mu = 0,1,2,3 \quad i = 1,2,3 \quad c = 1 \text{ in natural units .} \quad (2\text{-}1)$$

Contravariant components, and their siblings described below, are essential to relativity theory, and QFT is grounded in special relativity. To avoid confusion, whenever we want to raise a component to a power, we will use parenthesis, e.g., the contravariant z component of the position vector squared is $(x^3)^2$. From henceforth, we will use natural units, and not write c.

From special relativity, we know the differential proper time passed on an object (with $c=1$) is

$$(d\tau)^2 = (dt)^2 - dX_1 dX_1 - dX_2 dX_2 - dX_3 dX_3 . \quad (2\text{-}2)$$

If we define <u>covariant components</u> of the 4D position vector as

$$x_\mu = \begin{bmatrix} x_0 \\ x_1 \\ x_2 \\ x_3 \end{bmatrix} = \begin{bmatrix} t \\ -X_1 \\ -X_2 \\ -X_3 \end{bmatrix} = [t, -X_i]^T , \quad (2\text{-}3)$$

Covariant components have negative 3D Cartesian coordinates

then (2-2) becomes

$$(d\tau)^2 = dx^0 dx_0 + dx^1 dx_1 + dx^2 dx_2 + dx^3 dx_3 = \underbrace{dx^\mu dx_\mu}_{\substack{\text{summation} \\ \text{convention}}} , \quad (2\text{-}4)$$

where on the RHS, we have introduce the shorthand <u>Einstein summation convention</u>, in which repeated indices are summed, and which we will use throughout the book. If we do not wish to sum when repeated indices appear, we will underline the indices, e.g., $dx^{\underline{\mu}} dx_{\underline{\mu}}$ means no summation.

Repeated indices means summation

We can obtain (2-3) by means of a matrix operation on (2-1), i.e.,

$$x_\mu = g_{\mu\nu} x^\nu = \begin{bmatrix} 1 & 0 & 0 & 0 \\ 0 & -1 & 0 & 0 \\ 0 & 0 & -1 & 0 \\ 0 & 0 & 0 & -1 \end{bmatrix} \begin{bmatrix} x^0 \\ x^1 \\ x^2 \\ x^3 \end{bmatrix} = \begin{bmatrix} 1 & 0 & 0 & 0 \\ 0 & -1 & 0 & 0 \\ 0 & 0 & -1 & 0 \\ 0 & 0 & 0 & -1 \end{bmatrix} \begin{bmatrix} t \\ X_1 \\ X_2 \\ X_3 \end{bmatrix},$$
$$\underbrace{}_{g_{\mu\nu}}$$

(2-5)

Getting covariant components from contravariant ones

where the matrix $g_{\mu\nu}$ is known as the <u>metric tensor</u>. Its <u>inverse</u>, $g^{\mu\nu}$, has the exact same form,

Contravariant and covariant forms of the metric tensor

$$\delta_\alpha{}^\nu = g_{\alpha\mu} g^{\mu\nu} = \begin{bmatrix} 1 & 0 & 0 & 0 \\ 0 & -1 & 0 & 0 \\ 0 & 0 & -1 & 0 \\ 0 & 0 & 0 & -1 \end{bmatrix} \begin{bmatrix} 1 & 0 & 0 & 0 \\ 0 & -1 & 0 & 0 \\ 0 & 0 & -1 & 0 \\ 0 & 0 & 0 & -1 \end{bmatrix}.$$
$$\underbrace{}_{g^{\mu\nu}}$$

(2-6)

where $\delta_\alpha{}^\nu$ is the <u>Kronecker delta</u> (= 0 if row $\alpha \neq$ column ν; = 1 if $\alpha = \nu$) . With the metric tensor and its inverse, we can re-write (2-4) as

$$(d\tau)^2 = g_{\mu\nu} dx^\mu dx^\nu = g^{\mu\nu} dx_\mu dx_\nu \,.$$

(2-7)

Partial derivatives with respect to x^μ and x_μ, often designated by $\partial_\mu \phi = \phi_{,\mu}$ and $\partial^\mu \phi = \phi^{,\mu}$, are

Contravariant and covariant derivatives

$$\partial_\mu = \frac{\partial}{\partial x^\mu} = \left(\frac{\partial}{\partial t}, \frac{\partial}{\partial x^i} \right)^T = \left(\frac{\partial}{\partial t}, \frac{\partial}{\partial X_i} \right)^T \quad \text{and} \quad \partial^\mu = \frac{\partial}{\partial x_\mu} = \left(\frac{\partial}{\partial t}, \frac{\partial}{\partial x_i} \right)^T = \left(\frac{\partial}{\partial t}, -\frac{\partial}{\partial X_i} \right)^T. \quad (2\text{-}8)$$

Note the spatial parts of x^μ and ∂^μ have opposite signs.

Raising and lowering indices

In general (see Prob. 4), we can raise or lower indices of any 4D vector w^μ using the (covariant) metric tensor and its inverse, the contravariant metric tensor, via $w^\mu = g^{\mu\nu} w_\nu$ and $w_\mu = g_{\mu\nu} w^\nu$.

For a matrix (a tensor using two indices), rather than a column quantity (vector with one index), we can use $g^{\mu\nu}$ to raise (or $g_{\mu\nu}$ to lower) either index, or use $g^{\mu\nu}$ twice for both indices. For example, for the matrix (tensor) $M_{\alpha\beta}$, we would have $M^{\mu\nu} = g^{\mu\alpha} g^{\nu\beta} M_{\alpha\beta}$.

Quantities for a single particle will be written in lower case, e.g., p_μ is the 4-momentum for a particle; for a collection of particles, in upper case, e.g., P_μ is 4-momentum for a collection of particles. Density values will be in script form, e.g., $\mathcal{H}$ for Hamiltonian density.

Script → density

Further, as one repeatedly sums p_μ and x^μ in QFT relations, we will employ the common streamlined notation $p_\mu x^\mu = px$ (the 4D inner product of 4 momentum and 4D position vectors.)

$p_\mu x^\mu = px$

2.3 Classical vs Quantum Plane Waves

As we will be dealing throughout the book with quantum plane waves, the following quick review of them is provided.

Fig. 2-1 illustrates the analogy between classical and quantum waves. Pressure plane waves, for example, can be represented as planes of constant *real* numbers (pressures) propagating through space. Particle wave function plane waves can be represented as planes of constant *complex* numbers (thus, constant phase angle) propagating through space. Theoretically, the planes extend to infinity in the y and z directions. The lower parts of Fig. 2-1 plot the numerical values of the waves on each plane vs. spatial position at a given instant of time. The complex wave has two components to plot; the real wave, only one. Plane wave packets for both pressure and wave function waves can be built up by superposition of many pure sinusoids, like those shown. (Though, as we will see, QFT rarely has need for wave packets.)

Real vs. complex (quantum) plane waves

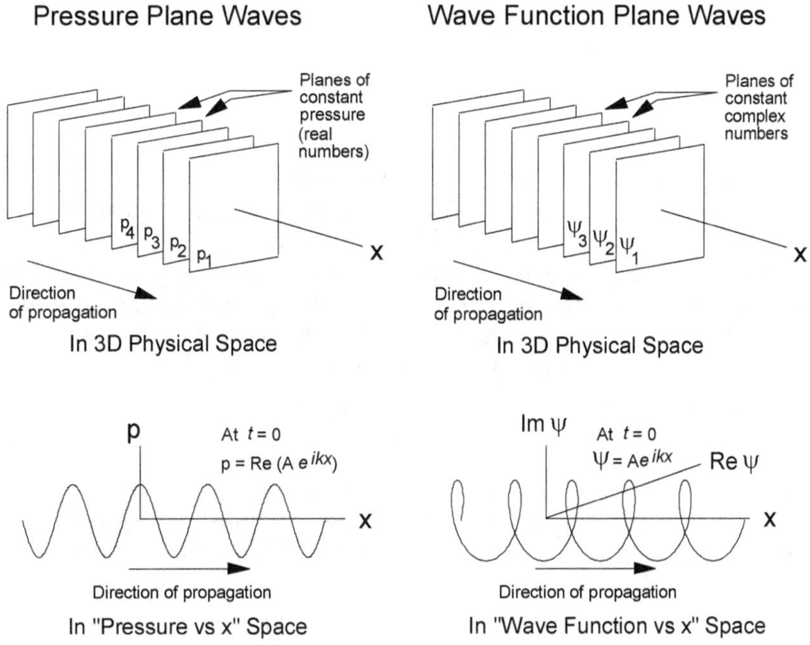

Pressure Plane Waves   Wave Function Plane Waves

Figure 2-1. Classical vs Quantum Plane Waves

2.4 Review of Variational Methods

2.4.1 Classical Particle Theory

Recall, from classical mechanics, that, given the <u>Lagrangian L</u> for a particle, which is the kinetic energy minus the potential energy,

$$L = T - V = \sum_{i=1}^{3} \tfrac{1}{2} m \left(\dot{x}^i \right)^2 - V \left(x^1, x^2, x^3 \right) = \frac{\mathbf{p}^2}{2m} - V \ , \qquad (2\text{-}9)$$

Definition of classical mechanics Lagrangian L

we can find the 3D equations of motion for the particle by the <u>Euler-Lagrange equation</u>, i.e.,

$$\frac{d}{dt} \left(\frac{\partial L}{\partial \dot{x}^i} \right) - \frac{\partial L}{\partial x^i} = 0 \ . \qquad (2\text{-}10)$$

Governing equation = Euler-Lagrange equation

This, with (2-9), readily reduces to Newton's 2nd law (with conservative force), $F^i = -\partial V / \partial x^i = m \ddot{x}^i$.

For a system of particles, we need only add an extra kinetic and potential energy term to (2-9) for each additional particle. For relativistic particles, we merely need to use relativistic kinetic and potential energy terms in (2-9), instead of Newtonian terms.

Recall also, that given the Lagrangian, we could find the Hamiltonian H, via the <u>Legendre transformation</u> (employing a Cartesian system where $x^i = x_i$ and $p^i = p_i$ [see Prob. 8]),

Legendre transformation $H \leftrightarrow L$

$$H = p_i \dot{x}^i - L, \quad \text{where} \quad p_i = \frac{\partial L}{\partial \dot{x}^i} = m \dot{x}^i \ \left(= p^i \text{ for Cartesian system} \right). \qquad (2\text{-}11)$$

p_i is the conjugate, or canonical, momentum of x^i. (Note that a contravariant component in the denominator is effectively equivalent to a covariant component in the entire entity, and vice versa.)

It is an <u>important point</u> that by knowing any one of H, L, or the equations of motion, we can readily deduce the other two using (2-9) through (2-11). That is, each completely describes the particle(s) and its (their) motion.

L, H, and equations of motion all tell us the same thing

<u>Equivalent entities</u>

Lagrangian $L \ \leftrightarrow \ $ equations of motion $\ \leftrightarrow \ $ Hamiltonian H

Hence, when we defined first quantization in Chap. 1 as i) keeping the classical Hamiltonian and ii) changing Poisson brackets to commutators, we could just as readily have used the Lagrangian L or the equations of motion [for $x^i(t)$] for i) instead. (Note that Poisson brackets are discussed on pg. 24 and summarized in Wholeness Chart 2-2 on pgs. 20 and 21.)

2.4.2 Pure Mathematics

We can apply the mathematical structure of the prior section to any kind of system, even some having nothing to do with physics. That is, if any system has a differential equation of motion (for example, an economic model), then one can find the Lagrangian for that system, as well as the Hamiltonian, the conjugate momentum, and more. So the mathematics derived for classical particles can be extrapolated and used to advantage in many other areas. Of course, one must then be careful in interpretation of the Hamiltonian, and similar quantities. The Hamiltonian, for example, will not, in general, represent energy, though many behavioral analogies (like conservation of H, etc.) will exist that can greatly aid in analyses of these other systems.

Variational math can be applied to many diverse areas in physics and elsewhere

2.4.3 Classical Field Theory

Classical field theory is analogous in many ways to classical particle theory. Instead of the Lagrangian L, we have the Lagrangian density $\mathcal{L}$. Instead of time t as an independent variable, we have $x^\mu = x^0, x^1, x^2, x^3 = t, x^i$ as independent variables. Instead of a particle described by $x^i(t)$, we have a field value described by $\phi(x^\mu)$ [or $\phi^r(x^\mu)$, where r designates different field types, or possibly, different spatial components of the same vector field (like **E** or **B** in electromagnetism).]

$$\text{Particle Theory} \rightarrow \text{Field Theory}$$
$$L, H, etc \rightarrow \mathcal{L}, \mathcal{H}, \text{ etc.} \qquad t \rightarrow x^\mu \qquad x^i(t) \rightarrow \phi^r(x^\mu)$$

Analogous entities in particle and field theories

From these correspondences in variables, we can intuit the analogous forms of (2-9) through (2-11) [though we will derive the Euler-Lagrange equation afterwards] for fields. Thus, the Lagrangian density, in terms of kinetic energy density and potential energy densities of the field, is

$$\mathcal{L} = \mathcal{T} - \mathcal{V} . \tag{2-12}$$

(Digressing here into the expressions for $\mathcal{T}$ and $\mathcal{V}$ in terms of the classical field ϕ would divert us away from our main purpose. In the next chapter we will see the form of these for a quantum field.)

Intuitive deduction of field relations from particle ones

The <u>Euler-Lagrange equation for fields</u> becomes

$$\boxed{\frac{\partial}{\partial x^\mu}\left(\frac{\partial \mathcal{L}}{\partial \phi^r_{,\mu}}\right) - \frac{\partial \mathcal{L}}{\partial \phi^r} = 0} . \tag{2-13}$$

The <u>Legendre transformation for the Hamiltonian density</u>, with π_r being the conjugate momentum density of the field ϕ^r, is

$$\boxed{\mathcal{H} = \pi_r \dot{\phi}^r - \mathcal{L} , \quad \text{where} \quad \pi_r = \frac{\partial \mathcal{L}}{\partial \dot{\phi}^r}} . \tag{2-14}$$

To see a real world example using (2-13), work through Prob. 6.

Compare (2-12) through (2-14) to (2-9) through (2-11), and note, that similar to particle theory, if we know any one of $\mathcal{L}$, $\mathcal{H}$ or the equations of motion, we can readily find the other two. That is, they are equivalent, and in our first assumption of second quantization (see Chap. 1), we could take any one of the three (not just $\mathcal{H}$ as we did in Chap. 1) as having the same form in quantum field theory as it did in classical field theory.

$\mathcal{L}$, $\mathcal{H}$, and eqs of motion all tell us the same thing

<u>Derivation of Euler-Lagrange Equation for Fields</u>

The fundamental assumption behind (2-13) is that the action of the field over an arbitrary 4D region Ω,

$$S = \int_T \underbrace{\int_V \mathcal{L}\left(\phi, \phi_{,\mu}\right) d^3\mathbf{x}}_{L} \, dt = \int_\Omega \mathcal{L}\left(\phi, \phi_{,\mu}\right) d^4x , \tag{2-15}$$

Formal derivation of Euler-Lagrange equation for fields

where $d^4x = d^3\mathbf{x}\,dt$ is an element of 4D volume, is stationary. More precisely, consider a virtual variation in ϕ of

$$\phi\left(x^\mu\right) \rightarrow \phi\left(x^\mu\right) + \delta\phi\left(x^\mu\right), \tag{2-16}$$

where the variation vanishes on the surface $\Gamma(\Omega)$ bounding the region Ω, i.e., $\delta\phi = 0$ on Γ. The "surface" here is actually three dimensional (rather than 2D), because it bounds a 4D region. This restriction on $\delta\phi$ is reasonable for a region Ω large enough so the field ϕ vanishes at its boundary.

For S to be stationary under the variation, we must have

$$\delta S = 0 . \tag{2-17}$$

Using (2-17) in (2-15), we have

$$\delta S = \int_\Omega \left\{ \frac{\partial \mathcal{L}}{\partial \phi} \delta\phi + \frac{\partial \mathcal{L}}{\partial \phi_{,\mu}} \delta\phi_{,\mu} \right\} d^4 x = \int_\Omega \left\{ \frac{\partial \mathcal{L}}{\partial \phi} \delta\phi + \underbrace{\frac{\partial \mathcal{L}}{\partial \phi_{,\mu}} \frac{\partial}{\partial x^\mu} \delta\phi}_{\text{term } Z} \right\} d^4 x . \tag{2-18}$$

With the last term on the RHS of (2-18), which we label "Z" here, re-written using

$$\frac{\partial}{\partial x^\mu} \left(\frac{\partial \mathcal{L}}{\partial \phi_{,\mu}} \delta\phi \right) = \left(\frac{\partial}{\partial x^\mu} \frac{\partial \mathcal{L}}{\partial \phi_{,\mu}} \right) \delta\phi + \underbrace{\frac{\partial \mathcal{L}}{\partial \phi_{,\mu}} \frac{\partial}{\partial x^\mu} \delta\phi}_{\text{term } Z} , \tag{2-19}$$

we can express (2-18) as

$$\delta S = \int_\Omega \left\{ \frac{\partial \mathcal{L}}{\partial \phi} \delta\phi - \left(\frac{\partial}{\partial x^\mu} \frac{\partial \mathcal{L}}{\partial \phi_{,\mu}} \right) \delta\phi \right\} d^4 x + \underbrace{\int_\Omega \frac{\partial}{\partial x^\mu} \left(\frac{\partial \mathcal{L}}{\partial \phi_{,\mu}} \delta\phi \right) d^4 x}_{= \int_\Gamma n_\mu \left(\frac{\partial \mathcal{L}}{\partial \phi_{,\mu}} \delta\phi \right) d^3 x} . \tag{2-20}$$

The last term in the above relation can, via the 4D version of Gauss's divergence theorem, be converted into an integral over the 3D "surface" Γ, as we show under the downward pointing bracket. In that integral, n_μ is the unit length 4D vector normal to the 3D surface Γ at every point on the surface, and it forms an inner product with the quantity in brackets by virtue of the summation over μ. Since we stipulated at the outset that $\delta\phi = 0$ on this surface, the last term in (2-20) must equal zero.

From (2-17), the first integral in (2-20),

$$\int_\Omega \underbrace{\left\{ \frac{\partial \mathcal{L}}{\partial \phi} - \left(\frac{\partial}{\partial x^\mu} \frac{\partial \mathcal{L}}{\partial \phi_{,\mu}} \right) \right\}}_{\text{must } = 0} \delta\phi \, d^4 x = 0 , \tag{2-21}$$

for any possible variation of ϕ, i.e., for any possible $\delta\phi$ everywhere within Ω. The only way this can happen is if the quantity inside the brackets equals zero. But this is just (2-13) for one field. A similar derivation can be made for each additional type of field, i.e., for different values of r in (2-13), and thus, we have proven (2-13).

<u>End of derivation</u>

2.4.4 Real vs. Complex Fields

In classical theory we typically deal with real fields, such as the displacement at every point in a solid or fluid, or the value of the **E** field in electrostatics. However, given our experience in NRQM, where complex wave functions were everywhere, so will we find that in QFT, quantum fields are commonly complex. Nothing in the above limited our derivation to real fields, so all of the relationships in this Sect. 2.4 are valid for complex fields, as well.

Classical field real; quantum fields usually complex

2.5 Classical Mechanics: An Overview

Wholeness Chart 2-2 is a summary of the key relations in all of classical physical theory (from the variational viewpoint.) The chart is intended <u>primarily as an overview</u> of past courses and as a lead in to quantum field theory, so a <u>detailed study of it is not really warranted at this time</u>. We have

Variational classical mechanics overview in Wholeness Chart 2-2

Wholeness Chart 2-2.

	Mathematically	**Non-relativistic Particle**
Independent variable(s)	t	t
Coordinates	$q_i = q_i(t)$, $i = 1,..,n$ (generalized)	$x^i = x^i(t)$, $i = 1, 2, 3$ (contravariant)
Lagrangian density	see Fields columns	not applicable for particle
Lagrangian	$L = L(q_i, \dot{q}_i, t)$	$L = L\left(x^i, \dot{x}^i, t\right) = \sum_i \tfrac{1}{2} m \left(\dot{x}^i\right)^2 - V\left(x^i, t\right)$
Action	$S = \int L\, dt$	as at left
Euler- Lagrange equation (From $\delta S = 0$.)	$\dfrac{d}{dt}\left(\dfrac{\partial L}{\partial \dot{q}_i}\right) - \dfrac{\partial L}{\partial q_i} = 0$	$\dfrac{d}{dt}\left(\dfrac{\partial L}{\partial \dot{x}^i}\right) - \dfrac{\partial L}{\partial x^i} = 0$
Equations of motion for chosen coordinates	use explicit form for L in Euler-Lagrange equation	$m\ddot{x}^i = -\dfrac{\partial V}{\partial x^i}$ usually V not function of t
Conjugate momentum density; total	see Fields columns; $p_i = \dfrac{\partial L}{\partial \dot{q}_i}$	n/a ; $p_i = \dfrac{\partial L}{\partial \dot{x}^i} = m\dot{x}^i \left(= p^i \text{ for Cartesian}\right)$
Physical momentum density; total	not relevant, purely math	n/a ; same as conjugate momentum
Alternative formulation	q_i, p_i and $L = L(q_i, p_i, t)$	x^i, p_i and $L = p^2/2m - V(x^i, t)$
Hamiltonian density; total	see Fields; $H = p_i \dot{q}_i - L$ (pure math)	n/a ; $H = p_i \dot{x}^i - L = p^2/2m + V$
Hamilton's Equations of Motion for conjugate variables	$\dot{p}_i = -\dfrac{\partial H}{\partial q_i} \qquad \dot{q}_i = \dfrac{\partial H}{\partial p_i}$	$\dot{p}_i = -\dfrac{\partial H}{\partial x^i} = -\dfrac{\partial V}{\partial x^i} \qquad \dot{x}^i = \dfrac{\partial H}{\partial p_i}$
Poisson Brackets, definition	for $u = u(q_i, p_i, t)$, $v = v(q_i, p_i, t)$ $\{u,v\} = \dfrac{\partial u}{\partial q_i}\dfrac{\partial v}{\partial p_i} - \dfrac{\partial u}{\partial p_i}\dfrac{\partial v}{\partial q_i}$	for $u = u(x^i, p_i, t)$, $v = v(x^i, p_i, t)$ $\{u,v\} = \dfrac{\partial u}{\partial x^i}\dfrac{\partial v}{\partial p_i} - \dfrac{\partial u}{\partial p_i}\dfrac{\partial v}{\partial x^i}$
Equations of motion in terms of Poisson brackets i) any variable ii) conjugate variables	i) for $v = H$ $\dfrac{du}{dt} = \{u, H\} + \dfrac{\partial u}{\partial t}$ ii) for i) plus $u = q_i$ or p_i $\dot{p}_i = \{p_i, H\} = -\dfrac{\partial H}{\partial q_i}$; $\dot{q}_i = \{q_i, H\} = \dfrac{\partial H}{\partial p_i}$	i) for $v = H$ $\dfrac{du}{dt} = \{u, H\} + \dfrac{\partial u}{\partial t}$ ii) for i) plus $u = x^i$ or p_i $\dot{p}_i = \{p_i, H\} = -\dfrac{\partial H}{\partial x^i}$; $\dot{x}^i = \{x^i, H\} = \dfrac{\partial H}{\partial p_i}$
Poisson Brackets for conjugate variables	$\{q_i, p_j\} = \delta_{ij}$ $\{q_i, q_j\} = \{p_i, p_j\} = 0$	$\{x^i, p_j\} = \delta^i{}_j$ $\{x^i, x^j\} = \{p_i, p_j\} = 0$

Summary of Classical (Variational) Mechanics

Non-relativistic Fields	Relativistic Particle	Relativistic Fields
$x^i, t \quad i = 1, 2, 3$	t	$x^\mu \quad \mu = 0, 1, 2, 3$
$\phi^r(x^i, t) \quad r = \text{field type} = 1, \dots, n$	$x^i = x^i(t), \quad i = 1, 2, 3$	$\phi^r(x^\mu) \quad r = \text{field type} = 1, \dots, n$
$\mathcal{L} = \mathcal{L}\left(\phi^r, \dot{\phi}^r, \partial_i \phi^r, x^i, t\right)$	not applicable for particle	$\mathcal{L} = \mathcal{L}\left(\phi^r, \partial_\mu \phi^r, x^\mu\right)$
$L = \int \mathcal{L} d^3x$	$L\left(x^i, v^i, t\right) = -m\sqrt{1-v^2} - V$	$L = \int \mathcal{L} d^3x$
$S = \int L dt = \int \mathcal{L} d^3x\, dt$	$S = \int L dt$	$S = \int L dt = \int \mathcal{L} d^3x\, dt$
$\dfrac{d}{dt}\left(\dfrac{\partial \mathcal{L}}{\partial \dot{\phi}^r}\right) + \dfrac{d}{dx^i}\left(\dfrac{\partial \mathcal{L}}{\partial \phi^r{}_{,i}}\right) - \dfrac{\partial \mathcal{L}}{\partial \phi^r} = 0$	$\dfrac{d}{dt}\left(\dfrac{\partial L}{\partial v^i}\right) - \dfrac{\partial L}{\partial x^i} = 0$	$\dfrac{\partial}{\partial x^\mu}\left(\dfrac{\partial \mathcal{L}}{\partial \phi^r{}_{,\mu}}\right) - \dfrac{\partial \mathcal{L}}{\partial \phi^r} = 0$
$\mathcal{L}$ above in Euler-Lagrange equation	$\dfrac{d}{dt}\left(\dfrac{\partial L}{\partial v^i}\right) = -\dfrac{\partial V}{\partial x^i}; \quad V\left(x^i, v^i\right)$	$\mathcal{L}$ above in Euler-Lagrange equation
$\pi_r = \dfrac{\partial \mathcal{L}}{\partial \dot{\phi}^r}; \quad \Pi_r = \int \pi_r d^3x$	n/a ; $p^i = \dfrac{\partial L}{\partial v^i} = \dfrac{mv^i}{\sqrt{1-v^2}} - \dfrac{dV}{dv^i}$	$\pi_r = \dfrac{\partial \mathcal{L}}{\partial \dot{\phi}^r}; \quad \Pi_r = \int \pi_r d^3x$
$\not{p}_i = \pi_r \dfrac{\partial \phi^r}{\partial x^i}; \quad p_i = \int \not{p}_i d^3x$	n/a ; = conjugate momentum	$\not{p}_i = \pi_r \dfrac{\partial \phi^r}{\partial x^i}; \quad p_i = \int \not{p}_i d^3x$
$\mathcal{L} = \mathcal{L}\left(\phi^r, \pi_r, \partial_i \phi^r, x^i, t\right)$	$L = L(x^i, p^i, t)$	$\mathcal{L} = \mathcal{L}\left(\phi^r, \pi_r, \partial_i \phi^r, x^i, t\right)$
$\mathcal{H} = \pi_r \dot{\phi}^r - \mathcal{L}; \quad H = \int \mathcal{H} d^3x$	n/a ; $H = p^i v^i - L = T + V$	$\mathcal{H} = \pi_r \dot{\phi}^r - \mathcal{L}; \quad H = \int \mathcal{H} d^3x$
same form as Relativistic Fields	$\dot{p}^i = -\dfrac{\partial H}{\partial x^i} = -\dfrac{\partial V}{\partial x^i} \quad \dot{x}^i = \dfrac{\partial H}{\partial p^i}$	$\dot{\pi}_r = -\dfrac{\delta \mathcal{H}}{\delta \phi^r} \qquad \dot{\phi}^r = \dfrac{\delta \mathcal{H}}{\delta \pi_r}$ where $\dfrac{\delta}{\delta \phi^r} = \dfrac{\partial}{\partial \phi^r} - \dfrac{\partial}{\partial x^i}\left(\dfrac{\partial}{\partial \phi^r{}_{,i}}\right)$
same form as Relativistic Fields	same form as Non-relativistic Particle, but different meaning for p^i	for $u = u(\phi^r, \pi_r, \partial_i \phi^r, t)$, $v = v(\phi^r, \pi_r, \partial_i \phi^r, t)$ $\{u, v\} = \left(\dfrac{\delta u}{\delta \phi^r}\dfrac{\delta v}{\delta \pi_r} - \dfrac{\delta u}{\delta \pi_r}\dfrac{\delta v}{\delta \phi^r}\right)\delta(\mathbf{x} - \mathbf{y})$
same form as Relativistic Fields	same form as Non-relativistic Particle	i) for $U = \int u\, dV$; $\{U, H\} = \iint \{u, \mathcal{H}\} d^3y\, d^3\mathbf{x}$ $\dot{U} = \dfrac{dU}{dt} = \{U, H\} + \dfrac{\partial U}{\partial t}$ ii) for $u = \pi_r$; $\dot{\Pi}_r = \{\Pi_r, H\}$
same form as Relativistic Fields	same form as Non-relativistic Particle	$\{\phi^r, \pi_s\} = \delta^r{}_s \delta(\mathbf{x} - \mathbf{y}); \{\phi^r, \phi^s\} = \{\pi_r, \pi_s\} = 0$

other fish to fry. I did say in the preface that we would focus on the essentials, and this chart is provided solely as i) a reference (which may aid some readers in studying for graduate oral exams), and ii) a lead in to technical details regarding Poisson brackets and second quantization.

The full theory behind Wholeness Chart 2-2 can be found in Goldstein (see Preface). The most important points regarding field theory, as represented in the chart, and which we will need to understand, are listed below.

Note that, due to subtleties in the theory, non-relativistic chart relationships are most easily, and best considered at this point, expressed in Cartesian coordinates, where $x^i \to X_i$ and $p_i = p^i$.

2.5.1 Key Concepts in Field Theory

1. Generalized coordinates do not have to be independent of each other, and the Lagrangian L can have second and/or higher coordinate derivatives. However, in most cases, including those of Wholeness Chart 2-2, the coordinates are independent and L only contains first derivatives.

   For us: q^i are independent of each other and only 1st derivatives in L, $\mathcal{L}$

2. The $x^i(t)$ for particles are not quite the same thing as the x^i for fields. The former are not independent variables, but functions of time t that represent the particle position at any given t.

   The latter are independent variables, and not functions of time, but fixed locations in space upon which the value for the field (and other things like energy density) depends. The field and related density type quantity values also depend on the other independent variable, time.

   $x^i(t)$ for particles; x^i independent of time for fields

3. Different values for the r label for fields can represent
   i) completely different fields, as well as
   ii) different components in spacetime of the same vector field.

   r label = different field types or different components of field

4. In general, the Hamiltonian does not have to represent energy, and can be simply a quantity which obeys all of the mathematical relations shown in the chart. However, in the application of analytical mechanics, it proves immensely useful if the Hamiltonian is, in fact, energy (or an energy operator.) Similarly, in general, the Lagrangian does not have to equal kinetic energy minus potential energy (i.e., T - V), and can simply be a quantity which gives rise via the Lagrange equation to the correct equation(s) of motion (called <u>field equations</u> for fields.)

   Fortunately, in field theory, the Lagrangian density can be represented as kinetic energy density minus potential energy density, and the Hamiltonian density turns out to be total energy density. These correspondences carry over to quantum field theory.

   In our work, always $L = T - V$; $H = T + V$

5. For fields,

$$\frac{\partial \phi}{\partial t} = \frac{d\phi}{dt} = \dot{\phi} \qquad (2\text{-}22)$$

This is generally not true for other quantities. For an explanation of this, see Box 2-1.

For fields, partial and total time derivatives are the same thing

---

Box 2-1. Time Derivatives and Fields

Any field, say ϕ, is a function of space and time, i.e., $\phi = \phi(x^i,t)$, where x^i is an independent variable representing a coordinate (non-moving) point in space upon which field quantities depend.

Note that the total time derivative is

$$\frac{d\phi}{dt} = \frac{\partial \phi}{\partial x^i}\frac{dx^i}{dt} + \frac{\partial \phi}{\partial t}\frac{dt}{dt}$$

But since x^i is an independent variable like time, and hence is not a function of time, its time derivative above is zero. Thus,

$$\frac{d\phi}{dt} = \frac{\partial \phi}{\partial t} = \dot{\phi}$$

So the partial time derivative and the total time derivative of a field are one and the same thing, and both are designated with a dot over the field.

Note that quantities other than fields do not, in general, have this property. (See the Poisson bracket blocks in the fields section of Wholeness Chart 2-2.) It is necessary, therefore, when talking about time derivatives of quantities other than the fields themselves, to specify precisely whether we mean the total or partial derivative with respect to time.

The conclusions reached here apply in both the relativistic and non-relativistic field cases.

6. There are two kinds of momenta, conjugate and physical. In some cases, these are the same, but in general they are not. For fields, each of these can be either total momentum or momentum density. Box 2-2 derives the relations between conjugate and physical momentum densities.

2 kinds of momenta. Each kind can be total or density

7. Key difference between the particle and field approaches.

For a single particle, particle position coordinates are the generalized coordinates and particle momentum components are its conjugate momenta. For fields, each field is itself a generalized coordinate and each field has its own conjugate momentum (density). As noted, this field conjugate momentum (density) is different from the physical momentum (density) that the field possesses.

Generalized coords Particle: x^i

Field: ϕ^r

Box 2-2. Conjugate and Physical Momentum Densities

The relationship between physical momentum density and conjugate momentum density for fields is not so intuitive. It can be derived by assuming our physical 3-momentum density $\not{p}^i$ obeys the classical field variational relation of the RHS of (B2-2.1). (This can be intuited from (2-11), except that there we used a Cartesian system where $p_i = p^i$, and here we use the relativistic Minkowski metric system, where $p_i = -p^i$.) If we divide the particle relation by volume, we get a density relation.

$$p_i = \frac{\partial L}{\partial \dot{x}^i} \quad \xrightarrow[\text{divide by particle volume}]{\text{for small particle in medium,}} \quad \not{p}_i = \frac{\partial \mathcal{L}}{\partial \dot{x}^i} \, . \qquad \text{(B2-2.1)}$$

For continuous media like a fluid, $\dot{x}^i$ is the velocity of the medium (field) at the point where $\not{p}_i$ is measured. We note carefully that our x^i here is the position coordinate of a point fixed relative to the field (fluid particle in our example) and thus is time dependent. (It is different from the same x^i symbol we use in field theory, which is an independent variable that does not depend on time.) Further, the total derivative $\dot{x}^i = dx^i/dt$ equals the partial derivative with respect to time $\partial x^i / \partial t$, since $x^i(t)$ in the present case is only a function of time.

Now take the conjugate momentum density relation for relativistic fields (2-14),

$$\pi_r = \frac{\partial \mathcal{L}}{\partial \dot{\phi}^r} \, , \qquad \text{(B2-2.2)}$$

and divide the RHS of (B2-2.1) by (B2-2.2),

$$\frac{\not{p}_i}{\pi_r} = \frac{\partial \mathcal{L}/\partial \dot{x}^i}{\partial \mathcal{L}/\partial \dot{\phi}^r} = \frac{\partial \dot{\phi}^r}{\partial \dot{x}^i} = \frac{\partial \phi^r / \partial t}{\partial x^i / \partial t} = \frac{\partial \phi^r}{\partial x^i} \quad \rightarrow \quad \not{p}_i = \pi_r \frac{\partial \phi^r}{\partial x^i} \quad \rightarrow \quad \not{p}^i = -\pi_r \frac{\partial \phi^r}{\partial x^i} \, . \qquad \text{(B2-2.3)}$$

The partial derivative of ϕ^r with respect to either of our definitions of x^i (time dependent as the moving position of a point fixed to the field, or time independent as coordinates fixed in space) is the same because by definition, partial derivative means we hold everything else (specifically time here) constant. Thus, the above relation holds in field theory when we consider the x^i as independent variables (coordinates fixed in space).

---

8. Note that it is common in QFT to refer to the field conjugate momentum density as simply the conjugate momentum, the Hamiltonian density as merely the Hamiltonian, and the Lagrangian density as the Lagrangian. This may be unfortunate, but you will learn to live with gleaning the exact sense of these terms from context.

The word "density" often dropped in field theory

9. (See Appendix A if you do not feel comfortable with the material discussed in this paragraph.) The relativistic particle summary, as outlined in Wholeness Chart 2-2, is not, in the strictest sense, formulated covariantly. It describes relativistic behavior, but position and momentum are (non-Lorentz covariant) three vectors, and the Lagrangian and Hamiltonian are not <u>world scalars</u> (world scalars are invariant under Lorentz transformation.) Alternative approaches are possible using proper time for the independent variable and <u>world vector</u> (four vector) quantities for generalized coordinates and conjugate momenta. (Goldstein and Jackson [see Preface] show two different ways to do this.) In those treatments the Lagrangian and Hamiltonian are world scalars though the Hamiltonian does not turn out to be total energy. The approach taken here has been chosen because, in it, we have the advantage of having a Hamiltonian that represents total energy. Further, the parallel between relativistic particles and the usual treatment of relativistic fields becomes much more transparent.

Several ways to formulate variational relativistic theory

10. Some comment is needed on the several different equations of motion that one runs into.

A differential equation of motion is generally an equation that contains derivative(s) with respect to time of some entity, and has as its solution that entity expressed as an explicit function of time (and for fields, space, as well.) For example, $F^i = m\ddot{x}^i$ is the equation of motion for a particle, with $x^i(t)$ as its solution. There are in general two kinds of entities for which we have equations of motion. One is the generalized coordinates themselves. The other is any function of those coordinates, generally expressed as u or v in the next to last row of Wholeness Chart 2-2. (The first class is really a special case of the second, where, for example, u might equal the generalized coordinate itself.)

Eqs of motion exist for i) generalized coordinates, and ii) functions of those coordinates

In Wholeness Chart 2-2, the equations of motion for generalized coordinates are expressed in three different but equivalent ways: the Lagrange equations formulation, the Hamilton's equations formulation, and the Poisson bracket formulation. These are all different expressions for describing the same behavior of the generalized coordinates of a given system via different differential equations. For any particular application, one of these formulations may have some advantage over the others.

The other class of equation of motion for a function of generalized coordinates, say u, can be expressed for the purely mathematical case (the others are analogous) as

$$\frac{du\left(q_i, p_i, t\right)}{dt} = \frac{\partial u}{\partial q_i}\dot{q}_i + \frac{\partial u}{\partial p_i}\dot{p}_i + \frac{\partial u}{\partial t}. \tag{2-23}$$

Using Hamilton's equations for the time derivatives of q_i and p_i yields

$$\frac{du}{dt} = \underbrace{\frac{\partial u}{\partial q_i}\frac{\partial H}{\partial p_i} - \frac{\partial u}{\partial p_i}\frac{\partial H}{\partial q_i}}_{\substack{\text{Possion bracket} \\ \text{definition for } u \text{ and } H}} + \frac{\partial u}{\partial t} = \{u, H\} + \frac{\partial u}{\partial t}, \tag{2-24}$$

Poisson bracket definition used in equation of motion

which is effectively the same equation of motion as (2-23), for the same coordinate u, expressed instead in terms of a Poisson bracket. See the first line of the next to last row block in Wholeness Chart 2-2.

Summary of Forms of Differential Equations of Motion

For generalized coordinates (all three below are equivalent)

1. Lagrangian into Euler-Lagrange equation

2. Hamilton's equations of motion

3. Poisson bracket notation for 2 above

Forms for differential equations of motion

For a function of those generalized coordinates (both below are equivalent)

1. Total time derivative expressed as partial derivatives (see (2-23), not shown in Wholeness Chart 2-2.)

2. Total time derivative expressed in terms of Poisson bracket notation (see (2-24), also shown in Wholeness Chart 2-2.)

11. (See Appendix A Sects. 2.9.3 and 2.9.4, if you do not feel at home with the concepts of this paragraph.) The field equations (equations of motion) for relativistic fields keep the exact same form in any inertial frame of reference[1], i.e., they are Lorentz invariant. Components of four vectors in any of the equations can change from frame to frame, but the relationship between these components expressed in the field equation must remain inviolate. Four vectors transform via the Lorentz transformation of course, and are termed Lorentz covariant. Four scalars (world scalars) are invariant under a Lorentz transformation and look exactly the same to any observer. (e.g., Rest mass m [or simply mass m as it is more commonly called in relativity] of a free

Lorentz invariance (scalars and form of equations) and covariance (vectors and tensors)

---

[1] To be completely accurate, this is true strictly for Einstein synchronization, the synchronization convention of Lorentz transformations. If you are not a relativity expert, please don't worry about this fine point.

particle is a four scalar, where $m^2 = p^\mu p_\mu$. Another observer in a different (primed) frame could measure a different four momentum p'_μ, but would find the same mass via $p'^\mu p'_\mu = m^2$.

Note the result of demanding that the Euler-Lagrange equation (i.e., the field equation) (2-13) be Lorentz invariant. We know that, within that equation, x^μ, ϕ^r, and derivatives of x^μ are Lorentz covariant or invariant. So, in order for the whole equation to be Lorentz invariant, the Lagrangian density $\mathcal{L}$ must be invariant, i.e., a world scalar.

$\mathcal{L}$ is a Lorentz invariant scalar

Since d^4x is also a <u>Lorentz (world or four-) scalar</u> (i.e., four volume is the same in any Lorentz coordinate system, just as 3D volume is the same in any Cartesian system), the action S (see Chart 2-2) must be a Lorentz scalar as well. Note though that the total Lagrangian L is *not* a four scalar since d^3x is not a four scalar. Neither is the Hamiltonian or the Hamiltonian density. To see this, do Prob. 9.

L, H, and $\mathcal{H}$ are not Lorentz scalars

<u>End of Key Concepts in Field Theory points</u>

2.6 Schrödinger vs Heisenberg Pictures

In quantum theory, there are different methods by which one can describe state and operator behavior that all result in the same measurable quantity. That is, the underlying math differs, but the predictions one would make for experimentally measurable dynamical variables remain the same.

These different, but equivalent, ways are called different <u>pictures</u> and apply in the same way to all branches of quantum theory (NRQM, RQM, QFT.) Most QM courses more elementary than this one use what is known as the <u>Schrödinger picture</u>, and that is, no doubt, what you unconsciously thought in terms of, when you did NRQM. We will review that, and then introduce what is called the <u>Heisenberg picture</u>, which helps immensely in QFT with developing theory and doing calculations. Note carefully, before we start, that these terms *do not* refer to the Schrödinger wave approach vs the Heisenberg matrix approach to QM. Everything we do will comprise the wave approach, not the matrix approach, but there are two distinct pictures within that approach, i.e.,

Different pictures in quantum theory

Schrödinger Wave Approach   Heisenberg Matrix Approach
   1. Schrödinger picture
   2. Heisenberg picture.

We will review the Schrödinger picture and develop the Heisenberg picture in terms of NRQM, though the final results will be applicable to any branch of QM, including QFT.

2.6.1 The Schrödinger Picture

In QM, one has i) states (wave functions, particles, kets, state vectors), and ii) operators (such as momentum, the Hamiltonian, and the like), which act on those states. The real world value corresponding to any such operator that one would expect to measure in an experiment, i.e., the average value over many trials, is called the <u>expectation value</u>. The expectation value for any operator is typically designated with a bar over the operator and is found via the statistical relationship (with normalized wave function ψ)

Operator expectation value = "expected" or mean measurement

$$\bar{\mathcal{O}} = \int \psi^\dagger \mathcal{O} \psi \, d^3x = \langle \psi | \mathcal{O} | \psi \rangle. \qquad (2\text{-}25)$$

Calculating expectation value

The time derivative of the expectation value (2-25) (being what we would expect to measure in experiment for the rate of change of the corresponding dynamical variable) is (see Appendix B, Section 2.10.1, if you are concerned about switching total derivatives for partial derivatives below)

$$\frac{d\bar{\mathcal{O}}}{dt} = \frac{d}{dt} \langle \psi | \mathcal{O} | \psi \rangle = \left\langle \frac{\partial \psi}{\partial t} \middle| \mathcal{O} \middle| \psi \right\rangle + \left\langle \psi \middle| \frac{\partial \mathcal{O}}{\partial t} \middle| \psi \right\rangle + \left\langle \psi \middle| \mathcal{O} \middle| \frac{\partial \psi}{\partial t} \right\rangle. \qquad (2\text{-}26)$$

Eq of motion of expectation value

In the Schrödinger picture, the solutions to the Schrödinger equation

$$i \frac{\partial \psi_S}{\partial t} = H \psi_S \quad \text{or} \quad i \frac{\partial}{\partial t} | \psi \rangle_S = H | \psi \rangle_S \qquad (2\text{-}27)$$

In S.P., NRQM eq of motion of state (Schrödinger eq)

are the states ψ_S (or $| \psi \rangle_S$), which are time dependent. The subscript S indicates the Schrödinger picture (S.P.). In that picture, the operators are usually not time dependent. For example, using the familiar momentum operator $p_1{}^S = i\partial /\partial x^1$ for the S.P. in the x^1 direction, with

$$\psi_S = Ae^{-i(Et-\mathbf{p}\cdot\mathbf{x})} = |\psi\rangle_S \qquad A^\dagger A = \frac{1}{V}, \tag{2-28}$$

An example

(2-25) is

$$\overline{p}_1 = \int A^\dagger e^{i(Et-p^ix^i)}\left(i\frac{\partial}{\partial x^1}\right)Ae^{-i(Et-p^ix^i)}d^3x = \int A^\dagger e^{i(Et+p_ix^i)}\left(i\frac{\partial}{\partial x^1}\right)Ae^{-i(Et+p_ix^i)}d^3x = {}_S\langle\psi|p_1^S|\psi\rangle_S, \tag{2-29}$$

where the state is time dependent, but the operator p_1^S is not. That is, since the latter has no t in it,

$$\frac{dp_1^S}{dt} = \frac{\partial p_1^S}{\partial t} = 0. \tag{2-30}$$

In S.P., NRQM eq of motion of momentum operator (i.e., p_1^S constant in time)

Equation (2-26) for p_1 is then

$$\frac{d\overline{p}_1}{dt} = \frac{d}{dt}\langle\psi|p_1^S|\psi\rangle = {}_S\left\langle\frac{\partial\psi}{\partial t}\middle|p_1^S\middle|\psi\right\rangle_S + {}_S\left\langle\psi\middle|\frac{\partial p_1^S}{\partial t}\middle|\psi\right\rangle_S + {}_S\left\langle\psi\middle|p_1^S\middle|\frac{\partial\psi}{\partial t}\right\rangle_S, \tag{2-31}$$

where we leave in the zero quantity of (2-30), because we will want to generalize this result to all operators, including those rare cases where S.P. operators are time dependent (such as the Hamiltonian when $V = V(t)$.) Using the Schrödinger equation (2-27) and its complex conjugate for the ket and bra time derivatives, respectively, in (2-31), we get

$$\frac{d\overline{p}_1}{dt} = {}_S\left\langle\psi\middle|\left(iHp_1^S + \underbrace{\frac{\partial p_1^S}{\partial t}}_{=0} - ip_1^S H\right)\middle|\psi\right\rangle_S = {}_S\langle\psi|-i\left[p_1^S, H\right]|\psi\rangle_S + {}_S\left\langle\psi\middle|\frac{\partial p_1^S}{\partial t}\middle|\psi\right\rangle_S. \tag{2-32}$$

Eq of motion of momentum expectation value

Recall the old NRQM adage that the expectation value of any operator without explicit time dependence that commutes with the Hamiltonian is conserved (its time derivative is zero.) Note that (2-27), (2-30), and (2-31)/(2-32) are equations of motion for the state, momentum operator, and momentum expectation value, respectively, in the Schrödinger picture. These are generalized to any state and operator in Wholeness Chart 2-4.

(2-32) generalized to any operator in Chart 2-4

Note further that the partial time derivative $\partial/\partial t$ in the Schrödinger equation (2-27) acting on the ket is equivalent to the full time derivative d/dt by the same logic as that in Box 2-1. That is, the ket, or wave function, here is mathematically the same as a classical field, functionally dependent on the independent variables, x^i and t. So, we can write the equation of motion for a state (i.e., the Schrödinger equation) with either a partial or total time derivative.

2.6.2 The Heisenberg Picture

The Schrödinger picture states and operators can be transformed to states and operators having different form via what is known as a <u>unitary transformation</u> (see Box 2-3). The particular unitary transformation (where U is a <u>unitary operator</u>) for this is

$$U = e^{-iHt} \qquad \left(= e^{-iHt/\hbar} \text{ in non-natural units}\right), \tag{2-33}$$

Transforming between Schrödinger and Heisenberg pictures

where states and operators transform as

$$\begin{array}{ll} U^\dagger|\psi\rangle_S = |\psi\rangle_H & U^\dagger\mathcal{O}^S U = \mathcal{O}^H \\[2mm] U|\psi\rangle_H = |\psi\rangle_S & U\mathcal{O}^H U^\dagger = \mathcal{O}^S. \end{array} \tag{2-34}$$

Note the effect of the first relation in (2-34) on our sample ket (2-28),

$$U^\dagger|\psi\rangle_S = e^{iHt}Ae^{-i(Et-\mathbf{p}\cdot\mathbf{x})} = e^{iEt}Ae^{-i(Et-\mathbf{p}\cdot\mathbf{x})} = Ae^{i\mathbf{p}\cdot\mathbf{x}} = |\psi\rangle_H. \tag{2-35}$$

We find that the state, which was time dependent in the S.P., is *time independent* in the Heisenberg picture (H.P.). This statement is generally true for any state. (Think through it, if you like, for a more general wave function state of several terms.)

Thus, the equation of motion for a state in the S.P. (2-27), becomes, in the H.P,

$$\frac{d|\psi\rangle_H}{dt} = 0. \tag{2-36}$$

In H.P., eq of motion of state (state is constant in time)

Now taking the time derivative of the second relation in the top row of (2-34), we have (see Appendix B, Section 2.10.2, if you are concerned about switching total to partial derivatives)

$$\frac{d}{dt}\left(U^{\dagger}\mathcal{O}^{S}U\right) = (iH)\underbrace{e^{iHt}\mathcal{O}^{S}e^{-iHt}}_{\mathcal{O}^{H}} + e^{iHt}\left(\frac{\partial\mathcal{O}^{S}}{\partial t}\right)e^{-iHt} + \underbrace{e^{iHt}\mathcal{O}^{S}e^{-iHt}}_{\mathcal{O}^{H}}(-iH)$$

defined as $\hat{\partial}\mathcal{O}^{H}/\partial t$ ←Note the hat on ∂ defintion

$$= \frac{d\mathcal{O}^{H}}{dt} = -i\left[\mathcal{O}^{H},H\right] + \underbrace{\frac{\hat{\partial}\mathcal{O}^{H}}{\partial t}}_{=0 \text{ in this book}} \quad .$$

(2-37) *In H.P., eq of motion of operator*

We will not be considering any operators that are time dependent in the S.P., so for us, the last term in (2-37) will always be zero. Nonetheless, even in this case, we see that in the H.P., an operator time derivative can be non-zero, and thus, the operator, time dependent.

Box 2-3. Unitary Transformations in Quantum Theories

A *unitary transformation* is called unitary because its operation on (transformation of) a state vector leaves the magnitude of the state vector unchanged, i.e., the state vector magnitude is multiplied by unity. It is the complex space analogue of an *orthogonal transformation* in Cartesian coordinate space, which, when acting on a (real number) vector in that space, rotates the vector but does not stretch or compact it. A unitary transformation can be thought of as "rotating" a (complex number) state vector in Hilbert space (the complex space where each coordinate axis is an eigenvector) without changing the "length" (magnitude) of the vector. In NRQM, the square of the absolute value of the state vector is the square of its "length", and this is the probability density for measuring the particle. This means a unitary transformation of a state vector leaves the probability of detecting the particle unchanged. A unitary transformation multiplies probability by unity.

Recall, from classical mechanics, that an orthogonal transformation represented by a real matrix **A** has an inverse equal to the transpose of that matrix, i.e., $\mathbf{A}^{-1} = \mathbf{A}^{T}$. In the complex space of state vectors, a unitary transformation U has an analogous form for its inverse, the complex conjugate transpose, i.e., $U^{-1} = U^{\dagger}$ and so $U^{\dagger}U = 1$. The following example may make this clearer.

Consider $U = e^{-iHt}$, where H is the (hermitian) Hamiltonian operator. By inspection one knows its magnitude in complex space is unity and so its action on a state vector would not change the length of that state vector (though phase would change by –Ht.) Also, by inspection, $U^{\dagger}U = 1$. So, U performs a unitary transformation.

Wholeness Chart 2-3. Unitary vs Orthogonal Transformations		
	3D Cartesian Space (Real)	**Hilbert Space** (Complex)
Magnitude conserving transformation	Orthogonal $\mathbf{A}$ = matrix	Unitary $U = e^{iX}$
Effect on vector	rotates in real space	"rotates" in complex space
Physical effect	vector length unchanged	probability unchanged
Inverse	$\mathbf{A}^{-1} = \mathbf{A}^{T}$	$U^{-1} = U^{\dagger}$

How an exponential operator works

Do a Taylor expansion of $U = e^{-iHt}$ above about t, when U is operating on an energy eigenstate., i.e.,

$$U\left|\psi_{E}\right\rangle = e^{-iHt}\left|\psi_{E}\right\rangle = \left(1 - itH - \tfrac{1}{2}t^{2}H^{2} + ...\right)\left|\psi_{E}\right\rangle = \left(1 - itE - \tfrac{1}{2}t^{2}E^{2} + ...\right)\left|\psi_{E}\right\rangle = e^{-iEt}\left|\psi_{E}\right\rangle$$

So an operator in the exponent has the same effect in the exponent as it would if acting in the usual non-exponential way on an eigenstate. This conclusion is readily generalized to any state.

Note: Although it is common to write $U = e^{-iHt}$, it is implied that H (if you think of it as $i\partial/\partial t$) does not act on t. To be proper, the t should be placed before the H, as we did in the expansion above, but it usually is not done that way.

Because $H (= H^S$ by definition) commutes with itself, U and $U^\dagger$ commute with H, so using $\mathcal{O}^S = H^S = H$ in the second relation on the top line of (2-34),

Hamiltonian H has same form in S.P. and H.P.

$$H = H^S = H^H . \tag{2-38}$$

Finally, for (2-32) expressed in terms of a general operator $(p_1{}^S \to \mathcal{O}^S)$, we find, after inserting $UU^\dagger = 1$ where needed, that

$$\frac{d\bar{\mathcal{O}}}{dt} = {}_S\langle\psi|UU^\dagger\left(-i\left[\mathcal{O}^S, H\right]\right)UU^\dagger|\psi\rangle_S + {}_S\langle\psi|UU^\dagger\frac{\partial\mathcal{O}^S}{\partial t}UU^\dagger|\psi\rangle_S$$

$$= {}_H\langle\psi|\left(-i\left[\mathcal{O}^H, H\right]\right)|\psi\rangle_H + {}_H\langle\psi|\frac{\hat{\partial}\mathcal{O}^H}{\partial t}|\psi\rangle_H . \tag{2-39}$$

From which we see that the equation of motion for the expectation value of an operator has the same form in both pictures. This means that whichever picture we choose to work in, although the states and operators will be different, the predictions for quantities we can measure (dynamical variables) will be the same. So we can choose whichever system is easier to work with mathematically. For NRQM, this was the S.P. For QFT, as we will see, it is the H.P.

Eq of motion of expectation value has same form in S.P. and H.P.

Wholeness Chart 2-4. Schrödinger vs. Heisenberg Picture Equations of Motion

	States	Operators	Expectation Values
Schrödinger Picture	Time dependent $$i\frac{d}{dt}|\psi\rangle_S = H|\psi\rangle_S$$ (Schrödinger eq)	Usually time independent $$\frac{d\mathcal{O}^S}{dt} = \underbrace{\frac{\partial\mathcal{O}^S}{\partial t} = 0}_{\text{usually}}$$	$$\frac{d\bar{\mathcal{O}}}{dt} = {}_S\langle\psi|\left(-i\left[\mathcal{O}^S, H\right]+\frac{\partial\mathcal{O}^S}{\partial t}\right)|\psi\rangle_S$$ $|\psi\rangle_S$ changes in time; $\mathcal{O}^S$ usually const in time
Transform via $U = e^{-iHt/\hbar}$ $\Downarrow$	$U^\dagger|\psi\rangle_S = |\psi\rangle_H$	$U^\dagger\mathcal{O}^S U = \mathcal{O}^H$	$\frac{d\bar{\mathcal{O}}}{dt}$ invariant under the transformation
Heisenberg Picture	Time independent $$\frac{d|\psi\rangle_H}{dt} = 0$$	Often time dependent $$\frac{d\mathcal{O}^H}{dt} = -i\left[\mathcal{O}^H, H\right]+\underbrace{\frac{\hat{\partial}\mathcal{O}^H}{\partial t}}_{\substack{\text{usually}\\=0}}$$	Same as Schrödinger picture above with sub and superscript $S \to H$ and $\partial\mathcal{O}^H = \hat{\partial}\mathcal{O}^H$ $|\psi\rangle_H$ const in time; $\mathcal{O}^H$ often changes in time
Hamiltonian		$H^H = H^S = H$	
Key Relation	In S.P., the state eq of motion	In H.P., the operator eq of motion	In both pictures, expectation value and its equation of motion are the same, equally key.

Continuation of Wholeness Chart 1-2. Comparison of Three Quantum Theories

	NRQM	RQM	QFT
Most advantageous picture to use	Schrödinger picture	Schrödinger picture	Heisenberg picture

2.6.3 Visualizing Schrödinger and Heisenberg Pictures

One can think of the S.P. as quantum waves (wave functions, states, or kets) moving and evolving in time, but operators as constant (generally) in time. The H.P., by contrast, can be thought of as quantum waves frozen in time (static wave functions or time independent kets), with operators being what move and evolve. Either way, the expectation value (2-40) (what we would measure on average over many measurements) is the same, and so is its equation of motion.

$$\bar{\mathcal{O}} = {}_S\langle\psi|\mathcal{O}^S|\psi\rangle_S = {}_H\langle\psi|\mathcal{O}^H|\psi\rangle_H .\qquad(2\text{-}40)$$

S.P.: particle waves move, operators (usually) do not. H.P.: waves frozen, operators evolve. Measured values same in both.

The philosophical lesson to be learned from this is that we can have different models of reality predicting the same real world phenomena. In this case, in one model the states are waves that move and evolve. In the other model, the states never change. But, both are valid predictors of the laws of nature we observe in the physical universe. Hence, we should be wary of accepting any given model of reality as a "true" picture of what nature is actually doing.

2.7 Quantum Theory: An Overview

Wholeness Chart 2-5, Summary of Quantum Mechanics, overviews the fundamental branches of quantum theory in much the same way that Wholeness Chart 2-2 overviews the fundamental branches of classical theory. These correspond to, and elaborate on, the bottom and top parts, respectively, of Wholeness Chart 1-1 in Chap. 1. (We will temporarily leave $\hbar$ in our relations even though, in our units, it equals one, so that you, the reader, can see precisely where it comes into those, rather key, relations.)

Chart 2-5 summarizes QM

Note particularly, that in Wholeness Chart 2-5, all relations and quantities are expressed in the Heisenberg picture. If it were expressed in the Schrödinger picture, then many quantities (i.e., operators) such as H, p_i, and the like would have to be expressed as expectation values. In the H.P., the equation of motion for an operator (see H.P. row in Wholeness Chart 2-4) has the same time dependence as the expectation value for that operator (the bra and ket are constant in time in the right most block in that row.) That is, in the H.P. the operator equation of motion is the same as that of the expectation value. And the state (ket) equation of motion, which was quite critical in the S.P. (it is the Schrödinger equation), becomes rather meaningless, as the state is constant in time. So we can ignore the states in the H.P. summary of Wholeness Chart 2-5 and write the equations of motion in terms of the operators.

Chart 2-5 is in terms of H.P.

2.7.1 Classical vs. Quantum: Much is the Same

Note that everything in the first 12 blocks in the NRQM and RQM columns of Chart 2-5 is the same as that in Chart 2-2, from the independent variables used through Hamilton's equations of motion. For example, the Hamiltonian H has the same form for a particle in quantum mechanics as it does for a classical particle. (Recall from Chap. 1, this was criterion number one for first quantization.)

First 12 rows: Classical NR particle of Chart 2-2 same as NRQM of Chart 2-5

2.7.2 Poisson Brackets vs. Commutators: Something is Different

However, note that the equation of motion for a dynamical variable, represented by u, changes from (2-24) in classical non-relativistic particle theory to

$$\frac{du}{dt} = \frac{-i}{\hbar}[u,H] + \frac{\hat{\partial}u}{\partial t}\qquad(2\text{-}41)$$

in NRQM in the Heisenberg picture. Equation(2-41), which you should have seen before in your NRQM studies, was discovered independently by early quantum theorists. Yet it was striking to everyone how closely it parallels its classical counterpart (2-24). The fundamental difference is that the Poisson brackets have become commutators (with a factor of $-i/\hbar$ in front.)

Last 2 rows: Classical NR particle has Poisson brackets; NRQM has commutators

Similarly, the Poisson bracket relations for conjugate variables in classical theory (last line, third column in Wholeness Chart 2-2) parallel the commutators (last line, third column of Wholeness Chart 2-5) discovered early on in the development of NRQM.

So, the classical non-relativistic particle and the NRQM theories mimic one another, with one significant difference. All relations remain effectively the same except that the commutators of quantum theory correspond to Poisson brackets of classical theory (times a factor of $-i/\hbar$.)

2.7.3 Quantization and the Correspondence Principle

According to the *correspondence principle*, in the macroscopic limit, our quantum relations must reduce to the usual classical relations. But in comparing the last two blocks in the third columns (NR particle and NRQM) of Wholeness Charts 2-2 and 2-5, this can only be true if

	Comments	**Non-relativistic Quantum Mechanics**
Independent variables through Hamilton's equations of motion		Same form as top 12 blocks of Wholeness Chart 2-2
Commutator brackets, definition		for $u = u(x^i, p_i, t)$, $v = v(x^i, p_i, t)$ $$[u,v] = uv - vu$$

Equations of motion in terms of commutator brackets
 	
  i) any dynamical variable
  ii) conjugate variables
	Correspondence principle:
	
	Classical $\rightarrow$ Quantum
	
	$$\{u,v\} \rightarrow \frac{-i}{\hbar}[u,v]$$
		i) for $v = H$ $\dfrac{du}{dt} = \dfrac{-i}{\hbar}[u,H] + \dfrac{\partial u}{\partial t}$
		
		ii) for i) plus $u = x^i$ or p_i
		
		$\dot{p}_i = \dfrac{-i}{\hbar}[p_i, H] = -\dfrac{\partial H}{\partial x^i}$; $\dot{x}^i = \dfrac{-i}{\hbar}[x^i, H] = \dfrac{\partial H}{\partial p_i}$

Uncertainty principle		$\left[x^i, p_j\right] = i\hbar\delta^i{}_j$ $\left[x^i, x^j\right] = \left[p_i, p_j\right] = 0$

$$\underbrace{\left\{x^i, p_j\right\}}_{\substack{\text{classical}\\\text{dynamic}\\\text{variables}}} = \delta^i{}_j = \underbrace{\frac{-i}{\hbar}\left[x^i, p_j\right]}_{\substack{\text{quantum}\\\text{operators}}}. \qquad \left(\begin{array}{l}\text{Cartesian system, where}\\ p_j = p^j = \text{3-momentum}\end{array}\right) \qquad (2\text{-}42)$$

So the correspondence principle provides us with a key part of our method for quantization. That is, in going from classical theory to NRQM, we must take

$$\left\{x^i, p_j\right\} = \delta^i{}_j \xrightarrow[\text{1st quantization}]{} \left[x^i, p_j\right] = i\hbar\delta^i{}_j \qquad \text{(Cartesian system)} \qquad (2\text{-}43)$$

Classical NR particle theory becomes NRQM if Poisson brackets converted to commutators

Of course, as noted in Chap. 1, we also keep the same form of the Hamiltonian (or equivalently, the Lagrangian) as we had classically.

2.7.4 Extrapolation to Field Theory

Shortly after understanding this, one gets the idea that perhaps the same thing can be done with field theory. So, we try it. We postulate the same first twelve rows for Wholeness Chart 2-5 as we had in Wholeness Chart 2-2, and the same sort of bracket correspondence for the other rows as in NRQM/RQM, and see where it takes us. Does it indeed lead to a good theory, one that predicts the phenomena we observe? Very quickly we find that it does, and that new theory has come to be called *quantum field theory*. This means for going from our classical theory of fields to the quantum theory of fields is called second quantization, i.e.,

We guess: Classical relativistic field theory should become QFT if Poisson brackets converted to commutators

$$\left\{\phi^r(\mathbf{x},t), \pi_s(\mathbf{y},t)\right\} = \delta^r{}_s\delta(\mathbf{x}-\mathbf{y}) \xrightarrow[\text{2nd quantization}]{} \left[\phi^r(\mathbf{x},t), \pi_s(\mathbf{y},t)\right] = i\hbar\delta^r{}_s\delta(\mathbf{x}-\mathbf{y}) \quad (2\text{-}44)$$

where again, we keep the same form of the Hamiltonian (or equivalently, the Lagrangian) as we had classically. That is, as we develop QFT, we will use the same independent variables, the same sense for the Hamiltonian density as an energy density, the same Legendre transformation, the same Euler-Lagrange equation into which we will plug our Lagrangian density, the same conjugate momenta definitions, etc.

The delta function in $\mathbf{x} - \mathbf{y}$ in (2-44) ensures that we are only considering the field and its conjugate momentum density at the same point in space. We will see the role this plays in the mathematical development of the theory later.

Both of the processes (2-43) and (2-44) are formally called <u>canonical quantization</u>. They are canonical because it is the canonically conjugate variables - the generalized coordinates and their conjugate momenta - which are the center of attention. The term quantization arises because the metamorphosis of brackets, in going from the classical to quantum realm, changes the Poisson bracket relation for the canonical variables into the commutator, which is the mathematical basis of

Summary of Quantum Mechanics (Heisenberg Picture)

Non-relativistic Quantum Fields	Relativistic QM	Quantum Field Theory
	Same form as top 12 blocks of Wholeness Chart 2-2	Same form as top 12 blocks of Wholeness Chart 2-2
No theory generally used.	Same form as Non-relativistic Quantum Mechanics section, but different meaning for p_i	for $u = u(\phi^r, \pi_r, \partial_i \phi^r, t)$, $v = v(\phi^r, \pi_r, \partial_i \phi^r, t)$ $[u,v] = uv - vu$
	See Non-relativistic Quantum Mechanics section	i) for $U = \int u\, dV$; $[U,H] = UH - HU$ $\dot{U} = \dfrac{dU}{dt} = [U,H] + \dfrac{\hat{\partial}U}{\partial t}$ ii) for $u = \pi_r$; $\dot{\Pi}_r = [\Pi_r, H]$
	See Non-relativistic Quantum Mechanics section	$\boxed{\left[\phi^r, \pi_s\right] = i\hbar \delta^r{}_s \delta(\mathbf{x} - \mathbf{y}); \left[\phi^r, \phi^s\right] = [\pi_r, \pi_s] = 0}$

the uncertainty principle. The uncertainty principle is often called <u>the quantum principle</u>, hence the name *quantization*.

Quantization then, in a nutshell, is a means for deducing the governing quantum equations from knowledge of the classical macroscopic ones. We will begin to use it in the next chapter to develop our theory.

Quantization is a means for deducing quantum theory from classical theory

2.8 Chapter Summary

The bottom right hand block of Wholeness Chart 2-5, Summary of Quantum Mechanics, contains the essence of this chapter (enclosed in box with bold border). A quantum field and its own conjugate momentum density do not commute, whereas all other pairings of fields and momentum density do commute. This is one postulate at the basis of QFT (see (2-44).) The other postulate comprises keeping the same form for the Lagrangian density (or equivalently, either the Hamiltonian density or the field equations of motion) as in the classical realm. These postulates are known as second quantization. (I guess we've said this enough. ☺)

Natural units and their relation to other types of units, summarized in Wholeness Chart 2-1 and Sect. 2.1.7, comprise another key concept in the chapter. In natural units, $c = \hbar = 1$ (dimensionless), and all quantities are expressed in units of powers of MeV.

Other fundamental concepts include certain field relations in the right most column of Wholeness Chart 2-2, which apply in the quantum realm. These are i) the Euler-Lagrange equation for fields, ii) the definition of conjugate momentum density, and iii) the Legendre transformation for fields. (Note that we will do virtually nothing with Hamilton's equations, so you need not worry about them.)

Unitary transformations, designated often by U, are quite important in QFT and are summarized in Box 2-3. When acting on a state vector, unitary transformations do not change the "length" (magnitude) in complex space of the state, the square of which is probability density. Thus, unitary transformations conserve probability. Importantly, $U^{-1} = U^\dagger$.

Quantum theories can be expressed in two different pictures, called the Schrödinger and Heisenberg pictures, summarized in Wholeness Chart 2-4. In the S.P., states are time dependent, but operators usually are not. The H.P. is the opposite. For it, states are static (fixed in time) and operators often time dependent. The key equation of motion in the S.P. is the state equation of motion (the Schrödinger equation). The key equation of motion in the H.P. is the operator equation of motion. (There is, since the state is constant, effectively, no H. P. state equation of motion.) The H.P. is closer to the classical perspective in that the focus in both is on dynamical variables/operators such as H, p_i, etc., which may vary in time. (And there is no state equation of motion in the classical world, since, for it, there is no such thing as a state.) QFT is easier to develop in the H.P., so we will be using it, rather than the S.P.

2.9 Appendix A: Understanding Contravariant and Covariant Components

The concepts of contravariant and covariant components presented in Sect. 2.2 should be somewhat familiar to those who have studied the prerequisite material delineated in the preface. However, oftentimes, even those who have already been exposed to these concepts still do not feel completely at home with them. For them, and for any newcomers to the subject, I hope the following brief introduction will help.

2.9.1 A Trick for Conveniently Finding 4D Vector Length

Contravariant and covariant components are simply tricks that allow us to represent vectors (and tensors) in a way that helps us carry out certain mathematical procedures, like finding the magnitude of a vector in curved space or the proper time passing on a particle in special relativity. In this book, we will not be dealing with curved space, so all of the applications of contravariant and covariant component theory herein will be for the simpler case of <u>Minkowski space</u> (flat, 4D space with Cartesian space coordinates plus time.) We will, for starters, want to be able to calculate proper time on a particle (decay time of a particle, for instance, depends on proper time, not the lab time we see as the particle whizzes by.)

Consider how we find the length l of a vector in a 3D Cartesian system with one end of the vector at the origin, i.e.,

$$\left(l\right)^2 = \left(X_1\right)^2 + \left(X_2\right)^2 + \left(X_3\right)^2 = X_i X_i \quad \left(= \sum_i X_i X_i , \text{ repeated indices mean summation.} \atop \text{See Sect. 2.2} \right)$$

$$= \begin{bmatrix} X_1 & X_2 & X_3 \end{bmatrix} \begin{bmatrix} X_1 \\ X_2 \\ X_3 \end{bmatrix} = \begin{bmatrix} X_1 & X_2 & X_3 \end{bmatrix} \begin{bmatrix} 1 & 0 & 0 \\ 0 & 1 & 0 \\ 0 & 0 & 1 \end{bmatrix} \begin{bmatrix} X_1 \\ X_2 \\ X_3 \end{bmatrix} = X_i \underbrace{\delta_{ij} X_j}_{\substack{\sum_j \delta_{ij} X_j \\ = X_i}} \tag{2-45}$$

where, with a future purpose in mind, we insert an identity matrix, represented in index notation by the Kronecker delta δ_{ij} ($= 0$ if row $i \neq$ column j; $= 1$ if $i = j$), on the RHS.

Now, imagine a spatially 4D Cartesian system, where the length of a 4D vector is

$$\left(l\right)^2 = \left(X_0\right)^2 + \left(X_1\right)^2 + \left(X_2\right)^2 + \left(X_3\right)^2 = X_\mu X_\mu$$

$$= \begin{bmatrix} X_0 & X_1 & X_2 & X_3 \end{bmatrix} \begin{bmatrix} X_0 \\ X_1 \\ X_2 \\ X_3 \end{bmatrix} = \begin{bmatrix} X_0 & X_1 & X_2 & X_3 \end{bmatrix} \begin{bmatrix} 1 & 0 & 0 & 0 \\ 0 & 1 & 0 & 0 \\ 0 & 0 & 1 & 0 \\ 0 & 0 & 0 & 1 \end{bmatrix} \begin{bmatrix} X_0 \\ X_1 \\ X_2 \\ X_3 \end{bmatrix} = X_\mu \underbrace{\delta_{\mu\nu} X_\nu}_{X_\mu} \tag{2-46}$$

Now consider the 4D spacetime of special relativity theory (SRT), and the "length" of a 4D vector we have in mind is the proper time τ on an object passing by us. The 0th coordinate is now time instead of a spatial X_0 coordinate. From SRT, we know

$$\left(c\tau\right)^2 = \left(ct\right)^2 - \left(X_1\right)^2 - \left(X_2\right)^2 - \left(X_3\right)^2 = \text{ (how to write as summed indices?)}$$

$$= \begin{bmatrix} ct & X_1 & X_2 & X_3 \end{bmatrix} \begin{bmatrix} ct \\ -X_1 \\ -X_2 \\ -X_3 \end{bmatrix} \qquad c = 1 \text{ in natural units} \tag{2-47}$$

Note that because of the minus signs in our "length" (= proper time) calculation in (2-47), we can't use the nice summation symbolism of the first lines of (2-45) and (2-46). That was only good if all of the terms in the summation had the same sign. Fine for purely spatial coordinates of any dimension. Not possible if we have both time and space in the same coordinate system.

But here is a clever idea. Let's define the column matrix of the second line in (2-47) as a different set of vector components, with minus signs in front of the X_i. We could designate it with primes, if we like, so

$$X'_\mu = \begin{bmatrix} ct \\ -X_1 \\ -X_2 \\ -X_3 \end{bmatrix} = \begin{bmatrix} X_0 \\ -X_1 \\ -X_2 \\ -X_3 \end{bmatrix} \quad \text{and} \quad X_\mu = \begin{bmatrix} ct \\ X_1 \\ X_2 \\ X_3 \end{bmatrix} = \begin{bmatrix} X_0 \\ X_1 \\ X_2 \\ X_3 \end{bmatrix}. \qquad (2\text{-}48)$$

With this newly defined representation of our 4D vector, and $ct = X_0$, we can represent our vector "length" of (2-47) as

$$\left(c\tau\right)^2 = \left(ct\right)^2 - \left(X_1\right)^2 - \left(X_2\right)^2 - \left(X_3\right)^2 = \left(X_0\right)^2 - \left(X_1\right)^2 - \left(X_2\right)^2 - \left(X_3\right)^2 = X_\mu X'_\mu. \qquad (2\text{-}49)$$

And thus, we have a neat shorthand way to write out a vector length in 4D spacetime.

Unfortunately, the primed notation is used in relativity and elsewhere to indicate a different coordinate system in a different frame. In relativity, this is usually a frame having velocity relative to the unprimed frame. In the present case, we are only working in a single coordinate system. So, a different symbolism has arisen for this case (i.e., for finding vector lengths in the same coordinate system). While it can take a little getting used to, the symbolism entails using no primes, but instead raising the indices for one of the component sets in (2-49), and keeping the indices lowered for the other. We also generally use non-capital letters for 4D position vectors, and capital letters (with subscript indices only) for 3D Cartesian components. Thus, by the convention chosen,

$$x^\mu = \begin{bmatrix} x^0 \\ x^1 \\ x^2 \\ x^3 \end{bmatrix} = X_\mu = \begin{bmatrix} ct \\ X_1 \\ X_2 \\ X_3 \end{bmatrix} \quad \text{and} \quad x_\mu = \begin{bmatrix} x_0 \\ x_1 \\ x_2 \\ x_3 \end{bmatrix} = X'_\mu = \begin{bmatrix} ct \\ -X_1 \\ -X_2 \\ -X_3 \end{bmatrix}. \qquad (2\text{-}50)$$

With the above convention, our 4D vector length (2-49) becomes

$$\left(c\tau\right)^2 = x^0 x_0 + x^1 x_1 + x^2 x_2 + x^3 x_3 = x^\mu x_\mu. \qquad (2\text{-}51)$$

Of course, this can lead to some confusion, as before this, we have always used a superscript solely for raising a quantity to a power. To avoid this confusion, we will have to remember to enclose entities in parentheses when we mean the superscript as a power, as we did on the LHS of (2-51). From now on, superscripts without parentheses will designate components, not powers. Be forewarned, however, that, unfortunately, authors may not always strictly adhere to this practice, and you may have to glean the meaning of a superscript from context. (This isn't so hard *after* you get accustomed to this notation, but it can be difficult *before* you do.)

For reasons beyond the scope of this discussion, $\underline{x^\mu}$ was designated as the <u>contravariant components</u> form, and $\underline{x_\mu}$ as the <u>covariant components</u> form, of the same physical vector. As a mnemonic, just remember that the raised index contravariant components are the 3D Cartesian coordinates plus ct. The lowered index covariant components include a minus sign for the 3D part.

Contravariant and covariant components also allow us to readily find the 4D length of any vector, not just the 4D position vector x^μ. For example, the <u>four-velocity of relativity u^μ</u> for an object is

$$u^\mu = \frac{dx^\mu}{d\tau} = \frac{d}{d\tau}\begin{bmatrix} x^0 & x^1 & x^2 & x^3 \end{bmatrix} = \begin{bmatrix} u^0 & u^1 & u^2 & u^3 \end{bmatrix}, \qquad (2\text{-}52)$$

where

$$u^i = \frac{dx^i}{d\tau} = \frac{dx^i}{\sqrt{1 - v^2/c^2}\,dt} = \frac{v^i}{\sqrt{1 - v^2/c^2}} = \gamma v^i \; ; \quad u^0 = \frac{dx^0}{d\tau} = c\frac{dt}{d\tau} = \frac{c}{\sqrt{1 - v^2/c^2}} = \gamma c, \qquad (2\text{-}53)$$

u^i here is the derivative of the spatial coordinate with respect to proper time on the object τ, v^i is that with respect to coordinate time t, γ is the usual Lorentz factor common in relativity, and we will henceforth often write vectors as rows, rather than columns, to save space. The 4D length $|u^\mu|$ is found from

$$(u)^2 = \left|u^\mu\right|^2 = u^\mu u_\mu = \frac{dx^\mu}{d\tau}\frac{dx_\mu}{d\tau} = \begin{bmatrix} u^0 & u^1 & u^2 & u^3 \end{bmatrix} \begin{bmatrix} u_0 \\ u_1 \\ u_2 \\ u_3 \end{bmatrix} = \begin{bmatrix} u^0 & u^1 & u^2 & u^3 \end{bmatrix} \begin{bmatrix} u^0 \\ -u^1 \\ -u^2 \\ -u^3 \end{bmatrix} \quad (2\text{-}54)$$

$$= \left(u^0\right)^2 - \left(u^1\right)^2 - \left(u^2\right)^2 - \left(u^3\right)^2 = \gamma^2\left(c^2 - \left(v^1\right)^2 - \left(v^2\right)^2 - \left(v^3\right)^2\right) = c^2,$$

the last part of which students of relativity may recognize as the correct expression for the square of the magnitude of the four-velocity.

The magnitude of the <u>4-momentum $p^\mu = mu^\mu$</u> is then found from

$$(p)^2 = \left|p^\mu\right|^2 = p^\mu p_\mu = m^2 u^\mu u_\mu = m^2 c^2 \quad \left(= m^2 \text{ in natural units}\right). \qquad (2\text{-}55)$$

(2-55) tells us that for (massless) photons $(p)^2 = 0$, even though $p^\mu \neq 0$. (See Prob. 13.) Note from (2-55) that $p^0 = \gamma mc = E/c$, where E is relativistic energy, and p^i = relativistic 3-momentum.

For any general vector w^μ, with upper case letters representing 3D Cartesian components, we have

$$w^\mu = \begin{bmatrix} w_0 & W_1 & W_2 & W_3 \end{bmatrix} \quad w_\mu = \begin{bmatrix} w_0 & -W_1 & -W_2 & -W_3 \end{bmatrix} \quad (w)^2 = \left|w^\mu\right|^2 = w^\mu w_\mu . \quad (2\text{-}56)$$

In addition, we will often use differential elements of 4 vectors, such as dx^μ, and the relations (2-56) hold for such differential 4 vectors, as well (which should be fairly obvious, as a differential of a vector is also a vector in its own right.)

2.9.2 The Metric

Note that we can use a certain matrix to convert from contravariant to covariant components,

$$x_\mu = \begin{bmatrix} x_0 \\ x_1 \\ x_2 \\ x_3 \end{bmatrix} = \begin{bmatrix} ct \\ -X_1 \\ -X_2 \\ -X_3 \end{bmatrix} = \begin{bmatrix} 1 & 0 & 0 & 0 \\ 0 & -1 & 0 & 0 \\ 0 & 0 & -1 & 0 \\ 0 & 0 & 0 & -1 \end{bmatrix} \begin{bmatrix} ct \\ X_1 \\ X_2 \\ X_3 \end{bmatrix} = \underbrace{\begin{bmatrix} 1 & 0 & 0 & 0 \\ 0 & -1 & 0 & 0 \\ 0 & 0 & -1 & 0 \\ 0 & 0 & 0 & -1 \end{bmatrix}}_{g_{\mu\nu}} \underbrace{\begin{bmatrix} x^0 \\ x^1 \\ x^2 \\ x^3 \end{bmatrix}}_{x^\nu} = g_{\mu\nu}x^\nu . \quad (2\text{-}57)$$

This matrix $g_{\mu\nu}$ represents what is called the <u>metric</u> (of the coordinate space, which in this case is <u>Minkowski coordinate space</u>.) It lowers a raised index. It has an inverse that turns out to have the same form as it does.

$$\underbrace{\begin{bmatrix} 1 & 0 & 0 & 0 \\ 0 & -1 & 0 & 0 \\ 0 & 0 & -1 & 0 \\ 0 & 0 & 0 & -1 \end{bmatrix}}_{g_{\mu\nu}} \underbrace{\begin{bmatrix} 1 & 0 & 0 & 0 \\ 0 & -1 & 0 & 0 \\ 0 & 0 & -1 & 0 \\ 0 & 0 & 0 & -1 \end{bmatrix}}_{\left(g_{\mu\nu}\right)^{-1}=g^{\nu\alpha}} = \underbrace{\begin{bmatrix} 1 & 0 & 0 & 0 \\ 0 & 1 & 0 & 0 \\ 0 & 0 & 1 & 0 \\ 0 & 0 & 0 & 1 \end{bmatrix}}_{\delta_\mu{}^\alpha} . \quad (2\text{-}58)$$

The inverse of the metric can be used to raise indices, i.e.,

$$x^\mu = \begin{bmatrix} x^0 \\ x^1 \\ x^2 \\ x^3 \end{bmatrix} = \underbrace{\begin{bmatrix} 1 & 0 & 0 & 0 \\ 0 & -1 & 0 & 0 \\ 0 & 0 & -1 & 0 \\ 0 & 0 & 0 & -1 \end{bmatrix}}_{g^{\mu\nu}} \underbrace{\begin{bmatrix} x_0 \\ x_1 \\ x_2 \\ x_3 \end{bmatrix}}_{x_\nu} = g^{\mu\nu}x_\nu . \quad (2\text{-}59)$$

When indices are repeated, they are summed, and even when they are not, they are only dummy indices symbolizing coordinate axes numbers. So, it really doesn't matter what particular Greek letter we take for a summed index. Hence, $g^{\nu\alpha}$ represents the same entity as $g^{\mu\nu}$.

$g_{\mu\nu}$ is sometimes called the <u>covariant metric</u>, and $g^{\mu\nu}$, the <u>contravariant metric</u>. The term *metric* used alone usually means $g_{\mu\nu}$.

Note that with the metric, we can write (2-51) as

$$\left(c\tau\right)^2 = x^\mu x_\mu = g_{\mu\nu}x^\mu x^\nu . \tag{2-60}$$

Prove (2-60) to yourself three ways: by substituting the RHS of (2-57) into the middle part of the above, by writing out (2-60) in matrix form, and by doing the summation of terms implied by the repeated indices.

Note that the particular metric form of the metric in (2-57) is specific to Minkowski coordinates, which is all we will use in this book. Other coordinate systems (like 4D having time and a spherical spatial coordinate system) would have other forms for $g_{\mu\nu}$. Note that in general relativity, you will find the Minkowski metric, which is commonly designated by $g_{\mu\nu}$ in QFT, to be designated by the symbol $\eta_{\mu\nu}$. In relativity, $g_{\mu\nu}$ usually refers to any general metric, not necessarily of form shown in (2-57). But in this book, the symbol $g_{\mu\nu}$ always equals $\eta_{\mu\nu}$, the Minkowski metric.

The metric in (2-60) plays a role in 4D spacetime similar to the role played by the identity matrix of (2-45) and (2-46) for Cartesian spaces (which are purely spatial, with no time axis.) In fact, for Cartesian systems, the identity matrix *is* the metric, so for any vector **v**, $v^i = v_i$. (Do Prob. 8 for more on this.)

The form of the metric tells us a lot, in fact virtually everything, about the coordinate space we are dealing with. It is, in a sense, the *signature* of the coordinate space.

2.9.3 Invariance and Covariance

The quantity $c\tau$ of (2-60) is an example of what is known as a <u>4D scalar</u> (or <u>world scalar</u> or <u>Lorentz scalar</u>.) It is the length of a vector (timelike here) in spacetime.

In 3D space, a vector length remains the same (invariant) if we change (transform) coordinate systems. The components of the vector are different in a rotated (primed) coordinate system (i.e., $X_i' \neq X_i$), but the length remains the same. $l^2 = X_i X_i = X_i' X_i'$. By definition, a scalar is measured the same by observers using any coordinate system. Scalars are <u>invariant</u> under transformation to a new coordinate system.

The quantity $c\tau$, or simply the proper time τ passed on an object, is the same for all observers, is invariant in 4D spacetime, and hence is a scalar. $\left(c\tau\right)^2 = x^\mu x_\mu = x'^\mu x_\mu'$, even though $x'^\mu \neq x^\mu$; $x_\mu' \neq x_\mu$. The term <u>Lorentz invariance</u> is commonly used for 4D scalars.

Other such scalars are the magnitudes of the 4-velocity of (2-54) [equal to c] and the 4-momentum of (2-55) [equal to mc.] Change the unprimed coordinate values in those relations to primed coordinates of another observer in another coordinate frame, and the magnitudes remain the same. We will soon encounter yet other such scalars.

As noted, the components of a vector change in different coordinate systems. This is true in 3D if we rotate to new coordinate axes. It is also true in 4D spacetime for coordinate systems in relative motion with respect to one another (unprimed vs primed coordinates). In both cases, the length of the vector remains the same. Objects which behave in this manner (e.g., vectors like x^μ, u^μ, and p_μ) are said to be <u>covariant</u> under transformation to a new coordinate system. For spacetime, the term <u>Lorentz covariance</u> is common.

Note that the same term "covariant", as opposed to "contravariant", is also used with respect to vector components, but the meaning there is different.

2.9.4 Invariance and the QFT Wave Equations

As we will see, beginning in Chap. 3, contravariant/covariant component notation will provide us with a very useful way of writing the relativistic wave equations of RQM and QFT (see first block of Wholeness Chart 1-2 in Chap. 1) and their solutions. Importantly, these forms of the wave equations are *invariant*. By this we mean that the numerical values of the vector components in the equations will change as the coordinate system changes, but the relations between the vector components will remain the same. In other words, the wave equation has the same form (it looks the same mathematically), whether we use unprimed or primed coordinates. The wave equation is invariant. This is the famous principle of relativity known as <u>Lorentz invariance of the laws of nature</u>. Different observers see different vector component values, but they find the same laws of nature governing the behavior of those components. This is a fundamental principle of special relativity theory, and since QFT is grounded in special relativity, it is a fundamental principle of QFT. Any valid relativistic quantum theory must obey Lorentz invariance. Its governing equations must be invariant.

Note that, with respect to equations, the term *Lorentz covariance (of equations)* is used in the literature interchangeably with *Lorentz invariance (of equations)*. While the *form* of the equations is invariant, the *vectors* in the equation are covariant. Hence, the practice of using either term.

2.9.5 Other Uses for This Stuff

We have only scratched the surface of the mathematics of metrics, contravariant components, and covariant components, formally called <u>differential geometry</u> (or <u>tensor analysis</u>, or in the old days, <u>Riemannian geometry</u>.) Their enormous power becomes more evident when one studies curved spaces, such as the surface of a sphere or the spacetime around a black hole. However, hopefully, this Appendix A provides some justification for their use, which is widespread in QFT.

2.10 Appendix B: Partial vs Total Derivatives

2.10.1 For Relations Like (2-26)

In equation (2-26), one might think that, according to the product differentiation rule, the factor $\frac{\partial \mathcal{O}^S}{\partial t}$ should be a total derivative, as in $\frac{d \mathcal{O}^S}{dt}$, rather than a partial derivative. That is, we would expect the equation to look like (2-61), where the second line comes from Box 2-1, pg. 22.

$$\frac{d\bar{\mathcal{O}}}{dt} = \frac{d}{dt}\langle \psi | \mathcal{O} | \psi \rangle = \left\langle \frac{d\psi}{dt} \Big| \mathcal{O} \Big| \psi \right\rangle + \left\langle \psi \Big| \frac{d\mathcal{O}}{dt} \Big| \psi \right\rangle + \left\langle \psi \Big| \mathcal{O} \Big| \frac{d\psi}{dt} \right\rangle$$
$$= \left\langle \frac{\partial \psi}{\partial t} \Big| \mathcal{O} \Big| \psi \right\rangle + \left\langle \psi \Big| \frac{d\mathcal{O}}{dt} \Big| \psi \right\rangle + \left\langle \psi \Big| \mathcal{O} \Big| \frac{\partial \psi}{\partial t} \right\rangle . \tag{2-61}$$

But as long as our operators are functions of x^i and t or their derivatives (where $x \neq x(t)$), using similar logic to that of Box 2.1, we can take the total time derivative of the operator in the bottom row of (2-61) as a partial time derivative. This is what we do in (2-26).

2.10.2 For Relations Like (2-37)

We can generalize. Consider an entity, call it $\tilde{\mathcal{O}}$, that is a function of fields which are in turn functions of x^i and t. We will temporarily assume $\tilde{\mathcal{O}}$ is a classical entity, and later extrapolate to quantum operators and quantum fields.

$$\tilde{\mathcal{O}} = f(x,t)\, g(x,t)\, h(x,t) \qquad x \neq x(t) \tag{2-62}$$

$\tilde{\mathcal{O}}$ in (2-62) is analogous to $\mathcal{O}^H = U^\dagger \mathcal{O}^S U$ in (2-37). f, g, and h are analogous to $U^\dagger$, $\mathcal{O}^S$ and U.

$$\frac{d\tilde{\mathcal{O}}}{dt} = \frac{d\tilde{\mathcal{O}}}{df}\frac{df}{dt} + \frac{d\tilde{\mathcal{O}}}{dg}\frac{dg}{dt} + \frac{d\tilde{\mathcal{O}}}{dh}\frac{dh}{dt} = gh\frac{df}{dt} + fh\frac{dg}{dt} + fg\frac{dh}{dt}$$
$$= gh\left(\frac{\partial f}{\partial t}\frac{dt}{dt} + \frac{\partial f}{\partial x}\frac{dx}{dt}\right) + fh\left(\frac{\partial g}{\partial t}\frac{dt}{dt} + \frac{\partial g}{\partial x}\frac{dx}{dt}\right) + fg\left(\frac{\partial h}{\partial t}\frac{dt}{dt} + \frac{\partial h}{\partial x}\frac{dx}{dt}\right) = gh\frac{\partial f}{\partial t} + fh\frac{\partial g}{\partial t} + fg\frac{\partial h}{\partial t} . \tag{2-63}$$

Note the equivalence of the ends of the first and second rows in (2-63). So, as long as $\tilde{\mathcal{O}}$ is a function of fields (as it typically is in QFT), the partial time and total time derivatives on the RHS can be interchanged in these sorts of expressions. This holds true as long as x is not a function of t.

If f, g, and h are operators (as in QFT), we have to more careful about the order above (always keeping the factor with f in it to the left of the factor with g, and g to the left of h.)

2.11 Problems

1. Pretend you are scientist in the pre MKS system days, with knowledge of Newton's laws. Units of meters for length, kilograms for mass, and seconds for time have been proposed. What units would force be measured in? Would it be appropriate to give the units for force the shortcut name "newton"? Could you have, alternatively, chosen units for other quantities than length, mass, and seconds as fundamental, and derived units for the remaining quantities? Could you have chosen the speed of sound as one of your basic units and selected it as equal to one and dimensionless? If so, and time in seconds was another basic unit, what units would length have?

2. The fine structure constant α in the Gaussian system (cgs with electromagnetism) is $e^2/4\pi\hbar c$, dimensionless, and approximately equal to 1/137. Without doing any calculations and without looking at Wholeness Chart 2-1, what are its algebraic expression, its dimensions, and its numerical value in natural units? Why can you find the dimensions and numerical value so easily? Does charge have dimensions in natural units? Without looking up the electron charge in Gaussian units, calculate the charge on the electron in natural units. (Answer: .303.)

3. Suppose we have a term in the Lagrangian density of form $m^2\phi^2$, where m has dimensions of mass. What is the dimension M, in natural units, of the field ϕ?

4. a) Derive $x^\alpha = g^{\alpha\beta}x_\beta$. [Hint: Use (2-5) and (2-6), or alternatively, use the matrix form of the contravariant metric tensor along with column vectors in terms of Cartesian coordinates] Note that this relation and (2-5) hold in general for any 4D vector, not just the position vector.

    b) Express $\partial^\mu\partial_\mu$ in terms of i) contravariant and covariant 4D components, and ii) in terms of time t and Cartesian coordinates X_i. The operation $\partial_\mu\partial^\mu = \partial^\mu\partial_\mu$ is called the <u>d'Alembertian</u> operator, and is the 4D Minkowski coordinates analogue of the 3D Laplacian operator $\partial_i\partial_i = \partial^i\partial^i$ of Cartesian coordinates.

    c) Then find $\partial^\mu\partial_\mu\,(x^\alpha x_\alpha)$, where physical length of the interval of x^α is $\sqrt{x^\alpha x_\alpha}$, i) by expressing all terms in t and X_i, and ii) solely using 4D component notation. (For the last part, note, from a), that $\partial x^\alpha / \partial x_\beta = g^{\alpha\beta}$ and from (2-5), $\partial x_\alpha / \partial x^\beta = g_{\alpha\beta}$.)

5. Obtain your answer to the following question by inspection of the final equation in Box 2-2, and then ask yourself whether or not your conclusion feels right intuitively.

    If ϕ^r were a sinusoid, how would the physical momentum density of a short wavelength wave compare to that of a longer one?

6. Consider a classical, non-relativistic field of dust particles in outer space that are so diluted they do not exert any measurable pressure on one another. There is no gravitational, or other, potential density, i.e., $\mathcal{V}(x^i) = 0$. The density of particles is $\rho(x^i)$, which for our purposes we can consider constant in time. The displacement of the field (movement of each dust particle at each point) from its initial position is designated by the field value $\phi^r(x^i)$. $r = 1,2,3$, here, as there is a component of displacement, measured in length units, in each of the three spatial directions. ϕ^r and x^i are both measures of length, but the x^i are fixed locations in space, whereas the ϕ^r are displacements of the particles, in three spatial directions, relative to their initial positions.

    What is the kinetic energy density in terms of the field displacement ϕ^r (actually, it is in terms of the time derivatives of ϕ^r and ϕ_r)? What is the Lagrangian density for the field? Use (2-13) to find the differential equation of motion for the displacement ϕ^r. You should get $\rho\ddot{\phi}_r = 0$. Is this just Newton's second law for a continuous medium with no internal or external force?

7. Without looking back in the chapter, write down the Euler-Lagrange equation for fields. This is a good thing to memorize.

8. In a 3D Cartesian coordinate system, the metric $g_{\mu\nu} = \delta_{\mu\nu}$, the Kronecker delta, where μ, ν take on only values 1,2,3. In that case, it is better expressed as $g_{ij} = \delta_{ij}$ Show that, in such a system, $x^i = x_i$, velocity $v^i = v_i$, and 3-momentum $p^i = p_i$.

9.   Why are the Hamiltonian and the Hamiltonian density not Lorentz scalars? If they are to represent energy and energy density, respectively, does this make sense? (Does the energy of an object or a system have the same value for all observers? Do you measure the same kinetic energy for a plane passing overhead as someone on board the plane would?) Energy is the zeroth component of the four momentum p_μ. Does one component of a four vector have the same value for everyone?

10.   (Do this problem only if you have extra time and want to understand relativity better.) Construct a column like those shown in Wholeness Chart 2-2 for the Relativistic Particle case, but do the entire summary in terms of relativistically covariant relationships. (That is, start with world (proper) time τ and fill in the boxes using 4D momentum, etc.) Keep it simple by treating only a free particle (no potential involved.)

11.   Consider the unitary operator $U = e^{-iHt}$, where H is the Hamiltonian, and a non-energy eigenstate ket, $\left|\psi\right> = C_1\left|\psi_{E_1}\right> + C_2\left|\psi_{E_2}\right>$. What is $U\left|\psi\right>$?

12.   Consider the unitary operator $U = e^{-iH(t-t_0)}$ and $\left|\psi_E\right> = \left|Ae^{-i(Et_0 - \mathbf{p} \cdot \mathbf{x})}\right>$, an energy eigenstate at time t_0. What is $U\left|\psi_E\right>$? Does U here act as a translator of the state in time? That is, does it have the effect of moving the state that was fixed in time forward in time, and turning it into a dynamic entity rather than a static one? If we operate on this new dynamic state with $U^\dagger$, would we turn it back into a static state? Is that not what we do when we operate on a Schrödinger picture state to turn it into a (static) Heisenberg picture state? (Earlier in the chapter we took $t_0 = 0$ to make things simpler.)

13.   (Problem added in revision of 2nd edition). Express the components p^μ of 4-momentum for a photon. Assume it is traveling in the x^1 direction. Use natural units where speed of light $c = 1$. (Hint: Use energy expressed in terms of frequency f and 3-momentum in terms of wave length λ.) Then show that even though $p^\mu \neq 0$, $(p)^2 = p^\mu p_\mu = 0$. (Hint: Use speed of light expressed in terms of frequency and wave length.) Does this make sense in light of (2-55), given what we know about the photon mass? Then express p^μ in terms of $\omega = 2\pi f$ and wave number $k = 2\pi/\lambda$ where $\hbar = h/2\pi$. ($= 1$ in natural units).

Part One
Free Fields

Like a bird on a wire,
like a drunk in midnight choir,
I have tried in my way to be free.
Leonard Cohen

Chapter 3

Scalars: Spin 0 Fields

*..if I look back at my life as a scientist and a teacher, I think the most important
and beautiful moments were when I say, "ah-hah, now I see a little better" ...
this is the joy of insight which pays for all the trouble one has had in this career.*

Victor F. Weisskopf
Quarks, Quasars, and Quandaries

3.0 Preliminaries

This chapter presents the most fundamental concepts in the theory of quantum fields and contains the very essence of the theory. Master this chapter, and you are well on your way to mastering that theory.

3.0.1 Background

Early efforts to incorporate special relativity into quantum mechanics started with the non-relativistic Schrödinger equation,

Seeking a relativistic quantum theory?

$$i\hbar \frac{\partial}{\partial t}\phi = H\phi \qquad \text{where } H = \frac{p^2}{2m} + V = -\frac{\hbar^2}{2m}\nabla^2 + V \,, \qquad (3\text{-}1)$$

Try relativistic Hamiltonian in Schrödinger equation

and attempted to find a relativistic, rather than non-relativistic, form for the Hamiltonian H.[1] One might guess that approach would lead to a valid relativistic Schrödinger equation. This is, in essence, true but there is one problem, as we will see below.

In special relativity, the 4-momentum vector is Lorentz covariant, meaning its length in 4D space is invariant. For a free particle (i.e., $V = 0$),

$$p^\mu p_\mu = m^2 c^2 = g_{\mu\nu}p^\mu p^\nu = \begin{bmatrix} E/c & p^1 & p^2 & p^3 \end{bmatrix}\begin{bmatrix} E/c \\ -p^1 \\ -p^2 \\ -p^3 \end{bmatrix} \rightarrow \frac{E^2}{c^2} = \mathbf{p}^2 + m^2 c^2 \,. \qquad (3\text{-}2)$$

Relativistic energy E

Changing dynamical variables over to operators (as happens in quantization), i.e.,

$$E \rightarrow H \qquad \text{and} \qquad p^i \rightarrow -i\hbar\partial_i \,, \qquad (3\text{-}3)$$

Relativistic $E \rightarrow$ relativistic operator H

one finds, from the RHS of (3-2),

---

[1] Actually, Schrödinger first attempted to find a wave equation that was relativistic and came up with what later came to be known as the Klein-Gordon equation, which we will study in this chapter. He discarded it because of problems discussed later on herein, and because it gave wrong answers for the hydrogen atom. Shortly thereafter, he deduced the non-relativistic Schrödinger equation we are familiar with. Some time afterwards, other researchers then tried to "relativize" that equation, as discussed herein.

$$H = \sqrt{-\hbar^2 c^2 \partial_i \partial_i + m^2 c^4} \ , \tag{3-4}$$

*Bad news:
Relativistic H has
square root of a
differential
operator*

seemingly the only form a relativistic Hamiltonian could take. Unfortunately, taking the square root of terms containing a derivative is problematic, and difficult to correlate with the physical world.

The solution to the problem of finding a <u>relativistic Schrödinger equation</u> has been found, however, and as we will see in the next three chapters, turns out to be different for different spin types. This was quite unexpected at first, but has since become a cornerstone of relativistic quantum theory. (See first row of Wholeness Chart 1-2 in Chap. 1, pg. 7.)

*But answer has
been found, as we
will see*

Particles with zero spin, such as π-mesons (pions) and the famous Higgs boson, are known as <u>scalars</u>, and are governed by one particular relativistic Schrödinger equation, deduced by (after Schrödinger, actually), and named after, Oscar Klein and Walter Gordon. Particles with ½ spin, such as electrons, neutrinos, and quarks, and known as <u>spinors</u>, by a different relativistic Schrödinger equation, discovered by Paul Dirac. And particles with spin 1, such as photons and the W's and Z's that carry the weak charge, and known as <u>vectors</u>, by yet another relativistic Schrödinger equation, discovered by Alexandru Proca. The Proca equation reduces, in the massless (photon) case, to Maxwell's equations.

*Each spin type
has its own
relativistic wave
equation*

We will devote a separate chapter to each of these three spin types and the wave equation associated with each. We begin in this chapter with scalars.

3.0.2 Chapter Overview

RQM first,

where we will look at

*RQM overview
(scalars)*

- deducing the Klein-Gordon equation, the first relativistic Schrödinger equation, using the relativistic H^2,
- solutions (which are states = wave functions) to the Klein-Gordon equation,
- probability density and its connection to the funny normalization constant in the solutions, and
- the problem with negative energies in the relativistic solutions.

Then QFT,

- using the classical relativistic $\mathcal{L}$ (Lagrangian density) for scalar fields, and the Legendre transformation to get $\mathcal{H}$ (Hamiltonian density),

*QFT overview
(scalars)*

- from $\mathcal{L}$ and the Euler-Lagrange equation, finding the same Klein-Gordon equation, with the same mathematical form for the solutions, but this time the solutions are fields, not states,
- from 2nd quantization, finding the commutation relations for QFT,
- determining relevant operators in QFT: $H = \int \mathcal{H} \, d^3x$, number, creation/destruction, etc.,
- showing this approach avoids negative energy states,
- seeing how the vacuum is filled with quanta of energy ½$\hbar\omega$,
- deriving other operators (probability density, 3-momentum, charge) and
- picking up relevant loose ends (scalars = bosons, Fock (multiparticle) space).

And then,

- seeing quantum fields in a different light, as harmonic oscillators.

With finally, and importantly,

- finding the Feynman propagator, the mathematical expression for virtual particles.

Free (no force) Fields

In this chapter, as well as Chaps. 4 (spin ½) and 5 (spin 1), we will deal only with fields/particles that are not interacting, i.e., feel no force = "free". Thus, we will take potential energy $V = 0$. In Chap. 7, which begins Part 2 of the book, we will begin to investigate interactions.

*We study free (no
interactions)
case first*

3.1 Relativistic Quantum Mechanics: A History Lesson

3.1.1 Two Possible Routes to RQM

Recall from Chaps. 1 and 2, that 1st quantization, for both non-relativistic and relativistic particle theories, entails i) using the classical form of the Hamiltonian as the quantum form of the

Hamiltonian, and ii) changing Poisson brackets to commutators. We recall also from Prob. 6 of Chap. 1 that non-commutation of dynamical variables means those variables are operators (because ordinary numbers commute.) For example,

$$\left[p^i, x^j \right] = -i\hbar\delta_i^{\ j} \xleftrightarrow{\text{equivalent}} p^i = -i\hbar\partial_i \qquad (3\text{-}5)$$

as the RHS above is the only form that satisfies the LHS, and it is an operator.

Non-commuting variables must be operators

One might expect that this is the route we would follow to obtain RQM, i.e., 1st quantization of relativistic classical particle theory. However, historically, it was done differently. That is, RQM was first extrapolated from NRQM, not from classical theory. As illustrated in Fig. 3-1, it can be done either way.

In this book, to save space and time, we will only show one of these paths, the historical one represented by the lowest arrow in Fig. 3-1.

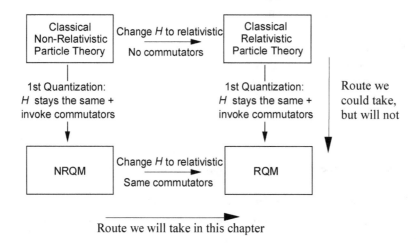

Figure 3-1. Different Routes to Relativistic Quantum Mechanics

3.1.2 Deducing the Klein-Gordon Equation

As we saw in Sect. 3.0.1, when we try to use a relativistic Hamiltonian in the Schrödinger equation, we have the problem of the partial derivative operator (see (3-4)) being under a square root sign. So, rather than use H, Klein and Gordon, in 1927, did the next best thing. They used H^2 instead. That is, they squared the operators (operate on each side twice rather than once) in the original Schrödinger equation (3-1) and thus from (3-2), obtained

Let's square operators on both sides of Schrödinger eq

$$\left(i\hbar\frac{\partial}{\partial t} \right)\left(i\hbar\frac{\partial}{\partial t} \right)\phi = H^2\phi = \left(\mathbf{p}_{oper}^2 c^2 + m^2 c^4 \right)\phi , \qquad (3\text{-}6)$$

which becomes from the square of (3-4)

Then use operator form for H^2

$$-\frac{\hbar^2}{c^2}\frac{\partial^2}{\partial t^2}\phi = \left(-\hbar^2\frac{\partial}{\partial X_i}\frac{\partial}{\partial X_i} + m^2 c^2 \right)\phi \ \rightarrow \ -\frac{\partial}{\partial x^0}\frac{\partial}{\partial x_0}\phi = \left(\frac{\partial}{\partial x^i}\frac{\partial}{\partial x_i} + \underbrace{\frac{m^2 c^2}{\hbar^2}}_{\mu^2} \right)\phi . \qquad (3\text{-}7)$$

Re-arranging, we have the <u>Klein-Gordon equation</u> (expressed in two equivalent ways with slightly different notation)

To get the Klein-Gordon equation

$$\left(\frac{\partial}{\partial x^\mu}\frac{\partial}{\partial x_\mu} + \mu^2 \right)\phi = 0 \quad \text{or} \quad \left(\partial_\mu\partial^\mu + \mu^2 \right)\phi = 0 , \quad \mu^2 = \frac{m^2 c^2}{\hbar^2}\left(= m^2 \text{ in nat. units} \right). \qquad (3\text{-}8)$$

As noted in Chap. 2, Prob. 4, the operation $\partial_\mu\partial^\mu = \partial^\mu\partial_\mu$ is called the <u>d'Alembertian</u> operator, and is the 4D Minkowski coordinates analogue of the 3D Laplacian operator $\partial_i\partial_i = \partial^i\partial^i$ of Cartesian coordinates.

In 1934, Pauli and Weisskopf [1] showed that the Klein-Gordon equation specifically describes a spin-0 (scalar) particle. This should become evident to us as we study the Dirac and Proca equations, for spin ½ and spin 1, later on, and compare them to the Klein-Gordon equation.

Klein-Gordon equation is specifically for scalars

3.1.3 The Solutions to the Klein-Gordon Equation

A solution set to (3-8), readily checked by substitution into (3-8) (which is good practice when using contravariant/covariant notation), is (where $E_n^2 - \mathbf{p}_n^2 = m^2$)

$$\phi(x) = \sum_{n=1}^{\infty} \frac{1}{\sqrt{2VE_n/\hbar}} \left(A_n e^{-\frac{i}{\hbar}(E_n t - \mathbf{p}_n \cdot \mathbf{x})} + \underbrace{B_n^\dagger e^{\frac{i}{\hbar}(E_n t - \mathbf{p}_n \cdot \mathbf{x})}}_{\text{absent in NRQM}} \right), \quad (3-9)$$

Solutions to Klein-Gordon equation (discrete)

where we will discuss the funny looking normalization factor in front, containing the volume V and the energy of the nth solution, later. The coefficients A_n and $B_n^\dagger$ are constants, and a complex conjugate form for the coefficient of the last term above, i.e., $B_n^\dagger$, is used because it will prove advantageous later.

This is a discrete set of solutions, typical for cases with waves constrained inside a volume V, though V can be taken as large as one wishes. Each discrete wavelength in the summation of (3-9) fits an integer number of times inside the volume V. Continuous (integral rather than sum) solutions, for waves not constrained inside a specific volume V, exist for (3-8) as well, but we are not concerned with them at this point.

Continuous solutions also exist

This solution set is also specifically for plane waves. We will not consider alternative solution forms for other wave shapes that would exist in problems with cylindrical or spherical geometries.

Only plane wave solutions here

The solution (3-9), because we are working in RQM, is a state, i.e., $\phi(x)$ above $= |\phi(x)\rangle$, for a single particle. Each individual term in the summation is an eigenstate. $\phi(x)$ is a general state superposition of eigenstates.

Solutions in RQM are states (particles)

Note that in NRQM, we only had terms in the counterpart to (3-9) that had the exponential form of $-i(E_n t - \mathbf{p}_n \cdot \mathbf{x})/\hbar$, because that was the only form that satisfied the non-relativistic Schrödinger equation. Because we are using the square of the relativistic Hamiltonian in RQM, we get additional solutions of exponential form $+i(E_n t - \mathbf{p}_n \cdot \mathbf{x})/\hbar$ that also solve the relativistic Klein-Gordon equation. You should do Prob. 1, at the end of the chapter, to justify the statements in this paragraph to yourself.

Relativistic form has extra set of solutions

With an aim towards using natural units, we note the following relations, where wave number $k_i = 2\pi/\lambda_i$ and we use the deBroglie relation $p^i = \hbar k^i$,

$$p_\mu = \begin{bmatrix} E/c \\ p_i \end{bmatrix} = \begin{bmatrix} E/c \\ -p^i \end{bmatrix} = \hbar k_\mu = \begin{bmatrix} \hbar\omega/c \\ -\hbar k^i \end{bmatrix} \xrightarrow{\text{nat. units}} p_\mu = \begin{bmatrix} E \\ -p^i \end{bmatrix} = k_\mu = \begin{bmatrix} \omega \\ -k^i \end{bmatrix}, \quad (3-10)$$

Relations for p_μ and k_μ

and recall the notation introduced in Chap. 2,

$$px = p_\mu x^\mu = Et - p^i x^i = Et - \mathbf{p} \cdot \mathbf{x} \qquad (= p^\mu x_\mu)$$

$$kx = k_\mu x^\mu = \omega t - k^i x^i = \frac{Et}{\hbar} - \frac{p^i x^i}{\hbar} = \frac{p_\mu}{\hbar} x^\mu \qquad (= k^\mu x_\mu) \quad (3-11)$$

Notation review

in nat. units → $E = \omega, \quad p_i = k_i, \quad p_\mu = k_\mu, \quad px = kx.$

It is then common to re-write (3-9) in natural units with the above notation. In doing so, we also switch the dummy summation variable n, which represents each individual wave in the summation, to the 3D vector quantity $\mathbf{k}$, representing the wave number and direction of each possible wave. For free fields, a given wave with wave number vector $\mathbf{k}$ has a particular energy (see (3-2) with $\mathbf{p} = \mathbf{k}$ in natural units), and we can designate that energy via either $E_\mathbf{k}$ or $\omega_\mathbf{k}$. It is common practice for scalars to use $\mathbf{k}$ (rather than $\mathbf{p}$) and $\omega_\mathbf{k}$ (rather than $E_\mathbf{p}$ or $E_\mathbf{k}$.)

[1] Pauli, W. and Weisskopf, V., Helv. Phys. Acta 7, 709 (1934). Translation in Miller, A. I., *Early Quantum Electrodynamics: A Source Book*, Cambridge U. Press, New York (1994)

The <u>Klein-Gordon equation solutions</u> (3-9) then become, in <u>natural units</u>

$$\phi(x) = \sum_{\mathbf{k}} \frac{1}{\sqrt{2V\omega_{\mathbf{k}}}} \left(A_{\mathbf{k}} e^{-ikx} + B_{\mathbf{k}}^{\dagger} e^{ikx} \right) \quad . \tag{3-12}$$

Natural units form of Klein-Gordon solutions

Except for Box 3-1, which reviews NRQM, we will henceforth, in this chapter, use natural units.

<u>Definition of Eigensolutions</u>

As noted previously, in RQM, the solution ϕ of (3-12) is that of a general (sum of eigenstates) single particle state. Each <u>eigenstate</u> has mathematical form (where we are going to omit the $2\omega_{\mathbf{k}}$ part here, because of what is coming)

Eigenstates of Klein-Gordon equation

$$\phi_{\mathbf{k},A} = \frac{e^{-ikx}}{\sqrt{V}} \quad \text{or} \quad \phi_{\mathbf{k},B^{\dagger}} = \frac{e^{ikx}}{\sqrt{V}}. \tag{3-13}$$

Each of these forms has what is called <u>unit norm</u>. That is, for $\phi_{\mathbf{k},A}$ (and similarly, for $\phi_{\mathbf{k},B^{\dagger}}$),

$$\int \phi_{\mathbf{k},A}^{\dagger} \phi_{\mathbf{k},A} d^3x = \frac{1}{V} \int_V e^{ikx} e^{-ikx} d^3x = 1, \tag{3-14}$$

Eigenstates have unit norm

or more generally, all such eigenstates are <u>orthonormal</u>, i.e., their inner products are

$$\int \phi_{\mathbf{k},A}^{\dagger} \phi_{\mathbf{k}',A} d^3x = \frac{1}{V} \int_V e^{ikx} e^{-ik'x} d^3x = \delta_{\mathbf{k}\mathbf{k}'}. \tag{3-15}$$

and are orthogonal

Similar relations to (3-15) exist for $\phi_{\mathbf{k},B^{\dagger}}$, and every $\phi_{\mathbf{k},A}$ is orthogonal to every $\phi_{\mathbf{k},B^{\dagger}}$. Work this out by doing Prob.2.

Relations (3-13) to (3-15) should look familiar from NRQM. There, (3-14) was the integral of the probability density for a particle in an eigenstate. In RQM, however, things are a little different, as we will see, and we use the term "unit norm" for the property displayed in (3-14).

Unit norm eigenstates were advantageous in NRQM, and they will be in QFT as well. That is the reason we omitted the $2\omega_{\mathbf{k}}$ part of our solutions (3-12) in forming our definitions (3-13). By so doing, the eigenstates then have unit norm, and things just turn out easier later on.

We defined eigenstates to have unit norm because it will be advantageous

3.1.4 Probability Density in RQM

We are going to investigate probability density in RQM, but first look over Box 3-1, and be sure you understand how probability density is derived in NRQM.

<u>Probability Density Using the Klein-Gordon Equation</u>

For RQM, we start with the Klein-Gordon equation rather than Schrödinger equation. First post-multiply it by $\phi^{\dagger}$, then subtract the complex conjugate equation post-multiplied by ϕ, i.e.,

Deduce RQM probability density using relativistic wave equation

$$\left\{ \frac{\partial^2}{\partial t^2} \phi = \left(\nabla^2 - \mu^2 \right) \phi \right\} \phi^{\dagger}$$

$$- \left\{ \frac{\partial^2}{\partial t^2} \phi^{\dagger} = \left(\nabla^2 - \mu^2 \right) \phi^{\dagger} \right\} \phi, \tag{3-16}$$

and note that $\mu^2 \phi^{\dagger} \phi - \mu^2 \phi \phi^{\dagger} = 0$. The LHS of the result can be replaced with the new LHS in (3-17) below, and the RHS with (3-18).

$$\underbrace{\frac{\partial^2 \phi}{\partial t^2} \phi^{\dagger} - \frac{\partial^2 \phi^{\dagger}}{\partial t^2} \phi}_{\text{LHS of result above}} + \underbrace{\frac{\partial \phi}{\partial t} \frac{\partial \phi^{\dagger}}{\partial t} - \frac{\partial \phi^{\dagger}}{\partial t} \frac{\partial \phi}{\partial t}}_{= 0} = \underbrace{\frac{\partial}{\partial t} \left(\frac{\partial \phi}{\partial t} \phi^{\dagger} - \frac{\partial \phi^{\dagger}}{\partial t} \phi \right)}_{\text{new LHS}} \tag{3-17}$$

$$\underbrace{\left(\nabla^2 \phi \right) \phi^{\dagger} - \left(\nabla^2 \phi^{\dagger} \right) \phi}_{\text{RHS of result above}} + \underbrace{\nabla \phi \cdot \nabla \phi^{\dagger} - \nabla \phi^{\dagger} \cdot \nabla \phi}_{= 0} = \underbrace{\nabla \cdot \left(\left(\nabla \phi \right) \phi^{\dagger} - \left(\nabla \phi^{\dagger} \right) \phi \right)}_{\text{new RHS}} \tag{3-18}$$

Box 3-1. Review of Non-Relativistic QM Probability Density

In non-relativistic quantum mechanics (NRQM), we encountered 1) the wave function solution to the Schrödinger equation Ψ, and 2) the particle probability density $\rho = \Psi^{\dagger}\Psi$ (or equivalently when Ψ is a scalar quantity, $\Psi^*\Psi$.) We review here the derivation of that relation for probability density.

Conserved quantities in field theory:

Recall the <u>continuity equation</u> of continuum mechanics and electromagnetism,

$$\frac{\partial \rho}{\partial t} + \nabla \cdot \mathbf{j} = 0 \qquad \left(\xrightarrow{\text{implies}} \int_V \rho \, d^3 x = \text{constant in time} \right), \qquad \text{(B3-1.1)}$$

where ρ is density (mass or charge density), $\mathbf{j}$ is the 3D current density (mass/area-sec or charge/area-sec), and V is all space, or at least large enough so that everywhere outside it, for all time, $\rho = 0$. V is fixed in space and time, whereas ρ can change in space and time inside V. Any conserved quantity (such as total mass M or total charge Q) obeys (B3-1.1).

The general procedure:

Use the governing quantum wave equation to deduce another equation having the form of the continuity equation (B3-1.1), and we will then know that ρ, whatever it turns out to be in that case, must represent a conserved quantity. Its integral over all space is constant in time. If we normalize ρ such that when integrated over all space, the result equals one, we can conjecture that ρ is the particle probability density (which when integrated over all space equals the probability that we will find the particle somewhere in all space, i.e., one.) Then throughout time, as our particle evolves, moves, and rearranges its probability density distribution, the total probability of finding it somewhere in space is always one. It turns out, from experiment, that the conjecture that this quantity ρ in NRQM equals probability density is true.

Probability Density Using the Schrödinger Equation:

First, pre-multiply the Schrödinger equation by the complex conjugate of the wave function, i.e.,

$$\Psi^{\dagger} \left\{ \frac{\partial}{\partial t}\Psi = \frac{1}{i\hbar}\left(-\frac{\hbar^2}{2M}\nabla^2 + V \right)\Psi \right\} \qquad \text{(B3-1.2)}$$

Then, post-multiply the complex conjugate of the Schrödinger equation by the wave function

$$\left\{ \frac{\partial}{\partial t}\Psi^{\dagger} = \frac{-1}{i\hbar}\left(-\frac{\hbar^2}{2M}\nabla^2 + V^{\dagger} \right)\Psi^{\dagger} \right\}\Psi \qquad \text{(B3-1.3)}$$

where the potential V is real so $V = V^{\dagger}$. Adding (B3-1.2) to (B3-1.3), we get

$$\Psi^{\dagger}\frac{\partial \Psi}{\partial t} + \frac{\partial \Psi^{\dagger}}{\partial t}\Psi = \Psi^{\dagger}\frac{1}{i\hbar}\left(-\frac{\hbar^2}{2M}\nabla^2 + V \right)\Psi + \left(\frac{-1}{i\hbar}\left(-\frac{\hbar^2}{2M}\nabla^2\Psi^{\dagger} + V^{\dagger}\Psi^{\dagger} \right) \right)\Psi \qquad \text{(B3-1.4)}$$

or

$$\frac{\partial\left(\Psi^{\dagger}\Psi \right)}{\partial t} = \underbrace{\frac{-\hbar}{2iM}\left(\Psi^{\dagger}\left(\nabla^2\Psi \right) - \left(\nabla^2\Psi^{\dagger} \right)\Psi \right)}_{\nabla \cdot \left[\Psi^{\dagger}(\nabla\Psi) - \left(\nabla\Psi^{\dagger} \right)\Psi \right]} + \underbrace{\frac{\Psi^{\dagger}V\Psi}{i\hbar} - \frac{V^{\dagger}\Psi^{\dagger}\Psi}{i\hbar}}_{=0 \text{ since } V^{\dagger}=V} \qquad \text{(B3-1.5)}$$

This is the same as the continuity equation (B3-1.1) if we take as our <u>probability density</u>

$$\rho = \Psi^{\dagger}\Psi , \qquad \text{(B3-1.6)}$$

and as our <u>probability current density</u> (sometimes just <u>probability current</u>)

$$\mathbf{j} = \frac{\hbar}{2iM}\left\{ \Psi^{\dagger}\left(\nabla\Psi \right) - \left(\nabla\Psi^{\dagger} \right)\Psi \right\} . \qquad \text{(B3-1.7)}$$

This is how the commonly used relation (B3-1.6) is found.

Equating the new LHS of (3-17) to the new RHS of (3-18), and to make future work easier, multiplying both sides by the constant i, gives the form of the continuity equation

Manipulations of the wave equation lead to an equation like the continuity equation

$$i\frac{\partial}{\partial t}\left(\frac{\partial \phi}{\partial t}\phi^\dagger - \frac{\partial \phi^\dagger}{\partial t}\phi\right) = i\nabla\cdot\left((\nabla\phi)\phi^\dagger - (\nabla\phi^\dagger)\phi\right) \;\rightarrow\; \frac{\partial\rho}{\partial t}+\nabla\cdot\mathbf{j}=0, \tag{3-19}$$

where probability density and the probability current for a Klein-Gordon particle are

From that, we deduce form of RQM probability density

$$\rho = j^0 = i\left(\frac{\partial\phi}{\partial t}\phi^\dagger - \frac{\partial\phi^\dagger}{\partial t}\phi\right),\text{ and} \tag{3-20}$$

$$\mathbf{j} = -i\left((\nabla\phi)\phi^\dagger - (\nabla\phi^\dagger)\phi\right) \qquad j^i = -i\left(\phi_{,i}\phi^\dagger - \phi^\dagger{}_{,i}\phi\right) = i\left(\phi^{,i}\phi^\dagger - \phi^{\dagger,i}\phi\right). \tag{3-21}$$

Importantly, and perhaps surprisingly, the relativistic form of the probability density (3-20) is _not_ the same as (B3-1.6), the NRQM probability density.

4 Currents

We introduce 4D notation for the scalar and 3D vector of (3-19) and define the scalar 4-current

$$j^\mu = \begin{bmatrix}\rho \\ \mathbf{j}\end{bmatrix} = \begin{bmatrix}\rho \\ j^i\end{bmatrix} = \begin{bmatrix}j^0 \\ j^i\end{bmatrix} = i\left(\phi^{,\mu}\phi^\dagger - \phi^{\dagger,\mu}\phi\right)\;. \tag{3-22}$$

4-current and 4D form of continuity equation

The 4D continuity equation form of (3-19) is then

$$\boxed{\frac{\partial j^\mu}{\partial x^\mu} = \partial_\mu j^\mu = j^\mu{}_{,\mu}=0}\,, \tag{3-23}$$

where we have shown three common notational ways to designate partial derivative. (3-23) tells us the _important fact_ that the 4-divergence _of the 4-current of any conserved quantity_ (total probability in this case) _is zero._

4-divergence of 4-current of conserved quantity always = 0

Probability for Klein-Gordon Discrete Solutions

For a single particle state in RQM, we are going to assume at first, for simplicity, that the solution (3-12), has only terms with coefficients $A_\mathbf{k}$, i.e., the general state ϕ contains no eigenstates shown with coefficients $B_\mathbf{k}^\dagger$. Probability density (3-20) is then (where primes do *not* denote derivatives with respect to spatial coordinates, merely different summation dummy variables)

Scalar probability density in terms of first Klein-Gordon solution set

$$\rho = \left(\sum_\mathbf{k}\frac{\omega_\mathbf{k}A_\mathbf{k}}{\sqrt{2\omega_\mathbf{k}}}\frac{e^{-ikx}}{\sqrt{V}}\right)\left(\sum_{\mathbf{k}'}\frac{A_{\mathbf{k}'}^\dagger}{\sqrt{2\omega_{\mathbf{k}'}}}\frac{e^{ik'x}}{\sqrt{V}}\right) + \left(\sum_{\mathbf{k}'}\frac{\omega_{\mathbf{k}'}A_{\mathbf{k}'}^\dagger}{\sqrt{2\omega_{\mathbf{k}'}}}\frac{e^{ik'x}}{\sqrt{V}}\right)\left(\sum_\mathbf{k}\frac{A_\mathbf{k}}{\sqrt{2\omega_\mathbf{k}}}\frac{e^{-ikx}}{\sqrt{V}}\right), \tag{3-24}$$

where the $\omega_\mathbf{k}$ and $\omega_{\mathbf{k}'}$ came from the time derivatives.

If we integrate ρ over the volume V (which is large enough to encompass the entire state), the result must equal 1. When we do so, all terms with $\mathbf{k}' \neq \mathbf{k}$ go to zero, so the $\omega_{\mathbf{k}'}\rightarrow\omega_\mathbf{k}$ and cancel out. The V term in the denominator cancels in the integration over the volume V, and the two terms result in a factor of 2 that cancels with the 2 in the denominator. The result is

Square of absolute value of coefficient $A_\mathbf{k}$ = probability of finding kth eigenstate

$$\int\rho d^3x = \sum_\mathbf{k}|A_\mathbf{k}|^2 = 1\,. \tag{3-25}$$

Thus $|A_\mathbf{k}|^2$ is the probability of measuring the $\mathbf{k}$th eigenstate, similar to what the coefficients of eigenstates represented in NRQM.

Difference from NRQM

Note that in RQM

Comparing probability in NRQM and RQM

$$\underbrace{\int \phi^\dagger\phi\, d^3x}_{\neq\rho} = \sum_\mathbf{k}\frac{(A_\mathbf{k})^2}{2\omega_\mathbf{k}}\neq 1 \quad\text{but}\quad \int\underbrace{i\left(\frac{\partial\phi}{\partial t}\phi^\dagger - \frac{\partial\phi^\dagger}{\partial t}\phi\right)}_{=\rho}d^3x = \sum_\mathbf{k}|A_\mathbf{k}|^2 = 1 \quad\text{(RQM)}, \tag{3-26}$$

whereas in NRQM, we had

$$\int \underbrace{\phi^{\dagger}\phi}_{=\rho} \, d^3x = \sum_{\mathbf{k}} |A_{\mathbf{k}}|^2 = 1 \qquad \text{(NRQM)} . \qquad (3\text{-}27)$$

<u>Normalization Factors</u>

Obtaining the RHS of (3-26) is the reason for the normalization factors $1/\sqrt{2\omega_{\mathbf{k}}V}$ used in the solution ϕ of (3-12) and (3-9). Those factors result in a total probability of one for a single particle and $|A_{\mathbf{k}}|^2$ as the probability for measuring the respective eigenstate. That is, the form of the relativistic field equation gave us the form of the probability density in (3-20) (and (3-26)), and the need to have total probability of unity gave us the normalization factors in the solutions.

*RQM normalization factors arise from need to have total probability = 1 and $|A_{\mathbf{k}}|^2$ = probability of **k**th state*

<u>Relativistic Invariance of Probability</u>

This total probability value of unity in (3-25) (and (3-26)) is a relativistic invariant (i.e., a world scalar.) If we change our frame, the energy spectrum (i.e., the $\omega_{\mathbf{k}}$ values) will change (kinetic energy for each energy-momentum eigenstate looks different). But these changes cancel out in the probability calculation, since the $\omega_{\mathbf{k}}$ cancel, and always result in a total probability of one for any frame. Further, the $A_{\mathbf{k}}$ here are constants that do not vary with frame, so the probability of finding any particular state is also independent of what frame the measurements are taken in.

Total probability and $A_{\mathbf{k}}$ are frame independent (relativistically invariant)

Note that this means the normalization factors chosen provide relativistic invariance of total probability, which we would not have had with any other choice.

3.1.5 Negative Energies in RQM

If we take our traditional operator form for H as $i\partial/\partial t$ and operate on one of our Klein-Gordon solution eigenstates of (3-12) and (3-13), we should get the energy eigenvalue $\omega_{\mathbf{k}}$. When we do this for the eigenstates with exponents in $-ikx$, all looks as expected.

$$H\phi_{\mathbf{k},A} = E_{\mathbf{k},A}\phi_{\mathbf{k},A} \;\rightarrow\; i\frac{\partial\phi_{\mathbf{k},A}}{\partial t} = i\frac{\partial}{\partial t}\frac{e^{-ikx}}{\sqrt{V}} = \omega_{\mathbf{k}}\frac{e^{-ikx}}{\sqrt{V}} = \omega_{\mathbf{k}}\phi_{\mathbf{k},A} = E_{\mathbf{k},A}\phi_{\mathbf{k},A} . \qquad (3\text{-}28)$$

However, when we do it for the eigenstates with exponents in $+ikx$, we have an "uh-oh", i.e.,

$$H\phi_{\mathbf{k},B^{\dagger}} = E_{\mathbf{k},B^{\dagger}}\phi_{\mathbf{k},B^{\dagger}} \;\rightarrow\; i\frac{\partial\phi_{\mathbf{k},B^{\dagger}}}{\partial t} = i\frac{\partial}{\partial t}\frac{e^{ikx}}{\sqrt{V}} = -\omega_{\mathbf{k}}\frac{e^{ikx}}{\sqrt{V}} = -\omega_{\mathbf{k}}\phi_{\mathbf{k},B^{\dagger}} = E_{\mathbf{k},B^{\dagger}}\phi_{\mathbf{k},B^{\dagger}} . \quad (3\text{-}29)$$

Half of our RQM eigenstates have negative energy

Since $\omega_{\mathbf{k}}$ is always a positive number, we have states with negative energies in RQM. We might have expected this, since we used the square of the Hamiltonian as the basis of RQM, and square roots typically have both positive and negative signs.

The bottom line: This is not an attribute of what a good theory has been expected to have, i.e., solely positive energies as we see in our world. As we will shortly see, QFT solved this dilemma (as well as others delineated in Chap. 1.)

3.1.6 Negative Probabilities in RQM

Do Prob. 3 to prove to yourself that a particle ϕ containing only eigenstates of the exponential form $+i(E_n t - \mathbf{p_n}\cdot\mathbf{x})/\hbar = ikx$ (i.e., those with coefficients $B_{\mathbf{k}}^{\dagger}$ in (3-12)) has total probability of being measured of -1. The extra states in RQM have physically untenable negative probabilities!

Time to move on to QFT.

Half of our RQM eigenstates have negative probability density

3.2 The Klein-Gordon Equation in Quantum Field Theory

3.2.1 States vs Fields

It should come as no surprise, to those who have read Chap. 1, that the fundamental scalar wave equation of RQM, the Klein-Gordon equation (3-8), is also the fundamental scalar wave equation of QFT, except that ϕ therein is considered a field, instead of a state. The word "field" in classical theory means an entity that, unlike a particle, is spread out, i.e., is a function of space (it has different values at different spatial locations) and typically also a function of time. The state ϕ of NRQM and RQM certainly fills that bill, but in quantum theory we don't use the word "field" for this, we use the word "state" (or "wave function" or "ket" or "particle".)

States & fields both spread out in space. But in quantum theories, "field" also means "operator"

The word "field" in quantum theory refers to a quantity that is spread out in space, but also, importantly, as we will soon see, is an operator in QFT. More properly, it is called a <u>quantum field</u> or an <u>operator field</u>, though the short term <u>field</u> is far more common. Confusingly, we use the same symbol ϕ in QFT for a field as we used for a state in NRQM and RQM.

Notation

In QFT, symbols such as ϕ, which are not part of a ket symbol, do not represent states, but fields. Unless otherwise explicitly noted, <u>in QFT notation</u>,

$|\phi\rangle$ symbolizes <u>a state</u> (particle)   and ϕ symbolizes <u>a field</u> (operator),

On the other hand, <u>in NRQM and RQM</u>, <u>both symbols</u> above represented the same thing, <u>a state</u>.

Notational difference between states and fields. In QFT, ϕ is not a state, but a field

We will understand these distinctions a little better later, but for now understand that formally, the Klein-Gordon equation in QFT is called a <u>field equation</u>, because its solution ϕ is a (quantum or operator) field. See the second and third rows of Wholeness Chart 1-2 in Chap. 1, pg. 7.

There are two common ways to derive this equation, which we present in the following two sections, plus a third, which is a good check on the theory and can be found in the Appendix A.

3.2.2 From RQM to QFT

Fig. 3-2 illustrates, schematically, the two basic routes to QFT. The quickest is at the bottom of the figure, for which we simply postulate that the solution ϕ of the Klein-Gordon equation (3-8) describes a field (instead of a particle). This is reasonable, since ϕ is a function of spatial location (and often time), i.e., it is a field in the formal mathematical sense.

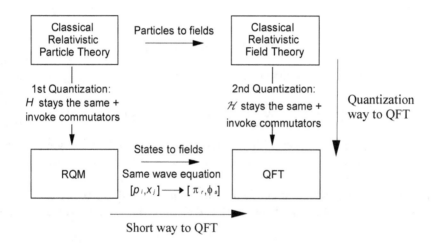

Two different routes to QFT

Figure 3-2. Different Routes to Quantum Field Theory

We then must apply the commutation relations for fields (see Chap. 2, pg. 31, Wholeness Chart 2-5, 6th column = 3rd column on right hand page), instead of the commutation relations for particle properties (same chart, 3rd column on left hand page). When we do this, and simply crank the mathematics, we obtain QFT. Because the QFT we then obtain describes the real world so well, it justifies the original postulate.

Short route: RQM → QFT. Similar math as 2nd quantization below

The formal mathematics are much the same as for the alternative route, illustrated on the RHS of Fig. 3-2, and treated in the next section.

3.2.3 From Classical Relativistic Fields to QFT

Classical Scalar Fields

The classical <u>Lagrangian density</u> for a free (no forces), real, relativistic scalar field ϕ has form

2nd quantization route: Classical fields → QFT

$$\mathcal{L}_0^0 = K\left(\partial_\alpha \phi \partial^\alpha \phi - \mu^2 \phi\phi\right) = K\left(\dot\phi\dot\phi + \partial_i \phi \partial^i \phi - \mu^2 \phi\phi\right) = K\left(\dot\phi\dot\phi - \underbrace{\partial_i \phi \partial_i \phi}_{\nabla\phi\cdot\nabla\phi} - \mu^2 \phi\phi\right), \quad (3\text{-}30)$$

Start with classical Lagrangian density for free scalar field

where ϕ, since it is a classical field, is real (not complex), μ is a constant to be determined by experiment, K is an arbitrary constant, the superscript "0" on $\mathcal{L}$ stands for scalar (with spin 0), and the subscript "0" means "free". This is not the place to do classical theory, so we will not derive (3-30) here. We do note in passing that (3-30) is a general result derived by insisting that ϕ and $\mathcal{L}$ are Lorentz invariants (i.e., world scalars – see Chap. 2 including appendix) and that the associated Euler-Lagrange equation is also Lorentz invariant in form. (3-30) is the only form that satisfies these conditions and results in a linear field equation (i.e., ϕ appears only to the first power.) A non-linear field equation might work, but is far more complicated. For free fields, we will find a linear equation works well.

Using the Legendre transformation, we can readily use (3-30) to find the Hamiltonian density, where $\pi_0{}^0$ is the field conjugate momentum,

$$\mathcal{H}_0^0 = \pi_0^0 \dot{\phi} - \mathcal{L}_0^0 = \underbrace{\frac{\partial \mathcal{L}_0^0}{\partial \dot{\phi}}}_{2K\dot{\phi}} \dot{\phi} - \mathcal{L}_0^0 = K\left(\dot{\phi}\dot{\phi} + \nabla\phi\cdot\nabla\phi + \mu^2\phi\phi\right). \tag{3-31}$$

Find Hamiltonian density from Legendre transformation

We may be tempted at this point to proceed with quantization, and simply use the $\mathcal{H}$ and $\mathcal{L}$ above along with the appropriate commutators. However, we know that in quantum mechanics most meaningful things are complex, not real. Quite the reverse of the macroscopic world we live in, and for which real fields of form ϕ generally apply.

Classical field taken as complex

So, we adopt one more postulate, which is that our field ϕ be complex. This means re-expressing our values for $\mathcal{H}$ and $\mathcal{L}$ in terms of a complex field, but such that $\mathcal{H}$ and $\mathcal{L}$ remain real (energy, and energy density $\mathcal{H}$, must be real numbers.) Doing this, where we choose to take $K=1$, yields the free, complex scalar field Lagrangian and Hamiltonian densities

$$\boxed{\mathcal{L}_0^0 = \left(\partial_\alpha \phi^\dagger\, \partial^\alpha \phi - \mu^2 \phi^\dagger \phi\right) = \left(\dot{\phi}^\dagger \dot{\phi} - \nabla\phi^\dagger\cdot\nabla\phi - \mu^2\phi^\dagger\phi\right)}, \text{ and} \tag{3-32}$$

$$\boxed{\mathcal{H}_0^0 = \frac{\partial\mathcal{L}_0^0}{\partial\dot{\phi}^r}\dot{\phi}^r - \mathcal{L}_0^0 = \underbrace{\frac{\partial\mathcal{L}_0^0}{\partial\dot{\phi}}}_{\pi_0^0 = \dot{\phi}^\dagger}\dot{\phi} + \underbrace{\frac{\partial\mathcal{L}_0^0}{\partial\dot{\phi}^\dagger}}_{\pi_0^{0\dagger} = \dot{\phi}}\dot{\phi}^\dagger - \mathcal{L}_0^0 = \dot{\phi}\dot{\phi}^\dagger + \nabla\phi^\dagger\cdot\nabla\phi + \mu^2\phi^\dagger\phi}. \tag{3-33}$$

Re-express Lagrangian and Hamiltonian densities in terms of complex fields

Take care to realize that ϕ and $\phi^\dagger$ are considered *separate fields in the summation over field types r*, and note the definitions of their respective conjugate momenta. That is, $\pi_0{}^0$ equals the complex conjugate of the time derivative of the field (not the time derivative of the field.) $\pi_0^{0\dagger}$ equals the time derivative of the field, not its complex conjugate.

If, as we progress, we find situations where real, rather than complex, fields are involved, we can simply deal with the special case of a complex field where the imaginary part is zero. Assuming a complex field above means we assumed the most general case.

Deriving the Klein-Gordon Field Equation

Substituting the Lagrangian density (3-32) into the Euler-Lagrange field equation,

$$\frac{\partial}{\partial x^\mu}\left(\frac{\partial\mathcal{L}}{\partial\phi^r{}_{,\mu}}\right) - \frac{\partial\mathcal{L}}{\partial\phi^r} = 0 \ . \tag{3-34}$$

Use $\mathcal{L}$ in Euler-Lagrange equation to get Klein-Gordon equation

yields the Klein-Gordon equation for fields, where again, the values $r=1,2$ signify, respectively, the field ϕ and its complex conjugate transpose $\phi^\dagger$ (also called the Hermitian conjugate) which, for scalars, is simply the complex conjugate,

$$\boxed{\begin{aligned}\left(\partial_\mu\partial^\mu + \mu^2\right)\phi &= \left(\square^2 + \mu^2\right)\phi = 0 \quad &\text{(a)}\\ \left(\partial_\mu\partial^\mu + \mu^2\right)\phi^\dagger &= \left(\square^2 + \mu^2\right)\phi^\dagger = 0. \quad &\text{(b)}\end{aligned}} \tag{3-35}$$

In the above, we have introduced the $\square^2$ symbol for the D'Alembertian, the 4D equivalent of the 3D Laplacian, $\nabla^2 = \partial^i\partial^i = \partial_i\partial_i = -\partial^i\partial_i$. (Note, some authors use $\square$ instead of $\square^2$.) We could, of course, also have obtained (3-35)(b) by taking the complex conjugate transpose of (3-35)(a), since everything inside the parentheses is real.

Recall from Chap. 2, that given any one of $\mathcal{H}$, $\mathcal{L}$, or the field equation, we can deduce any of the others (via the Legendre transformation and the Euler-Lagrange equation). So, knowing any one of these is equivalent to knowing any of the others, and our first postulate of 2nd quantization could have stipulated the same $\mathcal{L}$ in classical theory and QFT, or the same field equation, instead of $\mathcal{H}$.

The <u>discrete plane wave solutions</u> to (3-35) are the same as (3-12), and its Hermitian conjugate, from RQM, i..e.,[1]

$$\phi(x) = \sum_{\mathbf{k}} \frac{1}{\sqrt{2V\omega_{\mathbf{k}}}} a(\mathbf{k})e^{-ikx} + \sum_{\mathbf{k}} \frac{1}{\sqrt{2V\omega_{\mathbf{k}}}} b^{\dagger}(\mathbf{k})e^{ikx} \qquad (a)$$

$$= \qquad\qquad \phi^{+} \qquad + \qquad \phi^{-}$$

$$\phi^{\dagger}(x) = \sum_{\mathbf{k}} \frac{1}{\sqrt{2V\omega_{\mathbf{k}}}} b(\mathbf{k})e^{-ikx} + \sum_{\mathbf{k}} \frac{1}{\sqrt{2V\omega_{\mathbf{k}}}} a^{\dagger}(\mathbf{k})e^{ikx} \qquad (b)$$

$$= \qquad\qquad \phi^{\dagger+} \qquad + \qquad \phi^{\dagger-} \ .$$

(3-36)

Discrete plane wave solutions to Klein-Gordon field equation

Note the new symbolism for each of the solution forms. We use lower case coefficients in QFT because, as we will see, the coefficients play a much different role in QFT than they did in RQM, and we need to distinguish them.

The <u>continuous plane wave solutions</u> to (3-35) are

$$\phi(x) = \int \frac{d^{3}\boldsymbol{k}}{\sqrt{2(2\pi)^{3}\omega_{\mathbf{k}}}} a(\mathbf{k})e^{-ikx} + \int \frac{d^{3}\boldsymbol{k}}{\sqrt{2(2\pi)^{3}\omega_{\mathbf{k}}}} b^{\dagger}(\mathbf{k})e^{ikx} \qquad (a)$$

$$= \qquad\qquad \phi^{+} \qquad + \qquad \phi^{-}$$

$$\phi^{\dagger}(x) = \int \frac{d^{3}\boldsymbol{k}}{\sqrt{2(2\pi)^{3}\omega_{\mathbf{k}}}} b(\mathbf{k})e^{-ikx} + \int \frac{d^{3}\boldsymbol{k}}{\sqrt{2(2\pi)^{3}\omega_{\mathbf{k}}}} a^{\dagger}(\mathbf{k})e^{ikx} \qquad (b)$$

$$= \qquad\qquad \phi^{\dagger+} \qquad + \qquad \phi^{\dagger-} \ .$$

(3-37)

Continuous plane wave solutions to Klein-Gordon field equation

The continuous solutions represent waves that are not constrained to a specific volume. Wavelengths for such solutions do not have to fit an integer number of times inside a particular volume, and thus are not limited to discrete values.

Note also the shorthand notation for each of the four different solution sets underneath the brackets. You will see these symbols again and again, so you might want to consider making a copy of (3-36), pasting it above your desk, and doing memorization tests with yourself every day until they become ingrained in your consciousness. Try to remember that $\phi^{\dagger+}$ is *not* the complex conjugate of ϕ^{+}, for example, contrary to what you might expect. The + sign refers to a term with positive energy in the RQM sense (i.e., – sign before the energy in the exponent.) It might help to think that because $\dagger$ changes the sign of the imaginary part of every complex quantity, it also changes the sign of the symbol ϕ^{-}. So, $(\phi^{-})^{\dagger} = \phi^{\dagger+}$.

Learn the shorthand notation for the four types of solutions

---

[1] These solutions have the familiar $\pm i(\omega_{\mathbf{k}}t - \mathbf{k}\cdot\mathbf{x})$ form in the exponent, but functions of alternative form $\pm i(\omega_{\mathbf{k}}t + \mathbf{k}\cdot\mathbf{x})$ in the exponent also solve the Klein-Gordon equation. I call these alternative forms "supplemental solutions". These solutions, although not mathematically independent from the traditional solutions, result in different operators that can modify QFT in key ways. Supplemental solutions are discussed in R. D. Klauber, "A symmetry for Resolution of the Gauge Hierarchy Problem without SUSY, Null Higgs Condensate Energy, and Null Zero Point Energy" (2018) http://arxiv.org/abs/1802.03277 and "Mechanism for Vanishing Zero Point Energy" (2003) https://arxiv.org/abs/astro-ph/0309679.

However you do it, being able to readily recall the definitions of the symbols in (3-36) and (3-37) will help in the future.

<u>Finding μ^2</u>

Do Prob. 4 to prove to yourself that the value of μ^2, which appeared as an unknown constant in the theoretical determination of (3-30), has the same value it did in RQM, i.e.,

$$\boxed{\mu^2 = \frac{m^2 c^2}{\hbar^2} \quad (= m^2 \text{ in nat. units})}. \tag{3-38}$$

μ^2 has same value in QFT as in RQM

<u>Third Way to Klein-Gordon Equations: A Consistency Check</u>

Recall from Chap 2. and Wholeness Charts 2-2 and 2-5 (pgs. 20 and 30), that we could express the equations of motion for classical fields in terms of Poisson brackets in the former chart, and for Heisenberg picture quantum fields, in terms of commutators in the latter chart. The commutator-based equation of motion for a quantum field in the next to last box in the right hand column of Wholeness Chart 2-5 (reproduced below on the LHS of (3-39)) is in terms of the Hamiltonian and the field. For scalar fields, this equation of motion for ϕ should be essentially the same as the Klein-Gordon equation for ϕ. That is,

There is yet a third way to derive the field equation of motion, this time from a variational math relation

<u>Heisenberg Picture Field Equation of Motion</u>   <u>Klein-Gordon Field Equation</u>

$$\dot{\phi} = -i[\phi, H] \quad \xleftarrow{\text{should be same thing}} \quad (\partial_\mu \partial^\mu + \mu^2)\phi = 0 \quad (3\text{-}39)$$

In the Appendix A of this chapter, we show that this is indeed true, and thus our theory is self-consistent. It also proves that the Klein-Gordon field equation of QFT (3-35) (and (3-39)) derived above applies to the Heisenberg, not Schrödinger, picture.

This is a parallel path to do second quantization that is included in the route represented by the vertical arrow on the RHS of Fig. 3-2, but it uses a different, though related, part of the theory.

3.2.4 Summary Chart

All that we have done in this Sect. 3.2, and what we will do in the remainder of this chapter and the next two chapters, is summarized in Wholeness Chart 5-4 at the end of Chap. 5.

Note the summary is at the end of Chap. 5 because each column in it lists the key components in the development of QFT for one of the three spin types (spin 0, ½, and 1), and we won't be doing the latter two until Chaps. 4 and 5.

Be sure to use the summary wholeness chart, as you study this chapter and the next two

You can follow along in the chart, as we develop the theory for scalars, by reading the blocks in the Spin 0 column. You may want to stick a Post-It on that page as a book marker, so you can easily flip to it as you read along in this, and the following two, chapters.

3.3 Commutation Relations: The Crux of QFT

We will soon see how the commutation relations encompassed in the <u>second part of 2nd quantization</u>, found in the last box in the right hand column of Wholeness Chart 2-5 of Chap. 2, pg. 31, and reproduced in (3-40) below, lie at the root of, and structure, all of QFT. For scalars, they are

$$\left[\phi^r(\mathbf{x},t), \pi_s(\mathbf{y},t) \right] = \phi^r \pi_s - \pi_s \phi^r = i\delta^r{}_s \delta(\mathbf{x}-\mathbf{y}) \quad \left[\phi^r, \phi^s \right] = [\pi_r, \pi_s] = \left[\phi^r, \pi_s^\dagger \right] = 0. \tag{3-40}$$

2nd quantization commutation relations determine coefficient commutation

Note, in passing, that a complex conjugate of a field is considered a different field. In effect, if $\phi^r = \phi$, then $\phi^s = \phi^\dagger$, where $r \neq s$, so that $\left[\phi, \phi^\dagger\right] = 0$.

Of overriding importance in the theory, as we will see, are the following <u>coefficient commutation relations</u>, which we will derive below from the 2nd quantization postulate of (3-40).

$$\boxed{\left[a(\mathbf{k}), a^\dagger(\mathbf{k}') \right] = \left[b(\mathbf{k}), b^\dagger(\mathbf{k}') \right] = \delta_{\mathbf{k}\mathbf{k}'} \text{ (discrete)}; \quad = \delta(\mathbf{k}-\mathbf{k}') \text{ (continuous)}}. \tag{3-41}$$

Coefficient com rels 1) play fundamental role in QFT, and 2) imply coefficients are operators

The form of (3-41) should tell us immediately that the Klein-Gordon solution coefficients $a(\mathbf{k})$, $b(\mathbf{k})$, etc. in QFT of (3-36) and (3-37) are far different animals than the $A_{\mathbf{k}}$, $B_{\mathbf{k}}$, etc. in RQM of (3-12). The latter are merely numbers, which commute. We must, therefore, suspect that the $a(\mathbf{k})$, $b(\mathbf{k})$, etc. are operators, and as we will see, this suspicion will turn out to be correct.

Proof of coefficient commutation relations

To prove (3-41), start with (3-40) and take different spatial coordinates $\mathbf{x}$ and $\mathbf{y}$, but the same time coordinate t, for ϕ and π^0_0. This results in the equal time commutation relations

$$\left[\phi(\mathbf{x},t)\pi^0_0(\mathbf{y},t) - \pi^0_0(\mathbf{y},t)\phi(\mathbf{x},t)\right] = \left[\phi(\mathbf{x},t)\dot{\phi}^\dagger(\mathbf{y},t) - \dot{\phi}^\dagger(\mathbf{y},t)\phi(\mathbf{x},t)\right] = i\delta(\mathbf{x}-\mathbf{y}), \quad (3\text{-}42)$$

Proving coefficient commutation relations

which are only important at this point as a step in our proof. Then, plugging the discrete solutions (3-36) into the middle part of (3-42), where to save space we use the compressed notation $a_\mathbf{k} = a(\mathbf{k})$, etc., we get

$$\left(\sum_\mathbf{k}\frac{1}{\sqrt{2V\omega_\mathbf{k}}}a_\mathbf{k}e^{-i(\omega_\mathbf{k}t-\mathbf{k}\cdot\mathbf{x})} + \sum_\mathbf{k}\frac{1}{\sqrt{2V\omega_\mathbf{k}}}b^\dagger_\mathbf{k}e^{i(\omega_\mathbf{k}t-\mathbf{k}\cdot\mathbf{x})}\right)\left(\sum_{\mathbf{k}'}\frac{-i\omega_{\mathbf{k}'}}{\sqrt{2V\omega_{\mathbf{k}'}}}b_{\mathbf{k}'}e^{-i(\omega_\mathbf{k}t-\mathbf{k}'\cdot\mathbf{y})} + \sum_{\mathbf{k}'}\frac{i\omega_{\mathbf{k}'}}{\sqrt{2V\omega_{\mathbf{k}'}}}a^\dagger_{\mathbf{k}'}e^{i(\omega_\mathbf{k}t-\mathbf{k}'\cdot\mathbf{y})}\right)$$

$$-\left(\sum_{\mathbf{k}'}\frac{-i\omega_{\mathbf{k}'}}{\sqrt{2V\omega_{\mathbf{k}'}}}b_{\mathbf{k}'}e^{-i(\omega_\mathbf{k}t-\mathbf{k}'\cdot\mathbf{y})} + \sum_{\mathbf{k}'}\frac{i\omega_{\mathbf{k}'}}{\sqrt{2V\omega_{\mathbf{k}'}}}a^\dagger_{\mathbf{k}'}e^{i(\omega_\mathbf{k}t-\mathbf{k}'\cdot\mathbf{y})}\right)\left(\sum_\mathbf{k}\frac{1}{\sqrt{2V\omega_\mathbf{k}}}a_\mathbf{k}e^{-i(\omega_\mathbf{k}t-\mathbf{k}\cdot\mathbf{x})} + \sum_\mathbf{k}\frac{1}{\sqrt{2V\omega_\mathbf{k}}}b^\dagger_\mathbf{k}e^{i(\omega_\mathbf{k}t-\mathbf{k}\cdot\mathbf{x})}\right)$$

$$=\sum_\mathbf{k}\sum_{\mathbf{k}'}\frac{i\omega_{\mathbf{k}'}}{2V\sqrt{\omega_\mathbf{k}\omega_{\mathbf{k}'}}}\begin{pmatrix} -a_\mathbf{k}b_{\mathbf{k}'}e^{-i(\omega_\mathbf{k}+\omega_{\mathbf{k}'})t}e^{i(\mathbf{k}\cdot\mathbf{x}+\mathbf{k}'\cdot\mathbf{y})} + a_\mathbf{k}a^\dagger_{\mathbf{k}'}e^{-i(\omega_\mathbf{k}-\omega_{\mathbf{k}'})t}e^{i(\mathbf{k}\cdot\mathbf{x}-\mathbf{k}'\cdot\mathbf{y})} \\ -b^\dagger_\mathbf{k}b_{\mathbf{k}'}e^{i(\omega_\mathbf{k}-\omega_{\mathbf{k}'})t}e^{-i(\mathbf{k}\cdot\mathbf{x}-\mathbf{k}'\cdot\mathbf{y})} + b^\dagger_\mathbf{k}a^\dagger_{\mathbf{k}'}e^{i(\omega_\mathbf{k}+\omega_{\mathbf{k}'})t}e^{-i(\mathbf{k}\cdot\mathbf{x}+\mathbf{k}'\cdot\mathbf{y})} \\ +b_{\mathbf{k}'}a_\mathbf{k}e^{-i(\omega_\mathbf{k}+\omega_{\mathbf{k}'})t}e^{i(\mathbf{k}\cdot\mathbf{x}+\mathbf{k}'\cdot\mathbf{y})} + b_{\mathbf{k}'}b^\dagger_\mathbf{k}e^{i(\omega_\mathbf{k}-\omega_{\mathbf{k}'})t}e^{-i(\mathbf{k}\cdot\mathbf{x}-\mathbf{k}'\cdot\mathbf{y})} \\ -a^\dagger_{\mathbf{k}'}a_\mathbf{k}e^{-i(\omega_\mathbf{k}-\omega_{\mathbf{k}'})t}e^{i(\mathbf{k}\cdot\mathbf{x}-\mathbf{k}'\cdot\mathbf{y})} - a^\dagger_{\mathbf{k}'}b^\dagger_\mathbf{k}e^{i(\omega_\mathbf{k}+\omega_{\mathbf{k}'})t}e^{-i(\mathbf{k}\cdot\mathbf{x}+\mathbf{k}'\cdot\mathbf{y})} \end{pmatrix} = i\delta(\mathbf{x}-\mathbf{y}).$$

K-G solutions into equal time commutator

(3-43)

Using the math identity for the 3D Dirac delta function

$$\delta(\mathbf{x}-\mathbf{y}) = \frac{1}{V}\sum_{n=-\infty}^{+\infty}e^{-i\mathbf{k}_n\cdot(\mathbf{x}-\mathbf{y})} \quad \begin{pmatrix} \text{in our notation} = \frac{1}{V}\sum_\mathbf{k}e^{-i\mathbf{k}\cdot(\mathbf{x}-\mathbf{y})} \\ \\ \text{or equivalently, } \frac{1}{2V}\sum_\mathbf{k}\left(e^{-i\mathbf{k}\cdot(\mathbf{x}-\mathbf{y})} + e^{i\mathbf{k}\cdot(\mathbf{x}-\mathbf{y})}\right) \end{pmatrix} \quad (3\text{-}44)$$

Re-express Dirac delta function

on the RHS of the last row in (3-43), and matching terms, we see that all terms where $\mathbf{k}' \neq \pm\mathbf{k}$ must equal zero, since (3-44) has no terms in both $\mathbf{k}$ and $\mathbf{k}'$. These particular terms reduce to the following form, summed over $\mathbf{k}$ and $\mathbf{k}'$.

$$\frac{i\omega_{\mathbf{k}'}}{2V\sqrt{\omega_\mathbf{k}\omega_{\mathbf{k}'}}}\begin{pmatrix} \underbrace{(b_{\mathbf{k}'}a_\mathbf{k} - a_\mathbf{k}b_{\mathbf{k}'})}_{\text{must}=0}e^{-i(\omega_\mathbf{k}+\omega_{\mathbf{k}'})t}e^{i(\mathbf{k}\cdot\mathbf{x}+\mathbf{k}'\cdot\mathbf{y})} + \underbrace{(a_\mathbf{k}a^\dagger_{\mathbf{k}'} - a^\dagger_{\mathbf{k}'}a_\mathbf{k})}_{\text{must}=0}e^{-i(\omega_\mathbf{k}-\omega_{\mathbf{k}'})t}e^{i(\mathbf{k}\cdot\mathbf{x}-\mathbf{k}'\cdot\mathbf{y})} \\ +\underbrace{(b_{\mathbf{k}'}b^\dagger_\mathbf{k} - b^\dagger_\mathbf{k}b_{\mathbf{k}'})}_{\text{must}=0}e^{i(\omega_\mathbf{k}-\omega_{\mathbf{k}'})t}e^{-i(\mathbf{k}\cdot\mathbf{x}-\mathbf{k}'\cdot\mathbf{y})} + \underbrace{(b^\dagger_\mathbf{k}a^\dagger_{\mathbf{k}'} - a^\dagger_{\mathbf{k}'}b^\dagger_\mathbf{k})}_{\text{must}=0}e^{i(\omega_\mathbf{k}+\omega_{\mathbf{k}'})t}e^{-i(\mathbf{k}\cdot\mathbf{x}+\mathbf{k}'\cdot\mathbf{y})} \end{pmatrix} = 0$$

Terms where $\mathbf{k}' \neq \pm\mathbf{k}$

(3-45)

These must vanish, so their commutators must $= 0$

(All terms in summations with $\mathbf{k}' \neq \pm\mathbf{k}$ equal 0, as no terms on RHS in $\mathbf{k}$ and $\mathbf{k}'$)

So, all possible coefficient commutators with $\mathbf{k}' \neq \mathbf{k}$ or $-\mathbf{k}$ vanish. The remaining terms all have $\mathbf{k}' = \pm\mathbf{k}$, which means $\omega_\mathbf{k} = \omega_{\mathbf{k}'}$. Some of these have an exponential form $i(\omega_\mathbf{k} + \omega_{\mathbf{k}'})t$, and those terms give us a summation of terms over $\mathbf{k}$ having form, for each possible $\mathbf{k}'$, of

$$\frac{i\omega_\mathbf{k}}{2V\omega_\mathbf{k}}\begin{pmatrix} \underbrace{(b_\mathbf{k}a_\mathbf{k} - a_\mathbf{k}b_\mathbf{k})}_{\text{must}=0}\underbrace{e^{-i2\omega_\mathbf{k}t}}_{\neq 0}e^{i\mathbf{k}\cdot(\mathbf{x}+\mathbf{y})} + \underbrace{(b^\dagger_\mathbf{k}a^\dagger_\mathbf{k} - a^\dagger_\mathbf{k}b^\dagger_\mathbf{k})}_{\text{must}=0}e^{i2\omega_\mathbf{k}t}e^{-i\mathbf{k}\cdot(\mathbf{x}+\mathbf{y})} \quad (\leftarrow \mathbf{k}'=\mathbf{k}) \\ +\underbrace{(b_{-\mathbf{k}}a_\mathbf{k} - a_\mathbf{k}b_{-\mathbf{k}})}_{\text{must}=0}e^{-i2\omega_\mathbf{k}t}e^{i\mathbf{k}\cdot(\mathbf{x}-\mathbf{y})} + \underbrace{(b^\dagger_\mathbf{k}a^\dagger_{-\mathbf{k}} - a^\dagger_{-\mathbf{k}}b^\dagger_\mathbf{k})}_{\text{must}=0}e^{i2\omega_\mathbf{k}t}e^{-i\mathbf{k}\cdot(\mathbf{x}-\mathbf{y})} \quad (\leftarrow \mathbf{k}'=-\mathbf{k}) \end{pmatrix} = 0$$

Terms where $\mathbf{k}' = \pm\mathbf{k}$, i.e., those of form $\exp(i(\omega_\mathbf{k}+\omega_\mathbf{k})t)$

(3-46)

(All time dependent terms with $\mathbf{k}'=\pm\mathbf{k}$ equal 0, as no time dependence on RHS)

Commutators must $= 0$

For these terms, the coefficient commutators must vanish because the exponential in $\omega_\mathbf{k}$ varies in time, whereas there is no such variation on the RHS of the last row in (3-43).

The remaining terms have exponential form $i(\omega_\mathbf{k} - \omega_{\mathbf{k}'})t$ and $\mathbf{k}' = \pm \mathbf{k}$. Adding those terms for $\mathbf{k}' = \mathbf{k}$ with the terms for $\mathbf{k}' = -\mathbf{k}$ yields, with the relevant terms on the RHS of (3-43) (see 2nd row in parentheses of (3-44)) on the RHS below,

$$\frac{i}{2V}\left(\begin{array}{l} \underbrace{\left(a_\mathbf{k} a_\mathbf{k}^\dagger - a_\mathbf{k}^\dagger a_\mathbf{k}\right)}_{must\,=1} \underbrace{e^{-i(\omega_\mathbf{k}-\omega_\mathbf{k})t}}_{=1} e^{i\mathbf{k}\cdot(\mathbf{x}-\mathbf{y})} + \underbrace{\left(b_\mathbf{k} b_\mathbf{k}^\dagger - b_\mathbf{k}^\dagger b_\mathbf{k}\right)}_{must\,=1} e^{-i\mathbf{k}\cdot(\mathbf{x}-\mathbf{y})} \quad (\leftarrow \mathbf{k}'=\mathbf{k}) \\ + \underbrace{\left(a_\mathbf{k} a_{-\mathbf{k}}^\dagger - a_{-\mathbf{k}}^\dagger a_\mathbf{k}\right)}_{must\,=0} e^{i\mathbf{k}\cdot(\mathbf{x}+\mathbf{y})} + \underbrace{\left(b_{-\mathbf{k}} b_\mathbf{k}^\dagger - b_\mathbf{k}^\dagger b_{-\mathbf{k}}\right)}_{must\,=0} e^{-i\mathbf{k}\cdot(\mathbf{x}+\mathbf{y})} \quad (\leftarrow \mathbf{k}'=-\mathbf{k}) \end{array}\right) = \frac{i}{2V}\left(\begin{array}{l} e^{i\mathbf{k}\cdot(\mathbf{x}-\mathbf{y})} \\ + e^{-i\mathbf{k}\cdot(\mathbf{x}-\mathbf{y})}\end{array}\right) \text{(3-47)}$$

Remaining terms where $\mathbf{k}' = \pm \mathbf{k}$, i.e., those of form $exp(i(\omega_\mathbf{k} - \omega_\mathbf{k})t)$

Key commutators must = 1

(All time independent terms in summation with $\mathbf{k}' = \pm\mathbf{k}$ must equal RHS).

All terms with $(\mathbf{x} + \mathbf{y})$ in the exponents of the LHS must equal zero, as the RHS only has terms in $(\mathbf{x} - \mathbf{y})$. The LHS of (3-47) matches the RHS if each coefficient commutator in the first row equals one. Subtleties in justifying that as the only way to interpret (3-47) are shown in Appendix E.

The commutation relations for $a_\mathbf{k} a_{\mathbf{k}'}{}^\dagger$ and $b_\mathbf{k} b_{\mathbf{k}'}{}^\dagger$ in (3-45) to (3-47) are the same as (3-41). QED.

If you are ambitious, have extra time, and/or simply have to prove everything to yourself, do Prob. 7 to derive the continuous solution commutators of (3-41).

<u>End of coefficient commutation relations proof</u>

With the coefficient commutator relations in hand, we are finally ready to dive into the real core of QFT.

3.4 The Hamiltonian in QFT

We find the Hamiltonian by integrating the Hamiltonian density $\mathcal{H}$ over all space (a volume V containing the discrete solutions, which we can make as large as we like.) In QFT, we express $\mathcal{H}$ in terms of a complex field and substitute our field equation solutions.

$H = \int \mathcal{H}\,dV$

3.4.1 The Free Scalar Hamiltonian in Terms of the Coefficients

For a free scalar field $\mathcal{H} = \mathcal{H}_0{}^0$, as in (3-33), where we employ our discrete, plane wave solutions (3-36) we get

$H = \int \mathcal{H}\,dV$ in terms of the fields

$$H_0^0 = \int \mathcal{H}_0^0 d^3x = \int \left(\dot{\phi}\dot{\phi}^\dagger + \nabla\phi^\dagger \cdot \nabla\phi + \mu^2 \phi^\dagger\phi \right) d^3x =$$

$$\int \left(\sum_\mathbf{k} \frac{\partial}{\partial t}\frac{1}{\sqrt{2V\omega_\mathbf{k}}}\left(a(\mathbf{k})e^{-ikx} + b^\dagger(\mathbf{k})e^{ikx} \right) \right)\left(\sum_{\mathbf{k}'} \frac{\partial}{\partial t}\frac{1}{\sqrt{2V\omega_{\mathbf{k}'}}}\left(b(\mathbf{k}')e^{-ik'x} + a^\dagger(\mathbf{k}')e^{ik'x} \right) \right) d^3x \quad (3\text{-}48)$$

Deriving H in terms of the coefficients ↓

$$+ \int \left(-\partial_i \phi^\dagger \partial^i \phi + \mu^2\phi^\dagger\phi \right)d^3x.$$

The middle line of (3-48), i.e., the $\int \dot{\phi}\dot{\phi}^\dagger d^3x$ part, becomes

$$\int \left(\sum_\mathbf{k} \frac{i\omega_\mathbf{k}}{\sqrt{2V\omega_\mathbf{k}}}\left(-a(\mathbf{k})e^{-ikx} + b^\dagger(\mathbf{k})e^{ikx} \right) \right)\left(\sum_{\mathbf{k}'} \frac{i\omega_{\mathbf{k}'}}{\sqrt{2V\omega_{\mathbf{k}'}}}\left(-b(\mathbf{k}')e^{-ik'x} + a^\dagger(\mathbf{k}')e^{ik'x} \right) \right) d^3x. \quad (3\text{-}49)$$

or

$$\sum_\mathbf{k}\sum_{\mathbf{k}'} \frac{-\sqrt{\omega_\mathbf{k}}\sqrt{\omega_{\mathbf{k}'}}}{2V}\int \left(\begin{array}{l} a(\mathbf{k})b(\mathbf{k}')e^{-ikx}e^{-ik'x} - a(\mathbf{k})a^\dagger(\mathbf{k}')e^{-ikx}e^{ik'x} \\ -b^\dagger(\mathbf{k})b(\mathbf{k}')e^{ikx}e^{-ik'x} + b^\dagger(\mathbf{k})a^\dagger(\mathbf{k}')e^{ikx}e^{ik'x} \end{array}\right) d^3x. \quad (3\text{-}50)$$

The sum over $\mathbf{k}$ and $\mathbf{k}'$ is from negative infinity to positive infinity in the x, y, and z directions.

All terms in the integration in (3-50) result in zero except when $\mathbf{k}' = \mathbf{k}$ or $\mathbf{k}' = -\mathbf{k}$, because we are integrating orthogonal functions between their boundaries. (This is similar to $sin(2X)sin(4X)$ integrated with respect to X along a complete number of wavelengths, where here $\mathbf{k} = 2$ and $\mathbf{k}' = 4$.) Since the volume of integration in (3-50) equals V, we end up with

$$\int \dot{\phi}\dot{\phi}^\dagger d^3x = \sum_\mathbf{k} \frac{\omega_\mathbf{k}}{2}\left(-a(\mathbf{k})b(-\mathbf{k})e^{-2i\omega_\mathbf{k}t} + a(\mathbf{k})a^\dagger(\mathbf{k}) + b^\dagger(\mathbf{k})b(\mathbf{k}) - b^\dagger(\mathbf{k})a^\dagger(-\mathbf{k})e^{2i\omega_\mathbf{k}t} \right)$$

$$\qquad\qquad (3\text{-}51)$$

$$= \sum_\mathbf{k} \frac{(\omega_\mathbf{k})^2}{2\omega_\mathbf{k}}\left(-a(-\mathbf{k})b(\mathbf{k})e^{-2i\omega_\mathbf{k}t} + a(\mathbf{k})a^\dagger(\mathbf{k}) + b^\dagger(\mathbf{k})b(\mathbf{k}) - b^\dagger(-\mathbf{k})a^\dagger(\mathbf{k})e^{2i\omega_\mathbf{k}t} \right).$$

In the second row of (3-51), the sign change on the **k** in the first and last terms is justified since we are summing over all **k**, so for every term with **k** in it, there is another with − **k**. This modification will make things easier a bit later.

Following similar steps for the next term in (3-48) we get

$$-\int \partial_i \phi^\dagger \partial^i \phi d^3 x = \int \partial_i \phi^\dagger \partial_i \phi d^3 x$$

$$= \int \left(\sum_{\mathbf{k}} \frac{ik^i}{\sqrt{2V\omega_{\mathbf{k}}}} \left(b(\mathbf{k})e^{-ikx} - a^\dagger(\mathbf{k})e^{ikx} \right) \right) \left(\sum_{\mathbf{k'}} \frac{ik'^i}{\sqrt{2V\omega_{\mathbf{k'}}}} \left(a(\mathbf{k'})e^{-ik'x} - b^\dagger(\mathbf{k'})e^{ik'x} \right) \right) d^3 x \quad (3\text{-}52)$$

$$= \sum_{\mathbf{k}} \frac{\mathbf{k}^2}{2\omega_{\mathbf{k}}} \left(b(\mathbf{k})a(-\mathbf{k})e^{-2i\omega_{\mathbf{k}}t} + a^\dagger(\mathbf{k})a(\mathbf{k}) + b(\mathbf{k})b^\dagger(\mathbf{k}) + a^\dagger(\mathbf{k})b^\dagger(-\mathbf{k})e^{2i\omega_{\mathbf{k}}t} \right)$$

where we note that terms in the summation with both **k** and − **k** have an extra sign change since $k_i = -k'_i$ in the multiplication in the second line of (3-52).

Similarly, for the mass term in (3-48), we get (do Prob. 8 at the end of the chapter to prove it)

$$\int \mu^2 \phi^\dagger \phi d^3 x = \sum_{\mathbf{k}} \frac{\mu^2}{2\omega_{\mathbf{k}}} \left(b(\mathbf{k})a(-\mathbf{k})e^{-2i\omega_{\mathbf{k}}t} + b(\mathbf{k})b^\dagger(\mathbf{k}) + a^\dagger(\mathbf{k})a(\mathbf{k}) + a^\dagger(\mathbf{k})b^\dagger(-\mathbf{k})e^{2i\omega_{\mathbf{k}}t} \right). \quad (3\text{-}53)$$

Adding the final parts of (3-51), (3-52), and (3-53), and using $\mathbf{k}^2 + \mu^2 = (\omega_{\mathbf{k}})^2$ along with the coefficient commutation relations (3-41), we end up with

$$H_0^0 = \sum_{\mathbf{k}} \frac{\omega_{\mathbf{k}}}{2} \left(\underbrace{a(\mathbf{k})a^\dagger(\mathbf{k})}_{\text{use commutator}} + a^\dagger(\mathbf{k})a(\mathbf{k}) + b^\dagger(\mathbf{k})b(\mathbf{k}) + \underbrace{b(\mathbf{k})b^\dagger(\mathbf{k})}_{\text{use commutator}} \right)$$

$$= \sum_{\mathbf{k}} \omega_{\mathbf{k}} \left(a^\dagger(\mathbf{k})a(\mathbf{k}) + \tfrac{1}{2} + b^\dagger(\mathbf{k})b(\mathbf{k}) + \tfrac{1}{2} \right). \quad (3\text{-}54)$$

or simply

$$\boxed{H_0^0 = \sum_{\mathbf{k}} \omega_{\mathbf{k}} \left(N_a(\mathbf{k}) + \tfrac{1}{2} + N_b(\mathbf{k}) + \tfrac{1}{2} \right)}, \quad (3\text{-}55)$$

H = ∫ℋ dV in terms of the coefficients

where

$$\boxed{N_a(\mathbf{k}) = a^\dagger(\mathbf{k})a(\mathbf{k}) \qquad N_b(\mathbf{k}) = b^\dagger(\mathbf{k})b(\mathbf{k})} . \quad (3\text{-}56)$$

Expressions (3-55) and (3-56) lie at the heart of QFT, as we are about to see.

3.4.2 Number Operators

Consider what we must get if the Hamiltonian of (3-55) operates on a state (a ket) comprised of two free scalar particles, each in the same eigenstate of energy $\omega_{\mathbf{k}_1}$. We would expect that (multiparticle) state to have an energy eigenvalue equal to its total energy $2\omega_{\mathbf{k}_1}$, i.e.,

$$H_0^0 \left| 2\phi_{\mathbf{k}_1} \right\rangle = 2\omega_{\mathbf{k}_1} \left| 2\phi_{\mathbf{k}_1} \right\rangle . \quad (3\text{-}57)$$

But from (3-55), that means

$$\sum_{\mathbf{k}} \omega_{\mathbf{k}} \left(N_a(\mathbf{k}) + \tfrac{1}{2} + N_b(\mathbf{k}) + \tfrac{1}{2} \right) \left| 2\phi_{\mathbf{k}_1} \right\rangle = 2\omega_{\mathbf{k}_1} \left| 2\phi_{\mathbf{k}_1} \right\rangle . \quad (3\text{-}58)$$

How can we make sense of (3-58)? The answer is that it is not quite true, and that we can make sense of it all if, instead of (3-57) and (3-58), we consider

$$H_0^0 \left| 2\phi_{\mathbf{k}_1} \right\rangle = \sum_{\mathbf{k}} \omega_{\mathbf{k}} \left(\underbrace{N_a(\mathbf{k}) + \tfrac{1}{2}}_{\substack{a\ \text{type} \\ \text{particles}}} + \underbrace{N_b(\mathbf{k}) + \tfrac{1}{2}}_{\substack{b\ \text{type} \\ \text{particles}}} \right) \underbrace{\left| 2\phi_{\mathbf{k}_1} \right\rangle}_{\substack{a\ \text{type} \\ \text{particles}}} = \left(2\omega_{\mathbf{k}_1} + \sum_{\mathbf{k}} \omega_{\mathbf{k}} \left(\tfrac{1}{2} + \tfrac{1}{2} \right) \right) \left| 2\phi_{\mathbf{k}_1} \right\rangle, \quad (3\text{-}59)$$

H = ∫ℋ dV in terms of number operators

with the following interpretation.

$\underline{N_a(\mathbf{k})}$ = number operator with eigenvalue $n_a(\mathbf{k})$ = number of a particles with 3-mom $\mathbf{k}$ in the ket,

$\underline{N_b(\mathbf{k})}$ = number operator with eigenvalue $n_b(\mathbf{k})$ = number of b particles with 3-mom $\mathbf{k}$ in the ket,

and, the vacuum has ½ quantum of energy for each $\mathbf{k}$ for a particles, and also for b particles.

This might, at first, be considered a separate postulate, but if the $\mathcal{H}_0{}^0$ derived by 2nd quantization for quantum scalar fields is correct, this is the only possible interpretation of (3-59) that works. The part about the vacuum would be surprising to anyone who had not already heard that the vacuum is a seething caldron of virtual quanta. More on this shortly.

For $H = \int \mathcal{H}\,dV$ to work, the vacuum must be filled with ½ quanta

We also anticipate that the b type particles will be antiparticles, and the a types, normal particles. More on this later, as well.

Examples of number operators and kets

In light of the above, the following examples should be relatively straightforward. Note we designate b type particles with an overbar.

Example #1: 10 particle state

$$H_0^0 \left| 5\phi_{\mathbf{k}_1}, 2\phi_{\mathbf{k}_2}, \underbrace{3\bar{\phi}_{\mathbf{k}_3}}_{\substack{b \text{ type} \\ \text{particles}}} \right\rangle = \sum_{\mathbf{k}} \omega_{\mathbf{k}} \left(N_a(\mathbf{k}) + \tfrac{1}{2} + N_b(\mathbf{k}) + \tfrac{1}{2} \right) \left| 5\phi_{\mathbf{k}_1}, 2\phi_{\mathbf{k}_2}, 3\bar{\phi}_{\mathbf{k}_3} \right\rangle$$

$$= \left(5\omega_{\mathbf{k}_1} + 2\omega_{\mathbf{k}_2} + 3\omega_{\mathbf{k}_3} + \sum_{\mathbf{k}} \omega_{\mathbf{k}} \left(\tfrac{1}{2} + \tfrac{1}{2} \right) \right) \left| 5\phi_{\mathbf{k}_1}, 2\phi_{\mathbf{k}_2}, 3\bar{\phi}_{\mathbf{k}_3} \right\rangle. \tag{3-60}$$

and so we see that b type particles in our theory have positive energy. This resulted from our interpretation of H_0^0 in (3-59), i.e., from our postulate for the properties of $N_b(\mathbf{k})$ in (3-59) and the box following it. We know from experiment that antiparticles have positive energy, and this interpretation will lead to b particles filling the role of antiparticles. We will see later that b particles have opposite charge from the a particles, and so they fit nicely into our theory as antiparticles.

Every state has vacuum as part of it, so every state has huge vacuum energy

Example #2: Vacuum state

$$H_0^0 \underbrace{\left| 0 \right\rangle}_{\substack{\text{vacuum} \\ \text{state}}} = \sum_{\mathbf{k}} \omega_{\mathbf{k}} \left(N_a(\mathbf{k}) + \tfrac{1}{2} + N_b(\mathbf{k}) + \tfrac{1}{2} \right) \left| 0 \right\rangle = \underbrace{\sum_{\mathbf{k}} \omega_{\mathbf{k}} \left(\tfrac{1}{2} + \tfrac{1}{2} \right)}_{\text{infinite energy}} \left| 0 \right\rangle. \tag{3-61}$$

Note that every state is superimposed on the vacuum, so every state actually has infinite energy. We saw this in Example #1.

3.4.3 Zero Point (Vacuum) Energy

The infinite sum of ½ quanta in (3-61) represents the now famous perspective on the vacuum as being almost inconceivably crammed with energy, known as zero-point energy (ZPE.) In actuality, the sum, while enormous, is usually not considered infinite, but for reasons beyond the scope of our current discussion, to terminate at a very high level, known as the Planck scale.

Zero-point (vacuum) energy results from 2nd quantization postulate of non-commutation

It is important to recognize that this *vacuum energy arose from our postulate of 2nd quantization*, that a field and its conjugate momentum don't commute (see (3-40)). Because of this we got the coefficient commutation relations (3-41), and those were used in our derivation of the form of the Hamiltonian (see (3-54), which resulted in the appearance of ½$\omega_{\mathbf{k}}$ terms.

The field commutation relations of QFT are siblings to the particle commutation relations for NRQM and RQM. (See Wholeness Chart 2-5 in Chap. 2, pg. 30.) In the latter two theories, particle position and momentum do not commute, and this results in the renowned Heisenberg uncertainty relation between position and momentum. In QFT, the field and its conjugate momentum do not commute, implying a parallel uncertainty relationship between them. This, in this sense, may be partially behind the oft heard statement that the uncertainty principle is the cause of the zero-point energy. In Chap. 10, we cover another sense commonly meant.

Non-commutation leads to uncertainty. Part of reason for statements that uncertainty principle results in ZPE

Are the ½ Quanta Effervescent?

One often also hears that the ½ quanta "pop" in and out of the vacuum effervescently in particle/antiparticle pairs. I submit that this is a heuristic, at best, representation for the popular media. According to (3-61), there is no "popping", no evanescent physical reality alternated with nothingness, no pairing of particles at a creation event followed by a common mutual destruction event as one might see in a Feynman diagram (different from, but similar in nature to, what we saw in Fig. 1-1 of Chap. 1, pg. 2) Via our fundamental QFT relation for H_0^0, the ½ quanta are simply

Ruminations on vacuum effervescence

sitting in the vacuum. They may be virtual in some sense, and not real, but the QFT H operator does not suggest any intermittent sort of existence.

Further, our derivation of $H_0{}^0$ has been exclusively for free fields, where no interactions are included (more on interactions in later chapters). The particles (and antiparticles), for which $H_0{}^0$ determines the energy, do not interact with other particles or antiparticles. This means they can't create or destroy in pairs, since that is, above all, an interaction between the associated particles and antiparticles. So $H_0{}^0$ specifically does not measure the energy of such pairs and the ½ energy terms therein must be for free fields that are not "popping" in and out of the vacuum in pairs.

Some argue that experimental measurements of Casimir plate forces, the Lamb shift, and the anomalous magnetic moment of the electron demonstrate the existence of vacuum fluctuations. However, Casimir forces are generally computed by considering the ½ quanta such as those in (3-61) to be standing waves between the plates. That is, they do *not need*, in those analyses, to be "popping" in and out of the vacuum, but merely be sitting there continually. Further, the Casimir force can be computed without reference to zero-point energies at all, and thus may not be the conclusive proof for their existence it is widely taken to be. (See R. L. Jaffe, "Casimir Effect and the Quantum Vacuum", Phys. Rev. D72 021301(R) (2005) http://arxiv.org/abs/hep-th/0503158 .)

Casimir plates do not prove ZPE existence

The Lamb shift calculation involves so-called "vacuum fluctuations", but they are actually higher order corrections to propagators for interacting fields (which we study in later chapters), not the ½ quanta vacuum energy of free fields (which we study in this chapter.) The same is true of the QFT correction to the magnetic moment of the electron (the famed "anomalous magnetic moment".)

As a caveat, I note that the remarks in this sub-section entitled "Are the ½ Quanta Effervescent" reflect my personal position on vacuum energy. Many physicists believe quanta are continually "bubbling" in and out of existence in the vacuum. I simply have not seen a sound derivation of this in the literature, and don't believe it is supported by the formal derivation of ZPE.

Neither are Lamb shift nor anomalous magnetic moment solution due to free field ½ quanta (ZPE)

We will discuss this issue further when we get to interaction theory. (See Chap. 10.)

3.4.4 Positive Energy in QFT

Note that unlike RQM, all particles in QFT have positive energy. The QFT energy operator $H_0{}^0$ operating on states yields positive eigenvalues for both a and b types of particles. The RQM energy operator operating on states did not do that, as we saw in Sect. 3.1.5 (pg. 47).

$H = \int \mathcal{H}\, dV$ leads to positive energy for both QFT particle types

Continuation of Wholeness Chart 1-2. Comparison of Three Quantum Theories

	NRQM	**RQM**	**QFT**
Hamiltonian	$i\dfrac{\partial}{\partial t}$	$i\dfrac{\partial}{\partial t}$	$H = \int \mathcal{H} d^3x$
Sign of Energy E	positive	positive & negative	positive

3.4.5 Unit Norms and Orthogonality for Multiparticle States

Recall from NRQM, that it was advantageous to normalize states, i.e., change the constant in front of the ket such that the inner product of the state and its complex conjugate transpose (the bracket of the bra and ket) equaled unity. That is, we defined

$$\langle \phi_{\mathbf{k}} \| \phi_{\mathbf{k}} \rangle = \int_V \underbrace{\phi_{\mathbf{k}}{}^{\dagger}(\mathbf{x},t)\phi_{\mathbf{k}}(\mathbf{x},t)}_{\text{states}} d^3x = 1 \quad \text{and} \quad \langle \phi_{\mathbf{k}} \| \phi_{\mathbf{k}'} \rangle = \int_V \underbrace{\phi_{\mathbf{k}}{}^{\dagger}(\mathbf{x},t)\phi_{\mathbf{k}'}(\mathbf{x},t)}_{\text{states}} d^3x = 0, \quad \mathbf{k} \neq \mathbf{k}'. \quad (3\text{-}62)$$

As an aside, note that in (3-62), it is assumed that the ket is expressed in the position basis. For example, a plane wave momentum eigenstate in that basis would have form[1]

---

[1] The ket symbol $|\phi\rangle$ in general represents a particle state, but the form of the ket when we write it out mathematically, such as we did in the RHS of (3-63), changes with the basis we care to use. For example, we could express the ket in the momentum basis (in momentum space) instead of the position basis x; or in a number of other ways. Mathematically, (3-63) is really $|\phi\rangle_{x\,\text{basis}} = \langle x|\phi\rangle = Ae^{-i(Et-\mathbf{k}\cdot\mathbf{x})}$.

$$\left|\phi_{\mathbf{k}}\right\rangle = A e^{-i(Et - \mathbf{k}\cdot\mathbf{x})} \qquad \left(\left|\phi_{\mathbf{k}}\right\rangle \text{ here is expressed in the position basis}\right). \qquad (3\text{-}63)$$

In this book, unless otherwise stated, when we express $|\phi\rangle$ mathematically, we will assume the position basis as in (3-63). At such times, $|\phi\rangle$ will $= |\phi\rangle_{\text{x basis}}$. Now, back to the main point.

In NRQM and RQM, states are single particle states. In QFT, they are typically multiparticle, but we will also find it advantageous therein to normalize. So, we define our <u>symbols for multiparticle states</u> so that every such state is normalized (i.e., has <u>unit norm</u>.) For example, for a state comprising two a particles of 3-momentum $\mathbf{k}$, one of 3-momentum $\mathbf{k}'$, and five b particles of 3-momentum $\mathbf{k}''$, we would have (where the middle part is just a reminder of what we mean by the bracket notation[1])

$$\left\langle 2\phi_{\mathbf{k}}, \phi_{\mathbf{k}'}, 5\overline{\phi}_{\mathbf{k}''} \middle\| 2\phi_{\mathbf{k}}, \phi_{\mathbf{k}'}, 5\overline{\phi}_{\mathbf{k}''} \right\rangle = \int_V \left(\underbrace{2\phi_{\mathbf{k}}\phi_{\mathbf{k}'}5\overline{\phi}_{\mathbf{k}''}}_{\text{states}} \right)^{\dagger} \left(\underbrace{2\phi_{\mathbf{k}}\phi_{\mathbf{k}'}5\overline{\phi}_{\mathbf{k}''}}_{\text{states}} \right) d^3x = 1 \ . \qquad (3\text{-}64)$$

Multiparticle states in QFT have unit norm

Note that any (multiparticle) state is <u>orthogonal</u> to every other state that is not identical to it in particle types, particle numbers, and $\mathbf{k}$ values for each. For examples, where $\mathbf{k} \neq \mathbf{k}'$,

$$\left\langle 2\phi_{\mathbf{k}}, \phi_{\mathbf{k}'}, 5\overline{\phi}_{\mathbf{k}''} \middle\| \phi_{\mathbf{k}'}, 5\overline{\phi}_{\mathbf{k}''} \right\rangle = 0 \qquad \left\langle 2\phi_{\mathbf{k}} \middle\| \phi_{\mathbf{k}} \right\rangle = 0 \qquad \left\langle 5\phi_{\mathbf{k}''} \middle\| 5\overline{\phi}_{\mathbf{k}''} \right\rangle = 0 \qquad \left\langle \phi_{\mathbf{k}} \middle\| \phi_{\mathbf{k}'} \right\rangle = 0 \ . \qquad (3\text{-}65)$$

Different multiparticle states in QFT are orthogonal

Note on Notation

It is common practice in QFT to employ the bracket notation of the LHS of (3-64), and virtually never, the integral form shown between the equal signs. As noted earlier, in QFT, symbols such as $\phi_{\mathbf{k}}$, which are not part of a ket symbol, normally do not represent states, but operators/fields. When they play the role of states, as in (3-62), we must label them specifically, as there and in (3-64).

QFT virtually never uses ϕ_k to represent a state, nor an integral to represent an inner product of states

3.5 Expectation Values and the Hamiltonian

Note that the expectation value relation for an operator $\mathcal{O}$ in QFT for a single particle state is the same as that in the rest of quantum mechanics, i.e., provided $|\phi\rangle$ has unit norm,

$$\overline{\mathcal{O}} = \left\langle \phi \middle| \mathcal{O} \middle| \phi \right\rangle \ , \qquad (3\text{-}66)$$

Expectation value for single particle in QFT like those in NRQM and RQM

where we don't confuse the overbar (used here outside a bra or ket) for expectation (or average) value with its use inside a bra or ket, where it signifies b type particles. The expectation value is the average value we would measure over a large number of measurements of the state. If the particle is in an eigenstate of an observable (an operator), then every measurement of that observable for that state would be the same (the eigenvalue), and thus equal to the expectation value. An eigenstate of energy would measure the same value for energy upon every measurement. This is true for a single particle state, such as in (3-66).

It is also true for a multiparticle state, such as those we run into in QFT. For example, the multiparticle state in (3-60) is in an eigenstate of energy (the sum of the energies from each of the ten particles in the state.) Each particle therein has fixed mass plus a fixed momentum $\mathbf{k}$, and hence fixed total energy. Thus, the energy expectation value for the state in (3-60) is

$$\overline{H}_0^0 = \left\langle 5\phi_{\mathbf{k}_1}, 2\phi_{\mathbf{k}_2}, 3\overline{\phi}_{\mathbf{k}_3} \middle| H_0^0 \middle| 5\phi_{\mathbf{k}_1}, 2\phi_{\mathbf{k}_2}, 3\overline{\phi}_{\mathbf{k}_3} \right\rangle =$$

$$= \left\langle 5\phi_{\mathbf{k}_1}, 2\phi_{\mathbf{k}_2}, 3\overline{\phi}_{\mathbf{k}_3} \middle| \underbrace{\left(5\omega_{\mathbf{k}_1} + 2\omega_{\mathbf{k}_2} + 3\omega_{\mathbf{k}_3} + \sum_{\mathbf{k}} \omega_{\mathbf{k}}\left(\tfrac{1}{2} + \tfrac{1}{2}\right) \right)}_{\text{a number, not an operator, so can move outside}} \middle| 5\phi_{\mathbf{k}_1}, 2\phi_{\mathbf{k}_2}, 3\overline{\phi}_{\mathbf{k}_3} \right\rangle$$

$$= \left(5\omega_{\mathbf{k}_1} + 2\omega_{\mathbf{k}_2} + 3\omega_{\mathbf{k}_3} + \sum_{\mathbf{k}} \omega_{\mathbf{k}}\left(\tfrac{1}{2} + \tfrac{1}{2}\right) \right) \underbrace{\left\langle 5\phi_{\mathbf{k}_1}, 2\phi_{\mathbf{k}_2}, 3\overline{\phi}_{\mathbf{k}_3} \middle| 5\phi_{\mathbf{k}_1}, 2\phi_{\mathbf{k}_2}, 3\overline{\phi}_{\mathbf{k}_3} \right\rangle}_{=1}$$

$$= 5\omega_{\mathbf{k}_1} + 2\omega_{\mathbf{k}_2} + 3\omega_{\mathbf{k}_3} + \sum_{\mathbf{k}} \omega_{\mathbf{k}}\left(\tfrac{1}{2} + \tfrac{1}{2}\right). \qquad (3\text{-}67)$$

---

[1] For the purists, we note that (3-64) can be taken as an invariant relationship if we define our states properly. That is, we can define our multiparticle eigenstate with a factor of $1/\sqrt{V}$ in front, so the integrand in (3-64) yields a factor $1/V$. The integral over 3D space then yields a factor of V. V is non-invariant, but one in the numerator and one in the denominator cancel, leaving an invariant final result.

In general, for any operator $\mathcal{O}$, the expectation value for any multiparticle state is

$$\boxed{\bar{\mathcal{O}} = \langle \phi_1, \phi_2, \phi_3, \ldots | \mathcal{O} | \phi_1, \phi_2, \phi_3, \ldots \rangle} \ , \tag{3-68}$$

Expectation value for multiparticle state has same form as single particle state

where we will typically find the operator expressed in terms of number operators.

A concept that becomes important later on in QFT is that of the vacuum expectation value, or simply the VEV, whose symbol and mathematical expression are

$$\langle \mathcal{O} \rangle = \langle 0 | \mathcal{O} | 0 \rangle \ . \tag{3-69}$$

Expectation value for the vacuum, VEV

If you don't see it right away, do Prob. 9 to prove to yourself that the VEV of the free field scalar Hamiltonian is

$$\left\langle H_0^0 \right\rangle = \sum_{\mathbf{k}} \omega_{\mathbf{k}} \ . \tag{3-70}$$

Hamiltonian VEV

We expect to measure infinite (or enormous, if nature has a maximum $|\mathbf{k}|$) energy in the vacuum.

3.6 Creation and Destruction Operators

In this section, we will prove what is perhaps the most fundamental aspect of QFT, which we foreshadowed in Chap. 1, that the Klein-Gordon solution coefficients $a(\mathbf{k})$, $a^\dagger(\mathbf{k})$, $b(\mathbf{k})$, and $b^\dagger(\mathbf{k})$ are not numbers, but operators that create and destroy particles. Since certain combinations of them do not commute, we should expect them to be operators of some kind.

The coefficients are creation and destruction operators

3.6.1 Proving It

underline{Proof that $a(\mathbf{k})$ is a Particle Destruction Operator}

With the underline{notation $|n_{\mathbf{k}}\rangle$} denoting a multiparticle state of $n_{\mathbf{k}}$ a type particles (no b types for now), all with the same 4-momentum k^μ, what can we say about the state

Proving it.

$$a(\mathbf{k})|n_{\mathbf{k}}\rangle = |m_{\mathbf{k}}\rangle \ ? \tag{3-71}$$

To see, first operate on this state with our number operator $N_a(\mathbf{k}) = a^\dagger(\mathbf{k})\, a(\mathbf{k})$

$$N_a(\mathbf{k})|m_{\mathbf{k}}\rangle = N_a(\mathbf{k})a(\mathbf{k})|n_{\mathbf{k}}\rangle = \underbrace{a^\dagger(\mathbf{k})a(\mathbf{k})}_{\text{use commutator}} a(\mathbf{k})|n_{\mathbf{k}}\rangle \ . \tag{3-72}$$

Then, where noted above, use the commutation relations from (3-41), to find (3-72) equals

$$\left(a(\mathbf{k})a^\dagger(\mathbf{k}) - 1 \right)a(\mathbf{k})|n_{\mathbf{k}}\rangle = a(\mathbf{k})a^\dagger(\mathbf{k})a(\mathbf{k})|n_{\mathbf{k}}\rangle - a(\mathbf{k})|n_{\mathbf{k}}\rangle = a(\mathbf{k})N_a(\mathbf{k})|n_{\mathbf{k}}\rangle - a(\mathbf{k})|n_{\mathbf{k}}\rangle$$
$$= a(\mathbf{k})n_{\mathbf{k}}|n_{\mathbf{k}}\rangle - a(\mathbf{k})|n_{\mathbf{k}}\rangle = n_{\mathbf{k}}a(\mathbf{k})|n_{\mathbf{k}}\rangle - a(\mathbf{k})|n_{\mathbf{k}}\rangle = (n_{\mathbf{k}}-1)a(\mathbf{k})|n_{\mathbf{k}}\rangle = (n_{\mathbf{k}}-1)|m_{\mathbf{k}}\rangle. \tag{3-73}$$

So

$$N_a(\mathbf{k})|m_{\mathbf{k}}\rangle = (n_{\mathbf{k}}-1)|m_{\mathbf{k}}\rangle = m_{\mathbf{k}}|m_{\mathbf{k}}\rangle \qquad m_{\mathbf{k}} = n_{\mathbf{k}} - 1 \ . \tag{3-74}$$

Since the number operator operating on $|m_{\mathbf{k}}\rangle$ gives a number of particles one less than it did when operating on $|n_{\mathbf{k}}\rangle$, the operation of $a(\mathbf{k})$ on $|n_{\mathbf{k}}\rangle$ in (3-71) reduces the number of particles in the state by one. We conclude that $a(\mathbf{k})$ is a particle destruction operator.
End of Proof

Do Prob. 10, or at least part of it, to prove one or more of the last three relations below to yourself.

$$N_a(\mathbf{k})\left(a(\mathbf{k})|n_{\mathbf{k}}\rangle\right) = (n_{\mathbf{k}}-1)\left(a(\mathbf{k})|n_{\mathbf{k}}\rangle\right) \quad a(\mathbf{k})\text{ destroys an } a \text{ particle with momentum } \mathbf{k}$$

$$N_a(\mathbf{k})\left(a^\dagger(\mathbf{k})|n_{\mathbf{k}}\rangle\right) = (n_{\mathbf{k}}+1)\left(a^\dagger(\mathbf{k})|n_{\mathbf{k}}\rangle\right) \quad a^\dagger(\mathbf{k})\text{ creates an } a \text{ particle with momentum } \mathbf{k}$$

Summary of operator functions of coefficients

$$N_b(\mathbf{k})\left(b(\mathbf{k})|\bar{n}_{\mathbf{k}}\rangle\right) = (\bar{n}_{\mathbf{k}}-1)\left(b(\mathbf{k})|\bar{n}_{\mathbf{k}}\rangle\right) \quad b(\mathbf{k})\text{ destroys a } b \text{ particle with momentum } \mathbf{k} \tag{3-75}$$

$$N_b(\mathbf{k})\left(b^\dagger(\mathbf{k})|\bar{n}_{\mathbf{k}}\rangle\right) = (\bar{n}_{\mathbf{k}}+1)\left(b^\dagger(\mathbf{k})|\bar{n}_{\mathbf{k}}\rangle\right) \quad b^\dagger(\mathbf{k})\text{ creates a } b \text{ particle with momentum } \mathbf{k}.$$

underline{Creation operators} $a^\dagger(\mathbf{k})$ and $b^\dagger(\mathbf{k})$ are sometimes called underline{raising operators}, because they raise the number of particles in a state. underline{Destruction operators} $a(\mathbf{k})$ and $b(\mathbf{k})$ are sometimes called underline{lowering} operators, for what should be obvious reasons. States that have been operated on by a raising operator are sometimes called raised states; those by a lowering operator, lowered states.

3.6.2 Normalization Factors for Raised and Lowered States

When a raising operator operates on a ket, the resulting raised ket does not generally have unit norm (is not normalized.) Consider

$$a^\dagger(\mathbf{k})|n_\mathbf{k}\rangle = A|n_\mathbf{k}+1\rangle \, , \tag{3-76}$$

Raising and lowering particle number results in a factor in front of the new ket

where A is some constant (which is a number, not an operator, and could be complex). The original ket $|n_\mathbf{k}\rangle$ and the ket $|n_\mathbf{k}+1\rangle$ (without the constant A) in (3-76) have unit norm. (See (3-64) for one example.) Also, by taking the complex conjugate transpose of (3-76), we see the $a(\mathbf{k})$ acting leftward on the bra has the same raising effect as the $a^\dagger(\mathbf{k})$ acting on the ket,

$$\left(A|n_\mathbf{k}+1\rangle\right)^\dagger = \left(a^\dagger(\mathbf{k})|n_\mathbf{k}\rangle\right)^\dagger = \langle n_\mathbf{k}|a(\mathbf{k}) = \langle n_\mathbf{k}+1|A^\dagger \, . \tag{3-77}$$

Note that

$$\underbrace{\langle n_\mathbf{k}|a(\mathbf{k})}_{\langle n_\mathbf{k}+1|A^\dagger}a^\dagger(\mathbf{k})|n_\mathbf{k}\rangle = \langle n_\mathbf{k}+1|A^\dagger A|n_\mathbf{k}+1\rangle = A^\dagger A\underbrace{\langle n_\mathbf{k}+1\|n_\mathbf{k}+1\rangle}_{1} = A^\dagger A \, . \tag{3-78}$$

(3-78) also equals

$$\langle n_\mathbf{k}|\underbrace{a(\mathbf{k})a^\dagger(\mathbf{k})}_{\text{use commutator}}|n_\mathbf{k}\rangle = \langle n_\mathbf{k}|\underbrace{a^\dagger(\mathbf{k})a(\mathbf{k})}_{N_a(\mathbf{k})}+1|n_\mathbf{k}\rangle = \langle n_\mathbf{k}|n_\mathbf{k}+1|n_\mathbf{k}\rangle = n_\mathbf{k}+1 \, . \tag{3-79}$$

Equating the RHS's of (3-78) and (3-79), and for simplicity taking A as real (complex would also work, but be more complicated) yields

$$A = \sqrt{n_\mathbf{k}+1} \, . \tag{3-80}$$

From (3-76), we then have the first line in (3-81) below. Identical logic leads to the third line. Do Prob. 11 if you can't just accept the second and fourth lines without seeing for yourself how they are obtained.

$$\boxed{\begin{aligned} a^\dagger(\mathbf{k})|n_\mathbf{k}\rangle &= \sqrt{n_\mathbf{k}+1}\,|n_\mathbf{k}+1\rangle \\ a(\mathbf{k})|n_\mathbf{k}\rangle &= \sqrt{n_\mathbf{k}}\,|n_\mathbf{k}-1\rangle \\ b^\dagger(\mathbf{k})|\bar{n}_\mathbf{k}\rangle &= \sqrt{\bar{n}_\mathbf{k}+1}\,|\bar{n}_\mathbf{k}+1\rangle \\ b(\mathbf{k})|\bar{n}_\mathbf{k}\rangle &= \sqrt{\bar{n}_\mathbf{k}}\,|\bar{n}_\mathbf{k}-1\rangle \end{aligned}} \tag{3-81}$$

Factors arising from action of creation and destruction operators

Note that the above results are ultimately due to 2nd quantization. The non-commutation of fields and their conjugate momenta resulted in the coefficient commutation relations, which was a crucial part in the proof that the coefficients create and destroy states, as well as the derivation of the normalization constants shown above. *Second quantization turned the solution coefficients in RQM, which were merely constants, into creation and destruction operators in QFT.*

2nd quantization responsible for creation/ destruction operators

3.6.3 Annihilating the Vacuum

Note that the vacuum $|0\rangle$ has unit norm, like any other state, i.e.,

$$\langle 0\|0\rangle = 1 \, . \tag{3-82}$$

The vacuum state has unit norm

Note also that 0 (zero) is a number representing nothing, and is different from $|0\rangle$, the vacuum state, which actually is something. From (3-81), the action of a destruction operator on the vacuum results in zero. That is,

$$a(\mathbf{k})|0\rangle = \sqrt{0}\,|-1\rangle = 0 \, . \tag{3-83}$$

A destruction operator annihilates the vacuum, i.e., leaves 0

Don't worry about the funny looking ket in the middle (which is actually meaningless and not something you will ever see in the literature). The root of zero controls the final result.

In QFT lingo, one says "a lowering operator destroys (or annihilates) the vacuum".

3.6.4 Total Particle Number

For future use, we define the total particle number as the number of particles (i.e. *a* types) minus the number of antiparticles (*b* types). For scalars, the <u>total particle number operator</u> is

$$N(\phi) = \sum_{\mathbf{k}} \left(N_a(\mathbf{k}) - N_b(\mathbf{k}) \right). \tag{3-84}$$

Total particle number is number of particles minus number of antiparticles

Note the subtle difference in phraseology in that we commonly use the term "number of particles" as being equal to the number of particles *plus* the number of antiparticles. "Total particle number" on the other hand refers to a negative value for the number of antiparticles.

We will soon see that *b* particles have opposite charge from *a* particles, and thus, in many senses, represent their negatives. So, designating them as having negative total particle number seems reasonable.

3.6.5 $\phi(\mathbf{x})$ and $\phi^\dagger(\mathbf{x})$ as Operator Fields

Since the field solutions $\underline{\phi(x)}$ and $\underline{\phi^\dagger(x)}$ contain the operator coefficients, they are then also operators, or more properly, <u>operator fields</u> or <u>quantum fields</u>. As noted in Chap. 1 and earlier in the present chapter, this is often shortened in QFT to simply <u>fields</u>. At long last, we have proven our earlier statements about the solutions to the wave equation in QFT being operators (fields).

φ and φ† are operator fields since they contain coefficient operators

Note that for our field solutions of (3-36), ϕ acts as a total particle number lowering operator, because it destroys particles (via $a(\mathbf{k})$) and creates antiparticles (via $b^\dagger(\mathbf{k})$). The former decreases a positive total particle number, whereas the latter increases the magnitude of a negative total particle number. For $\phi^\dagger$, the situation is reversed: $a^\dagger(\mathbf{k})$ creates particles and $b(\mathbf{k})$ destroys antiparticles, both actions increasing the total particle number.

Thus, the <u>total particle lowering operator field</u> is (see (3-36) for full expression)

$$\phi = \underbrace{\phi^+}_{\substack{\text{destroys}\\\text{particles}}} + \underbrace{\phi^-}_{\substack{\text{creates}\\\text{anti-particles}}}, \tag{3-85}$$

φ is a total particle number lowering operator field

and the <u>total particle raising operator field</u> is

$$\phi^\dagger = \underbrace{\phi^{\dagger+}}_{\substack{\text{destroys}\\\text{anti-particles}}} + \underbrace{\phi^{\dagger-}}_{\substack{\text{creates}\\\text{particles}}}. \tag{3-86}$$

φ† is a total particle number raising operator field

When we originally saw the field solutions (3-36), it was suggested, as a mnemonic, that you make a copy of them and stick it over your desk. It would be good now to insert (3-85) and (3-86) into that copy and make them part of it, as we will be using those symbols and what they represent, over and over.

3.6.6 Normal Ordering

When the infinite sum of ½ quanta energy in (3-59) (or (3-61)) was first found, physicists wanted desperately to make it go away. The amount of energy involved should, via general relativity, curve the universe to such an enormous degree that the light emanating from your finger would be bent so much that it would never reach your eyes. But that isn't what happens in our world, so something isn't correct. In fact, if one cuts off the addition in (3-61) at the Planck scale, instead of at infinity, the difference in mass-energy level of the vacuum, between that summation and what is observed, is on the order of a factor of 10^{120}, the biggest discrepancy between theory and experiment in the history of science. (We show this in Chap. 10.)

One approach to solving ("hiding" may be a better word) this problem is something called normal ordering. <u>Normal ordering</u>, in any term, consists of moving all destruction operators to the right-hand side of that term. This has little impact for operators that commute, such as $a(\mathbf{k})$ and $b^\dagger(\mathbf{k})$, for example. That is, changing the term $a(\mathbf{k})b^\dagger(\mathbf{k})$ to $b^\dagger(\mathbf{k})a(\mathbf{k})$ is not an issue, since the factors commute, i.e., $a(\mathbf{k})b^\dagger(\mathbf{k}) = b^\dagger(\mathbf{k})a(\mathbf{k})$. However, the term $a(\mathbf{k})a^\dagger(\mathbf{k})$ is a different story, since $a(\mathbf{k})a^\dagger(\mathbf{k}) \neq a^\dagger(\mathbf{k})a(\mathbf{k})$.

Normal ordering puts all destruction operators in any term on the RHS

In particular, note the effect of normal ordering in our derivation of the number operator form of the Hamiltonian in (3-54). Instead of employing the commutator relations for the $a(\mathbf{k})a^\dagger(\mathbf{k})$ and $b(\mathbf{k})b^\dagger(\mathbf{k})$ terms as done in (3-54), we simply move all the destruction operators to the RHS, so those terms become $a^\dagger(\mathbf{k})a(\mathbf{k})$ and $b^\dagger(\mathbf{k})b(\mathbf{k})$. Thus, we never end up with the $\frac{1}{2}\omega_\mathbf{k}$ terms, the Hamiltonian is finite, just what we would originally have expected it to be, and the vacuum has zero energy, i.e.,

$$H_0^0 = \sum_\mathbf{k} \omega_\mathbf{k} \left(a^\dagger(\mathbf{k})a(\mathbf{k}) + b^\dagger(\mathbf{k})b(\mathbf{k}) \right) \quad (\text{normal ordered} = \text{what is observed}). \qquad (3\text{-}87)$$

Normal ordering makes vacuum energy go away

The Hamiltonian only has number operators yielding $n_\mathbf{k}\omega_\mathbf{k}$ energy for $n_\mathbf{k}$ particles, each having 3-momentum $\mathbf{k}$.

Although use of normal ordering became quite widespread, it suffers from a pretty fundamental problem. It violates the foundational postulate of non-commutation of certain operators, upon which all of QFT stands. Invoking normal ordering means assuming, in this one area of QFT, that $a(\mathbf{k})$ and $a^\dagger(\mathbf{k})$ (as well as $b(\mathbf{k})$ and $b^\dagger(\mathbf{k})$) commute! But they don't. And the fact that they don't is fundamental to every other part of QFT[1]. In normal ordering, we simply suspend commutation long enough to get a zero energy for the vacuum, then bring it back for the rest of the theory. It is not unreasonable to conclude that use of normal ordering for this purpose is questionable, at best[2].

Some argue that using normal ordering in this way is internally inconsistent

Note that normal ordering is often justified because particle/field behavior in our theories of classical mechanics, electromagnetism, and special relativity depends on energy difference, so we can take our reference as the vacuum energy level and all energies of interest are relative to that. In those theories, it is ΔE that is important, not E. So, why not assume our Hamiltonian represents $\Delta E = E - \infty$ instead of E? The answer, I and others submit, is because in general relativity, the theory depends on E. To be consistent with all of physics, we need H representing E, not ΔE.

Normal ordering of the Hamiltonian is also commonly used in other areas of QFT (pgs. 112, 208-209.) One is for determining charge on spin $\frac{1}{2}$ particles. Otherwise, the vacuum would have infinite charge. Since we have no theories where that can be simply subtracted, it is hard to justify use of normal ordering for charge derivation, and one's confidence in it is further eroded. (On the cited pages, we will see alternative derivations that do not use normal ordering.)

In any case, in spite of common use of normal ordering to "clean up" the Hamiltonian, the huge vacuum energy issue has not gone away in most physicists' eyes, and it remains a widely discussed, unsolved problem as of the year of this version of this book (though I offer a possible solution in the article cited in the footnote on page 50).

Paradoxically both 1) normal ordering is widely used to eliminate vacuum energy, and 2) vacuum energy is generally accepted as a fact

It may seem to you the reader that the entire issue is fraught with ambiguity, and that is probably a reasonable assessment. In spite of that, virtually everyone considers vacuum energy to be a reality.

3.6.7 The Observable Hamiltonian

One can distinguish observables, which in quantum theories are represented by operators, from the theoretically obtained expressions for the corresponding operators. That is, the $\frac{1}{2}\omega_\mathbf{k}$ terms in H_0^0 are not observed, so for reasons of practicality, we can consider the normal ordered Hamiltonian of (3-87) to be what we will call the <u>observable Hamiltonian</u>.

The normal ordered Hamiltonian is the observable Hamiltonian.

The relation (3-87), even though derived via the *ad hoc* and mathematically questionable normal ordering process, results in what is actually observed. This is why normal ordering persists.

Since we will not do a lot with the vacuum, we will find it more convenient and streamlined to use (3-87) for the Hamiltonian, except when we are specifically interested in vacuum energy. Just don't forget, as we do that, what the complete Hamiltonian, as derived via our theory, looks like.

3.7 Probability, Four Currents, and Charge Density

Probability in QFT is found in essentially the same way as we did for NRQM (see Box 3-1) and RQM (see Sect. 3.1.4, pg. 44.) That is, we use the governing wave equation and manipulate it to obtain a relationship like the continuity equation (3-19) (or (3-23) in 4D notation). The integral over all space of the quantity ρ in that relationship is conserved, and since total probability (of finding one or more particles) is also conserved, ρ has a good chance of being probability density. Experiment can confirm, or deny, that.

Probability in QFT found in similar way as in NRQM and RQM, i.e., from wave equation

---

[1] As an example of one such part, see Sect. 3.6.1, where commutation relations result in $a(\mathbf{k})$ and $b(\mathbf{k})$ being destruction operators, and $a^\dagger(\mathbf{k})$ and $b^\dagger(\mathbf{k})$ being creation operators.

[2] P. Teller, *An Interpretive Introduction to Quantum Field Theory*, Princeton University Press (1995). On page 130, with reference to normal ordering Teller states, "If, as appears to be the case, at this point one must use mathematically illegitimate tricks, concern is an appropriate response."

3.7.1 Four Currents, Operators, and Probability Density in QFT

In the present case, the solutions to the governing equation are operator fields, not states, so we would expect the resulting density ρ to be an operator density, rather than a numeric density. Our expectation will turn out to be true, as we see below.

Since our governing equation is the Klein-Gordon equation, and that is the same as in RQM, similar steps (3-16) to (3-20) can be followed. The result is the same 4-current relations as (3-22) and (3-23), except that ϕ and $\phi^\dagger$ are now operator fields (or simply, in QFT lingo "fields"),

Same wave equation as RQM → same form for probability 4-current

$$\boxed{j^\mu{}_{,\mu} = 0 \quad \text{with} \quad j^\mu = i\left(\phi^{,\mu}\phi^\dagger - \phi^{\dagger,\mu}\phi\right) \qquad j^\mu \text{ is an operator}}, \tag{3-88}$$

so

$$\rho = j^0 = i\left(\frac{\partial\phi}{\partial t}\phi^\dagger - \frac{\partial\phi^\dagger}{\partial t}\phi\right). \tag{3-89}$$

But now 4-current is an operator

Since (3-89) is an operator, we need its expectation value to find measurable probability density,

$$\bar\rho = \left\langle\phi_1,\phi_2,\phi_3,....\left|\rho\right|\phi_1,\phi_2,\phi_3,....\right\rangle. \tag{3-90}$$

So, need to find expectation value of probability operator

To evaluate (3-90), we need first to substitute our free field solutions (3-36) into (3-89). Do Prob. 12 to prove to yourself that if we restrict ourselves to particles in $\mathbf{k}$ eigenstates (which is typically the case in QFT), then this results in an effective density operator

$$\rho = \frac{1}{V}\sum_{\mathbf{k}}\left(a^\dagger(\mathbf{k})a(\mathbf{k}) - b^\dagger(\mathbf{k})b(\mathbf{k})\right) = \frac{1}{V}\sum_{\mathbf{k}}\left(N_a(\mathbf{k}) - N_b(\mathbf{k})\right). \tag{3-91}$$

Probability operator expressed in terms of number operators

3.7.2 Single Particle State

Let's now find the expectation value of ρ for a single a type particle state $|\phi_{\mathbf{k}'}\rangle$. We find all the number operators except the one for an a type particle with momentum $\mathbf{k}'$ yield zero, so

$$\bar\rho = \left\langle\phi_{\mathbf{k}'}\left|\rho\right|\phi_{\mathbf{k}'}\right\rangle = \left\langle\phi_{\mathbf{k}'}\left|\frac{1}{V}\sum_{\mathbf{k}}\left(N_a(\mathbf{k}) - N_b(\mathbf{k})\right)\right|\phi_{\mathbf{k}'}\right\rangle = \left\langle\phi_{\mathbf{k}'}\left|\frac{1}{V}\right|\phi_{\mathbf{k}'}\right\rangle = \frac{1}{V}. \tag{3-92}$$

Probability expectation value for single type a particle = 1, no surprise

For a plane wave, this is exactly our probability density, a flat distribution over the volume, whose integral over the volume equals one. So far, ρ looks like it could well be a probability distribution.

3.7.3 Multiparticle State

But now let's look at a multiparticle state.

$$\bar\rho = \left\langle 3\phi_{\mathbf{k}_1},\phi_{\mathbf{k}_2}\left|\rho\right|3\phi_{\mathbf{k}_1},\phi_{\mathbf{k}_2}\right\rangle = \left\langle 3\phi_{\mathbf{k}_1},\phi_{\mathbf{k}_2}\left|\frac{1}{V}\sum_{\mathbf{k}}\left(N_a(\mathbf{k}) - N_b(\mathbf{k})\right)\right|3\phi_{\mathbf{k}_1},\phi_{\mathbf{k}_2}\right\rangle$$

$$= \left\langle 3\phi_{\mathbf{k}_1},\phi_{\mathbf{k}_2}\left|\frac{4}{V}\right|3\phi_{\mathbf{k}_1},\phi_{\mathbf{k}_2}\right\rangle = \frac{4}{V}. \tag{3-93}$$

Probability expectation value for four type a particles = 4 > 1 !

When (3-93) is integrated over V, we get 4, the number of particles in the state! Since total probability is never greater than 1, our interpretation of ρ as a probability density seems to be in trouble for multiparticle states.

Partially for this reason, QFT rarely deals with probability densities for states. It concerns itself, instead, with *numbers* of particles (and antiparticles) in a state. Thus, the number operators play a major role. As we will see, this works well, and allows us to solve the kinds of problems in QFT we need to solve.

So, QFT deals with number of particles instead of probability

3.7.4 Antiparticles (Type b Particles)

Now consider the expectation value of ρ on a b type single particle state.

$$\bar\rho = \left\langle\bar\phi_{\mathbf{k}'}\left|\rho\right|\bar\phi_{\mathbf{k}'}\right\rangle = \left\langle\bar\phi_{\mathbf{k}'}\left|\frac{1}{V}\sum_{\mathbf{k}}\left(N_a(\mathbf{k}) - N_b(\mathbf{k})\right)\right|\bar\phi_{\mathbf{k}'}\right\rangle = \left\langle\bar\phi_{\mathbf{k}'}\left|\frac{-1}{V}\right|\bar\phi_{\mathbf{k}'}\right\rangle = -\frac{1}{V}. \tag{3-94}$$

Probability expectation value for single type b particle = −1 < 0 !!

So total probability of a b type particle would be negative! And for 3 such particles, it would be -3.

This was another tip to early researchers that the density they were dealing with was more readily related to charge density, and the b particles were antiparticles, with opposite charge (and charge density) from particles.

Led to conclusion that $\rho \propto$ charge probability density, and its spatial integral is total charge

Thus, we have started to see that the concept of probability density and the mathematics associated with it seem to lose some applicability in QFT. Particle number, however, takes on significance, as now antiparticles would simply be designated as having negative total particle numbers.

3.7.5 Charge Density Not Probability Density

If we multiply our four current operator (3-88) by the charge of a scalar particle q it behaves like a charge density operator, which we will designate by s^μ.

$$s^\mu_{\ ,\mu} = 0 \quad \text{with} \quad s^\mu = qj^\mu = iq\left(\phi^{,\mu}\phi^\dagger - \phi^{\dagger,\mu}\phi\right) , \tag{3-95}$$

so

Multiply 4-current operator by particle charge to get charge 4-current operator

$$\rho_{charge} = qj^0 = iq\left(\frac{\partial\phi}{\partial t}\phi^\dagger - \frac{\partial\phi^\dagger}{\partial t}\phi\right). \tag{3-96}$$

This makes sense, as charge would be distributed in parallel fashion to probability density, i.e., denser charge where the particle is more concentrated. Further, total charge using (3-93) multiplied by q would yield $4q$, the charge on the state. Similarly, the total charge on the state in (3-94) would be $-q$.

Thus, re-interpreting the operator ρ, as charge density, and the 3D part of the four current as charge current density is consistent. In actuality, it is demanded in order for our theory to agree with experiment. That empirical reality also forces us to accept b type particles as antiparticles.

Take care that in the future, we may use the symbol ρ as simply charge density, without a subscript. Since we will rarely, if ever, deal again with probability density, hopefully, there will be little confusion.

3.7.6 Caution in Evaluating Expectation Values of Density Operators

Some care must be taken in the evaluation of expectation values similar to that of (3-90). The bracket, expressed in the position basis, is an integration over space. But for operators with a spatial dependence such as ρ often has (and which is typical of charge, mass or any type of density), the spatial dependence in the operator is <u>not</u> included in the integration. That is, writing out the expectation value as an integral, we integrate over the $\mathbf{x}'$ of the state, but not the $\mathbf{x}$ of the operator.

Integration implied in expectation value is over ket $\mathbf{x}'$, not over operator $\mathbf{x}$

$$\left\langle\rho(\mathbf{x},t)\right\rangle = \left\langle\phi_\mathbf{k}(\mathbf{x}',t)\,|\,\rho(\mathbf{x},t)\,|\,\phi_\mathbf{k}(\mathbf{x}',t)\right\rangle = \int\phi^\dagger_{\mathbf{k},state}(\mathbf{x}',t)\rho(\mathbf{x},t)\phi_{\mathbf{k},state}(\mathbf{x}',t)d^3x' . \tag{3-97}$$

This was not evident in (3-92) and similar relations above, because there (for plane waves, specifically) the operator ρ was not a function of space.

The point in (3-97) generalizes to other types of operator functions that would be sandwiched inside a bra and a ket. We will run into these in the future.

3.7.7 The ϕ and $\phi^\dagger$ Normalization Constants Again

We just assumed, in all of our discussion so far, that the normalization constants in our solutions, $1/\sqrt{2\omega_\mathbf{k}V}$, that we derived in RQM, are also valid in QFT. Since our field solutions ϕ and $\phi^\dagger$ in QFT had the same form as the state solutions in RQM, and our 4-current $j^\mu = (\rho, \mathbf{j})$ in each case had the same form as well, this seems like a reasonable assumption. The assumption can be considered justified by our results in the above few sections. For example, (3-91) worked out as a correct form for density (probability or charge) only because of the form chosen for our constants. The square root of $2\omega_\mathbf{k}$ dropped out in getting (3-91) because of the two terms, each with two field factors multiplied, and the time derivatives in (3-89).

ϕ and $\phi^\dagger$ normalization constants from RQM work for QFT

We can therefore consider the results of the sections above as justification for the choice of normalization constants in the field solutions to the Klein-Gordon equation. All of so many other results, yet to be seen in our studies, will be further justification.

3.8 More on Observables

QFT, like the quantum theories studied before it, is interested in observable quantities, such as energy, 3-momentum, charge, and spin, which are represented in each of those theories by operators. The eigenvalues of those operators are what we measure. Expectation values of those operators are the averages of what we measure over many trials.

Regarding energy, we have already discussed the observable operator corresponding to it. See (3-87). Regarding spin, scalar particles have none, so we will put off discussion of particles that do have it to later chapters.

Finding operators other than H in terms of number operators

3.8.1 Charge Operator

Regarding charge, we need merely to integrate our charge density operator qj^0 of (3-96) and (3-91) over the entire volume, to get the <u>charge operator</u>

$$Q = \int s^0 d^3x = q \int j^0 d^3x = q \sum_{\mathbf{k}} \left(N_a(\mathbf{k}) - N_b(\mathbf{k}) \right). \qquad (3\text{-}98)$$

A typical multiparticle state is in a charge eigenstate with an eigenvalue of charge equal to the sum of the charges of all particles in the state. Hence, the eigenvalue equals the <u>charge expectation value</u>, since we will measure the same charge with each measurement. For a sample state,

Total charge operator yields total charge of a (multiparticle) state

$$\bar{Q} = \left\langle 7\phi_{\mathbf{k}_1}, \phi_{\mathbf{k}_2}, 5\bar{\phi}_{\mathbf{k}_3} \left| \underbrace{q \sum_{\mathbf{k}} \left(N_a(\mathbf{k}) - N_b(\mathbf{k}) \right)}_{Q} \right| 7\phi_{\mathbf{k}_1}, \phi_{\mathbf{k}_2}, 5\bar{\phi}_{\mathbf{k}_3} \right\rangle = 7q + q - 5q = +3q. \qquad (3\text{-}99)$$

Note, we derived (3-98) using (3-91). If you did Prob. 12, you saw that in deriving (3-91) we summed terms in ½ and − ½ that cancelled to net zero. In other words, (3-98) actually has a $+q/2$ and a $-q/2$ term for each $\mathbf{k}$. Thus, Q acting on the vacuum would sum up an infinite number of half charge quanta for both particles (positive charge) and anti-particles (negative charge), leaving zero total vacuum charge. (Thankfully, it does. If it didn't, our theory would be bound for the trash heap.)

Vacuum has zero charge

3.8.2 Three Momentum Operator

The three-momentum operator can be found using the relationship for physical momentum density at the bottom of Box 2-2 in Chap. 2, pg. 23, and integrating over the volume. (Also shown in the 9th block under the title in the RH column of Wholeness Chart 2-2, pg. 21.) That is,

$$p^i = \int \not{p}^i d^3x = -\int \pi_r \frac{\partial \phi^r}{\partial x^i} d^3x = -\int \left(\frac{\partial \mathcal{L}}{\partial \dot{\phi}} \frac{\partial \phi}{\partial x^i} + \frac{\partial \mathcal{L}}{\partial \dot{\phi}^\dagger} \frac{\partial \phi^\dagger}{\partial x^i} \right) d^3x . \qquad (3\text{-}100)$$

Substituting the Klein-Gordon solutions (3-36) and their conjugate momenta into (3-100), one obtains the <u>3-momentum operator</u> (do Prob. 13 to prove it)

$$\mathbf{P} = \sum_{\mathbf{k}} \mathbf{k} \left(N_a(\mathbf{k}) + N_b(\mathbf{k}) \right), \qquad (3\text{-}101)$$

Total 3-momentum operator yields total 3-momentum of a (multiparticle) state

which is pretty much what we would have expected. **P** operating on a multiparticle ket, with all particles in **k** eigenstates, would yield an eigenvalue equal to the number of particles and antiparticles with 3-momentum **k** multiplied by **k**. If this is less than obvious to you, do Prob. 14.

It is interesting that, similar to what happened to charge, we have ½ quanta in the vacuum with 3-momentum, but the total for the vacuum sums to zero. That is, in deriving (3-101), we get terms in the summation of ½ **k** + ½ **k** = **k** (one ½ quanta for each particle and one for each antiparticle), similar to what we had for energy. But unlike energy, this is a vector summation, and for every 3-momentum **k** in the sum, there is a 3-momentum − **k**, as well. The net is nil 3-momentum for the vacuum, which again, is a welcome result.

Vacuum has zero 3-momentum

So far in our theory, only energy has proved problematic in having a non-zero vacuum expectation value (VEV.)

3.8.3 The Four Momentum Operator

As discussed in the Appendix of Chap. 2, and elsewhere, the four momentum has energy in the 0th component (E/c in non-natural units) and 3-momentum for the other three components. Given (3-87) for $H_0^{\,0}$, and (3-101) for the free scalar field p^i, the <u>four momentum operator</u> is

$$\underbrace{P^{\mu} = K^{\mu}}_{\substack{\text{operators} \\ \text{here}}} = \underbrace{\begin{pmatrix} H \\ \mathbf{P} \end{pmatrix}}_{\substack{\text{for free} \\ \text{scalars}}} = \sum_{\mathbf{k}} \underbrace{\begin{pmatrix} \omega_{\mathbf{k}} \\ \mathbf{k} \end{pmatrix}}_{\substack{\text{usually what} \\ \text{we mean by} \\ \text{symbol } k^{\mu}}} \left(N_a\left(\mathbf{k}\right) + N_b\left(\mathbf{k}\right) \right), \tag{3-102}$$

4-momentum operator includes energy and 3-momentum

where we note that k^{μ} usually refers to the numeric (not operator) 4 vector ($\omega_{\mathbf{k}}, \mathbf{k}$).

3.9 Real Fields

So far, we have only dealt with complex fields. It is possible to have real fields, and in fact, we will see they play a key role in the theory. They turn out to be associated with neutral particles, which are, in fact, their own antiparticles.

To see this, look at a special case for our general field equation solutions (3-36) where $\phi = \phi^{\dagger}$, i.e., ϕ is real. In order for this to be true, we must have $a(\mathbf{k}) = b(\mathbf{k})$, and of course, $a^{\dagger}(\mathbf{k}) = b^{\dagger}(\mathbf{k})$. Thus, for this case,

$$\phi = \phi^{\dagger} = \sum_{\mathbf{k}} \frac{1}{\sqrt{2V\omega_{\mathbf{k}}}} a\left(\mathbf{k}\right) e^{-ikx} + \sum_{\mathbf{k}} \frac{1}{\sqrt{2V\omega_{\mathbf{k}}}} a^{\dagger}\left(\mathbf{k}\right) e^{ikx} \quad \left(\text{for } \phi \text{ real}\right), \tag{3-103}$$

which makes sense, since adding a complex number and its complex conjugate yields a real number.

In the charge operator (3-98), we would then have $N_a\left(\mathbf{k}\right) = N_b\left(\mathbf{k}\right)$ (i.e., $a^{\dagger}(\mathbf{k})a(\mathbf{k}) = b^{\dagger}(\mathbf{k})b(\mathbf{k})$), so charge would be zero for any such particle(s) state. Each b type particle operator (creation, destruction, charge, energy, etc.) will be the same operator as that for a particles. There is only one type of particle (for a real field), so that particle must be its own antiparticle.

Real (not complex) fields create and destroy neutral charge particles

Conclusion: Real fields are associated with charge-neutral particles, which are their own anti-particles.

Note on nomenclature: The term "real field" refers to the (operator) field solutions to the field equation (Klein-Gordon for scalars) that are not complex. The term "real particle" refers to a particle that is not virtual, but manifest and detectable.

Terminology difference between real fields and real particles

3.10 Characteristics of Klein-Gordon States

3.10.1 Bosons vs Fermions

Although at this point in your career, you should be familiar with bosons and fermions, and their behavior, Wholeness Chart 3-1 can serve as a refresher course. It should need no further comment.

Wholeness Chart 3-1. Bosons vs Fermions

	Bosons	**Fermions**
What role	typically forces	typically matter
Some examples	elementary: photons, Higgs composite: mesons	elementary: electrons, neutrinos, quarks composite: baryons (e.g., proton, neutron)
Behavior	can occupy same state	can't occupy same state
Spin	integer spin Scalars: spin 0 Vectors (e.g. photons): spin 1 Graviton: spin 2	half integer spin Spinors: spin ½ Gravitinos: spin 3/2

3.10.2 Klein-Gordon States are Bosons

Using the same scalar creation operator repeatedly on a state results in a raised state containing a number of the same particle with the same $\mathbf{k}$ (and thus the same energy and identical in all regards.) For example, using the creation operator of the first line of (3-81) acting first on the vacuum, then repeatedly on the newly created states, we have

$$a(\mathbf{k})^{\dagger}|0\rangle = |\phi_{\mathbf{k}}\rangle \;\rightarrow\; a(\mathbf{k})^{\dagger}|\phi_{\mathbf{k}}\rangle = \sqrt{2}\,|2\phi_{\mathbf{k}}\rangle \;\rightarrow\; a(\mathbf{k})^{\dagger}|2\phi_{\mathbf{k}}\rangle = \sqrt{3}\,|3\phi_{\mathbf{k}}\rangle \rightarrow ..... \quad (3\text{-}104)$$

Scalar kets can have > 1 particle in same state, so scalars are bosons, not fermions

We are not concerned, for this discussion, with the square root numeric coefficients, but with the fact that we can have multiparticle states with more than one individual particle in the same individual state.

This means Klein-Gordon states must be bosons. We sort of knew this because we were told that they have zero spin. But here we prove it.

As we will see in the next chapter, spinors do not have the characteristic displayed by (3-104).

3.10.3 Commutators with Scalars, Not Anti-Commutators

Let's see what happens if anti-commutators were used instead of commutators with the Klein-Gordon field equation solutions. That is, in the derivation of our number operator form of the Hamiltonian, in equation (3-54), try the <u>anti-commutators</u>

$$\left[a(\mathbf{k}), a^{\dagger}(\mathbf{k}') \right]_{+} = a(\mathbf{k})a^{\dagger}(\mathbf{k}') + a^{\dagger}(\mathbf{k}')a(\mathbf{k}) = \delta_{\mathbf{k}\mathbf{k}'}$$
$$\left[b(\mathbf{k}), b^{\dagger}(\mathbf{k}') \right]_{+} = b(\mathbf{k})b^{\dagger}(\mathbf{k}') + b^{\dagger}(\mathbf{k}')b(\mathbf{k}) = \delta_{\mathbf{k}\mathbf{k}'} . \qquad (3\text{-}105)$$

A minus sign is then introduced, such that instead of (3-55), we would get

$$H_0^0 = \sum_{\mathbf{k}} \omega_{\mathbf{k}} \left(\tfrac{1}{2} + \tfrac{1}{2} \right) , \qquad (3\text{-}106)$$

or for an <u>observable Hamiltonian</u> (ignoring vacuum energy)

$$H_0^0 = 0 \quad \text{and thus,} \quad H_0^0 \left| \phi_{\mathbf{k}_1} \right\rangle = 0 . \qquad (3\text{-}107)$$

Using anti-commutator instead of commutator → theory with faulty energy prediction (wrong)

Every real state would have zero energy, which is certainly not physically true. Therefore, we can only use commutators with spin 0 boson fields, and not anti-commutators.

We will find in the next chapter, that fermions in QFT are governed by anti-commutators in parallel fashion to the way in which bosons (scalars, at least, to this point in our studies) are governed by commutators. Just as anti-commutators can't work for bosons (scalars), as we saw above, we will also see later that commutators can't work for fermions.

Klein-Gordon scalars:
1) *must be bosons, and*
2) *can only use commutation relations*

3.11 Odds and Ends

3.11.1 Usefulness of 3-Momentum Discrete Eigenstates

As you will see in time, QFT can find real world experimental values, for things like scattering cross sections and decay half-lives, using only discrete **k** eigenstate forms for real states. These eigenstates, unlike wave packets, typically extend indefinitely in the **k** direction. But for experimental predictions, the particle states, which are actually wave packets, can be approximated to extremely high precision by such discrete **k** eigenstates.

Discrete form of solutions (not wave packets) suffice for extremely accurate real particle predictions

One exception is the propagator, the mathematical representation of virtual particles, which is best derived, as shown in Sect. 3.13, and most useful practically, via incorporation of the continuous (integral) form of the field equation solutions. This is, at least in part, because virtual particles are not constrained by boundary conditions to discrete **k** values and in certain cases must be integrated over all possible, continuous values of **k**.

Except, we will need continuous solutions to find Feynman propagator (for virtual particles)

3.11.2 Nevertheless, What about Non-eigen States?

In NRQM (and RQM) we commonly dealt with general states, i.e., non-eigen states, which were superpositions of two or more eigenstates. Granted, as noted above, that we can solve almost all QFT problems using **k** eigenstates, we might still ask "how does QFT compare in this regard to what we learned in NRQM?" It is a good question, troubling many students, no doubt, and not treated in any other text I am aware of.

A closely related question is "what is created or destroyed by the general solutions $\phi(x)$ (or $\phi^{\dagger}(x)$), which for discrete eigenstates, is a summation of terms, each containing a single particle eigenstate creation/destruction operator?" Does operation of $\phi^{\dagger}(x)$ on the vacuum, for instance, create an infinite number of single particles, or a single particle comprising an infinite number of momentum eigenstates? If the latter, what amplitudes (whose absolute values squared are probabilities) are assigned to each such eigenstate?

Creating a General Single Particle State (Discrete Solution Form)

To create a general single particle state, , we would need a creation operator of form

$$C = \sum_{\mathbf{k}} A_{\mathbf{k}} \mathbf{a}_{\mathbf{k}}^{\dagger} , \qquad (3\text{-}108)$$

Creation operator for single particle general (non-eigen) state, discrete case.

so that operation of C on the vacuum results in a sum of eigenstates,

$$C|0\rangle = \sum_{\mathbf{k}} A_{\mathbf{k}} \mathbf{a}_{\mathbf{k}}^{\dagger}|0\rangle = A_1 |\phi_1\rangle + A_2 |\phi_2\rangle + A_3 |\phi_3\rangle + ... = |\phi\rangle . \qquad (3\text{-}109)$$

In (3-108) and (3-109) $A_{\mathbf{k}}$ is a numerical coefficient, the square of the absolute value of which (for proper normalization of the ket eigenstates) equals the probability of finding the $\mathbf{k}$ eigenstate.

If only one term in C is used, then only one eigenstate with $|A_{\mathbf{k}}| = 1$ is created. If a more general state, comprising a sum of eigenstates, is created, then we are free to select the $A_{\mathbf{k}}$ as we please in order to create the particular general state we like, provided (for conservation of probability and correct normalization so total probability is unity)

$$\sum_{\mathbf{k}} | A_{\mathbf{k}} |^2 = 1 . \qquad (3\text{-}110)$$

Destroying a General Single Particle State (Discrete)

Note that the general single particle destruction operator

$$D = \sum_{\mathbf{k}} \mathbf{a}_{\mathbf{k}} \qquad (3\text{-}111)$$

Destruction operator for single particle general (non-eigen) state, discrete case.

acting on any single particle general state will lower that state to the vacuum. That is,

$$\left(\sum_{\mathbf{k}} \mathbf{a}_{\mathbf{k}} \right) |\phi\rangle = \left(\sum_{\mathbf{k}} \mathbf{a}_{\mathbf{k}} \right) \left(A_1 |\phi_1\rangle + A_2 |\phi_2\rangle + A_3 |\phi_3\rangle + ... \right)$$

$$= \left(\left(\sum_{\mathbf{k}} \mathbf{a}_{\mathbf{k}} \right) A_1 |\phi_1\rangle + \left(\sum_{\mathbf{k}} \mathbf{a}_{\mathbf{k}} \right) A_2 |\phi_2\rangle + \left(\sum_{\mathbf{k}} \mathbf{a}_{\mathbf{k}} \right) A_3 |\phi_3\rangle + ... \right) \qquad (3\text{-}112)$$

$$= A_1 \underbrace{\mathbf{a}_1 |\phi_1\rangle}_{|0\rangle} + A_1 \underbrace{\mathbf{a}_2 |\phi_1\rangle}_{0} + A_1 \underbrace{\mathbf{a}_3 |\phi_1\rangle}_{0} + ... + A_2 \underbrace{\mathbf{a}_1 |\phi_2\rangle}_{0} + A_2 \underbrace{\mathbf{a}_2 |\phi_2\rangle}_{|0\rangle} + 0 + ... = \underbrace{(A_1 + A_2 + ...)}_{\text{can normalize} = 1} |0\rangle .$$

Creating and Destroying Multi-particle State (Discrete)

Applying operators similar in form to (3-108) (with typically different values for $A_{\mathbf{k}}$ in each operator) twice in succession creates a two particle state where each particle is a single particle general state (i.e., each is a summation of momentum eigenstates.) Any number of such operators may be applied to create a state of any number of particles, each in a general (not necessarily eigen) state.

For multiparticle general states, repeatedly apply C and D operators

Applying (3-111) repeatedly will destroy one general state single particle upon each application.

What $\phi(x)$ and $\phi^{\dagger}(x)$ Create When Acting on the Vacuum

$\phi(x)$ acting on the vacuum will create a single general antiparticle state comprising a superposition of an infinite number of eigenstates, each with a constant coefficient in front of it, i.e.,

$$\phi(x)|0\rangle = \sum_{\mathbf{k}} \frac{1}{\sqrt{2V\omega_{\mathbf{k}}}} e^{-ikx} \underbrace{a(\mathbf{k})|0\rangle}_{0} + \sum_{\mathbf{k}} \frac{1}{\sqrt{2V\omega_{\mathbf{k}}}} e^{ikx} \underbrace{b^{\dagger}(\mathbf{k})|0\rangle}_{|\bar{\phi}_{\mathbf{k}}\rangle} = \sum_{\mathbf{k}} \underbrace{\frac{1}{\sqrt{2V\omega_{\mathbf{k}}}}}_{\text{constant}} \underbrace{e^{ikx}}_{\substack{\text{phase} \\ \text{factor}}} |\bar{\phi}_{\mathbf{k}}\rangle . \quad (3\text{-}113)$$

Similarly, $\phi^{\dagger}(x)$ acting on the vacuum will create a single particle general state comprising a superposition of particle eigenstates. For discrete solutions, these operations have little use in QFT.

Creating and Destroying Continuous Solution Forms of States

In analogous fashion, the continuous form of $\phi(x)$ (or $\phi^{\dagger}(x)$) acting on the vacuum yields a single antiparticle (particle) wave packet, i.e., an integral over all $\mathbf{k}$, rather than a sum. You don't need to know more now, but for the future, details are in this book's web site (address on pg. xvi, opposite pg. 1) at the link titled Non Eigen States, Wave Packets, and the Hamiltonian in QFT. A summary of that can be found in Appendix C of Chap. 10 herein. We will work more with these operations when we derive the Feynman propagator in Sect. 3.13.

3.11.3 c-numbers vs q-Numbers

The terms c-number and q-number were introduced by Paul Dirac to distinguish between *classical numbers* (real or complex), which commute, and *quantum operators*, which do not always commute. The term q-number can equally apply to the *eigenvalue* of a given quantum operator.

q-numbers are quantum numbers (operators or their eigenvalues);

Thus, the 3-momentum of a classical particle is a c-number. The 3-momentum of a quantum state, in a 3-momentum eigenstate, is a q-number.

Eigenstates are often labeled by their q-numbers (eigenvalues). For example, the n, l, and m numbers for electron levels in the hydrogen atom are quantum, or q, numbers. n represents the energy level number (which is simpler than specifying the energy itself); l, the angular momentum magnitude; and m, the z component of angular momentum. By specifying n, l, and m, one specifies the eigenstate of the electron in the atom.

c-numbers are classical numbers

3.11.4 Fock Space and Hilbert Space

As you should (hopefully) remember, a quantum state in NRQM is an abstract vector in an abstract vector space, analogous to a physical vector in 3D physical space. The same thing is true in RQM and QFT. This is summarized in Wholeness Chart 3-2.

In all quantum theories, basis vectors (which are typically eigenstates) are abstractions of the unit basis vectors along the 3D axes. A general state is a vector sum of certain amounts of each basis vector state. Operators in each kind of space act on the states in that space.

In NRQM and RQM, the states are single particle states and the abstract space they inhabit is called Hilbert space, which has a different single particle eigenstate as the basis vector of each "axis". The dimension of the Hilbert space for a given system is simply the number of linearly independent eigenstates in that system. This can, for many systems, be infinite.

In QFT, states are multiparticle, so the basis eigenstate of each "axis" is a multiparticle state. One "axis" basis vector might be an electron and a photon, each with particular 3-momentum. Another might be 2 photons and a positron with particular 3-momenta. Yet another might be an electron and photon like the first, except that at least one of them has different 3-momentum from the first. The multiparticle abstract state space of QFT is called Fock space, which is simply an extension of Hilbert space to multiparticle states.

Fock space is Hilbert space generalized to multiparticle states

Wholeness Chart 3-2. Physical, Hilbert, and Fock Spaces

	3D Physical Space	**Hilbert Space**	**Fock Space**
Character of a vector	Position vector in 3D	State vector $|\Psi\rangle$ in NRQM, RQM Single particle	State vector $|\phi_1, \phi_2, \dots\rangle$ in QFT Multi particle
Orthonormal basis vectors along "axes"	$\hat{\mathbf{i}}_1, \hat{\mathbf{i}}_2, \hat{\mathbf{i}}_3$	Normalized eigenvectors $|\Psi_1\rangle, |\Psi_2\rangle, |\Psi_3\rangle, |\Psi_4\rangle, \dots$	Normalized eigenvectors $\lvert 0\rangle, \lvert \phi_1\rangle, \lvert \phi_2\rangle, .. \lvert \phi_1, \phi_2\rangle, .. \lvert \phi_1, \phi_2, \phi_3\rangle, \dots$
Inner product	$\hat{\mathbf{i}}_i \cdot \hat{\mathbf{i}}_j = \delta_{ij}$	$\langle \Psi_r | \Psi_s \rangle = \delta_{rs}$	$\langle \phi_1 | \phi_1, \phi_2 \rangle = 0$; $\langle \phi_1, \phi_2 | \phi_1, \phi_2 \rangle = 1$; etc.
General state vector	$\mathbf{r} = x^1\hat{\mathbf{i}}_1 + x^2\hat{\mathbf{i}}_2 + x^3\hat{\mathbf{i}}_3$	$|\Psi\rangle = C_1|\Psi_1\rangle + C_2|\Psi_2\rangle + C_3|\Psi_3\rangle + \dots$	$|\Phi\rangle = C_1|\phi_1\rangle + C_2|\phi_2\rangle + \dots + C_{12}|\phi_1, \phi_2\rangle + C_{13}|\phi_1, \phi_3\rangle + \dots + C_{123}|\phi_1, \phi_2, \phi_3\rangle + \dots$
State vector & its components	point in 3D space = amount along each basis vector	"point" in Hilbert Space = amount of each single particle basis vector	"point" in Fock Space = amount of each multi particle basis vector
Operators	Matrices operate on vectors	Hamiltonian H, 3-momentum $\mathbf{P}$, etc operate on states	Hamiltonian H, $\mathbf{P}$, creation $a^\dagger(\mathbf{k})$, $b^\dagger(\mathbf{k})$, destruction $a(\mathbf{k}), b(\mathbf{k})$, charge Q operate on states

3.11.5 a(k) Destroys Any State without Single a Type Particle in k Eigenstate

Keep in mind (as we actually already did in the last row of (3-112)) that, for example,

$$\mathbf{a}_2\left|\phi_1, 4\phi_3, 7\overline{\phi}_2\right\rangle = 0, \tag{3-114}$$

Destruction operators of given kind and k *destroy any state not having like kind of particle in a* k *eigenstate*

which is zero, not the vacuum state. In general, a particular type particle destruction operator of given **k** acting on a state that has no particles of that type of the same 3-momentum **k** results in zero.

3.12 Harmonic Oscillators and QFT

One sometimes hears that particles in QFT can be considered to be harmonic oscillators. The reason for this can be seen with the aid of Wholeness Chart 3-3, which summarizes the states of the NRQM harmonic oscillator (RQM is similar, but more complicated) and particles in QFT.

One sees immediately that the energy levels of the two look very similar. Each level is $\hbar\omega$ above the one below it. (We keep the symbol $\hbar$ for this discussion, since it makes the NRQM summary look more familiar.) And strikingly, each also has a lowest level of energy, when $n = 0$ ($n_k = 0$ in QFT, to be precise), of ½ quantum (½$\hbar\omega$ or ½$\hbar\omega_k$.) More striking still, each has raising and lowering operators that raise and lower energy levels by $\hbar\omega$ (or $\hbar\omega_k$ for each extra particle in QFT.)

QFT particle states have similarities to harmonic oscillator energy states

These similarities led people to think in terms of QFT particles as harmonic oscillators. The vacuum was the lowest excited state. Each state above (in QFT, each additional particle) was simply a more excited state of the lowest state (in QFT, the vacuum state.) Operators acting on states raise or lower the number of particles, and thus the energy level, and so excite, or de-excite, the vacuum. Particles are just excitations of an underlying vacuum field.

Wholeness Chart 3-3. Quantum Harmonic Oscillator Compared to QFT Free States

	NRQM Harmonic Oscillator	**QFT Free States**
Energy Levels	$(n + \frac{1}{2})\hbar\omega$	$(n_k + \frac{1}{2})\hbar\omega_k$
Interpretation of n and n_k	single particle energy level 0,1,2,…	number of particles at $\hbar\omega_k$ energy
Interpretation of ω and ω_k	natural angular frequency of classical oscillator	angular frequency of particle of energy $\hbar\omega_k$
Lowest energy level	½ $\hbar\omega$	½ $\hbar\omega_k$
Interpretation of ↑	real particle in lowest state	vacuum, virtual particle
Raising operator	raises single particle energy one level	raises number of particles by one and thus, also raises energy one level
Lowering operator	lowers single particle energy one level	lowers number of particles by one and thus, also lowers energy one level
Wave form	Hermite polynomial	$e^{\pm ikx}$
Nature of wave form	real, non-sinusoid	complex, sinusoid
Motion	oscillates in one place	wave that moves
Spatial constraints	bound state, local region	unbound state, unlimited volume
Free or interaction	harmonic oscillator potential → force	free, no force

3.12.1 "Derivation" of QFT via Harmonic Oscillators

Some "derive" QFT from harmonic oscillator assumption

Some treatments actually introduce QFT via assuming states therein are harmonic oscillators. I submit this assumption can only be made after one already knows the form of the Hamiltonian (3-55) and the raising/lowering operators (3-75) as we derived them in Sect. 3.4 (pg. 53.) Otherwise, how could anyone understand they should simply assume the QFT states have energy levels similar to those of the harmonic oscillator?

I contend that assuming harmonic oscillator behavior in QFT is an unreasonable, and unfounded, assumption, but that starting with 2nd quantization (a parallel track to what was known to work in NRQM) is a reasonable assumption. However, the former approach is common.

3.12.2 Harmonic Oscillators Have Different Behavior than States

Note that the wave form for the harmonic oscillator is a Hermite polynomial, far different from the complex sinusoid of $e^{\pm ikx}$ that fields (and states) have in QFT. And, a harmonic oscillator doesn't move in space (other than up and down, or side to side, in one location), whereas waves (particles = states) do, i.e., they travel from place to place. Further, the free fields (and particles) we have been dealing with in QFT are unrestricted in space (for discrete solutions, volume V can be as large as the universe; for continuous solutions, there is no volume constraint), whereas harmonic oscillators are confined to a local region. Still further, harmonic oscillators are not free states like those we have treated, but feel force/interaction (due to the harmonic oscillator potential).[1]

But there are differences between QFT particle states and harmonic oscillator states

For discussion of a counter argument that might be made here for the vacuum, see Appendix B.

Note the caveats: This section and the one above it are my personal position on this matter, and may not be shared by many others. You, the reader, should make your own call.

Note also that one can still treat $e^{-i\omega_k t}$, the time factor alone of any term in (3-36), for fixed **x**, as a harmonic oscillator solution. This can help in analysis as shown, for example, in V. F. Mukhanov and S. Winitzki, *Introduction to Quantum Effects in Gravity*, (Cambridge 2007).

3.12.3 Vacuum Excitations = Real Particles

In spite of the foregoing, one can still think of real states as stable, excited states of the vacuum, since our raising operators can create a particle state from that vacuum, i.e.,

$$a^{\dagger}\left(\mathbf{k}\right)\left|0\right\rangle = \left|\phi_{\mathbf{k}}\right\rangle. \tag{3-115}$$

States can be thought of as excitations of the vacuum (the least excited, or ground, state)

The RHS above can be considered as the next highest state above the ground state (above the vacuum), and thus, an excited state of the vacuum. Considering such excited states specifically as *harmonic oscillator* excited states is a different matter.

3.13 The Scalar Feynman Propagator

The Feynman propagator, the mathematical formulation representing a virtual particle, such as the one represented by the wavy line in Fig. 1-1 of Chap. 1, pg. 2, is one of the toughest things, in my opinion, to learn and feel comfortable with in QFT. If you don't feel comfortable with it right away, don't worry about it. That is how virtually everyone feels. Over time, it will become more familiar, and if you are lucky and work hard, maybe even easy.

Feynman propagator not simple to understand

I have tried to take the derivation of the propagator one step at a time and emphasize what each step entails. Wholeness Chart 5-4 (at the end of Chap. 5) breaks these steps out clearly and should be used as an aid when studying the propagator derivation.

Use wholeness chart as you study the derivation

Propagators: NRQM vs QFT and Real vs Virtual Particles

Note that the propagator for real particles, which you may have studied in NRQM, is *not* the same as the Feynman propagator, which is explicitly for virtual particles in QFT. It may be confusing, but the Feynman propagator is often simply called, "the propagator". You will have to get used to discerning the difference from context.

In QFT, as we will see when we study interactions, a propagator for real particles is not generally needed, and we will not derive one here.

Feynman propagator for QFT virtual particles is different from propagator for real particles of NRQM & RQM

3.13.1 The Approach

The first part of QFT is a free particle theory (no interactions, as in this chapter and the next three). After this, interactions are introduced. In the course of deriving the interaction theory, a mathematical relationship arises that is called the Feynman propagator. Physically, it can be

We'll use the Feynman propagator when we get to interaction theory

---

[1] All of this harmonic oscillator business confused me greatly as a student. I simply could not understand how QFT states could possibly be essentially identical to harmonic oscillators. I was not confident enough to bring up the counter points mentioned herein, and they were never addressed. So, if you have seen and been confused by the harmonic oscillator approach, you are not alone.

visualized as representing a virtual particle that exists fleetingly and carries energy, momentum, and in some cases, charge from one real particle to another. Thus, it is the carrier, or mediator, of force (interaction.) See the virtual photon of Fig. 1-1 in Chap. 1, pg. 2.

It will help us pedagogically to derive the Feynman propagator now, rather than when we get to interactions. The derivation of interaction theory is fairly complicated and it will be easier, as we develop it, if we already know the mathematical relation for the Feynman propagator, rather than diverting our attention for several pages to derive it then.

But it's easier in the long run if we derive it here

Heuristically, it may help to consider the virtual particle as created at a particular spacetime point and destroyed at a later spacetime point, and this is how Feynman diagrams portray it. From this (heuristic) perspective the operator field $\phi^\dagger(y)$ can be considered to create a virtual scalar particle at event y (we used the symbol x_2 in Fig. 1-1), and the field operator $\phi(x)$ destroys that virtual particle at event x (x_1 in Fig. 1-1.) The scalar propagator incorporates these two field operators in a sort of "short-hand" way.

Note that the above "creation/destruction at a point" perspective can help initially in understanding the derivation of the propagator, but we caution that it will have to be modified and refined. We will save that to the end when, after digesting the derivation to follow, this modification will be easier to understand.

We will now derive a relationship for the propagator using the field operators acting on the vacuum, and will later see (Chap. 7) that this derived relationship arises naturally in the full mathematical development of the interaction theory.

3.13.2 Milestones in the Derivation

We develop the Feynman propagator in five distinct steps, starting with a physical interpretation. We represent that interpretation mathematically and then "massage" it in subsequent steps with more mathematics, until we obtain the form of the propagator that is most useful (in QFT interaction analysis).

Start with a physical visualization of the propagator and follow 5 distinct steps

The entire derivation is for continuous (not discrete) eigenstate solutions of the field equation (Klein-Gordon here), since the propagator represents a virtual particle in the vacuum and the vacuum is not confined to a volume V. We represent the scalar Feynman propagator with the symbol $i\Delta_F(x-y)$. (Including the imaginary factor i is common practice.)

Step 1: Math interpretation of the physical propagator

Step 1: Express the Feynman propagator $i\Delta_F$ as a mathematical representation of a particle or antiparticle created at one point in space and time in the vacuum and destroyed at another place and time.

Steps 2, 3, and 4: Math manipulation of relations for particle and anti-particle

Step 2: Express $i\Delta_F$ in terms of two commutators (one for particles and one for anti-particles).

Step 3: Express those two commutators as real integrals.

Step 4: Re-express those two real integrals as two contour (complex plane) integrals.

Step 5: Re-express the two contour integrals as a single integral over real, not complex, space, the form most suitable for analysis.

Step 5: Combining two integrals in complex space into one over real space

Step 1: The Feynman Propagator as the Vacuum Expectation Value of a Time Ordering Operator

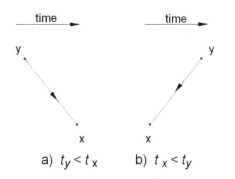

a) $t_y < t_x$ b) $t_x < t_y$

Figure 3-3. Creation & Destruction of Virtual Particle/Antiparticle

Fig 3-3a represents creation of a particle, which will be virtual, at y and destruction of it at x. Fig. 3-3b represents creation of an antiparticle at x and destruction of it at y. Virtual particles are never detected when real particles interact, so the same effect on the real particles could be realized by either of the processes in Fig. 3-3. That is, a virtual particle carrying charge from y to x would represent the same charge exchanges as a virtual antiparticle carrying opposite charge from x to y. Thus, we need a relationship for the propagator that includes both scenarios as possibilities.

That is, we need an operator that will create a particle first if $t_y < t_x$, but create an antiparticle first if $t_x < t_y$. Our Klein-Gordon solutions (3-37), (3-85), and

Step 1, first part, defining the time ordering operator T and seeing how it represents creation of either a virtual particle or antiparticle followed by its destruction

(3-86) provide the means for the desired creation and destruction operations. But these have to be arranged to provide us with the time ordering dependence of Fig. 3-3. To this end, consider the time ordering operator T, defined as follows.

If $t_y < t_x$, $\phi^\dagger(y)$ operates first (creates a particle) and is placed on the right, with $\phi(x)$ operating second (destroys the particle) and placed on the left.

$$\text{for } t_y < t_x \quad \text{(particle)} \qquad T\{\phi(x)\phi^\dagger(y)\} = \phi(x)\phi^\dagger(y). \qquad (3\text{-}116)$$

Of course, from (3-85), and (3-86), in (3-116), $\phi(x)$ also creates an antiparticle and $\phi^\dagger(x)$ also destroys an anti-particle, but we will see this effect ultimately drops out and does not play a role in the Feynman propagator.

If $t_x < t_y$, $\phi(x)$ operates first (creates an antiparticle) and is placed on the right with $\phi^\dagger(y)$ operating second (destroying the antiparticle) and placed on the left (where the effects of these operators for particles will drop out, as we will see.)

$$\text{for } t_x < t_y \quad \text{(anti-particle)} \qquad T\{\phi(x)\phi^\dagger(y)\} = \phi^\dagger(y)\phi(x). \qquad (3\text{-}117)$$

Step 1, second part, defining the transition amplitude as equal to the VEV of time ordering operator T

We now define what is called the transition amplitude, which equals the vacuum expectation value (VEV) of the above time ordering operator. It is an amplitude, similar to the amplitude of a wave function in NRQM, because, as we will shortly see, the square of its magnitude equals the probability density of it being observed. (As the square of the magnitude of the amplitude for a component of the wave function equals the probability of it being observed.)

This transition amplitude is

$$\langle 0|T\{\phi(x)\phi^\dagger(y)\}|0\rangle, \qquad (3\text{-}118)$$

We use the VEV because we will be interested in the expectation of finding a virtual particle traveling in the vacuum

which is the vacuum expectation value (VEV) of T, and this, as we will see below, represents both possible scenarios of Fig. 3-3. In wave mechanics for the position basis, the bracket above is an integration over all space. This is still true, but note carefully that the integration variable is over the space variable of the bra and ket (think $\mathbf{x}'$), but not the time ordering variables $\mathbf{x}$ and $\mathbf{y}$. (See Sect. 3.7.6, pg. 63.) In QFT notation, we tend to merely think of a bracket as equaling zero unless the bra and ket represent the same state.

To gain insight into (3-118), consider the transition amplitude operating on the vacuum when a virtual particle is propagated. Then, from (3-85), and (3-86), where an overbar in a state represents an antiparticle,

Gaining insight into the time ordering operator T acting on the vacuum

$$T\{\phi(x)\phi^\dagger(y)\}|0\rangle = \phi(x)\phi^\dagger(y)|0\rangle \qquad\qquad \text{for } t_y < t_x \quad \text{(particle)}$$

$$= \left(\underbrace{\phi^+(x)}_{\substack{\text{destroys}\\\text{particle}}} + \underbrace{\phi^-(x)}_{\substack{\text{creates}\\\text{antiparticle}}}\right)\left(\underbrace{\phi^{\dagger+}(y)|0\rangle}_{\substack{\text{destroys antiparticle,}\\\text{annihilates vacuum}}} + \underbrace{\phi^{\dagger-}(y)|0\rangle}_{\substack{\text{creates}\\\text{particle}}}\right) \qquad (3\text{-}119)$$

$$= \left(\phi^+(x) + \phi^-(x)\right)F(y)|\phi\rangle$$

$$= G(x)F(y)|0\rangle + H(x)F(y)|\bar{\phi}\phi\rangle.$$

G, F, and H are numeric factors that result from the creation and destruction operations (such as the normalization coefficients that are part of the field operators), which we will not express explicitly here. Thus, we have a general ket left, which in this case is part vacuum state, with the amplitude of the vacuum state part being GF, and the amplitude of the multiparticle state (scalar plus anti-scalar) part being HF. As we (hopefully) remember from NRQM, and which is true for all quantum theories, for appropriate normalization, the square of the magnitude of the amplitude of a state equals the probability of finding that state. Thus $(GF)^\dagger(GF)$ represents the probability of observing the vacuum state (no particles left after the transition.) To find the amplitude GF, we need only form an inner product of the last line of (3-119) with $\langle 0|$, i.e.,

Taking the inner product of the above $T|0\rangle$ with $\langle 0|$ to get the transition amplitude

$$\langle 0|T\{\phi(x)\phi^\dagger(y)\}|0\rangle = \langle 0|G(x)F(y)|0\rangle + \langle 0|H(x)F(y)|\bar{\phi}\phi\rangle$$

$$= G(x)F(y)\underbrace{\langle 0|0\rangle}_{=1} + H(x)F(y)\underbrace{\langle 0|\bar{\phi}\phi\rangle}_{=0} = G(x)F(y). \qquad (3\text{-}120)$$

Note in (3-119) that the $\phi^{\dagger-}(y)$ part of $\phi^\dagger(y)$, which created a particle, left the $F(y)$ factor, but the $\phi^{\dagger+}(y)$ part, which destroys an anti-particle or, in this case, annihilates the vacuum, resulted in zero. Also, the $\phi^+(x)$ part of $\phi(x)$, which destroyed a particle, left us with the $G(x)$ factor, but the $\phi^-(x)$ part created an anti-particle in the ket, and thus left zero (because the particle + antiparticle ket was orthogonal to the vacuum, the bra, leaving a bracket = 0.)

So, as we said above with reference to our original definition (3-116) of the time ordering operator, only the part of $\phi(x)$ that destroys a particle and the part of $\phi^\dagger(y)$ that creates a particle will be relevant.

In a similar way, the same time ordering operator can be used for antiparticle propagation (with time for x and y reversed) as in Fig 3-3b and (3-117). You can prove this by doing Prob. 17.

So, the VEV of the time ordering operator is an amplitude, the square of whose magnitude is the probability of the transition from the vacuum initially (represented by $|0\rangle$) to the vacuum finally (represented by $\langle 0|$). Actually, $|G(x)F(y)|^2$ is a probability density (to be precise, a double density), because it is a function of $\mathbf{x}$ and $\mathbf{y}$. That is, the location $\mathbf{y}$ where the virtual particle is created could be anywhere, and so could the location $\mathbf{x}$ where it is destroyed. We would need to integrate the probability density over all possible $\mathbf{x}$ and all possible $\mathbf{y}$ to get the actual probability, and this is what one does in interaction theory to calculate probabilities and cross sections.

The square of the absolute value of the transition amplitude is a probability density (for the transition to occur)

Given all of this, we can define our mathematical relationship for the processes shown in Fig. 3-3 as the VEV of the time ordering operator T. This is called, in honor of its discoverer, the <u>Feynman propagator $i\Delta_F$</u> (where we insert a factor of i because it makes things easier later on),

Redefine the transition amplitude as the Feynman propagator

$$\boxed{i\Delta_F(x-y) = \langle 0|T\{\phi(x)\phi^\dagger(y)\}|0\rangle}.$$ (3-121)

<u>Step 2: Expressing $i\Delta_F$ in Terms of Commutators</u>

Note for $t_y < t_x$, the case for <u>a virtual particle (not antiparticle)</u>, the Feynman propagator equals

Step 2, expressing Feynman propagator in terms of commutators

$$i\Delta_F(x-y) = \langle 0|\phi(x)\phi^\dagger(y)|0\rangle$$

$$= \langle 0|\phi^+(x)\underbrace{\phi^{\dagger+}(y)|0\rangle}_{=0} + \langle 0|\phi^+(x)\underbrace{\phi^{\dagger-}(y)|0\rangle}_{=(factor\,F)|\phi\rangle} + \langle 0|\phi^-(x)\underbrace{\phi^{\dagger+}(y)|0\rangle}_{=0} + \langle 0|\phi^-(x)\underbrace{\phi^{\dagger-}(y)|0\rangle}_{=(factor\,F)|\phi\rangle}$$ (3-122)

$$= \langle 0|\underbrace{\phi^+(x)\phi^{\dagger-}(y)|0\rangle}_{=(factor\,GF)|0\rangle} + \underbrace{(factor\,F)\langle 0|\phi^-(x)|\phi\rangle}_{=(factor\,HF)\langle 0|\bar\phi\phi\rangle=0} = \langle 0|\phi^+(x)\phi^{\dagger-}(y)|0\rangle.$$

Step 2, first part, finding Feynman propagator for virtual particle (not antiparticle) in terms of a commutator

where "factor" represents the non-operator quantities in each field operator term that are left unchanged when the creation and destruction coefficient operators act on a ket. Note the first "*factor*" in the under-bracket of the last line of (3-122) corresponds to GF in (3-120); in the second under-bracket "*factor*", to HF.

To the last part of (3-122), we can add zero in the form of

$$0 = \langle 0| - \phi^{\dagger-}(y)\underbrace{\phi^+(x)|0\rangle}_{=0}.$$ (3-123)

By adding a term equal to zero, we can use a commutator

Doing that, we find (3-122) becomes

$$i\Delta_F(x-y) = \langle 0|\phi^+(x)\phi^{\dagger-}(y) - \phi^{\dagger-}(y)\phi^+(x)|0\rangle = \langle 0|\left[\phi^+(x),\phi^{\dagger-}(y)\right]|0\rangle$$ (3-124)

In similar fashion, for $t_x < t_y$, the case for <u>a virtual antiparticle</u>, one finds, by doing Prob. 18, that

$$i\Delta_F(x-y) = \langle 0|\left[\phi^{\dagger+}(y),\phi^-(x)\right]|0\rangle.$$ (3-125)

Step 2, second part, expressing Feynman propagator for virtual antiparticle in terms of a commutator

In summary for Step 2, we have shown the Feynman propagator can be expressed in terms of commutators as

$$i\Delta_F(x-y) = \langle 0|\left[\phi^+(x),\phi^{\dagger-}(y)\right]|0\rangle \quad \text{if } t_y < t_x \quad \text{(virtual particle)}$$

$$= \langle 0|\left[\phi^{\dagger+}(y),\phi^-(x)\right]|0\rangle \quad \text{if } t_x < t_y \quad \text{(virtual anti-particle)}.$$ (3-126)

Step 2, summary

Step 3: Expressing Commutator Forms of $i\Delta_F$ as Integrals

Define the symbol $i\Delta^+$ as the commutator of the field type a solutions (for particle) of the first line of (3-126), i.e.,

$$i\Delta^+ (x - y) = \left[\phi^+ (x), \phi^{\dagger -} (y) \right],$$

(3-127)

where the solutions used on the RHS are the integral (continuous) form for the Klein-Gordon solutions (3-37). It is common usage to use a + sign to designate (3-127), rather than the letter a, which would be easier to remember. Just think "a type field" when you see +. Equation (3-127) is thus

$$i\Delta^+ (x - y) = \frac{1}{2(2\pi)^3} \iint \left[a(\mathbf{k}), a^\dagger (\mathbf{k}') \right] \frac{e^{-ikx} e^{ik'y}}{\sqrt{\omega_\mathbf{k} \omega_{\mathbf{k}'}}} d^3\mathbf{k} \, d^3\mathbf{k}'$$

$$= \frac{1}{2(2\pi)^3} \int \left(\int \frac{e^{ik'y}}{\sqrt{\omega_\mathbf{k} \omega_{\mathbf{k}'}}} \delta(\mathbf{k} - \mathbf{k}') d^3\mathbf{k}' \right) e^{-ikx} d^3\mathbf{k} ,$$

(3-128)

and hence,

$$i\Delta^+ (x - y) = \frac{1}{2(2\pi)^3} \int \frac{e^{-ik(x-y)}}{\omega_\mathbf{k}} d^3\mathbf{k} .$$

(3-129)

Similarly, where a minus sign stands for b type fields (since they are associated with antiparticles, the minus makes some sense), we define the commutator in the second line of (3-126),

$$i\Delta^- (x - y) = \left[\phi^{\dagger +} (y), \phi^- (x) \right] = \frac{1}{2(2\pi)^3} \iint \left[b(\mathbf{k}), b^\dagger (\mathbf{k}') \right] \frac{e^{ik'x} e^{-iky}}{\sqrt{\omega_\mathbf{k} \omega_{\mathbf{k}'}}} d^3\mathbf{k} \, d^3\mathbf{k}'$$

$$= \frac{1}{2(2\pi)^3} \int \frac{e^{ik(x-y)}}{\omega_\mathbf{k}} d^3\mathbf{k} ,$$

(3-130)

Thus, (note different authors may define the symbols $i\Delta^+$ and $i\Delta^-$ somewhat differently)

$$i\Delta^\pm (x - y) = \frac{1}{2(2\pi)^3} \int \frac{e^{\mp ik(x-y)}}{\omega_\mathbf{k}} d^3\mathbf{k} .$$

(3-131)

Note that though our earlier expressions for $i\Delta$ and $i\Delta^\pm$, contained operators that operated on the ket part of the VEV in (3-126), because the commutator of these operators in (3-128) (and similarly, in (3-130)) is a number, $i\Delta^\pm$ are simply numbers, not operators. Since the expectation value of a number is a number, $i\Delta_F$ of (3-126) is only that, a number. (To be precise it is a numeric *function*, not an operator *function*.) The bottom line is: We don't have to worry about operators, their effects, or VEV brackets any more, but can simply evaluate the Feynman propagator $i\Delta_F$ as a numeric mathematical relation.

Step 4: Expressing the Two Real Integrals $i\Delta^\pm$ as Contour Integrals

It will prove advantageous if we express (3-131) as contour integrals. Before doing so, we first review complex integral theory.

Consider the complex plane for a function f of the complex variable k_0, i.e., $f(k_0)$. Here, the symbol k_0 is not a pole (poles are usually designated with null subscript), but represents a complex number generalization of the zeroth component (the energy) of 4-momentum k. We concern ourselves with the particular case where k_0 takes on the real value $\omega_\mathbf{k}$.

From complex variable theory, we have the Cauchy integral formula.

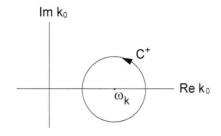

Figure 3-4. Contour Integral for Real, Positive Frequency

Step 3, first part, find $i\Delta^+$ = commutation relation for type a fields (particles) as an integral

Step 3, second part, find $i\Delta^-$ = commutation relation for type b fields (antiparticles)

Step 3, final part, combine above parts into one symbol $i\Delta^\pm$

Our VEV of operators expression of the propagator has become a simple numeric function

Review of integral in the complex plane

$$f\left(\omega_{\mathbf{k}}\right)=\frac{1}{i2\pi}\int_{C^+}\frac{f\left(k_0\right)}{k_0-\omega_{\mathbf{k}}}dk_0 \,.$$ (3-132)

Now, re-express (3-129) as a regular (not complex plane) integral as

Step 4, first part, express $i\Delta^+$ as a contour integral

$$i\Delta^+\left(x-y\right)=\frac{1}{\left(2\pi\right)^3}\int e^{i\mathbf{k}\cdot(\mathbf{x}-\mathbf{y})}\underbrace{\left\{\frac{e^{-i\omega_{\mathbf{k}}\left(t_x-t_y\right)}}{2\omega_{\mathbf{k}}}\right\}}_{f(\omega_{\mathbf{k}})}d^3\mathbf{k} \,,$$ (3-133)

where we take the bracketed quantity as equal to $f(\omega_{\mathbf{k}})$, and where

$$f\left(k_0\right)=\frac{e^{-ik_0\left(t_x-t_y\right)}}{k_0+\omega_{\mathbf{k}}} \quad \left(\text{so } f\left(\omega_{\mathbf{k}}\right)=\frac{e^{-i\omega_{\mathbf{k}}\left(t_x-t_y\right)}}{\omega_{\mathbf{k}}+\omega_{\mathbf{k}}} \text{ as above}\right).$$ (3-134)

We can then use (3-134) in (3-132) to re-express $f(\omega_{\mathbf{k}})$ in terms of a contour integral. Using this for the bracket in (3-133), we find (3-133) becomes

$$i\Delta^+\left(x-y\right)=\frac{1}{\left(2\pi\right)^3}\int e^{i\mathbf{k}\cdot(\mathbf{x}-\mathbf{y})}\left\{\frac{1}{i2\pi}\int_{C^+}\frac{f\left(k_0\right)}{k_0-\omega_{\mathbf{k}}}dk_0\right\}d^3\mathbf{k}$$

$$=\frac{1}{\left(2\pi\right)^3}\int e^{i\mathbf{k}\cdot(\mathbf{x}-\mathbf{y})}\left\{\frac{1}{i2\pi}\int_{C^+}\frac{e^{-ik_0\left(t_x-t_y\right)}}{\left(k_0-\omega_{\mathbf{k}}\right)\left(k_0+\omega_{\mathbf{k}}\right)}dk_0\right\}d^3\mathbf{k}$$ (3-135)

$$=\frac{-i}{\left(2\pi\right)^4}\int_{C^+}\frac{e^{-ik(x-y)}}{\left(k_0\right)^2-\left(\omega_{\mathbf{k}}\right)^2}d^4k \,.$$

where the integral notation now implies integration over four dimensions of the 4-momentum, with the 3-momentum part from $-\infty$ to $+\infty$ in real space and the energy part a contour integral in complex space. Note that the integral does not "blow up" because $k_0 \neq \omega_{\mathbf{k}}$ over the contour integral. We are using a mathematical trick that works, though it jars our usual understanding that, for real particles, the zeroth component of 4-momentum equals energy. k_0 has at this point become, for us, a variable that generally does not equal energy $\omega_{\mathbf{k}}$.

We modify (3-135) a little by noting what is always true mathematically for any four vector, and thus true for 4-momentum components,

Modifying terms in our result a little

$$k^2=\left(k_0\right)^2-\left(\mathbf{k}\right)^2 \quad \rightarrow \quad \left(k_0\right)^2=k^2+\left(\mathbf{k}\right)^2$$ (3-136)

and what is physically true relativistically for rest mass, energy, and 3-momentum (see (3-2)),

$$\omega_{\mathbf{k}}^2-\left(\mathbf{k}\right)^2=\mu^2 \quad \rightarrow \quad \omega_{\mathbf{k}}^2=\mu^2+\left(\mathbf{k}\right)^2 \,.$$ (3-137)

Substitute the RH expressions of (3-136) and (3-137) into the last line of (3-135) to get

$$i\Delta^+\left(x-y\right)=\frac{-i}{\left(2\pi\right)^4}\int_{C^+}\frac{e^{-ik(x-y)}}{k^2-\mu^2}d^4k \,.$$ (3-138)

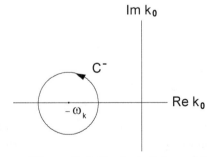

Im k_0

C^-

$-\omega_k$

Re k_0

Figure 3-5. Contour Integral for Real, Negative Frequency

For $i\Delta^-(x-y)$, we carry out similar steps except that the contour integral (still counter clock-wise [ccw], as in Fig. 3-4) is now about $-\omega_{\mathbf{k}}$. In Appendix C, we carry out these steps. When all is said and done, we find the only differences from (3-138) to be the sign and the contour, which is now about the negative frequency value and designated by C^-.

Step 4, second part, express $i\Delta^-$ as a contour integral

$$i\Delta^-\left(x-y\right) = \frac{i}{\left(2\pi\right)^4} \int_{C^-} \frac{e^{-ik(x-y)}}{k^2 - \mu^2} d^4k .$$ (3-139)

Step 5: Re-express $i\Delta_F$ in Most Convenient Form

Step 5, expressing Feynman propagator as integral over real, not complex space

We would like two things more: 1) express the propagator as a single function so we don't have to keep track (while we are integrating over spacetime and doing other things) of whether the virtual field is a particle or antiparticle (i.e., whether to use the Δ^+ or Δ^- function), and 2) have all our integrations over real numbers rather than deal with contour integrals.

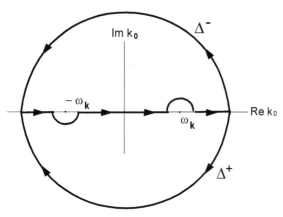

To do this, consider Fig. 3-6, where we have shown two contour integrals and the quantity being integrated in both cases is the integrand of (3-139). The top loop represents Δ^- (we have left out a factor of $\left(2\pi\right)^4$ in the figure) and encloses $-\omega_\mathbf{k}$ with a ccw path. The lower loop encloses $+\omega_\mathbf{k}$, but since it has a cw integration path, the result will have a sign change, and hence equals the ccw integration in (3-138). Thus, the lower loop represents Δ^+.

This means we can define the Feynman propagator Δ_F of (3-138) and (3-139) as proportional to the same integral over the two different loops of Fig. 3-6. We say "proportional" because we also have to include the concomitant integration over the 3D space

Figure 3-6. Contour Integrals for Δ^- and Δ^+

of $\mathbf{k}$ not shown in Fig. 3-6, as well as the various constants involved.

Two different contours for the Feynman propagator written with same integral, different meaning for path C_F

So we can then re-write the Feynman propagator of (3-138) and (3-139) with Fig. 3-6, as

$$i\Delta_F\left(x-y\right) = \frac{i}{\left(2\pi\right)^4} \int_{C_F} \frac{e^{-ik(x-y)}}{k^2 - \mu^2} d^4k ,$$ (3-140)

where the C_F on the integral defines which route (loop) we take in the plane of Fig. 3-6.

Now, consider enlarging the outer hemispheric parts of the two loops in Fig. 3-6, so they extend essentially to infinity. The value of the contour integrals over them will remain unchanged. But the k^2 value in the denominator of (3-140) will become so large that any contribution to the integral over those parts of the path will become negligible. (See Appendix D.) Thus, we can effectively take the integral (3-140) as extending only along the real axis from $-\infty$ to $+\infty$ as in Fig. 3-7.

Extending outer semicircle parts of contours to ∞

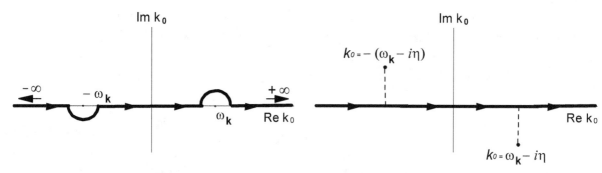

Fig. 3-7. Contour C_F for Δ_F   **Fig. 3-8. Contour and Displaced Poles for Δ_F**

We can further simplify by moving the poles an infinitesimal distance η off the real axis as shown in Fig. 3-8 and deform the contour so that it is all along the real axis. In the limit as $\eta \to 0$, the integral will have the same value, though we must now include this slight pole shift in the propagator expression (3-140). We do this by recalling from (3-136) and (3-137) that we used

Instead of integrating around poles on the axis, move poles slightly off the axis

$$k^2 - \mu^2 = \left(k_0\right)^2 - \left(\omega_{\mathbf{k}}\right)^2 \tag{3-141}$$

to obtain the denominator of (3-140), so we must temporarily restate (3-140) using the right hand side of (3-141), then shift the poles. Thus, (3-140) becomes

$$i\Delta_F\left(x-y\right) = \frac{i}{\left(2\pi\right)^4} \int_{-\infty}^{+\infty} \frac{e^{-ik(x-y)}}{\left(k_0\right)^2 - \left(\omega_{\mathbf{k}} - i\eta\right)^2} d^4k. \tag{3-142}$$

If we then use (3-141) again, ignore second order terms in η, and take $\varepsilon = 2\eta\,\omega_{\mathbf{k}}$, we have our <u>final result</u> for the Feynman scalar propagator

$$\boxed{i\Delta_F\left(x-y\right) = \frac{i}{\left(2\pi\right)^4} \int_{-\infty}^{+\infty} \frac{e^{-ik(x-y)}}{k^2 - \mu^2 + i\varepsilon} d^4k.} \tag{3-143}$$

Yields a single integral over real space representing both virtual particle & antiparticle, the most convenient form for the Feynman propagator

Note the advantages of this form. We now have a single mathematical relationship that automatically describes both a particle propagating from y to x *and* an antiparticle propagating from x to y. We also have done away with the cumbersome contour integrals in favor of a simple 4D integral over the entire real (not complex) 4-momentum space. In principle, we can evaluate this integral then take ε to zero after the integration is carried out.

Summary of Steps 1 to 5

Steps 1 to 4 for the virtual particle Feynman propagator were

Summary of propagator derivation

Steps 1 to 4

$$i\Delta_F\left(x-y\right) = \langle 0|T\{\phi(x)\phi^\dagger(y)\}|0\rangle = \langle 0|\phi(x)\phi^\dagger(y)|0\rangle \quad \text{if } t_y < t_x \text{ (particle)}$$

$$= \langle 0|\underbrace{\left[\phi^+(x),\phi^{\dagger-}(y)\right]}_{i\Delta^+(x-y),\text{ a number}}|0\rangle = \left[\phi^+(x),\phi^{\dagger-}(y)\right]\langle 0||0\rangle = \left[\phi^+(x),\phi^{\dagger-}(y)\right] \tag{3-144}$$

$$= i\Delta^+\left(x-y\right) = \frac{1}{2\left(2\pi\right)^3}\int\frac{e^{-ik(x-y)}}{\omega_{\mathbf{k}}}d^3\mathbf{k} = \frac{-i}{\left(2\pi\right)^4}\int_{C^+}\frac{e^{-ik(x-y)}}{k^2-\mu^2}d^4k \;,$$

and for the virtual anti-particle Feynman propagator,

$$i\Delta_F\left(x-y\right) = \langle 0|T\{\phi(x)\phi^\dagger(y)\}|0\rangle = \langle 0|\phi^\dagger(y)\phi(x)|0\rangle \quad \text{if } t_x < t_y \text{ (anti-particle)}$$

$$= \langle 0|\underbrace{\left[\phi^{\dagger+}(y),\phi^-(x)\right]}_{i\Delta^-(x-y),\text{ a number}}|0\rangle = \left[\phi^{\dagger+}(y),\phi^-(x)\right]\langle 0||0\rangle = \left[\phi^{\dagger+}(y),\phi^-(x)\right] \tag{3-145}$$

$$= i\Delta^-\left(x-y\right) = \frac{1}{2\left(2\pi\right)^3}\int\frac{e^{ik(x-y)}}{\omega_{\mathbf{k}}}d^3\mathbf{k} = \frac{i}{\left(2\pi\right)^4}\int_{C^-}\frac{e^{-ik(x-y)}}{k^2-\mu^2}d^4k \;.$$

The two contour integrals of (3-144) and (3-145) were combined in Step 5 to yield the single integral over real space of (3-143).

Step 5

3.13.3 Comments on the Propagator and Its Derivation

The Propagator and Interaction Theory

The derivation above was formulated with an eye to interaction theory. In that theory, amplitudes are derived for various kinds of interactions between various particles. The square of the magnitude of each amplitude turns out to be the probability of that particular interaction (transition) occurring. These transition amplitudes each depend on the initial real particles, the final real particles, and the virtual particle(s) that mediate the transition. It turns out that the factor in the amplitude representing the virtual particle contribution is identical to the Feynman propagator Δ_F as we defined it in the VEV of the time ordering operator (3-121). Thus, it is also equal to (3-143), so we can simply plug the RHS of (3-143) into the overall transition amplitude as part of our analysis.

Our definition of Feynman propagator here will pop up in our formal derivation of interaction theory

This is one reason we started with the relation $\phi\phi^\dagger$ to create and destroy a virtual scalar particle, rather than what one might initially expect, the simpler creation and destruction operator relation

$a(\mathbf{k})a^\dagger(\mathbf{k})$. Our heuristic approach was tailored to match what we knew would be coming in the mathematical development of interaction theory.

Feynman diags, & our derivation, imply creation/ destruction at a point, but more properly, waves created/destroyed & they spread out over space.

Meaning of Spacetime Points y and x

In Fig. 3-3a, we imply the virtual particle is created at y and destroyed at x. In Feynman diagrams virtual particles are depicted in this way, and at least one real incoming particle can be thought of as being destroyed at y, as in Fig. 1-1 of Chap. 1, pg.2, with a virtual particle created simultaneously at y. At x the virtual particle is destroyed, with the simultaneous creation of at least one outgoing real particle at x.

To be precise, it is more correct to think of the incoming, outgoing, and virtual particles as moving waves spread out in space. What we calculate for a given y and x is the probability density for the interaction as a function of the coordinates y and x. If $\mathbf{y}$ and $\mathbf{x}$ are closer, one would find the probability density for the interaction to occur is greater; if farther away, the probability density is less. Integrating over all $\mathbf{x}$ and $\mathbf{y}$ gives the total probability for observing the interaction.

We are really finding probability density as function of $\mathbf{x}$, $\mathbf{y}$

Momentum Space Form of the Propagator

From (3-143), we can readily write down the 4-momentum space form of the propagator, the Fourier transform of (3-143), which will be very useful,

Momentum space form of the propagator

$$\boxed{\Delta_F\left(k\right) = \frac{1}{k^2 - \mu^2 + i\varepsilon}}.$$ (3-146)

Green's Functions, Correlation Functions, and Propagators

Earlier version was physical space form

Feynman propagators have the form of functions known in mathematics as Green's functions, and you will sometimes see them referred to in that way. You may also see them referred to as correlation functions for free fields, because there is a correlation implied between events x and y.

Feynman propagator = Green function or correlation function

3.14 Chapter Summary

Scalars and Relativistic Quantum Mechanics (RQM)

Do Prob. 20 to create your own Wholeness Chart summary of scalars and RQM as in Sect. 3.1.

Scalars and Quantum Field Theory (QFT)

This part of the chapter is key. Know it, and you know most of the basic principles in QFT. Spin ½ and spin 1 field theory closely parallel that of scalars, so most of the conceptual battle is waged in this Chap. 3.

Free scalar QFT is summarized in the second column of Wholeness Chart 5-4 at the end of Chap. 5. If you can, more or less, reproduce that Wholeness Chart column without looking at it (that is, derive the essence of QFT), you have achieved something few have achieved.

QFT Grounded in 2nd Quantization

It is important to understand how the entire theory springs out of the two 2nd quantization postulates. All the operators (number, Hamiltonian, creation/destruction, 3-momentum, charge, etc) are a direct result of these postulates. So is the vacuum energy. Wholeness Chart 5-4 can help to make that transparent.

In particular, starting with the classical Lagrangian density (or Hamiltonian density), the commutation postulate gives us the rest of the theory, step-by-step, as illustrated below (where we use only a type particles to save space).

Although the steps shown below are specifically for scalars, the developments of QFT for spin ½ and spin 1 particles follow precisely the same conceptual steps.

Steps to QFT

$[\phi^r(\mathbf{x},t), \pi_s(\mathbf{y},t)] = i\delta^r{}_s\, \delta(\mathbf{x} - \mathbf{y}) \;\rightarrow\; [a(\mathbf{k}), a^\dagger(\mathbf{k}')] = \delta_{\mathbf{k}\mathbf{k}'} \;\rightarrow\; H_0^{\,0} \;\&\; \text{vacuum energy} \;\rightarrow$

$N_a(\mathbf{k}) = a^\dagger(\mathbf{k})a(\mathbf{k})$ as number operator $\;\rightarrow\; a^\dagger(\mathbf{k}),\, a(\mathbf{k})$ as creation/destruction operators

$$\downarrow \qquad\qquad\qquad\qquad \swarrow \quad \searrow$$

form of observable operators   the propagator   interaction theory – to be studied
(for real particles)   (for virtual particles)   (for real & virtual particles)

Kinds of Operators in QFT

In QFT there are two kinds of operators. One kind is the usual one from NRQM and RQM representing the dynamical variables of classical theory, such as the Hamiltonian (energy), the 3-momentum operator, charge, etc. The other kind comprises creation and destruction operators.

The first kind, when operating on an eigenstate, re-produces the original state multiplied by an eigenvalue. The second kind changes the state to another state (raising or lowering the number of particles in the state.) The second kind comprises the coefficients $a(\mathbf{k})$, $a^\dagger(\mathbf{k})$, $b(\mathbf{k})$, $b^\dagger(\mathbf{k})$, as well as the fields of which they are a part, ϕ and $\phi^\dagger$. Note that operators of this kind do not have eigenvalues, since their operation on a state changes that state, rather than re-producing it (times an eigenvalue), and hence they are generally *not* observable.

Wholeness Chart 3-4. Different Kinds of Operators in QFT

	Examples	Effect on Eigenstate	Observable?
Dynamical Variable Operators	H, $\mathbf{P}$, Q, $N_a(\mathbf{k})$	eigenvalue times original eigenstate	Yes
Raising and Lowering Operators	$a^\dagger(\mathbf{k})$, $a(\mathbf{k})$, $b^\dagger(\mathbf{k})$, $b(\mathbf{k})$	new eigenstate, one more/less particle	No
Fields	ϕ and $\phi^\dagger$	as above	No

Odds and Ends

For a summary of bosons vs fermions, and Fock space, see Wholeness Charts 3-1 (pg. 65) and 3-2 (pg. 68).

3.15 Appendix A: Klein-Gordon Equation from H.P. Equation of Motion

3.15.1 Background Math Needed for Delta Function Relation

From Arfken and Weber, *Mathematical Methods for Physicists*, 4th ed (Academic Press 1995), pg. 85,

$$\int \frac{d\delta(x'-a)}{dx'} f(x')dx' = -\int \frac{df(x')}{dx'}\delta(x'-a)dx' = -\frac{df(x)}{dx}\bigg|_{x=a}, \qquad (3\text{-}147)$$

where in our case we will have

$$x' \to \mathbf{x}' \quad a \to \mathbf{x} \quad f(x') \to \nabla'\phi(\mathbf{x}') \quad \frac{d\delta(x'-a)}{dx'} \to \nabla'\delta(\mathbf{x}'-\mathbf{x}), \qquad (3\text{-}148)$$

so that (3-147) becomes

$$\int \nabla'\delta(\mathbf{x}'-\mathbf{x})\cdot\nabla'\phi(\mathbf{x}',t)d^3x' = -\nabla\cdot\nabla\phi(\mathbf{x},t). \qquad (3\text{-}149)$$

3.15.2 Deriving the Scalar Field Equation

The Heisenberg equation of motion for any operator is

$$i\frac{d}{dt}\mathcal{O} = [\mathcal{O},H], \qquad (3\text{-}150)$$

and for a complex scalar field, with the aid of Box 2.1, pg. 22, this is

$$i\frac{\partial}{\partial t}\phi = [\phi,H]. \qquad (3\text{-}151)$$

Thus, using (3-33) for $\mathcal{H}$ to find $H = \int\mathcal{H}d^3x$, we have

$$i\frac{\partial}{\partial t}\phi(\mathbf{x},t) = [\phi(\mathbf{x},t),\int d^3x'\{\underbrace{\pi^\dagger\pi}_{\substack{\text{only non}\\\text{-zero result}}} + \nabla'\phi^\dagger\cdot\nabla'\phi + \mu^2\phi^\dagger\phi\}] \qquad (3\text{-}152)$$

where the quantities inside the integral are all functions of $\mathbf{x}'$ and t. Since $\phi(\mathbf{x},t)$ is not a function of $\mathbf{x}'$, we can evaluate the commutator inside the integral. The second and third terms inside the integral of (3-152) commute with ϕ, and thus drop out. Writing out the independent variable dependence only when needed for clarity, and using the field commutation relations for ϕ and π (reproduced below from Chap. 2, Wholeness Chart 2-5, pg. 31, last box in RH column) of

$$\left[\pi_s,\phi^r\right]=-i\delta^r{}_s\delta\left(\mathbf{x}'-\mathbf{x}\right),\tag{3-153}$$

in the second line below, where it says "subs", we find (3-152) becomes

$$i\frac{\partial}{\partial t}\phi(\mathbf{x},t)=\int d^3\mathbf{x}'\left[\phi(\mathbf{x},t),\pi^\dagger(\mathbf{x}',t)\pi(\mathbf{x}',t)\right]=\int d^3\mathbf{x}'\{\underbrace{\phi\pi^\dagger}_{\text{commute}}\pi-\pi^\dagger\underbrace{\pi\phi}_{\text{subs}}\}$$

$$=\int d^3\mathbf{x}'\{\underbrace{\pi^\dagger\phi\pi-\pi^\dagger\phi\pi}_{0}+\pi^\dagger(\mathbf{x}',t)i\delta(\mathbf{x}'-\mathbf{x})\}\tag{3-154}$$

$$=i\pi^\dagger(\mathbf{x},t).$$

Next, using (3-150) when the operator is the complex conjugate of the canonical momentum,

$$i\frac{\partial}{\partial t}\pi^\dagger(\mathbf{x},t)=[\underbrace{\pi^\dagger(\mathbf{x},t)}_{\substack{\text{function}\\\text{of }\mathbf{x}}},\int d^3\mathbf{x}'\{\underbrace{\underbrace{\pi^\dagger\pi}_{\substack{\to 0\text{ in}\\\text{commutator}}}+\nabla'\phi^\dagger\cdot\nabla'\phi+\mu^2\phi^\dagger\phi\}}_{\text{functions of }\mathbf{x}'}].\tag{3-155}$$

Note that $\nabla'\pi^\dagger(\mathbf{x},t)=0$, because the derivative of a function of $\mathbf{x}$ is with respect to a primed $\mathbf{x}''$, and we can move $\pi^\dagger$ inside and outside of any quantity the 3D spatial derivative operates on. We use this several times in what follows. We then focus on the second term in (3-155) and substitute (3-149) in the third line below where it says "use math relation above". That second term is

$$\int d^3\mathbf{x}'\{\underbrace{\pi^\dagger\nabla'\phi^\dagger}_{\nabla'(\pi^\dagger\phi^\dagger)}\cdot\nabla'\phi-\nabla'\phi^\dagger\cdot\underbrace{(\nabla'\phi)\pi^\dagger}_{\substack{\nabla'(\phi\pi^\dagger)\\\text{commute}}}\}=\int d^3\mathbf{x}'\{\underbrace{\nabla'(\pi^\dagger\phi^\dagger)}_{\substack{\text{use com}\\\text{relations}}}\cdot\nabla'\phi-\underbrace{\nabla'\phi^\dagger\cdot\nabla'(\pi^\dagger\phi)}_{\nabla'(\phi^\dagger\pi^\dagger)\cdot\nabla'\phi}\}$$

$$=\int d^3\mathbf{x}'\{\nabla'(\phi^\dagger\pi^\dagger)\cdot\nabla'\phi-\underbrace{\nabla'i\delta(\mathbf{x}'-\mathbf{x})\cdot\nabla'\phi}_{\text{use math relation above}}-\nabla'(\phi^\dagger\pi^\dagger)\cdot\nabla'\phi\}=i\nabla^2\phi(\mathbf{x},t).\tag{3-156}$$

By doing Prob. 6 at the end of the chapter, the reader can verify that evaluation of the third term in the RHS of (3-155), using similar (but simpler) steps, leads to

$$i\frac{\partial}{\partial t}\pi^\dagger(\mathbf{x},t)=i(\nabla^2-\mu^2)\phi(\mathbf{x},t).\tag{3-157}$$

Substituting the time derivative of (3-154) into (3-157), one gets the Klein-Gordon equation

$$\frac{\partial^2}{\partial t^2}\phi=(\nabla^2-\mu^2)\phi,\tag{3-158}$$

thus showing that the equation of motion of a scalar field in the Heisenberg picture, expressed in terms of commutation relations, is equivalent to the Klein-Gordon equation.

3.16 Appendix B: Vacuum Quanta and Harmonic Oscillators

One might argue that two vacuum quanta traveling waves of energy $\tfrac{1}{2}\hbar\omega_\mathbf{k}$ with 3-momenta $\mathbf{k}$ and $-\mathbf{k}$ could be superimposed to yield a vacuum standing wave, i.e., a harmonic oscillator (distributed in space). But the total energy of the standing wave would then be $\hbar\omega$, which is not the ground state of a quantum oscillator, and thus the parallel disappears. Further, the wave form would still be sinusoidal in nature, not that of a Hermite polynomial. Nor is any force/potential involved.

One might instead argue that the two traveling wave eigenstates are superimposed to comprise a general quantum state, wherein the probability of measuring each of the states is $\tfrac{1}{2}$. In this case, the expectation value for energy of the standing wave would be $\tfrac{1}{2}(\tfrac{1}{2}\hbar\omega_\mathbf{k})+\tfrac{1}{2}(\tfrac{1}{2}\hbar\omega_\mathbf{k})=\tfrac{1}{2}\hbar\omega_\mathbf{k}$. But, this violates (3-55), which tells us the energy must be $\hbar\omega_\mathbf{k}$. Thus, this interpretation is inconsistent with contemporary QFT. The points regarding lack of Hermite polynomial form and force/potential for QFT free states apply here, as well.

Still further, this logic implies the entire vacuum state is one general state comprised of all the ½ energy eigenstates. This would mean the expectation value for the energy of the vacuum is the average energy of all those eigenstates, not the sum of them. So, if one assumes (which is common) an upper limit on these energies of the Planck scale energy (see Appendix A of Chap. 10), the total vacuum energy of the universe could then not exceed the Planck energy, which is about that of a very small bit of dust and hardly anything to make a fuss about.

3.17 Appendix C: Propagator Derivation Step 4 for Real, Negative Frequency

We derive (3-139) from the scalar propagator derivation step 3, (3-131),

$$i\Delta^-\left(x-y\right)=\frac{1}{2\left(2\pi\right)^3}\int\frac{e^{ik(x-y)}}{\omega_{\mathbf{k}}}d^3\mathbf{k}=\frac{1}{2\left(2\pi\right)^3}\int\frac{e^{i\omega_{\mathbf{k}}\left(t_x-t_y\right)}}{\omega_{\mathbf{k}}}e^{-i\mathbf{k}\bullet(\mathbf{x}-\mathbf{y})}d^3\mathbf{k}$$

$$=\frac{1}{2\left(2\pi\right)^3}\int_{-\infty}^{\infty}\frac{e^{i\omega_{\mathbf{k}}\left(t_x-t_y\right)}}{\underbrace{\omega_{\mathbf{k}}}_{\substack{\text{even, same value}\\\text{for }\mathbf{k}\text{ or}-\mathbf{k}}}}\underbrace{\Big(\underbrace{cos\left(-\mathbf{k}\right)\bullet(\mathbf{x}-\mathbf{y})}_{\substack{\text{even, same value}\\\text{for }\mathbf{k}\text{ or}-\mathbf{k}}}+\underbrace{i\,sin\left(-\mathbf{k}\right)\bullet(\mathbf{x}-\mathbf{y})}_{\substack{\text{odd, integral of this}\\\text{contribution}=0\text{ for }\mathbf{k}\text{ or}-\mathbf{k}}}\Big)}\quad d^3\mathbf{k}\quad\text{(3-159)}$$

$$\underbrace{\quad}_{\text{integral same if take this}=cos\,\mathbf{k}\bullet(\mathbf{x}-\mathbf{y})+i\,sin\,\mathbf{k}\bullet(\mathbf{x}-\mathbf{y})=e^{i\mathbf{k}\bullet(\mathbf{x}-\mathbf{y})}}$$

$$=\frac{1}{\left(2\pi\right)^3}\int\underbrace{\frac{e^{i\omega_{\mathbf{k}}\left(t_x-t_y\right)}}{2\omega_{\mathbf{k}}}}_{=g\left(-\omega_{\mathbf{k}}\right)}e^{i\mathbf{k}\bullet(\mathbf{x}-\mathbf{y})}d^3\mathbf{k}\,.$$

From complex variable theory

$$g\left(-\omega_{\mathbf{k}}\right)=\frac{1}{i2\pi}\int_{C^-}\frac{g\left(k_0\right)}{k_0-\left(-\omega_{\mathbf{k}}\right)}dk_0\quad\text{with}\quad g\left(k_0\right)=\frac{e^{-ik_0\left(t_x-t_y\right)}}{-k_0+\omega_{\mathbf{k}}},\qquad\text{(3-160)}$$

which we can check is correct (i.e., equals the underbracket quantity in the last part of (3-159)) via

$$g\left(-\omega_{\mathbf{k}}\right)=\frac{e^{-i\left(-\omega_{\mathbf{k}}\right)\left(t_x-t_y\right)}}{-\left(-\omega_{\mathbf{k}}\right)+\omega_{\mathbf{k}}}\overset{checks}{=}\frac{e^{i\omega_{\mathbf{k}}\left(t_x-t_y\right)}}{2\omega_{\mathbf{k}}}.\qquad\text{(3-161)}$$

Putting the RH quantity in (3-160) into the LH quantity in (3-160), we have

$$g\left(-\omega_{\mathbf{k}}\right)=\frac{1}{i2\pi}\int_{C^-}\frac{1}{k_0+\omega_{\mathbf{k}}}\frac{e^{-ik_0\left(t_x-t_y\right)}}{-k_0+\omega_{\mathbf{k}}}dk_0=\frac{i}{2\pi}\int_{C^-}\frac{e^{-ik_0\left(t_x-t_y\right)}}{\left(k_0\right)^2-\left(\omega_{\mathbf{k}}\right)^2}dk_0\,.\qquad\text{(3-162)}$$

Using (3-162) in the last part of (3-159) results in

$$i\Delta^-\left(x-y\right)=\frac{i}{\left(2\pi\right)^4}\int_{C^-}\frac{e^{-ik(x-y)}}{\underbrace{\left(k_0\right)^2-\left(\omega_{\mathbf{k}}\right)^2}_{=k^2-\mu^2}}d^4k\,,\qquad\text{(3-163)}$$

where the underbracket part comes from (3-136) and (3-137). (3-163) is (3-139).

3.18 Appendix D: Enlarging the Integration Path of Fig. 3-6

The integral of Fig. 3-6 expressed in (3-140), can be written as

$$i\Delta_F\left(x-y\right)=\frac{i}{\left(2\pi\right)^4}\int_{C_F}\frac{e^{-ik_0\left(t_x-t_y\right)}e^{i\mathbf{k}\bullet(\mathbf{x}-\mathbf{y})}}{k^2-\mu^2}dk_0d^3k=\frac{i}{\left(2\pi\right)^4}\int_{C_F}\frac{e^{-i\left(\text{Re}k_0+i\text{Im}k_0\right)\left(t_x-t_y\right)}e^{i\mathbf{k}\bullet(\mathbf{x}-\mathbf{y})}}{k^2-\mu^2}dk_0\,d^3k$$

$$=\frac{i}{\left(2\pi\right)^4}\int_{C_F}\frac{e^{-i\left(\text{Re}k_0\right)\left(t_x-t_y\right)}e^{\left(\text{Im}k_0\right)\left(t_x-t_y\right)}e^{i\mathbf{k}\bullet(\mathbf{x}-\mathbf{y})}}{k^2-\mu^2}dk_0\,d^3k.$$

$$\text{(3-164)}$$

The factor with Re k_0 in the exponent oscillates and will be swamped by the denominator wherever $k_0^2 \to \infty$. But the factor with Imk_0 in the exponent is real, so we have to be careful about it.

For the lower half plane of Fig. 3-6, the integral $i\Delta_F(x-y)$ represents $\Delta^+(x-y)$, where $t_x > t_y$ and Imk_0 is negative. That means in (3-164) the factor with Imk_0 has a negative value in the exponent and will go to zero as Im $k_0 \to -\infty$.

For the upper half plane of Fig. 3-6, the integral $i\Delta_F(x-y)$ represents $\Delta^-(x-y)$, where $t_y > t_x$ and Imk_0 is positive. That means in that half of the plane that the factor with Imk_0 has a negative value in the exponent too and will go to zero as Im $k_0 \to +\infty$.

Thus the integral over the contour vanishes whenever $k_0^2 \to \infty$.

3.19 Appendix E: Justifying (3-47) Conclusions

Note that (3-47) is one term in a sum over $\mathbf{k}$, where for each term in $\mathbf{k}$ there is an additional one in $-\mathbf{k}$. Writing out two such terms and then summing them leads to

$$\left[a_{\mathbf{k}}, a_{\mathbf{k}}^\dagger\right]e^{i\mathbf{k}\cdot(\mathbf{x}-\mathbf{y})} + \left[b_{\mathbf{k}}, b_{\mathbf{k}}^\dagger\right]e^{-i\mathbf{k}\cdot(\mathbf{x}-\mathbf{y})} + \left[a_{\mathbf{k}}, a_{-\mathbf{k}}^\dagger\right]e^{i\mathbf{k}\cdot(\mathbf{x}+\mathbf{y})} + \left[b_{-\mathbf{k}}, b_{\mathbf{k}}^\dagger\right]e^{-i\mathbf{k}\cdot(\mathbf{x}+\mathbf{y})}$$

$$\left[a_{-\mathbf{k}}, a_{-\mathbf{k}}^\dagger\right]e^{-i\mathbf{k}\cdot(\mathbf{x}-\mathbf{y})} + \left[b_{-\mathbf{k}}, b_{-\mathbf{k}}^\dagger\right]e^{i\mathbf{k}\cdot(\mathbf{x}-\mathbf{y})} + \left[a_{-\mathbf{k}}, a_{\mathbf{k}}^\dagger\right]e^{-i\mathbf{k}\cdot(\mathbf{x}+\mathbf{y})} + \left[b_{\mathbf{k}}, b_{-\mathbf{k}}^\dagger\right]e^{i\mathbf{k}\cdot(\mathbf{x}+\mathbf{y})} \quad (3\text{-}165)$$

$$= e^{i\mathbf{k}\cdot(\mathbf{x}-\mathbf{y})} + e^{-i\mathbf{k}\cdot(\mathbf{x}-\mathbf{y})} + e^{-i\mathbf{k}\cdot(\mathbf{x}-\mathbf{y})} + e^{i\mathbf{k}\cdot(\mathbf{x}-\mathbf{y})} = 2e^{i\mathbf{k}\cdot(\mathbf{x}-\mathbf{y})} + 2e^{-i\mathbf{k}\cdot(\mathbf{x}-\mathbf{y})},$$

and thus,

$$\begin{array}{ll}
\left[a_{\mathbf{k}}, a_{\mathbf{k}}^\dagger\right] + \left[b_{-\mathbf{k}}, b_{-\mathbf{k}}^\dagger\right] = 2 & \qquad \left[b_{\mathbf{k}}, b_{\mathbf{k}}^\dagger\right] + \left[a_{-\mathbf{k}}, a_{-\mathbf{k}}^\dagger\right] = 2 \\[2mm]
\left[a_{\mathbf{k}}, a_{-\mathbf{k}}^\dagger\right] + \left[b_{\mathbf{k}}, b_{-\mathbf{k}}^\dagger\right] = 0 & \qquad \left[b_{-\mathbf{k}}, b_{\mathbf{k}}^\dagger\right] + \left[a_{-\mathbf{k}}, a_{\mathbf{k}}^\dagger\right] = 0
\end{array} \quad (3\text{-}166)$$

At this point, we could adopt a reasonable postulate that coefficients in (3-166) not having the same 3-momentum have zero commutators, and those that do have the same 3-momentum all have the same commutator values. That would give us (3-41) and lead to our present (good) theory of QFT.

If we were to be thorough, however, and repeat the process of (3-42) to (3-47) for other commutators, such as $\left[\phi, \phi^\dagger\right] = 0$, we would find other relations between coefficient commutators that would lead inevitably to (3-41). You can take my word for this, work it out yourself (which is tedious), or see it on the book website under Auxiliary Material (URL on pg. xvi, opposite pg. 1).

3.20 Problems

1. Substitute (3-9) into the non-relativistic Schrödinger equation (3-1), and also the relativistic Klein-Gordon equation (3-8), to prove to yourself that only terms with exponential form $-i(E_n t - \mathbf{p}_n\cdot\mathbf{x})/\hbar$ solve the Schrödinger equation, but all terms in (3-9) solve the Klein-Gordon equation. Do you see that the single time derivative in the former equation, and the second order time derivative in the latter, are responsible for this?

2. Prove that the orthonormality conditions (3-15) of states $\phi_{\mathbf{k},A}$ also apply to states $\phi_{\mathbf{k},B^\dagger}$.

3. Repeat steps (3-24) and (3-25) using the terms with coefficients $B_{\mathbf{k}}^\dagger$ in (3-12) instead of those with $A_{\mathbf{k}}$. You should find total probability is negative.

4. Express the Klein-Gordon equations (3-35) and their discrete solutions (3-36) in cgs units (i.e., with c and $\hbar \neq 1$) and plug the latter into the former to show that $\mu^2 = m^2 c^2/\hbar^2$.

5. Prove that the continuous solutions (3-37) solve the Klein-Gordon equations.

6. Show that the 3rd term in (3-155) of the Appendix equals $-i\mu^2\phi(\mathbf{x},t)$.

7. Derive the commutators for the continuous solutions to the Klein-Gordon field equation from the second postulate of 2nd quantization. (Warning: This problem may not be worth the significant investment in time needed.)

8.   Starting with the mass term in (3-48), derive (3-53).

9.   Find the VEV (vacuum expectation value) of the free field scalar Hamiltonian.

10. Show that $a^\dagger(\mathbf{k})$ creates an a type particle with 3-momentum $\mathbf{k}$, $b(\mathbf{k})$ destroys a b type particle with 3-momentum $\mathbf{k}$, and $b^\dagger(\mathbf{k})$ creates a b type particle with 3-momentum $\mathbf{k}$. Follow steps similar to those in (3-71) to (3-74).

11. Show $a(\mathbf{k})\left|n_\mathbf{k}\right\rangle = \sqrt{n_\mathbf{k}}\left|n_\mathbf{k}-1\right\rangle$. Does it follow in a heartbeat that $b(\mathbf{k})\left|\bar{n}_\mathbf{k}\right\rangle = \sqrt{\bar{n}_\mathbf{k}}\left|\bar{n}_\mathbf{k}-1\right\rangle$?

12. Substitute the free field solutions (3-36) to the Klein-Gordon equation into the probability density operator relation (3-89) and then insert that into (3-90) to find the effective probability density operator expressed in terms of number operators (3-91). It will help you in doing so to note that for any term where $\mathbf{k} \neq \mathbf{k}'$, the destruction and creation operators will cause the ket to be different from (orthogonal to) the bra, so the resulting term in the expectation value $\bar{\rho}$ will be zero. Hence, those terms can be ignored in determining an effective ρ.

   Note that the result you get is restricted to situations where all particles (in the ket) are in $\mathbf{k}$ eigenstates, which is almost invariably the case in QFT problems and applications. With particles in general (non $\mathbf{k}$ eigen) states, ρ becomes more complicated.

13. Using (3-100), the expression for 3-momentum in terms of the fields and their conjugate momenta, and the Klein-Gordon field equation solutions, prove (3-101), the number operator form of the 3-momentum operator useful for finding expectation values of 3-momentum. Hint: note that the expectation value $\langle \text{state}| a(\mathbf{k})\, a^\dagger(\mathbf{k}')|\text{state}\rangle$ for $\mathbf{k}$ not $= \mathbf{k}'$, is zero.

14. For the state $\left|2\phi_{\mathbf{k}_1}, 3\bar{\phi}_{\mathbf{k}_1}, \bar{\phi}_{\mathbf{k}_2}\right\rangle$, determine the expectation value of $\mathbf{P}$, the 3-momentum operator.

15. Show that for real (not complex) scalar fields, in order for π to be equal to $\dot{\phi}$, the constant K in the scalar Lagrangian density (3-30) must be ½. In general, in QFT, for real fields, we take K=½.

16. Show that if instead of the 2nd quantization, postulate #2 of commutator relations (3-40), we had anti-commutators between the field and its conjugate momentum, i.e.,

$$\left[\phi^r(\mathbf{x},t), \pi_s(\mathbf{y},t)\right]_+ = \phi^r \pi_s + \pi_s \phi^r = i\delta^r{}_s\delta(\mathbf{x}-\mathbf{y}) \qquad (3\text{-}167)$$

   then the coefficient commutators would be anti-commutator relations, i.e.,

$$\left[a(\mathbf{k}), a^\dagger(\mathbf{k}')\right]_+ = -\left[b(\mathbf{k}), b^\dagger(\mathbf{k}')\right]_+ = \delta_{\mathbf{k}\mathbf{k}'} . \qquad (3\text{-}168)$$

   (Hint: Just use opposite signs in (3-43) 2nd row and then in last two rows inside the bracket just before the last equal sign. Then, all commutators in (3-45) to (3-47) become anti-commutators.

17. Find the transition amplitude operating on the vacuum when a virtual anti-particle is propagated as shown in Fig. 3-3b. Use symbols for numeric factors resulting from creation and destruction operators acting on the vacuum and other states.

18. Prove (3-125).

19. Reproduce the essence, with the best detail you can muster, of the Spin 0 column in Wholeness Chart 5-4 without looking at it. That is, prove to yourself that you know how the free field part of QFT is developed.

20. Create your own Wholeness Chart summary of RQM, as presented in Sect. 3.1. Take each subsection heading of Sect. 3.1 as a block in the left-hand column of your chart. Put the main result(s) of that section in the block just to its right in the next column. In between main results insert blocks with short notes on how one gets from the material above to the result in the block below. If there are other comments you wish to add, put them in another column to the right of the others.

Chapter 4

Spinors: Spin ½ Fields

Niels Bohr: "What are you working on Mr. Dirac?"
Paul Dirac: "I'm trying to take the square root of something"

4.0 Preliminaries

While it may seem humorous to think of a physics Nobel laureate struggling over a square root problem, Dirac's meaning here was actually quite deep.

The quotes above purportedly came during a break at a 1927 conference Bohr and Dirac attended. Dirac later recalled that he continued on by saying he was trying to find a relativistic quantum theory of the electron, and Bohr commented, "But Klein has already solved that problem." Dirac then tried to explain he was not satisfied with the (Klein-Gordon) solution because it involved a 2nd order equation in time. That led to negative energy solutions, and he sought a 1st order equation like the non-relativistic Schrödinger equation. But the conference reconvened just then, and the discussion ended.

Dirac sought a first order relativistic Schrödinger equation

4.0.1 Background

Recall from Chap. 3, Sects. 3.0.1 (pg. 40) and 3.1.2 (pg. 42) that we had to use H^2 to develop our relativistic wave equation, because the relativistic Hamiltonian H entailed the operator $\partial_i \partial_i$ under a square root sign, and that had no meaning. Dirac wanted to find a meaningful H to use in a relativistic Schrödinger equation of form

$$H\psi = i\frac{\partial}{\partial t}\psi \,, \tag{4-1}$$

rather than

In other words, he sought a wave equation in H, not H²

$$H^2\phi = -\frac{\partial^2}{\partial t^2}\phi \qquad \text{(Klein-Gordon eq)} \,. \tag{4-2}$$

It is no secret that he succeeded, and his famous result, published in early 1928, is now known as the Dirac equation. We will study it in depth in this chapter.

It wasn't too long after Dirac's discovery of the correct form for (4-1), that people realized (4-2) actually describes scalars; and (4-1), spin ½ fermions, such as the electron. The mathematical nature of the <u>Dirac equation (4-1)</u> provided a good indication for this. That is, (4-1) turns out (as we will see) to be a matrix equation with H being a square matrix quantity and ψ, a column matrix.

His equation turned out to be specifically for spin ½ particles, not all particles

In NRQM, we represented up and down spin of particles via wave functions that had a two component column matrix "tacked on". $(1,0)^T$ represented spin up; and $(0,1)^T$, spin down. So, if ψ in (4-1) in RQM (and QFT) turns out to be a column matrix (and it does), then we could make a good bet that it will represent spinors, rather than scalars. We would be smart to make such a bet, as we would end up winning it.

The Dirac equation is a matrix equation

Interestingly, the column matrix solutions ψ to (4-1) turn out to have four components, rather than two. Given that the relativistic (scalar) solutions to the relativistic wave equation we found in Chap. 3 provided us with antiparticles, which essentially doubled our total number of fields/particles, this should not be too surprising. Four spin components is just what we need to represent particles with up or down spin (2 components) plus antiparticles with up or down spin (2 more components.)

4.0.2 Chapter Overview

Our approach to spin ½ fermions in this chapter will parallel that for spin 0 bosons. You may find it helpful to compare and contrast the bulleted material below with that of the Chapter Overview for scalars at the beginning of Chap. 3, pg. 41.

RQM first,

where we will look at

- the lack of a classical theory of fermions (no macroscopic fermionic behavior observed) and thus, being unable to use a classical H in 1st quantization,
- deducing the Dirac equation, a relativistic Schrödinger equation in H, not H^2,
- solutions (states in RQM) to the Dirac equation,
- probability density and its connection to the normalization constant in the solutions,
- negative energies and the Dirac equation solutions,
- how the Dirac solutions (unexpectedly at first) represent spin ½ particles, and
- spin and the spin operator acting on the solutions (which we didn't have with scalars).

Then QFT,

- noting the lack of classical, macroscopic fermionic fields and thus, being unable to use a classical $\mathcal{H}$ in 2nd quantization,
- assuming the RQM Dirac equation as the QFT field equation, with the same solution form,
- using the (Dirac) field equation to deduce the QFT $\mathcal{L}$ for spinors (the reverse route from the scalar case), and employing the Legendre transformation to get $\mathcal{H}$,
- assuming solution coefficients obey <u>anti-commutation</u> (instead of commutation) relations,
- determining relevant operators in QFT: $H = \int \mathcal{H} \, d^3x$, number, creation/destruction, etc.,
- showing this approach avoids real particle negative energy states,
- seeing how the vacuum is filled with spinor quanta of energy $- \frac{1}{2}\hbar\omega$,
- deriving other operators (probability density, 3-momentum, charge, spin), and
- showing spinors are fermions, and they won't work with commutation relations.

And then,

- finding the spinor Feynman propagator.

Free (no force) Fields

As in Chap. 3, we look herein only at free spinors.

4.1 Relativistic Quantum Mechanics for Spinors

4.1.1 No Classical Spinor Fields: Can We Quantize?

In Chap. 3, Wholeness Chart 3.1 (pg. 65), we recalled that, via the Pauli exclusion principle, fermions cannot occupy the same state within the same macro system. So, whereas photons (bosons) can occupy the same state and a lot of them can therefore reinforce one another to produce a macroscopic electromagnetic field, spinors (fermions) cannot do so. In other words, we have no classical macroscopic spinor fields to sense, interact with, and study experimentally. And thus, we have no classical theory of spinors.

First quantization started with the classical Hamiltonian (or equivalently, the Lagrangian) and used that as the quantum Hamiltonian (or quantum Lagrangian). But we have no classical spinor theory and thus no classical spinor Hamiltonian. Precisely parallel statements can be made for 2nd quantization. There is simply no classical theory with spinor Hamiltonian and Lagrangian densities.

So, we can't do 1st or 2nd quantization for spinors in the way it was advertised earlier, i.e., as THE way to obtain a good quantum theory. (My apologies for the false advertising, but you would have been confused at the time, otherwise.)

So how do we deduce a relativistic spinor quantum theory? We answer this question in the next section by showing how Dirac did it (though he was actually trying to do something else.)

Spinor theory development parallels scalar theory

RQM overview (spinors)

QFT overview (spinors)

Still only free particles/fields in this chapter

No classical theory for spin ½ particles/fields, because fermions can't occupy same state

So we can't do 1ˢᵗ or 2ⁿᵈ quantization

Dirac found another way

4.1.2 Dirac's Approach to RQM: Another History Lesson

Dirac's primary goal was a 1st order relativistic Schrödinger equation, and he postulated that if it existed, it must have the general form (where, as before, we use the ket form symbolism for the wave equation solution in particle quantum theory)

General form a 1st order RQM equation must have

$$i\frac{\partial}{\partial t}|\psi\rangle = H|\psi\rangle = \left(\boldsymbol{\alpha}\boldsymbol{\cdot}\mathbf{p} + \beta m\right)|\psi\rangle \, . \tag{4-3}$$

In (4-3), $\mathbf{p}$ is particle three momentum (an operator in quantum theories), and the vector α and the scalar β would have to be determined. Thus, the equation would be first order in the time derivative (and hopefully yield only positive energy solutions). Also, the relativistic free particle H would be a linear function of both $\mathbf{p}$ and mass m. The key question then is 'what are α and β?' in order for this equation to be true.

To find the answer, Dirac reasoned that H^2 and $|\psi\rangle$ must also satisfy the usual relativistic energy momentum relation (and therefore the Klein-Gordon equation)

Square of this equation must equal K-G eq

$$-\frac{\partial^2}{\partial t^2}|\psi\rangle = H^2|\psi\rangle = \left(\mathbf{p}^2 + m^2\right)|\psi\rangle \, . \tag{4-4}$$

Squaring the operators in (4-3) and inserting the results into (4-4), we get

$$-\frac{\partial^2}{\partial t^2}|\psi\rangle = H^2|\psi\rangle = \left(\alpha_i p_i + \beta m\right)\left(\alpha_j p_j + \beta m\right)|\psi\rangle$$

$$= \left(\alpha_i^2 p_i^2 + \underbrace{\overbrace{\alpha_i\alpha_j + \alpha_j\alpha_i}^{i > j \text{ here}}}_{\text{must}=0,} p_i p_j + \underbrace{\left(\alpha_i\beta + \beta\alpha_i\right)}_{\text{must}=0} p_i m + \beta^2 m^2 \right)|\psi\rangle, \tag{4-5}$$

This squaring restricts form of terms in general equation

where comparison with the RHS of (4-4) shows the bracketed quantities in the lower line above must equal zero. That comparison also shows that $\alpha_i^2 = 1$ and $\beta^2 = 1$. In summary, where <u>anti-commutators</u> are defined as $[\alpha_i, \alpha_j]_+ = \alpha_i\alpha_j + \alpha_j\alpha_i$,

$$\left[\alpha_i, \alpha_j\right]_+ = \left[\alpha_i, \beta\right]_+ = 0 \quad i \neq j \quad \alpha_1, \alpha_2, \alpha_3, \beta \text{ all anti-commute with each other,}$$

$$\left(\alpha_1\right)^2 = \left(\alpha_2\right)^2 = \left(\alpha_3\right)^2 = \left(\beta\right)^2 = 1 \text{ (the identity matrix)}. \tag{4-6}$$

The α_i, β thus must be matrices with certain properties

If α_i and β were numbers they would have to commute and could not possibly anti-commute. Hence, they can only be matrices. Since these matrices are operators operating on $|\psi\rangle$, then $|\psi\rangle$ itself must be a multicomponent object (i.e., a column matrix, at least.)

Using (4-6), one can show that the α_i and β matrices are traceless, hermitian, have ± 1 eigenvalues, and must have an even dimension of at least four. It will save time if you can simply accept these results. If not, then please prove them to yourself. I do note that I, myself, have never done so.

Choosing the minimum dimension case (four), Dirac and Pauli came up with a set of matrices which solve all of the above conditions (specifically (4-6)) which is now called the <u>standard (or Dirac-Pauli) representation</u>, and which we will study in some depth in this chapter. There are, however, other possible choices for α_i and β that satisfy the same conditions. Two of these, called the <u>Weyl and Majorana representations</u>, are also four dimensional and can be convenient for some advanced applications, but we will ignore them herein.

Dirac & Pauli found a set of 4X4 matrices that worked

Square matrices in a 4D space must be 4X4, and thus from (4-3), if $|\psi\rangle$ is a column matrix (a vector), it must have four components (a 4D vector). Take care to note that the 4D space we are talking about here is *not* the four-dimensional physical space of relativity theory, but an abstract space, often called <u>spinor space</u>.

The matrices Dirac and Pauli found for spinor space are

The 4D abstract space of the solutions is called spinor space

$$\beta = \begin{bmatrix} 1 & & & \\ & 1 & & \\ & & -1 & \\ & & & -1 \end{bmatrix} \quad \alpha_1 = \begin{bmatrix} & & & 1 \\ & & 1 & \\ & 1 & & \\ 1 & & & \end{bmatrix} \quad \alpha_2 = \begin{bmatrix} & & & -i \\ & & i & \\ & -i & & \\ i & & & \end{bmatrix} \quad \alpha_3 = \begin{bmatrix} & & 1 & \\ & & & -1 \\ 1 & & & \\ & -1 & & \end{bmatrix}, \quad (4\text{-}7)$$

Form of the α_i, β matrices

where blank components equal zero. (4-7) is commonly written using the 2X2 Pauli matrices σ_i as

$$\beta = \begin{bmatrix} I & 0 \\ 0 & -I \end{bmatrix} \quad \alpha_1 = \begin{bmatrix} 0 & \sigma_1 \\ \sigma_1 & 0 \end{bmatrix} \quad \alpha_2 = \begin{bmatrix} 0 & \sigma_2 \\ \sigma_2 & 0 \end{bmatrix} \quad \alpha_3 = \begin{bmatrix} 0 & \sigma_3 \\ \sigma_3 & 0 \end{bmatrix}, \quad (4\text{-}8)$$

where 0 represents the 2X2 null matrix.

Note that the Klein-Gordon equation can be considered the "square" of the Dirac equation and hence any solution $|\psi\rangle$ which solves the Dirac equation also solves the Klein-Gordon equation.

Solutions to Dirac equation also solve K-G equation

4.1.3 More Convenient Way to Express the Matrices

The Dirac equation can be expressed in a more convenient way by pre-multiplying (4-3) by β. To help when we do that, we define four matrices, called <u>Dirac matrices</u> or <u>gamma matrices</u>, as

Dirac matrices γ^μ, found from α_i and β, are better to work with

$$\gamma^0 = \beta \quad \gamma^1 = \beta\alpha_1 \quad \gamma^2 = \beta\alpha_2 \quad \gamma^3 = \beta\alpha_3 \,, \quad (4\text{-}9)$$

where you can do Prob. 2 to show these equal

$$\gamma^0 = \begin{bmatrix} 1 & & & \\ & 1 & & \\ & & -1 & \\ & & & -1 \end{bmatrix} \quad \gamma^1 = \begin{bmatrix} & & & 1 \\ & & 1 & \\ & -1 & & \\ -1 & & & \end{bmatrix} \quad \gamma^2 = \begin{bmatrix} & & & -i \\ & & i & \\ & i & & \\ -i & & & \end{bmatrix} \quad \gamma^3 = \begin{bmatrix} & & 1 & \\ & & & -1 \\ -1 & & & \\ & 1 & & \end{bmatrix}, (4\text{-}10)$$

or commonly, as

Form of Dirac matrices

$$\gamma^0 = \begin{bmatrix} I & 0 \\ 0 & -I \end{bmatrix} \quad \gamma^1 = \begin{bmatrix} 0 & \sigma_1 \\ -\sigma_1 & 0 \end{bmatrix} \quad \gamma^2 = \begin{bmatrix} 0 & \sigma_2 \\ -\sigma_2 & 0 \end{bmatrix} \quad \gamma^3 = \begin{bmatrix} 0 & \sigma_3 \\ -\sigma_3 & 0 \end{bmatrix}. \quad (4\text{-}11)$$

From henceforth, we will do virtually nothing with the α_i and β matrices, and focus on the γ^μ matrices, instead.

Note the <u>Hermiticity conditions</u> (which you can prove by doing Prob. 3),

Complex conjugate transpose relations for Dirac matrices

$$\gamma^{\mu\dagger} = \gamma^0 \gamma^\mu \gamma^0 \,. \quad (4\text{-}12)$$

4.1.4 The Dirac Equation Expressed with Dirac Matrices

Dirac's original 1st order equation (4-3) in terms of α and β, pre-multiplied by β, takes on the form

$$i \underbrace{\beta}_{\gamma^0} \frac{\partial}{\partial t} |\psi\rangle = \left(\underbrace{\beta\alpha_i}_{\gamma^i} p_i + \underbrace{\beta^2}_{I} m \right) |\psi\rangle = \left(-i\gamma^i \frac{\partial}{\partial x^i} + m \right) |\psi\rangle, \quad (4\text{-}13)$$

or rearranged as what is formally called the <u>Dirac equation</u>

$$\boxed{ \sum_{\eta=1}^{4} \left(\sum_{\mu=0}^{3} i \left(\gamma^\mu \right)_{\kappa\eta} \partial_\mu - m\delta_{\kappa\eta} \right) |\psi\rangle_\eta = 0 \qquad \kappa = 1, 2, 3, 4 }, \quad (4\text{-}14)$$

Dirac equation in terms of Dirac matrices & all indices written out

where we have written out the 4X4 spinor space indices in κ and η, and the summation signs, in order to make it explicitly clear what is going on in spinor space. Note that the Dirac equation is actually *four separate non-matrix equations*, one for each value of the index κ. And each of these equations entails a sum of matrix components (sum over μ), each post multiplied by one of the four components (in η index) of the column vector $|\psi\rangle$. Yes, it seems complicated. But also, yes, it works. And also, yes, it is considered beautiful by many.

Dirac equation is actually four non-matrix equations

You will get used to the complication in time. When you do, you should gain an appreciation for the beauty, as well. In the words of the equation's discoverer,

> *"The research worker, in his efforts to express the fundamental laws of Nature in mathematical form, should strive mainly for mathematical beauty. He should take simplicity into consideration in a subordinate way to beauty ... It often happens that the requirements of simplicity and beauty are the same, but where they clash, the latter must take precedence. "* [1]
>
> — Paul A. M. Dirac

You should do Prob. 4 to provide some practice with (4-14), and then note that the <u>common way to write the Dirac equation</u> is to hide the spinor space indices in κ and η, i.e.,

Common, short hand form of Dirac equation

$$\boxed{\left(i\gamma^{\mu}\partial_{\mu} - m\right)|\psi\rangle = 0} \,, \tag{4-15}$$

where you have to be vigilant to remember the implicit 4X4 spinor space matrix/column nature of (4-15) as expressed explicitly in (4-14).

<u>Another notation</u> commonly used, which is the most streamlined of all, is

Slash notation also very common in Dirac equation

$$\displaystyle{\not{\partial} = \gamma^{\mu}\partial_{\mu}} \quad \text{so, the Dirac equation} \rightarrow \left(i\not{\partial} - m\right)|\psi\rangle = 0\,. \tag{4-16}$$

We note in passing that

$$m \rightarrow \frac{mc}{\hbar} \quad \text{in non-natural units in the Dirac equation}\,. \tag{4-17}$$

4.1.5 Solutions to the Dirac Equation

We can write out (4-15) fully as

$$i\gamma^{\mu}\partial_{\mu}|\psi\rangle = i\left(\gamma^{0}\partial_{0} + \gamma^{1}\partial_{1} + \gamma^{2}\partial_{2} + \gamma^{3}\partial_{3}\right)|\psi\rangle = m|\psi\rangle = \tag{4-18}$$

$$i\left(\begin{bmatrix} 1 & & & \\ & 1 & & \\ & & -1 & \\ & & & -1 \end{bmatrix}\partial_{0} + \begin{bmatrix} & & & 1 \\ & & 1 & \\ & -1 & & \\ -1 & & & \end{bmatrix}\partial_{1} + \begin{bmatrix} & & & -i \\ & & i & \\ & i & & \\ -i & & & \end{bmatrix}\partial_{2} + \begin{bmatrix} & & 1 & \\ & & & -1 \\ -1 & & & \\ & 1 & & \end{bmatrix}\partial_{3}\right)\begin{vmatrix} \psi_{1} \\ \psi_{2} \\ \psi_{3} \\ \psi_{4} \end{vmatrix}\rangle \tag{4-19}$$

Writing out Dirac equation

$$= i\left(\begin{bmatrix} \partial_{0} & 0 & \partial_{3} & \partial_{1} - i\partial_{2} \\ 0 & \partial_{0} & \partial_{1} + i\partial_{2} & -\partial_{3} \\ -\partial_{3} & -\partial_{1} + i\partial_{2} & -\partial_{0} & 0 \\ -\partial_{1} - i\partial_{2} & \partial_{3} & 0 & -\partial_{0} \end{bmatrix}\begin{vmatrix} \psi_{1} \\ \psi_{2} \\ \psi_{3} \\ \psi_{4} \end{vmatrix}\rangle = m\begin{vmatrix} \psi_{1} \\ \psi_{2} \\ \psi_{3} \\ \psi_{4} \end{vmatrix}\rangle.$$

Note that the numeric subscripts on the ∂ symbols refer to derivatives with respect to time and space, whereas the numeric subscripts on the components of $|\psi\rangle$ refer to the respective components of the ket in spinor space.

(4-19) is a 4X4 matrix problem, for which we can try solutions of form $|\psi\rangle = |u_{\alpha}e^{\pm ikx}\rangle$, where u_{α} is a four-component spinor space column matrix. Doing this and carrying out the derivatives in (4-19), we end up with an 4X4 eigenvalue problem. This has four solutions $|\psi^{(n)}\rangle$, where $n = 1,2,3,4$, with each such solution having four spinor space components. We will not go through the tedium of doing this. Rather, I will simply provide the solutions, and you will do Prob. 5 to prove to yourself, by substitution, that they are indeed valid solutions to (4-19).

The <u>Dirac equation solutions</u> in the Dirac-Pauli (standard) representation are

---

[1] (Footnote added in 2018 version of book): The Dirac equation and much else in physics have beauty, to be sure. But many (e.g., S. Hossenfelder in *Lost in Math: How Beauty Leads Physics Astray* (Basic Books 2018)) are starting to question it as a guiding principle, since some very beautiful and elegant theories (e.g., grand unified and supersymmetry theories) have not found much confirmation in experiments. Younger readers should get to see how well this principle holds up in the next half century or so.

$$\left|\psi^{(1)}\right\rangle = \sqrt{\frac{E+m}{2m}}\underbrace{\begin{pmatrix} 1 \\ 0 \\ \dfrac{p^3}{E+m} \\ \dfrac{p^1+ip^2}{E+m} \end{pmatrix}}_{\text{spinor } u_1 = \text{part of solution in 4D spinor space}}\underbrace{e^{-ipx}}_{\substack{\text{4D}\\ \text{physical}\\ \text{space}\\ \text{part}}} = u_1 e^{-ipx} \qquad \left|\psi^{(2)}\right\rangle = \sqrt{\frac{E+m}{2m}}\underbrace{\begin{pmatrix} 0 \\ 1 \\ \dfrac{p^1-ip^2}{E+m} \\ \dfrac{-p^3}{E+m} \end{pmatrix}}_{\text{spinor } u_2} e^{-ipx} = u_2 e^{-ipx}$$

*Solutions to Dirac equation (discrete, plane waves, **p** eigenstates)*

(4-20)

$$\left|\psi^{(3)}\right\rangle = \sqrt{\frac{E+m}{2m}}\underbrace{\begin{pmatrix} \dfrac{p^3}{E+m} \\ \dfrac{p^1+ip^2}{E+m} \\ 1 \\ 0 \end{pmatrix}}_{\text{spinor } v_2} e^{ipx} = v_2 e^{ipx} \qquad \left|\psi^{(4)}\right\rangle = \sqrt{\frac{E+m}{2m}}\underbrace{\begin{pmatrix} \dfrac{p^1-ip^2}{E+m} \\ \dfrac{-p^3}{E+m} \\ 0 \\ 1 \end{pmatrix}}_{\text{spinor } v_1} e^{ipx} = v_1 e^{ipx}.$$

Yes, again, these are more complicated than solutions we have dealt with in the past, but you will get used to them with time. Note several things in (4-20). Any constant instead of $\sqrt{(E+m)/2m}$ would suffice, but that choice was made because things will work better later on with it, as we will see. The symbol E is always a positive number of magnitude equal to the energy. p^i is positive if it points in the positive direction of its respective axis. These are plane wave solutions. We have defined new symbols $\underline{u_r(\mathbf{p})}$ and $\underline{v_r(\mathbf{p})}$ (r =1,2), which are the column vectors multiplied by the constant shown, are functions only of **p** for a given m (since $E = \sqrt{\mathbf{p}^2+m^2}$), and go by the name spinors, or four-spinors. Note that the particles represented by the $\underline{|\psi^{(n)}\rangle}$ are also often called spinors. We will show shortly that r values represent different spin states (for example, u_1 represents spin up, and u_2 represents spin down in the particle at-rest system.) As you might expect, we will find the solutions containing $v_r(\mathbf{p})$ are associated with antiparticles; and those with $u_r(\mathbf{p})$, with particles. More on this later, but for now, take care to note the reverse order numbering on $v_{2,1}$ from $u_{1,2}$, which is customary.

Column vector parts of solutions called spinors

The solutions (4-20) are eigenstates of **p**, since every measurement of 3-momentum of the particles they represent would result in the value **p**. They are also eigenstates of energy, since, for given m, a free particle of 3-momentum **p** has a fixed E.[1]

Components with $E/c+ mc$ (for non-natural units) in the denominator are often called the small components, because due to the c in the denominator, for **p** of non-relativistic speeds, they are dwarfed by the components having just a 1 in them, which are often called the large components.

Inner Product of Spinors

We will find the inner product of each spinor in (4-20) with itself a valuable thing to know. For $u_1(\mathbf{p})$,

---

[1] Similar to that mentioned in the footnote on pg. 50 in Chap. 3 for scalars, there are supplemental forms for the solutions to the Dirac equation having exponents of $\pm i(E_\mathbf{p} t + \mathbf{p} \cdot \mathbf{x})$, instead of $\pm i(E_\mathbf{p} t - \mathbf{p} \cdot \mathbf{x})$, but these have been widely ignored. The possible impact on QFT of including these solution forms in the theory is discussed in the reference cited in the aforementioned footnote.

$$u_1^\dagger(\mathbf{p})u_1(\mathbf{p}) = \frac{E+m}{2m}\begin{pmatrix} 1 & 0 & \dfrac{p^3}{E+m} & \dfrac{p^1-ip^2}{E+m} \end{pmatrix}\begin{pmatrix} 1 \\ 0 \\ \dfrac{p^3}{E+m} \\ \dfrac{p^1+ip^2}{E+m} \end{pmatrix} = \frac{E+m}{2m}\left(1+\frac{\mathbf{p}^2}{(E+m)^2}\right)$$

Inner product of spinors with themselves

(4-21)

$$= \frac{E+m}{2m}\left(\frac{(E+m)^2+\mathbf{p}^2}{(E+m)^2}\right) = \frac{\overbrace{E^2+m^2}^{E^2-\mathbf{p}^2}+2Em+\mathbf{p}^2}{2m(E+m)} = \frac{2E^2+2Em}{2m(E+m)} = \frac{E}{m}.$$

By doing Prob. 6, you can feel comfortable with the general result (underline means no summation),

$$u_{\underline{r}}^\dagger(\mathbf{p})u_{\underline{r}}(\mathbf{p}) = v_{\underline{r}}^\dagger(\mathbf{p})v_{\underline{r}}(\mathbf{p}) = \frac{E}{m},$$ (4-22)

Spinor magnitude = $\sqrt{E/m}$

which we got because of our original choice of the constant in (4-20). (You may want to ruminate, by looking at (4-21), on how our spinor inner product would have been a little more complicated if we had chosen unity as our original constant in (4-20). We make our choices for arbitrary constants in order to conform with custom. They are what are commonly used.)

Orthogonality of Spinors

The inner product of u_1 and u_2 is

$$u_1^\dagger(\mathbf{p})u_2(\mathbf{p}) = \frac{E+m}{2m}\begin{pmatrix} 1 & 0 & \dfrac{p^3}{E+m} & \dfrac{p^1-ip^2}{E+m} \end{pmatrix}\begin{pmatrix} 0 \\ 1 \\ \dfrac{p^1-ip^2}{E+m} \\ \dfrac{-p^3}{E+m} \end{pmatrix}$$

Orthogonality of spinors

(4-23)

$$= \frac{E+m}{2m}\left(\frac{p^3\left(p^1-ip^2\right)-p^3\left(p^1-ip^2\right)}{(E+m)^2}\right) = 0.$$

By doing Prob. 7, you can prove to yourself that

$$u_r^\dagger(\mathbf{p})u_s(\mathbf{p}) = v_r^\dagger(\mathbf{p})v_s(\mathbf{p}) = 0 \qquad r \neq s .$$ (4-24)

We can combine (4-22), (4-24), and the first part of Prob. 8 into the general result

$$\boxed{u_r^\dagger(\mathbf{p})u_s(\mathbf{p}) = v_r^\dagger(\mathbf{p})v_s(\mathbf{p}) = \frac{E}{m}\delta_{rs} \qquad u_r^\dagger(\mathbf{p})v_s(-\mathbf{p}) = 0}.$$ (4-25)

General spinor inner product relations

Orthogonality of Eigensolutions

We can show that the eigensolutions (4-20) to the Dirac equation are orthogonal in the usual quantum mechanical way. For example, for first two solutions of (4-20)

$$\left\langle \psi^{(1)}\middle|\psi^{(2)}\right\rangle = \int \psi_{state}^{(1)\dagger}\psi_{state}^{(2)}d^3x = \int u_1^\dagger(\mathbf{p})e^{+ipx}u_2(\mathbf{p})e^{-ipx}d^3x = \underbrace{u_1^\dagger(\mathbf{p})u_2(\mathbf{p})}_{=0}\underbrace{\int e^{+ipx}e^{-ipx}d^3x}_{=V}.$$ (4-26)

For the first and third solutions,

$$\left\langle \psi^{(1)}\middle|\psi^{(3)}\right\rangle = \int u_1^\dagger(\mathbf{p})e^{+ipx}v_2(\mathbf{p})e^{+ipx}d^3x = \underbrace{u_1^\dagger(\mathbf{p})v_2(\mathbf{p})}_{=0 \text{ for } \mathbf{p}=0}\underbrace{\int e^{+ipx}e^{+ipx}d^3x}_{=0 \text{ for } \mathbf{p}\neq 0} = 0 .$$ (4-27)

By continuing with different solution pairs (do Prob. 9 for practice), one can prove that

$$\left\langle \psi^{(m)}\middle|\psi^{(n)}\right\rangle = 0 \text{ for } m \neq n .$$ (4-28)

Dirac eigen solutions are orthogonal

Given that we should know mathematically that eigenvector solutions in any eigenvalue problem are generally orthogonal, and that the Dirac equation represents an eigenvalue problem, this should not be too surprising.

General Solution to Dirac Equation

The most general solution to the Dirac equation is a sum (or integral) of all eigenstates, each having a (typically complex) coefficient representing the amount of that eigenstate in the total general state. The discrete plane wave general solution is then

$$\psi_{state} = |\psi\rangle = \sum_{r,\mathbf{p}} \sqrt{\frac{m}{VE_{\mathbf{p}}}} \left(C_r(\mathbf{p})u_r(\mathbf{p})e^{-ipx} + D_r^\dagger(\mathbf{p})v_r(\mathbf{p})e^{ipx} \right), \tag{4-29}$$

General solution = sum of eigenstate solutions

where $C_r(\mathbf{p})$ and $D_r^\dagger(\mathbf{p})$ are the coefficients. We will explain later the reason for the unusual normalization constant chosen in terms of mass, energy, and volume, though from (4-22) and the normalization we did in Chap. 3, you may be sensing what is coming.

4.1.6 The Adjoint Dirac Equation

Unlike the Klein-Gordon equation, the Dirac equation is a matrix equation. So, rather than a complex conjugate form of the wave equation, we need to consider taking a complex conjugate transpose of that equation, and of its solutions (4-20). But, as it turns out, the theory works more coherently if rather than $|\psi\rangle^\dagger = \langle\psi| = \psi^\dagger_{state}$, we define and use the adjoint

$$\bar{\psi}_{state} = \psi^\dagger_{state}\gamma^0 = |\psi\rangle^\dagger \gamma^0 = \langle\psi|\gamma^0 = \langle\bar{\psi}|, \tag{4-30}$$

Definition of adjoint

where an inner product between the row vector $|\psi\rangle^\dagger = \langle\psi| = \psi^\dagger_{state}$ and the gamma matrix are implied, and we hope the symbolism has not become unwieldy. You should do Prob. 10 to show yourself what the four row vectors $\langle\bar{\psi}^{(n)}|$ look like.

The adjoint Dirac equation is obtained by taking the complex conjugate transpose of the Dirac equation (4-15), post multiplying by γ^0, and using the Hermiticity conditions along with the adjoint definition. (Do this in Prob. 11.) The result is

$$i\partial_\mu \langle\bar{\psi}|\gamma^\mu + m\langle\bar{\psi}| = 0 \quad. \tag{4-31}$$

The adjoint Dirac equation

Simply by deriving (4-31), we have proven that the adjoint (4-30) solves the adjoint Dirac equation, as long as $|\psi\rangle$ solves the Dirac equation. By doing Prob. 12, you can justify it to yourself in a more "hands on" way.

Adjoint spinors are defined as the row vectors

$$\bar{u}_r = u_r^\dagger\gamma^0 \qquad \bar{v}_r = v_r^\dagger\gamma^0, \tag{4-32}$$

Adjoint spinor definitions

which, with (4-29) and (4-30), gives us the discrete plane wave adjoint general solution form

$$\bar{\psi}_{state} = \langle\bar{\psi}| = \sum_{r,\mathbf{p}} \sqrt{\frac{m}{VE_{\mathbf{p}}}} \left(D_r(\mathbf{p})\bar{v}_r(\mathbf{p})e^{-ipx} + C_r^\dagger(\mathbf{p})\bar{u}_r(\mathbf{p})e^{ipx} \right). \tag{4-33}$$

General adjoint solution

4.1.7 Probability Density for Dirac Fermions in RQM

Probability and the Four-Current Using the Dirac Equation

Recall from Chap. 3 (see Box 3-1, pg. 45, and Sect. 3.1.4, pg. 44) that, in any theory, we can typically use the governing equation (and often its complex conjugate transpose) to find a conserved quantity, which in quantum theory can be total probability. We did this in NRQM and for scalars in RQM. So, we try the same general approach for spin ½ particles, but note that researchers found early on that the adjoint Dirac equation was better for this purpose than simply the complex conjugate of the Dirac equation.

This is actually much simpler to do for the first order Dirac equation and its adjoint than it was for the second order Klein-Gordon equation, so we leave it as Prob. 13 for the reader. The result of that problem is

Dirac equation and its adjoint yield a 4-divergence of a 4-current = 0

$$\partial_\mu j^\mu = 0 \qquad j^\mu = (\rho, \mathbf{j}) = \bar{\psi}_{state}\, \gamma^\mu \psi_{state} = \left\langle \bar{\psi} \middle| \gamma^\mu \middle| \psi \right\rangle_{\text{not integ}}, \tag{4-34}$$

Form of the 4-current

where the subscript "not integ" means we are not integrating over space in the bracket shown, contrary to what the bracket symbol typically implies in the position basis. (4-34) means the total quantity

$$\int_V j^0 d^3x = \int_V \rho\, d^3x = \int_V \bar{\psi}_{state}\, \gamma^0 \psi_{state}\, d^3x = \left\langle \bar{\psi} \middle| \gamma^0 \middle| \psi \right\rangle$$

$$= \int_V \psi^\dagger_{state}\, \underbrace{\gamma^0 \gamma^0}_{I}\, \psi_{state}\, d^3x = \left\langle \psi \middle| \psi \right\rangle = Q' \tag{4-35}$$

The conserved quantity related to the 4-current

is conserved for V = all space, and ρ is the density value corresponding to Q'. Naturally, we would like ρ to be a probability density and Q' to be total probability, as we had in NRQM.

Probability for the Dirac Discrete Solutions

For a single particle state in RQM, we assume at first that the solution (4-29) has only terms with coefficients C_r (i.e., only has spinors of form u_r), i.e., the general state $|\psi\rangle$ contains no eigenstates with coefficients D_r (i.e., no spinors of form v_r). Our ρ of (4-34) is then

Is that conserved quantity probability?

$$\rho = \left(\sum_{r,\mathbf{p}} \sqrt{\frac{m}{VE_\mathbf{p}}} C^\dagger_r(\mathbf{p}) \underbrace{\bar{u}_r(\mathbf{p})\, e^{ipx}}_{u^\dagger_r(\mathbf{p})\gamma^0} \right) \gamma^0 \left(\sum_{r',\mathbf{p}'} \sqrt{\frac{m}{VE_{\mathbf{p}'}}} C_{r'}(\mathbf{p}') u_{r'}(\mathbf{p}') e^{-ip'x} \right)$$

$$= \left(\sum_{r,\mathbf{p}} \sqrt{\frac{m}{VE_\mathbf{p}}} C^\dagger_r(\mathbf{p}) u^\dagger_r(\mathbf{p}) e^{ipx} \right)\left(\sum_{r',\mathbf{p}'} \sqrt{\frac{m}{VE_{\mathbf{p}'}}} C_{r'}(\mathbf{p}') u_{r'}(\mathbf{p}') e^{-ip'x} \right). \tag{4-36}$$

If (4-36) is probability density, its integral over all space must equal 1. In such integration all terms with $\mathbf{p}' \neq \mathbf{p}$ go to zero due to the exponential term, and all remaining terms with $r' \neq r$ go to zero from the spinor orthogonality relation (4-24). We end up with

$$\int \rho\, d^3x = \sum_{r,\mathbf{p}} \frac{m}{VE_\mathbf{p}} \left(C^\dagger_r(\mathbf{p}) C_r(\mathbf{p}) \underbrace{u^\dagger_r(\mathbf{p}) u_r(\mathbf{p})}_{E_\mathbf{p}/m} \underbrace{\int e^{-ipx} e^{ipx} d^3x}_{V} \right) = \sum_{r,\mathbf{p}} C^\dagger_r(\mathbf{p}) C_r(\mathbf{p}) = \sum_{r,\mathbf{p}} |C_r(\mathbf{p})|^2 = 1. \tag{4-37}$$

For C_r type particles, the conserved quantity can be interpreted as probability

Thus, by taking ρ as probability density, $|C_r(\mathbf{p})|^2$ is the probability of measuring the particle with spin state r and 3-momentum eigenstate $\mathbf{p}$, similar to what the coefficients of eigenstates represented in NRQM, and for scalars in RQM (see Chap. 3, Sect. 3.1.4, pgs. 44-47).

Normalization Factors

(4-37) is the reason we used $\sqrt{m/VE_\mathbf{p}}$ as our normalization factor in solutions (4-20). A different such factor would not have resulted in unity on the RHS of (4-37).

We chose constant in solutions so conserved quantity would equal 1

4.1.8 Negative Energies for D Type Dirac Fermions

Do Prob. 14 to prove to yourself that a Dirac fermion represented by a solution with exponential form $-ipx$ (that has spinor u_r and coefficient $C_r(\mathbf{p})$), has positive energy; and one represented by the solution form ipx (spinor v_r and coefficient $D^\dagger_r(\mathbf{p})$), has negative energy.

D_r type particles have negative energy

Thus, Dirac did not solve one of the problems he originally set out to solve. Half of his solutions represent particles with negative energies, just like the Klein-Gordon solutions did.

4.1.9 Probability for D Type Dirac Fermions

Repeating the process of (4-36) for a D type particle instead of a C type, we find

$$\int \rho\, d^3x = \sum_{r,\mathbf{p}} \frac{m}{VE_\mathbf{p}} \left(D_r(\mathbf{p}) D^\dagger_r(\mathbf{p}) \underbrace{v^\dagger_r(\mathbf{p}) v_r(\mathbf{p})}_{E_\mathbf{p}/m} \underbrace{\int e^{-ipx} e^{ipx} d^3x}_{V} \right) = \sum_{r,\mathbf{p}} D_r(\mathbf{p}) D^\dagger_r(\mathbf{p}) = \sum_{r,\mathbf{p}} |D_r(\mathbf{p})|^2 = 1 \tag{4-38}$$

Probability for D_r type particle

and a total positive probability (unlike the negative probability in the scalar case) that we can set equal to 1.

4.1.10 A Great Use for the Dirac Approach: Spin

So, though Dirac's approach still resulted in negative energies, he did solve two of the problems he set out to solve (a first order equation and no negative probabilities.) Well beyond that, however, he provided physics with something of enormous value. His equation, and its solutions, allow us to model relativistic particles with spin ½, which could not be done before. Importantly, this turned out to give correct relativistic solutions for the hydrogen atom, which the Klein-Gordon equation did not.

Dirac approach of great value for spin ½

RQM Spin Operator

Before delving into relativistic spin, you should read over and understand Box 4-1, Review of Spin in NRQM. That box is summarized in the second column of Wholeness Chart 4-1, pg. 102.

In RQM, we have recently seen that each of the solutions Dirac came up with has a column matrix (spinor) built in. But it is a four, not two, component column matrix. It wasn't long before people realized that Dirac's four solutions (4-20) represented two kinds of particles, regular particles and antiparticles. The four components of each solution were different for each, and represented the four states of spin up and spin down for particles, and spin up and spin down for antiparticles.

The spin operator can be derived formally from tensor analysis of the energy-momentum and angular momentum tensors in 4D spacetime, but that is pretty complicated.[1] We will simply state the RQM spin operator Σ_i as follows, then compare it to the NRQM version of the spin operator, and work some examples, to justify the form we have assumed.

$$\Sigma_i = \frac{\hbar}{2}\begin{bmatrix} \sigma_i & 0 \\ 0 & \sigma_i \end{bmatrix} \rightarrow \Sigma_1 = \frac{\hbar}{2}\begin{bmatrix} & 1 & & \\ 1 & & & \\ & & & 1 \\ & & 1 & \end{bmatrix} \quad \Sigma_2 = \frac{\hbar}{2}\begin{bmatrix} & -i & & \\ i & & & \\ & & & -i \\ & & i & \end{bmatrix} \quad \Sigma_3 = \frac{\hbar}{2}\begin{bmatrix} 1 & & & \\ & -1 & & \\ & & 1 & \\ & & & -1 \end{bmatrix} . (4\text{-}39)$$

RQM spin operator

Note (4-39) is a 3D object in physical space (3 components i of the spin angular momentum), but each of the components in that space is itself a 4X4 matrix in relativistic 4D spinor space, rather than non-relativistic 2D spinor space of NRQM. In relativity, spin has three spatial components, as it did non-relativistically. But, relativistically, each component must act on a 4D column vector in spinor space. So, if we were to guess at a 4D matrix spin operator, then (4-39), where we formed 4D matrices from our 2D Pauli matrices, would be a good first guess. Compare (4-39) to (B4-1.1) of Box 4-1. And, as we will see, this guess turns out to be correct.

As in NRQM, we can choose the direction of our z axis however we like, and it was easier in NRQM if we lined it up in the direction of spin of our particle, so spin would be either up (plus z direction), or down (minus z direction). We'll do something similar in RQM, so we'll focus for the present on the z direction spin component operator Σ_3.

Dirac Fermion Spins: Stationary Particle Examples

Consider a Dirac particle which, for simplicity, is simply sitting in front of us and not moving. So $\mathbf{p}=0$ in our frame, and the solutions (4-20) become much simplified. What then, are their respective spins? For the first such solution, we have

First Dirac solution, if particle stationary, is spin up eigenstate of RQM spin operator

$$\Sigma_3\left|\psi^{(1)}\right\rangle = \frac{\hbar}{2}\begin{bmatrix} 1 & & & \\ & -1 & & \\ & & 1 & \\ & & & -1 \end{bmatrix}\underbrace{\sqrt{\frac{m+m}{2m}}\begin{pmatrix} 1 \\ 0 \\ 0 \\ 0 \end{pmatrix}}_{u_1}e^{-ipx} = \frac{\hbar}{2}\begin{pmatrix} 1 \\ 0 \\ 0 \\ 0 \end{pmatrix}e^{-ipx} = \frac{\hbar}{2}\left|\psi^{(1)}\right\rangle \quad \hbar=1 \text{ in n.u.} \quad (4\text{-}40)$$

and the spin is $+\hbar/2$, indicating that our first Dirac solution is for spin up.

---

[1] See F. Mandl and G. Shaw, *Quantum Field Theory*, 1st ed. (John Wiley 1984), Chap 2, pgs. 38-39 and Chap 4, pg. 65.

Box 4-1. Review of Spin in Non-relativistic Quantum Mechanics

NRQM Spin Operator

In NRQM our spin operator S_i was the Pauli matrices σ_i times the factor $\hbar/2$,

$$S_i = \frac{\hbar}{2}\sigma_i \;\rightarrow\; S_1 = \frac{\hbar}{2}\sigma_1 = \frac{\hbar}{2}\begin{bmatrix} 0 & 1 \\ 1 & 0 \end{bmatrix} \quad S_2 = \frac{\hbar}{2}\sigma_2 = \frac{\hbar}{2}\begin{bmatrix} 0 & -i \\ i & 0 \end{bmatrix} \quad S_3 = \frac{\hbar}{2}\sigma_3 = \frac{\hbar}{2}\begin{bmatrix} 1 & 0 \\ 0 & -1 \end{bmatrix} \qquad (B4\text{-}1.1)$$

Spin (non-orbital angular momentum) S_i in physical space acted like a vector with three components, as it does classically. A particle, or object, can have spin components in any of the three dimensions. In spinor space, however, spin was represented by a 2X2 matrix, one for each 3D component.

Eigenstates of the Spin Operator Components

z direction

Thus, a wave function (ket) with a column matrix representing spin up in the *z* direction $(1,0)^T$ had spin $\hbar/2$; and one in the down direction $(0,1)^T$ had spin $-\hbar/2$.

$$S_3\left|\psi_{\mathbf{p},up}\right\rangle = S_3\left(A(\mathbf{p})e^{-ipx}\begin{bmatrix} 1 \\ 0 \end{bmatrix}\right) = A(\mathbf{p})e^{-ipx}\frac{\hbar}{2}\begin{bmatrix} 1 & 0 \\ 0 & -1 \end{bmatrix}\begin{bmatrix} 1 \\ 0 \end{bmatrix} = A(\mathbf{p})e^{-ipx}\frac{\hbar}{2}\begin{bmatrix} 1 \\ 0 \end{bmatrix} = \frac{\hbar}{2}\left|\psi_{\mathbf{p},up}\right\rangle$$

$$\qquad (B4\text{-}1.2)$$

$$S_3\left|\psi_{\mathbf{p},down}\right\rangle = S_3\left(A'(\mathbf{p})e^{-ipx}\begin{bmatrix} 0 \\ 1 \end{bmatrix}\right) = -\frac{\hbar}{2}\left|\psi_{\mathbf{p},down}\right\rangle .$$

The spin up and spin down states are eigenstates of S_3, the *z* direction component of the spin operator S_i.

x direction

$(1,1)^T$ and $(1,-1)^T$ are readily shown to be eigenstates of the *x* component of the spin operator, i.e., S_1, via

$$S_1\left|\psi_{\mathbf{p},+x\,\mathrm{spin}}\right\rangle = S_1\left(B(\mathbf{p})e^{-ipx}\begin{bmatrix} 1 \\ 1 \end{bmatrix}\right) = B(\mathbf{p})e^{-ipx}\frac{\hbar}{2}\begin{bmatrix} 0 & 1 \\ 1 & 0 \end{bmatrix}\begin{bmatrix} 1 \\ 1 \end{bmatrix} = B(\mathbf{p})e^{-ipx}\frac{\hbar}{2}\begin{bmatrix} 1 \\ 1 \end{bmatrix} = \frac{\hbar}{2}\left|\psi_{\mathbf{p},+x\,\mathrm{spin}}\right\rangle$$

$$\qquad (B4\text{-}1.3)$$

$$S_1\left|\psi_{\mathbf{p},-x\,\mathrm{spin}}\right\rangle = B'(\mathbf{p})e^{-ipx}\frac{\hbar}{2}\begin{bmatrix} 0 & 1 \\ 1 & 0 \end{bmatrix}\begin{bmatrix} 1 \\ -1 \end{bmatrix} = B'(\mathbf{p})e^{-ipx}\frac{\hbar}{2}\begin{bmatrix} -1 \\ 1 \end{bmatrix} = -\frac{\hbar}{2}\left|\psi_{\mathbf{p},-x\,\mathrm{spin}}\right\rangle .$$

Note that we can construct the *x* component eigenstates from linear combinations of the *z* component eigenstates, i.e.

$$\begin{bmatrix} 1 \\ 1 \end{bmatrix} = \begin{bmatrix} 1 \\ 0 \end{bmatrix} + \begin{bmatrix} 0 \\ 1 \end{bmatrix} \qquad\qquad \begin{bmatrix} 1 \\ -1 \end{bmatrix} = \begin{bmatrix} 1 \\ 0 \end{bmatrix} - \begin{bmatrix} 0 \\ 1 \end{bmatrix} . \qquad\qquad (B4\text{-}1.4)$$

y direction

Do Prob. 15 to prove to yourself that $(i,-1)^T$ and $(i,1)^T$ are eigenstates of the *y* component of the spin operator, i.e., S_2, and that they can be constructed from linear combinations of the up and down (*z* direction) eigenstates.

Up and Down (z Direction) Eigenstates Span the 2D Spinor Space (and are Basis Vectors of that Space)

Any spin state (not necessarily pointed in one of the *x,y,* or *z* directions) as visualized in 3D physical space is composed of components in the *x,y,* and *z* directions of the 3D spin vector. This is a simple law of vector components.

Each of these three components for the *x,y,z* directions is represented in 2D spinor space as a different column matrix. But, importantly, all possible such column matrices can be expressed as linear combinations of the up and down (*z* direction) eigenstate column matrices. We showed this for the *x* direction in (B4-1.4). You showed it for *y*, if you did the problem suggested above. (We didn't normalize (B4-1.4). The normalized forms are $\left(1/\sqrt{2}\right)(1,1)^T$ and $\left(1/\sqrt{2}\right)(1,-1)^T$.)

Since the *x* and *y* direction spin states can be constructed via linear combination of the *z* direction eigenspin states, all possible spin states can be composed of various combinations of the *z* direction eigenspin states, i.e., of the up and down states $(1,0)^T$ and $(0,1)^T$. In mathematical language, we say $(1,0)^T$ and $(0,1)^T$ are basis vectors that span spinor space.

General Solution Includes All Possible Spin States

$$|\psi\rangle = \sum_{\mathbf{p}}\left(C_+(\mathbf{p})e^{-ipx}\begin{bmatrix} 1 \\ 0 \end{bmatrix} + C_-(\mathbf{p})e^{-ipx}\begin{bmatrix} 0 \\ 1 \end{bmatrix}\right) \qquad\qquad (B4\text{-}1.5)$$

The general solution (B4-1.5) includes all possible spin states. For example, for given $\mathbf{p}$, with $C_-(\mathbf{p}) = -C_+(\mathbf{p})$, spin is in the negative *x* direction (see (B4-1.3) above. Choosing coefficients correctly yields any given particle spin direction.

Bottom Line: If we develop our theory for up and down states, it will be applicable to all possible other spin states, too.

Now do Prob. 16 to prove to yourself that the other three solutions, for a stationary particle (or antiparticle) represent, respectively, spin down ($-\hbar/2$), spin up, and spin down states.

Other Dirac solutions, if particle stationary, are one spin up and two spin down states

We included the symbol $\hbar$ in our spin discussion so far, so that you could see the role it played in non-natural unit formulations, but from now on, we will assume natural units, such that $\hbar = 1$. And we will refer to spin magnitudes as ½ , which is more common than $\hbar/2$.

Dirac Fermion Spins: Moving Particles Discussion

As soon as we have a moving particle, things get more complicated, as we have to include non-zero p^i values in our solutions (4-20). This complication, which we didn't have in NRQM, is due to relativistic effects.

Due to relativity, a moving particle's spin is more complicated

Classical Relativistic Effects on a Spinning Object

Read Box 4-2 to gain some appreciation for the effect on a classical, macroscopic object's angular momentum direction when viewed from a frame in which the object has translational speed approaching that of light.

*Classically, relativistic translation alters **L***

Box 4-2. Classical Macroscopic Spinning Object Translating at Relativistic Speed

In 4D relativistic theory, angular momentum is a 2nd order tensor, but it can be treated simply as a vector formed from the integral over a rotating body of $dm(\mathbf{r} \times \mathbf{v}_t) = dm(\mathbf{r} \times (\boldsymbol{\omega} \times \mathbf{r}))$, where symbols should be obvious. When a macroscopic object like a spinning disk, as shown below, moves close to the speed of light, distances contract in the direction of the velocity, and this makes the plane of the disk appear to turn. (See figure below.)

The closer the disk gets to the speed of light, the more the disk surface appears in the observer's frame to align normal to the velocity direction. In the rest frame translating with the disk itself, the disk still appears aligned in the original way.

In the observer's frame, though, the angular momentum **L** appears to turn toward the direction of the velocity becoming **L′**. The greater the speed, the greater this turning. At light speed, **L′** and **v** become parallel.

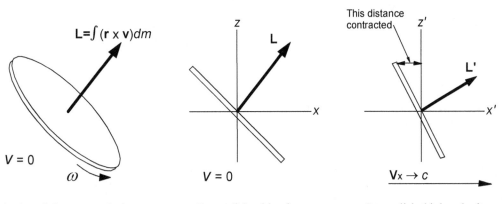

Spinning disk, no translation   Same disk, side view   Same disk, high velocity

For **L** in the opposite direction and the same positive x direction velocity, **L′** would point more and more towards the *negative x* direction, as speed increased.

Quantum Mechanics Experiments: Spin Magnitude Unchanged at Any Speed

We know from all the testing on high speed quantum particles like electrons, that spin (particle angular momentum) does not change in magnitude at all, no matter what speed the particle has (as measured from our reference frame). Electrons always have spin of $\hbar/2$ (½ in natural units), no matter what their velocity.

Experimentally, quantum particle spin magnitude always the same

Quantum mechanically, then, at high speed, a particle's angular momentum (spin) magnitude remains unchanged, but its direction appears to us in our frame to realign itself closer to that of the translational velocity vector. See Fig. 4-1. As velocity approaches c, the angular momentum (spin) approaches the velocity direction (or directly opposite direction.) For massless particles like photons, which travel at c, the spin is always aligned parallel with the velocity vector, either in the direction of velocity or in the opposite direction. Such a state is called a pure <u>helicity</u> state.

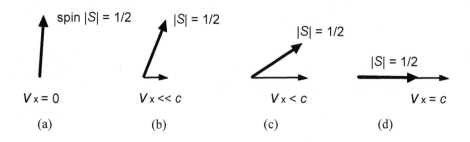

*As $v \to c$, quantum particle spin axis re-aligns closer to the **v** direction. For $v = c$, the particle is in a pure helicity state*

Figure 4-1. Effect of Transverse Velocity on Dirac Particle Spin

*Math form of Dirac spinor changes with **p** as physical spin direction re-aligns*

Mathematically, these kinds of relativistic complications are incorporated into the form of the spinors $u_r(\mathbf{p})$ and $v_r(\mathbf{p})$ (by their dependence on 3-momentum and thus ultimately, on velocity, as we will see below) and by how they are combined to form more general spin states. But it is nonetheless good to have some physical appreciation for why spin angular momentum direction can, for relativistic cases, depend upon linear momentum **p**.

Spinor Components Really Dependent on Velocity

Spinor form actually dependent on just velocity, not 3-momentum

With reference to the prior paragraph, note that the spinor components are actually dependent on particle velocity, rather than momentum, by the following logic.

Energy and momentum are expressed (in non-natural units to make it easier to understand)

$$E = \frac{mc^2}{\sqrt{1 - v^2/c^2}} \qquad p^i = \frac{mv^i}{\sqrt{1 - v^2/c^2}}, \qquad (4\text{-}41)$$

so in the coefficient and spinor components of the Dirac spinor (4-20) the mass m drops out. This leaves them a function solely of velocity (and thus in line with our knowledge, and with Box 4-2, that relativistic effects are dependent on velocity.)

Dirac Spinors: What Happens When **v** $\neq$ 0

Note what happens to the spin as seen by us, for an electron whose spin is represented solely by u_1, but has $p^1 \neq 0$, with $p^2 = p^3 = 0$ in our frame (the lab.)

$$\Sigma_3 \left| \psi^{(1)} \right\rangle = \frac{1}{2} \begin{bmatrix} 1 & & & \\ & -1 & & \\ & & 1 & \\ & & & -1 \end{bmatrix} \sqrt{\frac{E+m}{2m}} \begin{pmatrix} 1 \\ 0 \\ 0 \\ \dfrac{p^1}{E+m} \end{pmatrix} e^{-ipx} = \frac{1}{2} \sqrt{\frac{E+m}{2m}} \begin{pmatrix} 1 \\ 0 \\ 0 \\ \dfrac{-p^1}{E+m} \end{pmatrix} e^{-ipx} \neq \frac{1}{2} \left| \psi^{(1)} \right\rangle. \quad (4\text{-}42)$$

u_1 = spin up eigenstate if particle at-rest. u_1 not generally a spin up eigenstate for moving particle

So such a moving electron is not in a spin up eigenstate as seen in the lab. This relates to our discussion above about translation affecting the direction of the spin vector axis. Here, the mathematics bears this out. (We will have more to say on this shortly.) u_1 for a non-translating electron has spin up, but u_1 for an electron with high transverse velocity is not an up eigenstate.

Now consider u_1 representing an electron traveling in the z direction instead of the x direction.

$$\Sigma_3 \left| \psi^{(1)} \right\rangle = \frac{1}{2} \begin{bmatrix} 1 & & & \\ & -1 & & \\ & & 1 & \\ & & & -1 \end{bmatrix} \sqrt{\frac{E+m}{2m}} \begin{pmatrix} 1 \\ 0 \\ \dfrac{p^3}{E+m} \\ 0 \end{pmatrix} e^{-ipx} = \frac{1}{2} \sqrt{\frac{E+m}{2m}} \begin{pmatrix} 1 \\ 0 \\ \dfrac{p^3}{E+m} \\ 0 \end{pmatrix} e^{-ipx} = \frac{1}{2} \left| \psi^{(1)} \right\rangle. \quad (4\text{-}43)$$

But if particle motion is in direction of at-rest spin axis, it is also a spin eigenstate when moving

This electron, represented by u_1, is an up eigenstate as it moves, just as it was when it was at rest. Relativistically, this makes sense, as the plane of a spinning disk with **L** aligned in the direction of **p** would not appear to turn as **p** increased from zero to a relativistic value.

In general, boosts in the spin axis direction leave u_1, u_2, v_2 and v_1 in the same spin eigenstates as they would be at rest. Boosts in other directions take them out of those spin eigenstates. We will look a bit at the math behind this, for general particle states, shortly, but doing Probs. 17 and 18 now will enable you to get some feeling for how this happens with two particular Dirac particle states.

The At-Rest Coordinate System

The four solutions (4-20) (for given **p**) and their associated spinors $u_r(\mathbf{p})$ and $v_r(\mathbf{p})$ can be thought of as up/down spin eigenstates if the respective particles represented by each were decelerated to be at rest in the lab. We will call the coordinate system, fixed in the lab, in which a particle represented by $u_r(\mathbf{p})$ or $v_r(\mathbf{p})$ would be decelerated to be at rest, and its spin would be in the z direction, the at-rest coordinate system. Note this is a lab frame and is *different from the rest frame coordinate system*, which is fixed to the particle.

The at-rest coord system is the lab frame with particle decelerated to $\mathbf{v} = 0$ and z lab axis parallel to spin of decelerated particle

Dirac Fermion Spins: How to Think About Them

$u_1(\mathbf{p})$ and $v_2(\mathbf{p})$ in this system would then represent spin up stationary states; $u_2(\mathbf{p})$ and $v_1(\mathbf{p})$ represent spin down stationary states. We will virtually never work with electrons that are not moving, so our spinors will always look more complicated than the one in (4-40). But it can help us mentally as we work with such states, that are often not in a spin eigenstate in the lab, to think of them as spin eigenstates for the at-rest system, i.e., for a particle decelerated to $\mathbf{v} = 0$ in the lab.

Think of r index on spinors as representing z spin eigenstate for at-rest coord system

The Four-Spinors Span the 4D Spinor Space

In NRQM 2D spinor space, as we showed in Box 4-1, the two eigenspinors $(1,0)^T$ and $(0,1)^T$ (spin up and spin down, respectively) spanned the space of all possible spins (all possible spins can be constructed from linear combinations of those two eigenspin states, so they are basis vectors for 2D spinor space.) The general solution state (B4-1.5), a linear superposition of spin up and spin down eigenstates, contains all possible spin direction states. Each possible distinct general particle state has different coefficients C_+ and C_- for the eigenspin solutions summed to form that general state. This results in a different spin direction for the particle for each different set of coefficients. Thus, in NRQM, all possible spin directions can be represented by general solution (B4-1.5).

By analogy, we can surmise that the four Dirac spinors u_1, u_2, v_2 and v_1 of (4-20) span the RQM 4D spinor space of all possible spins and momenta, and thus, are basis vectors for that space. Our RQM general solution (4-29) contains within it all possible relativistic spin states.

More mathematically, we should know that a 4D space is spanned by four column vectors, each of four components, where these vectors are all independent of one another. Generally, the vector solutions of an eigenvalue problem, which is what the Dirac equation solutions are, are independent and complete, and thus we can conclude, span the space. They can be used as basis vectors.

The four Dirac spinors span the 4D spinor space of RQM

Further, the stationary particle case has spinors of form (ignoring the normalization factor in front) $(1,0,0,0)^T$, $(0,1,0,0)^T$, $(0,0,1,0)^T$, and $(0,0,0,1)^T$. These are obviously independent (and orthogonal). That independence is not changed by a boost (giving the particle a velocity relative to our frame.)

General RQM Solution Contains All Possible Spin Directions

Hence, similar to NRQM, our RQM general solution (4-29) for spinors contains all possible spin states within it. Different coefficients $C_1(\mathbf{p})$ and $C_2(\mathbf{p})$ will yield different spin states for C type particles. And different coefficients $D^\dagger_1(\mathbf{p})$ and $D^\dagger_2(\mathbf{p})$ will yield different spin states for D type particles.

RQM spinor general solution encompasses all possible spin directions

To see how this works, look at Fig. 4-1, and consider how each of the four states shown therein can be represented by their respective terms in the general particle state solution (4-29). In general, for $j = a,b,c,d$, the four states shown (for a C type particle) in Fig. 4-1 are

An example

$$\left| \psi_{(j)} \right\rangle = \sqrt{\frac{m}{VE_{\mathbf{p}_j}}} \left(C_1(\mathbf{p}_j) u_1(\mathbf{p}_j) + C_2(\mathbf{p}_j) u_2(\mathbf{p}_j) \right) e^{-ip_j x}. \tag{4-44}$$

State (a) there is effectively spin up with $\mathbf{p}_a = 0$, where from (4-20) and (4-29),

$$\left|\psi_{(a)}\right\rangle = \sqrt{\frac{m}{VE_{\mathbf{p}_a}}}C_1(0)u_1(0)e^{-ip_a x} = \sqrt{\frac{m}{VE_{\mathbf{p}_a}}}\sqrt{\frac{E_{\mathbf{p}_a}+m}{2m}}\underbrace{\begin{pmatrix}1\\0\\0\\0\end{pmatrix}}_{u_1(0)}e^{-ip_a x}, \tag{4-45}$$

which from (4-40) is an eigenstate of Σ_3. So, for state (a), (4-44) has effectively, $C_1 = 1$ and $C_2 = 0$.

For the last state (d), where the particle is traveling at the speed of light, (4-44) becomes an eigenstate of Σ_1 (see Prob. 18) with eigenvalue ½. That is,

$$\left|\psi_{(d)}\right\rangle = \sqrt{\frac{m}{VE_{\mathbf{p}_d}}}C_1(\infty)u_1(\infty)e^{-ip_d x} + \sqrt{\frac{m}{VE_{\mathbf{p}_d}}}C_2(\infty)u_2(\infty)e^{-ip_d x}$$

$$= \sqrt{\frac{m}{VE_{\mathbf{p}_d}}}\left(\underbrace{\sqrt{\frac{E_{\mathbf{p}_d}+m}{2m}}\begin{pmatrix}1\\0\\0\\1\end{pmatrix}}_{u_1(\infty)} + \underbrace{\sqrt{\frac{E_{\mathbf{p}_d}+m}{2m}}\begin{pmatrix}0\\1\\1\\0\end{pmatrix}}_{u_2(\infty)}\right)e^{-ip_d x} = \sqrt{\frac{m}{VE_{\mathbf{p}_d}}}\underbrace{\sqrt{\frac{E_{\mathbf{p}_d}+m}{2m}}\begin{pmatrix}1\\1\\1\\1\end{pmatrix}}_{\text{eigenstate of }\Sigma_1}e^{-ip_d x}, \tag{4-46}$$

where here, we must have $C_1 = 1$ and $C_2 = 1$ (in the normalized version, $C_1 = C_2 = 1/\sqrt{2}$) to get the proper eigenstate on the RHS of the lower row in (4-46).

For "in between" states (b) and (c), C_1 and C_2 would have other values. We won't get into how those values are determined here. We only want to make the following point.

The $u_{1,2}$ values are determined in a given problem solely by $\mathbf{p}$ (or equivalently, $\mathbf{v}$.) A particle has given $\mathbf{p}$, the $u_{1,2}$ are determined, and they serve as our spinor space basis vectors. Then, for a given spin alignment in our physical world problem, we have to choose the correct values for $C_{1,2}$ to mathematically represent that spin state, for the given $\mathbf{p}$. The spin basis vectors $u_{1,2}$ span spinor space, but we have to determine how much of each we need for their linear superposition to equal the spin state we are dealing with. Parallel logic holds for D type particles and $v_{2,1}$.

$\mathbf{p}$ determines $u_{1,2}$ and then spin is represented by correct linear combination of u_1 and u_2

So we see that although our development of the theory focused on only the two states of spin up and spin down (for stationary particle = at-rest frame particle) states, the theory is applicable to all possible spin direction states.

One State That is Impossible

Note that we can never have a relativistic state where the spin vector and $\mathbf{p}$ are at right angles.

Dirac Spinors Become NRQM Spinors in Low Speed Limit

Classical relativistic mechanics approaches Newtonian mechanics in the limit where speed is much less than that of light. In parallel fashion, we should expect that our RQM solutions approach NRQM solutions as $v \to 0$, as well. By doing Prob. 19, you can gain an understanding of how this does indeed happen with Dirac spinors. If it didn't, our solutions could not be correct, so this is one more check on the theory we have developed.

RQM spinors $\to$ NRQM spinors, as $v \to 0$.

Spinors vs Particle Spin Dependence on Velocity

As all of this relativistic spin stuff is a bit mind bending, the following heuristic description of it is offered to help in conceptualizing what is actually going on physically.

Fig. 4-2 illustrates spin for a Dirac fermion in the at-rest system (particle at-rest in the lab) and a boosted particle system in terms of its x and z components in the lab, and also in terms of its spinor u_1 and u_2 components.

Note that u_1 and u_2 actually exist in spinor space (they are spinor space basis vectors in that space), but they correspond to directions in physical space. For example, in the at-rest system, u_1 represents spin up and so can be visualized as a spatial vector that points in the $+z$ direction. Similarly, in the at-rest system, u_2 represents spin down, so can be visualized as a vector pointing in the $-z$ direction. (See Fig 4-2a.)

Representing the u_r as vectors is a heuristic oversimplification though, and in fact is not really correct, as operations like spinor addition work a little differently than vector addition. (See Winter[1].) However, temporarily visualizing them as such can aid in our understanding of how they and spin behave, relative to the at-rest coordinate system, for varying particle velocities.

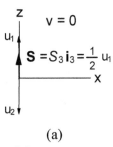

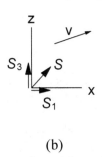

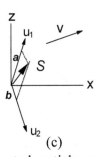

(a)	(b)	(c)
Particle at rest system = lab system	Boosted particle, spin x and z components in lab	Boosted particle, spin u_1 and u_2 components in lab

u_1 and u_2 can be visualized heuristically as directions in physical space

Figure 4-2. Heuristic Look at Spin in the At-Rest System and in a Particle Boosted System

In the particle boosted system (Fig 4-2b and c), u_1 and u_2, as we learned, no longer represent spin in the $+$ and $- z$ directions, but can be visualized as vectors pointing in other directions, where such directions depend on the particular boost velocity **v**.

So, both the spin direction of **S** and the spin space basis vectors u_1 and u_2, as visualized in physical space, change direction with boost velocity **v**. (Compare Fig. 4-2a to Fig. 4-2c.)

At-rest system

In the at-rest system, spin is aligned with the z axis, and in our example, it is the positive direction of the z axis. Spin is **S** ($= \mathbf{S}_3$) = ½ $\mathbf{u}_1$ and

$$\mathbf{S} = \mathbf{S}_3 = S_3\,\mathbf{i}_3 \qquad \text{or} \qquad \mathbf{S} = \tfrac{1}{2}\mathbf{u}_1 . \qquad (4\text{-}47)$$

Particle boosted system

In the boosted system, we can express the spin (no longer aligned with the z axis of the lab) as

$$\mathbf{S} = S_1\mathbf{i}_1 + S_3\mathbf{i}_3 \qquad \text{or} \qquad \mathbf{S} = a\mathbf{u}_1 + b\mathbf{u}_2 , \qquad (4\text{-}48)$$

(where u_1 and u_2 are no longer aligned with the z axis either).

*Dependence of u_1 and u_2 directions on **v** is different from spin direction dependence*

Note that when the particle is boosted, the directions of u_1 and u_2 change, and so does the direction of the spin **S**. However, they change a little differently. **S** for the at-rest particle is solely composed of u_1, whereas for the boosted particle it has components of both u_1 and u_2. A common new student mistake is to think that **S** and u_1 change in the same manner with **v**, and that the RHS of (4-47), if it holds at-rest, will hold for any velocity.

Again, the vector addition implied in Fig. 4-2c and the RHS of (4-48) is not quite correct for spinors, but hopefully, the underlying concept that the amount of each of u_1 and u_2 in **S** changes with velocity has become clear.

Summarizing Eigen Status of u_1 and u_2

$u_1(\mathbf{p})\,e^{-ipx}$ and $u_2(\mathbf{p})\,e^{-ipx}$ is each always an eigenstate of the Dirac equation (for any **p**).

$u_1(\mathbf{p})\,e^{-ipx}$ and $u_2(\mathbf{p})\,e^{-ipx}$ is each sometimes an eigenstate of z spin, i.e. of Σ_3 (for **p** = 0 or = $p^3\mathbf{i}_3$).

$u_1(\mathbf{p})\,e^{-ipx}$ and $u_2(\mathbf{p})\,e^{-ipx}$ are always basis vectors for any general state $|\psi\rangle$ (for any **p**).

$u_1(\mathbf{p})$ and $u_2(\mathbf{p})$ is each sometimes an eigenstate of z spin, i.e. of Σ_3 (for **p** = 0 or = $p^3\mathbf{i}_3$).

$u_1(\mathbf{p})$ and $u_2(\mathbf{p})$ are always basis vectors in 4D spinor space (for any **p**).

$u_1(\mathbf{p})$ and $u_2(\mathbf{p})$ change orientation, as visualized in physical space, as **p** changes.

Spin **S** (often written in relativity as Σ) changes direction with **p**, but differently than u_1 and u_2.

Summarizing the essence of u_1 and u_2

---

[1] Winter, Rolf G., *Quantum Physics,* Wadsworth (1979), Chap. 9.

Any general spin state u can be represented as a linear combination of u_1 and u_2 (for any $\mathbf{p}$).

$$u(\mathbf{p}) = C_1(\mathbf{p})\, u_1(\mathbf{p}) + C_2(\mathbf{p})\, u_2(\mathbf{p})$$

Any general particle state includes a spin part plus a spacetime part (for any given $\mathbf{p}$).

$$|\psi_{\mathbf{p}}\rangle = \sqrt{\frac{m}{2VE_{\mathbf{p}}}}\, u(\mathbf{p})\, e^{-ipx} = \sqrt{\frac{m}{2VE_{\mathbf{p}}}}\, (C_1(\mathbf{p})\, u_1(\mathbf{p})\, e^{-ipx} + C_2(\mathbf{p})\, u_2(\mathbf{p})\, e^{-ipx})$$

4.1.11 RQM Helicity Operator

We mentioned helicity briefly above in discussing how particles approaching the speed of light approach a state where their spin axis and the velocity vector $\mathbf{v}$ are parallel. It is possible, of course, for particles not traveling at the speed of light to have spin aligned with velocity (or any other direction.) With massless particles ($v = c$), however, the two must be aligned, in perfect helicity.

In general, if the spin axis (using the right-hand rule), of a particle is in the direction of $\mathbf{v}$ one says the particle has <u>positive helicity</u>. If spin points in the direction of $-\mathbf{v}$, the particle has <u>negative helicity</u>. The question we can ask then is "what is the helicity operator"?

Positive and negative helicity defined

The degree of helicity a particle has can be defined in terms of the angle between the spin vector and the velocity vector. It is maximum if that angle is zero. Zero helicity would exist if the angle were 90^0. The dot product of the spin vector with a unit vector in the $\mathbf{p}$ (or equivalently, the $\mathbf{v}$) direction behaves in just this way and has come to be the mathematical definition of helicity.

Helicity varies in degree, from max (full) to zero

Quantum mechanically, spin has three components, either for S_i in NRQM or Σ_i in RQM. These three components represent the three spatial components (x,y,z directions) of spin in physical space, though each component is a matrix in spinor space. So, our spin operator Σ in RQM plays the role of a 3-vector in physical space that points in the direction of spin. The inner product in physical space of the spin operator vector Σ and the unit vector in the $\mathbf{p}$ direction ($= \mathbf{p}/|\mathbf{p}|$) would then be the RQM <u>helicity operator</u>. That is,

Helicity operator = dot product of spin operator and $\mathbf{p}$

$$\Sigma_{\mathbf{p}} = \Sigma \cdot i_{\mathbf{p}} = \Sigma \cdot \frac{\mathbf{p}}{|\mathbf{p}|} = \Sigma_1 \frac{p^1}{|\mathbf{p}|} + \Sigma_2 \frac{p^2}{|\mathbf{p}|} + \Sigma_3 \frac{p^3}{|\mathbf{p}|}\,. \tag{4-49}$$

(4-49) is a 4X4 matrix in spinor space because each Σ_i is a 4X4 matrix. (4-49) is a scalar in physical space because it is the inner product of two vectors.

Helicity operator is a scalar in physical space, but a matrix in spinor space

<u>A Helicity Example</u>

Consider a case where a particle is in the first eigenstate of (4-20), which we know from prior work would be a spin up eigenstate if the particle were stationary. Take (4-49) with positive $p^3 \neq 0$, $p^1 = p^2 = 0$ and operate on that first eigenstate.

Example of positive helicity eigenstate

$$\Sigma \cdot \frac{\mathbf{p}}{|\mathbf{p}|} |\psi^{(1)}\rangle = \Sigma_3 \underbrace{\frac{p^3}{|\mathbf{p}|}}_{=1} |\psi^{(1)}\rangle = \frac{1}{2} \begin{bmatrix} 1 & & & \\ & -1 & & \\ & & 1 & \\ & & & -1 \end{bmatrix} \sqrt{\frac{E+m}{2m}} \begin{pmatrix} 1 \\ 0 \\ \dfrac{p^3}{E+m} \\ 0 \end{pmatrix} e^{-ipx}$$

$$= \frac{1}{2}\sqrt{\frac{E+m}{2m}} \begin{pmatrix} 1 \\ 0 \\ \dfrac{p^3}{E+m} \\ 0 \end{pmatrix} e^{-ipx} = \frac{1}{2}|\psi^{(1)}\rangle. \tag{4-50}$$

So, this state has a helicity eigenvalue of $+\frac{1}{2}$, which we should have known from non-mathematical considerations since $|\psi^{(1)}\rangle$ is an up spin ($+z$ direction) eigenstate as long as its velocity is in the z direction, as it is in this problem.

Note that if p^3 were negative ($-z$ direction),

$$\Sigma_3 \left| \psi^{(1)} \right\rangle = \frac{1}{2} \begin{bmatrix} 1 & & & \\ & -1 & & \\ & & 1 & \\ & & & -1 \end{bmatrix} \sqrt{\frac{E+m}{2m}} \begin{pmatrix} 1 \\ 0 \\ \frac{p^3}{E+m} \\ 0 \end{pmatrix} e^{-ipx} = \frac{1}{2} \left| \psi^{(1)} \right\rangle, \text{ and} \qquad (4\text{-}51)$$

Example of negative helicity eigenstate

$$\Sigma \cdot \frac{\mathbf{p}}{|\mathbf{p}|} \left| \psi^{(1)} \right\rangle = \Sigma_3 \underbrace{\frac{p^3}{|\mathbf{p}|}}_{\substack{= -1 \\ \text{since} \\ p^3 < 0}} \left| \psi^{(1)} \right\rangle = -\frac{1}{2}\sqrt{\frac{E+m}{2m}} \begin{pmatrix} 1 \\ 0 \\ \frac{p^3}{E+m} \\ 0 \end{pmatrix} e^{-ipx} = -\frac{1}{2} \left| \psi^{(1)} \right\rangle. \qquad (4\text{-}52)$$

Thus, from (4-51), the spin would still be up (+ ½ spin eigenvalue), but from (4-52), helicity would be negative (– ½ helicity eigenvalue).

In general, a + ½ helicity state for spinors means the spin is in the direction of **p**; a – ½ helicity eigenvalue means spin is in the direction of – **p**.[1]

A Second Helicity Example

Consider a Dirac particle in state $\left| \psi^{(1)} \right\rangle$ with positive $p^2 \neq 0$, $p^1 = p^3 = 0$. We would expect this is not a helicity eigenstate, since we know the spin, if velocity were zero, would be in the + z direction and here, the velocity is in the y direction. Let's see if the math tells us the same thing.

$$\Sigma \cdot \frac{\mathbf{p}}{|\mathbf{p}|} \left| \psi^{(1)} \right\rangle = \Sigma_2 \underbrace{\frac{p^2}{|\mathbf{p}|}}_{=1} \left| \psi^{(1)} \right\rangle = \frac{1}{2} \begin{bmatrix} & -i & & \\ i & & & \\ & & & -i \\ & & i & \end{bmatrix} \sqrt{\frac{E+m}{2m}} \begin{pmatrix} 1 \\ 0 \\ 0 \\ \frac{ip^2}{E+m} \end{pmatrix} e^{-ipx}$$

Example of state that is not a helicity eigenstate

$$= \frac{1}{2}\sqrt{\frac{E+m}{2m}} \begin{pmatrix} 0 \\ i \\ \frac{p^2}{E+m} \\ 0 \end{pmatrix} e^{-ipx} \neq \pm \frac{1}{2} \left| \psi^{(1)} \right\rangle. \qquad (4\text{-}53)$$

The lower line above is not proportional to the original state $\left| \psi^{(1)} \right\rangle$, so the state is not in a helicity eigenstate, as we suspected.

To gain more practice with helicity, do Prob. 20.

4.1.12 Summary of RQM for Dirac Particles

Do Prob. 21, which asks you to construct your own Wholeness Chart summary for RQM Dirac particles for all but spin and helicity.

Spin and helicity are summarized herein in Wholeness Chart 4-1 for both NRQM and RQM.

---

[1] Some authors define the helicity operator as $2\Sigma_{\mathbf{p}}$ so their helicity eigenvalues are ±1 instead of ± ½.

Wholeness Chart 4-1. Spin ½ Particle Spin Summary

	NRQM	RQM
Physical space	3D space plus time	4D spacetime
Spinor space dimensions	2D	4D
Wave function = ket (plane waves here)	physical space plus time part e^{-ikx}, and spinor space part (2D spinor)	physical spacetime parts e^{-ikx} and e^{+ikx}, and spinor space part (4D spinor)
Spinor forms = basis vectors (Dirac-Pauli representation for RQM)	$\chi_+ = \begin{bmatrix} 1 \\ 0 \end{bmatrix} \quad \chi_- = \begin{bmatrix} 0 \\ 1 \end{bmatrix}$	$u_1 = \sqrt{\dfrac{E+m}{2m}} \begin{pmatrix} 1 \\ 0 \\ \dfrac{p^3}{E+m} \\ \dfrac{p^1+ip^2}{E+m} \end{pmatrix} \quad u_2 = \sqrt{\dfrac{E+m}{2m}} \begin{pmatrix} 0 \\ 1 \\ \dfrac{p^1-ip^2}{E+m} \\ \dfrac{-p^3}{E+m} \end{pmatrix}$ similar forms for v_2 and v_1, here and below
Spinor form dependence	Independent of velocity $\mathbf{v}$	Depends on velocity $\mathbf{v}$ (on $\mathbf{p}$)
Spin operator	$S_i = \dfrac{\hbar}{2}\sigma_i \quad \sigma_i = $ 2X2 Pauli matrices	$\Sigma_i = \dfrac{\hbar}{2}\begin{bmatrix} \sigma_i & 0 \\ 0 & \sigma_i \end{bmatrix} \quad$ 4X4 matrices
Spin operator character	3D physical space angular momentum vector similar to linear momentum vector $\mathbf{p}$. Both are quantum operators. Spin vector components = matrix operators in spinor space.	
Spinors spin eigenstate?	Yes. Always	Generally, no. Yes, if $\mathbf{v} = 0$, or spin & x^3 aligned
Spin in directions of 3D axes (Linear combinations not normalized here)	For any velocity $\chi_+ \rightarrow +z\,\text{spin}; \ \chi_- \rightarrow -z\,\text{spin}$ $\chi_+ + \chi_- \rightarrow +x; \ \chi_+ - \chi_- \rightarrow -x$ $i\chi_+ - \chi_- \rightarrow +y; i\chi_+ + \chi_- \rightarrow -y$	For $\mathbf{v} = 0$, or spin and $\mathbf{v}$ both in direction indicated $u_1 \rightarrow +z\,\text{spin}; \ u_2 \rightarrow -z\,\text{spin}$ $u_1 + u_2 \rightarrow +x; \ u_1 - u_2 \rightarrow -x$ $iu_1 - u_2 \rightarrow +y; \ iu_1 + u_2 \rightarrow -y$
Spin operator eigenvalues	$\pm \hbar/2$ $+ (-)$ for given σ_i if spin in pos (neg) i axis direction	$\pm \hbar/2$ (½ in natural units) $+ (-)$ for given Σ_i if spin in pos (neg) i axis direction
General spin direction (Can normalize coefficients if desired.)	$\chi = C_+ \chi_+ + C_- \chi_-$ C_+, C_- determine spin direction. Choose them to fit problem spin, including spins not along x,y,z axes	$u = C_1(\mathbf{p})u_1(\mathbf{p}) + C_2(\mathbf{p})u_2(\mathbf{p})$ C_1, C_2 determine spin direction. Choose them to fit particular problem spin for any given $\mathbf{p}$ (including spins not along $\mathbf{p}$ or x,y,z axes.)
Helicity	$\sigma_\mathbf{p} = \sigma \cdot \dfrac{\mathbf{p}}{|\mathbf{p}|} = \sigma_i \dfrac{p^i}{p}$	$\Sigma_\mathbf{p} = \Sigma \cdot \dfrac{\mathbf{p}}{|\mathbf{p}|} = \Sigma_i \dfrac{p^i}{p}$
Spinors helicity eigenstates?	Generally no. Yes if spin and $\mathbf{p}$ aligned for χ_+ or χ_-	Generally no. Yes, if spin and $\mathbf{p}$ aligned for u_1 or u_2
Helicity eigenstates	Any general state where spin parallel to line of action of $\mathbf{p}$	
Helicity eigenvalues	$\pm \hbar/2$ (½ in natural units) $+$ if spin in $\mathbf{p}$ direction; $-$ if in $-\mathbf{p}$ direction	
General states summary	A state is in a spin eigenstate of S_i (Σ_i) if spin is aligned in $+$ or $-$ direction of ith axis. A state is in a helicity eigenstate if spin is aligned with $\mathbf{p}$ or $-\mathbf{p}$. Eigenstate or not, coefficients on the $\chi_{+,-}$ (or $u_{1,2}$) can be chosen to produce the mathematical state for the particular spin direction (and $\mathbf{p}$), and thus for the helicity, as well.	

4.2 The Dirac Equation in Quantum Field Theory

4.2.1 Summary Chart

All that we will do in the remainder of this chapter is summarized in Wholeness Chart 5-4 at the end of Chap. 5. I highly recommend following along, step by step, in that chart as we progress from here to the end of this chapter.

Be sure to use the summary wholeness chart, as you study this chapter

4.2.2 From RQM to QFT

Recall from Sect. 4.1.1, pg. 85, that because fermions cannot occupy the same state, they cannot reinforce one another to produce a macroscopic fermionic field. So, we have no classical spinor field theory. That is, we have no Hamiltonian density, Lagrangian density, nor wave equation (all of which are essentially equivalent), which we could use for quantizing and deriving a quantum spinor field theory.

No classical theory for spin ½ fields, so we can't use $\mathcal{H}$ or $\mathcal{L}$ for 2ⁿᵈ quantization postulate #1

But in our scalar theory we found the QFT wave equation for fields (in the Heisenberg picture) was identical in form to the RQM equation for states (in the Schrödinger picture). It is not a great leap of faith, therefore, to assume the same thing will hold true for spin ½ fermions. And, of course, as history and experiment have proven, it does.

So, we take a hint from scalar theory and take the Dirac RQM equation as our QFT field equation

Thus, the <u>Dirac equation for fields</u> (where we will, as with scalar fields, work in the Heisenberg picture), from (4-15) for states, is

$$\left(i\gamma^\mu \partial_\mu - m\right)\psi = 0 \ .$$ (4-54)

<u>Its eigensolutions</u> (for given **p**), identical in form to (4-20) but fields now instead of states, are

$$\psi^{(1)} = u_1 e^{-ipx} \quad \psi^{(2)} = u_2 e^{-ipx} \quad \psi^{(3)} = v_2 e^{ipx} \quad \psi^{(4)} = v_1 e^{ipx} \ ,$$ (4-55)

whose mathematical behavior we have already learned quite a bit about.

The <u>adjoint Dirac equation for fields</u>, from (4-31), is

And the RQM adjoint equation as our QFT adjoint equation

$$i\partial_\mu \overline{\psi}\gamma^\mu + m\overline{\psi} = 0 \ ,$$ (4-56)

with <u>adjoint eigensolutions</u>

$$\overline{\psi} = \psi^\dagger \gamma^0 \ \rightarrow \ \overline{\psi}^{(1)} = u_1^\dagger \gamma^0 e^{ipx} = \overline{u}_1 e^{ipx} \quad \overline{\psi}^{(2)} = \overline{u}_2 e^{ipx} \quad \overline{\psi}^{(3)} = \overline{v}_2 e^{-ipx} \quad \overline{\psi}^{(4)} = \overline{v}_1 e^{-ipx} \ .$$ (4-57)

4.2.3 Summary of General Plane Wave Solutions

For the Dirac and adjoint Dirac equation, the <u>general discrete plane wave solutions</u> are

$$
\begin{aligned}
\psi &= \sum_{r,\mathbf{p}} \sqrt{\frac{m}{VE_\mathbf{p}}}\left(c_r(\mathbf{p})u_r(\mathbf{p})e^{-ipx} + d_r^\dagger(\mathbf{p})v_r(\mathbf{p})e^{ipx}\right) \\
&= \qquad\qquad\quad \psi^+ \qquad + \qquad \psi^- \\
\overline{\psi} &= \sum_{r,\mathbf{p}} \sqrt{\frac{m}{VE_\mathbf{p}}}\left(d_r(\mathbf{p})\overline{v}_r(\mathbf{p})e^{-ipx} + c_r^\dagger(\mathbf{p})\overline{u}_r(\mathbf{p})e^{ipx}\right) \\
&= \qquad\qquad\quad \overline{\psi}^+ \qquad + \qquad \overline{\psi}^- \ ,
\end{aligned}
$$ (4-58)

General plane wave solutions to Dirac field equation

and the <u>general continuous plane wave solutions</u> are

$$\psi = \sum_r \sqrt{\frac{m}{(2\pi)^3}} \int \frac{d^3\mathbf{p}}{\sqrt{E_\mathbf{p}}}\left(c_r(\mathbf{p})u_r(\mathbf{p})e^{-ipx} + d_r^\dagger(\mathbf{p})v_r(\mathbf{p})e^{ipx}\right)$$

$$\overline{\psi} = \sum_r \sqrt{\frac{m}{(2\pi)^3}} \int \frac{d^3\mathbf{p}}{\sqrt{E_\mathbf{p}}}\left(d_r(\mathbf{p})\overline{v}_r(\mathbf{p})e^{-ipx} + c_r^\dagger(\mathbf{p})\overline{u}_r(\mathbf{p})e^{ipx}\right) .$$ (4-59)

Note that for fields we use lower case for the coefficients, as we did for fields in the scalar treatment. You may be anticipating that we want to distinguish them from the RQM coefficients (upper case) because we will find the QFT coefficients to be operators, rather than mere numbers. If so, you will turn out to be right, as we will see.

4.2.4 The Dirac Lagrangian, Conjugate Momentum, and Hamiltonian

From the Dirac equation (4-54), and trial and error, we can deduce the <u>Lagrangian (density) for free spinor fields</u> to be

$$\mathcal{L}_0^{1/2} = \overline{\psi}\left(i\gamma^\alpha \partial_\alpha - m\right)\psi \ , \tag{4-60}$$

which can be checked by plugging into the Euler-Lagrange equation,

$$\frac{\partial}{\partial x^\mu}\left(\frac{\partial \mathcal{L}}{\partial \phi^n_{,\mu}}\right) - \frac{\partial \mathcal{L}}{\partial \phi^n} = 0, \quad \text{with} \quad \phi^1 = \overline{\psi}; \ \phi^2 = \psi; \ \mathcal{L} = \mathcal{L}_0^{1/2} \ . \tag{4-61}$$

By doing Prob.22, you can prove to yourself that for $n = 1$ above, we get the Dirac equation, and for $n = 2$, we get the adjoint Dirac equation.

<u>Conjugate momenta</u> for ψ and $\overline{\psi}$ are

$$\pi^{1/2} = \frac{\partial \mathcal{L}_0^{1/2}}{\partial \psi_{,0}} = i\overline{\psi}\gamma^0 = i\psi^\dagger \gamma^0 \gamma^0 = i\psi^\dagger \qquad \overline{\pi}^{1/2} = \frac{\partial \mathcal{L}_0^{1/2}}{\partial \overline{\psi}_{,0}} = 0, \tag{4-62}$$

where the adjoint momentum on the RHS might be a little surprising, but is true.

The <u>Dirac Hamiltonian density</u> can be found from the Legendre transformation as

$$\mathcal{H}_0^{1/2} = \pi^{1/2}\dot{\psi} + \overline{\pi}^{1/2}\dot{\overline{\psi}} - \mathcal{L}_0^{1/2} = i\psi^\dagger \dot{\psi} - \mathcal{L}_0^{1/2} = i\underbrace{\psi^\dagger \gamma^0}_{\overline{\psi}}\gamma^0 \dot{\psi} - \mathcal{L}_0^{1/2}$$

$$= i\overline{\psi}\gamma^0\dot{\psi} - i\overline{\psi}\gamma^0\dot{\psi} - i\overline{\psi}\gamma^i\partial_i\psi + m\overline{\psi}\psi = -i\overline{\psi}\gamma^i\partial_i\psi + m\overline{\psi}\psi \ . \tag{4-63}$$

Dirac Lagrangian (density) $\mathcal{L}$ for free fields, deduced from Dirac equation

Dirac conjugate momenta

Dirac Hamiltonian density $\mathcal{H}$ from $\mathcal{L}$

4.3 Anti-commutation Relations for Dirac Fields

4.3.1 No Spinor Poisson Brackets: Try Assuming Coefficient Commutators

Since we don't have macroscopic spinor fields, and thus no associated Poisson brackets, we can't really carry out second quantization postulate #2, in which we took Poisson brackets over into commutators. But just as we took a hint in Sect. 4.2.2 from the scalar fields case for postulate #1 (i.e., we assumed that the QFT wave equation was the same as the RQM wave equation in both scalar and spin ½ cases), we can take a similar hint for postulate #2. That is, we can postulate that the Dirac solution coefficients $c_r(\mathbf{p})$ and $d_r(\mathbf{p})$ obey the same sort of commutation relations that the Klein-Gordon solution coefficients $a(\mathbf{k})$ and $b(\mathbf{k})$ did.

But, when early researchers did this, they soon found that such *commutation relations did not work for Dirac fields*. They did not produce a viable theory that matched the real world. (We will see this near the end of the chapter, but for now, just accept it.)

No classical spinor field Poisson brackets → guess that Dirac coeffs have commutation relations like scalars

That guess is wrong

4.3.2 Dirac Coefficient Anti-commutation Relations Do Work

However, it was soon found that <u>coefficient anti-commutation relations</u>, parallel in form to the scalar coefficient commutation relations, did work. These are

$$\boxed{\left[c_r(\mathbf{p}), c_s^\dagger(\mathbf{p}')\right]_+ = \left[d_r(\mathbf{p}), d_s^\dagger(\mathbf{p}')\right]_+ = \delta_{rs}\delta_{\mathbf{pp}'} \text{ (discrete)}; \ = \delta_{rs}\delta(\mathbf{p}-\mathbf{p}') \text{ (continuous)}.}$$

All other anti-commutators between coefficients equal zero. (4-64)

A new guess: Dirac coeffs obey anti-commutation relations.

We will thus use (4-64) as our postulate #2 of 2nd quantization for Dirac fields. See Fig. 4-3. And before too long, we will prove to ourselves that indeed, this postulate does give us a viable spinor field theory.

This guess is our postulate #2 for 2nd quantization of spinor fields

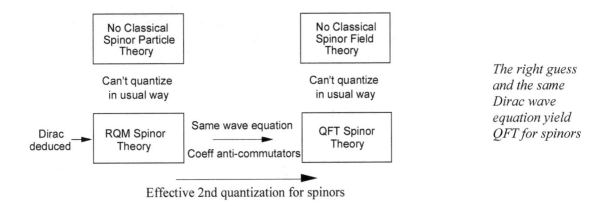

The right guess and the same Dirac wave equation yield QFT for spinors

Figure 4-3. The Route to QFT for Spin ½ Fields

4.4 The Dirac Hamiltonian in QFT

4.4.1 The Free Dirac Hamiltonian in Terms of the Coefficients

Similar to what we did for scalar fields, we find the Dirac Hamiltonian by integrating the Dirac Hamiltonian density of (4-63) over all space (a volume V containing the discrete solutions, which we can make as large as we like), i.e.,

$H = \int \mathcal{H} dV$

$$H_0^{1/2} = \int \mathcal{H}_0^{1/2} d^3 x = \int \left(-i\overline{\psi}\gamma^i \partial_i \psi + m\overline{\psi}\psi \right) d^3 x \ . \tag{4-65}$$

Parallel to what we did for scalars, we substitute the Dirac general solution (4-58) into (4-65),

Begin derivation of coefficient form of H

$$H_0^{1/2} = \int \left(-i\overline{\psi}\gamma^i \partial_i \psi + m\overline{\psi}\psi \right) d^3 x =$$

$$\int \left(\sum_{r,\mathbf{p}} \sqrt{\frac{m}{VE_{\mathbf{p}}}} (d_r(\mathbf{p})\overline{v}_r(\mathbf{p})e^{-ipx} + c_r^\dagger(\mathbf{p})\overline{u}_r(\mathbf{p})e^{ipx}) \right) \times$$

$$\left(-i\gamma^i \partial_i \right) \left(\sum_{s,\mathbf{p'}} \sqrt{\frac{m}{VE_{\mathbf{p'}}}} (c_s(\mathbf{p'})u_s(\mathbf{p'})e^{-ip'x} + d_s^\dagger(\mathbf{p'})v_s(\mathbf{p'})e^{ip'x}) \right) d^3 x \quad \text{(4-66)}$$

$$+ \int m \left(\sum_{r,\mathbf{p}} \sqrt{\frac{m}{VE_{\mathbf{p}}}} (d_r(\mathbf{p})\overline{v}_r(\mathbf{p})e^{-ipx} + c_r^\dagger(\mathbf{p})\overline{u}_r(\mathbf{p})e^{ipx}) \right) \times$$

$$\left(\sum_{s,\mathbf{p'}} \sqrt{\frac{m}{VE_{\mathbf{p'}}}} (c_s(\mathbf{p'})u_s(\mathbf{p'})e^{-ip'x} + d_s^\dagger(\mathbf{p'})v_s(\mathbf{p'})e^{ip'x}) \right) d^3 x \ .$$

The first of the two integrals above (i.e., the 2nd and 3rd lines of (4-66), which are multiplied together and represent the derivative term) becomes

$$\underset{+1}{\underbrace{(-i)i}} \int \left(\sum_{r,\mathbf{p}} \sqrt{\frac{m}{VE_{\mathbf{p}}}} (d_r(\mathbf{p})\overline{v}_r(\mathbf{p})e^{-i\left(E_{\mathbf{p}}t - p^i x^j\right)} \right) \left(\sum_{s,\mathbf{p'}} \sqrt{\frac{m}{VE_{\mathbf{p'}}}} c_s(\mathbf{p'})\gamma^i \underset{\underset{\partial_i}{\underbrace{}}}{p'^i} u_s(\mathbf{p'})e^{-i\left(E_{\mathbf{p'}}t - p'^i x^j\right)} \right) d^3 x$$

$$+ \int \left(\sum_{r,\mathbf{p}} \sqrt{\frac{m}{VE_{\mathbf{p}}}} d_r(\mathbf{p})\overline{v}_r(\mathbf{p})e^{-i\left(E_{\mathbf{p}}t - p^i x^j\right)} \right) \left(\sum_{s,\mathbf{p'}} \sqrt{\frac{m}{VE_{\mathbf{p'}}}} d_s^\dagger(\mathbf{p'})\gamma^i \left(-p'^i\right) v_s(\mathbf{p'})e^{i\left(E_{\mathbf{p'}}t - p'^i x^j\right)} \right) d^3 x$$

$$\tag{4-67}$$

$$+ \int \left(\sum_{r,\mathbf{p}} \sqrt{\frac{m}{VE_{\mathbf{p}}}} c_r^\dagger(\mathbf{p})\overline{u}_r(\mathbf{p})e^{i\left(E_{\mathbf{p}}t - p^i x^j\right)} \right) \left(\sum_{s,\mathbf{p'}} \sqrt{\frac{m}{VE_{\mathbf{p'}}}} c_s(\mathbf{p'})\gamma^i p'^i u_s(\mathbf{p'})e^{-i\left(E_{\mathbf{p'}}t - p'^i x^j\right)} \right) d^3 x$$

$$+ \int \left(\sum_{r,\mathbf{p}} \sqrt{\frac{m}{VE_{\mathbf{p}}}} c_r^\dagger(\mathbf{p})\overline{u}_r(\mathbf{p})e^{i\left(E_{\mathbf{p}}t - p^i x^j\right)} \right) \left(\sum_{s,\mathbf{p'}} \sqrt{\frac{m}{VE_{\mathbf{p'}}}} d_s^\dagger(\mathbf{p'})\gamma^i \left(-p'^i\right) v_s(\mathbf{p'})e^{i\left(E_{\mathbf{p'}}t - p'^i x^j\right)} \right) d^3 x \ .$$

In (4-67) above, all cases where the terms in the summations, when multiplied, don't give us $e^0 = 1$ for the spatial part of the exponent drop out, because an integral over all space of the oscillating function $e^{if(\mathbf{x})}$, where $f(\mathbf{x}) \neq 0$, is zero. So, in the first and last lines, only terms with $\mathbf{p} = -\mathbf{p}'$ will survive. And in the 2nd and 3rd lines, only terms in $\mathbf{p}' = \mathbf{p}$ will. Thus, (4-67), line-by-line, reduces (where we use $\mathbf{p}$ for all dummy indices) to (4-68). We assume, as in RQM, that the order of spinors and coefficients (such as c_r and d_r) can be interchanged at will, but we must preserve the order of spinor entities as it represent matrix/vector multiplication in spinor space.

$$+\int\left(\overbrace{\sum_{r,s,\mathbf{p}}\frac{m}{VE_{\mathbf{p}}}d_r(-\mathbf{p})\,\overline{v}_r(-\mathbf{p})\gamma^i p^i u_s(\mathbf{p})}^{\text{part used in Box 4-3}}\,c_s(\mathbf{p})e^{-i2E_{\mathbf{p}}t}\right)d^3x$$
will drop out when combined with equation below - see Box 4-3

$$+\int\left(\sum_{r,s,\mathbf{p}}\frac{m}{VE_{\mathbf{p}}}d_r(\mathbf{p})\overline{v}_r(\mathbf{p})\gamma^i\left(-p^i\right)v_s(\mathbf{p})d_s^\dagger(\mathbf{p})\right)d^3x$$

$$+\int\left(\sum_{r,s,\mathbf{p}}\frac{m}{VE_{\mathbf{p}}}c_r^\dagger(\mathbf{p})\overline{u}_r(\mathbf{p})\gamma^i p^i u_s(\mathbf{p})c_s(\mathbf{p})\right)d^3x \tag{4-68}$$

$$+\int\left(\sum_{r,s,\mathbf{p}}\frac{m}{VE_{\mathbf{p}}}c_r^\dagger(-\mathbf{p})\overline{u}_r(-\mathbf{p})\gamma^i\left(-p^i\right)v_s(\mathbf{p})d_s^\dagger(\mathbf{p})e^{i2E_{\mathbf{p}}t}\right)d^3x \quad.$$
will drop out when combined with equation below - similar to Box 4-3

In similar fashion, the last two lines of (4-66), representing the mass term in $H_0^{1/2}$, become

$$\int\left(\overbrace{\sum_{r,s,\mathbf{p}}\frac{m}{VE_{\mathbf{p}}}d_r(-\mathbf{p})\,\overline{v}_r(-\mathbf{p})m u_s(\mathbf{p})}^{\text{used in Box 4-3}}c_s(\mathbf{p})e^{-i2E_{\mathbf{p}}t}\right)d^3x$$
will drop out when combined with equation above - see Box 4-3

$$+\int\left(\sum_{r,s,\mathbf{p}}\frac{m}{VE_{\mathbf{p}}}d_r(\mathbf{p})\overline{v}_r(\mathbf{p})m v_s(\mathbf{p})d_s^\dagger(\mathbf{p})\right)d^3x$$

$$+\int\left(\sum_{r,s,\mathbf{p}}\frac{m}{VE_{\mathbf{p}}}c_r^\dagger(\mathbf{p})\overline{u}_r(\mathbf{p})m u_s(\mathbf{p})c_s(\mathbf{p})\right)d^3x \tag{4-69}$$

$$+\int\left(\sum_{r,s,\mathbf{p}}\frac{m}{VE_{\mathbf{p}}}c_r^\dagger(-\mathbf{p})\overline{u}_r(-\mathbf{p})m v_s(\mathbf{p})d_s^\dagger(\mathbf{p})e^{i2E_{\mathbf{p}}t}\right)d^3x.$$
will drop out when combined with equation above - similar to Box 4-3

The first row of (4-68) added to the first row of (4-69), as we do in (4-66), sum to zero as shown in Box 4-3. Similarly the last rows of (4-68) and (4-69) sum to zero as well. (Prove it in Prob. 23.)

Then we will look at the second lines of (4-68) and (4-69) but first need to recall (4-25), i.e.,

$$v_r^\dagger(\mathbf{p})v_s(\mathbf{p})=\overline{v}_r(\mathbf{p})\gamma^0 v_s(\mathbf{p})=\frac{E_{\mathbf{p}}}{m}\delta_{rs}=\frac{p_0}{m}\delta_{rs} \;\rightarrow\; \overline{v}_r(\mathbf{p})\gamma^0 p_0 v_s(\mathbf{p})=\frac{(p_0)^2}{m}\delta_{rs}=\frac{E_{\mathbf{p}}^2}{m}\delta_{rs}. \tag{4-70}$$

Adding the second lines of (4-68) and (4-69), and using (4-70) in the second line below, we find

$$\int\left(\sum_{r,s,\mathbf{p}}\frac{m}{VE_{\mathbf{p}}}d_r(\mathbf{p})\overline{v}_r(\mathbf{p})\left(-\gamma^i p^i+m\right)v_s(\mathbf{p})d_s^\dagger(\mathbf{p})\right)d^3x$$

$$=\int\left(\sum_{r,s,\mathbf{p}}\frac{m}{VE_{\mathbf{p}}}d_r(\mathbf{p})\left(\overline{v}_r(\mathbf{p})\left(\gamma^i p_i+m+\gamma^0 p_0\right)v_s(\mathbf{p})-\frac{E_{\mathbf{p}}^2}{m}\delta_{rs}\right)d_s^\dagger(\mathbf{p})\right)d^3x \tag{4-71}$$

$$=\underbrace{\frac{1}{V}\int d^3x}_{=1}\left(\sum_{r,s,\mathbf{p}}\frac{m}{E_{\mathbf{p}}}d_r(\mathbf{p})\left(\overline{v}_r(\mathbf{p})\left(\underbrace{\gamma^\mu p_\mu}_{\not{p}}+m\right)v_s(\mathbf{p})-\frac{E_{\mathbf{p}}^2}{m}\delta_{rs}\right)d_s^\dagger(\mathbf{p})\right).$$

Box 4-3. Terms That Drop Out of Derivation of Hamiltonian in Terms of Coefficients

To prove how the first row of (4-68) and (4-69), when summed, equal zero, we first need to derive a relationship involving the spinor $u_s(\mathbf{p})$.

Relationship for $u_s(\mathbf{p})$

Consider the Dirac equation and a single eigensolution to it having 3-momentum $\mathbf{p}$ and spin s,

$$\left(i\gamma^\mu \partial_\mu - m\right)\psi = \left(i\not\partial - m\right)\psi = 0 \quad \text{with} \quad \psi = c_{\underline{s}}(\mathbf{p})u_{\underline{s}}(\mathbf{p})e^{-ipx}. \tag{B4-3.1}$$

This results in

$$\left(\gamma^\mu p_\mu - m\right)c_{\underline{s}}(\mathbf{p})u_{\underline{s}}(\mathbf{p})e^{-ipx} = \left(\not p - m\right)c_{\underline{s}}(\mathbf{p})u_{\underline{s}}(\mathbf{p})e^{-ipx} = 0 \tag{B4-3.2}$$

Neither $c_s(\mathbf{p})$ nor the exponential equal zero, so the remaining factors must equal zero, and thus

$$\left(\gamma^\mu p_\mu - m\right)u_s(\mathbf{p}) = \left(\not p - m\right)u_s(\mathbf{p}) = 0. \tag{B4-3.3}$$

Re-expressing and Combining Terms in (4-68) and (4-69)

Now from the complex conjugate transpose of (4-25), where r and s are dummy variables and thus interchangeable, and the relation holds for any $\mathbf{p}$, including $-\mathbf{p}$,

$$u_r^\dagger(\mathbf{p})v_s(-\mathbf{p}) = 0 \;\rightarrow\; v_s^\dagger(-\mathbf{p})u_r(\mathbf{p}) = 0 \;\rightarrow\; v_r^\dagger(-\mathbf{p})u_s(\mathbf{p}) = 0$$

$$\rightarrow\; v_r^\dagger(-\mathbf{p})\gamma^0\gamma^0 u_s(\mathbf{p}) = 0 \;\rightarrow\; \bar{v}_r(-\mathbf{p})\gamma^0 u_s(\mathbf{p}) = 0 \;\rightarrow\; \bar{v}_r(-\mathbf{p})\gamma^0 p_0 u_s(\mathbf{p}) = 0. \tag{B4-3.4}$$

The middle part of the first line of (4-68) can then be written as

$$\bar{v}_r(-\mathbf{p})\gamma^i p^i u_s(\mathbf{p}) = \bar{v}_r(-\mathbf{p})\gamma^i \left(-p_i\right)u_s(\mathbf{p}) - \underbrace{\bar{v}_r(-\mathbf{p})\gamma^0 p_0 u_s(\mathbf{p})}_{=0 \text{ from (B4-3.4)}} = -\bar{v}_r(-\mathbf{p})\underbrace{\gamma^\mu p_\mu}_{\not p} u_s(\mathbf{p}). \tag{B4-3.5}$$

When we then use the RHS instead of the LHS of (B4-3.5) in the first line of (4-68) and add that to the first line of (4-69), as the terms are in the original summation of (4-66), we get

$$-\int\left(\sum_{r,s,\mathbf{p}}\frac{m}{VE_\mathbf{p}}d_r(-\mathbf{p})\left(\bar{v}_r(-\mathbf{p})\underbrace{\left(\not p - m\right)u_s(\mathbf{p})}_{=0 \text{ from (B4-3.3)}}\right)c_s(\mathbf{p})e^{-i2E_\mathbf{p}t}\right)d^3x = 0. \tag{B4-3.6}$$

Thus, the first and last lines of (4-68) and (4-69) drop out, as promised.

Now, if you did Prob. 23, you proved that

$$\left(\not p + m\right)v_s(\mathbf{p}) = 0, \tag{4-72}$$

which means (4-71) reduces to

$$\sum_{r,\mathbf{p}}\frac{m}{E_\mathbf{p}}d_r(\mathbf{p})\left(-\frac{E_\mathbf{p}^2}{m}\right)d_r^\dagger(\mathbf{p}) = -\sum_{r,\mathbf{p}}E_\mathbf{p}\underbrace{d_r(\mathbf{p})d_r^\dagger(\mathbf{p})}_{\substack{\text{use anti-}\\\text{commutator}}} = \sum_{r,\mathbf{p}}E_\mathbf{p}\left(\underbrace{d_r^\dagger(\mathbf{p})d_r(\mathbf{p})}_{\text{call this }\bar{N}_r(\mathbf{p})} - 1\right). \tag{4-73}$$

By doing Prob. 24, which essentially follows steps like (4-70) to (4-73) for the third lines of (4-68) and (4-69), one finds the sum of those third lines to be

$$\sum_{r,\mathbf{p}}E_\mathbf{p}\underbrace{c_r^\dagger(\mathbf{p})c_r(\mathbf{p})}_{\text{call this }N_r(\mathbf{p})}. \tag{4-74}$$

The only terms remaining in (4-66) are (4-73) and (4-74), so the free field Dirac Hamiltonian is

$$H_0^{1/2} = \sum_{r,\mathbf{p}} E_{\mathbf{p}} \left(N_r(\mathbf{p}) - \tfrac{1}{2} + \bar{N}_r(\mathbf{p}) - \tfrac{1}{2} \right),$$ (4-75)

End of derivation of H in terms of coefficients

where $\quad N_r(\mathbf{p}) = c_{\underline{r}}^\dagger(\mathbf{p}) c_{\underline{r}}(\mathbf{p}) \qquad \bar{N}_r(\mathbf{p}) = d_{\underline{r}}^\dagger(\mathbf{p}) d_{\underline{r}}(\mathbf{p}) \quad$ (underbars mean no summation). (4-76)

An Example

Now consider the Hamiltonian of (4-75) acting on a state with one c type Dirac particle in an eigenstate of 3-momentum $\mathbf{p}'$, another c particle with $\mathbf{p}''$, and a d type particle with $\mathbf{p}'''$. Note that Dirac d type particles in a ket are represented by an overbar, i.e., by $\bar{\psi}$.

An example: Comparing math results for H to real world energy

$$H_0^{1/2} \left| \psi_{r',\mathbf{p}'}, \psi_{r'',\mathbf{p}''}, \bar{\psi}_{r''',\mathbf{p}'''} \right\rangle = \underbrace{\sum_{r,\mathbf{p}} E_{\mathbf{p}} \left(N_r(\mathbf{p}) - \tfrac{1}{2} + \bar{N}_r(\mathbf{p}) - \tfrac{1}{2} \right)}_{\text{true mathematically}} \left| \psi_{r',\mathbf{p}'}, \psi_{r'',\mathbf{p}''}, \bar{\psi}_{r''',\mathbf{p}'''} \right\rangle$$

$$= \left(\underbrace{E_{\mathbf{p}'} + E_{\mathbf{p}''} + E_{\mathbf{p}'''}}_{\text{must be true physically}} + \underbrace{\sum_{r,\mathbf{p}} E_{\mathbf{p}} \left(-\tfrac{1}{2} - \tfrac{1}{2} \right)}_{\text{unexpected } -\infty \text{ energy}} \right) \left| \psi_{r',\mathbf{p}'}, \psi_{r'',\mathbf{p}''}, \bar{\psi}_{r''',\mathbf{p}'''} \right\rangle .$$ (4-77)

4.4.2 Dirac Number Operators

The only way this makes sense is if (4-77) is

$N_r(\mathbf{p})$ = number operator with eigenvalue $n_r(\mathbf{p})$ = underline{number of c particles of 3-mom $\mathbf{p}$}, spin r in the ket,

$\bar{N}_r(\mathbf{p})$ = number operator with eigenvalue $\bar{n}_r(\mathbf{p})$ = underline{number of d particles with $\mathbf{p}$ and spin r} in the ket,

and, the vacuum has $-$ ½ quantum of energy for each $\mathbf{p}$, r for c particles, and also for d particles.

Comparison yields number operators

4.4.3 Zero Point (Vacuum) Dirac Particle Energy

Note that

$$H_0^{1/2} \left| 0 \right\rangle = \sum_{r,\mathbf{p}} E_{\mathbf{p}} \left(N_r(\mathbf{p}) - \tfrac{1}{2} + \bar{N}_r(\mathbf{p}) - \tfrac{1}{2} \right) \left| 0 \right\rangle = \underbrace{\sum_{r,\mathbf{p}} E_{\mathbf{p}} \left(-\tfrac{1}{2} - \tfrac{1}{2} \right)}_{\text{infinite negative energy}} \left| 0 \right\rangle .$$ (4-78)

Vacuum appears filled with Dirac ½ quanta of negative energy

It is a striking fact that, according to our theory, the vacuum appears to be filled with an infinite number of Dirac c and d type virtual particle quanta, each of different frequency (with ½ $E_{\mathbf{p}}$ = ½ $\omega_{\mathbf{p}}$ in natural units), and each having *negative energy*.

Does this negative vacuum energy really exist? If one accepts that virtual scalars (and as we will see in Chap. 5, virtual photons) fill the vacuum with positive energy, then one has to accept what (4-78) is telling us. There appears to be no experimental evidence, however, unlike that often claimed for positive vacuum energy, that this is so. Additionally, it then becomes even harder to presume states in QFT are harmonic oscillator states, as no known such oscillator has a negative energy ground state.

Many, like myself, feel that something is still missing from the extant theory, and that we should keep an open mind with regard to what is happening in the vacuum.[1]

4.4.4 Positive Energy for Real Dirac Type d Particles in QFT

Consider a multiparticle state of *real* (not virtual, like in the vacuum above) d type particles, which in RQM, with $H = i\partial/\partial t$, had negative energies. Here, with $H = \int \mathcal{H} \, dV$ = (4-75) instead acting on a state with two d type particles with different 3-momentum, we find

---

[1] The article cited in the footnotes on pgs. 50 and 89 shows that by including the alternative forms of the solutions to the field equations in the theory, all ½ quanta terms in the vacuum, of both positive and negative energy, sum to zero.

$$H_0^{1/2}\left|\bar{\psi}_{r',\mathbf{p}'},\bar{\psi}_{r'',\mathbf{p}''}\right\rangle = \sum_{r,\mathbf{p}} E_{\mathbf{p}}\left(N_r(\mathbf{p}) - \tfrac{1}{2} + \bar{N}_r(\mathbf{p}) - \tfrac{1}{2}\right)\left|\bar{\psi}_{r',\mathbf{p}'},\bar{\psi}_{r'',\mathbf{p}''}\right\rangle$$

$$= \left(\underbrace{E_{\mathbf{p}'} + E_{\mathbf{p}''}}_{\substack{\text{real particle}\\\text{energy}}}\ \underbrace{-\infty}_{\substack{\text{from}\\\text{vacuum}}}\right)\left|\bar{\psi}_{r',\mathbf{p}'},\bar{\psi}_{r'',\mathbf{p}''}\right\rangle . \tag{4-79}$$

d type Dirac particles have positive energy in QFT, like antiparticles do in experiments

So, our theory produces only positive energy real particles, and we will expect to find the d type particles are antiparticles, with opposite charge of particles, but with the same (positive) energy.

4.4.5 Unit Norms and Orthogonality for Multiparticle Spinor States

Similar to what we had in QFT for scalars, we define our symbols for multiparticle states such that every such state is normalized, i.e., its inner product with itself is unity. That is, we choose the constant in front of the state (when it is expressed in wave function form) to get this result. This constant is hidden inside (it is implied, not shown) the ket symbol, when we use that to represent the state. For example, using the ket symbol, which is more common in QFT, we have

Orthonormality of spinor states similar to scalar states

$$\left\langle\psi_{r',\mathbf{p}'},\bar{\psi}_{r'',\mathbf{p}''}\middle\|\psi_{r',\mathbf{p}'},\bar{\psi}_{r'',\mathbf{p}''}\right\rangle = 1 . \tag{4-80}$$

And by their nature, each such multiparticle state is orthogonal to every other state that is not identical to it in particle types, particle numbers, $\mathbf{p}$, and spin r. If we expressed the following examples in terms of integrals over all space (inner products) of the wave function form of the states, we would find those integrals equal zero. Note states can contain both bosons and fermions.

$$\left\langle 4\phi_{\mathbf{k}'},\bar{\psi}_{r'',\mathbf{p}''}\middle\|3\phi_{\mathbf{k}'},\bar{\psi}_{r'',\mathbf{p}''}\right\rangle = 0 \qquad \underbrace{\left\langle\psi_{r',\mathbf{p}'},\bar{\psi}_{r'',\mathbf{p}''}\middle\|\psi_{r',\mathbf{p}'},\bar{\psi}_{r''',\mathbf{p}''}\right\rangle}_{r''\neq r'''} = 0 . \tag{4-81}$$

4.5 Expectation Values and the Dirac Hamiltonian

Parallel to scalars (see Chap. 3, Sect. 3.5, pgs. 57-58, for details), we find the expectation value of any operator for any multiparticle state, with ψ_j representing the jth state and can be either a c or d type particle, is

As are expectation values of operators

$$\bar{\mathcal{O}} = \left\langle\psi_1,\psi_2,\psi_3,....\middle|\mathcal{O}\middle|\psi_1,\psi_2,\psi_3,....\right\rangle . \tag{4-82}$$

4.6 Creation and Destruction Operators

It will probably not come as a big surprise that the $c_r(\mathbf{p})$ and $d_r(\mathbf{p})$ operators destroy Dirac particles, and their complex conjugates create Dirac particles. We prove this below.

The coefficients create and destroy Dirac particles

4.6.1 Proving It

Before going through the proof, we should note some relations we'll need that are a direct result of certain anti-commutation relations noted, but not shown, in (4-64). Two of these are

$$\left[c_r^\dagger(\mathbf{p}),c_r^\dagger(\mathbf{p})\right]_+ = \left[c_r(\mathbf{p}),c_r(\mathbf{p})\right]_+ = 0 . \tag{4-83}$$

The first of these gives us

$$c_r^\dagger(\mathbf{p})c_r^\dagger(\mathbf{p}) + c_r^\dagger(\mathbf{p})c_r^\dagger(\mathbf{p}) = 0 \rightarrow \left(c_r^\dagger(\mathbf{p})\right)^2 = 0 . \tag{4-84}$$

Do Prob. 25 to prove that from the second of (4-83) and similar relations for $d_r(\mathbf{p})$ and $d_r^\dagger(\mathbf{p})$,

$$\left(c_r(\mathbf{p})\right)^2 = 0 \qquad \left(d_r^\dagger(\mathbf{p})\right)^2 = 0 \qquad \left(d_r(\mathbf{p})\right)^2 = 0 . \tag{4-85}$$

Note, as proven when doing Prob. 25, that none of (4-84) and (4-85) would be true if the coefficients obeyed commutation, rather than anti-commutation, relations.

Proof that $c_r(\mathbf{p})$ is a Destruction Operator

We ask what state results from $c_r(\mathbf{p})$ acting on a single c type particle of spin r and 3-momentum $\mathbf{p}$ (no sum on r below),

Proving $c_r(\mathbf{p})$ destroys a single particle state

$$c_r(\mathbf{p})\left|\psi_{r,\mathbf{p}}\right\rangle = \left|?\right\rangle . \tag{4-86}$$

To see, operate on this unknown state with our number operator $N_r(\mathbf{p}) = c_r^\dagger(\mathbf{p})\,c_r(\mathbf{p})$.

$$N_r(\mathbf{p})|?\rangle = n_?|?\rangle = n_? \, c_r(\mathbf{p})\big|\psi_{r,\mathbf{p}}\big\rangle = \underbrace{c_r^\dagger(\mathbf{p})c_r(\mathbf{p})}_{\text{use anticommutator}}\, c_r(\mathbf{p})\big|\psi_{r,\mathbf{p}}\big\rangle$$

$$= \Big(1 - c_r(\mathbf{p})c_r^\dagger(\mathbf{p})\Big)c_r(\mathbf{p})\big|\psi_{r,\mathbf{p}}\big\rangle = c_r(\mathbf{p})\big|\psi_{r,\mathbf{p}}\big\rangle - c_r(\mathbf{p})\underbrace{c_r^\dagger(\mathbf{p})c_r(\mathbf{p})}_{N_r(\mathbf{p})}\big|\psi_{r,\mathbf{p}}\big\rangle. \qquad (4\text{-}87)$$

$$= c_r(\mathbf{p})\big|\psi_{r,\mathbf{p}}\big\rangle - c_r(\mathbf{p})\underbrace{n_r(\mathbf{p})}_{=1}\big|\psi_{r,\mathbf{p}}\big\rangle = \underbrace{(1-1)}_{n_?=0}\underbrace{c_r(\mathbf{p})\big|\psi_{r,\mathbf{p}}\big\rangle}_{|?\rangle}.$$

Since the number eigenvalue of $|?\rangle$ is zero, we conclude that $|?\rangle = |0\rangle$ and thus, from (4-86), that $c_r(\mathbf{p})$ destroyed the Dirac particle state $|\psi_{r,\mathbf{p}}\rangle$.

Note that, using (4-85), we can prove that $c_r(\mathbf{p})$ annihilates the vacuum, i.e.,

$$c_r(\mathbf{p})|0\rangle = c_r(\mathbf{p})\Big(c_r(\mathbf{p})\big|\psi_{r,\mathbf{p}}\big\rangle\Big) = \underbrace{\big(c_r(\mathbf{p})\big)^2}_{0}\big|\psi_{r,\mathbf{p}}\big\rangle = 0 . \qquad (4\text{-}88)$$

Proving $c_r(\mathbf{p})$ annihilates the vacuum

<u>End of Proof</u>

<u>Proof that $c_r^\dagger(\mathbf{p})$ is a Creation Operator</u>

What state results from $c_r^\dagger(\mathbf{p})$ acting on the vacuum?

$$c_r^\dagger(\mathbf{p})|0\rangle = |?\rangle . \qquad (4\text{-}89)$$

Start by operating on the above with the number operator.

Proving $c_r^\dagger(\mathbf{p})$ creates a single particle state from the vacuum

$$N_r(\mathbf{p})|?\rangle = n_?|?\rangle = \underbrace{N_r(\mathbf{p})}_{c_r^\dagger(\mathbf{p})c_r(\mathbf{p})}\, c_r^\dagger(\mathbf{p})|0\rangle = c_r^\dagger(\mathbf{p})\underbrace{c_r(\mathbf{p})c_r^\dagger(\mathbf{p})}_{\text{use anticommutator}}|0\rangle$$

$$= c_r^\dagger(\mathbf{p})\bigg(1 - \underbrace{c_r^\dagger(\mathbf{p})c_r(\mathbf{p})}_{N_r(\mathbf{p})}\bigg)|0\rangle = c_r^\dagger(\mathbf{p})(1-0)|0\rangle = c_r^\dagger(\mathbf{p})|0\rangle = |?\rangle . \qquad (4\text{-}90)$$

From the first and last parts of (4-90), we see that the state in question has $n_? = 1$ as the eigenvalue of $N_r(\mathbf{p})$ and thus, must be the single particle state $|?\rangle = |\psi_{r,\mathbf{p}}\rangle$. Hence, from (4-89), $c_r^\dagger(\mathbf{p})$ creates a single Dirac particle out of the vacuum.

<u>End of Proof</u>

4.6.2 The Surprise Result of $c_r^\dagger(\mathbf{p})$ Acting on a Single Particle State

From the parallels so far with scalar creation and destruction operators, one might expect that $c_r^\dagger(\mathbf{p})$ acting on a single particle state would create a two-particle state. But if we recall that two or more fermions cannot occupy the same state (same r and $\mathbf{p}$), it seems such a thing would have to be prohibited. Let's see if it is. Using (4-84), we get

Proving $c_r^\dagger(\mathbf{p})$ annihilates a state with type c particle $|\psi_{r,\,\mathbf{p}}\rangle$ in it

$$c_r^\dagger(\mathbf{p})\underbrace{\big|\psi_{r,\mathbf{p}}\big\rangle}_{c_r^\dagger(\mathbf{p})|0\rangle} = \underbrace{\big(c_r^\dagger(\mathbf{p})\big)^2}_{=\,0}|0\rangle = 0 . \qquad (4\text{-}91)$$

So action of $c_r^\dagger(\mathbf{p})$ on a state containing the Dirac particle $|\psi_{r,\mathbf{p}}\rangle$ destroys that state.

So, the theory we've developed tells us that <u>we cannot create (we cannot have) multiparticle states with more than one Dirac particle in a given single particle state</u>. Dirac particles obey the Pauli exclusion principle. This is a blessing. If it were otherwise, we would not have a valid fermionic field theory.

<u>Anti-commutators give us fermionic behavior</u>

As noted above, if our coefficients obeyed commutator relations (like scalar bosons do), we would not have gotten (4-84), and that is what caused the operation in (4-91) to prohibit creation of a two particle state with both particles being c type fermions having the same spin and 3-momentum. This is why we stated in Sect. 4.3 on pg. 104 that commutators would not work for Dirac particles and we had to use anti-commutators.

General rule:

    Coefficient commutation relations work for bosons and allow more than one identical single particle state to co-exist in the same multiparticle state.

    Coefficient anti-commutation relations work for fermions and do not allow more than one identical single particle state to co-exist in the same multiparticle state.

*Commutators →
bosonic behavior
Anti-commutators
→ fermionic
behavior*

4.6.3 Parallel Results for d Type Particles

    A parallel analysis for d type Dirac particles yields exactly parallel results as we had above for c types. $d_r^\dagger(\mathbf{p})$ and $d_r(\mathbf{p})$ operators create (from the vacuum) and destroy d type particle states. Application of $d_r(\mathbf{p})$ to a state containing a single d particle $\left|\overline{\psi}_{r,\mathbf{p}}\right\rangle$, results in the vacuum $|0\rangle$.

*Similar reasoning
and results for
d type Dirac
particles*

4.6.4 Total Particle Number

    As with scalars, total particle number is defined as the number of particles (i.e. c types) minus the number of antiparticles (d types). For spinors, the total particle number operator is

$$N(\psi) = \sum_{r,\mathbf{p}} \left(N_r(\mathbf{p}) - \overline{N}_r(\mathbf{p}) \right). \tag{4-92}$$

*Total particle
number is number
of particles minus
number of
antiparticles*

    Again, note the subtle difference in phraseology. "Number of particles" (which is different from "total particle number") equals the number of particles *plus* the number of antiparticles.

4.6.5 $\psi(x)$ and $\overline{\psi}(x)$ as Operator (or Quantum) Fields

    Thus, the total particle lowering operator field is

$$\psi = \sum_{r,\mathbf{p}} \sqrt{\frac{m}{VE_{\mathbf{p}}}} \left(c_r(\mathbf{p}) u_r(\mathbf{p}) e^{-ipx} + d_r^\dagger(\mathbf{p}) v_r(\mathbf{p}) e^{ipx} \right)$$

$$= \underbrace{\psi^+}_{\substack{\text{destroys} \\ \text{particles}}} + \underbrace{\psi^-}_{\substack{\text{creates} \\ \text{anti-particles}}}, \tag{4-93}$$

*ψ is a total
particle number
lowering
operator field*

and the total particle raising operator field is

$$\overline{\psi} = \sum_{r,\mathbf{p}} \sqrt{\frac{m}{VE_{\mathbf{p}}}} \left(d_r(\mathbf{p}) \overline{v}_r(\mathbf{p}) e^{-ipx} + c_r^\dagger(\mathbf{p}) \overline{u}_r(\mathbf{p}) e^{ipx} \right)$$

$$= \underbrace{\overline{\psi}^+}_{\substack{\text{destroys} \\ \text{anti-particles}}} + \underbrace{\overline{\psi}^-}_{\substack{\text{creates} \\ \text{particles}}}. \tag{4-94}$$

*$\overline{\psi}$ is a total
particle number
raising
operator field*

4.7 QFT Spinor Charge Operator and Four Current

4.7.1 Simple Deduction of Dirac Charge Operator

    From what we know about the number operators, and parallel to what we found for scalar fields, we can simply define our Dirac charge operator as

$$\boxed{Q = -e \sum_{r,\mathbf{p}} \left(N_r(\mathbf{p}) - \overline{N}_r(\mathbf{p}) \right)}, \tag{4-95}$$

*Spinor charge
operator*

where $-e$ is the charge on the electron. Note that, with this definition, d type particles will have a charge of $+e$, which would qualify them as antiparticles of the electron. Note the operation of (4-95) on a typical state.

$$-e \sum_{r,\mathbf{p}} \left(N_r(\mathbf{p}) - \overline{N}_r(\mathbf{p}) \right) \left| \psi_{r_1,\mathbf{p}_1}, \psi_{r_1,\mathbf{p}_2}, \overline{\psi}_{r_1,\mathbf{p}_1} \right\rangle = \underbrace{-e(1+1-1)}_{\text{tot charge} = -e} \left| \psi_{r_1,\mathbf{p}_1}, \psi_{r_1,\mathbf{p}_2}, \overline{\psi}_{r_1,\mathbf{p}_1} \right\rangle. \tag{4-96}$$

A state with two electrons and one positron has a total charge of $-e$.

4.7.2 The Dirac Charge Operator from the Four Current

But we could, instead, derive (4-95) more formally.

Spinor fields in QFT are governed by the same Dirac equation as spinor states were in RQM. And we used that equation in Sect. 4.1.7 (pg. 91) and Prob. 13 to find the conserved quantity Q' of (4-35) associated with the divergenceless four-current j^μ of (4-34). The exact same steps with fields yield the exact same expressions as they did for states, except now ψ and $\bar\psi$ are operator (quantum) fields with constants $C_r(\mathbf{p})$, $D_r(\mathbf{p})$ $\to$ operators $c_r(\mathbf{p})$, $d_r(\mathbf{p})$. Thus, we have the

A conserved 4-current spinor operator

$$\text{spinor 4-current operator} \quad j^\mu = (\rho, \mathbf{j}) = \bar\psi \gamma^\mu \psi \quad \text{with} \quad \partial_\mu j^\mu = 0, \tag{4-97}$$

and

$$\int_V j^0 d^3x = \int_V \rho\, d^3x = \int_V \bar\psi \gamma^0 \psi\, d^3x = \int_V \psi^\dagger \gamma^0 \gamma^0 \psi\, d^3x = \int_V \psi^\dagger \psi\, d^3x = Q'. \tag{4-98}$$

Derivation of Charge Operator from Divergenceless 4-Current

Substituting the solutions (4-93) and (4-94) post-multiplied by γ^0 into (4-98) yields

Deriving the spinor charge 4-current operator

$$Q' = \int_V \psi^\dagger \psi\, d^3x = \int_V \left(\sum_{r,\mathbf{p}} \sqrt{\frac{m}{VE_\mathbf{p}}} \left(d_r(\mathbf{p}) v_r^\dagger(\mathbf{p}) e^{-ipx} + c_r^\dagger(\mathbf{p}) u_r^\dagger(\mathbf{p}) e^{ipx} \right) \right) \times$$
$$\left(\sum_{s,\mathbf{p}'} \sqrt{\frac{m}{VE_{\mathbf{p}'}}} \left(c_s(\mathbf{p}') u_s(\mathbf{p}') e^{-ip'x} + d_s^\dagger(\mathbf{p}') v_s(\mathbf{p}') e^{ip'x} \right) \right) d^3x. \tag{4-99}$$

As we've seen before, all above products where $|\mathbf{p}| \neq \pm\, \mathbf{p}'$ result in exponentials that oscillate in space, and when integrated over all space, are zero. The $\mathbf{p} = \mathbf{p}'$ terms then are

$$\left(\sum_{r,s,\mathbf{p}} \frac{m}{VE_\mathbf{p}} \left(d_r(\mathbf{p}) d_s^\dagger(\mathbf{p}) \underbrace{v_r^\dagger(\mathbf{p}) v_s(\mathbf{p})}_{\frac{E_\mathbf{p}}{m}\delta_{rs}} + c_r^\dagger(\mathbf{p}) c_s(\mathbf{p}) \underbrace{u_r^\dagger(\mathbf{p}) u_s(\mathbf{p})}_{\frac{E_\mathbf{p}}{m}\delta_{rs}} \right) \right) \underbrace{\int_V d^3x}_{V}, \tag{4-100}$$

where the underbrackets are from (4-25). The $\mathbf{p}' = -\mathbf{p}$ terms are

$$\left(\sum_{r,s,\mathbf{p}} \frac{m}{VE_\mathbf{p}} \left(d_r(-\mathbf{p}) c_s(\mathbf{p}) \underbrace{v_r^\dagger(-\mathbf{p}) u_s(\mathbf{p})}_{=0} e^{-i2E_\mathbf{p}t} + c_r^\dagger(-\mathbf{p}) d_s^\dagger(\mathbf{p}) \underbrace{u_r^\dagger(-\mathbf{p}) v_s(\mathbf{p})}_{=0} e^{i2E_\mathbf{p}t} \right) \right) \int_V d^3x \tag{4-101}$$

where we again use (4-25) to show those terms go to zero. (4-99) then becomes

$$Q' = \int \rho\, d^3x = \sum_{r,\mathbf{p}} \left(c_r^\dagger(\mathbf{p}) c_r(\mathbf{p}) + \underbrace{d_r(\mathbf{p}) d_r^\dagger(\mathbf{p})}_{\substack{\text{use anti-}\\\text{commutator}}} \right) = \sum_{r,\mathbf{p}} \left(\underbrace{c_r^\dagger(\mathbf{p}) c_r(\mathbf{p})}_{N_r(\mathbf{p})} - \underbrace{d_r^\dagger(\mathbf{p}) d_r(\mathbf{p})}_{\bar N_r(\mathbf{p})} + 1 \right). \tag{4-102}$$

The "1" may look strange here, but recall we only need to satisfy the 4-divergence relation (4-97). So we can add a constant 4-vector to $j^\mu = (\rho, \mathbf{j})$ and the result will still have zero 4-divergence. We can also multiply j^μ by any constant and the 4-divergence of zero will still hold. So, define a new 4-vector, the spinor charge density 4-current operator s^μ (different from spinor 4-current operator j^μ),

The spinor charge 4-current operator (a density)

$$s^\mu = -e\left(j^\mu - \frac{1}{V}\sum_{r,\mathbf{p}} \begin{bmatrix} 1 \\ 1 \\ 1 \\ 1 \end{bmatrix} \right) \quad \to \quad \partial_\mu s^\mu = 0. \tag{4-103}$$

So, our operator for our new conserved quantity, which is the spinor charge operator, is (4-95), i.e.,

The spinor charge operator

$$Q = \int s^0 d^3x = \int -e\left(j^0 - \frac{1}{V}\sum_{r,\mathbf{p}} 1 \right) d^3x = -e\left(Q' - \sum_{r,\mathbf{p}} 1 \right) = -e \sum_{r,\mathbf{p}} \left(N_r(\mathbf{p}) - \bar N_r(\mathbf{p}) \right). \tag{4-104}$$

End of derivation

Be aware that other texts invoke normal ordering in the middle of (4-102), instead of the anti-commutator, to re-order the $d_r(\mathbf{p}) d_r^\dagger(\mathbf{p})$ term. By doing so, they do not get the "1" term on the RHS, and find $s^\mu = -e\,j^\mu$ instead of (4-103). As I've discussed earlier, normal ordering assumes anti-commutation of operators $d_r(\mathbf{p})$ and $d_r^\dagger(\mathbf{p})$ that, according to the fundamentals of the theory, do not anti-commute, so I favor the derivation shown above. I note further, in passing, that the approach referenced in the footnote on pg. 50 eliminates the pesky "1" that shows up on the RHS of (4-102).

4.8 Dirac Three Momentum Operator

4.8.1 Simple Deduction of Dirac Momentum Operator

Similar to what we did in Sect. 4.7.1, we can simply define our <u>Dirac 3-momentum operator</u> as

$$\mathbf{P} = \sum_{r,\mathbf{p}} \mathbf{p}\left(N_r(\mathbf{p}) + \bar{N}_r(\mathbf{p})\right), \qquad (4\text{-}105)$$

The spinor 3-momentum operator

since operation by (4-105) on any multiparticle state will yield a sum of the $\mathbf{p}$ values for all the single particles in that state. Do Prob. 26 as an example.

4.8.2 Dirac Three-Momentum Operator from the Conjugate Momentum

One can derive the 3-momentum operator more formally using the relation between 3-momentum density and conjugate momentum of the field from Box 2-2 (pg. 23) of Chap.2, i.e., (with sum on n index)

$$\not{p}^i = -\pi_n \frac{\partial \phi^n}{\partial x^i} \qquad n = 1, \phi^1 = \psi \; ; \; n = 2, \phi^2 = \bar{\psi}. \qquad (4\text{-}106)$$

Deriving the spinor 3-momentum operator

Then,

$$P^i = \int \not{p}^i d^3x = \underbrace{\sum_{r,p^i} p^i\left(N_r\left(p^i\right) + \bar{N}_r\left(p^i\right)\right)}_{\text{derived from field equation solutions}} \;\; \rightarrow \;\; \mathbf{P} = \sum_{r,\mathbf{p}} \mathbf{p}\left(N_r(\mathbf{p}) + \bar{N}_r(\mathbf{p})\right). \quad (4\text{-}107)$$

The derivation of the expression after the second equal sign above is similar to the kind of thing we have done before for H and Q, and we won't do it here. If you have trouble simply accepting (4-105), then please do Prob. 27, which asks you to derive (4-107).

4.9 Dirac Spin Operator in QFT

4.9.1 Deduction of Dirac QFT Spin Operator

From what we found for the Hamiltonian, charge and 3-momentum operators, one might guess that the Dirac spin operator in QFT would be

$$\text{(try assuming)} \;\; _{\text{QFT}}\Sigma = \Sigma \sum_{r,\mathbf{p}}\left(N_r(\mathbf{p}) + \bar{N}_r(\mathbf{p})\right) \;\; \rightarrow \;\; _{\text{QFT}}\Sigma_i = \Sigma_i \sum_{r,\mathbf{p}}\left(N_r(\mathbf{p}) + \bar{N}_r(\mathbf{p})\right), \quad (4\text{-}108)$$

Guessing the QFT spinor spin operator

where $\Sigma = \Sigma_i$ is the spin operator of RQM (See (4-39)). Thus, action of (4-108) for $i = 3$ on a sample state would probably give us the total spin in the z direction of the state, e.g.,

$$_{\text{QFT}}\Sigma_3 \left|\psi_{up,\mathbf{p}_1}, \psi_{down,\mathbf{p}_2}, \bar{\psi}_{up,\mathbf{p}_1}\right\rangle = \Sigma_3 \sum_{r,\mathbf{p}}\left(N_r(\mathbf{p}) + \bar{N}_r(\mathbf{p})\right)\left|\psi_{up,\mathbf{p}_1}, \psi_{down,\mathbf{p}_2}, \bar{\psi}_{up,\mathbf{p}_1}\right\rangle$$

$$= (1+1+1)\underbrace{\Sigma_3 \left|\psi_{up,\mathbf{p}_1}, \psi_{down,\mathbf{p}_2}, \bar{\psi}_{up,\mathbf{p}_1}\right\rangle}_{\text{but not a defined operation}}. \qquad (4\text{-}109)$$

But as noted in the underbracket above, our RQM spin operator is not defined for a multiparticle state. So, it is not evident how Σ_3 would act on states as found in QFT.

The guess isn't defined for multi-particle states

But what we can do is define our <u>QFT Dirac spin operator</u> as

$$_{\text{QFT}}\Sigma_i = \int_V \psi^\dagger \Sigma_i \psi\, d^3x \;\; \rightarrow \;\; _{\text{QFT}}\Sigma_3 = \int_V \psi^\dagger \Sigma_3 \psi\, d^3x, \qquad (4\text{-}110)$$

Definition of QFT spinor spin operator that will work

and note what we get when we substitute the general solutions to the Dirac equation for ψ and $\psi^\dagger$.

<u>For Type c Particles Only</u>

For simplicity, we will restrict ourselves to c type particles only, and later extend our results to include d types, as well. From (4-110), with the preceding superscript c to designate c type,

Examining our definition of QFT spinor spin operator

$$\prescript{c}{}{}_{\text{QFT}}\Sigma_3 = \int_V \left(\sum_{r,\mathbf{p}} \sqrt{\frac{m}{VE_\mathbf{p}}} c_r^\dagger(\mathbf{p}) u_r^\dagger(\mathbf{p}) e^{ipx} \right) \Sigma_3 \left(\sum_{s,\mathbf{p}'} \sqrt{\frac{m}{VE_{\mathbf{p}'}}} c_s(\mathbf{p}') u_s(\mathbf{p}') e^{-ip'x} \right) d^3x . \quad (4\text{-}111)$$

As we should be getting used to by now, all terms where $\mathbf{p} \neq \mathbf{p}'$ will go to zero in the integration, giving us

$$\prescript{c}{}{}_{\text{QFT}}\Sigma_3 = \left(\sum_{r,s,\mathbf{p}} \frac{m}{E_\mathbf{p}} c_r^\dagger(\mathbf{p}) c_s(\mathbf{p}) u_r^\dagger(\mathbf{p}) \Sigma_3 u_s(\mathbf{p}) \right) \frac{1}{V} \int_V d^3x = \left(\sum_{r,s,\mathbf{p}} \frac{m}{E_\mathbf{p}} u_r^\dagger(\mathbf{p}) \Sigma_3 u_s(\mathbf{p}) c_r^\dagger(\mathbf{p}) c_s(\mathbf{p}) \right). \quad (4\text{-}112)$$

For a single particle state of spin s, the c operators will destroy that state, then create ones of spins r, i.e.,

$$\prescript{c}{}{}_{\text{QFT}}\Sigma_3 \left| \psi_{s,\mathbf{p}} \right\rangle = \left(\sum_{r,s',\mathbf{p}'} \frac{m}{E_{\mathbf{p}'}} u_r^\dagger(\mathbf{p}') \Sigma_3 u_{s'}(\mathbf{p}') c_r^\dagger(\mathbf{p}') c_{s'}(\mathbf{p}') \right) \left| \psi_{s,\mathbf{p}} \right\rangle = \sum_r \frac{m}{E_\mathbf{p}} \underbrace{\left(u_r^\dagger(\mathbf{p}) \Sigma_3 u_s(\mathbf{p}) \right)}_{\text{a number}} \left| \psi_{r,\mathbf{p}} \right\rangle . \quad (4\text{-}113)$$

So, the expectation value of what we would measure for spin in the z direction for the given state with s spin would be

$$\left\langle \psi_{s,\mathbf{p}} \left| \prescript{c}{}{}_{\text{QFT}}\Sigma_3 \right| \psi_{s,\mathbf{p}} \right\rangle = \left\langle \psi_{s,\mathbf{p}} \left| \left(\sum_r \frac{m}{E_\mathbf{p}} u_r^\dagger(\mathbf{p}) \Sigma_3 u_s(\mathbf{p}) \left| \psi_{r,\mathbf{p}} \right\rangle \right) \right. = \sum_r \frac{m}{E_\mathbf{p}} \left\langle \psi_{s,\mathbf{p}} \left| (\text{a number}) \right| \psi_{r,\mathbf{p}} \right\rangle$$

$$= 0 \text{ for } r \neq s ; \qquad = \underbrace{\frac{m}{E_\mathbf{p}} u_r^\dagger(\mathbf{p}) \Sigma_3 u_r(\mathbf{p})}_{\text{the number}} \text{ for } r = s . \quad (4\text{-}114)$$

Hence the only way we can measure anything corresponding to this operator is if $r = s$. So, with no loss in measurement prediction capability (nor thus, in generality), we can drop terms in (4-112) that result in zero expectation values, leaving us with

$$\prescript{c}{}{}_{\text{QFT}}\Sigma_3 = \sum_{r,\mathbf{p}} \frac{m}{E_\mathbf{p}} u_r^\dagger(\mathbf{p}) \Sigma_3 u_r(\mathbf{p}) c_r^\dagger(\mathbf{p}) c_r(\mathbf{p}) = \sum_{r,\mathbf{p}} \frac{m}{E_\mathbf{p}} u_r^\dagger(\mathbf{p}) \Sigma_3 u_r(\mathbf{p}) N_r(\mathbf{p}) . \quad (4\text{-}115)$$

Now if $\left| \psi_{r,\mathbf{p}} \right\rangle$ is in a z direction eigenstate, say the up state, then (4-115) acting on it yields

$$\prescript{c}{}{}_{\text{QFT}}\Sigma_3 \left| \psi_{up,\mathbf{p}_1} \right\rangle = \left(\sum_{r,\mathbf{p}} \frac{m}{E_\mathbf{p}} u_r^\dagger(\mathbf{p}) \Sigma_3 u_r(\mathbf{p}) N_r(\mathbf{p}) \right) \left| \psi_{up,\mathbf{p}_1} \right\rangle = \frac{m}{E_{\mathbf{p}_1}} u_{up}^\dagger(\mathbf{p}_1) \underbrace{\Sigma_3 u_{up}(\mathbf{p}_1)}_{\frac{1}{2} u_{up}(\mathbf{p}_1)} \left| \psi_{up,\mathbf{p}_1} \right\rangle$$

Our definition of spin operator seems to work nicely for Σ_3

$$= \frac{1}{2} \frac{m}{E_{\mathbf{p}_1}} \underbrace{u_{up}^\dagger(\mathbf{p}_1) u_{up}(\mathbf{p}_1)}_{E_{\mathbf{p}_1}/m} \left| \psi_{up,\mathbf{p}_1} \right\rangle = \frac{1}{2} \left| \psi_{up,\mathbf{p}_1} \right\rangle . \quad (4\text{-}116)$$

And the eigenvalue of our operator is ½, representing an up eigenstate.

The key to all this is that the Σ_3 operator in (4-110) operates on the column spinor of ψ in (4-110), and the destruction and creation operators $c_r(\mathbf{p})$ and $d^\dagger_r(\mathbf{p})$ associated with that column spinor operate on the ket in question. So, the effect we get on a multiparticle state, due to the number operators $N_r(\mathbf{p})$ [see (4-115), applied in (4-116) to only a single particle state], is what we desired. i.e., the total spin in the i direction from all particles in the multiparticle state.

All of the above steps can be repeated analogously for Σ_1 and Σ_2 to yield the general result

It also works nicely in general, i.e. Σ_i works

$$\prescript{c}{}{}_{\text{QFT}}\Sigma_i = \sum_{r,\mathbf{p}} \frac{m}{E_\mathbf{p}} u_r^\dagger(\mathbf{p}) \Sigma_i u_r(\mathbf{p}) N_r(\mathbf{p}) . \quad (4\text{-}117)$$

For Both Type c and d Particles

The same steps we went through for c particles and fields above work for d types as well. Thus,

$$_{\text{QFT}}^{\ \ d}\Sigma_i = \left(\sum_{r,\mathbf{p}} \frac{m}{E_\mathbf{p}} v_r^\dagger(\mathbf{p}) \Sigma_i v_r(\mathbf{p}) \overline{N}_r(\mathbf{p}) \right). \tag{4-118}$$

Same operator works for d type particles, as well

Carrying out similar steps for both types yields the <u>QFT spin operator in terms of number operators</u>,

$$_{\text{QFT}}\Sigma_i = {}_{\text{QFT}}^{\ \ c}\Sigma_i + {}_{\text{QFT}}^{\ \ d}\Sigma_i = \sum_{r,\mathbf{p}} \frac{m}{E_\mathbf{p}} \left(u_r^\dagger(\mathbf{p}) \Sigma_i u_r(\mathbf{p}) N_r(\mathbf{p}) + v_r^\dagger(\mathbf{p}) \Sigma_i v_r(\mathbf{p}) \overline{N}_r(\mathbf{p}) \right). \tag{4-119}$$

Spin operator in terms of number operators

4.9.2 Formal Derivation of Dirac QFT Spin Operator

Angular momentum in relativity is, strictly speaking, a 4D skew symmetric tensor (the purely spatial components of which can be employed as the three components of a 3D pseudo-vector). That angular momentum tensor can be converted into an angular momentum density tensor, which can be quantized. This gets quite complicated, and we won't do it here, but will note that upon doing so, one finds the quantum field theory spin operator to be (4-110). And thus, with a quite formal derivation from general principles, one ends up with (4-119).

Spin operator can be derived more formally

4.10 QFT Helicity Operator

As noted before, helicity is simply the component of the spin along $\mathbf{p}$, i.e., the dot product of the 3D spin vector and the unit vector in the 3-momentum direction. So, the <u>QFT helicity operator</u> is

$$_{\text{QFT}}\Sigma_\mathbf{p} = \int \psi^\dagger \left(\Sigma \cdot \frac{\mathbf{p}}{|\mathbf{p}|} \right) \psi\, d^3x = \int \psi^\dagger \left(\Sigma_i \frac{p^i}{p} \right) \psi\, d^3x . \tag{4-120}$$

QFT helicity operator in terms of fields

In <u>terms of number operators</u>, from (4-119), this becomes

$$_{\text{QFT}}\Sigma_\mathbf{p} = \sum_{r,\mathbf{p}} \frac{m}{E_\mathbf{p}} \left(u_r^\dagger(\mathbf{p}) \Sigma_i \frac{p^i}{p} u_r(\mathbf{p}) N_r(\mathbf{p}) + v_r^\dagger(\mathbf{p}) \Sigma_i \frac{p^i}{p} v_r(\mathbf{p}) \overline{N}_r(\mathbf{p}) \right). \tag{4-121}$$

QFT helicity operator in terms of number operators

4.11 Odds and Ends

4.11.1 Inner and Outer Products in Spinor Space

For products of two fields, when the adjoint field is on the left and spinor indices are suppressed, an inner product is implied. Thus, where, as always, repeated indices mean summation,

$$\overline{\psi}\psi = \overline{\psi}_\beta \psi_\beta = \psi_\alpha^\dagger \gamma_{\alpha\beta}^0 \psi_\beta = \text{ a scalar quantity} . \tag{4-122}$$

Inner product of spinor fields = a number

When the adjoint field is on the right, an outer product (a tensor/matrix) is implied. For example,

$$\psi\overline{\psi} = \psi_\alpha \overline{\psi}_\beta = \psi_\alpha \psi_\delta^\dagger \gamma_{\delta\beta}^0 = X_{\alpha\beta} = \text{ a matrix quantity in spinor space} \tag{4-123}$$

Outer product of spinor fields = a 4X4 matrix

For spinor field anti-commutators, which for us, are almost always outer products, we mean

$$\left[\psi, \overline{\psi}\right]_+ = \left[\psi, \overline{\psi}\right]_{+\alpha\beta} = \psi_\alpha \overline{\psi}_\beta + \overline{\psi}_\beta \psi_\alpha = \left[\overline{\psi}, \psi\right]_+ = \left[\overline{\psi}, \psi\right]_{+\alpha\beta} \tag{4-124}$$

If these rules are broken under any special circumstances, we will write out the indices.

4.11.2 Dirac Matrices and Spinor Relations

There are a number of ways Dirac matrices and spinors may be manipulated, and many of these will prove convenient along the way, particularly when we get to interaction theory. One of these we have already seen: the hermiticity conditions (4-12).

Proving each and every one of these can eat up a lot of time, is not something I, the author, have ever done, and is not recommended. Proving one or two of them to yourself can, however, give you some confidence in all of them.

Appendix A of this chapter lists many of these relationships. Prob. 31 asks you to prove one of them. The rest can be treated similarly to the way we treat integral tables. I don't know anyone who has ever proven all of the relations in any integral table to her/himself, but every physicist I know commonly uses them with confidence.

Many ways to manipulate Dirac matrices summarized in Appendix A

4.11.3 Subtleties Regarding Spin

Even if you have worked hard to understand Dirac spinors, their dependence on 3-momentum, and the spin operator, you probably still feel a little insecure (a little "fuzzy") about them. If you don't, you probably either i) are a very rare student or ii) have not reflected too deeply on them.

Extant texts typically do far less to explain them than we have in this chapter, and seem to ignore key issues and questions, the analysis and answers for which would be highly enlightening.

The First Issue: General (Non-pure u_1 or u_2) States

One such example is how, in QFT, one might treat a particle that is not in a pure u_1 or u_2 state. Typically, one would expect the overwhelming majority of particles treated to not have their momenta and spin directions so perfectly aligned as to be representable by only one of u_1 or u_2. Yet in QFT, one almost invariably deals with kets such as $\left|\psi_{r,\mathbf{p}}\right\rangle$ or $\left|\psi_{r,\mathbf{p}},\psi_{r',\mathbf{p}'},\psi_{r'',\mathbf{p}''}\right\rangle$, for which, it is seemingly implied, the spin values r, r', and r'' equal 1 or 2. These are spin basis states, so it seems these kets cannot represent more general (non-basis) states.

How to handle non eigen spin states?

In interaction theory applications, for decay, the particle is analyzed at rest. And we can simply align the z axis of our chosen reference frame with the spin axis of the particle at rest. No problem.

In scattering, particles are typically high energy (approaching light speed) and so have their 3-momenta effectively aligned with their spins. So, we can simply choose our z axis for a given particle in that direction and things get simplified, because such a particle must then be in either a spin up u_1 state or a spin down u_2 state. Still further, in scattering, calculations are highly statistical, and certain averaging processes over all possible spin states take place. Those cover over and automatically account for more general, non-pure u_1 or u_2 spin basis states.

And yet, the theory still seems a little lacking, if, as it might seem, it only handles pure spin basis states.

The Second Issue: Multiple Particles and Different At-rest Systems

The solutions to the Dirac equation are expressed in terms of the spinors u_1, u_2, v_1 and v_2, but, it is important to realize that the p^i values used in the usual expressions (4-20) must be expressed in the at-rest coordinate system of the particle (decelerate the particle to rest in the lab) for which the lab z axis is taken parallel to the spin of the decelerated particle.

u_r and v_r expressed in at-rest system of each particular particle, not all in the same coord system

But if we have a multiparticle state, and we take our coordinate system as the at-rest system of one particle, this will usually not be the same as the at-rest system of any of the other particles. They will, in general, have different alignments of the z axis. So, we can't use the same coordinate system for all particles, or we won't represent the spin of those other particles correctly. But using more than one coordinate system makes computation a lot more difficult, of course.

The Full Resolutions of These Questions in Appendix B

Virtually everyone learns QFT without considering the second issue above, and you can certainly do so, as well. The first issue is contained (via spin averaging) in most treatments, but often unclearly. For those who really wish to get to the bottom of these matters, I have covered them in greater depth in Appendix B of this chapter. A summary of that appendix follows below.

Summary of Appendix B

The First Issue

1. Feynman diagrams in the literature may label a Dirac particle with given $\mathbf{p}$ and spin of subscript r. That implies r = either 1 or 2, and thus the state should be a pure basis state. But particles are rarely in a pure basis state, so the symbolism seems incomplete and unable to represent the full range of states in nature.

Summary of resolution of first issue

2. In the ket symbolism of this text, we will represent a general, not necessarily pure basis, state as $|\psi_{\mathbf{p}}\rangle$ and a pure basis state as $|\psi_{r,\mathbf{p}}\rangle$. In Appendix B, we show how the mathematical predictions for observables for $|\psi_{\mathbf{p}}\rangle$ work out.

3. In experiments, spin is typically not measured, but we may calculate a transition amplitude (see Chap. 1, pg. 3) for a given interaction for a particular basis state spin (r = 1 or 2) for each particle, and then use such basis state spin results to find general state results. Typically, this entails an averaging over all incoming particles of all possible spin states. Recall that quantum predictions

(expectation values, scattering, decay) of measurements are statistical in nature, so such averages work well.

The Second Issue

1. The usual standard representation forms of spinors u_1 and u_2 only represent actual spin if the particle at-rest coordinate system has spin parallel to the z axis direction. That is, p^i used in the usual forms of u_1 and u_2 must be measured in the at-rest system.

Summary of resolution of second issue

2. If we have two (or more) particles in a multiparticle state, each typically has a different at-rest coordinate system. So, we can either i) represent each particle with usual forms for u_1 and u_2 using p^i values for each measured in its own (different from the others) at-rest system, and thereby make computation very difficult, or ii) represent all particles by the usual forms of u_1 and u_2 using p^i values for all measured in the same coordinate system, but then, at best, only one particle would have its actual spin represented correctly.

3. We are typically concerned about expectation values, and such values are the same whether we use 2i) or 2ii). 2ii) is easier computationally because everything is done in the same coordinate system. It works because in taking expectation values, the "incorrectness" in spin representation in the ket is canceled by the "incorrectness" in spin representation in the bra.

4.11.4 Eigenstates of Three-Momentum Not So Problematic as Spin

Note that we can get away with considering only **p** eigenstate (basis state) particles in our kets, even though every free, real world particle is actually a wave packet (a general particle state which is an integral over different **p** eigenstates). This is because, in many cases, including those commonly treated in QFT, a particle eigenstate of certain **p** value (every measurement of 3-momentum would yield **p**) is an excellent approximation to a wave packet with 3-momentum expectation value **p** (the average of many measurements would be **p**).

p *eigenstates excellent approx of wave packet with* **p** *expectation value*

4.12 The Spinor Feynman Propagator

We will follow similar steps to derive the Feynman propagator for Dirac fields as we did for scalar fields in Chap. 3 (pgs. 70-78). Note that, as before, Wholeness Chart 5-4 at the end of Chap. 5 breaks these steps out clearly, and should be used as an aid when studying the spinor propagator derivation.

4.12.1 The Approach

Recall that the first part of QFT, including this chapter, deals with free particle/fields (no interactions). When we later introduce interactions, a mathematical relationship called the Feynman propagator will arise, which corresponds, physically, to a virtual particle that mediates force, i.e., it carries energy, momentum, and often charge from one real particle to another. It will make things easier in the long run, however, if we derive the Feynman propagator for spinors now, rather than when we get to interactions.

We'll use the spinor Feynman propagator when we get to interaction theory

But it's easier in the long run if we derive it here

Similar to what we did with scalars, we will, heuristically, consider the operator field $\bar{\psi}(y)$ to create a virtual Dirac particle at event y, and $\psi(x)$ to destroy that virtual particle at event x. The spinor propagator incorporates these two field operators.

Can heuristically think of the propagator as creating a virtual particle at y and destroying it at x

Note, that as we showed with scalars, while this "creation/destruction at a point" perspective helps in understanding the derivation of the propagator, the propagator really corresponds to a kind of probability density function in y and x. It represents the probability density (actually, the square of its magnitude represents probability density, though it is a bit more complicated as other factors are eventually involved) of a Dirac particle appearing at y and disappearing at x. It is a double density in that it is a function of both y and x, two independent variables, rather than one.

But it really correlates with a probability density in x and y

4.12.2 Milestones in the Derivation

We proceed in five distinct steps, precisely parallel to those we followed in Chap. 3 for scalars (pgs. 70-78). The entire derivation is for continuous (not discrete) eigenstate solutions of the field equation (Dirac equation here), as the propagator is not confined to a volume V. We represent the spinor Feynman propagator with the symbol $iS_F(x-y)$. This actually turns out to be a 4X4 matrix in spinor space, which should not be too surprising, since the Dirac equation and its solutions live in

Start with physical visualization of the propagator & follow 5 steps

Spinor propagator is a 4X4 *matrix*

4D spinor space. So, we also will, at times, use the symbol $iS_{F\alpha\beta}(x-y)$ for the spinor propagator where the α and β subscripts range from 1 to 4 and represent the matrix components.

Step 1: Math interpretation of the physical propagator

Step 1: Express the Feynman propagator iS_F as a mathematical representation of a spinor or anti-spinor created at one point in space and time in the vacuum and destroyed at another place and time.

Step 2: Express iS_F in terms of two anti-commutators (one for particles and one for anti-particles).

Steps 2, 3, 4: Math manipulation of relations for particle and anti-particle

Step 3: Express those two anti-commutators as real integrals.

Step 4: Re-express those two real integrals as two contour (complex plane) integrals.

Step 5: Re-express the two contour integrals as a single real integral, the form most suitable for analysis.

Step 5: Combining two complex integrals into a single real one.

4.12.3 The Derivation

Step 1: The Feynman Propagator as the VEV of a Time Ordering Operator

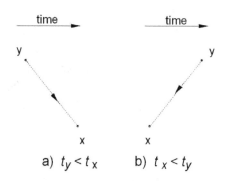

Figure 4-4. Creation & Destruction of Virtual Spinor/Anti-spinor

Fig. 4-4a represents creation of a virtual spinor at y and destruction of it at x. Figure 4-4b represents creation of a virtual anti-spinor at x and destruction of it at y. As explained in Chap. 3, we need a relationship for the propagator that includes both scenarios as possibilities.

That is, we need an operator that will create a particle first if $t_y < t_x$, but create an antiparticle first if $t_x < t_y$. Our Dirac equation solutions provide the means for the desired creation and destruction operations, but these have to be arranged to provide us with the time ordering dependence of Fig. 4-4. As we did with scalars, we consider the time ordering operator T, for spinors defined as follows.

Step 1, first part, defining the time ordering operator T and seeing how it represents creation of either a virtual particle or antiparticle followed by its destruction

If $t_y < t_x$, $\bar\psi(y)$ operates first (creates a particle) and is placed on the right, with $\psi(x)$ operating second (destroys the particle) and placed on the left.

$$\text{for } t_y < t_x \qquad T\{\psi(x)\bar\psi(y)\} = \psi(x)\bar\psi(y). \tag{4-125}$$

If $t_x < t_y$, $\psi(x)$ operates first (creates an antiparticle) and is placed on the right with $\bar\psi(y)$ operating second (destroying the antiparticle) and placed on the left.

$$\text{for } t_x < t_y \qquad T\{\psi(x)\bar\psi(y)\} = -\bar\psi(y)\psi(x). \tag{4-126}$$

where we note the minus sign on the right-hand side. This sign is arbitrary, as either a plus or minus will still result in an antiparticle created at x and destroyed at y. But choosing it as a minus here in our definition will make things work out to be more compact mathematically later.

Step 1, second part, defining the transition amplitude as equal to the VEV of the time ordering operator T

As with scalars, the transition amplitude equals the vacuum expectation value (VEV) of the above time ordering operator T, i.e.,

$$\langle 0|T\{\psi(x)\bar\psi(y)\}|0\rangle, \tag{4-127}$$

which, similar to that in Chap. 3, represents both possible scenarios of Fig. 4-4.

The VEV of T is the expectation of finding a virtual particle traveling in the vacuum

To gain insight into (4-127), review step 1 in the scalar propagator derivation of Chap. 3, pgs. 71-73 . This insight allows us to define our mathematical relationship for the processes shown in Fig. 4-4 as the VEV of the time ordering operator T. This is the spinor Feynman propagator S_F (where we include an extra factor of i because it makes the derivation simpler later on),

$$\boxed{iS_F(x-y) = \langle 0|T\{\psi(x)\bar\psi(y)\}|0\rangle}. \tag{4-128}$$

Redefine the transition amplitude as the Feynman propagator

Note that the RHS of (4-128) is an outer product in spinor space and thus the LHS is a matrix. We can write out the spinor indices α and β to make this clear,

$$iS_{F\alpha\beta}\left(x-y\right) = \left\langle 0\left|T\left\{\psi_\alpha\left(x\right)\bar{\psi}_\beta\left(y\right)\right\}\right|0\right\rangle \ . \tag{4-129}$$

We will write out spinor indices α and β when it aids in understanding

We will write out the spinor indices when it helps to illuminate a derivation, but otherwise will follow the common custom of suppressing them.

<u>Step 2: Expressing iS_F in Terms of Anti-Commutators</u>

Similar to step 2 of the propagator derivation in Chap. 3 (pg. 73), for $t_y < t_x$, the case for a virtual scalar (not anti-scalar), where the Feynman scalar propagator was expressed in terms of commutators, we can here express the <u>virtual spinor propagator</u> in terms of anti-commutators. To do this, first note that

Step 2, expressing Feynman propagator in terms of anti-commutators

$$iS_F\left(x-y\right) = \left\langle 0\left|T\left\{\psi\left(x\right)\bar{\psi}\left(y\right)\right\}\right|0\right\rangle = \left\langle 0\left|\psi\left(x\right)\bar{\psi}\left(y\right)\right|0\right\rangle \quad \text{if } t_y < t_x \ \text{(particle)}$$

$$= \underbrace{\left\langle 0\left|\psi^+\left(x\right)\bar{\psi}^+\left(y\right)\right|0\right\rangle}_{=0} + \underbrace{\left\langle 0\left|\psi^+\left(x\right)\bar{\psi}^-\left(y\right)\right|0\right\rangle}_{=(factor)|\psi\rangle} + \underbrace{\left\langle 0\left|\psi^-\left(x\right)\bar{\psi}^+\left(y\right)\right|0\right\rangle}_{=0} + \underbrace{\left\langle 0\left|\psi^-\left(x\right)\bar{\psi}^-\left(y\right)\right|0\right\rangle}_{=(factor)|\psi\rangle} \tag{4-130}$$

$$= \underbrace{\left\langle 0\left|\psi^+\left(x\right)\bar{\psi}^-\left(y\right)\right|0\right\rangle}_{=(factor)|0\rangle} + \underbrace{\left\langle 0\left|\psi^-\left(x\right)\right|\psi\right\rangle}_{=(factor)\left\langle 0|\bar{\psi}\psi\right\rangle=0} = \left\langle 0\left|\psi^+\left(x\right)\bar{\psi}^-\left(y\right)\right|0\right\rangle \ ,$$

Step 2, first part, expressing Feynman propagator for virtual spinor (not anti-spinor) in terms of an anti-commutator

where "factor" represents the non-operator quantities in each field operator term that are left unchanged when the creation and destruction coefficient operators act on a ket. To the last part of (4-130), we can add zero in the form of

$$0 = \left\langle 0\left|\bar{\psi}^-\left(y\right)\underbrace{\psi^+\left(x\right)}_{=0}\right|0\right\rangle . \tag{4-131}$$

By adding a term equal to zero, we can use an anti-commutator

And thus, (4-130) becomes (with spinor indices written out)

$$iS_{F\alpha\beta}\left(x-y\right) = \left\langle 0\left|\psi_\alpha^+\left(x\right)\bar{\psi}_\beta^-\left(y\right) + \bar{\psi}_\beta^-\left(y\right)\psi_\alpha^+\left(x\right)\right|0\right\rangle = \left\langle 0\left|\left[\psi_\alpha^+\left(x\right),\bar{\psi}_\beta^-\left(y\right)\right]_+\right|0\right\rangle \tag{4-132}$$

In similar fashion, for $t_x < t_y$, the case for <u>a virtual anti-spinor propagator</u>, one finds

$$iS_{F\alpha\beta}\left(x-y\right) = -\left\langle 0\left|\left[\bar{\psi}_\alpha^+\left(y\right),\psi_\beta^-\left(x\right)\right]_+\right|0\right\rangle . \tag{4-133}$$

Step 2, second part, expressing Feynman propagator for virtual anti-spinor in terms of an anti-commutator

In summary for Step 2, we have shown the spinor Feynman propagator can be expressed in terms of anti-commutators (which are outer products with spinor indices suppressed below) as

$$iS_F\left(x-y\right) = \left\langle 0\left|\left[\psi^+\left(x\right),\bar{\psi}^-\left(y\right)\right]_+\right|0\right\rangle \quad \text{if } t_y < t_x \ \text{(virtual spinor)}$$

$$= -\left\langle 0\left|\left[\bar{\psi}^+\left(y\right),\psi^-\left(x\right)\right]_+\right|0\right\rangle \quad \text{if } t_x < t_y \ \text{(virtual anti-spinor)}. \tag{4-134}$$

Step 2, summary

<u>Step 3: Expressing Anti-Commutator Forms of iS_F as Integrals</u>

Define the symbol iS^+ as the anti-commutator of the field type c solutions (for spinors, not anti-spinors) of the first line of (4-134), i.e.,

$$iS_{\alpha\beta}^+\left(x-y\right) = \left[\psi_\alpha^+\left(x\right),\bar{\psi}_\beta^-\left(y\right)\right]_+ = \left[\psi^+\left(x\right),\bar{\psi}^-\left(y\right)\right]_{+\alpha\beta} \ . \tag{4-135}$$

Step 3, first part, express iS^+ = anti-commutation relation for type c fields as an integral

The solutions used on the RHS above are the integral (continuous) form for the Dirac equation solutions (4-59). It is common usage to use a $+$ superscript on the LHS of (4-135) to designate c type fields, rather than the letter c, which would be easier to remember. Just think "c type field" when you see $+$ (and remember that c types create and destroy particles, not antiparticles). Equation (4-135) is thus

$$iS^+_{\alpha\beta}(x-y)$$

$$= \frac{m}{(2\pi)^3} \sum_r \sum_s \iint \left(c_r(\mathbf{p})c_s^\dagger(\mathbf{p}')u_{r\alpha}(\mathbf{p})\bar{u}_{s\beta}(\mathbf{p}') + c_s^\dagger(\mathbf{p}')c_r(\mathbf{p})\underbrace{\bar{u}_{s\beta}(\mathbf{p}')u_{r\alpha}(\mathbf{p})}_{=u_{r\alpha}(\mathbf{p})\bar{u}_{s\beta}(\mathbf{p}')} \right) \frac{e^{-ipx}e^{ip'y}}{\sqrt{E_\mathbf{p}E_{\mathbf{p}'}}} d^3\mathbf{p}\,d^3\mathbf{p}'$$

$$= \frac{m}{(2\pi)^3} \sum_r \sum_s \iint \underbrace{\left[c_r(\mathbf{p}),c_s^\dagger(\mathbf{p}') \right]_+}_{\delta_{rs}\delta(\mathbf{p}-\mathbf{p}')} \underbrace{u_r(\mathbf{p})\bar{u}_s(\mathbf{p}')}_{\substack{\text{outer product, 4X4} \\ \text{matrix in } \alpha \text{ and } \beta}} \frac{e^{-ipx}e^{ip'y}}{\sqrt{E_\mathbf{p}E_{\mathbf{p}'}}} d^3\mathbf{p}\,d^3\mathbf{p}' \qquad (4\text{-}136)$$

$$= \frac{m}{(2\pi)^3} \int \underbrace{\sum_r u_r(\mathbf{p})\bar{u}_r(\mathbf{p})}_{\substack{=\frac{\not{p}+m}{2m}\text{(Appendix A)}}} \frac{e^{-ip(x-y)}}{E_\mathbf{p}} d^3\mathbf{p} = \frac{1}{2(2\pi)^3} \int \underbrace{\left(\not{p}+m \right)}_{\substack{\alpha,\beta \text{ indices} \\ \text{suppressed}}} \frac{e^{-ip(x-y)}}{E_\mathbf{p}} d^3\mathbf{p},$$

where in the second line, since the index labeling means we take α first (row index) and β second (column index), we take $\bar{u}_{s\beta}u_{r\alpha} = u_{r\alpha}\bar{u}_{s\beta} = \left(u_r\bar{u}_s\right)_{\alpha\beta}$. In the last line we use relation (4-153) from Appendix A.

Similarly, where a minus sign superscript stands for d type fields (since they are associated with antiparticles, the minus makes some sense), and we use (4-154) of Appendix A,

Step 3, second part, express $iS^- =$ anti-commutation relation for type d fields as an integral

$$iS^-_{\alpha\beta}(x-y) = -\left[\bar{\psi}^+_\beta(y),\psi^-_\alpha(x) \right]_+ = \frac{-1}{2(2\pi)^3}\int \underbrace{\left(\not{p}-m \right)}_{\substack{\alpha,\beta \text{ indices} \\ \text{suppressed}}} \frac{e^{ip(x-y)}}{E_\mathbf{p}} d^3\mathbf{p} \qquad (4\text{-}137)$$

Note that though our earlier expressions for iS_F and $iS^\pm$, contained operators that operated on the ket part of the VEV in (4-134), because the anti-commutator of these operators in (4-136) (and similarly, in (4-137)) is a number, $iS^\pm$ are simply (matrix) numbers, not operators. Since the expectation value of a number is a number, iS_F of (4-134) is only that, a number (matrix number). (To be precise it is a numeric spinor space matrix *function*. But it is *not* an *operator* matrix function.) The bottom line: We don't have to worry about operators, their effects, or VEV brackets any more, but can simply evaluate the Feynman propagator iS_F as a numeric mathematical relation.

Our VEV of operators expression of the propagator has become a simple numeric function

It will help us in step 4, if we re-write (4-136) and (4-137), as (α, β suppressed)

Re-writing iS^+ and iS^- to help in next step

$$iS^+(x-y) = \left(i\gamma^\mu\frac{\partial}{\partial\left(x^\mu-y^\mu\right)} + m \right)\underbrace{\frac{1}{2(2\pi)^3}\int \frac{e^{-ip(x-y)}}{E_\mathbf{p}} d^3\mathbf{p}}_{i\Delta^+(x-y)}, \qquad (4\text{-}138)$$

$$iS^-(x-y) = \left(i\gamma^\mu\frac{\partial}{\partial\left(x^\mu-y^\mu\right)} + m \right)\underbrace{\frac{1}{2(2\pi)^3}\int \frac{e^{ip(x-y)}}{E_\mathbf{p}} d^3\mathbf{p}}_{i\Delta^-(x-y)}, \qquad (4\text{-}139)$$

where the $\Delta^\pm$ are taken from the Chap. 3 derivation of the scalar propagator, step 3 (pgs. 74), with $\mathbf{k} \to \mathbf{p}$, $\omega \to E$, $k \to p$. We can thus write both of the above anti-commutation relations as

Step 3, final part, combine above parts into one symbol $iS^\pm$

$$iS^\pm(x-y) = \left(i\gamma^\mu\frac{\partial}{\partial\left(x^\mu-y^\mu\right)} + m \right)i\Delta^\pm(x-y). \qquad (4\text{-}140)$$

Step 4: Expressing the Two $iS^\pm$ Integrals in Real Space as Contour Integrals

From Chap. 3, step 4, pgs. 74-76,

Step 4, express $iS^\pm$ as a contour integral

$$i\Delta^\pm(x-y) = \underbrace{\frac{1}{2(2\pi)^3}\int \frac{e^{\mp ik(x-y)}}{\omega_k} d^3\mathbf{k}}_{\substack{\text{integral over 3-} \\ \text{momentum space}}} = \underbrace{\frac{\mp i}{(2\pi)^4}\int_{C^\pm} \frac{e^{-ik(x-y)}}{k^2-\mu^2} d^4k}_{\substack{\text{integral in 4-momentum space,} \\ \text{energy part is contour integral}}}. \qquad (4\text{-}141)$$

Thus, re-writing $iS^{\pm}$ in terms of $i\Delta^{\pm}$, as we did in (4-140), allows us to evaluate $iS^{\pm}$ using what we already know from Chap. 3, i.e., (4-141). Taking $\mathbf{k} \to \mathbf{p}$, $\omega \to E$, $k \to p$, $\mu \to m$ in (4-141) and substituting the RHS of (4-141) into (4-140), we find

$$iS^{\pm}(x-y) = \left(i\gamma^{\mu} \frac{\partial}{\partial\left(x^{\mu} - y^{\mu}\right)} + m \right) \frac{\mp i}{(2\pi)^4} \int_{C^{\pm}} \frac{e^{-ip(x-y)}}{p^2 - m^2} d^4 p$$

$$= \frac{\mp i}{(2\pi)^4} \int_{C^{\pm}} \frac{\left(\not{p} + m\right)}{p^2 - m^2} e^{-ip(x-y)} d^4 p. \tag{4-142}$$

Step 5: Re-express S_F in Most Convenient Form

We would like two things more: 1) express the propagator as a single function so we don't have to keep track (while we are integrating over spacetime and doing other things) of whether the virtual field is a particle or antiparticle (i.e., whether to use the S^+ or S^- function), and 2) have all our integrations over real numbers rather than deal with contour integrals.

Step 5, expressing Feynman propagator as integral over real, not complex space

We did this for scalars in step 5 of the scalar propagator derivation to turn the RHS of (4-141) into a single integral in real (not complex) space, i.e.,

$$\Delta_F(x-y) = \frac{1}{(2\pi)^4} \int_{-\infty}^{+\infty} \frac{e^{-ik(x-y)}}{k^2 - \mu^2 + i\varepsilon} d^4 k, \tag{4-143}$$

Yields a single integral representing both virtual particle and antiparticle, most convenient form for Feynman propagator

where $\varepsilon \ll 1$. By direct analogy (we could do virtually the same steps over again with a different integrand), we can take the last line of (4-142) to get the final result for the <u>spinor Feynman propagator</u>

$$S_F(x-y) = \frac{1}{(2\pi)^4} \int_{-\infty}^{+\infty} \frac{e^{-ip(x-y)}\left(\not{p} + m\right)}{p^2 - m^2 + i\varepsilon} d^4 p. \tag{4-144}$$

Summary spinor Feynman propagator derivation

Summary of Steps 1 to 5

Steps 1 to 4 for the virtual spin ½ particle Feynman propagator were

$$iS_F(x-y) = \langle 0| T\{\psi(x)\bar{\psi}(y)\}|0\rangle = \langle 0|\psi(x)\bar{\psi}(y)|0\rangle \quad \text{if } t_y < t_x \text{ (particle)}$$

$$= \langle 0|\underbrace{\left[\psi^+(x), \bar{\psi}^-(y)\right]_+}_{iS^+_{\alpha\beta}(x-y), \text{ a number}}|0\rangle = \left[\psi^+(x), \bar{\psi}^-(y)\right]_+ \langle 0\|0\rangle = \left[\psi^+(x), \bar{\psi}^-(y)\right]_+ \tag{4-145}$$

Steps 1 to 4

$$= iS^+_{\alpha\beta}(x-y) = \frac{1}{2(2\pi)^3} \int \left(\not{p} + m\right) \frac{e^{-ip(x-y)}}{E_{\mathbf{p}}} d^3\mathbf{p} = \frac{-i}{(2\pi)^4} \int_{C^+} \frac{\left(\not{p} + m\right) e^{-ip(x-y)}}{p^2 - m^2} d^4 p.$$

For virtual spin ½ anti-particles, they were

$$iS_F(x-y) = \langle 0| T\{\psi(x)\bar{\psi}(y)\}|0\rangle = -\langle 0|\bar{\psi}(y)\psi(x)|0\rangle \quad \text{if } t_x < t_y \text{ (anti-particle)}$$

$$= -\langle 0|\left[\bar{\psi}^+(y), \psi^-(x)\right]_+|0\rangle = -\left[\bar{\psi}^+(y), \psi^-(x)\right]_+ \langle 0\|0\rangle = -\left[\bar{\psi}^+(y), \psi^-(x)\right]_+ = -\left[\psi^-(x), \bar{\psi}^+(y)\right]_+ \tag{4-146}$$

$$= iS^-(x-y) = \frac{-1}{2(2\pi)^3} \int \left(\not{p} - m\right) \frac{e^{ip(x-y)}}{E_{\mathbf{p}}} d^3\mathbf{p} = \frac{i}{(2\pi)^4} \int_{C^-} \frac{\left(\not{p} + m\right) e^{-ip(x-y)}}{p^2 - m^2} d^4 p.$$

The two contour integrals in the last lines of (4-145) and (4-146) were combined in Step 5 to yield the single integral over real space of (4-144).

Step 5

Momentum Space Form of the Propagator

From (4-144) we can readily write down the <u>4-momentum space form of the propagator</u>, i.e., its Fourier transform, as

$$S_F(p) = \frac{\not{p} + m}{p^2 - m^2 + i\varepsilon} = \left(\not{p} + m\right)\Delta_F(p). \tag{4-147}$$

Momentum space form of the propagator

4.13 Appendix A. Dirac Matrices and Spinor Relations

We first re-state (4-12) in order to have it grouped with other such relations.

Hermiticity conditions:

$$\gamma^{\mu\dagger} = \gamma^0 \gamma^\mu \gamma^0 \,. \tag{4-148}$$

Spinor relations from Dirac equation

From the Dirac equation, we find

$$\left(i\not\partial - m\right)\psi = 0 \rightarrow \left(i\not\partial - m\right)u_r\left(\mathbf{p}\right)e^{-ipx} = 0 \rightarrow \left(\not p - m\right)u_r\left(\mathbf{p}\right)e^{-ipx} = 0\,. \tag{4-149}$$

Multiplying (4-149) by e^{ipx} yields the top LH relation of (4-150). Doing the same thing with a negative frequency eigen solution yields the top RH relation. Similar procedures with the adjoint Dirac equation yield the bottom relations in (4-150).

$$\begin{align}
\left(\not p - m\right)u_r\left(\mathbf{p}\right) = 0 \qquad & \left(\not p + m\right)v_r\left(\mathbf{p}\right) = 0 \\
\overline{u}_r\left(\mathbf{p}\right)\left(\not p - m\right) = 0 \qquad & \overline{v}_r\left(\mathbf{p}\right)\left(\not p + m\right) = 0
\end{align} \tag{4-150}$$

Most General Form of Orthonormality Conditions

By plugging and chugging the appropriate spinors and gamma matrices, one can show the <u>most general inner product spinor relation</u>

$$\boxed{\overline{u}_r\left(\mathbf{p}\right)\gamma^\mu u_s\left(\mathbf{p}\right) = \overline{v}_r\left(\mathbf{p}\right)\gamma^\mu v_s\left(\mathbf{p}\right) = \frac{p^\mu}{m}\delta_{rs}.} \tag{4-151}$$

Note that for $\mu = 0$, (4-151) becomes

$$\overline{u}_r\left(\mathbf{p}\right)\gamma^0 u_s\left(\mathbf{p}\right) = u_r^\dagger\left(\mathbf{p}\right)\gamma^0\gamma^0 u_s\left(\mathbf{p}\right) = u_r^\dagger\left(\mathbf{p}\right)u_s\left(\mathbf{p}\right) = \frac{E}{m}\delta_{rs}\,, \tag{4-152}$$

which is (4-25), a special case of (4-151). For a different case, see Prob. 28.

Spinor outer product relations

The outer product of spinors is useful in the derivation of the Feynman propagator for spinors and in scattering calculations. The first of these is

$$u_r(\mathbf{p})\overline{u}_r\left(\mathbf{p}\right) = \frac{\not p + m}{2m} = \frac{\gamma^\mu p_\mu + m}{2m} = \frac{\gamma^\mu_{\alpha\beta}p_\mu + mI_{\alpha\beta}}{2m} = u_{r\,\alpha}\left(\mathbf{p}\right)\overline{u}_{r\,\beta}\left(\mathbf{p}\right). \tag{4-153}$$

(Summation on repeated indices $r = 1,2$, $\mu = 0,1,2,3$; outer product $\rightarrow$ 4X4 matrix in α,β) Similarly,

$$v_r(\mathbf{p})\overline{v}_r\left(\mathbf{p}\right) = \frac{\not p - m}{2m} = \frac{\gamma^\mu p_\mu - m}{2m} = \frac{\gamma^\mu_{\alpha\beta}p_\mu - mI_{\alpha\beta}}{2m} = v_{r\,\alpha}\left(\mathbf{p}\right)\overline{v}_{r\,\beta}\left(\mathbf{p}\right) \tag{4-154}$$

The above two relations can be proven by simple substitution using the spinors of (4-20), the Dirac matrices of (4-10), and $p^i = -p_i$. Prob. 29 asks you to do this for (4-153).

Dirac matrices sort of like 4-vector components

From (4-9), the Dirac matrices may be written symbolically as

$$\gamma^\mu = \left(\beta, \beta\alpha_i\right) \quad \text{shorthand for} \rightarrow \quad \gamma^0 = \beta \quad \gamma^i = \beta\alpha_i\,, \tag{4-155}$$

where γ^μ has four components and seems like it might be a kind of 4-vector in physical space. We will see in Chap. 6 that, in some ways, it behaves like this, even though each of its components is a 4X4 matrix in the separate abstract spinor space. (This is not unlike the Pauli matrices σ_i of NRQM, where each of the three matrices behaves like a 3-vector component in physical space, though it is a 2X2 matrix in abstract spin space.)

Dirac matrices anti-commutation relations:

Note that

$$\gamma^1\gamma^1 + \gamma^1\gamma^1 = 2\begin{bmatrix} & & & 1 \\ & & 1 & \\ & -1 & & \\ -1 & & & \end{bmatrix}\begin{bmatrix} & & & 1 \\ & & 1 & \\ & -1 & & \\ -1 & & & \end{bmatrix} = 2\begin{bmatrix} -1 & & & \\ & -1 & & \\ & & -1 & \\ & & & -1 \end{bmatrix} = -2I = 2g^{11}I. \quad (4\text{-}156)$$

Similarly, you can show by doing Prob. 30 that $\gamma^1\gamma^2 + \gamma^2\gamma^1 = 0 = 2g^{12}$ and $\gamma^0\gamma^0 + \gamma^0\gamma^0 = 2I = 2Ig^{00}$. Doing this for every combination of Dirac matrices (or more formally for all of them symbolically as in part ii of Prob. 30), we find

$$\left[\gamma^\mu, \gamma^\nu\right]_{+\alpha\beta} = \gamma^\mu_{\alpha\gamma}\gamma^\nu_{\gamma\beta} + \gamma^\nu_{\alpha\gamma}\gamma^\mu_{\gamma\beta} = 2g^{\mu\nu}I_{\alpha\beta} \qquad \text{shorthand} \rightarrow \quad \left[\gamma^\mu, \gamma^\nu\right]_+ = 2g^{\mu\nu}$$

$$\left[\gamma^\mu, \gamma^\nu\right]_+ = \gamma^\mu\gamma^\nu + \gamma^\nu\gamma^\mu = 2Ig^{\mu\nu} \qquad\qquad\qquad\qquad\qquad (4\text{-}157)$$

$$= 2g^{\mu\nu}\begin{bmatrix} 1 & & & \\ & 1 & & \\ & & 1 & \\ & & & 1 \end{bmatrix} = 2I\begin{bmatrix} 1 & & & \\ & -1 & & \\ & & -1 & \\ & & & -1 \end{bmatrix}^{\mu\nu} \quad \begin{pmatrix} I \text{ is 4X4 spinor space identity matrix} \\ g^{\mu\nu} \text{ is 4X4 spacetime metric matrix} \end{pmatrix}.$$

Since Dirac matrices multiply via matrix multiplication, for Dirac matrices of given μ and ν numbers, the result is a 4X4 matrix in spinor space, which turns out to be the 4X4 spinor space identity matrix multiplied by either 1, –1, or 0. $g^{\mu\nu}$, which takes on these three values in the appropriate manner, is thus a good symbolic way to remember these relations.

Note that (4-157) means that gamma matrices anti-commute only when $\mu \neq \nu$ (then, $\gamma^\mu\gamma^\nu = -\gamma^\nu\gamma^\mu$).

Dirac matrices with lower indices definition:

$$\gamma_\mu = g_{\mu\nu}\gamma^\nu. \qquad (4\text{-}158)$$

Contraction identities

From (4-157), relations (4-159) may be derived. (If you really want to take the time.) Note that all matrix multiplications result in other matrices, so a numeric value like "4" really symbolizes "4I" where I is the identity matrix in spinor space. Prob. 31 asks you to derive the first relation below.

$$\gamma_\lambda\gamma^\lambda = 4, \qquad\qquad \gamma_\lambda\gamma^\alpha\gamma^\lambda = -2\gamma^\alpha$$

$$\gamma_\lambda\gamma^\alpha\gamma^\beta\gamma^\lambda = 4g^{\alpha\beta} \qquad \gamma_\lambda\gamma^\alpha\gamma^\beta\gamma^\gamma\gamma^\lambda = -2\gamma^\gamma\gamma^\beta\gamma^\alpha \qquad (4\text{-}159)$$

$$\gamma_\lambda\gamma^\alpha\gamma^\beta\gamma^\gamma\gamma^\delta\gamma^\lambda = 2\left(\gamma^\delta\gamma^\alpha\gamma^\beta\gamma^\gamma + \gamma^\gamma\gamma^\beta\gamma^\alpha\gamma^\delta\right).$$

For 4-momentum (see Prob. 32),

$$\not{p}\not{p} = p^2. \qquad (4\text{-}160)$$

For a four vector A_μ, where "A slash" is defined by $\not{A} = \gamma^\alpha A_\alpha$, the following contraction relations hold.

$$\gamma_\lambda \not{A} \gamma^\lambda = -2\not{A} \qquad\qquad \gamma_\lambda \not{A}\not{B} \gamma^\lambda = 4AB$$

$$\gamma_\lambda \not{A}\not{B}\not{C} \gamma^\lambda = -2\not{C}\not{B}\not{A} \qquad \gamma_\lambda \not{A}\not{B}\not{C}\not{D} \gamma^\lambda = 2\left(\not{D}\not{A}\not{B}\not{C} + \not{C}\not{B}\not{A}\not{D}\right) \qquad (4\text{-}161)$$

Completeness relations

By subtracting (4-154) from (4-153), one can derive what are known as the completeness relations. Note that (4-162) entails summation on r and outer product multiplication.

$$\left(u_{r\alpha}(\mathbf{p})\bar{u}_{r\beta}(\mathbf{p}) - v_{r\alpha}(\mathbf{p})\bar{v}_{r\beta}(\mathbf{p})\right) = \delta_{\alpha\beta} \qquad (4\text{-}162)$$

Traces

$$\text{Tr}(UV) = \text{Tr}(VU) \qquad U,V \text{ are } n \times n \text{ matrices} \qquad (4\text{-}163)$$

For a product of an <u>odd number</u> of γ-matrices, $\qquad \text{Tr}\left(\gamma^\alpha \gamma^\beta .....\gamma^\mu \gamma^\nu\right) = 0 \qquad (4\text{-}164)$

For products of an <u>even number</u> of γ-matrices,

$$\text{Tr}\left(\gamma^\alpha \gamma^\beta\right) = 4g^{\alpha\beta} \qquad \text{Tr}\left[\gamma^\alpha,\gamma^\beta\right] = \text{Tr}\left[\gamma^\alpha \gamma^\beta - \gamma^\beta \gamma^\alpha\right] = 0$$

$$\text{Tr}\left(\gamma^\alpha \gamma^\beta \gamma^\gamma \gamma^\delta\right) = 4\left(g^{\alpha\beta}g^{\gamma\delta} - g^{\alpha\gamma}g^{\beta\delta} + g^{\alpha\delta}g^{\beta\gamma}\right) \qquad (4\text{-}165)$$

from which, one can find

$$\text{Tr}\left(A\!\!\!/\,B\!\!\!/\right) = 4(AB)$$

$$\text{Tr}\left(A\!\!\!/\,B\!\!\!/\,C\!\!\!/\,D\!\!\!/\right) = 4\{(AB)(CD) - (AC)(BD) + (AD)(BC)\}$$

$$\text{Tr}\left(A\!\!\!/_1 A\!\!\!/_2 ... A\!\!\!/_{2n}\right) = (A_1 A_2)\text{Tr}\left(A\!\!\!/_3 ... A\!\!\!/_{2n}\right) - (A_1 A_3)\text{Tr}\left(A\!\!\!/_2 A\!\!\!/_4 ... A\!\!\!/_{2n}\right)$$

$$+ ...... + (A_1 A_{2n})\text{Tr}\left(A\!\!\!/_2 A\!\!\!/_3 ... A\!\!\!/_{2n-1}\right). \qquad (4\text{-}166)$$

Also, $\qquad A\!\!\!/\,B\!\!\!/ = AB + \frac{1}{2}\left[\gamma^\alpha,\gamma^\beta\right]A_\alpha B_\beta = 2AB - B\!\!\!/\,A\!\!\!/, \qquad (4\text{-}167)$

with particular cases $\qquad A\!\!\!/\,A\!\!\!/ = A^2 \qquad A\!\!\!/\,B\!\!\!/ = -B\!\!\!/\,A\!\!\!/, \text{ if } AB = 0. \qquad (4\text{-}168)$

For <u>any product</u> of γ-matrices,

$$\text{Tr}\left(\gamma^\alpha \gamma^\beta ....\gamma^\mu \gamma^\nu\right) = \text{Tr}\left(\gamma^\nu \gamma^\mu ....\gamma^\beta \gamma^\alpha\right) \quad \rightarrow \quad \text{Tr}\left(A\!\!\!/_1 A\!\!\!/_2 ... A\!\!\!/_{2n}\right) = \text{Tr}\left(A\!\!\!/_{2n} ... A\!\!\!/_2 A\!\!\!/_1\right). \qquad (4\text{-}169)$$

4.14 Appendix B. Relativistic Spin: Getting to the Real Bottom of It All

I have not seen the following matter addressed in the literature, though I consider it to be fairly fundamental and something that should be understood clearly by all. I have deduced it myself in order to assuage gnawing uncertainties I held for quite some time about the theory. Thus, I caution the reader to be vigilant and note that any errors found herein are solely my own.

Consider **p** and spin both in the x direction as depicted in Fig. 4-1(d) (pg. 96). As shown about half way down the right column in Wholeness Chart 4-1 (pg. 102), this state is not represented by one of our (at-rest system up/down) basis state solutions u_1 or u_2, but by the general solution (non-basis state solution) $u = u_1 + u_2$ (not normalized here.) u in this case is an eigensolution for spin in the x direction, but not of the z direction. So, formally, the state shown in Fig. 4-1(d) would be represented as

$$\left|\psi_{x\text{spin},\mathbf{p}}\right\rangle = K\left|u_1(\mathbf{p})e^{-ipx} + u_2(\mathbf{p})e^{-ipx}\right\rangle, \qquad (4\text{-}170)$$

where K is a suitable constant providing correct normalization.

Consider as well, the state of Fig. 4-1(b), which likewise is not an eigenstate of the at-rest system up/down eigensolutions u_1 or u_2. It must have a general state like that shown in Wholeness Chart 4-1 two blocks below the block referred to in the previous paragraph, i.e.,

$$\left|\psi_{\nearrow\text{spin},\mathbf{p}}\right\rangle = \left|C_1' u_1(\mathbf{p})e^{-ipx} + C_2' u_2(\mathbf{p})e^{-ipx}\right\rangle, \qquad (4\text{-}171)$$

where for given **p**, the constants C_1' and C_2' must be determined such that spin will be in the direction shown in Fig. 4-1(b). Thus, if (4-171) is taken as our general state of any given Dirac particle with given **p**, then the particular orientation of the spin vector direction will lead to particular values of C_1' and C_2'. Hence, C_1' and C_2' are functions of both **p** and spin direction.

So, what is the issue? There are two of which I am aware.

4.14.1 The First Issue: General (Non-pure u_1 or u_2) States

In the QFT literature, kets are typically treated as eigenstates of **p** and basis states (up/down eigenstates in the at-rest system) spin r. A ket is represented as $\left|\psi_{r,\mathbf{p}}\right\rangle$ for a single particle; or for example, $\left| \psi_{r,\mathbf{p}}, \psi_{r',\mathbf{p}'}, \psi_{r'',\mathbf{p}''} \right\rangle$ for multiparticle states. They are generally not represented as (4-171). In other words, such kets cannot represent spin other than for particles in u_1 or u_2 basis spin states, and thus, seem incomplete.

4.14.2 Resolution of the First Issue

Consider the (non spin basis state) particle state

$$\left| \psi_{\text{n.s.b.},\mathbf{p}} \right\rangle = \underbrace{\left| C_1 \sqrt{\frac{m}{VE_{\mathbf{p}}}} u_1(\mathbf{p}) e^{-ipx} + C_2 \sqrt{\frac{m}{VE_{\mathbf{p}}}} u_2(\mathbf{p}) e^{-ipx} \right\rangle}_{\text{non spin basis state}}, \tag{4-172}$$

where $|C_1|^2$ is the probability of measuring the u_1 state, and $|C_2|^2$ is that of the u_2 state. The expectation value of any operator for this state is

$$\bar{\mathcal{O}} = \left\langle \psi_{\text{n.s.b.},\mathbf{p}} \left| \mathcal{O} \right| \psi_{\text{n.s.b.},\mathbf{p}} \right\rangle =$$

$$\left\langle C_1 \sqrt{\frac{m}{VE_{\mathbf{p}}}} u_1(\mathbf{p}) e^{-ipx} + C_2 \sqrt{\frac{m}{VE_{\mathbf{p}}}} u_2(\mathbf{p}) e^{-ipx} \left| \mathcal{O} \right| C_1 \sqrt{\frac{m}{VE_{\mathbf{p}}}} u_1(\mathbf{p}) e^{-ipx} + C_2 \sqrt{\frac{m}{VE_{\mathbf{p}}}} u_2(\mathbf{p}) e^{-ipx} \right\rangle. \tag{4-173}$$

Or, for each spin basis state being an eigenstate of $\mathcal{O}$, with $\mathcal{O}_{r=1,\mathbf{p}}$ and $\mathcal{O}_{r=2,\mathbf{p}}$ representing the operator eigenvalues for each spin basis state,

$$\bar{\mathcal{O}} = \left\langle C_1 \sqrt{\frac{m}{VE_{\mathbf{p}}}} u_1(\mathbf{p}) e^{-ipx} \left| \underbrace{\mathcal{O}_{r=1,\mathbf{p}}}_{\mathcal{O}\text{ eigenval}} \right| C_1 \sqrt{\frac{m}{VE_{\mathbf{p}}}} u_1(\mathbf{p}) e^{-ipx} \right\rangle + \left\langle C_1 \sqrt{\frac{m}{VE_{\mathbf{p}}}} u_1(\mathbf{p}) e^{-ipx} \left| \mathcal{O}_{r=2,\mathbf{p}} \right| C_2 \sqrt{\frac{m}{VE_{\mathbf{p}}}} u_2(\mathbf{p}) e^{-ipx} \right\rangle$$

$$+ \left\langle C_2 \sqrt{\frac{m}{VE_{\mathbf{p}}}} u_2(\mathbf{p}) e^{-ipx} \left| \mathcal{O}_{r=1,\mathbf{p}} \right| C_1 \sqrt{\frac{m}{VE_{\mathbf{p}}}} u_1(\mathbf{p}) e^{-ipx} \right\rangle + \left\langle C_2 \sqrt{\frac{m}{VE_{\mathbf{p}}}} u_2(\mathbf{p}) e^{-ipx} \left| \mathcal{O}_{r=2,\mathbf{p}} \right| C_2 \sqrt{\frac{m}{VE_{\mathbf{p}}}} u_2(\mathbf{p}) e^{-ipx} \right\rangle. \tag{4-174}$$

The first term in (4-174) is

$$\mathcal{O}_{r=1,\mathbf{p}} \int_V C_1^\dagger \sqrt{\frac{m}{VE_{\mathbf{p}}}} u_1^\dagger(\mathbf{p}) e^{ipx} C_1 \sqrt{\frac{m}{VE_{\mathbf{p}}}} u_1(\mathbf{p}) e^{-ipx} d^3x$$

$$= \mathcal{O}_{r=1,\mathbf{p}} C_1^\dagger C_1 \underbrace{u_1^\dagger(\mathbf{p}) u_1(\mathbf{p})}_{E_{\mathbf{p}}/m} \frac{m}{VE_{\mathbf{p}}} \int_V d^3x = \mathcal{O}_{r=1,\mathbf{p}} |C_1|^2 . \tag{4-175}$$

The second term in (4-174) is zero, i.e.,

$$\mathcal{O}_{r=2,\mathbf{p}} \int_V C_1^\dagger \sqrt{\frac{m}{VE_{\mathbf{p}}}} u_1^\dagger(\mathbf{p}) e^{ipx} C_2 \sqrt{\frac{m}{VE_{\mathbf{p}}}} u_2(\mathbf{p}) e^{-ipx} d^3x = \mathcal{O}_{r=2,\mathbf{p}} C_1^\dagger C_2 \underbrace{u_1^\dagger(\mathbf{p}) u_2(\mathbf{p})}_{=0} \frac{m}{VE_{\mathbf{p}}} \int_V d^3x = 0. \tag{4-176}$$

Similarly, as you can prove by doing Prob. 33, the third term = 0, and the fourth term = $\mathcal{O}_{r=2,\mathbf{p}}$ $|C_2|^2$. Then,

$$\bar{\mathcal{O}} = \mathcal{O}_{r=1,\mathbf{p}} |C_1|^2 + \mathcal{O}_{r=2,\mathbf{p}} |C_2|^2 , \tag{4-177}$$

where the probability of measuring the $r = 1$ basis state is $|C_1|^2$, that of the $r = 2$ basis state is $|C_2|^2$, and $|C_1|^2 + |C_2|^2 = 1$. This should look pretty familiar from NRQM, and it may be somewhat gratifying to find it here in RQM and QFT also.

If $\mathcal{O} = H$, the Hamiltonian, then our energy expectation value would be

$$\bar{E} = E_{\mathbf{p}} |C_1|^2 + E_{\mathbf{p}} |C_2|^2 = E_{\mathbf{p}} \left(|C_1|^2 + |C_2|^2 \right) = E_{\mathbf{p}}, \tag{4-178}$$

as it must be since the two different spin basis states in (4-172) have the same energy. Similar effects would be seen for other operators like $\mathbf{p}$ or Σ_i (although there are some tricky issues for Σ_i.)

The Resolution for Free Particles

We showed above that the mathematical predictions of measurable quantities for general, non-basis, free states work out the same as for basis states.

In this book and elsewhere, you may see a Feynman diagram or other figure of a Dirac particle with a given **p**, which is not necessarily in a spin basis state. It is common to represent such a ket without the subscript r, i.e.,

$$\left|\psi_{\text{n.s.b.,}\mathbf{p}}\right\rangle \quad \text{represented as} \quad \left|\psi_{\mathbf{p}}\right\rangle. \tag{4-179}$$

<u>The Resolution for Interacting Particles</u>

A transition amplitude (the mathematical representation of an interaction, whose magnitude squared is the probability of the interaction [see Chap. 1, pg. 3, and later chapters on interactions herein]) can be calculated for each basis spin state $\left|\psi_{r,\,\mathbf{p}}\right\rangle$ in a given interaction. These can be used, as we will see later on, to calculate transition amplitudes for more general states $\left|\psi_{\mathbf{p}}\right\rangle$.

In scattering experiments, spin is typically not detected, so an average of all possible incoming spin states is used. Our outcome predictions are generally of expectation value type form, which is itself an average, so things work out OK. We will get to the math of that with regard to interactions in later chapters.

4.14.3 The Second Issue: Multiple Particles and Different At-rest Systems

The second issue is more complicated. Suppose you had a particle such as that shown in Fig. 4-5(a), but you wished to express the state in a different (rotated), primed coordinate system where the x'^{3} axis is aligned with the spin direction. Note that it can be expressed in the unprimed coordinate system where the x^{3} axis is aligned with the at-rest system spin direction by finding the correct C_1 and C_2 in (4-171). But how would we express it in the primed system (as shown in Fig 4-5(b))?

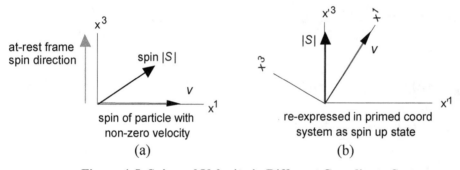

Figure 4-5. Spin and Velocity in Different Coordinate Systems

In the primed system, the spin is up, and so we figure our state must be an eigenstate of Σ_3 in that system. We might then consider that we need to find C_1 and C_2 in (4-171) such that this is true for 3-momentum aligned as shown (i.e., p'^{3} and $p'^{1} \neq 0$, $p'^{2} = 0$). Thus,

$$\left|\psi'_{\uparrow \text{spin},\mathbf{p}'}\right\rangle = \left|C_1 u_1(\mathbf{p}')e^{-ip'x'} + C_2 u_2(\mathbf{p}')e^{-ip'x'}\right\rangle = \left|C_1 \begin{pmatrix} 1 \\ 0 \\ \dfrac{p'^{3}}{E+m} \\ \dfrac{p'^{1}}{E+m} \end{pmatrix} e^{-ip'x'} + C_2 \begin{pmatrix} 0 \\ 1 \\ \dfrac{p'^{1}}{E+m} \\ \dfrac{-p'^{3}}{E+m} \end{pmatrix} e^{-ip'x'}\right\rangle. \tag{4-180}$$

Then operate on (4-180) with Σ_3,

$$\Sigma_3 \left| \psi'_{\uparrow\, spin, \mathbf{p'}} \right\rangle = \frac{1}{2} \begin{bmatrix} 1 & & & \\ & -1 & & \\ & & 1 & \\ & & & -1 \end{bmatrix} \left| \begin{pmatrix} C_1 \\ C_2 \\ \dfrac{C_1 p'^3 + C_2 p'^1}{E+m} \\ \dfrac{C_1 p'^1 - C_2 p'^3}{E+m} \end{pmatrix} e^{-ip'x'} \right\rangle$$

$$= \frac{1}{2} \left| \begin{pmatrix} C_1 \\ -C_2 \\ \dfrac{C_1 p'^3 + C_2 p'^1}{E+m} \\ \dfrac{-C_1 p'^1 + C_2 p'^3}{E+m} \end{pmatrix} e^{-ip'x'} \right\rangle \overset{?}{\underset{\substack{\text{see equation}\\ \text{(4-180) for} \to}}{=}} \frac{1}{2} \left| \psi'_{\uparrow\, spin, \mathbf{p'}} \right\rangle = \frac{1}{2} \left| \begin{pmatrix} C_1 \\ C_2 \\ \dfrac{C_1 p'^3 + C_2 p'^1}{E+m} \\ \dfrac{C_1 p'^1 - C_2 p'^3}{E+m} \end{pmatrix} e^{-ip'x'} \right\rangle . \tag{4-181}$$

If the equal sign with the question mark over it is truly an equal sign in (4-181), then the state is an eigenstate of spin up as shown in Fig. 4-5(b). But for that to be true, C_2 must = 0, and p'^1 must = 0. But in Fig. 4-5(b), $p'^1 \neq 0$.

Conclusion #1: We cannot use the traditional forms studied in this chapter for Dirac spinors u_1, u_2 and the spin operator Σ_3 to represent the particle state of Fig. 4-5(b). (That is, using those forms, there is no possible mathematical expression for a spin up state in a coordinate frame with the z direction aligned with spin and with velocity having a component transverse to that z direction.)

For Fig. 4-5(b), no lab frame with the z axis aligned with $\mathbf{S}_{v \neq 0}$ spin direction will work. It won't yield a z spin eigenvector $\Sigma_3 |\psi\rangle = \frac{1}{2}|\psi\rangle$. Changing $\mathbf{v}$ (without aligning it with $\mathbf{S}$) can cause any possible alignment of spin $\mathbf{S}_{v \neq 0}$ for a particle, except the original $\mathbf{S}_{v=0}$ spin direction. None of these other systems will work (give the proper eigenvector in the z direction). Therefore, for Fig. 4-5(b), only the $\mathbf{v} = 0$ system with the z axis aligned with the $\mathbf{S}_{v=0}$ spin direction does work.

Conclusion #2: The traditional forms for u_1, u_2, and Σ_3 (more generally, Σ_i) can only be used in a coordinate system for which the particle at-rest coordinate system spin is aligned with the z direction (positive or negative direction.)

Our Mistake Above: In the above example (4-181), we should have transformed u_1, u_2, and Σ_3 into their equivalent forms in the primed system, i.e., where T is the transformation in spinor space for a coordinate rotation transformation for Cartesian coordinates in physical space such as that shown in Fig. 4-5,

$$u'_1 = Tu_1 \qquad u'_2 = Tu_2 \qquad \Sigma'_i = T\Sigma_i T^{-1} \qquad \left| \psi'_{\uparrow\, spin, \mathbf{p'}} \right\rangle = T\left| \psi_{\nearrow\, spin, \mathbf{p}} \right\rangle, \tag{4-182}$$

where T is a 4X4 matrix in spinor space. (We won't determine its mathematical form here, as it would take us far afield of our present objective. It is presented in Chap. 6.)

Correcting that Mistake

So, if we were to go to the (considerable, at this point) trouble to deduce T for Fig. 4-5, we would find

$$\Sigma'_3 \left| \psi'_{\uparrow\, spin, \mathbf{p'}} \right\rangle = \frac{1}{2} \left| \psi'_{\uparrow\, spin, \mathbf{p'}} \right\rangle =$$

$$\underset{T\Sigma_3 T^{-1}}{\underbrace{\Sigma'_3}} \left| C_1 \underset{Tu_1}{\underbrace{u'_1(\mathbf{p'})}} e^{-ip'x'} + C_2 \underset{Tu_2}{\underbrace{u'_2(\mathbf{p'})}} e^{-ip'x'} \right\rangle = \frac{1}{2} \left| C_1 u'_1(\mathbf{p'}) e^{-ip'x'} + C_2 u'_2(\mathbf{p'}) e^{-ip'x'} \right\rangle, \tag{4-183}$$

where we would solve for the C_1 and C_2 values that satisfy (4-183). Then the ket in (4-183) with those particular values would represent the state shown in Fig. 4-5(b) in the primed coordinate system. But u_1 and u_2 would *not* look like (4-20). They would have other forms.

We did all of the above to emphatically demonstrate Conclusion #2 above. Keeping that in mind can help a lot when tackling relativistic spin problems.

<u>The Crux of the Second Issue</u>

The issue then, in QFT, is that in every case with two or more particles in different spin and **p** states in the same multiparticle ket, we have to use a different coordinate system for each particle, if we want to employ the usual forms for u_1, u_2, and Σ_i. We have to use the coordinate system for which, for that particular particle, the at-rest system spin axis is in a pure up or down state. So, each particle in a multiparticle state like $\left| \psi_{r,\mathbf{p}}, \psi_{r',\mathbf{p}'}, \psi_{r'',\mathbf{p}''} \right\rangle$ needs its own coordinate system, since each typically has a different orientation for its at-rest system spin direction. It seems we can't use just the lab frame system for all of them. But they are all part of the same problem/case (interaction or whatever) we are studying and trying to solve.

For example, consider two electrons in the same multiparticle state as shown in Fig. 4-6, each having different velocity and different at-rest system spin direction. We would represent the multiparticle ket as

$$\left| \psi_A, \psi_B \right\rangle = \left| \left(C_{A1} u_1 (\mathbf{p}_A) e^{-ip_A x_A} + C_{A2} u_2 (\mathbf{p}_A) e^{-ip_A x_A} \right) , \left(C_{B1} u_1 (\mathbf{p}_B) e^{-ip_B x_B} + C_{B2} u_2 (\mathbf{p}_B) e^{-ip_B x_B} \right) \right\rangle , (4\text{-}184)$$

where for given $\mathbf{p}_A$ we would determine C_{A1} and C_{A2} to yield the correct spin for electron A when it is moving as we would see in the lab, and carry out similar steps for the constants for electron B. u_1 and u_2 here have the standard form (4-20) we are familiar with. Component $p_A{}^i$ values would be for those of electron A in the at-rest system direction coordinate axes for electron A shown in Fig. 4-6. $p_B{}^i$ would be for components of 3-momentum of electron B in the coordinate at-rest system of electron B.

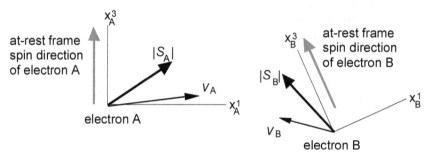

Figure 4-6. Two Electrons with Different p and Different At-Rest Spin Directions

The crux of our present issue then is that, if we take $x_A{}^i$ as our lab coordinates, they are fine for electron A, but they don't work for electron B. We need to use coordinate system $x_B{}^i$ to describe electron B if we want to use the standard forms for u_1 and u_2. On the other hand, solving problems is enormously easier if everything is done in the same coordinate system.

A parallel concern is the representation of a ket, where each particle is labeled in terms of its 3-momentum and spin as $\left| \psi_{r_A,\mathbf{p}_A}, \psi_{r_B,\mathbf{p}_B} \right\rangle$. The 3-momenta can't be compared directly (to see which direction each is going with respect to the other) because each particle's 3-momentum is expressed in a different coordinate system. And what the r spin label means on each is equally non-comparable. $r = 1$ for electron A means a different spin direction in the lab than $r = 1$ for electron B.

4.14.4 Resolution of the Second Issue

<u>Re-express Entire Multiparticle Ket in Terms of One Particle's At-Rest system</u>

To resolve this, first consider re-expressing the B electron part of (4-184) by transforming it to the at-rest coordinate system of electron A. To do this, keep in mind that both the A and B at-rest coordinate systems are at rest with respect to the lab, but rotated relative to one another. (Each particle at-rest system is found from decelerating the particle to be at rest in the lab, and aligning the at-rest system x^3 axis with the spin.) We can transform $\mathbf{p_B}$ ($= p_B{}^i$) to the at-rest coordinate system A using a simple rotation transformation on the $p_B{}^i$ components, i.e., re-express 3-momentum of electron B in coordinates of the at-rest system A. We will use the subscript symbol B/A to represent B electron components in the A at-rest coordinate system.

In addition to transforming $p_B{}^i$ to $p_{B/A}{}^i$, we need to transform the spinors u_r to the new coordinate system as well, analogous to what we did in (4-182), but the coordinate systems are somewhat different here. We haven't developed this, and won't here, but consider that if we did, the transformation in spinor space corresponding to a change in physical space from coordinate system B to A would be symbolized by T_{AB}. This represents changing the B electron components in the B electron at-rest coordinate system to those of the A electron at-rest coordinate system. Then (4-184) could be represented as

$$\left| \psi_A , \psi_{B/A} \right\rangle = \left| \psi_A , \left(C_{B1} \underbrace{T_{AB} u_1 \left(\mathbf{p}_{B/A} \right)}_{u_{1B/A}\left(\mathbf{p}_{B/A} \right)} e^{-ip_{B/A}x_{B/A}} + C_{B2} \underbrace{T_{AB} u_2 \left(\mathbf{p}_{B/A} \right)}_{u_{2B/A}\left(\mathbf{p}_{B/A} \right)} e^{-ip_{B/A}x_{B/A}} \right) \right\rangle =$$

$$\left| \left(\begin{array}{c} C_{A1} u_1 \left(\mathbf{p}_A \right) e^{-ip_A x_A} \\ + C_{A2} u_2 \left(\mathbf{p}_A \right) e^{-ip_A x_A} \end{array} \right) \left(\begin{array}{c} C_{B1} u_{1B/A} \left(\mathbf{p}_{B/A} \right) e^{-ip_{B/A}x_{B/A}} \\ + C_{B2} u_{2B/A} \left(\mathbf{p}_{B/A} \right) e^{-ip_{B/A}x_{B/A}} \end{array} \right) \right\rangle .$$

$\hspace{10cm}$ (4-185)

$u_{rB/A}$ is the rth spinor column matrix for the B electron expressed in the A electron at-rest system.

Now we have the entire multiparticle ket expressed in terms of a single coordinate system (4-185), but the column matrix components of $u_{rB/A}$ will not look like the familiar ones, so we cannot use the definitions (4-20) we have grown used to.

What We Measure is Key

As we should have learned well by now, the critical element in quantum theories is what we can measure when we do experiments, and that is reflected in the expectation value of the operator corresponding to the particular dynamical variable measured. We want to demonstrate that the expectation value of any dynamical variable, corresponding to operator $\mathcal{O}$, is the same no matter what system we express our particle state in. To keep things simple, consider measuring the expectation value of $\mathcal{O}$ for only the single B particle of (4-185) in the at-rest coordinate system A. The A particle measurement in the A particle at-rest coordinate system is trivial. The expectation value for $\mathcal{O}_{B/A}$ ($\mathcal{O}$ for the B particle measured in the A particle at-rest system), where $\mathcal{O}_{B/B}$ is $\mathcal{O}$ for the B particle expressed in the B at-rest system is thus

$$\overline{\mathcal{O}}_{B/A} = \underbrace{\left\langle \psi_{B/A} \middle| \mathcal{O}_{B/A} \middle| \psi_{B/A} \right\rangle}_{\substack{\text{all in coord system A,} \\ \text{spinors not usual form}}} = \left\langle \psi_{B/A} \middle| T_{AB} \underbrace{\mathcal{O}_{B/B}}_{\substack{\text{coord} \\ \text{sys B}}} T_{AB}^{-1} \middle| \psi_{B/A} \right\rangle$$

$\hspace{11cm}$ (4-186)

$$= \left\langle \begin{array}{c} C_{B1} T_{AB} u_1 \left(\mathbf{p}_{B/A} \right) e^{-ip_{B/A}x_{B/A}} \\ + C_{B2} T_{AB} u_2 \left(\mathbf{p}_{B/A} \right) e^{-ip_{B/A}x_{B/A}} \end{array} \middle| T_{AB} \mathcal{O}_{B/B} T_{AB}^{-1} \middle| \begin{array}{c} C_{B1} T_{AB} u_1 \left(\mathbf{p}_{B/A} \right) e^{-ip_{B/A}x_{B/A}} \\ + C_{B2} T_{AB} u_2 \left(\mathbf{p}_{B/A} \right) e^{-ip_{B/A}x_{B/A}} \end{array} \right\rangle .$$

In integral form, (4-186) is

$$\overline{\mathcal{O}}_{B/A} = \int_V \left(\begin{array}{c} C_{B1}^\dagger u^\dagger{}_1 \left(\mathbf{p}_{B/A} \right) \underbrace{T_{AB}^\dagger}_{T_{AB}^{-1}} e^{ip_{B/A}x_{B/A}} \\ + C_{B1}^\dagger u^\dagger{}_2 \left(\mathbf{p}_{B/A} \right) \underbrace{T_{AB}^\dagger}_{T_{AB}^{-1}} e^{ip_{B/A}x_{B/A}} \end{array} \right) T_{AB} \mathcal{O}_{B/B} T_{AB}^{-1} \left(\begin{array}{c} C_{B1} T_{AB} u_1 \left(\mathbf{p}_{B/A} \right) e^{-ip_{B/A}x_{B/A}} \\ + C_{B2} T_{AB} u_2 \left(\mathbf{p}_{B/A} \right) e^{-ip_{B/A}x_{B/A}} \end{array} \right) d^3x . \text{(4-187)}$$

Transformations of form T_{AB} are unitary, so their complex conjugate transposes equal their inverses, as noted in the underbrackets above. Every T_{AB} in (4-186) is pre-multiplied by its inverse and drops out. We thus get

$$\overline{\mathcal{O}}_{B/A} = \int_V \left(\begin{matrix} C^\dagger_{B1} u^\dagger{}_1\left(\mathbf{p}_{B/A}\right)e^{ip_{B/A}x_{B/A}} \\ + C^\dagger_{B1} u^\dagger{}_2\left(\mathbf{p}_{B/A}\right)e^{ip_{B/A}x_{B/A}} \end{matrix} \right) \mathcal{O}_{B/B} \left(\begin{matrix} C_{B1} u_1\left(\mathbf{p}_{B/A}\right)e^{-ip_{B/A}x_{B/A}} \\ + C_{B2} u_2\left(\mathbf{p}_{B/A}\right)e^{-ip_{B/A}x_{B/A}} \end{matrix} \right) d^3x =$$

$$\underbrace{\left\langle \begin{matrix} C_{B1} u_1\left(\mathbf{p}_{B/A}\right)e^{-ip_{B/A}x_{B/A}} \\ + C_{B2} u_2\left(\mathbf{p}_{B/A}\right)e^{-ip_{B/A}x_{B/A}} \end{matrix} \right|}_{\substack{\text{usual form of spinors with } \mathbf{p}_B \\ \text{components in A coord system}}} \mathcal{O}_{B/B} \underbrace{\left| \begin{matrix} C_{B1} u_1\left(\mathbf{p}_{B/A}\right)e^{-ip_{B/A}x_{B/A}} \\ + C_{B2} u_2\left(\mathbf{p}_{B/A}\right)e^{-ip_{B/A}x_{B/A}} \end{matrix} \right\rangle}_{\text{ditto of comment at left}} . \qquad (4\text{-}188)$$

Since the A at-rest system and the B at-rest system are both at rest with respect to each other, energy measured in one system of any particle is the same as that measured in the other system for the same particle. Thus, if $\mathcal{O}$ is the Hamiltonian, $\mathcal{O}_{B/B} = \mathcal{O}_{B/A}$ (energy of B particle is the same measured in B coordinate system and A coordinate system). If $\mathcal{O}$ is charge, the same thing is true.

Thus, for these operators, we can find the measurable quantities we want by using the ket (and bra) form of (4-188), i.e., employ the familiar form (4-20) for u_r of the B electron as a function of its 3-momentum components as measured in the A system.

For directional quantities like 3-momentum and spin, the argument is a tad subtler. Suppose we have chosen at-rest system A as our lab frame, and we want to know the expectation value of momentum for the B electron in the $x_A{}^3$ direction, i.e., $p_{B/A}3$ (operator symbol $P_{B/A}3$.) Well, in either system this is just the vertical component (in Fig. 4-6) of $m\mathbf{v}_B$, where $\mathbf{v}_B$ is shown in the figure. We can express this in either coordinate system as shown in Fig. 4-7.

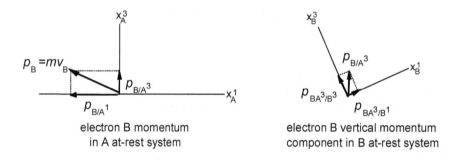

electron B momentum
in A at-rest system

electron B vertical momentum
component in B at-rest system

Figure 4-7. Vertical Component of 3-Momentum Expressed in Two Different At-Rest Systems

Thus, the vertical component of electron B's 3-momentum, expressed in each system, is simply

$$p_{B/A^3} \mathbf{i}_{A^3} = p_{BA^3/B^1} \mathbf{i}_{B^1} + p_{BA^3/B^3} \mathbf{i}_{B^3} \ \left(\text{number}\right)$$
$$\rightarrow \ P_{B/A^3} \mathbf{i}_{A^3} = P_{BA^3/B^1} \mathbf{i}_{B^1} + P_{BA^3/B^3} \mathbf{i}_{B^3} \ \left(\text{operator}\right) . \qquad (4\text{-}189)$$

Hence, we can use the operator form on the right side of the equal sign in the second part of (4-189) as the $\mathcal{O}_{B/B}$ operator in (4-188). We will find our expectation value for 3-momentum in the vertical direction to be

$$\overline{P}_{BA^3/B^1} \mathbf{i}_{B^1} + \overline{P}_{BA^3/B^3} \mathbf{i}_{B^3} = \left\langle \psi_{B/A} \left| P_{BA^3/B^1} \mathbf{i}_{B^1} + P_{BA^3/B^3} \mathbf{i}_{B^3} \right| \psi_{B/A} \right\rangle, \qquad (4\text{-}190)$$

which is the 3-momentum vertical direction component expectation value. But that is really the same thing as

$$\overline{P}_{B/A^3} \mathbf{i}_{A^3} = \overline{P}_{BA^3/B^1} \mathbf{i}_{B^1} + \overline{P}_{BA^3/B^3} \mathbf{i}_{B^3} = \underbrace{\left\langle \psi_{B/A} \left| P_{B/A^3} \mathbf{i}_{A^3} \right| \psi_{B/A} \right\rangle}_{\substack{\text{all in coord system A,} \\ \text{ket spinors of usual form}}} . \qquad (4\text{-}191)$$

Key Point for Vectors:

If we wish to know the expectation value of a vector quantity of any particle along the direction of any axis in the A coordinate system, we can simply find the expectation value in the same direction expressed in terms of the B coordinate system components. Thus, we can use the form of (4-188) for any directional quantity in the B system.

Finally, the Resolution of the Second Issue

So, the expectation value of any operator expressed in the A coordinate system, even if acting on a particle with an at-rest coordinate system different from A, can be found using (4-188). That is, it can be found using kets expressed in terms of the standard relations for $u_{1,2}$ and components of 3-momentum expressed in the A system.

This makes a lot of things easier, since we can use a single coordinate system throughout. Note, this works because doing so results in the same measured quantities (same expectation values), even though the spinor forms $u_{1,2}$ used are not, strictly speaking, the correct ones to describe the actual particle spin state. The "incorrectness" in the ket is canceled by the "incorrectness" in the bra, resulting in a correct expectation value.

4.14.5 Multiparticle States and the v_1 and v_2 Solutions

Although, to save space and make things simpler, we often dealt with single particle states in this appendix, the reasoning is equally applicable to the multiparticle states of QFT.

Also, all of the logic in this appendix applies in direct parallel fashion to the Dirac equation eigensolutions containing v_1 and v_2.

4.14.6 Summary of this Appendix B

See the end of Sect. 4.11.3 (pg. 116) for a summary of this appendix.

If you don't get all this, don't worry. Go with the summary results. Later, when you are a practicing physicist, if you have questions about spin stuff, come back to this.

4.15 Problems

1. Show that *i*) (4-7) solve the relations (4-6), and *ii*) they also fulfill the requirements of being traceless, hermitian (matrix equals its complex conjugate transpose), and β and α_1 have ± 1 eigenvalues. Note that eigenvectors for β are any constant times any of $(1,0,0,0)^T$, $(0,1,0,0)^T$, $(0,0,1,0)^T$, and $(0,0,0,1)^T$. Eigenvectors for α_1 are any constant times any of $(1,0,0,1)^T$, $(0,1,1,0)^T$, $(1,0,0,-1)^T$, and $(0,1,-1,0)^T$. Do you believe α_2 and α_3 would also fulfill all these relations and requirements, if we took the trouble to examine them?

2. Show that (4-9) equals (4-10) and (4-11).

3. Prove the Hermiticity conditions $\gamma^{\mu\dagger} = \gamma^0 \gamma^\mu \gamma^0$ by substitution, plugging, and chugging.

4. Use the form (4-14) of the Dirac equation (reproduced below), insert the Dirac gamma matrices (4-10), and show each term in the summation over μ. Write the whole thing as a matrix equation (which should be a sum of matrices where that sum is post-multiplied by the column matrix $|\psi\rangle$). Note that a different derivative is taken on the whole vector (column matrix) $|\psi\rangle$ for each Dirac gamma matrix. Note also that the resulting matrix equation has four rows, and thus, may be considered as four separate scalar equations, one for each index κ. Each such equation has all four components of $|\psi\rangle$ in it, although each such equation will have one component equal to zero.

$$\sum_{\eta=1}^{4} \left(\sum_{\mu=0}^{3} i\left(\gamma^\mu\right)_{\kappa\eta} \partial_\mu - m\delta_{\kappa\eta} \right) |\psi\rangle_\eta = 0 \qquad \kappa = 1,2,3,4 \qquad (4\text{-}14)$$

5. Pick one or more of the solutions (4-20) and show via substitution that each one you picked satisfies the Dirac equation (4-15).

6. Find the inner products of at least two of $u_2^\dagger(\mathbf{p})u_2(\mathbf{p})$, $v_2^\dagger(\mathbf{p})v_2(\mathbf{p})$, and $v_1^\dagger(\mathbf{p})v_1(\mathbf{p})$.

7. Find the inner product of $v_1^\dagger(\mathbf{p})v_2(\mathbf{p})$.

8. Show that the inner product of $u_1^\dagger(\mathbf{p})v_2(-\mathbf{p}) = 0$ and that $u_1^\dagger(\mathbf{p})u_2(-\mathbf{p}) = 0$.

9. Show that for the solutions (4-20), $\langle \psi^{(1)} | \psi^{(4)} \rangle = 0$ and $\langle \psi^{(3)} | \psi^{(4)} \rangle = 0$

10. Use (4-10) and (4-20) to show yourself what the four adjoints $\langle \bar{\psi}^{(n)} |$ look like.

11. Derive the adjoint Dirac equation (4-31).

12. Pick one of the adjoints $\langle \bar{\psi}^{(n)} |$ you found in Prob. 10, plug it into the adjoint Dirac equation, and chug the math to show it solves that equation.

13. Derive the 4-current related to the conserved quantity associated with Dirac particles that might be interpreted as probability. (Hint: Pre-multiply the Dirac equation by $\bar{\psi}_{state}$, post-multiply the adjoint Dirac equation by ψ_{state}, add the two, then group the resulting terms such that the four-derivative ∂_μ is taken with respect to a quantity in brackets. That quantity will be the four-current with zero four-divergence in spacetime.) You should get the four current j^μ of the RHS of (4-34) multiplied by the constant i, but the constant is irrelevant in $\partial_\mu j^\mu = 0$, so, we drop it.

14. Operate on the four solutions (4-20) with the Hamiltonian $H = i\partial/\partial t$ to show that those with coefficient $D_r^\dagger(\mathbf{p})$ have negative energies and those with $C_r(\mathbf{p})$ have positive energies.

15. Show that the NRQM states $(i,-1)^T e^{-ipx}$ and $(i,1)^T e^{-ipx}$ are eigenstates of the y component of the NRQM spin operator, i.e., S_2, and that they can be constructed from linear combinations of the NRQM up and down (z direction) eigenstates.

16. Take $\mathbf{p} = 0$ and operate on the last three of the solutions (4-20) with the relativistic spin operator Σ_3 to show that they are spin eigenstates of Σ_3 when they represent particles that are not moving.

17. Pick one or two, other than the first, of the solutions (4-20) and show that for $p^1 \neq 0$, $p^2 = p^3 = 0$, a particle, which would be in a z direction (up) spin eigenstate if $\mathbf{p}$ were zero, will not be in that up spin eigenstate. Then show, for the same solutions chosen, but with $p^1 = p^2 = 0$, $p^3 \neq 0$ as seen in the lab, that the particle is in the same spin eigenstate as it would be if $\mathbf{p}$ were zero.

18. Show that as a Dirac particle of spinor form $u = u_1 + u_2$ approaches the speed of light in the x direction (i.e., $p^1 \to \infty$, $p^2 \ll p^1$, and $p^3 \ll p^1$), then u approaches a spin state in the plus x direction (i.e., the spinor $u = u_1 + u_2$ is an eigenvector of Σ_1 of (4-39) with eigenvalue $+$ ½ (in natural units.) Then show that for $p^2 \approx p^1$ and $p^3 \approx p^1$, this is not true.

19. Take the relativistic solutions $|\psi^{(1)}\rangle$ and $|\psi^{(2)}\rangle$ of (4-20) for the Dirac equation in the non-relativistic limit, i.e., with $v \ll 1$ ($c=1$ in natural units), and show that they reduce to approximately $(1,0,0,0)^T e^{-ipx}$ and $(0,1,0,0)^T e^{-ipx}$. Do these look remarkably like the wave functions of a spin ½ particle you encountered in NRQM, i.e., spin up state $e^{-ipx}(1,0)^T$ and spin down state $e^{-ipx}(0,1)^T$?

Operate on each of the two low speed 4-spinor states you found with the 4D spin operator Σ_3 of (4-39). Now, operate on each of the two NRQM 2-spinor states above for up and down spin with ½$\hbar\sigma_3$. Do you get the same spin eigenstates and eigenvalues? Does the low speed RQM case for C type (i.e., having spinors u_r) particles parallel that of NRQM?

20. Take a Dirac particle in state $|\psi^{(2)}\rangle$ with $p^1 \neq 0$, $p^2 = p^3 = 0$. Without doing any math, is this in a helicity eigenstate or not? Operate on the state with the helicity operator and prove your answer mathematically. Then repeat the exercise for the same state with $p^3 \neq 0$, $p^1 = p^2 = 0$.

21. Construct your own Wholeness Chart summarizing Sect. 4.1, RQM for Spinors. You do not need to summarize spin and helicity, as that is done for you in Wholeness Chart 4-1.

22. Using (4-60), show that for $n = 1$ in (4-61), we get the Dirac equation; and for $n = 2$, we get the adjoint Dirac equation.

23. Prove that the last lines of (4-68) and (4-69) sum to zero. Hint: First use the Dirac equation with a single $\mathbf{p}$ eigensolution of spin s for a d type field to show that $(\not{p} + m)v_s(\mathbf{p}) = 0$. Then, use the RH relation of (4-25) to show $\bar{u}_r(-\mathbf{p})\gamma^0 p_0 v_s(\mathbf{p}) = 0$. Then, use that to re-write the last line of (4-68) in terms of $\not{p}$. Finally, combine that line with the last line of (4-69) and show that together they equal zero.

24. Show that third lines of (4-68) and (4-69) sum to (4-74).

25. Show that $\left(c_r(\mathbf{p})\right)^2 = 0$, $\left(d_r^\dagger(\mathbf{p})\right)^2 = 0$, and $\left(d_r(\mathbf{p})\right)^2 = 0$. Would any of these or (4-84) be true if the coefficients obeyed commutation relations, rather than anti-commutation relations?

26. Operate on the state $\left|\psi_{r_1,\mathbf{p}_1}, \psi_{r_2,\mathbf{p}_1}, \psi_{r_1,\mathbf{p}_3}, \bar{\psi}_{r_1,\mathbf{p}_1}\right\rangle$ with the $\mathbf{P}$ operator of (4-105) and show the eigenvalue from such operation equals the total 3-momentum of that state.

27. Derive the Dirac 3-momentum operator of (4-107).

28. Show that (4-151) holds for γ^1 and $u_1(\mathbf{p})$. (Hint: Simply insert the standard representation relations and matrix multiply.)

29. Prove (4-153), i.e., $u_r(\mathbf{p})\bar{u}_r(\mathbf{p}) = \dfrac{\not{p} + m}{2m}$, by substituting the spinors of (4-20), the Dirac matrices of (4-10), and $p^i = -p_i$.

30. Prove i) several of the anti-commutation relations (4-157) by plugging and chugging for different values of μ and ν, and/or ii) all of the anti-commutation relations (4-157) using (4-6). (Hint for part ii): For $i \neq j$, use the top line of (4-6), insert $\beta\beta = I$ in between the factors in each term, and take $\alpha_i\beta = -\beta\alpha_i$. For i and 0, pre-multiply $[\alpha_i,\beta]_+ = 0$ by β and substitute γ^μ values. For $i = j$, insert $\beta\beta$ between the factors and use $[\alpha_i,\beta]_+$. For 0 and 0, it is simple.)

31. Using the anti-commutation relations (4-157) and the definition $\gamma_\mu = g_{\mu\nu}\gamma^\nu$, prove that $\gamma_\mu\gamma^\mu = 4I$, where I is the 4X4 identity matrix in spinor space.

32. Prove (4-160). (Hint: Use the Dirac matrices anti-commutation relations (4-157) in
$$\not{p}\not{p} = p_\mu\gamma^\mu p_\nu\gamma^\nu = p_\mu p_\nu\gamma^\mu\gamma^\nu = p_0 p_0\gamma^0\gamma^0 + p_0 p_1\gamma^0\gamma^1 + ... + p_1 p_0\gamma^1\gamma^0 + ...)$$

33. Show that the third term of (4-174) equals 0, and the fourth term is $\mathcal{O}_{r=2,\mathbf{p}} |C_2|^2$.

Chapter 5

Vectors: Spin 1 Fields

"Three passions, simple but overwhelmingly strong, have governed my life: the longing for love, the search for knowledge, and unbearable pity for the suffering of mankind. These passions, like great winds, have blown me hither and thither, in a wayward course, over a great ocean ...

... I have wished to understand the hearts of men. I have wished to know why the stars shine. And I have tried to apprehend the Pythagorean power by which number holds sway above the flux. A little of this, but not much, I have achieved."

Excerpts from "What I Have Lived For"
by Bertrand Russell

5.0 Preliminaries

Few of you who have come this far in this book do not share, in some part, Russell's passion for knowledge. It is my hope that, as each of us lives his or her life, we can also share in his other two, most noteworthy, passions.

With regard to the "power by which number holds sway above the flux", few have more ably demonstrated that power in their work than James Clerk Maxwell. His casting of the phenomena of electricity and magnetism into one elegant and holistic mathematical structure will forever remain one of the monumental achievements in the history of mankind.

Although his famous Maxwell equations were formulated for a classical world, they play an equally fundamental role quantum mechanically, as we shall soon see.

5.0.1 Background

Maxwell first published his equations in 1864, well before the advent of special relativity, so they were framed in a distinctly 3D spatial plus time format. And that is how virtually every physics student first learns them. However, as QFT is a distinctly relativistic theory, we will need to work with Maxwell's equations in the more appropriate 4D format. This treatment of electromagnetism is typically reserved for graduate courses, after students have gained some level of comfort with the 3 + 1 dimensional approach. We will review the 4D approach, but hopefully, it is something that readers of this book have already been exposed to, as it serves as the bedrock for QFT of photons (massless spin 1 bosons).

Maxwell's equations in 4D format are the basis for QFT of photons

5.0.2 Chapter Overview

Our approach to spin 1 bosons (called <u>vectors</u>, for reasons that will become apparent) in this chapter is threefold, including i) a review of classical electromagnetic theory, ii) RQM for photons, and iii) QFT for photons. As we will see, the second and third of these bear striking parallel to comparable aspects of spin 0 boson theory, and this will help to make our work easier.

Spin 1 boson theory development parallels spin 0 boson theory

Vector bosons, like scalars, can be massive or massless, but since our focus in this book is on quantum electrodynamics (QED), where force is mediated by photons, we will be virtually exclusively concerned with photons, which are massless vector bosons.

The following bulleted points provide an overview of this chapter. You may find it helpful to compare and contrast the material below for RQM and QFT with that of the Chapter Overview for scalars at the beginning of Chap. 3.

A review of classical e/m first,

where we will look at

- the pre-relativistic version of Maxwell's equations, their (3D) vector and scalar potentials, and how they describe classical electromagnetic fields/waves,

- those same equations and e/m fields/waves represented covariantly (i.e., in special relativity) as a single equation for a 4D potential A^μ, and

- the classical relativistic Lagrangian density $\mathcal{L}$ for classical e/m fields.

Classical e/m overview

Then RQM for photons (of which e/m waves are made),

- deducing the quantum Maxwell equation in terms of the 4D potential A^μ by applying 1st quantization to classical theory,

- solutions $|A^\mu\rangle$ to that equation (the 4D potential will represent a photon mathematically), and

- noting that those solutions parallel the Klein-Gordon solutions, and so, much of what we learned for scalars can be carried over directly to vectors with $|\phi\rangle \rightarrow |A^\mu\rangle$.

RQM preview (photons = massless vectors)

Then QFT for photons,

- from 2nd quantization, finding the same Maxwell equation, with the same mathematical form for the solutions A^μ, but this time the solutions are quantum fields, not states,

- using the classical relativistic $\mathcal{L}$ for e/m fields and the Legendre transformation to get $\mathcal{H}$ (Hamiltonian density),

- from 2nd quantization, finding the photon field A^μ commutation relations for QFT,

- determining relevant QFT operators for photons by a short cut method: comparing to similar operators for scalars: H, number, creation/destruction, momentum, charge, etc., and

- finding the Feynman propagator for photons by analogy with the scalar propagator.

QFT preview (photons = massless vectors)

As in Chaps. 3 and 4, in this chapter, we will deal only with free particles/fields.

Only free photons in this chapter

5.1 Review of Classical Electromagnetism

5.1.1 Maxwell's Equations in 3D Plus Time Formulation

The Equations

With no charge density ρ_{charge} nor charge current $\mathbf{j}_{charge}$ present (meaning only free fields are involved, as ρ_{charge} and $\mathbf{j}_{charge}$ interact with the electric field $\mathbf{E}$ and the magnetic field $\mathbf{B}$), Maxwell's equations in a vacuum, in naturalized ($c = 1$, $\hbar = 1$), Heaviside-Lorentz units (i.e., units that are rationalized, meaning no factors of 4π appear in Maxwell's equations, and have $\varepsilon_0 = \mu_0 = 1$) are (in the formulation conceived by Oliver Heaviside)

$$\nabla \cdot \mathbf{E} = 0 \qquad \text{(a)}$$

$$\nabla \times \mathbf{B} = \frac{\partial \mathbf{E}}{\partial t} \qquad \text{(b)}$$

$$\nabla \cdot \mathbf{B} = 0 \qquad \text{(c)} \qquad\qquad (5\text{-}1)$$

$$\nabla \times \mathbf{E} = -\frac{\partial \mathbf{B}}{\partial t} . \qquad \text{(d)}$$

Sourceless Maxwell's equations in vacuum

These are commonly referred to as the sourceless Maxwell's equations because there are no sources ρ or $\mathbf{j}$ for the fields in the region in which the fields are being analyzed. (The sources for the fields would be elsewhere outside of the region in which we are interested.)

The Scalar and Vector Potentials: Making It Easier to Solve Maxwell's Equations

Maxwell's equations (5-1) are four coupled differential equations in two vector fields $\mathbf{E}$ and $\mathbf{B}$. To make them easier to solve, we would like, if we could, to re-express them in simpler form.

This can be done if we define a scalar potential $\Phi(\mathbf{x},t)$ (not to be confused with our scalar field ϕ of Chap. 3) and a vector potential $\mathbf{A}(\mathbf{x},t)$ so they solve

$$\mathbf{B} = \nabla \times \mathbf{A}, \qquad \mathbf{E} = -\nabla\Phi - \frac{\partial \mathbf{A}}{\partial t}, \qquad\qquad (5\text{-}2)$$

Potentials: a way to make Maxwell's equations easier to solve

Vector and scalar potentials Φ and $\mathbf{A}$ yield $\mathbf{E}$ and $\mathbf{B}$

then (5-1)(c) and (d) are solved automatically (identically) by Φ and $\mathbf{A}$. On the other hand, substitution of (5-2) into (5-1)(a) and (b) yields

*The two
equations Φ
and $\mathbf{A}$ must
solve,
equivalent of
Maxwell's eqs*

$$-\nabla^2\Phi - \frac{\partial}{\partial t}(\nabla\cdot\mathbf{A}) = 0 \qquad \text{(a)}$$

$$\underbrace{\nabla\times\nabla\times\mathbf{A}}_{\nabla(\nabla\cdot\mathbf{A})-\nabla^2\mathbf{A}} = -\nabla\frac{\partial\Phi}{\partial t} - \frac{\partial^2\mathbf{A}}{\partial t^2} \Rightarrow \frac{\partial^2\mathbf{A}}{\partial t^2} - \nabla^2\mathbf{A} = -\nabla\frac{\partial\Phi}{\partial t} - \nabla(\nabla\cdot\mathbf{A}), \quad \text{(b)}$$

(5-3)

which Φ and $\mathbf{A}$ must solve. If we can solve (5-3) for Φ and $\mathbf{A}$, then we can find the fields $\mathbf{E}$ and $\mathbf{B}$ from (5-2). So, instead of trying to solve the four equations (5-1) for two vector fields $\mathbf{E}$ and $\mathbf{B}$ (6 component unknowns), we can instead solve the two equations (5-3) for a scalar field Φ and a vector field $\mathbf{A}$ (3 components plus 1 scalar = 4 unknowns). Simpler, no?

Ambiguity in the Scalar and Vector Potentials

Before trying to solve (5-3) for Φ and $\mathbf{A}$, we should realize that Φ and $\mathbf{A}$ are not unique. Note that if we define other quantities (designated by primes) by

$$\Phi' = \Phi + \frac{\partial f}{\partial t}, \qquad \mathbf{A}' = \mathbf{A} - \nabla f, \qquad (5\text{-}4)$$

*Different forms
for scalar and
vector potentials
also yield
$\mathbf{E}$ and $\mathbf{B}$*

then these new quantities Φ' and $\mathbf{A}'$, used in place of Φ and $\mathbf{A}$ in (5-2), yield exactly the same $\mathbf{E}$ and $\mathbf{B}$, regardless of the form of f. That is,

$$\mathbf{B} = \nabla\times\mathbf{A}' = \nabla\times(\mathbf{A} - \nabla f) = \nabla\times\mathbf{A}$$

$$\mathbf{E} = -\nabla\Phi' - \frac{\partial\mathbf{A}'}{\partial t} = -\nabla\left(\Phi + \frac{\partial f}{\partial t}\right) - \frac{\partial}{\partial t}(\mathbf{A} - \nabla f) = -\nabla\Phi - \frac{\partial\mathbf{A}}{\partial t}.$$

(5-5)

Thus, if we have solutions Φ and $\mathbf{A}$ that solve (5-3) (i.e., that yield an $\mathbf{E}$ and $\mathbf{B}$ that solve Maxwell's equations (5-1)), then for any f, Φ' and $\mathbf{A}'$ of (5-4) also solve (5-1)(c) and (d) identically, and also solve (5-3). Φ' and $\mathbf{A}'$ thus yield the same $\mathbf{E}$ and $\mathbf{B}$ solutions to (5-1).

By way of example, put primes on Φ and $\mathbf{A}$ in (5-3) and plug in the primed values of (5-4),

*Those different
forms for scalar
and vector
potentials also
solve above
equations in Φ
and $\mathbf{A}$*

$$-\nabla^2\overbrace{\left(\Phi + \frac{\partial f}{\partial t}\right)}^{\Phi'} - \frac{\partial}{\partial t}\left(\nabla\cdot\overbrace{(\mathbf{A} - \nabla f)}^{\mathbf{A}'}\right) = 0 \qquad \text{(a)}$$

$$\frac{\partial^2}{\partial t^2}\overbrace{(\mathbf{A} - \nabla f)}^{\mathbf{A}'} - \nabla^2\overbrace{(\mathbf{A} - \nabla f)}^{\mathbf{A}'} = -\nabla\frac{\partial}{\partial t}\overbrace{\left(\Phi + \frac{\partial f}{\partial t}\right)}^{\Phi'} - \nabla\left(\nabla\cdot\overbrace{(\mathbf{A} - \nabla f)}^{\mathbf{A}'}\right). \qquad \text{(b)}$$

(5-6)

These equal

$$-\nabla^2\Phi - \nabla^2\frac{\partial f}{\partial t} - \frac{\partial}{\partial t}(\nabla\cdot\mathbf{A}) + \frac{\partial}{\partial t}\nabla^2 f = 0 \qquad \text{(a)}$$

$$\frac{\partial^2}{\partial t^2}\mathbf{A} - \frac{\partial^2}{\partial t^2}\nabla f - \nabla^2\mathbf{A} + \nabla^2\nabla f = -\nabla\frac{\partial}{\partial t}\Phi - \nabla\frac{\partial^2 f}{\partial t^2} - \nabla(\nabla\cdot\mathbf{A}) + \nabla^2\nabla f. \qquad \text{(b)}$$

(5-7)

And after cancelling terms (time derivatives and space derivatives commute), (5-7) are identical to (5-3).

Bottom line: If the potentials Φ and $\mathbf{A}$ yield $\mathbf{E}$ and $\mathbf{B}$ that solve Maxwell's equations, then so do Φ' and $\mathbf{A}'$ of (5-4), where f can be any function we like. We will, of course, want to choose an f that will make Maxwell's equations as easy as possible to solve.

Gauge Theory

The formal name for any theory formulated in terms of one or more potentials (two potentials, Φ and $\mathbf{A}$ here), where different potentials result in the same observable quantities ($\mathbf{E}$ and $\mathbf{B}$ here), is gauge theory. A gauge-invariant transformation changes the potential(s), also called gauge(s), from one form to another, but leaves the observables unchanged (invariant). For us, such a transformation (see (5-4)) can be written as

*Gauge theory:
transformations
of potential(s),
i.e., gauge(s),
yield same
observables*

$$\Phi \to \Phi + \frac{\partial f}{\partial t}, \qquad \mathbf{A} \to \mathbf{A} - \nabla f . \tag{5-8}$$

The primes helped us to understand how the potentials changed, but in the future we will often not use them, but simply talk of gauge transformations (transformations of the potentials) as expressed by (5-8). In our present case, we will understand that terminology to mean we can modify any gauge (potential) solutions Φ and $\mathbf{A}$ by the function f, in the manner shown in (5-8), and still have a viable solution (the same $\mathbf{E}$ and $\mathbf{B}$). Thus, for us the term "changing gauge" will mean changing f.

<u>Picking a Useful Gauge</u>

Of course, we want to pick an f that makes (5-3) easy to solve. Suppose that a solution $\mathbf{A}''$ exists for (5-3), but we don't know what it is. Then another solution $\mathbf{A}$ exists, where $\mathbf{A}$ is defined as

We want the particular gauge (potential) that makes solving easiest

$$\mathbf{A} = \mathbf{A}'' - \nabla f . \tag{5-9}$$

Let's pick f such that the following, known as the <u>Coulomb gauge</u>, is true.

Here, that is the Coulomb gauge

$$\nabla \cdot \mathbf{A} = 0 . \tag{5-10}$$

In other words, so that using (5-9) in (5-10) yields

$$\nabla \cdot \mathbf{A}'' - \nabla^2 f = 0 \quad \to \quad \nabla \cdot \mathbf{A}'' = \nabla^2 f . \tag{5-11}$$

In principle, we can solve (5-11) for f and use that in (5-9) to get our new solution $\mathbf{A}$ of (5-9). But we never have to actually do that. We can simply assume that a solution for f exists that will give us an $\mathbf{A}$ for which (5-10) holds true, and then use (5-10) in (5-3). When we do that, (5-3) becomes

$$\nabla^2 \Phi = 0 \qquad \text{(a)}$$
$$\frac{\partial^2 \mathbf{A}}{\partial t^2} - \nabla^2 \mathbf{A} = -\nabla \frac{\partial \Phi}{\partial t} . \qquad \text{(b)} \tag{5-12}$$

One solution to (5-12)(a) is $\Phi = 0$. Using that, (5-12)(b) becomes

The Coulomb gauge turns the equations in Φ & $\mathbf{A}$ into the wave equation in $\mathbf{A}$

$$\frac{\partial^2 \mathbf{A}}{\partial t^2} - \nabla^2 \mathbf{A} = 0 \quad \to \quad \frac{\partial^2}{\partial t^2}\mathbf{A} - \frac{\partial^2}{\partial x^i \partial x^i}\mathbf{A} = \frac{\partial^2}{\partial x^\mu \partial x_\mu}\mathbf{A} = \partial_\mu \partial^\mu \mathbf{A} = \Box^2 \mathbf{A} = 0 , \tag{5-13}$$

i.e., the wave equation (expressed above in various notations). This has the simple plane wave solution

$$\mathbf{A}(\mathbf{x},t) = \mathbf{A}_0 \, e^{\pm i(\omega t - \mathbf{k} \cdot \mathbf{x})} , \tag{5-14}$$

leading via (5-2) to

From that wave equation we get the wave $\mathbf{A}$, and from it, the waves $\mathbf{E}$ and $\mathbf{B}$

$$\mathbf{E} = -\underbrace{\nabla \Phi}_{\substack{\Phi=0 \\ \text{here}}} - \frac{\partial \mathbf{A}}{\partial t} = \mp i\omega \mathbf{A}_0 \, e^{\pm i(\omega t - \mathbf{k} \cdot \mathbf{x})} \qquad \text{(a)}$$

$$\mathbf{B} = \nabla \times \mathbf{A} = \mp i(\mathbf{k} \times \mathbf{A}_0) e^{\pm i(\omega t - \mathbf{k} \cdot \mathbf{x})} . \qquad \text{(b)} \tag{5-15}$$

(5-15) represents propagating electric and magnetic fields, perpendicular to one another and traveling in the third perpendicular direction $\mathbf{k}$. (Do Prob. 1 to show this.) That is, they represent an electromagnetic wave (photons).

Since we can always readily find $\mathbf{E}$ and $\mathbf{B}$ from $\mathbf{A}$ whenever we want, it is simplest to work with a single equation (5-13) and the single field $\mathbf{A}$, rather than multiple equations in $\mathbf{E}$ and $\mathbf{B}$. Thus, it is common practice to represent, and refer to, electromagnetic fields as $\mathbf{A}$.

<u>Bottom line</u>

If we pick our potential $\mathbf{A}$ such that it satisfies the Coulomb gauge (5-10), then solving Maxwell's equations becomes greatly simplified. That gauge lets us take $\Phi = 0$ and results in the single, well known, and easily solvable wave equation in $\mathbf{A}$, i.e.,

Form of Maxwell's eqs in Coulomb gauge = wave equation in $\mathbf{A}$

$$\boxed{\partial_\mu \partial^\mu \mathbf{A} = 0 \qquad \text{Maxwell's equations in Coulomb gauge (i.e., when } \nabla \cdot \mathbf{A} = 0)} . \tag{5-16}$$

We then find $\mathbf{E}$ and $\mathbf{B}$ from (5-2).

5.1.2 Maxwell's Equation in 4D (Covariant) Formulation

The Four-Vector Potential

All of the prior Sect. 5.1.1 is well and good, but the formulation is not relativistically covariant, and we will need such a formulation for investigating photon behavior in QFT. For that, let's begin by defining a 4D vector potential using Φ and $\mathbf{A}$ that (I, as the author, know ahead of time) will turn out to yield a suitable covariant e/m formulation,

$$A^{\mu}\left(\mathbf{x},t\right)=\left(\Phi\left(\mathbf{x},t\right),\mathbf{A}\left(\mathbf{x},t\right)\right)^{\mathrm{T}} \quad \text{or} \quad A^{\mu}\left(x\right)=\begin{pmatrix}\Phi\left(x\right)\\A^{1}\left(x\right)\\A^{2}\left(x\right)\\A^{3}\left(x\right)\end{pmatrix}, \tag{5-17}$$

The four-vector potential in covariant theory

where the notation (x) means $(\mathbf{x},t)$.

Then, let's define a field $F^{\mu\nu}(x)$ (which is a tensor field since it has two 4D indices μ and ν) that we can construct from (5-17) as

$$F^{\mu\nu}\left(x\right)=\partial^{\nu}A^{\mu}\left(x\right)-\partial^{\mu}A^{\nu}\left(x\right). \tag{5-18}$$

$F^{\mu\nu}$ tensor components (found from A^{μ}) equal E^{i} and B^{i}

Then consider (5-18), where $\mu=1$ and $\nu=2$ and we refer to (5-2),

$$F^{12}\left(x\right)=\underbrace{\partial^{2}}_{-\partial_{2}}A^{1}\left(x\right)-\underbrace{\partial^{1}}_{-\partial_{1}}A^{2}\left(x\right)=\partial_{1}A^{2}\left(x\right)-\partial_{2}A^{1}\left(x\right)$$

$$=\frac{\partial}{\partial x^{1}}A^{2}\left(x\right)-\frac{\partial}{\partial x^{2}}A^{1}\left(x\right)=\underbrace{\left(\nabla\times\mathbf{A}\left(x\right)\right)^{3}}_{x^{3}\text{ direction component}}=B^{3}\left(x\right). \tag{5-19}$$

For $\mu=2$ and $\nu=1$, in the second part of (5-19) the terms are reversed, so $F^{21}(x)=-B^{3}$. By doing Prob. 2, you can prove to yourself that

$$F^{31}\left(x\right)=-F^{13}\left(x\right)=B^{2}\left(x\right) \qquad F^{23}\left(x\right)=-F^{32}\left(x\right)=B^{1}\left(x\right). \tag{5-20}$$

Note from the next to last part of (5-19), and similarly with (5-20), that we are simply re-expressing the magnetic field part of (5-2) here in terms of four vector components of (5-17). This is the more "sophisticated" covariant way of doing the same thing we did for the 3D + time formulation.

For $\mu=0$ and $\nu=1$, where we use (5-2) again, and where we won't explicitly denote the x dependence,

$$F^{01}=\partial^{1}A^{0}-\partial^{0}A^{1}=\frac{\partial\Phi}{\partial x_{1}}-\frac{\partial A^{1}}{\partial t}=-\frac{\partial\Phi}{\partial x^{1}}-\frac{\partial A^{1}}{\partial t}=\underbrace{\left(-\left(\nabla\Phi\right)-\frac{\partial\mathbf{A}}{\partial t}\right)^{1}}_{x^{1}\text{ direction component}}=E^{1}. \tag{5-21}$$

Do Prob. 3 to prove to yourself that

$$F^{10}=-E^{1} \qquad F^{02}=-F^{20}=E^{2} \qquad F^{03}=-F^{30}=E^{3}$$

$$F^{00}=F^{11}=F^{22}=F^{33}=0. \tag{5-22}$$

Note from the next to last part of (5-21), and similarly with the first line of (5-22), that we are simply re-expressing the electric field part of (5-2) here in the more "sophisticated" terms of the four vector components of (5-17).

Summarizing the above, we find that the matrix representation of $F^{\mu\nu}(x)$ is

$$F^{\mu\nu}\left(x\right)=\begin{bmatrix}0 & E^{1} & E^{2} & E^{3}\\-E^{1} & 0 & B^{3} & -B^{2}\\-E^{2} & -B^{3} & 0 & B^{1}\\-E^{3} & B^{2} & -B^{1} & 0\end{bmatrix}, \tag{5-23}$$

$F^{\mu\nu}$ matrix components

where E^1 and B^1 represent what we designated by E_x and B_x in Cartesian coordinates before we worked with contravariant and covariant components, just as x^1 represents X_1 in Cartesian coordinates. Ditto for the other E^i and B^i. ($E^1 = -E_1 = E_x$, $B^1 = -B_1 = B_x$, $E^2 = -E_2 = E_y$, etc.)

This all jibes with what we know from the 3D+1 formulation. So, $F^{\mu\nu}(x)$ is a valid representation, in skew-symmetric matrix form, of the components of the electric field **E** and the magnetic field **B**, where those components are derived from what in 3D+1 formulation was the scalar potential Φ and the 3D vector potential A^i. But here, we have expressed those two potentials as the single 4D vector potential A^μ of (5-17).

Maxwell's Equations in Terms of $F^{\mu\nu}(x)$

Maxwell's equations can be expressed in terms of $F^{\mu\nu}(x)$, i.e., in terms of E^i and B^i, but we won't need to do that for our work, as we will be focusing on $A^\mu(x)$ and the equation governing that. Those who wish can find the equations in terms of $F^{\mu\nu}(x)$ in virtually any graduate level text on electromagnetism.

Four-Vector Potential Makes It Easier to Solve Maxwell's Equations in 4D

What we would like to do is re-express our equations (5-3) which are in terms of the 3D+1 scalar potential Φ and three-vector potential A^i in terms of the 4D four-vector potential $A^\mu(x)$. If we solve that equation for $A^\mu(x)$, then we have our electric and magnetic field solutions from (5-18) and (5-23).

4-vector makes solving Maxwell eqs easier

With the aid of (5-3), we can show that <u>Maxwell's equations for $A^\mu(x)$</u> are

$$\partial^\alpha \partial_\alpha A^\mu(x) - \partial^\mu\left(\partial_\nu A^\nu(x)\right) = 0 \ . \qquad (5\text{-}24)$$

Maxwell eqs in terms of A^μ

To show that, first take (5-24) with $\mu = 0$,

$$\underbrace{\frac{\partial^2}{\partial t^2}A^0(x)}_{\text{drops out}} + \frac{\partial^2}{\partial x^i \partial x_i}A^0(x) - \frac{\partial}{\partial t}\left(\underbrace{\frac{\partial}{\partial t}A^0(x)}_{\text{drops out}} + \frac{\partial}{\partial x^1}A^1(x) + \frac{\partial}{\partial x^2}A^2(x) + \frac{\partial}{\partial x^3}A^3(x)\right) \qquad (5\text{-}25)$$

$$= -\frac{\partial^2}{\partial x^i \partial x^i}A^0(x) - \frac{\partial}{\partial t}\frac{\partial}{\partial x^i}A^i(x) = -\nabla^2\Phi - \frac{\partial}{\partial t}(\nabla\cdot\mathbf{A}) = 0,$$

which is (5-3)(a). Doing Prob. 4 for $\mu = i$, we find that (5-24) equals (5-3)(b).

$$\partial^\alpha \partial_\alpha A^i(x) = \partial^i\left(\partial_\nu A^\nu(x)\right) \quad \Rightarrow \quad \frac{\partial^2\mathbf{A}}{\partial t^2} - \nabla^2\mathbf{A} = -\nabla\frac{\partial\Phi}{\partial t} - \nabla(\nabla\cdot\mathbf{A}). \qquad (5\text{-}26)$$

Thus, (5-24) is the 4D equivalent of (5-3). Solving either means we've solved Maxwell's equations.

Ambiguity in the Four-Vector Potential

Φ and A^i of A^μ are the same animals as those in the 3D + time formulation of (5-2), and yield **E** and **B** that are valid solutions of Maxwell's equations. So, we can incorporate the ambiguity we found via the gauge transformation of (5-8) directly into the components of A^μ.

*Like Φ and **A**, A^μ can take different forms that also solve Maxwell eqs*

That is, if we have a solution to Maxwell's equations $A^\mu(x)$, then we can transform that solution to another solution $A'^\mu(x)$, using the same function $f(x)$, i.e.,

$$A^\mu \to A'^\mu = \begin{pmatrix} \Phi' \\ A'^i \end{pmatrix} = \begin{pmatrix} \Phi + \dfrac{\partial f}{\partial t} \\ A^i - \dfrac{\partial}{\partial x^i}f \end{pmatrix} = \begin{pmatrix} \Phi + \dfrac{\partial f}{\partial t} \\ A^i + \dfrac{\partial}{\partial x_i}f \end{pmatrix} = \begin{pmatrix} \Phi + \partial^0 f \\ A^i + \partial^i f \end{pmatrix} = A^\mu + \partial^\mu f \qquad (5\text{-}27)$$

*The valid forms A^μ can take, i.e., gauge transformations that leave **E** and **B** invariant*

From the analysis of (5-4) through (5-7), we know that substitution of $A'^\mu(x)$ from (5-27) for $A^\mu(x)$ in (5-24) will result in all terms in f dropping out and give us (5-24) in terms of $A^\mu(x)$ back again. You can also do Prob. 5 to show yourself, via a fully covariant analysis, that this is true. Thus, both $A^\mu(x)$ and $A'^\mu(x)$ solve Maxwell's equation in terms of the four-potential.

Bottom line: If the four-potential $A^\mu(x)$ yields **E** and **B** that solve Maxwell's equations, then so does $A'^\mu(x)$ of (5-27), where f can be any function we like. As before, we will want to choose an f that will make Maxwell's equations in $A^\mu(x)$ as easy as possible to solve.

Picking a Useful 4D Gauge

The 3D Coulomb gauge of (5-10) only included A^i, i.e., only 3 components of A^μ, and thus it would not be a good gauge to use in covariant (4D) formulations, where we want four components of a four vector (which transforms covariantly – see Appendix of Chap. 2.) Note that, via analogy with (5-10), if we could have a gauge like the following, called the <u>Lorenz gauge</u>[1], then (5-24) would be greatly simplified.

The Lorenz gauge makes Maxwell eqs in A^μ easiest to solve

$$\boxed{\partial_\nu A^\nu(x) = 0}\ . \tag{5-28}$$

But can we have (5-28)? We can if using it leaves **E** and **B** unchanged. Well, the transformation (5-27) leaves them unchanged, as we proved. Let's assume we have a valid solution $A''^\mu(x)$ for Maxwell's equation. Then

$$A^\mu(x) = A''^\mu(x) + \partial^\mu f(x) \tag{5-29}$$

is also a solution to Maxwell's equations. But to make those solutions easier to solve we also want $A^\mu(x)$ to solve (5-28). Can we choose $A^\mu(x)$ so this is so?

We can always find an f (but we don't have to actually do it) that provides the Lorenz gauge

Plugging $A^\mu(x)$ of (5-29) into (5-28) yields

$$\partial_\mu A''^\mu(x) + \partial_\mu \partial^\mu f(x) = 0 \quad \rightarrow \quad \partial_\mu A''^\mu(x) = -\partial_\mu \partial^\mu f(x)\ . \tag{5-30}$$

So, knowing $A''^\mu(x)$ we can, in principle, solve (5-30) for $f(x)$, and for that particular $f(x)$, $A^\mu(x)$ will, in addition to solving Maxwell's equations, also solve (5-28). By doing the latter it will make our Maxwell equations in terms of the four-potential easier to solve.

We never need to actually solve for $f(x)$. We just need to know that we could solve for it, and so doing would give us a four potential $A^\mu(x)$ that solves the Lorenz gauge. We also never need to know what $A''^\mu(x)$ is. We just know that such a solution must exist. Knowing that $f(x)$ and $A''^\mu(x)$ exist if we wanted to find them is all that is necessary. Knowing that, allows us to know that (5-28) can be true, and that helps us immensely in finding the solution $A^\mu(x)$.

So, with (5-28), Maxwell's equations (5-24) become

$$\partial^\alpha \partial_\alpha A^\mu(x) = \left(\frac{\partial^2}{\partial t^2} + \frac{\partial^2}{\partial x^i \partial x_i}\right) A^\mu(x) = \left(\frac{\partial^2}{\partial t^2} - \frac{\partial^2}{\partial x^i \partial x^i}\right) A^\mu(x) = 0\ , \tag{5-31}$$

which is, once again, the well-known classical wave equation, this time in terms of $A^\mu(x)$.

Bottom line for 4D Formulation

If we pick our four-potential $A^\mu(x)$ such that it satisfies the Lorenz gauge (5-28), then solving Maxwell's 4D equations becomes greatly simplified. That particular choice of gauge results in the single, easily solvable wave equation in $A^\mu(x)$., i.e.,

With the Lorenz gauge, Maxwell's eqs take form of the wave equation

$$\boxed{\partial_\alpha \partial^\alpha A^\mu(x) = 0 \qquad \text{Maxwell's equations in Lorenz gauge (i.e., when } \partial_\mu A^\mu(x) = 0)}\ . \tag{5-32}$$

We then find **E** and **B** from (5-18) and (5-23)[2]. These will obviously behave like waves, since they are governed by the wave equation. They are electromagnetic waves, i.e., photons.

Summary of Classical Electromagnetism and Potentials

All of the preceding material in this chapter is summarized in Wholeness Chart 5-1 below.

---

[1] Named after Ludvig Lorenz, who conceived it, not Hendrik Lorentz of Lorentz transformation, Lorentz invariance, and Lorentz force fame. This is often, incorrectly, called the Lorentz gauge.

[2] Actually, there are many possible Lorenz gauges, all of which solve (5-28), since many different f solve (5-30). For example, any function equal to f plus a constant times x^1 solves (5-30). Because of this, some authors refer to (5-28) as the <u>Lorenz condition,</u> for which there are many possible Lorenz gauges.

Wholeness Chart 5-1. Summary of Classical Electromagnetism Potential Theory

	3D + time	4D
Starting point	Four Maxwell's equations in **E**, **B**	Four Maxwell's equations in **E**, **B**
	To make solving the equations above easier, we introduce the potentials below.	
Potentials	$\Phi(\mathbf{x},t)$, $\mathbf{A}(\mathbf{x},t)$	$A^\mu(x) = [\Phi(x), A^i(x)]^T$
E, **B** in terms of the potentials	$\mathbf{B} = \nabla \times \mathbf{A}$ $\mathbf{E} = -\nabla\Phi - \dfrac{\partial \mathbf{A}}{\partial t}$	$F^{\mu\nu}(x) = \partial^\nu A^\mu(x) - \partial^\mu A^\nu(x)$ $F^{01} = E^1, F^{02} = E^2, F^{12} = B^3$, etc.
Maxwell's equations in terms of potentials	$-\nabla^2\Phi - \dfrac{\partial}{\partial t}(\nabla \cdot \mathbf{A}) = 0$ $\dfrac{\partial^2 \mathbf{A}}{\partial t^2} - \nabla^2 \mathbf{A} = -\nabla\dfrac{\partial \Phi}{\partial t} - \nabla(\nabla \cdot \mathbf{A})$	$\partial^\alpha \partial_\alpha A^\mu(x) - \partial^\mu\left(\partial_\nu A^\nu(x)\right) = 0$
	Above, a reduced number of equations, easier to solve.	
Above even easier to solve if this were true	$\nabla \cdot \mathbf{A} = 0$ $\left(\partial_i A^i = 0\right)$ Coulomb gauge	$\partial_\nu A^\nu(x) = 0$ Lorenz gauge
	We can use the above because many different potentials solve Maxwell's equations, so we pick one of those that also solves these gauge equations. See below.	
Gauge transformations	$\Phi \to \Phi + \dfrac{\partial f}{\partial t}$, $\mathbf{A} \to \mathbf{A} - \nabla f$	$A^\mu(x) \to A^\mu(x) + \partial^\mu f(x)$
	If the left side of the arrow above solves Maxwell's equations, then so does the right side, for any f. We want an f that will make the Coulomb/Lorenz gauge true.	
If $\mathbf{A}''$ (A''^μ) solves Max eqs, then so does $\to$	$\mathbf{A}'' \to \mathbf{A}'' - \nabla f = \mathbf{A}$, $\Phi'' \to \Phi'' + \dfrac{\partial f}{\partial t}$	$A''^\mu(x) \to A''^\mu(x) + \partial^\mu f(x) = A^\mu(x)$
The f we need solves the gauge condition	$\nabla \cdot (\mathbf{A}'' - \nabla f) = 0 \;\to\; \nabla \cdot \mathbf{A}'' = \nabla^2 f$	$\partial_\mu\left(A''^\mu + \partial^\mu f\right) = 0 \;\to\; \partial_\mu A''^\mu = -\partial_\mu \partial^\mu f$
	We don't need to solve for f. We simply know at least one $\mathbf{A}$ (A^μ) exists that satisfies the gauge condition, and we assume our solution $\mathbf{A}$ (A^μ) is one of those.	
Maxwell's eqs then are	$\partial_\mu \partial^\mu \mathbf{A} = 0$ (for $\Phi = 0$)	$\partial_\alpha \partial^\alpha A^\mu(x) = 0$
	Solutions to above are waves. **E**, **B** are found from them and are waves, also.	

Solutions to the 4D Wave Equation in $A^\mu(x)$

The wave equation (5-32) is virtually identical to the Klein-Gordon equation which we studied in depth in Chap. 3 (pg. 49), except for three things: i) photons are massless ($\mu = 0$ in Klein-Gordon equation), ii) an electromagnetic wave is a classical world, measurable entity and thus is real, not complex, and iii) the solution $A^\mu(x)$ is a four-vector, not a scalar like ϕ. Hence, (by doing Prob. 6) you can show that if the photon field is real, i.e., $A^\mu(x) = A^{\dagger\mu}(x)$, then its <u>plane wave, discrete solution</u> has form

$$A^\mu(x) = \underbrace{\sum_{r,\mathbf{k}} \frac{1}{\sqrt{2V\omega_\mathbf{k}}} \varepsilon_r^\mu A_r(\mathbf{k}) e^{-ikx}}_{A^{\mu+}} + \underbrace{\sum_{r,\mathbf{k}} \frac{1}{\sqrt{2V\omega_\mathbf{k}}} \varepsilon_r^\mu A_r^\dagger(\mathbf{k}) e^{ikx}}_{A^{\mu-}}, \qquad (5\text{-}33)$$

Solution to Maxwell's eqs in Lorenz gauge

Solution is real, not complex

where $A_r(\mathbf{k})$ is a number, generally complex, and for each r, ε_r^μ is a four dimensional vector, which we can take, without loss of generality, to be unit length. Note from Prob. 6 how the form of (5-33), because it is real, differs from the Klein-Gordon solutions (which were complex) in not having a second set of coefficients, such as $B_r(\mathbf{k}) \neq A_r(\mathbf{k})$.

The Four Polarization Vectors ε_r^μ

The ε_r^μ, called <u>polarization vectors</u>, can take a little time getting used to. First, since we are dealing in spacetime, they must have four components. Second, to span a 4D space, we need four independent vectors. Hence, there must be four of them, each with four components. The μ superscript stands for the four components ($\mu = 0,1,2,3$). The subscript r stands for the four independent vectors ($r = 0,1,2,3$), which we will take to be orthogonal. In general, each independent vector ε_r^μ has components along each of the four axes in 4D. See Fig. 5-1(a).

Visualizing the polarization vectors

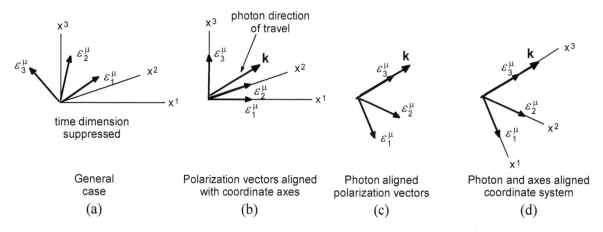

General case	Polarization vectors aligned with coordinate axes	Photon aligned polarization vectors	Photon and axes aligned coordinate system
(a)	(b)	(c)	(d)

Figure 5-1. Visualizing the Polarization Vectors in Different Alignments

However, to make things simpler, we can align our vectors ε_r^μ with our four coordinate axes, as in Fig. 5-1(b). When we do that, it is easy to recognize that in this special <u>polarization vectors aligned coordinate system</u>

$$\varepsilon_0^\mu = \left(1,0,0,0\right)^{\mathrm{T}} \quad \varepsilon_1^\mu = \left(0,1,0,0\right)^{\mathrm{T}} \quad \varepsilon_2^\mu = \left(0,0,1,0\right)^{\mathrm{T}} \quad \varepsilon_3^\mu = \left(0,0,0,1\right)^{\mathrm{T}}. \tag{5-34}$$

From (5-34), we easily recognize that the ε_r^μ are orthogonal. Using our 4D metric $g_{\mu\nu}$, we see that

$$\varepsilon_{\mu 0}\varepsilon_0^\mu = g_{\mu\nu}\varepsilon_0^\nu\varepsilon_0^\mu = 1, \ \ \varepsilon_{\mu 1}\varepsilon_1^\mu = g_{\mu\nu}\varepsilon_1^\nu\varepsilon_1^\mu = -1, \ \ \varepsilon_{\mu 2}\varepsilon_2^\mu = g_{\mu\nu}\varepsilon_2^\nu\varepsilon_2^\mu = -1, \ \ \varepsilon_{\mu 3}\varepsilon_3^\mu = g_{\mu\nu}\varepsilon_3^\nu\varepsilon_3^\mu = -1$$

$$\varepsilon_{\mu 0}\varepsilon_1^\mu = g_{\mu\nu}\varepsilon_0^\nu\varepsilon_1^\mu = 0, \ \ \varepsilon_{\mu 0}\varepsilon_2^\mu = g_{\mu\nu}\varepsilon_0^\nu\varepsilon_2^\mu = 0, \ \text{etc.} \tag{5-35}$$

So, in general, the <u>orthogonality conditions</u> for ε_r^μ are

$$\varepsilon_{\mu r}\varepsilon_s^\mu = g_{rs} = -\zeta_r\delta_{rs} \quad \text{where} \quad \zeta_0 = -1 \quad \zeta_1 = \zeta_2 = \zeta_3 = 1, \tag{5-36}$$

Orthogonality of polarization vectors

we use the metric g_{rs} as a mnemonic, and we introduce the symbol ζ_r, which will be of use later on. Note that (5-36) is also true in any other coordinate system, since (5-36) represent inner products between vectors ε_r^μ, and inner products are scalars, which are the same in any coordinate system. Also, physically, the inner product of two vectors can be expressed independently of a coordinate system as simply the product of the vector lengths times the cosine of the angle between them. So it is always the same regardless of the coordinate system we choose to work in.

Same orthogonality conditions in any coordinate system

Note we can choose to align our ε_3^μ vector with the $\mathbf{k}$, the direction of travel of the photon (regardless of the coordinate system alignment), as in Fig. 5-1(c). We deem this the <u>photon polarization vector alignment</u>, and for it, ε_3^μ is called the <u>longitudinal polarization</u> (vector), i.e., where bold indicates a 3D vector direction,

$$\varepsilon_3 = \mathbf{k} / |\mathbf{k}|. \tag{5-37}$$

For ε_3^μ in $\mathbf{k}$ direction: ε_3^μ is longitudinal, ε_1^μ and ε_2^μ are transverse

ε_1^μ and ε_2^μ are orthogonal to ε_3^μ, and for this alignment, are called the <u>transverse polarizations</u>.

$$\mathbf{k} \cdot \varepsilon_1 = \mathbf{k} \cdot \varepsilon_2 = 0. \tag{5-38}$$

Thus, we will expect $\varepsilon_1{}^\mu$ and $\varepsilon_2{}^\mu$ here to be in the same plane as the **E** and **B** vectors, since they are transverse to the direction of travel of an electromagnetic wave.

$\varepsilon_0{}^\mu$ is then scalar, or time-like

$\varepsilon_0{}^\mu$, for the photon aligned polarization vectors (Fig. 5-1(c) and (5-37)) and in the polarization vectors aligned coordinate system (Fig. 5-1(b) and (5-34)), points in the time (4th dimension) direction and in such systems is called the <u>time-like</u> or <u>scalar polarization</u>.

We could also choose to align all of the x^3 axis ($\mu = 3$ direction), $\varepsilon_3{}^\mu$, and the **k** vector, which can be very helpful. (See Fig. 5-1(d).) We will call this the <u>photon aligned coordinate system</u> (rather than the more cumbersome, but more correct, "photon-polarization vectors-axes aligned system"). It is easiest to visualize all of this $\varepsilon_r{}^\mu$ stuff in a coordinate system that has both its axes aligned with the $\varepsilon_r{}^\mu$ and also has $\varepsilon_3{}^\mu$ aligned in the **k** direction, as in Fig. 5-1(d)[1].

The photon-aligned system is easiest to work with

Physical View of **A**, **B**, **E**, and Polarization Vectors $\varepsilon_r{}^\mu$

Fig. 5-2 can give us a physical feeling for how the vector potential A^μ relates to the polarization vectors $\varepsilon_r{}^\mu$ and to **B** and **E**. In the figure, we work in the photon aligned coordinate system, suppress the time dimension, focus on just the 3-vector part **A** of A^μ, show only the $r = 1$ polarization state of A^μ for a particular **k**, and for simplicity (and with no loss of generality) take the coefficients to be real, i.e., $A_1(\mathbf{k}) = A_1{}^\dagger(\mathbf{k})$. That is, Fig. 5-2 depicts one term in the summation of (5-33),

Looking at one polarization state of A^μ

$$A^\mu(x) = \frac{1}{\sqrt{2V\omega_k}}\varepsilon_1{}^\mu A_1(\mathbf{k})e^{-ikx} + \frac{1}{\sqrt{2V\omega_k}}\varepsilon_1{}^\mu A_1(\mathbf{k})e^{ikx} = \sqrt{\frac{2}{V\omega_k}}\varepsilon_1{}^\mu A_1(\mathbf{k})\cos kx, \quad (5\text{-}39)$$

Math form of that polarization state

where $\varepsilon_1{}^\mu = (0,1,0,0)^T$ in this coordinate system. This is <u>linear polarization</u>, in which the fields **A**, **B**, and **E** do not rotate. (In circular polarization they rotate about the **k** vector.)

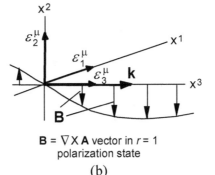

Pictorial form of that polarization state

The **A** vector in $r = 1$ polarization state

(a)

*This is linear polarization with no rotation of fields about **k***

B = ∇ X **A** vector in $r = 1$ polarization state

(b)

Figure 5-2. Physical Look at Vector Potential and Its Curl in $r = 1$ Polarization State

For Fig. 5-2, **A** is aligned in the x^1 direction (the $\varepsilon_1{}^\mu$ basis vector direction.) It is a sinusoidal wave (note cosine in (5-39)) moving in the $+x^3$ direction (the **k** or $\varepsilon_3{}^\mu$ direction), so the field **A** is a transverse vector field varying in space (and propagating over time). **B** = ∇ x **A**, so **B** is orthogonal to the plane of the **A** field, and as should be obvious from the figure, is aligned in the x^2 direction ($\varepsilon_2{}^\mu$ direction.)

*For **A** field in x^1 direction polarization state, **B** field is in x^2 direction*

From

$$\nabla \times \mathbf{E} = \nabla \times \left(-\nabla\Phi - \frac{\partial \mathbf{A}}{\partial t}\right) = -\nabla \times \frac{\partial \mathbf{A}}{\partial t} = -\frac{\partial}{\partial t}(\nabla \times \mathbf{A}) = -\frac{\partial \mathbf{B}}{\partial t}, \quad (5\text{-}40)$$

*And **E** field is in x^1 direction*

which is simply one of Maxwell's equations in **E** and **B**, we see that **B** is orthogonal to the plane of the **E** wave. We don't show **E** in Fig. 5-2, but it would be in the x^1 direction ($\varepsilon_1{}^\mu$ direction.)

---

[1] It is a subtle point, but we can take the components of $\varepsilon_r{}^\mu$ to be real as long as we are dealing with linear polarization, but if we wish to describe circular or elliptic polarization, we would require complex polarization vectors and concomitant modifications to the treatment herein.

For **A** aligned with the $\varepsilon_2{}^\mu$ direction in Fig. 5-2, **B** would be perpendicular to it (x^1 direction) and **E** would be parallel. Note that an **A** in any direction in the plane orthogonal to **k** can be constructed by a superposition of the $\varepsilon_1{}^\mu$ and $\varepsilon_2{}^\mu$ states, and so our solution (5-33) is the most general solution for A^μ.

*For **A** in x^2 direction polarization, **B** in x^1 direction, **E** in x^2 direction*

5.1.3 The Classical Electromagnetic Lagrangian

Since we are dealing with electromagnetic *fields*, the Lagrangian *density*, rather than the Lagrangian is relevant. There is actually more than one possible Lagrangian density that leads to a correct classical theory of electromagnetism. The simplest of these, first proposed by Fermi, turns out to be suitable for quantization (later in this chapter) and, is, for free fields (subscript 0 below),

$$\mathcal{L}_0^{e/m} = -\tfrac{1}{2}\left(\partial_\nu A_\mu(x)\right)\left(\partial^\nu A^\mu(x)\right). \tag{5-41}$$

e/m Lagrangian density for Lorenz gauge

(5-41) can be verified by inserting it into the Euler-Lagrange field equation

$$\frac{\partial}{\partial x^\nu}\left(\frac{\partial \mathcal{L}}{\partial \phi^n{}_{,\nu}}\right) - \frac{\partial \mathcal{L}}{\partial \phi^n} = 0, \quad \text{with} \quad \phi^n = A_\mu \; ; \; \mathcal{L} = \mathcal{L}_0^{e/m}, \tag{5-42}$$

to get (5-32). The minus sign in (5-41) is not really needed here, but is the extant convention, since in the development of QFT, things then turn out more conveniently. (5-41) works for the Lorenz gauge.

In Chap. 11 we will look at another possible Lagrangian (density) that yields Maxwell's equations and proves more useful in interaction theory. For now, however, (5-41) works well for free fields and is simpler to use while we begin to get our feet wet with QFT electromagnetic theory.

5.2 Relativistic Quantum Mechanics for Photons

5.2.1 Brief History of the Photon

Although Maxwell's equations were discovered in the mid 1800s, the solution form was wavelike, and the entity that solution represented was considered to be a wave. So, at that time and for some time thereafter, it was assumed it could not be particle-like. However, in 1899, Planck, and then in 1905, Einstein, explained certain electromagnetic phenomena in terms of indivisible packets, or quanta of electromagnetic energy, that came to be known as photons. In the ensuing years, the concept of wave/particle duality gained acceptance, and today we know that photons, like all other particles such as the electron, possess both wave and particle properties. This, of course, played a major role in the development of quantum theory.

Knowing that electromagnetic quanta (photons) existed, researchers in the early 1900s soon realized that each photon had an energy of $\hbar\omega$ and a 3-momentum of $\hbar\mathbf{k}$, and that they could be treated like massless particles that obey special relativity.

The advent of the Pauli exclusion principle, with the realization that fermions and bosons differ, led to the understanding that many bosons could coalesce into a macroscopic, classical field. The electromagnetic fields classical researchers had worked with were each simply a collection of photons.

Maxwell's eqs for classical e/m wave apply to individual photon

So, if Maxwell's equations described a classical field, they had to also apply to each of the individual photons making up that macroscopic field. The solution to Maxwell's equations $A^\mu(x)$ could then be used as either a representation of a macroscopic classical field, or as a representation of a "classical", single photon. True, that this "classical" photon had some strange properties (e.g., wave/particle duality), but without invoking commutation relations (and thereby quantizing the theory), the photon, and the theory surrounding it, were still essentially non-quantum.

5.2.2 1st Quantization and RQM for Photons

The top RHS of Fig. 5-3 illustrates the manner discussed above in which one can deduce a (somewhat) classical, relativistic electromagnetic particle theory for photons from the classical, relativistic theory of electromagnetic fields (Maxwell's theory.)

Given that theory, we could then follow the usual steps for 1st quantization and a RQM theory of photons. However, there is a quicker and easier way.

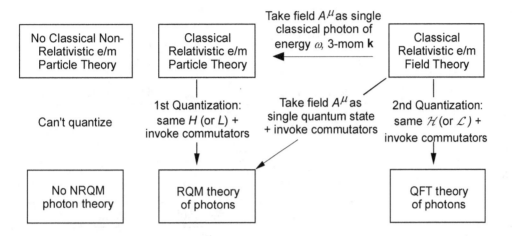

Ways from classical e/m theory to RQM and QFT

Figure 5-3. Quantizing the Electromagnetic Field

Recall that the Hamiltonian H, the Lagrangian L, and the governing wave equation are but different forms for representing the same thing. From any one of these, we can deduce the other two. We know that the governing equation for photons is Maxwell's equation (in the four-potential $A^\mu(x)$ form), so we can simply use that equation for quantization instead of going to all the trouble of finding $H = \int \mathcal{H}\,d^3x$, or $L = \int \mathcal{L}\,d^3x$. That is, we follow the diagonal arrow in Fig. 5-3 to quantize our classical field theory into a quantum particle (RQM) theory of photons.

Simplest route to RQM is diagonal arrow above

Before looking more closely at that, note, as represented on the LHS of Fig. 5-3, that because electromagnetic fields/photons are relativistic, there can be no non-relativistic theory of them, classical or quantum.

<u>RQM Wave Equation for Photons</u>

The first step in 1st quantization comprises taking the same e/m wave equation we had classically, (5-32), and thus, the same solution form for the state $|A^\mu(x)\rangle$ as for the classical $A^\mu(x)$, (5-33), which we repeat below.

1st quantization, first step: Same wave equation in RQM as in classical e/m

$$\partial_\alpha \partial^\alpha A^\mu_{state}(x) = \partial_\alpha \partial^\alpha \left| A^\mu \right\rangle = 0 . \tag{5-43}$$

$$\left| A^\mu \right\rangle = \sum_{r,\mathbf{k}} \frac{1}{\sqrt{2V\omega_\mathbf{k}}} \varepsilon^\mu_r(\mathbf{k}) A_r(\mathbf{k}) e^{-ikx} + \sum_{r,\mathbf{k}} \frac{1}{\sqrt{2V\omega_\mathbf{k}}} \varepsilon^\mu_r(\mathbf{k}) A_r^\dagger(\mathbf{k}) e^{ikx} . \tag{5-44}$$

<u>Polarization, **B, E** Unrelated to Spin</u>

When students learn of circular polarization states, where the transverse **E** and **B** states rotate around the **k** vector direction as the e/m wave propagates, they often confuse that rotation with photon spin. The two are unrelated. Classical angular momentum increases with rotation rate, and the rate of the rotation of **A** (and thus of **E** and **B**) increases with $\omega_\mathbf{k}$ in (5-39), i.e., with the energy of the photon. But every photon, regardless of energy, has the same spin 1 value. Further, as we saw on pg. 143, we can have linear polarization where the **E** and **B** do not rotate around the **k** vector, but the photon still had spin 1.

Circular polarization is not related to spin

Bottom line: Don't confuse circular polarization with spin. They are different.

<u>Polarization, Not Spin, States are Basis States for Photons</u>

In electromagnetism and the quantum theories derived from it, we formulate the theory with polarization states rather than spin states. Photon spin is always in the + or − **k** direction and comprises only two possible states for given **k**. Polarization vectors have four possible states, mutually orthogonal in 4D space, and thus can serve as basis states, whereas spin (for photons) cannot.

Polarization states, not spin states, used as photon basis vectors

We compare and contrast spinor spin basis states with photon polarization basis states in Wholeness Chart 5-2.

Wholeness Chart 5-2. Comparing Spinor and Polarization Basis States

	Dirac Particles	Photons
Wave equation	Dirac equation	Maxwell equation in $\lvert A^\mu \rangle$
For simplicity	Only consider c_r solutions	Only consider a_r solutions
Single eigensolution to wave equation = basis vector for general states	$\lvert \psi_{r,\mathbf{p}} \rangle = \sqrt{\dfrac{m}{VE_\mathbf{p}}} C_r(\mathbf{p}) u_r(\mathbf{p}) e^{-ipx}$	$\lvert A^\mu_{r,\mathbf{k}} \rangle = \dfrac{1}{\sqrt{2V\omega_\mathbf{k}}} A_r(\mathbf{k}) \varepsilon^\mu_r(\mathbf{k}) e^{-ikx}$
Basis vector (one for each r value for given $\mathbf{p}$ or $\mathbf{k}$)	In spinor space, $u_r(\mathbf{p})$ $r = 1,2$	In physical spacetime, $\varepsilon_r{}^\mu(\mathbf{k})$ $r = 0,1,2,3$
r value related?	Symbol r used in both cases, but has different meanings.	
Are basis vectors spin eigenstates?	Generally not. Yes, for $\mathbf{p} = 0$ or in x^3 direction.	No. Although $\varepsilon_3{}^\mu$ is, if aligned with $\mathbf{k}$. Other $\varepsilon_r{}^\mu$ would not be.
What are basis vectors related to?	Spin	Photon polarization (direction of A^μ in space)

RQM Commutation Relations for Photons

The second step in 1st quantization is taking Poisson brackets over into commutators (with the factor of i.) Note, for example, what this means for 3-momentum and position and one of the eigenstates from the first summation set in (5-44).

$$\left[x^1, p^1 \right] = i \quad \rightarrow \quad \left[x^1, p^1 \right] \left| A^\mu \right\rangle = i \left| A^\mu \right\rangle = i \frac{1}{\sqrt{2V\omega_\mathbf{k}}} \varepsilon_r^\mu A_r(\mathbf{k}) e^{-ikx} \quad \rightarrow \quad \underbrace{p^1}_{\text{operator}} = -i \frac{\partial}{\partial x^1} \quad (5\text{-}45)$$

1st quantization, second step: Poisson brackets to commutators

The commutation relations mean the dynamical variables in classical theory become operators in quantum theory, as we have seen before. From (5-45),

$$\underbrace{-i \frac{\partial}{\partial x^1}}_{p^1 \text{ operator}} \left| A^\mu \right\rangle = -i \frac{\partial}{\partial x^1} \left(\frac{1}{\sqrt{2V\omega_\mathbf{k}}} \varepsilon_r^\mu A_r(\mathbf{k}) e^{-i\left(\omega t - p^i x^j \right)} \right) = \underbrace{p^1}_{\substack{p^1 \\ \text{number}}} \left| A^\mu \right\rangle . \qquad (5\text{-}46)$$

The Nomenclature "Vector" for Spin 1 Particles

Note now, why the term "vector" is used for spin 1 bosons. Because $A^\mu(x)$, which represents that boson, is a four vector.

5.2.3 Same Problems for RQM for Photons as Seen Before

It should come as little surprise that RQM for photons suffers from the same deficiencies we saw before for RQM for scalar and spinor particles. By doing Prob. 7, you can see that, for example, we have negative energies for half of our solutions (5-44).

Photon RQM problems like those for scalars and spinors

And so, as before, in order to get a full and satisfactory relativistic theory, we need to move on to the QFT of photons.

5.3 The Maxwell Equation in Quantum Field Theory

5.3.1 Summary Chart

All that we will do in the remainder of this chapter is summarized in Wholeness Chart 5-4 on pg. 156 at the end of the chapter. As in prior chapters, I highly recommend following along, step by step, in that chart as we progress through QFT for photons.

Be sure to use the summary wholeness chart, as you study this chapter

5.3.2 From Classical Electromagnetism to QFT

The RHS of Fig. 5-3 shows the steps to QFT for photons via 2nd quantization. The first step is taking the classical Lagrangian density (or Hamiltonian density or field equations) directly over to QFT. The photon field $A^\mu(x)$ is real, so we don't have to convert our real classical field to a complex field.

The classical Lagrangian density (5-41) equals the QFT Lagrangian density. We repeat it here, but change the superscript from "e/m" to "1", indicating a spin 1 boson.

2nd quantization, first step: Use classical Lagrangian density

$$\mathcal{L}_0^1 = -\tfrac{1}{2}\left(\partial_\nu A_\mu(x)\right)\left(\partial^\nu A^\mu(x)\right). \tag{5-47}$$

From (5-47) and the Euler-Lagrange equation, one obtains the QFT field equation (which equals (5-32)) of

Same Lagrangian yields same field (wave) equation as classically

$$\partial_\alpha\partial^\alpha A^\mu(x) = 0, \tag{5-48}$$

where $A^\mu(x)$ is a quantum field, not a quantum state. The discrete plane wave solution of (5-48) is

$$A^\mu(x) = \underbrace{\sum_{r,\mathbf{k}}\frac{1}{\sqrt{2V\omega_\mathbf{k}}}\varepsilon_r^\mu(\mathbf{k})a_r(\mathbf{k})e^{-ikx}}_{A^{\mu+}} + \underbrace{\sum_{r,\mathbf{k}}\frac{1}{\sqrt{2V\omega_\mathbf{k}}}\varepsilon_r^\mu(\mathbf{k})a_r^\dagger(\mathbf{k})e^{ikx}}_{A^{\mu-}}, \tag{5-49}$$

Plane wave solutions to field equation

where we change the coefficients from capital to lower case letters, which we are using to indicate operators. We anticipate these coefficients, in QFT, will turn out to be creation and destruction operators, as scalar and spinor field theory coefficients were.

The continuous plane wave solutions, which one can check by doing Prob. 8, are

$$A^\mu(x) = \underbrace{\sum_r\int\frac{d^3k}{\sqrt{2(2\pi)^3\omega_\mathbf{k}}}\varepsilon_r^\mu(\mathbf{k})a_r(\mathbf{k})e^{-ikx}}_{A^{\mu+}} + \underbrace{\sum_r\int\frac{d^3k}{\sqrt{2(2\pi)^3\omega_\mathbf{k}}}\varepsilon_r^\mu(\mathbf{k})a_r^\dagger(\mathbf{k})e^{ikx}}_{A^{\mu-}}. \tag{5-50}$$

5.3.3 The Parallel with Scalars

Note how (5-48), (5-49), and (5-50) are virtually identical to the parallel relations for scalars (massless scalars with $\mu = 0$ in this case). (See Chap. 3, pgs. 49-50.) They differ only in that the latter are not vectors and are complex, not real. (5-47) is identical to the scalar free Lagrangian density, except for those differences and the factor of $-\tfrac{1}{2}$.

Photon relations very similar to scalars

These similarities will allow us to take many relations we spent much time deriving in scalar (and spinor) theory directly over to vector theory, simply by changing ϕ to $A^\mu(x)$. You will probably want to put a book marker in Chap. 3 in the QFT development pages there, as we will be referring to, and simply lifting, many of those derived relations for use in this chapter with photons.

5.3.4 Conjugate Momentum and Hamiltonian Density

From (5-47), the photon conjugate momentum is

$$\pi_\mu^1 = \frac{\partial \mathcal{L}_0^1}{\partial \dot{A}^\mu} = \frac{\partial}{\partial \dot{A}^\mu} \left(-\tfrac{1}{2} \underbrace{(\partial_0 A_\nu)}_{\dot{A}_\nu} \underbrace{(\partial^0 A^\nu)}_{\dot{A}^\nu} - \underbrace{\tfrac{1}{2}(\partial_i A_\nu)(\partial^i A^\nu)}_{\text{no } \dot{A}^\nu, \dot{A}_\nu, \text{ so drops out}} \right) .$$

$$= -\tfrac{1}{2} \left(\left(\underbrace{\frac{\partial}{\partial \dot{A}^\mu} \dot{A}_\nu}_{g_{\mu\nu}} \right) \dot{A}^\nu + \dot{A}_\nu \left(\underbrace{\frac{\partial}{\partial \dot{A}^\mu} \dot{A}^\nu}_{\delta_{\mu\nu}} \right) \right) = -\dot{A}_\mu .$$

(5-51)

Conjugate momentum for A^μ

If you don't see the $g_{\mu\nu}$ part above right away, do Prob. 9, in which I lead you through the derivation. From (5-51), the Hamiltonian density is

$$\mathcal{H}_0^1 = \pi_\mu^1 \dot{A}^\mu - \mathcal{L}_0^1 = -\dot{A}_\mu \dot{A}^\mu + \tfrac{1}{2} \dot{A}_\mu \dot{A}^\mu + \tfrac{1}{2}(\partial_i A_\mu)(\partial^i A^\mu) = -\tfrac{1}{2} \dot{A}_\mu \dot{A}^\mu + \tfrac{1}{2}(\partial_i A_\mu)(\partial^i A^\mu)$$

$$= -\tfrac{1}{2} \dot{A}_\mu \dot{A}^\mu - \tfrac{1}{2}(\partial^i A_\mu)(\partial^i A^\mu) = -\tfrac{1}{2}(\partial^\nu A_\mu)(\partial^\nu A^\mu).$$

(5-52)

Hamiltonian density in terms of A^μ

5.3.5 The Underlying Reason Why A^μ is Real

In Sect. 5.1.2 on pg. 141 we gave a heuristic reason why the solutions to the photon wave equation (5-48) were real. The quantum reason is that photons are their own anti-particles.

Recall from Chap. 3, Sect. 3.9 on pg. 65, that if a particle is its own anti-particle, then its field solution must be real, not complex. So, A^μ is real. (A particle that is its own anti-particle must be chargeless, but a chargeless particle, e.g., a neutrino, does not have to be its own antiparticle.)

We also showed, in Prob. 15 of Chap. 3, that in the mathematical treatment of real fields, a factor of ½ arises in the Lagrangian density that is not in the Lagrangian density for complex fields.

Particles that are their own anti-particles have real, not complex solutions. Photon is one of these.

5.4 Commutation Relations for Photon Fields

The second step in 2nd quantization comprises taking the Poisson brackets of classical field theory over to commutators (with a factor of i) in QFT. In scalar theory, we showed these commutators were

2nd quantization, second step: Poisson brackets to commutators

$$\left[\phi^r(\mathbf{x},t), \pi_s(\mathbf{y},t) \right] = \phi^r \pi_s - \pi_s \phi^r = i\delta_{s}^r \delta(\mathbf{x}-\mathbf{y}),$$

(5-53)

and from these (in a 1½ page proof), we showed the coefficients in the solutions must then obey

$$\left[a(\mathbf{k}), a^\dagger(\mathbf{k}') \right] = \left[b(\mathbf{k}), b^\dagger(\mathbf{k}') \right] = \delta_{\mathbf{kk}'} \text{ (discrete);} \quad = \delta(\mathbf{k}-\mathbf{k}') \text{ (continuous).}$$

(5-54)

Rather than go through the tedium of a virtually identical proof, from the parallels between scalar bosons and vector bosons, we will simply extrapolate from the above. Thus, from (5-53)

$$\left[A^\mu(\mathbf{x},t), \pi_\nu^1(\mathbf{y},t) \right] = i\delta_{\nu}^\mu \delta(\mathbf{x}-\mathbf{y}) \rightarrow \left[A^\mu(x), \pi^{\nu 1}(y) \right] = ig^{\mu\nu} \delta(\mathbf{x}-\mathbf{y}),$$

(5-55)

which leads (from (5-54)) to the <u>photon coefficient commutation relations</u>

Photon coefficient commutation relations

$$\boxed{\begin{array}{ll} \left[a_r(\mathbf{k}), a_s^\dagger(\mathbf{k}') \right] = \zeta_r \delta_{rs} \delta_{\mathbf{kk}'} & \text{(discrete)} \qquad \zeta_0 = -1, \zeta_{1,2,3} = 1 \\ \qquad\qquad = \zeta_r \delta_{rs} \delta(\mathbf{k}-\mathbf{k}') & \text{(continuous)} \end{array}}$$

(5-56)

All other commutators, such as $[a_r(\mathbf{k}), a_s(\mathbf{k})]$, equal zero for any r and s, as with scalars. The ζ_r part arises from the $g^{\mu\nu}$ of (5-55), which was not present for scalars. This is zero except when $\mu = \nu$, and then the spatial parts have – sign and the time part +. But an additional minus sign is introduced into (5-55) and (5-56) by (5-51).

5.5 The QFT Hamiltonian for Photons

5.5.1 The Free Photon Hamiltonian in Terms of the Coefficients

By inserting the field solutions (5-49) into the expression for the Hamiltonian density $\mathcal{H}_0^1$ (5-52) and integrating over all space we get H_0^1, the full <u>photon Hamiltonian</u>, in terms of the coefficients. Again, this is virtually the same thing we did for scalars, so we carry over that result there to here.

Hamiltonian in terms of number operator

$$H_0^1 = \sum_{\mathbf{k},r} \omega_{\mathbf{k}} \left(N_r(\mathbf{k}) + \tfrac{1}{2} \right) \qquad (5\text{-}57)$$

$$N_r(\mathbf{k}) = \zeta_r a_r^\dagger(\mathbf{k}) a_r(\mathbf{k}) \quad \text{the number operator for photons} \qquad (5\text{-}58)$$

Photon number operator

5.5.2 Zero Point Vacuum Energy Again

As you can see from (5-57), the photon energy expectation value for the vacuum includes a summing of ½ $\omega_{\mathbf{k}}$ over every photon energy and polarization r to infinity, as we had for scalars (and similarly for spinors). All of the comments made with regard to this in prior chapters are equally applicable here.

Photon vacuum energy like that for scalars

5.5.3 Positive Energy for Photons in QFT

We do end up with positive energies for all real (meaning not virtual as opposed to a real vs complex sense) photons from (5-57), though given the $\zeta_0 = -1$ in the summation of (5-58), it seems like there could be photons with negative energy, so this is not obvious at this point. We resolve this issue in Sect. 5.8.3.

Photon QFT avoids problems of RQM like negative energy

5.6 Other Photon Operators in QFT

In similar fashion to what we did for scalars and spinors, we can derive all other operators. Again, to avoid the tedium, we simply state these as follows.

<u>Photon creation and destruction operators</u>

$$a_r^\dagger(\mathbf{k})|n_{\mathbf{k},r}\rangle = \sqrt{n_{\mathbf{k},r}+1}\,|n_{\mathbf{k},r}+1\rangle \qquad a_r^\dagger(\mathbf{k})\text{ creates a photon with momentum }\mathbf{k}\text{, polarization }r$$
$$a_r(\mathbf{k})|n_{\mathbf{k},r}\rangle = \sqrt{n_{\mathbf{k},r}}\,|n_{\mathbf{k},r}-1\rangle \qquad a_r(\mathbf{k})\text{ destroys a photon with momentum }\mathbf{k}\text{, polarization }r \qquad (5\text{-}59)$$

Photon creation and destruction operators

<u>Total photon particle number</u>

$$N\left(A^\mu\right) = \sum_{\mathbf{k},r} N_r(\mathbf{k}) \qquad (5\text{-}60)$$

Other photon operators

<u>Total particle number lowering and raising</u>

$$A^{\mu+} \quad \text{particle lowering operator field (contains } a_r(\mathbf{k}))$$
$$A^{\mu-} \quad \text{particle raising operator field (contains } a_r^\dagger(\mathbf{k})) . \qquad (5\text{-}61)$$

<u>Four-current operator</u>

$$j^\mu = -i\left(A_\alpha{}'^{\mu\dagger} A^\alpha - A_\alpha{}'^\mu A^{\alpha\dagger} \right) \;\; = 0 \;\text{ for photons since } A_\alpha{}'^{\mu\dagger} = A_\alpha{}'^\mu . \qquad (5\text{-}62)$$

Photon 4-current identically zero

If we were to take j^0 of (5-62) as our probability operator, as early researchers expected it to be, then we would have zero probability of ever finding a photon. But since the photon has zero charge, and thus, zero charge density, those researchers concluded (5-62) must be a charge density operator.

<u>Charge operator</u>

The integral of j^0 of (5-62) over all space to get charge is, of course, zero. The charge operator for photons is identically zero.

So, photon charge operator identically zero

Three-momentum operator

$$\boxed{\mathbf{P} = \sum_{\mathbf{k},r} \mathbf{k} N_r(\mathbf{k})} \quad .$$

(5-63)

Spin operator not important for photons

Spin for photons is always aligned with the $\mathbf{k}$ direction and always of magnitude 1. It is therefore not a key, unknown variable, like it was for spin ½ particles. So little attention is put on it in QFT, and one typically does not use a photon spin operator in QFT calculations. For completeness, however, we can state it as

$$\mathbf{S}_\gamma = \sum_{\mathbf{k},r} \frac{\mathbf{k}}{|\mathbf{k}|} N_{r,\pm}(\mathbf{k}) \qquad N_{r,\pm}(\mathbf{k}) = +N_r(\mathbf{k}) \text{ for } \mathbf{k} \text{ direction spin}$$

(5-64)

$$N_{r,\pm}(\mathbf{k}) = -N_r(\mathbf{k}) \text{ for } -\mathbf{k} \text{ direction spin}.$$

If we needed it, (5-64) would give us the total spin angular momentum of a collection of photons.

Helicity operator not important for photons, either

Since spin for photons is always aligned with the $\mathbf{k}$ direction, and the helicity operator for a single photon is simply the inner product of the unit vector in the $\mathbf{k}$ direction and spin $\mathbf{S}$, the helicity operator is simply

$$\text{Helicity operator}_\gamma = \sum_{\mathbf{k},r} N_{r,\pm}(\mathbf{k}) \quad .$$

(5-65)

One virtually never uses the helicity operator for photons, however, since they are always in helicity eigenstates, and also since adding up helicities for a collection of photons, has no real meaning or purpose.

5.7 The Photon Propagator

The photon propagator is derived in much the same fashion as the scalar propagator. We start by defining the propagator as the time ordered operator like we did in Chaps. 3 (pg. 72) and 4 (pg. 118),

$$T\{A^\mu(x)A^\nu(y)\} = A^\mu(x)A^\nu(y) \qquad \text{for } t_y < t_x \quad (\text{particle = photon})$$

(5-66)

$$= A^\nu(y)A^\mu(x) \qquad \text{for } t_x < t_y \quad (\text{anti-particle = photon}).$$

But for photons, the particle is its own anti-particle, and thus A^μ can either create or destroy a photon. We also have four-vector components μ and ν, so we will expect the final form of the photon propagator to have those components as well.

Following the same 5 steps as in Chap. 3 (pgs. 71-77), and also Chap. 4 (pgs. 118-121), we arrive at the photon propagator

$$\boxed{D_F^{\mu\nu}(x-y) = \frac{-g^{\mu\nu}}{(2\pi)^4} \int \frac{e^{-ik(x-y)}}{k^2 + i\varepsilon} d^4k \quad \text{in physical space}}$$

(5-67)

$$\boxed{D_F^{\mu\nu}(k) = \frac{-g^{\mu\nu}}{k^2 + i\varepsilon} \quad \text{in four-momentum space}} .$$

(5-68)

5.8 More on Quantization and Polarization

5.8.1 Problem with the Lorenz Gauge

We started all of the above by assuming we could simply use Maxwell's equations under the Lorenz gauge (and hence use the corresponding Lagrangian as well) carried over directly to QFT. That implies that $\partial_\mu A^\mu$ of the Lorenz gauge would be considered an operator in QFT that is identically equal to zero. Unfortunately, that is not strictly true, because, as we are about to see, the Lorenz condition, employed in the direct way, is incompatible with the commutation relations.

Consider the commutator (the bottom line of all the algebra below to (5-71) is that the RHS $\neq 0$)

$$\underbrace{[\partial_\mu A^\mu(x)}_{\substack{= 0 \text{ in} \\ \text{Lorenz gauge}}}, A^\nu(y)] = \left(\sum_{r,\mathbf{k}} \frac{-ik_\mu}{\sqrt{2V\omega_\mathbf{k}}} \varepsilon_r^\mu(\mathbf{k}) a_r(\mathbf{k}) e^{-ikx} + \sum_{r,\mathbf{k}} \frac{ik_\mu}{\sqrt{2V\omega_\mathbf{k}}} \varepsilon_r^\mu(\mathbf{k}) a_r^\dagger(\mathbf{k}) e^{ikx} \right) \times$$

$$\left(\sum_{s,\mathbf{k}'} \frac{1}{\sqrt{2V\omega_{\mathbf{k}'}}} \varepsilon_s^\nu(\mathbf{k}') a_s(\mathbf{k}') e^{-ik'y} + \sum_{s,\mathbf{k}'} \frac{1}{\sqrt{2V\omega_{\mathbf{k}'}}} \varepsilon_s^\nu(\mathbf{k}') a_s^\dagger(\mathbf{k}') e^{ik'y} \right)$$

(5-69)

$$- \left(\sum_{s,\mathbf{k}'} \frac{1}{\sqrt{2V\omega_{\mathbf{k}'}}} \varepsilon_s^\nu(\mathbf{k}') a_s(\mathbf{k}') e^{-ik'y} + \sum_{s,\mathbf{k}'} \frac{1}{\sqrt{2V\omega_{\mathbf{k}'}}} \varepsilon_s^\nu(\mathbf{k}') a_s^\dagger(\mathbf{k}') e^{ik'y} \right) \times$$

$$\left(\sum_{r,\mathbf{k}} \frac{-ik_\mu}{\sqrt{2V\omega_\mathbf{k}}} \varepsilon_r^\mu(\mathbf{k}) a_r(\mathbf{k}) e^{-ikx} + \sum_{r,\mathbf{k}} \frac{ik_\mu}{\sqrt{2V\omega_\mathbf{k}}} \varepsilon_r^\mu(\mathbf{k}) a_r^\dagger(\mathbf{k}) e^{ikx} \right),$$

which due to commutators that equal zero becomes

$$\underbrace{[\partial_\mu A^\mu(x)}_{\substack{= 0 \text{ in} \\ \text{Lorenz gauge}}}, A^\nu(y)] = \sum_{s,\mathbf{k}'} \left(\begin{array}{l} \sum_{r,\mathbf{k}} \dfrac{-ik_\mu}{\sqrt{2V\omega_\mathbf{k}}} \dfrac{1}{\sqrt{2V\omega_{\mathbf{k}'}}} \varepsilon_r^\mu(\mathbf{k}) \varepsilon_s^\nu(\mathbf{k}') \underbrace{\left[a_r(\mathbf{k}), a_s^\dagger(\mathbf{k}') \right]}_{\zeta_r \delta_{rs} \delta_{\mathbf{k}\mathbf{k}'}} e^{-ikx+ik'y} \\ + \sum_{r,\mathbf{k}} \dfrac{ik_\mu}{\sqrt{2V\omega_\mathbf{k}}} \dfrac{1}{\sqrt{2V\omega_{\mathbf{k}'}}} \varepsilon_r^\mu(\mathbf{k}) \varepsilon_s^\nu(\mathbf{k}') \underbrace{\left[a_r^\dagger(\mathbf{k}), a_s(\mathbf{k}') \right]}_{-\zeta_r \delta_{rs} \delta_{\mathbf{k}\mathbf{k}'}} e^{ikx-ik'y} \end{array} \right). \quad (5\text{-}70)$$

*Lorenz gauge
says something
should be zero,
but commutation
relations yield
non-zero*

So that (5-69) and (5-70) finally become

$$\underbrace{[\partial_\mu A^\mu(x)}_{\substack{= 0 \text{ in} \\ \text{Lorenz gauge}}}, A^\nu(y)] = \sum_{r,\mathbf{k}} \frac{-ik_\mu}{2V\omega_\mathbf{k}} \zeta_r \varepsilon_r^\mu(\mathbf{k}) \varepsilon_r^\nu(\mathbf{k}) \left(e^{-ik(x-y)} + e^{ik(x-y)} \right)$$

(5-71)

$$= \sum_{r,\mathbf{k}} \frac{-ik_\mu}{V\omega_\mathbf{k}} \zeta_r \, \varepsilon_r^\mu(\mathbf{k}) \varepsilon_r^\nu(\mathbf{k}) \cos\left(k(x-y) \right) \neq 0.$$

The LHS of (5-71) is zero if the Lorenz condition holds. But the RHS is not zero, according to the rules for commutators we established earlier. Of different solutions to this problem, possibly the best was found by Gupta and Bleuler, as described in the following section.

5.8.2 The Gupta-Bleuler Resolution

The Weak Lorenz Condition

Gupta and Bleuler[1] replaced the Lorenz condition (5-28) with the weaker condition

$$\partial_\mu A^{\mu+}(x) |\Psi\rangle = 0, \quad (5\text{-}72)$$

*An alternative to
the Lorenz gauge
by Gupta-Bleuler*

where the $|\Psi\rangle$ represents a ket of any number of photons. The operator in the weak Lorenz condition contains only destruction coefficient operators of form $a_r(\mathbf{k})$. The adjoint of (5-72) is

$$\langle \Psi | \partial_\mu A^{\mu-}(x) = 0. \quad (5\text{-}73)$$

So, the expectation value of the Lorenz condition equals zero, i.e.,

*Expectation value
for Lorenz
condition should
match classical
value of zero*

$$\overline{\partial_\mu A^\mu(x)} = \langle \Psi | \partial_\mu A^\mu(x) | \Psi \rangle = \langle \Psi | (\underbrace{\partial_\mu A^{\mu-}(x)}_{=0} + \underbrace{\partial_\mu A^{\mu+}(x)}_{=0}) | \Psi \rangle = 0. \quad (5\text{-}74)$$

Classically, when we make macroscopic measurements, we measure expectation values (to high precision with large numbers of quanta, typically). So the Lorenz condition still holds classically

---

[1] K. Bleuler, Eine neue Methode zur Behandlung der longitudinalen und skalaren Photonen, *Helvetica Phys. Acta* 23 (1950), 567–586. S.N. Gupta, Theory of longitudinal photons in quantum electrodynamics, *Proc. Phys. Soc.* Sect.A. 63 (1950), 681–691.

and the Maxwell's equation in terms of A^μ we have been using holds in this limit. Thus, predictions from our theory should correspond to actual measurements in experiments.

Actually, using the full Lorenz condition (5-28) instead of (5-72) results in an expectation value of the Lorenz condition of zero as well. This is because the action of the Lorenz condition on the ket must result in a ket different from the bra of the expectation value relation. That is,

$$\overline{\partial_\mu A^\mu(x)} = \langle\Psi|\partial_\mu A^\mu(x)|\Psi\rangle = \langle\Psi|\underbrace{\partial_\mu A^{\mu+}(x)|\Psi\rangle}_{\substack{|\Psi\text{ minus a photon}\rangle \\ \text{orthogonal to }|\Psi\rangle}} + \langle\Psi|\underbrace{\partial_\mu A^{\mu-}(x)|\Psi\rangle}_{\substack{|\Psi\text{ plus a photon}\rangle \\ \text{orthogonal to }|\Psi\rangle}} = 0 . \quad (5\text{-}75)$$

Expectation value for Lorenz condition in both approaches is zero. Both good from that perspective.

5.8.3 Meaning of the Weak Lorenz Condition

<u>The Constraint on Scalar and Longitudinal Photons</u>

To understand (5-72), substitute the solutions (5-49) for $A^{\mu+}$ and consider the photon aligned coordinate system of Fig. 5-1(d), where (5-34) and (5-37) hold.

What does the weak Lorenz condition mean in terms of operator coefficients?

$$\partial_\mu A^{\mu+}(x) = \sum_\mu \partial_\mu\left(\sum_{r,\mathbf{k}}\frac{1}{\sqrt{2V\omega_\mathbf{k}}}\varepsilon_r^\mu a_r(\mathbf{k})e^{-ikx}\right) = \sum_{r,\mathbf{k},\mu}\frac{1}{\sqrt{2V\omega_\mathbf{k}}}\varepsilon_r^\mu a_r(\mathbf{k})\partial_\mu e^{-ikx}$$

$$= \sum_\mathbf{k}\frac{-i}{\sqrt{2V\omega_\mathbf{k}}}\left(\begin{array}{l} \underset{=1}{\underbrace{\varepsilon_0^0}} a_0(\mathbf{k})\omega_\mathbf{k} + \underset{=0}{\underbrace{\varepsilon_0^1}} a_0(\mathbf{k})\underset{=0}{\underbrace{k_1}} + \underset{=0}{\underbrace{\varepsilon_0^2}} a_0(\mathbf{k})\underset{=0}{\underbrace{k_2}} + \underset{=0}{\underbrace{\varepsilon_0^3}} a_0(\mathbf{k})k_3 + \\[2mm] \underset{=0}{\underbrace{\varepsilon_1^0}} a_1(\mathbf{k})\omega_\mathbf{k} + \underset{=1}{\underbrace{\varepsilon_1^1}} a_1(\mathbf{k})\underset{=0}{\underbrace{k_1}} + \underset{=0}{\underbrace{\varepsilon_1^2}} a_1(\mathbf{k})\underset{=0}{\underbrace{k_2}} + \underset{=0}{\underbrace{\varepsilon_1^3}} a_1(\mathbf{k})k_3 + \\[2mm] r=2 \text{ row is zero as above for } r=1 \qquad + \\[2mm] \underset{=0}{\underbrace{\varepsilon_3^0}} a_3(\mathbf{k})\omega_\mathbf{k} + \underset{=0}{\underbrace{\varepsilon_3^1}} a_3(\mathbf{k})\underset{=0}{\underbrace{k_1}} + \underset{=0}{\underbrace{\varepsilon_3^2}} a_3(\mathbf{k})\underset{=0}{\underbrace{k_2}} + \underset{=1}{\underbrace{\varepsilon_3^3}} a_3(\mathbf{k})\underset{\substack{=-k^3 \\ =-\omega_\mathbf{k}}}{\underbrace{k_3}} \end{array}\right)e^{-ikx} . \quad (5\text{-}76)$$

In the very last term of (5-76) we invoke the relativistic relation, in natural units, $m^2 = E^2 - \mathbf{p}^2 = (\omega_\mathbf{k})^2 - \mathbf{p}^2 = (\omega_\mathbf{k})^2 - (k^3)^2$, where $m = 0$. So $k^3 = \omega_\mathbf{k} = -k_3$. Then, for every $\mathbf{k}$, (5-76) acting on $|\Psi\rangle$ becomes

$$\partial_\mu A^{\mu+}(x)|\Psi\rangle = 0 \quad\rightarrow\quad \left(a_3(\mathbf{k}) - a_0(\mathbf{k})\right)|\Psi\rangle = 0, \quad \text{all } \mathbf{k} . \quad (5\text{-}77)$$

(5-77) is a constraint on the linear combinations of longitudinal and scalar photons for each value of $\mathbf{k}$ that may be present in a state. It places no restriction on the transverse photons that may be in that state.

For use in the next section, we re-write (5-77) and its adjoint as

$$\left(a_3(\mathbf{k}) - a_0(\mathbf{k})\right)|\Psi\rangle = 0 \quad\rightarrow\quad a_3(\mathbf{k})|\Psi\rangle = a_0(\mathbf{k})|\Psi\rangle \quad\overset{\text{adjoint}}{\rightarrow}\quad \langle\Psi|a_3^\dagger(\mathbf{k}) = \langle\Psi|a_0^\dagger(\mathbf{k}) . \quad (5\text{-}78)$$

Relation between scalar and longitudinal coefficients operating on a state

<u>The Expectation Value of Energy of an Allowed State</u>

We can see the effect of the constraint on the RHS of (5-77) (imposed by the weak Lorenz condition of the LHS of (5-77)) by calculating the expectation value of the energy of an allowed state $|\Psi\rangle$. From (5-57) and (5-58), where we ignore the ½ quanta of the vacuum,

$$\overline{H_0^1} = \langle\Psi|H_0^1|\Psi\rangle = \langle\Psi|\sum_\mathbf{k}\sum_{r=0}^3\omega_\mathbf{k}\left(N_r(\mathbf{k})\right)|\Psi\rangle = \langle\Psi|\sum_\mathbf{k}\sum_{r=0}^3\omega_\mathbf{k}\left(\zeta_r a_r^\dagger(\mathbf{k})a_r(\mathbf{k})\right)|\Psi\rangle =$$

$$\langle\Psi|\sum_\mathbf{k}\omega_\mathbf{k}\left(-a_0^\dagger(\mathbf{k})a_0(\mathbf{k})|\Psi\rangle + a_1^\dagger(\mathbf{k})a_1(\mathbf{k})|\Psi\rangle + a_2^\dagger(\mathbf{k})a_2(\mathbf{k})|\Psi\rangle + a_3^\dagger(\mathbf{k})a_3(\mathbf{k})|\Psi\rangle\right). \quad (5\text{-}79)$$

We re-write this using (5-78)

$$\overline{H_0^1} = \sum_{\mathbf{k}} \omega_{\mathbf{k}} \begin{pmatrix} \underbrace{-\langle\Psi|a_0^{\dagger}(\mathbf{k})\,a_0(\mathbf{k})|\Psi\rangle}_{-\langle\Psi|a_3^{\dagger}(\mathbf{k})\,a_3(\mathbf{k})|\Psi\rangle} + \langle\Psi|a_1^{\dagger}(\mathbf{k})\,a_1(\mathbf{k})|\Psi\rangle \\ +\langle\Psi|a_2^{\dagger}(\mathbf{k})\,a_2(\mathbf{k})|\Psi\rangle + \langle\Psi|a_3^{\dagger}(\mathbf{k})\,a_3(\mathbf{k})|\Psi\rangle \end{pmatrix} \qquad (5\text{-}80)$$

$$= \sum_{\mathbf{k}} \omega_{\mathbf{k}} \left(\langle\Psi|a_1^{\dagger}(\mathbf{k})\,a_1(\mathbf{k})|\Psi\rangle + \langle\Psi|a_2^{\dagger}(\mathbf{k})\,a_2(\mathbf{k})|\Psi\rangle \right).$$

Or

$$\overline{H_0^1} = \sum_{\mathbf{k}} \omega_{\mathbf{k}} \sum_{r=1}^{2} \langle\Psi|a_r^{\dagger}(\mathbf{k})\,a_r(\mathbf{k})|\Psi\rangle , \qquad (5\text{-}81)$$

That relation leads to scalar and longitudinal terms cancelling in the Hamiltonian expectation

And that leads to a Hamiltonian expectation that effectively sums over only transverse states

which means the only contribution to the energy expectation value is from transverse photons. The scalar energy expectation value is negative, but it is always cancelled by a positive longitudinal energy expectation value of the same magnitude.

A similar effect occurs, in parallel manner, for any observable, since each such observable (e.g., 3-momentum $\mathbf{P}$ of (5-63)) incorporates the number operator $N_r(\mathbf{k})$ of (5-79) above. So the same steps would be followed for the expectation value of any other operator as is done in (5-79) to (5-81).

And other operator expectations also effectively sum over only transverse states

This jibes with classical e/m theory, since we know that $\mathbf{E}$ and $\mathbf{B}$ in a classical electromagnetic wave are always perpendicular to $\mathbf{k}$. And from the definition of our potential $A^{\mu} = (\Phi, \mathbf{A})$ of either (5-2) or (5-18), we know that $\mathbf{E}$ points in the same direction as $\mathbf{A}$. Thus, classically, $\mathbf{A}$ is always orthogonal to the wave propagation direction $\mathbf{k}/|\mathbf{k}|$.

This matches what we observe classically, only transverse states

We have never measured any real longitudinal photon states. Nor have we ever done so for scalar (time-like) states. The Gupta-Bleuler approach, in the classical limit of expectation values, yields what we observe in our universe, only transverse, observable photons. Very elegant, in my humble opinion.

Comments on Scalar and Longitudinal Waves

Note that we have not proven that scalar and longitudinal photons do not exist, only that, if they do, they cancel one another in finding expectation values for any classical entity. Expectation values relate to real (not virtual) particles. We might then, still expect that scalar and longitudinal photon states would play a role in the Feynman propagator, which represents a virtual, and unobservable, particle. In fact, they do. In interaction theory, calculations of effects from particles mediating force (like photons in quantum electrodynamics) must include all four polarization states.

Real scalar and longitudinal states not measurable, but they play a role in propagator (as virtual particle states)

Finally, we note a subtle point that you should not worry about too much at this time. That is, it can be shown that the allowed mixtures of longitudinal and scalar photons is equivalent to a gauge transformation between two potentials, both of which are in Lorenz gauges.

Comment on Lorenz Gauge

You may be wondering "if we got the simplest form of Maxwell's equations based on the Lorenz condition holding for A^{μ}, but it isn't true, how can we do 2nd quantization and assume we have the same wave equation (= same Lagrangian)?"

Redefine 2nd quantization as same but coefficient constraint as additional condition. It works.

The simplest answer is that by doing so, the theory works. Predictions match experiment. From a theoretical standpoint, we could just include this as part of 2nd quantization, with the added condition that, acting on a state, the scalar and longitudinal coefficient operators are related in such a way as to cancel out their effects in finding expectation values.

5.8.4 Summary of Weak Lorenz Condition

Wholeness Chart 5-3 below summarizes the full vs weak Lorenz condition approaches.

Wholeness Chart 5-3. Gupta-Bleuler Weak Lorenz Condition Overview

	Full Lorenz Condition	Gupta-Bleuler Weak Lorenz Condition
The Constraint	$\partial_\mu A^\mu(x) = 0$	$\partial_\mu A^{\mu+}(x)\lvert\Psi\rangle = 0$
Expectation Value of Lorenz Condition Given the Constraint	$\langle\Psi\lvert\partial_\mu A^\mu(x)\lvert\Psi\rangle = 0$	$\langle\Psi\lvert\partial_\mu A^\mu(x)\lvert\Psi\rangle =$ $\langle\Psi\lvert\partial_\mu\left(A^{\mu+}(x) + A^{\mu-}(x)\right)\lvert\Psi\rangle = 0$
	Either condition OK classically. Yields same macroscopic Lorenz gauge.	
Theory derived from constraint + 2nd quantization	commutation relations, number operator, observables operators	same commutators, number operator, observables operators as at left
Problem with constraint?	Yes. $[\underbrace{\partial_\mu A^\mu(x)}_{=0}, A^\nu(y)] = \underbrace{V^\nu(x-y)}_{\text{numeric vector}} \neq 0$ Commutators from 2nd quantization in above yield different result than constraint.	No. $[\underbrace{\partial_\mu A^\mu(x)}_{\neq 0}, A^\nu(y)] = \underbrace{V^\nu(x-y)}_{\substack{\text{same} \\ \text{numeric vector}}} \neq 0$ Commutators from 2nd quantization in above yield same result as constraint.
Operator result of constraint, in photon aligned coordinate system	$a_3(\mathbf{k})\lvert\Psi\rangle \neq a_0(\mathbf{k})\lvert\Psi\rangle$	$a_3(\mathbf{k})\lvert\Psi\rangle = a_0(\mathbf{k})\lvert\Psi\rangle$
Effect on observable operators expectation value	$\langle\Psi\lvert\sum_{r=0}^{3} N_r(\mathbf{k})\lvert\Psi\rangle \neq \langle\Psi\lvert\sum_{r=1}^{2} N_r(\mathbf{k})\lvert\Psi\rangle$	$\langle\Psi\lvert\sum_{r=0}^{3} N_r(\mathbf{k})\lvert\Psi\rangle = \langle\Psi\lvert\sum_{r=1}^{2} N_r(\mathbf{k})\lvert\Psi\rangle$
Result	longitudinal and scalar (timelike) photon states no different from transverse → predict they are observed, but they aren't	longitudinal and scalar (timelike) photon states always cancel → predict none observed and they aren't

5.9 Photon Polarization Issues Similar to Spinor Spin Issues

One might raise several issues relating to polarization states, some similar to what we saw for spinors at the end of Chap. 4.

Other Coordinate Systems for a Given **k** Direction

First, the treatment in Sect. 5.8 employed only the photon aligned coordinate system. In other coordinate systems, the polarization vectors $\varepsilon_r{}^\mu$ would not be represented so simply, and the analysis would have been markedly more difficult and less transparent. In the most general coordinate system, the $\varepsilon_r{}^\mu$ would then each have up to four non-zero components. And $\varepsilon_1{}^\mu$, for example, would not by itself represent a pure transverse state. Any of the transverse, longitudinal, or scalar states would be represented by a superposition of different amounts of each of the $\varepsilon_r{}^\mu$.

Coordinate systems other than photon aligned system can work, but more complicated

Though the analysis would be more complicated, the final result would remain the same. Only transverse states, regardless of the coordinate system chosen to express them in, would remain observable.

How to Handle Polarization States Not in x^1 or x^2 Directions

The treatment so far assumed we only had two transverse polarization states in the x^1 and x^2 directions. But what about a state where **A** is aligned at 45° between these axes, or in any other direction? Such a state is a linear superposition of the two transverse states.

Non-eigen state kets can be handled. See Chap. 4 for details.

We will not enter into a detailed discussion of this here, but simply note that the answer parallels that for "The First Issue" for spinors found in Chap. 4, pg. 116. In short, in experiments we typically average over polarization states and not actually measure those states. This averaging uses the x^1 and x^2 direction polarization states as basis vectors, with each state within the ensemble of states having different amounts of each. The average of all such composite states is used in predicting experimental results. For experiments that might monitor polarization states, the expectation values for non-axis aligned polarization states can be calculated for such states in similar ways to what we discussed for spin ½ particles.

In the ket symbolism in this book, we will represent a photon state with polarization in an axis direction as $|\gamma_{r,\mathbf{k}}\rangle$ and a more general polarization state as $|\gamma_{\mathbf{k}}\rangle$.

How to Handle Multiple Photons with Different **k** Directions

We have used a preferred coordinate system, the photon aligned coordinate system, where the x^3 axis is aligned with the **k** vector of the photon. This makes interpretation and analysis much simpler. But what happens when we have more than one photon, each with a different **k** direction? Then, at most, we have simple expressions, like those we have seen in this chapter, for only one such photon (the one with its **k** aligned with the lab x^3.)

*Multiple photons, all with different **k** directions, and thus different photon aligned systems, can be handled via coordinate transformations. See Chap. 4 for details.*

This question parallels "The Second Issue" of Chap. 4, pg. 117, and like the prior issue above, we will not go into great detail on it here. In short, one can determine quantities for a given photon in its photon aligned coordinate system then transform those quantities to the (different alignment) lab system. In the process of the transformation, things like the expressions for, and interpretation of, ε^μ get a lot more complicated. But one can still crank the math, typically tedious, and get those expressions.

Expectation values in any coordinate system will turn out to be the same. So one can pick the coordinate system one likes for a given photon (the photon aligned system is best, obviously) to calculate an expectation value, then transform that value to lab coordinates.

5.10 Where to Next?

We have achieved a milestone in the study of QFT, the coverage of the fundamentals for the three spin types. Knowing this well will serve us well in interaction theory, to come after we investigate symmetry and its importance in QFT, in the next chapter.

Knowledge of free fields will serve us well in study of interacting fields

5.11 Summary Chart

Chaps. 3, 4, and 5 are summarized in Wholeness Chart 5-4 that follows.

QED/FIELD THEORY OVERVIEW: PART 1

Wholeness Chart 5-4. From Field Equations to Propagators and Observables
Heisenberg Picture, Free Fields

	__Spin 0__	__Spin ½__	__Spin 1__
Classical Lagrangian density, free	$\mathcal{L}_0^0 = K\left(\partial_\alpha \phi \partial^\alpha \phi - \mu^2 \phi\phi\right)$	None. Macroscopic spinor fields not observed.	$\mathcal{L}_0^1 = \underbrace{\frac{\mu^2}{2} A^\mu A_\mu}_{\mu=0} - \frac{1}{2}\left(\partial_\nu A_\mu\right)\left(\partial^\nu A^\mu\right)$ for photons
2nd quantization, Postulate #1	Bosons: Quantum field $\mathcal{L}$ (or equivalently, $\mathcal{H}$) same as classical, fields are complex, and $K=1$. Spinors: Dirac eq from RQM with states → fields. Deduce $\mathcal{L}$ from Dirac eq; $\mathcal{H}$ from Legendre transf.		
QFT Lagrangian density, free	$\mathcal{L}_0^0 = \left(\partial_\alpha \phi^\dagger \partial^\alpha \phi - \mu^2 \phi^\dagger \phi\right)$	$\mathcal{L}_0^{1/2} = \overline{\psi}\left(i\partial\!\!\!/ - m\right)\psi \qquad \partial\!\!\!/ = \gamma^\alpha \partial_\alpha$	As above for classical.
	$\mathcal{L}$ ↑ into the Euler-Lagrange equation yields ↓		
Free field equations	$\left(\partial_\alpha \partial^\alpha + \mu^2\right)\phi = 0$ $\left(\partial_\alpha \partial^\alpha + \mu^2\right)\phi^\dagger = 0$	$(i\gamma^\alpha \partial_\alpha - m)\psi = 0$ $(i\partial_\alpha \overline{\psi}\gamma^\alpha + m\overline{\psi}) = 0 \qquad \overline{\psi} = \psi^\dagger \gamma^0$	$\left(\partial_\alpha \partial^\alpha + \mu^2\right)A^\mu = 0 \quad$ photon $\mu = 0$ $A^{\mu\dagger} = A^\mu$ for chargeless (photon)
Conjugate momenta	$\pi_0^0 = \frac{\partial \mathcal{L}_0^0}{\partial \dot{\phi}} = \dot{\phi}^\dagger; \; \pi_0^{0\dagger} = \frac{\partial \mathcal{L}_0^0}{\partial \dot{\phi}^\dagger} = \dot{\phi}$	$\pi^{1/2} = i\psi^\dagger; \; \overline{\pi}^{1/2} = 0$	$\pi_\mu^1 = -\dot{A}_\mu$
Hamiltonian density	$\mathcal{H}_0^0 = \pi_0^0 \dot{\phi} + \pi_0^{0\dagger}\dot{\phi}^\dagger - \mathcal{L}_0^0$ $= \left(\dot{\phi}\dot{\phi}^\dagger + \nabla\phi^\dagger \cdot \nabla\phi + \mu^2 \phi^\dagger \phi\right)$	$\mathcal{H}_0^{1/2} = \pi^{1/2}\dot{\psi} - \mathcal{L}_0^{1/2}$	$\mathcal{H}_0^1 = \pi_\mu^1 \dot{A}^\mu - \mathcal{L}_0^1$
Free field solutions	$\phi = \phi^+ + \phi^-$ $\phi^\dagger = \phi^{\dagger+} + \phi^{\dagger-}$	$\psi = \psi^+ + \psi^-$ $\overline{\psi} = \overline{\psi}^+ + \overline{\psi}^-$	$A^\mu = A^{\mu+} + A^{\mu-}$ (photon)
Discrete eigenstates (Plane waves, constrained to volume V)	$\phi(x) = \sum_{\mathbf{k}} \frac{1}{\sqrt{2V\omega_\mathbf{k}}}(a(\mathbf{k})e^{-ikx}$ $\qquad\qquad + b^\dagger(\mathbf{k})e^{ikx})$ $\phi^\dagger(x) = \sum_{\mathbf{k}} \frac{1}{\sqrt{2V\omega_\mathbf{k}}}(b(\mathbf{k})e^{-ikx}$ $\qquad\qquad + a^\dagger(\mathbf{k})e^{ikx})$	$\psi = \sum_{r,\mathbf{p}} \sqrt{\frac{m}{VE_\mathbf{p}}}(c_r(\mathbf{p})u_r(\mathbf{p})e^{-ipx}$ $\qquad\qquad + d_r^\dagger(\mathbf{p})v_r(\mathbf{p})e^{ipx})$ $\overline{\psi} = \sum_{r,\mathbf{p}} \sqrt{\frac{m}{VE_\mathbf{p}}}(d_r(\mathbf{p})\overline{v}_r(\mathbf{p})e^{-ipx}$ $\qquad\qquad + c_r^\dagger(\mathbf{p})\overline{u}_r(\mathbf{p})e^{ipx})$	$A^\mu =$ $\sum_{r,\mathbf{k}} \frac{1}{\sqrt{2V\omega_\mathbf{k}}}(\varepsilon_r^\mu(\mathbf{k})a_r(\mathbf{k})e^{-ikx}$ $\qquad\qquad + \varepsilon_r^\mu(\mathbf{k})a_r^\dagger(\mathbf{k})e^{ikx})$
Continuous eigenstates (Plane waves, no volume constraint)	$\phi(x) = \int \frac{d\mathbf{k}}{\sqrt{2(2\pi)^3 \omega_\mathbf{k}}}(a(\mathbf{k})e^{-ikx}$ $\qquad\qquad + b^\dagger(\mathbf{k})e^{ikx})$ $\phi^\dagger(x) = \int \frac{d\mathbf{k}}{\sqrt{2(2\pi)^3 \omega_\mathbf{k}}}(b(\mathbf{k})e^{-ikx}$ $\qquad\qquad + a^\dagger(\mathbf{k})e^{ikx})$	$\psi = \sum_r \sqrt{\frac{m}{(2\pi)^3}}\int \frac{d^3\mathbf{p}}{\sqrt{E_\mathbf{p}}}(c_r(\mathbf{p})u_r(\mathbf{p})e^{-ipx}$ $\qquad\qquad + d_r^\dagger(\mathbf{p})v_r(\mathbf{p})e^{ipx})$ $\overline{\psi} = \sum_r \sqrt{\frac{m}{(2\pi)^3}}\int \frac{d^3\mathbf{p}}{\sqrt{E_\mathbf{p}}}(d_r(\mathbf{p})\overline{v}_r(\mathbf{p})e^{-ipx}$ $\qquad\qquad + c_r^\dagger(\mathbf{p})\overline{u}_r(\mathbf{p})e^{ipx})$ spinor indices on u_r, v_r, and ψ suppressed. $r = 1,2$.	$A^\mu =$ $\sum_r \int \frac{d\mathbf{k}}{\sqrt{2(2\pi)^3 \omega_\mathbf{k}}}(\varepsilon_r^\mu(\mathbf{k})a_r(\mathbf{k})e^{-ikx}$ $\qquad\qquad + \varepsilon_r^\mu(\mathbf{k})a_r^\dagger(\mathbf{k})e^{ikx})$ $r = 0,1,2,3$ (4 polarization vectors)

2nd quantization Postulate #2	Bosons: $\left[\phi^r(\mathbf{x},t),\pi_s(\mathbf{y},t)\right]=\left[\phi^r\pi_s-\pi_s\phi^r\right]=i\delta^r{}_s\delta(\mathbf{x}-\mathbf{y})$, ϕ^r = any field, other commutators = 0. Spinors: Coefficient anti-commutation relations parallel coefficient commutation relations for bosons.		
	Bosons: using conjugate momenta expressions in ↑ yields ↓		
Equal time commutators (intermediate step only)	$\left[\phi(\mathbf{x},t),\dot{\phi}^\dagger(\mathbf{y},t)\right]=i\delta(\mathbf{x}-\mathbf{y})$	Not needed for spinor derivation.	$\left[A^\mu(\mathbf{x},t),\dot{A}^\nu(\mathbf{y},t)\right]$ $=-ig^{\mu\nu}\delta(\mathbf{x}-\mathbf{y})$
	Bosons: Using free field solutions in ↑ with 3D Dirac delta function (e.g., for discrete solutions, $\delta(\mathbf{x}-\mathbf{y})=\dfrac{1}{2V}\sum\limits_{n=-\infty}^{+\infty}\left(e^{-i\mathbf{k}_n\cdot(\mathbf{x}-\mathbf{y})}+e^{i\mathbf{k}_n\cdot(\mathbf{x}-\mathbf{y})}\right)$), and matching terms, yields the coefficient commutators ↓.		
Coefficient commutators	$\left[a(\mathbf{k}),a^\dagger(\mathbf{k}')\right]=\left[b(\mathbf{k}),b^\dagger(\mathbf{k}')\right]$	$\left[c_r(\mathbf{p}),c_s^\dagger(\mathbf{p}')\right]_+=\left[d_r(\mathbf{p}),d_s^\dagger(\mathbf{p}')\right]_+$	$\left[a_r(\mathbf{k}),a_s^\dagger(\mathbf{k}')\right]$
discrete	$=\delta_{\mathbf{k}\mathbf{k}'}$	$=\delta_{rs}\delta_{\mathbf{p}\mathbf{p}'}$	$=\zeta_r\delta_{rs}\delta_{\mathbf{k}\mathbf{k}'}$ $\quad\zeta_0=-1,\ \zeta_{1,2,3}=1$
continuous	$=\delta(\mathbf{k}-\mathbf{k}')$	$=\delta_{rs}\delta(\mathbf{p}-\mathbf{p}')$	$=\zeta_r\delta_{rs}\delta(\mathbf{k}-\mathbf{k}')$
Other coeffs	All other commutators = 0	All other anti-commutators = 0	All other commutators = 0
The Hamiltonian Operator			
	Substituting the free field solutions into the free Hamiltonian density $\mathcal{H}_0$, integrating $H_0=\int\mathcal{H}_0\,d^3x$, and using the coefficient commutators ↑ in the result, yields ↓. Acting on states with H_0 yields number operators.		
H_0	$\sum\limits_{\mathbf{k}}\omega_{\mathbf{k}}\left(N_a(\mathbf{k})+\tfrac{1}{2}+N_b(\mathbf{k})+\tfrac{1}{2}\right)$	$\sum\limits_{\mathbf{p},r}E_{\mathbf{p}}\left(N_r(\mathbf{p})-\tfrac{1}{2}+\bar{N}_r(\mathbf{p})-\tfrac{1}{2}\right)$	$\sum\limits_{\mathbf{k},r}\omega_{\mathbf{k}}\left(N_r(\mathbf{k})+\tfrac{1}{2}\right)$
Number operators	$N_a(\mathbf{k})=a^\dagger(\mathbf{k})a(\mathbf{k})$ $N_b(\mathbf{k})=b^\dagger(\mathbf{k})b(\mathbf{k})$	$N_r(\mathbf{p})=c_r^\dagger(\mathbf{p})c_r(\mathbf{p})$ $\bar{N}_r(\mathbf{p})=d_r^\dagger(\mathbf{p})d_r(\mathbf{p})$	$N_r(\mathbf{k})=\zeta_r\,a_r^\dagger(\mathbf{k})a_r(\mathbf{k})$
Creation and Destruction Operators			
	Evaluating $N_a(\mathbf{k})\,a(\mathbf{k})\,|n_\mathbf{k}\rangle$ (similar for other particle types) with ↑ and the coefficient commutators yields ↓		
creation	$a^\dagger(\mathbf{k}),\ b^\dagger(\mathbf{k})$	$c_r^\dagger(\mathbf{p}),\ d_r^\dagger(\mathbf{p})$	$a_r^\dagger(\mathbf{k})$
destruction	$a(\mathbf{k}),\ b(\mathbf{k})$	$c_r(\mathbf{p}),\ d_r(\mathbf{p})$	$a_r(\mathbf{k})$
Normaliz factors lowering	$a(\mathbf{k})|n_k\rangle=\sqrt{n_k}\,|n_k-1\rangle$	$c_r(\mathbf{p})|\psi_{r,\mathbf{p}}\rangle=|0\rangle$	as with scalars
raising	$a^\dagger(\mathbf{k})|n_k\rangle=\sqrt{n_k+1}\,|n_k+1\rangle$	$c_r^\dagger(\mathbf{p})|0\rangle=|\psi_{r,\mathbf{p}}\rangle$	as with scalars
tot particle num	$N(\phi)=\sum\limits_{\mathbf{k}}\left(N_a(\mathbf{k})-N_b(\mathbf{k})\right)$	$N(\psi)=\sum\limits_{\mathbf{p},r}\left(N_r(\mathbf{p})-\bar{N}_r(\mathbf{p})\right)$	$N(A^\mu)=\sum\limits_{\mathbf{k},r}N_r(\mathbf{k})$
tot particle num: lowering	$\phi=\phi^++\phi^-$	$\psi=\psi^++\psi^-$	$A^{\mu+}$
raising	$\phi^\dagger=\phi^{\dagger+}+\phi^{\dagger-}$	$\bar{\psi}=\bar{\psi}^++\bar{\psi}^-$	$A^{\mu-}$

Four Currents and Probability			
Four currents (operators) $j^\mu,_\mu = 0$	$j^\mu = (\rho, \mathbf{j}) = -i\left(\phi^{\dagger},^\mu \phi - \phi,^\mu \phi^\dagger\right)$	$j^\mu = (\rho, \mathbf{j}) = \bar{\psi}\gamma^\mu \psi$	$j^\mu = -i\left(A_\alpha,^{\mu\dagger} A^\alpha - A_\alpha,^\mu A^{\alpha\dagger}\right)$ $= 0$ for photons $\left(A_\alpha^{\dagger} = A_\alpha\right)$
	Emphasis in field theory is usually on the number of particles ($N(\mathbf{k})$ operator), and particle probability densities are rarely used. For completeness, however, and to make the connection with quantum mechanics, they are included below. (Antiparticles would have negative values of those below!)		
Single particle probability density (not operator)	$\bar{\rho}(\mathbf{x},t) =$ $\langle\phi(\mathbf{x}',t)\vert j^0(\mathbf{x},t)\vert\phi(\mathbf{x}',t)\rangle$ Note integration over $\mathbf{x}'$, not $\mathbf{x}$ For type a plane wave, $\bar{\rho} = \dfrac{1}{V}$	As at left, but with Dirac j^0 above.	= 0 for chargeless particles.
Charge, not probability	Scalar type b particle $\rightarrow$ negative ρ. Photons $\rightarrow \rho = 0$. Led to conclusion that j^0 is really proportional to *charge* probability density.		
Observables			
	Observable operators like total energy, three momentum, and charge are found by integrating corresponding density operators over all 3-space. (For spin ½, electrons assumed below with $q = -e$)		
H	$P_0 = \sum_{\mathbf{k}} \omega_{\mathbf{k}}\left(N_a(\mathbf{k}) + N_b(\mathbf{k})\right)$	$P_0 = \sum_{\mathbf{p},r} E_{\mathbf{p}}\left(N_r(\mathbf{p}) + \bar{N}_r(\mathbf{p})\right)$	$P_0 = \sum_{\mathbf{k},r} \omega_{\mathbf{k}} N_r(\mathbf{k})$
$P_i = 3$-momentum	$\mathbf{P} = \sum_{\mathbf{k}} \mathbf{k}\left(N_a(\mathbf{k}) + N_b(\mathbf{k})\right)$	$\mathbf{P} = \sum_{\mathbf{p},r} \mathbf{p}\left(N_r(\mathbf{p}) + \bar{N}_r(\mathbf{p})\right)$	$\mathbf{P} = \sum_{\mathbf{k},r} \mathbf{k}\, N_r(\mathbf{k})$
s^μ	$qj^\mu = q(\rho, \mathbf{j})$	$q\left(j^\mu - (\text{constant})\right) \rightarrow \partial_\mu s^\mu = 0$	0 for photons
Q	$\int s^0 d^3x =$ $q\sum_{\mathbf{k}}\left(N_a(\mathbf{k}) - N_b(\mathbf{k})\right)$	$\int s^0 d^3x =$ $-e\sum_{\mathbf{p},r}\left(N_r(\mathbf{p}) - \bar{N}_r(\mathbf{p})\right)$	0 for photons
Spin operator for RQM states and QFT fields	N/A	$\Sigma = \Sigma_i = \dfrac{1}{2}\begin{bmatrix} \sigma_i & 0 \\ 0 & \sigma_i \end{bmatrix}$ $i = 1,2,3$ $\sigma_i = $ 2D Pauli matrices	magnitude = 1 for photons,
Helicity operator for RQM states and QFT fields	N/A	$\dfrac{\Sigma \cdot \mathbf{p}}{\vert\mathbf{p}\vert}$	helicity eigenstates
Spin operator for QFT states	N/A	$\int \psi^\dagger \Sigma \psi \, d^3x$	magnitude = 1 for photons,
Helicity operator for QFT states	N/A	$\int \psi^\dagger \left(\dfrac{\Sigma \cdot \mathbf{p}}{\vert\mathbf{p}\vert}\right)\psi \, d^3x$	helicity eigenstates

Bosons, Fermions, and Commutators

Operations on states with creation, destruction, and number operators above yield the properties below.

Properties of states:	$n_a(\mathbf{k}) = 0,1,2,\ldots,\infty$ So, spin 0 states bosonic.	$n_r(\mathbf{p}) = 0,1$ only So, spin ½ states fermionic.	$n_r(\mathbf{k}) = 0,1,2,\ldots,\infty$ So, spin 1 states bosonic.
Bosons can only employ commutators Fermions can only employ anti-commutators	If anti-commutators used instead of commutators with Klein-Gordon equation solutions, then observable (not counting vacuum energy) Hamiltonian operator would have form $H_0^{\;0} = 0$ and $H_0^{\;0}|\phi_{\mathbf{k}}\rangle = 0$,i.e., all scalar particles would have zero energy. Hence, we cannot use anticommutators with spin 0 bosons.	Commutators lead to 2 or more identical particle states co-existing in same multiparticle state. Anti-commutators lead to only one given single particle state per multi-particle state. Therefore, commutators cannot be used with spin ½ fermions. This is further proof that we need commutators with bosons.	Same as spin 0.

The Feynman Propagator

Creation and destruction of free particles (& antiparticles) and their propagation visualized below.

Feynman diagrams	a) $t_y < t_x$ b) $t_x < t_y$	a) $t_y < t_x$ b) $t_x < t_y$	a) $t_y < t_x$ b) $t_x < t_y$
Step 1 Time ordered operator T	If $t_y < t_x$, $T\{\phi(x)\phi^\dagger(y)\} = \phi(x)\phi^\dagger(y)$, i.e., the $\phi^\dagger(y)$ operates first, and should be placed on the right. If $t_x < t_y$, $T\{\phi(x)\phi^\dagger(y)\} = \phi^\dagger(y)\phi(x)$, i.e, the $\phi(x)$ operates first, and should be placed on the right. Note that $\phi(x)$ commutes with $\phi^\dagger(y)$ for $x \neq y$. [Fermion fields anti-commute.]		
Transition amplitude (double density in x and y)	$\langle 0|T\{\phi(x)\phi^\dagger(y)\}|0\rangle = i\Delta_F$	$\langle 0|T\{\psi_\alpha(x)\bar{\psi}_\beta(y)\}|0\rangle = iS_{F\alpha\beta}$	$\langle 0|T\{A^\mu(x)A^\nu(y)\}|0\rangle = iD_F^{\;\mu\nu}$
	The above vacuum expectation values (transition amplitudes) represent both 1) creation of a particle at y, destruction at x, and 2) creation of an antiparticle at x, destruction at y $\Big\}$ transition amplitude = Feynman propagator		
Step 2 Propagator in terms of two commutators	By adding a term equal to zero to the Feynman propagator above, it can be expressed as vacuum expectation values (VEVs) of two commutators (anti-commutators for fermions)		
	$i\Delta_F(x-y) =$ $\langle 0|\big[\phi^+(x),\phi^{\dagger-}(y)\big]|0\rangle \; t_y < t_x$ $\langle 0|\big[\phi^{\dagger+}(y),\phi^-(x)\big]|0\rangle \; t_x < t_y$	$iS_{F\alpha\beta}(x-y) =$ $\langle 0|\big[\psi_\alpha^+(x),\bar{\psi}_\beta^-(y)\big]_+|0\rangle \; t_y < t_x$ $-\langle 0|\big[\bar{\psi}_\beta^+(y),\psi_\alpha^-(x)\big]_+|0\rangle \; t_x < t_y$	$iD_F^{\;\mu\nu}(x-y) =$ $\langle 0|\big[A^{\mu+}(x),A^{\nu-}(y)\big]|0\rangle \; t_y < t_x$ $\langle 0|\big[A^{\nu+}(y),A^{\mu-}(x)\big]|0\rangle \; t_x < t_y$

Step 3 As 3-momentum integrals	With the coefficient commutation relations, the above two commutators/anti-commutators (for each spin type) can be expressed as two integrals over 3-momentum space		
Definition of symbols for commutators	$\left[\phi^+(x),\phi^{\dagger-}(y)\right]=i\Delta^+(x-y)$ $\left[\phi^{\dagger+}(y),\phi^-(x)\right]=i\Delta^-(x-y)$	$\left[\psi^+_\alpha(x),\overline{\psi}^-_\beta(y)\right]_+=iS^+_{\alpha\beta}(x-y)$ $-\left[\overline{\psi}^+_\beta(y),\psi^-_\alpha(x)\right]_+=iS^-_{\alpha\beta}(x-y)$	$\left[A^{\mu+}(x),A^{\nu-}(y)\right]=iD^{\mu\nu+}(x-y)$ $\left[A^{\nu+}(y),A^{\mu-}(x)\right]=iD^{\mu\nu-}(x-y)$

$$i\Delta^\pm=\frac{1}{2(2\pi)^3}\int\frac{e^{\mp ik(x-y)}}{\omega_k}d^3\mathbf{k}\qquad iS^\pm=\frac{\pm1}{2(2\pi)^3}\int\frac{(\displaystyle{\not}p\pm m)e^{\mp ip(x-y)}}{E_\mathbf{p}}d^3\mathbf{p}\qquad iD^{\mu\nu\pm}=-g^{\mu\nu}i\Delta^\pm$$

Δ^+, S^+, $D^{\mu\nu+}$ represent particles; Δ^-, S^-, $D^{\mu\nu-}$ represent anti-particles. Symbols $S^\pm=S^\pm{}_{\alpha\beta}$

Although fields such as ϕ are operators, because of their coefficient commutation relations, each integral above is a number, not an operator. The expectation value of a number X is simply the same number X. $\langle0|X|0\rangle=X\langle0|0\rangle=X$. So, the Feynman propagator will also be simply a number (no brackets needed.)

Step 4

As contour integrals

Contour integral theory (integration in the complex plane) permits the above two integrals (for each spin type) over real 3-momentum space to be expressed as contour integrals.

$$i\Delta^\pm=\frac{\mp i}{(2\pi)^4}\int_{C^\pm}\frac{e^{-ik(x-y)}}{k^2-\mu^2}d^4k\qquad iS^\pm=\frac{\mp i}{(2\pi)^4}\int_{C^\pm}\frac{(\displaystyle{\not}p+m)e^{-ip(x-y)}}{p^2-m^2}d^4p\qquad iD^{\mu\nu\pm}=\frac{\mp ig^{\mu\nu}}{(2\pi)^4}\int_{C^\pm}\frac{e^{-ik(x-y)}}{k^2\underbrace{-\mu^2}_{photon=0}}d^4k$$

Step 5

As one integral

Taking certain limits with contour integrals in the complex plane yields a single form for the Feynman propagator that works for any time ordering and will prove more convenient.

in physical space

$$\Delta_F(x-y)=\frac{1}{(2\pi)^4}\int\frac{e^{-ik(x-y)}}{k^2-\mu^2+i\varepsilon}d^4k\qquad S_{F\alpha\beta}(x-y)=\frac{1}{(2\pi)^4}\int\frac{(\displaystyle{\not}p+m)e^{-ip(x-y)}}{p^2-m^2+i\varepsilon}d^4p\qquad D_F^{\mu\nu}(x-y)=\frac{-g^{\mu\nu}}{(2\pi)^4}\int\frac{e^{-ik(x-y)}}{k^2+i\varepsilon}d^4k$$

in momentum space

$$\Delta_F(k)=\frac{1}{k^2-\mu^2+i\varepsilon}\qquad S_{F\alpha\beta}(p)=\frac{\displaystyle{\not}p+m}{p^2-m^2+i\varepsilon}\qquad D_F^{\mu\nu}(k)=\frac{-g^{\mu\nu}}{k^2+i\varepsilon}$$

5.12 Appendix: Completeness Relations

Completeness Relations for Polarization Vectors

There are some relations between the polarization vectors $\varepsilon_r{}^\mu$ that are a little less transparent than the orthogonality conditions (5-36), but which may be used at some points in QFT. Whereas for the orthogonality conditions we sum on the spacetime index μ, if instead, we sum on the spinor index r, we get other relations called the <u>completeness relations</u> (which we state first, then describe how to derive)

$$\sum_{r=0}^{3} \zeta_r \varepsilon_r^{\mu} \varepsilon_r^{\nu} = -g^{\mu\nu} \quad \text{where} \quad \zeta_0 = -1 \quad \zeta_1 = \zeta_2 = \zeta_3 = 1. \tag{5-82}$$

It is easiest to derive these using a coordinate system aligned with the ε_r^{μ}, such as those of Fig. 5-1(b) or (d), pg. 142, where (5-34) and (5-35) hold. Since (5-82) is a tensor equation (i.e., it has two spacetime indices μ and ν), transformation to another Lorentz coordinate system (such as by rotating the spatial axes) will leave the relation unchanged. ((5-82) is valid no matter how we choose to align the axes.)

The completeness relations (5-82) can be proven by substitution of the values in (5-34) directly into (5-82). Doing this for two or three sets of values for μ and ν, should convince you of its validity.

Note that it is not as easy to visualize the meaning of (5-82) as it is for the orthogonality relations (5-36). Simply knowing that (5-82) is valid mathematically should suffice.

We also had completeness relations for spinors. (See Chap. 4, pg. 123.)

5.13 Problems

1. From the Coulomb gauge defining equation $\nabla \cdot \mathbf{A} = 0$, and $\mathbf{A}$ of (5-14), show that $\mathbf{A}$ and the wave momentum $\mathbf{k}$ are perpendicular. From (5-15) show how $\mathbf{E}$ and $\mathbf{B}$ are perpendicular to each other and also to $\mathbf{k}$.

2. Use (5-18) and (5-2) to show that $F^{31}(x) = -F^{13}(x) = B^2(x)$. If you feel you need to, also show that $F^{23}(x) = -F^{32}(x) = B^1(x)$.

3. Use (5-18) and (5-2) to show that $F^{02} = -F^{20} = E^2$. If you feel you need to, also show that $F^{03} = -F^{30} = E^3$. Is it obvious that $F^{00} = F^{11} = F^{22} = F^{33} = 0$?

4. Show that for $\mu = i$, (5-24) equals (5-3)(b).

5. Show that substitution of $A'^{\mu}(x)$ from (5-27) for $A^{\mu}(x)$ in (5-24) will result in all terms in f dropping out and give (5-24) in terms of $A^{\mu}(x)$ back again.

6. Show that $A^{\mu}(x) = \sum_{r,\mathbf{k}} \dfrac{1}{\sqrt{2V\omega_{\mathbf{k}}}} \varepsilon_r^{\mu} A_r(\mathbf{k}) e^{-ikx} + \sum_{r,\mathbf{k}} \dfrac{1}{\sqrt{2V\omega_{\mathbf{k}}}} \varepsilon_r^{\mu} B_r^{\dagger}(\mathbf{k}) e^{ikx}$ is a solution of the wave equation in $A^{\mu}(x)$. Then show that if $A^{\mu}(x)$ is real, i.e., $A^{\mu}(x) = A^{\dagger\mu}(x)$, then $B_r^{\dagger}$ must $= A_r^{\dagger}$. Then note that the two terms in the solution are complex conjugates of one another. Does the sum of a complex number and its complex conjugate always equal a real number?

7. Using the RQM Hamiltonian operator $H = i\partial/\partial t$, show that each of the eigensolutions of (5-44) with coefficients $A_r^{\dagger}(\mathbf{k})$ has negative energy.

8. Show that (5-50) solves (5-48).

9. Start with the four-vector $w_{\mu} = g_{\mu\nu} w^{\nu}$ and take the partial derivative $\partial/\partial w^{\nu}$ of both sides to show that $\partial w_{\mu}/\partial w^{\nu} = g_{\mu\nu}$. w_{μ} can, of course, be any four-vector, such as $\dot{A}_{\nu}$.

Chapter 6

Symmetry, Invariance, and Conservation for Free Fields

"The time has come", the walrus said, "to speak of many things,
of symmetries, Lagrangians, and changeless transformings."

Re-rendering of Lewis Carroll
by R. Klauber

6.0 Preliminaries

My apologies to Lewis Carroll for the liberties taken with his great work, but the Jabberwockian, oxymoron-like phrase "changeless transforming" will come to have deep significance for us. We will find it central to our understanding of symmetry in general, and more specifically, in our study of quantum field theory.

6.0.1 Background

Symmetry is one of the most aesthetically captivating and philosophically meaningful concepts known to mankind. Rooted originally in the arts, it has evolved and re-emerged in our modern age as a unified and holistic structural basis for all of science.

But if so, what then, particularly in mathematical terms, is it? If, in a work of art, it is a quality, perhaps somewhat abstract and related closely to feeling and emotion, how does it relate to physics? Can it be defined precisely?

We begin our answer to these questions after the chapter preview below.

6.0.2 Chapter Overview

First, an introduction to symmetry,
  where we will look at
  * a simple definition of symmetry without math,
  * examples of symmetry, and
  * a mathematical definition of symmetry.

A simple definition of symmetry with examples

Then, symmetry in classical physics, including
  * laws of nature symmetric under Lorentz transformation, i.e., laws are invariant in spacetime (same for all inertial observers)
  * symmetry in the Lagrangian $L \to$ a related quantity is conserved

Symmetry in classical mechanics

Then, symmetry in quantum field theory, including
  * field equations symmetric under Lorentz transformation, i.e., they are invariant in spacetime (same for all inertial observers)
  * symmetry in the Lagrangian density $\mathcal{L} \to$ a related quantity is conserved

  * symmetry, gauges, and gauge theories

Symmetry in QFT

Free vs interacting fields

We will deal primarily with free particles and fields in this chapter, but the principles will apply in general, as we shall see when we investigate interactions.

Symmetry principles apply to free and interacting cases, but only free in this chapter

6.1 Introduction to Symmetry

6.1.1 Symmetry Simplified

Each of us has some intuitive feel for what symmetry is, though most might, at least at first, have some difficulty coming up with a very precise definition. Certainly, snowflakes have symmetry, and so do cylinders and beach balls. A map of New York probably does not. Just what exactly is it that we sense about an object that causes us to deem it symmetric?

To see what that certain something is, imagine yourself looking at a real-life version of the cylinder depicted in the figure below. Then imagine closing your eyes for a moment, and during the time you can't see, someone else rotates the cylinder about the vertical axis shown in the figure. When you open your eyes is there any way you could tell that the rotation had taken place? The answer, of course, is no, but what does that mean?

It means that even though something changed (the rotational position of the cylinder), something else remained unchanged. The form we perceive, the wholeness that is the cylinder, looks exactly the same. The act of moving or "transforming" the cylinder simultaneously exhibits the qualities of both change (transformation) and non-change (invariance).

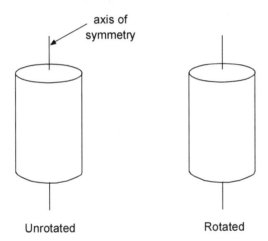

Figure 6-1. Symmetry of a Cylinder

Symmetry is the propensity for non-change with change

So what then is *symmetry*? It is simply the propensity *for non-change with change, for invariance under transformation*. In many cases, such as this one, it is a relationship between the whole and the parts in which the whole can exhibit changelessness while the component parts change. In virtually every case, it involves superficial change with more profound non-change.

Symmetry manifests to greater or lesser degrees. A sphere, for instance, has more symmetry than a cylinder because it possesses innumerable (rather than only one) possible axes about which it could be rotated and still appear the same. A snowflake has even less symmetry than a cylinder, since there are only six discrete positions into which it could be rotated where no change could be discerned. A baseball glove has no symmetry whatever. There are absolutely no ways it could be rotated (not counting multiples of $360°$) without looking distinctly different.

Different degrees of non-change with change mean different degrees of symmetry

Symmetry extends beyond rotation. Consider an infinite length horizontal line. Translate it 10 meters to the right. It still looks the same. It has translational symmetry. Consider the human body where the right half is reflected to the left, and the left half reflected to the right. It still looks the same (to good approximation.) To high degree, our bodies have mirror, or reflection, symmetry.

There are continuous symmetries, like the cylinder of Fig. 6-1, a sphere, or the infinite straight line discussed above. For these, transformation is continuous. And there are discrete symmetries,

Different kinds of symmetry: rotational, translational, reflection

like the snowflake, an infinite picket fence, or any reflection symmetry. For these, the transformation only maintains an invariant quality in certain discrete positions.

Symmetries can be continuous or discrete

Extrapolating these ideas beyond mere geometry and rotation, we can begin to understand why symmetry is considered so meaningful and fascinating. Non-change with change permeates many diverse phenomena. In many works of visual art, such as those of Escher or Indian mandalas, this principle is evident. In architecture, it has been pervasive throughout the ages. In music, the refrain, typically the essence of a song, remains the same, while other lyrics change. And that certain something we sense in the work of a great master is typically there throughout all of his or her individual pieces. We know that a Bach sonata, even if we have never heard it before, is by Bach. We know a Picasso painting, even if we have never seen it before, is by Picasso. We sense symmetry.

Symmetry plays a major role in the arts and elsewhere

6.1.2 Symmetry Mathematically

In mathematical terms, the rotations, translations, and reflections we discussed in the previous section are known as <u>transformations</u>. Any transformation, by definition, is a change of something. If the transformation is symmetric, something else remains unchanged, or in math terms, <u>invariant</u>. Not all transformations are symmetric, of course. We will look at some mathematical examples below, but first we need to note one more thing.

Mathematically, symmetry comprises invariance under transformation

The transformation depicted in Fig. 6-1 can be understood either as a rotation of the cylinder in one direction while we remain fixed (an <u>active transformation</u>, by name), or alternatively, as a rotation of our viewing frame of reference in the other direction while the cylinder remains fixed (a <u>passive transformation</u>). The same thing is true for snowflakes, the translation of a straight line, and more. Transformations typically involve a *change of perspective*, a change in the relationship between the observer and the thing being observed.[1]

Transformation is change of object with observer fixed or vice versa.

Mathematically, when we change our position of observation, it is equivalent to using a new, different reference frame and coordinate system, oriented differently from, and/or displaced relative to, the original. So a transformation can be viewed simply as a change of coordinate system, and this is often represented as a shifting from unprimed to primed coordinates. We will focus on this (passive transformation) interpretation, the most common one in physics, and most relevant to QFT.

Changing observer = changing coordinate system, most useful interpretation

Example #1

So how about some simple examples? For starters, see Fig. 6-2, where on the left-hand side we show the function

$$f\left(x^1, x^2\right) = \left(x^1\right)^2 + \left(x^2\right)^2.$$

(6-1)

Example of a function symmetric under rotation transformation

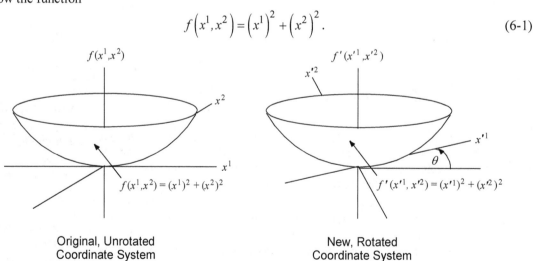

Original, Unrotated
Coordinate System

New, Rotated
Coordinate System

Figure 6-2. Example of a Function Symmetric Under Coordinate Transformation

---

[1] As an aside, with regard to the Schrödinger and Heisenberg pictures of Chap. 2 (see Wholeness Chart 2-4, pg. 28), the changing state vector with a non-changing operator of the former is analogous to a 3D vector rotating via an active transformation with a non-changing observer. Operators, like *H* for example, reflect the dynamical variables an observer observes. The Heisenberg picture, with changing operators and non-changing ("frozen") state vectors, is analogous to a passive 3D transformation, where the observer's viewpoint changes, but the vector does not.

We then change to a coordinate system rotated relative to the first, where our transformation from the first set of coordinates to the second is

$$x'^1 = x^1 \cos\theta + x^2 \sin\theta \qquad x'^2 = -x^1 \sin\theta + x^2 \cos\theta, \qquad (6\text{-}2)$$

with the inverse transformation being

$$x^1 = x'^1 \cos\theta - x'^2 \sin\theta \qquad x^2 = x'^1 \sin\theta + x'^2 \cos\theta. \qquad (6\text{-}3)$$

In matrix form, these are

$$\begin{bmatrix} x'^1 \\ x'^2 \end{bmatrix} = \underbrace{\begin{bmatrix} \cos\theta & \sin\theta \\ -\sin\theta & \cos\theta \end{bmatrix}}_{T} \begin{bmatrix} x^1 \\ x^2 \end{bmatrix} \qquad \begin{bmatrix} x^1 \\ x^2 \end{bmatrix} = \underbrace{\begin{bmatrix} \cos\theta & -\sin\theta \\ \sin\theta & \cos\theta \end{bmatrix}}_{T^{-1}=T^{\mathrm{T}}} \begin{bmatrix} x'^1 \\ x'^2 \end{bmatrix}, \qquad (6\text{-}4)$$

2D rotation transformation

where we designate the transformation by T, whose inverse is its own transpose.

Substituting (6-3) into (6-1) to express our function in the new system primed coordinates yields

$$f\left(x^1,x^2\right) = \left(x^1\right)^2 + \left(x^2\right)^2 =$$

$$f'\left(x'^1,x'^2\right) = \left(x'^1 \cos\theta - x'^2 \sin\theta\right)^2 + \left(x'^1 \sin\theta + x'^2 \cos\theta\right)^2 \qquad (6\text{-}5)$$

$$= \left(x'^1\right)^2 \left(\cos^2\theta + \sin^2\theta\right) + \left(x'^2\right)^2 \left(\cos^2\theta + \sin^2\theta\right) = \left(x'^1\right)^2 + \left(x'^2\right)^2 = f\left(x'^1,x'^2\right).$$

Function has same form in original or primed coordinates → it is symmetric under the transformation

The function has exactly the same form in both coordinate systems, exactly the same form whether we express it in terms of the unprimed or primed coordinates. Given Fig. 6-2, this should not be much of a surprise.

The prime on f' is used to indicate it has, in general, different functional form from f, which is the case for non-symmetric functions. But since the function f here is symmetric under the transformation, the functional form of f and f' is the same, so we drop the prime. This can be more easily understood with the following example.

Example #2

Consider the function

$$g\left(x^1,x^2\right) = \left(x^2\right)^2. \qquad (6\text{-}6)$$

Example of function not symmetric under rotation transformation

Express (6-6) in the primed coordinate system by substituting (6-3) into it, and we get

$$g = \left(x^2\right)^2 = \left(x'^1 \sin\theta + x'^2 \cos\theta\right)^2 = \left(x'^1\right)^2 \sin^2\theta + \left(x'^2\right)^2 \cos^2\theta + 2x'^1 x'^2 \sin\theta\cos\theta \neq \left(x'^2\right)^2. \ (6\text{-}7)$$

Thus, g has different form in the two systems and is *not* symmetric under the transformation T.

$$g\left(x^1,x^2\right) = g'\left(x'^1,x'^2\right) \neq g\left(x'^1,x'^2\right) \quad \text{but} \quad f\left(x^1,x^2\right) = f'\left(x'^1,x'^2\right) = f\left(x'^1,x'^2\right). \qquad (6\text{-}8)$$

The transformed form of g, represented by g', has the same value at the same physical point, but it is not the same form in terms of the primed coordinates as g was in terms of the unprimed coordinates. But f', the transformed form of f, did have the same form in terms of both sets of coordinates, and thus, we dropped the prime on f on the RHS of (6-8).

In spite of its non-symmetry under rotation, g is symmetric under a different kind of transformation, the translation to a coordinate system which is displaced relative to the first along the x^1 axis, i.e., $x^1 \to x'^1 = x^1 + $ constant, or

But same function is symmetric under a translation transformation

$$\begin{bmatrix} x'^1 \\ x'^2 \end{bmatrix} = \begin{bmatrix} x^1 \\ x^2 \end{bmatrix} + \begin{bmatrix} K \\ 0 \end{bmatrix} \qquad K = \text{constant}. \qquad (6\text{-}9)$$

Substitution of (6-9) into (6-6) yields $g'(x'^1, x'^2) = (x'^2)^2$, having the same form in both systems.

If a coordinate is missing from f, then f is symmetric with respect to change of that coordinate

Lessons from the Examples

From Example #2, we can deduce the general rule that if a coordinate is missing in a given function, that function is invariant under a transformation solely in the direction of that coordinate (and also under multiplication of the coordinate by a constant, which will be less important for us.) The function is symmetric with respect to that transformation.

In both examples and in general, the value of a particular function at a given physical point in space is the same under any transformation, symmetric or not. The new coordinates are simply a new way to designate that particular point with different numbers, but it is the same point in space, and hence must have the same numeric value for the function there. If f or g were a physical entity, like pressure, simply changing our coordinates would not change the value of the pressure at any given point in space, even though the numbers describing that point's location are different.

Value of a scalar function at a physical point stays same under any transformation

So under *any* transformation of coordinate axes, the value at a physical point of every possible scalar function is invariant. Under a *symmetry* transformation the form of the function also is invariant. Under a non-symmetry transformation, the form of the function looks different in terms of the new coordinates, and we represent that functional difference with a prime on the function label.

Form of a scalar function stays same under a symmetry transformation

Scalars are Invariant, Vectors are Covariant

Consider a 2D position vector in physical space represented in the unprimed coordinates of Example #1 by $x^i = (x^1, x^2)$. Under the rotation transformation T, this becomes $x'^i = (x'^1, x'^2) \neq (x^1, x^2)$. A different (i.e., non-invariant) set of coordinates represents the exact same vector. But it is the same vector at the same physical location, and in fact, has the same length in each coordinate system equal to

Vector components change under transformation

$$|\mathbf{x}| = |x^i| = \sqrt{\left(x^1\right)^2 + \left(x^2\right)^2} = \sqrt{\left(x'^1\right)^2 + \left(x'^2\right)^2} = |x'^i|. \tag{6-10}$$

So the scalar value at the point (equal to the length of the position vector at that point) is the same in both systems, but the coordinate values are not.

But vector length & direction in physical space unchanged for any coord system

It is generally true of every vector $\mathbf{v}$, not just the position vector shown here, that its physical, measurable length (a scalar value) remains unchanged under any coordinate transformation, but its component values change. This is called covariance. Scalar values are invariant under coordinate transformation; vector components are covariant. (Don't confuse this use of the word "covariant" with our use of the terms covariant and contravariant coordinates.)

Vectors are covariant under coordinate transformation

Parallel to scalars, if the vector components remain unchanged under a given transformation, then that transformation is a symmetry transformation, i.e., $v'^j(x'^i) = v^j(x'^i)$. One example is the **E** field around a point charge, which points radially outward from the point, described in a coordinate system with origin at the point. Rotating to a new coordinate system, we find the same functional dependence of the **E** field on the new coordinates. See Prob. 7.

Vector transformation symmetric if components unchanged

All of these conclusions are valid for any dimension space, and in particular for our purposes, the 4D spacetime of relativity theory. They are also valid for systems of generalized coordinates, not just Cartesian like those shown here, and for both particles and fields. Probs. 1 through 6 and Wholeness Chart 6-1 can help you gain more comfort with these concepts.

All of above true for 4D and other spaces, as well

Wholeness Chart 6-1. Symmetry Summary

	Non-Symmetric Transformation	**Symmetric Transformation**
Coordinate values change?	Yes	Yes
Scalar value at a physical point the same?	Yes	Yes
Form of function invariant?	No	Yes
Vector magnitude and direction at a physical point the same?	Yes	Yes
Vector components invariant?	No	Yes
Vector components vary covariantly?	Yes	No, invariant
General rule: If a function h is not a function of the jth coordinate x^j, then h is symmetric under the transformation $x^j \rightarrow x^j + $ constant		

6.2 Symmetry in Classical Mechanics

6.2.1 Invariance of the Laws of Nature

Symmetry turns out to play an extraordinary role in the physics of our creation. Albert Einstein, in possibly the most far reaching of any scientific discovery, provided the first insight into the universe's innate symmetry. He showed, via his theories of relativity, that even though the visible world of changing objects appears different at different places, in different times, to different observers, the physical laws of nature governing those objects remain invariant regardless of when, where, or how they are perceived. The laws of physics, acting on a subtler, more holistic level of creation, exhibit changelessness in the midst of change and are said to be *symmetric* throughout spacetime.

Einstein showed laws of nature symmetric (invariant) in spacetime

We review this discovery by Einstein and the classical mechanics leading up to it below. This should not be new material for most readers, but since it forms a good part of the foundation for understanding the ramifications of symmetry in physics, I provide the following overview, which many readers are probably well versed enough in to skim, or skip, over.

6.2.2 Brief History of Einstein's Insight into Symmetry

Newton's Laws: Invariant under Galilean Transformations

Newton's laws of motion are trivially symmetric under a rotation transformation because $\mathbf{F}=m\mathbf{a}$ (from which the other two laws can be derived) is a vector equation, and as we showed above, any vector maintains its magnitude and direction unchanged in physical space under a passive rotation transformation. So, the $\mathbf{F}$ and $\mathbf{a}$ vectors will still be aligned with each other in any new coordinate system and still have the same proportionality constant of m. There will be new coordinates for each, but the same equation relating those coordinates will hold. If $F^i = ma^i$, then $F'^i = ma'^i$. Similar results hold for a translation transformation, which you can show by doing Prob. 8.

$\mathbf{F}=m\mathbf{a}$ invariant under rotation and translation

Another type of transformation involves changing from one coordinate system to another where the transformed system has a constant velocity relative to the first of $\mathbf{v}$. Any fixed coordinate value in the original system appears to move in the $-\mathbf{v}$ direction relative to the second. This, as most readers should know, is called a <u>Galilean transformation</u> for the 3D plus time of classical mechanics and is

The (3D) Galilean transformation is to coordinate system having constant velocity relative to original system

$$\begin{bmatrix} x^1 \\ x^2 \\ x^3 \end{bmatrix} \rightarrow \begin{bmatrix} x'^1 \\ x'^2 \\ x'^3 \end{bmatrix} = \begin{bmatrix} x^1 - v^1 t \\ x^2 - v^2 t \\ x^3 - v^3 t \end{bmatrix} \quad \text{or} \quad \mathbf{x} \rightarrow \mathbf{x} - \mathbf{v}t. \quad (6\text{-}11)$$

In Newtonian/Galilean mechanics, time does not change from one system to the other. It is invariant and thus labeled by t in both systems above.

Newton's second law is invariant under this transformation because of the second order time derivative in x^i. That is,

$$\begin{bmatrix} F^1 \\ F^2 \\ F^3 \end{bmatrix} = m \begin{bmatrix} \ddot{x}^1 \\ \ddot{x}^2 \\ \ddot{x}^3 \end{bmatrix} \rightarrow = m \begin{bmatrix} \ddot{x}'^1 \\ \ddot{x}'^2 \\ \ddot{x}'^3 \end{bmatrix} = m \frac{d^2}{dt^2} \begin{bmatrix} x^1 - v^1 t \\ x^2 - v^2 t \\ x^3 - v^3 t \end{bmatrix} = m \begin{bmatrix} \ddot{x}^1 \\ \ddot{x}^2 \\ \ddot{x}^3 \end{bmatrix} = \begin{bmatrix} F^1 \\ F^2 \\ F^3 \end{bmatrix} \quad \text{or} \quad \mathbf{F} = m\mathbf{a} \rightarrow \mathbf{F} = m\mathbf{a}. \quad (6\text{-}12)$$

$\mathbf{F}=m\mathbf{a}$ invariant under Galilean transformation

Prior to Maxwell's appearance on the scene, it was generally assumed that all laws of nature were invariant under Galilean transformations.

Maxwell's Laws: Invariant under a Different Kind of Transformation

However, with the publication of Maxwell's equations in 1864, it was realized that his laws of nature, in contrast, do not transform symmetrically under a Galilean transformation. If one invoked (6-11) in the coordinates of Maxwell's equations, the result was a set of equations of different form, quite unlike the behavior that we saw in (6-12). That exercise is fairly involved and would lead us too far afield, so we won't get into it here.

Maxwell's eqs NOT invariant under Galilean transformation

It was, however, realized that Maxwell's equations were invariant under a different transformation between coordinate systems in relative motion. This transformation is 4D with time and space transformations, rather than simply 3D spatial, and is called the <u>Lorentz transformation</u>, after its discoverer. It is, where we lose no generality by restricting relative velocity to a single coordinate direction (since Maxwell's laws are symmetric under rotation), and where we write out c in non-natural units this one time,

Maxwell's eqs ARE invariant under Lorentz transformation

$$\begin{bmatrix} x^0 \\ x^1 \\ x^2 \\ x^3 \end{bmatrix} \rightarrow \begin{bmatrix} x'^0 \\ x'^1 \\ x'^2 \\ x'^3 \end{bmatrix} = \begin{bmatrix} \gamma\left(x^0 - \frac{v}{c}x^1\right) \\ \gamma\left(x^1 - \frac{v}{c}x^0\right) \\ x^2 \\ x^3 \end{bmatrix} = \underbrace{\begin{bmatrix} \gamma & -\gamma\frac{v}{c} & 0 & 0 \\ -\gamma\frac{v}{c} & \gamma & 0 & 0 \\ 0 & 0 & 1 & 0 \\ 0 & 0 & 0 & 1 \end{bmatrix}}_{\Lambda} \begin{bmatrix} x^0 \\ x^1 \\ x^2 \\ x^3 \end{bmatrix} \qquad \gamma = \frac{1}{\sqrt{1 - \frac{v^2}{c^2}}} \quad v = v^1 . \quad (6\text{-}13)$$

The (4D) Lorentz transformation

Again, because it would lead us away from our present tasks, we will not go through the lengthy process of showing that Maxwell's equations retain the same form under (6-13). Note that we will designate the Lorentz transformation with the symbol Λ.

Einstein's Resolution

Einstein and others wanted all equations of nature to display invariance under the same transformation. There were two major sets of laws at the time, Newton's for mechanics and Maxwell's for electromagnetism. But they didn't transform to moving coordinates in the same way. Something had to give.

Scientists wanted all laws symmetric under same transformation

Einstein intuited that the speed of light must be the same for all observers, whether they are fixed relative to one another or have relative constant velocity. This was quite a radical insight and turned out to be true. Since the wave solution to Maxwell's equations yielded a speed of e/m waves (light) of c, that meant those equations must yield the same result in any coordinate system that was not accelerating (nor in a gravitational field). To do this, they must have the same form in all such coordinate systems. The only transformation that did that was Lorentz's.

Einstein figured c = same for all observers

This meant Maxwell's eqs same for all

Einstein took that "to the bank". He knew it meant that in order for the equations of mechanics to also be symmetric under the Lorentz transformation, they must be modified. Newton's laws were not the exact truth, but only a very good approximation under normal human conditions to more accurate and precise laws.

So, Lorentz transfn must be correct one for e/m and mechanics laws

We won't write Einstein's law of mechanics here, but refer interested readers to any textbook on relativity[1]. The point is that his reformulation of mechanics 1) is invariant in form under Lorentz transformations, and 2) reduces to high accuracy, for objects moving at speeds much less than c, to Newton's 2nd law.

Einstein reformulated mechanics to be invariant under Lorentz transfn

We summarize these results in Wholeness Chart 6-2.

That reformulation is special relativity

Wholeness Chart 6-2. Galilean vs Lorentz Transformations

	Galilean transformation	Lorentz transformation
Newton's laws symmetric?	Yes	No
Maxwell's laws symmetric?	No	Yes
Einstein law of mechanics symmetric?	No	Yes
Valid for any speed?	No	Yes
Valid at low speed?	Yes, approximately	Yes

Special relativity mechanics become classical mechanics at $v \ll c$ ($v \ll 1$ in natural units)

---

[1] For example, Hartle, James B., *Gravity*, Pearson (2003), Chap. 5

Einstein carried this idea further in the development of his general theory of relativity. In very general terms, the same concept holds. At any given point in spacetime, the laws of nature, expressed as relationships between physical entities (like scalars, vectors, and tensors) are invariant in form. However, the Lorentz transformation is specifically for differences in velocity in non-accelerating, non-gravitational, systems. All of our work in this text will assume acceleration and gravitational effects are zero or small enough to be ignored.

Our work special relativity, not general relativity

The bottom line: All laws of nature are symmetric (invariant) under Lorentz transformation[1]. They are the same for all observers in relative constant velocity motion.

Laws of nature symmetric under Lorentz transformation

6.2.3 More with Lorentz Transformations

Index notation for Lorentz transformations

The relation (6-13) expressed in shorthand notation for the position vector, and the corresponding transformations for a four vector V^μ and a 4D tensor $T^{\mu\nu}$ (for those familiar with tensors) can be written as

$$x'^\mu = \Lambda^\mu{}_\nu x^\nu \qquad V'^\mu\left(x'^\alpha\right) = \Lambda^\mu{}_\nu V^\nu\left(x^\alpha\right) \qquad T'^{\mu\nu}\left(x'^\alpha\right) = \Lambda^\mu{}_\delta \Lambda^\nu{}_\gamma T^{\delta\gamma}\left(x^\alpha\right). \qquad (6\text{-}14)$$

Note that Λ^{-1}, the inverse of Λ, can be obtained by taking $\mathbf{v} \rightarrow -\mathbf{v}$ in (6-13) since each coordinate system seems to be going in the opposite direction with respect to the other. Λ^{-1} will transform x'^μ back into x^μ.

Length of any four-vector invariant

Recall from (6-10) that the length of a vector in 3D is unchanged under a coordinate system transformation, i.e., the length is a scalar and thus invariant. The same thing is true in 4D for four-vectors. Recall further, from Chap. 2, particularly the appendix Sect. 2.9.3, pg. 35, that the length of any four-vector, symbolized by w^μ, is the square root of the 4D inner product, i.e., of

Length of any 4D vector invariant under Lorentz transformation

$$w_\mu w^\mu = w_0 w^0 + w_1 w^1 + w_2 w^2 + w_3 w^3 = w^0 w^0 - w^1 w^1 - w^2 w^2 - w^3 w^3 = \text{scalar invariant} \quad (6\text{-}15)$$

and that this is the same for any observer in any inertial coordinate system. This applies to any vector, be it a position vector like x^μ, the differential of position dx^μ, the four-velocity u^μ of the Chap. 2 appendix, the four-potential A^μ, the partial derivative ∂^μ, or any other.

Do Prob. 9 to show that under a Lorentz transformation (6-15) is invariant and therefore is Lorentz invariant. We often call such scalars in 4D world scalars or Lorentz scalars.

4D Laplacian derivative invariant

Similarly, just as $x_\mu x^\mu$ is Lorentz invariant, so is the differential length squared $dx_\mu dx^\mu$, as well as its partial differential sibling $\partial_\mu \partial^\mu$. Putting the latter into the denominator, we get the corresponding 4D Laplacian derivative

4D Laplacian invariant under Lorentz transformation

$$\frac{\partial}{\partial x^\mu}\frac{\partial}{\partial x_\mu} = \partial_\mu \partial^\mu = \frac{\partial}{\partial x^0}\frac{\partial}{\partial x^0} - \frac{\partial}{\partial x^1}\frac{\partial}{\partial x^1} - \frac{\partial}{\partial x^2}\frac{\partial}{\partial x^2} - \frac{\partial}{\partial x^3}\frac{\partial}{\partial x^3} = \text{scalar invariant derivative}, \quad (6\text{-}16)$$

which, as noted, is Lorentz invariant itself. So, where X represents any quantity,

$$\frac{\partial}{\partial x^\mu}\frac{\partial}{\partial x_\mu} = \frac{\partial}{\partial x'^\mu}\frac{\partial}{\partial x'_\mu} \quad \rightarrow \quad \frac{\partial}{\partial x^\mu}\frac{\partial}{\partial x_\mu}X = \frac{\partial}{\partial x'^\mu}\frac{\partial}{\partial x'_\mu}X. \qquad (6\text{-}17)$$

---

[1] I don't want to confuse readers, but most specialists in relativity would bring up caveats here. For one, in systems with clock synchronization done under a different convention than Einstein's, the laws of nature actually do take different form. The Lorentz transformation assumes Einstein synchronization. We will stick with that, the simplest, most efficient, synchronization and with the most widely used transformation of Lorentz. Our statements with regard to symmetries thereunder will hold true in general for our work and are widely accepted as valid criteria, which good theories should meet.

However, a second caveat involves research being done at the time of this writing that questions whether Lorentz symmetry needs to be upheld in certain very advanced theories of elementary particles.

Please do not worry about these things now. You can do so, when and if your work leads you into these other areas.

In general, any time we sum over pairs of indices, even if the factors in the summation are partial derivatives, we get a scalar invariant as a result.

Most general transformation between 4D coordinates

The most general transformation we could have in spacetime would comprise 1) a 4D translation (translating our coordinate axes in space, time, or both), 2) a rotation in space, and 3) a Lorentz transformation to a frame with different relative velocity. (We ignore reflection.)

The rotation in 3D is the same as we should have seen in classical mechanics and thus, for our purposes, is not of great interest. It does allow us to rotate our 3D axes, however, so that the relative velocity between our original and transformed systems is along the x^1 axes of both. This lets us use the Lorentz transformation in its simplest form, the Λ of (6-13). With this form we state the general transformation between coordinate systems, known as the Poincaré transformation as

$$x'^{\mu} = \Lambda^{\mu}{}_{\nu}\left(x^{\nu} + a^{\nu}\right) \qquad a^{\nu} = \text{constant four vector}. \qquad (6\text{-}18)$$

Poincaré transformation

Our laws of nature invariably involve partial derivatives with respect to time and space, i.e, with respect to infinitesimal differences in 4D position. These differences between coordinates in any given coordinate system are unchanged by displacing the coordinates by any constant a^{ν}. So the laws of nature will not be changed by a translation of the 4D coordinate system by a^{ν}. The laws are the same at any place x^i, and at any time x^0.

Laws of nature symmetric under Poincaré transformation

Bottom line: Thus, you may hear it said sometimes that the laws of nature are invariant with respect to Poincaré transformations.

6.2.4 Other Kinds of Symmetry

There are other kinds of symmetry, other than that of Lorentz symmetry in spacetime. You should have studied this, at least to some degree, in classical mechanics, but we will review the essence of it here.

Other kinds of symmetry exist

Symmetry of the Lagrangian Implies a Conserved Quantity

Consider the Euler-Lagrange equation for a particle in Newtonian mechanics

$$\frac{d}{dt}\left(\frac{\partial L}{\partial \dot{x}^i}\right) - \frac{\partial L}{\partial x^i} = 0 \qquad L = T - V \qquad p_i = \frac{\partial L}{\partial \dot{x}^i}. \qquad (6\text{-}19)$$

If the Lagrangian L is not an explicit function of the spatial coordinate x^i, then $\partial L/\partial x^i = 0$ on the LHS above. Thus, the time derivative of p_i is zero.

$$\frac{d}{dt}\left(\frac{\partial L}{\partial \dot{x}^i}\right) = \frac{dp_i}{dt} = 0 \quad \text{with} \quad \frac{\partial L}{\partial x^i} = 0 \quad \text{when} \quad L \neq L\left(x^i\right). \qquad (6\text{-}20)$$

Hence p_i is constant and thus, conserved. This makes sense since the only source for spatial dependence in L is the potential energy V. The gradient of potential energy $-\partial V/\partial x^i$ is force on the particle. If we have no V dependence on x^i, then there is no force in the x^i direction, and momentum p_i is constant. For example, if V is a function of x^1, but not of x^2 or x^3, then p_1 is not conserved, but p_2 and p_3 are.

$L \neq L(x^i)$ means p_i conserved

Note this means the Lagrangian is symmetric. As we saw on pg. 166, whenever a function is not dependent on a coordinate, it is symmetric in that coordinate. Changing the coordinate via a translation to a new coordinate value makes no change in the function.

Bottom line: If the Lagrangian is symmetric in a coordinate, then the conjugate momentum for that coordinate is conserved. This is true not just for the coordinates x^i, but for any generalized coordinates q^i (We reviewed generalized coordinates in Chap. 2. $q^i = x^i$ is just a special case).

L symmetric in generalized coord q^i means conjugate momentum p_i conserved

This should be review for you, but if you feel you need more practice with symmetry of L and conservation, do Probs. 10 and 11.

Symmetry of the Lagrangian Density

The prior section dealt with the Lagrangian of a particle, and similar effects arise for the Lagrangian density $\mathcal{L}$ of classical field theory. We will not, however, delve into that here, but simply move on to symmetries of $\mathcal{L}$ in quantum field theory and what they imply there.

Similar effect with Lagrangian density

6.3 Transformations in Quantum Field Theory

6.3.1 Scalars, Vectors, and Tensors

We have discussed, in the classical mechanics section above, how scalars and vectors transform under the Lorentz (and Poincaré) transformation. The same conclusions carry over to the fields of QFT. However, there are no spinors in classical theory, so we didn't discuss their transformation properties there, but we need to in QFT.

Quantum scalars, vectors transform in same way as classical

6.3.2 Spinor Transformations

When we transform our coordinate system either by a Lorentz transformation (boost = change in velocity) or a rotation of our coordinates (change in angle), we ask how a spinor field will transform. We know world scalars maintain their same value at an event and vectors transform according to (6-14). For spinors, we seek a matrix which is four by four in spinor space and which represents what happens to a spinor under a Lorentz transformation and/or a rotation of coordinates. That is, we seek D in

Spinors only in QM

$$\psi'\left(x'^{\mu}\right) = D\psi\left(x^{\mu}\right) \xrightarrow[\text{written out}]{\text{with spinor indices}} \psi'_{\alpha}\left(x'^{\mu}\right) = D_{\alpha\beta}\psi_{\beta}\left(x^{\mu}\right). \qquad (6\text{-}21)$$

Spinors have their own transformation

Deriving D can take pages, is quite complicated, and would take us far afield from our present direction, so I will just state it. Interested readers can find this derivation in certain texts or online[1]. The spinor transformation under Lorentz and rotation transformations is

$$D = e^{-i(\mathbf{L}\cdot\Theta+\mathbf{M}\cdot\mathbf{Q})} \qquad L^k = -\tfrac{i}{2}\varepsilon_{ij}{}^k\gamma^i\gamma^j, \quad \Theta^k = \left(\theta^1,\theta^2,\theta^3\right), \quad M^k = \tfrac{i}{2}\gamma^0\gamma^k, \quad Q^k = \left(v^1,v^2,v^3\right), (6\text{-}22)$$

where Θ^k represents rotation angles of the primed system with respect to the unprimed system; Q^k is a three vector of the boost velocities; and $\varepsilon_{ij}{}^k$ is zero unless i,j,k are all different, 1 if their order is 1,2,3 or 2,3,1 or 3,1,2, and -1 for other orders. (In Chap. 4, pg. 127, (4-182), our transformation T equals D of (6-22) where $\mathbf{v} = 0$ [i.e., $Q^k = 0$].)

Spinors have their own transformation

It is probably not beneficial, at this point, to worry too much more about (6-22). If, at some time in the future, your work takes you in a direction where you need to understand this transformation better, then you can study it more extensively then.

Note that in formal mathematical language, the set of all possible Lorentz transformations (all possible $\mathbf{v}$) is known as the <u>Lorentz group</u>. When the Lorentz group acts on the coordinate system, it changes what our spinors look like in the new system and this change is represented by D. So, D is called a <u>representation of the Lorentz group</u>. It "represents" that group in spinor space.

Spinor transformation is a representation of the Lorentz group

We note, again without proof here due to complexities involved, that

$$\bar{\psi}\psi = \text{world scalar} \qquad \bar{\psi}\gamma^{\mu}\psi = \text{transforms like four vector}. \qquad (6\text{-}23)$$

Spinor objects that transform like world scalars and vectors

The first part of Wholeness Chart 6-3 summarizes Lorentz transformations for scalars, vectors, spinors and tensors (for those experienced with tensors).

6.4 Lorentz Symmetry of the Lagrangian Density

As reviewed in Chap. 2, Sect. 2.5.1, point 11, pgs. 24-25, the Lagrangian density is a Lorentz scalar, in the sense that it has the same value at any event (4D point) as seen from any inertial coordinate system. But there is a deeper symmetry that $\mathcal{L}$ has as well. Its functional form, in terms of the fields of which it is composed, is also the same in any inertial coordinate system. It is a symmetric *function*, in addition to being a symmetric *value* at every (4D) point.

$\mathcal{L}$ symmetric under Lorentz transformation

We conclude this because of Einstein's postulate that the laws of nature (the field equation here) are invariant in form under Lorentz transformation. The Euler-Lagrange equation for fields, which is another form of the field equation, is a law of nature and must therefore be invariant in form as well. But since $\mathcal{L}$ is inserted into the invariant Euler-Lagrange equation to get the invariant field equation, $\mathcal{L}$ itself must be invariant in functional form. $\mathcal{L}$ is the same function of the fields in any inertial coordinate system.

---

[1] For examples, David Tong's lecture http://www.damtp.cam.ac.uk/user/dt281/qft/four.pdf, Sidney Coleman's notes www.quantumfieldtheory.info/Sydney_Coleman_QFT_lecture_notes.pdf pg. 125, or *Physics of Symmetry* by Jakob Schwichtenberg (2nd ed, Springer 2018), pgs. 72-75, 84-85.

This is summarized, with a concrete example from scalar field theory, in the last three rows of Wholeness Chart 6-3 below.

Note that, as discussed in the referenced page above, and the associated problem thereto, though $\mathcal{L}$ is a world scalar, L is not. Neither is the Hamiltonian H nor the Hamiltonian density $\mathcal{H}$. That is, none of L, H, and $\mathcal{H}$ have the same value when measured in coordinate systems having relative velocity to one another.

Wholeness Chart 6-3. Summary of Effect of Lorentz Transformation on Fields

	x^μ system	x'^μ system	Comment
	$x^\mu \longrightarrow$ Lorentz transformation $\Lambda \longrightarrow x'^\mu$		Shorthand symbols defined: $x'^\mu = \Lambda x^\mu$ x'^μ and x^μ represent the same event
Scalar field	$S(x^\mu)$	$S'(x'^\mu) = S(x^\mu)$ always $S'(x'^\mu) = S(x'^\mu)$ if sym form	✓ If S is symmetric function under Λ, S' has same functional form as S
Vector field	$V^\alpha(x^\mu)$	$V'^\alpha(x'^\mu) = \Lambda V^\alpha(x^\mu)$ always $V'^\alpha(x'^\mu) = V^\alpha(x'^\mu)$ if sym	$V'^\alpha V'_\alpha = V^\alpha V_\alpha$ invariant, V^α covariant 2 vectors, $V'^\alpha W'_\alpha = V^\alpha W_\alpha$ invariant $\leftarrow$ If V^α components sym under Λ
Tensor field	$T^{\alpha\beta}(x^\mu)$	$T'^{\alpha\beta}(x'^\mu) = \Lambda\Lambda T^{\alpha\beta}(x^\mu)$ always $T'^{\alpha\beta}(x'^\mu) = T^{\alpha\beta}(x'^\mu)$ if sym	$T'^{\alpha\beta} T'_{\alpha\beta} = T^{\alpha\beta} T_{\alpha\beta}$ invar, $T^{\alpha\beta}$ covar Other invariants exist such as trace $T^\alpha{}_\alpha$ $\leftarrow$ If $T^{\alpha\beta}$ components sym under Λ
Spinor field	$\psi(x^\mu)$	$\psi'(x'^\mu) = D\psi(x^\mu)$ $\bar{\psi}\psi$ invariant $\left(\bar{\psi}\gamma^\alpha\psi\right)' = \Lambda^\alpha{}_\beta \bar{\psi}\gamma^\beta\psi$	D = Lorentz group rep for spinors $\bar{\psi}\psi$ transforms like world scalar $\bar{\psi}\gamma^\alpha\psi$ transforms like 4-vector
Law of nature	$(\partial^\alpha\partial_\alpha + m)\,\phi(x^\mu) = 0$	$(\partial'^\alpha\partial'_\alpha + m)\,\phi'(x'^\mu) = 0$	Same form under Λ Example is Klein-Gordon field equation
Euler-Lagrange equation	$\dfrac{\partial}{\partial x^\alpha}\left(\dfrac{\partial\mathcal{L}}{\partial\phi^r{}_{,\alpha}}\right) - \dfrac{\partial\mathcal{L}}{\partial\phi^r} = 0$	$\dfrac{\partial}{\partial x'^{\alpha'}}\left(\dfrac{\partial\mathcal{L}'}{\partial\phi'^r{}_{,\alpha'}}\right) - \dfrac{\partial\mathcal{L}'}{\partial\phi'^r} = 0$	Same form under Λ
Lagrangian density	$\mathcal{L}(\phi^r(x^\mu))$ $\mathcal{L} = \partial_\alpha\phi^\dagger\,\partial^\alpha\phi - \mu^2\phi^\dagger\phi$	$\mathcal{L}'(\phi'^r(x'^\mu)) = \mathcal{L}(\phi'^r(x'^\mu))$ $\mathcal{L} = \partial'_\alpha\phi'^\dagger\,\partial'^\alpha\phi' - \mu^2\phi'^\dagger\phi'$	For Euler-Lagrange eq to be covariant (same form under Λ), $\mathcal{L}$ must keep same form under Λ, as well Example is Klein-Gordon $\mathcal{L}$

6.5 Other Symmetries of the Lagrangian Density: Noether's Theorem

6.5.1 Example of a Different Kind of Symmetry

There are other ways the Lagrangian density can be symmetric, other than under Lorentz transformations. For example, consider the scalar Lagrangian (density)

A different kind of symmetry for $\mathcal{L}$

$$\mathcal{L}_0^0 = \left(\partial_\nu\phi^\dagger\partial^\nu\phi - \mu^2\phi^\dagger\phi\right) \qquad (6\text{-}24)$$

where we introduce a transformation that changes the phase angle of the solution

$$\phi \rightarrow \phi' = \phi e^{-i\alpha} \tag{6-25}$$

Under this transformation, $\mathcal{L}$ is symmetric

where α is a real constant (and has nothing to do with the Lorentz sub/superscript index α). No change is made to x^μ of which ϕ is a function.

What does such a transformation do to the Lagrangian? (We will start dropping the word "density", as is common practice.) Note from (6-25) that $\phi = \phi' \, e^{i\alpha}$, and plugging that into (6-24) we have

$$\mathcal{L}_0^0\left(\phi^\dagger,\phi\right) = \left(\partial_\nu \phi^\dagger \partial^\nu \phi - \mu^2 \phi^\dagger \phi\right) \xrightarrow{\phi \rightarrow \phi' = \phi e^{-i\alpha}} \mathcal{L}_0^0 = \left(\partial_\nu \underbrace{\phi'^\dagger e^{-i\alpha}}_{\phi^\dagger} \partial^\nu \underbrace{\phi' e^{i\alpha}}_{\phi} - \mu^2 \underbrace{\phi'^\dagger e^{-i\alpha}}_{\phi^\dagger} \underbrace{\phi' e^{i\alpha}}_{\phi}\right)$$

$$= \left(\partial_\nu \phi'^\dagger \partial^\nu \phi' - \mu^2 \phi'^\dagger \phi'\right) = \mathcal{L}_0^0\left(\phi'^\dagger,\phi'\right). \tag{6-26}$$

So the Lagrangian is unchanged in form under the transformation. The transformed Lagrangian has the same form whether in terms of ϕ or ϕ'. Thus, the law of nature derived from the Lagrangian, the Klein-Gordon equation in this case, also looks the same in terms of ϕ or ϕ'. And so, all predictions for measurements using either solution will be the same.

6.5.2 Internal vs External Symmetries

Poincaré transformations (Lorentz plus 4D translation) and 3D rotations involve changes to our physical coordinates x^μ of our external world and are called <u>external transformations.</u>

External vs internal symmetries

Transformations like (6-25) have nothing to do with x^μ, but instead function in hidden spaces, behind the scene, like Hilbert or Fock space. They are called <u>internal transformations.</u>

In both cases, if something remains the same under the transformation, we have a symmetry (external or internal).

Note that the transformation (6-25) amounts to a rotation in the complex plane, which is an internal space. We will see repeatedly, as we delve further into QFT, that internal transformations often amount to what can be visualized in some cases as akin to rotations, and in others, as reflections, in non-physical, abstract spaces like Fock space.

Internal transformations like rotations and reflections in abstract space

6.5.3 Noether's Theorem

A general theorem to cover all types of transformations, but most useful for internal transformations, was discovered by Emmy Noether and bears her name. It plays an extremely important role in QFT, and in words, can be stated like this.

<u>Noether's theorem</u> (without math): A symmetry in the Lagrangian density implies an associated quantity is conserved.

Noether's theorem in words

This is reminiscent of the symmetry in the Lagrangian L of classical particle theory with respect to the generalized coordinate q^i (x^i in a Cartesian system) (see pg. 170) implying the conjugate momentum p_i is conserved.

<u>Proof of Noether's Theorem:</u>

If $\mathcal{L}$ is symmetric in some parameter α, then it is unchanged when α changes, i.e., its derivative with respect to α is zero.

Proof of Noether's theorem

$$\mathcal{L} = \mathcal{L}\left(\phi^r, \phi^r_{,\mu}\right) \text{ symmetric in } \alpha, \text{ then } \rightarrow \quad \frac{\partial \mathcal{L}}{\partial \alpha} = 0 = \underbrace{\frac{\partial \mathcal{L}}{\partial \phi^r}}_{\substack{\text{use Euler-} \\ \text{Lagrange} \\ \text{equation}}} \frac{\partial \phi^r}{\partial \alpha} + \frac{\partial \mathcal{L}}{\partial \phi^r_{,\mu}} \frac{\partial \phi^r_{,\mu}}{\partial \alpha} \tag{6-27}$$

Using $\partial \mathcal{L}/\partial \phi^r$ from the Euler-Lagrange field equation

$$\frac{\partial}{\partial x^\mu}\left(\frac{\partial \mathcal{L}}{\partial \phi^r_{,\mu}}\right) - \frac{\partial \mathcal{L}}{\partial \phi^r} = 0 \tag{6-28}$$

in (6-27), yields

$$0 = \left(\frac{\partial}{\partial x^\mu} \frac{\partial \mathcal{L}}{\partial \phi^r{}_{,\mu}} \right) \frac{\partial \phi^r}{\partial \alpha} + \frac{\partial \mathcal{L}}{\partial \phi^r{}_{,\mu}} \frac{\partial \phi^r{}_{,\mu}}{\partial \alpha} = \frac{\partial}{\partial x^\mu} \left(\frac{\partial \mathcal{L}}{\partial \phi^r{}_{,\mu}} \frac{\partial \phi^r}{\partial \alpha} \right) - \frac{\partial \mathcal{L}}{\partial \phi^r{}_{,\mu}} \underbrace{\frac{\partial}{\partial x^\mu} \frac{\partial \phi^r}{\partial \alpha}}_{\frac{\partial \phi^r{}_{,\mu}}{\partial \alpha}} + \frac{\partial \mathcal{L}}{\partial \phi^r{}_{,\mu}} \frac{\partial \phi^r{}_{,\mu}}{\partial \alpha} . \quad (6\text{-}29)$$

The last two terms cancel, leaving

$$\frac{\partial}{\partial x^\mu} \underbrace{\left(\frac{\partial \mathcal{L}}{\partial \phi^r{}_{,\mu}} \frac{\partial \phi^r}{\partial \alpha} \right)}_{j^\mu} = 0 \quad \rightarrow \quad \partial_\mu j^\mu = 0 \quad \rightarrow \quad \int\limits_{\substack{\text{all} \\ \text{space}}} j^0 d^3 x = Q' = \text{constant in time} . \quad (6\text{-}30)$$

The first two expressions above are simply the continuity equation for the quantity j^μ. That means the $\mu = 0$ component of j^μ is a density value that when integrated over all space is conserved. <u>End of proof.</u>

<u>Noether's theorem</u> (mathematically): If the Lagrangian density $\mathcal{L}(\phi^r, \phi^r{}_{,\mu})$ is symmetric in form with respect to a transformation in ϕ^r which is a function of parameter α, i.e., $\phi^r(x^\mu) \rightarrow \phi^r(x^\mu, \alpha)$, then the four current (using $\phi^r(x^\mu, \alpha)$)

$$j^\mu \left(\phi^r, \phi^r{}_{,\nu} \right) = \frac{\partial \mathcal{L}}{\partial \phi^r{}_{,\mu}} \frac{\partial \phi^r}{\partial \alpha} \quad \text{(sum on } r\text{)} \quad (6\text{-}31)$$

has zero four-divergence, $\partial_\mu j^\mu = 0$. Thus, its zeroth component j^0 integrated over all space is conserved, as is $q j^0$ integrated over all space, where q is a constant[1].

Noether's theorem mathematically

6.5.4 Applying to Our Example

Let's use (6-31) with our example of Sect. 6.5.1 above. Our symmetry transformation is (6-25). That is, we showed there that the scalar Lagrangian (6-24) is invariant in form under (6-25). But now we want to know what exactly is conserved under this symmetry.

From (6-24), we find the terms for the first factor on the RHS of (6-31) (note that summation over r in (6-31) has $r = 1$ for ϕ and $r = 2$ for $\phi^\dagger$)

Applying Noether's theorem

$$\frac{\partial \mathcal{L}_0^0}{\partial \phi_{,\mu}} = \frac{\partial}{\partial \phi_{,\mu}} \left(\phi^\dagger{}_{,\nu} \phi^{,\nu} - \mu^2 \phi^\dagger \phi \right) = \frac{\partial}{\partial \phi_{,\mu}} \left(\phi^\dagger{}_{,\nu} \phi^{,\nu} \right) = \underbrace{\frac{\partial \phi^\dagger{}_{,\nu}}{\partial \phi_{,\mu}}}_{0} \phi^{,\nu} + \phi^\dagger{}_{,\nu} \frac{\partial \phi^{,\nu}}{\partial \phi_{,\mu}} = \phi^\dagger{}_{,\nu} g^{\nu\mu} = \phi^{\dagger,\mu}$$

$$\frac{\partial \mathcal{L}_0^0}{\partial \phi^\dagger{}_{,\mu}} = \frac{\partial}{\partial \phi^\dagger{}_{,\mu}} \left(\phi^\dagger{}_{,\nu} \phi^{,\nu} - \mu^2 \phi^\dagger \phi \right) = \frac{\partial}{\partial \phi^\dagger{}_{,\mu}} \left(\phi^\dagger{}_{,\nu} \phi^{,\nu} \right) = \frac{\partial \phi^\dagger{}_{,\nu}}{\partial \phi^\dagger{}_{,\mu}} \phi^{,\nu} + \underbrace{\frac{\partial \phi^{,\nu}}{\partial \phi^\dagger{}_{,\mu}}}_{0} \phi^\dagger{}_{,\nu} = \delta^\mu_\nu \phi^{,\nu} = \phi^{,\mu} .$$

$$(6\text{-}32)$$

We find the second factor on the RHS of (6-31) from the transformation relation (6-25) $\phi \rightarrow \phi e^{-i\alpha}$,

$$\frac{\partial \phi(x^\eta, \alpha)}{\partial \alpha} = \frac{\partial}{\partial \alpha} \phi(x^\eta) e^{-i\alpha} = -i\phi(x^\eta) e^{-i\alpha}$$

$$\frac{\partial \phi^\dagger(x^\eta, \alpha)}{\partial \alpha} = \frac{\partial}{\partial \alpha} \phi^\dagger(x^\eta) e^{i\alpha} = i\phi^\dagger(x^\eta) e^{i\alpha} .$$

$$(6\text{-}33)$$

---

[1] There are subtleties to Noether's theorem, which are beyond the scope of our work, so we won't discuss them here. Noether originally introduced two theorems and there is a related third one. See Noether, E., *Nachr. d. Konig. Gesellsch., d. Wiss. zu Gottingen, Math-phys. Klasse* (1918), 235-257 (English translation: Tavel, M. A., 'Noether's theorem', *Transport Theory and Statistical Physics* 1(3) (1971), 183-207) and Brading, K., and Brown, H., arXiv:hep-th/0009058.

Using (6-32) and (6-33) in (6-31) we find

$$j^\mu\left(\phi^r,\phi^r_{,\nu}\right) = \frac{\partial\mathcal{L}}{\partial\phi_{,\mu}}\frac{\partial\phi}{\partial\alpha} + \frac{\partial\mathcal{L}}{\partial\phi^\dagger_{,\mu}}\frac{\partial\phi^\dagger}{\partial\alpha} = -i\underbrace{\phi^{\dagger,\mu}(x^\eta,\alpha)}_{\phi^{\dagger,\mu}(x^\eta)e^{i\alpha}}\phi(x^\eta)e^{-i\alpha} + i\underbrace{\phi^{,\mu}(x^\eta,\alpha)}_{\phi^{,\mu}(x^\eta)e^{-i\alpha}}\phi^\dagger(x^\eta)e^{i\alpha}$$

We get same four current as we found for charge in Chap 3

(6-34)

$$= i\left(\phi^{,\mu}(x^\eta)\phi^\dagger(x^\eta) - \phi^{\dagger,\mu}(x^\eta)\phi(x^\eta)\right).$$

This is identical to the scalar four-current, with zero four-divergence and conserved Q'. (See (6-30) as in Chap. 3 (3-21) and (3-23), pg. 46). In Chap. 3, we found this using the Klein-Gordon equation. Here, we found it using Noether's theorem. (Richard Feynman once said that a good physicist should be able to find the same result via different paths.)

From the ensuing discussion in Chap. 3, we learned that in RQM, j^0 can be interpreted for a particle as probability density and qj^0, where q is charge on a single particle, as charge density. So in RQM, $\int q\,j^0 d^3x = \int s^0 d^3x = q$ is conserved, as charge must be. Obviously, if $\partial_\mu j^\mu = 0$, then so does $\partial_\mu\left(q\,j^\mu\right) = \partial_\mu s^\mu = 0$.

6.5.5 Charge Operator in QFT

As we learned in Chaps. 3, 4, and 5, in QFT entities like j^μ are a little different in the sense that they are really operators that operate on states. Indeed, if we follow the steps we did in Chap. 3 in exactly the same way (use (6-34) to find j^0, plug that into the RHS of (6-30), and multiply by an arbitrary constant q equal to the charge on one particle), we find

$$Q = qQ' = q\int j^0 d^3x = \int s^0 d^3x = q\sum_{\mathbf{k}}\left(N_a(\mathbf{k}) - N_b(\mathbf{k})\right). \tag{6-35}$$

What is really physically conserved is the charge of the multiparticle state, i.e.,

$$Q\left|n_1\phi_1,n_2\phi_2,...\overline{n}_1\overline{\phi}_1,...\right\rangle = q\sum_{\mathbf{k}}\left(N_a(\mathbf{k}) - N_b(\mathbf{k})\right)\left|n_1\phi_1,n_2\phi_2,...\overline{n}_1\overline{\phi}_1,...\right\rangle$$

When we say an operator is conserved, we really mean the associated physical value is

$$= \underbrace{q\left(n_1 + n_2 + .. - \overline{n}_1 - ..\right)}_{\text{conserved}}\left|n_1\phi_1,n_2\phi_2,...\overline{n}_1\overline{\phi}_1,...\right\rangle. \tag{6-36}$$

Above, we simply stated the measured charge is conserved. To prove it, consider the following.

Proof that "conservation of an operator" derivation means conserved measured quantity

The state (ket) in (6-36) is an eigenstate of the charge operator Q, and in fact, every state with a given number of particles is a charge eigenstate. That is, if we measure the total charge of a given multiparticle state, we will get a certain number. If we imagined we had an exact duplicate of that multiparticle state at the same moment in time, and then measured its charge, we would get the same number again. Repeating this duplication and measurement, we would always get the same number eigenvalue for total charge. That, of course, is the characteristic of an eigenstate. A general state superposition of eigenstates would sometimes measure one eigenvalue and sometimes another.

Proof that conserved operator means physical value is conserved.

The average of (imagined) repeated measurements of the same state at the same moment in time is, of course, the expectation value of the quantity measured. For an eigenstate, then, this average, the expectation value, is the same as the eigenvalue that is measured each time.

$$\overline{Q} = \left\langle n_1\phi_1,n_2\phi_2,...\overline{n}_1\overline{\phi}_1,...\left|Q\right|n_1\phi_1,n_2\phi_2,...\overline{n}_1\overline{\phi}_1,...\right\rangle$$

$$= \left\langle n_1\phi_1,n_2\phi_2,...\overline{n}_1\overline{\phi}_1,...\left|q\sum_{\mathbf{k}}\left(N_a(\mathbf{k}) - N_b(\mathbf{k})\right)\right|n_1\phi_1,n_2\phi_2,...\overline{n}_1\overline{\phi}_1,...\right\rangle$$

(6-37)

$$= \left\langle n_1\phi_1,n_2\phi_2,...\overline{n}_1\overline{\phi}_1,...\left|q\left(n_1 + n_2 + .. - \overline{n}_1 - ..\right)\right|n_1\phi_1,n_2\phi_2,...\overline{n}_1\overline{\phi}_1,...\right\rangle$$

$$= q\left(n_1 + n_2 + .. - \overline{n}_1 - ..\right)\underbrace{\left\langle n_1\phi_1,n_2\phi_2,...\overline{n}_1\overline{\phi}_1,...\right.\left|n_1\phi_1,n_2\phi_2,...\overline{n}_1\overline{\phi}_1,...\right\rangle}_{=1} = q\left(n_1 + n_2 + .. - \overline{n}_1 - ..\right).$$

So, if we ask, how the expectation value of an eigenstate changes over time, we are asking how the measured eigenvalue changes over time. We are asking if the time derivative of (6-37) is zero. If it is, $\overline{Q}$ is conserved. Thus,

$$\frac{d\overline{Q}}{dt} = \underbrace{\left(\frac{d}{dt}\langle n_1\phi_1, n_2\phi_2, \dots \overline{n}_1\overline{\phi}_1, \dots \right|}_{= 0 \text{ in Heisenberg picture}} Q \left| n_1\phi_1, n_2\phi_2, \dots \overline{n}_1\overline{\phi}_1, \dots \right\rangle$$

$$+ \left\langle n_1\phi_1, n_2\phi_2, \dots \overline{n}_1\overline{\phi}_1, \dots \right| \underbrace{\frac{dQ}{dt}}_{= 0} \left| n_1\phi_1, n_2\phi_2, \dots \overline{n}_1\overline{\phi}_1, \dots \right\rangle \qquad (6\text{-}38)$$

$$+ \left\langle n_1\phi_1, n_2\phi_2, \dots \overline{n}_1\overline{\phi}_1, \dots \right| Q \underbrace{\left(\frac{d}{dt}\left| n_1\phi_1, n_2\phi_2, \dots \overline{n}_1\overline{\phi}_1, \dots \right\rangle\right)}_{= 0 \text{ in Heisenberg picture}} = 0.$$

Saying an operator is conserved means its expectation value is

Note, from Chap. 2 Wholeness Chart 2-4, pg. 28, that states do not have time dependence in the Heisenberg picture, the picture that we employ for QFT free fields. The middle line above is zero because we showed in (6-30) that the operator Q is conserved ($Q = qQ'$), and thus its time derivative is zero. So, the total time derivative of $\overline{Q}$ is zero. (The same conclusion would be reached in the Schrödinger picture, but it would be a little more complicated to derive.)

End proof

Bottom line: The expectation value (expected measurement) of a conserved operator is conserved. If the state measured is in an eigenstate, any measurement at any time will yield the same eigenvalue.

For eigenstate of an operator, expectation value = eigenvalue

So when we cavalierly say in QFT that Q is conserved, remember that Q is really an operator, which it is difficult to think of as being conserved, and that the real thing conserved is the numeric result of operating on a ket with Q. Keep in mind, however, that in QFT, virtually everyone speaks of the operator itself as being conserved.

So, eigenvalue is conserved if operator is "conserved"

As the particles represented by the ket of (6-36) move through the universe, each time we operate on that ket with Q, we will get the same number eigenvalues, the same charge.

Similarly, j^μ is also an operator and its zero 4-divergence really means that the corresponding component numeric values for the physical particles represented by the ket it would act on, have zero 4-divergence. E.g,, the operation of $\partial_\mu j^\mu = 0$ on that ket would always yield zero times the ket.

We have been dealing strictly with free particles, but we will soon find, and Noether's theorem will help us to do it, that interacting particles conserve total charge as well. This is something we know already is true in the physical world, of course, but our theory would hardly be worth anything if we didn't find the same thing there.

6.5.6 More on Symmetry and Noether's Theorem

Spinors and Vectors

It should come as little surprise that spinor and vector four currents, giving rise to conserved charge, such as we found in Chaps. 4 and 5, can be derived from Noether's theorem, as well. You can prove that to yourself, if you really need to, by doing Probs. 12 and 13.

Spinors and vectors similar to above example for scalars

Other test for conservation: commuting with Hamiltonian

You may recall that in NRQM, a dynamical variable was conserved if its operator commuted with the Hamiltonian. That is, for an operator $\mathcal{O}$ with dynamical variable numeric value $\mathcal{O}$,

$$[H, \mathcal{O}] = 0 \quad \text{means} \quad \frac{d\mathcal{O}}{dt} = 0. \qquad (6\text{-}39)$$

Does $[H,Q] = 0$ mean conservation of Q, as it did in NRQM?

You may, at some point, have wondered why this wasn't used in the development of QFT. It is a good question. So, does this test for conservation hold in QFT as well?

To answer, consider our scalar charge operator (6-35) and the scalar Hamiltonian operator from Chap. 3 expressed in terms of number operators. Number operators commute, for example,

$$N_a(\mathbf{k})N_b(\mathbf{k}')\left| n_\mathbf{k}\phi_\mathbf{k}, \overline{n}_{\mathbf{k}'}\overline{\phi}_{\mathbf{k}'}\right\rangle = N_a(\mathbf{k})\overline{n}_{\mathbf{k}'}\left| n_\mathbf{k}\phi_\mathbf{k}, \overline{n}_{\mathbf{k}'}\overline{\phi}_{\mathbf{k}'}\right\rangle = n_\mathbf{k}\overline{n}_{\mathbf{k}'}\left| n_\mathbf{k}\phi_\mathbf{k}, \overline{n}_{\mathbf{k}'}\overline{\phi}_{\mathbf{k}'}\right\rangle$$

$$= \overline{n}_{\mathbf{k}'}n_\mathbf{k}\left| n_\mathbf{k}\phi_\mathbf{k}, \overline{n}_{\mathbf{k}'}\overline{\phi}_{\mathbf{k}'}\right\rangle = N_b(\mathbf{k}')N_a(\mathbf{k})\left| n_\mathbf{k}\phi_\mathbf{k}, \overline{n}_{\mathbf{k}'}\overline{\phi}_{\mathbf{k}'}\right\rangle. \qquad (6\text{-}40)$$

Number operators commute

So

$$[H, Q] = \left[\sum_\mathbf{k}\omega_\mathbf{k}\left(N_a(\mathbf{k}) + N_b(\mathbf{k})\right), q\sum_\mathbf{k}\left(N_a(\mathbf{k}) - N_b(\mathbf{k})\right)\right] = 0 \qquad (6\text{-}41)$$

H and Q commute. And charge q is conserved. We conclude that this method [LHS of (6-39)] for determining whether or not a quantity is conserved is valid in QFT, as it was in NRQM.

Yes, $[H,Q]= 0$ means conservation of Q in QFT, too.

We caution, however, that no one (at least in my experience) in QFT seems to use (6-39) to do so, and no text I know of shows it. Noether's theorem comprises the standard in that regard to which everyone generally adheres. But since one so often has the experience in learning QFT of wondering where some basic principle of NRQM went to in this new and very different theory, I felt it good to provide this discussion of it.

But $[H,Q]= 0$ almost never used in QFT

The various ways to determine a conserved quantity are listed in Wholeness Chart 6-4.

Other uses

Noether's theorem is used repeatedly throughout QFT, and we will eventually see it can tell us whether weak and strong charges are conserved, as well. There are still other uses for symmetry as we will see when we get to interactions.

Noether's theorem has wide range of applications

Wholeness Chart 6-4. Ways to Determine If a Quantity is Conserved

	1st Method	**2nd Method**	**3rd Method**
Steps	Manipulating wave equation and its complex conjugate	Noether's theorem	Operator Q' commutation with H
Result	Four current with zero divergence, $\partial_\mu j^\mu = 0$	Four current with zero divergence, $\partial_\mu j^\mu = 0$	$[Q',H] = 0$
Meaning	$\int j^0 d^3x = Q'$ conserved	$\int j^0 d^3x = Q'$ conserved	Q' conserved
Application	$Q = qQ'$ = electric charge (conserved)		
Other applications	Could be used for weak and strong charge conservation, but not common	Weak and strong charge conservation, energy and 3 momentum conservation	As at left.

6.6 Symmetry, Gauges, and Gauge Theory

6.6.1 A Simple Example and Definitions

You have probably heard that quantum electromagnetic, weak, and strong force theories are called <u>gauge theories</u>. So are other theories you are already familiar with, such as the classical gravitational and electromagnetic field theories. See Chap. 5, pgs. 135-141. As a very simple example, consider an electrostatic field potential $\Phi(\mathbf{x})$ where $\mathbf{E}$, the force on a particle per unit charge, is

Simple example of a classical gauge field transformation

$$\mathbf{E} = -\nabla\Phi = -\nabla(\Phi+C) = -\nabla\Phi' \qquad \Phi' = \Phi + C \qquad C = \text{constant}. \qquad (6\text{-}42)$$

Our measurable $\mathbf{E}$ is the same for Φ or Φ'. So $\mathbf{E}$ is symmetric under the transformation $\Phi\to\Phi'$. We call Φ (or Φ') our <u>gauge field</u>; and $\Phi\to \Phi' = \Phi + C$, the <u>gauge transformation</u>. Each different configuration Φ' is a different <u>gauge</u> of the gauge field. That is, for each different value of C, we have a different gauge (for the same gauge field.)

In Chap. 5 this got more complicated for electrodynamics, where we also had a vector potential (gauge vector field) $\mathbf{A}$.

Definitions

<u>Gauge invariance</u> (or <u>gauge symmetry</u>) is the property of a field theory in which different configurations of the underlying fundamental, but unobservable, field(s) result in identical observable properties.

Definitions related to gauge theory

The unobservable field, often a potential field, is called the <u>gauge field</u>.

A <u>gauge transformation</u> changes the gauge field from one configuration to another.

Each different configuration of the gauge field is a different <u>gauge</u>.

A theory having gauge invariance (symmetry) is called a <u>gauge theory</u>.

6.6.2 Free Quantum Field Theory and Gauges

Recall from Wholeness Chart 3-4 at the end of Chap. 3 (pg.79) that the fields such as ϕ, ψ, and A^μ are themselves not observable. They cannot be measured directly. (We prove that they have zero expectation value in Chap. 7.) But properties of the fields like energy, momentum, and charge are measurable. Our dynamical variable operators, which include number operators, reflect this. They typically have non-zero expectation values.

Quantum gauge theories

Note that under the internal transformation (6-25), repeated below,

$$\phi \to \phi' = \phi\, e^{-i\alpha}, \qquad (6\text{-}43)$$

Scalar gauge transformation

the Lagrangian (6-26) remained invariant. Thus, the Klein-Gordon field equation derived from that Lagrangian is invariant, i.e., ϕ' solves the K-G equation as well as ϕ. All our dynamical variable operators are ultimately derived from the Lagrangian, so they too will be the same for ϕ'. As one example, see the 4-current of (6-34) in which we effectively substitute ϕ' for ϕ and get the same result for j^μ and Q.

Free QFT is a gauge theory, because $\mathcal{L}$, and thus measurables, unchanged under gauge transfm

The transformation (6-43) is a gauge transformation of the underlying, unobservable field ϕ. The theory of free scalar quantum fields is a gauge theory, because all measurable quantities are unchanged under the gauge transformation. By doing Prob. 16, you can show the same thing is true for free Dirac spinor fields.

Note that the (internal) gauge transformation (6-43) is simply a change in phase of the field. This is similar to NRQM, where we may recall that we could change the phase of the wave function, but observables like probability density, energy, and momentum remained unchanged. A solution to the Schrödinger equation could have any constant phase factor and still be a solution predicting the same measurable results.

Transfm here is a phase change Observables unchanged like phase change in NRQM

Thus, gauge symmetries are internal symmetries. (See Sect. 6.5.2, pg. 173.)

Gauge symmetry is an internal symmetry

More formal definition

We can also say that a <u>gauge theory</u> is a type of field theory in which the Lagrangian (density) is invariant under a continuous (not discrete) transformation.

6.7 Chapter Summary

We have seen that for

Symmetry and transformations in general

- symmetry is the propensity for non-change with superficial change
- mathematically, symmetry is invariance under transformation
- Wholeness Chart 6-1 compares and contrasts symmetric and non-symmetric transformations

    Scalar value at a point is always invariant. Scalar function form invariant only under symmetry transformation

    Vector components at a point vary co-variantly. Vector length and direction in physical space same under any transformation. Vector function form invariant only under symmetry transformation

    A scalar or vector function that is not a function of a coordinate x^j is symmetric with respect to a displacement in the j coordinate direction.

Transformations in classical mechanics

- the laws of nature are symmetric under Lorentz transformation, i.e., invariant in spacetime
- symmetry of the classical Lagrangian L under a translation transformation of a generalized coordinate q^j (often x^j) means the conjugate momentum p_j of that coordinate is conserved
- similar effects for classical Lagrangian density $\mathcal{L}$

Transformations in QFT (see Wholeness Chart 6-3)

- scalar and vector quantum fields transform like classical ones did; spinors do not exist classically, but have their own form for QFT transformations
- symmetry of the QFT Lagrangian density $\mathcal{L}$ under Lorentz transformation means field equation (law of nature) is invariant in form for different inertial observers

- Noether's theorem: If $\mathcal{L}$ is symmetric under a change of a parameter, then there is an associated quantity that is conserved
- there are three ways to determine if a quantity is conserved (see Wholeness Chart 6-4), though Noether's theorem method is the most useful and covers widest range of cases.
- a gauge theory is a field theory for which the Lagrangian (and thus all measurables) remains invariant under a transformation of the underlying unmeasurable gauge field
- a gauge symmetry is an internal symmetry; a Lorentz symmetry is an external symmetry

6.8 Problems

1. Is the function $F = 2(x^1)^2 + (x^2)^2$ symmetric under rotation in the x^1-x^2 plane? Guess first, then prove (or disprove) your answer by expressing F in terms of a rotated set of coordinates x'^1-x'^2, i.e,, as $F'(x'^1, x'^2)$, where θ is the angle of rotation between the two coordinate systems.

2. In Prob. 1, at the point $(x^1, x^2) = (1,2)$, F has the value 6. If we transform to the rotated coordinate system x'^1-x'^2 with $\theta = 45°$, what are the coordinates of that same physical point in space in that coordinate system? Using your expression F' for F in terms of x'^1 and x'^2, show that $F'(x'^1, x'^2)$ at that physical point equals 6, as well.

3. Without doing any calculations, is the function $G = (x^1)^2 + (x^2)^2 + (x^3)^2$ symmetric under rotation in 3D space? Is $H = (x^1)^2 + 3(x^2)^2 + (x^3)^2$?

4. Is the function $J = (x^1)^2 + (x^3)^2$ symmetric under the translation $x^2 \rightarrow x'^2 = x^2 + a$, where a is a constant? Is it symmetric under $x^3 \rightarrow x'^3 = x^3 + a$?

5. Is the differential equation $\partial_i x^i = 3$ symmetric under the translation $x^2 \rightarrow x'^2 = x^2 + a$, where a is a constant? Is it symmetric under $x^2 \rightarrow x'^2 = x^2 + (x^2)^2$?

6. Consider the position vector $(x^1, x^2) = (3,4)$. This vector's length is 5, and for the x^1 axis horizontal, its angle with the horizontal is $53°$. What are this vector's position coordinates in the x'^1-x'^2 coordinate system of Prob. 1? What is its length? Calculate it. What is its angle with the horizontal? What is its angle with respect to the x'^1 axis? Express your answer in terms of θ.

7. On page 166 we briefly discussed the spherical symmetry of the electric field around a point charge. It is easier mathematically to consider the symmetry of the simpler case of an infinitely long line of uniformly distributed charge. This radiates an electric field **E** in a coordinate system with x^3 axis aligned with the line of charge of components (where ϕ below is the relevant cylindrical coordinate system angle)

$$\begin{bmatrix} E^1 \\ E^2 \\ E^3 \end{bmatrix} = \frac{E_0}{r} \begin{bmatrix} \cos\phi \\ \sin\phi \\ 0 \end{bmatrix} = \frac{E_0}{r} \begin{bmatrix} x^1/r \\ x^2/r \\ 0 \end{bmatrix} = \frac{E_0}{\left(x^1\right)^2 + \left(x^2\right)^2} \begin{bmatrix} x^1 \\ x^2 \\ 0 \end{bmatrix}.$$

Express E^i and x^i above in the primed coordinate system of Fig. 6-2 on page 164 using (6-3) to show that

$$\begin{bmatrix} E'^1 \\ E'^2 \\ E'^3 \end{bmatrix} = \frac{E_0}{\left(x'^1\right)^2 + \left(x'^2\right)^2} \begin{bmatrix} x'^1 \\ x'^2 \\ 0 \end{bmatrix},$$

and thus, that the vector field components E^i in this case are symmetric under the rotation transformation of Fig. 6-2. If you feel ambitious, repeat the analysis for the **E** field around a point charge.

8. Show that $F^i = m\ddot{x}^i$ is symmetric under the transformation $x^i \rightarrow x^i + a^i$, where a^i is a constant for each i.

9.   Transform the components in (6-15) by the Lorentz transformation (6-13) and show that $w_\mu\, w^\mu = w'_\mu\, w'^\mu$.

10.   For a particle attached to a spring confined to move in one dimension, the potential energy $V =$ ½ $k\, (x^1)^2$. Use this to find the Lagrangian of this system. Is this Lagrangian symmetric with respect to translations of x^1? Is momentum conserved in the x^1 direction? Find the equation of motion for the system using the Lagrangian approach. What does the momentum equal? Is this Lagrangian symmetric in x^2? Is momentum conserved in the x^2 direction? Does this make sense physically?

11.   For a disk attached to a spring confined to rotate in the plane of the disk about an axis, the potential energy is $V =$ ½ $k\theta^2$, where θ is the angle of rotation. I is the mass moment of inertia about the axis. What is the Lagrangian of this system? Is this Lagrangian symmetric in θ? Is angular momentum conserved? Find the equation of motion for the system using the Lagrangian approach. What does the angular momentum equal? If there were no spring, would the Lagrangian be symmetric in θ? Would angular momentum be conserved?

12.   Show that the Lagrangian density for free Dirac fermions (see Chap. 4 or Wholeness Chart 5-4 at the end of Chap. 5)) is symmetric under the transformation $\psi \rightarrow \psi\, e^{-i\alpha}$. Use Noether's theorem and the same transformation to show that for Dirac particles, $j^\mu = (\rho, \mathbf{j}) = \bar{\psi}\gamma^\mu\psi$ where $\partial_\mu j^\mu = 0$.

13.   Show that for photons $j^\mu = 0$. Assume temporarily that A^μ is complex, so we can write the Lagrangian as $\mathcal{L}_0^{e/m} = -\frac{1}{2}\big(\partial_\nu A_\mu(x)\big)^\dagger \big(\partial^\nu A^\mu(x)\big)$. Use Noether's theorem with the transformation $A^\mu \rightarrow A^\mu e^{-i\alpha}$, to obtain j^μ with $\partial_\mu j^\mu = 0$. Then, show that by taking A^μ as real, we must have $j^\mu = 0$.

14.   Show that the total (not density) 3-momentum k^i for free scalars is conserved. Use our knowledge that the conjugate momentum for x^i is k_i, the total (not density) 3-momentum (expressed in covariant components), and it is conserved if L is symmetric (invariant) under the coordinate translation transformation $x^i \rightarrow x'^i = x^i + \alpha^i$, where α^i is a constant 3D vector. Then, show the same result via commutation of the three-momentum operator of Chap. 3 (see Wholeness Chart 5-4, pg. 158) with the Hamiltonian. (Solution is posted on book website. See URL on pg. xvi, opposite pg. 1.)

15.   Use the transformation $x^0 \rightarrow x'^0 = x^0 + \alpha$ for free scalars to show that energy $\omega_\mathbf{k}$ is conserved. Note that the conjugate momentum for time is energy. Is it immediately obvious that you will get the same results from commutation of the energy operator with the Hamiltonian? (Tricky wording here?)

16.   Show that the Hamiltonian density for free Dirac fermions is symmetric under the same transformation as in Prob. 12.

17.   Is Dirac's field theory a gauge theory? What is the gauge field? Give an example of one of its gauges. What is the gauge transformation? Is this an external or internal symmetry?

Part Two

Interacting Fields

*"The mystery was, where did the electron come from when a nucleus suffered a beta
decay? Fermi's answer was that the electron .. is* created *in the act of decay,
through an interaction of the field of the electron with the fields of [other particles].
...QFT gave rise to a new view not only of particles but also of the forces among them."*
Steven Weinberg
The Search for Unity: Notes for a History of QFT
MIT Press (1977)

Chapter 7

Interactions: The Underlying Theory

> *"Physics is very muddled again at the moment; it is much too hard
> for me anyway, and I wish I were a movie comedian or something
> like that and had never heard anything about physics!"*
> Wolfgang Pauli
> In a letter to R. Kronig, 25 May 1925

> *"Like every other branch of learning, the study of the rainbow is a giant
> onion. Each cook merely succeeds in removing another layer, and then,
> after a short blush of satisfaction, some iconoclast points out that there
> is at least one more layer to be removed before the core is attained."*
> Raymond L. Lee, Jr. and Alistair B. Fraser
> *The Rainbow Bridge: Rainbows in Art, Myth,
> and Science* (PSU Press, 2001), p. 276.

7.0 Preliminaries

Within a few decades, Pauli's 1925 consternation was well resolved by the development of QFT and concomitant experimental discoveries. As Steven Weinberg noted in the quote on the prior page, the new theory changed physicists' world view by providing a mathematical structure that framed forces (interactions) not merely in terms of exchanges in energy and momentum between particles, but also as the means by which particles are created, destroyed, and transmigrated from one type to another. The layer of Nature troubling Pauli and others had been peeled back and exposed.

7.0.1 Background

However, even within that layer, there were sub-layers. There were different kinds of interactions the standard model (SM) of QFT was eventually able to describe. These comprise the electromagnetic, weak, and strong forces.

The first of these succumbed to theory by the 1940-50s in the form of quantum electrodynamics (QED), the primary subject of the remainder of this volume. The weak and strong forces, the subjects of what may one day, if I have the stamina, be a second volume following this one, came later, but by the 1970s were fundamental, fairly well understood, facets of QFT.

Gravity, the fourth known interaction, seems to comprise, from a quantum perspective, a substantial layer of knowledge all to itself. In some way, virtually everyone agrees, it must fall under the umbrella of a theoretic structure encompassing it and the other three forces. However, at the time of this writing, a few years from the 100th anniversary of Pauli's expression of frustration, we still do not have a complete theory of quantum fields that includes all four interaction types.

7.0.2 Chapter Overview

In this chapter, after first discussing fundamental aspects of electromagnetic interactions in classical theory and RQM, we lay the general groundwork for interactions of all three interaction types in the SM of QFT. In the next chapter, we will narrow that more general approach to electromagnetism specifically and develop QED.

First, we review classical e/m with source terms (i.e. with interactions)
- Maxwell's equations in 3D + 1 formulation in $\mathbf{E}$, $\mathbf{B}$ including sources ρ_{charge}, $\mathbf{j}_{charge}$
- Maxwell's equations in 4D formulation in A^μ including 4D source $-ej^\mu = (\rho_{charge}, \mathbf{j}_{charge})$
- The classical interaction e/m Lagrangian (i.e., including $-ej^\mu$)

Classical e/m interaction theory

Then, we consider RQM for e/m interactions, specifically
- Photons represented by A^μ in Maxwell's 4D equation including interaction term $-ej^\mu$
- The quantum interaction e/m Lagrangian (i.e., including $-ej^\mu$)
- The Dirac equation in ψ modified to include e/m interaction from A^μ
- An example: Interaction Dirac equation solves the relativistic hydrogen atom problem

Relativity, e/m and QM combined in RQM interaction theory

And then, interactions in QFT, including
- The Interaction Picture (a third kind of picture beyond the S.P. and the H.P.)
- The S-matrix (scattering matrix) and the S (scattering) operator
- Dyson's expansion of the S operator
- Wick's theorem applied to Dyson's S operator expansion

Interactions in QFT

Free vs interacting fields

In all prior chapters we have worked solely with free fields/particles. In the remainder of the book we deal solely with interacting fields/particles.

Interacting, not free, fields from here on.

7.1 Interactions in Relativistic Quantum Mechanics

7.1.1 Maxwell's Equations with Sources

In Chap. 5, Eq. (5-1), we showed the source free (no interactions) Maxwell field equations in a vacuum, in naturalized Heaviside-Lorentz units. If we include the charge density ρ_{charge} and current density $\mathbf{j}_{charge}$, these equations, in 3D + time format, describe interactions for $\mathbf{E}$ and $\mathbf{B}$ with electrically charged sources.

$$\nabla \cdot \mathbf{E} = \rho_{charge} \qquad (a)$$

$$\nabla \times \mathbf{B} - \frac{\partial \mathbf{E}}{\partial t} = \mathbf{j}_{charge} \qquad (b)$$

$$\nabla \cdot \mathbf{B} = 0 \qquad (c)$$

$$\nabla \times \mathbf{E} + \frac{\partial \mathbf{B}}{\partial t} = 0 \; . \qquad (d)$$

(7-1)

Maxwell's equations with source terms

By introducing the potentials Φ and $\mathbf{A}$, where

$$\mathbf{B} = \nabla \times \mathbf{A}, \qquad \mathbf{E} = -\nabla\Phi - \frac{\partial \mathbf{A}}{\partial t}, \qquad (7\text{-}2)$$

Scalar and vector potentials

one gets for (7-1)(a) and (b),

$$-\nabla^2 \Phi - \frac{\partial}{\partial t}(\nabla \cdot \mathbf{A}) = \rho_{charge} \qquad (a)$$

$$\frac{\partial^2 \mathbf{A}}{\partial t^2} - \nabla^2 \mathbf{A} + \nabla \frac{\partial \Phi}{\partial t} + \nabla(\nabla \cdot \mathbf{A}) = \mathbf{j}_{charge}, \qquad (b)$$

(7-3)

Maxwell's equations in terms of scalar and vector potentials

and $0 = 0$ for both of (7-1)(c) and (d).

Re-writing (7-3) in terms of the 4-potential A^μ and the 4-current j^μ, we can do Prob. 1 to show

$$\partial^\alpha \partial_\alpha A^\mu(x) - \partial^\mu \left(\partial_\nu A^\nu(x)\right) = -ej^\mu(x) \quad \text{with} \quad A^\mu = \left(\Phi, A^i\right), \; -ej^\mu = \left(\rho_{charge}, \mathbf{j}_{charge}\right). \quad (7\text{-}4)$$

Maxwell equation in terms of 4 potential

If, as we did in the sourceless case, we employ the Lorenz gauge condition

$$\partial_\nu A^\nu (x) = 0 \,, \qquad\qquad (7\text{-}5)$$

the <u>Maxwell 4D interaction equation</u> in terms of the four potential becomes

$$\partial^\alpha \partial_\alpha A^\mu (x) = -e\, j^\mu. \qquad\qquad (7\text{-}6)$$

Maxwell equation in Lorenz gauge

For the free field case of Chap. 5, we had the same equation with the electric charge source 4-current $-e.j^\mu = 0$.

7.1.2 The Classical Electromagnetic Interaction Lagrangian

The full electromagnetic Lagrangian (density), including interactions, must give rise to (7-6) when substituted into the Euler-Lagrange field equation

$$\frac{\partial}{\partial x^\nu}\left(\frac{\partial \mathcal{L}}{\partial \phi^n{}_{,\nu}}\right) - \frac{\partial \mathcal{L}}{\partial \phi^n} = 0, \quad \text{with} \quad \phi^n = A_\mu; \mathcal{L} = \mathcal{L}^{e/m}\,. \qquad (7\text{-}7)$$

Euler-Lagrange equation

By doing Prob. 2, one can prove that the <u>full electromagnetic field classical Lagrangian</u> is

$$\mathcal{L}^{e/m} = \underbrace{-\tfrac{1}{2}\big(\partial_\nu A_\mu (x)\big)\big(\partial^\nu A^\mu (x)\big)}_{\mathcal{L}_0^{e/m}} + \underbrace{e\, j^\mu (x) A_\mu (x)}_{\mathcal{L}_I^{e/m}}\,, \qquad (7\text{-}8)$$

Classical e/m free plus interaction $\mathcal{L}$

where "0" and "I" subscripts denote the free and interaction parts, respectively, of the Lagrangian.

7.1.3 Electromagnetic Interactions in Relativistic Quantum Mechanics

Relations (7-6) and (7-8) hold for classical electromagnetism where $-e j^\mu$ is the classical electric charge 4-current. In quantization, we assume the quantum form of the Lagrangian (density or total) is the same as the classical, and thus, so would be the resulting wave equation. In RQM, we would then consider A^μ to represent the quantum photon state (ket, wave function). Thus, (7-8) represents the RQM electromagnetic Lagrangian (density).

But then, quantum mechanically, how should one interpret $-e j^\mu$? In Chap. 4 (eq. (4-34) to (4-37) pg. 92), we saw that the probability 4-current for an electron in RQM, where ψ_{state} represents the electron wave function state, is

$$j^\mu = (\rho, \mathbf{j}) = \bar{\psi}_{state}\, \gamma^\mu \psi_{state} \quad \text{where} \quad \partial_\mu j^\mu = 0\,. \qquad (7\text{-}9)$$

RQM electron probability density

It seems natural to assume charge density varies directly with probability density (denser regions of the particle itself would be regions of higher measured charge density). Thus, we can assume

$$-e j^\mu = -e\bar{\psi}_{state}\, \gamma^\mu \psi_{state} \qquad\qquad (7\text{-}10)$$

RQM electron charge density

where the total charge of the electron would be (see Chap. 4, (4-37))

$$-e\int j^0\, d^3x = -e\,. \qquad\qquad (7\text{-}11)$$

Using (7-6), (7-8), and (7-10), we can then represent the <u>RQM interaction wave equation for a photon</u> as

$$\partial^\alpha \partial_\alpha A^\mu_{state} = -e\, \bar{\psi}_{state}\, \gamma^\mu \psi_{state} \qquad\qquad (7\text{-}12)$$

RQM photon interaction wave equation

with the corresponding <u>RQM e/m interaction Lagrangian for a photon</u> as

$$\mathcal{L}^{e/m} = \underbrace{-\tfrac{1}{2}\big(\partial_\nu A_\mu{}_{state}\big)\big(\partial^\nu A^\mu_{state}\big)}_{\mathcal{L}_0^{e/m}} + \underbrace{e\, \bar{\psi}_{state}\, \gamma^\mu \psi_{state} A_\mu{}_{state}}_{\mathcal{L}_I^{e/m}}\,. \qquad (7\text{-}13)$$

RQM photon $\mathcal{L}$

(7-12) governs the behavior of a photon (A^μ_{state}) in the presence of an electron (ψ_{state}).

7.1.4 The Electromagnetic Interaction Dirac Equation

In Chap. 4, we studied the free Dirac equation, which described an electron not interacting with any electromagnetic particle (i.e., photon). In Chap. 5, we studied Maxwell's equation for a free photon, not interacting with any electrons (or positrons), i.e., (7-12) with the RHS = 0. In the prior section we developed Maxwell's equation for a photon interacting with an electron, i.e., (7-12).

What we need to develop, in order to complete the picture, is the full Dirac equation describing the electron interacting with a photon.

The Full Quantum Electromagnetic Lagrangian

To this end, consider that $\mathcal{L}_0^{e/m}$ of (7-13) represents the free photon part of the full e/m Lagrangian. If we assume $\mathcal{L}_I^{e/m}$ represents the e/m interaction part for both the photon and the electron (since it contains factors of both), then we need only add the free electron contribution to (7-13) to get a Lagrangian containing all terms relevant to photons, electrons, and interactions between them. From Chap. 4, (eq. (4-60), pg. 104), we know this term to be

$$\mathcal{L}_0^{1/2} = \overline{\psi}_{state}\left(i\gamma^\mu\partial_\mu - m\right)\psi_{state} \ . \tag{7-14}$$

Free electron term in $\mathcal{L}$

Thus, the full e/m Lagrangian is

$$\mathcal{L}^{1/2,1} = \underbrace{-\frac{1}{2}\left(\partial_\nu A_{\mu \atop state}\right)\left(\partial^\nu A^\mu_{state}\right)}_{\mathcal{L}_0^1 = \mathcal{L}_0^{e/m}} + \underbrace{\overline{\psi}_{state}\left(i\gamma^\mu\partial_\mu - m\right)\psi_{state}}_{\mathcal{L}_0^{1/2}} + \underbrace{e\,\overline{\psi}_{state}\,\gamma^\mu\psi_{state}A_{\mu \atop state}}_{\mathcal{L}_I^{1/2,1} = \mathcal{L}_I^{e/m}} \tag{7-15}$$

Full e/m quantum $\mathcal{L}$

where we change superscript notation to reflect the spins of the particles in each term of (7-15).

The Full Dirac Equation

To find the interaction form of the Dirac equation, we use (7-15) in (7-7) with $\phi^{n=1} = \overline{\psi}$. The result (do Prob. 3) is

$$\left(i\gamma^\mu\partial_\mu - m\right)\psi_{state} = -e\,\gamma^\mu\psi_{state}A_{\mu \atop state} \tag{7-16}$$

Full (including e/m interaction) Dirac equation

As an aside, using (7-15) in (7-7) with $\phi^{n=2} = \psi$ results in the adjoint full Dirac equation.

Clarification

Some readers may be concerned at this point that we have used the Lagrangian *density* methodology, which is normally reserved for quantum and classical fields, to develop the full Dirac equation for quantum states (corresponding to particles, not fields). One would expect to use the total Lagrangian L (integration of $\mathcal{L}$ over all space) instead of $\mathcal{L}$, since (7-16) is a wave equation for interacting states, not fields. This may be a little confusing.

The presentation above is somewhat historical, as similar logic was used to deduce the interaction Dirac equation, prior to the full development of QFT. Given that the classical e/m equation in A^μ is a field, not particle, equation, we, and early researchers, had little choice.

However, even in the context of 1st (particle) and 2nd (field) quantization as we have come to understand them, the issue is not such a big one. This is because we did not employ commutation relations for fields A^μ and ψ, analogous to Poisson bracket relations, in the above development. It is the adoption of commutation relations for those fields that turns them into creation and destruction operators quantum mechanically. We did not do that, so A^μ and ψ remain as states, not quantum fields, in the above treatment. Of course, for RQM, we would still have commutation relations for dynamical variable operators, such as p_x and X, though we would not have them for A^μ and ψ.

A^μ, ψ commutators not employed above, so they are states here, not quantum operators

7.1.5 The Relativistic Hydrogen Atom: Applying the Full Dirac Equation

We made one assumption in determining (7-15) and thus (7-16), and this was that $\mathcal{L}_I^{e/m} = \mathcal{L}_I^{1/2,1}$ completely represented the interaction between A^μ and ψ, for both, not simply one, of them. If we apply (7-16) to a real world problem, and predict the measured results, we would have good grounds to accept that assumption and (7-16) as valid.

Early researchers did exactly that by applying (7-16) to the hydrogen atom and found excellent agreement with experiment. They found it predicted relativistic effects not accounted for by the non-relativistic Schrödinger equation. We do not repeat this rather extensive analysis, but outline the basic steps involved and the final results below. Interested readers can find the full treatment in Merzbacher, E., *Quantum Mechanics*, 2nd ed., Chap. 22 (John Wiley, 1970) or Itzykson, C. and Zuber, J.B., *Quantum Field Theory*, Chap. 2 (McGraw-Hill, 1980).

Early researchers showed full Dirac equation worked for H atom

<u>Fundamental Steps: Relativistic Hydrogen Atom Analysis</u>

For the analysis of the relativistic hydrogen atom, there are <u>two assumptions</u>.

Steps for relativistic H atom analysis in RQM

1) The full (interaction) Dirac equation (7-16) governs.

2) The e/m potential exerted by the proton nucleus and felt by the bound electron is solely a Coulomb (static, no moving charge sources = no magnetic fields, so **A**=0) potential with 4-potential

Assume full Dirac equation and Coulomb potential

$$A^\mu_{state} = (\Phi, 0) \qquad \Phi = \frac{Ze}{4\pi r}, \tag{7-17}$$

where Ze is the charge of the nucleus ($Z = 1$ for hydrogen) and r is the radial distance from the assumed point nucleus.

Then, carry out <u>the analysis</u>.

Solve full Dirac equation in spherical coords

3) Express the interaction Dirac equation (7-16) in spherical coordinates.

4) Solve (7-16) (the spinor space indices are hidden, so this is really a matrix equation) for ψ_{state} using (7-17).

5) Find the energy eigenvalues, i.e., the electron energy levels, of the eigenstate solutions ψ_{state}.

With <u>the result</u>

6) The fine structure formula for electron energy levels found via the analysis correctly describes the relativistic effects on the measured hydrogen atom spectral line distribution.

Energy eigenvalues found describe relativistic fine structure correctly

7.1.6 RQM Interactions Summary

Do Prob. 4 to construct a wholeness chart summarizing electromagnetic interactions in RQM.

7.2 Interactions in Quantum Field Theory

7.2.1 The QFT Electromagnetic Interaction Wave Equations

In Chaps. 3, 4, and 5, we saw that for each spin type, the Schrödinger picture wave equation for a state and the Heisenberg picture wave equation for the associated quantum field had the same form. In the former case, the wave equation solution was a state, i.e., a particle wave function. In the latter, the wave equation solution was a quantum field, i.e., an operator that created and destroyed states. For free scalars, this equation was the Klein-Gordon equation; for spinors, it was the (no interactions) Dirac equation; and for massless vectors (photons), it was Maxwell's equation (sourceless, in terms of A^μ).

It would seem natural, therefore, to assume the same thing is true for interactions. (See lower part of Fig. 7-1.) And thus, our <u>interacting spinor field and photon wave equations in the Heisenberg picture</u> should simply be (7-12) and (7-16) for fields, i.e.,

QFT e/m interacting fields equations

$$\boxed{\partial^\alpha \partial_\alpha A^\mu = -e\bar{\psi}\gamma^\mu\psi} \tag{7-18}$$

$$\boxed{(i\gamma^\mu \partial_\mu - m)\psi = -e\gamma^\mu A_\mu \psi} \tag{7-19}$$

where the order of ψ and A_μ in (7-19) is unimportant, even though they are operators, since different type fields commute. The <u>associated Lagrangian</u> (7-15) <u>for the ψ and A^μ operator fields</u> is

QFT e/m interacting Lagrangian density

$$\boxed{\mathcal{L}^{1/2,1} = -\frac{1}{2}\left(\partial_\nu A_\mu\right)\left(\partial^\nu A^\mu\right) + \underbrace{\bar{\psi}\left(i\gamma^a \partial_\alpha - m\right)\psi}_{\mathcal{L}_0^{1/2}} + \underbrace{e\bar{\psi}\gamma^\mu A_\mu\psi}_{\mathcal{L}_I^{1/2,1}}} \tag{7-20}$$

Alternatively, of course, we could go through the steps of 2nd quantization similar to what we did in earlier chapters, only this time it would include interactions. (See RHS of Fig. 7-1.) This approach would not be much different from what we did in Sects. 7.1.3 and 7.1.4 (adopting the classical field $\mathcal{L}$ as the quantum field $\mathcal{L}$) except that we would also invoke field commutation relations for the bosons and anti-commutation relations for the fermions. By either approach we obtain (7-18) through (7-20) with the fields A^μ and ψ being operators that create and destroy states.

Two routes to QED

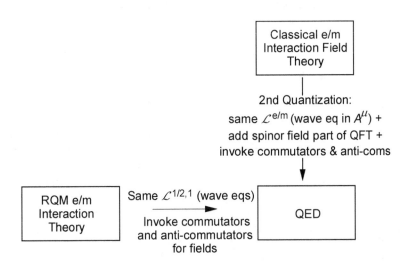

Figure 7-1. Two Paths to Quantum Electrodynamics

*QFT can solve
simple problems
like H atom as
RQM did*

7.2.2 The Naïve vs Realistic Approaches to Solving Interaction Wave Equations

For the hydrogen atom, we saw in RQM that one could use (7-19) (in its (7-16) form for states) with a simple Coulomb potential in the zeroth component of A^μ and readily obtain a good solution. We could do the same thing in QFT, the only difference being that our solutions ψ and $\bar{\psi}$ would destroy and create electron states (which were the same H atom bound states found in RQM).

*QFT solves more
difficult cases, too*

But in QFT, we seek to do much more than this. We will need more general interactions between electrons, positrons, and photons where A^μ cannot be represented so simply and where it is not independent of ψ. In general, A^μ and ψ will depend on one another. That makes (7-18) and (7-19) into non-linear coupled partial differential equations, which are notoriously difficult to solve. Solving them in closed form, for all but the simplest cases like (7-17), is essentially impossible.

*Naïve to think we
can just solve
coupled QFT
PDEs directly*

So, whereas, we might naïvely consider that, for any particular problem, all we need to do is solve (7-18) and/or (7-19) for the given boundary and initial conditions, the reality is not quite so simple. We must seek other, non-closed form, solution avenues.

*To solve the PDEs
in QED:
i) perturbation
ii) trick =
Interaction Picture*

Modern day computers can help in providing numerical solutions, but early researchers in QFT did not have such things. Also, the route those researchers did take provides considerable insight into the inner workings of the theory. That route, for the QFT e/m interaction theory known as quantum electrodynamics (QED), was forged in large part by Richard Feynman, Freeman Dyson, Julian Schwinger, and Sin-Itiro Tomonaga. It involves two things,

      i) perturbation theory, and
      ii) a trick known as the Interaction Picture.

The first of these, for QFT, is expressed in terms of the second, as described in the next section.

7.3 The Interaction Picture

7.3.1 Review of Heisenberg and Schrödinger Pictures

Recall from Chap. 2, Sect. 2.6, pgs. 25-29 (summarized in Wholeness Chart 2-4, which is repeated as the top half of Wholeness Chart 7-1 below) that in the Schrödinger Picture (S.P.), states are time dependent and operators are usually time independent. In the Heisenberg Picture (H.P.) it is reversed. States are time independent and operators are often time dependent. The key equation in the S.P. is the state equation of motion (top left block in NRQM is the Schrödinger equation, and in spinor RQM, the Dirac equation), since the operator is usually unchanging. The key equation in the H.P. is the operator equation of motion, since the state is unchanging (frozen in time, as it were). Both pictures make the same predictions for measured values, e.g., expectation values.

*Review of the
S.P. and H. P.*

Wholeness Chart 7-1. Comparing Schrödinger, Heisenberg, and Interaction Pictures

Schrödinger vs. Heisenberg Pictures (from Chap. 2 with $|\Psi\rangle$ = generic state)

	States	Operators	Expectation Values
Schrödinger Picture	Time dependent $$i\frac{d}{dt}|\Psi\rangle_S = H|\Psi\rangle_S$$	Usually time independent $$\frac{d\mathcal{O}^S}{dt} = \underbrace{\frac{\partial \mathcal{O}^S}{\partial t} = 0}_{\text{usually}}$$	$$\frac{d\bar{\mathcal{O}}}{dt} = {}_S\langle\Psi|\left(-i\left[\mathcal{O}^S,\underbrace{H^S}_{=H}\right] + \frac{\partial \mathcal{O}^S}{\partial t}\right)|\Psi\rangle_S$$ $\|\psi\rangle_S$ changes in time; $\mathcal{O}^S$ usually const
Transform via $U = e^{-iHt}$ $\Downarrow$	$U^\dagger|\Psi\rangle_S = |\Psi\rangle_H$	$U^\dagger \mathcal{O}^S U = \mathcal{O}^H$	$\dfrac{d\bar{\mathcal{O}}}{dt}$ invariant under the transformation
Heisenberg Picture	Time independent $$\frac{d|\Psi\rangle_H}{dt} = 0$$	Often time dependent $$\frac{d\mathcal{O}^H}{dt} = -i\left[\mathcal{O}^H, H\right] + \underbrace{U^\dagger\frac{\partial \mathcal{O}^S}{\partial t}U}_{\substack{\text{defined} = \frac{\partial \mathcal{O}^H}{\partial t} \\ \text{usually} = 0}}$$	Like S.P. above, but scripts $S \to H$ $$\frac{d\bar{\mathcal{O}}}{dt} = {}_H\langle\Psi|\left(-i\left[\mathcal{O}^H,\underbrace{H^H}_{=H}\right] + \frac{\partial \mathcal{O}^H}{\partial t}\right)|\Psi\rangle_H$$ $\|\Psi\rangle_H$ const in time; $\mathcal{O}^H$ often not const
Hamiltonian	$H^H = H^S = H$ (for free fields, $H = H_0$, but in general, H includes free plus interaction parts)		
Key Relation	In S.P., the state eq of motion	In H.P., the operator eq of motion	Expectation value and its eq of motion are the same and equally key in both pictures

Schrödinger vs. Interaction Pictures

	States	Operators	Expectation Values
Schrödinger Picture	Same as top row		
Symbol defs	$H^S = H_0 + H_I^S$ $(H_0 = H_0{}^S)$	$H^I = H_0{}^I + H_I{}^I$	
Transform via $U_0 = e^{-iH_0t}$ $\Downarrow$	$U_0^\dagger|\Psi\rangle_S = |\Psi\rangle_I$	$U_0^\dagger \mathcal{O}^S U_0 = \mathcal{O}^I$	$\dfrac{d\bar{\mathcal{O}}}{dt}$ invariant under the transformation
Interaction Picture	Time dependent on interaction part H_I^I $$i\frac{d}{dt}|\Psi\rangle_I = H_I^I|\Psi\rangle_I$$	Time dependent on $H_0^I = H_0 = H_0{}^S$ $$\frac{d\mathcal{O}^I}{dt} = -i\left[\mathcal{O}^I, \underbrace{H_0}_{H_0^I}\right] + \underbrace{U_0^\dagger\frac{\partial \mathcal{O}^S}{\partial t}U_0}_{\substack{\text{defined} = \frac{\partial \mathcal{O}^I}{\partial t} \\ \text{usually} = 0}}$$	Like S.P. above, but scripts $S \to I$ $$\frac{d\bar{\mathcal{O}}}{dt} = {}_I\langle\Psi|\left(-i\left[\mathcal{O}^I, H^I\right] + \frac{\partial \mathcal{O}^I}{\partial t}\right)|\Psi\rangle_I$$ $\|\Psi\rangle_I$ and $\mathcal{O}^I$ both change in time
Hamiltonian	$H_0{}^I = H_0 = H_0{}^S$ $H_I^I = U_0^\dagger H_I^S U_0 \neq H_I^S$ $\to$ $H^I = H_0 + H_I{}^I$		
Key Relation	Both state and operator eqs of motion important in I.P.		Expectation value same as S.P., H.P.

Clarification

Distinguishing between the partial vs. total time derivative of an operator can sometimes be a bit confusing. It is related to the explicit vs implicit time dependence of the operator.

For example, the potential energy Φ of a charged particle near a capacitor depends on the charge on the capacitor (and typically the distance from it.) That charge could be i) dependent solely on the electrical dynamics of a freely oscillating circuit to which the capacitor is attached and not a direct function of time *per se*, or ii) driven by a voltage source across the capacitor whose voltage varies

Distinguishing between total and partial time derivative of an operator

(typically sinusoidally) with time. In both cases, the charge (and thus the potential energy nearby) changes in time. In the second case, it is an explicit function of time; in the first, it is implicit.

Mathematically, the potential, and the charge q, in each case can be expressed as

i) potential $\Phi_i \propto q = q(R,L,C,i,\dot{i})$ Φ_i implicit function of time (i varies in time)

ii) potential $\Phi_e \propto q = q(t) = CV(t)$ Φ_e explicit function of time,   (7-21)

where R and L are circuit resistance and inductance, C is capacitance, and i is current. Thus,

i) $\dfrac{\partial \Phi_i}{\partial t} = 0$, though $\dfrac{d\Phi_i}{dt}$ often $\neq 0$ Φ_i implicit function of time

ii) $\dfrac{\partial \Phi_e}{\partial t} \neq 0$, and so $\dfrac{d\Phi_e}{dt}$ generally $\neq 0$ Φ_e explicit function of time.   (7-22)

Φ_i can change in time, because it depends on current i and that varies with time (i is the solution to the unforced RLC circuit), but is not explicitly tied to the passage of time t.

Of course, we could have a potential at a point in space due to both a time dependent voltage source capacitor plus another capacitor in a freely oscillating circuit. In that case,

iii) potential $\Phi \propto q = q(R,L,C,i,\dot{i},t)$ Φ explicit function of time.   (7-23)

The partial and total time derivative behaviors for (7-23) are the same as (7-22) *ii)*.

As an example, the S.P. Hamiltonian operator H^S is comprised of K.E. plus P.E., where the latter could be like Φ above. Φ is, in the vast majority of real world situations, not an explicit function of time, and since K.E. is not an explicit function of time, H^S is then not an explicit function of time either. All S.P. operators in all cases considered in this book will not be explicit functions of time.

From the top row, operators column of Wholeness Chart 7-1, we note that S.P. operators only change in time (total time derivative) if they are explicit functions of time. But, in this book, we will always have $\mathcal{O}^S \neq \mathcal{O}^S(t)$, and so any $\mathcal{O}^S$ we deal with will be constant in time. Its total time derivative will be zero.

$\mathcal{O}$ as a Dynamical Variable Operator

If our operator $\mathcal{O}$ is a dynamical variable operator like H, **P**, etc., then in all our cases in this book, since the S.P. total time derivative of $\mathcal{O}^S$ is zero, there is only a trivial equation of motion for $\mathcal{O}^S$ in the S.P. In the H.P. (see 3rd row, operators column) the operator has a non-zero equation of motion if the operator in the H.P. does not commute with H. If it commutes, the operator in the H.P. is conserved.

$\mathcal{O}$ as a Quantum Field

In the Appendix of Chap. 3, we showed that the operator equation of motion for free relativistic scalar fields ($\mathcal{O}^H = \phi$) in the H.P. (3rd row, operators column) actually reduces to the Klein-Gordon equation. It can be shown that the same general form reduces, for spinors, to the Dirac equation, and for vectors, to the Maxwell equation.

Expectation Values for $\mathcal{O}$ as a Dynamical Variable Operator

The expectation value of a dynamical variable operator is the same in the S.P. and the H.P., as one can see (with $|\Psi\rangle$ representing, generically, any particular state) via

$$\bar{\mathcal{O}} = {}_S\langle\Psi|\mathcal{O}^S|\Psi\rangle_S = {}_S\langle\Psi|I\mathcal{O}^S I|\Psi\rangle_S = \underbrace{{}_S\langle\Psi|U}_{{}_H\langle\Psi|}\underbrace{U^\dagger\mathcal{O}^S U}_{\mathcal{O}^H}\underbrace{U^\dagger|\Psi\rangle_S}_{|\Psi\rangle_H} = {}_H\langle\Psi|\mathcal{O}^H|\Psi\rangle_H . \quad (7\text{-}24)$$

In similar fashion, the time derivative of the expectation value of such an operator can also be seen to be the same in either picture. (Do Prob. 5 to show this.)

Expectation Values for $\mathcal{O}$ as a Quantum Field

Note that the expectation value for the Klein-Gordon scalar field operator ϕ, for a particular state of two scalar particles, each of 3-momentum **k'** is, in the Heisenberg picture (see Chap. 3, relation (3-81), pg. 59, for coefficients arising from creation/destruction operators)

Side notes (right margin):

If operator only directly dependent on quantities (that may vary in time), then partial time derivative = 0

If operator directly dependent on time (a direct function of t), then partial time derivative $\neq 0$

In most cases, quantum operators not an explicit function of time

In this book, all partial time derivatives of operators = 0

Operator can be a dynamical variable

Or it can be a quantum field, where H.P. eq of motion is K-G, Dirac, or Maxwell

Expectation value of dynamical variable same in S.P. and H.P.

$$\overline{\phi} = \left\langle 2\phi_{\mathbf{k}'} \middle| \phi \middle| 2\phi_{\mathbf{k}'} \right\rangle = \left\langle 2\phi_{\mathbf{k}'} \middle| \left(\sum_{\mathbf{k}} \frac{1}{\sqrt{2V\omega_{\mathbf{k}}}} a(\mathbf{k}) e^{-ikx} + \sum_{\mathbf{k}} \frac{1}{\sqrt{2V\omega_{\mathbf{k}}}} b^{\dagger}(\mathbf{k}) e^{ikx} \right) \middle| 2\phi_{\mathbf{k}'} \right\rangle$$

$$= 0 + 0 + ... + \frac{\sqrt{2} e^{-ik'x}}{\sqrt{2V\omega_{\mathbf{k}'}}} \underbrace{\left\langle 2\phi_{\mathbf{k}'} \middle\| \phi_{\mathbf{k}'} \right\rangle}_{=0} + 0 + ... + \frac{e^{ik_1 x}}{\sqrt{2V\omega_{\mathbf{k}_1}}} \underbrace{\left\langle 2\phi_{\mathbf{k}'} \middle\| 2\phi_{\mathbf{k}'}, \overline{\phi}_{\mathbf{k}_1} \right\rangle}_{=0} \qquad (7\text{-}25)$$

$$+ \frac{e^{ik_2 x}}{\sqrt{2V\omega_{\mathbf{k}_2}}} \underbrace{\left\langle 2\phi_{\mathbf{k}'} \middle\| 2\phi_{\mathbf{k}'}, \overline{\phi}_{\mathbf{k}_2} \right\rangle}_{=0} + ... + \frac{e^{ik'x}}{\sqrt{2V\omega_{\mathbf{k}'}}} \underbrace{\left\langle 2\phi_{\mathbf{k}'} \middle\| 2\phi_{\mathbf{k}'}, \overline{\phi}_{\mathbf{k}'} \right\rangle}_{=0} + ... = 0,$$

where the zeros in the second row result from the effect of each $a(\mathbf{k})$ destroying (making equal to zero) any state not containing a scalar particle of 3-momentum $\mathbf{k}$. When the $a(\mathbf{k}')$ operator acts, it reduces the ket by one particle with resulting bracket equaling zero. Each of the creation operators $b^{\dagger}(\mathbf{k})$ creates an antiparticle in the ket, so the bra and ket are orthogonal, and the resulting bracket is zero.

The result is zero expectation value for ϕ, and this would be true for whatever state, including the vacuum, that we choose. This effectively means that we would measure nothing if we tried to measure the quantum field ϕ. The field itself is unmeasurable. Hence, the RH column of Wholeness Chart 7-1 would be all zeros, as fields have no expectation values that could vary in time.

Expectation value of a quantum field is zero

This contrasts with dynamical variable operators, such as energy, momentum, and charge, which for any given state, do have expectation values and are indeed measurable. Thus, the RH column of Wholeness Chart 7-1 relates specifically to dynamical variables.

Expectation value of dyn variable is measurable; of quantum field, not measurable

In general (do Prob. 6 for practice), the expectation value for any quantum field, including spinor and vector fields, is zero[1].

Pick the Easiest Picture to Do What We Want to Do

The S.P. was easier to use in NRQM and RQM. The H.P. was easier to use in QFT for free particles, and that is what we employed in our development of QFT in Chaps. 3, 4, and 5.

Use the picture that makes things easiest

7.3.2 The Third Kind of Picture

It turns out that a third picture, the <u>Interaction Picture</u> (I.P.) is easier to use for interactions in QFT. For one reason, as we are soon to find out, it facilitates use of perturbation theory in place of trying to solve the coupled, non-linear, partial differential equations (7-18) and (7-19).

Easiest picture to use for interactions is a 3rd kind, the Interaction Picture

Additionally, the I.P. allows us to analyze interacting fields using all the results of our free QFT development from Chaps. 3, 4, and 5, so, we won't have to go through similar lengthy steps for interacting fields. We get this huge benefit by breaking the Hamiltonian into two parts.

Breaking the Hamiltonian into Free and Interaction Parts

The Hamiltonian (total, not density) for e/m interactions in the S.P. can be expressed, from the Lagrangian density (7-15) and the Legendre transformation, as (with ϕ^r generically representing any quantum field)

Two parts of the Hamiltonian; the free part and the interaction part

$$\underbrace{H^S}_{H} = \underbrace{H^{1/2,1}}_{\text{just e/m}} = \underbrace{\int \left(\pi_r \dot{\phi}^r - \mathcal{L}_0^1 - \mathcal{L}_0^{1/2} \right) d^3x}_{H_0^S = H_0 \text{ (free part)}} \underbrace{- \int \mathcal{L}_I^{1/2,1} d^3x}_{H_I^S \text{ (interaction part)}}, \qquad (7\text{-}26)$$

where, for simplicity, we will take the symbol H as the S.P. Hamiltonian (as we did in the S.P. vs. H.P. treatment) and H_0 as the S.P. free part of the Hamiltonian. H_I^S represents the interaction part of the Hamiltonian in the S.P. Thus, (7-26) is simply

$$H = H_0 + H_I^S . \qquad (7\text{-}27)$$

Note, for future reference, that for all cases in this book,

$$\mathcal{H}_I = -\mathcal{L}_I \quad \text{and} \quad H_I = -L_I . \qquad (7\text{-}28)$$

---

[1] Exceptions to this exist in more advanced areas of QFT, but please do not worry about them for now.

<u>Using Only the Free Part of the Hamiltonian to Transform to the Interaction Picture</u>

The transformation from the S.P. to the I.P. is similar to the transformation from the S.P. to the H.P. (see Wholeness Chart 7-1, 2nd row, LH column) except that we only use the free field part of the full Hamiltonian without the interaction part. Thus, the <u>transformation from the S.P. to the I.P.</u> is

$$U_0 = e^{-iH_0 t} .$$
(7-29)

Use only the free part to transform to the I.P.

where U_0 is a unitary operator (see Box 2-3 of Chap. 2, pg. 27), where

$$U_0^\dagger |\Psi\rangle_S = |\Psi\rangle_I ,$$
(7-30)

The I.P. transformation

and where subscripts "S" and "I" on generic states $|\Psi\rangle$ indicate the S.P. and I.P., respectively. For operators, where superscripts "S" and "I" represent the S.P. and I.P., respectively,

$$U_0^\dagger \mathcal{O}^S U_0 = \mathcal{O}^I .$$
(7-31)

Transforming states

Transforming operators

<u>Parts of the Hamiltonian Expressed in the I.P.</u>

For the free part of the Hamiltonian operator $H_0 = H_0{}^S$, we see that

$$H_0^I = U_0^\dagger H_0^S U_0 = U_0^\dagger H_0 U_0 = e^{iH_0 t} H_0 e^{-iH_0 t} = H_0 e^{iH_0 t} e^{-iH_0 t} = H_0 ,$$
(7-32)

because H_0 commutes with itself. (See expansion of e^{-iHt} example in Box 2-3.) Thus,

$$H_0 = H_0^S = H_0^I .$$
(7-33)

By doing Prob. 7, one can see that this equality generally does not hold for the interaction part,

$$H_I^I = U_0^\dagger H_I^S U_0 \underbrace{\neq H_I^S}_{\text{generally}} ,$$
(7-34)

and thus, we will represent the interaction picture Hamiltonian as

$$H^I = H_0 + H_I^I .$$
(7-35)

Expressing parts of H in I.P.

7.3.3 Equations of Motion in the I.P.

The <u>I.P. equation of motion for operators</u> is

$$\frac{d\mathcal{O}^I}{dt} = \frac{d}{dt}\left(U_0^\dagger \mathcal{O}^S U_0\right) = \frac{dU_0^\dagger}{dt}\mathcal{O}^S U_0 + U_0^\dagger \underbrace{\frac{\partial \mathcal{O}^S}{\partial t} U_0}_{\text{defined} = \frac{\partial \mathcal{O}^I}{\partial t}} + U_0^\dagger \mathcal{O}^S \frac{dU_0}{dt} =$$
(7-36)

$$\frac{de^{iH_0 t}}{dt}\mathcal{O}^S e^{-iH_0 t} + e^{iH_0 t}\mathcal{O}^S \frac{de^{-iH_0 t}}{dt} + \frac{\partial \mathcal{O}^I}{\partial t} = iH_0 \underbrace{e^{iH_0 t}\mathcal{O}^S e^{-iH_0 t}}_{\mathcal{O}^I} - \underbrace{e^{iH_0 t}\mathcal{O}^S e^{-iH_0 t}}_{\mathcal{O}^I} iH_0 + \frac{\partial \mathcal{O}^I}{\partial t} ,$$

or

$$\frac{d\mathcal{O}^I}{dt} = -i\left[\mathcal{O}^I, H_0\right] + \underbrace{\frac{\partial \mathcal{O}^I}{\partial t}}_{=0 \text{ in this book}} .$$
(7-37)

Operator eq of motion in I.P.

where the last term, throughout this book, is zero, because we only deal with operators for which $\partial \mathcal{O}^S / \partial t = 0$ (see first line of (7-36)).

Thus, the equation of motion for operators in the I.P. depends only on the free part of the Hamiltonian.

The <u>I.P. equation of motion for states</u> is (found by doing Prob. 8, which every reader should do)

$$i\frac{d}{dt}|\Psi\rangle_I = H_I^I |\Psi\rangle_I .$$
(7-38)

State eq of motion in I.P.

And hence, the equations of motion for states in the I.P. depends only on the interaction part of the Hamiltonian.

The <u>I.P. equation of motion for expectation values</u> is (found by doing Prob. 9)

$$\frac{d\bar{\mathcal{O}}}{dt} = {}_I\langle\Psi|\left(-i\left[\mathcal{O}^I,H^I\right]+\frac{\partial\mathcal{O}^I}{\partial t}\right)|\Psi\rangle_I .$$ (7-39)

Expectation value eq of motion

The above relations (7-27) to (7-38) are summarized in the bottom half of Wholeness Chart 7-1.

7.3.4 Big Benefit from the I.P.

Note (from Wholeness Chart 7-1) that (7-37) has the same form as the operator equation of motion in the H.P., except that we have H_0 in the I.P. and H in the H.P.

We can use all operator results from free field in H.P. for interacting fields in I.P.

This means if we were working in the H.P. for the special case of free fields, where $H = H_0$, then that H.P. special case equation of motion for an operator would be the same as the I.P. general case equation of motion for the operator. Hence, we can take all results we obtained for operator behavior in the H.P. free field development of Chaps. 3, 4, and 5 and use them, as is, in the I.P. general case development for interacting fields.

For quantum field operators (such as $\mathcal{O}^I = \phi$, ψ, or A^μ), (7-37) has identical form to the H.P. equation of motion for fields where $H = H_0$. Thus from our note earlier on pg. 189 under "$\mathcal{O}$ as a Quantum Field", (7-37) reduces to the Klein-Gordon equation for scalars, the free Dirac equation for spinor fields, and the free Maxwell equation for photons. Quantum fields in the I.P. behave just like the free quantum fields we have already studied.

All quantum fields are same in I.P. as free case in H.P. Same field equations, same solutions

If you are now saying "great" to yourself, then you recognize the value of the I.P.

7.3.5 Expectation Values in the I.P.

Expectation values remain the same in the I.P. as they were in the S.P. (and H.P.). This should not be too hard to prove to yourself by doing Prob. 10. Measured values remain the same, but the underlying math differs. A familiar refrain in studies of the quantum realm.

Expectation values same in I.P. as in S.P. and H.P.

7.3.6 Visualizing States in the I.P.

Consider a single particle scalar plane wave state expressed in the coordinate basis. In the S.P., it looks like, where K is a normalization factor and sub/superscript meaning should be obvious,

The I.P. transformation takes the H_0 dependence out of states

$$|\phi\rangle_S = \phi_{state}^S = Ke^{-iEt+i\mathbf{k}\cdot\mathbf{x}} = Ke^{-iE_0t-iE_It+i\mathbf{k}\cdot\mathbf{x}} .$$ (7-40)

Note that E_I is a number, a measured value if we were to measure, and thus is the same in any picture. Transform the state (7-40) to the I.P.,

$$U_0^\dagger|\phi\rangle_S = e^{iH_0t}\phi_{S,state} = K\underbrace{e^{iH_0t}e^{-iE_0t-iE_It+i\mathbf{k}\cdot\mathbf{x}}}_{\text{see Box 2-3 expansion}} == Ke^{iE_0t}e^{-iE_0t-iE_It+i\mathbf{k}\cdot\mathbf{x}} = Ke^{-iE_It+i\mathbf{k}\cdot\mathbf{x}} = |\phi\rangle_I .$$ (7-41)

So we see that the state in the I.P. varies in time only with the interaction energy, in accord with (7-38). (Use (7-41) in (7-38)). The operator $U_0^\dagger$ takes out the H_0 dependence of the ket.

7.3.7 Bottom Line for the I.P.

In the I.P.

- the state equation of motion depends on only the interaction Hamiltonian H_I^I,

The I.P, and its advantages, in a nutshell

- operator equations of motion depend on only the free Hamiltonian H_0, thus, importantly,

- the operator equations of motion in the I.P. are the same as the operator equations of motion in the H.P. for *free* fields (i.e., for $H^H = H_0$ with $H_I^H = 0$), so all operator relations derived in Chaps. 3, 4, and 5 are valid in the I.P.,

- meaning the free field case Klein-Gordon, Dirac, and Maxwell equations (of motion) from the H.P. are the same as those in the interacting case in the I.P., and so

- quantum fields ϕ, ψ, and A^μ in the I.P. (the solutions to the field equations of motion) are the same as the *free* quantum fields solutions in the H.P. developed in Chaps. 3, 4, and 5.

So in the I.P.,

- we only need to solve the state equation of motion (7-38).

Wholeness Chart 7-2 summarizes the above and provides examples from the three pictures for states and operators.

Wholeness Chart 7-2: Examples from the Three Pictures

	States	Operators	Expectation Values
Schrödinger Picture	$i\dfrac{d}{dt}\lvert\Psi\rangle_S = H\lvert\Psi\rangle_S$ Time dependent Examples Schrödinger eq in NRQM Klein-Gordon, Dirac, Maxwell eqs in RQM	$\dfrac{d\mathcal{O}^S}{dt}=\dfrac{\partial\mathcal{O}^S}{\partial t}=0$ Always time indep in this book No operator eq of motion to solve	$\dfrac{d\bar{\mathcal{O}}}{dt}={}_S\langle\Psi\rvert-i\big[\mathcal{O}^S,H\big]\rvert\Psi\rangle_S$ $\lvert\psi\rangle_S$ changes in time; $\mathcal{O}^S$ always const in this book
Heisenberg Picture	$\lvert\Psi\rangle_H$ = constant Time independent Like frozen ripple on icy pond. No state eq of motion to solve.	$\dfrac{d\mathcal{O}^H}{dt}=-i\big[\mathcal{O}^H,H\big]$ Time dependent if $[\mathcal{O}^H,H]\neq 0$ Examples Interacting field eqs of QFT: For $\mathcal{O}^H=\phi\rightarrow$ Klein-Gord interact eq For $\mathcal{O}^H=\psi\rightarrow$ Dirac interact eq For $\mathcal{O}^H=A^\mu\rightarrow$ Maxwell interact eq, Dyn variable opers, $\mathcal{O}^H=\mathbf{P},H,$ etc	$\dfrac{d\bar{\mathcal{O}}}{dt}={}_H\langle\Psi\rvert-i\big[\mathcal{O},^H H\big]\rvert\Psi\rangle_H$ $\lvert\Psi\rangle_H$ const in time; $\mathcal{O}^H$ often changes in time Same $\dfrac{d\bar{\mathcal{O}}}{dt}$ as S.P.
Key point		If $H_I^H=0$ (i.e., $H=H_0$) then above equations are free field equations.	
Interaction Picture	$i\dfrac{d}{dt}\lvert\Psi\rangle_I = H_I^I\lvert\Psi\rangle_I$ Time dependent if $H_I^I\neq 0$	$\dfrac{d\mathcal{O}^I}{dt}=-i\big[\mathcal{O}^I,H_0\big]$ Time dependent if $[\mathcal{O}^I,H_0]\neq 0$ Examples Free field eqs of QFT (If $H=H_0$, $\mathcal{O}^I=\mathcal{O}^H$) For $\mathcal{O}^I(=\mathcal{O}^H)=\phi\rightarrow$ Klein-Gord eq For $\mathcal{O}^I(=\mathcal{O}^H)=\psi\rightarrow$ Dirac eq For $\mathcal{O}^I(=\mathcal{O}^H)=A^\mu\rightarrow$ Maxwell eq Free field dyn variable opers of QFT	$\dfrac{d\bar{\mathcal{O}}}{dt}={}_I\langle\Psi\rvert-i\big[\mathcal{O}^I,H^I\big]\rvert\Psi\rangle_I$ $\lvert\Psi\rangle_I$ changes in time if $H_I^I\neq 0$; $\mathcal{O}^I$ changes in time if $[\mathcal{O}^I,H_0]\neq 0$ Same $\dfrac{d\bar{\mathcal{O}}}{dt}$ as S.P. and H.P.
Key points	In I.P. we have to solve state equation of motion (if there are interactions).	If we use I.P., we can use all operator relations from free field QFT (Chaps. 3, 4, and 5) in interacting field QFT. (As an aside, I.P. is same as H.P. if there are no interactions, since $H=H_0$)	

7.3.8 A Subtle Point

I don't recommend that the newcomer get sidetracked thinking too much on the following at this point in her/his studies. We include it here to be complete.

Our commutation relations for fields involve the conjugate momentum (again, ϕ^r represents a generic field)

$$\pi_r=\frac{\partial\mathcal{L}}{\partial\dot{\phi}^r}\quad\text{for free fields}\rightarrow\quad\pi_r=\frac{\partial\mathcal{L}_0}{\partial\dot{\phi}^r}\ . \tag{7-42}$$

In Chaps. 3, 4, and 5 we dealt exclusively with free fields and thus, exclusively with the free part of the Lagrangian $\mathcal{L}_0$. In this chapter, we are dealing with the full Lagrangian, $\mathcal{L} = \mathcal{L}_0 + \mathcal{L}_I$. If $\mathcal{L}_I$ contains no time derivatives of fields, then the RH and LH equations in (7-42) are the same. In QED, this is true. See $\mathcal{L}_I^{1/2,1}$ of (7-20).

A subtle point relevant to weak and strong interactions

Thus, all of the relations we developed earlier involving commutators (including creation and destruction operator properties, number operators, Hamiltonian and charge operators in terms of number operators, etc.) will still hold in QED.

For weak interactions, the interaction Lagrangian does contain derivatives, so the equality of the RH and LH sides of (7-42) does not hold. However, by making similar assumptions to those of this chapter for weak interactions, one can obtain a viable theory.

7.3.9 Operator Relations We Can Use in the I.P.

Thus, if we stay in the I.P., we can use

1. free field operator solutions ϕ, ψ, and A^μ of Part I for interactions
2. free field operator creation and destruction properties for interactions
3. free field number operators for interactions
4. free field observables operators for interactions
5. free field Feynman propagators for interactions

Operator results from free field theory we can use in interaction theory if we use the Interaction Picture

7.3.10 Notation Change

From now on, we will be working solely in the I.P., so we will tend to drop the I superscripts (on operators) and subscripts (on states) indicating the I.P., except for H_I^I and $\mathcal{H}_I^I$.

Will generally drop "I" script indicating I.P. from here on

7.3.11 "Eigenstates" in QFT

The word <u>eigenstate</u> in QFT generally means a state of one or more particles for which each particle has a specific, definite 3-momentum, and additionally, for each fermion, a specific, definite spin state and for each photon, a specific, definite polarization. Such multiparticle states are eigenstates of the 3-momentum operator and the Hamiltonian (since a given particle energy level is readily deduced from its 3-momentum). In the I.P., these operators would be the free field operators H_0 and P_0. Such eigenstates are also eigenstates of number and charge operators, as we have developed them in Chaps. 3, 4, and 5. That is, each eigenstate has a definite number of each type of particle and a definite charge (equal to the sum of charges on the individual particles).

"Eigenstate" = each particle in state has definite 3-momentum

Eigenstate also has definite total energy, num of particles, & charge

7.3.12 Summary Wholeness Chart

Wholeness Chart 8-4 at the end of Chap. 8 summarizes the interaction picture results and all material between here and the end of Chap. 8, i..e, from operators, states, and propagators in the I.P. to Feynman Rules for QED interactions.

I strongly recommend keeping that chart by your side as you work through the rest of this, and all of the next, chapters.

Use Wholeness Chart at end of Chap. 8 as study rest of this chapter.

7.4 The S Operator and the S Matrix

In particle theory we have many different interactions that can occur. Many different incoming particles can interact with one another. And for every set of incoming particles, there are multiple final (and often different) sets of outgoing particles.

By way of example, one such set of incoming particles is the electron and positron of Fig. 1-1 in Chap. 1, pg. 2. In that figure, the outgoing particles are a muon and an anti-muon. But another possibility comprises the intermediary virtual photon mutating back into an outgoing electron and positron. Or an outgoing tau (also called a tauon or tau lepton) and anti-tau.

As another example, we could have two incoming electrons that scatter off one another by exchanging a virtual photon resulting in two outgoing electrons (with many different possible individual momenta). Or an incoming electron and a photon scattering off one another (Compton scattering) to result in an outgoing electron and photon (with different individual momenta).

These different interactions have different probabilities to occur. So how do we keep track of each and every possible interaction, or more particularly, how do we keep track of the individual probabilities for each to occur? The answer, at least in principle, is something called the <u>S Matrix</u>, shorthand for <u>Scattering Matrix</u> (since each of these interactions is considered a scattering process).

*We keep track
of probabilities
of different
interactions via
the S Matrix*

7.4.1 The S Matrix

Consider a column vector for which each component represents a different initial (incoming) quantum eigenstate (typically multiparticle), as on the RH side of (7-43).

$$
\begin{bmatrix} e_{r^1,\mathbf{p}^1},e^+_{r^2,\mathbf{p}^2} \\ e_{r^1,\mathbf{p}^1},e^+_{r^2,\mathbf{p}^3} \\ e_{r^1,\mathbf{p}^2},e^+_{r^2,\mathbf{p}^3} \\ \vdots \\ e_{r^2,\mathbf{p}^1},e_{r^2,\mathbf{p}^2} \\ \vdots \\ e_{r^1,\mathbf{p}^1},\gamma_{r^2,\mathbf{k}^1} \\ \vdots \end{bmatrix}_f
=
\begin{bmatrix} S_{11} & S_{12} & S_{13} & \cdots & S_{1i} & \cdots & \cdots & \cdots \\ S_{21} & S_{22} & S_{23} & \cdots & S_{2i} & \cdots & \cdots & \cdots \\ S_{31} & S_{32} & S_{33} & \cdots & S_{3i} & \cdots & \cdots & \cdots \\ \cdots & \cdots & \cdots & \cdots & \cdots & \cdots & \cdots & \cdots \\ S_{f1} & S_{f2} & S_{f3} & \cdots & S_{fi} & \cdots & \cdots & \cdots \\ \cdots & \cdots & \cdots & \cdots & \cdots & \cdots & \cdots & \cdots \\ \cdots & \cdots & \cdots & \cdots & \cdots & \cdots & \cdots & \cdots \\ \cdots & \cdots & \cdots & \cdots & \cdots & \cdots & \cdots & \cdots \end{bmatrix}
\begin{bmatrix} e_{r^1,\mathbf{p}^1},e^+_{r^2,\mathbf{p}^2} \\ e_{r^1,\mathbf{p}^1},e^+_{r^2,\mathbf{p}^3} \\ e_{r^1,\mathbf{p}^2},e^+_{r^2,\mathbf{p}^3} \\ \vdots \\ e_{r^2,\mathbf{p}^1},e_{r^2,\mathbf{p}^2} \\ \vdots \\ e_{r^1,\mathbf{p}^1},\gamma_{r^2,\mathbf{k}^1} \\ \vdots \end{bmatrix}_i
\qquad (7\text{-}43)
$$

$\underbrace{\qquad}_{\text{Final Eigenstates}}$ $\underbrace{\qquad\qquad\qquad}_{\text{S Matrix}}$ $\underbrace{\qquad}_{\text{Initial Eigenstates}}$

Each component of the column vector on the LH represents a different final (outgoing) eigenstate.

*Each S_{fi} of S
Matrix is the
transition
amplitude
between an initial
eigenstate and a
final eigenstate*

The square of the absolute value of each component of the S Matrix connecting a given initial and final state equals the probability of that transition taking place. For example, the probability of the first initial state above, of an electron and positron with particular spins and momenta, interacting to become the second final state, of an electron and positron with somewhat different particular spins and momenta, is

$$
S^{\dagger}_{21}S_{21} = |S_{21}|^2 = \text{probability of 1st eigenstate transitioning to 2nd}. \qquad (7\text{-}44)
$$

*$|S_{fi}|^2$ is probability
of transition from
initial eigenstate
$|i\rangle$ to final
eigenstate $|f\rangle$*

Thus, if we recall the meaning of "transition amplitude" from Chap. 1, each component of the S Matrix, S_{fi} is a <u>transition amplitude</u> for a particular reaction (scattering event, transition, interaction) between particles in particular eigenstates.

<u>Our job</u> from here on out: Learn how to calculate S_{fi} for any given interaction.

Note that i) the S Matrix is huge, infinite really, as it not only includes all possible types of interactions, but all possible momenta magnitudes and directions for each type (all possible eigenstates), ii) many (most) of the components of (7-43) equal zero, since many interactions (such as $e,\gamma \rightarrow e,e^+$) are impossible (for the example, because charge is not conserved), and iii) the diagonal components effectively represent no interaction taking place, i.e., we have the same exact final state as the initial state. For example, $|S_{33}|^2$ represents the probability that the particles in the 3rd (typically multiparticle) final eigenstate will be measured to be unchanged from the initial ones.

*S Matrix
represents all
interactions and
has mostly zero
comps, with
diagonal comps
for $|i\rangle = |f\rangle$*

7.4.2 The S Operator

You should now go back and read pgs. 2-3 in Chap.1. Note that (1-6) is the transition probability for the initial state of Fig. 1-1 to transition/scatter into the final state. It is the square of the absolute value of the particular component of the S Matrix connecting those two states.

For simplicity at this point, consider having final $e^- e^+$ in Fig. 1-1 instead of final $\mu^- \mu^+$. $e^- e^+ \rightarrow e^- e^+$ is known as <u>Bhabha scattering</u>, and Fig. 1-1 with the above change represents one type of Bhabha scattering. Note that we got to (1-6) from (1-1), which is similar to (7-45) below. In both, operators act on a ket and step-by-step convert that initial ket state into the final ket state.

*Review of
scattering
example from
Chap. 1*

$$
\text{Trans amplit} = \left\langle e^+ e^- \left| \iint K \left(\overline{\psi} A \psi\right)_{x_1} \left(\overline{\psi} A \psi\right)_{x_2} dx_1 dx_2 \right| e^+ e^- \right\rangle = \left\langle e^+ e^- \left| \underset{\substack{Bhabha}}{S_{operator}} \right| e^+ e^- \right\rangle \quad (7\text{-}45)
$$

$\underbrace{\qquad\qquad\qquad\qquad\qquad}$
operators that convert
initial ket into final ket
= same state as bra

one operator
symbol=several
sub-operators

*S_{Bhabha} operator
changes initial
eigenstate ket to
final eigenstate
ket*

In Chap. 1, we alluded to, but for simplicity effectively ignored, the integration shown in (7-45). (We will see how it comes in along with certain constants and subtleties not shown in (7-45).) You can begin to suspect that the ψ and A operators shown in (7-45) are the fields we have been dealing with for several chapters now. As in (1-1) to (1-6), these operators destroy the initial particles, generate a virtual photon that propagates until it is destroyed, at which point the final particles are created. The bra represents the final state. When, after all the operator fields in (7-45) have operated on the ket, the ket has changed to match the bra, and there are no more operators left, only a number sandwiched between the bra and ket, which we call S_{Bhabha}. The bra and ket inner product =1, so we are left with the number S_{Bhabha} as our transition amplitude (our associated S Matrix component).

Thus, where we express the spin and momenta explicitly,

$$\text{Trans amplit} = \left\langle e^+_{r^3,\mathbf{p}^3} \, e^-_{r^4,\mathbf{p}^4} \right| \underbrace{S_{\substack{operator \\ Bhabha}}}_{\substack{comprised\ of \\ sub\ operators}} \left| e^+_{r^1,\mathbf{p}^1} \, e^-_{r^2,\mathbf{p}^2} \right\rangle$$

$$= \left\langle e^+_{r^3,\mathbf{p}^3} \, e^-_{r^4,\mathbf{p}^4} \right| \underbrace{S_{Bhabha}}_{\substack{number\ left \\ after\ opers\ act}} \left| e^+_{r^3,\mathbf{p}^3} \, e^-_{r^4,\mathbf{p}^4} \right\rangle \qquad (7\text{-}46)$$

$$= S_{Bhabha} \underbrace{\left\langle e^+_{r^3,\mathbf{p}^3} \, e^-_{r^4,\mathbf{p}^4} \middle| e^+_{r^3,\mathbf{p}^3} \, e^-_{r^4,\mathbf{p}^4} \right\rangle}_{=1} = S_{fi} \ \text{for this interaction.}$$

Scattering transition amplitude S_{Bhabha} found from operator $S_{\substack{operator, \\ Bhabha}}$, initial ket, and final bra

So, for any interaction, we seek an operator $S_{oper,fi}$ that sandwiched between the initial state ket and the final state bra, gives us a number that equals the transition amplitude (i.e., the relevant S_{fi}.)

That is, for the operator $S_{oper,fi}$

$$S_{fi} = \left\langle f \middle| S_{oper,fi} \middle| i \right\rangle \qquad (\text{no sum on } i \text{ or } f). \qquad (7\text{-}47)$$

Note that, $S_{oper,fi}$ takes the initial eigenstate into the final eigenstate.

$$S_{oper,fi} \middle| i \right\rangle = S_{fi} \middle| f \right\rangle \qquad (\text{no sum on } i \text{ or } f). \qquad (7\text{-}48)$$

Transition amplitude S_{fi} for any given $|i\rangle$ and $|f\rangle$ found from similar operator acting on $|i\rangle$

7.4.3 General (Non-eigen) States

Beyond the single final eigenstate of (7-48), we seek a most general form of S_{oper}, such that the same S_{oper} will give us the appropriate S_{fi} regardless of the initial and final states. That is, one single operator S_{oper} that can be used for every possible initial ket and final bra of (7-47), but gives us different (correct) S_{fi} numbers for each different interaction. This is what we will soon determine.

Note that a given initial state of certain particle types (such as an electron and positron) can result in any of several possible final state particle types (electron and positron, muon and anti-muon, or tau and anti-tau). Further each of these final particle combinations can have different individual momenta from that of the initial particles (though the total momentum would be the same as initially), and thus there are many possible final eigenstates even for the same final particle types.

We seek a single operator S_{oper} that includes all such operators, i.e., that is good for any $|i\rangle$ or $|f\rangle$

So, for any given $|i\rangle$, there are many possible $|f\rangle$, each with a different probability of resulting from the interaction. The final general state is a sum of eigenstates each having a certain probability of being detected when the state is measured (when general state collapses to an eigenstate). Symbolically, where C_f is the amplitude of the normalized fth final eigenstate,

$$\underbrace{|F\rangle}_{\substack{general \\ final\ state}} = \sum_f C_f |f\rangle = \underbrace{S_{oper}}_{\substack{yields \\ all\ final \\ eigstates}} |i\rangle = \sum_f S_{oper,fi} |i\rangle = \underbrace{\left(S_{oper,1i} + S_{oper,2i} + ... \right)}_{S_{oper}} |i\rangle$$

$$= \underbrace{S_{oper,1i} |i\rangle}_{\substack{1st\ final \\ eigenstate}} + \underbrace{S_{oper,2i} |i\rangle}_{\substack{2nd\ final \\ eigenstate}} + \underbrace{S_{oper,3i} |i\rangle}_{\substack{3rd\ final \\ eigenstate}} + ... = S_{1i} |f=1\rangle + S_{2i} |f=2\rangle + S_{3i} |f=3\rangle + ... \qquad (7\text{-}49)$$

$$= \sum_f S_{fi} |f\rangle .$$

S_{oper} = sum of operators, each of which converts $|i\rangle$ into a different $|f\rangle$

Amplitude of each eigenstate $|f\rangle$ is transition amplitude S_{fi}, = S Matrix comp

From comparison of the last line with the second part of the first line, we see that S_{fi} is the amplitude of the normalized fth final state, and thus we have the

probability of measuring final state $|f\rangle$ given initial state $|i\rangle = C_f^\dagger C_f = S_{fi}^\dagger S_{fi} = |S_{fi}|^2$. $\qquad$ (7-50)

$$(\text{no sum on } f)$$

Since the total probability of finding some final eigenstate upon measuring is one, then

$$\sum_f |S_{fi}|^2 = 1 \qquad \text{conservation of probability} \qquad (7\text{-}51)$$

(7-51) is called the <u>conservation of probability</u> relation, because our probability of measuring the initial state before the interaction was 1, and the probability of measuring some one of the final states is 1. (We can't measure nothing after the interaction. We must, with a probability of 1, measure something and all possibilities are included in the general final state.)

7.4.4 Finding the S Matrix from the S Operator

From (7-49), we can find any given S_{fi} component of the S Matrix, such as the $f = 2$ final state,

$$\langle f = 2 || F\rangle = \langle f = 2 | \sum_f S_{fi} | f\rangle = \langle f = 2 | \underbrace{\left(S_{1i} | f=1\rangle + S_{2i} | f=2\rangle + S_{3i} | f=3\rangle + ..\right)}_{\left(S_{oper,1i} + S_{oper,2i} + ...\right) | i\rangle = S_{oper} | i\rangle} \qquad (7\text{-}52)$$

$$= \langle f = 2 | S_{2i} | f = 2\rangle + \left(\text{many zero terms}\right) = S_{2i} = \langle f = 2 | S_{oper} | i\rangle.$$

In general,

$$\boxed{S_{fi} = \langle f | S_{oper} | i\rangle} \quad . \qquad (7\text{-}53)$$

7.5 Finding the S Operator

You may now be thinking that since we have the creation and destruction operators we need already from our theory so far, we can simply use them in relations like (7-45). To some degree, you may have a point, but there is a formal theory giving us the form of S_{oper}, and we shall develop it here. For one reason, we would not know the appropriate constants to use in a relation like (7-45) without the full theory. Additionally, there are subtler issues we have yet to bring up which that theory addresses, but our simplified approach does not.

7.5.1 The S Operator from the State Equation of Motion

We can find the S_{oper} we seek from the state equation of motion, the only thing we haven't already solved for in the I.P. formulation. In the I.P., our state equation of motion, where $|\Psi\rangle$ represents a generic state (multiparticle typically), is

$$i\frac{d}{dt}|\Psi(t)\rangle_I = H_I^I |\Psi(t)\rangle_I . \qquad (7\text{-}54)$$

In terms of the notation in (7-54), and with subscripts i and f indicating initial and final, respectively,

$$|F\rangle = \sum_f S_{fi} | f\rangle = |\Psi(t_f)\rangle_I = S_{oper}(t_f, t_i) |\Psi(t_i)\rangle_I = S_{oper}(t_f, t_i) |i\rangle, \qquad (7\text{-}55)$$

where we note the S_{oper} must be a function of the initial and final times. Taking our final time t_f as time t in (7-54) (the state at time t is the final state in effect that has evolved by time t), we have

$$|\Psi(t)\rangle_I = S_{oper}(t, t_i) |\Psi(t_i)\rangle_I . \qquad (7\text{-}56)$$

Using (7-56) in (7-54) yields

$$i\frac{d}{dt}\left(S_{oper} |\Psi(t_i)\rangle_I\right) = H_I^I \left(S_{oper} |\Psi(t_i)\rangle_I\right). \qquad (7\text{-}57)$$

This becomes

$$i\frac{dS_{oper}}{dt}\Big|\Psi(t_i)\rangle_I + iS_{oper}\underbrace{\frac{d}{dt}\Big|\Psi(t_i)\rangle_I}_{\substack{= 0 \text{ as ket not}\\ \text{function of } t}} = H_I^I S_{oper}\Big|\Psi(t_i)\rangle_I,\qquad(7\text{-}58)$$

and thus the <u>differential equation for S_{oper}</u> is

$$\boxed{i\frac{dS_{oper}}{dt} = H_I^I S_{oper}}\ .\qquad(7\text{-}59)$$

Differential equation for S_{oper}

This has the solution

$$S_{oper} = \underset{\substack{\text{must}\\ =1}}{K}\ e^{-i\int_{t_i}^{t} H_I^I dt'},\qquad(7\text{-}60)$$

where we note that K in front must equal unity, since, for $t = t_i$ in (7-60), $S_{oper} = K$, but to agree with (7-56), with $t = t_i$, S_{oper} must $= I$, the identity operator.

We caution the reader that we are being a bit cavalier mathematically here, since H_I^I is an operator and not a c number. Doing it in this manner should help make things easier to understand at this point, but after finishing Sect. 7.6, you should read Appendix B, to see precisely what we mean by (7-60).

Our final form for S_{oper} with $t=t_f$ in (7-60) is (note the incorporation of the Hamiltonian density)

$$\boxed{S_{oper} = e^{-i\int_{t_i}^{t_f} H_I^I dt} = e^{-i\int_{t_i}^{t_f}\int_V \mathcal{H}_I^I d^4x}}\ .\qquad(7\text{-}61)$$

Solution = S_{oper}

In principle (it isn't so simple practically, as we will see), if we know H_I^I, we know S_{oper}, and we can find the S matrix from (7-53)

$$S_{fi} = \langle f|S_{oper}|i\rangle = \langle f|\underbrace{S_{oper}|i\rangle}_{\substack{\text{1 eig}\\ \text{state}}}\underset{\substack{|F\rangle=\text{sum}\\ \text{of eig states}}}{} = \langle f|e^{-i\int_{t_i}^{t_f}\int_V \mathcal{H}_I^I d^4x}\underbrace{|\Psi(t_i)\rangle}_{\substack{i\text{th eigen}\\ \text{state here}}}.\qquad(7\text{-}62)$$

Importantly, note that because (7-61) is a unitary operator (see Box 2-3 of Chap. 2, pg. 27) the norm ("length" in abstract Fock space) of $|F\rangle$ remains the same as that of $|i\rangle$. If $|i\rangle$ was normalized to unit norm, then so must $|F\rangle$ be, i.e.,

If initial state had unit norm,

$$\text{If } \langle i|i\rangle = 1, \quad \text{then} \quad 1 = \langle i|I|i\rangle = \langle i|S_{oper}^\dagger S_{oper}|i\rangle = \langle F|F\rangle = 1.\qquad(7\text{-}63)$$

because S_{oper} is unitary, final general state has unit norm

If each $|f\rangle$ is normalized to have unit norm (which for us it always is), then from (7-63) and (7-49), the conservation of probability relation (7-51) must hold.

7.5.2 Ways to Visualize the S Operator

There are two ways we can visualize the operation of S_{oper} on a state.

<u>Physical Space Example</u>

Consider our Chap. 1 example of an initial e^- and e^+ annihilating one another to become a virtual photon, which in turn mutates into a final μ^- and μ^+. Each possible outgoing state could have different momenta for the two particles (with same total momentum). Each such state would be a different possible (multiparticle) final eigenstate. Each would, in general, have a different probability of being measured from the other final eigenstates. Of course, we could also have final states of fermion pairs e^-e^+ or $\tau^-\tau^+$, each with many different possible particle momenta. So,

$$\overbrace{\Big|e^-_{r^1,\mathbf{p}^1}, e^+_{r^2,\mathbf{p}^2}\Big\rangle}^{|i=1\rangle} \text{ via } S_{oper} \Rightarrow \underbrace{S_{11}\overbrace{\Big|e^-_{r^1,\mathbf{p}^1}, e^+_{r^2,\mathbf{p}^2}\Big\rangle}^{|f=1\rangle} + S_{21}\overbrace{\Big|e^-_{r^1,\mathbf{p}^3}, e^+_{r^2,\mathbf{p}^4}\Big\rangle}^{|f=2\rangle} + ...}_{\text{electron-positron final states}} + \underbrace{S_{\mu1}\Big|\mu^-_{r^1,\mathbf{p}^1}, \mu^+_{r^2,\mathbf{p}^2}\Big\rangle}_{\text{muon-antimuon final state}} + ...\ (7\text{-}64)$$

Physical space view of action of $S_{oper} \rightarrow$ one state turns into another

S_{oper} converts the initial eigenstate into a final general state sum of eigenstates where the square of the absolute magnitude of the coefficient of each final eigenstate, S_{fi}, equals the probability of measuring that final eigenstate.

<u>Fock Space Example</u>

In Fock space (see Chap. 3, Wholeness Chart 3-2, pg. 68), where every eigenstate can be visualized as a separate axis in an infinite (abstract) dimensional space, the S_{oper} can be visualized as a sort of abstract "rotation" in that space. The initial state vector $|i\rangle$ is "rotated" by the S_{oper} into a new vector with components along the eigenstate basis axes.

In Fig. 7-2, we illustrate this for a final general state composed of only two component eigenstates (because we can't draw any more axes on a 2-D sheet of paper).

This should seem reasonable, as (7-61) has the form of a function $e^{-i\theta}$ where θ is the angle a complex quantity (a phasor) in complex space is rotated through. Also, from (7-63), we see the norm (abstract "length" = 1) of the state vector is unchanged under the operation S_{oper}, i.e., we have a different final vector but it has the same magnitude. This is what happens in rotation.

By close analogy, a complex number can be represented as $z = Re^{i\phi}$ with length R. Operating on z with $e^{-i\theta}$ is $e^{-i\theta}z = e^{-i\theta}Re^{i\phi} = Re^{-i\theta+i\phi}$, which has the same length R, but is rotated through θ.

Fock space view of action of S_{oper} → like a "rotation" of the state vector

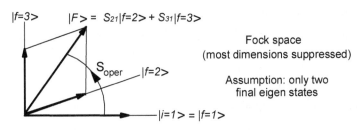

Figure 7-2. The Action of the S Operator Visualized in Fock Space

7.5.3 What is H_I^I in the I.P. of QFT?

The interaction Hamiltonian H_I^I is different for different interaction types, i.e., different kinds of forces. It is one thing for electromagnetic interactions, a different thing for weak interactions, and yet another thing for strong interactions. Our present task is to identify it for electromagnetism.

From Fig. 7-1 and Wholeness Chart 7-2, recall we seem to have discovered a general rule that the interaction Lagrangian (or wave equation, or Hamiltonian) for states in the S.P. of RQM is the same as that for fields in the I.P. of QFT. For free fields, the H.P. and I.P. are the same, so this is consistent with our prior work.

Using this rule for RQM e/m theory and QFT e/m theory (which is <u>quantum electrodynamics</u> or QED), we got (7-20). From that, we can find the e/m interaction Hamiltonian $_{QED}H_I^I$ via

Interaction Hamiltonian density different for e/m, weak, strong forces

$$\mathcal{H}_I^{1/2,1} = -\mathcal{L}_I^{1/2,1} = -e\bar{\psi}(x)\gamma^\mu A_\mu(x)\psi(x) \qquad _{QED}H_I^I = \int_V \mathcal{H}_I^{1/2,1}d^3x, \qquad (7\text{-}65)$$

Form of interaction Hamiltonian density for e/m. i.e., QED

and for QED, this can be used in (7-61) and (7-62).

In the I.P., fields (operators) behave like free fields, as we have seen. Given this, we take the fields ψ, $\bar{\psi}$, and A^μ as free fields (even though they are part of the interaction Hamiltonian).

We leave further exploration of (7-65) and the development of QED to the next chapter.

7.5.4 Scattering Volume V Same for All Particles and Taken ≈ ∞

The question then arises as to what volume V, and what times t_i and t_f, we are referring to when we find a given S_{fi} from (7-62). Won't the probability of an interaction rise if two initial particles are together for a longer period of time? And will the volume over which a particle (wave) is spread affect the probability of it interacting with another particle (wave)?

First, the volume V we <u>integrate</u> over in (7-65) and (7-62) is the same volume we use in the normalization factors $1/\sqrt{2V\omega_k}$ and $\sqrt{m/VE_p}$ for the fields A^μ and ψ in (7-65). (See Wholeness Chart 5-4, pg. 156.) All particle-waves are assumed to occupy that same volume V and thus the Hamiltonian density will be generally non-zero over all V.

Of course, almost every real-world situation differs from this. Real world interacting particles do not typically occupy the exact same volume. In scattering experiments, for example, target matter may be a fixed object, but a stream of particles is directed toward it. A target particle wave is essentially confined to the object, but the other particle wave is directed toward the object and typically not only extends outside the object's surface, but only spends part of the time with any portion of it passing through the object's volume.

Volume V:
i) same for all
fields,
ii) equals
integration vol,
iii) taken → ∞
to simplify math

Full resolution of this issue, which is often a source of significant consternation for students, is a topic best discussed in Chap. 17 (Scattering), after we have learned more theory. For now, we need to simply accept that the S_{oper} acts on a multiparticle state wherein all particles are confined to the same volume V for the same time period. And the probability for any final multiparticle state to be measured inside V at the end of the time period is $|S_{fi}|^2$.

Further, it turns out, as we will see in Chap. 8 and beyond, that the mathematics involved in finding S_{fi} are made enormously simpler, instead of being quite intractable, by taking $V \to \infty$. So, in developing the theory, we will take V as extremely large. This again, does not seem to correlate with the real world, as particles/fields do not extend to infinity, or even effectively to infinity. However, we will find, again in Chap. 17, that results obtained by making this assumption can actually be used to successfully predict real world experimental results.

We will be able to use results assuming V→ ∞ to predict experiment with V finite

7.5.5 Scattering Time Duration Taken ≈ ∞

The probability of finding a particular final state inside a particular 3D region should change with time. This is similar to particle decay. As time goes on, the probability of a decay being measured increases. There generally are, of course, a number of different possible final states, with different probabilities of being measured, at any given time.

Consider the situation for which $t_i \to -\infty$ and $t_f \to \infty$. In that case, we would have a sort of equilibrium of final states, in the sense that S_{oper} would have had enough time to operate such that each possible final eigenstate probability would be fixed, and no longer changing with time.

Infinite time of interaction yields equilibrium of final states

We would then have a certain probability for each particular final state to be measured. From the initial multiparticle state $|i=1\rangle$, we would have probability $|S_{21}|^2$ of measuring the 2nd final state, $|S_{31}|^2$ of measuring the 3rd final state, etc. after infinite time had passed.

We will find that, similar to 3D volume, considering the integration time for the operation of S_{oper} to be very large, effectively infinite, will be advantageous in developing the mathematics of QFT. Fortunately, again as we will eventually see, the results using this assumption can be used to accurately predict real world experimental outcomes.

Taking integration time $(t_f - t_i) \to \infty$ simplifies math and can lead to correct prediction for experiment for finite time duration

Thus, we define an <u>infinite spacetime</u> S_{oper} by the <u>symbol S</u> as

$$S = \underset{V \to \infty}{S_{oper}} \left(t_f \to \infty, t_i \to -\infty \right) = e^{-i\int_{-\infty}^{\infty} \mathcal{H}_I^I d^4x} . \qquad (7\text{-}66)$$

S_{oper} for infinite 4D spacetime region = S symbol

7.6 Expanding S_{oper}

S_{oper} can be expanded (in a Taylor series like $e^x = 1 + x + x^2/2! + x^3/3! + \ldots$) as

$$S_{oper}\left(t_f, t_i\right) = e^{-i\int_{t_i}^{t_f} H_I^I(t)dt} = I - i\int_{t_i}^{t_f} H_I^I\left(t_1\right)dt_1 - \frac{1}{2!}T\left\{\left(\int_{t_i}^{t_f} H_I^I\left(t_1\right)dt_1\right)\left(\int_{t_i}^{t_f} H_I^I\left(t_2\right)dt_2\right)\right\} + \ldots$$

S_{oper} expanded in Taylor series

$$(7\text{-}67)$$

$$= \sum_{n=0}^{\infty} \frac{(-i)^n}{n!}\int_{t_i}^{t_f} \ldots\ldots\int_{t_i}^{t_f} T\left\{H_I^I\left(t_1\right)H_I^I\left(t_2\right)\ldots H_I^I\left(t_n\right)\right\} dt_1 dt_2 \ldots dt_n .$$

If the H_I^I above were numeric functions of time, it wouldn't really matter what order, with respect to the t_n, we carry out the integrations in (7-67). However, since they are comprised of operators (see (7-65), for example) that act on a ket state to their right, as in (7-55) and (7-62), we have to be sure that at each point in the integration, the time-wise earliest operators are acting first. The operation of the operators must occur in the order they occur in nature, in time sequence. So, the earliest operator must operate on the ket first, i.e., it must be furthest to the right. The next earliest to operate must be second from the right, etc.

Earlier in time operators in that expansion operate first

This means that at each point in the n dimensional space where each axis is a different t_n, the H_I^I dependent on the earliest time of the t_n should act first, the H_I^I dependent on the next earliest of the t_n should act next, etc. See Fig. 7-3 for a two dimensional illustration of this, where we are integrating over the t_1–t_2 plane.

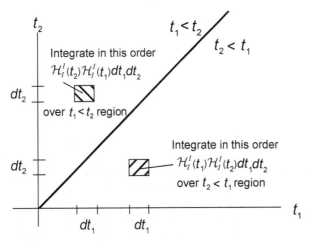

Figure 7-3. Time Ordering of Integrand in t_n Space (All but two dimensions suppressed)

The time ordering in the integrand is indicated by operator T

We indicate this ordering with the time ordering operator symbol T (see Chap. 3, pg. 72) in the integrands of (7-67). The order of the integrand operators is rearranged as we integrate over all t_n dimensions, such that the operators are time ordered at every point $(t_1, t_2, \dots t_n)$.

See Appendix B to clarify some subtle mathematical issues in (7-60), (7-61), and (7-67).

Taking our integration limits to infinity in both space and time, as in (7-66), turns (7-67) into the time ordered infinite spacetime S_{oper}, i.e., the Dyson expansion of the S operator,

Dyson expansion of S, where integration is over infinite 4D space

$$ S = \sum_{n=0}^{\infty} \frac{(-i)^n}{n!} \int_{-\infty}^{\infty} \dots \int_{-\infty}^{\infty} T\left\{ \mathcal{H}_I^I(t_1) \mathcal{H}_I^I(t_2) \dots \mathcal{H}_I^I(t_n) \right\} d^4x_1 d^4x_2 \dots d^4x_n \qquad . \qquad (7\text{-}68) $$

<u>SPECIAL NOTE:</u> (7-68) is the <u>usual definition for</u> the term "S operator" as used in QFT. Infinite time and space, along with time ordering, are assumed for the more general case of arbitrary time and space limits, in what we have been calling S_{oper}. So a big difference between S_{oper} and S is that time integration and volume of interaction for the latter are always infinite. From now on, when we use the term "S operator", unless otherwise stated, we will mean (7-68).

<u>2nd NOTE:</u> Similarly, the S matrix elements S_{fi} in the literature usually correspond to the values for infinite time duration and infinite volume of the interaction. For us, they could represent either case, of finite time/volume or infinite time/volume, but from here on, unless otherwise specified, they will correspond to the infinite time/volume case.

Also note the symbols we can use for the terms in (7-68)

Symbols $S^{(n)}$ used for terms in Dyson expansion

$$ S = \underbrace{I}_{S^{(0)}} \underbrace{-i \int_{-\infty}^{\infty} \mathcal{H}_I^I(x_1) d^4x_1}_{S^{(1)}} \underbrace{-\frac{1}{2!} \int_{-\infty}^{\infty} \int_{-\infty}^{\infty} T\left\{ \mathcal{H}_I^I(x_1) \mathcal{H}_I^I(x_2) \right\} d^4x_1 d^4x_2}_{S^{(2)}} + \dots = \sum_{n=0}^{\infty} S^{(n)} \quad .(7\text{-}69) $$

7.7 Wick's Theorem Applied to Dyson Expansion

7.7.1 Background for Wick's Theorem

Time ordering is hard mathematically; normal ordering is easier

Time ordering, as we have in (7-68) (or equivalently, (7-69)), is cumbersome to handle mathematically, because we can't keep the same order of operators throughout the integration over time. Fortunately, we can convert S into non-time ordered form, where the order of operators does not change during integration, via a handy theorem developed by Gian-Carlo Wick.

Wick's theorem can convert Dyson time order to normal order

Wick's theorem converts time ordered products of operators into normal ordered products of operators (see Chap. 3, pgs. 60-61) and some things called "contractions".
<u>Definition:</u>
For generic fields A and B (either of which could be ϕ, ψ, A^μ, $\phi^\dagger$, etc) where

$$A = \underbrace{A^+}_{\substack{\text{destruc} \\ \text{oper}}} + \underbrace{A^-}_{\substack{\text{creation} \\ \text{oper}}} = A^d + A^c \qquad B = \underbrace{B^+}_{\substack{\text{destruc} \\ \text{oper}}} + \underbrace{B^-}_{\substack{\text{creation} \\ \text{oper}}} = B^d + B^c ,$$
(7-70)

a <u>contraction</u> (note the square edged under bracket symbol) is defined here (we have chosen the simplest of several equivalent ways to define the contraction) as

*"Contraction"
defined*

$$\underline{A(x_1)B(x_2)} = \left[A^+(x_1), B^-(x_2) \right]_{\mp} = \left[A^d(x_1), B^c(x_2) \right]_{\mp} \qquad \text{if } t_2 < t_1$$

$$= \pm \left[B^+(x_2), A^-(x_1) \right]_{\mp} = \pm \left[B^d(x_2), A^c(x_1) \right]_{\mp} \qquad \text{if } t_1 < t_2 .$$
(7-71)

The plus sign subscript implies anti-commutation, which is used if both A and B are fermionic. The $\pm$ in front of the relations on the 2nd row takes a plus sign for commutation. a minus for anti-commutation.

Note that for fields, as we have seen from the beginning, all commutators/anti-commutators of (7-71) are zero unless $A = \phi$ and $B = \phi^\dagger$, or $A = \psi$ and $B = \bar\psi$, etc. For the non-zero relations, let's first review our derivation of the propagator.

<u>Review of the Propagator</u>

In Chap. 3, (3-144) and (3-145), pg. 77, we summarized the scalar Feynman propagator derivation first four steps as, for a particle (where $x \to x_1$ and $y \to x_2$),

$$i\Delta_F(x_1 - x_2) = \langle 0 | T\{\phi(x_1)\phi^\dagger(x_2)\} | 0 \rangle = \langle 0 | \phi(x_1)\phi^\dagger(x_2) | 0 \rangle \quad \text{if } t_2 < t_1 \text{ (particle)}$$

$$= \langle 0 | \underbrace{\left[\phi^+(x_1), \phi^{\dagger-}(x_2) \right]}_{\text{a number}} | 0 \rangle = \underbrace{\left[\phi^+(x_1), \phi^{\dagger-}(x_2) \right]}_{\text{same number}} = \underbrace{i\Delta^+(x_1 - x_2)}_{\leftarrow \text{ symbol for}}$$
(7-72)

$$= \frac{1}{2(2\pi)^3} \int \frac{e^{-ik(x_1 - x_2)}}{\omega_{\mathbf{k}}} d^3\mathbf{k} = \frac{-i}{(2\pi)^4} \int_{C^+} \frac{e^{-ik(x_1 - x_2)}}{k^2 - \mu^2} d^4k ,$$

*Review of
propagator
derivation
for $t_2 < t_1$*

and for an anti-particle,

$$i\Delta_F(x_1 - x_2) = \langle 0 | T\{\phi(x_1)\phi^\dagger(x_2)\} | 0 \rangle = \langle 0 | \phi^\dagger(x_2)\phi(x_1) | 0 \rangle \quad \text{if } t_1 < t_2 \text{ (anti-particle)}$$

$$= \langle 0 | \underbrace{\left[\phi^{\dagger+}(x_2), \phi^-(x_1) \right]}_{\text{a number}} | 0 \rangle = \underbrace{\left[\phi^{\dagger+}(x_2), \phi^-(x_1) \right]}_{\text{same number}} = \underbrace{i\Delta^-(x_1 - x_2)}_{\leftarrow \text{symbol for}}$$
(7-73)

$$= \frac{1}{2(2\pi)^3} \int \frac{e^{ik(x_1 - x_2)}}{\omega_{\mathbf{k}}} d^3\mathbf{k} = \frac{i}{(2\pi)^4} \int_{C^-} \frac{e^{-ik(x_1 - x_2)}}{k^2 - \mu^2} d^4k .$$

*Review of
propagator
derivation
for $t_1 < t_2$*

The fifth and final derivation step combined the final parts of (7-72) and (7-73) into a single relation (see (3-143), pg. 77)

$$i\Delta_F(x_1 - x_2) = \frac{i}{(2\pi)^4} \int_{-\infty}^{+\infty} \frac{e^{-ik(x_1 - x_2)}}{k^2 - \mu^2 + i\varepsilon} d^4k .$$
(7-74)

*Review of
propagator
derivation for
either time
order*

<u>End of Propagator Review</u>

Note that (7-71) has the same form as the middle term of the middle row for both (7-72) and (7-73). Thus, whenever a contraction (7-71) of scalar fields is non-zero, it is a scalar Feynman propagator. Similar results hold for spinor and photon fields.

Thus, <u>special cases of contractions</u> (the only non-zero cases) are

*Contractions are
zero except when
they equal a
Feynman
propagator*

$$\underline{\phi(x_1)\phi^\dagger(x_2)} = \underline{\phi^\dagger(x_2)\phi(x_1)} = i\Delta_F(x_1 - x_2)$$

$$\underline{\psi_\alpha(x_1)\bar\psi_\beta(x_2)} = -\underline{\bar\psi_\beta(x_2)\psi_\alpha(x_1)} = iS_{F\alpha\beta}(x_1 - x_2)$$
(7-75)

$$\underline{A^\mu(x_1)A^\nu(x_2)} = iD_F^{\mu\nu}(x_1 - x_2).$$

Review of Normal Ordering

Recall from our earlier work in Chap. 3 (Sect. 3.6.6, pgs. 60-61) that normal ordering consists of placing all destruction operators to the right-hand side inside a given term. I noted in that chapter that I consider the application of normal ordering there to eliminate the ½ quanta terms in the vacuum *ad hoc*, but in the present application, it arises naturally as a part of the mathematics of Wick's theorem, as we are about to see. Symbolically, we use the symbol N to indicate normal ordering, i.e.,

$$N(ABCD...) = \text{all destruction operators placed to right of all creation operators} . \quad (7\text{-}76)$$

Review of normal ordering, placing destruction operators all on right

As part of our definition of normal ordering, note that the <u>switching</u> places of two <u>adjacent fermionic fields</u> gives rise to a <u>sign change</u>. This can be justified, in part, because fermionic fields obey anti-commutation relations. For example[1], given $[C,D]_+ = 0$, $CD = -DC$.

Interchanging two adjacent fermions in normal ordering changes sign

Non-adjacent Contraction Operators

We will encounter terms like $A(x_1)B(x_2)C(x_3)$, where an operator (B in this example) is sandwiched between two operators that form a contraction. Because contractions are Feynman propagators they are c-numbers and not operators (even though they can be expressed, as in the LHS of (7-75), in terms of two operator fields.) Hence in a term they share with operator fields, contractions can be placed anywhere without affecting the term. That is, contractions commute with operators. For example,

$$A(x_1)B(x_2)C(x_3) = A(x_1)C(x_3)B(x_2) = B(x_2)A(x_1)C(x_3). \quad (7\text{-}76)+1$$

2nd edition, this section inserted: Non-adjacent contraction operators

However, in what is about to come, it will prove efficient for our symbolism if we define <u>normal ordering of terms including contractions</u> so that the following holds true.

For B a destruction operator $\;\rightarrow\; N\left\{A(x_1)B^d(x_2)C(x_3)\right\} = \pm A(x_1)C(x_3)B^d(x_2)$

minus sign for C and B^d
both fermionic

$$(7\text{-}76)+2$$

For B a creation operator $\;\;\rightarrow\; N\left\{A(x_1)B^c(x_2)C(x_3)\right\} = \pm B^c(x_2)A(x_1)C(x_3)$

minus sign for A and B^c
both fermionic

Definition of normal ordering for non-adjacent contraction operators

In other words, normal ordered terms with contractions in them will take on whatever sign would occur if we treated the contraction as two operators rather than a c-number, and exchanged operators to put creation operators on the LHS and destruction operators on the RHS of the contraction.

This definition extends to terms with more than one operator (such as B above) and more than one contraction. Simply make the successive exchanges needed to get all operators originally sandwiched inside the contraction outside of it, and count the sign changes needed for those exchanges. Only fermions exchanged with adjacent fermions entail a sign change.

Notation Modification

We will henceforth use subscripts in place of parenthesis arguments for fields, i.e.,

$$A(x_1) \rightarrow A_{x_1} \qquad A(x_1)B(x_1)C(x_1) \rightarrow (ABC)_{x_1} \qquad \psi(x_1)\gamma^\mu A_\mu(x_1)\bar{\psi}(x_1) \rightarrow \left(\psi\gamma^\mu A_\mu\bar{\psi}\right)_{x_1} . \quad (7\text{-}77)$$

Streamlining notation

---

[1] One might wonder about the cases where we switch positions of particular adjacent fermion operator fields whose anti-commutator does not equal zero. Things still work out if we follow the same procedure for this case, as well, but it is more complicated. We address this issue in Sect. 7.8.

7.7.2 Wick's Theorem

We only state Wick's theorem here and leave the justification of it for Sect. 7.8. The theorem turns a time ordered product into a series of normal ordered terms and contractions, which, as we noted, helps us in calculating interaction probabilities, because we can keep the operators in the same order as we integrate over time.

Wick's Theorem is

Wick's theorem → turning time ordering into normal ordering plus contractions

Justification to follow

$$T\left\{(AB...)_{x_1} ....(AB...)_{x_n}\right\} = N\left\{(AB...)_{x_1} ....(AB...)_{x_n}\right\}$$
$$+ N\left\{\underline{(AB..)_{x_1}(AB..)_{x_2}} ...\right\} + N\left\{\underline{(AB..)_{x_1}(AB..)_{x_2}} ..\right\} + ... + N\left\{... \underline{(A...Z)_{x_{n-1}}(A...Z)_{x_n}}\right\}$$
$$+ N\left\{(ABC...)_{x_1}(ABC...)_{x_2} ...\right\} + N\left\{(ABC...)_{x_1}(ABC...)_{x_2} ...\right\} + ...$$
$$+ \text{(all normal ordered terms with three non-equal times contractions)}$$
$$+ \text{ etc.}$$

(7-78)

Note that there are no contractions between operators operating at the same time. We say there are no "equal times contractions" in Wick's theorem. We will discuss equal times contractions in Sect. 7.8.4

No contractions for operators operating at same time

Note on Time Ordering: Just as in normal ordering, if we switch positions of adjacent fermions in time ordering, we get a sign change as well.

As an aside, aren't you glad now that we derived the Feynman propagator in Chap 3 (and Chaps. 4 and 5), so we didn't have to detour here through all those pages of sticky math before getting to (7-71), (7-75), and (7-78).

7.7.3 Wick's Theorem and QED

In QFT, we take a given $\mathcal{H}_I{}^I$ (from e/m, weak, or strong theory) in Dyson's expansion (7-69), then convert that via Wick's theorem (7-78) to manageable form. In the next chapter we will apply that procedure to electromagnetism by using (7-78) to re-express the integrands of (7-69) where

$\mathcal{H}_I{}^I$ → Dyson expansion → Wick's theorem

$$\left(AB...\right)_{x_1} = \mathcal{H}_I^I\left(x_1\right) = -e\left(\bar{\psi}\gamma^\mu A_\mu \psi\right)_{x_1} \quad \left(AB...\right)_{x_2} = \mathcal{H}_I^I\left(x_2\right) = -e\left(\bar{\psi}\gamma^\mu A_\mu \psi\right)_{x_2} \quad \text{etc.} \quad (7\text{-}79)$$

Form of $\mathcal{H}_I{}^I$ used in QED

The result will be QED.

7.8 Justifying Wick's Theorem

7.8.1 A Key Issue

In both time ordering and normal ordering, the assumption is made that we can simply interchange operator order (with a sign change for two fermions). This assumes commutation (two bosons or a boson and a fermion) or anti-commutation (two fermions) always holds.

BUT, with N and T, how can we simply commute (or anti-commute) operators that don't commute (anti-commute)?

Yet, as I noted in Chap. 3, when normal ordering was discussed there, doing so flies in the face of the very foundation of QFT itself, i.e., that certain operators don't commute (or for fermions, anti-commute). In fact, even in this chapter, our definition of the contraction (7-71) implies non-zero commutation/anti-commutation for certain fields. So, it may seem at first blush, that in the customary approach using Wick's theorem (7-78), we blithely assume it is non-zero in some places and zero in others.

7.8.2 So What is Really Going On?

The Short Answer

If we start with any order of operators $A_1B_2C_3D_4...$, we can re-order them any way we like, as long as, when we do, we are sure to use appropriate commutation and (for two adjacent fermions) anti-commutation relations, which can be non-zero in some cases. The result for any such procedure equals the same thing as the originally ordered relation, i.e.,

Because if we use the full commutation (anti-commutation) relations to re-order we derive Wick's theorem

$$\begin{pmatrix} \text{Original} \\ \text{order} \end{pmatrix} = A_1 B_2 C_3 D_4 \ldots = \begin{pmatrix} \text{Re-ordered timewise} \\ \text{using com/anti-com rels} \end{pmatrix} = \begin{pmatrix} \text{Re-ordered, destruc opers on} \\ \text{right using com/anti-com rels} \end{pmatrix}. \quad (7\text{-}80)$$

If we define the <u>symbol T_c</u> to represent time ordering via substitution of appropriate commutation (anti-commutation for fermions) relations, and the <u>symbol N_c</u> to represent an ordering of operators with all destruction operators to the right of all construction operators via the same substitution of commutators/anti-commutators method, then we can express (7-80) as

$$A_1 B_2 C_3 D_4 \ldots = T_c \left\{ A_1 B_2 C_3 D_4 \ldots \right\} = N_c \left\{ A_1 B_2 C_3 D_4 \ldots \right\}. \quad (7\text{-}81)$$

Re-ordering a term with T_c and N_c (using com/anti-com relations) doesn't change the term

We state (justification to follow) that this reduces to Wick's theorem (7-78) with the <u>time ordering operator T</u> simply a time ordering of operators assuming all commutators and anti-commutators are zero (so switching positions of adjacent bosons is always OK, switching positions of adjacent fermions always simply results in a sign change, and switching positions of a fermion and boson is always OK, as well.) Similarly, the <u>normal ordering operator N</u> entails the same switching method to get destruction operators all to the right of all creation operators. Thus, Wick's theorem for operators $A_1 B_2 C_3 D_4 \ldots$ is

$$T\left\{ A_1 B_2 C_3 D_4 \ldots \right\} = N\left\{ A_1 B_2 C_3 D_4 .. \right\} + N\left\{ A_1 B_2\ C_3 D_4 \ldots \right\} + N\left\{ A_1 B_2 C_3\ D_4 .. \right\} + \ldots$$

$$+ N\left\{ A_1 B_2\ C_3 D_4 .. \right\} + N\left\{ A_1 B_2 C_3 D_4 .. \right\} + \ldots \quad (7\text{-}82)$$

But such re-ordering gives rise to the Wick theorem expressed via T and N operations

+ (all normal ordered terms with three contractions) + etc.

In other words, for Wick's theorem application, we assume for bosonic fields A_1 and B_2, that $A_1 B_2 = B_2 A_1$, even if the two fields don't commute. Similarly, for two fermionic fields C_3 and D_4, we assume $C_3 D_4 = -D_4 C_3$, even if the two don't anti-commute. The theorem is structured so that the result is the same as one would get by using the method of (7-81).

<u>A Boson Example</u>

To help make this clearer, consider the simple example of only two fields in (7-81) where we take $A_1 = \phi(x_1)$ and $B_2 = \phi^\dagger(x_2)$. We choose these because we know they don't commute, i.e., $[\phi(x_1), \phi^\dagger(x_2)] \neq 0$, since the operators $a(\mathbf{k})$ and $a^\dagger(\mathbf{k})$ (and $b(\mathbf{k})$ and $b^\dagger(\mathbf{k})$) are involved, and they don't commute. So, $\phi(x_1)\phi^\dagger(x_2) \neq \phi^\dagger(x_2)\phi(x_1)$, or with the original symbols, $A_1 B_2 \neq B_2 A_1$. So, one would presume, if the theory is consistent, that we can't simply make the exchange $A_1 B_2 \rightarrow B_2 A_1$, as if they did commute.

Start to justify Wick's theorem with an example: two bosons we are familiar with

With this choice of fields, (7-81) becomes (see the first page of Wholeness Chart 5-4 at the end of Chap. 5 and note that if you haven't locked it into your memory yet that superscript minus sign on a field designates construction, and a positive sign designates destruction, then it is time to do so)

Time to lock into memory that script $+$ on a field means destruction, $-$ means construction

$$A_1 B_2 = \phi(x_1)\phi^\dagger(x_2) = \left(\phi^+(x_1) + \phi^-(x_1) \right)\left(\phi^{\dagger+}(x_2) + \phi^{\dagger-}(x_2) \right)$$

$$= \phi^+(x_1)\phi^{\dagger+}(x_2) + \phi^+(x_1)\phi^{\dagger-}(x_2) + \phi^-(x_1)\phi^{\dagger+}(x_2) + \phi^-(x_1)\phi^{\dagger-}(x_2). \quad (7\text{-}83)$$

Thus, let's set this equal to both the right and left hand sides of an equation, as in (7-80), then re-order the LHS with lower times further to the right and the RHS with destruction operators all on the right. We <u>use the commutation relations</u> and do not assume, as in (7-82) that the fields simply commute.

$$T_c\left\{ A_1 B_2 \right\} = N_c\left\{ A_1 B_2 \right\} \rightarrow T_c\left\{ \phi(x_1)\phi^\dagger(x_2) \right\} = N_c\left\{ \phi(x_1)\phi^\dagger(x_2) \right\} \quad \text{or} \downarrow$$

$$T_c\left\{ \underbrace{\phi^+(x_1)\phi^{\dagger+}(x_2) + \phi^+(x_1)\phi^{\dagger-}(x_2) + \phi^-(x_1)\phi^{\dagger+}(x_2) + \phi^-(x_1)\phi^{\dagger-}(x_2)}_{\text{re-order with earlier time on right using commutation relations}} \right\} = \quad (7\text{-}84)$$

Re-express same two boson relation using T_c and N_c

$$N_c\left\{ \underbrace{\phi^+(x_1)\phi^{\dagger+}(x_2)}_{\text{2 destruction operators}} + \underbrace{\phi^+(x_1)\phi^{\dagger-}(x_2)}_{\substack{\text{only term without destruction} \\ \text{operator on right}}} + \underbrace{\phi^-(x_1)\phi^{\dagger+}(x_2)}_{\substack{\text{destruction operator} \\ \text{on right}}} + \underbrace{\phi^-(x_1)\phi^{\dagger-}(x_2)}_{\text{2 creation operators}} \right\}.$$

<u>Case 1</u> $t_2 < t_1$

For $t_2 < t_1$, the LHS (2nd row) of (7-84) is already in time order. The only term in the RHS (bottom row) that doesn't have destruction operators on the right is the second term. Those two fields don't commute, i.e.,

That relation for $t_2 < t_1$

$$\phi^+(x_1)\phi^{\dagger -}(x_2) - \phi^{\dagger -}(x_2)\phi^+(x_1) = \left[\phi^+(x_1), \phi^{\dagger -}(x_2)\right] \underbrace{= i\Delta^+(x_1 - x_2)}_{\substack{\text{an aside, don't worry} \\ \text{about for now}}} \neq 0, \quad (7\text{-}85)$$

where we reference the middle row of (7-72) for the meaning of the $i\Delta^+$ symbol. Some readers report it is easier to follow the derivation the first time through if one ignores the Feynman propagator equivalence and focuses on the commutator form.

Finding $\phi^+(x_1)\phi^{\dagger -}(x_2)$ from (7-85) and using it for the 2nd term in the bottom row of (7-84), we find (7-84) becomes

$$\underbrace{\phi^+(x_1)\phi^{\dagger +}(x_2) + \phi^+(x_1)\phi^{\dagger -}(x_2) + \phi^-(x_1)\phi^{\dagger +}(x_2) + \phi^-(x_1)\phi^{\dagger -}(x_2) =}_{=\phi(x_1)\phi^\dagger(x_2) = A_1 B_2 \text{ in time order}}$$

$$\phi^+(x_1)\phi^{\dagger +}(x_2) + \underbrace{\phi^{\dagger -}(x_2)\phi^+(x_1) + \left[\phi^+(x_1), \phi^{\dagger -}(x_2)\right]}_{\text{only term we changed}} \qquad (7\text{-}86)$$

$$\underbrace{+ \phi^-(x_1)\phi^{\dagger +}(x_2) + \phi^-(x_1)\phi^{\dagger -}(x_2)}_{\substack{\text{all non-contraction terms have destruction operators, if any, on the right,} \\ \text{re-ordered using non-zero commutation relations}}}$$

From (7-72) again, and the top row of (7-71),

A contraction term arises from this procedure

$$\left[\phi^+(x_1), \phi^{\dagger -}(x_2)\right] = \overset{\text{ignore for now}}{i\Delta^+(x_1 - x_2)} = \overset{\text{ignore for now}}{i\Delta_F(x_1 - x_2)} = \underbrace{\phi(x_1)\overline{\phi^\dagger(x_2)}}_{\text{for } t_2 < t_1} \qquad (7\text{-}87)$$

(7-86) was obtained by holding fast to the commutation relations of QFT. Using (7-87), and noting that a contraction is just a number, so normal ordering is not an issue, we can re-write (7-86) as

Result for this case: a relation in T and N plus a contraction term

$$\left. \begin{array}{l} \underbrace{T\{\phi(x_1)\phi^\dagger(x_2)\}}_{\substack{\text{time ordered assuming} \\ \text{all operators commute}}} = \underbrace{N\{\phi(x_1)\phi^\dagger(x_2)\}}_{\substack{\text{normal ordered assuming} \\ \text{all operators commute}}} + \underbrace{\left[\phi^+(x_1), \phi^{\dagger -}(x_2)\right]}_{\substack{\text{a number, no operators, so} \\ \text{essentially already normal ordered}}} \\[2em] = N\{\phi(x_1)\phi^\dagger(x_2)\} + \phi(x_1)\overline{\phi^\dagger(x_2)} \end{array} \right\} \; t_2 < t_1. \quad (7\text{-}88)$$

where the N normal ordering and T time ordering in (7-88) means we assume for that ordering process that all commutators are zero so we can simply switch orders of adjacent bosons.

<u>Case 2</u> $t_1 < t_2$

Now, take (7-84) where $t_1 < t_2$. The RHS will remain the same and equals the last two lines of (7-86). We do need, however, to re-express every term in the middle line of (7-84).

$$T_c\left\{ \underbrace{\phi^+(x_1)\phi^{\dagger +}(x_2) + \phi^+(x_1)\phi^{\dagger -}(x_2) + \phi^-(x_1)\phi^{\dagger +}(x_2) + \phi^-(x_1)\phi^{\dagger -}(x_2)}_{\text{re-order with earlier time } t_1 \text{ on right using commutation relations}} \right\} =$$

That same relation for $t_1 < t_2$

$$T_c\left\{ \underbrace{\phi^+(x_1)\phi^{\dagger +}(x_2)}_{\substack{\text{2 destruction} \\ \text{operators commute}}} + \underbrace{\phi^+(x_1)\phi^{\dagger -}(x_2)}_{\substack{\text{need to use non-zero} \\ \text{commutation relation}}} + \underbrace{\phi^-(x_1)\phi^{\dagger +}(x_2)}_{\substack{\text{need to use non-zero} \\ \text{commutation relation}}} + \underbrace{\phi^-(x_1)\phi^{\dagger -}(x_2)}_{\substack{\text{2 creation} \\ \text{operators commute}}} \right\} = \qquad (7\text{-}89)$$

$$\underbrace{N_c\{\phi(x_1)\phi^\dagger(x_2)\}}_{\text{same as } t_2 < t_1 \text{ case}}$$

Note that for the first and last terms in the middle row, the fields commute. We can re-express the 2nd term using (7-85). We note the fields in the 3rd term do not commute, i.e.,

$$\phi^-(x_1)\phi^{\dagger+}(x_2) = \phi^{\dagger+}(x_2)\phi^-(x_1) - \left[\phi^{\dagger+}(x_2),\phi^-(x_1)\right]. \tag{7-90}$$

Substituting (7-90) and (7-85) into the middle row of (7-89) and the RHS of (7-88) into the lowest row, we find (7-89) becomes

$$T_c\left\{\phi(x_1)\phi^\dagger(x_2)\right\} = \phi^{\dagger+}(x_2)\phi^+(x_1) + \underbrace{\phi^{\dagger-}(x_2)\phi^+(x_1) + \left[\phi^+(x_1),\phi^{\dagger-}(x_2)\right]}_{\text{2nd term}}$$

$$+\underbrace{\phi^{\dagger+}(x_2)\phi^-(x_1) - \left[\phi^{\dagger+}(x_2),\phi^-(x_1)\right]}_{\text{3rd term}} + \phi^{\dagger-}(x_2)\phi^-(x_1) \tag{7-91}$$

$$= N_c\left\{\phi(x_1)\phi^\dagger(x_2)\right\} = N\left\{\phi(x_1)\phi^\dagger(x_2)\right\} + \left[\phi^+(x_1),\phi^{\dagger-}(x_2)\right].$$

The bracket commutators in the first and last rows cancel. The negative bracket in the middle line can be moved to the RHS (the normal order side), yielding

$$\phi^{\dagger+}(x_2)\phi^+(x_1) + \phi^{\dagger-}(x_2)\phi^+(x_1) + \phi^{\dagger+}(x_2)\phi^-(x_1) + \phi^{\dagger-}(x_2)\phi^-(x_1)$$

$$= N\left\{\phi(x_1)\phi^\dagger(x_2)\right\} + \left[\phi^{\dagger+}(x_2),\phi^-(x_1)\right] \qquad \text{for } t_1 < t_2. \tag{7-92}$$

So (where we use the middle line of (7-73) for the Feynman propagator equivalence that we are ignoring on the first look at this derivation), from the bottom row of (7-71),

$$\left[\phi^{\dagger+}(x_2),\phi^-(x_1)\right] = \overbrace{i\Delta^-(x_1-x_2)}^{\text{ignore for now}} = \overbrace{i\Delta_F(x_1-x_2)}^{\text{ignore for now}} = \underbrace{\phi(x_1)\phi^\dagger(x_2)}_{\text{for } t_1<t_2}, \tag{7-93}$$

A contraction term arises in this case too

and (7-92) becomes

$$\underbrace{T\left\{\phi(x_1)\phi^\dagger(x_2)\right\}}_{\substack{\text{time ordered assuming}\\\text{all operators commute}}} = \underbrace{N\left\{\phi(x_1)\phi^\dagger(x_2)\right\}}_{\substack{\text{normal ordered assuming}\\\text{all operators commute}}} + \underline{\phi(x_1)\phi^\dagger(x_2)} \quad \text{for } t_1 < t_2. \tag{7-94}$$

Same result for this case of reverse time order

Comparing (7-94) and (7-88), we see that the same relation holds for either ordering.

A Fermion Example

By doing Probs. 12 and 13, you can prove to yourself that the same general relation

$$T\left\{\psi(x_1)\bar\psi(x_2)\right\} = N\left\{\psi(x_1)\bar\psi(x_2)\right\} + \underline{\psi(x_1)\bar\psi(x_2)} \quad \text{for } t_1 < t_2 \text{ and } t_2 < t_1 \tag{7-95}$$

Same result for two fermions

holds for two fermions as well.

One Boson and One Fermion Example

By doing Prob. 14, you can prove to yourself, knowing that the commutator of a boson and fermion is always zero, that

$$T\left\{\phi(x_1)\bar\psi(x_2)\right\} = N\left\{\phi(x_1)\bar\psi(x_2)\right\} + \underbrace{\phi(x_1)\bar\psi(x_2)}_{=0} \quad \text{for } t_1 < t_2 \text{ and } t_2 < t_1. \tag{7-96}$$

Same result for one fermion and one boson

Other Combinations

Photon and scalar fields commute, as do neutrino and electron fields. So the general relation (7-96) holds for all other combinations of fields, where the fields are of different types.

Same result for different type bosons/fermions

Bottom Line

We have proven Wick's theorem for two fields of any types for any time order. For generic fields A_1 and B_2, it is

$$T\{A_1 B_2\} = N\{A_1 B_2\} + A_1 B_2 \quad \text{for any time order}. \tag{7-97}$$

Result: Wick's theorem proven for two fields

7.8.3 More Than Two Fields

Hopefully, we have gained some comfort with Wick's theorem by working through the examples with only two fields. And perhaps, we can simply accept that the mathematicians have proven Wick's theorem formally, for us. Much like integral tables, which we employ regularly without proving each relation we use, we can simply accept Wick's theorem, apply it to our work at hand, and move on.

Three fields treated, and extended by induction to more fields, in Appendix A

Those wishing to dig deeper and understand a bit better can read Appendix A where we extend the above type of analysis to three fields. In that appendix, we then use induction to justify Wick's theorem for any number of fields.

For those who feel the need for a formal proof, see the original article "The Evaluation of the Collision Matrix" by G. C. Wick (Phys. Rev. **80**, 268, 1950), or any of "Notes on Wick's Theorem in Many-Body Theory" by Luca Guido Molinari (wwwteor.mi.infn.it/~molinari/NOTES/Wick.pdf), *Quantum Field Theory for Mathematicians* by R. Ticciati (Cambridge University Press 1999, pg. 85-87), and *Field Quantization* by W. Greiner and J. Reinhardt (Springer, 1966, pg. 231-233).

7.8.4 The Issue of Equal Time Operators

<u>Turning Our Different Times Relation into One with Some Equal Times</u>

Readers may have noticed that (7-82) seemed to be stated for fields $A_1 B_2 C_3 D_4...$ where times t_1 of field A_1, t_2 of field B_2, t_3 of C_3, etc. are all different (none are the same time.) In contrast, our statement of Wick's theorem (7-78) was more general in the sense that it has several fields at the same time, such as $A_1 B_1 C_1 D_1... = (ABCD...)_{x_1}$ all at the same time, and $A_2 B_2 C_2 D_2... = (ABCD...)_{x_2}$ all at the same, but different from t_1, time.

Operators at same time seemingly not treated in above

We can generalize (7-82) by taking, for example, $t_1 = t_2$, so that $A_1 B_2 \rightarrow A_1 B_1$. That is, wherever we have different fields in (7-82), we can just assume some have equal times. We should thus be able to derive Wick's theorem (7-78) entailing more than one field at the same time from our relation (7-82). We would find (7-82) then looks like

But can generalize by taking $t_2 = t_1$, for example

$$T\{A_1 B_1 C_1 .. F_2 G_2 ..\} = N\{A_1 B_1 C_1 .. F_2 G_2 ..\} + N\left\{ A_1 B_1 C_1 .. F_2 G_2 .. \right\} + N\left\{ A_1 B_1 C_1 .... F_2 G_2 .. \right\} + ..$$

$$+ N\left\{ A_1 B_1 C_1 .. F_2 G_2 .. \right\} + N\left\{ A_1 B_1 C_1 .. F_2 G_2 .. \right\} + ... \tag{7-98}$$

+ (all normal ordered terms with three contractions) + etc.,

which with slightly different notation looks a lot like Wicks' theorem (7-78).

<u>The Fly in the Ointment</u>

The one difference between (7-98) and Wicks' theorem (7-78) is that the latter (7-78) has no equal times contractions, whereas the former (7-98) does. How do we resolve this?

But then we get equal times contractions, which are not in Wick's theorem

<u>Resolving the Fly in the Ointment</u>

i) The Traditional Resolution: Normal Ordering in Interaction Hamiltonian

In traditional QFT, we apply Wick's theorem using $\mathcal{H}_I^I$, which for QED takes the form (7-79), which we repeat below.

Many treatments resolve this by assuming $\mathcal{H}_I^I$ is already normal ordered

$$(AB...)_{x_1} = \mathcal{H}_I^I(x_1) = -e\left(\bar{\psi} \gamma^\mu A_\mu \psi \right)_{x_1} \qquad (AB...)_{x_2} = \mathcal{H}_I^I(x_2) = -e\left(\bar{\psi} \gamma^\mu A_\mu \psi \right)_{x_2} \quad \text{etc.} \tag{7-99}$$

In that approach it is common to assume the fields in each of $\mathcal{H}_I^I(x_1)$, $\mathcal{H}_I^I(x_2)$, etc are already normal ordered. That is,

$$\mathcal{H}_I^I(x_1) = -eN\left\{ \bar{\psi} \gamma^\mu A_\mu \psi \right\}_{x_1} \qquad \mathcal{H}_I^I(x_2) = -eN\left\{ \bar{\psi} \gamma^\mu A_\mu \psi \right\}_{x_2} \quad \text{etc.} \tag{7-100}$$

If that is so, then all equal time contractions on the RHS of (7-98) are zero, since each is arrived at by re-ordering the fields $\psi, \bar{\psi}, A_\mu$ for each $\mathcal{H}_I^I(x_i)$ so they are normal ordered. But if they are already normal ordered, no such re-ordering is required, and we have no equal times contractions.

But, as I've said before, normal ordering assumes all fields commute (or for fermions, anti-commute), and since QFT is grounded in, and only exists because of, non-commutation (non-anti-commutation) relations, there seems to be an inconsistency. So, I prefer the following resolution.

ii) Another Resolution without Invoking Normal Ordering in $\mathcal{H}_I{}^I$:

Consider three fields (like $\bar{\psi}, A_\mu, \psi$) operating at the same time that we will label A_1, B_1, C_1, each composed of a construction plus a destruction operator. Superscripts c,d imply construction and destruction, respectively. With our N_c and T_c reordering, one such component of $A_1 B_1 C_1$ yields

$$N_c\left\{A_1^c B_1^d C_1^d\right\} = A_1^c B_1^d C_1^d = N\left\{A_1^c B_1^d C_1^d\right\} = T_c\left\{A_1^c B_1^d C_1^d\right\} = A_1^c B_1^d C_1^d = T\left\{A_1^c B_1^d C_1^d\right\}$$
$$T\left\{A_1^c B_1^d C_1^d\right\} = N\left\{A_1^c B_1^d C_1^d\right\} \quad \rightarrow \text{ no equal time commutator (contraction) in Wick theorem.} \tag{7-101}$$

But this may contradict basic postulates of QFT

Other resolution for equal times contractions = 0

Another component, where we note that if all operators operate at the same time, we can time re-order them any way we like (as long as we include the proper commutation relation), yields

$$N_c\left\{A_1^d B_1^c C_1^d\right\} = B_1^c A_1^d C_1^d + \left[A_1^d, B_1^c\right]C_1^d = N\left\{A_1^d B_1^c C_1^d\right\} + \left[A_1^d, B_1^c\right]C_1^d$$
$$T_c\left\{A_1^d B_1^c C_1^d\right\} = \underbrace{B_1^c A_1^d C_1^d}_{\substack{\text{can time order}\\\text{any way we like}}} + \underbrace{\left[A_1^d, B_1^c\right]C_1^d}_{\substack{\text{as long as we}\\\text{include commutator}}} = T\left\{B_1^c A_1^d C_1^d\right\} + \left[A_1^d, B_1^c\right]C_1^d$$
$$= T\left\{A_1^d B_1^c C_1^d\right\}$$

$\left.\begin{array}{l}\text{first line}\\\text{equal}\\\text{to second}\end{array}\right\}$ (7-102)

$$T\left\{A_1^d B_1^c C_1^d\right\} = N\left\{A_1^d B_1^c C_1^d\right\} \quad \rightarrow \text{ no equal time commutator (contraction) in Wick theorem.}$$

For simplifying choice for order of equal time factors, equal times contractions cancel out, and we can leave them out of Wick's theorem

Repeating for each component of $A_1 B_1 C_1$, we can always choose the time order T we want, since all operators operate at the same time. With the right choice, we get a commutation relation on one side of the equation that cancels with one on the other. Parallel logic holds for fermions/anti-commutators.

This choice of time ordering results in the simplest form for our theory (always the preferred starting point in any theory development) and as we shall see, correctly predicts experiment.

Thus, $N\{A_1 B_1 C_1\} = T\{A_1 B_1 C_1\} \rightarrow$ no equal time commutator (contraction) in Wick theorem. (7-103)

To those who might contend that the above is simply sleight-of-hand use of normal ordering, I reply that equal-time commutators, if included anyway, can either be exorcised from QED via procedures beyond the scope of this book and/or lead to non-physical situations. For example, conservation of 4-momentum in certain associated interactions would only be possible for particles having zero energy, i.e., for particles that do not exist, and thus can be ignored. We will see this in the appendix of Chap. 8.

If we included them anyway, they would lead to non-physical situations (or could otherwise be cleansed from the theory of QED)

The bottom line: Equal-time contractions don't play a role in Wick's theorem (7-78) for QED.

7.8.5 Summary of Wick's Theorem

To get Wick's theorem, we start with a series of operator fields, operating in arbitrary order and set it equal to itself, i.e., $A_1 B_2 C_3 D_4 \ldots = A_1 B_2 C_3 D_4 \ldots$

On the LHS, we then re-arrange operator fields using commutation/anti-commutation relations such that earlier times are to the right of later times. We herein use the symbol T_c to represent this re-ordering procedure. The final result of the LHS equals the original LHS expression, since at each step, we simply substituted equivalent relations for the original pair of adjacent operators.

Summary of how Wick theorem arises from commutator/anti-commutator relations

On the RHS, we re-arrange operator fields using commutation/anti-commutation relations such that destruction operators are all to the right of creation operators. We herein use the symbol N_c to represent this re-ordering procedure. The final result of the RHS equals the original RHS expression. Thus, the final RHS equals the final LHS.

The final result of these operations is the same as employing Wick's theorem (7-78).

In Wick's theorem, the time ordering operation T re-orders operators with earlier times to the right of later times, but assumes we can switch orders of adjacent operators as if they commuted (or for two fermions, anti-commuted). Similarly, the normal ordering N operator re-orders with destruction operators all on the right, but assumes we can switch orders of adjacent operators as if they commuted (or for two fermions, anti-commuted).

Using the T_c and N_c procedures, we find contractions arising in the final result. Using the T and N operations, we insert those same contractions, as designated in Wick's theorem.

7.8.6 Applying Wick's Theorem

In QFT, we use Wick's theorem to re-express the Dyson expansion (7-69) in a more suitable form, more amenable to calculating S matrix values (and thus probabilities). In Dyson's expansion, we assume the T operator, not the T_C procedure, re-orders the integrands.

$\mathcal{H}_I{}^I$ in S of Dyson expansion in terms of T operator → S in terms of N and contractions in Wick's theorem

7.9 Comment on Normal Ordering of the Hamiltonian Density

Many practitioners of QFT, and many textbooks, employ normal ordering of the Hamiltonian density, both the free and interaction parts, as standard operating procedure. In carrying out such ordering, it is assumed that all possible pairs of operator fields have zero commutation (or anti-commutation) relations.

As I've noted in this chapter (interaction Hamiltonian) and in Chap. 3 (free Hamiltonian), I consider this procedure to be inconsistent with the foundational basis of QFT, which postulates the non-zero value of commutation (anti-commutation) of certain operator fields.

However, as a counter argument, it is possible that Nature has organized things such that the Hamiltonian density is actually normal ordered, "out of the gate" as it were, on the micro level. On the macro level, with many quantum fields forming a macroscopic field, the normal ordering issue is not relevant, as macroscopic fields, to high order commute. This can be seen by re-writing a typical commutation relation in non-natural units, e.g.,

$$\left[\phi^r\left(\mathbf{x},t\right), \pi_s\left(\mathbf{y},t\right) \right] = \phi^r \pi_s - \pi_s \phi^r = i\hbar \delta^r{}_s \delta\left(\mathbf{x}-\mathbf{y}\right). \qquad (7\text{-}104)$$

Non-zero commutators (anti-commutators) values are very small, ≈ 0 at macroscopic scales

In our day-to-day measurement systems, $\hbar$ is extremely small (1.055×10^{-34} joule-sec) and so this commutator (7-104) is, from a macroscopic perspective, effectively zero. (7-104) is one of the postulates of second quantization, from which all of our other non-zero commutation relations, such as those for $a(\mathbf{k})$ and $a^\dagger(\mathbf{k})$, are derived. Hidden in all of those relations, when expressed in natural units, was the fact that, from a human sized point of view, the commutators are effectively zero.

So from human perspective, all fields effectively commute (anti-commute)

And thus, for our macroscopic Hamiltonian density, we can have any order for the fields involved. The usual such order, the one we assume has the same quantum form when we quantize, is not normal ordered. But it could be, and we simply are unaware of it, because macroscopically the normal ordered and non-normal ordering Hamiltonian densities are essentially the same thing.

So $\mathcal{H}$ could be normal ordered at micro scale, but our classical theory formulation would be blind to it & have evolved in non-normal ordered form

Thus, I concede, though I am not comfortable with it, that Nature might indeed have normal ordered Hamiltonian densities at the quantum level. As I have shown in the theoretical developments in this book, however, such normal ordering is not required to develop a consistent theory.

The one possible exception to this is the prediction of a virtually infinite number ½ quanta in the vacuum as we discussed in Chap. 3. Normal ordering of the free Hamiltonian density there eliminates those pesky things, which have yet to be unequivocally measured in experiment. In a footnote in Chap. 3 (pg. 50), I referenced an article by me in which this issue may be resolved without recourse to normal ordering.

But a viable QFT can be developed without recourse to normal ordering of the Hamiltonian density

<u>Bottom line</u>: QFT can be developed without imposing normal ordering on the fields in the Hamiltonian density. It can be argued that such normal ordering is inconsistent with the commutation relation postulates upon which QFT is founded. However, it is possible that Nature may incorporate normal ordering at the quantum level, though our classical theories, from which we induce QFT, are formulated without it.

7.10 Chapter Summary

In Prob. 4, you were asked to summarize the first part of this chapter on RQM interaction theory.

The theme of the remainder of the chapter, QFT interaction theory underlying principles, is summarized in the first two pages or so of Wholeness Chart 8-4 at the end of Chap. 8. That comprises an overview of the interaction picture, the S operator, the S matrix, Dyson's expansion of the S operator, and Wick's theorem.

This chapter summary at end of Chap. 8

Note that the principles developed in this chapter are applicable to the electromagnetic, weak, and strong interactions. For the rest of this volume, we apply these principles solely to electromagnetism. That application gives us quantum electrodynamics (QED).

This chapter: general principles of interaction theory applicable to e/m, weak, strong forces

7.11 Appendix A: Justifying Wick's Theorem via Induction

7.11.1 Three Generic Fields

We used specific fields ϕ and ψ we are familiar with in Sect. 7.8.2 in place of the more generic notation A_1 and B_2. However, we now switch to the more general notation, so that we can do one example with generic fields and the results will be applicable to any types of fields.

Consider three such fields A_1, B_2, and C_3, where we reorder the RH and LH sides of the identity

$$\underbrace{A_1B_2C_3}_{\substack{\text{Re-ordered timewise} \\ \text{using com \& anti-} \\ \text{com relations}}} = \underbrace{A_1B_2C_3}_{\substack{\text{Re-ordered, destruc} \\ \text{opers on right using} \\ \text{com \& anti-com rels}}} \quad \text{i.e.,} \rightarrow \quad T_c\{A_1B_2C_3\} = N_c\{A_1B_2C_3\}. \tag{7-105}$$

Justifying Wick's theorem for three fields

As with two fields, re-order using commutator (anti-commutator) relations

<u>Case 1</u>, $t_3 < t_2 < t_1$

For this case, the LHS of (7-105) is in time order and doesn't need to be changed.

$$T_c\{A_1B_2C_3\} = A_1B_2C_3. \tag{7-106}$$

Start with a case for a particular time order

The first term after the equal sign in (7-105) becomes

$$A_1B_2C_3 = N_c\{A_1B_2C_3\} = N_c\left\{\left(A_1^d + A_1^c\right)\left(B_2^d + B_2^c\right)\left(C_3^d + C_3^c\right)\right\}$$

$$= N_c\left\{\left(A_1^d B_2^d + A_1^d B_2^c + A_1^c B_2^d + A_1^c B_2^c\right)\left(C_3^d + C_3^c\right)\right\} \tag{7-107}$$

$$= N_c\left\{\left(A_1^d B_2^d \overbrace{\pm B_2^c A_1^d + \left[A_1^d, B_2^c\right]_{\mp}} + A_1^c B_2^d + A_1^c B_2^c\right)\left(C_3^d + C_3^c\right)\right\}.$$

where we substituted the commutation/anti-commutation relation for $A_1^d B_2^c$, with the minus sign in the last row for anti-commutation (when both fields are fermions). Now multiply the C_3 field on the RHS of the last row of (7-107).

$$A_1B_2C_3 = N_c\{A_1B_2C_3\} = N_c\left\{A_1^d B_2^d C_3^d \pm B_2^c A_1^d C_3^d + \underbrace{\left[A_1^d, B_2^c\right]_{\mp}}_{\text{a } c \text{ number}} C_3^d + A_1^c B_2^d C_3^d + A_1^c B_2^c C_3^d\right\}$$

$$+ N_c\left\{A_1^d B_2^d C_3^c \pm B_2^c A_1^d C_3^c + \underbrace{\left[A_1^d, B_2^c\right]_{\mp}}_{\text{a } c \text{ number}} C_3^c + A_1^c B_2^d C_3^c + A_1^c B_2^c C_3^c\right\}. \tag{7-108}$$

The upper row of (7-108) already has destruction operators all on the right side. The lower row needs re-ordering.

$$A_1B_2C_3 = N_c\{A_1B_2C_3\} = A_1^d B_2^d C_3^d \pm B_2^c A_1^d C_3^d + A_1^c B_2^d C_3^d + A_1^c B_2^c C_3^d + \left[A_1^d, B_2^c\right]_{\mp} C_3^d$$

$$+ N_c\left\{A_1^d \underbrace{B_2^d C_3^c}_{\substack{\pm C_3^c B_2^d \\ +\left[B_2^d, C_3^c\right]_{\mp}}} \pm B_2^c \underbrace{A_1^d C_3^c}_{\substack{\pm C_3^c A_1^d \\ +\left[A_1^d, C_3^c\right]_{\mp}}} + A_1^c \underbrace{B_2^d C_3^c}_{\substack{\pm C_3^c B_2^d \\ +\left[B_2^d, C_3^c\right]_{\mp}}}\right\} + A_1^c B_2^c C_3^c + \left[A_1^d, B_2^c\right]_{\mp} C_3^c. \tag{7-109}$$

Re-arranging (7-109), with boxes numbering terms to keep track of them, yields

$$A_1B_2C_3 = N_c\{A_1B_2C_3\}$$

$$= \boxed{1}\,A_1^d B_2^d C_3^d \pm \boxed{2}\,B_2^c A_1^d C_3^d + \boxed{3}\,A_1^c B_2^d C_3^d + \boxed{4}\,A_1^c B_2^c C_3^d + \boxed{5}\,\left[A_1^d, B_2^c\right]_{\mp} C_3 \pm N_c\left\{\underbrace{A_1^d C_3^c}_{\substack{\boxed{6}\,\pm C_3^c A_1^d \\ +\left[A_1^d, C_3^c\right]_{\mp} \\ \boxed{7}}} B_2^d\right\} \tag{7-110}$$

$$+ \boxed{8}\,A_1^d \left[B_2^d, C_3^c\right]_{\mp} \pm \boxed{9}\,B_2^c C_3^c A_1^d \pm \boxed{10}\,B_2^c \left[A_1^d, C_3^c\right]_{\mp} \pm \boxed{11}\,A_1^c C_3^c B_2^d + \boxed{12}\,A_1^c \left[B_2^d, C_3^c\right]_{\mp} + \boxed{13}\,A_1^c B_2^c C_3^c.$$

Re-grouping once again, we have

$$A_1B_2C_3 = A_1^d \overset{\boxed{1}}{B_2^d} C_3^d \pm \overset{\boxed{2}}{B_2^c A_1^d} C_3^d + A_1^c \overset{\boxed{3}}{B_2^d} C_3^d + A_1^c B_2^c \overset{\boxed{4}}{C_3^d}$$

$$\pm C_3^c \overset{\boxed{6}}{A_1^d B_2^d} \pm B_2^c \overset{\boxed{9}}{C_3^c A_1^d} \pm A_1^c \overset{\boxed{11}}{C_3^c B_2^d} + A_1^c B_2^c \overset{\boxed{13}}{C_3^c} \tag{7-111}$$

$$+ \underbrace{\overset{\boxed{5}}{\left[A_1^d, B_2^c\right]_{\mp}}}_{A_1B_2 \text{ for } t_2 < t_1} C_3 + A_1 \underbrace{\overset{\boxed{8}+\boxed{12}}{\left[B_2^d, C_3^c\right]_{\mp}}}_{B_2C_3 \text{ for } t_3 < t_2} \pm B_2^c \underbrace{\overset{\boxed{10}}{\left[A_1^d, C_3^c\right]_{\mp}}}_{A_1C_3 \text{ for } t_3 < t_1} \pm \underbrace{\overset{\boxed{7}}{\left[A_1^d, C_3^c\right]_{\mp}}}_{A_1C_3 \text{ for } t_3 < t_1} B_2^d .$$

where the signs in the last two terms depend on whether fermion switching has been involved to get to each term. We can re-write the last two terms, where the sign changing under fermion switch will be taken into account automatically by the N operator (see (7-76) + 2) as

$$\pm B_2^c \underbrace{\left[A_1^d, C_3^c\right]_{\mp}}_{A_1C_3 \text{ for } t_3 < t_1} \pm \underbrace{\left[A_1^d, C_3^c\right]_{\mp}}_{A_1C_3 \text{ for } t_3 < t_1} B_2^d = \pm B_2^c \underbrace{A_1 C_3}_{} \pm \underbrace{A_1 C_3}_{} B_2^d = N\left\{A_1 \underbrace{B_2 C_3}_{}\right\}. \tag{7-112}$$

So finally, where terms noted in under brackets refer to (7-111), and we use (7-112)

$$\downarrow \; A_1B_2C_3 \text{ re-arranged using full commutation relations } \downarrow$$

$$T\left\{A_1B_2C_3\right\} = \tag{7-113}$$

$$\underbrace{N\left\{A_1B_2C_3\right\}}_{\text{1st 8 terms}} + \underbrace{N\{A_1\underbrace{B_2C_3}\}}_{\text{1st term in last row}} + \underbrace{N\{\underbrace{A_1}B_2\underbrace{C_3}\}}_{\text{2nd term in last row}} + \underbrace{N\{A_1\underbrace{B_2}C_3\}}_{\text{last 2 terms}} \text{ for } t_3 < t_2 < t_1$$

Other cases other than $t_3 < t_2 < t_1$

For cases where $t_3 < t_2 < t_1$ is not true, we would have the same result for the RHS of (7-105), i.e., it would equal the RHS of (7-113). But the time ordered side would be ordered using the T_c operator (using the commutator/anti-commutator relations when switching field positions.) As with the two field case, we would get commutator/anti-commutator relations on the time ordered side that would cancel with identical relations on the N_c ordered side. You can do Prob. 15 to prove the case for $t_2 < t_3 < t_1$, and then, if you wish, play around with other time sequences to prove it in general. The final result is that (7-113) holds for any time sequence for any three fields.

$$T\left\{A_1B_2C_3\right\} = N\left\{A_1B_2C_3\right\} + N\{A_1\underbrace{B_2C_3}\} + N\{\underbrace{A_1}B_2\underbrace{C_3}\} + N\{A_1\underbrace{B_2}\underbrace{C_3}\} \text{ for any time order} . \tag{7-114}$$

7.11.2 Wick's Theorem via Induction

Comparing (7-114) for three fields to (7-97) for two fields, we can see a pattern emerging. If we were to carry out one more example with four fields, we would see additional types of terms entailing two contractions. This pattern would be fully reflected by (7-82), and thus by Wick's theorem in full, (7-78).

7.12 Appendix B: Operators in Exponentials and Time Ordering

7.12.1 Math Reference

For what follows we recall (hopefully, you have seen this relation before) the <u>Baker-Campbell-Hausdorff</u> formula, where, for A and B as operators,

$$e^A e^B = e^{A+B+\frac{1}{2}[A,B]+\frac{1}{12}[A,[A,B]]-\frac{1}{12}[B,[A,B]]-\frac{1}{24}[B,[A,[A,B]]]+\ldots} . \tag{7-115}$$

If A and B commute, or are c numbers, we get the familiar simple addition of exponents result.

Contractions arise, as with two fields, but more of them

Signs of various terms depend on fermion/fermion exchanges needed to get the terms

Those exchanges (their signs) taken into account by the N operation symbol meaning

Result: Wick's theorem for three fields, this time order case

Same result for all time orders = Wick's theorem for three fields

A pattern emerges for more fields. That pattern is Wick's theorem

7.12.2 Solution of Differential Equation with Operator

Consider (7-59), which we repeat below for convenience,

$$i\frac{dS_{oper}}{dt} = H_I^I S_{oper},$$ (7-59)

and consider what we might naïvely expect to be the solution, (7-60),

$$S_{oper} = e^{-i\int_{t_i}^{t} H_I^I dt'}.$$ (7-60)

Then, the LHS of (7-59) can be found via (with (7-65) in the last line below)

$$i\frac{dS_{oper}(t,t_i)}{dt} = i\lim_{\Delta t \to 0}\frac{1}{\Delta t}\left(S_{oper}(t+\Delta t,t_i) - S_{oper}(t,t_i)\right)$$

$$= i\lim_{\Delta t \to 0}\frac{1}{\Delta t}\left(e^{-i\int_{t_i}^{t+\Delta t} H_I^I(t')dt'} - e^{-i\int_{t_i}^{t} H_I^I(t')dt'}\right)$$

$$= i\lim_{\Delta t \to 0}\frac{1}{\Delta t}\left(e^{-i\int_{t}^{t+\Delta t} H_I^I(t')dt' -i\int_{t_i}^{t} H_I^I(t')dt'} - e^{-i\int_{t_i}^{t} H_I^I(t')dt'}\right)$$ (7-116)

where in QED, $H_I^I(t')=\int \mathcal{H}_I^{1/2,1}(t')d^3x$ and $\mathcal{H}_I^{1/2,1}(t') = -e\bar{\psi}(x')\gamma^\mu\psi(x')A_\mu(x')$.

<u>For Commuting Variables in the Exponent</u>

If the first term in the 3rd line of (7-116) contained only c numbers or commuting operators in H_I^I, then we would have, via (7-115),

$$i\frac{dS_{oper}(t,t_i)}{dt} = i\lim_{\Delta t \to 0}\frac{1}{\Delta t}\underbrace{\left(e^{-i\int_{t}^{t+\Delta t} H_I^I(t')dt'}e^{-i\int_{t_i}^{t} H_I^I(t')dt'} - e^{-i\int_{t_i}^{t} H_I^I(t')dt'}\right)}_{\text{only if } H_I^I \text{ at different} \atop \text{times commute}}$$

$$= i\lim_{\Delta t \to 0}\frac{1}{\Delta t}\left(e^{-i\int_{t}^{t+\Delta t} H_I^I(t')dt'} - 1\right)e^{-i\int_{t_i}^{t} H_I^I(t')dt'}$$

$$= i\lim_{\Delta t \to 0}\frac{1}{\Delta t}\left(e^{-i\int_{t}^{t+\Delta t} H_I^I(t')dt'} - 1\right)S_{oper}(t,t_i)$$ (7-117)

$$= i\lim_{\Delta t \to 0}\frac{1}{\Delta t}\left(1 - iH_I^I(t)\Delta t - \frac{1}{2}\left(H_I^I(t)\Delta t\right)^2 +... - 1\right)S_{oper}(t,t_i)$$

$$= i\left(-iH_I^I(t)\right)S_{oper}(t,t_i) = H_I^I(t)S_{oper}(t,t_i),$$

which is the same as (7-59).

But, the first term in the 3rd line of (7-116) contains operators in H_I^I, that do not commute at different times (during the integration over time process), and so, it seems, due to (7-115), we cannot conclude that (7-60) is a solution of (7-59).

However, we can still make (7-60) meaningful by attaching a particular interpretation to the symbolism, as we show below.

For Commuting or Non-commuting Variables in the Exponent

To find a viable solution to (7-59) that is good regardless of the implications of (7-115), we start by noting that (7-59) can be solved with

$$S_{oper}\left(t,t_i\right)=1-i\int_{t_i}^{t}H_I^I\left(t_1\right)S_{oper}\left(t_1,t_i\right)dt_1 , \qquad (7\text{-}118)$$

though this assumes we know the form of $S_{oper}(t_1, t_i)$. But we can use (7-118) repeatedly to solve (7-59) via iteration. That is, plug

$$S_{oper}\left(t_1,t_i\right)=1-i\int_{t_i}^{t_1}H_I^I\left(t_2\right)S_{oper}\left(t_2,t_i\right)dt_2 \qquad (7\text{-}119)$$

into (7-118), then express $S_{oper}(t_2, t_i)$ in terms of integration over t_3 using (7-118), etc. The resulting infinite series looks like

$$\overbrace{}^{\text{See LHS of Fig. 7-4}}$$
$$S_{oper}\left(t,t_0\right)=1-i\int_{t_0}^{t}H_I^I\left(t_1\right)dt_1+(-i)^2\int_{t_0}^{t}H_I^I\left(t_1\right)\left(\int_{t_0}^{t_1}H_I^I\left(t_2\right)dt_2\right)dt_1$$
$$+(-i)^3\int_{t_0}^{t}H_I^I\left(t_1\right)\left(\int_{t_0}^{t_1}H_I^I\left(t_2\right)\left(\int_{t_0}^{t_2}H_I^I\left(t_3\right)dt_3\right)dt_2\right)dt_1+... \qquad (7\text{-}120)$$

where no assumption need be made about the commutation properties of factors in the integrands. In this case, for the bottom line term above, we must integrate over t_3 first, then t_2, then t_1. In the term with an integration over t_n, the order is $t_n, t_{n-1}, ... t_2, t_1$. The LHS of Fig. 7-4 is a graphic representation of the integration regions involved in the double integral term in (7-120).

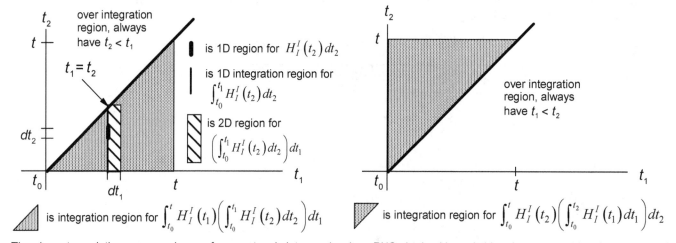

The above two relations are equal, even for operators in integrands, since RHS obtained by switching dummy variables $t_1 \leftrightarrow t_2$ in LHS.

Figure 7-4. Regions of Integration Related to Double Integration Term in S_{oper} Expansion

Fig. 7-4 should be relatively self explanatory (a picture is worth a thousand words). From the last line in that figure, where we simply switch dummy variables in the last term of the first line below,

$$\int_{t_0}^{t}\int_{t_0}^{t} T\left\{H_I^I(t_1)H_I^I(t_2)\right\}dt_2 dt_1 = \underbrace{\int_{t_0}^{t}\left(\int_{t_0}^{t_1} H_I^I(t_1)H_I^I(t_2)dt_2\right)dt_1}_{t_2<t_1 \text{ region}}$$

$$+ \underbrace{\int_{t_0}^{t}\left(\int_{t_0}^{t_2} H_I^I(t_2)H_I^I(t_1)dt_1\right)dt_2}_{t_1<t_2 \text{ region}}$$

$$= \underbrace{\int_{t_0}^{t}\left(\int_{t_0}^{t_1} H_I^I(t_1)H_I^I(t_2)dt_2\right)dt_1}_{t_2<t_1 \text{ region}} \qquad (7\text{-}121)$$

$$+ \underbrace{\int_{t_0}^{t}\left(\int_{t_0}^{t_1} H_I^I(t_1)H_I^I(t_2)dt_2\right)dt_1}_{t_2<t_1 \text{ region}}$$

$$= 2\int_{t_0}^{t}\left(\int_{t_0}^{t_1} H_I^I(t_1)H_I^I(t_2)dt_2\right)dt_1 .$$

Thus, for the double integration term in (7-120), we can substitute

$$\int_{t_0}^{t}\left(\int_{t_0}^{t_1} H_I^I(t_1)H_I^I(t_2)dt_2\right)dt_1 = \frac{1}{2}\int_{t_0}^{t}\int_{t_0}^{t} T\left\{H_I^I(t_1)H_I^I(t_2)\right\}dt_2 dt_1 \qquad (7\text{-}122)$$

For the triple integration term in the bottom row of (7-120), we get

$$\int_{t_0}^{t}\int_{t_0}^{t}\int_{t_0}^{t} T\left\{H_I^I(t_1)H_I^I(t_2)H_I^I(t_3)\right\}dt_3 dt_2 dt_1$$

$$= \int_{t_0}^{t} H_I^I(t_1)\left(\int_{t_0}^{t_1} H_I^I(t_2)\left(\int_{t_0}^{t_2} H_I^I(t_3)dt_3\right)dt_2\right)dt_1$$

$$+ \int_{t_0}^{t} H_I^I(t_1)\left(\int_{t_0}^{t_1} H_I^I(t_3)\left(\int_{t_0}^{t_3} H_I^I(t_2)dt_2\right)dt_3\right)dt_1$$

$$+ \int_{t_0}^{t} H_I^I(t_2)\left(\int_{t_0}^{t_2} H_I^I(t_1)\left(\int_{t_0}^{t_1} H_I^I(t_3)dt_3\right)dt_1\right)dt_2$$

$$+ \int_{t_0}^{t} H_I^I(t_2)\left(\int_{t_0}^{t_2} H_I^I(t_3)\left(\int_{t_0}^{t_3} H_I^I(t_1)dt_1\right)dt_3\right)dt_2 \qquad (7\text{-}123)$$

$$+ \int_{t_0}^{t} H_I^I(t_3)\left(\int_{t_0}^{t_3} H_I^I(t_1)\left(\int_{t_0}^{t_1} H_I^I(t_2)dt_2\right)dt_1\right)dt_3$$

$$+ \int_{t_0}^{t} H_I^I(t_3)\left(\int_{t_0}^{t_3} H_I^I(t_2)\left(\int_{t_0}^{t_2} H_I^I(t_1)dt_1\right)dt_2\right)dt_3$$

$$= 6\int_{t_0}^{t} H_I^I(t_1)\left(\int_{t_0}^{t_1} H_I^I(t_2)\left(\int_{t_0}^{t_2} H_I^I(t_3)dt_3\right)dt_2\right)dt_1 ,$$

or

$$\int_{t_0}^{t} H_I^I(t_1)\left(\int_{t_0}^{t_1} H_I^I(t_2)\left(\int_{t_0}^{t_2} H_I^I(t_3)dt_3\right)dt_2\right)dt_1$$

$$= \frac{1}{3!}\int_{t_0}^{t}\int_{t_0}^{t}\int_{t_0}^{t} T\left\{H_I^I(t_1)H_I^I(t_2)H_I^I(t_3)\right\}dt_3 dt_2 dt_1 \qquad (7\text{-}124)$$

Repeating the procedure for higher integration number terms, we end up with (7-67) (with t_f there equal to t here).

Thus, we need to interpret the exponentials in (7-60), (7-61), (7-66), and (7-67) as being defined by (7-67), i.e., as implying time ordering in the expansion. Only by doing this can we avoid the issues that non-commutation would bring in via the Baker-Campbell-Hausdorff formula, (7-115).

Note that some authors use the time ordering symbol T to indicate this, as in

$$S_{oper}\left(t_f, t_i\right) = Te^{-i\int_{t_i}^{t_f} H_I^I(t)dt}. \tag{7-125}$$

We don't do that in this book, but we need to keep in mind that by (7-61) we mean (7-67).

7.13 Problems

1. Show by taking each value of $\mu = 0,1,2,3$, that (7-4) is equivalent to (7-3).

2. Use (7-8) in (7-7) to get (7-6).

3. Use the full e/m quantum Lagrangian (7-15) in the Euler-Lagrange equation (7-7) with $\phi^{n=1} = \bar{\psi}$ to get the full Dirac equation (7-16).

4. Create your own wholeness chart summarizing, in less than a page, Sect. 7.1 on Interactions in RQM.

5. From the equation of motion for the expectation value of an operator in the S.P. (top row, RH column of Wholeness Chart 7-1) derive the same equation of motion in the H.P. (3rd row, RH column of Wholeness Chart 7-1.) (Hint: Simply insert $I = UU^\dagger$ in appropriate places in the S.P. equation of motion.)

6. Show that the vacuum expectation value (VEV) for the photon field is zero, i.e., $\langle 0| A^\mu |0\rangle = 0$. Pick any state you like and show that the expectation value of A^μ for that state is zero.

7. Show that $H_I^I = U_0^\dagger H_I^S U_0 = H_I^S$ only if H_I^S commutes with H_0. The last part of Box 2-3 in Chap. 2 (pg. 27) may help.

8. Start with the S.P. state equation of motion (top left box in Wholeness Chart 7-1) and use U_0 to derive the I.P. state equation of motion (7-38). (Hint: Insert $I = U_0 U_0^\dagger$ before the kets, express $H = H^S = H_0 + H_I^S$, and pre-multiply the entire equation by $U_0^\dagger$.)

9. Start with the S.P. expectation value equation of motion (top right box in Wholeness Chart 7-1) and use U_0 to derive the I.P. expectation value equation of motion (7-39).

10. From the S.P. expectation value of an operator, show that it equals the I.P. expectation value, i.e. show (using (7-24) as a guide) that

$$\bar{\mathcal{O}} = {}_S\langle\Psi|\mathcal{O}^S|\Psi\rangle_S = {}_I\langle\Psi|\mathcal{O}^I|\Psi\rangle_I.$$

11. Show that a commutator relation between two operators in both the H.P. and the I.P. is the same as the commutator relation between the same two operators expressed in the S.P., if that commutator equals a constant.

12. Use the full anti-commutation relations to re-order $\psi(x_1)\bar{\psi}(x_2)$ two ways, first with earlier times on the right assuming $t_2 < t_1$, then with destruction operators on the right. Each of these relations is equal to the original relation $\psi(x_1)\bar{\psi}(x_2)$, and therefore they are equal to each other. Show that Wick's theorem (7-95) and (7-82) holds for $t_2 < t_1$, where the two operator fields are ordered using the T and N operators (which assume all fields commute/anti-commute.) Hint: Near the end, use (4-145) of Chap. 4 and (7-75).

13. Repeat Prob. 12 for $t_1 < t_2$. Hint: Near the end, use (4-146) of Chap. 4 and (7-75).

14. Repeat Probs. 12 and 13 for the two fields $\phi(x_1)\bar{\psi}(x_2)$.

15. In a manner similar to what we did in (7-106) to (7-113) for three fields for the case $t_3 < t_2 < t_1$, show that the same relation (7-113) holds for three fields for $t_2 < t_3 < t_1$. Remember that in both cases, when we place destruction operators on the right using commutation/anti-commutation relations, we must have the same RH side, as in (7-111), since it is the same thing regardless of time order.

Chapter 8

QED: Quantum Field Interaction Theory Applied to Electromagnetism

"At the Pocono Manor Inn that spring day in 1948, Feynman
introduced his diagrams ... By all indications, [it] was a flop."

David Kaiser, Kenji Ito, and Karl Hall
Spreading the Tools of Theory
Soc. Stud. Sci. 34/6, 879–922

And so he felt that his lecture was a complete failure, "a hopeless presentation".

S. S. Schweber recalling taped interview
with Richard Feynman on Nov 13, 1980

8.0 Preliminaries

The Pocono Manor conference, a gathering of luminaries including Oppenheimer, Bohr, Dirac, Wigner, Schwinger and others, was focused on the major problem of the day – how to handle those notorious infinities that crept into the exact (not approximate, as in the present chapter) calculations of transition amplitudes. (See Chaps. 9, 12, 13, 14, and 15 herein.) As part of the demonstration of his solution to the issue, Feynman introduced his now famous Feynman diagrams, a simple, symbolic, short-hand way to organize the complicated mathematical expressions involved, and also his Feynman rules, which associated certain mathematical quantities with each aspect of a given diagram.

No one understood him, and several were skeptical. Some even wondered whether Feynman, himself, knew what he was doing. Yet, before too long after that conference, the utility of the diagrams became apparent, and their usage spread far and wide.

The lecture Feynman gave that day used Feynman diagrams as a tool for a greater end, elimination of the infinities that plagued the theory. And what he did that day worked. It worked so well that it earned him a share (along with Schwinger and Tomonaga, who did the same thing in a different way) of the Nobel Prize in 1965. On that, he had this to say.

"The work I have done has, already, been adequately rewarded and recognized. Imagination reaches out repeatedly trying to achieve some higher level of understanding, until suddenly I find myself momentarily alone before one new corner of nature's pattern of beauty and true majesty revealed. That was my reward."

Richard P. Feynman
Nobel banquet speech

8.0.1 Background

In the last chapter, we showed how the S operator transforms an initial state into a final state, and how the transition amplitude S_{fi} can be found from the S operator. We also solved for the form of the S operator and then expanded that (via the Dyson expansion) into an infinite series of terms involving time ordering of operators and integration over all of space and time. Finally, we saw how Wick's theorem could convert the time ordering into more manageable normal ordered form.

8.0.2 Chapter Overview

In this chapter, we will take the result of Wick's theorem and apply it to electromagnetism to develop quantum electrodynamics (QED). Note that for this chapter, we will only find approximate values for S_{fi}. We will do this by noting that the higher order terms in the expansion of the S operator for QED should be small, and so restrict our analysis here to second and lower order terms.

We will

Steps in applying QFT interaction theory to e/m to second order approximation

- Use $\mathcal{H}_I^I$ for QED in the Dyson-Wick's expansion of the S operator
- Examine only lower order terms $S^{(0)}$, $S^{(1)}$, $S^{(2)}$ in the expansion
- Show how $S^{(0)}$ implies no transition, and $S^{(1)}$ is not a real physical process and irrelevant
- Analyze all possible terms in $S^{(2)}$, the second order portion of the S operator
- Use Feynman diagrams to represent individual interactions
- Calculate transition amplitudes for typical QED (e.g., Bhabha) scattering processes
- Detail the short cut method to find transition amplitudes, i.e., Feynman rules
- Examine mixed lepton (μ and τ particles in addition to electrons/positrons) interactions
- Define inelastic vs. elastic scattering (interactions)
- Discuss how to add amplitudes
- Look at attraction and repulsion in light of QED

Note that from now on we will be dealing exclusively with QFT, and not RQM, because the latter cannot handle the general case of mutation of particles from one type to another.

8.0.3 Summary Chart

All that we will do in this chapter is summarized in Wholeness Chart 8-4 at the end of the chapter. As in prior chapters, I highly recommend following along, step by step, in that chart as we progress through this chapter.

Be sure to use the summary wholeness chart, as you study this chapter

8.1 Dyson-Wick's Expansion for QED Hamiltonian Density

Recall from Chap. 7, pg. 201, (7-68) and (7-69), the Dyson expansion of the S operator

S operator generically

$$ S = I - i\int_{-\infty}^{\infty} \mathcal{H}_I^I(x_1)d^4x_1 \; -\frac{1}{2!}\int_{-\infty}^{\infty}\int_{-\infty}^{\infty} T\left\{\mathcal{H}_I^I(x_1)\mathcal{H}_I^I(x_2)\right\}d^4x_1 d^4x_2 \; +..... . \qquad (8\text{-}1) $$

For the interaction Hamiltonian density in (8-1) we use the relation discovered to work for RQM, because we have learned that Hamiltonians for RQM expressed in the Schrödinger Picture, as a rule, take the same form for QFT expressed in the Heisenberg Picture. We are working in the Interaction Picture, for which operators (such as the Hamiltonian density) take the same form in the I.P. as in the H.P. So, for electromagnetic interactions between electrons, positrons, and photons, the quantum Hamiltonian density takes the form, where, with $A_\mu\gamma^\mu = A\!\!\!/$ (similar to $\partial_\mu\gamma^\mu = \partial\!\!\!/$),

Interaction Hamiltonian density for QED

$$ \mathcal{H}_I^I = -\mathcal{L}_I^I = -e\overline{\psi}A_\mu\gamma^\mu\psi = -e\overline{\psi}A\!\!\!/\psi = -e\left(\overline{\psi}^+ + \overline{\psi}^-\right)\left(A^+ + A^-\right)\left(\psi^+ + \psi^-\right), \qquad (8\text{-}2) $$

And plugging (8-2) into (8-1) gives us

S operator for QED interaction Hamiltonian density = sum of terms

$$ S = \underbrace{I}_{S^{(0)}} \underbrace{+ie\int_{-\infty}^{\infty}\left(\overline{\psi}A\!\!\!/\psi\right)_{x_1}d^4x_1}_{S^{(1)}} \underbrace{-\frac{1}{2!}e^2\int_{-\infty}^{\infty}\int_{-\infty}^{\infty} T\left\{\left(\overline{\psi}A\!\!\!/\psi\right)_{x_1}\left(\overline{\psi}A\!\!\!/\psi\right)_{x_2}\right\}d^4x_1 d^4x_2}_{S^{(2)}} + ... = \sum_{n=0}^{\infty} S^{(n)}. \quad (8\text{-}3) $$

Note that for each $S^{(n)}$, we have a factor in front of e^n. The fine structure constant α in the Gaussian system is $e^2/4\pi\hbar c$, dimensionless, and approximately equal to 1/137. Recall from our work in Chap. 2 (Prob. 2 there) that a dimensionless quantity is dimensionless in any system of units, and has the same value in any system of units. Thus, in natural units

charge e value in natural units

$$ \alpha = \frac{e^2}{4\pi} \approx \frac{1}{137} \;\; \rightarrow \; e \approx \sqrt{\frac{4\pi}{137}} \approx .303 . \qquad (8\text{-}4) $$

Hence, in (8-3), each term has a dimensionless factor in front of approximately $.303/n$ that of the prior term (the n comes from the factorial in the denominator of each term). Thus, higher order terms should be significantly smaller than lower order terms and, in principle, we can approximate (8-3) by taking only the first few terms. In this chapter, we only deal with $S^{(0)}$, $S^{(1)}$, and $S^{(2)}$.

In part, due to extra factor of e/n in each term in S expansion, higher order terms smaller

We can then use Wick's theorem from Chap. 7 to re-express each time ordered term of the S operator (8-3) in normal ordered plus contractions form.

8.1.1 $S^{(0)}$ Term

The first term in (8-3) needs no re-arranging, as it is solely the identity operator. $S^{(0)} = I$.

Zeroth order term $S^{(0)} = I$

8.1.2 $S^{(1)}$ Term

The second term in (8-3) has factors operating all at the same time t_1, and so can be considered time ordered. Wick's theorem for this case reduces to

$$T\left\{(AB...)_{x_1}\right\} = N\left\{(AB...)_{x_1}\right\}. \tag{8-5}$$

So, in terms of normal ordering,

$$S^{(1)} = ie \int_{-\infty}^{\infty} N\left\{\bar{\psi} A \psi\right\}_{x_1} d^4 x_1 . \tag{8-6}$$

First order term $S^{(1)}$

8.1.3 $S^{(2)}$ Term

The third term in (8-3), with integrals over two events, is time ordered. For that, we use Wick's theorem expressed for only two spacetime events, i.e.,

$$T\left\{(AB...)_{x_1} (AB...)_{x_2}\right\} = N\left\{(AB...)_{x_1} (AB...)_{x_2}\right\}$$
$$+ N\left\{(AB...)x_1(AB...)x_2\right\} + N\left\{(AB...)x_1(AB...)x_2\right\} + ...$$
$$+ N\left\{(ABC...)x_1(ABC...)x_2\right\} + N\left\{(ABC...)x_1(ABC...)x_2\right\} + ... \tag{8-7}$$

+ (all normal ordered terms with three non-equal times contractions) + etc.

We then re-express the $n = 2$ term in (8-3) via (8-7) with $A = \bar{\psi}$, $B = A$, and $C = \psi$ to yield

$$S^{(2)} = -\frac{1}{2!} e^2 \iint d^4 x_1 d^4 x_2 \times$$

$$\begin{pmatrix} N\left\{(\bar{\psi} A \psi)_{x_1} (\bar{\psi} A \psi)_{x_2}\right\} \\ + N\left\{(\bar{\psi} A \psi)_{x_1} (\bar{\psi} A \psi)_{x_2}\right\} + N\left\{(\bar{\psi} A \psi)_{x_1} (\bar{\psi} A \psi)_{x_2}\right\} + N\left\{(\bar{\psi} A \psi)_{x_1} (\bar{\psi} A \psi)_{x_2}\right\} \\ + N\left\{(\bar{\psi} A \psi)_{x_1} (\bar{\psi} A \psi)_{x_2}\right\} + \left(\text{other terms where single contraction} = 0\right) \\ \text{contraction} = 0 \\ + N\left\{(\bar{\psi} A \psi)_{x_1} (\bar{\psi} A \psi)_{x_2}\right\} + N\left\{(\bar{\psi} A \psi)_{x_1} (\bar{\psi} A \psi)_{x_2}\right\} + N\left\{(\bar{\psi} A \psi)_{x_1} (\bar{\psi} A \psi)_{x_2}\right\} \\ + \left(\text{other terms where at least one of two contractions} = 0\right) \\ + (\bar{\psi} A \psi)_{x_1} (\bar{\psi} A \psi)_{x_2} + \left(\text{terms where contractions} = 0\right). \end{pmatrix} \tag{8-8}$$

Second order term $S^{(2)}$ has many sub-terms

8.1.4 Physical Interpretation of the $S^{(n)}$

Recall the S operator transforms an initial (multi-particle) eigenstate state $|i\rangle$ into a general final state $|F\rangle$, which is a superposition of all the possible final eigenstates $|f\rangle$ that the initial state could transition into. The amplitude of the fth state in the summation of states composed by $|F\rangle$ is simply $\langle f\|F\rangle$. The square of the absolute magnitude of that amplitude is the probability of measuring $|f\rangle$. We call that the transition amplitude for $|i\rangle$ becoming $|f\rangle$. Mathematically, the transition amplitude is

Each term in S contributes to transforming initial state to final state

Thus, each term in S produces a term in the transition amplitude S_{fi}

$$S_{fi} = \left\langle f \,|\, F \right\rangle = \left\langle f \,\Big|\, \sum_{f'} S_{f'i} \,\Big|\, f' \right\rangle = \left\langle f \,|\, S \,|\, i \right\rangle = \left\langle f \,\big|\, S^{(0)} + S^{(1)} + S^{(2)} + ... \,\big|\, i \right\rangle = S_{fi}^{(0)} + S_{fi}^{(1)} + S_{fi}^{(2)} + ... , \quad (8\text{-}9)$$

so, we have to evaluate the impact of each $S^{(n)}$ on the initial state $|i\rangle$. We have to determine what states, with what amplitudes, each term in S turns a particular initial state into.

8.2 $S^{(0)}$ Physically

Consider the operation of the $S^{(0)}$ term in the S operator on a state, say an electron and a photon in momentum eigenstates. We find.

$S^{(0)}$ yields zero unless final state = initial state, then $S_{fi}^{(0)} = 1$

$$S_{fi}^{(0)} = \left\langle f \,|\, S^{(0)} \,|\, i \right\rangle = \left\langle f \,\big|\, S^{(0)} \,\big|\, e_{\mathbf{p},r}^-, \gamma_{\mathbf{k},s} \right\rangle = \left\langle f \,|\, I \,\big|\, e_{\mathbf{p},r}^-, \gamma_{\mathbf{k},s} \right\rangle = \left\langle f \,\big|\, e_{\mathbf{p},r}^-, \gamma_{\mathbf{k},s} \right\rangle .$$

$$= 1 \text{ if } \left\langle f \right| = \left\langle e_{\mathbf{p},r}^-, \gamma_{\mathbf{k},s} \right|, \quad 0 \text{ otherwise.} \qquad (8\text{-}10)$$

This is the contribution to S_{fi} from the zeroth term in the S operator expansion for two particles that don't interact. Each particle has the same final momentum and spin/polarization state that it had originally. They passed by each other, but didn't interact. The identity matrix in the first line of (8-10) simply reproduced the initial state. We can symbolize this by writing

$$\left| e_{\mathbf{p},r}^-, \gamma_{\mathbf{k},s} \right\rangle \rightarrow \left| e_{\mathbf{p},r}^-, \gamma_{\mathbf{k},s} \right\rangle . \qquad (8\text{-}11)$$

There are no virtual particles exchanged and no change in 4-momentum.

8.3 $S^{(1)}$ Physically

Consider the operation of the $S^{(1)}$ term in the S operator on an initial state

$S^{(1)}$ term, expressed as creation and destruction field operators, is actually 8 sub-terms

$$S^{(1)} |i\rangle = (-i)\int d^4 x_1 N\{-e\overline{\psi} A \psi\}_{x_1} |i\rangle = ie \int d^4 x_1 N\Big\{ e\left(\overline{\psi}^+ + \overline{\psi}^-\right)\left(A^+ + A^-\right)\left(\psi^+ + \psi^-\right)\Big\}_{x_1} |i\rangle. \quad (8\text{-}12)$$

If we multiply out the factors in (8-12), we will have eight different sub-terms contributing to the $S^{(1)}$ term in S. We will label these sub-terms as $S_j^{(1)}$ where $j = 1, 2, ..., 8$.

8.3.1 A Look at One of the Eight Sub-Terms

Let us examine one of these where our initial state is an electron and positron of particular momenta and spin states, i.e.,

Examining one of the 8 terms in $S^{(1)} \rightarrow$ what it is physically

$$S_1^{(1)} \left| e_{\mathbf{p}_1,r_1}^-, e_{\mathbf{p}_2,r_2}^+ \right\rangle = ie \int d^4 x_1 N\left\{ \overline{\psi}^+ A_\mu^- \gamma^\mu \psi^+ \right\}_{x_1} \underbrace{\left| e_{\mathbf{p}_1,r_1}^-, e_{\mathbf{p}_2,r_2}^+ \right\rangle}_{|i\rangle} = ie \int d^4 x_1 \underbrace{\left\{ A_\mu^- \overline{\psi}^+ \gamma^\mu \psi^+ \right\}_{x_1}}_{\text{normal ordered}} \left| e_{\mathbf{p}_1,r_1}^-, e_{\mathbf{p}_2,r_2}^+ \right\rangle \quad (8\text{-}13)$$

Substituting the expressions for the photon and spinor fields (see Wholeness Chart 5-4 at the end of Chap. 5), we find (8-13) becomes

Chugging the math

$$S_1^{(1)} \left| e_{\mathbf{p}_1,r_1}^-, e_{\mathbf{p}_2,r_2}^+ \right\rangle = ie \int d^4 x_1 \left(\sum_{s,\mathbf{k}} \sqrt{\frac{1}{2V\omega_{\mathbf{k}}}} \varepsilon_{\mu,s}(\mathbf{k}) a_s^\dagger(\mathbf{k}) e^{ikx_1} \right)\left(\sum_{r',\mathbf{p}'} \sqrt{\frac{m}{VE_{\mathbf{p}'}}} d_{r'}(\mathbf{p}') \overline{v}_{r'}(\mathbf{p}') e^{-ip'x_1} \right)\gamma^\mu \times$$

$$\left(\sum_{r'',\mathbf{p}''} \sqrt{\frac{m}{VE_{\mathbf{p}''}}} c_{r''}(\mathbf{p}'') u_{r''}(\mathbf{p}'') e^{-ip''x_1} \right)\left| e_{\mathbf{p}_1,r_1}^-, e_{\mathbf{p}_2,r_2}^+ \right\rangle . \qquad (8\text{-}14)$$

Destruction operators $d_{r'}$ and $c_{r''}$ will destroy the ket (i.e., make it equal to zero) for all terms in the sum except when i) $\mathbf{p}' = \mathbf{p}_2$ and $r' = r_2$, and when ii) $\mathbf{p}'' = \mathbf{p}_1$ and $r'' = r_1$. Those will reduce the ket to the vacuum state by destroying the electron and positron we started out with. Thus, we have

$$S_1^{(1)} \left| e_{\mathbf{p}_1,r_1}^-, e_{\mathbf{p}_2,r_2}^+ \right\rangle =$$

$$ie \int d^4 x_1 \left\{ \left(\sum_{s,\mathbf{k}} \sqrt{\frac{1}{2V\omega_{\mathbf{k}}}} \varepsilon_{\mu,s}(\mathbf{k}) a_s^\dagger(\mathbf{k}) e^{ikx_1} \right)\frac{m}{V}\sqrt{\frac{1}{E_{\mathbf{p}_1}E_{\mathbf{p}_2}}} \overline{v}_{r_2}(\mathbf{p}_2) e^{-ip_2 x_1} \gamma^\mu u_{r_1}(\mathbf{p}_1) e^{-ip_1 x_1} \right\} |0\rangle . \quad (8\text{-}15)$$

Each term of the remaining sum in (8-15) creates a photon with different momentum and polarization states. So, (8-15) becomes (note sum over all photon states)

$$S_1^{(1)}\left|e^-_{\mathbf{p}_1,r_1},e^+_{\mathbf{p}_2,r_2}\right\rangle =$$

$$ie\int d^4x_1\left\{\sum_{s,\mathbf{k}}\sqrt{\frac{1}{2V\omega_{\mathbf{k}}}}\varepsilon_{\mu,s}(\mathbf{k})e^{ikx_1}\frac{m}{V}\sqrt{\frac{1}{E_{\mathbf{p}_1}E_{\mathbf{p}_2}}}\overline{v}_{r_2}(\mathbf{p}_2)e^{-ip_2x_1}\gamma^\mu u_{r_1}(\mathbf{p}_1)e^{-ip_1x_1}\right\}\left|\gamma_{\mathbf{k},s}\right\rangle. \quad (8\text{-}16)$$

Suppose $|\gamma_{\mathbf{k}1,s1}\rangle$ is our final state of a single photon. For this final state, note that

$$\underbrace{\left\langle\gamma_{\mathbf{k}_1,s_1}\right|}_{\langle f|}S_1^{(1)}\underbrace{\left|e^-_{\mathbf{p}_1,r_1},e^+_{\mathbf{p}_2,r_2}\right\rangle}_{|i\rangle}=S_{1,fi}^{(1)}=\text{transition amplitude for Feynman diagram of Fig.8-1}, \quad (8\text{-}17)$$

and from (8-16) and (8-17), where all terms having different bra and ket states drop out,

$$S_{1,fi}^{(1)}=ie\int d^4x_1\underbrace{\left\{\sqrt{\frac{1}{2V\omega_{\mathbf{k}_1}}}\varepsilon_{\mu,s_1}(\mathbf{k}_1)e^{ik_1x_1}\frac{m}{V}\sqrt{\frac{1}{E_{\mathbf{p}_1}E_{\mathbf{p}_2}}}\overline{v}_{r_2}(\mathbf{p}_2)e^{-ip_2x_1}\gamma^\mu u_{r_1}(\mathbf{p}_1)e^{-ip_1x_1}\right\}}_{\text{a number, no operators}}\underbrace{\left\langle\gamma_{\mathbf{k}_1,s_1}\middle|\middle|\gamma_{\mathbf{k}_1,s_1}\right\rangle}_{=1}$$

$$=ie\frac{m}{\sqrt{2V^3}}\sqrt{\frac{1}{\omega_{\mathbf{k}_1}E_{\mathbf{p}_1}E_{\mathbf{p}_2}}}\varepsilon_{\mu,s_1}(\mathbf{k}_1)\overline{v}_{r_2}(\mathbf{p}_2)\gamma^\mu u_{r_1}(\mathbf{p}_1)\underbrace{\int e^{i(k_1-p_2-p_1)x_1}d^4x_1}_{(2\pi)^4\delta^{(4)}(k_1-p_2-p_1)} \quad (8\text{-}18)$$

$$=ie(2\pi)^4\delta^{(4)}(k_1-p_2-p_1)\sqrt{\frac{1}{2V\omega_{\mathbf{k}_1}}}\sqrt{\frac{m}{VE_{\mathbf{p}_1}}}\sqrt{\frac{m}{VE_{\mathbf{p}_2}}}\varepsilon_{\mu,s_1}(\mathbf{k}_1)\overline{v}_{r_2}(\mathbf{p}_2)\gamma^\mu u_{r_1}(\mathbf{p}_1).$$

Result of the math: transition amplitude contribution from this one part of $S^{(1)}$

Since $\mathbf{p}_2$, $\mathbf{p}_1$, r_2, and r_1 are known, we can calculate (8-18) (though with all the gamma matrices and spinors involved it would be tedious). When we do scattering cross section calculations near the end of the book, we will integrate over momenta, so delta functions will come into play there.

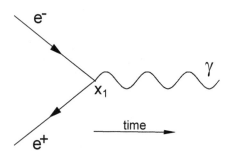

Figure 8-1. Single Vertex Interaction Feynman Diagram

First important note: The Dirac delta function arising in our calculation (see (8-18)) ensures that the outgoing 4-momentum of the final state photon equals the incoming total 4-momentum of the two initial state particles. This, we will see, is a general principle that holds for all transition amplitudes, throughout QFT. Outgoing 4-momentum for any interaction <u>vertex</u> (three particles interacting at a point in a Feynman diagram) equals incoming 4-momentum.

Dirac delta function in transition amplitude ensures outgoing 4-momentum equals incoming

And this conforms with what we know transpires in our world. Total 3-momentum and total energy, in a system without external forces acting on it, are conserved.

Process represented in Fig. 8-1

<u>Second important note</u>: The interaction represented mathematically by (8-18), and pictorially by Fig. 8-1, is not physically viable and does not occur. To understand why, see Box 8-1.

This interaction cannot occur physically because it is "off-shell"

8.3.2 The Other Seven Sub-Terms

Probs. 1 and 2 look at two of the other seven sub-terms in the $S^{(1)}$ term of S (see RHS of (8-12)). Doing those problems and Prob. 3, we see that each of those sub-terms represents a single vertex Feynman diagram, that each conserves 4-momentum, and that each is not physically viable. Interactions they represent are i) $\gamma\to e^-e^+$; ii) $e^-\gamma\to e^-$; iii) $\gamma e^-e^+\to$ vacuum, and more. (Prob. 3.)

There are seven other parts of $S^{(1)}$, representing different types of vertices

8.3.3 Each Sub-Term Yields Zero for All but One Case

Note that the sub-term $S_2^{(1)}$ of Prob. 1 applied to the incoming electron and positron state of Fig. 8-1 yields zero.

Each part of $S^{(1)}$ yields zero except the one part where field operators match states

$$S_2^{(1)}\left|e^-_{\mathbf{p}_1,r_1},e^+_{\mathbf{p}_2,r_2}\right\rangle=ie\int d^4x_1 N\{\overline{\psi}^-\mathcal{A}^+\psi^+\}x_1\left|e^-_{\mathbf{p}_1,r_1},e^+_{\mathbf{p}_2,r_2}\right\rangle=0 \quad (8\text{-}19)$$

because $\mathcal{A}^+$ destroys any state (turns it into zero) that does not have a photon in it. Similarly, all sub-terms except $S_1^{(1)}$ will turn an incoming electron/positron state to zero. So, we get

Box 8-1. "Off the Mass Shell" – Why a Single Vertex Interaction is Not Physical

Relativistic 4-Momentum Relation

Recall that, for a single particle,

$$m^2 = E^2 - \mathbf{p}^2 = p^\mu p_\mu \quad = 0 \text{ for photons .} \tag{B8-1.1}$$

For a system of particles (like our initial electron and positron state of Fig. 8-1), we determine an invariant mass m_{sys} (which is the same for all observers) from (B8-1.2) (where E_{sys} is the sum of energies of the individual particles, and $\mathbf{p}_{sys}$ is the sum of 3-momenta of those particles),

$$E_{sys}^2 - \mathbf{p}_{sys}^2 = m_{sys}^2 , \tag{B8-1.2}$$

and this is non-zero for the initial multiparticle state (system) of Fig. 8-1. (Note that m_{sys} can be different for systems having the same particles (i.e., same sum of individual rest masses), where those particles have different energies.)

Applied to the Interaction of Fig. 8-1

From the delta function in (8-18), we know that the outgoing 4-momentum for the photon of Fig. 8-1 equals the incoming total 4-momentum of the electron and positron, i.e.,

$$p_i^\mu = \begin{pmatrix} E_1 + E_2 \\ \mathbf{p}_1 + \mathbf{p}_2 \end{pmatrix} = p_f^\mu = \begin{pmatrix} \omega_{\mathbf{k}_1} \\ \mathbf{k}_1 \end{pmatrix} . \tag{B8-1.3}$$

From, (B8-1.2) and (B8-1.3), for the incoming state, we have

$$p_i^\mu p_{i\,\mu} = (E_1 + E_2)^2 - (\mathbf{p}_1 + \mathbf{p}_2) \cdot (\mathbf{p}_1 + \mathbf{p}_2) = m_{sys}^2 \neq 0 . \tag{B8-1.4}$$

For the outgoing state, the photon, we have

$$p_f^\mu p_{f\,\mu} = \underbrace{(E_1 + E_2)^2}_{\omega_{\mathbf{k}_1}} - \underbrace{(\mathbf{p}_1 + \mathbf{p}_2)}_{\mathbf{k}_1} \cdot \underbrace{(\mathbf{p}_1 + \mathbf{p}_2)}_{\mathbf{k}_1} = m_{sys}^2 \neq 0 . \tag{B8-1.5}$$

But (B8-1.5) must equal zero for a photon. That it doesn't equal zero means we can't produce a real photon as shown in Fig. 8-1 and thus the reaction pictured there cannot take place. It is prohibited by the relativistic law of nature (B8-1.1).

Virtual Particles Not So Limited

We will soon see that virtual particles are not limited by (B8-1.1) and (B8-1.2). They are never measured, but only exist fleetingly. Thus, an interaction like $e^-e^+ \to \gamma \to e^-e^+$ (Bhabha scattering of Fig. 8-2, pg. 221) where γ is a virtual photon and, as we will see, does not obey (B8-1.1) and (B8-1.2), is possible. Virtual particles have certain freedoms to ignore some laws that real particles must obey. They do this so fleetingly that we can never measure this behavior. However, the initial state and the final state (e^-e^+ and e^-e^+ in this example) can be measured and must obey (B8-1.2).

The photon of Fig. 8–1, contrary to the photon in Bhabha scattering, is not interacting again and so it is real. It is not fleeting and virtual.

Off the Mass Shell

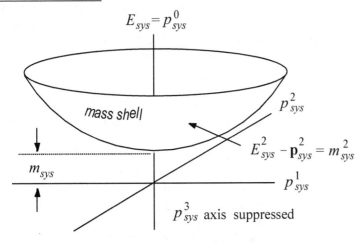

A plot of (B8-1.2) looks like the figure shown, i.e., a shell upon which real particle energies and 3-momenta values must lie.

The surface in the figure is called the <u>mass shell</u>. Real particles must be "on the mass shell" (or simply "on-shell"). Virtual particles can be, and generally are, "off the mass shell" (or simply "off-shell").

$m_{sys} = 0$ for photons, so the photon mass shell touches the origin. The photon of Fig. 8-1 is off-shell for its own mass $m_{sys} = m = 0$ (because, due to 4-momentum conservation, it is on the mass shell shown at left of the incoming particles state).

Do Prob. 18 for more insight on this.

$$S_{fi} = \langle f | S^{(1)} | i \rangle = \langle \gamma | \underbrace{S_1^{(1)} + S_2^{(1)} + \ldots + S_8^{(1)}}_{\text{all yield zero}} | e^-_{\mathbf{p}_1, r_1}, e^+_{\mathbf{p}_2, r_2} \rangle = \underbrace{\langle \gamma |}_{\substack{\text{on-shell} \\ \text{photon}}} \underbrace{S_1^{(1)} | e^-_{\mathbf{p}_1, r_1}, e^+_{\mathbf{p}_2, r_2} \rangle}_{\text{off-shell photon}} = 0 \quad (8\text{-}20)$$

because the only ket left is an off-shell photon with $k^\mu = p_1^\mu + p_2^\mu$, and that is a different state from, and thus orthogonal to, any real final state photon, which cannot have this value for k^μ. Thus, the transition amplitude for Fig. 8-1 is zero. Similar logic for all single vertex interactions means we can simply ignore $S^{(1)}$ from here on.

So we can ignore $S^{(1)}$

8.3.4 Conclusions

$S^{(1)}$ gives rise to single vertex interactions, which are not physical, i.e., they do not occur.

We can represent all eight sub-terms (vertex interactions) of the $S^{(1)}$ term via

$$S^{(1)} = ie \int d^4 x_1 N \left\{ \bar{\psi} A^\mu \gamma_\mu \psi \right\}_{x_1} \qquad \text{where} \qquad (8\text{-}21)$$

$\bar{\psi} = \bar{\psi}^+ + \bar{\psi}^-$ destroys spinor anti-particles and creates spinor particles

$\psi = \psi^+ + \psi^-$ destroys spinor particles and creates spinor anti-particles

$A^\mu = A^{\mu+} + A^{\mu-}$ destroys photons and creates photons.

$S^{(1)}$ gives rise to all possible single vertex interactions

None of these can occur physically

It is noteworthy that we have finally reached the point in QFT where we see the transmutation of particles (e^+ and e^- into a photon in Fig. 8-1) first discussed in Chap. 1. More of this below.

8.4 $S^{(2)}$ Physically

If you guessed that since $S^{(0)}$ has no vertices, and $S^{(1)}$ represented single vertex interactions, that $S^{(2)}$ will represent two vertex interactions (like the scattering of Fig. 1-1, pg. 2), you guessed right. Let's see how that happens by examining (8-8) term by term.

$S^{(2)}$ gives rise to all possible two vertex interactions

8.4.1 The No Propagator Term $S_A^{(2)}$

The first term in (8-8) is

$$S_A^{(2)} = -\frac{1}{2!} e^2 \iint d^4 x_1 d^4 x_2 N \left\{ \left(\bar{\psi} A \psi \right)_{x_1} \left(\bar{\psi} A \psi \right)_{x_2} \right\}. \qquad (8\text{-}22)$$

This represents two independent processes like $S^{(1)}$ (for example, Fig. 8-1). The two processes do not interact with one another and each behaves as if the other did not exist. There is no virtual particle (Feynman propagator) linking them. Think of two separate single vertex Feynman diagrams, each being one of the eight different types you found in Prob. 3. Neither of these can occur, as seen in Sect. 8.3, so $S_A^{(2)}$ does not represent a real physical process and is ignored in QFT.

$S_A^{(2)}$, the first sub-term in $S^{(2)}$ is like two separate $S^{(1)}$ terms and thus cannot occur

8.4.2 The Photon Propagator Term $S_B^{(2)}$

Consider the second term in (8-8) acting on an initial state

$$S_B^{(2)} | i \rangle = -\frac{1}{2!} e^2 \iint d^4 x_1 d^4 x_2 N \left\{ (\bar{\psi} A \psi)_{x_1} (\bar{\psi} A \psi)_{x_2} \right\} | i \rangle. \qquad (8\text{-}23)$$

Remember the contraction is the Feynman propagator, so (8-23) represents an interaction having a photon virtual particle. Also, the propagator is a number not an operator, so the only operators at play in (8-23) are spin ½ creation and destruction operators. Thus, the only initial states (8-23) could destroy would be electron/positron states; and the only final states it could create would be electron/positron states. We call these types of interactions four external lepton interactions.

$S_B^{(2)}$ sub-term in $S^{(2)}$ has a photon propagator and creates/destroys positrons and electrons

Let's examine what (8-23) would do to an incoming state composed of one electron and one positron with an outgoing state also composed of one electron and one positron. Note that we will not write out symbols for e^+ and e^- when it is obvious what the particles are. The time forward arrow represents e^- and the time backward arrow, its anti-particle e^+.

In Fig. 8-2, both scenarios have the same incoming and outgoing particle states, which are all that we can measure. The internal virtual particle interaction is not measurable, so there is no way we can tell which of the two interactions in Fig. 8-2 gave us the Bhabha scattering. (As an aside, our example in Chap. 1 ($e^- + e^+ \to \mu^- + \mu^+$) only had a diagram like that on the left hand side of Fig. 8-2. That case was chosen there to keep things simple.) Bhabha scattering actually entails both types of interaction, i) an annihilation of e^- with an e^+ followed by a creation of the same two types of

$S_B^{(2)}$ can represent Bhabha scattering

particles, and ii) one of the incoming particles emitting a virtual photon which is then absorbed by the other particle. In both cases, we can have the same initial and final particle states, i.e., the same incoming momenta and spins for both and the same outgoing momenta and spins for both.

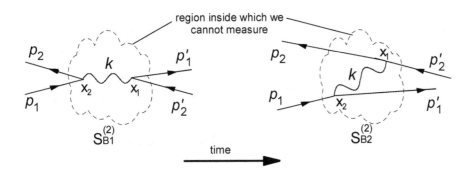

Bhabha scattering can occur in two ways, represented by two different aspects of $S_B^{(2)}$

Figure 8-2 Bhabha Scattering Can Occur in Two Ways

When we calculate the transition amplitude for Bhabha scattering, we have to add the contributions from each way in which it can take place. Thus, for the Fig. 8-2 interactions,

Bhabha transition amplitude is sum of transition amplitudes of the two different ways

$$S_{Bhabha} = S_{B1}^{(2)} + S_{B2}^{(2)}. \tag{8-24}$$

The First Type of Bhabha Scattering

Consider the first type of Bhabha scattering with the S operator acting on the incoming state,

$$S_B^{(2)}\left|e^-_{\mathbf{p}_1,r_1}, e^+_{\mathbf{p}_2,r_2}\right\rangle = -\tfrac{1}{2!}e^2 \iint d^4x_1 d^4x_2 N\left\{(\overline{\psi}A\psi)_{x_1}(\overline{\psi}A\psi)_{x_2}\right\}\left|e^-_{\mathbf{p}_1,r_1}, e^+_{\mathbf{p}_2,r_2}\right\rangle. \tag{8-25}$$

Calculating the transition amplitude for the first way

At x_2, the ψ^+ part of ψ_{x_2} would destroy the ket electron; the $\overline{\psi}^+$ part of $\overline{\psi}_{x_2}$ would destroy the ket positron; and we would be left with the vacuum $|0\rangle$. The propagator would then create a virtual photon at x_2 and propagate it to x_1 where it would be destroyed. Then, the ψ^- part of ψ_{x_1} would create a positron at x_1; and the $\overline{\psi}^-$ part of $\overline{\psi}_{x_1}$ would create an electron there.

We find the transition amplitude for the first type of Bhabha scattering represented by $S_{B1}^{(2)}$ via

$$S_{B1}^{(2)} = \left\langle e^-_{\mathbf{p}_1',r_1'}, e^+_{\mathbf{p}_2',r_2'}\right| \underbrace{S_B^{(2)}}_{\substack{\text{only opers for} \\ \text{LH of Fig 8-2}}} \left|e^-_{\mathbf{p}_1,r_1}, e^+_{\mathbf{p}_2,r_2}\right\rangle = \frac{-e^2}{2}\left\langle e^-_{\mathbf{p}_1',r_1'}, e^+_{\mathbf{p}_2',r_2'}\right| \times$$

$$\iint d^4x_1 d^4x_2 \left((\overline{\psi}^- A \psi^-)_{x_1}(\overline{\psi}^+ A \psi^+)_{x_2} + N\left\{(\overline{\psi}^+ A \psi^+)_{x_1}(\overline{\psi}^- A \psi^-)_{x_2}\right\}\right)\left|e^-_{\mathbf{p}_1,r_1}, e^+_{\mathbf{p}_2,r_2}\right\rangle, \tag{8-26}$$

in which we can think of the second term above as annihilation at x_1 and creation at x_2 instead of the other way around. Note that (8-26) drops terms in $S_B^{(2)}$ that correspond to the RH side of Fig. 8-2. When we normal order the second term, $\psi^+(x_1)$ is switched once with $\overline{\psi}^-(x_2)$, introducing a minus sign, then switched with $\psi^-(x_2)$, introducing a second minus sign and resulting in no total sign change. The propagator is just a number, so it can be moved anywhere without effect (though care has to be taken with keeping the correct spinor multiplication order).

Carrying out similar switching for $\overline{\psi}^+(x_1)$ with $\overline{\psi}^-(x_2)$ and then $\psi^-(x_2)$, we end up with

$$S_{B1}^{(2)} = \frac{-e^2}{2}\left\langle e^-_{\mathbf{p}_1',r_1'}, e^+_{\mathbf{p}_2',r_2'}\right| \iint d^4x_1 d^4x_2 \left((\overline{\psi}^- A \psi^-)_{x_1}(\overline{\psi}^+ A \psi^+)_{x_2} + (\overline{\psi}^- A \psi^-)_{x_2}(\overline{\psi}^+ A \psi^+)_{x_1}\right)\left|e^-_{\mathbf{p}_1,r_1}, e^+_{\mathbf{p}_2,r_2}\right\rangle$$

$$= -e^2\left\langle e^-_{\mathbf{p}_1',r_1'}, e^+_{\mathbf{p}_2',r_2'}\right| \iint d^4x_1 d^4x_2 (\overline{\psi}^- A \psi^-)_{x_1}(\overline{\psi}^+ A \psi^+)_{x_2}\left|e^-_{\mathbf{p}_1,r_1}, e^+_{\mathbf{p}_2,r_2}\right\rangle. \tag{8-27}$$

where we simply exchanged dummy variables in the second term of the first line to get the second line. Substituting the field equation solutions for the operators at x_2 in (8-27), we have

$$S_{B1}^{(2)} = -e^2 \left\langle e^-_{\mathbf{p}_1',r_1'}, e^+_{\mathbf{p}_2',r_2'} \right| \iint d^4x_1 d^4x_2 \left(\overline{\psi}^- \gamma^\mu \psi^- \right)_{x_1} iD_{F\mu\nu}(x_1 - x_2) \times$$

$$\left(\sum_{r',\mathbf{p}'} \sqrt{\frac{m}{VE_{\mathbf{p}'}}} d_{r'}(\mathbf{p}')\overline{v}_{r'}(\mathbf{p}')e^{-ip'x_2} \right)\gamma^\nu \left(\sum_{r'',\mathbf{p}''} \sqrt{\frac{m}{VE_{\mathbf{p}''}}} c_{r''}(\mathbf{p}'')u_{r''}(\mathbf{p}'')e^{-ip''x_2} \right) \left| e^-_{\mathbf{p}_1,r_1}, e^+_{\mathbf{p}_2,r_2} \right\rangle. \quad (8\text{-}28)$$

Note where the gamma matrices show up in (8-28) and that all terms where the destruction operators don't match the ket particle types, momenta, and spins result in zero. The terms that do match up turn the ket into the vacuum. So, (8-28) becomes

$$S_{B1}^{(2)} = -e^2 \left\langle e^-_{\mathbf{p}_1',r_1'}, e^+_{\mathbf{p}_2',r_2'} \right| \iint d^4x_1 d^4x_2 \left(\overline{\psi}^- \gamma^\mu \psi^- \right)_{x_1} iD_{F\mu\nu}(x_1 - x_2) \times$$

$$\frac{m}{V}\sqrt{\frac{1}{E_{\mathbf{p}_1}E_{\mathbf{p}_2}}}\overline{v}_{r_2}(\mathbf{p}_2)\gamma^\nu u_{r_1}(\mathbf{p}_1)e^{-ip_2 x_2}e^{-ip_1 x_2}\left|0\right\rangle. \quad (8\text{-}29)$$

Substituting the propagator relation we derived at the end of Chap. 5 and re-arranging, we find

$$S_{B1}^{(2)} = -e^2 \left\langle e^-_{\mathbf{p}_1',r_1'}, e^+_{\mathbf{p}_2',r_2'} \right| \int \left(\overline{\psi}^- \gamma^\mu \psi^- \right)_{x_1} \times$$

$$\left(\int \frac{-ig_{\mu\nu}}{(2\pi)^4} \int \frac{e^{-ik(x_1-x_2)}}{k^2 + i\varepsilon}d^4k \frac{m}{V}\sqrt{\frac{1}{E_{\mathbf{p}_1}E_{\mathbf{p}_2}}}\overline{v}_{r_2}(\mathbf{p}_2)\gamma^\nu u_{r_1}(\mathbf{p}_1)e^{-ip_2 x_2}e^{-ip_1 x_2}d^4x_2 \right)d^4x_1\left|0\right\rangle$$

$$= e^2 \left\langle e^-_{\mathbf{p}_1',r_1'}, e^+_{\mathbf{p}_2',r_2'} \right| \int \left(\overline{\psi}^- \gamma^\mu \psi^- \right)_{x_1} ig_{\mu\nu}\overline{v}_{r_2}(\mathbf{p}_2)\gamma^\nu u_{r_1}(\mathbf{p}_1) \times$$

$$\frac{m}{V}\sqrt{\frac{1}{E_{\mathbf{p}_1}E_{\mathbf{p}_2}}}\left(\int \frac{1}{k^2+i\varepsilon}e^{-ikx_1}\frac{1}{(2\pi)^4}\underbrace{\left(\int e^{ikx_2}e^{-ip_2 x_2}e^{-ip_1 x_2}d^4x_2 \right)}_{(2\pi)^4\delta^{(4)}(k-p_2-p_1)}d^4k \right)d^4x_1\left|0\right\rangle. \quad (8\text{-}30)$$

The delta relation that results above tells us that $k = p_1 + p_2$, the virtual photon 4-momentum equals the sum of the incoming particles 4-momenta. Again, we see conservation of energy and 3-momentum at a vertex, a consistently recurring theme. And again, we see that the photon virtual particle is off-shell, i.e., $k^2 \neq 0$.

The delta relation picks the single value $k = p_1 + p_2$ out of the integration over k. So now,

$$S_{B1}^{(2)} = e^2 \left\langle e^-_{\mathbf{p}_1',r_1'}, e^+_{\mathbf{p}_2',r_2'} \right| \frac{m}{V}\sqrt{\frac{1}{E_{\mathbf{p}_1}E_{\mathbf{p}_2}}}\left(\int \left(\overline{\psi}^- \gamma^\mu \psi^- \right)_{x_1}\frac{e^{-i(p_2+p_1)x_1}}{(p_2+p_1)^2+i\varepsilon}d^4x_1 \right)ig_{\mu\nu}\overline{v}_{r_2}(\mathbf{p}_2)\gamma^\nu u_{r_1}(\mathbf{p}_1)\left|0\right\rangle. \quad (8\text{-}31)$$

Substituting the relations for the spinor creation field operators at x_1 into (8-31) yields

$$S_{B1}^{(2)} = e^2 \frac{m}{V}\sqrt{\frac{1}{E_{\mathbf{p}_1}E_{\mathbf{p}_2}}}\left\langle e^-_{\mathbf{p}_1',r_1'}, e^+_{\mathbf{p}_2',r_2'} \right| \left(\int \frac{e^{-i(p_2+p_1)x_1}}{(p_2+p_1)^2+i\varepsilon}\left(\sum_{r,\mathbf{p}} \sqrt{\frac{m}{VE_{\mathbf{p}}}}c_r^\dagger(\mathbf{p})\overline{u}_r(\mathbf{p})e^{ipx_1} \right)\gamma^\mu \times \right.$$

$$\left. \left(\sum_{r',\mathbf{p}'} \sqrt{\frac{m}{VE_{\mathbf{p}'}}}d_{r'}^\dagger(\mathbf{p}')v_{r'}(\mathbf{p}')e^{ip'x_1} \right)d^4x_1 \right)ig_{\mu\nu}\overline{v}_{r_2}(\mathbf{p}_2)\gamma^\nu u_{r_1}(\mathbf{p}_1)\left|0\right\rangle. \quad (8\text{-}32)$$

The creation operators will turn the ket into a sum of eigenstates, but only the state in that sum that matches the bra will provide a non-zero result (from the inner product of the bra and ket equaling zero unless they are the same state.) Thus, (8-32) becomes

$$S_{B1}^{(2)} = e^2 \left(\frac{m}{V} \right)^2 \sqrt{\frac{1}{E_{\mathbf{p}_1}E_{\mathbf{p}_2}}}\sqrt{\frac{1}{E_{\mathbf{p}_1'}E_{\mathbf{p}_2'}}}\left\langle e^-_{\mathbf{p}_1',r_1'}, e^+_{\mathbf{p}_2',r_2'} \right| \left(\int e^{-i(p_2+p_1)x_1}e^{i(p_1'+p_2')x_1}d^4x_1 \right) \times$$

$$\overline{u}_{r_1'}(\mathbf{p}_1')\gamma^\mu v_{r_2'}(\mathbf{p}_2')\frac{ig_{\mu\nu}}{(p_2+p_1)^2+i\varepsilon}\overline{v}_{r_2}(\mathbf{p}_2)\gamma^\nu u_{r_1}(\mathbf{p}_1)\left| e^-_{\mathbf{p}_1',r_1'}, e^+_{\mathbf{p}_2',r_2'} \right\rangle. \quad (8\text{-}33)$$

The integral over x_1 is another delta function telling us that $p_2' + p_1' = p_2 + p_1$, the total outgoing 4-momentum equals the total incoming 4-momentum, which, as we discovered earlier, equals the 4-momentum of the virtual photon propagator. Energy and 3-momentum are conserved at every vertex. And the final, real particle state is on-shell.

The result of all this, where we note that the fraction factor in (8-33) with the $i\varepsilon$ as part of the denominator is simply (up to a sign) the Feynman propagator in momentum space, is

$$\underbrace{S_{B1}^{(2)}}_{\substack{\text{transition}\\\text{amplitude}}} = \sqrt{\frac{m}{VE_{\mathbf{p}_1}}}\sqrt{\frac{m}{VE_{\mathbf{p}_2}}}\sqrt{\frac{m}{VE_{\mathbf{p}_1'}}}\sqrt{\frac{m}{VE_{\mathbf{p}_2'}}}\underbrace{\left\langle e_{\mathbf{p}_1',r_1'}^-,e_{\mathbf{p}_2',r_2'}^+\Big|\Big|e_{\mathbf{p}_1',r_1'}^-,e_{\mathbf{p}_2',r_2'}^+\right\rangle}_{=1}(2\pi)^4\,\delta^{(4)}\left(p_1+p_2-\left(p_1'+p_2'\right)\right)$$

$$\times\underbrace{\left(-e^2\right)\bar{u}_{r_1'}(\mathbf{p}_1')\gamma^\mu v_{r_2'}(\mathbf{p}_2')iD_{F\mu\nu}\left(k=p_1+p_2\right)\bar{v}_{r_2}(\mathbf{p}_2)\gamma^\nu u_{r_1}(\mathbf{p}_1)}_{\text{Feynman amplitude }\mathcal{M}_{B1}^{(2)}}. \tag{8-34}$$

Result: transition amplitude for the first way

You may wish to do Prob. 4 now to compare the Chap. 1 transition amplitude mock "calculation" (for final $\mu^-\,\mu^+$ instead of $e^-\,e^+$) to the precise steps (8-25) to (8-34). Note one major difference. Operation, right to left is in *normal*, *not time* order (as implied earlier to simplify things).

Unlike implied in Chap. 1 heuristic example, normal, not time, order

Note the definition of the <u>Feynman amplitude</u> for this interaction, $\mathcal{M}_{B1}^{(2)}$ (don't confuse with the transition amplitude of which it is a part). Using that, we compress (8-34) into

$$S_{B1}^{(2)} = \left(\prod_{\mathbf{p}}^{\text{all ext fermions}}\sqrt{\frac{m}{VE_{\mathbf{p}}}}\right)(2\pi)^4\,\delta^{(4)}\left(p_1+p_2-\left(p_1'+p_2'\right)\right)\mathcal{M}_{B1}^{(2)}. \tag{8-35}$$

Feynman amplitude is part of transition amplitude

The Second Type of Bhabha Scattering

Note in Fig. 8-2, that on the RHS, either x_2 or x_1 could occur first, and our math takes care of both cases automatically. We start with the same fundamental relation (8-25) we had for the first type of Bhabha scattering,

$$S_B^{(2)}\left|e_{\mathbf{p}_1,r_1}^-,e_{\mathbf{p}_2,r_2}^+\right\rangle = -\frac{e^2}{2!}\iint d^4x_1 d^4x_2\,N\left\{(\overline{\psi}\slashed{A}\psi)_{x_1}(\overline{\psi}\slashed{A}\psi)_{x_2}\right\}\left|e_{\mathbf{p}_1,r_1}^-,e_{\mathbf{p}_2,r_2}^+\right\rangle, \tag{8-36}$$

Calculating the transition amplitude for the second way

but this time, we pick the creation and destruction operators corresponding to the second type of Bhabha scattering. That is, instead of (8-27), we have

$$S_{B2}^{(2)} = \left\langle e_{\mathbf{p}_1',r_1'}^-,e_{\mathbf{p}_2',r_2'}^+\right|\underbrace{S_B^{(2)}}_{\substack{\text{only opers for}\\\text{RH of Fig 8-2}}}\left|e_{\mathbf{p}_1,r_1}^-,e_{\mathbf{p}_2,r_2}^+\right\rangle = \frac{-e^2}{2}\left\langle e_{\mathbf{p}_1',r_1'}^-,e_{\mathbf{p}_2',r_2'}^+\right|\times$$

$$\iint d^4x_1 d^4x_2\,N\left\{(\overline{\psi}^+A_\mu\gamma^\mu\psi^-)_{x_1}(\overline{\psi}^-A_\nu\gamma^\nu\psi^+)_{x_2}+(\overline{\psi}^-A_\mu\gamma^\mu\psi^+)_{x_1}(\overline{\psi}^+A_\nu\gamma^\nu\psi^-)_{x_2}\right\}\left|e_{\mathbf{p}_1,r_1}^-,e_{\mathbf{p}_2,r_2}^+\right\rangle, \tag{8-37}$$

which, if we switch dummy integration variables again, the first and second terms above are equivalent. That eliminates the ½ factor and leaves (where we now show spinor indices)

$$S_{B2}^{(2)} = -e^2\left\langle e_{\mathbf{p}_1',r_1'}^-,e_{\mathbf{p}_2',r_2'}^+\right|\iint d^4x_1 d^4x_2\,N\left\{(\overline{\psi}_\alpha^+A_\mu\gamma_{\alpha\beta}^\mu\psi_\beta^-)_{x_1}(\overline{\psi}_\delta^-A_\nu\gamma_{\delta\eta}^\nu\psi_\eta^+)_{x_2}\right\}\left|e_{\mathbf{p}_1,r_1}^-,e_{\mathbf{p}_2,r_2}^+\right\rangle. \tag{8-38}$$

This can be visualized as the incoming electron destroyed at x_2 with the outgoing electron created at x_2 and a photon propagator starting at x_2. The incoming positron is destroyed at x_1 with the outgoing positron created at x_1 and the photon propagator ending at x_1. The reverse situation, where x_1 and x_2 are interchanged has been included by the factor of two enveloped in going from (8-37) to (8-38).

Normal ordering of (8-38), where we now see the value of writing out spinor indices, gives us

$$S_{B2}^{(2)} = e^2\left\langle e_{\mathbf{p}_1',r_1'}^-,e_{\mathbf{p}_2',r_2'}^+\right|\iint d^4x_1 d^4x_2\left((\overline{\psi}_\delta^-)_{x_2}(A_\mu\gamma_{\alpha\beta}^\mu\psi_\beta^-)_{x_1}(\overline{\psi}_\alpha^+)_{x_1}(A_\nu\gamma_{\delta\eta}^\nu\psi_\eta^+)_{x_2}\right)\left|e_{\mathbf{p}_1,r_1}^-,e_{\mathbf{p}_2,r_2}^+\right\rangle. \tag{8-39}$$

Note carefully that we have to put (8-39) not simply in any normal order, but in the same normal order as the bottom row of (8-27). That is, both transition sub amplitudes must have, in order from the right side moving leftward, operators performing e^- destruction, e^+ destruction, e^+ creation, and e^- creation. If we are to add transition amplitudes as in (8-24), the operations on the kets in each case must be in the same order. As it turns out, doing this results in the necessary sign to give us the correct total transition amplitudes, such as (8-24) for the present case of Bhabha scattering.

Inserting destruction field operator solutions and the propagator symbol into (8-39), we get

$$S_{B2}^{(2)} = e^2\left\langle e_{\mathbf{p}_1',r_1'}^-,e_{\mathbf{p}_2',r_2'}^+\right|\iint d^4x_1 d^4x_2(\overline{\psi}_\delta)_{x_2}(\psi_\beta)_{x_1}iD_{F\mu\nu}(x_1-x_2)\gamma_{\alpha\beta}^\mu\times$$

$$\left(\sum_{r',\mathbf{p}'}\sqrt{\frac{m}{VE_{\mathbf{p}'}}}d_{r'}(\mathbf{p}')\bar{v}_{r',\alpha}(\mathbf{p}')e^{-ip'x_1}\right)\gamma_{\delta\eta}^\nu\left(\sum_{r'',\mathbf{p}''}\sqrt{\frac{m}{VE_{\mathbf{p}''}}}c_{r''}(\mathbf{p}'')u_{r'',\eta}(\mathbf{p}'')e^{-ip''x_2}\right)\left|e_{\mathbf{p}_1,r_1}^-,e_{\mathbf{p}_2,r_2}^+\right\rangle. \tag{8-40}$$

We carry out the remaining steps below with little comment. I have tried not to skip steps, so hopefully you can follow them without too much difficulty.

$$S_{B2}^{(2)} = e^2 \left\langle e_{\mathbf{p}_1',r_1'}^-, e^+{}_{\mathbf{p}_2',r_2'} \left| \iint d^4x_2 d^4x_1 (\overline{\psi}_\delta)_{x_2} (\psi_{\overline{\beta}})_{x_1} \gamma_{\alpha\beta}^\mu \times \right. \right.$$

$$\left. \left(\frac{-ig_{\mu\nu}}{(2\pi)^4} \int \frac{e^{-ik(x_1-x_2)}}{k^2+i\varepsilon} d^4k \right) \frac{m}{V} \sqrt{\frac{1}{E_{\mathbf{p}_1}E_{\mathbf{p}_2}}} \overline{v}_{r_2,\alpha}(\mathbf{p}_2) \gamma_{\delta\eta}^\nu u_{r_1,\eta}(\mathbf{p}_1) e^{-ip_2 x_1} e^{-ip_1 x_2} \left| 0 \right\rangle \right. \quad (8\text{-}41)$$

$$= e^2 \left\langle e_{\mathbf{p}_1',r_1'}^-, e^+{}_{\mathbf{p}_2',r_2'} \left| \iint d^4x_2 d^4x_1 \left(\sum_{r,\mathbf{p}} \sqrt{\frac{m}{VE_{\mathbf{p}}}} c_r^\dagger(\mathbf{p}) \overline{u}_{r,\delta}(\mathbf{p}) e^{ipx_2} \right) \left(\sum_{r',\mathbf{p}'} \sqrt{\frac{m}{VE_{\mathbf{p}'}}} d_{r'}^\dagger(\mathbf{p}') v_{r',\beta}(\mathbf{p}') e^{ip'x_1} \right) \times \right. \right.$$

$$\left. \gamma_{\alpha\beta}^\mu \left(\frac{-ig_{\mu\nu}}{(2\pi)^4} \int \frac{e^{-ik(x_1-x_2)}}{k^2+i\varepsilon} d^4k \right) \frac{m}{V} \sqrt{\frac{1}{E_{\mathbf{p}_1}E_{\mathbf{p}_2}}} \overline{v}_{r_2,\alpha}(\mathbf{p}_2) \gamma_{\delta\eta}^\nu u_{r_1,\eta}(\mathbf{p}_1) e^{-ip_2 x_1} e^{-ip_1 x_2} \left| 0 \right\rangle \right. \quad (8\text{-}42)$$

$$= e^2 \frac{m^2}{V^2} \sqrt{\frac{1}{E_{\mathbf{p}_1}E_{\mathbf{p}_2}E_{\mathbf{p}_1'}E_{\mathbf{p}_2'}}} \left\langle e_{\mathbf{p}_1',r_1'}^-, e^+{}_{\mathbf{p}_2',r_2'} \left| \iint d^4x_2 d^4x_1 \, \overline{u}_{r_1',\delta}(\mathbf{p}_1') e^{ip_1'x_2} v_{r_2',\beta}(\mathbf{p}_2') e^{ip_2'x_1} \gamma_{\alpha\beta}^\mu \times \right. \right.$$

$$\left. \left(\frac{-ig_{\mu\nu}}{(2\pi)^4} \int \frac{e^{-ik(x_1-x_2)}}{k^2+i\varepsilon} d^4k \right) \overline{v}_{r_2,\alpha}(\mathbf{p}_2) \gamma_{\delta\eta}^\nu u_{r_1,\eta}(\mathbf{p}_1) e^{-ip_2 x_1} e^{-ip_1 x_2} \left| e_{\mathbf{p}_1',r_1'}^-, e^+{}_{\mathbf{p}_2',r_2'} \right\rangle + \left(\begin{array}{c} \text{terms with} \\ \text{bra} \neq \text{ket} \end{array} \right). \quad (8\text{-}43)$$

$$= e^2 \frac{m^2}{V^2} \sqrt{\frac{1}{E_{\mathbf{p}_1}E_{\mathbf{p}_2}E_{\mathbf{p}_1'}E_{\mathbf{p}_2'}}} \left\langle e_{\mathbf{p}_1',r_1'}^-, e^+{}_{\mathbf{p}_2',r_2'} \left\| e_{\mathbf{p}_1',r_1'}^-, e^+{}_{\mathbf{p}_2',r_2'} \right\rangle \overline{v}_{r_2,\alpha}(\mathbf{p}_2) \gamma_{\alpha\beta}^\mu v_{r_2',\beta}(\mathbf{p}_2') \overline{u}_{r_1',\delta}(\mathbf{p}_1') \gamma_{\delta\eta}^\nu u_{r_1,\eta}(\mathbf{p}_1) \times \right.$$

$$\int \left(\left(\frac{-ig_{\mu\nu}}{(2\pi)^4} \int \left(\underbrace{\int e^{ikx_2} e^{ip_1'x_2} e^{-ip_1 x_2} d^4x_2}_{(2\pi)^4 \delta^{(4)}(k-(p_1-p_1'))} \right) \frac{e^{-ikx_1}}{k^2+i\varepsilon} d^4k \right) e^{ip_2'x_1} e^{-ip_2 x_1} \right) d^4x_1 \; + 0 + 0 + ... \quad (8\text{-}44)$$

Note the delta function means the virtual photon four-momentum k must equal the loss in four-momentum from the electron labeled #1 in the RHS of Fig. 8-2. Continuing, we have

$$S_{B2}^{(2)} = e^2 \frac{m^2}{V^2} \sqrt{\frac{1}{E_{\mathbf{p}_1}E_{\mathbf{p}_2}E_{\mathbf{p}_1'}E_{\mathbf{p}_2'}}} \overline{v}_{r_2}(\mathbf{p}_2) \gamma^\mu v_{r_2'}(\mathbf{p}_2') \overline{u}_{r_1'}(\mathbf{p}_1') \gamma^\nu u_{r_1}(\mathbf{p}_1) \times$$

$$\left(\frac{-ig_{\mu\nu}}{(p_1-p_1')^2+i\varepsilon} \right) \left(\underbrace{\int e^{-i(p_1-p_1')x_1} e^{ip_2'x_1} e^{-ip_2 x_1} d^4x_1}_{(2\pi)^4 \delta^{(4)}(p_1+p_2-(p_1'+p_2'))} \right). \quad (8\text{-}45)$$

And thus, finally,

$$S_{B2}^{(2)} = \left(\prod_{\mathbf{p}}^{\text{all ext fermions}} \sqrt{\frac{m}{VE_{\mathbf{p}}}} \right) (2\pi)^4 \delta^{(4)} \left(p_1 + p_2 - (p_1' + p_2') \right) \mathcal{M}_{B2}^{(2)}, \quad (8\text{-}46)$$

Result: transition amplitude for the second way

where

$$\mathcal{M}_{B2}^{(2)} = e^2 \overline{v}_{r_2}(\mathbf{p}_2) \gamma^\mu v_{r_2'}(\mathbf{p}_2') i D_{\mu\nu}(p_2'-p_2) \overline{u}_{r_1'}(\mathbf{p}_1') \gamma^\nu u_{r_1}(\mathbf{p}_1). \quad (8\text{-}47)$$

The Complete Bhabha Scattering Transition Amplitude

The total transition amplitude for 2nd order ($n = 2$) Bhabha scattering, with (8-46) and (8-35), is

$$S_{Bhabba} = S_{B1}^{(2)} + S_{B2}^{(2)} = \left(\prod_{\mathbf{p}}^{\text{all ext fermions}} \sqrt{\frac{m}{VE_{\mathbf{p}}}} \right) (2\pi)^4 \delta^{(4)} \left(p_1 + p_2 - (p_1' + p_2') \right) \left(\mathcal{M}_{B1}^{(2)} + \mathcal{M}_{B2}^{(2)} \right), (8\text{-}48)$$

Complete transition amplitude = sum of those for both ways

where we see that once again the incoming total 4-momentum equals the outgoing total 4-momentum, and that from the intermediate steps, 4-momentum is conserved at every vertex.

"Whew…" you are saying, with some justification, "that was a lot of algebra". And we haven't even calculated the gamma matrices and spinor multiplications or the sums over μ and ν (which we will do in calculating cross sections near the end of the book). Typically, each interaction we determine the transition amplitude for involves at least this much algebra and calculation.

Feynman Rules to the Rescue

Fortunately, that is where Feynman rules, which you may have heard of, come in. They allow us, with an associated Feynman diagram as a guide, to simply write out transition amplitudes like (8-35) and (8-46), fairly straightforwardly, with much less work.

Soon we will see how Feynman rules cut all this algebra

Before looking at Feynman rules, however, it will strengthen our understanding if I do another transition amplitude the long way for you here, and you do one for yourself as an exercise. My example will be Compton scattering (next section), and yours will be Møller scattering (Fig. 8-4).

Møller Scattering, Another Application of $S_B^{(2)}$

Note that Møller scattering (to be analyzed in Sect. 8.4.4) entails two initial electrons and two final electrons (four external fermions) with a photon virtual particle transmitting the force and so will be described by (8-23), as well. Compton scattering (see Fig. 8-3), on the other hand, entails an initial photon and an electron, a final photon and electron, and an electron propagator. That cannot be handled with (8-23). But it can be handled by other terms in $S^{(2)}$, as discussed in the next section.

8.4.3 The Fermion Propagator Term $S_C^{(2)}$

Consider the third and fourth terms (second and third terms in 2nd row inside bracket) in (8-8) acting on an initial state.

$$S_C^{(2)}|i\rangle = -\tfrac{1}{2!}e^2 \iint d^4x_1 d^4x_2 \left(N\left\{ (\overline{\psi}A\psi)_{x_1}(\overline{\psi}A\psi)_{x_2} \right\} + N\left\{ (\overline{\psi}A\psi)_{x_1}(\overline{\psi}A\psi)_{x_2} \right\} \right)|i\rangle . \quad (8\text{-}49)$$

The first term in (8-49) actually equals the second term as we can see from Box 8-2, and thus, we can re-express (8-49) as

$$S_C^{(2)}|i\rangle = -e^2 \iint d^4x_1 d^4x_2 N\left\{ (\overline{\psi}A\psi)_{x_1}(\overline{\psi}A\psi)_{x_2} \right\}|i\rangle . \quad (8\text{-}50)$$

The contraction here is a fermion propagator and the incoming particles that could be destroyed by (8-50) could comprise an electron and a photon. Effectively, the electron and photon could scatter off one another as in Fig. 8-3. This is called <u>Compton scattering</u>. Note from Fig. 8-3, that it can occur in two different ways, i.e., have the same real particles in and out.

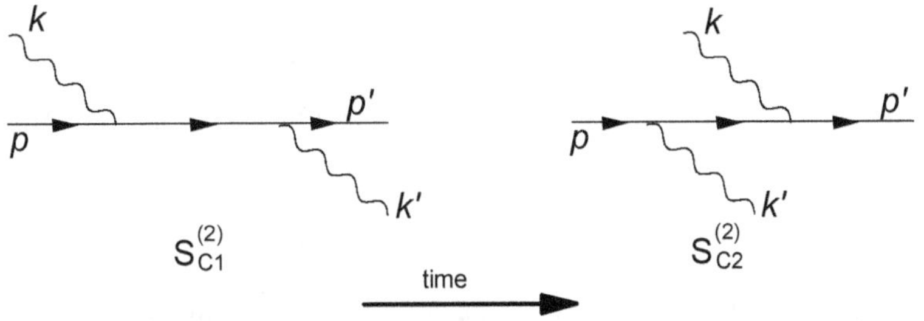

Figure 8-3. Compton Scattering Can Occur in Two Ways

It might be better if we labeled the LH vertex on each side of Fig. 8-3 with x_2, and the RH with x_1, as it would be easier to track the analysis. But, it is common practice to not write in these labels in Feynman diagrams, and we leave them out here, so the reader can get accustomed to the practice.

Remember all we can measure are the incoming and outgoing particles. We cannot measure the virtual particle, i.e., we cannot measure the actual interaction. So, that interaction could occur in either of the ways we label above as $S_{C1}^{(2)}$ or $S_{C2}^{(2)}$. That is, either i) an electron could absorb a photon (equivalent to an electron and a photon being destroyed and a virtual electron being created at x_2), and later emit a photon (equivalent to the virtual electron being destroyed and both a real electron and a real photon being created at x_1); or ii) an electron could emit a photon (equivalent to an electron being destroyed and a real photon along with a virtual electron being created at x_2), and later absorb a photon (equivalent to the virtual electron and a real photon being destroyed while a real electron is created at x_1).

Note that only the $S_C^{(2)}$ terms of all the $n = 2$ terms will result in destruction of an initial electron and photon ket, and then creation of a final electron and photon ket. The S matrix transition amplitude for second order Compton scattering is thus (with incoming particles unprimed, outgoing primed)

$S_B^{(2)}$ can also handle Møller scattering, as we will see later

$S_C^{(2)}$ has two sub-terms which have fermion propagators and can create/ destroy electrons and photons

These two sub-terms are equivalent mathematically and can be combined into one expression

$S_C^{(2)}$ can describe Compton scattering

We can only measure incoming and outgoing real particles, not virtuals

So interaction could occur in either of two ways, and we don't know which

$S_C^{(2)}$ is only $S^{(2)}$ term that will describe Compton scattering

---

Box 8-2. Showing the First Term in $S_C^{(2)}$ Equals the Second Term

Normal Ordered Commutator/Anti-commutator Equals Zero

Consider normal ordering the commutator of two boson fields which do not commute, such as

$$N\underbrace{\left[\phi^+(x_1),\phi^{\dagger-}(x_2)\right]}_{\neq 0} = N\left\{\phi^+(x_1)\phi^{\dagger-}(x_2)-\phi^{\dagger-}(x_2)\phi^+(x_1)\right\} = \phi^{\dagger-}(x_2)\phi^+(x_1)-\phi^{\dagger-}(x_2)\phi^+(x_1)=0 . \quad \text{(B8-2.1)}$$

Similarly, as you can prove by doing Prob. 5 part a), the normal ordering of an anti-commutator of two fermion fields, which do not anti-commute, equals zero as well.

Adjacent Boson or Fermion Fields Inside a Normal Ordered Product

Thus, we can simply exchange order of adjacent bosons inside any normal ordered product, even if they don't commute and even if they are part of a contraction. Since this is true for the parts ϕ^+, ϕ^-, $\phi^{\dagger+}$, and $\phi^{\dagger-}$ of ϕ and $\phi^\dagger$, it is true more generally for ϕ and $\phi^\dagger$. A parallel statement can be made for ψ and $\bar\psi$. Thus, for example,

$$N\left\{\phi(x_1)\phi^\dagger(x_2)\psi(x_3)\bar\psi(x_4)...\right\} = N\left\{\left(\phi^\dagger(x_2)\phi(x_1)+\underbrace{\left[\phi(x_1),\phi^\dagger(x_2)\right]}_{\substack{\text{drops out due to}\\\text{normal ordering}}}\right)\psi(x_3)\bar\psi(x_4)...\right\}$$

$$\text{(B8-2.2)}$$

$$= N\left\{\phi^\dagger(x_2)\phi(x_1)\psi(x_3)\bar\psi(x_4)...\right\}.$$

Similarly, we can exchange the order of adjacent fermion fields, along with a sign change, inside any normal ordered product, as you can prove by doing Prob. 5, part b). Fermions and adjacent bosons always commute, as we know.

The First Term in $S_C^{(2)}$

With the above results, we can re-order factors in the first term of $S_C^{(2)}$. Note there are two sign changes after the first equal sign below (meaning no sign change) because fermion/fermion exchanges are made twice.

$$N\left\{(\bar\psi_\alpha A_\mu \gamma^\mu_{\alpha\beta}\psi_\beta)_{x_1}(\bar\psi_\delta A_\nu \gamma^\nu_{\delta\eta}\psi_\eta)_{x_2}\right\} = N\left\{(\bar\psi_\delta)_{x_2}(\bar\psi_\alpha A_\mu \gamma^\mu_{\alpha\beta}\psi_\beta)_{x_1}(A_\nu \gamma^\nu_{\delta\eta}\psi_\eta)_{x_2}\right\} =$$

$$N\left\{(\bar\psi_\delta A_\nu \gamma^\nu_{\delta\eta})_{x_2}(\bar\psi_\alpha A_\mu \gamma^\mu_{\alpha\beta}\psi_\beta)_{x_1}(\psi_\eta)_{x_2}\right\} = N\left\{(\bar\psi_\delta A_\nu \gamma^\nu_{\delta\eta})_{x_2}(\bar\psi_\alpha)_{x_1}(\psi_\eta)_{x_2}(-A_\mu \gamma^\mu_{\alpha\beta}\psi_\beta)_{x_1}\right\} = \quad \text{(B8-2.3)}$$

$$N\left\{(\bar\psi_\delta A_\nu \gamma^\nu_{\delta\eta}(-\psi_\eta))_{x_2}(-\bar\psi_\alpha A_\mu \gamma^\mu_{\alpha\beta}\psi_\beta)_{x_1}\right\} = N\left\{(\bar\psi_\delta A_\nu \gamma^\nu_{\delta\eta}\psi_\eta)_{x_2}(\bar\psi_\alpha A_\mu \gamma^\mu_{\alpha\beta}\psi_\beta)_{x_1}\right\}.$$

In the integral of (8-49), the x_1 and x_2 are dummy variables so we can make the exchange $x_1 \leftrightarrow x_2$.

$$S_{Compton} = \langle f|S|i\rangle = \left\langle e^-_{\mathbf{p}',s'},\gamma_{\mathbf{k}',r'}\left|\sum_n S^{(n)}\right|e^-_{\mathbf{p},s},\gamma_{\mathbf{k},r}\right\rangle \approx \left\langle e^-_{\mathbf{p}',s'},\gamma_{\mathbf{k}',r'}\left|S_C^{(2)}\right|e^-_{\mathbf{p},s},\gamma_{\mathbf{k},r}\right\rangle$$

$$= \left\langle e^-_{\mathbf{p}',s'},\gamma_{\mathbf{k}',r'}\left|(-e^2)\iint d^4x_1 d^4x_2 N\left\{(\bar\psi A \psi)_{x_1}(\bar\psi A \psi)_{x_2}\right\}\right|e^-_{\mathbf{p},s},\gamma_{\mathbf{k},r}\right\rangle \quad \text{(8-51)}$$

Transition amplitude for Compton scattering

$$= -e^2 \left\langle e^-_{\mathbf{p}',s'},\gamma_{\mathbf{k}',r'}\left|\iint d^4x_1 d^4x_2 N\left\{\left(\bar\psi^+ +\bar\psi^-\right)_{x_1}\left(A^+ + A^-\right)_{x_1} \times \right.\right.\right.$$

$$\left.\left.\left.\left(iS_F(x_1-x_2)\right)\left(A^+ + A^-\right)_{x_2}\left(\psi^+ + \psi^-\right)_{x_2}\right\}\right|e^-_{\mathbf{p},s},\gamma_{\mathbf{k},r}\right\rangle . \quad \text{(8-52)}$$

After the operators raise and lower the ket, only two terms in (8-52) remain (i.e., have identical bra and ket). They are

$$S_{Compton} = -e^2 \left\langle e^-_{\mathbf{p}',s'},\gamma_{\mathbf{k}',r'}\left|\iint d^4x_1 d^4x_2 N\{\underbrace{\bar\psi^-_{x_1}A^-_{x_1}\left(iS_F(x_1-x_2)\right)A^+_{x_2}\psi^+_{x_2}}_{\text{will result in }S_{C1}^{(2)}\text{ term}}\right.\right.$$

One part for each way Compton scattering can occur

$$+ \underbrace{\bar\psi^-_{x_1}A^+_{x_1}\left(iS_F(x_1-x_2)\right)A^-_{x_2}\psi^+_{x_2}}_{\text{will result in }S_{C2}^{(2)}\text{ term}}\}\left|e^-_{\mathbf{p},s},\gamma_{\mathbf{k},r}\right\rangle . \quad \text{(8-53)}$$

Deriving transition amplitude contribution from first way Compton scattering can occur

First type of Compton Scattering

Continuing with only the first term in (8-53), we have (with Dirac indices explicitly shown)

$$\langle f | S_{C1}^{(2)} | i \rangle = \left\langle e_{\mathbf{p}',s'}^{-}, \gamma_{\mathbf{k}',r'} \left| S_{C1}^{(2)} \right| e_{\mathbf{p},s}^{-}, \gamma_{\mathbf{k},r} \right\rangle = \left\langle e_{\mathbf{p}',s'}^{-}, \gamma_{\mathbf{k}',r'} \right| \left(-e^2\right) \iint d^4x_1 d^4x_2 \times$$

$$N \left\{ \begin{array}{l} \left(\sum_{s'',\mathbf{p}''} \sqrt{\dfrac{m}{VE_{\mathbf{p}''}}} c_{s''}^{\dagger}(\mathbf{p}'') \overline{u}_{s'',\alpha}(\mathbf{p}'') e^{ip''x_1} \right) \left(\sum_{r'',\mathbf{k}''} \sqrt{\dfrac{1}{2V\omega_{\mathbf{k}''}}} a_{r''}^{\dagger} \varepsilon_{\mu,r''}(\mathbf{k}'') e^{ik''x_1} \gamma_{\alpha\beta}^{\mu} \right) \times \\[2mm] \dfrac{1}{(2\pi)^4} \int d^4q\, iS_{F\beta\delta}(q) e^{-iq(x_1-x_2)} \times \\[2mm] \left(\sum_{r''',\mathbf{k}'''} \sqrt{\dfrac{1}{2V\omega_{\mathbf{k}'''}}} a_{r'''} \varepsilon_{\nu,r'''}(\mathbf{k}''') e^{-ik'''x_2} \gamma_{\delta\eta}^{\nu} \right) \left(\sum_{s''',\mathbf{p}'''} \sqrt{\dfrac{m}{VE_{\mathbf{p}'''}}} c_{s'''}(\mathbf{p}''') u_{s''',\eta}(\mathbf{p}''') e^{-ip'''x_2} \right) \end{array} \right\} \left| e_{\mathbf{p},s}^{-}, \gamma_{\mathbf{k},r} \right\rangle. \quad (8\text{-}54)$$

The quantity to be normal ordered happens to already be normal ordered (all destruction operators are on the RHS). And each term in the summations of the last two factors before the ket will destroy the ket (leave zero), except for terms that match the ket particles' momenta and spins, and those will leave the vacuum. So,

$$\left\langle e_{\mathbf{p}',s'}^{-}, \gamma_{\mathbf{k}',r'} \left| S_{C1}^{(2)} \right| e_{\mathbf{p},s}^{-}, \gamma_{\mathbf{k},r} \right\rangle = \left\langle e_{\mathbf{p}',s'}^{-}, \gamma_{\mathbf{k}',r'} \right| \left(-e^2\right) \iint d^4x_1 d^4x_2 \left(\sum_{s'',\mathbf{p}''} \sqrt{\dfrac{m}{VE_{\mathbf{p}''}}} c_{s''}^{\dagger}(\mathbf{p}'') \overline{u}_{s'',\alpha}(\mathbf{p}'') e^{ip''x_1} \right) \times$$

$$\left(\sum_{r'',\mathbf{k}''} \sqrt{\dfrac{1}{2V\omega_{\mathbf{k}''}}} a_{r''}^{\dagger} \varepsilon_{\mu,r''}(\mathbf{k}'') e^{ik''x_1} \gamma_{\alpha\beta}^{\mu} \right) \dfrac{1}{(2\pi)^4} \int d^4q\, iS_{F\beta\delta}(q) e^{-iq(x_1-x_2)} \times \quad (8\text{-}55)$$

$$\left(\sqrt{\dfrac{1}{2V\omega_{\mathbf{k}}}} \varepsilon_{\nu,r}(\mathbf{k}) e^{-ikx_2} \right) \gamma_{\delta\eta}^{\nu} \left(\sqrt{\dfrac{m}{VE_{\mathbf{p}}}} u_{s,\eta}(\mathbf{p}) e^{-ipx_2} \right) |0\rangle.$$

Each term in the summations in the first and second rows above will create a particle in the ket, and for each of these, only numbers sandwiched between the ket and bra will remain. The numbers can be moved outside the respective bras and kets, and thus, the bra and ket in each case will form an inner product multiplied by a number.

$$\left\langle e_{\mathbf{p}',s'}^{-}, \gamma_{\mathbf{k}',r'} \left| S_{C1}^{(2)} \right| e_{\mathbf{p},s}^{-}, \gamma_{\mathbf{k},r} \right\rangle = \left\langle e_{\mathbf{p}',s'}^{-}, \gamma_{\mathbf{k}',r'} \right| \left(-e^2\right) \iint d^4x_1 d^4x_2 \times$$

$$\sum_{s'',\mathbf{p}''} \sum_{r'',\mathbf{k}''} \left\{ \begin{array}{l} \left(\sqrt{\dfrac{m}{VE_{\mathbf{p}''}}} \overline{u}_{s'',\alpha}(\mathbf{p}'') e^{ip''x_1} \right) \left(\sqrt{\dfrac{1}{2V\omega_{\mathbf{k}''}}} \varepsilon_{\mu,r''}(\mathbf{k}'') e^{ik''x_1} \gamma_{\alpha\beta}^{\mu} \right) \times \\[2mm] \dfrac{1}{(2\pi)^4} \int d^4q\, iS_{F\beta\delta}(q) e^{-iq(x_1-x_2)} \times \\[2mm] \left(\sqrt{\dfrac{1}{2V\omega_{\mathbf{k}}}} \varepsilon_{\nu,r}(\mathbf{k}) e^{-ikx_2} \right) \gamma_{\delta\eta}^{\nu} \left(\sqrt{\dfrac{m}{VE_{\mathbf{p}}}} u_{s,\eta}(\mathbf{p}) e^{-ipx_2} \right) \end{array} \right\} \left| e_{\mathbf{p}'',s''}^{-}, \gamma_{\mathbf{k}'',r''} \right\rangle. \quad (8\text{-}56)$$

$$\underbrace{\phantom{\left(\sqrt{\dfrac{1}{2V\omega_{\mathbf{k}}}} \varepsilon_{\nu,r}(\mathbf{k}) e^{-ikx_2} \right) \gamma_{\delta\eta}^{\nu} \left(\sqrt{\dfrac{m}{VE_{\mathbf{p}}}} u_{s,\eta}(\mathbf{p}) \right)}}_{\text{numbers}}$$

Pulling the numbers outside the bras and kets, we have

$$\left\langle e_{\mathbf{p}',s'}^{-}, \gamma_{\mathbf{k}',r'} \left| S_{C1}^{(2)} \right| e_{\mathbf{p},s}^{-}, \gamma_{\mathbf{k},r} \right\rangle = \sum_{s'',\mathbf{p}''} \sum_{r'',\mathbf{k}''} \underbrace{\left\langle e_{\mathbf{p}',s'}^{-}, \gamma_{\mathbf{k}',r'} \middle| e_{\mathbf{p}'',s''}^{-}, \gamma_{\mathbf{k}'',r''} \right\rangle}_{\delta_{\mathbf{p}'\mathbf{p}''}\delta_{s's''}\delta_{\mathbf{k}'\mathbf{k}''}\delta_{r'r''}} \left(-e^2\right) \iint d^4x_1 d^4x_2 \times$$

$$\left(\sqrt{\dfrac{m}{VE_{\mathbf{p}''}}} \overline{u}_{s'',\alpha}(\mathbf{p}'') e^{ip''x_1} \right) \left(\sqrt{\dfrac{1}{2V\omega_{\mathbf{k}''}}} \varepsilon_{\mu,r''}(\mathbf{k}'') e^{ik''x_1} \gamma_{\alpha\beta}^{\mu} \right) \times \quad (8\text{-}57)$$

$$\dfrac{1}{(2\pi)^4} \int d^4q\, iS_{F\beta\delta}(q) e^{-iq(x_1-x_2)} \left(\sqrt{\dfrac{1}{2V\omega_{\mathbf{k}}}} \varepsilon_{\nu,r}(\mathbf{k}) e^{-ikx_2} \right) \gamma_{\delta\eta}^{\nu} \left(\sqrt{\dfrac{m}{VE_{\mathbf{p}}}} u_{s,\eta}(\mathbf{p}) e^{-ipx_2} \right).$$

$$= 0 + 0 + ... + \underbrace{\left\langle e^-_{\mathbf{p}',s'}, \gamma_{\mathbf{k}',r'} \mid e^-_{\mathbf{p}',s'}, \gamma_{\mathbf{k}',r'} \right\rangle}_{=1} \left(-e^2 \right) \iint d^4x_1 d^4x_2 \times$$

$$\sqrt{\frac{m}{VE_{\mathbf{p}''}}} \bar{u}_{s',\alpha}(\mathbf{p}') e^{ip'x_1} \sqrt{\frac{1}{2V\omega_{\mathbf{k}'}}} \varepsilon_{\mu,r'}(\mathbf{k}') e^{ik'x_1} \gamma^\mu_{\alpha\beta} \frac{1}{(2\pi)^4} \int d^4q iS_{F\beta\delta}(q) e^{-iq(x_1-x_2)} \times \qquad (8\text{-}58)$$

$$\sqrt{\frac{1}{2V\omega_{\mathbf{k}}}} \varepsilon_{\nu,r}(\mathbf{k}) e^{-ikx_2} \gamma^\nu_{\delta\eta} \sqrt{\frac{m}{VE_{\mathbf{p}}}} u_{s,\eta}(\mathbf{p}) e^{-ipx_2} + 0 + 0 + ...$$

Re-arranging factors in the above yields

$$\left\langle f \mid S^{(2)}_{C1} \mid i \right\rangle = \left\langle e^-_{\mathbf{p}',s'}, \gamma_{\mathbf{k}',r'} \mid S^{(2)}_{C1} \mid e^-_{\mathbf{p},s}, \gamma_{\mathbf{k},r} \right\rangle =$$

$$\left(-e^2 \right) \left(\frac{m}{VE_{\mathbf{p}'}} \right)^{1/2} \left(\frac{1}{2V\omega_{\mathbf{k}'}} \right)^{1/2} \left(\frac{1}{2V\omega_{\mathbf{k}}} \right)^{1/2} \left(\frac{m}{VE_{\mathbf{p}}} \right)^{1/2} \bar{u}_{s',\alpha}(\mathbf{p}') \varepsilon_{\mu,r'}(\mathbf{k}') \gamma^\mu_{\alpha\beta} \varepsilon_{\nu,r}(\mathbf{k}) \gamma^\nu_{\delta\eta} u_{s,\eta}(\mathbf{p}) \times \;(8\text{-}59)$$

$$\frac{1}{(2\pi)^4} \int d^4q iS_{F\beta\delta}(q) \left\{ \int d^4x_1 e^{-iqx_1} e^{ip'x_1} e^{ik'x_1} \int d^4x_2 e^{iqx_2} e^{-ikx_2} e^{-ipx_2} \right\}$$

Noting that

$$\int d^4x_1 e^{ix_1(p'+k'-q)} \int d^4x_2 e^{ix_2(q-p-k)} = (2\pi)^4 \delta^{(4)}\left(p'+k'-q \right)(2\pi)^4 \delta^{(4)}\left(q-(p+k) \right), \quad (8\text{-}60)$$

we see 4-momentum is conserved at every vertex, and incoming 4-momentum equals outgoing,

$$q = p + k = p' + k'. \qquad (8\text{-}61)$$

The last line of (8-59), with (8-60), becomes

$$(2\pi)^4 iS_{F\beta\delta}(p+k). \qquad (8\text{-}62)$$

The resulting expression for (8-59) is

$$\left\langle f \mid S^{(2)}_{C1} \mid i \right\rangle = (2\pi)^4 \delta^{(4)}\left(p'+k'-p-k \right) \left(\frac{m}{VE_{\mathbf{p}'}} \right)^{1/2} \left(\frac{m}{VE_{\mathbf{p}}} \right)^{1/2} \left(\frac{1}{2V\omega_{\mathbf{k}'}} \right)^{1/2} \left(\frac{1}{2V\omega_{\mathbf{k}}} \right)^{1/2} \mathcal{M}^{(2)}_{C1}, \;(8\text{-}63)$$

Result for first way of Compton scattering

where the Feynman amplitude is

$$\mathcal{M}^{(2)}_{C1} = -e^2 \bar{u}_{s',\alpha}(\mathbf{p}') \varepsilon_{\mu,r'}(\mathbf{k}') \gamma^\mu_{\alpha\beta} iS_{F\beta\delta}(q=p+k) \varepsilon_{\nu,r}(\mathbf{k}) \gamma^\nu_{\delta\eta} u_{s,\eta}(\mathbf{p}). \qquad (8\text{-}64)$$

Second Type of Compton Scattering

Similarly, for the second term in (8-53), one gets the same relation for $\left\langle f \mid S^{(2)}_{C2} \mid i \right\rangle$ as (8-63) except that $\mathcal{M}^{(2)}_{C1}$ is replaced with

$$\mathcal{M}^{(2)}_{C2} = -e^2 \bar{u}_{s',\alpha}(\mathbf{p}') \varepsilon_{\mu,r}(\mathbf{k}) \gamma^\mu_{\alpha\beta} iS_{F\beta\delta}(q=p-k') \varepsilon_{\nu,r'}(\mathbf{k}') \gamma^\nu_{\delta\eta} u_{s,\eta}(\mathbf{p}) \qquad (8\text{-}65)$$

Transition amplitude contribution from second way for Compton scattering

The Complete Compton Scattering Transition Amplitude

Hence, the full second order Compton transition amplitude, including both cases of Fig. 8-3, is

Compton's $S_{fi} = \left\langle f \mid S \mid i \right\rangle_{Comp} = S_{Compton} = \left\langle f \mid S^{(2)}_C \mid i \right\rangle = \left\langle f \mid S^{(2)}_{C1} \mid i \right\rangle + \left\langle f \mid S^{(2)}_{C2} \mid i \right\rangle$

$$= (2\pi)^4 \delta^{(4)}\left(p'+k'-p-k \right) \left(\frac{m}{VE_{\mathbf{p}'}} \right)^{1/2} \left(\frac{m}{VE_{\mathbf{p}}} \right)^{1/2} \left(\frac{1}{2V\omega_{\mathbf{k}'}} \right)^{1/2} \left(\frac{1}{2V\omega_{\mathbf{k}}} \right)^{1/2} \left(\mathcal{M}^{(2)}_{C1} + \mathcal{M}^{(2)}_{C2} \right), \quad (8\text{-}66)$$

Complete transition amplitude = sum of contributions from both ways

or

$$S_{Compton} = \left(\prod_{\mathbf{p}''}^{\substack{\text{all external} \\ \text{fermions}}} \sqrt{\frac{m}{VE_{\mathbf{p}''}}} \right) \left(\prod_{\mathbf{k}''}^{\substack{\text{all external} \\ \text{bosons}}} \sqrt{\frac{1}{2V\omega_{\mathbf{k}''}}} \right) (2\pi)^4 \delta^{(4)}\left(p'+k'-p-k \right) \left(\mathcal{M}^{(2)}_{C1} + \mathcal{M}^{(2)}_{C2} \right). \; (8\text{-}67)$$

Note the similarity with the Bhabha scattering transition amplitude (8-48), and once again, we will soon learn to do this sort of thing in much shorter fashion using Feynman rules.

8.4.4 Returning to the Photon Propagator Term $S_B^{(2)}$

After reading this section, you should do Prob. 6 to derive the transition amplitude for <u>Møller scattering</u> (Fig. 8-4). Note that, similar to Compton and Bhabha scattering, there are two ways for it to occur in which the outgoing electrons have the same individual momenta (and spins). And since we can only measure the incoming and outgoing particles, we have no way of knowing which of the two may have occurred. We thus must add the two amplitudes to get the total amplitude.

In Fig. 8-4, we place time on the vertical, rather than horizontal, axis. Feynman diagrams can be expressed in either fashion. Here we show the alternative, so you get used to both.

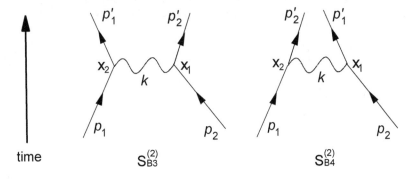

Figure 8-4. Møller Scattering Can Occur in Two Ways

In finding the result, take care to note that either x_1 or x_2 can occur first and this is already accounted for in the Dyson time ordering converted, via Wick's theorem, to normal ordering.

A separate issue is that in our expression for the transition amplitude, the p_2 electron could be destroyed at x_2 instead of x_1 (and vice versa for the p_1 electron). If you work the math of Prob. 6, you will see these are two different terms, each of which contributes to the amplitude. However, they have the same form if we simply switch dummy integration variables $x_2 \leftrightarrow x_1$ in one of them. Hence, we will get two equal terms for the LHS of Fig. 8-4, and two equal terms for the RHS. This allows us to use only one of them for the LHS and one for the RHS, but multiply the transition amplitude expression by 2. This effectively eliminates the ½ in the factor $-e^2/2$.

The final answer you should get for Prob. 6, the second order Møller scattering amplitude, is

$$S_{Møller} = \left(\prod_{\mathbf{p}}^{\substack{\text{all external} \\ \text{fermions}}} \sqrt{\frac{m}{VE_{\mathbf{p}}}} \right) (2\pi)^4 \, \delta^{(4)} \left(p_1' + p_2' - p_1 - p_2 \right) \left(\mathcal{M}_{B3}^{(2)} + \mathcal{M}_{B4}^{(2)} \right) \tag{8-68}$$

$$\mathcal{M}_{B3}^{(2)} = + e^2 \bar{u}_{r_1'}(\mathbf{p}_1') \gamma^\mu u_{r_1}(\mathbf{p}_1) i D_{F\mu\nu} \left(k = p_1 - p_1' \right) \bar{u}_{r_2'}(\mathbf{p}_2') \gamma^\nu u_{r_2}(\mathbf{p}_2)$$

$$\mathcal{M}_{B4}^{(2)} = - e^2 \bar{u}_{r_2'}(\mathbf{p}_2') \gamma^\mu u_{r_1}(\mathbf{p}_1) i D_{F\mu\nu} \left(k = p_1 - p_2' \right) \bar{u}_{r_1'}(\mathbf{p}_1') \gamma^\nu u_{r_2}(\mathbf{p}_2). \tag{8-69}$$

The interaction probability is $|S_{Møller}|^2$, so the overall sign of $\mathcal{M}_{B3}^{(2)} + \mathcal{M}_{B4}^{(2)}$ is unimportant for determining it. The relative sign difference between them is important, because that will affect the absolute magnitude of their sum. If one of (8-69) is positive, the other must be negative. Note that in the process of finding $\mathcal{M}_{B3}^{(2)}$, we have to normal order. In so doing, we put the destruction operators $\psi^+(\mathbf{p}_1)$ and $\psi^+(\mathbf{p}_2)$ at the end. But they could be ordered there as $\psi^+(\mathbf{p}_1)\psi^+(\mathbf{p}_2)$, or as $\psi^+(\mathbf{p}_2)\psi^+(\mathbf{p}_1)$. Either would be correct normal order, but the sign of the resulting amplitude would be different. The key is that in finding $\mathcal{M}_{B4}^{(2)}$, we have to normal order in the same order as we did for $\mathcal{M}_{B3}^{(2)}$.

Beyond these considerations, by convention, $\mathcal{M}_{B3}^{(2)}$ is taken as positive, since (as we will see in Chap. 16) this leads to the correct classical sign convention when we derive classical potential field theory from QFT. Without noting it, we have already adopted the sign convention for Bhabha scattering that will have similar ramifications for classical theory.

8.4.5 The Electron/Positron Closed Loop Term $S_D^{(2)}$

The first two terms of the 4th line inside the brackets of (8-8) are

$$S_D^{(2)} = -\frac{1}{2!}e^2 \iint d^4x_1 d^4x_2 \left(N\left\{ (\bar\psi A \psi)_{x_1} (\bar\psi A \psi)_{x_2} \right\} + N\left\{ (\bar\psi A \psi)_{x_1} (\bar\psi A \psi)_{x_2} \right\} \right). \qquad (8\text{-}70)$$

$S_D^{(2)}$ comprises two terms

In the manner detailed in Box 8-2, we can reorder either of these by switching adjacent fields (and introducing a minus sign each time we switch fermions). Doing this to the first term makes it look like the second term (an even number of fermion/fermion switches means no sign change), except that the coordinates x_1 and x_2 are interchanged. But we know these are just dummy variables so we can switch $x_2 \leftrightarrow x_1$ and the first term then looks just like the second. Thus (8-70) becomes

$$S_D^{(2)} = -e^2 \iint d^4x_1 d^4x_2 N\left\{ (\bar\psi A \psi)_{x_1} (\bar\psi A \psi)_{x_2} \right\}, \qquad (8\text{-}71)$$

These two terms are equivalent mathematically and can be combined into one term

which is represented by the Feynman diagram of Fig. 8-5.

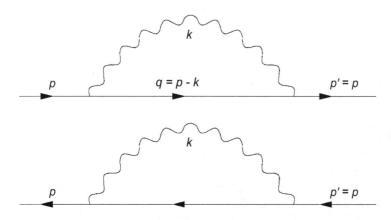

This one term can describe both an electron closed loop and a positron closed loop

Figure 8-5. Electron Closed Loop (top) and Positron Closed Loop (bottom)

Note (8-71) can represent destruction of an electron, propagation of both electron and photon virtual particles, and then the creation of another electron. It can also represent destruction of a positron, propagation of both positron and photon virtual particles, then the creation of another positron. We will focus on the electron case, since electrons are more common in our world and since the results exactly parallel those for the positron case.

The closed loop diagrams are also called self-energy diagrams, for reasons which will become apparent in Chaps. 9 and 12.

Note the following, soon to be seen, quite important fact about closed loops. In diagrams without closed loops, all virtual particle 4-momenta are precisely pinned down at each vertex. See, for example, Fig. 8-2 on pg. 221, where the virtual photon has 4-momentum equal to the sum of the incoming 4-momenta, i.e., $k = p_1 + p_2$. In Fig. 8-5, with closed loops, however, one of the virtual particles' 4-momenta can be anything. That is, for any value of k (the virtual photon 4-momentum), q (the virtual electron 4-momentum) simply takes on the value $q = p - k$. The sum total of k and q must equal p, but this does not determine k and q separately.

Closed loop diagrams are also called self-energy diagrams

One might expect we would need to integrate the expression for the transition amplitude of Fig. 8-5 over all values of k, i.e., over all values of k^0 from $-\infty$ to $+\infty$, and all values of $\mathbf{k}$ from $-\infty$ to $+\infty$ along all three spatial directions. As we are about to see mathematically in the following steps (presented without comment), that expectation would be correct.

4-momenta k and q can vary in closed loop as long as they sum to p

$$S_{e\,loop} = \left\langle e_{\mathbf{p}',s'}^- \left| S_D^{(2)} \right| e_{\mathbf{p},s}^- \right\rangle = \left\langle e_{\mathbf{p}',s'}^- \left| -e^2 \iint d^4x_1 d^4x_2 \, (\bar\psi^- A_\mu \gamma^\mu \psi)_{x_1} (\bar\psi A_\nu \gamma^\nu \psi^+)_{x_2} \right| e_{\mathbf{p},s}^- \right\rangle$$

$$= \left\langle e_{\mathbf{p}',s'}^- \left| -e^2 \iint d^4x_1 d^4x_2 \, i D_{F\mu\nu}(x_1 - x_2)(\bar\psi^-)_{x_1} \gamma^\mu i S_F(x_1 - x_2) \gamma^\nu (\psi^+)_{x_2} \right| e_{\mathbf{p},s}^- \right\rangle \qquad (8\text{-}72)$$

Deriving electron closed loop transition amplitude

$$= -e^2 \left\langle e^-_{\mathbf{p}',s'} \left| \iint d^4x_1 d^4x_2 \, iD_{F\mu\nu}\left(x_1 - x_2\right)\left(\sum_{\mathbf{p}'',s''} \sqrt{\frac{m}{VE_{\mathbf{p}''}}}\,\overline{u}_{s''}(\mathbf{p}'')c^\dagger_{s''}(\mathbf{p}'')e^{ip''x_1}\right)\gamma^\mu \times \right. \right.$$

$$\left. \left. iS_F\left(x_1 - x_2\right)\gamma^\nu \sqrt{\frac{m}{VE_\mathbf{p}}}\,u_s(\mathbf{p})e^{-ipx_2}\right|0\right\rangle \qquad (8\text{-}73)$$

$$= -e^2 \left\langle e^-_{\mathbf{p}',s'} \left| \iint d^4x_1 d^4x_2 \, iD_{F\mu\nu}\left(x_1 - x_2\right)\sqrt{\frac{m}{VE_{\mathbf{p}'}}}\sqrt{\frac{m}{VE_\mathbf{p}}}\,\overline{u}_{s'}(\mathbf{p}')e^{ip'x_1}\gamma^\mu iS_F\left(x_1 - x_2\right)\gamma^\nu u_s(\mathbf{p})e^{-ipx_2}\right| e^-_{\mathbf{p}',s'}\right\rangle$$

$$= -e^2 \iint d^4x_1 d^4x_2 \left(\frac{-ig_{\mu\nu}}{\left(2\pi\right)^4}\int \frac{e^{-ik(x_1-x_2)}}{k^2 + i\varepsilon}d^4k\right)\frac{m}{V}\sqrt{\frac{1}{E_{\mathbf{p}'}E_\mathbf{p}}}\,\overline{u}_{s'}(\mathbf{p}')\gamma^\mu \times$$

$$\left(\frac{1}{\left(2\pi\right)^4}i\int \frac{\left(\not{q}+m\right)e^{-iq(x_1-x_2)}}{q^2 - m^2 + i\varepsilon}d^4q\right)\gamma^\nu u_s(\mathbf{p})e^{ip'x_1}e^{-ipx_2}. \qquad (8\text{-}74)$$

Re-arranging by moving factors into helpful positions with regard to the integration signs yields

$$S_{e\,loop} = -e^2 \iint d^4q\,d^4k \left(\frac{1}{\left(2\pi\right)^4}\right)\underbrace{\left(\frac{-ig_{\mu\nu}}{k^2 + i\varepsilon}\right)}_{iD_{\mu\nu}(k)}\underbrace{\left(\int e^{-ikx_1}e^{-iqx_1}e^{ip'x_1}d^4x_1\right)}_{(2\pi)^4\delta^{(4)}\left(q-\left(p'-k\right)\right)}\underbrace{\left(\int e^{ikx_2}e^{iqx_2}e^{-ipx_2}d^4x_2\right)}_{(2\pi)^4\delta^{(4)}\left(q-\left(p-k\right)\right)} \times$$

$$\frac{m}{V}\sqrt{\frac{1}{E_{\mathbf{p}'}E_\mathbf{p}}}\,\overline{u}_{s'}(\mathbf{p}')\gamma^\mu \frac{1}{\left(2\pi\right)^4}\underbrace{i\left(\frac{\left(\not{q}+m\right)}{q^2 - m^2 + i\varepsilon}\right)}_{iS_F(q)}\gamma^\nu u_s(\mathbf{p}). \qquad (8\text{-}75)$$

With the Dirac delta functions, integration over q leaves $p' - k = p - k$, meaning $p' = p$, and thus,

$$S_{e\,loop} = -e^2 \delta^{(4)}\left(p - p' = 0\right)\frac{m}{VE_\mathbf{p}}\int d^4k\, iD_{F\mu\nu}\left(k\right)\overline{u}_{s'}(\mathbf{p})\gamma^\mu iS_F\left(p-k\right)\gamma^\nu u_s(\mathbf{p}). \qquad (8\text{-}76)$$

The result integrates over all k in loop keeping $q = p - k$

You can show that only the $s' = s$ spin state in (8-76) survives by doing Prob. 7, or you can just take my word for it. Thus, the electron loop transition amplitude is

$$S_{e\,loop} = \overbrace{\prod_{\mathbf{p}'}}^{\substack{\text{all external}\\\text{fermions}}}\sqrt{\frac{m}{VE_{\mathbf{p}'}}}\left(2\pi\right)^4\delta^{(4)}\left(p - p'\right)\mathcal{M}_{e\,loop}, \qquad (8\text{-}77)$$

Final expression for electron closed loop transition amplitude

where

$$\mathcal{M}_{e\,loop} = \frac{-e^2}{\left(2\pi\right)^4}\int d^4k\, iD_{F\mu\nu}\left(k\right)\overline{u}_s(\mathbf{p})\gamma^\mu iS_F\left(p-k\right)\gamma^\nu u_s(\mathbf{p}). \qquad (8\text{-}78)$$

And so, yes, as we conjectured, we do have to integrate over all values of k. Note that for this integration, because the virtual particles can be off-shell, the energy $k^0 = \omega$ is not constrained by the relativistic relation $m^2 = \omega^2 - \mathbf{k}^2$ (which, of course, equals zero for photons). So the integration over k^0 is independent of the integration over $\mathbf{k}$.

In the integration over 4D k space, ω is not constrained by $\mathbf{k}$ because virtual photon is off-shell

8.4.6 The Photon Closed Loop Term $S_E^{(2)}$

Like the electron, the photon has a closed loop interaction, as well, which arises from the last term in the fourth line inside the brackets of (8-8), i.e.,

$$S_E^{(2)} = -\frac{1}{2!}e^2 \iint d^4x_1 d^4x_2\, N\left\{(\overline{\psi}\not{A}\psi)_{x_1}(\overline{\psi}\not{A}\psi)_{x_2}\right\}. \qquad (8\text{-}79)$$

$S_E^{(2)}$ describes photon closed loop

As represented by the Feynman diagram of Fig. 8-6, this has a real incoming and outgoing photon connected by electron and positron virtual particles forming a closed loop. (The overbar symbol is sometimes used in Feynman diagrams for virtual anti-particles. It is just a symbol. The overbar here is not related to adjoint fields.) The photon closed loop (or self-energy) diagram is also known as a <u>vacuum polarization loop</u> diagram because, in it, the chargeless photon sitting in the vacuum is polarized, i.e., it splits into a particle with plus (pole) charge and a particle with negative (pole) charge.

Photon closed loop also called self-energy loop or vacuum polarization loop

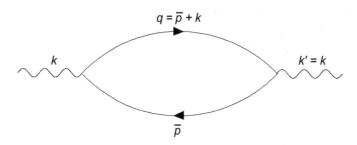

$$q = \bar{p} + k$$

$$k \qquad\qquad k' = k$$

$$\bar{p}$$

Figure 8-6. Photon Closed Loop

Using the middle row of (7-75) from Chap. 7, we can re-express (8-79) as

$$S_E^{(2)} = -\frac{1}{2!}e^2 \iint d^4x_1 d^4x_2 \left(-iS_{F\eta\alpha}(x_2 - x_1)\right)\gamma^\mu_{\alpha\beta}iS_{F\beta\delta}(x_1 - x_2)\gamma^\nu_{\delta\eta}N\left\{\left(A^+_\mu + A^-_\mu\right)_{x_1}\left(A^+_\nu + A^-_\nu\right)_{x_2}\right\}$$ *Re-expressing* $S_E^{(2)}$

$$= \frac{1}{2!}e^2 \iint d^4x_1 d^4x_2 iS_{F\eta\alpha}(x_2 - x_1)\gamma^\mu_{\alpha\beta}iS_{F\beta\delta}(x_1 - x_2)\gamma^\nu_{\delta\eta} \times \qquad\qquad (8\text{-}80)$$

$$\left(\underbrace{\left(A^+_\mu\right)_{x_1}\left(A^+_\nu\right)_{x_2}}_{\text{will go to zero}} + \left(A^-_\nu\right)_{x_2}\left(A^+_\mu\right)_{x_1} + \left(A^+_\mu\right)_{x_1}\left(A^-_\nu\right)_{x_2} + \underbrace{\left(A^-_\mu\right)_{x_1}\left(A^-_\nu\right)_{x_2}}_{\text{will go to zero}}\right).$$

The first and last terms above will go to zero when we sandwich them between single photon bra and ket states. We note again that the x_1 and x_2 are dummy integration variables and can be switched without changing the result. Similarly, μ and ν are dummy variables and so can be interchanged. Hence, the middle two terms are the same mathematically. Thus, we will find *Deriving the transition amplitude for photon closed loop*

$$S_{\gamma \atop loop} = \left\langle \gamma_{k',r'}\left|S_E^{(2)}\right|\gamma_{k,r}\right\rangle = \left\langle \gamma_{k',r'}\left|\frac{1}{2!}e^2 \iint d^4x_1 d^4x_2 iS_{F\eta\alpha}(x_2 - x_1)\gamma^\mu_{\alpha\beta}iS_{F\beta\delta}(x_1 - x_2)\gamma^\nu_{\delta\eta} \times\right.\right.$$

$$\left.\left.\left(\left(A^-_\mu\right)_{x_1}\left(A^+_\nu\right)_{x_2} + \left(A^-_\mu\right)_{x_1}\left(A^+_\nu\right)_{x_2}\right)\right|\gamma_{k,r}\right\rangle \qquad\qquad (8\text{-}81)$$

$$= \left\langle \gamma_{k',r'}\left|e^2 \iint d^4x_1 d^4x_2 iS_{F\eta\alpha}(x_2 - x_1)\gamma^\mu_{\alpha\beta}iS_{F\beta\delta}(x_1 - x_2)\gamma^\nu_{\delta\eta}\left(A^-_\mu\right)_{x_1}\left(A^+_\nu\right)_{x_2}\right|\gamma_{k,r}\right\rangle.$$

We can note a general rule, that we have seen over and over, that the factorials in the denominators of terms in the S operator expansion disappear in the transition amplitude.

Substituting the field solution $\left(A^+_\nu\right)_{x_2}$ into (8-81) leaves the vacuum state. $\left(A^-_\mu\right)_{x_1}$ acting on that is

$$S_{\gamma \atop loop} = e^2 \iint d^4x_1 d^4x_2 iS_{F\eta\alpha}(x_2 - x_1)\gamma^\mu_{\alpha\beta}iS_{F\beta\delta}(x_1 - x_2)\gamma^\nu_{\delta\eta} \times$$

$$\left(\sum_{k'',r''}\sqrt{\frac{1}{2V\omega_{k''}}}\varepsilon_{\mu,r''}(k'')e^{ik''x_1}\underbrace{\left\langle\gamma_{k',r'}\right\|\left.\gamma_{k'',r''}\right\rangle}_{\delta_{k'k''}\delta_{r'r''}}\right)\sqrt{\frac{1}{2V\omega_k}}\varepsilon_{\nu,r}(k)e^{-ikx_2} \qquad (8\text{-}82)$$

$$= \frac{-e^2}{2V\sqrt{\omega_k \omega_{k'}}}\iint d^4x_1 d^4x_2 \left(\frac{1}{(2\pi)^4}\int \frac{(\bar{p} + m)_{\eta\alpha}e^{-i\bar{p}(x_2 - x_1)}}{\bar{p}^2 - m^2 + i\varepsilon}d^4\bar{p}\right)\gamma^\mu_{\alpha\beta} \times$$

$$\qquad\qquad (8\text{-}83)$$

$$\left(\frac{1}{(2\pi)^4}\int \frac{(\slashed{q} + m)_{\beta\delta}e^{-iq(x_1 - x_2)}}{q^2 - m^2 + i\varepsilon}d^4q\right)\gamma^\nu_{\delta\eta}\varepsilon_{\mu,r'}(k')\varepsilon_{\nu,r}(k)e^{ik'x_1}e^{-ikx_2}.$$

$$= \frac{-e^2}{2V\sqrt{\omega_{\mathbf{k}}\omega_{\mathbf{k}'}}} \iint d^4\overline{p}\,d^4q \left(\frac{1}{(2\pi)^4} \underbrace{\int \frac{e^{i(\overline{p}-q)x_1} e^{ik'x_1} d^4x_1}{(2\pi)^4 \delta^{(4)}\big(q-(k'+\overline{p})\big)}}_{} \right) \underbrace{\frac{(\overline{\slashed{p}}+m)_{\eta\alpha}}{\overline{p}^2-m^2+i\varepsilon}}_{S_{F\eta\alpha}(\overline{p})} \gamma^\mu_{\alpha\beta} \underbrace{\frac{(\slashed{q}+m)_{\beta\delta}}{q^2-m^2+i\varepsilon}}_{S_{F\beta\delta}(q)} \gamma^\nu_{\delta\eta} \times$$

$$\varepsilon_{\mu,r'}(\mathbf{k}')\varepsilon_{\nu,r}(\mathbf{k}) \left(\frac{1}{(2\pi)^4} \underbrace{\int \frac{e^{i(q-\overline{p})x_2} e^{-ikx_2} d^4x_2}{(2\pi)^4 \delta^{(4)}\big(q-(k+\overline{p})\big)}}_{} \right). \qquad (8\text{-}84)$$

Part of the transition amplitude is a trace of a spinor space matrix

From the Dirac delta functions and the integration over q, $k' = k$, so

$$= \frac{-e^2}{2V\omega_{\mathbf{k}}} \delta^{(4)}(k-k') \int d^4\overline{p} \ \underbrace{S_{F\eta\alpha}(\overline{p})\gamma^\mu_{\alpha\beta} S_{F\beta\delta}(\overline{p}+k)\gamma^\nu_{\delta\eta}}_{\text{trace in spinor space of matrix, } M^{\mu\nu}_{\eta\eta}} \ \varepsilon_{\mu,r'}(\mathbf{k})\varepsilon_{\nu,r}(\mathbf{k}). \qquad (8\text{-}85)$$

Our final result for the photon loop transition amplitude, with Tr indicating the trace in spinor space, is

$$\mathcal{M}_{\substack{\gamma \\ loop}} = \frac{-e^2}{(2\pi)^4} \left\{ \mathrm{Tr} \int d^4\overline{p}\, S_F(\overline{p})\gamma^\mu S_F(\overline{p}+k)\gamma^\nu \right\} \varepsilon_{\mu,r'}(\mathbf{k})\varepsilon_{\nu,r}(\mathbf{k})$$

$$S_{\substack{\gamma \\ loop}} = \left(\prod_{\mathbf{k}}^{\substack{\text{all ext} \\ \text{bosons}}} \sqrt{\frac{1}{2V\omega_{\mathbf{k}}}} \right) (2\pi)^4 \delta^{(4)}(k-k') M_{\gamma\,loop}. \qquad (8\text{-}86)$$

Final result: transition amplitude for photon closed loop

<u>Point to Note:</u> Closed loop diagrams invariably lead to integrations from $-\infty$ to $+\infty$ over all four axes of 4-momentum space. We saw this already for the electron and photon closed loops.

8.4.7 Subtle, Important, and Confusing Point

If you look closely at Fig. 8-6, you will see that the 4-momenta don't seem to sum to zero at the vertex, as we have seen in every prior example. Physically, they actually do, but because of some wrinkles in our mathematics (about to be explained) we draw our Feynman diagrams so that positive 4 momentum for virtual particles is interpreted in the direction of the arrows (which for antiparticles go in the opposite direction of time.)

4-momentum for virtual antiparticle is tricky, due to math manipulations

Note in (8-81) that, if we consider the LH vertex in Fig. 8-6 to be x_2, and the RH vertex to be x_1, then the inner contraction in (8-79) can be considered to create an electron at x_2 and propagate it to x_1, where it is destroyed. The outer contraction in (8-79), on the other hand, would then create a positron at x_2 and propagate it to x_1, where it is destroyed. However, the outer contraction (positron) is expressed mathematically in the same form as the inner contraction (electron) except that the argument of S_F is $(x_2 - x_1)$, rather than $(x_1 - x_2)$. In effect, the sense of this is that the electron travels from x_2 to x_1, whereas the positron travels from x_1 to x_2. This is not what happens physically, but our mathematical shortcuts, which changed things to make the math simpler, makes the final math expression look this way.

Math manipulations simplified expressions but have effect of changing sign in virtual anti-particle 4-momentum math

Hence, for the virtual positron, the math looks like the four momentum is carried from x_1 to x_2 (in the opposite direction of time). Thus, we end up with a delta function in (8-84) that tells us $q = k + \overline{p}$, rather than $q = k - \overline{p}$, which we might expect. The former is what we get via the math and going in the direction of the arrows in the Feynman diagram. The latter is what is really happening physically, going in the direction of time.

Another way to look at this is that the $\overline{p}$ used in Fig. 8-6 and in our mathematical expressions is 4-momentum traveling backward in time, so it has a built in minus sign. The *physical value* we would measure in our world where time goes forward is $-\overline{p}$.

If p is physical 4-momentum of virtual anti-fermion, then $-p$ is 4-momentum used in Feynman diagrams and associated math

This built in minus sign on 4-momenta for virtual antiparticle fermions *does not hold* true for i) virtual or real photons nor ii) *real antiparticle fermions*. Recall our Bhabha scattering with an incoming electron/positron pair of Fig. 8-2, pg. 221 with the virtual photon $k = p_1 + p_2$ of (8-34). The incoming positron momentum was positive in the direction of time and equal to p_2. The virtual photon momentum was k and equal to the sum of the momenta of the electron and positron as we would measure them with instruments (with time going forward).

Wholeness Chart 8-1. Keeping Four-momenta Signs Straight

	Four Momenta		
Real Particles	**True, Physical Value**	**Value in Transition Amplitudes**	**Labeled in Feynman Diagrams**
electron	p_e	p_e	p_e
positron	p_p	p_p	p_p
photon	k	k	k
Virtual Particles			
electron	p_e	p_e	p_e
positron	p_p	$-p_p$	$-p_p$
photon	k	k	k

To make this clearer (hopefully), consider Compton scattering of a positron, as in Fig. 8-7.

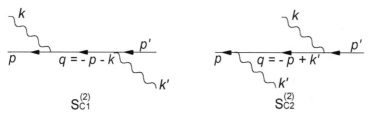

An example: positron Compton scattering

Figure 8-7. Positron Compton Scattering Feynman Diagrams

If you like, you can do Prob. 8 to show the transition amplitude for the LH diagram interaction is (8-67) with Feynman amplitude

$$\mathcal{M}_{C1}^{(2)} = e^2 \overline{v}_{s\alpha}(\mathbf{p}) \varepsilon_{\mu,r}(\mathbf{k}) \gamma_{\alpha\beta}^{\mu} i S_{F\beta\delta} (q = -p-k) \varepsilon_{\nu,r'}(\mathbf{k}') \gamma_{\delta\eta}^{\nu} v_{s'\eta}(\mathbf{p}') .$$ (8-87)

This strange sign change applies only to virtual anti-particles, not real anti-particles

Note the real positrons have positive momenta, as in the real world, whereas the virtual positron, for our Feynman diagram and our math, has momentum equal to the negative of its true value.

All of this is summarized in Wholeness Chart 8-1.

8.4.8 The Vacuum Bubble Term $S_F^{(2)}$

The final non-zero term in (8-8) is

$$S_F^{(2)} = -\frac{1}{2!} e^2 \iint d^4 x_1 d^4 x_2 \left(\overline{\psi} A \psi\right)_{x_1} \left(\overline{\psi} A \psi\right)_{x_2} .$$ (8-88)

$S_F^{(2)}$ describes vacuum bubble

By now, we should be able to guess right off that the Feynman diagram for (8-88) looks like Fig. 8-8.

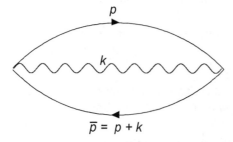

Figure 8-8. Vacuum Bubble Feynman Diagram

By doing Prob.9, you can show that (8-88) equals

$$S_F^{(2)} = -\delta^{(4)}(0)\frac{e^2}{(2\pi)^4}\iint S_{F\eta\alpha}(p+k)\gamma_{\alpha\beta}^\mu S_{F\beta\delta}(p)\gamma_{\delta\eta}^\nu iD_{F\mu\nu}(k)d^4k\,d^4p \quad . \qquad (8\text{-}89)$$

Transition amplitude for vacuum bubble

<u>Note</u>: The relation (8-89) implies conservation of 4-momentum at each vertex (when we realize that $\overline{p}$ as shown really means $-\overline{p} = -p - k$ physically, going forward in time).

4-momentum conserved at vertex

The vacuum bubble means we start with zero 4-momentum in the vacuum, we end with zero four-momentum in the vacuum, and in between we have a sum total 4-momentum of zero for all the virtual particles mediating the interaction.

Of course, this means either one or two of the virtual particles in Fig. 8-8 must have negative energy, as well as 3-momentum going in the opposite direction of its travel (for time moving forward). This, of course, is unheard of for real particles, but virtual particles, as we have seen with off-shell analyses, do not obey all the laws of nature we are used to seeing for real particles.

So, at least one virtual particle in bubble must have negative energy

And at least one has $\mathbf{p}$ (or $\mathbf{k}$) in opposite direction of its v

Since vacuum bubbles do not interact with real particles, they make no change in the manifest universe, and therefore, for all intents and purposes, they do not exist (at least phenomenologically.)

Vacuum bubbles don't interact with manifest universe

Similarly, negative energy pops up in other loop integrations, as well. For example, in the top part of Fig. 8-5, pg. 230, for k having large energy ($k^0 > p^0$), $q^0 = p^0 - k^0$ will be negative. And of course, we integrate the loop over all energy values from $k^0 = -\infty$ to $+\infty$, anyway, so obviously, virtual particles can have negative energies. Additionally, in similar fashion, a virtual particle can have a 3-momentum vector of opposite direction to its direction of travel.

8.5 The Shortcut Method: Feynman Rules

You may have started noticing certain patterns arising in our transition amplitude calculations. For one, there always seems to be a spinor of form $u_r(\mathbf{p})$ arising for every incoming electron, a spinor $\overline{v}_r(\mathbf{p})$ for every incoming positron, a spinor $\overline{u}_r(\mathbf{p})$ for every outgoing electron, etc. There is a propagator expressed in momentum space for every virtual particle, such as $iD_{F\alpha\beta}(k)$ for each virtual photon. For every second order interaction, there is a factor of e^2. For every vertex there is a γ^μ matrix. For first order terms, there is always one vertex; for second order, there are always two. And there always seems to be a Dirac delta function ensuring the incoming total momentum equals the outgoing total momentum.

Feynman rules: the quick and easy way to find transition amplitudes

Well, in the late 1940s, Richard Feynman noticed these patterns, as well. And from them, he devised a set of rules, <u>Feynman rules</u>, that allow us to simply write down the transition amplitude by looking at the associated Feynman diagram. Given how much work we've seen it takes to determine one transition amplitude, this is welcomed relief from all that tedium.

Although we do not show justification for it here, these rules apply to any order, not just second order, i.e., not just to $S^{(2)}$, but to $S^{(n)}$, where n is any number. We will not look at higher order interactions until the next chapter.

8.5.1 Topologically Distinct Feynman Diagrams

The rules themselves apply to what are termed <u>topologically different Feynman diagrams</u>. These are different from one another in ways other than simply changing of the labeling of vertices.

Changing $x_1 \leftrightarrow x_2$ leaves same topology, does not change Feynman diagram for purpose of Feynman rules

For example, in the LH diagram of Fig. 8-4 on pg. 229, if we switch the vertex labeling such that $x_1 \leftrightarrow x_2$, the diagram is not topologically different. The two diagrams with different labeling are, for the purposes of Feynman rules, one and the same. Similarly, if we interchange the vertex event labels for the RH diagram of Fig. 8-4, i.e., $x_1 \leftrightarrow x_2$, the new diagram is not topologically different.

However, the LH and RH diagrams of Fig. 8-4 are topologically distinct. In the LH diagram, the final particle labeled p'_1 shares a vertex with the initial particle labeled p_1. In the RH diagram, on the other hand, the final particle labeled p'_2 shares a vertex with the initial particle labeled p_1. So, the RH and LH diagrams are topologically different, and each makes its own contribution to the total amplitude for the interaction (same incoming and outgoing individual particle properties.)

8.5.2 Feynman Rules

Feynman's rules, for electrons and positrons in QED (there are more evolved versions for weak and strong interactions), are shown in Box 8-3.

Feynman's rules

Box 8-3. Feynman Rules for QED

A. The S matrix element (the transition amplitude) for a given interaction is

$$S_{fi} = \delta_{fi} + \left((2\pi)^4 \delta^{(4)}\left(P_f - P_i\right) \left(\prod^{\substack{\text{all external} \\ \text{bosons}}} \sqrt{\frac{1}{2V\omega_{\mathbf{k}}}} \right) \left(\prod^{\substack{\text{all external} \\ \text{fermions}}} \sqrt{\frac{m}{VE_{\mathbf{P}}}} \right) \right) \mathcal{M} \qquad \mathcal{M} = \sum_{n=1}^{\infty} \mathcal{M}^{(n)} \qquad \text{(B8-3.1)}$$

where P_f is the total 4-momentum of all final particles, P_i is the total 4-momentum of all initial particles, and the contribution $\mathcal{M}^{(n)}$ comes from the nth order perturbation term of the S operator, $S^{(n)}$.

B. The Feynman amplitude $\mathcal{M}^{(n)}$ is obtained from all of the topologically distinct, connected (i.e., all lines connected to one another in a given diagram) Feynman diagrams which contain n vertices. The contribution to each $\mathcal{M}^{(n)}$ is obtained by the following.

1. For each vertex, include a factor $ie\gamma^{\mu}$.

2. For each internal photon line, labeled by 4-momentum k, include a factor $iD_{F\mu\nu}\left(k\right) = i\dfrac{-g_{\mu\nu}}{k^2 + i\varepsilon}$

3. For each internal fermion line, labeled by 4-momentum p, write a factor $iS_F\left(p\right) = i\dfrac{\not{p} + m}{p^2 - m^2 + i\varepsilon}$

4. For each external line, write one of the following spinor factors, where $\mathbf{p}$ and $\mathbf{k}$ indicate basis states of corresponding 3-momenta, r represents spin state for fermions and polarization state for photons, [see (4-93) and (4-94), pg. 111]

        a) for each initial electron: $u_r(\mathbf{p})$
        b) for each final electron: $\bar{u}_r(\mathbf{p})$
        c) for each initial positron: $\bar{v}_r(\mathbf{p})$
        d) for each final positron: $v_r(\mathbf{p})$
        e) for each initial photon: $\varepsilon_{r,\mu}(\mathbf{k})$
        f) for each final photon: $\varepsilon_{r,\mu}(\mathbf{k})$

5. The spinor factors (γ matrices, S_F functions, spinors) for each fermion line are ordered so that, reading from right to left, they occur in the same sequence as following the fermion line in the direction of its arrows through the vertex. (Order is important as it conveys spinor matrix multiplication order when we do not show spinor indices.)

6. The four-momenta at each vertex are conserved (same total after as before).

7. For each closed loop of internal fermions only (without photons inside the loop itself, like what we call a "photon loop" which internally has an electron and a positron), take the trace (in spinor space) of the resulting matrix and multiply by a factor of (-1).

8. For each 4-momentum q which is not fixed by 4-momentum conservation, carry out the integration

$$\frac{1}{(2\pi)^4} \int d^4 q \quad \text{One such integration for each closed loop (fermion/fermion or fermion/photon loop).}$$

9. Multiply the expression by (-1) for each interchange of neighboring fermion operators (each associated with a particular spinor factor) which would be required to place the expression in appropriate normal order. "Appropriate", when we are adding sub amplitudes, means each sub amplitude must be in the same, not just any, normal order of destruction and creation operators.

8.5.3 Applying Feynman Rules

Let's use these rules with some of the interactions we've already calculated transition amplitudes for the long way.

Bhabha Scattering (Second Order)

Using Fig. 8-2 on pg. 221, let's apply Feynman rules to the LH diagram. Following the fermion arrows for incoming particles, we place the electron spinor $u_r(\mathbf{p})$ on the far right. Then, to the left of that, we place an $ie\gamma^{\nu}$ matrix factor, representing the vertex on the left. Next, to the left of that, we add the positron spinor $\bar{v}_r(\mathbf{p})$

An example: seeing how Feynman rules eliminate mountains of algebra

$$\overline{v}_{r_2}(\mathbf{p}_2)ie\gamma^v u_{r_1}(\mathbf{p}_1) \quad \leftrightarrow \quad \overline{v}_{r_2\alpha}(\mathbf{p}_2)ie\gamma^v_{\alpha\beta}u_{r_1\beta}(\mathbf{p}_1) \tag{8-90}$$

The spinor indices are usually hidden, but on the RHS of (8-90), we have written them out for clarity. This is why the order of placement of spinor quantities is important.

Then, following the LH diagram of Fig. 8-2, place the Feynman photon propagator in momentum space to the left of (8-90) followed by the appropriate spinor for the final positron (following the arrows), the vertex gamma matrix factor, and the spinor for the final electron.

$$\mathcal{M}_{B1}^{(2)} = \overline{u}_{r_1'}(\mathbf{p}_1')ie\gamma^\mu v_{r_2'}(\mathbf{p}_2')iD_{\mu\nu}(k=p_2+p_1)\overline{v}_{r_2}(\mathbf{p}_2)ie\gamma^v u_{r_1}(\mathbf{p}_1) \tag{8-91}$$

The final step for the LH diagram simply comprises using the above with (B8-3.1) of Box 8-3,

$$S_{B1}^{(2)} = \left(\prod_{\mathbf{p}}^{\substack{\text{all external} \\ \text{fermions}}} \sqrt{\frac{m}{VE_{\mathbf{p}}}}\right)(2\pi)^4\delta^{(4)}\left(p_1+p_2-(p_1'+p_2')\right)\times \tag{8-92}$$

$$\left(-e^2\overline{u}_{r_1'}(\mathbf{p}_1')\gamma^\mu v_{r_2'}(\mathbf{p}_2')iD_{\mu\nu}(k=p_1+p_2)\overline{v}_{r_2}(\mathbf{p}_2)\gamma^v u_{r_1}(\mathbf{p}_1)\right).$$

Of course, we could have done all of this in one step, but we took several lines to make our first use of Feynman rules clearer. Compare the work involved getting (8-92) to the two pages of algebra we used getting the same relation (8-34). You don't need a PhD in physics to realize the advantages of this approach over the earlier one.

For the RH diagram of Fig. 8-2, let's follow a similar procedure. The first vertex (bottom of figure) yields

$$\overline{u}_{r_1'}(\mathbf{p}_1')ie\gamma^v u_{r_1}(\mathbf{p}_1). \tag{8-93}$$

Then, we include the propagator and the second vertex terms to get (8-94). Note we had to include a factor of $(-1)^3$ in (8-94) because the operator associated with the $\overline{u}$ spinor would have to be moved leftward past two other operators (spinors), followed by the operator associated with the $\overline{v}$ spinor being moved rightward past one operator (spinor), to obtain the same operator (spinor) order as (8-91).

$$\mathcal{M}_{B2}^{(2)} = -\overline{v}_{r_2}(\mathbf{p}_2)ie\gamma^\mu v_{r_2'}(\mathbf{p}_2')iD_{\mu\nu}(p_2-p_2')\overline{u}_{r_1'}(\mathbf{p}_1')ie\gamma^v u_{r_1}(\mathbf{p}_1), \tag{8-94}$$

which equals (8-47). And thus, we get (8-48)

$$S_B^{(2)} = S_{B1}^{(2)} + S_{B2}^{(2)} = \left(\prod_{\mathbf{p}}^{\substack{\text{all external} \\ \text{fermions}}} \sqrt{\frac{m}{VE_{\mathbf{p}}}}\right)(2\pi)^4\delta^{(4)}\left(p_1+p_2-(p_1'+p_2')\right)\left(\mathcal{M}_{B1}^{(2)}+\mathcal{M}_{B2}^{(2)}\right). \tag{8-95}$$

Other Interactions

You should now do Probs. 10 to 14 to gain practice using Feynman rules to find transition amplitudes.

Finishing those, you have achieved a major milestone in your personal history of studying physics. Knowing how to use Feynman diagrams and apply Feynman rules comprises the very essence of QFT. We have refinements and applications to add, surely, but you now own the core of the theory.

8.6 Points to Be Aware of

Some additional things to know

The following points are things that have not been made clear yet, but that you should understand. Some of them often confuse students (and longer-term practitioners) of QFT, but hopefully, this will not be the case for you after reading this section.

Comparing Typical Perturbation Theory with QED

QED an application of perturbation theory

QED is often called a perturbation theory, which it essentially is, although it differs in some ways from the classic textbook type perturbation problem. To see the similarities and differences between the two, look at Wholeness Chart 8-2 below, which should be fairly self-explanatory.

Wholeness Chart 8-2. Comparing Typical Perturbation Theory to QED

Typical Perturbation Problem		QED
Governing equation Readily solvable if it didn't have a certain small term		Governing equation (interaction picture) $i\dfrac{d}{dt}\lvert\Psi\rangle = H_I^I\lvert\Psi\rangle$ where $\lvert\Psi\rangle = S_{oper}\lvert i\rangle$
Solving governing equation		
Separate into 2 equations		$S_{oper} = e^{-i\int_{t_i}^{t} H_I^I\,dt'} \rightarrow S = e^{-i\int_{-\infty}^{+\infty} H_I^I\,dt'} = S^{(0)} + S^{(1)} + S^{(2)} + ...$ $\lvert f\rangle = S\lvert i\rangle$
Solvable part: Governing equation minus small part that makes it hard to solve	Perturbation part : Small term part of governing equation that makes it hard to solve	
Closed form solution to solvable eq, f_{closed}	Solution to perturbation term eq, f_{pert} generally approximate, often a power series	$\lvert f\rangle_{free} = S^{(0)}\lvert i\rangle = I\lvert i\rangle$ free state, closed part | $\lvert f\rangle_{pert} = \left(S^{(1)} + S^{(2)} + ...\right)\lvert i\rangle$ perturbation part
Total (approx) solution $f = f_{closed} + f_{pert}$		$\lvert f\rangle = \lvert f\rangle_{free} + \lvert f\rangle_{pert} = \left(I + S^{(1)} + S^{(2)} + S^{(3)} + ...\right)\lvert i\rangle$
f_{pert} has 2nd, 3rd, etc order terms. More terms = more accurate solution.		Using up to $S^{(2)}$ = 2nd order solution as in this chapter. More terms = more accurate solution.

Each Interaction Term in $\mathcal{L} \rightarrow$ Vertex (for a Spacetime Event)

Note by looking over our development of each term in (8-8) in Sects. 8.1 through 8.4.8, or better by glancing through Wholeness Chart 8-4 at the end of this chapter (pg. 248), you can convince yourself that each term in $\mathcal{H}_I^I$, or equivalently, in $\mathcal{L}_I^I$ (since $\mathcal{L}_I^I = -\mathcal{H}_I^I$) results in a vertex interaction comprising the particles generated by the particular fields occurring in that term. Here so far, in QED, the only term in $\mathcal{L}_I^I$ is $\bar{\psi}\slashed{A}\psi$, so this gives rise to a vertex with two Dirac fermions and a photon.

Looking at $\mathcal{L}_I^1$, one can tell what vertices look like because each term represents a vertex

A first order term in S gives rise to one such vertex; a second order term in S gives rise to two such vertices; etc. But the nature of each vertex is determined by the form of the term in the Lagrangian (density). For example, we have no vertices with three Dirac fermions, or two Dirac fermions and two photons, etc. We only have vertices where each has two spin ½ particles and one photon. We can tell this right off, simply by examining the Lagrangian.

In more advanced theories (including the extension of QED of Sect. 8.7 below, as well as weak and strong interaction theory) we will have other terms in the Lagrangian (which end up in the Hamiltonian and thus in S). Just by looking at the form of these terms, we will know immediately what types of vertex interactions that theory gives rise to. The Lagrangian, we will find over and over, has a whole lot of information packed into it, and being able to read that information readily can help us in many ways to be more efficient and save time.

Not a big deal now, but in more advanced theories knowing this is a big help

Waves Not Particles → Not Lines in Spacetime

Feynman diagrams show particle tracks as lines in spacetime, and this implies they are point-like objects. But in reality, as we know, particles are waves spread through space. So, keep in mind that Feynman diagrams are a little deceptive in this regard. They are good tools, but only symbols of a more complex reality.

States in Feynman diagrams look like point particles moving, but really waves spread out in space

If you like, you can think of the lines on a Feynman diagram representing incoming, outgoing, and propagated waves that collapse at a point (a vertex) when interacting, much like the wave function collapses in NRQM when measured. However, this, too, is not really accurate, and I will probably be criticized for suggesting you think of it so, so don't take it literally.

Volumes V in our Relations

In the S operator, $\mathcal{H}_I^{\,I} = -e\bar{\psi}A\!\!\!/\psi$ is integrated over all space, to infinity in all directions, but the volume V in the factor in front of each field solution ($\sqrt{m/VE_\mathbf{p}}$ for fermions, for example) is not infinite. Yet when we integrated over all space (see the last line of (8-30), for example) to get our transition amplitudes, we assumed each field extended to infinity. That gave us the delta functions in the transition amplitudes. But that would mean the V in the field solution denominators would be infinite, and thus the solutions themselves equal to zero.

In the present QED approach, V is volume that all particles occupy together throughout interaction

What we are really doing, in effect, is to take V very large, but not actually infinite. In that case, the integrals over V in our transition amplitude derivations (again see the last line of (8-30) are only approximately equal to delta functions. But, if we take V large enough, which we do, the approximation of each such integral is, for all intents and purposes, equal to a delta function.

V taken as very large, effectively infinity. Makes math easier

Further, by taking the same V in our factors in front for all particles (positrons, electrons, and photons), we are assuming that all particles (all waves) are spread equally throughout the same V.

These assumptions may seem limiting at this point, because we commonly deal with particle waves that do not extend to infinity. In addition, it seems an elongated particle wave passing through another localized (in a small region) particle wave would not share the same volume V, and thus our analysis method would not work in this case. We will resolve these issues in Chap. 17, where we show how to get meaningful real-world answers using this methodology. For now, simply think of all fields (and thus the particles they create and destroy) in a given calculation as occupying the same, extremely large but not quite infinite, volume V.

We will see later how this can be used for scattering calculations with finite V and where particles pass thru one another

Waves Interact Over Volume, Not at Vertex Point

Also, Feynman diagrams imply, for example, that an incoming particle sheds a virtual propagator particle at a particular spacetime point (vertex). But recall that we integrate over all spacetime to get a final transition amplitude relation in terms of momenta, with no spacetime coordinates involved.

So a Feynman diagram simply represents a single differential event (at each vertex), which is part of an integration over all spacetime. It represents one differential part of a transition amplitude density (analogous to probability density), which in principle varies from event to event. By integrating over all events (all spacetime) we get the total transition amplitude.

Vertices seem to indicate an interaction of point particles at a point, but symbolic for interaction over region of spacetime by waves

So what we really have, unlike the diagrams seem to show, is waves spread out over a region V which can interact with one another. For example, for Bhabha scattering (first case of annihilation) an electron wave and a positron wave simultaneously occupy the same V with some amplitude density for making the transition, which is a function of x_1 and x_2. We integrate over all x_1 and x_2 to get the transition amplitude and take the square of the absolute value of that to get probability of the interaction occurring.

Hence, physically, it can be visualized as the electron and positron wave collapsing with a photon propagator wave emerging, then the virtual photon wave gives rise to an outgoing electron wave and positron wave, which each fill the volume V precisely. The interaction is of waves over a region, not of point-like objects at points.

Delta Function Clarification

One might wonder about the delta functions in our transition amplitudes, such as $\delta^{(4)}\left(p_1 + p_2 - \left(p_1' + p_2'\right)\right)$ in (8-35), since they are either infinite or zero, and we are supposed to be calculating probabilities between 0 and 1. Well, it turns out that, in scattering, we will integrate over momenta, and when we do, the delta functions will pick out only the momenta where final total value equals initial total. For now, just live with the delta function as part of our transition amplitude.

Delta functions to be integrated over momenta in scattering calculations, so don't worry that they equal infinity at a point.

Can Virtual Photons Travel Faster than Light?

If we think of virtual particles as appearing at x_2 and disappearing at x_1, as they appear in Feynman diagrams, then since the physical locations of x_1 and x_2 can be further apart than light can travel in the time between the two events, one could conclude that virtual particles can travel faster than light.

Virtual particles appear to travel faster than light

However, if we consider virtual particles (and real particles) to be waves that interact over an entire region, this interpretation may not be quite accurate. Nevertheless, in our transition amplitude integrations over all spacetime, we do find contributions from events x_1 outside the light cone of event x_2. So, perhaps it is not so incorrect to say that interactions may occur, in part, at speed > c.

But really fields transferring energy and momentum (yet for some regions of interaction still > c)

Recall that wave function collapse in NRQM happens instantaneously at all points. So, events on these waves that are outside one another's light cones somehow join the collapse, even though no signal could travel fast enough to trigger them to do so.

We are verging on philosophy here. (Nothing wrong with that.) The bottom line, for our present purposes, is that the math works. It predicts experiment.

New Terms Subtract Out of 1s on Diagonal of S operator

This is a subtle point that troubled me for years, so I include it here. Don't worry about it too much for now, but if it ever troubles you in the future, return and read the following.

A subtle point, not to spend too much time on now

Note that $S^{(0)} = I$ is a unitary operator. (Its matrix determinant is 1.) It has 1s on the diagonal and zeros elsewhere. With no interactions, the probability of an incoming state leaving as the exact same state would be 1.

One might think that adding $S^{(n)}$ in $S = S^{(0)} + S^{(1)} + S^{(2)}$ … terms would add to the off diagonal components, while the original 1s remain on the diagonal. Thus, the probability for no interaction for any initial state would remain 1, while we had non-zero values for interactions that left different final states. And so the total probability of getting any of these final states (including the original state unchanged) would be the sum of the individual probabilities, and this would be greater than 1.

However, the higher order contributions can add negative components to the S matrix diagonal, thereby reducing the probability for the final state being the same as the initial state. And thus, the total probability for finding any one of the many final states can equal 1.

General, not Basis, Spin Particle States

Note that all of our transition amplitude calculations entailed specific spin basis states, such as u_1, u_2, v_2, etc. One could ask, what then, do we do for a particle general spin state, such as $u = a_1 u_1 + a_2 u_2$?

Finding transition amplitudes for general, not basis, spin states.

The answer is this. First, we find the transition amplitude S_{fi1} for u_1 and the transition amplitude S_{fi2} for u_2. Then, we find the transition amplitude for the general state u as

$$S_{fi} = a_1 S_{fi1} + a_2 S_{fi2} \,. \tag{8-96}$$

Angular Momentum is Conserved

A final (generally, multiparticle) state will have the same total angular momentum (total spin in our cases) as the incoming state, but individual particles within the state can have different spin states than they came in with. We showed the conservation of angular momentum principle in Prob. 7 for the special case single electron state of Fig. 8-5 (pg. 230). The γ^μ matrices help take care of this automatically for us.

Total angular momentum conserved in interactions.

γ^μ matrices help the math work out.

Switching Identical Fermions Changes Transition Amplitude Sign

This too, is a subtle point that you may wish to hold off on until some time in the future. The choice is yours.

Another subtle point – switching fermion labels switches sign of transition amplitude

Recall from NRQM, that for a multiparticle state, if we switched labels (switched positions in physical space) of the fermions, the multiparticle wave function changed sign. A corresponding thing happens in QFT.

Note in the Møller scattering of Fig. 8-4 on pg. 229, the difference between the two ways the scattering can occur boils down to the electron labeled p'_1 being interchanged with the electron labeled p'_2. And note that the transition amplitudes for the two different ways have opposite signs.

This turns out to be a general rule in QFT. Interchanging labels on indistinguishable fermions changes the sign of the resulting transition amplitude.

Attraction vs Repulsion in QED

After all this work getting to our present state of understanding of QFT, you may be wondering (as I did at a similar stage), how the theory handles repulsion vs. attraction, how it handles like charges vs unlike charges. In classical theory, the force law in the two cases had opposite signs. If

QFT is about forces (interactions), then surely it must be able to describe this trivial example, no?

QFT concerned with probabilities of interactions, but in Chap. 16, we derive Coulomb force law.

One, somewhat evasive, answer is that in QFT we are usually not dealing with force laws, but with probabilities of interactions occurring, e.g.., the chance of repulsion (or attraction) occurring between two particles. We are typically not much concerned with whether it really is attraction or repulsion. For either case, one gets a positive probability value.

However, in Sect. 8.10 we take a non-mathematical, descriptive look at how QFT would describe the interaction between two charged macroscopic objects. And in Chap. 16, we formally deduce the Coulomb potential for repulsion and attraction from QFT.

Our Formalism Appropriate for Scattering, Not Bound States

Our approach good for scattering, not bound states

The QFT formalism we are developing is not appropriate for bound states, but is well suited for scattering. It handles probabilities for transient interactions well, but a different tack is needed for quasi-stable states such as the hydrogen atom.[1] Introductory QFT courses rarely address bound states, and we will abide by this tradition. First things (and the most useful things), first.

8.7 Including Other Charged Leptons in QED

Including muons and taus in QED

So far, we have dealt only with charged leptons of the electron/positron type. As physicists have learned from experiment, there are two more families of leptons, the muon/anti-muon (symbols μ^- and μ^+) and the tau/anti-tau particle (symbols τ^- and τ^+) families. Electrons, muons, and taus all have −1 charge. Positrons, anti-muons and anti-taus all have +1 charge. Electrons and positrons have the same mass. Muons and anti-muons have the same mass, which is about 200 times the electron mass. Taus and anti-taus have the same mass, which is about 170 times the muon mass.

Of course, muons and taus are extremely rare and not generally seen in our world. However, they do turn up in high energy particle accelerators, so we need to be able to handle them in our theory. One might guess we simply need to add extra terms to our Hamiltonian density (Lagrangian density), similar in form to the electron/positron/photon term used so far, for each of the muon and tau families. This guess would be correct, and the new interaction Hamiltonian that correctly describes all three families is

The mixed lepton QED interaction Hamiltonian

$$\mathcal{H}_I^{\,I} = -\mathcal{L}_I^{\,I} = -e\sum_{l=1}^{3}\bar{\psi}_l A^\mu \gamma_\mu \psi_l = -e\sum_{l=1}^{3}\left(\bar{\psi}_l^{\,+} + \bar{\psi}_l^{\,-}\right)\left(\mathcal{A}^+ + \mathcal{A}^-\right)\left(\psi_l^{\,+} + \psi_l^{\,-}\right), \qquad (8\text{-}97)$$

where the summation over l represents a separate Hamiltonian density for electrons/positrons (l = 1), muons/anti-muons (l = 2), and taus/anti-taus (l = 3). (Quarks are also electrically charged, but we don't consider quarks in this volume).

No vertices that mix lepton families allowed

Note that with (8-97) we will now have vertices with fermions of the same family interacting with a photon. Just as we have seen $e^-e^+A^\mu$, $e^-e^-A^\mu$ and the like vertices, we will now also have $\mu^-\mu^+A^\mu$, $\mu^-\mu^-A^\mu$ and the like vertices, and $\tau^-\tau^+A^\mu$, $\tau^-\tau^-A^\mu$ and the like vertices.

Note that we don't have families mixing at a vertex. That is, we don't have $e^-\mu^+A^\mu$ or $\mu^-\tau^-A^\mu$ type vertex terms in (8-97), which means, for example, that in Fig. 8-9 we could not have a μ^- (or τ^-) in place of the e^- on the LHS. No electron/positron along with a μ (or τ) at the same vertex.

8.7.1 An Interaction Involving More than One Lepton Family

A mixed lepton interaction example

In Fig. 8-9, we see the process of Chap. 1, pgs. 2-3, involving a muon/anti-muon pair being produced by an electron and positron. This is like our Bhabha scattering, but with final muons instead of electron/positron.

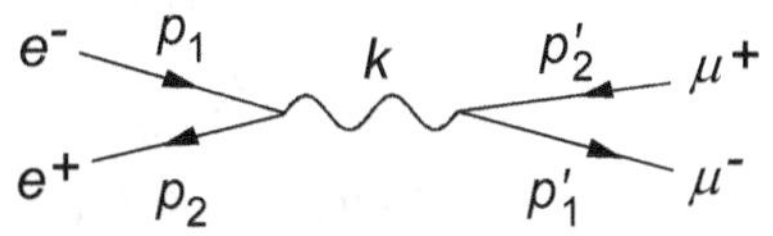

Figure 8-9. The Process $e^+ + e^- \rightarrow \mu^+ + \mu^-$

[1] See Mandl & Shaw, *Quantum Field Theory,* 2nd ed (Wiley 2010), 187-191, S. Schweber, *An Introduction to Relativistic Quantum Field Theory* Silvan (Dover 2005), 705-720 and Berestetskii et al, *Quantum Electrodynamics* (Butterworth,1982), 101, 343-347, 371-376, 552-559.

We could, if we wished, deduce the transition amplitude for Fig. 8-9 from basic principles. However, it is much simpler to just take our result for Bhabha scattering (via first way of Fig. 8-2, pg. 221), assuming creation operators for muons instead of those for electrons and positrons. That is, from (8-35) and (8-34), we get

Transition amplitude for our mixed lepton interaction example

$$S^{(2)}_{e^-e^+ \to \mu^- \mu^+} = \left(\prod_{\mathbf{p}}^{\substack{\text{all external} \\ \text{particles}}} \sqrt{\frac{m}{VE_{\mathbf{p}}}} \right) (2\pi)^4 \delta^{(4)}\left(p_1 + p_2 - \left(p_1' + p_2' \right) \right) \mathcal{M}^{(2)}_{e^-e^+ \to \mu^- \mu^+}. \tag{8-98}$$

with $\mathcal{M}^{(2)}_{e^-e^+ \to \mu^- \mu^+} = -e^2 \bar{u}_{r_1' \ l=2}(\mathbf{p}_1') \gamma^\mu v_{r_2' l=2}(\mathbf{p}_2') i D_{\mu\nu}\left(k = p_1 + p_2 \right) \bar{v}_{r_2 \ l=1}(\mathbf{p}_2) \gamma^\nu u_{r_1 \ l=1}(\mathbf{p}_1).$ (8-99)

Note that there is no contribution to the amplitude here like the 2nd way for Bhabha scattering to occur, as shown on the RHS of Fig. 8-2. We chose this example in Chap. 1 because it is simpler.

8.7.2 Feynman Rules for Multiple Families

From the example above, we can glean that our Feynman rules can be generalized for three families of leptons as follows.

Feynman rules for mixed lepton interactions

1) Obtain the Feynman amplitude assuming all leptons are electrons/positrons.

2) For lines representing other lepton "flavors" (another word for "family") replace spinors and/or propagators with those representing the other flavors.

Note that the only difference in the form of the result will be the masses.

8.7.3 Elastic vs Inelastic Scattering

The term inelastic scattering (or inelastic interaction) refers to the particles involved changing into different types of particles with different masses (recall we mean "rest mass" by the term "mass"), such as the scattering shown in Fig. 8-9. The outgoing particles are different than the incoming particles, and hence some kinetic energy must be converted into mass (or vice versa depending on the particular interaction).

Inelastic scatter = particle masses change

The total amount of energy (in the form of mass plus kinetic energy) must stay the same. But if the total mass changes, then that change must be compensated for by an equivalent opposite change in kinetic energy.

Elastic scattering (or elastic interaction), on the other hand, implies we have the same type final particles (with the same masses) as the initial particles. (See Figs. 8-2 and 8-4 on pgs. 221 and 229, for examples.) No mass is created out of, or destroyed to yield, kinetic energy. All energy exchange is purely kinetic, and this is a characteristic of classical elastic interactions, hence the name.

Elastic scatter = same particles out as in, masses unchanged, only K.E. transferred

8.7.4 Meaning of the Word "Force" Changes a Bit

Prior to QFT, the meaning of the word "force" was limited to that of a "push or pull" exerted by one body on another body. This is depicted in the RH diagram in Fig. 8-2, pg. 221 where the electron and positron exert force on one another, and neither particle changes its identity. In that diagram, the force is mediated by the virtual photon, which carries energy and momentum from one particle to the other, thereby changing the motion of each.

However, in QFT, "force" is extended to include the action of the virtual particle in the LH diagram of Fig. 8-2, as well. There, in addition to being the carrier of energy and momentum, the virtual particle mediates a transmutation of particle types. The electron and positron turn into a virtual photon. This, like the RH diagram, represents elastic scattering because the final particle types are the same as the initial.

The term "force" extends to include particle transmutation interactions

In Fig. 8-9 above, however, the electromagnetic force (mediated by the virtual photon) changes the initial particle types into different ones, i.e., results in inelastic scattering. The point is we now use the term "force" in a more general way, to involve any type, elastic or inelastic, of interaction.

8.8 When to Add Amplitudes and When to Add Probabilities

For two different Feynman diagrams, given the same initial state in each, when the final particles are also in the same state (not just particle type, but also same momenta and spins) in each, we add the amplitudes from the diagrams and take the square of the absolute value of that sum to get total probability of finding the state with those final particle states.

Add amplitudes when incoming and outgoing states are the same, but different ways exist for it to occur

We can't distinguish between the two processes because we can't measure what is happening with the virtual particles involved. So, the total amplitude is the sum of the amplitudes of the indistinguishable interactions. The probability is the square of the absolute value of that.

When the final states from two different Feynman diagrams (having the same initial state) are different (as with $e^-e^+ \rightarrow e^-e^+$ and $e^-e^+ \rightarrow \mu^-\mu^+$, or $e^-e^+ \rightarrow e^-e^+$ and $e^-e^+ \rightarrow e'^-e'^+$ where primes indicate different momenta), then we take the square of the absolute value of each individual amplitude to get the probability of each final state occurring. The probability of an interaction occurring at all that would yield any one of the final states is the sum of these probabilities.

When outgoing states are different, add probabilities of each way for total probability of observing any one among the final states

In this case, we can distinguish between the two processes, since the final measurable results are distinguishable. So each has its probability of occurring independent of the other. The total probability of any of the different final states being found is the sum of the probabilities of the different final states.

8.9 Wave Packets and Complex Sinusoids

Particles are typically of wave packet form, as in the LH side of Fig. 8-10. Yet in our analyses, they parallel the form of the field operators that create and destroy them, i.e., complex sinusoids e^{ipx}, as in the RH side of Fig. 8-10. (Fig. 8-10 shows only the real components of complex waves.)

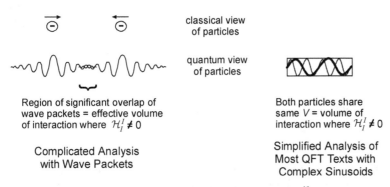

classical view of particles

quantum view of particles

Region of significant overlap of wave packets = effective volume of interaction where $\mathcal{H}_I^I \neq 0$

Both particles share same V = volume of interaction where $\mathcal{H}_I^I \neq 0$

Complicated Analysis with Wave Packets

Simplified Analysis of Most QFT Texts with Complex Sinusoids

*Particles really wave packets, but our analysis simplifies to **k** eigenstates*

Figure 8-10. The Complicated Wave Packet vs Simplified e^{ikx} Solution Form Analyses

Obviously, wave packets are more complicated to deal with and analyze than sinusoids. But, as it turns out, we can make excellent predictions for scattering and decay phenomena with the simpler complex sinusoids, and virtually all introductory QFT texts deal exclusively with them.

Keep in mind that in our analyses, all incoming and outgoing particles occupy the same volume. Their field operators all have the same V in their normalization factors. See the RH of Fig. 8-10. However, two particles approaching each other from opposite directions, as in the LH of Fig. 8-10, are more accurately modeled as wave packets.

*For wave packets or **k** eigenstates, Hamiltonian only non-zero where all particle waves non-zero*

Our interaction Hamiltonian $\mathcal{H}_I^I = -e\bar{\psi}A^\mu\gamma_\mu\psi$ is only non-zero where the fields $\bar{\psi},\psi$ and A^μ are all non-zero (or effectively so). For the wave packet model of those fields, this only becomes significantly different from zero where the wave packets of the particles overlap. Of course, the wave packets actually extend to infinity, but outside of a certain region they are so diminished that one can consider them zero.

In contrast, for the complex sinusoid model, all fields are uniformly distributed throughout the same volume V (which we take as very large) and zero elsewhere. Again, this model works for the most significant experiments, and after the discussion of Sect. 8.10 below, we will henceforth in this book confine our attention to the mathematically far simpler model of the RH side of Fig. 8-10.

8.10 Looking Closer at Attraction and Repulsion

8.10.1 Two Elementary Particles Attracting or Repelling

As promised, we now look at a more or less heuristic overview of repulsion and attraction, given what we have learned about QFT.

Heuristic look at attraction and repulsion in QED

Repulsion

Consider two like charged particles approaching one another, such as the electrons in the left side of Fig. 8-11. If they are going to repel one another, as we know they do, the virtual particles transmitting the electromagnetic force must have certain properties.

For the electron coming in from the left (with velocity toward the upper right) the virtual photon, as it leaves that real particle, must kick that particle back so its velocity has a component in the negative horizontal direction. In order to do that, the virtual photon must carry 3-momentum **k** in its own direction of travel. When the electron coming in from the right accepts that virtual photon 3-momentum, its velocity must receive a kick in the positive horizontal direction. Hence, the two electrons appear to repel one another.

*In repulsion case, virtual particle **k** is in same direction as its velocity **v***

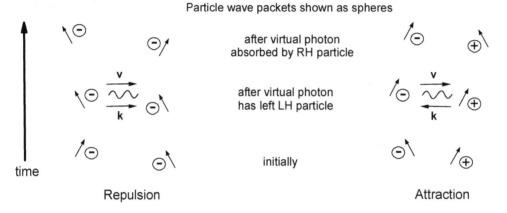

Figure 8-11. Virtual Particle Momenta for Real Particle Repulsion and Attraction

Attraction

For attraction, as in the right side of Fig. 8-11, things are not quite so intuitive. If the electron on the LH moving away is to be attracted, it must pick up a momentum (velocity) component in the positive horizontal direction. The only way it can do that, upon shedding a virtual photon, is if that virtual photon has 3-momentum pointing to the left. That is, the virtual photon in this case carries 3-momentum in the opposite direction of its velocity. This certainly seems strange from a classical mechanics perspective, but it is the only way for attraction to occur.

*In attraction case, virtual particle **k** is in opposite direction of **v***

When the positron on the RH moving away absorbs the virtual photon, it picks up 3-momentum in the negative horizontal direction. Hence, it picks up an additional velocity component in that direction, too. The electron and positron are pulled closer together.

Summary of Two Particle Attraction and Repulsion

Once again, we see that virtual particles disobey some basic physical laws with which we have grown familiar. For them, 3-momentum can actually be in the opposite direction of velocity. In fact, for attraction, this sort of behavior is essential. More generally, 3-momentum in other cases can be in other directions not parallel to virtual particle velocity (where we define that velocity in terms of the length vector between emission and absorption events divided by the time between them.)

Of course, in reality, we are looking at particles which are field-like, in the sense that they are spread out in space. An interaction is more like an interaction of one particle field with another, and during this interaction, 3-momentum of appropriate direction is transferred. Again, the Feynman diagrams tend to make us think of point-like particles, rather than field-like particles.

Again, virtual particles disobey some traditional laws for real particles

In all cases, however, we do find total 3-momentum and total energy, including all real and virtual particles, are conserved. These conservation laws still hold.

8.10.2 Charged Approaching Macroscopic Objects Attracting and Repelling

Now consider two macroscopic charged objects approaching one another along the same line of action. Arrows for virtual particles in Fig. 8-12 represent velocity **v** (not 3-momentum **k**).

Charged macro bodies approaching one another

Repulsion

On the left side of Fig. 8-12, both objects are negatively charged. Many virtual photons are emitted and absorbed by the many charged particles on each macroscopic body. These virtuals have 3-momentum in the direction of their motion. As the bodies get closer together, the wave packets of the real particles in each macroscopic body overlap more and more, resulting in more and more virtuals being emitted and absorbed. The result is stronger repulsive force felt by each body.

Like charges approaching: virtual 3-momentum **k** *in* **v** *direction*

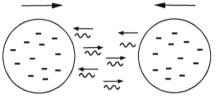

The closer they get, the more the real wave packets of the objects overlap → the more virtuals exchanged → the stronger the force

Individual virtuals have **k** in direction Individual virtuals have **k** in opposite
of **v** and positive energy   direction of **v** and negative energy

Repulsion   Attraction

Figure 8-12. Virtual Particle 3-Momenta and Energy for Macro Repulsion and Attraction

Note that the macroscopic bodies lose speed as they approach one another, and thereby their respective energies decrease. Classically we interpret this lost energy as a gain in the classical electric field energy between the particles. The field potential is of form

$$V_{\substack{classical \\ repulsion}} \propto \frac{q_1 q_2}{r} > 0 \,, \tag{8-100}$$

so that the total energy, where *KE* is kinetic energy and subscripts indicate the bodies, is

$$E_{total} = KE_1 + V_{\substack{classical \\ repulsion}} + KE_2 = \text{constant} \,. \tag{8-101}$$

As *r* gets smaller, *KE* of the macro bodies decreases, but the electric field potential energy increases.

In QFT, we would say the kinetic energies of the macro bodies plus all the energy of the virtuals is conserved,

$$E_{total} = KE_1 + E_{virtuals} + KE_2 = \text{constant} \,. \tag{8-102}$$

And so, we can see that what is interpreted classically as classical field energy, is, in the quantum realm, energy of virtual particles,

Like charges approaching, virtual particles have positive energy

$$V_{\substack{classical \\ repulsion}} = E_{virtuals} > 0 \,. \tag{8-103}$$

In similar fashion, 3-momentum of the macro bodies plus all the virtuals is conserved.

Attraction

Similar to what we saw in Fig. 8-11, in Fig. 8-12 attraction involves virtual particles with 3-momenta in the opposite directions of their velocities. This conserves total 3-momentum and provides attractive behavior, where the objects' speeds towards one another increase.

Opposite charges: virtual **k** *opposite direction of* **v**

Similar to repulsion, the classical energy relation for attraction (see right side of Fig. 8-12) is

$$E_{total} = KE_1 + V_{\substack{classical \\ attraction}} + KE_2 = \text{constant} \,, \tag{8-104}$$

where the attraction field potential energy is negative

$$V_{\substack{classical \\ attraction}} \propto \frac{q_1 q_2}{r} < 0 \,. \tag{8-105}$$

As the macro objects approach each other, they speed up, so their kinetic energy increases. But the classical field energy decreases (becomes more negative) to compensate and keep total energy constant.

Given that we must have the same energy relation, in terms of virtual particles, as (8-102), this must mean that the virtual particles energy must be negative.

$$V_{classical \atop attraction} = E_{virtuals} < 0 . \qquad (8\text{-}106)$$

Opposite charges approaching, virtual particles have negative energy

Strange, but necessary. Virtual particles involved in attraction between approaching bodies must have negative energies and 3-momenta in the opposite direction of their motion.

Summary of Attraction and Repulsion for Approaching Objects

The classical field energy equals the virtual particle total energy. As the bodies get closer, the number of virtuals exchanged increases, making the macroscopic force seem greater. The virtual particle energies and 3-momenta are what is required to yield this behavior in the large.

8.10.3 Negative Energies Again?

It is interesting to note that we seem to have found our way back to negative energy. Recall, that finding negative energies in RQM was one of the driving forces to develop QFT. Ironically, after all that development, it turns up anew.

Negative energy not a big concern because virtual, not real, particles possess it

However, the key difference is that now it is only virtual particles which can have negative energies. It is still prohibited for real particles, the ones we can measure, and therefore does not contradict our classical reality, as RQM seemed to do.

8.10.4 Charged Objects Receding from One Another

Do Prob. 17 to prove to yourself that charged objects receding from one another along the same line of action exchange virtual photons with the same energy and 3-momentum characteristics as those of approaching objects.

Wholeness Chart 8-3. Summary of Virtual Photon Properties for 1D Attraction and Repulsion

For receding objects, virtual particles same characteristics as approaching objects

	Virtual Particles in Repulsion		**Virtual Particles in Attraction**	
	3-momentum	**Energy**	**3-momentum**	**Energy**
Approaching and Receding Objects	**v** direction	positive	− **v** direction	negative

There are additional subtleties in the above that we will discuss later in Chap. 17.

8.11 The Degree of the Propagator Contribution to the Transition Amplitude

Note the form of the momentum space photon propagator,

$$iD_{F\mu\nu}(k) = -i\frac{g_{\mu\nu}}{k^2 + i\varepsilon} , \qquad (8\text{-}107)$$

which appears in any transition amplitude having a virtual photon. Remember that the virtual photon is off-shell, so the value of k^2 does not have to be zero, as with a real photon.

Propagator contribution greater, the closer virtual particle is to on-shell

For values of $|k| \rightarrow 0$, this takes on a higher value than for larger $|k|$. ($|k|$ is the length in 4D of the photon four-momentum vector, i.e., $|k| = \sqrt{k^2} = \sqrt{\omega_{\mathbf{k}}^2 - \mathbf{k}^2}$) So, the closer a virtual photon is to being on-shell, the greater its contribution is to the transition amplitude, and thus, the higher the probability of that particular interaction where that value of 4-momentum is transferred from one real particle to another. This does not necessarily mean less strong interactions (smaller changes in 4-momentum of the real particles involved) are more likely, because k^2 can be small when both $\omega_{\mathbf{k}}^2$ and $\mathbf{k}^2$ are high. If the latter are higher, the interaction is stronger.

8.12 Summary of Where We Have Been: Chaps. 7 and 8

Wholeness Chart 8-4 on the next four pages summarizes all of interaction theory to here, i.e., Chaps. 7 and 8. However, the following is a briefer summary of how we got to this point.

Overview of our progress

Summary of the Ground We Have Covered

Free field theory via H_0 → Interaction Picture → State evolution via $|\Phi(t)\rangle = S_{oper}|i\rangle$ (we seek S_{oper})

      ↓   ↓   ↓

    Free fields   State eq of motion via H_I^I → use $|\Phi\rangle$ above in state eq
                (H_I^I in terms of free fields)

→ solution $S_{oper}(t)$ in terms of H_I^I, time ordered → $S = S_{oper}(t=\infty)$ → Wick's theorem

→ S normal ordered + contractions → transition amplitude $S_{fi} = \langle f|S|i\rangle$ → algebra → S_{fi}

→ probability of interaction = $|S_{fi}|^2$

Or the Shortcut (Based on the Above) Instead

Feynman rules → S_{fi} → $|S_{fi}|^2$

8.13 Appendix: Returning to Equal-Times-Commutators and Wick's Theorem

Recall that at the end of Sect. 7.8.4 on pg. 209, we noted that had we included equal-times contractions in Wick's theorem, it would have given rise to non-physical situations. Now that we have digested the current chapter, we can investigate just what that means.

Consider if the following term, containing an equal-times contraction, were in the S operator.

$$S_X^{(2)} = -e^2 \int d^4x_1 d^4x_2 N \left\{ \left(\bar{\psi}A\psi\right)_{x_1} \left(\bar{\psi}A\psi\right)_{x_2} \right\} \tag{8-108}$$

(8-108) represents, in one case, an initial real electron and positron at x_2 annihilated into a virtual photon that then travels to x_1. There it becomes a very strange virtual fermion, one that begins and ends at the same spacetime point, and thus can hardly be said to exist. Essentially, there are no real final particles, i.e., the incoming particles disappear.

We could plug all the mathematical expressions for the fields and propagators into (8-108), in the same manner we have done repeatedly in this chapter to evaluate the transition amplitude

$$S_{EqTim} = \langle 0|S_X^{(2)}|e^-_{\mathbf{p}_1,r_1}, e^+_{\mathbf{p}_2,r_2}\rangle. \tag{8-109}$$

We get the same results via the Feynman rules short cut, so (8-109) becomes

$$S_{EqTim} = \sqrt{\frac{m}{VE_{\mathbf{p}_1}}} \sqrt{\frac{m}{VE_{\mathbf{p}_2}}} (2\pi)^4 \delta^{(4)}(p_1 + p_2) \mathcal{M}_{EqTim}. \tag{8-110}$$

From the delta function, we see that the only time the transition amplitude is non-zero is when the energies of the two incoming fermions total zero. That means each incoming fermion must have zero energy, i.e., neither can exist and the interaction doesn't happen.

For other interactions with equal-times contractions, either similar kinds of issues result, or for QED, the effect can be renormalized away (not treated in this text, but similar to procedures we will cover in Chap. 14.)

Thus, we can conclude, as stated in Chap. 7, that equal-times contractions, if included in our theory, would give rise to non-physical situations and/or can be otherwise exorcised from the theory. Physical reality and our theory match if equal-times contractions are not part of the application of Wick's theorem to QED, as we noted in Chap. 7.

QED/FIELD THEORY OVERVIEW: PART 2

Wholeness Chart 8-4. From Operators and Propagators to Feynman Rules

⇩ **INTERACTING FIELDS** ⇩

	In theory, the non-linear coupled partial differential interaction fields equations can be solved simultaneously to get interacting fields solutions and hence complete descriptions of all interactions. In practice this has not been possible and the following perturbation scheme has been developed.

Interaction Picture Approach

Interaction Picture $H^I = H_0^I + H_I^I$	Motion of <u>states</u> governed by H_I^I : $\qquad\qquad i\dfrac{d}{dt}\big|\Psi\big\rangle_I = H_I^I \big|\Psi\big\rangle_I$ Motion of <u>operators</u> governed by $H_0^I = H_0 \qquad \dfrac{dO^I}{dt} = -i\big[O^I, H_0\big]$ $\phi,\ \psi,\ A^\mu$ are operators, so depend on H_0 only. Further, for them the above operator equation reduces to the free field equations on the first page of Part 1 of this chart. (See end of Chap. 5.)
Results of Interaction Picture	For interactions, if employ I.P., we can use 1. free field operator solutions $\phi,\ \psi,\ A_\mu$ of Part 1 as I.P. fields solutions 2. free field operator creation and destruction properties 3. free field number operators 4. free field observables operators 5. free field Feynman propagators 6. state equations of motion in H_I^I to determine change in state in time (i.e., interactions)
H_I^I	Spatial integral of $\mathcal{H}_I^I = -\mathcal{L}_I^{\,I}$ (with operators taken as free field solutions) $= H_I^I$. e.g., for QED $\ H_I^I = \int \mathcal{H}_I^I d^3\mathbf{x}$ with $\mathcal{H}_I^I = -e\,\overline{\psi}\,\gamma^\mu \psi\, A_\mu$
New notation	Drop script "I" on states and operators other than $\mathcal{H}_I^I$ and H_I^I.
S matrix	$|S_{fi}|^2$ is probability of ith eigenstate (often multiparticle) transitioning to fth eigenstate S_{fi} is transition amplitude.
S_{oper}	General scattered state: $\qquad \big|\Psi(t_f)\big\rangle = S_{oper}\big|\Psi(t_i)\big\rangle \quad \big(\big|F\big\rangle = S_{oper}\big|i\big\rangle$ typically$\big)$ $|i\rangle$ = an initial eigenstate. $|f\rangle$ = a final eigenstate. $|\Psi(t)\rangle = |F\rangle$ = sum of eigenstates usually
Finding S matrix from S_{oper}	$S_{fi} = \big\langle f\big|S_{oper}\big|i\big\rangle = \big\langle f\,\big|\,\Psi(t_f)\big\rangle\ $ so $\ \big|\Psi(t_f)\big\rangle = \displaystyle\sum_f |f\rangle S_{fi}$ Conservation of probability (not particles) is $\displaystyle\sum_f |S_{fi}|^2 = 1$
S_{oper} equation of motion	I.P. state eq of motion at top for $|\Psi(t)\rangle = S_{oper}|\Psi(t_i)\rangle$ yields $\quad i\dfrac{dS_{oper}}{dt} = H_I^I S_{oper}$
Solution to S_{oper} equation of motion	$S_{oper} = e^{-i\int_{t_i}^{t_f} H_I^I dt} = e^{-i\int_{t_i}^{t_f}\int_V \mathcal{H}_I^I d^4 x}$ (time ordered), so $\ \big|\Psi(t)\big\rangle = S_{oper}\big|\Psi(t_i)\big\rangle = e^{-i\int_{t_i}^{t} H_I^I dt'}\big|\Psi(t_i)\big\rangle$
S_{oper} for all space and time $= S$	$S = S_{oper}\underset{V\to\infty}{\big(t_f \to \infty, t_i \to -\infty\big)} = e^{-i\int_{-\infty}^{\infty}\mathcal{H}_I^I d^4 x} \qquad S$ = "S operator" terminology in QFT typically

S matrix for S operator	For ∞ space and time, $S_{fi} = \langle f|S|i \rangle = \langle f|\Psi(t=\infty) \rangle$ so $|\Psi(t=\infty)\rangle = \sum\limits_f |f\rangle S_{fi}$
Dyson expansion	Below is Dyson expansion of S solution above
exact: 2nd order:	$S = I \underbrace{}_{S^{(0)}} \underbrace{-i\int_{-\infty}^{\infty}\mathcal{H}_I^I(x_1)d^4x_1}_{S^{(1)}} \underbrace{-\frac{1}{2!}\int_{-\infty}^{\infty}\int_{-\infty}^{\infty}T\{\mathcal{H}_I^I(x_1)\mathcal{H}_I^I(x_2)\}d^4x_1d^4x_2}_{S^{(2)}} + ... = \sum\limits_{n=0}^{\infty}S^{(n)}$ $= \sum\limits_{n=0}^{\infty}\frac{(-i)^n}{n!}\int_{-\infty}^{\infty}.....\int_{-\infty}^{\infty}d^4x_1d^4x_2...d^4x_nT\{\mathcal{H}_I^I(x_1)\mathcal{H}_I^I(x_2)...\mathcal{H}_I^I(x_n)\}$ To this point treatment is exact. Perturbation arises from using fewer than ∞ terms in the above. $S \cong I + (-i)\int_{-\infty}^{\infty}d^4x_1\mathcal{H}_I^I(x_1) + \frac{(-i)^2}{2!}\int_{-\infty}^{\infty}d^4x_1d^4x_2T\{\mathcal{H}_I^I(x_1)\mathcal{H}_I^I(x_2)\}$
Contractions of operators	Definition: $\underset{\sqcup}{AB} = \langle 0|T\{AB\}|0 \rangle$ = Feynman propagators if A and B are certain fields. Special (only non-zero) cases: $\overline{\phi(x_1)\phi^{\dagger}(x_2)} = \overline{\phi^{\dagger}(x_2)\phi(x_1)} = i\Delta_F(x_1-x_2)$ $\overline{\psi_\alpha(x_1)\overline{\psi}_\beta(x_2)} = -\overline{\overline{\psi}_\beta(x_2)\psi_\alpha(x_1)} = iS_{F\alpha\beta}(x_1-x_2)$ $\overline{A^\mu(x_1)A^\nu(x_2)} = iD_F^{\mu\nu}(x_1-x_2).$
Wick's Theorem	$T\{(AB...)_{x_1}.......(AB...)_{x_n}\} = N\{(AB...)_{x_1}.......(AB...)_{x_n}\}$ $+N\{\,\underline{(AB..)_{x_1}(AB..)_{x_2}}\,...\} + N\{\,\underline{(AB..)_{x_1}(AB..)_{x_2}}\,..\} + \begin{pmatrix}\text{all other normal ordered non} \\ \text{equal time contractions}\end{pmatrix}$ $+N\{\,(ABC...)_{x_1}(ABC...)_{x_2}\,...\} + \begin{pmatrix}\text{all other normal ordered non} \\ \text{equal time double contractions}\end{pmatrix}$ $+\,\left(\text{all normal ordered non equal time triple contractions}\right)$ $+$ etc.
QED interactions	Above basic principles of interaction theory applied to QED below
$\mathcal{H}_I^I$ for QED	$\mathcal{H}_I^I = -\mathcal{L}_I^I = -e\overline{\psi}\gamma^\mu A_\mu\psi = -e\left(\overline{\psi}^+ + \overline{\psi}^-\right)\left(A^+ + A^-\right)\left(\psi^+ + \psi^-\right)$ (for electrons and positrons)
Dyson expansion for QED	Using above in Dyson expansion of S operator along with the extended Wick's theorem to turn the time ordered integrand into a normal ordered integrand yields the QED S operator $S = \sum\limits_{n=0}^{\infty}S^{(n)} = S^{(0)} + S^{(1)} + S^{(2)} + S^{(3)} + ... \begin{pmatrix}\text{higher} \\ \text{order terms}\end{pmatrix}$ where $S^{(0)}, S^{(1)}, S^{(2)}$ shown below

	Operator	Matrix Elements
$S^{(0)}$	$= I$ no transition of particles $|f\rangle = |i\rangle$	$S_{fi}^{(0)} = \langle f|S^{(0)}|i \rangle$ typical process: $|e^-,\gamma\rangle \rightarrow |e^-,\gamma\rangle$ no virtual particles, no 4-momentum change

$S^{(1)}$	$= (-i)\int d^4x_1 N\left\{-e\bar{\psi}\gamma^\mu A_\mu\psi\right\}_{x_1}$ = all three external particles interaction terms = 8 terms in all, but these processes are <u>not real physical processes</u> (because resultant particle(s) off-shell)	Typical non-physical process:
$S^{(2)}$ $\quad S_A^{(2)}$	$= \dfrac{-e^2}{2!}\int d^4x_1 d^4x_2 N\left\{\left(\bar{\psi}A\psi\right)_{x_1}\left(\bar{\psi}A\psi\right)_{x_2}\right\}$ No real physical processes.	Two processes like $S^{(1)}$ above going on independently.
$\quad S_B^{(2)}$	$= \dfrac{-e^2}{2!}\int d^4x_1 d^4x_2 N\left\{\left(\bar{\psi}A\psi\right)_{x_1}\left(\bar{\psi}A\psi\right)_{x_2}\right\}$ = all four external leptons interaction terms	$S_{Bfi}^{(2)} = \left\langle f\middle|S_B^{(2)}\middle|i\right\rangle$ typical process (Bhabha scattering): $\left|e^-,e^+\right\rangle \rightarrow \left|e^-,e^+\right\rangle$
$\quad S_C^{(2)}$	$= \dfrac{-e^2}{2!}\int d^4x_1 d^4x_2 N\left\{\left(\bar{\psi}A\psi\right)_{x_1}\left(\bar{\psi}A\psi\right)_{x_2}\right\}$ $\qquad + N\left\{\left(\bar{\psi}A\psi\right)_{x_1}\left(\bar{\psi}A\psi\right)_{x_2}\right\}$ $= -e^2\int d^4x_1 d^4x_2 N\left\{\left(\bar{\psi}A\psi\right)_{x_1}\left(\bar{\psi}A\psi\right)_{x_2}\right\}$ = all two external leptons, two external photons interaction terms.	$S_{Cfi}^{(2)} = \left\langle f\middle|S_C^{(2)}\middle|i\right\rangle$ typical process (Compton scattering): $\left|i\right\rangle = \left|e^-,\gamma\right\rangle \rightarrow \left|f\right\rangle = \left|e^-,\gamma\right\rangle$ with virtual electron mediating scatter.
$\quad S_D^{(2)}$	$= -e^2\int d^4x_1 d^4x_2 N\left\{\left(\bar{\psi}A\psi\right)_{x_1}\left(\bar{\psi}A\psi\right)_{x_2}\right\}$ = two external leptons terms (lepton self-energy)	$S_{Dfi}^{(2)} = \left\langle f\middle|S_D^{(2)}\middle|i\right\rangle$ electron and positron loops
$\quad S_E^{(2)}$	$= \dfrac{-e^2}{2!}\int d^4x_1 d^4x_2 N\left\{\left(\bar{\psi}A\psi\right)_{x_1}\left(\bar{\psi}A\psi\right)_{x_2}\right\}$ = two external photons term (photon self-energy)	$S_{Efi}^{(2)} = \left\langle f\middle|S_E^{(2)}\middle|i\right\rangle$ photon loop
$\quad S_F^{(2)}$	$= \dfrac{-e^2}{2!}\int d^4x_1 d^4x_2 N\left\{\left(\bar{\psi}A\psi\right)_{x_1}\left(\bar{\psi}A\psi\right)_{x_2}\right\}$ = no external particles term	$S_{Ffi}^{(2)} = \left\langle f\middle|S_F^{(2)}\middle|i\right\rangle$ vacuum bubble
$S^{(3)}, S^{(4)}$, etc.	Higher order terms. Ignored for now.	

Sample probability determination	Compton scattering, two ways: (Assumption: Particles are plane waves in box of volume V.) See Chap. 8, Sect. 8.4.3, pg. 225 for derivation of the following Compton's $S_{fi} = S_{Compton} = \langle f \vert S \vert i \rangle_{Comp} = \langle f \vert S_{C1}^{(2)} + S_{C2}^{(2)} \vert i \rangle$ $$= \left(\prod_{\mathbf{p}''}^{\substack{\text{all external} \\ \text{fermions}}} \sqrt{\frac{m}{VE_{\mathbf{p}''}}} \right) \left(\prod_{\mathbf{k}''}^{\substack{\text{all external} \\ \text{bosons}}} \sqrt{\frac{1}{2V\omega_{\mathbf{k}''}}} \right) (2\pi)^4 \, \delta\left(p' + k' - p - k\right)\left(\mathcal{M}_{C1}^{(2)} + \mathcal{M}_{C2}^{(2)} \right).$$ where $\quad \mathcal{M}_{C1}^{(2)} = -e^2 \bar{u}_{s',\alpha}(\mathbf{p}') \varepsilon_{\mu,r'}(\mathbf{k}') \gamma_{\alpha\beta}^{\mu} iS_{F\beta\delta}(q = p+k) \varepsilon_{v,r}(\mathbf{k}) \gamma_{\delta\eta}^{v} u_{s,\eta}(\mathbf{p})$ $\quad\quad\quad\quad \mathcal{M}_{C2}^{(2)} = -e^2 \bar{u}_{s',\alpha}(\mathbf{p}') \varepsilon_{\mu,r}(\mathbf{k}) \gamma_{\alpha\beta}^{\mu} iS_{F\beta\delta}(q = p-k') \varepsilon_{v,r'}(\mathbf{k}') \gamma_{\delta\eta}^{v} u_{s,\eta}(\mathbf{p})$ Probability of Compton scattering $= \left\vert \langle f \vert S \vert i \rangle_{Comp} \right\vert^2$
Adding amplitudes	When two or more diagrams have the same external states in and out, add amplitudes for each contributing diagram, then square the absolute value of the result to get probability. For probability that any of two or more outcomes (different external states out) may occur from the same external states in, square absolute value of individual amplitudes first and then add.
2 ways to calculate probability	1) Go through tedious derivation like Chap. 8, Sect. 8.4.3 for each interaction 2) Use short cut of Feynman rules (listed in Box 8-3, pg. 236.)
	All three lepton types treated below.
Mixed lepton $\mathcal{H}_I^I$	$$\mathcal{H}_I^I = -\mathcal{L}_I^I = -e\sum_l \bar{\psi}_l A_\mu \gamma^\mu \psi_l = -e\sum_l \left(\bar{\psi}_l^+ + \bar{\psi}_l^- \right)\left(A^+ + A^- \right)\left(\psi_l^+ + \psi_l^- \right)$$

Mixed lepton S operator	Each $\bar{\psi} A \psi$ term in S expression above for single lepton type replaced by $\sum_l \bar{\psi}_l A \psi_l$ term.
	
	$$S = \sum_{n=0}^{\infty} \frac{(ie)^n}{n!} \int_{-\infty}^{\infty} ..... \int_{-\infty}^{\infty} d^4x_1 ... d^4x_n \sum_{l_1=1}^{3} .... \sum_{l_n=1}^{3} T\left\{ \left(\bar{\psi}_{l_1} A \psi_{l_1} \right)_{x_1} ... \left(\bar{\psi}_{l_n} A \psi_{l_n} \right)_{x_n} \right\}$$
	
	= terms like previous blocks for e^-, e^+ +
	    " " " " " muons +
	    " " " " " taus +
	
	terms mixing lepton types (but not mixed at a vertex).

Typical interaction	$e^- + e^+ \to \mu^- + \mu^+$ (with photon mediating.)
Mixed lepton summary	1) Obtain the Feynman amplitude assuming all leptons are electrons/positrons. 2) For lines in the Feynman diagram representing other lepton flavors, replace spinors and/or propagators with those representing the other flavors.

8.14 Problems

1. Show that the sub-term $S_2^{(1)} = ie \int d^4x_1 N\{\bar{\psi}^- A^+ \psi^+\}_{x_1}$ represents destruction of an initial electron and an initial photon, followed by creation of another electron. Draw the Feynman diagram for this process. Evaluate the integral to the point where you can show that outgoing 4-momentum for the final electron equals incoming 4-momentum of the initial multi-particle state.

2. Show that the sub-term $S_3^{(1)} = ie \int d^4x_1 N\{\bar{\psi}^- A^+ \psi^-\}_{x_1}$ represents destruction of an initial photon followed by creation of an electron and a positron. Draw the Feynman diagram for this process. Evaluate the integral to the point where you can show that outgoing 4-momentum for the multi-particle final state equals incoming 4-momentum of the initial photon.

    Is this a real physical process that could actually occur? Prove your answer using the mathematics of the mass shell.

    Show that sub-term $S_3^{(1)}$ applied to the incoming state of Fig. 8-1 results in zero.

3. Draw Feynman diagrams for all eight possible sub-terms of $S^{(1)}$. Label each with the related mathematical expression of the sub-term represented. (You don't have to evaluate the integrals.)

4. Compare the mathematical steps of this chapter (8-25) to (8-34) for calculating S_{B1}, the transition amplitude for the annihilation type of Bhabha scattering, to the heuristic steps for the transition amplitude of the interaction on pgs. 2-3 of Chap. 1. (Note that we used μ^- and μ^+ as the final particles there instead of e^- e^+, so that we would have only one relevant Feynman diagram, i. e., the LHS of Fig. 8-2. The RHS of the figure would not exist for e^- $e^+ \to \mu^- \mu^+$, since the final particles are different particles from the initial particles.) Identify which factors in the treatment in this chapter correspond to the factors K, K_2, K_γ, and K_1 in Chap. 1. What did we leave out of the Chap. 1 treatment (to make it simple)?

5. a) Show $N\left\{\left[\bar{\psi}^+(x_1), \psi^-(x_2)\right]_+\right\} = 0$, even though the anti-commutator shown does not equal

    zero.

    b) Using a), show that $N\{\phi(x_1)\phi^\dagger(x_2)\psi(x_3)\bar{\psi}(x_4)...\} = N\{-\phi(x_1)\phi^\dagger(x_2)\bar{\psi}(x_4)\psi(x_3)...\}$.

6. Derive the transition amplitude for Møller scattering (see Fig. 8-4, pg. 229), which equals (8-68) with (8-69). Do this the long way, i.e., without Feynman rules. Note that to get correct signs one has to normal order both sub amplitudes in the same normal order. "Same" here refers to the momentum/spinor values of the operators, not their x_1 and x_2 dependency. Note also that this is a fairly lengthy problem. (See comments in the two paragraphs following (8-69).)

7. Prove that $s'=s$ in (8-76). Recall that $iS_F(p-k) = \dfrac{(p_\alpha - k_\alpha)\gamma^\alpha + m}{(p-k)^2 - m^2 + i\varepsilon}$, and from Chap. 4

    Appendix A (pg. 122), that $\left[\gamma^\alpha, \gamma^\nu\right]_+ = 2g^{\alpha\nu}$, $(\not{p} - m)u_r(\mathbf{p}) = 0$, $g_{\mu\nu}\gamma^\mu\gamma^\nu = \gamma_\nu\gamma^\nu = 4$, and

    $\bar{u}_r(\mathbf{p})\gamma^\mu u_s(\mathbf{p}) = \dfrac{p^\mu}{m}\delta_{rs}$. (Hint: Start with the integrand of (8-76) and use the gamma matrices

    anti-commutation relations to switch the order of γ^α and γ^ν. Then use the relation after that above to get certain terms to drop out.)

8. Derive the transition amplitude for positron Compton scattering of the LHS Fig. 8-7, pg. 234. Do this the long way, without Feynman rules.

9. Find the transition amplitude relation, in terms of propagators, for the vacuum bubble diagram of Fig. 8-8, pg. 234. Do this the long way, without Feynman rules.

10. Use Feynman rules (pg. 236) to find the transition amplitude (8-67) for Compton scattering of electrons.

11. Use Feynman rules to find the transition amplitude (8-68) for Møller scattering of electrons.

12. Use Feynman rules to find the transition amplitude (8-77) for the electron/photon closed loop of an electron.

13. Use Feynman rules to find the transition amplitude (8-86) for the photon closed loop.

14. Use Feynman rules to find the transition amplitude (8-89) for an electron/positron/photon vacuum bubble.

15. Draw the two Feynman diagrams for the interaction $e^+ e^- \rightarrow \gamma \gamma$. Using Feynman's rules, write out the amplitude for one of the diagrams.

16. Draw the two Feynman diagram for the interaction $\gamma \gamma \rightarrow e^+ e^-$. Using Feynman's rules, write out the amplitude for one of the diagrams.

17. Carry out an analysis parallel to that of Sect. 8.10.2 for two macroscopic charged bodies receding from one another along the same line of action. Prove to yourself that for repulsion the virtual particles have positive energy; and that for attraction, the virtual particles have negative energy.

18. (Added in revision of 2018.) Draw the mass shell for the photon. Suppress the k^2 and k^3 dimensions, to make it easier. That is, plot E vs k^1. Does it touch the origin? Are the sides of the shell, for a photon, straight lines? Does the mass shell for a massive particle like that shown in Box 8-1 approach that of a photon asymptotically for very high E (speed approaching c)?

Chapter 9

Higher Order Corrections

""The last function of reason is to recognize that there
are an infinity of things which surpass it."
Blaise Pascal

9.0 Background

In Chap. 8, we investigated the role of the second order term $S^{(2)}$ in the S operator expansion (where $S = S^{(0)} + S^{(1)} + S^{(2)} + S^{(3)} + \dots$), and how using only that term in the state equation of evolution $|f\rangle = S |i\rangle$ should give us a good approximation to the transition amplitude for a given interaction. That is, $S_{fi} = \langle f | S | i \rangle \approx \langle f | S^{(2)} | i \rangle$. This is because $S^{(0)}$ and $S^{(1)}$ don't give rise to real changes in particles, and because, the reasoning goes, higher order terms are significantly smaller than $S^{(2)}$.

$S^{(2)}$ developed in Chap. 8

When we add in these higher order terms, the additional terms in the transition amplitude are called <u>higher order corrections</u> to the transition amplitude.

9.0.1 Higher Order Corrections

As we are about to find out, higher order corrections are problematic in two ways. For one, they significantly complicate the analysis. That, however, is a minor problem relative to the second.

$S^{(n)}$ for $n > 2$ developed in this chapter

In a fashion quite unexpected from a mathematically naïve perspective, higher order corrections, as we will see, turn our nice, neat, finite transition amplitudes of Chap. 8 into monstrous, difficult-to-tame infinite numbers, about which you have no doubt heard. For instance, using higher order terms in our Compton scattering calculation results in a transition amplitude of infinity. Since the square of the absolute value of the transition amplitude is the probability of the interaction occurring, and since probability must lie between zero and one, "Houston, we have a problem."

Problem: parts of these higher order terms are infinite

9.0.2 Loop Diagrams

Actually, as I have purposely refrained from mentioning prior to here, infinities crop up even in some second order transition amplitudes. These are specifically from the loop diagrams we saw in Fig. 8-5 pg. 230 (electron closed loop) and Fig. 8-6 pg. 232 (photon closed loop), and also a loop associated with each vertex. More on this later.

We will see loop diagrams of Chap 8 have infinite amplitudes

Non-loop second order terms (such as those associated with the diagrams we examined in Chap. 8 for Bhabha, Compton, and Møller scattering) are free from the infinity problem, fortunately.

9.0.3 Chapter Overview

The main topic we want to cover in this chapter is

We'll examine higher order corrections

- higher order corrections due to terms $S^{(3)}$, $S^{(4)}$, $S^{(5)}$, ... in the S operator expansion

In doing so, we will see that

- diagrams with loops have infinite transition amplitudes,
- loops are common components of higher order Feynman diagrams.

Further, we will note

- the mathematics for why terms that should be finite are infinite, and
- that we will not show the infinity problem solution (renormalization) until Part Three.

9.1 Higher Order Correction Terms

In Chap. 8, we showed the S operator for QED as (see (8-3) pg. 215)

$$S = \sum_{n=0}^{\infty} S^{(n)} = \underbrace{I}_{S^{(0)}} + \underbrace{ie\int \left(\overline{\psi} A \psi\right)_{x_1} d^4x_1}_{S^{(1)}} \underbrace{-\tfrac{1}{2!}e^2 \iint T\left\{\left(\overline{\psi} A \psi\right)_{x_1} \left(\overline{\psi} A \psi\right)_{x_2}\right\} d^4x_1 d^4x_2}_{S^{(2)}} +$$

$$\underbrace{-\tfrac{i}{3!}e^3 \iiint T\left\{\left(\overline{\psi} A \psi\right)_{x_1} \left(\overline{\psi} A \psi\right)_{x_2} \left(\overline{\psi} A \psi\right)_{x_3}\right\} d^4x_1 d^4x_2 d^4x_3 + ...}_{S^{(3)}} \qquad (9\text{-}1)$$

Dyson expansion of S, including higher order terms

We evaluated the $S^{(0)}$, $S^{(1)}$, and $S^{(2)}$ terms of this in Chap. 8 using Wick's theorem to express each term as normal ordering operators with contractions rather than time ordering, as in (9-1). We will now evaluate the $S^{(3)}$ and $S^{(4)}$ terms, and draw general conclusions about terms of even higher order.

Note that $S^{(3)}$ is a third order term with respect to e, the electron charge, and in that context $S^{(4)}$ is fourth order. We will use this nomenclature for the time being, but refine it later to reflect more common usage employed in the literature.

9.1.1 Third Order in e Correction Terms

We will use Wick's theorem (9-2) below to re-express $S^{(3)}$ of (9-1).

$$T\left\{(AB...)_{x_1} ...(AB...)_{x_n}\right\} = N\left\{(AB...)_{x_1} ...(AB...)_{x_n}\right\}$$

$$+ N\left\{\underbrace{(AB..)_{x_1}(AB..)_{x_2}}\; ...\right\} + N\left\{(AB..)_{x_1} \underbrace{(AB..)_{x_2} ..}\right\} + ... + N\left\{...\underbrace{(A...Z)_{x_{n-1}}(A...Z)_{x_n}}\right\}$$

$$+ N\left\{\underbrace{(ABC...)_{x_1}(ABC...)_{x_2}}\; ...\right\} + N\left\{\underbrace{(ABC...)_{x_1}(ABC...)_{x_2} ...}\right\} + ... \qquad (9\text{-}2)$$

Wick's theorem re-expresses any term in S as normal ordered terms & contractions

+ (all normal ordered terms with three non-equal times contractions) + etc.

Using (9-2) for $S^{(3)}$ of (9-1), we find

$$S^{(3)} = -\tfrac{i}{3!}e^3 \iiint T\left\{\left(\overline{\psi} A \psi\right)_{x_1} \left(\overline{\psi} A \psi\right)_{x_2} \left(\overline{\psi} A \psi\right)_{x_3}\right\} d^4x_1 d^4x_2 d^4x_3$$

$$= -\tfrac{i}{3!}e^3 \iiint \Bigg(N\left\{\left(\overline{\psi} A \psi\right)_{x_1} \left(\overline{\psi} A \psi\right)_{x_2} \left(\overline{\psi} A \psi\right)_{x_3}\right\} +$$

$$+ N\left\{\underbrace{(\overline{\psi} A \psi)_{x_1} (\overline{\psi} A \psi)_{x_2}} \left(\overline{\psi} A \psi\right)_{x_3}\right\} + N\left\{(\overline{\psi} A \psi)_{x_1} \underbrace{(\overline{\psi} A \psi)_{x_2} (\overline{\psi} A \psi)_{x_3}}\right\}$$

$$+ N\left\{\underbrace{(\overline{\psi} A \psi)_{x_1} (\overline{\psi} A \psi)_{x_2} (\overline{\psi} A \psi)_{x_3}}\right\} + ...$$

$$+ N\left\{\underbrace{(\overline{\psi} A \psi)_{x_1} (\overline{\psi} A \psi)_{x_2} (\overline{\psi} A \psi)_{x_3}}\right\} + N\left\{\underbrace{(\overline{\psi} A \psi)_{x_1} (\overline{\psi} A \psi)_{x_2} (\overline{\psi} A \psi)_{x_3}}\right\} + ...$$

$$(9\text{-}3)$$

Using Wick theorem on $S^{(3)}$ term

+ (all normal ordered terms with three non-equal times contractions) + etc. $\Big) d^4x_1 d^4x_2 d^4x_3$

Most of the terms in (9-3) are zero. For example, for the single contraction terms, the first two shown are zero because, recall, contractions of a field with itself are zero, as are its contractions with a different field type (such as lepton with boson). So we can immediately drop out all terms in (9-3) except those comprising contractions of a field with its own complex conjugate (the adjoint for fermions; the field itself for photons, as the photon field is real, so it is its own complex conjugate). Each of the many terms left represents its own interaction reflected in its own Feynman diagram.

Most terms in $S^{(3)}$ are zero

Remaining terms represent different Feynman diagrams

Of the remaining terms, let's look at the third single contraction term of (9-3), with integrand

$$N\left\{\underbrace{(\overline{\psi} A \psi)_{x_1} (\overline{\psi} A \psi)_{x_2}} \left(\overline{\psi} A \psi\right)_{x_3}\right\}, \qquad (9\text{-}4)$$

Looking at one term with single contraction

which is not zero via the above reasoning. One of the Feynman diagrams for this would look like Fig. 9-1. (This is an unconnected diagram, as we discussed in Chap. 8. That is, not all lines are connected to one another in a single network.)

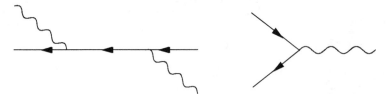

Figure 9-1. Feynman Diagram for One of the Single Contraction Terms of $S^{(3)}$

Note that in Fig. 9-1 we have a single vertex interaction, acting alone, which we learned in Chap.8 is not physically possible, because it has at least one real (not virtual) particle off shell. We conclude that any possible term in $S^{(3)}$ that has a vertex factor $\left(\bar{\psi}A\psi\right)_{x_i}$ alone, unconnected to a contraction, is not physical and can be ignored.

This further reduces the total number of terms we need to be concerned about in (9-3) considerably. Let's look at one of these, the three contractions term with integrand

$$N\left[\left(\bar{\psi}A\psi\right)_{x_1}\left(\bar{\psi}A\psi\right)_{x_2}\left(\bar{\psi}A\psi\right)_{x_3}\right]. \qquad (9\text{-}5)$$

A typical Feynman diagram for this looks like Fig. 9-2.

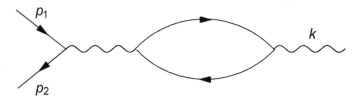

Figure 9-2. Feynman Diagram for One of the Three Contraction Terms of $S^{(3)}$

Note that the net result of Fig. 9-2 is similar to what we saw in Fig. 8-1, pg. 218, repeated below as Fig. 9-3, where a real electron and a real positron transmute into a single off-shell photon.

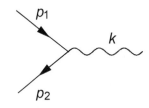

Figure 9-3. Feynman Diagram with Equivalent Incoming and Outgoing 4-Momenta as in Fig. 9-2

In both the Fig. 9-2 and 9-3 cases, we have conservation of energy and 3-momentum at each vertex. So, for the same incoming electron/positron state, we will get the same off-shell photon. Hence the process of Fig. 9-2 is not real, and (9-5) can be ignored.

<u>An aside:</u> We reiterate what we said in Sect. 8.3.3, pgs. 219-220, about non-physical states. Mathematically, the operators involved in Fig. 9-3 create an off-shell photon ket, which has different 4-momentum than an on-shell (real world) single photon bra can ever have. The bra and ket don't match, so their inner product yields zero. Hence, the mathematics behind Fig. 9-3 will always yield a zero transition amplitude value, and we can simply ignore that process. <u>End of aside.</u>

In similar fashion, we can show that every term in (9-3) is un-physical and contributes zero to the transition amplitude. As another example, by doing Prob. 1, you can prove to yourself that four contraction terms in (9-3) are not physical.

<u>Conclusion #1:</u> $S^{(3)}$ plays no role in QED and can be ignored.

Note that all Feynman diagrams for $S^{(3)}$ have three vertices. And recall that all such diagrams for $S^{(2)}$ had two vertices; $S^{(1)}$ had only one vertex; and $S^{(0)}$ had none. Vertices represent events x_n and, we had three events in $S^{(3)}$. (See (9-1).)

Further, $S^{(3)}$ represented *all* possible Feynman diagrams with three vertices, without missing any. As we saw, there are many of these (which are all, in this case, either zero or non-physical.)

Conclusion #2: $S^{(n)}$ represents all possible Feynman diagrams with exactly *n* vertices.

Each term in $S^{(n)}$ has n vertices

Rather than going to all the trouble of expanding $S^{(3)}$ via Wick's theorem and examining each possible term, we could instead have simply noted that $S^{(3)}$ had three vertices and drawn all of the three vertex Feynman diagrams. Then we could simply apply Feynman's rules to each of those diagrams, without worrying about the math of contractions and all the rest. This general approach holds true for any order term, i.e., for any number of vertices/events.

Conclusion #3: We can simplify the process of finding transition amplitudes for terms in any $S^{(n)}$ by just drawing all possible Feynman diagrams with *n* vertices and applying Feynman's rules to each.

Let's use shortcut Feynman diagrams and rules to evaluate $S^{(n)}$ terms

From Box 9-1, we see that for $S^{(3)}$, we will get 3! = 6 identical terms corresponding to six diagrams that are topologically similar (i.e., from Chap. 8, Sect. 8.5.1, pg. 235, those which look the same when we exchange any two event labels with each other, such as x_1 for x_3.) Thus, when these six terms are added together, the 3! in the denominator of $S^{(3)}$ cancels out. This has already been taken into account in Feynman's rules, so we don't have to even worry about the factorial factor.

Similar logic holds for any $S^{(n)}$ where *n*! identical terms will arise from topologically similar diagrams, yielding a single term without *n*! in the denominator representing a single topologically distinct diagram. And again, Feynman's rules, which apply to each topologically different diagram, already take this into account.

Conclusion #4: Feynman's rules ignore the factorial in the denominator of each term in (9-1) because those rules are applied to each topologically distinct diagram.

Feynman's rules take n! factor of Wick expansion into account

9.1.2 Fourth Order in e Correction Terms

We are going to use Conclusion #3 above (keeping in mind Conclusion #4) to evaluate

$$S^{(4)} = \frac{1}{4!} e^4 \iiint \int T \left\{ \left(\bar{\psi} A \psi \right)_{x_1} \left(\bar{\psi} A \psi \right)_{x_2} \left(\bar{\psi} A \psi \right)_{x_3} \left(\bar{\psi} A \psi \right)_{x_4} \right\} d^4 x_1 d^4 x_2 d^4 x_3 d^4 x_4 \qquad (9\text{-}6)$$

$S^{(4)}$ term in Dyson expansion

of (9-1). That is, we can simply draw all <u>connected</u> (no vertices sitting alone, but all lines and vertices in the same network) Feynman diagrams with four vertices. These will be many. But they will fall into groups, where all diagrams in a group have the same incoming and outgoing particles. Let's look at some examples. (We'll ignore muons and taus for the time being.)

Let's jump right to Feyn diags for one type of 4th order interaction

Bhabha Scattering Group of Contributions Coming from $S^{(4)}$

Consider the $S_{B1}^{(2)}$ contribution to the Bhabha scattering transition amplitude of the LH of Fig. 8-2, pg. 221. It represents the contribution from an electron/positron pair incoming and annihilating, and an electron/positron pair outgoing, where there are 2 vertices. (Mathematically, the lowest order term of significance for this interaction.)

That type is Bhabha scattering

From (9-6), we get the next highest order contribution to the Bhabha scattering transition amplitude, $S_{B1}^{(4)}$. It is represented by all four-vertex topologically distinct Feynman diagrams with the same incoming and outgoing states. These are shown as the last 11 diagrams of Figure 9-4.

Our program amounts to finding the transition amplitudes for every diagram of Fig. 9-4 and adding them together to get the total transition amplitude for the interaction with 4th order in *e* accuracy. Note that the first diagram in Fig. 9-4 is 2nd order, 3rd order terms are all zero (as we showed earlier), and all the remaining diagrams in the figure are 4th order.

We'll use Feynman rules to get amplit for each diagram in Fig. 9-4, then add them for total amplit to 4th order

Other Types of Scattering

For Compton, Møller and other types of scattering, we would carry out similar steps. We would draw all connected Feynman diagrams with two and four vertices having the same incoming and outgoing states. We use Feynman rules to determine and calculate the transition amplitude for each of these, add them, and take the square of the absolute value of the result to determine the probability of that particular type of scattering occurring (to 4th order in *e* accuracy.)

Similar steps for other kinds of scattering

Box 9-1. Where the $n!$ Goes in Feynman Rules

Using Wick's theorem for $S^{(3)}$ of (9-3), we found one term in the integrand was (9-5), repeated below and reflected in Fig. 9-2, where the vertices are labeled x_3, x_2, x_1 going from left to right in the figure, but from right to left in the integrand,

$$N\left\{ (\overline{\psi}A\psi)_{x_1} (\overline{\psi}A\psi)_{x_2} (\overline{\psi}A\psi)_{x_3} \right\}. \tag{B9-1.1}$$

There is another term of the form

$$N\left\{ (\overline{\psi}A\psi)_{x_1} (\overline{\psi}A\psi)_{x_2} (\overline{\psi}A\psi)_{x_3} \right\}. \tag{B9-1.2}$$

Recall from Box 8-2, pg. 226, that inside normal ordered terms, we can simply exchange adjacent bosons and have the same result, whether the bosons are part of a contraction or not. We can do the same thing with fermions except that we have to multiply by –1 when we switch positions of adjacent fermion fields. A boson and fermion that are adjacent can simply be switched.

Thus, we can re-order (B9-1.2) (noting that fermion switches occur twice for each fermion, so the net result is no sign change), as

$$(B9\text{-}1.2) = N\left\{ (\overline{\psi}A\psi)_{x_1} (\overline{\psi}A\psi)_{x_3} (\overline{\psi}A\psi)_{x_2} \right\}. \tag{B9-1.3}$$

But, since (B9-1.1) and (B9-1.3) are integrands in an integration over x_1, x_2, and x_3, which are dummy variables, then they result in the same thing after being integrated. And thus, we can switch x_2 with x_3 without changing that result. So, (B9-1.3) is equal to (B9-1.1), meaning (B9-1.2) is equal to (B9-1.1).

Effectively, we have shown that i) switching dummy event variables x_2 and x_3 has no effect on the final result, and ii) switching dummy event variables turns one term in the Wick's expansion, (B9-1.1), into another equivalent term, (B9-1.2). That is, (B9-1.1) = (B9-1.2) = (B9-1.3).

In similar fashion, we can get additional terms in the Wick's expansion by simply permuting the event variables. From the order of (B9-1.1) of $x_1x_2x_3$, we got $x_1x_3x_2$ in (B9-1.3) representing (B9-1.2). Four other terms will result from event orders $x_2x_3x_1$, $x_2x_1x_3$, $x_3x_2x_1$, and $x_3x_1x_2$, resulting in $6 = 3!$ terms. These six terms can be added together as one term times 6, which cancels the 3! in the denominator of (9-3). (If you do Prob. 2, you can find another one of those six terms.)

As we have learned, switching dummy variables in a Feynman diagram yields what is termed a topologically similar diagram. Hence, if we use the diagram of Fig. 9-2, without labeling the vertices, it can represent all 6 terms. Thus, Fig. 9-2 represents all topologically similar diagrams, and Feynman rules apply to it, without a 3!.

In general then, for any order n, Feynman rules are applied to each topologically distinct diagram. They are not applied separately to topologically similar diagrams.

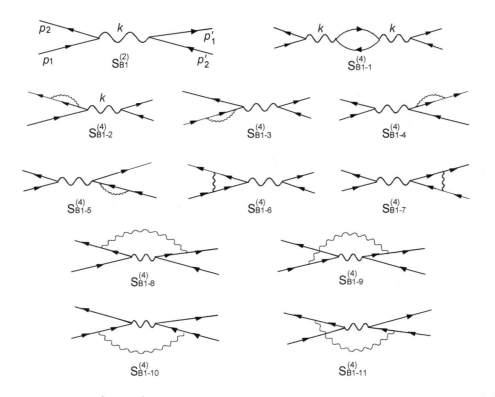

Figure 9-4. e^2 and e^4 Order Contributions to First Kind of Bhabha Scattering

One $S^{(2)}$ term and eleven $S^{(4)}$ terms for first kind of Bhabha scattering

9.1.3 THE Problem

Everything seems great to this point. And this is what early researchers in the field must have felt. A few calculations (simple in concept, but extensive and algebra intensive when you get down to it) and *voila*, we have our probability of the interaction occurring.

Unfortunately, some of the terms in our example of Fig. 9-4 (and in fact, for any e^4 order transition amplitude contribution for any interaction) turn out, when calculated, to be infinite! This is the famous, infamous problem that we discussed at the beginning of this chapter and Chap. 8.

Unfortunately, some terms are infinite, as we will see below

Let's look at one example to see how the infinity arises.

The Photon Loop Diagram

The second diagram in the top row of Fig. 9-4 has a photon loop (made up of a virtual electron and a virtual positron). We show it in Fig. 9-5 with the momenta labeled, as well as the spacetime index at each vertex (to help keep track when using Feynman's rules).

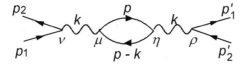

Figure 9-5. The $S^{(4)}_{B1-1}$ Term Feynman Diagram for Bhabha Scattering

4th order Bhabha scattering term with a photon loop part

By doing Prob. 3, you can show that the Feynman amplitude for this is

$$
\mathcal{M}^{(4)}_{B1-1} =
$$
$$
\frac{-e^4}{(2\pi)^4} \bar{u}_{r_1'}(\mathbf{p}_1')\gamma^\rho v_{r_2'}(\mathbf{p}_2') D_{F\rho\eta}(k) \left\{ \mathrm{Tr}\int S_F(p-k)\gamma^\eta S_F(p)\gamma^\mu d^4 p \right\} D_{F\mu\nu}(k)\bar{v}_{r_2}(\mathbf{p}_2)\gamma^\nu u_{r_1}(\mathbf{p}_1). \tag{9-7}
$$

Feynman amplitude for this term has an integral over 4 momentum p

Note from our analysis in Chap. 8 (see Fig. 8-6 and accompanying text on pgs. 231-233) that there are two equivalent ways to evaluate the loop integral of Fig. 9-5. In Chap. 8 we took the fermion leg in the loop as $\bar{p}+k$ instead of p, the anti-fermion leg as $\bar{p}$ instead of $p-k$, and we integrated over $\bar{p}$ instead of p. Both integrals equal the same thing, merely differing in choice of

We used different dummy variables in Chap. 8, but loop integral result is the same

dummy variables. While we may use one or the other, to keep things as easy as possible for new learners, we will try to use the representation of (9-7) and Fig. 9-5 exclusively from now on.

Every momentum value in (9-7) is fixed except p, which must be integrated over 4D momentum space from $+\infty$ to $-\infty$ along all four axes. And we know all of the factors are finite except for the integral, which we don't know yet. Let's estimate the integral value by assuming parts of the integral where any component p^μ is large could be the problematic portions, if there are any. That is, we'll specifically investigate whether the integral blows up for large values of p^μ.

The integral of (9-7) is (top row below), and is estimated by (bottom row below),

Approximating the integral for large p

$$\int S_F(p)\gamma^\mu S_F(p-k)\gamma^\eta d^4p = \int \frac{\not{p}+m}{p^2-m^2+i\varepsilon}\gamma^\mu \frac{\not{p}-\not{k}+m}{(p-k)^2-m^2+i\varepsilon}\gamma^\eta d^4p$$

$$\underrightarrow{\text{for contribution}\atop\text{from large }p} \approx \int \frac{p_\nu}{p^2}\gamma^\nu\gamma^\mu\gamma^\sigma\gamma^\eta \frac{p_\sigma}{p^2}d^4p \xrightarrow[\text{also ignoring spacetime}\atop\text{indices to simplify}]{\text{ignoring }\gamma\text{ matrices}\atop\text{to estimate simply}} \int \frac{pp}{p^4}d^4p \approx \int \frac{d^4p}{p^2}.$$

$$(9\text{-}8)$$

To evaluate the last part of (9-8), consider that when integrating over a 2D surface, 3D volume, or 4D region, the following hold true for radial coordinate r, where $d^n r$ represents the region of integration for an n dimensional space[1].

Evaluation trick: re-express integral over d^4p as over dp

$$dA = d^2r = \underbrace{2\pi r}_{\substack{\text{1D circumf}\\\text{of circle in}\\\text{2D space}}} dr \qquad dV = d^3r = \underbrace{4\pi r^2}_{\substack{\text{2D surf of}\\\text{sphere in}\\\text{3D space}}} dr \qquad dR_{4D} = d^4r = \underbrace{2\pi^2 r^3}_{\substack{\text{3D vol} =\\\text{hypersurface of}\\\text{4D hypersphere}\\\text{in 4D space}}} dr \;. \quad (9\text{-}9)$$

Applying the last expression in (9-9) with $r \to p$ to the last part of (9-8), (9-8) becomes

Integral estimated to diverge with p^2 at high p

$$\int_{-\infty}^\infty \frac{d^4p}{p^2} = \int_0^\infty 2\pi^2 \frac{p^3}{p^2} dp = 2\pi^2 \int_0^\infty p\,dp = \pi^2 p^2\,\big|_0^\infty \;, \qquad (9\text{-}10)$$

which diverges with the square of p as p gets large, and is called <u>quadratically divergent</u>.

Note the photon loop portion of Fig. 9-5 results in the divergent integral. The calculation we did was independent of the rest of the diagram (rest of the transition amplitude calculation).

Any term with photon loop will have divergent integral

This procedure of estimating the degree of divergence is called <u>power counting</u>, <u>naïve power counting</u>, or <u>superficial power counting</u>. It is called naïve/superficial because, as we will see in later chapters, the actual degree of divergence can be less than this estimate. Power counting tells us the maximum degree of divergence an integral may have. For example, the photon loop actually diverges with the log of p at high p. More on this later.

Power counting only indicates max degree of divergence

<u>Conclusion</u>: Whenever we have a photon loop, in any scattering case, we will get a factor of infinity in our transition amplitude. Naïve power counting indicates the divergence may be proportional to p^2, for large p.

Photon loop max divergence with p^2, for large p

9.1.4 Other Diagrams

Other diagrams in Fig. 9-4 also lead to infinite transition amplitudes. Each of the terms $S_{B1\text{-}2}^{(4)}$ to $S_{B1\text{-}5}^{(4)}$ has a lepton loop, and each, as we show below, results in a factor of infinity.

<u>Lepton Loop Diagrams</u>

Consider $S_{B1\text{-}3}^{(4)}$, for example. All factors in the Feynman amplitude that do not involve an integration will be finite. The loop has a photon propagator and a fermion propagator, so we can do a quick estimate of its potential for divergence, similar to what we did above for the photon loop.

Lepton loops have divergent integrals

---

[1] To be precise, we should do something called a <u>Wick rotation</u> transformation here first, which converts our relations from Minkowski space to Euclidean 4D space. Readers seeing this material for the first time should not worry too much about this, as the final estimates and conclusions we draw are valid in both cases. Wick rotation takes $g_{\mu\nu} \to -\delta_{\mu\nu}$ with $x^0 \to ix^0$. Multiplying by i in complex space is like rotating by $90°$, hence the "rotation" part of the nomenclature. We cover Wick rotation in Chap. 15.

For k as the virtual photon 4-momentum in the loop, and p_1 for the incoming electron, the virtual electron will have 4-momentum $p_1 - k$. Thus, the loop contribution to the transition amplitude can be estimated to diverge via

$$\int S_F\left(p_1 - k\right) D_{F\mu\nu}\left(k\right) d^4 k = \int \frac{\not{p}_1 - \not{k} + m}{\left(p_1 - k\right)^2 - m^2 + i\varepsilon}\frac{g_{\mu\nu}}{k^2 + i\varepsilon} d^4 k \qquad (9\text{-}11)$$

$$\xrightarrow[\text{large } k]{\substack{\text{ignoring }\gamma\text{ matrices}\\ \text{to estimate simply}\\ \text{also ignoring spacetime}\\ \text{indices to simplify}}} \approx \int \frac{k}{k^4} d^4 k = 2\pi^2 \int \frac{1}{k^3} k^3 dk = 2\pi^2 k \big|_0^\infty.$$

Power counting → max lepton loop integral divergence with p, for large p

Power counting for the lepton loop indicates it diverges with the first power of k as $k \to \infty$. We call this <u>linear divergence</u>. Again, the actual degree of divergence might be less.

<u>Vertex Loop Diagrams</u>

Do Prob. 4 to prove to yourself that the vertex loop on the LH side of the $S^{(4)}_{B1\text{-}6}$ diagram (in this book, we deem a vertex with two lepton propagators and a photon propagator as shown, a <u>vertex loop</u>, though that terminology is not common elsewhere) may diverge with the natural log of k. That is, you will find

$$2\pi^2 \int_0^\infty \frac{dk}{k} = 2\pi^2 \ln k \big|_0^\infty. \qquad (9\text{-}12)$$

Power counting → vertex loop max divergence with ln k, for large k

Power counting indicates the vertex loop has a <u>logarithmic divergence</u> as k gets large.

Note that we are not really considering the origin ($k = 0$) part of (9-12) because the denominator in the integral really has extra terms (m, k, $i\varepsilon$) that we dropped early on because we were focused on large k values. For small k values, the $k = 0$ part of (9-12) is not a valid estimate, and we will discuss this more in later chapters.

<u>Other Diagrams</u>

Consider the $S^{(4)}_{B1\text{-}8}$ diagram in Fig. 9-4, which we show below with appropriate momenta and spacetime labels. Note there are no <u>simple loops</u> (i.e., photon loop, fermion loop, or vertex loop as shown above) in Fig. 9-6, though there is what we might call a "loop". However, the term <u>loop</u> in QFT, generally, and in this book, always, refers only to one of the three simple loops shown above.

Another term in $S^{(4)}$

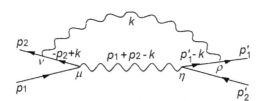

Figure 9-6. The $S^{(4)}_{B1\text{-}8}$ Term Feynman Diagram for Bhabha Scattering

By doing Prob. 5, you can use Feynman's rules to find the Feynman amplitude for Fig. 9-6,

$$\mathcal{M}^{(4)}_{B1\text{-}8} = \frac{1}{\left(2\pi\right)^4} e^4 \int d^4 k\, \overline{u}_{r'}\left(\mathbf{p}_1'\right)\gamma^\rho S_F\left(p_1' - k\right)\gamma^\eta v_{s'}\left(\mathbf{p}_2'\right) \times$$

$$D_{F\eta\mu}\left(p_1 + p_2 - k\right) D_{F\rho\nu}\left(k\right)\overline{v}_s\left(\mathbf{p}_2\right)\gamma^\nu S_F\left(-p_2 + k\right)\gamma^\mu u_r\left(\mathbf{p}_1\right). \qquad (9\text{-}13)$$

Feynman amplitude for this term

Once again, to determine the maximum degree of divergence of the integral in (9-13) for large k, we can ignore the gamma matrices, the summations on spacetime indices inside the fermion propagators, and the mass m inside those propagators, as well. Thus, (9-13), for $k \to \infty$ becomes

$$\mathcal{M}^{(4)}_{B1\text{-}8} \to \int S_F\left(p_1' - k\right) D_F\left(p_1 + p_2 - k\right) D_F\left(k\right) S_F\left(-p_2 + k\right) d^4 k$$

$$= \int \frac{\not{p}_1' - \not{k} + m}{\left(p_1' - k\right)^2 - m^2 + i\varepsilon}\frac{1}{\left(p_1 + p_2 - k\right)^2 + i\varepsilon}\frac{1}{k^2 + i\varepsilon}\frac{-\not{p}_2 + k + m}{\left(-p_2 + k\right)^2 - m^2 + i\varepsilon} d^4 k \qquad (9\text{-}14)$$

The integral in this term converges for large k

$$\xrightarrow[\text{for large } k]{} \approx \int \frac{1}{k}\frac{1}{k^2}\frac{1}{k^2}\frac{1}{k} d^4 k = 2\pi^2 \int \frac{1}{k^6} k^3 dk = -\pi^2 \frac{1}{k^2} \text{ for large } k.$$

So, this Feynman amplitude integral converges in the large k region of the integration.

For $k \to 0$, the middle line of (9-14) reduces to

$$\mathcal{M}^{(4)}_{B1-8} \xrightarrow{\text{for small } k} \int \frac{\not{p}_1' + m}{p_1'^2 - m^2 + i\varepsilon} \frac{1}{\left(p_1 + p_2\right)^2 + i\varepsilon} \frac{1}{k^2 + i\varepsilon} \frac{-\not{p}_2 + m}{p_2^2 - m^2 + i\varepsilon} d^4 k \qquad (9\text{-}15)$$

And also for $k \to 0$

$$\approx \int \left(\text{constant}\right) \frac{1}{k^2} d^4 k \approx 2\pi^2 \int \frac{1}{k^2} k^3 dk = \pi^2 k^2 \text{ for small } k.$$

Hence, the integral in (9-13) is finite.

In similar fashion, the amplitudes for the last three diagrams of Fig. 9-4 are finite, as well. Prob. 6 will give you some practice in proving this.

In similar fashion the other terms for Fig. 9-4 are finite

9.1.5 Extrapolations, Generalizations, and Comments

We have seen, with particular examples, how integrals arising in our transition amplitudes can diverge and lead to non-finite values for those amplitudes. We need to extrapolate and generalize these results.

The Other Part of the Bhabha Scattering Amplitude

All of the above pertained only to the LH side of Fig. 8-2, pg. 221. We would have to do similar analysis of the RH side of that figure to include the other way Bhabha scattering can occur. We would then add all of the amplitudes for all of the diagrams (all of those in Fig. 9-4 plus all of those of 2nd and 4th order for the RH side of Fig. 8-2) to get the total amplitude (leaving aside for the moment the issue of many of those terms being infinite).

For complete Bhabha amplitude, need to add in other way it can occur

Other Kinds of Scattering

For Compton, Møller and other kinds of scattering, we would have to follow parallel steps. Draw all of the appropriate Feynman diagrams for each, corresponding to terms of order e^2 and e^4, write down, using Feynman's rules, the amplitudes for each, and evaluate the integrals inside each of those separate relations. In principle, the sum of all those separate amplitudes for one type of scattering (e.g. Compton) would be added to get the total amplitude for that type of scattering (to fourth order in e accuracy.) See Prob. 7.

Same steps for Compton, Møller, other scattering types

However, just as we found above for Bhabha scattering, we would find that in each case, the photon, the electron, and vertex loops would all lead to factors of infinity in the transition amplitude. And each would have the same type of divergence at high 4-momentum in their integrals. See Prob. 8.

In all cases, photon, fermion, vertex loops are infinite, all else finite

All other terms would be finite[1]. Only the loop factors are unbounded.

Higher Order Terms

It turns out (we won't show it, but you can probably intuit it from our earlier work), that all higher order terms in e^n, where n is odd, drop out.

$S^{(n)}$ with n odd all yield zero and can be ignored

The remaining terms, where n is even, must be considered. In doing so (we won't do it here), one would find that again, the only infinite factors in any diagram come from the 3 (simple) loops.

$S^{(n)}$ with even $n >$ 4 done in similar fashion & loops yield ∞ there too

Common Nomenclature

Since we only have terms with factors of e to an even power, and since the fine structure constant (in natural Heaviside-Lorentz units) is

$$\alpha = \frac{e^2}{4\pi}, \qquad (9\text{-}16)$$

we can replace our e^2 and e^4 factors in our amplitudes with $4\pi\alpha$ and $(4\pi)^2\alpha^2$, respectively.

Common to consider $S^{(2)}$ 1st order in α, $S^{(4)}$ 2nd order in α

What we have been calling 2nd order in e, is typically called 1st order in α. 4th order in e is called 2nd order in α. Particle physicists seem to prefer using α to e in this context, though both can be found in the literature.

Also, high momentum divergences are referred to as <u>ultraviolet divergences</u> (ultraviolet waves have higher energy). We will also encounter (in later chapters) divergences as $k \to 0$, which are commonly called <u>infrared divergences</u>.

Ultraviolet (high energy) and infrared (low energy) divergences

---

[1] Actually, there are cases, such as a Compton diagram with three fermion propagators as edges of a "triangle", that diverge. But, in such cases, there are two such diagrams with integrals of equal magnitude and opposite sign. So, they cancel and do not together contribute an infinite value to the total amplitude.

<u>Why Only the Loops Cause Problems?</u>

The loops cause us problems because they are simple, in the sense that they typically have a small number of propagators (two for photon and fermion loops, three for a vertex loop). The propagators all have 4-momentum factors in their denominators. Photon propagators have k^2 in the denominator; fermions effectively have k (k in numerator, k^2 in denominator). (In some cases, the integration variable is p instead of k.)

When we multiply the propagators together, at high k, we get approximately $1/k^n$ in the integrand, i.e.,

$$\int \frac{1}{k^n} d^4k = 2\pi^2 \int \frac{1}{k^n} k^3 dk = 2\pi^2 \int k^{3-n} dk . \tag{9-17}$$

For $n > 4$, the integral is bounded for high k. But the number n depends on the number and type of propagators involved. Photon propagators have an effective extra k in the denominator compared to fermion propagators. And the more propagators in general, the higher n.

Look at the last four diagrams of Fig. 9-4. Each has four propagators with the dummy variable k appearing in each propagator. So, we would expect a larger value for n than, say, the second diagram in Fig. 9-4, which has fewer propagators carrying the dummy integration variable k.

In general, we can conclude that when a sufficient number of propagators are part of a single integration (all propagators carry the same dummy integration variable k [or p]), the integral will converge at high k.

Simple loops, with fewer factors in integration variable k (or p) in denominator, diverge

9.1.6 The Solution to the Infinity Problem

As you are aware from the beginning remarks of Chap. 8, the problem of the infinities in QED was solved in the late 1940s by Feynman, Schwinger, and Tomonaga. We will explore that solution in Chaps. 12 to 15.

For now, we note that the solution entails a redefining of e and m such that the infinite values are enveloped in those symbols and thus removed from our calculations. That results in finite transition amplitudes. If this sounds complicated, it is. That is why we save it for later on.

The procedure for removing the infinities in this way is called <u>renormalization</u>. In effect, we "renormalize" (redefine) values of certain quantities to give us finite answers.

You may find it heartening that after renormalization, we can simply use our 2nd order in e (1st order in α) results from Chap. 8 to get real world answers. Renormalization is not easy to understand, but the net result makes our physics much easier in the long run. We can do it with what we've learned already.

We will study renormalization, the solution to the infinity problem, in Chaps. 12 to 15

Renormalization will allow us to use the results of Chap. 8 to get complete scattering solutions

9.1.7 Is Something "Rotten in Denmark" Mathematically?

One thing is fishy in all of this infinity business. Recall that in Chap. 7, in our development of the S operator, we found that S_{oper} was a unitary operator. That is,

$$S_{oper} = e^{-i\int_{t_i}^{t_f} H_I^I dt} = e^{-i\int_{t_i}^{t_f} \int_V \mathcal{H}_I^I d^4x} \quad (\text{time ordered in expansion}). \tag{9-18}$$

The operation of S_{oper} on an initial state turned that state into a final state, i.e.,

$$\underbrace{|F\rangle}_{\substack{\text{general}\\\text{final state}}} = \sum_f C_f |f\rangle = S_{oper}|i\rangle = \sum_f S_{fi} \underbrace{|f\rangle}_{\substack{\text{final}\\\text{eigstates}}} . \tag{9-19}$$

$|i\rangle$ is defined as having unit norm. If S_{oper} is a unitary operator, then it merely "rotates" in Fock space the initial state $|i\rangle$ into the final general state $|F\rangle$. That means the magnitude of $|F\rangle$ is the same as that of $|i\rangle$

$$\langle i|i\rangle = 1 = \langle F|F\rangle \tag{9-20}$$

The final eigenstates $|f\rangle$ all are defined as having unit norms. Thus, the sum of the squares of the absolute values of the coefficients in (9-19) must equal unity.

$$\sum_f |S_{fi}|^2 = 1 \quad \text{conservation of probability under action of } S_{oper}. \tag{9-21}$$

Something is funny, because S_{oper} is a unitary operator

Which means no coefficient S_{fi} can have magnitude > 1

But we just showed that S_{fi} → ∞ at e^4 order

We made a big deal of this property back in Chap. 7. But, as we've found out in this chapter, some S_{fi} are infinite. So obviously the sum in (9-21) can't be correct. What have we done wrong?

Haag's Theorem

A theorem by Haag[1] states that a unitary operator $U(t_f, t_i)$ acting between finite times t_i and t_f will no longer remain unitary in the limit of $t_i \to -\infty$ or $t_f \to \infty$. In fact, it becomes very badly behaved.

Haag's theorem says a unitary operator becomes non-unitary for infinite times

In the development of interaction theory, one takes, as we did, the integration limits in (9-18) to infinity. We defined the symbol S to represent S_{oper} in this limit.

$$S = \underset{V \to \infty}{S_{oper}} \left(t_f \to \infty, t_i \to -\infty \right) = e^{-i\int_{-\infty}^{\infty} \mathcal{H}_I^I d^4 x} \tag{9-22}$$

S, via Haag's theorem, is not unitary, as S_{oper} is. So, our S matrix elements for S in

And our S is S_{oper} at infinite times

$$|F\rangle = \sum_f C_f |f\rangle = S|i\rangle = \sum_f S_{fi} |f\rangle \quad \left(S_{fi} \text{ here are for } S = S_{oper} \text{ in } \infty \text{ limit case} \right) \tag{9-23}$$

cannot be expected to satisfy (9-21). And they don't. Many of the S_{fi} are infinite.

In visualizing the action of S on the initial state (see Fig. 7-2, pg. 199), the magnitude ("length") of the state vector is not maintained equal to 1 as S "rotates" the initial state in Fock space. The state vector can be visualized as stretched in length (to infinity). Its magnitude is no longer finite, since some of its components have magnitudes $|S_{fi}|^2$ that are unbounded.

What Does This Mean?

Mathematicians and philosophers of science (and some physicists) have had, and continue to have, significant debates over the impact of Haag's theorem on QFT. Some consider it makes the foundation of the theory shaky, and thus, the entire theory suspect.

Some worry about this, but in practice QFT extraordinarily successful

The bottom line is that QFT, despite some seeming rough edges for the mathematical purists, predicts reality with extraordinary precision. The theory works. To most who use it, that is validation enough.

It all works, of course, because there ultimately is a solution to the infinity problem, as we will see in Chaps. 12 to 15.

9.1.8 Chapter Summary: QED Higher Order Corrections

- Higher order corrections to a given interaction transition amplitude comprise adding the $S^{(n)}$ terms for $n > 2$ to $S^{(2)}$ and finding

Overview of higher order corrections

$$S_{fi} = I_{fi} + \overset{(1)}{S_{fi}} + \overset{(2)}{S_{fi}} + \overset{(3)}{S_{fi}} + \overset{(4)}{S_{fi}} + ... = \langle f | I + \overset{(1)}{S} + \overset{(2)}{S} + \overset{(3)}{S} + \overset{(4)}{S} + ... | i \rangle$$

- All terms with odd n yield zero and can be ignored
- Simple photon, fermion, and vertex loops lead to unbounded transition amplitudes
- All else in transition amplitude calculations are finite
- There is a solution to the infinite amplitude problem, which we will soon study
- Short cut to find an nth order correction for a given interaction (e.g. Compton):
    - i) Draw all possible topologically distinct, connected Feynman diagrams with n vertices
    - ii) Use Feynman's rules to calculate the transition amplitude for each diagram
    - iii) Add all of the transition amplitudes found to get the total $S_{fi}^{(n)}$ for the interaction

---

[1] See P. Teller, *An Interpretive Introduction to Quantum Field Theory*, Princeton (1995), 115-116, 122-124 and R. Ticciati, *Quantum Field Theory for Mathematicians*, Cambridge, (1999), 84.

Wholeness Chart 9-1. Loop Corrections

Our Term	Other Common Terminology	Diagram	Potential divergence as $k, p \to \infty$	Nomenclature, Comments
photon loop	photon self-energy, vacuum polarization, closed fermion loop		p^2	(naïve) quadratic divergence
fermion loop	fermion self-energy		k	(naïve) linear divergence
vertex loop	vertex correction, vertex modification		$\ln k$	(naïve) logarithmic divergence
lepton triangle	triangle graph		p	Arises in Compton scattering. which we didn't do. Cancels with another having same magnitude, opposite sign

9.2 Problems

1. Write down one of the terms with four contractions in (9-3), then prove to yourself that it is not physical. From what you did, is it apparent that any four-contraction term in (9-3) is non-physical?

2. Write down one of the three contraction terms in (9-3), other than (B9-1.1) and (B9-1.2) of Box 9-1 on pg. 258, that has two fermion contractions and one photon contraction. Re-arrange it by commuting/anti-commuting fields until it looks like (B9-1.1), except with event labels x_i are in different order. In an integral over all three x_i, can we just change the variable labels to look identical to (B9-1.1)?

3. Find the Feynman amplitude (9-7) for Fig. 9-5 on pg. 259 using Feynman's rules.

4. Prove, using the power counting method, that the vertex loop integral on the LH side of the $S_{B1-6}^{(4)}$ diagram in Fig. 9-4 on pg. 259 potentially diverges with the natural log of k for large k. Do you think virtually the same analysis would show the vertex loop integral on the RH side of $S_{B1-7}^{(4)}$ in Fig. 9-4 potentially diverges with $\ln k$ for large k?

5. Use Feynman's rules to show the Feynman amplitude for Fig. 9-6 on pg. 261 equals (9-13).

6. Prove that, for high k, diagram $S_{B1-10}^{(4)}$ of Fig. 9-4 on pg. 259 results in a finite Feynman amplitude.

7.   Draw all the topologically distinct Feynman diagrams for Compton scattering for 2nd and 4th order in e. See Fig. 8-3 on pg. 225. You should have 20 in all, two 2nd order and 18 4th order. (Hint: For each side of Fig. 8-3 one obtains seven 4th order diagrams by inserting photon loops, fermion loops, and vertex loop corrections. For each side there is another 4th order diagram having virtual photons passing from the incoming fermion to the outgoing fermion. That totals 16 4th order diagrams. There are two more that each have an internal fermion triangle with the two external photons and one virtual photon, each such photon at one of the three vertices of the triangle.)

8.   For Prob. 7, pick one of the diagrams with a fermion loop for the incoming electron and show that the loop integral potentially diverges linearly with k for large k, where k is the four-momentum of the photon propagator inside the loop.

Chapter 10

The Vacuum Revisited

"There might be more than you can see."

It's Not My Time
Three Doors Down

10.0 Background

10.0.1 Vacuum "Fluctuations"

The term "vacuum fluctuations" has several meanings in the literature, one of which is related to the higher order corrections of the last chapter. Other uses of it have little to do with them.

In my experience, although passed off as a seemingly simple concept in articles written for lay audiences, the term "vacuum fluctuations" is not only commonly misused in those articles, but from anything but a superficial perspective, invariably and inordinately confusing. At least it can be so for someone just getting grounded in QFT, who tries to relate such renditions of the term to the fundamentals of that theory.

Term "vacuum fluctuations" confusing as it has several meanings

I also feel obliged to pass on that I have found a number of established physicists seemingly at a loss to explain the so-called vacuum fluctuations to me in terms of those fundamentals. I write this chapter in hopes of clarifying this issue, as best I can, for newcomers and perhaps, for others.

10.0.2 Chapter Overview

We will see how the term "vacuum fluctuations" can refer to any of the distinctly different

- evanescent particle pair creation and destruction of lay literature fame,
- ½ quanta expectation value for the vacuum state $|0\rangle$ (as in Chaps. 3, 4, 5),
- vacuum bubbles with three virtual particles (as in Chap. 8), and
- higher order correction virtual particles (as in Chap. 9).

Possible meanings for "vacuum fluctuations"

Then, we will

- compare the above four cases theoretically, and
- compare them to experiment.

Compare them to theory and experiment

Following that, we will

- review the uncertainty principle as applied to the first two cases, and
- analyze how wave packet theory in QFT relates to those cases.

Examine QFT uncertainty and wave packets

10.1 Vacuum Fluctuations: The Theory

In discussing the theoretical side of vacuum fluctuations, we begin with "the story", the much circulated description of the vacuum, of which you have no doubt heard. We then review relevant parts of QFT, which we studied in earlier chapters, and compare those to the story.

The well told story: particle-antiparticle pairs popping in and out of the vacuum

10.1.1 The Story

A vacuum fluctuation is typically described as particle pair creation and destruction in the vacuum. A particle and its anti-particle pop into existence out of the vacuum, as shown in Fig. 10-1, and then, quickly (presumably before they can be measured) annihilate one another. Total charge is conserved since there was zero charge before the pair was created, total charge zero when both exist, and then zero charge once again after they mutually destruct.

Energy for the pair, which are both typically considered to have positive energy, is "borrowed" from the vacuum according to the uncertainty principle for energy and time. Similarly, 3-momenta of the two particles do not have to cancel one another, as total 3-momentum is also borrowed from the vacuum via the uncertainty principle for 3-momentum and space.[1]

Energy and momentum "borrowed" for short time via uncertainty principle

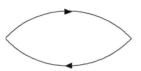

Figure 10-1. Pair Production Vacuum Fluctuation via "The Story"

The fluctuation does not exist long enough to be measured, so we never can detect it using our instruments. That is, for energy variation ΔE away from zero of significant enough magnitude to be measured, the duration of the existence of that energy (the time variation Δt) is unimaginably small, since $\hbar$ is so small (in everyday measuring units $\hbar = 1.0546 \times 10^{-34}$ joule-sec). This is the impact of the uncertainty relation for energy and time, the LH relation below.

$$\Delta E \Delta t \gtrsim \frac{\hbar}{2} \qquad \Delta p^i \Delta x^i \gtrsim \frac{\hbar}{2}. \qquad (10\text{-}1)$$

Similarly, for 3-momentum variation of appreciable magnitude (RH relation above), the variation from zero of the length measurement of the particle in the direction of the 3-momentum is extremely small and typically below the threshold of detection via instruments. Also, a large 3-momentum variation entails a large energy variation, and thus an extremely short time interval for the existence of that variation, meaning it goes undetected. Because the particle/anti-particle pairs are not detectable, they are considered to be virtual, not real, particles.

Enormous number of pairs always being created and destroyed

Particle/anti-particle pairs of very high energy (and 3-momenta) exist for extremely fleeting moments of time (in regions of extremely tiny size). The vacuum is everywhere presumed to be a boiling cauldron of particle/anti-particle pairs, of all possible energies and momenta, popping continuously in and out of existence.

If this were true, then at very small time and distance scales, energy and 3-momenta values for the pairs would become so large as to exceed the mass-energy density needed to create black holes. Thus, microscopic black holes would be continually appearing and disappearing in the vacuum as a result of the creation and destruction of the most energetic pairs. One could imagine this as being like bubbling foam, and the phrase coined by John Wheeler[2], "<u>quantum foam</u>", has caught on.

At Planck time and distance scale, particle energy so large, tiny black holes form

The dimensions and energy levels of the quantum foam bubbles are those corresponding to mass-energy, time, and distance scales at which microscopic black holes would form. This scale is called the <u>Planck scale</u>, i.e., approximately

$$t_P = 5.39 \times 10^{-44} \text{ sec} \qquad m_P = 2.18 \times 10^{-8} \text{ kg} \qquad l_P = 1.62 \times 10^{-35} \text{ m}.$$

So many tiny black holes bubbling in and out → quantum foam

Renditions of the pair production/quantum foam story are often accompanied with comments that all of this is a result of QFT. So, let's review what we have learned so far in QFT about particle creation and destruction, and the properties of the vacuum.

10.1.2 Quantum Field Theory Phenomena that May be Relevant

The question we would like to answer is "where in QFT, if anywhere, can the vacuum fluctuation pair production phenomenon be found?" To start our quest for that answer, let's delineate the phenomena we have learned about that relate either to the vacuum or to virtual particle creation and destruction.

Does the pair production story "jibe" with QFT?

<u>Free Fields Half Quanta in the Vacuum</u>

In Chap. 3, Sects. 3.4.1 to 3.4.3 on pgs. 53-55, we found our theory predicts that for scalar fields, the vacuum is filled with free scalar particles and antiparticles, one for each possible energy level. Each particle's energy level is $\frac{1}{2}\hbar\omega_k$ ($= \frac{1}{2}\omega_k$ in natural units), i.e., they are half quanta in the

1st candidate from QFT: ½ quanta of vacuum, the ZPE

---

[1] This process is not to be confused with strong field pair production, which is not a purely vacuum process, but entails a virtual photon from a strong e/m field turning into an electron/positron pair.

[2] J.A. Wheeler, On the Nature of Quantum Geometrodynamics, *Annal. Phys.* **2**, 604-614 (1957). For a lay person rendition, see J.A. Wheeler, *Geons, Black Holes, and Quantum Foam*, Norton (1998).

sense of what we normally consider quanta to be. In principle, the fields range in energy from zero to infinity, although many consider the Planck energy to be a realistic upper limit cutoff. The energy from these half quanta is commonly called "zero-point energy" (ZPE).

In Chap. 4, we saw similar ½ quanta for fermions, but those had negative energies. In Chap. 5, we obtained similar results for photons as for scalars, i.e., an energy of $\frac{1}{2}\hbar\omega_{\mathbf{k}}$ for each $\mathbf{k}$ and r polarization state. These fermion and photon energies are also encompassed by the term ZPE.

ZPE fermions have negative energy

Interacting Fields Vacuum Bubbles

In Chap. 8, we saw how the interaction part of the theory predicts vacuum bubbles of a lepton, anti-lepton, and photon, such as that in Fig. 8-8, repeated below as Fig. 10-2.

2ⁿᵈ QFT candidate: three (virtual) particle vacuum bubbles

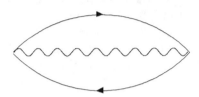

Figure 10-2. Vacuum Bubble for Interacting Fields

Interacting Fields Virtual Particle Higher Order Corrections

Any interaction has a simple, first order in α, version, and higher order corrections. For example, in Fig. 9-4, pg. 259, we show Feynman diagrams for the lowest order term (upper left, α order) and eleven second order (α^2 order) terms. In the second order diagrams, the extra virtual particles are sometimes called vacuum fluctuations. For examples, the photon loop (or closed fermion loop) in the middle of the $S_{B1-1}^{(4)}$ diagram, the positron loop (virtual positron and virtual photon) of $S_{B1-2}^{(4)}$, and the longer virtual photon in $S_{B1-11}^{(4)}$ would all, in this context, be considered vacuum fluctuations. Higher order (α^3, α^4, ..) corrections would result in a plethora of such virtual particles.

3ʳᵈ QFT candidate: virtual particles from higher order corrections

Use of the Term "Vacuum Fluctuations"

The term "vacuum fluctuations" can be used to describe each of the above phenomena. That is, it can mean i) the story of pair production/annihilation and quantum foam, ii) half quanta fields, iii) three virtual particles vacuum bubbles, and iv) virtual particles in interactions. Our overriding question is whether i) is the same as any of ii), iii), or iv).

"Vacuum fluctuations" can mean any of above four cases

Comparing Each QFT Phenomenon with The Story

We will now take a separate section to compare each of the above phenomena found in QFT to the story and determine which, if any, of them corresponds to the vacuum pair production scenario of Sect. 10.1.1.

We will now compare them

Please note carefully that I have not seen elsewhere the material presented below in Sects. 10.1.3 to 10.1.6. Neither have I seen a complete summary of related experimental evidence as shown in Sect.10.2. The conclusions drawn in these sections are my own, arrived at after many years of pondering the vacuum fluctuation issue and searching for relevant analyses/reviews.

Given this, I ask two things. 1) Please consider that the views expressed are not embraced by many physicists (I don't believe they have been considered by many), and you must formulate your own position on the matter. 2) If any reader knows of other suitable summaries of this material, either agreeing or disagreeing with the material/position presented herein, please notify me via the website for this book. (See URL on pg. xvi, opposite pg. 1.)

10.1.3 Free Field Half Quanta (ZPE)

The reader should now re-read Sect. 3.4.3 Zero Point (Vacuum) Energy, pgs. 55-56, and Sect. 3.6.6 Normal Ordering, pgs. 60-61, to review key concepts regarding the free field ½ quanta.

1ˢᵗ: ZPE vs the pair popping story

First, we note that the ZPE quanta are derived from free field theory and therefore no interactions are involved. In the story of Fig. 10-1, however, a particle and antiparticle are created via a vertex interaction and then destroyed via an interaction at a second vertex. It does not seem, therefore, that the free ZPE fields can form particle/anti-particle vacuum bubbles because they don't interact with one another. They are free, not interacting, particles. There are no Feynman diagrams associated with them.

ZPE are from free field theory so shouldn't have interacting pairs

Second, we re-iterate what was noted in Sect. 3.4.3. In the QFT derivation of half quanta vacuum energy, those half quanta appear to simply be steadily "sitting" in the vacuum, and not "popping in and out" of it. There is no apparent mechanism whereby they exist part of the time, but not all of the time. In fact, the free Hamiltonian operating on the vacuum state (see (3-61), pg. 55) shows that the vacuum quanta are eigenstates of energy. Energy eigenstates have the same energy no matter when they are measured. They exist continually.

Via theory, ZPE seem steady and not "popping" in and out

Third, ZPE is not borrowed according to QFT, it just exists in the vacuum.

ZPE not borrowed, but simply "there"

Fourth, fermions have negative energy vacuum ½ quanta, as we saw in Chap. 4. There is no uncertainty principle for negative energy (although one could probably consider postulating one). And how would mini black holes form from negative energy? And if one posits (see pg. 61) we can simply use the infinite vacuum energy as a baseline and deal with $\Delta E = E - \infty$, then in interactions, we have one baseline ($+\infty$) for bosons and another ($-\infty$) for fermions[1]. Does that make sense?

Negative ZPE: no uncertainty principle and total vacuum energy < 0

Finally, there are more known fermions than bosons in nature, so according to QFT, the total ZPE should be negative, the opposite of that promoted in the pair popping story. In this light, tales of a vacuum filled with (positive) energy seem, to put it kindly, strange.

10.1.4 Interacting Fields Vacuum Bubbles

In QED we only have vertices with three particles. There are no two particle vertices as proposed in the pair production scenario of Fig. 10-1.

2nd: vac bubbles vs pair popping

No two particle vertices in standard QFT

The pair production scenario entails a temporary "borrowing" of energy (and momentum) from the vacuum via the uncertainty principle, meaning a net non-zero sum of the virtual pair particles' energies. But we learned in interaction theory that the sum of all incoming energy (and 3-momentum) at any vertex equals the total outgoing energy (and 3-momentum). In Fig. 10-1 there is zero incoming 4-momentum, so the total 4-momentum of the pair must also be zero. Thus, the story of vacuum pair production appears to be in conflict with QFT.

Vertex energy conservation → zero total energy in pair scenario, no borrowing

Quantum foam presumably results from large energies being found over short time and distance scales, resulting in the formation of myriads of tiny black holes. But, via QFT vertex conservation, there must be no net energy resulting from the pair production of Fig. 10-1, and so it does not seem it could be a source of the energy needed to form the black holes.

Zero total energy per pair → no mini black holes

Further, the three particle vertices of Fig. 10-2 result in no net energy for the three virtual particles, so they can not result in tiny black holes.

Nor can there be mini black holes from three-particle bubbles

Thus, the vacuum bubbles of Fig. 10-2 cannot, via standard QFT, play a role in the vacuum pair production story. Nor can they interact with other particles, since, they have no external legs.

10.1.5 Interacting Fields Virtual Particles

The virtual particles in higher order contributions to interactions, such as depicted in Fig. 9-4, pg. 259, have nothing to do with the vacuum *per se*, but with the particles (both real and virtual) off of which they "hang". They are manifestations springing from entities other than the vacuum.

3rd: Higher order virtual particles vs pair popping

Nevertheless, agreement between certain experiments (see below) and QFT predictions using higher order corrections (also called "<u>radiative corrections</u>") are sometimes cited as demonstrative of vacuum fluctuations. This would not seem appropriate.

Higher order correction virtual particles not found alone in vacuum, but with real particles

By any measure, virtual particles participating in interactions between real particles cannot be considered the mechanism for vacuum fluctuation pair production.

10.1.6 Conclusion: Theory and Vacuum Fluctuations

The story of vacuum pair production/destruction commonly called "vacuum fluctuations" is not consonant with the fundamental theory of QFT/QED, in particular, not with ZPE quanta, vacuum bubbles, nor higher order correction virtual particles.

There are no pair creation/ destruction Feynman diagrams

<u>Bottom line:</u> There are no Feynman diagrams like Fig. 10-1, and none for the ½ quanta.

10.2 Vacuum Fluctuations and Experiment

In this section, we describe certain experiments that are used to justify the existence of vacuum fluctuations and how they relate to the theoretical side of QFT, as we understand it.

Vacuum pair production story vs experiment

---

[1] Pointed out to the author in 2015 by Anna Pearson, then an Oxford University PhD candidate.

10.2.1 Casimir Plates

Two flat plates brought close together experience a small attractive force at very small separation distances. This effect was first predicted by Dutch physicists Hendrik B. G. Casimir and Dirk Polder in 1948. The attractive force has been attributed to ZPE, in heuristic and very simple terms, because the vacuum quantum waves outside the plates presumably exert greater force than the vacuum quantum waves between the plates. However, there are other interpretations.

> ... the Casimir effect is often invoked as decisive evidence that the zero point energies of quantum fields are "real". On the contrary, Casimir effects can be formulated and Casimir forces can be computed without reference to zero point energies.... The Casimir force is simply the (relativistic, retarded) van der Waals force between the metal plates.... So, the concept of zero point fluctuations is a heuristic and calculational aid in the description of the Casimir effect, but not a necessity.... No known phenomenon, including the Casimir effect, demonstrates that zero point energies are "real".[1]

There are two key things to note.

1) While the Casimir effect can be calculated by assuming ZPE half quanta, the same result can also be calculated another way without using them at all. It thus does not prove their existence, contrary to what is often claimed.

Casimir experiment not proof ZPE quanta exist

2) In the Casimir calculation that does employ ZPE, the quanta are assumed to be continuously existing *standing waves, not particle pairs popping in and out of existence.*

Even if it were, no "popping" of particles needed

Conclusions: Casimir plate experiments do not provide proof of the existence of ZPE ½ quanta in any form. Even if one were, nevertheless, to consider such quanta responsible for the Casimir effect, there is no evidence those quanta are paired and evanescent (pop in and out of existence).

10.2.2 Lamb Shift

The Lamb shift is a small difference between the two energy levels $^2S_{1/2}$ and $^2P_{1/2}$ of the hydrogen atom, which according to RQM, should have the same energies. QFT, in its QED form, predicts this shift, and that prediction was one of the great early successes of the theory.

The Lamb shift calculation is long and difficult[2]. It is often described as taking vacuum fluctuations into account in order to obtain the correct result. However, in actuality, these "vacuum fluctuations" are really the radiative, or higher order, corrections (extra virtual particles in Feynman diagrams) of Sect. 10.1.5. These corrections to the Coulomb potential of the hydrogen atom (in diagrams, extra virtual photons, electrons, and positrons) yield the correct energy levels.

Lamb shift due to higher order corrections of real particle interactions

It has nothing to do with the vacuum

Conclusion: The Lamb shift does not prove vacuum pair production/destruction.

10.2.3 Anomalous Magnetic Moment of the Electron

The Dirac equation in RQM leads to a prediction of the magnetic dipole moment of the electron that is slightly different from that measured in experiment. The measured value was called the anomalous magnetic moment.

Using QED with radiative corrections, the anomalous magnetic moment value was accurately calculated[3]. This, too, is sometimes attributed to vacuum fluctuations. However, here again, it is higher order (radiative) corrections having nothing to do with the vacuum that are invoked.

Anomalous mag moment from higher order corrections of real interactions

Conclusion: The anomalous magnetic moment does not prove vacuum pair production/destruction.

10.2.4 The Fulling-Davies-Unruh Effect

If the vacuum is filled with ZPE, then as Stephen Fulling (1973), Paul Davies (1975) and Bill Unruh (1976) described, an accelerating observer would find the ZPE to look like black-body

It has nothing to do with the vacuum

---

[1] R. L. Jaffe, "Casimir Effect and the Quantum Vacuum", *Phys. Rev. D72* 021301(R) (2005) http://arxiv.org/abs/hep-th/0503158 .

[2] C. Itzykson and J.B. Zuber, *Quantum Field Theory* (McGraw-Hill 1985). Sect. 7-3-2, 358-365.

[3] See Chap. 16 of this book.

radiation, whereas the non-accelerating observer would not[1]. This effect became commonly known as the Unruh effect, though it is more appropriate to use all three names when referring to it.

The Fulling-Davies-Unruh effect is not directly related to vacuum ½ quanta, vacuum bubbles, or radiative corrections, but to the difference between vacua in inertial and non-inertial frames[2]. The accelerated observer measuring the inertial vacuum detects particles not observed by the inertial observer. From this perspective, the effect seems quite unrelated to any concepts considered herein.

Related to non-inertial frame vacuum, not elementary QFT

Some experimenters believe they have detected the Fulling–Davies–Unruh effect, but as of the date of this book, the claimed observations are controversial and under dispute. (See cited Wikipedia URL for the latest on this controversy.)

Controversial if Unruh effect has been measured

Conclusions: There is no incontrovertible proof that Fulling–Davies–Unruh radiation exists. Even if it is confirmed, it would not prove pair popping or that ZPE quanta exist.

Even if measured, not proof of pair "popping"

10.2.5 Measured Vacuum Energy

As noted in Chap. 3, ZPE calculations[3], assuming a Planck scale maximum allowable value (rather than infinity) for the ½ quanta, predict a positive vacuum energy density for bosons on the order of 10^{74} GeV^4 (natural units), whereas the observed value is $\leq 10^{-47}$ GeV. (See Appendix A.) This is the famous largest discrepancy between theory and experiment in the history of science.

Measured vs theory ZPE differ by $>10^{120}$ (for bosons)

Usually going unmentioned in such discussions is that the total vacuum energy for all known bosons and fermions would be *negative* and of this order, as discussed at the end of Sect. 10.1.3

Conclusion: The observed vacuum energy density does not support the existence of ZPE or any other form of vacuum fluctuations.

For both bosons and fermions, theoretic total ZPE is negative

10.2.6 Experimental Evidence Conclusion

At the time of this text version, there is no experimental evidence irrefutably demonstrating the existence of virtual particle/anti-particle pairs popping in and out of the vacuum and unrelated to interactions between real particles. Further, there is no such irrefutable evidence for ZPE ½ quanta in any form, including continuously existing waves. (See Appendix F, added in 2018 text version.)

No experimental evidence proving ZPE, let alone pair "popping"

Bottom line: No known experiment proves vacuum particle/anti-particle pair production.

10.3 Further Considerations of Uncertainty Principle

10.3.1 Uncertainty Principle and Commutation Relations

Non-relativistic Quantum Mechanics

Recall from NRQM that the uncertainty principle is a direct result of non-commutation of certain operator pairs. In short, most elementary quantum mechanics courses prove that for any operators P and Q, the relation (10-2) below, where the < > brackets indicate expectation value and Δ indicates standard deviation, holds. Note that ΔP and ΔQ are not operators, so we can switch their order.

Brief review of how non-commutation results in uncertainty

$$(\Delta P)(\Delta Q) \geq \tfrac{1}{2}\left|i\left\langle [P,Q] \right\rangle\right|. \tag{10-2}$$

For position and momentum, we know that

$$\left[x^i, p^j\right] = i\hbar\delta_{ij} \;\rightarrow\; \left[p^j, x^i\right] = -i\hbar\delta_{ij}. \tag{10-3}$$

Taking $Q = x^i$ and $P = p^j$, (10-3) into (10-2) yields the position/momentum uncertainty principle,

$$(\Delta p^j)(\Delta x^i) \geq \frac{\hbar}{2}\delta_{ij} \quad \text{equivalent to} \quad (\Delta x^i)(\Delta p^j) \geq \frac{\hbar}{2}\delta_{ij}. \tag{10-4}$$

So non-commutation of operators means an uncertainty principle exists for those operators.

The LHS of (10-4) is generally equal to, or only a little larger than, the RHS. For Gaussian wave packets, for example, the RHS equals the LHS.

---

[1] See http://en.wikipedia.org/wiki/Unruh_effect.

[2] See V.F. Mukhanov and S. Winitzki, *Introduction to Quantum Effects in Gravity*, Cambridge (2010), Chap. 8.

[3] S. Weinberg, "The Cosmological Constant Problem", *Reviews of Modern Physics,* **61**, 1, 1-23 (Jan 1989). See also, Appendix A of this chapter.

Relativistic Quantum Mechanics

Similar logic results in a similar relation as (10-4) for RQM.

Quantum Field Theory

For 1st quantization (particles), we used (10-3). (See Chap. 1, pg. 4.) For 2nd quantization (fields),

$$\left[\phi_r\left(\mathbf{x},t\right),\pi_S\left(\mathbf{y},t\right)\right]=i\hbar\delta_{rs}\delta\left(\mathbf{x}-\mathbf{y}\right), \qquad (10\text{-}5)$$

Non-commutation in field theory leads to field uncertainty

where ϕ_r represents a (bosonic) field, and π_s represents the field conjugate momentum. Substituting $Q = \phi_r$ and $P = \pi_S$ in (10-2), we get

$$\left(\Delta\phi_r\left(\mathbf{x},t\right)\right)\left(\Delta\pi_S\left(\mathbf{y},t\right)\right)\geq\frac{\hbar}{2}\delta_{rs}\delta\left(\mathbf{x}-\mathbf{y}\right), \qquad (10\text{-}6)$$

an <u>uncertainty principle for fields</u>.

Recall from Chap. 3 (or the Wholeness Chart at the end of Chap. 5) that (10-5) gives rise to the commutation relations for creation and destruction operators, such as $[a(\mathbf{k}),a^\dagger(\mathbf{k}')] = \delta_{\mathbf{k}\mathbf{k}'}$. Those commutation relations are used in the derivation of H_0 in terms of number operators (Chap. 3, (3-54) to (3-56), pgs. 54-55) and in so doing, give rise to the ½ quanta energy terms in H_0.

Non-commutation in field theory also leads to ZPE

So, in QFT, the commutation relation (10-5) gives rise to i) the uncertainty principle for fields (10-6) and also to ii) the ½ quanta in the vacuum. This is one reason why the ½ quanta are often said to be the result of the quantum uncertainty principle.

One reason ZPE sometimes said due to uncertainty

10.3.2 Uncertainty Principle for Fields and Measurement

As we noted at the end of Chap. 3 in Wholeness Chart 3-4, pg. 79 and also showed in Chap. 7 pgs. 189-190, fields are not observable. They have zero expectation values. The same is true of their conjugate momenta. This might lead us to believe that (10-6) is essentially meaningless with regard to measurement in the real world, since we can't measure the quantities it describes.

But QFT fields not measurable anyway, so uncertainty not relevant

10.3.3 Uncertainty Principle for Particles and Measurement

Almost the entirety of this book, like almost all introductory texts on QFT, deals with particles in pure momentum eigenstates, represented by $\mathbf{k}$ or $\mathbf{p}$. Recall we have noted that the 3-momentum eigenstates are good enough approximations to real particles (which are invariably wave packets and not momentum eigenstates) to yield highly accurate answers for scattering problems.

Consider uncertainty in QFT for particles, not fields

Pure $\mathbf{k}$ (or $\mathbf{p}$) particle states extend across the entire region of 3D space in which measurements could take place (which could be the entire universe). Whenever a given such particle momentum (or energy) is measured, it will have the same value each time. That is, there is no variation in momentum (or energy).

This applies not simply to real particles, but also to the presumed ½ quanta of the vacuum. Each such quantum has definite $\mathbf{k}$ value and is an eigenstate, not a general state (not a superposition of eigenstates.) Thus, there would be no variation in its energy or momentum upon measurement (assuming it could ever be measured.)

In typical QFT, ZPE particles are pure $\mathbf{k}$ states. No momentum or energy uncertainty, anyway.

Further, both large and small $|\mathbf{k}|$ values would be comprised by waves that extend spatially to infinity. Each would have definite momentum and completely indefinite location. But microscopic black holes need highly energetic particles to be packed into small regions. The ½ quanta do not seem to satisfy this requirement.

10.3.4 Uncertainty Principle and Ground States

In NRQM, you probably saw how the ground state of a system such as the hydrogen atom, or the harmonic oscillator, could be deduced, approximately, from the uncertainty principle.[1] In such cases, the estimated ground state energy (lowest possible energy state of the system) was found to be non-zero. This seems to be the source for referencing the uncertainty principle in the pair popping and the vacuum ½ quanta cases.

However, the hydrogen atom and harmonic oscillator are bound state systems, i.e., particles experience a force via a potential. They are not free systems. Free systems in NRQM do not have to have a non-zero energy ground state. Free particle energy ground states can be zero (for zero

---

[1] See, for examples, R.G. Winter, *Quantum Physics* (Wadsworth, 1979), pgs. 18-19, or S. Gasiororwicz, *Quantum Physics* (Wiley, 1974), pgs. 37-40.

velocity, where we ignore the mass-energy equivalence in NRQM, and have wave functions of form e^{-ikx}, which have $\Delta x \to \infty$). That is, the lowest energy eigenstate for free particles has zero energy.

Both the pair popping of lay literature and the ½ quanta of QFT represent free systems. If the uncertainty principle does not lead to a specific, discrete non-zero ground state for free systems of energy eigenstate particles in NRQM, then how can one use that particular argument to support non-zero ground state free particles in QFT?[1]

In fact, the usual treatment of uncertainty in state energy and momentum to determine the ground state gives one a zero ground state for free particles in any quantum theory, including QFT. Significantly, the vacuum states in NRQM and RQM have zero energy.

For a free particle wave packet (see Sect. 10.4 below for more), the uncertainty principle implies that any of a range of values for energy could be found upon measurement. But this range is continuous. That is, there is no specific, discrete ground state energy, such as $\frac{1}{2}\hbar\omega_k$, for free wave packets. So, again, why should we expect the uncertainty principle to provide one in the pair popping scenario or QFT, if it doesn't in NRQM or RQM? Its use in the latter two theories is the supposed justification for applying it in the former one.

Specifically, the vacuum has $\Delta x^i \to \infty$. $(\Delta x^i)(\Delta p^i) \geq \hbar/2$, so we can have $(\Delta p^i) = 0$, and Δp^i = exactly 0. One might surmise from this that a zero energy vacuum does not violate uncertainty.

10.3.5 Uncertainty Principle Conclusions

> Bottom line: In standard QFT, particularly with particles in **k** eigenstates, vacuum fluctuations do not appear to arise from an uncertainty principle for fields or particles.

10.4 Wave Packets

One could counter much of the foregoing Sect. 10.3 with the argument that definite **k** states are not precise representatives of real world particles, which are really wave packets, and so our entire development of QFT is simply an approximation. That is, the ½ quanta are more probably wave packets of indefinite **k**, with an expectation (mean) value for **k** of $\bar{\mathbf{k}}$ and a standard deviation $\Delta\mathbf{k}$ about that mean. Then, all this talk of uncertainty in **k** and **x** would start to have meaning.

But wave packet approach to QFT would have uncertainty

To examine this, we would need to re-develop our entire theory for the continuous solutions, rather than discrete solutions, to the QFT wave equations. (See Chap. 3, (3-36) and (3-37) on pg. 50.) This would take an entire chapter, or more, and would lead us astray from more immediate goals. However, in Appendix C, I summarize important steps in this development and reference the book website, where detailed development of continuous solutions is presented. I do not recommend study of that appendix on one's first sojourn into QFT. I do recommend it once one has gained a solid footing in the theory of the discrete eigen solutions, which comprise almost the entire book.

So, examine QFT for wave packets

The <u>Hamiltonian operator for discrete solutions of the scalar field</u> is

$$H_0^0 = \sum_{\mathbf{k}} \omega_{\mathbf{k}}\left(N_a(\mathbf{k}) + \tfrac{1}{2} + N_b(\mathbf{k}) + \tfrac{1}{2}\right)$$

Hamiltonian for discrete solution fields

$N_a(\mathbf{k}), N_b(\mathbf{k})$ = number of real particles, dimensionless   (10-7)

$\tfrac{1}{2}$ = number of vacuum particles, dimensionless.

The corresponding <u>Hamiltonian operator for continuous solutions of the scalar field</u> is (where we write $\iiint d^3k$ [which is the same as $\int d\mathbf{k}$] as $\int d^3k$ to save space yet make clear we mean triple integration over momentum space; and use $\delta(0)$ as a short form for $\delta^3(0)$)

Hamiltonian for continuous solution fields, i.e., wave packet particles

$$H_0^0 = \int \omega_{\mathbf{k}}\left(\mathcal{N}_a(\mathbf{k}) + \tfrac{1}{2}\delta(0) + \mathcal{N}_b(\mathbf{k}) + \tfrac{1}{2}\delta(0)\right)d^3k$$

$\mathcal{N}_a(\mathbf{k}), \mathcal{N}_b(\mathbf{k})$ = (num real particles, all **x** space) / unit **k** space vol, dimensions $1/M^3$

$\tfrac{1}{2}$ = (num vacuum particles per **x** space vol) / unit **k** space vol, dimensions $1/M^3$ (10-8)

$\delta(0)$ = infinite volume of universe

---

[1] Those wishing to counter with the argument that free states in QFT are harmonic oscillators should see Chap. 3, Sect. 3.12, pgs. 69-70.

The units in the parenthetical part of the integrand of (10-8) are natural units with momentum cubed in the denominator. Momentum, in natural units, has the dimension of mass since velocity in those units is dimensionless. $\mathcal{N}_a(\mathbf{k})$ and $\mathcal{N}_b(\mathbf{k})$ for the continuous solutions are <u>number density operators</u>, rather than just number operators as in the discrete case.

The mathematical form in position space of a scalar wave packet ket of a particle (not anti-particle) in QFT is

$$|\phi\rangle = \left| \int \frac{A(\mathbf{k}')e^{-ik'x}}{\sqrt{(2\pi)^3}}\, d^3k' \right\rangle , \qquad (10\text{-}9)$$

Single particle QFT state in position space

Where $A(\mathbf{k}')$ is a *coefficient* (*not an operator*) that defines the shape of the wave packet. Typically, $\left|A(\mathbf{k}')\right|$ has a Gaussian shape as in Fig. 10-3a

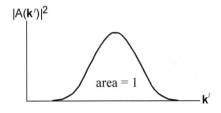
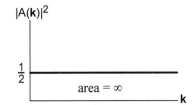

a) Typical shape of $|A(\mathbf{k}')|^2$ for particle wave   b) Shape of $|A(\mathbf{k})|^2$ for vacuum

Figure 10-3. Shape of $|A(\mathbf{k}')|^2$ Coefficient in k Space

$|A(\mathbf{k}')|^2$ represents the wave packet density per unit volume in $\mathbf{k}'$ space at a given $\mathbf{k}'$ for a single particle wave packet. By doing Prob. 1, you can show that for $|\phi\rangle$ having unit norm, then

$$\int \left|A(\mathbf{k}')\right|^2 d^3k' = 1 . \qquad (10\text{-}10)$$

Coefficient property for unit norm state

Wholeness Chart 10-2 in Appendix C, by comparison with discrete solutions, may make this a bit clearer.

When (10-8) acts on (10-9), we get

$$H_0^0|\phi\rangle = \int \omega_{\mathbf{k}} \left(\mathcal{N}_a(\mathbf{k}) + \tfrac{1}{2}\delta(0) + \mathcal{N}_b(\mathbf{k}) + \tfrac{1}{2}\delta(0) \right) d^3k \left| \int \frac{A(\mathbf{k}')e^{-ik'x}}{\sqrt{(2\pi)^3}}\, d^3k' \right\rangle$$

$$= \int \omega_{\mathbf{k}} \left(\left|A(\mathbf{k})\right|^2 + \tfrac{1}{2}\delta(0) + \tfrac{1}{2}\delta(0) \right) d^3k \left| \int \frac{A(\mathbf{k}')e^{-ik'x}}{\sqrt{(2\pi)^3}}\, d^3k' \right\rangle . \qquad (10\text{-}11)$$

Continuous field Hamiltonian acting on single particle wave packet state

At each $\mathbf{k}'$ inside the ket, the number operator $\mathcal{N}_a(\mathbf{k})$ will pick up the number density of particles at that value of $\mathbf{k} = \mathbf{k}'$. Had our ket state comprised two wave packet particles of exactly the same particle state (same $A(\mathbf{k}')$ distribution function), we would have obtained

$$H_0^0|2\phi\rangle = \int \omega_{\mathbf{k}} \left(\mathcal{N}_a(\mathbf{k}) + \tfrac{1}{2}\delta(0) + \mathcal{N}_b(\mathbf{k}) + \tfrac{1}{2}\delta(0) \right) d^3k \left| 2\int \frac{A(\mathbf{k}')e^{-ik'x}}{\sqrt{(2\pi)^3}}\, d^3k' \right\rangle$$

$$= \int \omega_{\mathbf{k}} \left(2\left|A(\mathbf{k})\right|^2 + \delta(0) \right) d^3k \left| 2\int \frac{A(\mathbf{k}')e^{-ik'x}}{\sqrt{(2\pi)^3}}\, d^3k' \right\rangle . \qquad (10\text{-}12)$$

Continuous field Hamiltonian acting on two particle wave packet state

However, for simplicity, we are not going to consider multiparticle states anymore, just single particle ones.

We now ask "what is the expectation value of energy for the single particle wave packet state of (10-11)?" The answer is

$$\bar{E} = \langle\phi|H_0^0|\phi\rangle = \left\langle \int \frac{A(\mathbf{k}'')e^{-ik''x}}{\sqrt{(2\pi)^3}}\,d^3k'' \left| \int \omega_\mathbf{k}\left(\mathcal{N}_a(\mathbf{k})+\mathcal{N}_b(\mathbf{k})+\delta(0)\right)d^3k \right| \int \frac{A(\mathbf{k}')e^{-ik'x}}{\sqrt{(2\pi)^3}}\,d^3k' \right\rangle$$

$$= \int \omega_\mathbf{k}\left(\left|A(\mathbf{k})\right|^2+\delta(0)\right)d^3k \underbrace{\left\langle \int \frac{A(\mathbf{k}'')e^{-ik''x}}{\sqrt{(2\pi)^3}}\,d^3k'' \left\| \int \frac{A(\mathbf{k}')e^{-ik'x}}{\sqrt{(2\pi)^3}}\,d^3k' \right. \right\rangle}_{=1}$$

(10-13)

$$= \int \omega_\mathbf{k}\left(\left|A(\mathbf{k})\right|^2+\tfrac{1}{2}\delta(0)+\tfrac{1}{2}\delta(0)\right)d^3k = \underbrace{\int\left|A(\mathbf{k})\right|^2\omega_\mathbf{k}d^3k}_{\substack{\text{real particle expectation}\\ \text{energy} = \bar{\omega}}} + \underbrace{\delta(0)\int \tfrac{1}{2}\omega_\mathbf{k}d^3k}_{\substack{\text{vacuum particle}\\ \text{expectation energy}}} + \underbrace{\delta(0)\int \tfrac{1}{2}\omega_\mathbf{k}d^3k}_{\substack{\text{vacuum anti-particle}\\ \text{expectation energy}}}.$$

We get two parts to our answer. One is the expectation value of the wave packet particle energy $\bar{\omega}$. The other is the energy of the vacuum, infinite and reminiscent of the discrete solutions states summation, which we repeat below for a single particle state for comparison.

$$\bar{E} = \langle\phi_{\mathbf{k}'}|H_0^0|\phi_{\mathbf{k}'}\rangle = \langle\phi_{\mathbf{k}'}|\sum_\mathbf{k}\omega_\mathbf{k}\left(N_a(\mathbf{k})+\tfrac{1}{2}+N_b(\mathbf{k})+\tfrac{1}{2}\right)|\phi_{\mathbf{k}'}\rangle = \langle\phi_{\mathbf{k}'}|\omega_{\mathbf{k}'}+\sum_\mathbf{k}\omega_\mathbf{k}\left(\tfrac{1}{2}+\tfrac{1}{2}\right)|\phi_{\mathbf{k}'}\rangle$$

$$= \left(\omega_{\mathbf{k}'}+\sum_\mathbf{k}\omega_\mathbf{k}\left(\tfrac{1}{2}+\tfrac{1}{2}\right)\right)\langle\phi_{\mathbf{k}'}\|\phi_{\mathbf{k}'}\rangle = \underbrace{\omega_{\mathbf{k}'}}_{\substack{\text{real}\\\text{particle}\\\text{energy}}} + \underbrace{\sum_\mathbf{k}\tfrac{1}{2}\omega_\mathbf{k}}_{\substack{\text{vacuum}\\\text{particle}\\\text{energy}}} + \underbrace{\sum_\mathbf{k}\tfrac{1}{2}\omega_\mathbf{k}}_{\substack{\text{vacuum}\\\text{anti-particle}\\\text{energy}}}.$$

(10-14)

10.4.1 An Important Point

Note in (10-13) that the particle state, being a wave packet, has an envelope $A(\mathbf{k})$, which is effectively only non-zero over a limited range of $\mathbf{k}$ values. And the integral (10-10) means its amplitude even over that range is finite.

On the other hand, the two vacuum energy contributions of (10-13) (last two terms) have no such limiting envelope. In fact, the amplitude of the effective coefficient for each of those terms is $1/\sqrt{2}$ per unit $\mathbf{x}$ space volume, over the entire range of $\mathbf{k}$. For the vacuum, we have $A_\infty(\mathbf{k}) = 1/\sqrt{2}$ in the next to last term of (10-13), and coefficient $B_\infty(\mathbf{k}) = 1/\sqrt{2}$ in the last term. See Fig. 10-3b.

So, according to (10-13), we can think of the vacuum as two single particle wave packets (particle and anti-particle) in every $\mathbf{x}$ space unit volume, each with constant amplitude = $1/\sqrt{2}$ over all $\mathbf{k}$ in $\mathbf{k}$ space. But such a constant amplitude cannot be normalized like (10-10), and each vacuum particle then has a numerical probability of being measured of infinity. (Recall, $|A(\mathbf{k})|^2$ is probability density per unit $\mathbf{k}$.) This is clearly mathematically inconsistent.

In other words, there appears to be no interpretation whereby the vacuum is comprised, in a meaningful way, of individual wave packet particle states. Without wave packet states, we have no uncertainty principle, no greater energy/momentum expectation value for shorter time and distance scales.

10.4.2 If the Vacuum Had Wave Packet ½ Quanta

If the vacuum were filled with ½ quanta wave packets having $\bar{\omega}_{\bar{\mathbf{k}}}/2$ energy expectation values, then each such wave packet would need its own envelope $A_j(\mathbf{k})$, where j labels the wave packet. In other words, our expected energy would, instead of (10-13), look like

$$\bar{E} = \langle\phi|H_0^0|\phi\rangle = \underbrace{\int\left|A(\mathbf{k})\right|^2\omega_\mathbf{k}d^3k}_{\substack{\text{real particle}\\ \text{expectation energy} = \bar{\omega}}} + \sum_j^\infty \underbrace{\tfrac{1}{2}\int\left|A_j(\mathbf{k})\right|^2\omega_\mathbf{k}d^3k}_{\substack{\text{vacuum expectation energy}\\ \bar{\omega}_j/2 \text{ for } j\text{th particle}}} + \sum_j^\infty \underbrace{\tfrac{1}{2}\int\left|A_j(\mathbf{k})\right|^2\omega_\mathbf{k}d^3k}_{\substack{\text{vacuum expectation energy}\\ \bar{\omega}_j/2 \text{ for } j\text{th anti-particle}}}.$$

(10-15)

The vacuum contribution, for wave packet ½ quanta, would look like the RHS of (10-15), rather than the RHS of the last row of (10-13). If we have ½ quanta wave packets in the vacuum, then each such ½ quantum needs a wave packet form, and that means an envelope $A_j(\mathbf{k}) \neq$ constant for all $\mathbf{k}$. It would need $A_j(\mathbf{k})$ obeying (10-10).

10.4.3 Wave Packet Conclusions:

1. QFT does not appear to predict separate wave packets in the vacuum, but a seeming single particle packet (and a similar single anti-particle packet) per unit **x** space volume of envelope $A_j(\mathbf{k}) = 1/\sqrt{2}$.

2. Thus, the uncertainty principles for energy/time and momentum/position do not appear applicable to the vacuum as described by standard QFT.

> Bottom line: QFT wave packet analysis does not appear to support vacuum fluctuations.

QFT seems to predict a single, weird looking wave packet in each unit volume of the vacuum, not many usual type packets

10.5 Further Considerations

10.5.1 Vacuum Fluctuations May Yet Exist

Of course, all of the above does not mean that vacuum fluctuations do not exist. The uncertainty principle leads one to think that they may well proliferate in the vacuum. Many of the world's top physicists are, in fact, convinced that they do in some form. But, the precise mechanism by which it would occur, if it does, is not obvious in standard QFT.

Vacuum fluctuations may still exit.

10.5.2 More Advanced Theories

As of the date of this text, we have no viable theory of quantum gravity. Perhaps, in that theory, when it is finally developed, there will be vacuum fluctuations, driven by the uncertainty principle. Further, superstring (more properly M-theory) theorists regularly consider vacuum fluctuations of the strings. The competing theory of loop quantum gravity considers spacetime comprised of small "quanta" of geometry linked in ways that lead to microscopic behavior similar to that of spacetime foam. So, perhaps one of these theories, or a third known as "twistor theory", will one day put a firm foundation under vacuum fluctuations.

> Bottom line: It is possible more advanced theories can prescribe vacuum fluctuations.

Vacuum fluctuations may be valid part of advanced theories

10.6 Chapter Summary

Wholeness Chart 10-1, along with the following bottom line statement, summarize the present chapter. See also, Appendix F, which was added to the 2018 version of the text.

> Bottom line for this chapter: According to standard QFT, only three particle virtual bubbles can truly be called "vacuum fluctuations", and they have zero net energy so cannot contribute to vacuum energy.

Only vacuum bubbles are truly vacuum fluctuations, but their energy = 0

10.7 Addenda

10.7.1 Hidden in the Theory is a Way for ½ Quanta to Disappear

I noted in Chap. 3 (footnote on pg. 50) that there are little recognized alternative solutions to the QFT field equations that are not used in the standard renditions of the theory. The traditional solutions have the familiar $\pm\, i\,(\omega_{\mathbf{k}}\, t - \mathbf{k}\cdot\mathbf{x})$ form in the exponent. The alternative solutions to the field equations have form $\pm\, i\,(\omega_{\mathbf{k}}\, t + \mathbf{k}\cdot\mathbf{x})$. I call these supplemental solutions.

When the supplemental solutions are included in the theory (see footnote citation), one finds boson energy terms $-\omega_{\mathbf{k}}/2$ arising in the vacuum that cancel the contributions from the traditional solutions, leaving a net vacuum expectation energy of zero. A similar cancellation effect occurs for fermions. This, of course, is closer to the observed value.

Author's research: If include alternative solutions to field equations in QFT, ZPE disappear

10.7.2 Caveats Again

As I said earlier, much in this chapter comprises my own ruminations on the vacuum fluctuations subject. The reader should consider this in drawing her/his own conclusions.

Wholeness Chart 10-1. Comparison of Vacuum Fluctuation Scenarios

	Pair Production	½ Quanta, Zero Point Energy (ZPE)	Virtual Bubbles	Radiative Corrections
Basic Description	Particle and anti-particle pairs continually popping in and out of vacuum	Quanta of $\frac{1}{2}\omega_k$ sitting in vacuum, particles and anti-particles	Three virtual particles arise from and dissolve into vacuum	Higher order virtual particle corrections to lowest order interaction
Chapter in this book	This chapter.	Chaps. 3,4,5	Chap. 8	Chap. 9
How Proposed to Arise?	Uncertainty principle for states	2nd quantization	Interaction terms in Hamiltonian	Interaction terms in Hamiltonian
Typical Feynman diagram		None		
Does such a Feynman diagram exist in QFT?	No	N/A	Yes	Yes
Free or Interaction Theory?	Not in QFT, free or interacting	Free	Interacting	Interacting
Interacts with rest of universe?	Yes, it is claimed	No (part of free theory)	No	Yes
Positive, negative, or zero vacuum energy?	Positive, borrowed via uncertainty principle	Bosons positive, fermions negative, total negative	Zero. Negative virtual energy cancels positive	No contribution to vacuum energy
Mathematics behind it	Non-commutation p_x and $x \rightarrow$ uncertainty in p_x and x	Non-commutation π and $\phi \rightarrow \frac{1}{2}\omega_k$ in vacuum for all k	Interaction term $e\bar{\psi}\gamma^\mu A_\mu \psi$ in $\mathcal{H}$	Same as at left
Uncertainty? Measure?	Can measure p_x, x in principle	Can't measure π, ϕ even in principle	Can't measure	Measure indirectly
Exist continually or evanescently?	Evanescent	Continually exist (no mechanism for otherwise)	Evanescent	Evanescent
Arise in vacuum alone?	Yes	Yes	Yes	No
Experiments prove?				
Casimir plates	No, despite claims	No	No	Possibly
Lamb shift	No	No	No	Yes
Anomalous magnetic moment	No	No	No	Yes
Fulling-Davis-Unruh	No	Inconclusive	No	No
Yields known vacuum energy?	No	No	Close, at the least	N/A
Wave packet theory supports?	No	Ill-defined. Does not appear supported by QFT math.	Yes	Yes

10.8 Appendix A: Theoretical Value for Vacuum Energy Density

10.8.1 The Cut-off Method

Given (10-14), the vacuum energy density of the ZPE in a rectangular solid shape volume V is

$$\bar{\rho}_v = \frac{\bar{E}_v}{V} = \frac{1}{V}\langle 0|H_0^0|0\rangle = \frac{1}{V}\langle 0|\sum_{\mathbf{k}}\omega_{\mathbf{k}}\left(\frac{1}{2}+\frac{1}{2}\right)|0\rangle = \frac{1}{V}\sum_{\mathbf{k}}\omega_{\mathbf{k}} = \frac{1}{l_1 l_2 l_3}\sum_{\mathbf{k}}\omega_{\mathbf{k}} = \sum_n \frac{\omega_{\mathbf{k}_n}}{l_1 l_2 l_3}, \quad (10\text{-}16)$$

where l_i is the length of the ith side of volume V inside of which the particle/waves are found, and we have used slightly different labeling in the last expression to suit our immediate needs. Boundary conditions on V give the wavelength of the nth wave ($n = 1, 2,..$) in the x^1 direction as $\lambda_{n\,1} = l_1/n$, with similar relations for the other two directions. So, the wave number $k_{n\,1}$ of the nth wave is

$$k_{n1} = \frac{2\pi}{\lambda_{n\,1}} = \frac{2\pi n}{l_1}. \quad (10\text{-}17)$$

Defining $\Delta k_1 = k_{(n+1)\,1} - k_{n\,1}$ we have, from (10-17),

$$\Delta k_1 = \frac{2\pi(n+1)}{l_1} - \frac{2\pi\,n}{l_1} = \frac{2\pi}{l_1} \quad \text{so that} \quad \frac{1}{l_1} = \frac{\Delta k_1}{2\pi}.. \quad (10\text{-}18)$$

Similar results hold for l_2 and l_3. Thus, (10-16) (with the relativistic expression for energy) is

$$\bar{\rho}_v = \sum_n \frac{\omega_{\mathbf{k}_n}}{l_1 l_2 l_3} = \sum_n \omega_{\mathbf{k}_n}\frac{\Delta k_1}{2\pi}\frac{\Delta k_2}{2\pi}\frac{\Delta k_3}{2\pi} = \sum_n \sqrt{m^2 + \mathbf{k}_n{}^2}\frac{\Delta k_1}{2\pi}\frac{\Delta k_2}{2\pi}\frac{\Delta k_3}{2\pi}. \quad (10\text{-}19)$$

For l_1 very large, from (10-18), $\Delta k_1 \rightarrow dk_1$, with similar expressions for large l_2 and l_3. In the limit of large l_i (large volume V), (10-19) becomes

$$\bar{\rho}_v = \int_{-\infty}^{\infty}\sqrt{m^2+\mathbf{k}^2}\frac{dk_1}{2\pi}\frac{dk_2}{2\pi}\frac{dk_3}{2\pi} = \int_{-\infty}^{\infty}\frac{1}{(2\pi)^3}\sqrt{m^2+\mathbf{k}^2}\,d^3k \quad (10\text{-}20)$$

Using (9-9) on pg. 260, we find this becomes

$$\bar{\rho}_v = \int_0^{\infty}\frac{1}{(2\pi)^3}\sqrt{m^2+k^2}\,4\pi k^2 dk = \frac{1}{2\pi^2}\int_0^{\infty}\sqrt{m^2+k^2}\,k^2 dk. \quad (10\text{-}21)$$

This is obviously infinite, unless we take an upper limit cutoff, typically considered the Planck scale mass (energy). This is because a particle with energy of the Planck mass is assumed to have an associated Compton wavelength (size of the particle) so small that its associated mass-energy forms a microscopic black hole. Such particles would instantaneously collapse, so smaller size (larger energy) particles may not be able to exist in our universe. Given such logic, with Λ = Planck mass $\gg m$, we find (10-21) is

$$\bar{\rho}_v = \frac{1}{2\pi^2}\int_0^{\Lambda}\sqrt{m^2+k^2}\,k^2 dk \approx \frac{1}{2\pi^2}\int_0^{\Lambda}k^3 dk = \frac{\Lambda^4}{8\pi^2} \quad (10\text{-}22)$$

In natural units, the Planck mass $\approx 1.22\times 10^{19}$ GeV. Thus, we get the theoretical value

$$\bar{\rho}_v \approx \frac{\left(1.22\times 10^{19}\right)^4}{8\pi^2}\text{GeV}^4 = 2.80\times 10^{74}\text{ GeV}^4. \quad (10\text{-}23)$$

Note length in natural units is GeV^{-1}, so an energy per unit volume would be measured in GeV^4. The experimental value for the upper limit on energy density of the vacuum is

$$\bar{\rho}_v \le 10^{-47}\text{ GeV}^4, \quad (10\text{-}24)$$

a discrepancy between theory and experiment by a factor of more than 10^{120}. Yikes.

Note that in this scenario, the vacuum particles are complex sinusoidal waves extending across the universe from one end to the other (just like the real particles in this scenario).

10.8.2 Other Methods of Calculating Vacuum Energy

As pointed out by J. Martin [Everything you always wanted to know about the cosmological constant problem (but were afraid to ask), *C. R. Physique*, **13**, 566–665 (2012)], the cutoff method is not Lorentz invariant, since the energy Λ is different in different frames, and therefore, though simple in concept, is not valid. Martin uses a Lorentz invariant evaluation of (10-20) and arrives at a vacuum energy density "only" 10^{55} times greater than that observed.

10.9 Appendix B: Symmetry Breaking, Mass Terms, and Vacuum Pairs

This appendix is not for newcomers to QFT, but for veterans familiar with electroweak interactions and symmetry breaking.

In QED, mass terms are part of the free Hamiltonian. So, their effect shows up when we determine the energy expectation value for the Hamiltonian for any given state, including the vacuum state (VEV of the Hamiltonian). Thus, the mass terms end up contributing as part of the ½ energy quanta of the vacuum and, as we saw, do not have Feynman diagram type interactions.

In electro-weak theory with symmetry breaking, however, mass terms arise from coupling of the Bose and Fermi fields with the Higgs field. When the Higgs field symmetry breaks, the Higgs field gets a VEV, and massless fields get masses. Since these mass terms, such as

$$m\overline{\psi}\psi \tag{10-25}$$

arise in the interaction (not the free) part of $\mathcal{H}$ and $\mathcal{L}$, they result in interactions of the type shown in the Feynman diagrams of Fig. 10-4.

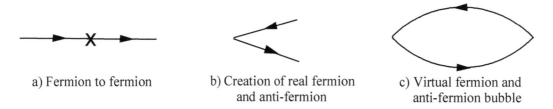

| a) Fermion to fermion | b) Creation of real fermion and anti-fermion | c) Virtual fermion and anti-fermion bubble |

Figure 10-4. Three Types of Interactions from Symmetry Breaking Mass Terms

<u>For Fig. 10-4b</u>

One could then posit that we do indeed have pairs of particles "popping out" of the vacuum, as in Fig. 10-4b, which represents the first order term in the amplitude. Similarly, there would be destructions of fermion and anti-fermion pairs as well (not shown). Key points to be made in this regard are

• The energies and momenta of the particles in Fig. 10-4b must still sum to zero, so there is no vacuum energy contribution.

• The pairs of Fig. 10-4b are real, not virtual. Real particles cannot have negative energy. But since total energy should sum to zero, one of the particles must have negative energy. Hence, we can conclude that Fig. 10-4b does not represent a real physical process and cannot occur.

• The particle pairs so produced presumably would not yield the Casimir plate quantitative result. The ½ quanta generating that result had different mathematical factors than would result from the particles of Fig. 10-4b.

• The pairs do not arise in the vacuum alone, but as a result of the Higgs field. In essence, Fig. 10-4b really has a Higgs field source (which might be visualized as a pre-symmetry breaking Higgs particle coming in from the left.) These pairs are *not* pure vacuum pairs.

• At high energy, particles are massless and terms of form (10-25) do not exist. That is, they are replaced by a Higgs field interacting with the fermion and anti-fermion as discussed in the prior point. In other words, there are no interactions like that of Fig. 10-4b at high energy, so no tiny black holes (posited to result from high energy vacuum particles) could form.

• The probability of interaction occurring (pair formation) is not a function of e/m coupling α, nor weak coupling, nor strong coupling, but of the Higgs coupling.

<u>For Fig. 10-4c</u>

Using the Dyson-Wick expansion for the amplitude, we would find second order terms of form

$$m^2 \int d^4x_1 d^4x_2 N \left\{ \left(\overline{\psi}\psi\right)_{x_1} \left(\overline{\psi}\psi\right)_{x_2} \right\} \tag{10-26}$$

represented by Fig. 10-4c.

- The energies and momenta of the particles in Fig. 10-4c must still sum to zero, so the total loop energy is zero, and there is no vacuum energy contribution.

- One of the virtual particles must have negative energy. If one insisted on applying an uncertainty principle, then it must be applied for negative energy as well. Hence, any "borrowed energy" fluctuations about zero employed to produce Fig. 10-4c must be both positive and negative. The sum of all such fluctuations must be zero, and that means zero vacuum energy.

- The last four bullets for Fig. 10-4b apply here as well.

These points lead to the conclusions that neither of the pair productions of Fig. 10-4b and c is the vacuum pair popping production commonly referred to in the literature, and they make no contribution to vacuum energy.

For Higgs Vacuum Energy Remnant

To be complete in our discussion of vacuum energy, we need to consider the vacuum energy generated by the spontaneous breaking of the electroweak symmetry via the Higgs mechanism. That process leaves an energy density remnant in the vacuum, known as the Higgs condensate energy density.

But the Higgs condensate energy simply sits (essentially statically) in the vacuum unconnected to other particles. It is not related to particle-antiparticle pair popping in and out of existence via the uncertainty principle.

10.10 Appendix C: Comparison of QFT for Discrete vs Continuous Solutions

The overview in Wholeness Chart 10-2 (below and on the following pages) is presented without explanatory text (which can be found at the website for this book listed on pg. xvi, opposite pg. 1). Extensive study of it may be warranted for QFT veterans but is not recommended for newcomers.

Wholeness Chart 10-2. Discrete vs Continuous Versions of QFT
(Only Scalars Shown)

	Discrete	**Continuous**
Field Equations Solutions	$\phi(x) = \sum_{\mathbf{k}} \dfrac{1}{\sqrt{2V\omega_{\mathbf{k}}}}\left(a(\mathbf{k})e^{-ikx} + b^\dagger(\mathbf{k})e^{ikx}\right)$ $\phi^\dagger(x) = \sum_{\mathbf{k}} \dfrac{1}{\sqrt{2V\omega_{\mathbf{k}}}}\left(b(\mathbf{k})e^{-ikx} + a^\dagger(\mathbf{k})e^{ikx}\right)$	$\phi(x) = \int \dfrac{d^3k}{\sqrt{2(2\pi)^3\omega_{\mathbf{k}}}}\left(a(\mathbf{k})e^{-ikx} + b^\dagger(\mathbf{k})e^{ikx}\right)$ $\phi^\dagger(x) = \int \dfrac{d^3k}{\sqrt{2(2\pi)^3\omega_{\mathbf{k}}}}\left(b(\mathbf{k})e^{-ikx} + a^\dagger(\mathbf{k})e^{ikx}\right)$
Coefficient commutators	$\left[a(\mathbf{k}),a^\dagger(\mathbf{k}')\right]=\left[b(\mathbf{k}),b^\dagger(\mathbf{k}')\right]=\delta_{\mathbf{kk}'}$	$\left[a(\mathbf{k}),a^\dagger(\mathbf{k}')\right]=\left[b(\mathbf{k}),b^\dagger(\mathbf{k}')\right]=\delta(\mathbf{k}-\mathbf{k}')$
$\mathcal{H}_0^0$	$\pi_0^0\dot{\phi}+\pi_0^{0\dagger}\dot{\phi}^\dagger-\mathcal{L}_0^0=\left(\dot{\phi}\dot{\phi}^\dagger+\nabla\phi^\dagger\cdot\nabla\phi+\mu^2\phi^\dagger\phi\right)$	as at left, and in terms of number operators $\int\omega_{\mathbf{k}}\left(\dfrac{\mathcal{N}_a(\mathbf{k})}{V}+\dfrac{1}{2}+\dfrac{\mathcal{N}_b(\mathbf{k})}{V}+\dfrac{1}{2}\right)d^3k \quad V\to\delta(0)$
H_0^0	$\sum_{\mathbf{k}}\omega_{\mathbf{k}}\left(N_a(\mathbf{k})+\tfrac{1}{2}+N_b(\mathbf{k})+\tfrac{1}{2}\right)$ $N_a(\mathbf{k})=a^\dagger(\mathbf{k})\,a(\mathbf{k}),\; N_b(\mathbf{k})=b^\dagger(\mathbf{k})\,b(\mathbf{k})$	$\int\omega_{\mathbf{k}}\left(\mathcal{N}_a(\mathbf{k})+\tfrac{1}{2}\delta(0)+\mathcal{N}_b(\mathbf{k})+\tfrac{1}{2}\delta(0)\right)d^3k$ $\mathcal{N}_a(\mathbf{k})=a^\dagger(\mathbf{k})\,a(\mathbf{k}),\; \mathcal{N}_b(\mathbf{k})=b^\dagger(\mathbf{k})\,b(\mathbf{k})$
Operator Units	$N_a(\mathbf{k})$, number of real particles, unitless, M^0 $\tfrac{1}{2}$, number of vacuum particles, unitless $a(\mathbf{k})$, $a^\dagger(\mathbf{k})$, unitless Similar for $N_b(\mathbf{k})$, $b(\mathbf{k})$, $b^\dagger(\mathbf{k})$,	$\mathcal{N}_a(\mathbf{k})$, (num real particles)/(k space vol), M^{-3} $\tfrac{1}{2}$, (num vac particles)/(k vol) /(x vol), M^{-6} $a(\mathbf{k})$, $a^\dagger(\mathbf{k})$, $M^{-3/2}$ Similar for $\mathcal{N}_b(\mathbf{k})$, $b(\mathbf{k})$, $b^\dagger(\mathbf{k})$,

Single Particle (No Anti-particle) State Relations		
Eigenstate Creation	$a^\dagger(\mathbf{k})\|0\rangle = \|\phi_\mathbf{k}\rangle = \left\|\dfrac{e^{-ikx}}{\sqrt{V}}\right\rangle$ Eigenstate at one point in **k** space, spread over volume V in **x** space.	$a^\dagger(\mathbf{k})\|0\rangle = \|\phi(\mathbf{k})\rangle = \left\|\dfrac{e^{-ikx}}{\sqrt{(2\pi)^3}}\right\rangle$ Eigenstate at one point in **k** space, spread over universe in **x** space. $V \to \infty$
General State Creation, C is General State Creation Operator	$\|\phi\rangle = C\|0\rangle = \sum_\mathbf{k} A_\mathbf{k} a^\dagger(\mathbf{k})\|0\rangle$ $= \left\|\dfrac{1}{\sqrt{V}}\left(A_{\mathbf{k}_1} e^{-ik_1 x} + A_{\mathbf{k}_2} e^{-ik_2 x} + ..\right)\right\rangle = \left\|\sum_\mathbf{k} A_\mathbf{k} \dfrac{e^{-ikx}}{\sqrt{V}}\right\rangle$ Coefficient $A_\mathbf{k}$ unitless. State units $l^{-3/2} = M^{3/2}$	$\|\phi\rangle = C\|0\rangle = \int A(\mathbf{k}) a^\dagger(\mathbf{k}) d^3 k \|0\rangle$ $= \left\|\int A(\mathbf{k}) \dfrac{e^{-ikx}}{\sqrt{(2\pi)^3}} d^3 k\right\rangle$ Coefficient $A(\mathbf{k})$ units $l^{3/2} = M^{-3/2}$. State units $l^{-3/2} = M^{3/2}$
Creation Operator C	$C = \sum_\mathbf{k} A_\mathbf{k} a^\dagger(\mathbf{k})$	$C = \int A(\mathbf{k}) a^\dagger(\mathbf{k}) d^3 k$
State Norms	$\langle\phi\|\phi\rangle = 1$	$\langle\phi\|\phi\rangle = 1$
Coefficient Properties	$\sum_\mathbf{k} \|A_\mathbf{k}\|^2 = 1$	$\int \|A(\mathbf{k})\|^2 d^3 k = 1$
For an Eigenstate	Only one $A_\mathbf{k}$, with $\|A_\mathbf{k}\| = 1$. $\|A_\mathbf{k}\| / \sqrt{V} \to 1/\sqrt{V}$	Not very meaningful.
$N_a(\mathbf{k})$ Acting on General State	$N_a(\mathbf{k})\left\|\sum_{\mathbf{k}'} A_{\mathbf{k}'} \dfrac{e^{-ik'x}}{\sqrt{V}}\right\rangle$ $= \|A_{\mathbf{k}'}\|^2 \delta_{\mathbf{kk}'}\left\|\sum_{\mathbf{k}'} A_{\mathbf{k}'} \dfrac{e^{-ik'x}}{\sqrt{V}}\right\rangle = \|A_\mathbf{k}\|^2\left\|\sum_{\mathbf{k}'} A_{\mathbf{k}'} \dfrac{e^{-ik'x}}{\sqrt{V}}\right\rangle$	$\mathcal{N}_a(\mathbf{k})\left\|\int A(\mathbf{k}') \dfrac{e^{-ik'x}}{\sqrt{(2\pi)^3}} d^3 k'\right\rangle$ $= \|A(\mathbf{k}')\|^2 \delta(\mathbf{k}-\mathbf{k}')\left\|\int A(\mathbf{k}') \dfrac{e^{-ik'x}}{\sqrt{(2\pi)^3}} d^3 k'\right\rangle$
Eigenstate Energy Expectation Value	$\bar{E} = \left\langle\phi_{\mathbf{k}'}\right\|\sum_\mathbf{k}\omega_\mathbf{k}\left(N_a(\mathbf{k})+\tfrac{1}{2}+N_b(\mathbf{k})+\tfrac{1}{2}\right)\left\|\phi_{\mathbf{k}'}\right\rangle$ $= \left(\omega_{\mathbf{k}'} + \tfrac{1}{2}\sum_\mathbf{k}\omega_\mathbf{k} + \tfrac{1}{2}\sum_\mathbf{k}\omega_\mathbf{k}\right)\langle\phi_{\mathbf{k}'}\|\|\phi_{\mathbf{k}'}\rangle$ $= \omega_{\mathbf{k}'} + \sum_\mathbf{k}\omega_\mathbf{k}$	$\bar{E} = \left\langle\phi(\mathbf{k}')\right\|\int\omega_\mathbf{k}(\mathcal{N}_a(\mathbf{k})+\tfrac{1}{2}\delta(0)$ $+ \mathcal{N}_b(\mathbf{k})+\tfrac{1}{2}\delta(0))d^3 k\left\|\phi(\mathbf{k}')\right\rangle$ $= \int\omega_\mathbf{k}\left(\delta(\mathbf{k}-\mathbf{k}')+\tfrac{1}{2}\delta(0)+\tfrac{1}{2}\delta(0)\right)d^3 k\langle\phi\|\|\phi\rangle$ $= \omega_{\mathbf{k}'} + \delta(0)\int\omega_\mathbf{k} d^3 k$
General State Energy Expectation Value	$\bar{E} = \left\langle\sum_{\mathbf{k}'} A_{\mathbf{k}'}\dfrac{e^{-ik'x}}{\sqrt{V}}\right\|\sum_\mathbf{k}\omega_\mathbf{k}(N_a(\mathbf{k})+\tfrac{1}{2}$ $+ N_b(\mathbf{k})+\tfrac{1}{2})\left\|\sum_{\mathbf{k}'} A_{\mathbf{k}'}\dfrac{e^{-ik'x}}{\sqrt{V}}\right\rangle$ $= \left(\sum_{\mathbf{k}'}\omega_{\mathbf{k}'} A_{\mathbf{k}'}^\dagger A_{\mathbf{k}'}\delta_{\mathbf{kk}'} + \tfrac{1}{2}\sum_\mathbf{k}\omega_\mathbf{k} + \tfrac{1}{2}\sum_\mathbf{k}\omega_\mathbf{k}\right)\langle\phi\|\phi\rangle$ $= \sum_\mathbf{k}\|A_\mathbf{k}\|^2\omega_\mathbf{k} + \sum_\mathbf{k}\omega_\mathbf{k} = \bar{\omega} + \sum_\mathbf{k}\omega_\mathbf{k}$	$\bar{E} = \left\langle\int A(\mathbf{k}')\dfrac{e^{-ik'x}}{\sqrt{(2\pi)^3}}d^3 k'\right\|\int\omega_\mathbf{k}(\mathcal{N}_a(\mathbf{k})+\tfrac{1}{2}\delta(0)$ $+ \mathcal{N}_b(\mathbf{k})+\tfrac{1}{2}\delta(0))d^3 k\left\|\int A(\mathbf{k}')\dfrac{e^{-ik'x}}{\sqrt{(2\pi)^3}}d^3 k'\right\rangle$ $= \int\omega_\mathbf{k}\left(\|A(\mathbf{k}')\|^2\delta(\mathbf{k}-\mathbf{k}')+\delta(0)\right)d^3 k\langle\phi\|\|\phi\rangle$ $= \int\|A(\mathbf{k})\|^2\omega_\mathbf{k} d^3 k + \delta(0)\int\omega_\mathbf{k} d^3 k = \bar{\omega} + \delta(0)\int\omega_\mathbf{k} d^3 k$

Multi-particle (without Anti-particles) State Relations	
Multi Eigen Particles Creation	$a^\dagger(\mathbf{k}_1)a^\dagger(\mathbf{k}_2)...\|0\rangle = \|\phi_{\mathbf{k}_1},\phi_{\mathbf{k}_2},...\rangle$	$a^\dagger(\mathbf{k}_1)a^\dagger(\mathbf{k}_2)...\|0\rangle = \|\phi(\mathbf{k}_1),\phi(\mathbf{k}_2),...\rangle$
Multi General Particles Creation, C_q **is** q**th Particle Creation Operator**	$\|\phi_q,\phi_r,...\rangle = (C_q C_r ..)\|0\rangle$ $= \left(\sum_{\mathbf{k}} A_{q\mathbf{k}}a^\dagger(\mathbf{k})\right)\left(\sum_{\mathbf{k}} A_{r\mathbf{k}}a^\dagger(\mathbf{k})\right)(...)\|0\rangle$ $= \left\|\sum_{\mathbf{k}} A_{q\mathbf{k}}\frac{e^{-ikx_q}}{\sqrt{V}},\sum_{\mathbf{k}} A_{r\mathbf{k}}\frac{e^{-ikx_r}}{\sqrt{V}},...\right\rangle$	$\|\phi_q,\phi_r,...\rangle = (C_q C_r ..)\|0\rangle$ $= \left(\int A_q(\mathbf{k})a^\dagger(\mathbf{k})d^3k\right)\left(\int A_r(\mathbf{k})a^\dagger(\mathbf{k})d^3k\right)(..)\|0\rangle$ $= \left\|\int A_q(\mathbf{k})\frac{e^{-ikx_q}}{\sqrt{(2\pi)^3}}d^3k,\int A_r(\mathbf{k})\frac{e^{-ikx_r}}{\sqrt{(2\pi)^3}}d^3k,...\right\rangle$
Normalized Creation Operator C_q	$C_q = \sum_{\mathbf{k}} A_{q\mathbf{k}}a^\dagger(\mathbf{k})$	$C_q = \int A_q(\mathbf{k})a^\dagger(\mathbf{k})d^3k$
State Norms	$\langle\phi_q,\phi_r,...\|\phi_q,\phi_r,...\rangle = 1$	$\langle\phi_q,\phi_r,...\|\phi_q,\phi_r,...\rangle = 1$
Coefficient Properties	$\sum_{\mathbf{k}}\|A_{q\mathbf{k}}\|^2 = 1,\quad \sum_{\mathbf{k}}\|A_{r\mathbf{k}}\|^2 = 1,$ etc.	$\int\|A_q(\mathbf{k})\|^2 d^3k = 1,\quad \int\|A_r(\mathbf{k})\|^2 d^3k = 1,$ etc.
$N_a(\mathbf{k})$ **Acting on Multi General Particles State**	$N_a(\mathbf{k})\|\phi_q,2\phi_r,...\rangle$ $= N_a(\mathbf{k})\left\|\sum_{\mathbf{k}'} A_{q\mathbf{k}'}\frac{e^{-ik'x_q}}{\sqrt{V}},2\sum_{\mathbf{k}'} A_{r\mathbf{k}'}\frac{e^{-ik'x_r}}{\sqrt{V}},...\right\rangle$ $= \left(\|A_{q\mathbf{k}'}\|^2\delta_{\mathbf{k}\mathbf{k}'} + 2\|A_{r\mathbf{k}'}\|^2\delta_{\mathbf{k}\mathbf{k}'}+..\right)\|\phi_r,2\phi_r,...\rangle$ $= \left(\|A_{q\mathbf{k}}\|^2 + 2\|A_{r\mathbf{k}}\|^2 + ..\right)\|\phi_r,2\phi_r,...\rangle$	$\mathcal{N}_a(\mathbf{k})\|\phi_q,2\phi_r,...\rangle = \mathcal{N}_a(\mathbf{k})\times$ $\left\|\int A_q(\mathbf{k}')\frac{e^{-ik'x_q}}{\sqrt{(2\pi)^3}}d^3k',2\int A_r(\mathbf{k}')\frac{e^{-ik'x_r}}{\sqrt{(2\pi)^3}}d^3k',..\right\rangle$ $= (\|A_q(\mathbf{k}')\|^2\delta(\mathbf{k}-\mathbf{k}')$ $+ 2\|A_r(\mathbf{k}')\|^2\delta(\mathbf{k}-\mathbf{k}')+..)\|\phi_r,2\phi_r,...\rangle$
Multi Eigen Particles Energy Expectation Value	$\bar{E} = \langle\phi_{\mathbf{k}_1},2\phi_{\mathbf{k}_2},...\|\sum_{\mathbf{k}}\omega_{\mathbf{k}}(N_a(\mathbf{k})+\tfrac{1}{2}+$ $N_b(\mathbf{k})+\tfrac{1}{2})\|\phi_{\mathbf{k}_1},2\phi_{\mathbf{k}_2},...\rangle$ $= (\omega_{\mathbf{k}_1} + 2\omega_{\mathbf{k}_2} + ..$ $+\tfrac{1}{2}\sum_{\mathbf{k}}\omega_{\mathbf{k}} + \tfrac{1}{2}\sum_{\mathbf{k}}\omega_{\mathbf{k}})\langle\phi_{\mathbf{k}_1},2\phi_{\mathbf{k}_2},...\|\phi_{\mathbf{k}_1},2\phi_{\mathbf{k}_2},...\rangle$ $= \omega_{\mathbf{k}_1} + 2\omega_{\mathbf{k}_2} + ... + \sum_{\mathbf{k}}\omega_{\mathbf{k}}$	$\bar{E} = \langle\phi(\mathbf{k}_1),2\phi(\mathbf{k}_2)...\|\int\omega_{\mathbf{k}}(\mathcal{N}_a(\mathbf{k})+\tfrac{1}{2}\delta(0)$ $+\mathcal{N}_b(\mathbf{k})+\tfrac{1}{2}\delta(0))d^3k\|\phi(\mathbf{k}_1),2\phi(\mathbf{k}_2),...\rangle$ $= (\omega_{\mathbf{k}_1} + 2\omega_{\mathbf{k}_2} + ... + \delta(0)\int\omega_{\mathbf{k}}d^3k)\times$ $\langle\phi(\mathbf{k}_1),2\phi(\mathbf{k}_2),.\|\phi(\mathbf{k}_1),2\phi(\mathbf{k}_2),.\rangle$ $= \omega_{\mathbf{k}_1} + 2\omega_{\mathbf{k}_2} + ... + \delta(0)\int\omega_{\mathbf{k}}d^3k$
Multi General Particles Energy Expectation Value	$\bar{E} = \langle\phi_q,2\phi_r,...\|\sum_{\mathbf{k}}\omega_{\mathbf{k}}(N_a(\mathbf{k})+\tfrac{1}{2}$ $+N_b(\mathbf{k})+\tfrac{1}{2})\|\phi_q,2\phi_r,...\rangle$ $= (\sum_{\mathbf{k}}\|A_{q\mathbf{k}}\|^2\omega_{\mathbf{k}} + 2\sum_{\mathbf{k}}\|A_{r\mathbf{k}}\|^2\omega_{\mathbf{k}} + ...$ $+\tfrac{1}{2}\sum_{\mathbf{k}}\omega_{\mathbf{k}} + \tfrac{1}{2}\sum_{\mathbf{k}}\omega_{\mathbf{k}})\langle\phi_q,2\phi_r,...\|\phi_q,2\phi_r,...\rangle$ $= \bar{\omega}_q + 2\bar{\omega}_r + ... + \sum_{\mathbf{k}}\omega_{\mathbf{k}}$	$\bar{E} = \langle\phi_q,2\phi_r,...\|\int\omega_{\mathbf{k}}(\mathcal{N}_a(\mathbf{k})+\tfrac{1}{2}\delta(0)$ $+\mathcal{N}_b(\mathbf{k})+\tfrac{1}{2}\delta(0))d^3k\|\phi_q,2\phi_r,...\rangle$ $= (\int\omega_{\mathbf{k}}\|A_q(\mathbf{k})\|^2 d^3k + 2\int\omega_{\mathbf{k}}\|A_r(\mathbf{k})\|^2 d^3k + ...$ $+\delta(0)\int\omega_{\mathbf{k}}d^3k)\langle\phi_q,2\phi_r,...\|\phi_q,2\phi_r,...\rangle$ $= \bar{\omega}_q + 2\bar{\omega}_r + ... + \delta(0)\int\omega_{\mathbf{k}}d^3k$

Note			In the energy expectation derivation for the continuous case, one finds a Dirac delta function squared in the vacuum energy part. This is undefined mathematically. By some perspectives, its evaluation leaves a vacuum term of energy $\omega(\mathbf{k}=0)$ which equals μ (one particle mass). An alternative perspective is shown above.

10.11 Appendix D: Free Fields and "Pair Popping" Re-visited

One possible issue some might raise with this chapter needs to be addressed. That is, by using the Interaction Picture (I.P.), where operator fields are free and particle behavior is described by the interaction Hamiltonian, are we somehow obscuring some physics? Recall we derived Feynman diagrams from only the interaction term in the Hamiltonian. If we included the free Hamiltonian in such a derivation, would we possibly find Feynman diagrams producing and destroying particle/antiparticle pairs?

Specifically, $\mathcal{H}_0$ has creation and destruction operators paired together in the same terms. For scalar fields, in the Heisenberg Picture (H.P.), we found

$$\mathcal{H}_0^0 = \pi_0^0 \dot{\phi} + \pi_0^{0\dagger} \dot{\phi}^\dagger - \mathcal{L}_0^0 = \left(\dot{\phi}\dot{\phi}^\dagger + \nabla\phi^\dagger \cdot \nabla\phi + \mu^2 \phi^\dagger \phi \right) = \left(\pi_0^{0\dagger}\pi_0^0 + \nabla\phi^\dagger \cdot \nabla\phi + \mu^2 \phi^\dagger \phi \right) \quad (10\text{-}27)$$

$$\text{with} \quad \phi(x) = \sum_{\mathbf{k}} \frac{1}{\sqrt{2V\omega_{\mathbf{k}}}} \left(a(\mathbf{k})e^{-ikx} + b^\dagger(\mathbf{k})e^{ikx} \right). \quad (10\text{-}28)$$

So, we might end up with a creation ($a^\dagger(\mathbf{k})$ or $b^\dagger(\mathbf{k})$) and a destruction ($a(\mathbf{k})$ or $b(\mathbf{k})$) operator in the same term in H_0^0. In particular, if we had terms containing factors of $a^\dagger(\mathbf{k})b^\dagger(\mathbf{k})$ or $a^\dagger(\mathbf{k})b^\dagger(-\mathbf{k})$, we might expect creation of a particle and an antiparticle at the same event in the vacuum.

To see how states might change, let's use the Schrödinger picture (S.P.), in which operators do not change in time, but states do. In the S.P. for free fields, scalar states are governed by

$$i\frac{d}{dt}|\Phi\rangle = H_0^0 |\Phi\rangle, \quad (10\text{-}29)$$

where $|\Phi\rangle$ is in general a multi-scalar particle state and H_0^0 is the same in the H.P. as the S.P. (See Wholeness Chart 2-4, pg. 28.) Using H_0^0, specifically the RHS of (10-27), along with (10-28) and its associated quantities, we can follow the same steps as we did in Chap. 8 to determine the evolution of the state $|\Phi\rangle$. (See Wholeness Chart 8-4, pgs. 248-251.) This leads us to

$$\left| \Phi(t_f) \right\rangle = S \left| \Phi(t_i) \right\rangle = e^{-i\int_{-\infty}^{\infty} H_0^0 dt'} \left| \Phi(t_i) \right\rangle \qquad S_{fi} = \left\langle \Phi(t_f) \middle| S \middle| \Phi(t_i) \right\rangle = \left\langle f \middle| S \middle| i \right\rangle. \quad (10\text{-}30)$$

With the Dyson expansion, we find

$$S = \underbrace{I}_{S^{(0)}} \underbrace{-i\int_{-\infty}^{\infty} \mathcal{H}_0^0(x_1)d^4x_1}_{S^{(1)}} \underbrace{-\frac{1}{2!}\int_{-\infty}^{\infty}\int_{-\infty}^{\infty} T\left\{\mathcal{H}_0^0(x_1)\mathcal{H}_0^0(x_2)\right\}d^4x_1 d^4x_2}_{S^{(2)}} + \dots\dots . \quad (10\text{-}31)$$

For the $S^{(1)}$ term, the integration over all space was carried out in Chap. 3, Sect. 3.4.1, pgs. 53-54. In that derivation, we saw all terms containing factors $a^\dagger(\mathbf{k})b^\dagger(-\mathbf{k})$, $a^\dagger(\mathbf{k})b^\dagger(\mathbf{k})$, $a(\mathbf{k})b(-\mathbf{k})$, and $a(\mathbf{k})b(\mathbf{k})$ dropped out. That is, no terms remain that create a particle/anti-particle pair at the same event in the vacuum. Ditto for destruction of such a pair.

For the $S^{(2)}$ term, at any given point in time, we can integrate $H_0^0(x_1)$ over $d^3\mathbf{x_1}$ without regard to the integration of $H_0^0(x_2)$ over $d^3\mathbf{x_2}$. For that integration, we would get the same result as for $S^{(1)}$, i.e., no terms with factors creating or destroying a particle/anti-particle pair. The same result would hold for $S^{(n)}$ for any n.

The transition amplitude S_{fi} would, therefore, not contain any terms creating/destroying such pairs. And so, we would have no Feynman diagrams representing such a thing.

Further, by reviewing the above cited section of Chap. 3, one can see that the ½ quanta terms, commonly considered "vacuum fluctuations" come from the $a(\mathbf{k})a^\dagger(\mathbf{k})$ and $b(\mathbf{k})b^\dagger(\mathbf{k})$ terms and the coefficient commutation relations. Even if we chose to use these terms directly, without employing the commutation relations, the $a(\mathbf{k})a^\dagger(\mathbf{k})$ term is not coupled to the $b(\mathbf{k})b^\dagger(\mathbf{k})$term so both terms together would not represent a vertex in a Feynman diagram. In that interpretation, one might think of the $a(\mathbf{k})a^\dagger(\mathbf{k})$ as representing creation of a particle and destruction of the same particle at the same event, i.e., nothing would happen as time evolves. No evanescence. No pair popping.

In summary, for <u>free fields</u>

- Terms in the free Hamiltonian density containing two creation operators that might create a particle/antiparticle pair at an event drop out of the full (not density) Hamiltonian.

- The only terms surviving in the full Hamiltonian have creation and destruction operators paired. These would create and destroy the same particle at the same event, i.e., nothing would effectively happen.

We conclude that the free field components of the Hamiltonian do not lead to particle/antiparticle pairs popping in and out of the vacuum.

10.12 Appendix E: Considerations for Finite Volume Interactions

All of the foregoing material in this chapter related to "standard" QFT, in which fields/particles are considered to extend over infinite volume V and infinite time T. That assumption, as we will see in Part 4 of this book, leads to accurate real-world predictions for real world fields/particles of finite extensions in V and T.

In developing our theory, this assumption gave rise to Dirac delta functions (see (8-30), pg. 222) because we integrated over unbounded space and time. These Dirac delta functions, arising in each transition amplitude, led to strict conservation of 4-momentum at every vertex. Had V and T been finite instead of unbounded, integration would not have led to Dirac delta functions, and so one might question if, with finite V and T, the resulting relation would lead to uncertainty in outgoing 4-momentum. Presumably, for large V and T, the relation would approximate a Dirac delta function implying approximate, but not exact, conservation of 4-momenta. And thus, smaller V and T would mean 4-momenta would be less constrained to be conserved.

This would give rise to an uncertainty in outgoing 4-momentum at any vertex for which the fields did not have infinite extension in V and T. Smaller V and T means greater uncertainty in 3-momentum and energy, respectively, and this correlates with the familiar uncertainty principle.

To examine this more closely, consider the Dirac delta function shown in (8-30), pg. 222, where $k = P_f$ is the 4-momentum leaving the vertex and $P_i = p_1 + p_2$ is the incoming 4-momentum,

$$(2\pi)^4 \, \delta^{(4)}\left(P_f - P_i\right) = \int_{-\infty}^{\infty} e^{i\left(P_f - P_i\right)x_2} d^4 x_2 \,. \tag{10-32}$$

Now consider the RHS of (10-32) integrated over finite, instead of infinite, V and T, where, to keep things simple, we use the 1D correlate of the 4D integral, and represent that with the symbol I,

$$\int_{-V/2,-T/2}^{V/2,T/2} e^{i\left(P_f - P_i\right)x_2} d^4 x_2 \xrightarrow{\ 1D\ } \int_{-L/2}^{L/2} e^{i\left(P_f - P_i\right)x_2} dx_2 = I\left(P_f - P_i\right) \tag{10-33}$$

The integral is easy to evaluate, and I is found to be

$$I\left(P_f - P_i\right) = \frac{e^{i\left(P_f - P_i\right)x_2}}{i\left(P_f - P_i\right)}\Bigg|_{-L/2}^{L/2} = \frac{\cos\left(P_f - P_i\right)x_2 + i\sin\left(P_f - P_i\right)x_2}{i\left(P_f - P_i\right)}\Bigg|_{-L/2}^{L/2} = 2\frac{\sin\left(\left(P_f - P_i\right)\frac{L}{2}\right)}{\left(P_f - P_i\right)}. \tag{10-34}$$

In the development of NRQM, RQM, and QFT (see (3-24) to (3-25), pg. 46 and Sect. 3.4.1, pgs. 53-54), we typically assume

$$P_i = \frac{2\pi n_i}{L} \quad P_f = \frac{2\pi n_f}{L} \quad \rightarrow \quad P_f - P_i = \frac{2\pi\left(n_f - n_i\right)}{L} \qquad n_i, n_f \text{ integers}, \tag{10-35}$$

because (10-35) results in orthogonal functions of e^{iPx} and zero values for quantities like probability density in NRQM for particles at $L/2$ and $-L/2$, as well as certain terms in the probability of RQM and in the Hamiltonian of QFT that must be zero. (See above references.)

<u>n_i and n_f as integers</u>

For (10-35) in (10-34), we find

$$I\left(P_f - P_i\right) = 2\frac{sin\left(\dfrac{2\pi\left(n_f - n_i\right)}{L}\dfrac{L}{2}\right)}{\dfrac{2\pi\left(n_f - n_i\right)}{L}} = L\frac{sin\left(\pi\left(n_f - n_i\right)\right)}{\pi\left(n_f - n_i\right)}.$$ (10-36)

Due to the numerator, this is zero except for $n_f = n_i$. Then

$$I\left(P_f - P_i\right)_{n_f = n_i} = \lim_{n_f \to n_i}\left(L\frac{sin\left(\pi\left(n_f - n_i\right)\right)}{\pi\left(n_f - n_i\right)}\right) = L\frac{\pi\left(n_f - n_i\right)}{\pi\left(n_f - n_i\right)} = L$$ (10-37)

So, I is zero, except when $n_f = n_i$, i.e., when $P_f = P_i$. That behaves like a Dirac delta function for argument $P_f \neq P_i$. However when $P_f = P_i$, I is not ∞, as a Dirac delta function is, as long as L is finite.

Looking again at our transition amplitude calculation in (8-30), pg. 222, we see that the finite L (V there for 3D case; V and T, for 4D) will still leave us with a zero value unless $P_f = P_i$ ($k = p_1 + p_2$ there.) The value of the transition amplitude will change because we now have L (V for 3D, VT for 4D) as finite when $P_f = P_i$, but other values for P_f are prohibited (have zero probability of occurring.)[1]

Bottom line: For n_f and n_i as integers, and finite volume and time, we still must have strict 4-momentum conservation at a vertex. That is, there is no uncertainty principle at play giving rise to evanescent energy and 3-momentum "popping in and out" of the vacuum.

<u>n_f and n_i as non-integers</u>

If, however, n_f and n_i could be non-integers, then I of (10-36) can have non zero values when $n_f \neq n_i$ (and thus when $P_f \neq P_i$). Analogous results hold for 4D, so for finite V and T, we could have non-zero probability (due to a non-zero value in the RHS of (10-32)) for $P_f \neq P_i$ and not have strict conservation of 4-momentum[2].

Bottom line: For n_f and n_i as non-integers, and finite volume and time, we do not have strict 4-momentum conservation at a vertex. That is, there would be an uncertainty principle of sorts at play, which could give rise to evanescent energy and 3-momentum "popping in and out" of the vacuum. For infinite volume and time, strict conservation exists.

<u>Impact of n_f and n_i as non-integers on various kinds of "vacuum fluctuations"</u>

If non-integer values for n_f and n_i manifest in nature, then the following may be surmised for each type of "vacuum fluctuation" in QFT.

"Pair Popping"

The functional form of the transition amplitude and thus questions involving the Dirac delta function found therein are not relevant to the pair popping story, as there are no transition amplitudes having vertices with only two (not three, as for vacuum bubbles) particles. (See "Virtual Bubbles" section below.)

---

1 In the limit where $L \to \infty$, (10-37) becomes $2\pi \delta\left(P_f - P_i\right)$. When $V \to \infty$, we get the 3D Dirac delta function, and for $T, V \to \infty$, the 4D relation.

2 Additional analysis, which we mention but not do here, leads to the conclusion that for n_i and n_f as non-integers, we do get a Dirac delta function in (10-33) (and (10-32) as $L \to \infty$ ($V, T \to \infty$).

Zero Point Energy

The non-integer n_f and n_i condition would not modify anything we have said herein about the ZPE ½ quanta, as they represent free fields, with no vertices, i.e., no interactions. However, it does relate to virtual vacuum bubbles and radiative corrections, which are manifestations of interacting fields. (See Wholeness Chart 10-1, pg. 278, and below.)

Virtual Bubbles

A 3-particle virtual bubble has zero initial 4-momentum, but as noted above for finite V and T, it could then, after the first vertex, have a non-zero total 4-momentum (solely for non-integer n_f and n_i). And this then starts to look like the pair popping scenario (even though there are three, not two particles.)

However, we have seen that negative energy virtual particles are as likely as positive energy ones. So, the sum total energy of the bubble could be positive or negative. The sum over large numbers of such bubbles would be effectively zero energy. In other words, even for small values of V and T, there would be no net global contribution to the energy of the vacuum from virtual bubbles. It is conceivable, however, that tiny black holes could exist for positive energy bubbles, and possibly "white holes", we could call them, for the negative ones. We could have quantum foam, but zero total vacuum energy.

Radiative Corrections

As noted, radiative corrections do not arise alone in the vacuum and make no direct contribution to vacuum energy. This is true for finite, or infinite, V and T. Additionally, variations in energy from uncertainty at each vertex would go in both directions (positive and negative) and cancel globally, over many interactions.

BUT remember

Integer values for n_f and n_i in (10-35) seem to be required by nature. If this were not true, we would not have orthogonal functions as our solutions to the RQM/QFT wave equations and certain derivations, such as that for the number operator form of the Hamiltonian, would no longer be valid.

Bottom line:

Thus, vacuum energy, carried by particles popping in and out of the vacuum (for virtual 3 particle bubbles), appears inconsistent with the rest of our theory. To my knowledge, this issue (regarding non-integer n_f and n_i in transition amplitudes) has not been explored in great depth and might make a good research topic for someone. If any reader does pursue this, please apprise me of the results (via the website for this book, the address of which is found on pg. xvi, opposite pg. 1.)

10.13 Appendix F: Vacuum Fluctuations Update (Added in 2018 Text Revision)

There are several experimental results and two theoretical papers related to vacuum fluctuations, as well as another well-known phenomenon often linked to the vacuum, that were not originally covered in this chapter. One of the experiments (actually a cosmological observation) was done in 2012, and I was unaware of it when I wrote the book (2013, 1st edition). The others and the theoretical articles have only been made public in the three years before the 2018 revision of the text. I review these in chronological order below and supply links to the original articles. I then briefly address spontaneous radiation emission, often linked in the past to vacuum fluctuations.

Note that this appendix is posted on the book website (see pg. xvi, opposite pg. 1 for URL) with live links for websites cited below.

10.13.1 Cosmological Observations of Photons and Neutrinos (2012-2018)

Tiny scattering effects from Planck-scale quantum foam on photons propagating over billions of light-years should be cumulative and lead to detectable dispersion of those photons when they arrive on Earth. Lack of such dispersion would support the notion that vacuum fluctuations do not exist. A 2012 analysis of gamma ray bursts (GRBs) by Nemiroff et al[1] implied no Planck-scale quantum

---

[1] R. J. Nemiroff, R. Connolly, J. Holmes, and A. B. Kostinski1, "Bounds on Spectral Dispersion from Fermi-detected Gamma Ray Bursts" *Phys. Rev. Lett.* **108** (23): 231103 (2012). https://arxiv.org/abs/1109.5191.

foam. Popular accounts include "Cosmic race ends in a tie" by R. Cowen, *Nature*. (10 January 2012) and "Spacetime: A smoother brew than we knew" (January 2013) [1]. Other research, such as that by Vasileiou et al[2] also indicate smooth spacetime at the Planck-scale.

However, later work by Xu and Ma[3] and Amelino-Camelia et al[4] seem to contradict this, as they suggest evidence that cosmological photons and neutrinos may disperse. However, none of these results, either for or against spacetime foam, is statistically ironclad.

10.13.2 Usual Analysis of Casimir Plate Effect May be Faulty (2016 - 2017)

Nikolić[5] notes, among other points, that typical analyses of the Casimir effect use a Hamiltonian that has implicit dependence on matter fields and illegitimately treat it as if the dependence were explicit. He contends the true origin of the Casimir force is the van der Waals force.

10.13.3 Vacuum Fluctuations Experiment (2017)

An experimental group at the University of Konstanz[6] claimed the first direct detection of ZPE fluctuations in a laboratory experiment. Their technical article is quite difficult for a non-specialist in nonlinear optics to understand, so I have written a pedagogic introduction[7] to their work on the book website. Note that in that article I question whether ZPE fluctuations have really been detected and provide reasons why they may not have been. The result is controversial.

10.13.4 Spontaneous Emission

As early as 1913, A. Einstein and O. Stern[8] noted that a zero-point energy term had to be added to the classical theory to obtain the Planck radiation spectrum formula. Subsequent research, cited and summarized by P. W. Milonni[9], extended that perspective to spontaneous emission of radiation from an atom. It appeared that a vacuum contribution was needed to help "jiggle" an orbiting electron and "stimulate" it to jump down an energy level, thereby emitting e/m radiation.

However, Milonni, probably the leading expert on the subject, has noted that, similar to the Casimir plates case, there are different ways to carry out the calculations, and in at least one of them, no vacuum contribution is needed. He says (Milonni 1988), " .. the effects usually attributed to vacuum field fluctuations may instead be attributed to radiation reaction."

He goes on to say

> *"..radiation reaction nevertheless offers a valid basis for understanding spontaneous emission, provided the radiation reaction field is handled properly as a quantum-mechanical operator.*

---

[1] www.nature.com/news/cosmic-race-ends-in-a-tie-1.9768; https://phys.org/news/2013-01-spacetime-smoother-brew-knew.html .

[2] Vasileiou, V., Granot, J., Piran, T. and Amelino-Camelia, G., "A Planck-scale limit on spacetime fuzziness and stochastic Lorentz invariance violation", *Nat. Phys. Lett.* **11**, 344-346, April 2015 www.nature.com/articles/nphys3270 .

[3] Xu, H. and Ma, B.Q., "Light Speed Variation from GRB 160509A", *Phys. Lett. B* **760** (2016) 602 https://arxiv.org/abs/1607.08043

[4] Amelino-Camelia, G., D'Amico, G., Rosati, G. and Loret, N., "In vacuo-dispersion features for GRB neutrinos and photons", *Nat. Astron.* 1, 0139 (2017) https://arxiv.org/abs/1612.02765.

[5] H. Nikolić, "Proof that Casimir forces do not originate from vacuum energy", *Phys. Lett. B* **761** (2016) 197-202. https://arxiv.org/abs/1605.04143, and "Is zero-point energy physical? A toy model for Casimir-like effect", *Ann. of Phys*, **383** (2017) 181-195 https://arxiv.org/abs/1702.03291.

[6] Riek, C., Sulzer, P., Seeger, M., Moskalenko, A.S., Burkard, G., Seletskiy, D.V., and Leitenstorfer, A.. "Subcycle quantum electrodynamics", *Nature* **541**, 376-379 (19 Jan 2017) https://arxiv.org/abs/1611.06773.   Popular accounts include "Traffic Jam in Empty Space" http://www.uni-konstanz.de/en/university/news-and-media/current-announcements/news/news-in-detail/verkehrsstau-im-nichts/ and "Physicists observe weird quantum fluctuations of empty space – maybe" www.sciencemag.org/news/2015/10/physicists-observe-weird-quantum-fluctuations-empty-space-maybe.

[7] Klauber, R., "Vacuum Fluctuations Detection: A Pedagogic Overview of the University of Konstanz Experiment (Nov 2017) www.quantumfieldtheory.info/pedagog_U_Konstanz.pdf .

[8] Einstein, A., and Stern, O. *Ann. Phys.* **40**, 551.

[9] Milonni. P./W., "Different Ways of Looking at the Electromagnetic Vacuum", *Physica Scripta*, T21, 102-109 (1988) and *The Quantum Vacuum: An Introduction to Quantum Electrodynamics*, (Academic Press, 1994).

.. It was shown in the case of spontaneous emission that the physical interpretation suggested by quantum electrodynamics is more or less a consequence of the way we choose to order <u>commuting</u> (underlining added) atomic and field operators.

.. The level shifts and widths can be attributed exclusively to radiation reaction ..., or to linear combinations of the two.

.. There is no ordering that attributes the radiative decay of a level entirely to the vacuum field.

..Furthermore this picture (of the vacuum contribution) offers no explanation as to why there is no spontaneous <u>absorption</u> (underlining added) from the vacuum field."

Note that it is the order of operators that commute which changes the relative contributions of the ZPE and radiation reaction. In all the work we have done, the order of commuting operators is unimportant. It is the order of non-commuting operators that impacts our results, and about which we need to take special care. Here, Milonni tells us, the order of *commuting* operators affects the degree to which we can attribute spontaneous emission to ZPE or radiation reaction effects. For a certain order, there is no vacuum contribution. For another order, the ZPE quanta play a part, and the radiation reaction plays a part. There is no ordering for which the effect is entirely attributable to the vacuum. For all orderings, the final result is the same. But the attribution of cause varies.

Hence, like we have seen in other cases, most notably the Casimir effect, the experimentally verified result can be determined theoretically without recourse to vacuum fluctuations.

Still further, if the vacuum plays a role in spontaneous emission, why is there no spontaneous absorption by it?

10.13.5 ZPE and Experimental Measurement

If ZPE fluctuations really impact the physical world, we should be able to detect them directly. Yet, a detector picks up the non-vacuum contribution, but nothing from the vacuum.

As noted by Jaynes[1]

"It seems to me that, if you say radiation is "real," you ought to mean by that, that it can be detected by a real detector. But an optical pyrometer sees only the Planck term, and not the zero-point term, in black body-radiation.

It is a supple ontology which supposes that vacuum fluctuations are just real enough to shift the hydrogen 2s level by 4 microvolts; but not real enough to be seen by our eyes, although in the optical band they correspond to a flux of over 100 kilowatts/cm².
Nevertheless, the dark-adapted eye, looking for example at a faint star, can see real radiation of the order of 10^{-15} watts/cm²."

10.14 Problem

1. Show that for the single particle state $|\phi\rangle$, which can be expressed in function form as

$$\phi_{state} = \int \frac{A(\mathbf{k}')e^{-ik'x}}{\sqrt{(2\pi)^3}} \, d^3k'$$ having unit norm, i.e. $\langle\phi|\phi\rangle = 1$, that $\int |A(\mathbf{k}')|^2 d^3k' = 1$. Hint: The

bra $\langle\phi|$ can be expressed in function form as $\phi_{state}^{\dagger} = \int \frac{A^{\dagger}(\mathbf{k}'')e^{ik''x}}{\sqrt{(2\pi)^3}} \, d^3k''$, and the norm implies

integration of the bra times the ket over $\mathbf{x}$, with $\int e^{i(\mathbf{k}''-\mathbf{k}')\cdot\mathbf{x}} \, d^3x = (2\pi)^3 \delta^{(3)}(\mathbf{k}'' - \mathbf{k}')$.

---

[1] Jaynes, E. T., *Coherence and Quantum Optics IV*, edited by L. Mandel and E. Wolf (Plenum Press, New York, 1978), http://bayes.wustl.edu/etj/articles/electrodynamics.today.pdf , pgs. 5-6.

Chapter 11

Symmetry, Invariance, and Conservation for Interacting Fields

Brahma

If the red slayer think he slays,
 Or if the slain think he is slain,
They know not well the subtle ways
 I keep, and pass, and turn again.

Far or forgot to me is near;
 Shadow and sunlight are the same;
The vanished gods to me appear;
 And one to me are shame and fame.

They reckon ill who leave me out;
 When me they fly, I am the wings;
I am the doubter and the doubt,
 And I the hymn the Brahmin sings.

The strong gods pine for my abode,
 And pine in vain the sacred Seven;
But thou, meek lover of the good!
 Find me, and turn thy back on heaven.

A poem about symmetry
by Ralph Waldo Emerson

11.0 Preliminaries

I was tempted to say nothing about the above poem. I was also tempted to give readers a problem challenging them to find the symmetry in it. I decided to leave it with the following.

The word "keep" in the last line of the first stanza can mean maintain or stay the same. "pass" is equivalent to translate, and "turn", to rotate. Something more profound stays the same, while something more superficial is translated and/or rotated. Sound familiar?

There is yet more symmetry in the poem, but poems tend to be less effective with detailed explanation, so I leave it at that.

11.0.1 Background

We saw the relevance of symmetry for free fields in Chap. 6. In the present chapter, we will explore its relevance for interacting fields. We will find that the same conservation principles and laws we derived earlier for free fields also hold for interactions. In addition, we will find a new kind of symmetry relevant to interactions, which is laden with profundity for QFT.

11.0.2 Review

Recall that symmetry is the propensity for non-change with superficial change. Mathematically, it entails a change in perspective (coordinates by which the observer measures), while the thing being observed remains the same. It is invariance under transformation (under coordinate change).

*Symmetry =
invariance under
transformation*

A scalar function f is symmetric if it has the same form in terms of coordinates of different systems, i.e., where x^i and x'^i represent two different coordinate systems,

$$f\left(x^1, x^2, x^3\right) = f\left(x'^1, x'^2, x'^3\right). \tag{11-1}$$

We saw a real-world example where the laws of nature for free particles/fields (such as the Klein-Gordon equation) are symmetric under a Poincaré (Lorentz plus 4D translation) transformation. They have the same form under a spatial translation, a translation in time, a rotation, and/or a boost.

We saw further, that for free fields/particles, symmetries are related to conservation laws. For example, via Noether's theorem, we showed that a symmetry in the free field Lagrangian (density) meant that an associated quantity (charge in our example in Chap. 6) is conserved.

Symmetry of the first example is <u>external symmetry</u>, because it entails transformation of spacetime coordinates (i.e., coordinates of the physical, external world). That of the second, leading to charge conservation, is <u>internal symmetry</u>, because it entails a transformation in an abstract, non-physical space (complex space of $e^{i\alpha}$ for charge).

You may wish to read over Chap. 6 at this point to refresh your memory on these things.

11.0.3 Chapter Overview

We will first look at

- a slightly different, alternative form for the Lagrangian, which will prove beneficial to our discussion of symmetry in this chapter.

Then, we will look at symmetry in QFT when interactions are involved, specifically

- field equations symmetric under Lorentz transformation, i.e., they are invariant in spacetime (same for all inertial observers)

- symmetry in the full Lagrangian density (including interaction terms) $\mathcal{L} \to$ a related quantity is conserved (via <u>global symmetry</u>, which is what we have used up to here)

- symmetry, gauges, and interactions related to <u>local symmetry</u>, which we will define

Free vs interacting fields

We will deal primarily with interacting particles and fields in this chapter.

11.1 A Helpful Modification to the Lagrangian

11.1.1 The Lagrangian We Have Been Using

Recall the QED full Lagrangian we have been using has form

$$\mathcal{L}^{1/2,1} = \underbrace{-\tfrac{1}{2}\left(\partial_\nu A_\mu\right)\left(\partial^\nu A^\mu\right)}_{\mathcal{L}_0^1} + \underbrace{\bar\psi\left(i\gamma^\mu\partial_\mu - m\right)\psi}_{\mathcal{L}_0^{1/2}} + \underbrace{e\bar\psi\,\gamma^\mu\psi A_\mu}_{\mathcal{L}_I^{1/2,1}} \;. \tag{11-2}$$

When (11-2) is used in the Euler-Lagrange equation

$$\frac{\partial}{\partial x^\mu}\left(\frac{\partial\mathcal{L}}{\partial\phi^r{}_{,\mu}}\right) - \frac{\partial\mathcal{L}}{\partial\phi^r} = 0 \;. \tag{11-3}$$

with $\mathcal{L} = \mathcal{L}^{1/2,1}$ and $\phi^r = \bar\psi$, we obtain the QED field equation governing interactions of the Dirac fermion field with the electromagnetic field, i.e.,

$$\left(i\gamma^\mu\partial_\mu - m\right)\psi = -e\,\gamma^\mu\psi A_\mu \;. \tag{11-4}$$

With $\phi^r = A_\mu$ in (11-3), we obtain the governing equation for the electromagnetic field interacting with the Dirac field, i.e.,

$$\partial^\alpha\partial_\alpha A^\mu = -e\bar\psi\,\gamma^\mu\psi \;. \tag{11-5}$$

The terms on the RHS of both (11-4) and (11-5) are source terms reflecting the interaction. (Zero on the RHS yields the free field equation in each case.)

11.1.2 An Alternative Lagrangian that Also Works

There is some latitude in the form we choose for the Lagrangian. As a trivial example, we can add a constant to it and still get the same field equations (11-4) and (11-5). We can construct a less trivial, and more useful, Lagrangian with the aid of the definition in Chap. 5 of $F^{\mu\nu}$, repeated below.

$$F^{\mu\nu}(x) = \partial^{\nu}A^{\mu}(x) - \partial^{\mu}A^{\nu}(x) = \begin{bmatrix} 0 & E^1 & E^2 & E^3 \\ -E^1 & 0 & B^3 & -B^2 \\ -E^2 & -B^3 & 0 & B^1 \\ -E^3 & B^2 & -B^1 & 0 \end{bmatrix}. \qquad (11\text{-}6)$$

Now, if we redefine our QED Lagrangian by changing only the first term in (11-2), we get

$$\begin{aligned} \mathcal{L}^{1/2,1} &= -\tfrac{1}{4}F^{\mu\nu}F_{\mu\nu} + \bar{\psi}\left(i\gamma^{\mu}\partial_{\mu} - m\right)\psi + e\bar{\psi}\gamma^{\mu}\psi A_{\mu} \\ &= -\tfrac{1}{4}\left(\partial^{\nu}A^{\mu} - \partial^{\mu}A^{\nu}\right)\left(\partial_{\nu}A_{\mu} - \partial_{\mu}A_{\nu}\right) + \bar{\psi}\left(i\gamma^{\mu}\partial_{\mu} - m\right)\psi + e\bar{\psi}\gamma^{\mu}\psi A_{\mu} \\ &= -\tfrac{1}{4}\left(\partial^{\nu}A^{\mu}\partial_{\nu}A_{\mu} - \partial^{\nu}A^{\mu}\partial_{\mu}A_{\nu} \underbrace{-\partial^{\mu}A^{\nu}\partial_{\nu}A_{\mu}}_{=\,2\text{nd term}} + \underbrace{\partial^{\mu}A^{\nu}\partial_{\mu}A_{\nu}}_{=\,1\text{st term}}\right) + \bar{\psi}\left(i\gamma^{\mu}\partial_{\mu} - m\right)\psi + e\bar{\psi}\gamma^{\mu}\psi A_{\mu}. \end{aligned} \qquad (11\text{-}7)$$

A second, alternative form for the Lagrangian

Or, finally,

$$\mathcal{L}^{1/2,1} = -\tfrac{1}{2}\left(\partial^{\nu}A^{\mu}\partial_{\nu}A_{\mu} - \partial^{\nu}A^{\mu}\partial_{\mu}A_{\nu}\right) + \bar{\psi}\left(i\gamma^{\mu}\partial_{\mu} - m\right)\psi + e\bar{\psi}\gamma^{\mu}\psi A_{\mu}. \qquad (11\text{-}8)$$

We will, for the Lorenz gauge condition (which we have assumed throughout[1]), obtain the same interaction field equations (11-4) and (11-5). You should now do Probs. 1 and 2 to prove this to yourself.

Conclusion: We can use either form (11-2) or (11-8) for our Lagrangian and have the same interaction field theory results. We will shortly find that (11-8) serves our purposes better in this chapter.

Both forms work, but 2nd is advantageous in this chapter

So, why did I stick to form (11-2) throughout the book up to this point? Because it is easier to understand at first, and I didn't want to complicate things for you at an early learning stage. Also, you could not appreciate the advantages of using the second, more complicated form (11-8) until we got to this chapter. Additionally, you will become more adept at the theory by spending time with both formulations, which we do in this book.

Note, in passing, that in the alternative form of (11-8), we have, in terms of the electric and magnetic fields,

$$-\tfrac{1}{4}F^{\mu\nu}F_{\mu\nu} = -\tfrac{1}{2}\left(-\mathbf{E}^2 + \mathbf{B}^2\right). \qquad (11\text{-}9)$$

11.1.3 New Symbolism

From here on out, we will use the streamlined symbol $\mathcal{L}$ for the full QED Lagrangian, which we have until here designated by $\mathcal{L}^{1/2,1}$. We will use $\mathcal{L}_0$ for the free part of the full QED $\mathcal{L}$; and $\mathcal{L}_I$ for the interaction part.

Hereafter, symbol $\mathcal{L}$ means $\mathcal{L}^{1/2,1}$

---

[1] And which, for these purposes, is effectively the same as the Gupta-Bleuler weak Lorenz condition of Chap. 5.

11.2 External Symmetry for Interacting Fields

11.2.1 Lorentz Symmetry of Full $\mathcal{L}$

Note that both of the full QED Lagrangian forms above are invariant under a Lorentz transformation. We can see this the short-hand way by just recalling from Chap. 6 that $\bar{\psi}\gamma^\mu\psi$ behaves under Lorentz transformation like a 4-vector. Hence every term in either (11-2) or (11-8) is a Lorentz scalar (world scalar), since a contraction of two 4-vectors is a Lorentz scalar. Such scalars are invariant in form under Lorentz transformations, and so (where primes indicate another coordinate system obtained from the first system via Lorentz transformation and x is shorthand for 4D position [usually x^α])

Full Lagrangian $\mathcal{L}$ is Lorentz invariant

$$\mathcal{L}\big(\bar{\psi}(x),\psi(x),A_\mu(x)\big) = \mathcal{L}\big(\bar{\psi}'(x'),\psi'(x'),A_\mu'(x')\big). \qquad (11\text{-}10)$$

$\mathcal{L}$ is symmetric under Lorentz transformation. It has external symmetry.

If you want to spend the time, you can prove this the long way by applying the same Lorentz transformation to every factor of each term in (11-8) (or (11-2)) that has a coordinate μ or ν.

11.2.2 Lorentz Symmetry of Laws of Nature for Interactions

Using the primed form of (11-10) in the Euler-Lagrange equation (11-3) having primed fields and spacetime coordinates, we would obtain the same field equations (11-4) and (11-5), except that all fields and coordinates would be primed. Hence, the interaction field equations (the laws of nature for interactions) would take the same form in any two coordinate systems obtained from one another via Lorentz transformation.

So, interaction laws of nature (field equations) are Lorentz invariant

The interaction field equations, like the Lagrangian, have external symmetry.

11.2.3 External Symmetry of the Interaction Probability

In a good theory, the probability of an interaction occurring should not vary with the observer measuring it. That is, the transition amplitude (whose absolute value squared is the probability) should remain invariant under Lorentz transformation.

Note that like every term in (11-8) (and (11-2)), $\mathcal{L}_I$ (which is the same in both (11-2) and (11-8)) is itself invariant under Lorentz transformation. It takes the same form with fields and coordinates primed, as with them unprimed. Thus, $\mathcal{H}_I = -\mathcal{L}_I$ is also invariant. The S operator is a function of $\mathcal{H}_I$,

i.e., $S = e^{-i\int_{-\infty}^{\infty}\mathcal{H}_I^I d^4x}$, so S is also invariant (d^4x is a Lorentz scalar). Thus, where primes on states indicate the same initial and final states as in the unprimed frame but seen in the primed frame (though the same state will be seen with different energy and 3-momenta from a different Lorentz frame), and Λ indicates a generic Lorentz transformation representation,

$$S_{fi} = \langle f|S|i\rangle = \langle f|S_{fi}|f\rangle = S_{fi}\langle f|f\rangle$$

$$S_{fi} = \langle f|\underbrace{\Lambda^{-1}\Lambda}_{I} S \underbrace{\Lambda^{-1}\Lambda}_{I}|i\rangle = \langle f|\Lambda^{-1}\underbrace{\Lambda S \Lambda^{-1}}_{S'=S}\Lambda|i\rangle = \langle f'|S_{f'i'}|f'\rangle = S_{f'i'}\langle f'|f'\rangle = S_{f'i'}. \qquad (11\text{-}11)$$

Interaction probability is Lorentz invariant

In the above, $S' = S$ means the operator S' has the same form in terms of primed fields as S has for unprimed fields. Thus S' creates and destroys the same particle states in the ket as S does. Each such particle state simply looks different (different 3-momentum and energy) in the primed and unprimed systems.

So, the transition amplitude S_{fi} for a particular initial multi-particle state scattering into a particular final multi-particle state is the same as seen from two different Lorentz frames. Thus, $|S_{fi}|^2$ is the same as well. The probability is invariant under Lorentz transformation. It has external symmetry.

In summary,

Why probability is Lorentz invariant

$$\mathcal{L}_I \text{ sym} \rightarrow \mathcal{H}_I \text{ sym} \rightarrow S \text{ sym} \rightarrow S_{fi} \text{ sym} \rightarrow |S_{fi}|^2 \text{ sym.}$$

11.3 Internal Symmetry and Conservation for Interactions

11.3.1 Review of Noether's Theorem Applied to Free Fields

For Free Scalar Field

Recall from Chap. 6, that the free scalar Lagrangian was invariant under a transformation in phase. That is, for the transformation

Review of Noether's theorem for free fields

$$\phi \rightarrow \phi' = e^{-i\alpha}\phi \quad \text{(implies also } \phi^\dagger \rightarrow \phi'^\dagger = \phi^\dagger e^{i\alpha}) \tag{11-12}$$

where α is a real constant, the Lagrangian (with μ^2 completely different from the index μ)

For scalars

$$\mathcal{L}_0^0\left(\phi^\dagger, \phi, \phi^\dagger_{,\mu}, \phi^{,\mu}\right) = \phi^\dagger_{,\mu}\phi^{,\mu} - \mu^2\phi^\dagger\phi \tag{11-13}$$

remains the same function of primed fields as it was of unprimed fields,

$$\mathcal{L}_0^0\left(\phi^\dagger, \phi, \phi^\dagger_{,\mu}, \phi^{,\mu}\right) = \mathcal{L}_0^0\left(\phi'^\dagger, \phi', \phi'^\dagger_{,\mu}, \phi'^{,\mu}\right). \tag{11-14}$$

We proved this in Chap. 6, but it should be relatively easy for you to do it again here. So the Lagrangian (11-13) is symmetric under the transformation (11-12).

We then used <u>Noether's theorem</u>, re-stated below for convenience.

If the Lagrangian density $\mathcal{L}(\phi^r, \phi^r_{,\mu})$ is symmetric in form with respect to a transformation in ϕ^r that is a function of a parameter α, i.e., $\phi^r(x^\eta) \rightarrow \phi^r(x^\eta, \alpha)$, then the four current

$$j^\mu\left(\phi^r, \phi^r_{,\nu}\right) = \frac{\partial\mathcal{L}}{\partial\phi^r_{,\mu}}\frac{\partial\phi^r}{\partial\alpha} \quad \text{(sum on } r) \tag{11-15}$$

(using $\phi^r(x^\eta, \alpha)$) has zero four-divergence, $\partial_\mu j^\mu = 0$. Thus, its zeroth component j^0 integrated over all space is conserved, as is $q_\phi j^0$ integrated over all space, where q_ϕ is a constant, which can be taken as the charge on a single scalar particle.

Applying Noether's theorem with $\phi^{r=1} = \phi$ and $\phi^{r=2} = \phi^\dagger$ to (11-12) and (11-13), we found

$$j_\phi^\mu\left(\phi^r, \phi^{r,\mu}\right) = i\left(\phi^{,\mu}(x^\eta)\phi^\dagger(x^\eta) - \phi^{\dagger,\mu}(x^\eta)\phi(x^\eta)\right) \quad \text{total charge } Q_\phi = q_\phi\int j^0 d^3x \quad \text{conserved}. \tag{11-16}$$

For Free Dirac Fermion Fields

In Chap. 6, if you did Prob. 12, you showed that under the transformation

For Dirac spinor fields

$$\psi \rightarrow \psi' = e^{-i\alpha}\psi \quad \text{(implies also } \overline{\psi} \rightarrow \overline{\psi}' = \overline{\psi}e^{i\alpha}), \tag{11-17}$$

the free Dirac Lagrangian is symmetric, i.e.,

$$\mathcal{L}_0^{1/2} = \overline{\psi}\left(i\partial\!\!\!/ - m\right)\psi == \overline{\psi}'\left(i\partial\!\!\!/ - m\right)\psi', \tag{11-18}$$

and thus, via Noether's theorem with $\phi^{r=1} = \psi$, $\phi^{r=2} = \overline{\psi}$, and $-e$ the charge on an electron,

$$j_\psi^\mu = (\rho, \mathbf{j}) = \overline{\psi}\gamma^\mu\psi \quad \text{total charge conserved, since } Q_\psi = -e\int j^0 d^3x \quad \text{conserved}. \tag{11-19}$$

For Free Photon Fields

In Chap. 6, if you did Prob. 13, you showed that under the transformation

For photons

$$A_\mu \rightarrow A'_\mu = e^{-i\alpha}A_\mu, \tag{11-20}$$

the free electromagnetic Lagrangian of Chap. 6,

$$\mathcal{L}_0^1 = -\tfrac{1}{2}\left(\partial_\nu A_\mu\right)\left(\partial^\nu A^\mu\right) \tag{11-21}$$

is not symmetric, and there is no 4-current for photons, i.e.,

$$j_{A_\mu}^\mu = 0 \quad \rightarrow \quad \text{total charge } Q_{A_\mu} = 0 \quad \text{conserved}. \tag{11-22}$$

Similar results can be shown for the free e/m Lagrangian in (11-8). (See Prob. 3.)

Aside: Comment on Gauges

Note that the internal transformations (11-12) and (11-17) are gauge transformations because we transformed the fields, yet all the things we can measure remain the same. We can't measure the actual field, as we've noted earlier (Chap. 3, pg. 79, and Chap. 7, pgs. 189-190). The field equations, the Lagrangian, and the Hamiltonian don't change, and thus the number, energy, 3-momentum, charge operators we derive from them don't change under the transformation either. The latter operators are observables. They have expectation values for any given (multi-particle, in general) state.

Internal symmetries are gauge symmetries

The fields ϕ, ψ, A_μ are gauge fields. Each different α we choose for ϕ or ψ is a different gauge, and recall, we can use different gauges for A_μ, too (though we are working in the Lorenz gauge throughout this book.) Thus, we have gauge invariance, or gauge symmetry.

11.3.2 Noether's Theorem Applied to Interactions

Deriving the Interaction Four-Current and Conserved Quantity

The derivation in Chap. 6 of Noether's theorem (11-15) was for a very general case, and not restricted solely to free fields. Thus, we can apply it to the QED (full) Lagrangian, including both free and interacting terms for Dirac spinor fields and photon fields.

Noether's theorem also applies to interactions

We start with the transformation (11-17) and the QED $\mathcal{L}$ (11-2), or (alternatively) (11-8),

$$\mathcal{L} = \mathcal{L}^{1/2,1} = -\tfrac{1}{2}\left(\partial_\nu A_\mu\right)\left(\partial^\nu A^\mu\right) + \bar{\psi}\left(i\gamma^\mu \partial_\mu - m\right)\psi + e\bar{\psi}\gamma^\mu\psi A_\mu$$

$$\text{or} = -\tfrac{1}{2}\underbrace{\left(\partial^\nu A^\mu \partial_\nu A_\mu - \partial^\nu A^\mu \partial_\mu A_\nu\right)}_{\tfrac{1}{4}F^{\mu\nu}F_{\mu\nu}} + \bar{\psi}\left(i\gamma^\mu \partial_\mu - m\right)\psi + e\bar{\psi}\gamma^\mu\psi A_\mu. \tag{11-23}$$

Full Lagrangian is symmetric under internal transformation

By doing Prob. 4, you can show that (11-23) (both versions) is symmetric under (11-17). (This should be relatively easy, because the transformation (11-17) does not affect any factors of A_μ or its derivatives in (11-23).)

Now, we need only apply Noether's theorem (11-15), with $\phi^{r=1}= \psi$, $\phi^{r=2} =\bar{\psi}$, and $\phi^{r=3} = A^\nu$ to find a conserved quantity. So, our conserved four-current is

So, there is a conserved current

$$j^\mu = \frac{\partial \mathcal{L}}{\partial \psi_{,\mu}}\frac{\partial \psi}{\partial \alpha} + \frac{\partial \mathcal{L}}{\partial \bar{\psi}_{,\mu}}\frac{\partial \bar{\psi}}{\partial \alpha} + \frac{\partial \mathcal{L}}{\partial A^\nu_{,\mu}}\frac{\partial A^\nu}{\partial \alpha}, \tag{11-24}$$

and with (11-23), and (11-17),

$$\frac{\partial \mathcal{L}}{\partial \psi_{,\mu}} = \bar{\psi}i\gamma^\mu \qquad \frac{\partial \psi}{\partial \alpha} = -i\psi \qquad \frac{\partial \mathcal{L}}{\partial \bar{\psi}_{,\mu}} = 0 \qquad \frac{\partial A^\nu}{\partial \alpha} = 0, \tag{11-25}$$

$$j^\mu = \bar{\psi}\gamma^\mu\psi \qquad Q' = -e\int j^0 d^3x \ \text{ conserved}, \tag{11-26}$$

That current, and its associated charge are the same as in free field theory

the same result as we had for the free Dirac field (11-19). This should not be too surprising as any interactions Dirac particles have are with chargeless photons, which cannot possibly change the total charge.

Gauge Symmetry for Interactions

Just as noted above for free fields, we have a gauge symmetry here for interacting fields. ψ and A_μ are the gauge fields. Different α represent different gauges. The QED Lagrangian is symmetric under the gauge transformation and thus so are the Hamiltonian, the field equations, and all the operators derived from them.

Interacting fields, like free fields, are gauge fields

The Charge Operator in Terms of Number Operators

Q' in (11-26) is an operator of the same form as we had in the free field theory of Dirac particles. In the interaction picture we use free fields, so we will derive the same final result for the charge operator, in terms of number operators, as we had in Chap. 4. (See (4-104), pg. 113.) This is

$$Q = -e\sum_{\mathbf{p},r}\left(N_r\left(\mathbf{p}\right) - \bar{N}_r\left(\mathbf{p}\right)\right), \tag{11-27}$$

Charge operator Q same as in free theory

which is constant in time due to (11-26).

How the Q Operator Works During Interactions

Many QFT courses/texts get to (11-26) and conclude that because of it, charge is conserved in interactions. However, what is measured in interactions is a function not only of the charge operator, but of the states, as well. We need to be concerned with the change in time of the total charge eigenvalue of a (generally multi-particle) state, which is what one would measure.

Similar to what we discussed in Chap. 6 (Sect. 6.5.5, pg. 175), when we say charge is conserved, it really means the multi-particle state that we begin with keeps the same total charge as time evolves. The individual particles themselves may mutate into other types of particles, but the total charge of all the particles remains the same.

Thus, we ask how the eigenvalue of Q changes as a state evolves (i.e., as particles change from those in $|i\rangle$ to those in $|f\rangle$). That is, for the transition

$$Q|i\rangle = q_{i\,tot}|i\rangle \xrightarrow{\text{state transition}} Q|f\rangle = QS|i\rangle = q_{f\,tot}|f\rangle \,, \qquad (11\text{-}28)$$

Q acting on initial and final states yields same charge

does $q_{i\,tot} = q_{f\,tot}$?

The answer lies in the $QS|i\rangle$ expression above. We know first, that the operator Q doesn't change during the transition because it is conserved. If Q commutes with S, then

$$Q|f\rangle = q_{f\,tot}|f\rangle = QS|i\rangle = SQ|i\rangle = Sq_{i\,tot}|i\rangle = q_{i\,tot}S|i\rangle = q_{i\,tot}|f\rangle \rightarrow q_{f\,tot} = q_{itot}\,, \quad (11\text{-}29)$$

So measured charge is conserved during interactions

and total charge remains unchanged during the transition.

It is not immediately obvious that Q and S commute. The proof is fairly extensive. We show they do commute in the Appendix, and thus, total charge is conserved.

Of course, we already know, from our experience with Feynman diagrams in earlier chapters, that charge is conserved during interactions. In all vertices in Feynman diagrams, the events where particles mutate into other particles, we saw that particles only evolve into other particles in ways that conserve charge. That is, we didn't have any vertices where charge is not conserved, such as two electrons being created from an incoming photon, or a positron and a photon being created from an incoming photon. None of these, or similar charge conservation violating, vertices resulted from our theory. But in (11-29), we prove it explicitly.

We already knew this from Feynman diagrams, but here formally proven

If you really want to reinforce your appreciation of this point, do Prob. 5, which asks you to show the time rate of change of the expectation value for charge Q is zero.

11.4 Global vs Local Transformations and Symmetries

Up to this point in this book, we have dealt exclusively with simpler kinds of symmetries in which any change introduced to a field via a transformation in the field entailed the same change everywhere in spacetime. For example, the internal transformation (11-17) changed the field by the phase angle α, and such change in phase was the same for the field everywhere in space, for all time. As another example, the external Lorentz transformation transformed the field by the same boost velocity everywhere at all points in space, for all time. Such symmetries are called global symmetries.

Global transformation = same change everywhere in spacetime

In contrast, we could have an internal transformation where $\alpha = \alpha(x^\mu)$, where the phase angle change is not constant but a function of 3D position and time,

$$\psi \rightarrow \psi' = e^{-i\alpha(x^\mu)}\psi \,. \qquad (11\text{-}30)$$

Local transformation = different change at different points in spacetime

If the Lagrangian is invariant under such a transformation, we say it has local symmetry. The (local) transformation has symmetry at local points everywhere, but it is not the same transformation at every point.

Similarly, for the external Lorentz transformation, one could imagine it being different at various points in a fluid, where we want to transform to the rest coordinate system at every point in the fluid. The velocity at different points of the fluid would be different, so our transformation to the local rest frame would have a different boost velocity at each such point. This would be an example of an external local transformation. We will not deal further in this book with external (Lorentz) local transformations.

Local or global change can be internal or external

Recall that if the Lagrangian is symmetric under a given transformation, the transformation can be referred to as a gauge transformation, because nothing we measure will change, even though the representation of the underlying (gauge) field changes.

If $\mathcal{L}$ invariant under any of these, it is symmetric, and we have a gauge theory

We are about to investigate what it means for the Lagrangian to be symmetric under a local symmetry. We will find it plays a profound role in interaction QFT, i.e., in the gauge theory of interacting quantum fields.

Wholeness Chart 11-1. Types of Transformations

	External Transformation		Internal Transformation	
	Global	**Local**	**Global**	**Local**
Where in this book	All external transformations up to this section	Not treated	All internal transformations up to this section	Material following this section
Characterized by	Lorentz, same boost at all x^μ	Lorentz, different boost at different x^μ	Change in field, same at all x^μ	Change in field, different at different x^μ
Example	$A_\nu \to A'_\nu = \Lambda^\eta_\nu A_\eta$	$A_\nu \to A'_\nu = \Lambda^\eta_\nu(x^\mu) A_\eta$	$\psi \to \psi' = e^{-i\alpha}\psi$	$\psi \to \psi' = e^{-i\alpha(x^\mu)}\psi$
Symmetric under above?	If $\mathcal{L}$ invariant	If $\mathcal{L}$ invariant	If $\mathcal{L}$ invariant	If $\mathcal{L}$ invariant

11.5 Local Symmetry and Interaction Theory

11.5.1 Local Transformation of the Lagrangian

We saw in Chap. 6 that the QED Lagrangian was symmetric under the global internal transformation (11-17) (next to last column example of Wholeness Chart 11-1) with α = constant. This was true for the free Lagrangian $\mathcal{L}_0$ (see (11-18)) and the full (including interactions) Lagrangian $\mathcal{L}$ (see (11-23), which is repeated below), as we saw in Prob. 4.

Investigating if local internal transformations yield symmetric Lagrangian

$$\mathcal{L} = \mathcal{L}^{1/2,1} = \underbrace{-\frac{1}{2}\left(\partial_\nu A_\mu\right)\left(\partial^\nu A^\mu\right) + \overline{\psi}\left(i\gamma^\nu \partial_\nu - m\right)\psi}_{\mathcal{L}_0} + \underbrace{e\overline{\psi}\gamma^\nu\psi A_\nu}_{\mathcal{L}_I}$$

$$\text{or} = \underbrace{-\frac{1}{2}\left(\partial^\nu A^\mu \partial_\nu A_\mu - \partial^\nu A^\mu \partial_\mu A_\nu\right)}_{\frac{1}{4}F^{\mu\nu}F_{\mu\nu}} + \overline{\psi}\left(i\gamma^\nu \partial_\nu - m\right)\psi + e\overline{\psi}\gamma^\nu\psi A_\nu. \tag{11-31}$$

<u>Question</u>: Under the local internal transformation (11-30) (last column example in Wholeness Chart 11-1) is the Lagrangian symmetric?

<u>Part 1</u>: For $\mathcal{L}_0$?

Looking at just the free Lagrangian part of (11-31), both choices for the free photon part have no ψ field components, so those parts are unchanged under (11-30). So let's look at the fermionic part.

$$\mathcal{L}_0^{1/2} = \overline{\psi}\left(i\gamma^\nu \partial_\nu - m\right)\psi \xrightarrow{\psi \to e^{-i\alpha(x^\mu)}\psi} \overline{\psi}e^{i\alpha(x^\mu)}\left(i\gamma^\nu \partial_\nu - m\right)\psi e^{-i\alpha(x^\mu)}$$

$$= \overline{\psi}e^{i\alpha(x^\mu)}i\gamma^\nu\left(\partial_\nu\psi\right)e^{-i\alpha(x^\mu)} + \underbrace{\overline{\psi}e^{i\alpha(x^\mu)}i\gamma^\nu\psi\partial_\nu e^{-i\alpha(x^\mu)}}_{} - \overline{\psi}e^{i\alpha(x^\mu)}m\psi e^{-i\alpha(x^\mu)} \tag{11-32}$$

$$= \underbrace{\overline{\psi}\left(i\gamma^\nu \partial_\nu - m\right)\psi}_{\mathcal{L}_0^{1/2}} + \overline{\psi}\gamma^\nu\psi\partial_\nu\alpha(x^\mu) \neq \mathcal{L}_0^{1/2}.$$

Under $\psi \to e^{-i\alpha(x^\mu)}\psi$, free Lagrangian not symmetric

The free Lagrangian $\mathcal{L}_0$ is not symmetric under this local transformation.

<u>Part 2</u>: How about for $\mathcal{L} = \mathcal{L}_0 + \mathcal{L}_I$?

Let's look at the full Lagrangian (11-31) for the same transformation. Again, we know the photon part in the $\mathcal{L}_0$ part of $\mathcal{L}$ is invariant because it contains no components of ψ. So, we will look at the other parts of $\mathcal{L}$. We can use our results from (11-32) directly as part of the relation below.

$$\mathcal{L}_0^{1/2} + \mathcal{L}_I = \bar{\psi}\left(i\gamma^\nu \partial_\nu - m\right)\psi + e\bar{\psi}\gamma^\nu \psi A_\nu \xrightarrow{\ \psi \to e^{-i\alpha(x^\mu)}\psi\ }$$

$$\bar{\psi}\left(i\gamma^\nu \partial_\nu - m\right)\psi + \bar{\psi}\gamma^\nu \psi \partial_\nu \alpha(x^\mu) + e\bar{\psi}e^{i\alpha(x^\mu)}\gamma^\nu e^{-i\alpha(x^\mu)}\psi A_\nu \qquad (11\text{-}33)$$

$$= \underbrace{\bar{\psi}\left(i\gamma^\nu \partial_\nu - m\right)\psi}_{\mathcal{L}_0^{1/2}} + \bar{\psi}\gamma^\nu \psi \partial_\nu \alpha(x^\mu) + \underbrace{e\bar{\psi}\gamma^\nu \psi A_\nu}_{\mathcal{L}_I} \neq \mathcal{L}_0^{1/2} + \mathcal{L}_I .$$

Under $\psi \to e^{-i\alpha(x^\mu)}\psi$, full Lagrangian not symmetric

The full Lagrangian $\mathcal{L}$ is not symmetric under this local transformation, either.

Question: Can we, in addition to the transformation on ψ, also transform A_μ in some way such that under both transformations, the full Lagrangian is symmetric?

How about both $\psi \to e^{-i\alpha(x^\mu)}\psi$ and $A_\nu \to A_\nu - (1/e)\partial_\nu\alpha(x^\mu)$?

Answer: Yes. If we take a clue from classical theory and transform the photon field in the manner we did there (Wholeness Chart 5-1, pg. 141), where Maxwell's equations (and thus the classical $\mathcal{L}$) were symmetric under $A_\nu \to A_\nu + \partial_\nu f(x^\mu)$. If we take that gauge transformation with the specific form $f = -\alpha/e$, then

$$A_\nu \to A_\nu' = A_\nu - \frac{1}{e}\partial_\nu \alpha(x^\mu), \qquad (11\text{-}34)$$

and as we will see, (11-34) will make one of our two alternative forms for $\mathcal{L}$ symmetric.

Proof:

Taking our full Lagrangian with the first form of the free photon Lagrangian of (11-31), we find (changing our dummy indices to avoid confusion with the μ index in $\alpha(x^\mu)$)

$$\mathcal{L} = \mathcal{L}_0^1 + \mathcal{L}_0^{1/2} + \mathcal{L}_I = -\frac{1}{2}\left(\partial_\beta A_\nu\right)\left(\partial^\beta A^\nu\right) + \bar{\psi}\left(i\gamma^\nu \partial_\nu - m\right)\psi + e\bar{\psi}\gamma^\nu\psi A_\nu$$

$$\xrightarrow[A_\nu \to A_\nu - \frac{1}{e}\partial_\nu\alpha(x^\mu)]{\psi \to e^{-i\alpha(x^\mu)}\psi} \underbrace{-\frac{1}{2}\left(\partial_\beta A_\nu\right)\left(\partial^\beta A^\nu\right)}_{\mathcal{L}_0^1} + \frac{1}{2e}\left(\partial_\beta \partial_\nu \alpha(x^\mu)\right)\partial^\beta A^\nu + \frac{1}{2e}\left(\partial_\beta A_\nu\right)\left(\partial^\beta \partial^\nu \alpha(x^\mu)\right)$$

Under both, first form of full Lagrangian not symmetric

$$\qquad (11\text{-}35)$$

$$-\frac{1}{2e^2}\left(\partial_\beta \partial_\nu \alpha(x^\mu)\right)\left(\partial^\beta \partial^\nu \alpha(x^\mu)\right) + \underbrace{\bar{\psi}\left(i\gamma^\nu \partial_\nu - m\right)\psi}_{\mathcal{L}_0^{1/2}} + \underbrace{\bar{\psi}\gamma^\nu\psi\partial_\nu\alpha(x^\mu)}_{\text{cancels}}$$

$$+ \underbrace{e\bar{\psi}\gamma^\nu\psi A_\nu}_{\mathcal{L}_I} - \underbrace{\bar{\psi}\gamma^\nu\psi\partial_\nu\alpha(x^\mu)}_{\text{cancels}} \neq \mathcal{L} .$$

So, the first form of our Lagrangian is not symmetric.

But, if we take our full Lagrangian with the second form of the free photon Lagrangian of (11-31), we find

$$\mathcal{L} = \mathcal{L}_0^1 + \mathcal{L}_0^{1/2} + \mathcal{L}_I = \underbrace{-\frac{1}{2}\left(\partial_\beta A_\nu \partial^\beta A^\nu - \partial_\nu A_\beta \partial^\beta A^\nu\right)}_{-\frac{1}{4}F_{\nu\beta}F^{\nu\beta} = \mathcal{L}_0^1} + \bar{\psi}\left(i\gamma^\nu \partial_\nu - m\right)\psi + e\bar{\psi}\gamma^\nu\psi A_\nu$$

$$\xrightarrow[A_\nu \to A_\nu - \frac{1}{e}\partial_\nu\alpha(x^\mu)]{\psi \to e^{-i\alpha(x^\mu)}\psi} \underbrace{-\frac{1}{2}\left(\partial_\beta A_\nu \partial^\beta A^\nu - \partial_\nu A_\beta \partial^\beta A^\nu\right)}_{\mathcal{L}_0^1}$$

Under both, second form of full Lagrangian is symmetric

$$+\underbrace{\frac{1}{2e}\left(\partial_\beta\partial_\nu\alpha(x^\mu)\right)\partial^\beta A^\nu}_{\text{cancels }\downarrow} +\underbrace{\frac{1}{2e}\left(\partial_\beta A_\nu\right)\left(\partial^\beta\partial^\nu\alpha(x^\mu)\right)}_{\text{cancels }\downarrow} -\underbrace{\frac{1}{2e^2}\left(\partial_\beta\partial_\nu\alpha(x^\mu)\right)\left(\partial^\beta\partial^\nu\alpha(x^\mu)\right)}_{\text{cancels }\downarrow} \qquad (11\text{-}36)$$

$$-\frac{1}{2e}\left(\partial_\nu\partial_\beta\alpha(x^\mu)\right)\partial^\beta A^\nu -\frac{1}{2e}\left(\partial_\nu A_\beta\right)\left(\partial^\beta\partial^\nu\alpha(x^\mu)\right) +\frac{1}{2e^2}\left(\partial_\nu\partial_\beta\alpha(x^\mu)\right)\left(\partial^\beta\partial^\nu\alpha(x^\mu)\right)$$

$$+\underbrace{\bar{\psi}\left(i\gamma^\nu\partial_\nu - m\right)\psi}_{\mathcal{L}_0^{1/2}} + \underbrace{\bar{\psi}\gamma^\nu\psi\partial_\nu\alpha(x^\mu)}_{\text{cancels term after next}} + \underbrace{e\bar{\psi}\gamma^\nu\psi A_\nu}_{\mathcal{L}_I} - \underbrace{\bar{\psi}\gamma^\nu\psi\partial_\nu\alpha(x^\mu)}_{\text{cancels term before prior}} = \mathcal{L} .$$

Bottom line: By carrying out the set of local transformations $\psi \to \psi' = e^{-i\alpha(x^\mu)}\psi$ and $A_\nu \to A_\nu' = A_\nu - (1/e)\partial_\nu\alpha(x^\mu)$, and using $\mathcal{L}_0^1 = -\frac{1}{4}F_{\nu\beta}F^{\nu\beta}$ for the free photon field Lagrangian (second form above), the full QED Lagrangian $\mathcal{L}$ remains invariant.

This is a primary reason for preferring the second form (11-8) for the interaction Lagrangian. It gives us a symmetric interaction theory, which has great significance in QED (as we see in the next section) and further important ramifications for more advanced areas of QFT, such as weak and strong interactions.

A good reason, as we will see, to prefer the second form of $\mathcal{L}$

11.5.2 The Meaning of Local Symmetry for Interaction Theory

The Free Lagrangian and the Set of Both Local Transformations

Note that using the set of local transformations $\psi \to \psi' = e^{-i\alpha(x^\mu)}\psi$ and $A_\nu \to A'_\nu = A_\nu - (1/e)\partial_\nu\alpha(x^\mu)$, while leaving the second form of the full QED Lagrangian $\mathcal{L}$ invariant, does not do so for the free Lagrangian. To see this for the second form of the Lagrangian, drop the interaction terms in (11-36), and we get

$$\mathcal{L}_0 = \mathcal{L}_0^1 + \mathcal{L}_0^{1/2} \xrightarrow[A_\nu \to A_\nu - \frac{1}{e}\partial_\nu\alpha(x^\mu)]{\psi \to e^{-i\alpha(x^\mu)}\psi}$$

$$\underbrace{-\frac{1}{2}\left(\partial_\beta A_\nu \partial^\beta A^\nu - \partial_\nu A_\beta \partial^\beta A^\nu\right)}_{\mathcal{L}_0^1} + \underbrace{\bar{\psi}\left(i\gamma^\nu\partial_\nu - m\right)\psi + \bar{\psi}\gamma^\nu\psi\partial_\nu\alpha(x^\mu)}_{\mathcal{L}_0^{1/2}} \neq \mathcal{L}_0. \quad (11\text{-}37)$$

Under both, neither form of free Lagrangian is symmetric

By doing Prob. 6, you can show that the first form of the Lagrangian with $\mathcal{L}_0^1 = -\frac{1}{2}\left(\partial_\beta A_\nu\right)\left(\partial^\beta A^\nu\right)$ is also not symmetric under the local transformation set.

Summary

Wholeness Chart 11-2. Summary of Global and Local Internal Symmetry for $\mathcal{L}$ and $\mathcal{L}_0$

Summary of all this to help keep it straight

	1st Form of $\mathcal{L}_0{}^1$		2nd Form of $\mathcal{L}_0{}^1$	
	$\mathcal{L}_0$	$\mathcal{L}$	$\mathcal{L}_0$	$\mathcal{L}$
Global transformation	$\psi \to \psi' = e^{-i\alpha}\psi$			
Symmetric?	Yes	Yes	Yes	Yes
Local transformation	$\psi \to \psi' = e^{-i\alpha(x^\mu)}\psi$			
Symmetric?	No	No	No	No
Local transformation set	$\psi \to \psi' = e^{-i\alpha(x^\mu)}\psi$ and $A_\nu \to A_\nu - (1/e)\partial_\nu\alpha(x^\mu)$			
Symmetric?	No	No	No	Yes

Note: For global transformation set, α = constant in "Local transformation set" above, and we get the same as "Global transformation" row above

Significance of Local Symmetry

It turns out that the only way we can have a locally symmetric theory is if that theory includes all of the criteria of the last block in the RH column above. In particular, it must use the particular transformations shown, and no others. There appears to be no other way.

Thus, if we require our theory to have local symmetry, it must

1) use the second form for our Lagrangian with the $F_{\mu\nu}F^{\mu\nu}$ term for free photons,
2) use the particular local transformation set for ψ and A_ν shown above, and
3) use the full Lagrangian, including the specific interaction term of form $e\bar{\psi}\gamma^\nu\psi A_\nu$.

This has led to the following general rule for QFT.

If we require local symmetry, it forces us to add a particular interaction term to $\mathcal{L}_0$

That particular interaction term is exactly what is needed for a correct theory of interactions

General rule for QFT: If we start with the free Lagrangian and require it to be locally symmetric, then it can only be so if we add to it the particular interaction term(s) that actually describe(s) interactions in the real world.

This is a general rule that can be used not only in QED, but in more advanced theories. For examples, in weak and strong interaction quantum field theories, one can insist on local symmetry and then deduce the particular form the interaction terms in $\mathcal{L}$ must have to yield that symmetry. This can be a guide for developing correct theories of nature.

Alternative Way to View Significance of Local Symmetry

In the above sub-section, we concluded that imposing local symmetry on our theory forces us to insert the correct interaction term into the Lagrangian. Conversely, we can look at it in reverse.

We found a correct interaction theory that describes nature in earlier chapters (plus the next four chapters on renormalization). When we investigated that theory, we found it was symmetric (under the local transformations (11-30) and (11-34)). We then must conclude that nature herself is symmetric. It is her "nature", so to speak.

Conversely, we could say instead that nature loves symmetry, because it arises naturally in all our theories

> **Alternative general rule:** Nature just seems to love symmetry. Without looking for it, symmetry arises naturally in every theory we have.

Second Form for the Lagrangian Used Henceforth

Since the second form of the Lagrangian using the $F_{\mu\nu}F^{\mu\nu}$ term works for local symmetry, but the first form doesn't, we will henceforth use only that second form.

From here on, we use the 2nd form of $\mathcal{L}$ exclusively

11.5.3 Things to Be Aware of

Note the following.

Things to keep in mind

- Global symmetry is a special case of local symmetry where α = constant.

- Noether's theorem applies locally, as well as globally, as nothing in our derivation of it restricted our symmetry transformation to global cases.

Noether good for local symmetries, too, and yields same current

- Thus, we can derive a conserved current (the same conserved current) from global symmetry or local symmetry using the second form of the full Lagrangian.

- Symmetry under transformation of one or more fields means the fields are gauge fields (different α in the QED case, means different gauges).

- You may hear the term <u>Lie group</u> in this context. A Lie group is the set of all possible continuous transformations of the fields in a gauge theory. In our case, for the field ψ, it is the set of $e^{-i\alpha(x^\mu)}$ for all possible $\alpha(x^\mu)$. This group is called a <u>U(1) group</u>, where the U means unitary and the 1 means it is a 1X1 matrix transformation. (U(n) would entail an nXn matrix.)

Lie groups

- QED is generally regarded as the first, and simplest, physical gauge theory.

- The interaction term in the Lagrangian is often written in other texts as the RHS below

Interaction term commonly written in terms of 4-current

$$e\overline{\psi}\gamma^\nu\psi A_\nu = e\, j^\nu A_\nu = -s^\nu A_\nu\,. \tag{11-38}$$

- We will see in Chaps. 13, 14, and 15 that local symmetry is essential to renormalization. Without gauge invariance, we don't have a viable, finite theory. Symmetry is not just beautiful, appealing, and philosophically elegant. It is essential.

11.5.4 Local Symmetry Requires Massless Photon

Note that if the photon had mass, we would have a term in the Lagrangian like $m^2 A^\nu A_\nu$. The rest of the Lagrangian is symmetric under the local transformation shown in Wholeness Chart 11-2, pg. 295. So, for the $\mathcal{L}$ to be symmetric if the photon had mass, the photon mass term alone would have to be symmetric. But it is not, as we can see from (11-39) below.

Photon mass term not symmetric under local transformation

$$m^2 A^\nu A_\nu \xrightarrow[A_\nu \to A_\nu - \frac{1}{e}\partial_\nu\alpha(x^\mu)]{\psi \to e^{-i\alpha(x^\mu)}\psi} m^2 A^\nu A_\nu - \frac{2}{e}A^\nu\partial_\nu\alpha\left(x^\mu\right) + \frac{1}{e^2}\partial^\nu\alpha\left(x^\mu\right)\partial_\nu\alpha\left(x^\mu\right) \neq m^2 A^\nu A_\nu \,. \tag{11-39}$$

So if we insist on a symmetric Lagrangian (i.e., on gauge invariance), then the photon must be massless. We emphasize this here because the concept plays a major role in advanced QFT.

Requiring local symmetry of $\mathcal{L}$ means photon must be massless

> Gauge invariance of $\mathcal{L}$ means the photon is massless.
> Conversely: If photon had mass, $\mathcal{L}$ not gauge invariant.

11.5.5 Local Symmetry Also Yields Charge Conservation

By doing Prob. 7, you can show that under the full local transformation (11-30) and (11-34), Noether's theorem gives us the same result as (11-26). We obtained (11-26) via a global transformation on the full (free plus interaction) Lagrangian. Doing Prob. 7 illustrates that the full local symmetry transformation on the full Lagrangian yields the same conservation relation. Thus, by either the global or the local symmetry approach, we get the same conserved quantity (charge).

11.6 Minimal Substitution

Note that if we define something called the <u>gauge covariant derivative</u>[1] as

$$D_v = \partial_v - ieA_v,$$ (11-40)

then we can find the interaction Lagrangian by substituting (11-40) for ∂_v into the free Lagrangian. Specifically,

$$\mathcal{L}_0 = \mathcal{L}_0^1 + \mathcal{L}_0^{1/2} = \underbrace{-\frac{1}{2}\left(\partial_\beta A_v \partial^\beta A^v - \partial_v A_\beta \partial^\beta A^v\right)}_{-\frac{1}{4}F_{v\beta}F^{v\beta}=\mathcal{L}_0^1} + \bar{\psi}\left(i\gamma^v \partial_v - m\right)\psi$$

$$\xrightarrow{\partial_v \to D_v = \partial_v - ieA_v} -\frac{1}{2}\left(D_\beta A_v D^\beta A^v - D_v A_\beta D^\beta A^v\right) + \bar{\psi}\left(i\gamma^v D_v - m\right)\psi$$

$$= -\frac{1}{2}\left(\partial_\beta A_v \partial^\beta A^v - \partial_v A_\beta \partial^\beta A^v\right)$$

$$+ \frac{ie}{2}A_\beta A_v \partial^\beta A^v + \frac{ie}{2}\partial_\beta A_v A^\beta A^v \underbrace{- \frac{ie}{2}A_v A_\beta \partial^\beta A^v}_{\text{cancels 1st term at left}} \underbrace{- \frac{ie}{2}\partial_v A_\beta A^\beta A^v}_{\text{cancels 2nd term at left}}$$

$$\underbrace{- \frac{e^2}{2}A_\beta A_v A^\beta A^v}_{\text{cancels with next term}} + \frac{e^2}{2}A_v A_\beta A^\beta A^v + \bar{\psi}\left(i\gamma^v \partial_v - m\right)\psi + e\bar{\psi}\gamma^v \psi A_v$$ (11-41)

$$= \mathcal{L}_0 + \mathcal{L}_I = \mathcal{L}$$

Gauge covariant derivative defined

Minimal substitution = substituting gauge covariant derivative into $\mathcal{L}_0$ to get $\mathcal{L}$

This process of substituting the gauge covariant derivative for the usual derivative in the free Lagrangian is called <u>minimal substitution</u>.

By doing Prob. 8, you can show that carrying out minimal substitution on the first form of our free Lagrangian does not yield the correct interaction Lagrangian. This is yet another reason why the second form is preferred.

The chapter summary begins on the next page, in order to keep it on a single page.

---

[1] The term "gauge covariant derivative" is different from, and should not be confused with, the "covariant derivative" of general relativity (and more generally, tensor analysis). It is unfortunate that similar nomenclature is used for the two cases. In the QFT we are studying, spacetime is flat (Minkowski space) and we use a Lorentz metric, so the covariant derivative of GR has Christoffel symbol of zero. This is made more confusing by the common dropping of the word "gauge" in QFT when referring to the gauge covariant derivative. You often have to glean the meaning from context.

11.7 Chapter Summary

We have seen the following.

- An alternative form for the Lagrangian that is more suitable for interactions

$$\mathcal{L} = -\tfrac{1}{4} F^{\mu\nu} F_{\mu\nu} + \bar{\psi}\left(i\gamma^\mu \partial_\mu - m\right)\psi + e\bar{\psi}\gamma^\mu\psi A_\mu$$

- The (external) Lorentz symmetry of the full Lagrangian (including interaction term)
- The external symmetry of interaction probability (same probability for all observers)
- Charge conservation in interactions as a result of Noether's theorem
- Global and local symmetries for internal and external transformations

    See Wholeness Chart 11-1 summary, pg. 293
- Local symmetry applied to interaction theory

    See Wholeness Chart 11-2 summary, pg. 295
- From the above, we have two perspectives, depending on whether we want to consider the chicken or the egg to come first.

    1st perspective: Requiring local symmetry for the Lagrangian means we must add a term to the free Lagrangian (only second form works) that turns out to be the correct form for the interaction term in the full Lagrangian. Local symmetry dictates the form of the interaction.

    2nd perspective: After deriving the correct interaction theory from other principles unrelated to symmetry, we find the theory is locally symmetric. Nature seems to love symmetry.
- Requiring the QED Lagrangian to be gauge invariant restricts the photon to being massless.
- Local gauge invariance is essential for renormalizing our theory. (To be seen in future chapters.)
- Substituting the gauge covariant derivative $\partial_\nu \rightarrow D_\nu = \partial_\nu - ieA_\nu$ into the free Lagrangian (only second form works) is called minimal substitution and results in the correct form of the interaction Lagrangian.

Wholeness Chart 11-3. Summary of Symmetry Effects for Interactions

	External Symmetry in Full $\mathcal{L}$		Internal Symmetry in Full $\mathcal{L}$	
Type	Global	Local	Global	Local
Result	Laws of nature (field equations) invariant (same for all observers)	Not treated herein.	Conserved charge exists	i) Must add correct interaction term to $\mathcal{L}_0$ to make $\mathcal{L}$ symmetric → correct theory arises ii) Conserved charge exists

11.8 Appendix: Showing [Q, S] = 0

Note from our definition of the S operator,

$$S = e^{-i\int \mathcal{H}_I^I d^4x} \quad \text{with} \quad \mathcal{H}_I^I = -e\bar{\psi}\gamma^\mu\psi A_\mu \quad \text{(with time ordering in expansion)}, \quad (11\text{-}42)$$

that any operator such as Q will commute with S, if it commutes with $\mathcal{H}_I^I$ (because the expansion of S is a time ordered series having factors of $\mathcal{H}_I^I$ in each term of that series). Thus, we need to show that

$$Q = -e\sum_{\mathbf{p},r}\left(N_r(\mathbf{p}) - \bar{N}_r(\mathbf{p})\right) = -e\sum_{\mathbf{p},r}\left(c_r^\dagger(\mathbf{p})c_r(\mathbf{p}) - d_r^\dagger(\mathbf{p})d_r(\mathbf{p})\right) \quad (11\text{-}43)$$

commutes with

$$\mathcal{H}_I^I = -e\,\overline{\psi}\,\gamma^\mu\psi\,A_\mu = -e\sum_{\mathbf{p},r}\sqrt{\frac{m}{VE_\mathbf{p}}}\left(d_r(\mathbf{p})\overline{v}_r(\mathbf{p})e^{-ipx} + c_r^\dagger(\mathbf{p})\overline{u}_r(\mathbf{p})e^{ipx}\right)\times$$

$$\gamma^\mu\sum_{\mathbf{p'},r'}\sqrt{\frac{m}{VE_{\mathbf{p'}}}}\left(c_{r'}(\mathbf{p'})u_{r'}(\mathbf{p'})e^{-ip'x} + d_{r'}^\dagger(\mathbf{p'})v_{r'}(\mathbf{p'})e^{ip'x}\right)A_\mu. \tag{11-44}$$

In (11-44) we have not shown the explicit form for A_μ because all the creation and destruction operators within it, $a^\dagger{}_s(\mathbf{k})$ and $a_s(\mathbf{k})$, commute with all the creation and destruction operators on the RHS of (11-43). So, those factors of A_μ in the commutation $[Q, \mathcal{H}_I^I]$ will all commute and we don't have to worry about them.

To streamline everything, let's just look at one term in $[Q, \mathcal{H}_I^I]$, where all factors have the same $\mathbf{p}$ and r in all the summations of (11-43) and (11-44). Other terms will follow in parallel, or more trivial, fashion. To streamline even further, let's drop the e, square root factors, and A_μ factor, as they play no role in determining whether $[Q, \mathcal{H}_I^I] = 0$, or not. Thus, with no summation on repeated r here,

$$\left[Q, \mathcal{H}_I^I\right]\xrightarrow[\text{no sum on }r]{\text{one term of summations}}$$

$$\propto \Big(\underbrace{c_r^\dagger(\mathbf{p})c_r(\mathbf{p})}_{N_r(\mathbf{p})} - \underbrace{d_r^\dagger(\mathbf{p})d_r(\mathbf{p})}_{\overline{N}_r(\mathbf{p})}\Big)\underbrace{\left(d_r(\mathbf{p})\overline{v}_r(\mathbf{p})e^{-ipx} + c_r^\dagger(\mathbf{p})\overline{u}_r(\mathbf{p})e^{ipx}\right)}_{\text{relevant parts of }\overline{\psi}}\times$$

$$\gamma^\mu\underbrace{\left(c_r(\mathbf{p})u_r(\mathbf{p})e^{-ipx} + d_r^\dagger(\mathbf{p})v_r(\mathbf{p})e^{ipx}\right)}_{\text{relevant parts of }\psi} \tag{11-45}$$

$$-\underbrace{\left(d_r(\mathbf{p})\overline{v}_r(\mathbf{p})e^{-ipx} + c_r^\dagger(\mathbf{p})\overline{u}_r(\mathbf{p})e^{ipx}\right)}_{\text{relevant parts of }\overline{\psi}}\times$$

$$\gamma^\mu\underbrace{\left(c_r(\mathbf{p})u_r(\mathbf{p})e^{-ipx} + d_r^\dagger(\mathbf{p})v_r(\mathbf{p})e^{ipx}\right)}_{\text{relevant parts of }\psi}\Big(\underbrace{c_r^\dagger(\mathbf{p})c_r(\mathbf{p})}_{N_r(\mathbf{p})} - \underbrace{d_r^\dagger(\mathbf{p})d_r(\mathbf{p})}_{\overline{N}_r(\mathbf{p})}\Big).$$

Before carrying out the multiplication of (11-45), recall that

$$\left[c_r(\mathbf{p}), c_r^\dagger(\mathbf{p})\right]_+ = 1 \qquad \left[c_r(\mathbf{p}), c_r(\mathbf{p})\right]_+ = \left[c_r^\dagger(\mathbf{p}), c_r^\dagger(\mathbf{p})\right]_+ = 0$$

$$\left[c_r(\mathbf{p}), d_r(\mathbf{p})\right]_+ = \left[c_r^\dagger(\mathbf{p}), d_r^\dagger(\mathbf{p})\right]_+ = \left[c_r(\mathbf{p}), d_r^\dagger(\mathbf{p})\right]_+ = \left[c_r^\dagger(\mathbf{p}), d_r(\mathbf{p})\right]_+ = 0. \tag{11-46}$$

From the RHS of the top row of (11-46), as we saw in Chap. 4 ((4-84) and (4-85), pg. 109),

$$c_r(\mathbf{p})c_r(\mathbf{p}) = 0 \qquad c_r^\dagger(\mathbf{p})c_r^\dagger(\mathbf{p}) = 0 \qquad (\text{no sum on } r). \tag{11-47}$$

Now let's expand the 2nd and 3rd rows part of (11-45) using X symbols to represent pure numbers (which have no effect on commutation or anti-commutation). For the $N_r(\mathbf{p})$ part of that,

$$c_r^\dagger(\mathbf{p})c_r(\mathbf{p})\Big(d_r(\mathbf{p})c_r(\mathbf{p})\underbrace{\overline{v}_r(\mathbf{p})\gamma^\mu u_r(\mathbf{p})e^{-ipx}e^{-ipx}}_{X_{dc}} + d_r(\mathbf{p})d_r^\dagger(\mathbf{p})\underbrace{\overline{v}_r(\mathbf{p})\gamma^\mu v_r(\mathbf{p})e^{-ipx}e^{ipx}}_{X_{dd^\dagger}}$$

$$+ c_r^\dagger(\mathbf{p})c_r(\mathbf{p})\underbrace{\overline{u}_r(\mathbf{p})\gamma^\mu u_r(\mathbf{p})e^{ipx}e^{-ipx}}_{X_{c^\dagger c}} + c_r^\dagger(\mathbf{p})d_r^\dagger(\mathbf{p})\underbrace{\overline{u}_r(\mathbf{p})\gamma^\mu v_r(\mathbf{p})e^{ipx}e^{ipx}}_{X_{c^\dagger d^\dagger}}\Big) \tag{11-48}$$

$$= c_r^\dagger(\mathbf{p})c_r(\mathbf{p})\Big(d_r(\mathbf{p})c_r(\mathbf{p})X_{dc} + d_r(\mathbf{p})d_r^\dagger(\mathbf{p})X_{dd^\dagger} + c_r^\dagger(\mathbf{p})c_r(\mathbf{p})X_{c^\dagger c} + c_r^\dagger(\mathbf{p})d_r^\dagger(\mathbf{p})X_{c^\dagger d^\dagger}\Big).$$

Our objective is to move the $c_r^\dagger(\mathbf{p})c_r(\mathbf{p}) = N_r(\mathbf{p})$ factor to the right hand side of the above using the anti-commutation relations (11-46). If we do that, we will be able to show the 2nd and 3rd rows of (11-45) (i.e., (11-48)) cancel with the 4th and 5th rows of (11-45), and thus, that Q and $\mathcal{H}_I^I$ commute.

We are now going to drop all $\mathbf{p}$ and r notation to save space, and hopefully, make things easier to follow, and not harder.

The last row of (11-48) (2nd and 3rd rows of (11-45) for $N = c^\dagger c$) then becomes

$$= \underbrace{c^\dagger c d c}_{\substack{-c^\dagger d cc \\ =0}} X_{dc} + \underbrace{c^\dagger c d d^\dagger}_{\substack{-c^\dagger d c d^\dagger \\ = c^\dagger d d^\dagger c}} X_{dd^\dagger} + c^\dagger c c^\dagger c X_{c^\dagger c} + \underbrace{c^\dagger c c^\dagger d^\dagger}_{\substack{= c^\dagger d^\dagger c c^\dagger \\ = c^\dagger d^\dagger (1 - c^\dagger c)}} X_{c^\dagger d^\dagger})$$

$$= c^\dagger d d^\dagger c X_{dd^\dagger} + c^\dagger c c^\dagger c X_{c^\dagger c} - c^\dagger d^\dagger c^\dagger c X_{c^\dagger d^\dagger} + c^\dagger d^\dagger X_{c^\dagger d^\dagger} \tag{11-49}$$

$$= d d^\dagger c^\dagger c X_{dd^\dagger} + c^\dagger c c^\dagger c X_{c^\dagger c} + \underbrace{c^\dagger c^\dagger}_{=0} d^\dagger c X_{c^\dagger d^\dagger} + c^\dagger d^\dagger X_{c^\dagger d^\dagger}$$

$$= \left(d d^\dagger X_{dd^\dagger} + c^\dagger c X_{c^\dagger c} \right) c^\dagger c + c^\dagger d^\dagger X_{c^\dagger d^\dagger} .$$

The last row of (11-48) (2nd and 3rd rows of (11-45) for $\bar{N} = d^\dagger d$) then becomes

$$= -d^\dagger \underbrace{dd}_{=0} c X_{dc} - d^\dagger \underbrace{dd}_{=0} d^\dagger X_{dd^\dagger} - \underbrace{d^\dagger d c^\dagger c}_{\substack{d^\dagger c^\dagger cd \\ = c^\dagger c d^\dagger d}} X_{c^\dagger c} - \underbrace{d^\dagger d c^\dagger d^\dagger}_{\substack{-d^\dagger c^\dagger dd^\dagger \\ = -d^\dagger c^\dagger (1 - d^\dagger d)}} X_{c^\dagger d^\dagger}$$

$$= -c^\dagger c d^\dagger d X_{c^\dagger c} + d^\dagger c^\dagger X_{c^\dagger d^\dagger} - \underbrace{d^\dagger c^\dagger d^\dagger d}_{\substack{-c^\dagger d^\dagger d^\dagger d \\ =0}} X_{c^\dagger d^\dagger} \tag{11-50}$$

$$= -c^\dagger c X_{c^\dagger c} d^\dagger d + d^\dagger c^\dagger X_{c^\dagger d^\dagger} .$$

Adding (11-49) and (11-50), we get the total of the 2nd and 3rd rows of (11-45),

$$= \left(d d^\dagger X_{dd^\dagger} + c^\dagger c X_{c^\dagger c} \right) c^\dagger c - c^\dagger c X_{c^\dagger c} d^\dagger d . \tag{11-51}$$

The 4th and 5th rows of (11-45) for $N = c^\dagger c$ are

$$= -dc \underbrace{c^\dagger c}_{1 - cc^\dagger} X_{dc} - dd^\dagger c^\dagger c X_{dd^\dagger} - c^\dagger c c^\dagger c X_{c^\dagger c} - \underbrace{c^\dagger d^\dagger c^\dagger}_{-c^\dagger c^\dagger d^\dagger = 0} c X_{c^\dagger d^\dagger}$$

$$= -dc X_{dc} + d \underbrace{cc}_{=0} c^\dagger X_{dc} - dd^\dagger c^\dagger c X_{dd^\dagger} - c^\dagger c c^\dagger c X_{c^\dagger c} = -dc X_{dc} - \left(dd^\dagger X_{dd^\dagger} + c^\dagger c X_{c^\dagger c} \right) c^\dagger c. \tag{11-52}$$

The 4th and 5th rows of (11-45) for $\bar{N} = d^\dagger d$ are

$$= +dc \underbrace{d^\dagger d}_{1 - dd^\dagger} X_{dc} + d \underbrace{d^\dagger d^\dagger}_{=0} dX_{dd^\dagger} + c^\dagger c d^\dagger dX_{c^\dagger c} + c^\dagger \underbrace{d^\dagger d^\dagger}_{=0} dX_{c^\dagger d^\dagger}$$

$$= dc X_{dc} - \underbrace{dcdd^\dagger}_{-ddcd^\dagger = 0} X_{dc} + c^\dagger c d^\dagger dX_{c^\dagger c} = dc X_{dc} + c^\dagger c X_{c^\dagger c} d^\dagger d. \tag{11-53}$$

The total of 4th and 5th rows of (11-45) is (11-52) plus (11-53), or

$$= -\left(dd^\dagger X_{dd^\dagger} + c^\dagger c X_{c^\dagger c} \right) c^\dagger c + c^\dagger c X_{c^\dagger c} d^\dagger d . \tag{11-54}$$

The total for the 2nd to 5th rows of (11-45), which is proportional to one term in $[Q, \mathcal{H}_I^I]$ is (11-51) plus (11-54), and this equals zero. Thus,

$$\left[Q, \mathcal{H}_I^I \right] \xrightarrow{\text{one term of summations}} = 0 . \tag{11-55}$$

All other terms follow in similar, or more trivial, fashion, yielding

$$\left[Q, \mathcal{H}_I^I \right] = 0 \tag{11-56}$$

And thus, since S is a function of $\mathcal{H}_I^I$, then

$$[Q, S] = 0 \quad \text{Q.E.D.} \tag{11-57}$$

11.9 Problems

1. Similar to Prob. 9 at the end of Chap. 5, show that for any 4-vector A^α, $\partial A^{\alpha,\beta} / \partial A_{\mu,\nu} = g^{\alpha\mu} g^{\beta\nu}$. (Hint: Start with $A^{\alpha,\beta} = g^{\alpha\mu} g^{\beta\nu} A_{\mu,\nu}$.) Then show that $\partial A^{\alpha,\beta} / \partial A^{\mu,\nu} = \delta^\alpha{}_\mu \delta^\beta{}_\nu$. (Hint: Start with $A^{\alpha,\beta} = \delta^\alpha{}_\mu \delta^\beta{}_\nu A^{\mu,\nu}$.) You should then also be able to just accept that $\partial A_{\alpha,\beta} / \partial A_{\mu,\nu} = \delta^\mu{}_\alpha \delta^\nu{}_\beta$ and $\partial A_{\alpha,\beta} / \partial A^{\mu,\nu} = g_{\alpha\mu} g_{\beta\nu}$, but if not, prove them.

2. Use the results from Prob. 1, the Lorenz gauge condition $\partial_\nu A^\nu = 0$, and the Lagrangian (11-8) to show that (11-8) yields the interaction field equations (11-4) and (11-5). Hint: For (11-5), the bookkeeping can be easier if you start by changing the dummy indices in (11-8) from μ and ν to α and β, then, use the Euler-Lagrange equation (11-3) with dummy index $\mu \to \nu$ and $\phi^r = A_\mu$.

3. Show the free electromagnetic part of the Lagrangian of (11-8) is not symmetric under the transformation (11-20).

4. Show that both versions of $\mathcal{L}$ in (11-23) are symmetric under transformation (11-17).

5. Show that the time rate of change of the expectation value for charge Q is zero. Use Wholeness Chart 7-2, pg. 193 as a guide. From the Interaction Picture row, middle column, note that because $dQ/dt = 0$ (from Noether's theorem), then $[Q, H_0] = 0$. Then use the last column, same row, relation to determine the time rate of change of the expectation value of Q is zero. Note $H^I = H_0 + H_I^I$. Note also that if, as we proved in the Appendix, and used in (11-29), $[Q,S] = 0$, then because S is a function of H_I^I, we must also have $[Q, H_I^I] = 0$.

6. Show that under the local transformation set $\psi \to \psi' = e^{-i\alpha(x^\mu)}\psi$ and $A_\nu \to A'_\nu = A_\nu - (1/e)\partial_\nu\alpha(x^\mu)$, $\mathcal{L}_0 = -\frac{1}{2}\left(\partial_\beta A_\nu\right)\left(\partial^\beta A^\nu\right) + \bar\psi\left(i\gamma^\nu\partial_\nu - m\right)\psi$, the first form of the free QED Lagrangian, is not symmetric. (Short cut: Just delete the interaction terms from (11-35).

7. Using Noether's theorem, derive the conserved current for the full QED Lagrangian (second form) under the continuous, local transformation set of Prob. 6. (Hint: It will save you a lot of time if you realize that $\partial A^\nu/\partial \alpha = 0$, since the transformation for A_ν, while it contains $\partial_\nu\alpha$, contains no α.) Is this the same as we derived for the full Lagrangian under the global transformation where α = constant?

8. Use the first form of the free Lagrangian as shown in Prob. 6 and minimal substitution of the gauge covariant derivative $D_\nu = \partial_\nu - ieA_\nu$ to show that the result is not the correct form for the full Lagrangian (free plus interaction).

9. Show that the gauge covariant derivative $D_\nu = \partial_\nu - ieA_\nu$ acting on ψ transforms under the local internal transformation set $\psi \to \psi' = e^{-i\alpha(x^\mu)}\psi$ and $A_\nu \to A'_\nu = A_\nu - (1/e)\partial_\nu\alpha(x^\mu)$ as if it were a fermion field. That is, show $D_\nu\psi \to D'_\nu\psi' = e^{-i\alpha(x^\mu)}D_\nu\psi$. Note that $D'_\nu = \partial_\nu - ieA'_\nu$.

10. Show that under the transformations of Prob. 9, the field equations (11-4) and (11-5), which are respectively $\left(i\gamma^\nu\partial_\nu - m\right)\psi = -e\gamma^\nu\psi A_\nu$ and $\partial^\alpha\partial_\alpha A^\nu = -e\bar\psi\gamma^\nu\psi$, hold for ψ' and A'_ν. Hint: Both A_ν and A'_ν must satisfy the Lorenz condition $\partial_\nu A^\nu = \partial_\nu A'^\nu = 0$ in order for the second field equation (11-5) to hold for both, so therefore we must have $\partial_\nu\partial^\nu\alpha = 0$.

This page intentionally left blank.

Part Three

Renormalization:
Taming Those Notorious Infinities

"Happy is he who gets to know the reasons for things."
Virgil

Chapter 12

Overview of Renormalization

"To infinity and beyond."
Buzz Lightyear

12.0 Preliminaries

The quote from *Toy Story* cartoon character Buzz may seem humorous, but it is a fairly accurate description of the renormalization procedure in QFT. In Chap. 9, we saw how including higher order terms in our interaction calculations led to infinite transition amplitudes. In this and the subsequent chapters of Part Three, we will see how we can go beyond such infinities to obtain meaningful, finite transition amplitudes.

Renormalization makes infinity finite

Renormalization is probably the most difficult aspect of QFT to understand. Given the complexities of the theory, this is a strong statement. The Dirac equation and the Feynman propagator are difficult to master, but renormalization beats those, at least in my opinion. I will try herein to make it as easy to comprehend and as transparent as possible.

12.0.1 Background

Renormalization is an extensive study, and one can make a career out of it. Certainly, careers, and great reputations, have been made by finding ways to do it. More than one Nobel Prize has been awarded for it.

History of Renormalization

As noted in the introduction to Chap. 8, the transition amplitude infinities problem was the hottest topic in physics in the late 1940s. Three people, Feynman, Schwinger, and Tomonaga, independently solved the problem and jointly received the Nobel Prize for it in 1965. Some feel Freeman Dyson should have been included, as he put a sound mathematical foundation under the procedures developed by the other three. These procedures comprised renormalization for QED.

Nobel prizes for renormalization

History repeated itself for weak interaction theory, which was plagued by the same, though even more complicated, type of problem. In the 1970s, Gerardus 't Hooft and Martinus Veltman found a way to remove the infinites, a way to renormalize electroweak interactions. They were awarded the Nobel Prize for this in 1999.

For QED, weak, and strong theories

In 1982, Ken Wilson won the prize for what is called lattice gauge theory, a means for taming the infinities in non-perturbative QFT, such as those found in strong interactions. By contrast, the approaches of Feynman, Schwinger, Tomonaga, 't Hooft, and Veltman were all perturbative.

David Gross, H. David Politzer, and Frank Wilczek, with the help of Wilson's work, solved a problem rooted in (non-perturbative) strong interaction renormalization called quark asymptotic freedom and received the Nobel Prize for that in 2004.

For quantum gravity, the first things tried were renormalization techniques similar to what worked for e/m and weak theories. Despite massive efforts by many within the particle theorist community, no one has been able to make those viable. Those approaches seem dramatically unable to remove the infinities cropping up in the gravitational transition amplitudes.

Renormalization major block to consistent quantum gravity theory

Gravity, in the usual QFT sense, has simply been non-renormalizable. Alternative approaches, transcending the traditional one, include super-string theory (more correctly M-theory), loop quantum gravity, and twistor theory. These have had varying degrees of success, though, and as of this writing, the jury is still out on them.

Renormalization's Present Status

Despite its considerable success with standard QFT (sans gravity), not everyone has been happy with renormalization, including some people, like Richard Feynman, who played major roles in its development. Some consider the underlying mathematics suspect and posit that there must be a better way.

Renormalization works, but some question its foundations

The fact remains that renormalization broke a once seemingly intractable impasse, and allowed QFT to move forward and make extraordinarily accurate predictions of experimental results. Understanding it is essential for anyone wishing to become a field theorist.

12.0.2 Chapter Overview

In this chapter, we will take the simplest possible look at renormalization. We will crystallize the essence of it all, hopefully in a transparent way, before diving into the extensive mathematics of the process in the next three chapters.

This chapter: simplest possible overview

For renormalization, we will

- explain the root of the term,

- briefly describe a mathematical process we will use called "regularization",

- use the example of Bhabha scattering to illustrate the fundamental issues involved,

- see how higher order terms lead to factors that modify the charge, mass, propagators, and vertices (we'll use sub/superscript "Mod" on iD_F, iS_F, γ^μ as shorthand for the latter),

- express the redefinition of charge e and mass m that underlies renormalization,

- show how $\alpha = e^2/4\pi$, our QED coupling constant, is not really constant, but varies with the energy level of our interaction, and

- derive expressions for finding α as a function of one energy level, given another.

12.1 Whence the Term "Renormalization"?

Recall that in Chap. 7 we started with an initial multi-particle state $|i\rangle$ and via S_{oper} transformed that state into a final general state $|\Psi(t_f)\rangle$ comprised of a set of possible final eigenstates $|f\rangle$. (See Wholeness Chart 8-4 at the end of Chap. 8.) That is,

$$S_{oper}|i\rangle = |\Psi(t_f)\rangle = \sum_f |f\rangle S_{fi} \qquad S_{fi} = \langle f|S_{oper}|i\rangle \qquad \sum_f |S_{fi}|^2 = 1, \qquad (12\text{-}1)$$

where each possible final multi-particle eigenstate $|f\rangle$ has an amplitude S_{fi} (= the transition amplitude to that state), and $|S_{fi}|^2$ is the probability of measuring that final eigenstate. The last relation in (12-1) expresses the fact that $|\Psi(t_f)\rangle$ is normalized.

Renormalization "resets" our amplitudes so final general state is normalized again

However, when we took the limits

$$S = S_{oper} \underset{V\to\infty}{\left(t_f \to \infty, t_i \to -\infty\right)}, \qquad (12\text{-}2)$$

and evaluated our result at higher orders, we found

$$|S_{f'i}|^2 \to \infty, \quad |S_{f''i}|^2 \to \infty, \text{ etc., and } \quad \sum_f |S_{fi}|^2 \to \infty. \qquad (12\text{-}3)$$

In other words, via the last relation in (12-3), our final general state $|\Psi(t_f)\rangle$ is no longer normalized.

The process we are about to undertake will change our mathematics back so the individual amplitudes are finite, the last relation of (12-1) holds once again, and thus $|\Psi(t_f)\rangle$ will be normalized again. We will have renormalized our final result, hence the name "renormalization".

12.2 A Brief Mathematical Interlude: Regularization

As we saw in Chap. 9, internal loops in propagators and vertices can result in integrals that do not converge. In the process of renormalization, we will need to recast these integrals in a manageable form (one that is not infinite), at least during part of our renormalization process. During the renormalization process, we can then play some tricks that cause the troublesome integrals to "drop out" of the final result. We can then, at the end, restore them to their rightful, infinite value, but our final result will no longer diverge.

This process of temporarily rendering the infinite integrals as finite is called <u>regularization</u>. We illustrate the simplest way to regularize with the following example.

One way to regularize divergent integrals

Consider the divergent integral

$$\int_{-\infty}^{\infty} x^2 dx = \frac{1}{3} x^3 \Big|_{-\infty}^{\infty} \to \infty . \tag{12-4}$$

We can regularize (12-4), as follows

$$\int_{-\Lambda}^{\Lambda} x^2 dx = \frac{1}{3} x^3 \Big|_{-\Lambda}^{\Lambda} \to \frac{2}{3} \Lambda^3 \qquad \text{later take } \Lambda \to \infty . \tag{12-5}$$

Now imagine that (12-4) occurs in some analysis we are doing. We express it temporarily as the LHS of (12-5). Imagine further that in this analysis we have (12-4) multiplied by $1/\Lambda^3$. We would find the Λ factors cancel, leaving a finite number result. At that point, we could take the limit of $\Lambda \to \infty$ and restore the original physics of our analysis. But at that point, it doesn't matter, because there no longer is a Λ in our equations, and we have a finite final result.

The process of renormalization is not quite so simple, but it is analogous. During renormalization, as we will see, we do a similar thing with our divergent loop integrals as we have done here. That is, we regularize the troublesome integrals, employ a clever trick, and then find the final result no longer diverges even when we take $\Lambda \to \infty$.

12.3 A Renormalization Example: Bhabha Scattering

12.3.1 Setting Up the Total Amplitude Calculation to Order α^2

Fig. 12-1 is a duplicate of Fig. 9-4. which we have already seen in Chap. 9. It represents all of the possible first order and second order (in α) Feynman diagrams for the first kind of Bhabha scattering. The second kind of Bhabha scattering is not shown. (Fig. 8-2, pg. 221 shows both kinds.) The first order in α (second order in e) diagram (the upper left one in Fig. 12-1) is called a <u>tree diagram</u>. (In mathematics the term is used to refer to a diagram in which lines branch out from points without forming any closed loops.)

Diagrams of order e^2 are tree diagrams and have two vertices

Note that the energy involved in the interaction of Fig. 12-1 is k^0, the energy of the virtual photon in the tree diagram, which equals the sum of the energies of the incoming particles. This number, the energy level of a given interaction, is generally represented by the symbol k (or p for p^0 for leptons) in renormalization mathematics, even though k (and p) generally represents 4-momenta in Feynman diagrams and loop integrals. Often, the incoming particles are approaching one another from opposite directions, so 3-momenta effectively cancel leaving the virtual photon with lots of energy, but not much 3-momentum. (Remember it is off-shell, so this can happen.) In fact, in the center of mass frame, which is commonly used in QFT, total 3-momentum is always zero. (See Prob. 1.) Hence, using k to represent k^0 is not so far from reality. (See Prob. 2.)

Energy of an interaction = sum of incoming particle energies

k and p = four-momentum in diagrams, but sometimes = interaction energy

Recall the transition amplitude for a given interaction, to all orders n, is

Transition amplitude to all orders

$$S_{fi} = \delta_{fi} + \left((2\pi)^4 \delta^{(4)} \left(P_f - P_i \right) \left(\prod^{\substack{\text{all external} \\ \text{bosons}}} \sqrt{\frac{1}{2V\omega}} \right) \left(\prod^{\substack{\text{all external} \\ \text{fermions}}} \sqrt{\frac{m}{VE}} \right) \right) \mathcal{M} \qquad \mathcal{M} = \sum_{n=1}^{\infty} \mathcal{M}^{(n)} \tag{12-6}$$

Everything in (12-6) except the Feynman amplitude $\mathcal{M}$ and the component amplitudes $\mathcal{M}^{(n)}$, is the same for the same incoming and outgoing particles. As evidenced by Fig. 12-1, there can be many such diagrams with the same incoming and outgoing particles, i.e., many diagrams for the same interaction type (Bhabha scattering in this case).

Same initial and final states for each term in sum

For the Feynman amplitude of a given interaction type, $\mathcal{M}^{(n)}$ is the sum of the amplitudes from all diagrams of order n. If n is the order to which e is raised (that is, e^n is found in the amplitude), then all diagrams of order n would have n vertices.

In e^n and $\mathcal{M}^{(n)}$, n is number of vertices in a diagram

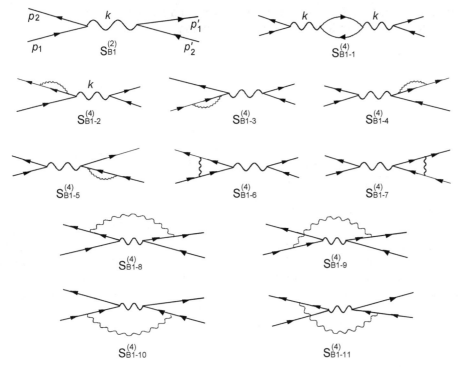

All 1ˢᵗ type Bhabha to 2ⁿᵈ and 4ᵗʰ order in e

All two and four vertex diagrams

Figure 12-1. e^2 and e^4 Contributions to First Kind of Bhabha Scattering

If we calculated the transition amplitudes for each diagram in Fig. 12-1, we would then have to add them to get the total amplitude (to order e^4, i.e., to order α^2) for the first type of Bhabha scattering. We would also have to add in all such contributions from the second type of Bhabha scattering. That is,

$$\mathcal{M}_{\substack{\text{Bhabha} \\ \text{total to} \\ \text{order } e^4}} = \underbrace{\mathcal{M}_{B1}^{(2)} + \sum_{i=1}^{11} \mathcal{M}_{B1-i}^{(4)}}_{\substack{\text{1st type of Bhabha scattering,} \\ \text{shown in Fig. 12-1}}} + \underbrace{\mathcal{M}_{B2}^{(2)} + \sum_{j=1}^{11} \mathcal{M}_{B2-j}^{(4)}}_{\substack{\text{2nd type of Bhabha scattering,} \\ \text{not shown in Fig. 12-1}}} \qquad (12\text{-}7)$$

Add amplitudes of all diagrams to order e^4 for total amplitude to same order

Integrals in higher order diagrams are dastardly. We'll use final result in this chapter.

As you might expect, given the dastardly unwieldy integrals involved for each separate diagram, evaluating (12-7) is an enormous, and very complicated, amount of work. The next three chapters are devoted to doing just that. For now, we will simply state the end result of all that work. However, before doing so, we need to introduce a change in our symbols.

It turns out that when we include higher than tree level amplitudes in our transition amplitude, the effective coupling e^2 (or equivalently, α) changes. That is, e (or α) appears to have a different value, according to whether or not we include the higher order parts of the amplitude. A similar thing occurs for the mass m in the lepton propagators.

We'll see that these integrals affect the value of e and m

Because of this, we will want to take the symbols e and m to represent what we would actually measure in experiment (for which nature would automatically include all the higher order contributions). Thus, we will redefine the symbols e and m that we have been using throughout the book so far as e_0 and m_0. Those latter symbols will represent charge and mass as they would be found if only tree level diagrams played any role in nature. We will call these the bare charge and bare mass, respectively, since they are not "dressed up" with contributions from additional Feynman diagrams beyond the tree level diagrams.

So we change symbols.

In practice, we can never measure bare charge or bare mass because nature always includes the higher order corrections in real world interactions. But the concept of such bare quantities will serve us well in our analyses.

Important Symbol Changes

e in all prior work will from henceforth be re-labeled as e_0 (<u>bare charge</u>)

α in all prior work will from henceforth be re-labeled as α_0 (<u>bare coupling constant</u>)

m in all prior work will from henceforth be re-labeled as m_0 (<u>bare mass</u>)

e_0, α_0, and m_0 mean tree level values and replace prior symbols e, α, and m

Thus, our tree level calculation in Chap. 8 (8-34), pg. 223, of the Feynman amplitude for the first diagram of Fig. 12-1 now looks the same except for these changes in notation, i.e.,

$$\mathcal{M}_{B1}^{(2)} = -e_0^2 \bar{u}_{r_1'}(\mathbf{p}_1')\gamma^\mu v_{r_2'}(\mathbf{p}_2') i D_{F\mu\nu}(k) \bar{v}_{r_2}(\mathbf{p}_2)\gamma^\nu u_{r_1}(\mathbf{p}_1). \tag{12-8}$$

Similarly, the Feynman amplitude for the top RH diagram in Fig. 12-1, which we found in Chap. 9 (9-7), pg. 259, is

$$\mathcal{M}_{B1-1}^{(4)} =$$

$$-\frac{e_0^4}{(2\pi)^4}\bar{u}_{r_1'}(\mathbf{p}_1')\gamma^\rho v_{r_2'}(\mathbf{p}_2') D_{F\rho\eta}(k)\left(\mathrm{Tr}\int \underbrace{S_F(p)\gamma^\mu S_F(p-k)}_{\text{with } m_0 \text{ not } m}\gamma^\eta d^4 p\right) D_{F\mu\nu}(k)\bar{v}_{r_2}(\mathbf{p}_2)\gamma^\nu u_{r_1}(\mathbf{p}_1). \tag{12-9}$$

That is, in every amplitude calculation at every level, we use e_0 instead of e, and m_0 instead of m. All of our work assumed the same values for these quantities at every order. We then add all of these together as in (12-7).

Regularization: integrals to ∞ will temporarily depend on parameter Λ, which eventually will go to ∞

Further, in a manner similar to that discussed in Sect. 12.2, we will want to regularize any integrals that do not converge when integrated to infinity. In effect, integrate them to some finite value, which we will call Λ, instead of infinity. After carrying out all the algebra and calculus using Λ, we then will take $\Lambda \to \infty$ and see if we get a meaningful answer. The hope, which we will see is realized, is that we can do that in such a way that our answer is meaningful, i.e., our transition amplitude remains finite and is correct.

12.3.2 Result of the Calculation

For now, we show the method assuming we can evaluate these integrals

As stated earlier, finding (12-7) takes many pages, and we will do that in Chaps. 13, 14, and 15. For now, we will show how the addition goes, then simply state the final result.

Note that in adding (12-8) to (12-9), the spinors and gamma matrices stay the same. They represent the same incoming (destroyed) and outgoing (created) particles, which will be same for all diagrams to any order in the Bhabha scattering of Fig. 12-1. So, adding (12-8) and (12-9) gives us (where we remind the reader that the trace is on spinor indices, which are hidden)

$$\mathcal{M}_{\substack{\text{Bhabha}\\ \text{1st type}\\ \text{2 terms only}}} = -e_0^2 \bar{u}_{r_1'}(\mathbf{p}_1')\gamma^\mu v_{r_2'}(\mathbf{p}_2')\left\{ iD_{F\mu\nu}(k)\right.$$

Example of adding tree level (e^2 order) diagram to one e^4 order diagram

$$+ e_0^2 D_{F\mu\eta}(k)\underbrace{\frac{1}{(2\pi)^4}\left(\mathrm{Tr}\int \underbrace{S_F(p)\gamma^\rho S_F(p-k)}_{\text{with } m_0 \text{ not } m}\gamma^\eta d^4 p\right) D_{F\rho\nu}(k)}_{\substack{\text{call this } iX^{\eta\rho}(k,\Lambda) \text{ with}\\ \text{integration to } \Lambda, \text{ not } \infty}}\left.\right\} \bar{v}_{r_2}(\mathbf{p}_2)\gamma^\nu u_{r_1}(\mathbf{p}_1) \tag{12-10}$$

$$= -e_0^2 \bar{u}_{r_1'}(\mathbf{p}_1')\gamma^\mu v_{r_2'}(\mathbf{p}_2')\left\{\underbrace{iD_{F\mu\nu}(k)}_{\text{tree diagram}} + \underbrace{iD_{F\mu\eta}(k)e_0^2 X^{\eta\rho}(k,\Lambda)D_{F\rho\nu}(k)}_{\text{photon self energy diagram}}\right\}\bar{v}_{r_2}(\mathbf{p}_2)\gamma^\nu u_{r_1}(\mathbf{p}_1),$$

where the big job in (12-10) would be to evaluate $X^{\eta\rho}(k,\Lambda)$. When we evaluate that, we find (see Chap. 14) the photon self-energy diagram term ends up making two changes to the amplitude as compared to the original tree diagram amplitude. (See (12-11) below.) That is, we get first a term that is a function of e_0^2, k, and Λ, and second, a modified form for the photon propagator. This modified propagator (modified to 2nd order in α) is only a function of k, but a different function of k than the original, non-loop, propagator.

$$\mathcal{M}_{\substack{\text{Bhabha}\\ \text{1st type}\\ \text{2 terms only}}} = -e_0^2\left\{1+\underbrace{\left(\text{function of } e_0^2, k, \Lambda\right)}_{\substack{\text{from photon}\\ \text{self energy term}}}\right\}\bar{u}_{r_1'}(\mathbf{p}_1')\gamma^\mu v_{r_2'}(\mathbf{p}_2')\ \underbrace{iD_{F\mu\nu}^{2nd\ Mod}(k)}_{\substack{\text{from photon}\\ \text{self energy term}}}\ \bar{v}_{r_2}(\mathbf{p}_2)\gamma^\nu u_{r_1}(\mathbf{p}_1). \tag{12-11}$$

Example of adding tree level and all e^4 order diagrams

The process of (12-10) and (12-11) is, however, only the beginning. We need to add in the other 10 diagrams of Fig. 12-1. Carrying out the exceptionally unwieldy integrations involved (see Chap. 15), we would find, where b_n is a "constant" we will discuss shortly, $F(k^2)$ is a finite valued function of k^2, $G^V(p_1, p_2)$ is a finite valued function of p_1 and p_2, and $k = p_1 + p_2 = p'_1 + p'_2$,

$$\mathcal{M}_{\substack{\text{Bhabha} \\ \text{1st type} \\ \text{total to } e^4}} = -\underbrace{e_0^2 \left\{ 1 + e_0^2 2b_n \, ln\frac{k}{\Lambda} + \mathcal{O}\left(e_0^4\right) \right\}}_{e^2(k,\Lambda)} \bar{u}_{r_1'}(\mathbf{p}_1') \underbrace{\left\{ \gamma^\mu + G^\mu\left(p_1',p_2'\right) \right\}}_{\substack{\gamma^\mu_{Mod} \\ 2nd}} v_{r_2'}(\mathbf{p}_2')$$

$$\underbrace{i D_{F\mu\nu}\left(k\right)\left\{1 - F\left(k^2\right)\right\}}_{\substack{D_{F\mu\nu}^{2nd}(k) \\ Mod}} \times \bar{v}_{r_2}(\mathbf{p}_2) \underbrace{\left\{ \gamma^\nu + G^\nu\left(p_1,p_2\right) \right\}}_{\substack{\gamma^\nu_{Mod} \\ 2nd}} u_{r_1}(\mathbf{p}_1) \qquad (12\text{-}12)$$

Yields finite modified propagator and vertices plus unbounded e^2

$$= -\underbrace{e^2(k,\Lambda)}_{\substack{\text{unbounded for} \\ e_0 \neq 0, \, \Lambda \to \infty}} \bar{u}_{r_1'}(\mathbf{p}_1') \underbrace{\gamma^\mu_{\substack{Mod \\ 2nd}}\left(p_1',p_2'\right)}_{finite} v_{r_2'}(\mathbf{p}_2') \, i \underbrace{D_{F\mu\nu}^{\substack{2nd \\ Mod}}(k)}_{finite} \bar{v}_{r_2}(\mathbf{p}_2) \underbrace{\gamma^\nu_{\substack{Mod \\ 2nd}}\left(p_1,p_2\right)}_{finite} u_{r_1}(\mathbf{p}_1).$$

The quantity $e^2(k,\Lambda)$ is dependent on k, the energy level of the interaction. For e_0 a non-zero constant and $\Lambda = \infty$, e is unbounded. All other terms are finite.

Note that the last line of (12-12) is identical with our tree level relation (12-8), except that

$$e_0 \to e(k,\Lambda), \; iD_{F\mu\nu}\left(k\right) \to iD_{F\mu\nu}^{\substack{2nd \\ Mod}}\left(k\right), \; \gamma^\nu \to \gamma^\nu_{\substack{Mod \\ 2nd}}\left(p_1,p_2\right), \text{ and } \gamma^\mu \to \gamma^\mu_{\substack{Mod \\ 2nd}}\left(p_1',p_2'\right).$$

Finding the modified propagator and vertex relations entails calculating the functions F and G^μ, which is anything but simple, and which we hold off on for now.

12.3.3 Renormalizing the Infinity Problem to e^4 Order Only

Note though that we still haven't resolved the infinity problem, since via

$$e^2\left(k,\Lambda\right) = e_0^2 \left(1 + e_0^2 2b_n \, ln\frac{k}{\Lambda} + \mathcal{O}\left(e_0^4\right) \right), \qquad (12\text{-}13)$$

for $\Lambda \to \infty$, $e^2(k) \to -\infty$, and thus our amplitude (12-12) $\to \infty$. This is where something radical is required, i.e., renormalization. Instead of $e^2(k)$ going to negative infinity in the limit, we assume that the right side of (12-13) in the limit of $\Lambda \to \infty$ corresponds to the (finite) value of the measured charge squared $e^2(k)$ at energy level k. That is, we assume

We redefine e to be the finite, measured value

$$\lim_{\Lambda \to \infty} e^2\left(k,\Lambda\right) = \text{finite value equal to (measured charge)}^2 \text{ at energy level } k. \qquad (12\text{-}14)$$

or
$$\lim_{\Lambda \to \infty} e_0^2 \left(1 + e_0^2 2b_n \, ln\frac{k}{\Lambda} + \mathcal{O}\left(e_0^4\right) \right) = \underbrace{\lim_{\Lambda \to \infty} e_0^2}_{= e_0^2} + \lim_{\Lambda \to \infty} e_0^4 2b_n \, ln\frac{k}{\Lambda} + \lim_{\Lambda \to \infty} \mathcal{O}\left(e_0^6\right)$$

$$\qquad (12\text{-}15)$$

$$= e^2\left(k\right) = \text{measured charge squared}.$$

But for this redefinition, we must have $e_0 = 0$

Now the only way to obtain a finite value for the measured charge squared $e^2(k)$ is to somehow cancel the minus infinity from the natural logarithm in the second term on the right side. This can only be done by having e_0 equal to zero, but in such a way that when we restore our theory to what it really is by taking $\Lambda \to \infty$, we get the bottom row of (12-15).

Taking our bare charge as zero, no doubt, seems strange. Essentially, it means fermions have no inherent charge, but that all charge is a result of higher order interactions. This may appear weird, but it has one major benefit. It works.

It also means (done precisely in Sect. 12.8) that measured charge varies with energy level k, and if we use the measured value of charge at energy level k, then we can compute our amplitude (12-12), including all contributions up to e^4 order, just from that relation alone. We need to know $F(k^2)$ and $G^\nu(p_1, p_2)$, in addition to $e(k)$, of course, and that is not simple. But it can be done.

e is a function of k, interaction energy level

In essence, though, given the specific calculations we have yet to carry out, we have renormalized our infinite amplitude so that now it is finite.

Bottom line: If we use our tree level amplitude relation (12-8) for the first type of Bhabha scattering, but instead of using the bare charge e_0, we use the measured charge at the energy level of the interaction $e(k)$, and we use the modified propagator and vertex relations, we will get the correct, finite amplitude (correct to order e^4, or equivalently α^2, here.)

12.3.4 The Second Type of Bhabha Scattering to e^4 Order

It turns out that if we did a similar analysis for all diagrams of the 2nd type Bhabha scattering (Fig. 8-2, pg. 221), we would end up with a parallel result. We would find the same amplitude algebraically as our tree level 2nd type amplitude, but with the same form $e(k, \Lambda)$ of (12-13) in place of e_0, the same form $iD_{F\mu\nu}^{\overset{Mod}{2nd}}(k)$ in place of $iD_{F\mu\nu}(k)$, and the same form $\gamma_{\overset{Mod}{2nd}}^{\mu}$ in place of γ^{μ}.

Repeating the process for the 2nd type of Bhabha scattering

That is (look at (8-47) pg. 224 for comparison), and where now $k = p_1 - p'_1 = p'_2 - p_2$,

$$\mathcal{M}_{\substack{\text{Bhabha} \\ \text{2nd type} \\ \text{total to } e^4}} = e_0^2 \left\{ 1 + e_0^2 2b_n \ln \frac{k}{\Lambda} + \mathcal{O}\left(e_0^4\right) \right\} \bar{v}_{r_2}(\mathbf{p}_2) \left\{ \gamma^\mu + G^\mu\left(p_2, p'_2\right) \right\} \times \tag{12-16}$$

$$v_{r'_2}(\mathbf{p}'_2) iD_{F\mu\nu}(k) \left\{ 1 - F\left(k^2\right) \right\} \bar{u}_{r'_1}(\mathbf{p}'_1) \left\{ \gamma^\nu + G^\nu\left(p'_1, p_1\right) \right\} u_{r_1}(\mathbf{p}_1)$$

$$= e^2(k) \bar{v}_{r_2}(\mathbf{p}_2) \gamma_{\overset{Mod}{2nd}}^{\mu}\left(p_2, p'_2\right) v_{r'_2}(\mathbf{p}'_2) iD_{F\mu\nu}^{\overset{Mod}{2nd}}(k) \bar{u}_{r'_1}(\mathbf{p}'_1) \gamma_{\overset{Mod}{2nd}}^{\nu}\left(p'_1, p_1\right) u_{r_1}(\mathbf{p}_1).$$

We get the same results for $e(k)$ and modified propagator and vertices

So, by making the same assumptions of (12-15), we can again use the measured value for charge at the energy level of the interaction along with the modified propagator and vertices, and just go with the same form as our tree level amplitude relation.

12.3.5 Total Bhabha Scattering to e^4 Order

The total (for both types) Bhabha scattering to e^4 order is the sum of (12-12) and (12-16),

$$\mathcal{M}_{\substack{\text{Bhabha} \\ \text{total} \\ \text{total to } e^4}} = \mathcal{M}_{\substack{\text{Bhabha} \\ \text{1st type} \\ \text{total to } e^4}} + \mathcal{M}_{\substack{\text{Bhabha} \\ \text{2nd type} \\ \text{total to } e^4}} = \underbrace{\mathcal{M}_{\substack{B1 \\ mod, 2nd}}^{(2)} + \mathcal{M}_{\substack{B2 \\ mod, 2nd}}^{(2)}}_{\substack{\text{tree level but with } e_0 \to e(k), \\ D_{F\mu\nu} \to D_{F\mu\nu}^{\overset{Mod}{2nd}}, \; \gamma^\mu \to \gamma_{\overset{Mod}{2nd}}^{\mu}}}, \tag{12-17}$$

Total Bhabha to e^4 order adds 1st and 2nd type amplitudes to e^4 order

where the notation with superscript (2) implies the 2nd order in e form of the amplitudes, but the modified versions of e, photon propagator, and vertex gamma matrix, as noted in the underbracket.

<u>Bottom line:</u> The substitution of $e(k)$ for e_0, $iD_{F\mu\nu}^{\overset{Mod}{2nd}}(k)$ for $iD_{F\mu\nu}(k)$, and $\gamma_{\overset{Mod}{2nd}}^{\mu}(p_1, p_2)$ for γ^μ in our tree level total amplitude, including both types of Bhabha scattering, yields the correct, finite Bhabha scattering amplitude (to order e^4 here).

12.4 Higher Order Contributions in Bhabha Scattering

12.4.1 Adding All Bhabha Amplitudes to All Orders

It turns out that if we add all relevant diagrams (an infinite number) to all order e^n, for $n \to \infty$ (see Fig. 12-2), we get a total Bhabha amplitude relation of form (12-18).

Bhabha scattering to all orders: add all relevant diagrams of all numbers of vertices

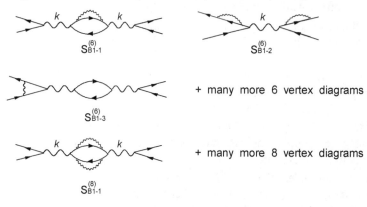

+ many more 10 and higher vertex diagrams

Figure 12-2. e^6 and Higher Contributions to First Kind of Bhabha Scattering

$$\underset{\substack{\text{total}}}{\mathcal{M}_{\text{Bhabha}}} = \underset{\substack{\text{1st type} \\ \text{all orders}}}{\mathcal{M}_{\text{Bhabha}}} + \underset{\substack{\text{2nd type} \\ \text{all orders}}}{\mathcal{M}_{\text{Bhabha}}}$$

$$= -e_0^2 \underbrace{\left\{1 + e_0^2\, 2b_n\, ln\frac{k}{\Lambda} + \left(e_0^4\text{ term}\right) + ...\right\}}_{e^2(k,\Lambda)} \bar{u}_{r_1'}(\mathbf{p}_1')\underbrace{\left\{\gamma^\mu + G^\mu\left(p_1',p_2'\right) + \binom{\text{higher}}{\text{order}}\right\}}_{\text{finite} = \gamma^\mu_{Mod}} v_{r_2'}(\mathbf{p}_2') \times$$

Modified finite propagators and vertices again, but different than e^4 order relations

$$i\underbrace{D_{F\mu\nu}\left(p_1 + p_2\right)\left\{1 - F\left(k^2\right) - \binom{\text{higher}}{\text{order}}\right\}}_{\text{finite} = D^{Mod}_{F\mu\nu}(k=p_1+p_2)} \bar{v}_{r_2}(\mathbf{p}_2)\underbrace{\left\{\gamma^\nu + G^\nu\left(p_1,p_2\right) + \binom{\text{higher}}{\text{order}}\right\}}_{\text{finite} = \gamma^\nu_{Mod}} u_{r_1}(\mathbf{p}_1)$$

(12-18)

$$+ e_0^2 \underbrace{\left\{1 + e_0^2\, 2b_n\, ln\frac{k}{\Lambda} + \left(e_0^4\text{ term}\right) + ...\right\}}_{e^2(k,\Lambda)} \bar{v}_{r_2}(\mathbf{p}_2)\underbrace{\left\{\gamma^\mu + G^\mu\left(p_2,p_2'\right) + \binom{\text{higher}}{\text{order}}\right\}}_{\text{finite} = \gamma^\mu_{Mod}} v_{r_2'}(\mathbf{p}_2') \times$$

Modified $e^2(k)$ again, but different than $e^2(k)$ for e^4 order diagrams

$$i\underbrace{D_{F\mu\nu}\left(p_2 - p_2'\right)\left\{1 - F\left(k^2\right) - \binom{\text{higher}}{\text{order}}\right\}}_{\text{finite} = D^{Mod}_{F\mu\nu}(k=p_2-p_2')} \bar{u}_{r_1'}(\mathbf{p}_1')\underbrace{\left\{\gamma^\nu + G^\nu\left(p_1',p_1\right) + \binom{\text{higher}}{\text{order}}\right\}}_{\text{finite} = \gamma^\nu_{Mod}} u_{r_1}(\mathbf{p}_1).$$

New symbols include higher order terms, not present at e^4 order

Where we use the symbols for e^2, $D^{Mod}_{F\mu\nu}$, and γ^μ_{Mod} without the "2nd" script to note that these include all higher order terms that were not present in the symbols for the e^4 case of Sect. 12.2.

12.4.2 Renormalizing e^2 to All Orders in Bhabha Scattering

We can then renormalize $e^2(k,\Lambda)$ of (12-18) in a similar way as we did in (12-14), i.e.,

Renormalize $e^2(k)$ in similar way as e^4 order relation

$$e^2(k) = e_0^2\left(1 + e_0^2\, 2b_n\, ln\frac{k}{\Lambda} + \left(e_0^4\text{ term}\right) + \left(e_0^6\text{ term}\right) + ...\right) = \text{(charge measured)}^2 \text{ at energy } k. \quad (12\text{-}19)$$

We again have e_0, our bare charge, equal to zero, but in such a way that (with $e_0^2 = 0$ in (12-19))

$$\underset{\Lambda\to\infty}{\text{limit}}\, e_0^2\left(1 + e_0^2\, 2b_n\, ln\frac{k}{\Lambda} + \left(e_0^4\text{ term}\right) + \left(e_0^6\text{ term}\right) + ...\right) = e^2(k). \quad (12\text{-}20)$$

12.4.3 Total Bhabha Scattering to All Orders

Thus, we can express (12-18) as

$$\underset{\substack{\text{total}}}{\mathcal{M}_{\text{Bhabha}}} = \underset{\substack{\text{1st type} \\ \text{all orders}}}{\mathcal{M}_{\text{Bhabha}}} + \underset{\substack{\text{2nd type} \\ \text{all orders}}}{\mathcal{M}_{\text{Bhabha}}}$$

Total Bhabha scattering to all orders

$$= -e^2(k)\, \bar{u}_{r_1'}(\mathbf{p}_1')\gamma^\mu_{Mod}\left(p_1',p_2'\right)v_{r_2'}(\mathbf{p}_2')\, iD^{Mod}_{F\mu\nu}\left(p_1+p_2\right)\bar{v}_{r_2}(\mathbf{p}_2)\gamma^\nu_{Mod}\left(p_1,p_2\right)u_{r_1}(\mathbf{p}_1) \quad (12\text{-}21)$$

$$+ e^2(k)\, \bar{v}_{r_2}(\mathbf{p}_2)\gamma^\mu_{Mod}\left(p_2,p_2'\right)v_{r_2'}(\mathbf{p}_2')\, iD^{Mod}_{F\mu\nu}\left(p_2-p_2'\right)\bar{u}_{r_1'}(\mathbf{p}_1')\gamma^\nu_{Mod}\left(p_1',p_1\right)u_{r_1}(\mathbf{p}_1).$$

Use tree diags, substitute modified relations to get correct amplitude

<u>Bottom line:</u> We can use our tree diagrams (both types) and simply substitute the measured charge $e(k)$ for e_0, $iD^{Mod}_{F\mu\nu}$ for $iD_{F\mu\nu}$, and γ^μ_{Mod} for γ^μ to get the finite, correct total amplitude for Bhabha scattering (to all orders here).

12.4.4 Running QED Coupling "Constant"

We can of course, with $e_0^2 = 4\pi\alpha_0$, re-express (12-20) as

$$\underset{\Lambda\to\infty}{\text{limit}}\, \alpha_0\left(1 + \alpha_0\, 8\pi b_n\, ln\frac{k}{\Lambda} + \left(\alpha_0^2\text{ term}\right) + \left(\alpha_0^3\text{ term}\right) + ...\right) = \alpha(k), \quad (12\text{-}22)$$

Re-expressing $e^2(k)$ as $\alpha(k)$

and that is how theorists tend to prefer it, actually. The fine structure "constant" α is more commonly referred to in QFT, as the <u>electromagnetic coupling "constant"</u>.

Note the word "constant" is now a bit of a misnomer, as we are finding both α and e are not really constants, but functions of k. In fact, since they "run" with k, α is often called the <u>running coupling constant</u>.

$\alpha(k)$ called "running coupling constant"

12.5 Same Result for Any Interaction

12.5.1 The Great Blessing

As we will see in forthcoming chapters, by analyzing different interactions, we can show that the same relation (12-19) holds for every type of interaction, and that the same modifications for the propagator and the vertex relations do as well. That is a great blessing. We can simply take tree level diagrams, find the amplitude for those, but use the measured charge $e(k)$ instead of the bare charge e_0 and the modified propagators and vertex instead of the bare propagators and bare vertex.

Fortunately, all our modified relations are the same for any interaction

12.5.2 But We'll Need to Calculate Some Things

Thus, renormalization makes calculation of interaction probabilities both finite and correct. That is great. The not so great part is deducing the mathematical forms such as $iD_{F\mu\nu}^{2nd\ Mod}(k)$, $\gamma_{Mod\ 2nd}^{\mu}(p_1, p_2)$ and $e(k)$, which proves that it works.

We'll still need to actually determine the modified quantities

Of particular interest to theorists is $e(k)$ and how it varies with k, the energy of the interaction. Knowing this variation, we can calculate what charge (or equivalently, what coupling "constant") we would measure at any energy level. We evaluate $e(k)$ in Sect. 12.8.

The evaluation of the modified propagator and vertex relations we save for the next three chapters.

12.6 We Also Need to Renormalize Mass

There is one more wrinkle to the renormalization business. Just as the bare charge we have been using led to infinite amplitudes, so does lepton mass. And just as we had to renormalize the charge, so we will also have to do with mass.

We also have to renormalize mass

In Sects. 12.2 to 12.5, we dealt with Bhabha scattering, where the tree level propagator was a photon, having zero mass, and to make things easier in that discussion, we ignored any renormalization effect on lepton mass in the second order fermion loop. But now that we have a rudimentary understanding, we need to refine that approach to make it completely correct.

For example, in Compton scattering, we have a tree level fermion propagator, where the mass is non-zero. The presence of fermion propagators in amplitude calculations leads to additional infinite terms, which we need to address.

To begin this, recall the momentum space fermion propagator has form (see Prob. 4 for RHS)

$$S_F(p) = \frac{(\not{p} + m_0)}{p^2 - m_0^2 + i\varepsilon} = \frac{1}{\not{p} - m_0 + i\varepsilon}, \qquad (12\text{-}23)$$

where we are using our new notation m_0 for mass.

12.6.1 Renormalized Mass m vs Bare Mass m_0 to Order e^4

When we add all possible interactions, with all possible Feynman amplitudes to order e^4, we find that we can extend our trick of using the tree level amplitudes with modified charge, propagators, and vertices as long as we also modify mass. Specifically, when we include all four vertex diagrams (all up to order e^4) in our amplitude calculation for a given interaction like Compton scattering, and we limit loop integrations to a very large energy Λ, rather than infinity, we find the propagator part of the amplitude takes on the form, with the function $H(p)$ a finite function of energy p,

Fermion propagator in terms of m_0

$$S_F(p) \xrightarrow[\text{to order } e^4]{\text{all diagrams}} \underbrace{\frac{1}{\not{p} - m_0 - \delta m + i\varepsilon}(1 - H(p))}_{\text{finite}} = S_F^{2nd\ Mod}(p) \qquad \delta m = e_0^2 \frac{3m}{8\pi^2} ln\frac{\Lambda}{m} + \mathcal{O}(1). \quad (12\text{-}24)$$

Term δm is infinite for $\Lambda \to \infty$

$\mathcal{O}(1)$ indicates unit order. The δm term and the $(1 - H(p))$ factor arise from the additional diagrams of order e^4. We can then define our propagator using

Redefine $m = m_0 + \delta m$

$$S_F^{2nd\ Mod}(p) = \frac{1}{\not{p} - m + i\varepsilon}(1 - H(p)) \qquad m = m_0 + \delta m, \qquad (12\text{-}25)$$

as measured (rest) mass

where now we consider m our observed mass. The bare mass m_0, like the bare charge e_0, is never observed, since we never observe an interaction composed solely of a tree level contribution.

Note from (12-24), as we take $\Lambda \to \infty$ to restore the actual theory as it is, $\delta m \to \infty$. If m is the finite mass we observe, then we must have the bare mass $m_0 = -\infty$. The difference between the positive infinity of δm and the negative infinity of m_0 leaves the finite quantity we measure, m.

$$m_0 = -\infty$$

$$\lim_{\Lambda \to \infty} m = \lim_{\Lambda \to \infty} m_0 + \lim_{\Lambda \to \infty} \delta m = m_0 + \lim_{\Lambda \to \infty} \delta m = -\infty + \infty = m. \tag{12-26}$$

This makes fermion propagator finite

Surely, this sounds strange. Stranger even than assuming our bare charge is zero. Yet, it works. This is yet one more example of the weirdness of the quantum realm. We can make calculations that reflect the physical world, but interpreting the underlying reality leaves us scratching our heads.

12.6.2 Renormalized Mass To All Orders

If we were to sum all the higher order diagrams, the effect we would find would be similar to (12-25), except that δm would include higher order terms. That is

Renormalize mass to all orders in same manner

$$S_F^{Mod}(p) = \frac{1}{\not{p} - m + i\varepsilon} \underbrace{\left(1 - H(p) - \binom{\text{higher}}{\text{order}} \right)}_{\text{finite}} \qquad m = m_0 + \delta m$$

$$\delta m = e_0^2 \frac{3m}{8\pi^2} \ln\frac{\Lambda}{m} + \mathcal{O}(1) + \left(\text{terms in } e_0^4, \Lambda \right) + \left(\text{higher order} \right) \tag{12-27}$$

$$m_0 = -\infty \qquad \lim_{\Lambda \to \infty} \delta m = \infty \qquad m = \text{finite, measured mass.}$$

12.6.3 No Energy Dependence in Renormalized Mass

Note that renormalized mass m is not a function of the energy level of the interaction, whereas renormalized charge $e(k)$ is a function of energy level, k (symbol p also used for this energy).

Unlike $e^2(k)$, renormalized mass not dependent on k

12.7 The Total Renormalization Scheme

To find the correct, finite transition amplitude, the absolute value of which squared equals the true probability of a particular interaction occurring, follow these steps.

1. Write down the tree level amplitude with bare quantities e_0 and m_0.

2. Replace e_0 and m_0 with renormalized values $e(k)$ and m, where k is the energy level of the interaction. m is the measured particle (rest) mass. $e(k)$ is measured charge at energy level k.

Summary of steps for renormalization

3. Replace propagators with modified propagators.

4. Replace the vertex relations with modified vertex relations.

5. If you need to find $e(k)$ (perhaps because the interaction is at some energy level for which we don't have experimental values for charge), see Sect. 12.8.

6. For the form of the modified propagators and vertices, see the next three chapters.

12.8 Express e(k) as e(p) or Other Symbol for Energy

Note that instead of expressing renormalized charge e as a function of the symbol k, it is more common to use the symbol p. That is, you are more likely to see $e(p)$ in the literature than $e(k)$. Also, it is common to use yet other symbols for interaction energy level, such as μ, with charge expressed as $e(\mu)$. In any case, whatever symbol is used for the argument for e, we understand that symbol represents energy level.

We can use p, μ, or other symbol for interaction energy

12.8.1 The Number bn

<u>Deducing the Form of b_n</u>

Taking $k \to p$ in (12-19), and using the expansion relation $\sqrt{1 + 2x} = 1 + x + (\text{higher order in } x)$, where x = all terms except the first in (12-19), we have

$$e(p) = e_0 \left(1 + e_0^2 b_n \ln\frac{p}{\Lambda} + \left(\text{higher order in } e_0^2 \right) \right), \tag{12-28}$$

Determining b_n

where we need to know what b_n stands for.

The first thing we have to know about $e(p)$ is that, as we will see eventually, it turns out to depend only on the photon propagator contribution. We will show in Chap. 14 that $e(p)$ is affected by all relevant diagrams for a given interaction, but contributions from all but the photon propagator cancel out.

For example, for the first type of Bhabha scattering in Fig. 12-1, pg. 307, all the contributions that would affect $e(p)$ from all diagrams, except the upper right hand one, cancel out. In that figure, the effect on $e(p)$ of diagrams $S_{B1-2}^{(4)}$ and $S_{B1-3}^{(4)}$ cancels with the effect from diagram $S_{B1-6}^{(4)}$. We will see how this works later on. For now, just accept it. Similarly, the effect on $e(p)$ from other diagrams in Fig. 12-1 cancels with their sibling diagrams, and when all is said and done, we are left with an $e(p)$ that is solely dependent on the photon propagator. (Note that the modified propagators and vertex relations do not have this kind of cancellation effect, only $e(p)$ does.)

Only photon propagator contributes to $e^2(p)$

Other effects cancel one another

So, given that (12-28) arises from the photon propagator of the $S_{B1-1}^{(4)}$ diagram of Fig. 12-1, we might guess that the second term in that relation has a lot to do with the photon self-energy (also called the photon loop or the closed fermion loop). It does.

Consider, however, that we also can have additional diagrams like those of $S_{B1-1}^{(4)}$, except that our closed fermion loop could be made of a muon/anti-muon pair instead of an electron/positron pair. Or it could be a tau/anti-tau pair. It could even be a quark/anti-quark pair (which we won't study in this book, but which are governed by the e/m force in addition to the weak and strong forces.) Examples of quark/anti-quark pairs are up quark/anti-up quark, down quark/anti-down quark, charmed quark/anti-charmed quark, etc.

Each different charged particle type contributes a loop to photon self energy

Still further, the closed loop could actually be the charged boson/anti-boson pair W^+ and W^- of weak interactions. These other particle types need to be taken into account in our analysis[1].

Each of these different particle type pairs makes its own contribution to $e(p)$ in (12-28). Each is actually an additional Feynman diagram that must be added in. This is what b_n takes into account. It adds in these extra diagrams. In deriving (12-28), we find

$$b_n = \sum_{a=1}^{n} \frac{1}{12\pi^2} \lambda_a Q_a^2 \quad \lambda_a = \text{number of possible pair types}, \quad Q_a = \text{charge in units of } e_0 . \quad (12\text{-}29)$$

b_n takes different particle loop types into account

For $a = 1$, we would have particle/anti-particle pairs of electron/positron, muon/anti-muon, tau/anti-tau, each having associated particle charge $Q_1 = 1$ (1 unit of e_0 per particle), and $\lambda_1 = 3$ (3 families of particles). For quarks of charge magnitude $(2/3)e_0$, we would take $a = 2$ with $Q_2 = 2/3$ and $\lambda_2 = 9$, because there are three such quarks (up, charmed, and top, each having an associated anti-quark), each of which comes in three different colors, making a total of nine such pairs. For quarks of charge magnitude $e_0/3$ (down, strange, and bottom quarks, each having an associated anti-quark), we would have $Q_3 = 1/3$ and $\lambda_3 = 9$, each coming in three different possible colors, making a total of nine such pairs. For W^+ and its anti-particle W^-, $Q_4 = 1$, and $\lambda_4 = 1$.

Thus, for all known charged particles, (12-29) would be

$$b_n = \frac{1}{12\pi^2} \Big\{ \underbrace{3(1)^2}_{\text{leptons}} + \underbrace{3\cdot 3\Big(\frac{2}{3}\Big)^2}_{\substack{\text{charge } 2/3 \\ \text{quarks}}} + \underbrace{3\cdot 3\Big(\frac{1}{3}\Big)^2}_{\substack{\text{charge } 1/3 \\ \text{quarks}}} + \underbrace{1(1)^2}_{W^+,W^-} \Big\} = \frac{1}{12\pi^2} 9 = \frac{3}{4\pi^2} . \quad (12\text{-}30)$$

b_n including all charged particle types

The Dependence of b_n on Energy Level

One thing to be cautious about is that a given particle/anti-particle pair can only arise if the energy of the interaction (incoming particles) is sufficient to create a given pair. For example, the tau has a mass of 1.78 GeV. If our interaction energy k in Fig. 12-3 were only 1 GeV, we would not be able to create tau/anti-tau pairs (needing two times 1.78 GeV). So we would have to leave them

---

[1] We have not, of course, studied quarks, W bosons, and their distinguishing characteristics, though you may have heard of them elsewhere. For now, simply accept the charge and λ_a numbers we use.

out of the (12-29) relation. Thus, the lepton pair number λ_1 would equal 2, because we could still create electron/positron pairs (.511 MeV each) and muon/anti-muon pairs (.106 GeV each). Similar logic applies to each quark pair and the weak bosons W^+ and W^-. (See Prob. 5.)

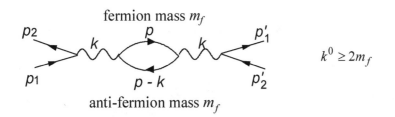

$k^0 \geq 2m_f$

Loop only occurs if for given particle i, total loop energy $\geq 2m_i$

Figure 12-3. Bhabha Scattering Photon Propagator Self Energy from One Particular Fermion at e^4 Order

We need an incoming energy equal to, or greater than, twice the mass of the particular particle type in the loop, or we will never have that loop[1]. So (12-30) was really for high energy interactions, specifically those with total incoming energy greater than twice the top quark mass (the heaviest known charged elementary particle.)

For future reference, see the table of charged particle masses below. Note that for quarks bound inside hadrons, like protons and neutrons, the large binding energy involved manifests as effective extra mass.

Table of charged particle masses

Table of Electrically Charged Particle Masses

Lepton	Q_a	Mass	**Quark**	Q_a	Mass for virtual loops	Effective, in proton	**Boson**	Mass
e -	−1	.511 MeV	up	+2/3	3 ± 2 MeV	.336 GeV	W^+, W^-	80.39 ± .02 GeV
μ	−1	.106 GeV	down	− 1/3	6 ± 2 MeV	.336 GeV		
τ	−1	1.777 GeV	charmed	+2/3	1.25 ± .1 GeV			
			strange	− 1/3	100 ± 20 MeV			
			top	+2/3	172.5 ± 1 GeV			
			bottom	− 1/3	4.25 ± .1 GeV			
1 eV = 1.60 X 10^{-12} erg = 1.78 X 10^{-33} gm					M = 10^6 G = 10^9			

12.8.2 Deriving e (p) , Given e (μ)

Evaluating (12-28), re-written below as (12-31), is problematic if we consider that, in reality, $\Lambda \to \infty$. But what we can do is evaluate $e(p)$ given a reference measurement at another energy level μ, that is, given $e(\mu)$. To see this, consider

Evaluating e(p), if e(μ) known at other energy level μ

$$e(p) = e_0 \left(1 + e_0^2 b_n \, ln\frac{p}{\Lambda} + \left(\text{higher order in } e_0^2\right)\right) \qquad (12\text{-}31)$$

$$e(\mu) = e_0 \left(1 + e_0^2 b_n \, ln\frac{\mu}{\Lambda} + \left(\text{higher order in } e_0^2\right)\right) \qquad (12\text{-}32)$$

Dividing (12-31) by (12-32) we find

---

[1] There may not seem to be a theoretical imperative for this assumption, since a virtual particle in a loop leg can have negative energy, which would thus be less than m. However, experiment supports the assumption, and an extensive evaluation of the loop integral for Fig. 12-3 (which would be unwieldy to say the least, given divergence/renormalization issues involved) should support it.

$$\frac{e(p)}{e(\mu)} = \frac{e_0\left(1 + e_0^2 b_n \, ln\frac{p}{\Lambda} + \left(\text{higher order in } e_0^2\right)\right)}{e_0\left(1 + e_0^2 b_n \, ln\frac{\mu}{\Lambda} + \left(\text{higher order in } e_0^2\right)\right)} = 1 + e_0^2 b_n \, ln\frac{p}{\Lambda} - e_0^2 b_n \, ln\frac{\mu}{\Lambda} + \left(\begin{array}{c}\text{higher order}\\ \text{in } e_0^2\end{array}\right)$$

$$= 1 + e_0^2 b_n \, ln\, p - e_0^2 b_n \, ln\, \Lambda - e_0^2 b_n \, ln\, \mu + e_0^2 b_n \, ln\, \Lambda + \left(\text{higher order in } e_0^2\right) \qquad (12\text{-}33)$$

$$= 1 + e_0^2 b_n \, ln\frac{p}{\mu} + \left(\text{higher order in } e_0^2\right).$$

Now, from (12-32), we have

$$e_0 = \frac{e(\mu)}{1 + e_0^2 b_n \, ln\frac{\mu}{\Lambda} + \left(\text{higher order in } e_0^2\right)} = e(\mu)\left(1 - e_0^2 b_n \, ln\frac{\mu}{\Lambda} + \left(\text{higher order in } e_0^2\right)\right). \quad (12\text{-}34)$$

Squaring (12-34) and putting it into the last line of (12-33) yields

$$\frac{e(p)}{e(\mu)} = 1 + \left(e^2(\mu)\left(1 - 2e_0^2 b_n \, ln\frac{\mu}{\Lambda} + \left(\text{higher order in } e_0^2\right)\right)\right)b_n \, ln\frac{p}{\mu} + \left(\text{higher order in } e_0^2\right)$$

$$= 1 + e^2(\mu)b_n \, ln\frac{p}{\mu} - e^2(\mu)2e_0^2 b_n \, ln\frac{\mu}{\Lambda}b_n \, ln\frac{p}{\mu} + \left(\text{higher order in } e_0^2\right) \qquad (12\text{-}35)$$

$$= 1 + e^2(\mu)b_n \, ln\frac{p}{\mu} + \left(\text{higher order in } e_0^2, e^2(\mu)\right).$$

Finally, we end up with

$$e(p) = e(\mu)\left(1 + e^2(\mu)b_n \, ln\frac{p}{\mu} + \left(\text{higher order}\right)\right). \qquad (12\text{-}36)$$

Expression for e(p) in terms of e(μ)

So, if we know $e(\mu)$ from experiment, we can calculate (to high accuracy if higher order terms are small, which they are) $e(p)$. We never have to take $\Lambda \to \infty$.

Of course, we will have to take care with b_n, as it might change over the region between energy μ and p. For such a transition, when a new fermion type of mass m_f would come into play, we could calculate $e(p')$ from (12-36) at the energy level $p' = 2m_f$, from our base of μ. Then take into account the added fermion loop by changing b_n to include the extra loop. Then calculate $e(p)$ from our new base p', where p is above $p' = 2m_f$, but below the level where the next loop would come into play.

In terms of the running coupling constant, we can re-write (12-36) as

Expression for α(p) in terms of α(μ)

$$\boxed{\alpha(p) = \alpha(\mu)\left(1 + \alpha(\mu)8\pi b_n \, ln\frac{p}{\mu} + \left(\text{higher order}\right)\right)}. \qquad (12\text{-}37)$$

Fig. 12-4 illustrates how the coupling constant changes as a function of p (the log of p, actually). Note the slope change at each point where the energy is great enough to create a new type of particle pair. The theoretical value predicted by (12-37) at 100 GeV matches what has been found in experiment.

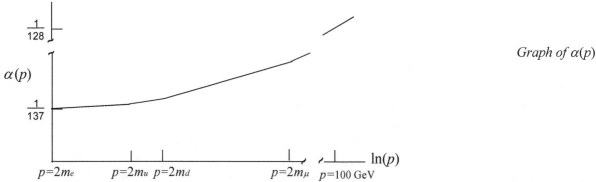

Graph of α(p)

Figure 12-4. QED Coupling Constant Dependence on Log of Interaction Energy

Note that because of the de Broglie relation, where 3-momentum is proportional to the inverse of wavelength, we have an effective inverse relation between particle energy (which depends on 3-momentum) and distance (i.e., particle wavelength). Higher energy particles have higher 3-momentum and shorter wavelengths[1]. Thus, our horizontal scale above could also be expressed as a distance scale, with lower values on the right end and higher values on the left. You will often see it expressed this way in the literature.

Energy level and distance inversely related

For all distances larger than sub-atomic, we can take $\alpha = \alpha_0 = 1/137$.

12.8.3 Renormalization Group Equation

Using (12-37), one can derive a differential equation for the behavior of the running QED coupling constant. First take the partial derivative of (12-37) with respect to μ and ignore the negligible contributions from higher order.

Deriving differential equation for $\alpha(p)$

$$\underbrace{\frac{\partial}{\partial\mu}\alpha(p)}_{0} = \frac{\partial}{\partial\mu}\alpha(\mu) + \left(\frac{\partial}{\partial\mu}\alpha^2(\mu)\right)8\pi b_n \ln\frac{p}{\mu} - \alpha^2(\mu)8\pi b_n \underbrace{\frac{\partial}{\partial\mu}\ln\mu}_{1/\mu} + \alpha^2(\mu)8\pi b_n \underbrace{\frac{\partial}{\partial\mu}\ln p}_{0} \quad (12\text{-}38)$$

Now evaluate (12-38) at $\mu = p$.

$$0 = \left(\frac{\partial}{\partial\mu}\alpha(\mu)\right)_{\mu=p} + \left(\left(\frac{\partial}{\partial\mu}\alpha^2(\mu)\right)8\pi b_n \underbrace{\ln\frac{p}{\mu}}_{ln1=0}\right)_{\mu=p} - \left(\alpha^2(\mu)8\pi b_n\frac{1}{\mu}\right)_{\mu=p} \quad (12\text{-}39)$$

This gives us

$$\left(\mu\frac{\partial}{\partial\mu}\alpha(\mu)\right)_{\mu=p} = \left(\alpha^2(\mu)8\pi b_n\right)_{\mu=p}, \quad (12\text{-}40)$$

or finally, what is called the <u>renormalization group equation (RGE)</u> (12-41) for the QED coupling constant, where the <u>RHS</u> is called the <u>beta function</u>.

$$\boxed{p\frac{\partial}{\partial p}\alpha(p) = \alpha^2(p)8\pi b_n = \beta(p,b_n)} \quad . \quad (12\text{-}41)$$

RGE for $\alpha(p)$

This is a differential equation expression for the evolution of the QED coupling constant with energy level p. By integrating it numerically or otherwise, one can find α at any energy level. From Fig. 12-4, we see the partial derivative in (12-41) is positive, and thus the beta function is positive. If the beta function were zero, then the QED coupling constant, and the theory itself, would be (energy) <u>scale invariant</u>. As an aside, the beta function in weak and strong interactions is negative.

It may seem like a lot of work to use the differential form (12-41) and integrate, when we have the algebraic form (12-37) we can just use directly. However, in more advanced theories of quantum fields, RGEs play an important role, so we provide an introduction to them here.

By doing Prob. 8, you can prove the corresponding RGE for $e(p)$ to yourself.

$$p\frac{\partial}{\partial p}e(p) = e^3(p)b_n. \quad (12\text{-}42)$$

RGE for $e(p)$

12.9 Things You May Run Into

<u>Screening</u>

What Fig. 12-4 is telling us is that as we probe closer to a charged particle like an electron (meaning higher energy, shorter wavelength, probing particles), the effective coupling constant (and thus, the charge) increases. This is a direct result of the mathematics of the theory, but it is often represented heuristically as a <u>screening</u> effect.

---

[1] This, of course, assumes particles are wave packets where a shorter packet means higher energy and momentum. In practice, particles in interactions are wave packets and not pure 3-momentum eigenstates as we assume during most of our QFT calculations. This actually all works out, without inconsistencies, as we hopefully will understand a bit better in Part 4 of this book.

In screening, a real electron emanates virtual photons, which mediate the e/m force and continually form higher order particle/anti-particle virtual loops. These loops have a positively charged particle leg and a negatively charged particle leg. The positively charged leg is attracted toward the real electron, while the negatively charged leg is repelled. Presumably a probe test particle is "screened" from "seeing" the full effect of the real electron charge by the particle/anti-particle loops. When the test particle is closer, there are fewer loops to screen it, so it sees a greater charge on the real electron.

In screening scenario, loops "screen" charge

I have thought to fair depth about this description and been unable to make the presumed principle work in a satisfactory manner, so I am not a fan of it. For me, the coupling constant dependence on energy/distance is simply something that falls out of the math and does not seem readily visualized via the screening scenario or any other I am aware of. I mention screening here because you, the reader, will probably run across it at some point.

Screening description is heuristic at best

Landau Pole

The QED coupling constant appears to go to infinity as $p \to \infty$. See Fig. 12-4. This is different from the infinity cropping up from higher order contributions. Lev Landau pointed this out, and the behavior is named after him, i.e., α is said to be infinite at the <u>Landau pole</u>. (Recall a pole is a number that causes a relation to go unbounded.) At energy levels above the Planck scale, however, everyone expects physics to be different, so the Landau pole is not a big concern.

α has Landau pole at ∞, but physics different beyond Planck scale

"Bare Charge" Terminology

We have taken the term bare charge for our symbol e_0, but note that that term is often used to mean the charge one would measure by getting infinitesimally close to a charged fermion. (For the screening viewpoint, this would correspond to being inside the screening virtual loops, and thus be the "bare", unscreened charge. In renormalization, on the other hand, e_0 is the charge one would have with only the lowest order contribution to the interaction.) Such smaller distance scales approaching zero correspond in Fig. 12-4 to momentum (and thus energy) levels approaching infinity. At such scales, α, and thus e, approach infinity. Hence, it is commonly said that bare charge of an electron is infinite (and this might imply a contradiction with what was said after (12-15).) It really depends on which of two ways you wish to use the label "bare charge".

"Bare charge" terminology can mean either
1) *e_0, or*
2) *$e(p)$ as $p \to \infty$*

12.10 Adiabatic Hypothesis

Prior to interaction, particles are not truly free of interactions, since they can have self-energy interactions like the incoming leptons do for $S_{B1-2}^{(4)}$ to $S_{B1-5}^{(4)}$ of Fig. 12-1, pg. 307. That is, there is really no time that interaction is not occurring. That is, there are no "bare" particles. All real particles are always "dressed" with loops like those in the cited diagrams.

Bare particles incoming and outgoing in theory are not so in reality

However, when we compute amplitudes, we consider the incoming particles (i.e., the ket $|i\rangle$) to be "bare", which one can visualize, in a sense that isn't really true, as the incoming lines in the cited diagrams prior to the loops forming. So mathematically, we think in terms of "bare" incoming particles that acquire "dressing" as they enter the interaction arena. In Fig. 12-1, we can think of the incoming particles as i) first, bare, then ii) adding self-energy loops, then iii) interacting with the exchange of a virtual photon (plus the higher order activity).

In doing this, we are using what is called the <u>adiabatic hypothesis</u>. "Adiabatic" in thermodynamics means a system does not exchange heat with the environment. Radiation is a form of heat. So, by assuming there are no self-energy radiative loops, we assume an adiabatic situation exists prior to the forming of the loops. For this, we must assume the coupling constant α (or equivalently, e) is turned off, so particles cannot self-interact. That is, we consider its effect on our interactions modified by a function of time, $e(p) \to f(t)e(p)$, where $f(t)$ is shown in Fig. 12-5.

Adiabatic hypothesis presumes bare particles exist before they become "dressed"

Prior to $-T$, the particles are bare (i.e., as $|i\rangle$). After $-T$, they become "dressed" due to self-interaction, but are not close enough yet to interact with one another. At $t = -\tau$, they become close enough to interact with one another, which continues until τ. The dressed particles then move apart and cease interacting, but remain dressed until T. At T, self-interaction ceases leaving the outgoing particles as bare, once again (i.e., as $|f\rangle$).

This is all fiction, of course, due in part to the fact that bare particles are never seen in the physical world and can never be measured. But it allows us to use our mathematical expressions which take $|i\rangle$ and $|f\rangle$ to be bare particles. After all our calculations are done, we take $-T \to -\infty$ and $T \to \infty$, to restore our fiction to reality, and obtain correct answers from our theory.

After finished calculating with adiabatic hypothesis, take $T \to \infty$

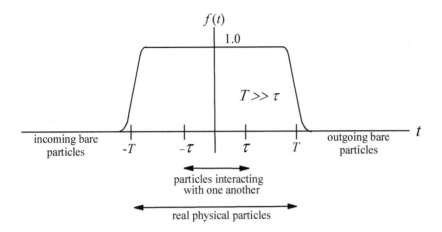

Figure 12-5. The Adiabatic Hypothesis

Some readers may be concerned that the time of particles interacting with one another, $-\tau < t < \tau$, was taken in earlier work with the S operator to be of infinite duration. In the adiabatic hypothesis, we can take τ to infinity, as well, but still much less than T.

In practice, we won't need to think about the adiabatic hypothesis, again. It simply provides justification for the manner in which we carry out our mathematics, which due to the presence of the bare initial and final particles might otherwise cause some concern.

No need to think about adiabatic hypothesis after this. Just calculate.

12.11 Regularization Revisited

As discussed briefly in Sect. 12.2, a process by which unbounded integral relations are tamed to yield finite results is called <u>regularization</u>. Typically, the regularization process yields a finite result as long as we take a given parameter Λ as finite, even though in reality it is infinite. That is, the result becomes infinite when $\Lambda \to \infty$. To solve this, one redefines quantities (as we did for e and m) such that when we take $\Lambda \to \infty$, we get a finite, final result.

Regularization helps tame infinite integrals. Gives result in terms of Λ instead of ∞.

There are a number of ways by which one can regularize the unbounded relations involving integrals that arise when we include higher order corrections in our amplitudes. We looked at one such way in Sect. 12.2. We will study others in Chap. 15.

Several ways to do it

In this chapter, we simply assumed we had already used a regularization process to get our amplitude in terms of a parameter Λ. (See (12-12), for example.) The regularization gave us

$$e^2\left(p,\Lambda\right)= e_0^2\left\{1+e_0^2\, 2b_n\, ln\frac{p}{\Lambda}+\mathcal{O}\left(e_0^4\right)\right\} \qquad (12\text{-}43)$$

as a factor in our amplitude(s). We then re-defined $e_0 = 0$, but such that when $\Lambda \to \infty$, $e(p)$ became the physical (measured in experiment) electric charge "constant" at energy level p.

We then redefine quantities so when take $\Lambda \to \infty$, we get valid results

The process of regularization is applied to the full propagator integral, and in so doing, also gives us the forms of the modified propagators and vertices. But those are finite and less problematic. Thus, regularizing the propagator integrals gave us unbounded parts (for $\Lambda \to \infty$) as in (12-43) and also finite parts, as in the modified propagators and vertices.

<u>Summary</u>: Regularization of an unbounded integral typically yields a finite part and another part that goes to infinity as a parameter $\Lambda \to \infty$. Since the theory calls for Λ to be infinite, we find a trick, such as redefining quantities, so that the latter part becomes finite when $\Lambda \to \infty$.

Regularization can give us finite and infinite (for $\Lambda\to\infty$) terms. We need a trick, like redefinition of quantities, to take care of the infinity.

12.12 Where We Stand

12.12.1 Can You Do Renormalization Problems?

If you understand the principles of this chapter, you can, in essence, do QFT, including the effects of higher order diagrams. However, to do this (to second order for example), you would need someone else to provide you with the relations for $\gamma^{\mu}_{Mod,2nd}$ $D^{Mod,2nd}_{F\mu\nu}$ and $S^{Mod,2nd}_{F}$ (see (12-12)), which we have not yet derived. You would also have to assume our relations for $e\left(p\right)$ and m of (12-31) and (12-27) are correct, which we also have yet to derive.

You can now solve scattering problems to any order, but need to use relations we haven't derived yet

So, yes, you can do scattering problems to higher order, simply by using tree level amplitudes and substituting physical charge $e(p)$ of (12-36) for e_0, physical mass m for m_0, modified propagators for bare propagators, and modified vertex relations for bare vertex ones.

12.12.2 Where Do We Go From Here?

In this chapter, we have laid out an overview of renormalization, which should be a solid foundation upon which to build. In Chaps. 13, 14, and 15 we get down to the mathematics underlying it all. This, as anyone who has studied it will attest, is messy, and that may be the understatement of the month for you. If you keep the fundamentals of this chapter in mind, as we progress through the coming mathematics, then hopefully, it will be easier going than it otherwise would have been.

We'll derive those relations in the next three chapters

12.13 Chapter Summary

12.13.1 The Renormalization Procedure

The process of renormalization to a given order n in α (i.e., in e^2) comprises the following.

Summary of how we renormalize

1. Draw all relevant Feynman diagrams with 2 to $2n$ vertices.

2. Add all the related Feynman amplitudes.

3. Evaluate all the propagator integrals in the amplitudes via regularization, yielding quantities e and m dependent on e_0 and m_0 respectively, and a parameter Λ, wherein e and m become infinite when $\Lambda \to \infty$. Other quantities will be obtained that are finite and not dependent on Λ. These other quantities modify propagators and vertex relations.

4. Redefine e_0 and m_0 such that, as $\Lambda \to \infty$, e and m equal the finite, physical values for the QED charge constant and fermion particle (rest) mass. e is dependent on interaction energy level p and is expressed as $e(p)$.

12.13.2 Solving Scattering Problems to Order n

To find a transition amplitude for a given interaction, do the following (a recap of Sect. 12.7).

Summary of how to solve problems given our renormalization

1. Write down the tree level amplitude with bare quantities e_0 and m_0.

2. Replace e_0 and m_0 with renormalized values $e(p)$ and m, where p is the energy level of the interaction, m is the measured particle (rest) mass, and $e(p)$ is measured charge at energy level p.

3. Replace propagators with modified propagators.

4. Replace the vertex relations with modified vertex relations.

12.13.3 Relations for α(p), e(p) and m

<u>Algebraic Relations</u>

Expressions for $e(p)$ and $\alpha(p)$, given $e(\mu)$ and $\alpha(\mu)$

$$\alpha(p) = \alpha(\mu)\left(1 + \alpha(\mu)8\pi b_n \, ln\frac{p}{\mu} + (\text{higher order})\right).$$

$$e(p) = e(\mu)\left(1 + e^2(\mu)b_n \, ln\frac{p}{\mu} + (\text{higher order})\right) \qquad b_n = \sum_{a=1}^{n}\frac{1}{12\pi^2}\lambda_a Q_a^2 \qquad (12\text{-}44)$$

λ_a = number of possible particle pair types, Q_a = charge in units of e_0

Expression for m

$$m = m_0 + \delta m \qquad \delta m = e_0^2\frac{3m}{8\pi^2}\,ln\frac{\Lambda}{m} + ...$$

$$m_0 = -\infty \qquad \underset{\Lambda\to\infty}{\text{limit}}\,\delta m = \infty \qquad m = \text{finite, measured mass.} \qquad (12\text{-}45)$$

<u>Differential Relations: Renormalization Group Equations</u>

Differential expressions for $e(p)$ and $\alpha(p)$ at energy point p

$$p\frac{\partial}{\partial p}\alpha(p) = \alpha^2(p)8\pi b_n \qquad (12\text{-}46)$$

$$p\frac{\partial}{\partial p}e(p) = e^3(p)b_n \qquad (12\text{-}47)$$

12.14 Problems

1.   In case your freshman physics is a little hazy, find the center of mass x_m of two masses m_1 and m_2, located respectively at x_1 and x_2. Then take the time derivative of $(m_1 + m_2)x_m$ to get the total 3-momentum of the system. If we are in another, primed, coordinate frame that is fixed to the COM of the system of particles, what is the velocity of the COM of the system in that primed coordinate frame? What is the total system 3-momentum as measured in the primed coordinate frame, i.e., in the COM frame?

2.   Show that for Bhabha scattering of the type in top two figures of Fig. 12-1, where k^0 is the energy of the virtual photon, and $\mathbf{k}$ is its 3-momentum, that $(k^0)^2 \geq \mathbf{k}^2$. (Hint: Use the relation (B8-1.2) of Box 8-1, pg. 219, i.e., for a system of particles $E_{sys}^2 - \mathbf{p}_{sys}^2 = m_{sys}^2$, where m_{sys} is generally not the total system mass, but an invariant quantity we determine knowing the total system energy and 3-momentum. E_{sys} and $\mathbf{p}_{sys}$ are the same both before and after the vertex interaction where the electron-positron pair mutate into a photon.)
     Note that if the photon were on shell, we would have $(k^0)^2 = \mathbf{k}^2$. If we were in the COM frame, we must have $\mathbf{k} = 0$.

3.   Does it make sense to you that for our mass renormalization, we can have
$$\lim_{\Lambda \to \infty} \delta m = (\text{constant})\lim_{\Lambda \to \infty} e_0^2 \, ln\frac{\Lambda}{m} = \infty \text{ and } e_0^2 = 0\,,$$ whereas in our charge renormalization, we
have $(\text{different constant})\lim_{\Lambda \to \infty} e_0^4 \, ln\frac{k}{\Lambda} = e^2(k) = \text{finite and } e_0^2 = 0$? Why?

4.   From the form of the fermion propagator $S_F(p) = \dfrac{(\not{p} + m_0)}{p^2 - m_0^2 + i\varepsilon}$, derive the equivalent form

$S_F(p) = \dfrac{1}{\not{p} - m_0 + i\varepsilon}$.   Hint: You   need   use $(\not{p})^2 = \not{p}\not{p} = p_\mu \gamma^\mu p_\nu \gamma^\nu = \gamma^\mu \gamma^\nu p_\mu p_\nu =$

$\frac{1}{2}\left(\gamma^\mu \gamma^\nu p_\mu p_\nu + \gamma^\nu \gamma^\mu p_\nu p_\mu\right) = \frac{1}{2}\left[\gamma^\mu, \gamma^\nu\right]_+ p_\mu p_\nu = g^{\mu\nu} p_\mu p_\nu = p^2$ from Chap. 4 Appendix A.

When you are all through, note that we have carried a 4X4 identity matrix through at every step. That is, when we multiply two gamma matrices together (in spinor space) we get a 4X4 matrix result. In terms of the Chap. 4 relation cited, with spinor matrices written out, it looks like $(\not{p})^2 = p_\mu \gamma_{\alpha\beta}^\mu p_\nu \gamma_{\beta\delta}^\nu = p^2 I_{\alpha\delta}$. The first two terms in the denominator of the second relation for

the propagator then are actually $\not{p} - m_0 = p_\mu \gamma_{\alpha\delta}^\mu - m_0 I_{\alpha\delta}$.

5.   From the Table of Fermion Masses on pg. 315, determine b_n for a scattering experiment at .9 GeV. (Answer: $1/(3\pi^2)$.) Determine b_n for an experiment at 1.4 GeV. (Hint: This takes no work.) Determine b_n for an experiment at 100 GeV. (Answer: $5/(9\pi^2)$.)

6.   Use (12-37) to calculate the QED coupling constant at 5.5 MeV. Assume the up quark mass is exactly 3 MeV and the down quark mass is exactly 6 MeV. Use 1/137 for $\alpha(\mu = 2m_e)$.

7.   Determine what the QED coupling constant will be at 6.0 MeV. (You need to assume the evaluation is at a virtually immeasurable amount less than 6.0 MeV.) Then determine what the coupling constant is at 12.0 MeV (i.e., immeasurably below 12.0). Use the same assumptions as Prob. 6.

8.   Following similar steps to what we did to find (12-41), derive (12-42).

Chapter 13

Renormalization Toolkit

My mind to me a kingdom is,
Such present joys therein I find
That it excels all other bliss
That earth affords or grows by kind.

Sir Edward Dyer

13.0 Preliminaries

13.0.1 Our Approach

In the last chapter we saw an overview perspective of renormalization. In this chapter, we will provide some basic, mostly background, tools one needs in order to do renormalization. These include certain mathematical relations, fundamental QFT identities, and the like. With this toolkit, in the next chapter we will then (finally) dive directly into the process of renormalization itself. Throughout this and the next chapter, we will not, however, evaluate any of the troublesome unbounded integrals discussed earlier. We save that for the chapter after next, which is devoted to regularization (taming infinite integrals by making them temporarily finite.)

Renormalization chapters' content summary

In summary,

Chap. 12: Overview of renormalization in simplest possible terms, with limited math.

Chap. 13: Assembling the tools we'll need to undertake renormalization.

Chap. 14: Putting it all together, i.e., actually renormalizing QED (using symbols for the infinite integrals evaluated via methods of Chap. 15).

Chap. 15: Regularization, expressing unbounded integrals in a form suitable for renormalization.

13.0.2 Chapter Overview

In this chapter, we will

- define shorthand symbols for the three divergent integrals,

- explore certain math relations and QFT identities needed for renormalization,

- detail the dependence of renormalization on gauge invariance,

- note changes to $\mathcal{L}$ and Feynman diagrams by replacing bare mass m_0 with physical mass m,

- see how part of one divergent integral equals part of another divergent integral, and

- express the propagators, vertex, and external particle relations to 2nd order in terms of the divergent integrals shorthand symbols.

This chapter → building the foundation for renormalization

13.1 The Three Key Integrals

13.1.1 Shorthand Expression of the Integrals

In Chap. 9 we saw how the problematic parts of the second order in α amplitude corrections comprised photon self-energy, electron self-energy, and vertex loop integrals, which all diverged to infinity. As noted in that chapter, and as we shall see further, in calculating radiative corrections to lowest order of *any* process, these same three divergent integrals occur and no others. Consequently, if we can resolve issues with these integrals, we can calculate 2nd order (in α) corrections readily.

Applying Feynman's rules in Fig. 13-1, excluding the incoming and outgoing particles, yields the amplitudes under each diagram, where the symbols are defined in (13-1) to (13-3).

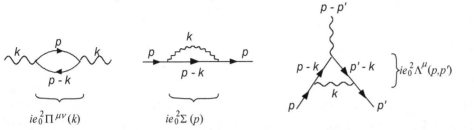

Three key corrections in renormalization

Figure 13-1. Photon Self-energy, Fermion Self-energy, and Vertex Loop Correction

To aid us in the future, we will represent the integrals in these amplitudes, respectively and apart from the ie_0^2 factors, by the symbols $\Pi^{\mu\nu}(k)$, $\Sigma(p)$, and $\Lambda^\mu(p, p')$.[1]

The 2nd order photon self-energy integral

$$\Pi^{\mu\nu}(k) = \frac{-i}{(2\pi)^4}\text{Tr}\int iS_F(p)\gamma^\mu iS_F(p-k)\gamma^\nu d^4p \tag{13-1}$$

The divergent integrals associated with those three key corrections

The 2nd order fermion self-energy integral

$$\Sigma(p) = \frac{i}{(2\pi)^4}\int iD_{F\alpha\beta}(k)\gamma^\alpha iS_F(p-k)\gamma^\beta d^4k \tag{13-2}$$

The 2nd order vertex loop correction integral

$$\Lambda^\mu(p,p') = \frac{-1}{(2\pi)^4}\int iD_{F\alpha\beta}(k)\gamma^\alpha iS_F(p'-k)\gamma^\mu iS_F(p-k)\gamma^\beta d^4k \tag{13-3}$$

The following mnemonics may help you keep these symbols straight. Π and photon both begin with "p", and $\Pi^{\mu\nu}$ represents photon self-energy. Σ represents fermion self-energy, and Sigma begins with "S", the label we use for the fermion propagator S_F. Λ^μ has an inverted V shape and represents the vertex correction, which begins with "v". Additionally, two-thirds of a vertex diagram even has a V shape when viewed at the appropriate angle.

Memorization tricks

As alluded to before, these integrals are evaluated by regularizing them. Regularization entails making otherwise divergent integrals temporarily convergent via judicious use of a <u>parameter</u>, which we represent by Λ (<u>not to be confused with Λ^μ of (13-3)</u>). Use of Λ allows the integral to be evaluated, with the final result expressed in terms of Λ. Then, in the final result at the end, one takes the limit of Λ (in the simplest way to regularize, $\Lambda \to \infty$) to obtain the real-world expression of the divergent integral.

Regularize (see Chap. 15) to express integrals in terms of finite Λ, then $\Lambda \to \infty$

When one carries out such regularization on (13-1) to (13-3), in each case, one finds a result with at least one term that is finite and at least one other that is finite for finite Λ, but infinite when we take the limit of Λ (where again, for the simplest kind of regularization, $\Lambda \to \infty$). We represent these different terms by the symbols shown in the relations below.

Symbols to express various terms obtained when regularize

$$\Pi^{\mu\nu}(k) = -\underbrace{g^{\mu\nu}k^2 A'(k,\Lambda)}_{\infty \text{ for } \Lambda \to \infty} - \underbrace{g^{\mu\nu}k^2\Pi_c(k^2)}_{\text{finite}} \tag{13-4}$$

$$\Sigma(p) = \underbrace{A(\Lambda,m)}_{-\infty \text{ for } \Lambda \to \infty} + \underbrace{(\not p - m)B(\Lambda)}_{\infty \text{ for } \Lambda \to \infty} + \underbrace{(\not p - m)\Sigma_c(\not p - m)}_{\text{finite}} \tag{13-5}$$

$$\Lambda^\mu(p,p') = \underbrace{L(\Lambda)\gamma^\mu}_{\infty \text{ for } \Lambda \to \infty} + \underbrace{\Lambda_c^\mu(p,p')}_{\text{finite}} \tag{13-6}$$

---

[1] Different authors use different symbols for these integrals. Peskin and Schroeder, for example, add a subscript "2" to indicate second order (no higher loops in each drawing of Fig. 13-1) and use the symbol we have to include all orders. Mandl and Shaw, on the other hand, use these symbols as we do.

Note the <u>subscript "c"</u> represents the convergent part of the integral. Note also that we have taken some latitude in using the physical mass m in (13-5). We should, in principle, use the bare mass m_0, but doing so makes the subsequent analysis far more difficult. With foresight (which you don't have yet), one knows that expressing (13-5) as we have shown it is valid.

The $\Sigma_c\left(\not{p}-m\right)$ in (13-5) actually stands for a series of terms, summed over i, of form $K_i\left(\not{p}-m\right)^i$. So, we can actually think of the expression for $\Sigma(p)$ as a Taylor expansion in terms of $\left(\not{p}-m\right)$. Note that we could instead expand around $\left(\not{p}-m_0\right)$. With foresight, we choose the former, and things eventually will work out easier that way. Realizing this will help us in a derivation to come. (13-4) and (13-6) can be considered as Taylor expansions, as well.

$\Sigma(p)$, $\Pi^{\mu\nu}(k)$, and $\Lambda^\mu(p,p')$ can be considered as Taylor expansions.

13.1.2 Full Expression of the Integrals

Evaluating (13-1) to (13-3) takes many pages of unwieldy mathematics and comprises most of the usual development of the renormalization process. In the process of doing this, most students get quite lost for long periods, and many never really gain a sound understanding of the underlying essence of renormalization.

In an attempt to ameliorate this experience for you the reader, we will in this chapter simply assume that we have integral tables available that we can use to readily express (13-1) to (13-3), or equivalently, (13-4) to (13-6). We can then, in relatively short order, get to the bottom of all this renormalization business. With that under our belts, in Chap. 15, we can return to show how those integrals are actually evaluated using regularization.

Thus,

$$\Pi^{\mu\nu}\left(k\right)=g^{\mu\nu}k^2\underbrace{2b_n\ln\frac{k}{\Lambda}}_{-A'\left(k,\Lambda\right)}+\underbrace{g^{\mu\nu}k^2\frac{1}{2\pi^2}\int_0^1 z\left(1-z\right)\ln\left(z\left(1-z\right)-m^2/k^2\right)dz}_{-\Pi_c(k^2)}. \qquad (13\text{-}7)$$

Explicitly expressing some of the terms found after regularization

$$\Sigma\left(p\right)=\underbrace{-\frac{3m}{8\pi^2}\ln\frac{\Lambda}{m}}_{A\left(\Lambda,m\right)}+\underbrace{\left(\not{p}-m\right)\frac{1}{8\pi^2}\ln\Lambda}_{B\left(\Lambda\right)}+\underbrace{\left(\not{p}-m\right)\Sigma_c\left(\not{p}-m\right)}_{\text{too complicated to express here}} \qquad (13\text{-}8)$$

$$\Lambda^\mu\left(p,p'\right)=\underbrace{\frac{1}{8\pi^2}\gamma^\mu\ln\Lambda}_{L\left(\Lambda\right)\gamma^\mu}+\underbrace{\Lambda_c^\mu(p,p')}_{\substack{\text{too complicated}\\\text{to express here}}} \qquad (13\text{-}9)$$

If your first response is angst at looking at the finite integrals, the rightmost terms of (13-7) to (13-9), remember that we are actually cutting to the chase and taking the simplest approach. Going through the steps of actually deriving these expressions can raise one's angst to some power significantly greater than one.

For now, we will explore the theory of renormalization by using symbols for tough integrals

There are ways to evaluate the finite integrals, but that does not concern us for the present. We will simply use the symbols $\Pi_c(k^2)$, $\Sigma_c(\not{p}-m)$, and $\Lambda_c^\mu(p,p')$ in what follows.

Although we can see from the above that $B=L$, we will soon show from general principles that the two are equal and that because of this, they end up cancelling one another in any amplitude calculation one would make for any possible interaction. This is actually an amazing blessing, for if it were not true, renormalization would be even more difficult, perhaps impossible, and we might not to this day have been able to renormalize the theory. More on this shortly.

We will find terms with B and L drop out

Note that the estimates of the integrals (13-7) to (13-9) we found via power counting in Chap. 9 were actually conservative and naïve. (See Wholeness Chart 9-1, pg. 265.) From those estimates, we expected (13-7) to diverge quadratically, but in actuality it is logarithmically divergent. (Λ here corresponds with our integration variable p limit in the photon loop of Chap. 9. Λ is an energy level, as was p there.)

Actual divergence $\leq$ superficial power counting estimate

In general, when we estimate an integral's degree of divergence via naïve power counting, the actual degree of divergence will be that of our estimate or less.

13.2 Relations We'll Need

13.2.1 Auxiliary Relations

As we saw in Appendix A of Chap. 4,

$$\left(\gamma^\nu p_\nu - m\right)u_r\left(\mathbf{p}\right) = 0 \qquad \bar{u}_r\left(\mathbf{p}\right)\left(\gamma^\nu p_\nu - m\right) = 0 \tag{13-10}$$

Some relations we'll need

$$\left[\gamma^\mu,\gamma^\nu\right]_+ = \gamma^\mu\gamma^\nu + \gamma^\nu\gamma^\mu = 2g^{\mu\nu}. \tag{13-11}$$

(13-11) is used in the last step of (13-12) below.

$$\gamma^\mu\gamma^\nu = \tfrac{1}{2}\left(\left[\gamma^\mu,\gamma^\nu\right]_+ + \left[\gamma^\mu,\gamma^\nu\right]\right) = g^{\mu\nu} + \tfrac{1}{2}\left[\gamma^\mu,\gamma^\nu\right] \tag{13-12}$$

For A and B any two operators, which need not commute (proof follows),

$$\frac{1}{A-B} = \frac{1}{A} + \frac{1}{A}B\frac{1}{A} + \frac{1}{A}B\frac{1}{A}B\frac{1}{A} + ... \tag{13-13}$$

<u>Proof</u>

Post multiply (13-13) by $A - B$.

$$1 = \frac{1}{A-B}\left(A-B\right) = \frac{1}{A}\left(A-B\right) + \frac{1}{A}B\frac{1}{A}\left(A-B\right) + \frac{1}{A}B\frac{1}{A}B\frac{1}{A}\left(A-B\right) + ...$$

$$= 1 \underbrace{- \frac{B}{A} + \frac{B}{A}}_{=0} \underbrace{- \frac{B}{A}\frac{B}{A} + \frac{B}{A}\frac{B}{A}}_{=0} \underbrace{- \frac{B}{A}\frac{B}{A}\frac{B}{A} + \frac{B}{A}\frac{B}{A}\frac{B}{A}}_{=0} + 0 + ... \tag{13-14}$$

<u>End of proof</u>

13.2.2 Gordon's Identity

Consider relations like (13-10) for a different momentum $\mathbf{p}'$ and spin state r',

$$\left(\gamma^\nu p'_\nu - m\right)u_{r'}\left(\mathbf{p}'\right) = 0 \qquad \bar{u}_{r'}\left(\mathbf{p}'\right)\left(\gamma^\nu p'_\nu - m\right) = 0. \tag{13-15}$$

Deriving Gordon's identity

Multiply the first of (13-10) by $\bar{u}_{r'}\left(\mathbf{p}'\right)\gamma^\mu$ on the left; then multiply the second of (13-15) by $\gamma^\mu u_r\left(\mathbf{p}\right)$ on the right; then add the two, to get

$$2m\bar{u}_{r'}\left(\mathbf{p}'\right)\gamma^\mu u_r\left(\mathbf{p}\right) = \bar{u}_{r'}\left(\mathbf{p}'\right)\left(\gamma^\mu\gamma^\nu p_\nu + \gamma^\nu\gamma^\mu p'_\nu\right)u_r\left(\mathbf{p}\right). \tag{13-16}$$

Using (13-12) in (13-16), we find

$$2m\bar{u}_{r'}\left(\mathbf{p}'\right)\gamma^\mu u_r\left(\mathbf{p}\right) = \bar{u}_{r'}\left(\mathbf{p}'\right)\left(p_\nu\left(g^{\mu\nu} + \tfrac{1}{2}\left[\gamma^\mu,\gamma^\nu\right]\right) + p'_\nu\left(g^{\nu\mu} + \tfrac{1}{2}\left[\gamma^\nu,\gamma^\mu\right]\right)\right)u_r\left(\mathbf{p}\right)$$

$$= \bar{u}_{r'}\left(\mathbf{p}'\right)\left(p_\nu\left(g^{\mu\nu} + \tfrac{1}{2}\left[\gamma^\mu,\gamma^\nu\right]\right) + p'_\nu\left(g^{\mu\nu} - \tfrac{1}{2}\left[\gamma^\mu,\gamma^\nu\right]\right)\right)u_r\left(\mathbf{p}\right). \tag{13-17}$$

Re-arranging and dividing by $2m$, we have <u>Gordon's Identity</u> (important, but don't memorize)

$$\boxed{\bar{u}_{r'}\left(\mathbf{p}'\right)\gamma^\mu u_r\left(\mathbf{p}\right) = \frac{p^\mu + p'^\mu}{2m}\bar{u}_{r'}\left(\mathbf{p}'\right)u_r\left(\mathbf{p}\right) + \frac{p_\nu - p'_\nu}{4m}\bar{u}_{r'}\left(\mathbf{p}'\right)\left[\gamma^\mu,\gamma^\nu\right]u_r\left(\mathbf{p}\right).} \tag{13-18}$$

13.2.3 A Key Identity

A crucial ingredient in renormalization is a relationship between the electron self-energy integral $\Sigma(p)$ and the vertex loop integral for the special case where $p = p'$, $\Lambda^\mu(p, p)$, which we will first state, then prove. It is called by some the <u>Ward identity</u>, but we will call it (for reasons to become obvious) the <u>original Ward identity</u>. It is

$$\boxed{\frac{\partial\Sigma(p)}{\partial p_\mu} = \Lambda^\mu\left(p,p\right).} \tag{13-19}$$

Original Ward identity

You might think the proof could simply be done by taking the derivative of $\Sigma(p)$ with respect to p^μ, but that entails taking derivatives of matrix relations in S_F and ends up with matrix multiplications where the ordering of those is ambiguous. We get around that issue by first finding the derivative of S_F a different way, as in steps (13-20) to (13-22) below.

Proof of the original Ward identity

From
$$\left(S_F\left(p\right)\right)^{-1} = \not{p} - m \tag{13-20}$$

we find
$$0 = \frac{\partial\left(1\right)}{\partial p_\eta} = \frac{\partial}{\partial p_\eta}\left(\left(S_F\left(p\right)\right)\left(S_F\left(p\right)\right)^{-1}\right) = \frac{\partial}{\partial p_\eta}\left(\left(S_F\left(p\right)\right)\left(\not{p} - m\right)\right)$$

$$= \frac{\partial S_F\left(p\right)}{\partial p_\eta}\left(\not{p} - m\right) + S_F\left(p\right)\frac{\partial}{\partial p_\eta}\left(\not{p} - m\right) = \frac{\partial S_F\left(p\right)}{\partial p_\eta}\left(S_F\left(p\right)\right)^{-1} + S_F\left(p\right)\gamma^\eta. \tag{13-21}$$

Or
$$\frac{\partial S_F\left(p\right)}{\partial p_\eta}\left(S_F\left(p\right)\right)^{-1} = -S_F\left(p\right)\gamma^\eta \;\rightarrow\; \frac{\partial S_F\left(p\right)}{\partial p_\eta} = -S_F\left(p\right)\gamma^\eta S_F\left(p\right). \tag{13-22}$$

Taking $p_\eta \rightarrow p_\eta - k_\eta$, (13-22) becomes

$$\frac{\partial S_F\left(p-k\right)}{\partial\left(p_\eta - k_\eta\right)} = -S_F\left(p-k\right)\gamma^\eta S_F\left(p-k\right). \tag{13-23}$$

Then, with (13-23) used in the second line below, we have

$$\frac{\partial\Sigma\left(p\right)}{\partial p_\mu} = \frac{\partial}{\partial p_\mu}\frac{i}{\left(2\pi\right)^4}\int iD_{F\alpha\beta}\left(k\right)\gamma^\alpha iS_F\left(p-k\right)\gamma^\beta d^4k$$

$$= \frac{i}{\left(2\pi\right)^4}\int iD_{F\alpha\beta}\left(k\right)\gamma^\alpha i\frac{\partial S_F\left(p-k\right)}{\partial\left(p_\eta - k_\eta\right)}\underbrace{\frac{\partial\left(p_\eta - k_\eta\right)}{\partial p_\mu}}_{=\delta_\eta^\mu}\gamma^\beta d^4k = \frac{i}{\left(2\pi\right)^4}\int iD_{F\alpha\beta}\left(k\right)\gamma^\alpha i\left(-S_F\left(p-k\right)\gamma^\mu S_F\left(p-k\right)\right)\gamma^\beta d^4k$$

$$= \frac{-1}{\left(2\pi\right)^4}\int iD_{F\alpha\beta}\left(k\right)\gamma^\alpha iS_F\left(p-k\right)\gamma^\mu iS_F\left(p-k\right)\gamma^\beta d^4k = \Lambda^\mu\left(p,p\right). \tag{13-24}$$

End of proof

13.2.4 The Ward Identities

Ward's name is associated with an additional set of identities, which play a key role in renormalization, and also in scattering calculations. They are called <u>Ward identities</u>, but to distinguish them from (13-19), we called the earlier relation the "original Ward identity". We derive the Ward identities in this section, but before that, we need a bit of background information.

Gauge Invariance Means Amplitude Invariance

Local gauge invariance means our Lagrangian $\mathcal{L}$ is symmetric in form under the transformations

$$\psi \rightarrow \psi' = e^{-i\alpha(x)}\psi \qquad A_v \rightarrow A_v' = A_v - \frac{1}{e}\partial_v\alpha\left(x\right), \tag{13-25}$$

where the numeric (not operator) field $\alpha(x)$ is our gauge (and is not the QED coupling constant). Since $\mathcal{L}\,(= \mathcal{L}_0 + \mathcal{L}_I)$, is unchanged in form, then each of $\mathcal{L}_I$ and $\mathcal{L}_0$ retains the same functional form, as well. (See (11-36), pg. 294.) That is, under a symmetry transformation of the full $\mathcal{L}$, even though $\mathcal{L}_I$ alone is not symmetric in its own right, in combination with $\mathcal{L}_0$, the transformation yields two terms $\mathcal{L}_I$ and $\mathcal{L}_0$ that are identical in form to the pre-transformation terms $\mathcal{L}_I$ and $\mathcal{L}_0$.

And thus, our transition amplitude must also be the same in form, as depicted symbolically in

$\mathcal{L}$ sym $\rightarrow$ $\mathcal{L}_I$ unchanged $\rightarrow$ $\mathcal{H}_I$ unchanged $\rightarrow$ S unchanged $\rightarrow$ S_{fi} unchanged $\rightarrow$ $|S_{fi}|^2$ unchanged.

Effectively, we can say that if $\mathcal{L}$ is symmetric under (13-25), then so is the amplitude S_{fi}. For the S operator,

$$S\left(\psi, A_\mu\right) = S\left(\psi', A_\mu'\right), \tag{13-25}+1$$

i.e., it has the same functional form in terms of unprimed or primed (transformed) fields.

With $\mathcal{M}$, our Feynman amplitude, the transition amplitude, as we found in Chap. 8, is

$$S_{fi} = \delta_{fi} + \left(\left(2\pi\right)^4\delta^{(4)}\left(P_f - P_i\right)\left(\prod_{\substack{\text{all external}\\\text{bosons}}}\sqrt{\frac{1}{2V\omega_{\mathbf{k}}}}\right)\left(\prod_{\substack{\text{all external}\\\text{fermions}}}\sqrt{\frac{m}{VE_{\mathbf{P}}}}\right)\right)\mathcal{M} \qquad \mathcal{M} = \sum_{n=1}^\infty\mathcal{M}^{(n)}. \tag{13-26}$$

Under the gauge transformation, the incoming and outgoing four-momenta P_i and P_f are unchanged, as are the volume V and the external particle energies, ω and E. Thus, $\mathcal{M}$ is gauge invariant if S_{fi} is.

Note that the gauge invariance applies to the total Feynman amplitude for *all* diagrams for given incoming and outgoing states. For example, in Bhabha scattering there are two ways for it to occur. (See Chap. 8, Fig. 8-2, pg. 221.) That is, for a given order in $e_0{}^n$, $\mathcal{M}^{(n)} = \mathcal{M}_{B1}^{(n)} + \mathcal{M}_{B2}^{(n)}$, where $\mathcal{M}_{B1}^{(n)}$ and $\mathcal{M}_{B2}^{(n)}$ each have many sub diagrams for $n \geq 2$. The point is that $\mathcal{M}^{(n)}$ is gauge invariant, but the individual $\mathcal{M}_{B1}^{(n)}$ and $\mathcal{M}_{B2}^{(n)}$ need not be.

Recognize that if $\mathcal{L}$ is gauge invariant, then $\mathcal{H}_I$ remains the same under any such gauge, and each term in our S operator expansion (each term contains n factors of $\mathcal{H}_I$) does also. Thus, for each order of interaction n, $S^{(n)}$ is effectively gauge invariant. Hence, so are $S_{fi}^{(n)}$ and $\mathcal{M}^{(n)}$.

An Example

Consider the initial photon of the LHS of Fig. 13-1 to be a real photon (rather than virtual, i.e., rather than a photon propagator). The self-energy Feynman amplitude of the real photon is

$$\mathcal{M}_{\gamma\,\text{self}}^{(2)} = \varepsilon_{r'\mu}(\mathbf{k}')\underbrace{\left\{\frac{1}{(2\pi)^4}\text{Tr}\int S_F(p)ie_0\gamma^\mu S_F(p-k)ie_0\gamma^\nu d^4p\right\}}_{\mathcal{M}_{\gamma\,\text{self}}^{(2)\mu\nu} = ie_0^2\,\Pi^{\mu\nu}(k)}\varepsilon_{r\nu}(\mathbf{k}) = \varepsilon_{r'\mu}(\mathbf{k}')\varepsilon_{r\nu}(\mathbf{k})\mathcal{M}_{\gamma\,\text{self}}^{(2)\mu\nu}, \quad(13\text{-}27)$$

where we represent the part of the interaction that does not include the interaction photon contributions as $\mathcal{M}_{\gamma\,\text{self}}^{(2)\mu\nu}$. Of course, we know that $\mathbf{k}' = \mathbf{k}$ and $r' = r$, but that is not important for present purposes and we want to generalize, so we leave in the primes.

The point is that for every interaction having one or more external photons, we can represent the Feynman amplitude in two factors, one for the photon polarization state(s) and one for the rest, where the latter has spacetime indices (which are summed with those on the polarization vectors).

Generalization

For any interaction having one or more photons as initial or final particle(s), we can represent the gauge invariant Feynman amplitude for any order n as

$$\mathcal{M}_{fi}^{(n)} = \varepsilon_{r_1\mu}(\mathbf{k}_1)\varepsilon_{r_2\nu}(\mathbf{k}_2)\varepsilon_{r_3\eta}(\mathbf{k}_3)...\mathcal{M}_{fi}^{(n)\mu\nu\eta...}(\mathbf{k}_1,\mathbf{k}_2,\mathbf{k}_3,...), \quad(13\text{-}28)$$

where we again note that (13-28) is gauge invariant only when the amplitude includes the sub amplitudes for every diagram having the same incoming and outgoing states.

Ward Identities

As we prove below, gauge invariance leads to the <u>Ward identities</u>

$$k_{1\mu}\mathcal{M}_{fi}^{(n)\mu}(\mathbf{k}_1,\mathbf{k}_2,..) = k_{2\nu}\mathcal{M}_{fi}^{(n)\nu}(\mathbf{k}_1,\mathbf{k}_2,..) = k_{1\mu}k_{2\nu}\mathcal{M}_{fi}^{(n)\mu\nu}(\mathbf{k}_1,\mathbf{k}_2,..) = ... = 0 \quad(13\text{-}29)$$

Proof of Ward Identities

The gauge transformation of (13-25) means $\partial_\nu\alpha(x)$ must satisfy Maxwell's wave equation (where since A_ν is real, $\alpha(x)$ should be real), since, if our Maxwell equation has form

$$\partial^\beta\partial_\beta A_\nu(x) = 0 \quad(13\text{-}29)+1$$

under (13-25), this becomes

$$\partial^\beta\partial_\beta\left(A_\nu' + \tfrac{1}{e}\partial_\nu\alpha\right) = 0 \quad\rightarrow\quad \partial^\beta\partial_\beta A_\nu' + \tfrac{1}{e}\partial^\beta\partial_\beta\partial_\nu\alpha = 0. \quad(13\text{-}29)+2$$

If we require (which we do, as our theory is built upon Maxwell's equation in this form)

$$\partial^\beta\partial_\beta A_\nu' = 0, \quad(13\text{-}29)+3$$

then from (13-29)+2, we see $\partial_\nu\alpha(x)$ must satisfy Maxwell's equation (LHS below).

$$\partial^\beta\partial_\beta\partial_\nu\alpha = 0 \quad\text{or}\quad \partial_\nu\partial^\beta\partial_\beta\alpha = 0 \xrightarrow[\text{solution also solves}]{\text{one possible }\alpha} \partial^\beta\partial_\beta\alpha = 0, \quad(13\text{-}29)+4$$

and $\alpha(x)$ can have essentially the same functional form as A^μ, the solution to (13-29)+1.[1]

For a real photon field described in the Lorenz gauge by the eigen state plane wave

$$A_\mu = \sum_{r,\mathbf{k}} \frac{1}{\sqrt{2V\omega_\mathbf{k}}} \left(\varepsilon_{\mu r}(\mathbf{k}) a_r(\mathbf{k}) e^{-ikx} + \varepsilon_{\mu r}(\mathbf{k}) a_r^\dagger(\mathbf{k}) e^{ikx} \right) \tag{13-30}$$

a useful form (one particular gauge) for $\alpha(x)$ is

$$\alpha(x) = \sum_{r,\mathbf{k}} \frac{1}{\sqrt{2V\omega_\mathbf{k}}} \left(\tilde{\alpha}_r(\mathbf{k}) e^{-ikx} + \tilde{\alpha}_r^\dagger(\mathbf{k}) e^{ikx} \right) \qquad \tilde{\alpha}_r(\mathbf{k}),\ \tilde{\alpha}_r^\dagger(\mathbf{k}) \text{ numbers, not operators.} \tag{13-31}$$

Thus, for this case, the photon gauge transformation of (13-25) becomes

$$A_\mu \to A'_\mu = A_\mu - \frac{1}{e} \partial_\mu \alpha(x) = A_\mu - \frac{1}{e} \sum_{r,\mathbf{k}} \frac{1}{\sqrt{2V\omega_\mathbf{k}}} \left(-ik_\mu \tilde{\alpha}_r(\mathbf{k}) e^{-ikx} + ik_\mu \tilde{\alpha}_r^\dagger(\mathbf{k}) e^{ikx} \right)$$

$$= A_\mu - \frac{1}{e} \sum_{r,\mathbf{k}} k_\mu \underbrace{\frac{1}{\sqrt{2V\omega_\mathbf{k}}} \left(-i\tilde{\alpha}_r(\mathbf{k}) e^{-ikx} + i\tilde{\alpha}_r^\dagger(\mathbf{k}) e^{ikx} \right)}_{\text{call this } \tilde{\alpha}_{r,\mathbf{k}}(x)} = A_\mu - \frac{1}{e} \underbrace{\sum_{r,\mathbf{k}} k_\mu \tilde{\alpha}_{r,\mathbf{k}}(x)}_{=\ \partial_\mu \alpha(x)} \tag{13-32}$$

$$= \sum_{r,\mathbf{k}} \frac{1}{\sqrt{2V\omega_\mathbf{k}}} \left(\left(\varepsilon_{\mu r}(\mathbf{k}) a_r(\mathbf{k}) + \frac{1}{e} ik_\mu \tilde{\alpha}_r(\mathbf{k}) \right) e^{-ikx} + \left(\varepsilon_{\mu r}(\mathbf{k}) a_r^\dagger(\mathbf{k}) - \frac{1}{e} ik_\mu \tilde{\alpha}_r^\dagger(\mathbf{k}) \right) e^{ikx} \right).$$

Consider a typical term of S expressed in factors of $\mathcal{H}_I$, for example, the Compton scattering term for $n = 2$, (pg. 225), under the symmetry transformation (13-25), with (13-25)+1,

$$S_C^{(2)}(\psi, A_\mu) = S_C^{(2)}(\psi', A'_\mu) = -e^2 \iint d^4x_1 d^4x_2 N\left\{ (\bar\psi A \psi)_{x_1} (\bar\psi A \psi)_{x_2} \right\}$$

$$\xrightarrow[\text{not positron}]{\text{for electron}} S_{C,e^-}^{(2)} = -e^2 \iint d^4x_1 d^4x_2 N\left\{ (\bar\psi A_\mu \gamma^\mu)_{x_1} \left[\psi_{x_1}^+, \bar\psi_{x_2}^- \right]_+ (A_\nu \gamma^\nu \psi)_{x_2} \right\} \xrightarrow{\text{under transf}}$$

$$= -e^2 \iint d^4x_1 d^4x_2 N\left\{ (\bar\psi_{x_1} e^{i\alpha(x_1)} \left(A_{\mu x_1}\gamma^\mu - \frac{1}{e}\partial_\mu\alpha(x_1)\gamma^\mu \right) \left[e^{-i\alpha(x_1)}\psi_{x_1}^+, \bar\psi_{x_2}^- e^{i\alpha(x_2)} \right]_+ \left(A_{\nu x_2}\gamma^\nu - \frac{1}{e}\partial_\nu\alpha(x_2)\gamma^\nu \right) e^{-i\alpha(x_2)}\psi_{x_2} \right\}$$

$$= -e^2 \iint d^4x_1 d^4x_2 N\left\{ (\bar\psi_{x_1} \left(A_{\mu x_1}\gamma^\mu - \frac{1}{e}\partial_\mu\alpha(x_1)\gamma^\mu \right) \left[\psi_{x_1}^+, \bar\psi_{x_2}^- \right]_+ \left(A_{\nu x_2}\gamma^\nu - \frac{1}{e}\partial_\nu\alpha(x_2)\gamma^\nu \right)\psi_{x_2} \right\}$$

$$= \underbrace{-e^2 \iint d^4x_1 d^4x_2 N\left\{ (\bar\psi A_\mu \gamma^\mu)_{x_1} \left[\psi_{x_1}^+, \bar\psi_{x_2}^- \right]_+ (A_\nu \gamma^\nu \psi)_{x_2} \right\}}_{S_{C,e^-}^{(2)}}$$

$$- e^2 \iint d^4x_1 d^4x_2 N\left\{ (\bar\psi A_\mu \gamma^\mu)_{x_1} \left[\psi_{x_1}^+, \bar\psi_{x_2}^- \right]_+ \left(-\frac{1}{e}\partial_\nu\alpha(x_2)\gamma^\nu \right)\psi_{x_2} \right\}$$

$$- e^2 \iint d^4x_1 d^4x_2 N\left\{ \bar\psi_{x_1} \left(-\frac{1}{e}\partial_\mu\alpha(x_1)\gamma^\mu \right) \left[\psi_{x_1}^+, \bar\psi_{x_2}^- \right]_+ (A_\nu \gamma^\nu \psi)_{x_2} \right\}$$

$$- e^2 \iint d^4x_1 d^4x_2 N\left\{ \bar\psi_{x_1} \left(-\frac{1}{e}\partial_\mu\alpha(x_1)\gamma^\mu \right) \left[\psi_{x_1}^+, \bar\psi_{x_2}^- \right]_+ \left(-\frac{1}{e}\partial_\nu\alpha(x_2)\gamma^\nu \right)\psi_{x_2} \right\} \tag{13-32}+1$$

Using our definition of $\tilde{\alpha}_{r,\mathbf{k}}$ from (13-32), this becomes

$$S_{C,e^-}^{(2)}(\psi, A_\mu) = S_{C,e^-}^{(2)}(\psi, A_\mu) + e\iint d^4x_1 d^4x_2 N\left\{ (\bar\psi A_\mu \gamma^\mu)_{x_1} \left[\psi_{x_1}^+, \bar\psi_{x_2}^- \right]_+ \left(\sum_{r,\mathbf{k}} k_\nu \tilde{\alpha}_{r,\mathbf{k}}(x_2)\gamma^\nu \right)\psi_{x_2} \right\}$$

$$+ e\iint d^4x_1 d^4x_2 N\left\{ \bar\psi_{x_1} \left(\sum_{r,\mathbf{k}} k_\mu \tilde{\alpha}_{r,\mathbf{k}}(x_1)\gamma^\mu \right) \left[\psi_{x_1}^+, \bar\psi_{x_2}^- \right]_+ (A_\nu \gamma^\nu \psi)_{x_2} \right\} \tag{13-32}+2$$

$$- \iint d^4x_1 d^4x_2 N\left\{ \bar\psi_{x_1} \left(\sum_{r',\mathbf{k}'} k'_\mu \tilde{\alpha}_{r',\mathbf{k}'}(x_1)\gamma^\mu \right) \left[\psi_{x_1}^+, \bar\psi_{x_2}^- \right]_+ \left(\sum_{r,\mathbf{k}} k_\nu \tilde{\alpha}_{r,\mathbf{k}}(x_2)\gamma^\nu \right)\psi_{x_2} \right\}.$$

We can see immediately, because identical terms appear on both sides of (13-32)+2, that the last three terms must sum to zero. More on that shortly.

---

[1] Alternatively, our theory was developed in the Lorenz gauge (see pg. 141), i.e., $\partial^\mu A_\mu = 0$, so we need $\partial^\mu A'_\mu = 0$ also. Thus, $0 = \partial^\mu A'_\mu = \partial^\mu \left(A_\mu - (1/e)\partial_\mu\alpha \right) = \partial^\mu A_\mu - (1/e)\partial^\mu\partial_\mu\alpha = 0 - (1/e)\partial^\mu\partial_\mu\alpha$. To keep the Lorenz gauge under the transformation, $\partial^\mu\partial_\mu\alpha = 0$, which is the same as $\partial^\beta\partial_\beta\alpha = 0$.

For now, using our symbol $iS^+(x_1 - x_2)$ for the electron propagator, we find

$$S_{C,e^-}^{(2)} = S_{C,e^-}^{(2)} + e\iint d^4x_1 d^4x_2 N\left\{\left(\overline{\psi}A_\mu\gamma^\mu\right)_{x_1} iS^+(x_1 - x_2)\left(\sum_{r,\mathbf{k}} k_\nu \tilde{\alpha}_{r,\mathbf{k}}(x_2)\gamma^\nu\right)\psi_{x_2}\right\}$$

$$+ e\iint d^4x_1 d^4x_2 N\left\{\overline{\psi}_{x_1}\left(\sum_{r,\mathbf{k}} k_\mu \tilde{\alpha}_{r,\mathbf{k}}(x_1)\gamma^\mu\right)iS^+(x_1 - x_2)\left(A_\nu\gamma^\nu\psi\right)_{x_2}\right\} \qquad (13\text{-}32)+3$$

$$- \iint d^4x_1 d^4x_2 N\left\{\overline{\psi}_{x_1}\left(\sum_{r',\mathbf{k}'} k_\mu' \tilde{\alpha}_{r',\mathbf{k}'}(x_1)\gamma^\mu\right)iS^+(x_1 - x_2)\left(\sum_{r,\mathbf{k}} k_\nu \tilde{\alpha}_{r,\mathbf{k}}(x_2)\gamma^\nu\right)\psi_{x_2}\right\}.$$

As long as we are restricting ourselves to electron Compton scattering, and not positron Compton scattering, we can express the electron propagator S^+ as the full propagator S (where the last three terms below sum to zero)

$$S_C^{(2)} = S_C^{(2)} + \underbrace{\sum_{r,\mathbf{k}} k_\nu \, e\iint d^4x_1 d^4x_2 \, \tilde{\alpha}_{r,\mathbf{k}}(x_2) N\left\{\left(\overline{\psi}A_\mu\gamma^\mu\right)_{x_1} iS(x_1 - x_2)\gamma^\nu\psi_{x_2}\right\}}_{S_{C1}^{(2)\nu}\ (\text{1st way, no initial photon})}$$

$$+ \underbrace{\sum_{r,\mathbf{k}} k_\mu \, e\iint d^4x_1 d^4x_2 \, \tilde{\alpha}_{r,\mathbf{k}}(x_1) N\left\{\overline{\psi}_{x_1}\gamma^\mu iS(x_1 - x_2)\left(A_\nu\gamma^\nu\psi\right)_{x_2}\right\}}_{S_{C2}^{(2)\mu}\ (\text{2nd way, no initial photon})} \qquad (13\text{-}32)+4$$

$$- \sum_{r',\mathbf{k}'} k_\mu' \sum_{r,\mathbf{k}} k_\nu \underbrace{\iint d^4x_1 d^4x_2 \, \tilde{\alpha}_{r',\mathbf{k}'}(x_1)\tilde{\alpha}_{r,\mathbf{k}}(x_2) N\left\{\overline{\psi}_{x_1}\gamma^\mu iS(x_1 - x_2)\gamma^\nu\psi_{x_2}\right\}}_{S_C^{(2)\,\mu\nu}\ (\text{both ways, no initial or final photon})}\;.$$

The two terms in the sum $\Sigma k_\nu S_{C1}^{(2)\nu} + \Sigma k_\mu S_{C2}^{(2)\mu}$ have the same external particles, so that sum must equal zero independently of $\Sigma k_\nu \Sigma k_\mu' S_C^{(2)\mu\nu}$, which has different external particles.

Recall that to find the amplitude $S_{fi} = S_{Compton}$ for Compton scattering (to 2nd order on the RHS below), we carry out steps, as we did in Chap. 8, to evaluate

$$S_{Compton} = \langle f|S|i\rangle = \langle e_{\mathbf{p}',s'}^-, \gamma_{\mathbf{k}',r'}|\sum_n S^{(n)}|e_{\mathbf{p},s}^-, \gamma_{\mathbf{k},r}\rangle \rightarrow S_{Compton}^{(2)} = \langle e_{\mathbf{p}',s'}^-, \gamma_{\mathbf{k}',r'}|S_C^{(2)}|e_{\mathbf{p},s}^-, \gamma_{\mathbf{k},r}\rangle \qquad (13\text{-}32)+5$$

When we did that, we found

$$S_{Compton}^{(2)} = \sqrt{\frac{m}{VE_{\mathbf{p}'}}}\sqrt{\frac{m}{VE_{\mathbf{p}}}}\sqrt{\frac{1}{2V\omega_{\mathbf{k}'}}}\sqrt{\frac{1}{2V\omega_{\mathbf{k}}}}(2\pi)^4 \delta^{(4)}(p'+k'-p-k)\mathcal{M}_{Compton}^{(2)}$$

$$\mathcal{M}_{Compton}^{(2)} = \mathcal{M}_{C1}^{(2)} + \mathcal{M}_{C2}^{(2)} \qquad \mathcal{M}_{C1}^{(2)} = -e^2 \overline{u}_{s'}(\mathbf{p}')\varepsilon_{\mu,r'}(\mathbf{k}')\gamma^\mu iS_F(q=p+k)\varepsilon_{\nu,r}(\mathbf{k})\gamma^\nu u_s(\mathbf{p}) \qquad (13\text{-}32)+6$$

$$\mathcal{M}_{C2}^{(2)} = -e^2 \overline{u}_{s'}(\mathbf{p}')\varepsilon_{\mu,r}(\mathbf{k})\gamma^\mu iS_F(q=p-k')\varepsilon_{\nu,r'}(\mathbf{k}')\gamma^\nu u_s(\mathbf{p}),$$

which results from the $S_C^{(2)}$ term in (13-32)+4. Doing a similar thing with the 2nd and 3rd terms on the RHS of (13-32)+4, where our initial state lacks the photon of (13-32)+6, we get (see Appendix)

$$\overbrace{\langle e_{\mathbf{p}',s'}^-, \gamma_{\mathbf{k}',r'}|\sum_{r,\mathbf{k}} k_\nu S_{C1}^{(2)\nu} + \sum_{r,\mathbf{k}} k_\mu S_{C2}^{(2)\mu}|e_{\mathbf{p},s}^-\rangle}^{=0} \rightarrow$$

$$0 = \frac{1}{e}\sqrt{\frac{m}{VE_{\mathbf{p}'}}}\sqrt{\frac{1}{2V\omega_{\mathbf{k}'}}}\sqrt{\frac{1}{2V\omega_{\mathbf{k}}}}\sqrt{\frac{m}{VE_{\mathbf{p}}}}(2\pi)^4 \delta^{(4)}(p'+k'-p-k)i\underbrace{k_\nu \mathcal{M}_{Compton}^{(2)\nu}}_{\text{must} = 0}\left(\tilde{a}_r(\mathbf{k}) + \tilde{a}_r^\dagger(-\mathbf{k})\right) \qquad (13\text{-}32)+7$$

$$\mathcal{M}_{Compton}^{(2)\nu} = \mathcal{M}_{C1}^{(2)\nu} + \mathcal{M}_{C2}^{(2)\nu} \qquad \mathcal{M}_{C1}^{(2)\nu} = -e^2 \, \varepsilon_{\mu,r'}(\mathbf{k}')\overline{u}_{s'}(\mathbf{p}')\gamma^\mu iS_F(p'+k')\gamma^\nu u_s(\mathbf{p})$$

$$\mathcal{M}_{C2}^{(2)\nu} = -e^2 \, \varepsilon_{\mu,r'}(\mathbf{k}')\overline{u}_{s'}(\mathbf{p}')\gamma^\nu iS_F(p'+k')\gamma^\mu u_s(\mathbf{p})$$

Thus (where the RHS of (13-33) follows from similar analysis of the last term in (13-32)+4),

$$k_\nu\left(\mathcal{M}_{C1}^{(2)\nu} + \mathcal{M}_{C2}^{(2)\nu}\right) = k_\nu \mathcal{M}_{Compton}^{(2)\nu} = 0 \qquad\qquad k_\mu' k_\nu \mathcal{M}_{Compton}^{(2)\mu\nu} = 0\,. \qquad (13\text{-}33)$$

If we take $\mathbf{k} \to \mathbf{k}_1$ and $\mathbf{k}' \to \mathbf{k}_2$, then (13-32)+7 and (13-33) above equal (13-29) for Compton scattering.

One should be able to visualize a similar result from any amplitude with fermion propagators and external fermions and photons. For the $\psi \to \psi'$ of (13-25), the $e^{-i\alpha(x)}$ and $e^{i\alpha(x)}$ factors will always cancel. The external $A_\nu \to A'_\nu$, due to the α part, will always leave a series of terms in the S operator expansion of form similar to those in (13-32)+5 (with appropriately more such terms when there are more factors of A_ν.) And these terms must sum to zero because a term like $S^{(2)}_{Ce^-}\left(\psi, A_\mu\right)$ in (13-32)+2 occurs on each side of the relationship resulting from the transformation.

For a photon propagator, we have (with similar results for $iD^{\mu\nu-}\left(x_1 - x_2\right)$)

$$iD^{\mu\nu+}\left(x_1 - x_2\right) = \left[A^{\mu+}_{x_1}, A^{\nu-}_{x_2} \right] \to \left[\left(A'^{\mu+}_{x_1} + \underbrace{\frac{1}{e}\partial^\mu\alpha^+\left(x_1\right)}_{\text{a number}} \right)\left(A'^{\nu-}_{x_2} + \underbrace{\frac{1}{e}\partial^\nu\alpha^-\left(x_2\right)}_{\text{a number}} \right) \right] = \left[A'^{\mu+}_{x_1}, A'^{\nu-}_{x_2} \right] \quad (13\text{-}34)$$

Thus, any photon propagator in any amplitude keeps the same form under the transformation, so we get no extra terms from it, as in (13-32)+4, that must equal zero.

And so, we have proven the Ward identities (13-29) using local gauge invariance (which manifested in (13-25)+1, the starting point of our proof). QED.

End of Ward identities proof

<u>End of proof</u>

Note that (13-19) is a relation for $n = 2$ order between the fermion loop and the vertex loop, whereas (13-29) is good at any order for any amplitude involving at least one external photon.

<u>Additional Identities</u>

There are yet other identities called <u>Ward-Takahashi identities</u>, of which (13-29) is a special case, but we will not treat those here. In Ward-Takahashi identities, the $k_{i\mu}$ are not restricted to represent external photons, but can be off shell (propagators), and the RHS of (13-29) is, for internal photons, not zero. The Ward identities are the Ward-Takahashi identities for real photons.

Ward identities a special case of Ward-Takahashi identities

<u>The Process</u>

For any amplitude relation of the form on the LHS of (13-35) below, the RHS, representing the <u>Ward identities</u>, is true. That is, we simply replace the polarization vector by the associated four-momentum and the result equals zero.

How to apply Ward identities

$$\boxed{\mathcal{M}^{(n)}_{fi}\left(\mathbf{k}_1,...,\mathbf{k}_j,...\right) = \varepsilon_{r_j\,\mu}\mathcal{M}^{(n)\,\mu}_{fi}\left(\mathbf{k}_1,...,\mathbf{k}_j,...\right) \quad \to \quad k_{j\,\mu}\mathcal{M}^{(n)\,\mu}_{fi}\left(\mathbf{k}_1,...,\mathbf{k}_j,...\right) = 0} \quad .(13\text{-}35)$$

<u>The Message</u>

Local gauge invariance leads to both charge conservation and the Ward identities. All three are different ways of saying the same thing. Each implies the other two.

Gauge invariance & Ward identities the same thing

$$\text{charge conservation} \quad \leftrightarrow \quad \text{local gauge invariance} \quad \leftrightarrow \quad \text{Ward identities.}$$

13.3 Ward Identities, Renormalization, and Gauge Invariance

Consider the scattering of light by light shown in Fig. 13-2. Two incoming photons scatter via fermion virtual particles to yield two outgoing photons. This is called <u>photon-photon scattering</u>, or <u>light-by-light scattering,</u> or less commonly, Delbrück scattering. Occasionally, it is referred to as a "four photon vertex", but this is misleading as there are really four vertices, not a single one with four photons connected directly to it.

Application of Ward identities in renormalization for photon-photon scattering case

Light-by-light scattering does not occur in classical electromagnetism, but does so in QFT due to higher order corrections. Classical electromagnetism contains only terms linear in the photon field A^μ and corresponds to our tree level diagrams. However, via the Dyson-Wicks expansion in QFT, we have terms contributing to the scattering amplitude beyond tree level, at second and higher order, which effectively make non-linear contributions.

Fig. 13-2 represents one way four external photons can scatter at second order in α. There are other ways the same states can scatter at second order, and Prob. 2 asks you to draw the Feynman diagrams for at least three other possibilities. Note that in Fig. 13-2 we have depicted a certain time order (from left to right) for the vertices in order to make the internal line four-momenta easy to determine. Depicting a different vertex time order (such as the upper left vertex before, rather than after, the lower left vertex) does *not* give us a different Feynman diagram, nor a different (sub) amplitude. See Prob. 3.

Since we have no two photon vertices in QED, there can be no first order scattering of light by light. This means the lowest possible order at which photons scatter is second order, described by $\gamma\gamma \to \gamma\gamma$.

If there were photon-photon scattering at tree level, then the process would be fairly common, and in practice, we would encounter light scattering off of light with fair regularity. But we don't. Light-light scattering is, in fact, extremely rare, as one might expect from a Feynman diagram representation of it such as Fig. 13-2.

Light-by-light scattering is non-classical, QFT 2nd order effect, and very rare

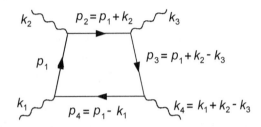

Figure 13-2. One Way for Photon-photon Scattering

Using Feynman rules, the second order amplitude for the photon-photon scattering of Fig. 13-2, where we distinguish that diagram from its sibling diagrams (second order, same incoming and outgoing photons, like you explored in Prob. 2) with the subscript (a), is

2nd order amplitude for one way of photon-photon scatter

$$\mathcal{M}_{\substack{\gamma\gamma\to\gamma\gamma \\ (a)}} = \varepsilon_\mu(\mathbf{k}_4)\varepsilon_\nu(\mathbf{k}_3)\varepsilon_\rho(\mathbf{k}_2)\varepsilon_\sigma(\mathbf{k}_1) \times$$

$$\underbrace{\frac{-e_0^4}{(2\pi)^4}\text{Tr}\int \frac{1}{\not{p}_4 - m + i\varepsilon}\gamma^\mu \frac{1}{\not{p}_3 - m + i\varepsilon}\gamma^\nu \frac{1}{\not{p}_2 - m + i\varepsilon}\gamma^\rho \frac{1}{\not{p}_1 - m + i\varepsilon}\gamma^\sigma d^4 p_1}_{\mathcal{M}^{\mu\nu\rho\sigma}_{\substack{\gamma\gamma\to\gamma\gamma \\ (a)}}} \quad (13\text{-}36)$$

$$= \varepsilon_\mu(\mathbf{k}_4)\varepsilon_\nu(\mathbf{k}_3)\varepsilon_\rho(\mathbf{k}_2)\varepsilon_\sigma(\mathbf{k}_1)\mathcal{M}^{\mu\nu\rho\sigma}_{\substack{\gamma\gamma\to\gamma\gamma \\ (a)}}.$$

Each propagator factor in (13-36) contains a factor of p_1, so in the same manner we estimated maximum possible divergences in Chap. 9, we have

Power counting estimate → log divergence

$$\mathcal{M}^{\mu\nu\rho\sigma}_{\substack{\gamma\gamma\to\gamma\gamma \\ (a)}} \propto \int_{-\infty}^\infty \frac{d^4 p_1}{(p_1)^4} \propto \int_0^\infty \frac{(p_1)^3 dp_1}{(p_1)^4} = ln\, p_1 \Big|_0^\infty, \quad (13\text{-}37)$$

a logarithmic divergence. This is a superficial estimate because, as we saw with the photon self-energy integral, the actual order of divergence may be less than this. (13-37) is a possible order of divergence for the photon-photon scattering amplitude, but it might, when the integral is actually evaluated, be less. It should not be more.

We can, however, show that the second order total photon-photon scattering amplitude does not diverge via another consideration. From our Ward identities (13-35), where letter subscripts represent different sub-amplitudes contributing to the total second order amplitude, we have

But Ward identities prove the total amplitude is not divergent

$$k_{1\mu}\mathcal{M}^{\mu\nu\rho\sigma}_{\gamma\gamma\to\gamma\gamma} = 0 \qquad \mathcal{M}^{\mu\nu\rho\sigma}_{\gamma\gamma\to\gamma\gamma} = \mathcal{M}^{\mu\nu\rho\sigma}_{\substack{\gamma\gamma\to\gamma\gamma \\ (a)}} + \mathcal{M}^{\mu\nu\rho\sigma}_{\substack{\gamma\gamma\to\gamma\gamma \\ (b)}} + \mathcal{M}^{\mu\nu\rho\sigma}_{\substack{\gamma\gamma\to\gamma\gamma \\ (c)}} + \dots \quad (13\text{-}38)$$

For $k_{1\mu}$ being arbitrary (any possible components), this can only be true if $\mathcal{M}^{\mu\nu\rho\sigma}_{\gamma\gamma\to\gamma\gamma}$ is finite.

Therefore, the amplitude for light-light scattering does not diverge.

So, the Ward identities mean light-by-light scattering has finite amplitude and by being true, allow the theory to be renormalized. In general, full renormalization of QFT is achievable only if the Ward identities are true. And they are only true because we have a gauge invariant (symmetric)

Renormalization possible due to Ward identities, which are due to gauge invariance

330   Chapter 13. Renormalization Toolkit

theory. Thus, we are able to renormalize our theory because it is gauge invariant. This is a point that is hard to overemphasize.

Conclusions:

Each of the following being true implies the others are also.

- The theory (specifically the Lagrangian) is locally gauge invariant (locally symmetric).
- A quantity (charge) is conserved.
- The theory has the correct interactions. (They result from demanding local gauge invariance).
- The Ward identities hold.
- The theory is renormalizable.

If the above are true, then

- the gauge boson (photon) is massless. (See Chap. 11, Sect. 11.5.4, pg. 296.)

Gauge invariance ↔ charge conserved ↔ correct $\mathcal{H}_I$ ↔ Ward identities ↔ QED is renormalizable

Note that we need a massless gauge boson (no mass term in the Lagrangian) to be able to renormalize, but other terms in the Lagrangian might still lead to irreconcilably infinite amplitudes. A massless photon is necessary, but not sufficient, for renormalization.

All of above → photon is massless

If you have, in the past, wondered what the great fuss about symmetry was all about, you should now be beginning to appreciate its role in modern theories of the quantum field. We seem to need it to have any viable theory at all.

Symmetry (gauge invariance) at the foundation of QFT

13.4 Changes in the Theory with m Instead of m₀

13.4.1 Counterterms in the Lagrangian and Hamiltonian

Given, as we saw in the last chapter, that we need to use the measured mass m, rather than the bare mass m_0, we need to investigate what that means for the QED Lagrangian

$$\mathcal{L} = \underbrace{-\tfrac{1}{4}F^{\mu\nu}F_{\mu\nu} + \bar{\psi}\left(i\gamma^\mu\partial_\mu - m_0\right)\psi}_{\mathcal{L}_0} + \underbrace{e_0\bar{\psi}\gamma^\mu\psi A_\mu}_{\mathcal{L}_I}.$$ (13-39)

We can get the Lagrangian into a form with m by substituting $m_0 = m - \delta m$ into (13-39). This yields

Re-expressing $\mathcal{L}$ in terms of physical mass leads to mass counterterm in $\mathcal{L}_I$

$$\mathcal{L} = \underbrace{-\tfrac{1}{4}F^{\mu\nu}F_{\mu\nu} + \bar{\psi}\left(i\gamma^\mu\partial_\mu - m\right)\psi}_{\text{mass renormalized }\mathcal{L}_0} + \underbrace{e_0\bar{\psi}\gamma^\mu\psi A_\mu \overbrace{+ \delta m\bar{\psi}\psi}^{\text{mass counter term}}}_{\text{mass renormalized }\mathcal{L}_I}.$$ (13-40)

Recall that $\mathcal{L}_I = -\mathcal{H}_I$ and that each term in $\mathcal{L}_I$ (or equivalently, $\mathcal{H}_I$) represents an interaction (a vertex, typically, in the sense that it gives rise to a corresponding vertex in a Feynman diagram). The first term in the mass renormalized $\mathcal{L}_I$ is simply the same term we had before mass renormalization with m here instead of m_0 then. It represents the vertex interaction of two fermions and a photon that we have become familiar with. But now we have another term in the (mass renormalized) interaction Lagrangian with just two fermions (and no photon) field, which is called the mass counterterm of the Lagrangian.

13.4.2 Counterterm Feynman Diagram

If we went through our entire derivation of Chaps. 7 and 8 again using (13-40), we would find this extra term, the mass counterterm, appearing at every step of our expansion of the S operator in terms of (the new renormalized) $\mathcal{H}_I$. The net result would be an extra interaction involving the same incoming and an outgoing fermion.

Thus, if we want to use a Lagrangian with the measured mass instead of the bare mass (which is more practical), then we can use (13-40). But that means that if we want to consider free fermions of mass m, then we can no longer consider them simply as in the Feynman diagram of the LHS of Fig. 13-3. We must also include a fermion self-interaction diagram (with an X in it) as in the RHS of Fig. 13-3, representing the mass counterterm. This extra diagram has an incoming fermion and an outgoing fermion just like a truly free (non-self-interacting) fermion diagram (as in the LHS of the figure) would have. So, we must add the two contributions. This means the old diagram with m_0 is now replaced by the equivalent of two diagrams, one involving m and one involving δm.

Mass counterterm manifests as its own Feynman diagram contribution

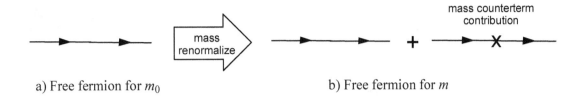

Figure 13-3. Equivalent Free Fermion Feynman Diagrams

Each term in $\mathcal{H}_I = -\mathcal{L}_I$ shows up in our Dyson-Wicks expansion with an i in front of it, so the Feynman rule for the mass counterterm will have one in it, as well. Thus, our Feynman rule #10 (we had nine before) for that term will be

10. For each mass counterterm diagram, add a term to the amplitude with a factor equal to $i\delta m$.

Mass counterterm contribution to Feynman rules →

iδm extra term

13.4.3 Dirac Equation with m Instead of m_0

The free part of (13-40) looks like the free part of the Lagrangian we have been dealing with since Part I of this book, except that m now takes the place of m_0. Since the Dirac equation is derived by placing the free Lagrangian into the Euler-Lagrange equation for fields, we would end up with the same form of Dirac equation in the mass renormalized theory, except that now our mass is represented by m. Thus, with mass re-normalized, the Dirac equation is

With physical mass m, have same form of Dirac equation

$$(i\gamma^\alpha \partial_\alpha - m)\psi = 0. \tag{13-41}$$

Hence, relations (13-10) remain true in the renormalized theory. So do all other operator relations with mass in them, since operators in the I.P. are the same as operators in free field H.P. theory (as in Chaps. 3 to 6.) That is, the free Lagrangian of (13-40) has the same form in m that all of our earlier work had for that form in m_0. As an example relation, the Gordon identity of (13-18) is true for measured mass m.

So all prior relations for m_0 remain the same for m

13.5 Showing the B in Fermion Loop Equals the L in Vertex Correction

For what we wish to do now, we need to express $\Lambda^\mu(p, p')$ in its most general possible form. Such a form must contain all possible related entities in the theory having components μ, and these are simply γ^μ, p^μ, and p'^μ. Thus,

Showing B=L using Gordon's and original Ward identities

$$\Lambda^\mu\left(p, p'\right) = a\gamma^\mu + b_1 p^\mu + b_2 p'^\mu \qquad \text{(most general possible form)}. \tag{13-42}$$

Start with most general form for Λ^μ

The values of a, b_1, and b_2 in (13-42) are such that (13-42) equals (13-9).

$\Lambda^\mu(p, p')$ represents a vertex, as shown in the RHS of Fig. 13-1. If the external photon in that vertex had $k^\mu = 0$, there would effectively be no photon, and then $p = p'$. In other words, (13-42) with $p = p'$ would represent a free fermion, with the vertex loop in Fig. 13-1 becoming a fermion self-energy loop.

Express Λ^μ for p = p'

Thus, where $p = p'$ and $b = b_1 + b_2$, (13-42) becomes

$$\Lambda^\mu\left(p, p\right) = a\gamma^\mu + bp^\mu \ \rightarrow \ \bar{u}_r\left(\mathbf{p}\right)\Lambda^\mu\left(p, p\right)u_r\left(\mathbf{p}\right) = \bar{u}_r\left(\mathbf{p}\right)a\gamma^\mu u_r\left(\mathbf{p}\right) + \bar{u}_r\left(\mathbf{p}\right)bp^\mu u_r\left(\mathbf{p}\right). \tag{13-43}$$

Note in Gordon's identity (13-18) that when $p' = p$, the last term in that identity is zero, i.e.,

Apply Gordon's identity

$$\bar{u}_r\left(\mathbf{p}\right)\gamma^\mu u_r\left(\mathbf{p}\right) = \frac{p^\mu + p^\mu}{2m}\bar{u}_r\left(\mathbf{p}\right)u_r\left(\mathbf{p}\right) = \frac{p^\mu}{m}\bar{u}_r\left(\mathbf{p}\right)u_r\left(\mathbf{p}\right). \tag{13-44}$$

Multiplying (13-44) by m and putting it into the last term of (13-43), we get

$$\bar{u}_r\left(\mathbf{p}\right)\Lambda^\mu\left(p, p\right)u_r\left(\mathbf{p}\right) = \bar{u}_r\left(\mathbf{p}\right)a\gamma^\mu u_r\left(\mathbf{p}\right) + \bar{u}_r\left(\mathbf{p}\right)bp^\mu u_r\left(\mathbf{p}\right)$$

$$= a\bar{u}_r\left(\mathbf{p}\right)\gamma^\mu u_r\left(\mathbf{p}\right) + mb\bar{u}_r\left(\mathbf{p}\right)\gamma^\mu u_r\left(\mathbf{p}\right). \tag{13-45}$$

So,

$$\bar{u}_r\left(\mathbf{p}\right)\Lambda^\mu\left(p, p\right)u_r\left(\mathbf{p}\right) = L\bar{u}_r\left(\mathbf{p}\right)\gamma^\mu u_r\left(\mathbf{p}\right) \qquad \text{where } L = a + mb. \tag{13-46}$$

(13-46) tells us that for the free particle (i.e., $p = p'$),

$$\Lambda^\mu\left(p,p\right)=L\left(\Lambda\right)\gamma^\mu \tag{13-47}$$

Gives us the free particle part of Λ^μ

and thus from (13-6),

$$\Lambda^\mu\left(p,p'\right)=\underbrace{L\left(\Lambda\right)\gamma^\mu}_{\substack{\text{free particle}\\\text{part}}}+\underbrace{\Lambda_c^\mu\left(p,p'\right)}_{\substack{=0\text{ for free}\\\text{particle, i.e., } p=p'}}. \tag{13-48}$$

From the original Ward identity (13-19) with (13-5) and (13-46),

$$\frac{\partial\Sigma\left(p\right)}{\partial p_\mu}=\Lambda^\mu\left(p,p\right)\quad\rightarrow\quad \bar{u}_r\left(\mathbf{p}\right)\frac{\partial\Sigma\left(p\right)}{\partial p_\mu}u_r\left(\mathbf{p}\right)=\bar{u}_r\left(\mathbf{p}\right)\Lambda^\mu\left(p,p\right)u_r\left(\mathbf{p}\right)$$

$$=\bar{u}_r\left(\mathbf{p}\right)\frac{\partial}{\partial p_\mu}\left(A+\left(p_\nu\gamma^\nu-m\right)B+\left(p_\nu\gamma^\nu-m\right)\Sigma_c\underbrace{\left(\not{p}-m\right)}_{\substack{\text{argument,}\\\text{not factor}}}\right)u_r\left(\mathbf{p}\right)=\bar{u}_r\left(\mathbf{p}\right)L\gamma^\mu u_r\left(\mathbf{p}\right) \tag{13-49}$$

$$=\bar{u}_r\left(\mathbf{p}\right)\gamma^\mu B u_r\left(\mathbf{p}\right)+\bar{u}_r\left(\mathbf{p}\right)\gamma^\mu\underbrace{\Sigma_c\left(\not{p}-m\right)u_r\left(\mathbf{p}\right)}_{=0}+\underbrace{\bar{u}_r\left(\mathbf{p}\right)\left(\not{p}-m\right)}_{=0}\frac{\partial}{\partial p_\mu}\Sigma_c\left(\not{p}\right)u_r\left(\mathbf{p}\right).$$

Certain terms drop out

where we used (13-10) to get terms to drop out. Note that $\Sigma_c\left(\not{p}-m\right)$ is part of an expansion in $\not{p}-m$, so each term in it has $\not{p}-m$ raised to a power. The factor of $\not{p}-m$ on the right in any such term acts on $u_r(\mathbf{p})$, and from (13-10) results in zero. Thus, the 2nd line RHS and the 3rd line of (13-49) give us

Leaving B=L

$$B\,\bar{u}_r\left(\mathbf{p}\right)\gamma^\mu u_r\left(\mathbf{p}\right)=L\,\bar{u}_r\left(\mathbf{p}\right)\gamma^\mu u_r\left(\mathbf{p}\right)\quad\rightarrow\quad B=L. \tag{13-50}$$

This result will play a vital role in renormalization, as we shall see in the next chapter.

13.6 Re-expressing 2nd Order Corrected Propagators, Vertex, and External Lines

2nd order expressions in terms of our new symbols

We now wish to express our modified propagators, modified vertex relation, and modified real particle relations in terms of the symbols A', Π_c, A, B, Σ_c, L, and Λ_c^μ of (13-4) to (13-6).

13.6.1 The 2nd Order Photon Propagator

Fig. 13-4 shows how the Feynman diagram for the photon propagator at first order becomes two diagrams at second order.

2nd order photon propagator

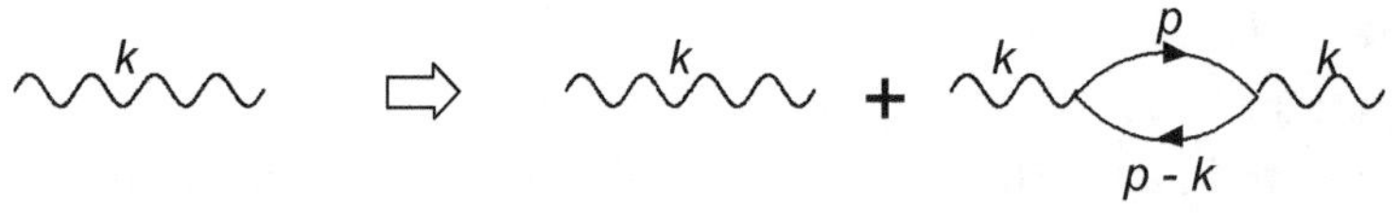

Figure 13-4. Photon Propagator Self-energy Correction to 2nd Order in α

This means the 2nd order photon propagator becomes

$$iD_{F\alpha\beta}\left(k\right)\;\Rightarrow\;iD_{F\alpha\beta}^{2nd}=iD_{F\alpha\beta}\left(k\right)+iD_{F\alpha\mu}\left(k\right)ie_0^2\Pi^{\mu\nu}\left(k\right)iD_{F\nu\beta}\left(k\right). \tag{13-51}$$

The RHS of (13-51), with (13-4) is thus

$$iD_{F\alpha\beta}^{2nd}\left(k\right)=-\frac{ig_{\alpha\beta}}{k^2+i\varepsilon}+\frac{-ig_{\alpha\mu}}{k^2+i\varepsilon}ie_0^2 g^{\mu\nu}\left(-k^2 A'\left(k,\Lambda\right)-k^2\Pi_c(k^2)\right)\underbrace{\frac{-ig_{\nu\beta}}{k^2+i\varepsilon}}_{\approx k^2}$$

2nd order photon propagator in terms of our symbols

$$=\frac{-ig_{\alpha\beta}}{k^2+i\varepsilon}+\frac{-ig_{\alpha\mu}}{k^2+i\varepsilon}e_0^2\delta^\mu_{\ \beta}\left(-A'\left(k,\Lambda\right)-\Pi_c(k^2)\right)=\underbrace{\frac{-ig_{\alpha\beta}}{k^2+i\varepsilon}}_{iD_{F\alpha\beta}(k)}\left(1-e_0^2 A'\left(k,\Lambda\right)-e_0^2\Pi_c(k^2)\right). \tag{13-52}$$

Compare (13-51) (and thus, (13-52)) with the modified photon propagator we discussed qualitatively in Chap. 12 and showed in (12-10), pg. 308.

13.6.2 The 2nd Order Fermion Propagator

Fig. 13-5 depicts Feynman diagrams for two different ways we can represent the 2nd order fermion propagator, depending on whether we wish to work with m_0 or m. (Sect. 13.4.2 pg. 330.)

2nd order fermion propagator

a) for m_0

b) for m

Two ways to represent with Feynman diagrams: for m0 and m

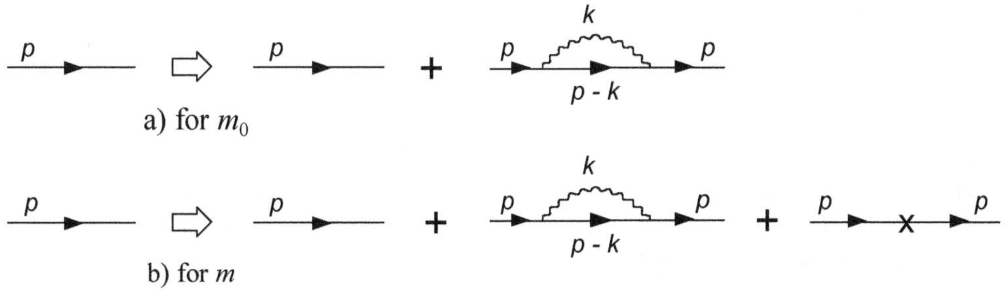

Figure 13-5. Fermion Propagator Self-energy Correction to Diagrams of 2nd Order in α

We will work with a) here first, then leave it as Prob. 4 for you to find the same thing for b). Thus, the 2nd order fermion propagator is

First way treated here. Other as a problem

$$iS_F(p) \Rightarrow iS_F^{2nd}(p) = iS_F(p) + iS_F(p)ie_0^2\Sigma(p)iS_F(p). \qquad (13\text{-}53)$$

This becomes, where we use operator relation (13-13) to get the second line below (don't confuse the A and B symbols used for generic operators in (13-13) with the totally different $A(\Lambda, m)$ and $B(\Lambda)$ symbols used in $\Sigma(p)$ of (13-5)),

$$iS_F^{2nd}(p) = i\underbrace{\frac{1}{\slashed{p} - m_0 + i\varepsilon}}_{A\text{ operator}} + i\underbrace{\frac{1}{\slashed{p} - m_0 + i\varepsilon}}_{A\text{ operator}}\underbrace{\left(-e_0^2\Sigma(p)\right)}_{B\text{ operator}}\underbrace{\frac{1}{\slashed{p} - m_0 + i\varepsilon}}_{A\text{ operator}}$$

$$= \frac{i}{A - B} + \binom{\text{higher}}{\text{order}} = \frac{i}{\slashed{p} - m_0 + e_0^2\Sigma(p) + i\varepsilon} + \binom{\text{higher}}{\text{order}} \qquad (13\text{-}54)$$

Converting $m_0 - e_0^2A$ to m

$$= \frac{i}{\slashed{p} - m_0 + \underbrace{e_0^2 A(\Lambda, m)}_{-\delta m} + e_0^2\left(\slashed{p} - m\right)B(\Lambda) + e_0^2\left(\slashed{p} - m\right)\Sigma_c\left(\slashed{p} - m\right) + i\varepsilon} + \binom{\text{higher}}{\text{order}}.$$

$$\underbrace{}_{-m}$$

Recall that $A(\Lambda, m)$ is unbounded, actually $A(\Lambda, m) \to -\infty$, as we take the parameter $\Lambda \to \infty$. But we can define measured mass $m = m_0 + \delta m$, with $\delta m = -e_0^2 A$ and end up with m alone in the denominator. As $\Lambda \to \infty$, then $\delta m \to \infty$. But if we assume $m_0 = -\infty$, then in the limit, $m_0 + \delta m = \infty - \infty = m$ remains finite. Thus (13-54), our renormalized fermion propagator to 2nd order, is

$$iS_F^{2nd}(p) = \frac{i}{\slashed{p} - m + e_0^2\left(\slashed{p} - m\right)B(\Lambda) + e_0^2\left(\slashed{p} - m\right)\Sigma_c\left(\slashed{p} - m\right) + i\varepsilon} + \binom{\text{higher}}{\text{order}}$$

$$= \frac{i\left(1 + e_0^2 B(\Lambda) + e_0^2\Sigma_c\left(\slashed{p} - m\right)\right)^{-1}}{\left(\slashed{p} - m\right) + i\varepsilon\left(1 + e_0^2 B(\Lambda) + e_0^2\Sigma_c\left(\slashed{p} - m\right)\right)^{-1}} + \binom{\text{higher}}{\text{order}}$$

2nd order fermion propagator in terms of our symbols

$$= \frac{i\left(1 - e_0^2 B(\Lambda) - e_0^2\Sigma_c\left(\slashed{p} - m\right)\right)}{\left(\slashed{p} - m\right) + i\varepsilon\left(1 - e_0^2 B(\Lambda) - e_0^2\Sigma_c\left(\slashed{p} - m\right)\right)} + \binom{\text{different}}{\text{higher order}} \qquad (13\text{-}55)$$

$$= \underbrace{\frac{i}{\slashed{p} - m + i\varepsilon}}_{iS_F(p)}\left(1 - e_0^2 B(\Lambda) - e_0^2\Sigma_c\left(\slashed{p} - m\right)\right) + \binom{\text{yet different}}{\text{higher order}}.$$

We have just shown, to 2nd order, how to renormalize fermion mass, as we described it qualitatively in the previous chapter. We have also obtained the renormalized fermion propagator to 2nd order. Compare (12-25) of Chap. 12, pg. 312, to the last row of (13-54) and (13-55).

13.6.3 2ⁿᵈ Order Incoming and Outgoing Particles

2ⁿᵈ order real particle relations

Although we haven't focused on it prior to this, we also have to include higher order corrections to the incoming and outgoing particles, not just the propagators and vertices. As examples of these, see the external fermion 2nd order corrections depicted in Fig. 12-1 (pg. 307) labeled $S_{B1-2}^{(4)}$ to $S_{B1-5}^{(4)}$.

For external fermions and photons, we can consider these corrections as if the first and second diagrams of Fig. 13-1 (pg. 323) had incoming and outgoing real particles (not propagators). So for a real fermion, the radiative correction to 2nd order diagram looks like Fig. 13-6, where we use dots at the beginnings of each diagram to indicate real (not virtual) particles. In Fig. 13-6, we are taking the approach of Fig. 13-5b), where renormalized mass is assumed.

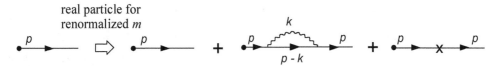

External fermion 2ⁿᵈ order correction Feynman diagrams

Figure 13-6. Real Fermion Self-energy Correction to Diagrams of 2ⁿᵈ Order in α

The initial fermion contribution to the amplitude, upon renormalization, becomes (where we note that in the last two diagrams of Fig. 13-6, the last line is really an internal line in the overall Feynman diagram for an entire reaction, so must be represented by a propagator)

$$u_r(\mathbf{p}) \Rightarrow u_r^{2nd}(\mathbf{p}) = u_r(\mathbf{p}) + \frac{i}{\not{p}-m+i\varepsilon}\,ie_0^2\Sigma(p)u_r(\mathbf{p}) + \frac{i}{\not{p}-m+i\varepsilon}(i\delta m)u_r(\mathbf{p})$$

$$= \left(1 - \frac{e_0^2 A(\Lambda,m) + e_0^2(\not{p}-m)B(\Lambda) + e_0^2(\not{p}-m)\Sigma_c(\not{p}-m) + \delta m}{\not{p}-m+i\varepsilon}\right)u_r(\mathbf{p}). \tag{13-56}$$

2ⁿᵈ order modification from substitution of relations for each diagram

Since $\delta m = -e_0^2 A(\Lambda,m)$, two terms will cancel above. Also, $e_0^2\Sigma_c(\not{p}-m)$ is an expansion having terms of $(\not{p}-m)$ to various powers. So each of these terms will lead to a factor of $(\not{p}-m)u_r(\mathbf{p})$ in (13-56). From (13-15), we know that each of these terms will equal zero. Thus, (13-56) becomes

Certain terms drop out

$$u_r(\mathbf{p}) \Rightarrow \left(1 - \frac{e_0^2(\not{p}-m)B(\Lambda)}{\not{p}-m+i\varepsilon}\right)u_r(\mathbf{p}) \qquad \text{indeterminate and naive}. \tag{13-57}$$

Direct substitution leads to indeterminate relation

We must be wary with (13-57) for two reasons. First the fermion is real and on shell, so we might at first consider that $(\not{p}-m)u_r(\mathbf{p})=0$ makes the second term in (13-57) equal to zero. However, we also have a factor of $\not{p}-m$ in the denominator, which would make us think the second term would be $e_0^2 B(\Lambda)$. We thus do not know really what that second term is, i.e., it is indeterminate.

Second, we have not considered, as discussed in Sect. 12.10 pg. 318 of last chapter, that the incoming fermion is initially bare, but becomes dressed via self-interactions. That is, we would need to incorporate the adiabatic hypothesis, whereby $\mathcal{H}_I$ is turned off initially, but then is turned on well before the particle begins to interact with any other particle.

The formal procedure for doing this is a bit complicated and can be found in Mandl and Shaw[1]. We will instead employ a simpler and more intuitive means to deduce the appropriate initial fermion renormalization from our renormalization (13-55) of the fermion propagator.

Can be solved formally, but complicated

---

[1] F. Mandl and G. Shaw, *Quantum Field Theory*, 1ˢᵗ ed, (Wiley) 1984, pgs. 192-193.

We'll deduce it a simpler way using the fermion 2nd order propagator

Note, from (13-55) that the renormalization factor by which we multiplied the bare propagator is, to second order,

$$1 - e_0^2 B(\Lambda) - e_0^2 \Sigma_c \left(\not{p} - m\right).\tag{13-58}$$

However, for the real particle, the Σ_c term acting on $u_r(\mathbf{p})$ goes to zero, as we noted above.

But keeping in mind that Σ_c drops out for real fermion

Now, the fermion propagator, as we showed in our derivation of Chap. 4, Sect. 4.12.3 pgs. 118-119 (or see Wholeness Chart 5-4 at the end of Chap. 5), is actually equal to the VEV of the time-ordering operator $T\left\{\psi_\alpha\left(x\right)\bar{\psi}_\beta\left(y\right)\right\}$, i.e.,

$$iS_{F\alpha\beta}\left(x-y\right) = \left\langle 0\left|T\left\{\psi_\alpha\left(x\right)\bar{\psi}_\beta\left(y\right)\right\}\right|0\right\rangle.\tag{13-59}$$

(13-59) has a spinor factor associated with each field. In essence, for the fermion (not anti-fermion) propagator, there is a factor of $u_r\left(\mathbf{p}\right)\bar{u}_r\left(\mathbf{p}\right)$. To renormalize the propagator, we multiply it by (13-58). So, in effect, each of the two spinors $u_r\left(\mathbf{p}\right)$ and $\bar{u}_r\left(\mathbf{p}\right)$ is multiplied by the square root of (13-58).

Two fields in the propagator, one in the external line → relation we want is square root of propagator relation

In (13-56) (and diagrammatically, Fig. 13-6) there is only one such spinor factor. So, we can surmise that the real particle renormalization is the square root of (13-58), where we drop the Σ_c term. This leads to

$$u_r\left(\mathbf{p}\right) \Rightarrow u_r^{2nd}\left(\mathbf{p}\right) = \left(1 - e_0^2 B(\Lambda)\right)^{1/2} u_r\left(\mathbf{p}\right) \approx \left(1 - \tfrac{1}{2}e_0^2 B(\Lambda)\right)u_r\left(\mathbf{p}\right) \quad \text{real } e^- \text{ renormalization}.\tag{13-60}$$

In similar fashion,

$$\bar{u}_r\left(\mathbf{p}\right) \Rightarrow \bar{u}_r^{2nd}\left(\mathbf{p}\right) \approx \left(1 - \tfrac{1}{2}e_0^2 B(\Lambda)\right)\bar{u}_r\left(\mathbf{p}\right)$$

$$v_r\left(\mathbf{p}\right) \Rightarrow v_r^{2nd}\left(\mathbf{p}\right) \approx \left(1 - \tfrac{1}{2}e_0^2 B(\Lambda)\right)v_r\left(\mathbf{p}\right)\tag{13-61}$$

$$\bar{v}_r\left(\mathbf{p}\right) \Rightarrow \bar{v}_r^{2nd}\left(\mathbf{p}\right) \approx \left(1 - \tfrac{1}{2}e_0^2 B(\Lambda)\right)\bar{v}_r\left(\mathbf{p}\right),$$

2nd order relations for fermion external lines in terms of our symbols

and for real photons,

$$\varepsilon_\mu\left(\mathbf{k}\right) \Rightarrow \varepsilon_\mu^{2nd}\left(\mathbf{k}\right) \approx \left(1 - \tfrac{1}{2}e_0^2 A'(\Lambda)\right)\varepsilon_\mu\left(\mathbf{k}\right).\tag{13-62}$$

2nd order relation for photon external line in terms of our symbols

Relations (13-60) to (13-62) are often called <u>external line renormalizations</u>.

13.6.4 The 2nd Order Vertex

2nd order vertex relation

The vertex modification to second order is depicted in Fig. 13-7.

The modified vertex contribution becomes (13-63). (Note from (13-3) that there is a minus sign built into Λ^μ, which was incorporated to give us a plus sign before the last term in (13-63), but which gives us an otherwise unexpected minus sign in the middle part of that equation.)

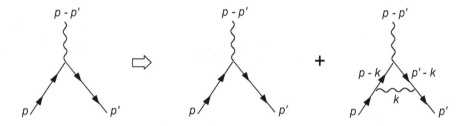

2nd order vertex Feynman diagrams

Figure 13-7. Vertex Correction to Diagrams of 2nd Order in α

$$ie_0\gamma^\mu \Rightarrow ie_0\gamma_{2nd}^\mu\left(p,p'\right) = ie_0\left(\gamma^\mu - \left(ie_0\right)^2\Lambda^\mu\left(p,p'\right)\right) = ie_0\left(\gamma^\mu + e_0^2\Lambda^\mu\left(p,p'\right)\right).\tag{13-63}$$

Add amplitude contributions from each diagram

Using (13-48), this becomes

$$ie_0\gamma^\mu \Rightarrow ie_0\gamma_{2nd}^\mu\left(p,p'\right) = \underbrace{ie_0\left\{\gamma^\mu\left(1 + e_0^2 L(\Lambda)\right) + e_0^2\Lambda_c^\mu\left(p,p'\right)\right\}}_{ie_0\gamma^\mu}.\tag{13-64}$$

2nd order vertex relation in terms of our symbols

13.7 Chapter Summary

The first part of Wholeness Chart 14-4 at the end of Chap. 14 summarizes the steps we took in this chapter to obtain our propagators, vertex, and external particle relations to second order.

See Wholeness Chart at end of Chap. 14

A summary of the relevance of local symmetry and key relations from this chapter follow.

Local gauge invariance implications

Each of the following being true implies the others are also.

Summarizing implications of local gauge invariance

- The theory (Lagrangian) is locally gauge invariant (locally symmetric).
- A quantity (charge) is conserved.
- The theory has the correct interactions.
- The Ward identities hold.
- The theory is renormalizable.

If the above are true, then

- the gauge boson (photon) must be massless.

Divergent integrals

Summary of key relations

$$\text{photon self energy}\quad \Pi^{\mu\nu}(k) = \underbrace{-g^{\mu\nu}k^2 A'(k,\Lambda)}_{\infty \text{ for } \Lambda \to \infty} - \underbrace{g^{\mu\nu}k^2 \Pi_c(k^2)}_{\text{finite}} \qquad (13\text{-}65)$$

$$\text{fermion self energy}\quad \Sigma(p) = \underbrace{A(\Lambda,m)}_{\infty \text{ for } \Lambda \to \infty} + \underbrace{(\not p - m)B(\Lambda)}_{\infty \text{ for } \Lambda \to \infty} + \underbrace{(\not p - m)\Sigma_c(\not p - m)}_{\text{finite}} \qquad (13\text{-}66)$$

$$\text{vertex correction}\quad \Lambda^{\mu}(p,p') = \underbrace{L(\Lambda)\gamma^{\mu}}_{\infty \text{ for } \Lambda \to \infty} + \underbrace{\Lambda_c^{\mu}(p,p')}_{\text{finite}} \qquad (13\text{-}67)$$

Operator relation

$$\frac{1}{A-B} = \frac{1}{A} + \frac{1}{A}B\frac{1}{A} + \frac{1}{A}B\frac{1}{A}B\frac{1}{A} + \dots \qquad (13\text{-}68)$$

Gordon's identity

$$\bar{u}_{r'}(\mathbf{p}')\gamma^{\mu}u_r(\mathbf{p}) = \frac{p^{\mu}+p'^{\mu}}{2m}\bar{u}_{r'}(\mathbf{p}')u_r(\mathbf{p}) + \frac{p_{\nu}-p'_{\nu}}{4m}\bar{u}_{r'}(\mathbf{p}')\left[\gamma^{\mu},\gamma^{\nu}\right]u_r(\mathbf{p}) \qquad (13\text{-}69)$$

Original Ward identity

$$\frac{\partial \Sigma(p)}{\partial p_{\mu}} = \Lambda^{\mu}(p,p) \quad \to \quad B = L \text{ in (13-66) and (13-67)}. \qquad (13\text{-}70)$$

Ward identities

$$\varepsilon_{r_j\nu}\mathcal{M}_{fi}^{(n)\mu..\nu..\eta...}\left(\mathbf{k}_1,...,\mathbf{k}_j,...\right) \quad \to \quad k_{j\nu}\mathcal{M}_{fi}^{(n)\mu..\nu..\eta...}\left(\mathbf{k}_1,...,\mathbf{k}_j,...\right) = 0. \qquad (13\text{-}71)$$

Replacing m_0 with m

$$m = m_0 + \delta m \quad \Rightarrow \quad \mathcal{L} = \underbrace{-\tfrac{1}{4}F^{\mu\nu}F_{\mu\nu} + \bar{\psi}\left(i\gamma^{\mu}\partial_{\mu} - m\right)\psi}_{\text{mass renormalized } \mathcal{L}_0} + \underbrace{e_0\,\bar{\psi}\,\gamma^{\mu}\psi A_{\mu} + \overbrace{\delta m\bar{\psi}\psi}^{\text{mass counter term}}}_{\text{mass renormalized } \mathcal{L}_I} \qquad (13\text{-}72)$$

We must add to any fermion line in a Feynman diagram → ——►—✕—►——

Additional <u>Feynman rule #10</u>: For each mass counterterm diagram, add a term to the amplitude with a factor equal to $i\delta m$.

Corrections to 2nd order

$$iD_{F\mu\nu}(k) \Rightarrow iD_{F\mu\nu}^{2nd}(k) = \left(iD_{F\mu\nu}(k)\right)\left(1 - e_0^2 A'(k,\Lambda) - e_0^2\Pi_c(k^2)\right). \qquad (13\text{-}73)$$

$$iS_F(p) \Rightarrow iS_F^{2nd}(p) = \left(iS_F(p)\right)\left(1 - e_0^2 B(\Lambda) - e_0^2\Sigma_c(\not{p}-m)\right) \qquad (13\text{-}74)$$

$$u_r(\mathbf{p}) \Rightarrow u_r^{2nd}(\mathbf{p}) = \left(1 - \tfrac{1}{2}e_0^2 B(\Lambda)\right)u_r(\mathbf{p}) \qquad \bar{u}_r(\mathbf{p}) \Rightarrow \bar{u}_r^{2nd}(\mathbf{p}) = \left(1 - \tfrac{1}{2}e_0^2 B(\Lambda)\right)\bar{u}_r(\mathbf{p})$$

$$v_r(\mathbf{p}) \Rightarrow v_r^{2nd}(\mathbf{p}) = \left(1 - \tfrac{1}{2}e_0^2 B(\Lambda)\right)v_r(\mathbf{p}) \qquad \bar{v}_r(\mathbf{p}) \Rightarrow \bar{v}_r^{2nd}(\mathbf{p}) = \left(1 - \tfrac{1}{2}e_0^2 B(\Lambda)\right)\bar{v}_r(\mathbf{p}) \quad (13\text{-}75)$$

$$\varepsilon_\mu(\mathbf{k}) \Rightarrow \varepsilon_\mu^{2nd}(\mathbf{k}) = \left(1 - \tfrac{1}{2}e_0^2 A'(\Lambda)\right)\varepsilon_\mu(\mathbf{k})$$

$$ie_0\gamma^\mu \Rightarrow ie_0\gamma_{2nd}^\mu(p,p') = \underbrace{ie_0\left\{\gamma^\mu\left(1 + e_0^2 L(\Lambda)\right) + e_0^2\Lambda_c^\mu(p,p')\right\}}_{ie_0\gamma^\mu} \qquad (13\text{-}76)$$

13.8 Appendix: Finding Ward Identities for Compton Scattering

Since we worked through every step of finding quite a number of amplitudes the long way in Chap. 8, we will be briefer here. If you have trouble at any point, please refer to the detailed derivation of the Compton scattering transition amplitude on pgs. 225-228, which closely parallels the following.

$$\left\langle e_{\mathbf{p}',s'}^-, \gamma_{\mathbf{k}',r'} \left| \left(\sum_{r,\mathbf{k}} k_\nu S_{C1}^{(2)\nu} + \sum_{r,\mathbf{k}} k_\mu S_{C2}^{(2)\mu}\right) \right| e_{\mathbf{p},s}^- \right\rangle = 0 \quad \text{because} \quad \sum_{r,\mathbf{k}} k_\nu S_{C1}^{(2)\nu} + \sum_{r,\mathbf{k}} k_\mu S_{C2}^{(2)\mu} = 0 \quad (13\text{-}77)$$

For Compton scattering the first way (LH of Fig. 8-3, pg. 225) without the incoming photon, the part of (13-77) with $S_{C1}^{(2)\nu}$ becomes

$$\left\langle e_{\mathbf{p}',s'}^-, \gamma_{\mathbf{k}',r'} \left| \sum_{r,\mathbf{k}} k_\nu S_{C1}^{(2)\nu} \right| e_{\mathbf{p},s}^- \right\rangle$$

$$= \left\langle e_{\mathbf{p}',s'}^-, \gamma_{\mathbf{k}',r'} \left| e \iint d^4x_1 d^4x_2\, N\left\{\left(\bar{\psi}A_\mu\gamma^\mu\right)_{x_1} iS(x_1-x_2)\left(\sum_{r,\mathbf{k}} k_\nu \tilde{\alpha}_{r,\mathbf{k}}(x_2)\right)\gamma^\nu\psi_{x_2}\right\} \right| e_{\mathbf{p},s}^- \right\rangle$$

$$= \left\langle e_{\mathbf{p}',s'}^-, \gamma_{\mathbf{k}',r'} \left| e \iint d^4x_1 d^4x_2 \left(\sum_{s'',\mathbf{p}''}\sqrt{\frac{m}{VE_{\mathbf{p}''}}}c_{s''}^\dagger(\mathbf{p}'')\bar{u}_{s''}(\mathbf{p}'')e^{ip''x_1}\right) \times \right. \qquad (13\text{-}78)$$

$$\left(\sum_{r'',\mathbf{k}''}\sqrt{\frac{1}{2V\omega_{\mathbf{k}''}}}a_{r''}^\dagger(\mathbf{k}'')\varepsilon_{\mu,r''}(\mathbf{k}'')e^{ik''x_1}\gamma^\mu\right)\frac{1}{(2\pi)^4}\int d^4q\, iS_F(q)e^{-iq(x_1-x_2)} \times$$

$$\left(i\sum_{r,\mathbf{k}} k_\nu\sqrt{\frac{1}{2V\omega_{\mathbf{k}}}}\left(-\tilde{a}_r(\mathbf{k})e^{-ikx_2} + \tilde{a}_r^\dagger(\mathbf{k})e^{ikx_2}\right)\right)\gamma^\nu\left.\left(\sum_{s''',\mathbf{p}'''}\sqrt{\frac{m}{VE_{\mathbf{p}'''}}}c_{s'''}(\mathbf{p}''')u_{s'''}(\mathbf{p}''')e^{-ip'''x_2}\right)\right| e_{\mathbf{p},s}^- \right\rangle$$

$$= \sum_{s'',\mathbf{p}''}\sum_{r'',\mathbf{k}''}\underbrace{\left\langle e_{\mathbf{p}',s'}^-, \gamma_{\mathbf{k}',r'} \right\| e_{\mathbf{p}'',s''}^-, \gamma_{\mathbf{k}'',r''}\right\rangle}_{\delta_{\mathbf{p}'\mathbf{p}''}\delta_{s's''}\delta_{\mathbf{k}'\mathbf{k}''}\delta_{r'r''}} e \iint d^4x_1 d^4x_2 \left(\sqrt{\frac{m}{VE_{\mathbf{p}''}}}\bar{u}_{s''}(\mathbf{p}'')e^{ip''x_1}\right)\left(\sqrt{\frac{1}{2V\omega_{\mathbf{k}''}}}\varepsilon_{\mu,r''}(\mathbf{k}'')e^{ik''x_1}\gamma^\mu\right)$$

$$\times\frac{1}{(2\pi)^4}\int d^4q\, iS_F(q)e^{-iq(x_1-x_2)}\left(i\sum_{r,\mathbf{k}} k_\nu\sqrt{\frac{1}{2V\omega_{\mathbf{k}}}}\left(-\tilde{a}_r(\mathbf{k})e^{-ikx_2} + \tilde{a}_r^\dagger(\mathbf{k})e^{ikx_2}\right)\right)\gamma^\nu\left(\sqrt{\frac{m}{VE_{\mathbf{p}}}}u_s(\mathbf{p})e^{-ipx_2}\right)$$

$$(13\text{-}79)$$

$$= e\sqrt{\frac{m}{VE_{\mathbf{p}'}}}\sqrt{\frac{1}{2V\omega_{\mathbf{k}'}}}\sqrt{\frac{m}{VE_{\mathbf{p}}}}\,\varepsilon_{\mu,r'}(\mathbf{k}')\bar{u}_{s'}(\mathbf{p}')\gamma^{\mu}\frac{1}{(2\pi)^4}\int d^4 q\, iS_F(q)\gamma^{\nu}u_s(\mathbf{p})\times$$

$$\left(i\sum_{r,\mathbf{k}}k_\nu\sqrt{\frac{1}{2V\omega_{\mathbf{k}}}}\left(-\tilde{a}_r(\mathbf{k})e^{-ikx_2}+\tilde{a}_r^\dagger(\mathbf{k})e^{ikx_2}\right)\right)\times \tag{13-80}$$

$$\left\{\int d^4 x_1 e^{-iqx_1}e^{ip'x_1}e^{ik'x_1}\int d^4 x_2 e^{iqx_2}e^{-ipx_2}\right\}$$

$$= e\sqrt{\frac{m}{VE_{\mathbf{p}'}}}\sqrt{\frac{1}{2V\omega_{\mathbf{k}'}}}\sqrt{\frac{m}{VE_{\mathbf{p}}}}\,\varepsilon_{\mu,r'}(\mathbf{k}')\bar{u}_{s'}(\mathbf{p}')\gamma^{\mu}\frac{1}{(2\pi)^4}\int d^4 q\, iS_F(q)\gamma^{\nu}u_s(\mathbf{p})\times$$

$$\left(\sum_{r,\mathbf{k}}\sqrt{\frac{1}{2V\omega_{\mathbf{k}}}}\int d^4 x_1 e^{-iqx_1}e^{ip'x_1}e^{ik'x_1}\,ik_\nu\begin{pmatrix}-\tilde{a}_r(\mathbf{k})\int d^4 x_2 e^{iqx_2}e^{-ipx_2}e^{-ikx_2}\\ +\tilde{a}_r^\dagger(\mathbf{k})\int d^4 x_2 e^{iqx_2}e^{-ipx_2}e^{+ikx_2}\end{pmatrix}\right) \tag{13-81}$$

$$= e\sqrt{\frac{m}{VE_{\mathbf{p}'}}}\sqrt{\frac{1}{2V\omega_{\mathbf{k}'}}}\sqrt{\frac{m}{VE_{\mathbf{p}}}}\,\varepsilon_{\mu,r'}(\mathbf{k}')\bar{u}_{s'}(\mathbf{p}')\gamma^{\mu}\frac{1}{(2\pi)^4}\int d^4 q\, iS_F(q)\gamma^{\nu}u_s(\mathbf{p})\times$$

$$(2\pi)^4\delta^{(4)}(q-p'-k')i\left(\sum_{r,\mathbf{k}}\sqrt{\frac{1}{2V\omega_{\mathbf{k}}}}k_\nu\begin{pmatrix}-\tilde{a}_r(\mathbf{k})(2\pi)^4\delta^{(4)}(q-p-k)\\ +\tilde{a}_r^\dagger(\mathbf{k})(2\pi)^4\delta^{(4)}(q-p+k)\end{pmatrix}\right)$$

$$= \frac{1}{e}\sqrt{\frac{m}{VE_{\mathbf{p}'}}}\sqrt{\frac{1}{2V\omega_{\mathbf{k}'}}}\sqrt{\frac{m}{VE_{\mathbf{p}}}}(2\pi)^4\underbrace{e^2\,\varepsilon_{\mu,r'}(\mathbf{k}')\bar{u}_{s'}(\mathbf{p}')\gamma^{\mu}iS_F(p'+k')\gamma^{\nu}u_s(\mathbf{p})}_{=\,-\mathcal{M}_{C1}^{(2)\nu}}\times \tag{13-82}$$

$$\left(i\sum_{r,\mathbf{k}}\sqrt{\frac{1}{2V\omega_{\mathbf{k}}}}k_\nu\begin{pmatrix}\underbrace{-\tilde{a}_r(\mathbf{k})\delta^{(4)}(p'+k'-p-k)}_{\substack{=\,0\text{ except when }k=p'+k'-p,\\ \text{the value in full Compton scattering}}}\\[2mm] \underbrace{+\;\tilde{a}_r^\dagger(\mathbf{k})\delta^{(4)}(p'+k'-p+k)}_{\substack{=\,0\text{ except when }k=-p'-k'+p,\text{ negative}\\ \text{of the value in full Compton scattering}}}\end{pmatrix}\right)$$

$$= -\frac{1}{e}\sqrt{\frac{m}{VE_{\mathbf{p}'}}}\sqrt{\frac{1}{2V\omega_{\mathbf{k}'}}}\sqrt{\frac{1}{2V\omega_{\mathbf{k}}}}\sqrt{\frac{m}{VE_{\mathbf{p}}}}(2\pi)^4\sum_r\delta^{(4)}(p'+k'-p-k)i\mathcal{M}_{C1}^{(2)\nu}\left(-k_\nu\tilde{a}_r(\mathbf{k})-k_\nu\tilde{a}_r^\dagger(-\mathbf{k})\right)$$

$$= \frac{1}{e}\sqrt{\frac{m}{VE_{\mathbf{p}'}}}\sqrt{\frac{1}{2V\omega_{\mathbf{k}'}}}\sqrt{\frac{1}{2V\omega_{\mathbf{k}}}}\sqrt{\frac{m}{VE_{\mathbf{p}}}}(2\pi)^4\sum_r\delta^{(4)}(p'+k'-p-k)ik_\nu\mathcal{M}_{C1}^{(2)\nu}\left(\tilde{a}_r(\mathbf{k})+\tilde{a}_r^\dagger(-\mathbf{k})\right) \tag{13-83}$$

<u>For Compton scattering the second way</u> (RH of Fig. 8-3), the part of (13-77) with $S_{C\,2}^{(2)\mu}$, after similar evaluation, yields, with the sub amplitude $\mathcal{M}_{C\,2}^{(2)\nu}$ as shown in (13-32)+7,

$$\left\langle e_{\mathbf{p}',s'}^-,\gamma_{\mathbf{k}',r'}\left|\sum_{r,\mathbf{k}}k_\mu S_{C\,2nd}^{(2)\mu}\right|e_{\mathbf{p},s}^-\right\rangle$$

$$= \frac{1}{e}\sqrt{\frac{m}{VE_{\mathbf{p}'}}}\sqrt{\frac{1}{2V\omega_{\mathbf{k}'}}}\sqrt{\frac{1}{2V\omega_{\mathbf{k}}}}\sqrt{\frac{m}{VE_{\mathbf{p}}}}(2\pi)^4\delta^{(4)}(p'+k'-p-k)ik_\nu\mathcal{M}_{C\,2}^{(2)\nu}\sum_r\left(\tilde{a}_r(\mathbf{k})+\tilde{a}_r^\dagger(-\mathbf{k})\right). \tag{13-84}$$

(13-83) and (13-84) summed equal the LHS of (13-77), so their sum equals zero. To do this, the coefficient of $\sum_r\tilde{a}_r(\mathbf{k})$, which is arbitrary, in that sum must vanish (as must the coefficient of $\sum_r\tilde{a}_r^\dagger(-\mathbf{k})$). The only way this can happen is if

$$k_\nu\left(\mathcal{M}_{C1}^{(2)\nu}+\mathcal{M}_{C\,2}^{(2)\nu}\right)=k_\nu\mathcal{M}_{Compton}^{(2)\nu}=0. \tag{13-85}$$

13.9 Problems

1. Show that $\left(\not{p}+m\right)v_r\left(\mathbf{p}\right)=0$. (Hint: Follow steps like we did to get (13-10).)

2. Draw at least three ways, other than that shown in Fig. 13-2, for which the incoming same two photon state scatters at second order into the same outgoing two photon state.

3. Re-draw the Feynman diagram of Fig. 13-2 with the upper left vertex occurring before the lower left vertex. Label the internal line four-momenta. Show by writing out the Feynman amplitude for this diagram using Feynman rules, that the amplitude you get is the same as we got in (13-36) for Fig. 13-2. (Hint: re-express (13-36) with p_2, p_3, and p_4 in terms of p_1. Then express your new diagram where all propagator factors are in terms of p_1. Remember that for anti-particle internal lines, the four-momentum has opposite sign from physical reality. See Wholeness Chart 8-1, pg. 234.) Realize that the diagram you drew for this problem is not one of the answers for Prob. 2.

4. Show that by using part b) of Fig. 13-5 for Feynman diagrams to 2nd order of fermion self-energy, you obtain (13-55). Hint: In (13-54) take $m_0 \rightarrow m$ and $ie_0^2\Sigma\left(p\right) \rightarrow ie_0^2\Sigma\left(p\right)+i\delta m$.

5. Show (13-62) using similar logic to what we used for (13-60). Note that in (13-52), the $e_0^2\Pi_c(k^2)$ term is an expansion with terms in k^2 to various powers, but that for a real photon $k^2 = 0$.

This page intentionally left blank.

Chapter 14

Renormalization: Putting It All Together

*Life is full of infinite absurdities, which, strangely enough,
do not even need to appear plausible, since they are true.*
Luigi Pirandello

14.0 Preliminaries

The previous two chapters served as background for this one. The first of those entailed a simplified overview of renormalization, and the second, a set of mathematical tools we will use in this chapter to carry out the actual, formal steps involved in renormalization. We note again that we will employ symbols for the divergent integrals and for the complicated convergent integrals encountered in doing this. These symbols were defined in the prior chapter, and we will assume familiarity with what they stand for. The actual evaluation of the integrals will be carried out in the next chapter.

Prior two chapters laid foundation for this one

14.0.1 Terminology

You may run into the phrases "on-shell renormalization" and "minimal subtraction renormalization" The on-shell renormalization scheme (also known as the physical scheme) is what we do in this book. The term refers to the fact that in this scheme one typically uses free fields, which the interaction picture we employ incorporates, and such fields (for associated external particles) are on-shell. The minimal subtraction renormalization scheme (or MS scheme), developed independently in 1973 by 't Hooft and Weinberg, treats the counterterms in the Lagrangian differently than the on-shell scheme. We will not do anything herein with the MS scheme.

Our approach is called "on-shell renormalization"

14.0.2 Second Order vs All Orders Renormalization

In the prior chapter we focused on 2nd order (in α) radiative corrections to tree level amplitudes. In this chapter, we will use the results of that chapter to i) carry out renormalization to 2nd order and then ii) sketch out the procedure and note the final results for renormalization to all orders.

*First, renormalize to 2nd order
Then: to all orders*

14.0.3 Chapter Overview

Specifically, in this chapter, we will first

- examine 2nd order (in α) corrections to Compton scattering, specifically the divergent loop integrals contributions,

We'll use example of Compton scattering, then generalize

- summarize the prior two chapters results for these divergent integrals,

- deduce the Compton amplitude to 2nd order,

- renormalize that amplitude so that it is finite via $m_0 \rightarrow m$ and $e_0 \rightarrow e$, and

- generalize our results for all possible types of QED scattering.

We will then

- repeat the above procedure for corrections to all orders.

Wholeness Charts 14-4 and 14-5 at the end of this chapter summarize this.

14.1 Renormalization Example: Compton Scattering

We begin by considering an example, Compton scattering. In addition to the two tree level (1st order in α_0) ways for this to occur that we saw in Chap. 8, there are additional 2nd order in α_0 (4th order in e_0) amplitude contributions we need to include. The Feynman diagrams for all of these are shown in Fig. 14-1 where the superscript on the S operators indicates number of vertices, i.e., order in factors of e_0, rather than α_0. (As an aside, this figure is the solution to Chap. 9, Prob. 7.)

Compton scattering to second order in α

Associated Feynman diagrams

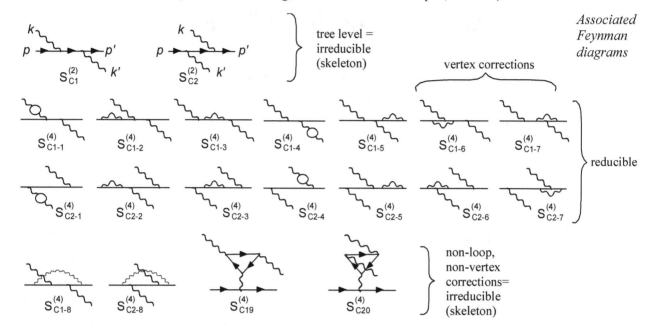

Figure 14-1. Compton Scattering: Feynman Diagrams Including 2nd Order Corrections

The first row in Fig. 14-1 represents the two 1st order sub-amplitudes we have seen before. The second row includes all the 2nd order corrections that can be made to the first diagram of the first row via loop and vertex corrections, including those to external particles as well as the propagator. The third row includes all similar corrections for the second diagram in the first row.

In addition to these, there are four other relevant diagrams, shown in the last row, which do not have propagator or vertex loops. The first two of these show virtual photons linking the incoming and outgoing fermions. The last two depict a situation we hinted at in Chap. 9, but did not show explicitly there. That is, we can have internal fermion triangles as shown, with the same incoming and outgoing particle states.

We will discuss the last four diagrams of Fig. 14-1 first, but before that, we need to define two classes of Feynman diagrams, which are handled differently while renormalizing. The first, represented by the middle two rows in Fig. 14-1 contain propagator/vertex loops and are called <u>reducible diagrams</u>. The second class, represented in the first and last rows, are called <u>irreducible (or skeleton) diagrams</u> and contain no propagator or vertex loops. The word "reducible" is used because reducible diagrams can be reduced to irreducible diagrams by removing the loops. For example, if we take the loop out of any diagram in row two, we get the 1st diagram of row one.

<u>Caution</u>: Some authors define these terms somewhat differently. See footnote, pg. 354.

Reducible vs irreducible diagrams

14.1.1 The Incoming to Outgoing Linking Virtual Photon Contribution

The first diagram in the last row, via Feynman rules with k'' as the virtual photon four momentum, has the amplitude

Amplitude for C1-8 diagram

$$\mathcal{M}_{C1-8}^{(4)} = \frac{(ie_0)^4}{(2\pi)^4} \int d^4 k'' \, \bar{u}_{r'}(\mathbf{p}') \gamma^\alpha i S_F(p-k''+k-k') \varepsilon_\beta(\mathbf{k}') \gamma^\beta \times$$

$$i S_F(p-k''+k) \gamma^\eta \varepsilon_\eta(\mathbf{k}) i S_F(p-k'') i D_{F\alpha\delta}(k'') \gamma^\delta u_r(\mathbf{p}).$$

$$(14-1)$$

Isolate the integral part and power count to determine the maximum possible divergence to get

$$\mathcal{M}_{C1-8}^{(4)} \xrightarrow[\text{ignoring } \gamma^{\mu}\text{s}]{\text{integral only}} \int S_F\left(p-k''+k-k'\right)S_F\left(p-k''+k\right)S_F\left(p-k''\right)D_F\left(k''\right)d^4k''$$

$$= \int \frac{1}{\left(p-k''+k-k'\right)-m+i\varepsilon} \frac{1}{\left(p-k''+k\right)-m+i\varepsilon} \frac{1}{\left(p-k''\right)-m+i\varepsilon} \frac{1}{k''^2+i\varepsilon} d^4k'' \qquad (14\text{-}2)$$

Power counting shows C1-8 contribution is finite

$$\xrightarrow{\text{for large } k''} \approx \int \frac{1}{k''}\frac{1}{k''}\frac{1}{k''}\frac{1}{\left(k''\right)^2}d^4k'' = 2\pi^2 \int \frac{1}{\left(k''\right)^5}\left(k''\right)^3 dk'' = -2\pi^2\frac{1}{k''} \text{ for large } k''.$$

So this contribution is finite (before even considering renormalization). By doing Prob. 2, you can show that the second diagram in the last row of Fig. 14-1 also leads to a finite contribution to the total amplitude. These results parallel what we discussed in Chap. 9, pgs. 261-262, "Other Diagrams" section.

So is similar C2-8 diagram

After determining the contributions from the other diagrams, more problematic since most involve divergent integrals, we should rightfully add in the $\mathcal{M}_{C1-8}^{(4)}$ and $\mathcal{M}_{C2-8}^{(4)}$ contributions. However, we will find that these contributions are negligible compared to those in the 2nd and 3rd rows of Fig. 14-1 and can, at second order, be ignored.

C1-8, C2-8 contributions will be negligible

Bottom line: The amplitudes for the first two diagrams in the last row of Fig. 14-1 are finite and negligible at 2nd order compared to the other diagrams.

14.1.2 The Triangle Diagrams Contribution

Note that the last diagram of Fig. 14-1 can be expressed in a different, but completely equivalent, way. That is, as we saw in Chap. 13, Prob. 3, the order of vertex events can be shown in different time order (time progresses from left to right here) without changing the mathematical expression of the associated Feynman amplitude. For the last diagram in Fig. 14-1, switching time ordering of the last to occur vertex with the first to occur vertex is shown in Fig. 14-2. The RHS of that figure is identical mathematically to the LHS.

Changing order of events in diagram does not change amplitude

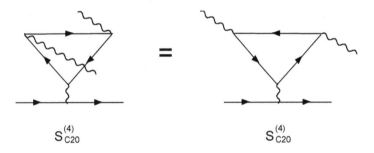

$$S_{C20}^{(4)} \qquad\qquad\qquad S_{C20}^{(4)}$$

So these diagrams are equivalent

Figure 14-2. Equivalent Triangle Diagrams

Thus, the last diagram of Fig. 14-1 can be re-drawn to look like the next to last diagram in the same figure, except that the directions of the fermion propagator arrows are reversed. A virtual fermion in one figure is a virtual anti-fermion in the other, and vice versa.

Sign reversal on propagator if arrow reversed

Recall from our original derivation of the Dirac particle propagator (see Chap. 4 (4-125) and (4-126), pg. 118) that the propagator for an anti-fermion has the opposite sign of the fermion. So, switching fermion and anti-fermion propagators introduces a minus sign into the total amplitude for each such switch. There are three switches in going from the next to last diagram in Fig. 14-1 to the RHS of Fig. 14-2. Since $(-1)^3 = -1$, the amplitudes for the two diagrams are the same except for sign. So, when added, they cancel exactly, and we don't have to consider them anymore.

So, triangle diagrams of Fig. 14-1 cancel

This procedure holds true for any fermion triangle in any interaction. This can be generalized according to Furry's theorem.

Furry's theorem: Diagrams containing an all fermion sided polygon having an odd number of sides always occur in pairs, and the contributions of such pairs to the total amplitude cancel out.

Diagrams with odd number sided fermion polygons always cancel

Bottom line for Compton: The triangle diagrams in Fig. 14-1 can be ignored. With the results of Sect. 14.1.1, this means all skeleton (irreducible) diagrams for Compton yield finite amplitudes.

Skeleton graphs have finite amplitudes

14.1.3 The Divergent Contributions

The remaining 2nd order in α diagrams in Fig. 14-1 are reducible, i.e., they all contain photon self-energy, fermion self-energy, or vertex corrections, which all yield divergent integrals, as we saw in earlier chapters. As we contended in those earlier chapters as a general principle, the only divergent integrals we will have to deal with, and renormalize, in this specific case of Compton scattering, are those three. This result can be generalized to any interaction.[1] Much of the remainder of this chapter will deal with incorporating these three integrals into our renormalization process.

<u>Bottom line</u>: Only reducible diagrams yield divergent integrals and thus, divergent amplitudes.

Only reducible diagrams have divergent amplitudes

True in general

14.1.4 The Second Order Total Amplitude

The total amplitude to 2nd order (in α_0) is then

$$\underset{\substack{\text{all ways} \\ \text{total to } e_0^4}}{\mathcal{M}_{\text{Compton}}} = \underbrace{\mathcal{M}_{C1}^{(2)} + \mathcal{M}_{C2}^{(2)}}_{\substack{\text{tree level, finite,} \\ \text{irreducible}}} + \underbrace{\sum_{i}^{i=7} \mathcal{M}_{C1-i}^{(4)} + \sum_{i}^{i=7} \mathcal{M}_{C2-i}^{(4)}}_{\substack{\text{contain divergent integrals,} \\ \text{reducible}}} + \underbrace{\mathcal{M}_{C1-8}^{(4)} + \mathcal{M}_{C2-8}^{(4)}}_{\substack{\text{finite,} \\ \text{irreducible}}} + \underbrace{\mathcal{M}_{C19}^{(4)} + \mathcal{M}_{C20}^{(4)}}_{\substack{\text{cancel out,} \\ \text{irreducible}}} . (14\text{-}3)$$

Adding amplitudes for all diagrams in Fig. 14-1 gives total 2nd order amplitude

14.2 Renormalizing 2nd Order Divergent Amplitudes

14.2.1 Steps of Renormalization

In the prior chapter, we have already done quite a bit of work towards our goal of renormalizing our divergent amplitudes to make them finite. This work is summarized in columns (I) to (VIII) of Wholeness Chart 14-4 at the end of this chapter (starting on pg. 368). Those columns deal directly, and independently, with the self-energy and vertex correction terms only. Starting in column (IX), which is new material, we use those results to renormalize the divergent sub-amplitudes of (14-3).

We review the prior chapter results of columns (I) to (VIII) immediately below.

Wholeness Chart 14-4 summarizes steps of 2nd order renormalization

Columns (I) to (VIII) cover previous chapter

(I) Feynman Diagram Modification

Column (I) of Wholeness Chart 14-4 shows the modifications at 2nd order to the propagator, external particle line, and vertex portions of Feynman diagrams. (See Chap. 13.)

Note that for the fermion self-energy, for completeness, we show both the bare fermion mass m_0 case (A) and the renormalized mass m case (B). However, in terms of developing the theory, we are getting a little ahead of ourselves, as strictly speaking, we cannot know the m case (B) until we proceed through to column (VIII) in the fermion row. So, for blocks (II) to (VII) in the fermion row, we will stick strictly to m_0 case (A) in our fermion propagators iS_F and think in terms of the Feynman diagrams for m_0 case (A). Once we know the theory, of course, we can use either case (A) or case (B), as we see fit.

For the external line correction, we only show the renormalized m case (B), simply because the derivation is easier in that case. But to do that, if we are developing our theory from scratch, keep in mind that we need to proceed through the fermion row to mass renormalization before starting on the external line correction derivation.

Column (I) shows diagrams to 2nd order for propagators, external particles, and vertex

(II) Propagator, Leg, Vertex Modification

Column (II) shows the additions one must make to the propagator, external line, and vertex to the mathematical representations of the associated Feynman diagrams. Here we use the symbols for the divergent integrals introduced at the beginning of Chap. 13, $\Pi^{\mu\nu}(k)$, $\Sigma(p)$, and $\Lambda^{\mu}(p, p')$.

Column (II) shows math for column (I) with symbols for loops

(III) Loop Integral

Here we explicitly express the integrals $\Pi^{\mu\nu}(k)$, $\Sigma(p)$, and $\Lambda^{\mu}(p,p')$ in terms of bare propagators.

Column (III) defines loop symbols

(IV) Most General Form of the Loop Integrals

We did not spend a great deal of time in Chap. 13 on the most general form possible for the self-energy integrals. We did examine this closely for the vertex correction in Sect. 13.5, pgs. 331-332, and we alluded to it for the fermion self-energy near the end of the same section. We note that in the process of evaluating the integrals, certain parts of the photon expression in this and the next

Column (IV) shows most general form of loop integral

---

[1] For a proof, see F. Mandl & G. Shaw, *Quantum Field Theory*, 1st Ed. (Wiley 1984), pgs. 216-218.

column do not appear in the final result. (See the next chapter.) Terms in the most general expression can, but do not need to, occur in any given specific case.

As noted in Chap. 13, we can express the general form for $\Sigma(p)$ in terms of $\left(\not{p}-m_0\right)$ or $\left(\not{p}-m\right)$. The development can be done with either, as the expression is general in nature and not constrained to either choice. However, using the form with m makes the development easier, so that is what we do here. This may be a little confusing, as we continue to express the fermion propagator, in subsequent column (VI), in terms of m_0. The propagator is a specific, previously derived relation, but $\Sigma(p)$ at this point is still generic.

We do not spend much time on these general forms at 2nd order, as via regularization we obtain the specific forms needed. However, they will help us later in renormalizing to higher orders.

(V) Series Expansion of the Loop Integrals

The general forms of column (IV) are here expressed as Taylor series expansions. The photon relation is expanded in powers of k^2; the fermion one, in powers of $\left(\not{p}-m\right)$. These expansions will be found (Chap. 15) to contain divergent and convergent terms, the latter labeled with subscript "c".

Column (V) contains series expansion form of loop integrals

(VI) Putting Expansions (V) into Propagator, Leg, Vertex Modifications (II)

Here we insert the loop integrals of column (V) into the 2nd order correction relations for the propagators and vertex of column (II). For the fermion case (14-6), we use the operator relation derived in Chap. 13,

Column (VI) puts expansions of (V) into math of (II)

$$\frac{1}{A-B}=\frac{1}{A}+\frac{1}{A}B\frac{1}{A}+\frac{1}{A}B\frac{1}{A}B\frac{1}{A}+.... \quad A,B \text{ any operators} \begin{pmatrix} \text{not the } A,B \text{ symbols in} \\ \text{propagator expansions} \end{pmatrix}, \quad (14\text{-}4)$$

with operators $A=\not{p}-m_0$ and $B=ie_0^2\Sigma(p)$ of column (V). Note the denominator for $iS_F^{2nd}(p)$ is an approximation, since we only used the first two terms in the RHS of (14-4) to obtain it. We sweep this approximation under the rug in our analysis by using the superscript notation "2nd", meaning higher order terms are ignored.

The external line relations, like all the others in this column, were derived in Chap. 13.

To second order, we then have from column (VI)

$$iD_{F\alpha\beta}^{2nd}(k)=\underbrace{\frac{-ig_{\alpha\beta}}{k^2+i\varepsilon}}_{iD_{F\,\alpha\beta}(k)}\big(\underbrace{1}_{\substack{\text{1st}\\\text{order}}}\underbrace{-e_0^2A'-e_0^2\Pi_c}_{\substack{\text{2nd order (loop)}\\\text{contribution}}}\big), \quad (14\text{-}5)$$

Evaluate fermion propagator using operator relation (14-4) above

$$iS_F^{2nd}(p)\approx\frac{i}{\underbrace{\not{p}-m_0+e_0^2A}_{-m}+e_0^2\left(\not{p}-m\right)B+e_0^2\left(\not{p}-m\right)\Sigma_c+i\varepsilon}$$

Renormalize mass in the process

$$=\frac{i}{\not{p}-m+e_0^2\left(\not{p}-m\right)B+e_0^2\left(\not{p}-m\right)\Sigma_c+i\varepsilon}=\frac{i}{\left(\not{p}-m\right)\left(1+e_0^2B+e_0^2\Sigma_c\right)+i\varepsilon} \quad (14\text{-}6)$$

$$\approx\frac{i}{\left(\not{p}-m+i\varepsilon\right)\left(1+e_0^2B+e_0^2\Sigma_c\right)}\approx\underbrace{\frac{i}{\not{p}-m+i\varepsilon}}_{iS_F(p)}\big(\underbrace{1}_{\substack{\text{1st}\\\text{order}}}\underbrace{-e_0^2B-e_0^2\Sigma_c}_{\substack{\text{2nd order (loop)}\\\text{contribution}}}\big),$$

$$u_r^{2nd}(\mathbf{p})=\left(1-e_0^2B\right)^{1/2}u_r(\mathbf{p})=\left(Z_f^{2nd}\right)^{1/2}u_r(\mathbf{p})\approx\big(\underbrace{1}_{\substack{\text{1st}\\\text{order}}}\underbrace{-\tfrac{1}{2}e_0^2B}_{\substack{\text{2nd order}\\\text{(loop)}}}\big)u_r(\mathbf{p})\approx\frac{1}{\left(1+e_0^2B\right)^{1/2}}u_r(\mathbf{p})$$

Get relations for all quantities with tree level and 2nd order parts

$$\text{same for } \bar{u}_r^{2nd}(\mathbf{p}),\ v_r^{2nd}(\mathbf{p}),\ \bar{v}_r^{2nd}(\mathbf{p}) \quad (14\text{-}7)$$

$$\varepsilon_\mu^{2nd}(\mathbf{k})=\left(1-e_0^2A'\right)^{1/2}\varepsilon_\mu(\mathbf{k})=\left(Z_\gamma^{2nd}\right)^{1/2}\varepsilon_\mu(\mathbf{k})\approx\big(\underbrace{1}_{\substack{\text{1st}\\\text{order}}}\underbrace{-\tfrac{1}{2}e_0^2A'}_{\substack{\text{2nd order}\\\text{(loop)}}}\big)\varepsilon_\mu(\mathbf{k}),$$

and

$$\gamma^{\mu}_{2nd}\left(p,p'\right)=\gamma^{\mu}\left(1+e_0^2 L\right)+e_0^2 \Lambda_c^{\mu}=\underbrace{\gamma^{\mu}}_{\substack{\text{1st}\\\text{order}}}\underbrace{+\gamma^{\mu}e_0^2 L+e_0^2 \Lambda_c^{\mu}}_{\substack{\text{2nd order (loop)}\\\text{contribution}}}. \tag{14-8}$$

(VII) The Infinite Part of (VI) as $\Lambda \to \infty$

As noted before, we will evaluate the unbounded integrals in the next chapter using regularization. In that, we determine the value of the integrals in terms of a parameter, typically symbolized as Λ (different from $\Lambda^{\mu}(p,p')$), whereby in the final result, as $\Lambda \to \infty$, the integral $\to \infty$. This column assumes we already know the result of this evaluation (as if we had simply looked it up in an integral table).

Note that we proved in Chap. 13 that $B = L$, even though both are unbounded. As we will soon see, because they are equal, they drop out of our relations, and we will not need to actually determine their precise form.

Column (VII) shows the result of regularizing the infinite parts, as if we took them from integral tables

(VIII) Fermion Mass *m* Renormalization

This column is only applicable to fermions. It was actually already shown in the first line of (14-6) above.

Here we fold the unbounded value for A in the denominator of $iS_F^{2nd}\left(p\right)$ into the finite value for observed, physical mass m by assuming the unobserved, bare mass m_0 is also unbounded. This leaves us with $\not{p} - m$ in the denominator instead of $\not{p} - m_0 + e_0^2 A(\Lambda) = \not{p} - m_0 - \delta m$. Thus, the fermion mass is renormalized.

With this relationship, we can re-express our Lagrangian in terms of m. That is, in the original Lagrangian, we substitute $m_0 = m - \delta m$. This results in an extra term in the Lagrangian with a factor of δm. This extra term leads to an extra Feynman diagram, and this is Case (B) in column (I). The term containing δm is known as the mass counterterm.

This extra Feynman diagram, i.e., Case (B), was used as the starting point for external fermion line renormalization in column (I).

Column (VIII) summarizes mass renormalization

The effect of using m, instead of m0, in $\mathcal{L}$ summarized in Column (I), Case B

End of review

Columns (IX) to (XIII) in Wholeness Chart 14-4 delineate the final steps in renormalization to 2nd order, which we elaborate on below.

End of material from prior chapters

↓ New material

(IX) Definitions

In this column we define shorthand symbols for quantities we will see repeatedly. The abbreviation h.o. stands for "higher order". The subscripts indicate association with photons (γ), fermions (f), or vertices (V). Other authors typically use subscript numbers 3, 2, and 1, respectively, for these, but I feel the alternative symbols used here make it easier to keep track of which relation is related to which entity.

$$Z_{\gamma}^{2nd} = 1 - e_0^2 A' \tag{14-9}$$

$$Z_f^{2nd} = 1 - e_0^2 B = \frac{1}{1+e_0^2 B} \quad \left(\text{ignore h.o. terms on RHS, at 2nd order}\right) \tag{14-10}$$

$$Z_V^{2nd} = 1 + e_0^2 L \tag{14-11}$$

We note that because $B = L$, then

$$Z_V^{2nd} = 1/Z_f^{2nd}. \tag{14-12}$$

Column (IX) defines symbols that include the unbounded quantities

(X) Amplitude Determination

We will now examine assembling a second order (in α) amplitude, given the foundation we have built for ourselves in Chaps. 12 and 13, and in this chapter. We start with our example from earlier in the chapter of Compton scattering.

Compton Scattering Amplitude

We now want to express the tree level and divergent sub-amplitudes summation of (14-3) for Compton scattering in a convenient form.

Column (X) summarizes using the above to assemble an amplitude (see below)

Look at tree level plus divergent parts of Compton amplitude to 2nd order

$$\underbrace{\mathcal{M}_{C1}^{(2)} + \mathcal{M}_{C2}^{(2)}}_{\text{tree level, finite}} + \underbrace{\sum_i^{i=7} \mathcal{M}_{C1-i}^{(4)} + \sum_i^{i=7} \mathcal{M}_{C2-i}^{(4)}}_{\text{contain divergent integrals}} \qquad (14\text{-}13)$$

Let's begin by examining just the first way for Compton scattering, which we have denoted with subscripts $C1$. It may help to follow the diagrams in Fig. 14-1 as we progress.

Start with adding sub-amplitudes of first way of Compton scattering

$$\mathcal{M}_{C1}^{(2)} + \sum_i^{i=7} \mathcal{M}_{C1-i}^{(4)} = \mathcal{M}_{C1}^{(2)} + \mathcal{M}_{C1-1}^{(4)} + \mathcal{M}_{C1-2}^{(4)} + \dots + \mathcal{M}_{C1-7}^{(4)}. \qquad (14\text{-}14)$$

From using Feynman's rules or referring back to (8-64), pg. 228, we have

$$\mathcal{M}_{C1}^{(2)} = -e_0^2 \, \bar{u}_{s'}\left(\mathbf{p}'\right) \varepsilon_{\mu,r'}\left(\mathbf{k}'\right) \gamma^\mu i S_F\left(p+k\right) \varepsilon_{\nu,r}\left(\mathbf{k}\right) \gamma^\nu u_s\left(\mathbf{p}\right). \qquad (14\text{-}15)$$

Tree level amplitude

Then, for the remaining seven amplitudes of (14-14), where we use the 2nd order (loop) relations of (14-5) to (14-8), we have

Sub-amplitude for each 2nd order (loop) contribution

$$\mathcal{M}_{C1-1}^{(4)} = -e_0^2 \bar{u}_{s'}\left(\mathbf{p}'\right) \varepsilon_{\mu,r'}\left(\mathbf{k}'\right) \gamma^\mu i S_F\left(p+k\right) \underbrace{\left(-\tfrac{1}{2} e_0^2 A'\right) \varepsilon_{\nu,r}\left(\mathbf{k}\right)}_{\substack{\text{2nd order (loop) external} \\ \text{photon contribution}}} \gamma^\nu u_s\left(\mathbf{p}\right) \qquad (14\text{-}16)$$

$$\mathcal{M}_{C1-2}^{(4)} = -e_0^2 \bar{u}_{s'}\left(\mathbf{p}'\right) \varepsilon_{\mu,r'}\left(\mathbf{k}'\right) \gamma^\mu i S_F\left(p+k\right) \varepsilon_{\nu,r}\left(\mathbf{k}\right) \gamma^\nu \underbrace{\left(-\tfrac{1}{2} e_0^2 B\right) u_s\left(\mathbf{p}\right)}_{\substack{\text{2nd order (loop) external} \\ \text{electron contribution}}} \qquad (14\text{-}17)$$

$$\mathcal{M}_{C1-3}^{(4)} = -e_0^2 \bar{u}_{s'}\left(\mathbf{p}'\right) \varepsilon_{\mu,r'}\left(\mathbf{k}'\right) \gamma^\mu \underbrace{i S_F\left(p+k\right)\left(-e_0^2 B - e_0^2 \Sigma_c\right)}_{\substack{\text{2nd order (loop) fermion} \\ \text{propagator contribution}}} \varepsilon_{\nu,r}\left(\mathbf{k}\right) \gamma^\nu u_s\left(\mathbf{p}\right) \qquad (14\text{-}18)$$

$$\mathcal{M}_{C1-4}^{(4)} = -e_0^2 \bar{u}_{s'}\left(\mathbf{p}'\right) \underbrace{\left(-\tfrac{1}{2} e_0^2 A'\right) \varepsilon_{\mu,r'}\left(\mathbf{k}'\right)}_{\substack{\text{2nd order (loop) external} \\ \text{photon contribution}}} \gamma^\mu i S_F\left(p+k\right) \varepsilon_{\nu,r}\left(\mathbf{k}\right) \gamma^\nu u_s\left(\mathbf{p}\right) \qquad (14\text{-}19)$$

$$\mathcal{M}_{C1-5}^{(4)} = -e_0^2 \underbrace{\left(-\tfrac{1}{2} e_0^2 B\right) \bar{u}_{s'}\left(\mathbf{p}'\right)}_{\substack{\text{2nd order (loop) external} \\ \text{electron contribution}}} \varepsilon_{\mu,r'}\left(\mathbf{k}'\right) \gamma^\mu i S_F\left(p+k\right) \varepsilon_{\nu,r}\left(\mathbf{k}\right) \gamma^\nu u_s\left(\mathbf{p}\right) \qquad (14\text{-}20)$$

$$\mathcal{M}_{C1-6}^{(4)} = -e_0^2 \bar{u}_{s'}\left(\mathbf{p}'\right) \varepsilon_{\mu,r'}\left(\mathbf{k}'\right) \gamma^\mu i S_F\left(p+k\right) \varepsilon_{\nu,r}\left(\mathbf{k}\right) \underbrace{\left(\gamma^\nu e_0^2 L + e_0^2 \Lambda_c^\nu\right)}_{\substack{\text{2nd order (vertex loop)} \\ \text{contribution}}} u_s\left(\mathbf{p}\right) \qquad (14\text{-}21)$$

$$\mathcal{M}_{C1-7}^{(4)} = -e_0^2 \bar{u}_{s'}\left(\mathbf{p}'\right) \varepsilon_{\mu,r'}\left(\mathbf{k}'\right) \underbrace{\left(\gamma^\mu e_0^2 L + e_0^2 \Lambda_c^\mu\right)}_{\substack{\text{2nd order (vertex loop)} \\ \text{contribution}}} i S_F\left(p+k\right) \varepsilon_{\nu,r}\left(\mathbf{k}\right) \gamma^\nu u_s\left(\mathbf{p}\right) \qquad (14\text{-}22)$$

Adding (14-15) through (14-22) as in (14-14), and noting that A', B, and L are scalars whereas Σ_c and Λ_c^μ have hidden spinor indices (and must therefore stay in the same location relative to other spinor type entities)[1], we get

Keep in mind: Σ_c & $\Lambda_c{}^\mu$ are spinor quantities. A', B, L are scalars

---

[1] Actually, to be precise, B is a scalar multiplied by the identity matrix in spinor space. So, B commutes with all spinor space entities (spinors or gamma matrices) and can be treated just like a scalar.

$$\mathcal{M}_{C1}^{(2)} + \sum_{i}^{i=7} \mathcal{M}_{C1-i}^{(4)} = \left(1 - \tfrac{1}{2}e_0^2 A' - \tfrac{1}{2}e_0^2 B - e_0^2 B - \tfrac{1}{2}e_0^2 A' - \tfrac{1}{2}e_0^2 B + e_0^2 L + e_0^2 L\right) \times$$

$$\underbrace{\left(-e_0^2 \bar{u}_{s'}\left(\mathbf{p}'\right)\varepsilon_{\mu,r'}\left(\mathbf{k}'\right)\gamma^{\mu} iS_F\left(p+k\right)\varepsilon_{\nu,r}\left(\mathbf{k}\right)\gamma^{\nu} u_s\left(\mathbf{p}\right)\right)}_{\mathcal{M}_{C1}^{(2)}}$$

Add all of above sub-amplitudes

$$-e_0^2 \bar{u}_{s'}\left(\mathbf{p}'\right)\varepsilon_{\mu,r'}\left(\mathbf{k}'\right)\gamma^{\mu} iS_F\left(p+k\right)\left(-e_0^2 \Sigma_c\right)\varepsilon_{\nu,r}\left(\mathbf{k}\right)\gamma^{\nu} u_s\left(\mathbf{p}\right) \qquad (14\text{-}23)$$

$$-e_0^2 \bar{u}_{s'}\left(\mathbf{p}'\right)\varepsilon_{\mu,r'}\left(\mathbf{k}'\right)\gamma^{\mu} iS_F\left(p+k\right)\varepsilon_{\nu,r}\left(\mathbf{k}\right)e_0^2 \Lambda_c^{\nu} u_s\left(\mathbf{p}\right)$$

$$-e_0^2 \bar{u}_{s'}\left(\mathbf{p}'\right)\varepsilon_{\mu,r'}\left(\mathbf{k}'\right)e_0^2 \Lambda_c^{\mu} iS_F\left(p+k\right)\varepsilon_{\nu,r}\left(\mathbf{k}\right)\gamma^{\nu} u_s\left(\mathbf{p}\right).$$

where the last three lines are finite (and so is the factor shown in the second row). Ignoring higher order terms and using our newly defined symbols, we can re-write (14-23) as

Use 2nd order approximation, ignoring higher order terms, to re-arrange above addition

$$\mathcal{M}_{C1}^{(2)} + \sum_{i}^{i=7} \mathcal{M}_{C1-i}^{(4)} \approx$$

$$\underbrace{\left(1 - \tfrac{1}{2}e_0^2 A'\right)}_{\left(Z_\gamma^{2nd}\right)^{1/2}} \underbrace{\left(1 - \tfrac{1}{2}e_0^2 B\right)}_{\left(Z_f^{2nd}\right)^{1/2}} \underbrace{\left(1 - e_0^2 B\right)}_{Z_f^{2nd}} \underbrace{\left(1 - \tfrac{1}{2}e_0^2 A'\right)}_{\left(Z_\gamma^{2nd}\right)^{1/2}} \underbrace{\left(1 - \tfrac{1}{2}e_0^2 B\right)}_{\left(Z_f^{2nd}\right)^{1/2}} \underbrace{\left(1 + e_0^2 L\right)}_{Z_V^{2nd}} \underbrace{\left(1 + e_0^2 L\right)}_{Z_V^{2nd}} \times \qquad (14\text{-}24)$$

Now the relation has certain factors that we defined in prior column

$$\left(-e_0^2 \bar{u}_{s'}\left(\mathbf{p}'\right)\varepsilon_{\mu,r'}\left(\mathbf{k}'\right)\underbrace{\left(\gamma^{\mu} + e_0^2 \Lambda_c^{\mu}\right)}_{\substack{\gamma^{\mu}_{e_0\,Mod} \\ 2nd}}\underbrace{iS_F\left(p+k\right)\left(1 - e_0^2 \Sigma_c\right)}_{\substack{iS_F^{2nd} \\ e_0\,Mod \\ (p+k)}}\varepsilon_{\nu,r}\left(\mathbf{k}\right)\underbrace{\left(\gamma^{\nu} + e_0^2 \Lambda_c^{\nu}\right)}_{\substack{\gamma^{\nu}_{e_0\,Mod} \\ 2nd}}u_s\left(\mathbf{p}\right)\right).$$

We define quantities in the last line of (14-24) that will be useful. To help keep notation straight, note that the "Mod" script designates only the convergent parts (first and second order) of (14-5) to (14-8).

Using these new symbols, we can express (14-24) more succinctly as

Re-expressing using new Z symbols and "Mod" scripts

$$\mathcal{M}_{C1}^{(2)} + \sum_{i}^{i=7} \mathcal{M}_{C1-i}^{(4)} \approx \left(Z_\gamma^{2nd}\right)\left(Z_f^{2nd}\right)^2\left(Z_V^{2nd}\right)^2 \times$$

$$\underbrace{\left(-e_0^2 \bar{u}_{s'}\left(\mathbf{p}'\right)\varepsilon_{\mu,r'}\left(\mathbf{k}'\right)\gamma^{\mu}_{\substack{e_0\,Mod \\ 2nd}}iS_F^{\substack{e_0\,Mod \\ 2nd}}\left(p+k\right)\varepsilon_{\nu,r}\left(\mathbf{k}\right)\gamma^{\nu}_{\substack{e_0\,Mod \\ 2nd}}u_s\left(\mathbf{p}\right)\right)}_{\mathcal{M}_{C1}^{(2)} \atop e_0\,Mod,2nd}, \qquad (14\text{-}25)$$

where the last line is a modified parallel to our tree level amplitude. That is, it is the same form as the first order (in α) expression but with modified expressions for the vertices and propagator. The RHS of the top row contains divergent quantities, the bottom row, all convergent quantities.

From (14-12), the fermion and vertex Z factors in (14-25) cancel out leaving

Fermion and vertex Z factors cancel leaving only Z_γ^{2nd} in front

$$\mathcal{M}_{C1}^{(2)} + \sum_{i}^{i=7} \mathcal{M}_{C1-i}^{(4)} \approx Z_\gamma^{2nd} \mathcal{M}_{C1 \atop e_0\,Mod,2nd}^{(2)}. \qquad (14\text{-}26)$$

Note that the fermion and vertex factors cancel because $B = L$, which resulted from the original Ward identity.[1] This, as noted before is a great blessing, as we no longer have to worry about the infinities introduced by higher order fermion and vertex corrections. They drop out of the theory.

You can show that we get a similar relation to (14-26) for the 2nd way for Compton scattering (indicated with the subscript C2) by doing Prob. 3. This relation is

Fermion and vertex Z factors cancel in second way for Compton, too

---

[1] Note that instead of factoring, as we did in going from (14-23) to (14-24), we could have simply added the terms in the first line of (14-23) that have Bs and Ls in them. All such terms would have cancelled with one another, leaving no B or L infinities to deal with. We did the factoring in (14-24) instead, because that is how you will find most other authors do it, and hopefully there will be less confusion when making comparisons between texts.

$$\mathcal{M}_{C2}^{(2)} + \sum_{i}^{i=7} \mathcal{M}_{C2-i}^{(4)} \approx Z_{\gamma}^{2nd} \, \mathcal{M}_{C2}^{(2)} \Big|_{\substack{e_0 \, Mod, 2nd}} . \qquad (14\text{-}27)$$

That is, the fermion and vertex Z factors drop out there as well. Additionally, our vertex and propagator modifications to the tree level amplitude are the same as those in (14-24), though the tree level amplitude itself for the 2nd way for Compton scattering has somewhat different form.

<u>Møller Scattering Example</u>

Note that in (14-24) we get a factor of Z_f^{2nd} for the fermion propagator, one of Z_V^{2nd} for every

Møller scattering gets Z factors in parallel fashion

vertex, one $\left(Z_f^{2nd}\right)^{1/2}$ for each external fermion, and one $\left(Z_{\gamma}^{2nd}\right)^{1/2}$ for each external photon. We also

get $iS_F^{2nd}\Big|^{e_0 \, Mod}$ replacing the tree level iS_F, and a $\gamma_{\substack{e_0 \, Mod \\ 2nd}}^{\mu}$ replacing each tree level γ^{μ}.

In Møller scattering, shown in Fig. 14-3, the same things occur, but additionally, for the photon propagator we get the replacement

But here we have photon propagator modification

$$iD_{F\mu\nu}(k) \Rightarrow iD_{F\mu\nu}^{2nd}\Big|^{e_0 \, Mod}(k) = iD_{F\mu\nu}(k)\left(1 - e_0^2 \Pi_c\right). \qquad (14\text{-}28)$$

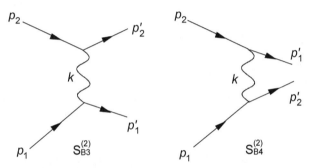

Figure 14-3. Møller Scattering (Two Ways to Occur at Tree Level)

By adding diagrams to 2nd order in α, the amplitude for the first way of Møller scattering, in parallel with (14-24), turns out to have Z factors for each of the four external fermions, the photon propagator, and the two vertices, in addition to having modified photon propagator and modified vertex. The amplitude is thus, with the subscript B3 symbolism of Chap. 8 (8-68),

Fermion and vertex Z factors cancel in first way for Møller, just like in Compton

$$\mathcal{M}_{B3}^{(2)} + \sum_{i}^{i=7} \mathcal{M}_{B3-i}^{(4)} \approx \left(Z_{\gamma}^{2nd}\right)\left(Z_f^{2nd}\right)^{1/2}\left(Z_f^{2nd}\right)^{1/2}\left(Z_f^{2nd}\right)^{1/2}\left(Z_f^{2nd}\right)^{1/2} Z_V^{2nd} Z_V^{2nd} \times$$

$$\underbrace{\left(e_0^2 \bar{u}_{r_1'}(\mathbf{p}_1') \gamma_{\substack{e_0 \, Mod \\ 2nd}}^{\mu} u_{r_1}(\mathbf{p}_1) iD_{F\mu\nu}^{2nd}\Big|^{e_0 \, Mod}(k) \bar{u}_{r_2'}(\mathbf{p}_2') \gamma_{\substack{e_0 \, Mod \\ 2nd}}^{\nu} u_{r_2}(\mathbf{p}_2) \right)}_{\mathcal{M}_{B3}^{(2)}\Big|_{e_0 \, Mod, 2nd}} \qquad (14\text{-}29)$$

$$= Z_{\gamma}^{2nd}\left(Z_f^{2nd}\right)^2 \left(Z_V^{2nd}\right)^2 \mathcal{M}_{B3}^{(2)}\Big|_{e_0 \, Mod, 2nd} = Z_{\gamma}^{2nd} \, \mathcal{M}_{B3}^{(2)}\Big|_{e_0 \, Mod, 2nd} .$$

For the second way for Møller scattering to occur, in similar fashion, we get

Same thing in second way for Møller scattering

$$\mathcal{M}_{B4}^{(2)} + \sum_{i}^{i=7} \mathcal{M}_{B4-i}^{(4)} \approx Z_{\gamma}^{2nd} \, \mathcal{M}_{B4}^{(2)}\Big|_{e_0 \, Mod, 2nd} . \qquad (14\text{-}30)$$

<u>Generalizing to Any Scattering Amplitude</u>

It turns out that in every second order amplitude calculation, the fermion and vertex factors in the divergent terms cancel and leave the photon factor Z_{γ}^{2nd} as our only divergent quantity. Additionally, in every such amplitude we merely substitute the 2nd order modified relations of

Fermion and vertex Z factors cancel in every type amplitude

(14-24) and (14-28) for propagators and vertices in the tree level amplitude to get the modified amplitude.

By doing Prob. 4, you can show that this same thing is true for Bhabha scattering.

Bottom line: In the divergent parts of our second order amplitude, for any type of QED scattering, the divergent fermion factor(s) will cancel with the divergent vertex factor(s), leaving only a photon divergent factor Z_γ^{2nd}. The remainder of the amplitude can be found by substituting (finite) modifications to the propagator and vertex relations for those in the tree level amplitude.

(XI) Cancellation of Z Factors

In this column we simply note the first sentence of the Bottom line summary above.

Only Z factor remaining in every type of amplitude is the photon one

(XII) Charge Renormalization

For Compton scattering, first way, we have

$$\mathcal{M}_{C1}^{(2)} + \sum_{i}^{i=7} \mathcal{M}_{C1-i}^{(4)} \approx Z_\gamma^{2nd} \mathcal{M}_{C1}^{(2)}{}_{e_0\,Mod,2nd}$$

$$= \left(1 - e_0^2 A'\right) e_0^2 \left(-\bar{u}_{s'}\left(\mathbf{p}'\right) \varepsilon_{\mu,r'}\left(\mathbf{k}'\right) \gamma^\mu_{\substack{e_0\,Mod\\2nd}} iS_F^{\substack{e_0\,Mod\\2nd}}\left(p+k\right) \varepsilon_{v,r}\left(\mathbf{k}\right) \gamma^v_{\substack{e_0\,Mod\\2nd}} u_s\left(\mathbf{p}\right) \right). \tag{14-31}$$

We now renormalize charge, as we saw in Chap. 12.

$$e_0^2\left(1 - e_0^2\,\underset{\infty}{\underline{A'}}\right) = e_0^2 Z_\gamma^{2nd} = e^2 = \text{measured charge squared, or}$$

$$e_0\left(1 - \tfrac{1}{2}e_0^2\,\underset{\infty}{\underline{A'}}\right) \approx e_0\left(Z_\gamma^{2nd}\right)^{1/2} \approx e \ = \text{measured charge.} \tag{14-32}$$

We renormalize charge, same for every case, by including Z_γ^{2nd} in definition of physical charge

The bare charge e_0 is zero, but e, due to the infinite quantity A', is finite and what we measure. Using the value for A' of column (VII), we find e is a function of the energy level at which the charge measurement is made. (See Chap. 12.)

$$\lim_{\Lambda\to\infty} e_0^2\left(1 + e_0^2\, 2b_n\, ln\frac{k}{\Lambda} + \mathcal{O}\left(e_0^4\right)\right) = \underbrace{\lim_{\Lambda\to\infty} e_0^2}_{= e_0^2} + \lim_{\Lambda\to\infty} e_0^4\, 2b_n\, ln\frac{k}{\Lambda} + \lim_{\Lambda\to\infty} \mathcal{O}\left(e_0^6\right) = e^2\left(k\right). \tag{14-33}$$

Just use e^2 in place of e_0^2 at front of amplitude for every interaction from now on

Moreover, every reaction we encounter will have the factor $\left(1 - e_0^2 A'\right)e_0^2$ in the 2nd order amplitude, as does (14-31).

Bottom line: So, from now on we can simply replace $\left(1 - e_0^2 A'\right)e_0^2$ with e^2 in every QED amplitude calculation to 2nd order, and our total amplitude to that order will be finite. We need to keep in mind that e depends on energy level (modestly via a log dependence).

(XIII) Substitution in Tree Amplitude to Get Amplitude to 2nd Order

One might expect we could then simply substitute the modified relations defined in the last row of (14-24) and (14-28), i.e.,

$$iS_F^{\substack{e_0\,Mod\\2nd}}\left(p+k\right) = iS_F\left(p+k\right)\left(1 - e_0^2 \Sigma_c\right)$$

$$\gamma^\mu_{\substack{e_0\,Mod\\2nd}} = \gamma^\mu + e_0^2 \Lambda_c^\mu \tag{14-34}$$

$$iD_{F\mu v}^{\substack{e_0\,Mod\\2nd}}\left(k\right) = iD_{F\mu v}\left(k\right)\left(1 - e_0^2 \Pi_c\right)$$

But propagator and vertex mods still have e_0^2 at this point

into (14-31) and other amplitudes, and all would be well. The issue is the e_0 found in each, which as we just saw, we now define as equal to zero. In such case, the relations of (14-34) would be none other than our tree level relations. However, experiment tells us (as we will see two chapters hence) that doing so does not give us the correct amplitude.

The resolution is this. The parts of (14-34) with e_0^2 factors arose from 2nd order corrections to first order propagators and vertices. These are actually themselves modified by 3rd and higher order (in α) corrections. That is, all that we have done here to correct a tree level Compton (or other) scattering amplitude to the next highest order, must also be done for these parts. See all but one of the sub-diagrams needed for this correction in Fig. 14-4. (Prob. 5 asks you to find the last one.)

But propagator and vertices have even higher order corrections not taken into account yet

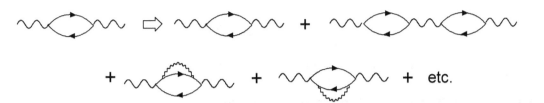

Figure 14-4. Next Higher Order Corrections to Photon Propagator

This process parallels what we have done in going from 1st to 2nd order, and in doing it, one finds a further correction to e_0^2 like that of (14-32). That is, a factor of Z_γ^{2nd} arises in front of the e_0^2 in (14-34). Thus, we can use the same charge renormalization of (14-32) in (14-34). We incorporate that, by defining modified propagators and vertices without e_0 script notation,

When we take corrections beyond 2nd order into account, e_0 $\to$ e in these modifications

$$iS_F^{\overset{Mod}{2nd}}\left(p+k\right)=iS_F\left(p+k\right)\left(1-e^2\Sigma_c\right)$$

$$\gamma^\mu_{\underset{2nd}{Mod}}=\gamma^\mu+e^2\Lambda_c^\mu \tag{14-35}$$

$$iD_{F\mu\nu}^{\overset{Mod}{2nd}}\left(k\right)=iD_{F\mu\nu}\left(k\right)\left(1-e^2\Pi_c\right).$$

So we redefine the modifications with e instead of e_0

We also define a <u>new symbol for the modified amplitude</u> $\underset{Mod,2nd}{\mathcal{M}_{C1}^{(2)}}$, now using physical charge throughout, including relations (14-35) instead of (14-34). Compare below with (14-31)

$$\mathcal{M}_{C1}^{(2)}+\sum_i^{i=7}\underset{e_0\,Mod,2nd}{\mathcal{M}_{C1-i}^{(4)}}\approx Z_\gamma^{2nd}\,\mathcal{M}_{C1}^{(2)}$$

$$=\underbrace{\left(1-e_0^2A'\right)e_0^2}_{e^2}\left(-\overline{u}_{s'}\left(\mathbf{p}'\right)\varepsilon_{\mu,r'}\left(\mathbf{k}'\right)\gamma^\mu_{\underset{2nd}{Mod}}iS_F^{\overset{Mod}{2nd}}\left(p+k\right)\varepsilon_{\nu,r}\left(\mathbf{k}\right)\gamma^\nu_{\underset{2nd}{Mod}}u_s\left(\mathbf{p}\right)\right)=\underset{Mod,2nd}{\mathcal{M}_{C1}^{(2)}}. \tag{14-36}$$

Similar definitions hold for the second way of Compton scattering (with $C1 \to C2$ above) and any other type of scattering (e.g., above $C1 \to B1$, the first way of Bhabha scattering).

<u>Bottom line:</u> To re-normalize any (otherwise unbounded) interaction amplitude to 2nd order in α, in the tree level amplitude make the substitutions

Renormalizing the originally unbounded sub-amplitudes means $e_0 \to e$ & propagators and vertices $\to$ mod

- $e_0 \Rightarrow e(k)$

- $iS_F, \gamma^\mu, iD_{F\mu\nu} \Rightarrow$ (14-35) above.

14.2.2 Two Routes from Tree to Renormalized Amplitude

The route we took here to renormalization comprised adding all of the sub-amplitudes for each diagram in Fig. 14-1, i.e., adding all of (14-15) through (14-22). This is the proper method of finding amplitudes that include higher orders, as we learned in Chap. 8 that whenever we have multiple Feynman diagrams with the same incoming and outgoing states, we add the amplitudes from each diagram to get the total amplitude.

However, it is common to do something somewhat different, and it may be good to understand the difference, so you are not confused when comparing treatments by other authors. The alternative approach entails simply substituting the relations (14-5) to (14-8) directly into the tree level amplitude in place of iS_F, γ^μ, $iD_{F\mu\nu}$, u_r, $\overline{u}_r$, v_r, $\overline{v}_r$ and ε_μ .

There are two, essentially equivalent, routes to find higher order amplitudes

In the first approach (above) we add all amplitudes from all relevant diagrams, where each diagram is either first or second (or higher) order. In the alternative approach, the "short cut" route (below), one simply substitutes, in the single tree level amplitude, the expression for each propagator, vertex, and external particle that includes first and second (and higher if considered) corrections. That is, one substitutes the relations (14-5) to (14-8) into the tree level amplitude in place of the tree level values for each propagator, vertex, and external particle.

Because of the 2nd (and higher) order approximations involved, the second, short cut approach is essentially the same as the first, formal approach. But it is the first that is actually the precise way, via the rules of QFT, to construct an amplitude to higher order.

These two approaches are summarized in Wholeness Chart 14-1.

Summary of the two routes

Wholeness Chart 14-1. Two Routes to Renormalization

__Formal, proper route__ (above)	__Short cut route__ (see Sect. 14.2.3 below)
	(This way actually derived from formal route and is an approximation of that route.)
Add tree + higher order amplitudes.	**Use only tree amplitude and substitute.**
See (14-15) to (14-22).	To tree amplitude, simply substitute relations
Use only higher order contribution in each higher order amplitude.	(14-5) to (14-8), which include 1st and higher order parts, for each propagator, each vertex, and each external particle.
E.g., for diagram with fermion propagator self-energy use fermion propagator	E.g., for tree fermion propagator substitute
$$iS_F\left(p\right)\left(-e_0^2 B - e_0^2 \Sigma_c\right).$$	$$iS_F^{2nd}\left(p\right) = iS_F\left(p\right)\left(1 - e_0^2 B - e_0^2 \Sigma_c\right).$$
↓	↓
Obtain summation amplitude as in (14-23)	

Re-arrange amplitude, via approximations, to have factors such as

$$Z_f^{2nd} = \left(1 - e_0^2 B\right) \text{ and } iS_F^{\overset{e_0 Mod}{2nd}} = \left(1 - e_0^2 \Sigma_c\right) iS_F$$

as in (14-24).

↓

Cancel Z_f^{2nd} and Z_V^{2nd} factors leaving $Z_\gamma^{2nd} \mathcal{M}_{\underset{e_0 Mod, 2nd}{Generic}}^{(2)}$

↓

Renormalize via $Z_\gamma^{2nd} e_0^2 = e^2$

14.2.3 The Alternative Route (Short Cut Method)

As noted above, instead of adding all of the first and second order diagram sub-amplitudes as we did in (14-13) to (14-23) to get (14-24), we can instead get (14-24) (to high approximation) by simply substituting (14-5) to (14-8) directly into the tree level amplitude (14-15).

$$\mathcal{M}_{C1}^{(2)} = -e_0^2 \; \bar{u}_{s'} \; (\mathbf{p}') \varepsilon_{\mu,r'} (\mathbf{k}') \; \gamma^\mu \; i \; S_F \left(p+k\right) \varepsilon_{\nu,r} (\mathbf{k}) \; \gamma^\nu \; u_s \; (\mathbf{p}). \tag{14-37}$$

$$\;\;\;\;\;\;\;\;\;\; \underset{\bar{u}_{s'}^{2nd}}{\downarrow} \;\;\; \underset{\varepsilon_{\mu,r'}^{2nd}}{\downarrow} \;\;\; \underset{\gamma_{2nd}^\mu}{\downarrow} \;\; \underset{S_F^{2nd}}{\downarrow} \;\;\;\;\; \underset{\varepsilon_{\nu,r}^{2nd}}{\downarrow} \;\;\; \underset{\gamma_{2nd}^\nu}{\downarrow} \; \underset{u_s^{2nd}}{\downarrow}$$

$$\rightarrow -e_0^2 \left(1 - \tfrac{1}{2}e_0^2 B\right) \bar{u}_{s'} (\mathbf{p}') \left(1 - \tfrac{1}{2}e_0^2 A'\right) \varepsilon_{\mu,r'} (\mathbf{k}') \times$$

$$\left(\gamma^\mu + \gamma^\mu e_0^2 L + e_0^2 \Lambda_c^\mu\right) iS_F \left(p+k\right) \left(1 - e_0^2 B - e_0^2 \Sigma_c\right) \times \tag{14-38}$$

$$\left(1 - \tfrac{1}{2}e_0^2 A'\right) \varepsilon_{\nu,r} (\mathbf{k}) \left(\gamma^\nu + \gamma^\nu e_0^2 L + e_0^2 \Lambda_c^\nu\right) \left(1 - \tfrac{1}{2}e_0^2 B\right) u_s (\mathbf{p}),$$

which to 2nd order is

$$\approx \underbrace{\left(1-\tfrac{1}{2}e_0^2 A'\right)}_{\left(Z_\gamma^{2nd}\right)^{1/2}}\underbrace{\left(1-\tfrac{1}{2}e_0^2 B\right)}_{\left(Z_f^{2nd}\right)^{1/2}}\underbrace{\left(1-e_0^2 B\right)}_{Z_f^{2nd}}\underbrace{\left(1-\tfrac{1}{2}e_0^2 A'\right)}_{\left(Z_\gamma^{2nd}\right)^{1/2}}\underbrace{\left(1-\tfrac{1}{2}e_0^2 B\right)}_{\left(Z_f^{2nd}\right)^{1/2}}\underbrace{\left(1+e_0^2 L\right)}_{Z_V^{2nd}}\underbrace{\left(1+e_0^2 L\right)}_{Z_V^{2nd}}\times$$

$$\left(-e_0^2 \bar{u}_{s'}\left(\mathbf{p}'\right)\varepsilon_{\mu,r'}\left(\mathbf{k}'\right)\underbrace{\left(\gamma^\mu + e_0^2 \Lambda_c^\mu\right)}_{\substack{\gamma_{e_0\,Mod}^\mu \\ 2nd}}\underbrace{iS_F\left(p+k\right)\left(1-e_0^2\Sigma_c\right)}_{\substack{iS_F^{2nd}\,(p+k)}}\varepsilon_{V,r}\left(\mathbf{k}\right)\underbrace{\left(\gamma^\nu + e_0^2\Lambda_c^\nu\right)}_{\substack{\gamma_{e_0\,Mod}^\nu \\ 2nd}}u_S\left(\mathbf{p}\right)\right). \qquad (14\text{-}39)$$

This is the same as (14-24), so

$$(14\text{-}39) \approx \mathcal{M}_{C1}^{(2)} + \sum_{i}^{i=7}\mathcal{M}_{C1-i}^{(4)}\,, \qquad\qquad (14\text{-}40)$$

and we can use either method (route) to get the Compton first way amplitude to 2nd order. Similar logic leads to use of either route for all types of amplitude higher order corrections. The second, alternative route is generally much quicker and easier.

14.3 The Total Amplitude to 2nd Order

From (14-3), the total Compton scattering amplitude to second order is then the finite value

$$\mathcal{M}_{\substack{\text{Compton}\\ \text{all ways}\\ \text{total to }e_0^4}} = \underbrace{\mathcal{M}_{C1}^{(2)} + \mathcal{M}_{C2}^{(2)}}_{\text{tree level, finite}} + \underbrace{\sum_{i}^{i=7}\mathcal{M}_{C1-i}^{(4)} + \sum_{i}^{i=7}\mathcal{M}_{C2-i}^{(4)}}_{\text{contain divergent integrals}} + \underbrace{\mathcal{M}_{C1-8}^{(4)} + \mathcal{M}_{C2-8}^{(4)}}_{\text{finite using }e_0^4} + \underbrace{\mathcal{M}_{C19}^{(4)} + \mathcal{M}_{C20}^{(4)}}_{\text{cancel out}}$$

$$(14\text{-}41)$$

$$= \underbrace{\overbrace{\mathcal{M}_{\substack{C1\\ Mod,2nd}}^{(2)} + \mathcal{M}_{\substack{C2\\ Mod,2nd}}^{(2)}}^{\text{finite using }e^2}} + \underbrace{\overbrace{\mathcal{M}_{\substack{C1-8\\ Mod,2nd}}^{(4)} + \mathcal{M}_{\substack{C2-8\\ Mod,2nd}}^{(4)}}^{\text{finite using }e^4,\ \text{negligible}}}\,.$$

The total amplitude for Compton scattering to 2nd order in α_0

The sub-amplitudes $\mathcal{M}_{C1-8}^{(4)}$ and $\mathcal{M}_{C2-8}^{(4)}$ originally had factors of $e_0{}^4$ (see (14-1)), but in similar manner as discussed for column (XIII) in the prior section, higher order corrections modify this to e^4. They are then of higher order in e than the other terms in the last line of (14-41), which are of order e^2, and can, at 2nd order, be ignored.

In the original sub-amplitudes that were finite, $e_0 \to e$ as well, due to higher order corrections

Thus, for any type of interaction in general, the renormalized amplitude to 2nd order is

These are then higher order in e and negligible at 2nd order

$$\mathcal{M}_{\substack{\text{Generic}\\ \text{all ways}\\ \text{total to }e_0^4}} = \underbrace{\sum_{j}^{\#\,\text{ways}}\mathcal{M}_{\text{Gen}\,j}^{(2)}}_{\text{tree level}} + \underbrace{\sum_{j,i}\mathcal{M}_{\text{Gen}\,j-i}^{(4)}}_{\text{contain divergent integrals}} + \underbrace{\sum_{j,k}\mathcal{M}_{\text{Gen}\,j-k}^{(4)}}_{\text{finite using }e_0^4} + \underbrace{\text{triangle\ \ diagrams}}_{\text{cancel out}}$$

$$(14\text{-}42)$$

$$= \underbrace{\overbrace{\sum_{j}^{\#\,\text{ways}}\mathcal{M}_{\substack{\text{Gen}\,j\\ Mod,2nd}}^{(2)}}^{\text{finite using }e^2}} + \underbrace{\overbrace{\sum_{j,k}\mathcal{M}_{\substack{\text{Gen}\,j-k\\ Mod,2nd}}^{(4)}}^{\text{finite using }e^4,\ \text{negligible}}}\,.$$

The total amplitude for any type scattering to 2nd order in α_0

Not only is (14-42) then finite, but its substitution in the transition amplitude S_{fi} equation (13-26), pg. 326, yields (to 2nd order accuracy) the actual probability $|S_{fi}|^2$ for the interaction to occur. More on this in Part Four of this book.

14.4 Renormalization to Higher Orders: Our Approach

As you have seen, it has taken us two and a half fair sized chapters to develop renormalization to second order. As you may imagine, developing it to higher orders takes even more space, time, and effort, as the number of Feynman graphs, and their complexity, increases dramatically and rapidly at each higher order.

Renormalization to nth order

Thus, in this book, we will not show every necessary step in so doing in detail, but rather provide a streamlined, systematic overview of the procedure. This procedure parallels what we did for the second order case and so, hopefully, will be relatively transparent and make sense logically.

We will outline the procedure here without extensive detail

The understanding gained by this approach should provide a reasonable foundation for i) moving on to applications in Part Four and thereafter drawing this book to a close, and ii) studying higher order renormalization in detail when, and if, the need arises in your career.

Many other texts provide in depth development of higher order renormalization. The purpose of this text is to get you grounded in the fundamentals of QFT, and not get bogged down in complications along the way. It is easier grinding through the details after one knows basic theoretical principles, rather than while one is trying to assimilate them.

14.5 Higher Order Renormalization Example: Compton Scattering

Compton scattering to nth order

We begin, parallel to what we did at second order, by considering an example, Compton scattering, now to arbitrarily high order n. In addition to the two first order and 18 second order Feynman diagrams of Fig. 14-1 (pg. 340), we have many additional diagrams, the actual number of which depends on the order n being considered.

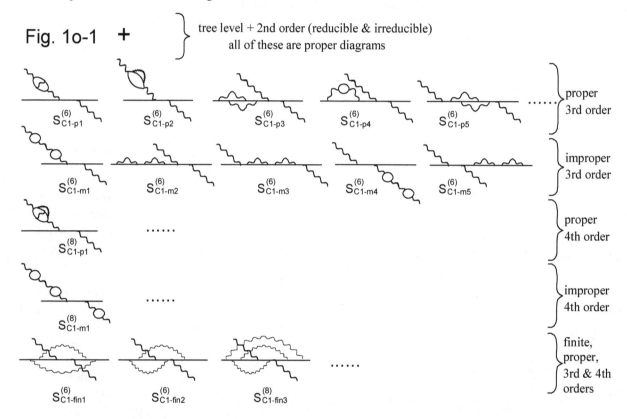

Figure 14-5. Compton Scattering Showing Some Higher Order Feynman Diagrams

The first row, represented by Fig. 14-1 includes all the 1st and 2nd order diagrams we have seen before. The second and third rows show certain (not all) 3rd order diagrams. The fourth and fifth rows show certain (not even close to all) 4th order diagrams. Rows two to five contain diagrams whose amplitudes diverge. The sixth row illustrates some (not all) of the finite amplitude diagrams for 3rd and 4th order. These do not contain any of the divergent loop diagrams. The last row indicates we also have to consider the second way for Compton scattering to occur plus orders higher than 4th, if we are including them. We don't even include triangle, pentangle, etc diagrams because, due to Furry's theorem, they all cancel out.

In Fig. 14-5 we introduce another category for distinguishing Feynman diagrams, proper vs improper. Note that <u>improper diagrams</u> have at least one virtual particle or external line from the tree diagram replaced with consecutive, side-by-side loop corrections, whereas <u>proper diagrams</u> do not. This distinction will help us in renormalizing amplitudes to 3rd and higher orders. In improper

Proper vs improper diagrams

graphs, at least one propagator or external line from the tree graph becomes a composite of two or more proper propagators aligned sequentially. An improper propagator can be split into proper propagators. See Wholeness Charts 14-2 (2nd and 3rd columns, next to last row) and 14-3 below.

Wholeness Chart 14-2. Types of Feynman Diagrams

Graph Type	Examples	Counter Examples	Definition	Use	Chapter
Connected	connected	same diagram unconnected	All lines connected to all others	Feynman rules apply only to connected diagrams	8
Topologically Distinct	topologically distinct	β / α / α / β topologically similar	Changing vertex label does not yield topologically distinct diagram	Eliminates $1/n!$ factor of Dyson-Wicks amplitude expansion in Feynman rules	8 (pg. 235) 9 (pg. 257)
Irreducible (Skeleton)	or irreducible (skeleton graph)	reducible	Irreducible = no self-energy or vertex mod parts. Reducible = self-energy or vertex mod part(s). If remove all, becomes irreducible.	Irreducible are finite. Reducible are ∞ (before renormalization).	14
Proper	proper diagrams	if cut here improper diagrams	Cannot split into proper graphs by cutting internal line. (Improper propagators are composites of adjacent proper ones.)	Renormalization for order > 2nd.	14
Primitively Divergent (not in this text, included because in other texts)	if cut here / primitive divergence / get this / convergent	if cut here / non-primitive divergence / get this / still divergent	Divergent graph which is converted to a convergent graph if any internal line is cut (replaced by 2 external lines)	All divergent graphs made of one or more primitive divergences. Used to prove that only reducible diagrams diverge.	Not in this text

Wholeness Chart 14-2 is provided to organize and summarize the various types of Feynman diagrams we have dealt with and are now dealing with. The last row depicts yet another category called primitively divergent diagrams, which we will not discuss herein. We include them for completeness and because you will probably run into them elsewhere. Primitively divergent diagrams are used to prove that only reducible diagrams diverge. See footnote on page 342.

Wholeness Chart 14-3. Comparing Certain Types of Feynman Diagrams[1]

Irreducible (Skeleton) (no loops, finite)		Reducible (loops, infinite)	
Tree Diagram	Non-tree Diagrams	Non-composite	Composite
Proper	Proper	Proper	Improper
Examples			

14.5.1 The nth Order Total Compton Amplitude

The total amplitude to nth order in α_0 is then

$$\mathcal{M}_{\substack{\text{Compton}\\ \text{all ways}\\ \text{total to } e_0^{2n}}} = \underbrace{\mathcal{M}_{C1}^{(2)} + \mathcal{M}_{C2}^{(2)}}_{\substack{\text{tree level, finite,}\\ \text{irreducible}}} + \underbrace{\sum_{k=2}^{n}\left(\sum_i \mathcal{M}_{C1-i}^{(2k)} + \sum_i \mathcal{M}_{C2-i}^{(2k)}\right)}_{\substack{\text{contain divergent integrals,}\\ \text{reducible}}} + \underbrace{\sum_{k=2}^{n}\left(\sum_i \mathcal{M}_{C1-\text{fin}\,i}^{(2k)} + \sum_i \mathcal{M}_{C2-\text{fin}\,i}^{(2k)}\right)}_{\substack{\text{finite,}\\ \text{irreducible}}} \quad (14\text{-}43)$$

Adding amplitudes for all diagrams in Fig. 14-5 gives total nth order amplitude

For $n = 2$, this reduces to (14-41).

14.6 Renormalizing nth Order Divergent Amplitudes

14.6.1 Steps of nth Order Renormalization

The steps for renormalizing to arbitrary order n are summarized in Wholeness Chart 14-5, which you should follow along with as we proceed in detail through each step below. These steps will parallel those for 2nd order renormalization, summarized in Wholeness Chart 14-4, just above 14-5.

Wholeness Chart 14-4 summarizes steps of nth order renormalization

(I) Feynman Diagram Modification

Column (I) of Wholeness Chart 14-5 shows some of the modifications at nth order to the propagator, external particle line, and vertex portions of Feynman diagrams. We have split them into proper and improper diagrams because it will prove helpful in step (II).

For the fermion case, we show only the bare fermion mass m_0 case (A) for two reasons: i) because we don't have enough space to also show the physical mass m case (B), and ii) because, like the 2nd order case, we use bare mass in the development of the propagator in order to renormalize mass and replace m_0 with m in step (VIII). As for 2nd order, in case (A) we have no mass counterterms in $\mathcal{L}$ and no additional Feynman diagrams. For case (B) we will have a renormalized mass and a counterterm in $\mathcal{L}$ with a factor of δm. This counterterm gives rise to additional diagrams having an X in fermion propagator lines.

In the external line block, we use the results from the fermion analysis, i.e., we use the physical mass m case (B). We could use either case (A) or (B) here and get the same results. But taking what we learned from the fermion step (VIII) and using it directly for external fermions is a lot easier. My general philosophy is that if two ways produce the same results, the easier is the better, and more efficient, way.

Column (I) shows diagrams to nth order for propagators, external particles, and vertex

(II) Propagator, Leg, Vertex Modification

Column (II) shows the additions one must make to the propagator, external line, and vertex to the mathematical representations of the associated Feynman diagrams. Let's look at each, one at a time.

Column (II) has nth order propagators, external particles, and vertex

---

[1] Peskin & Schroeder define irreducible as what we call proper reducible, and reducible as what we call improper reducible. For a single particle, they use the symbol 1PI (one particle irreducible) for what we would call one particle proper reducible. Mandl & Shaw use the definitions we have here.

<u>nth Order Photon Propagator</u>

Fig. 14-6 depicts tree level plus some other Feynman diagrams, including certain divergent corrections, to nth order for the photon propagator.

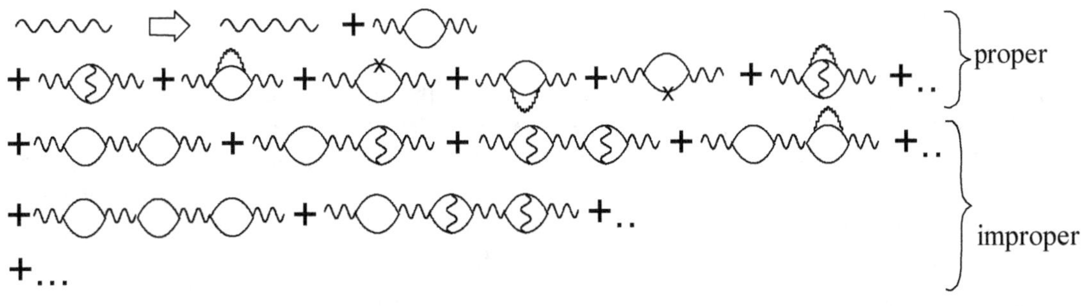

nth order photon propagator

Figure 14-6. Photon Propagator: Some of the Higher Order Divergent Corrections

Note the physical fermion mass δm corrections represented by the X symbol in Fig. 14-6 as we discussed in Chap. 13 and repeated in this chapter in step (VIII) for the 2nd order case. The nth order case, as we will see shortly, has a similar physical mass correction diagram/amplitude.

We define the sum of amplitudes from the last diagram of the first row plus all the diagrams of the second row in Fig. 14-6, i.e., all the proper diagrams except tree level, as

$$ie_0^2\Pi_{nth}^{\mu\nu}\left(k\right)=\text{sum of 2nd to }n\text{th order proper diagrams for photon propagator}. \quad (14\text{-}44)$$

More specifically, where we indicate the symbol $\Pi^{\mu\nu}(k)$ for the simple loop divergent integral introduced at the beginning of Chap. 13,

$$\Pi_{nth}^{\mu\nu}\left(k\right)=\underbrace{\frac{-i}{\left(2\pi\right)^4}\text{Tr}\int iS_F\left(p\right)\gamma^\mu iS_F\left(p-k\right)\gamma^\nu d^4p}_{\Pi^{\mu\nu}(k)\ \text{(last diagram, 1st row)}}\ +\begin{pmatrix}\text{sum of Fig. 14-6 2nd}\\\text{row contributions}\end{pmatrix}. \quad (14\text{-}45)$$

Then, in correcting the propagator from tree level to nth order, we have, with reference to Fig. 14-6,

$$iD_{F\,\alpha\beta}\left(k\right)\ \Rightarrow\ iD_{F\,\alpha\beta}^{nth}\left(k\right)$$

$$=\overbrace{\underbrace{iD_{F\,\alpha\beta}\left(k\right)}_{\text{tree photon propagator}}+\underbrace{iD_{F\,\alpha\mu}\left(k\right)ie_0^2\Pi_{nth}^{\mu\nu}\left(k\right)iD_{F\,\nu\beta}\left(k\right)}_{\text{last diagram of 1st row plus 2nd row}}}^{\text{proper diagrams}}$$

$$+\underbrace{iD_{F\,\alpha\mu}\left(k\right)ie_0^2\Pi_{nth}^{\mu\nu}\left(k\right)iD_{F\,\nu\delta}\left(k\right)ie_0^2\Pi_{nth}^{\delta\eta}\left(k\right)iD_{F\,\eta\beta}\left(k\right)}_{\text{3rd row}} \quad (14\text{-}46)$$

$$+\underbrace{iD_{F\,\alpha\mu}\left(k\right)ie_0^2\Pi_{nth}^{\mu\nu}\left(k\right)iD_{F\,\nu\delta}\left(k\right)ie_0^2\Pi_{nth}^{\delta\eta}\left(k\right)iD_{F\,\eta\rho}\left(k\right)ie_0^2\Pi_{nth}^{\rho\zeta}\left(k\right)iD_{F\,\zeta\beta}\left(k\right)}_{\text{4th row}}$$

$$+\left(\text{diagrams in higher rows to nth order}\right).$$

To elaborate, consider we can represent the sum of terms in (14-45) symbolically as

$$\Pi_{nth}^{\mu\nu}\left(k\right)=X^{\mu\nu}+Y^{\mu\nu}+Z^{\mu\nu}+..., \quad (14\text{-}47)$$

where $X^{\mu\nu}$ represents the contribution from the first non-tree level proper diagram of Fig. 14-6 ($\Pi^{\mu\nu}(k)$ in this case), $Y^{\mu\nu}$ represents that of the diagram after that, etc. Then in the 3rd row of (14-46) we would have terms of form

$$iD_{F\,\alpha\mu}\left(k\right)ie_0^2\Pi_{nth}^{\mu\nu}\left(k\right)iD_{F\,\nu\delta}\left(k\right)ie_0^2\Pi_{nth}^{\delta\eta}\left(k\right)iD_{F\,\eta\beta}\left(k\right)=$$

$$iD_{F\,\alpha\mu}\left(k\right)ie_0^2X^{\mu\nu}iD_{F\,\nu\delta}\left(k\right)ie_0^2X^{\delta\eta}iD_{F\,\eta\beta}\left(k\right)+iD_{F\,\alpha\mu}\left(k\right)ie_0^2X^{\mu\nu}iD_{F\,\nu\delta}\left(k\right)ie_0^2Y^{\delta\eta}iD_{F\,\eta\beta}\left(k\right)+... \quad (14\text{-}48)$$

$$+\ iD_{F\,\alpha\mu}\left(k\right)ie_0^2X^{\mu\nu}iD_{F\,\nu\delta}\left(k\right)ie_0^2Z^{\delta\eta}iD_{F\,\eta\beta}\left(k\right)\ +....$$

Thus, the 3rd row of (14-46) comprises contributions represented by all the (improper) diagrams in the 3rd row of Fig. 14-6. Similar logic holds for the 4th row of (14-46) and the 4th row of Fig. 14-6, and so on, to the nth order diagrams. $iD_{F\alpha\beta}^{nth}(k)$ of (14-46) therefore represents the photon propagator corrected to nth order as shown in Fig. 14-6.

nth Order Fermion Propagator

We follow a similar procedure for the fermion propagator as we did for the photon propagator. Fig. 14-7 shows a few of the proper and the improper corrections to nth order, where we assume bare mass m_0 (because here we need to derive the relation for physical mass m, as we did in the 2nd order case). *nth order fermion propagator*

We define the sum of amplitudes from the last diagram of the first row plus all the diagrams of the second row in Fig. 14-7, i.e., all the proper diagrams except tree level, as

$$ie_0^2 \Sigma_{nth}(p) = \text{sum of 2nd to } n\text{th order proper diagrams for fermion propagator} \qquad (14\text{-}49)$$

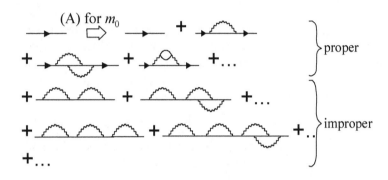

Figure 14-7. Fermion Propagator: Some of the Higher Order Divergent Corrections

More specifically, where we indicate the symbol $\Sigma(p)$ for the simple loop divergent integral introduced at the beginning of Chap. 13,

$$\Sigma_{nth}(p) = \underbrace{\frac{i}{(2\pi)^4} \int iD_{F\alpha\beta}(k)\, \gamma^\alpha iS_F(p-k)\gamma^\beta d^4k}_{\Sigma(p)} + \begin{pmatrix} \text{sum of Fig. 14-7 2nd} \\ \text{row contributions} \end{pmatrix} \qquad (14\text{-}50)$$

Then, similar to the photon case, in correcting the fermion propagator from tree level to nth order, we have

$$iS_F(p) \overset{\text{(A) for } m_0}{\Rightarrow} iS_F^{nth}(p)$$

$$= \overbrace{\underbrace{iS_F(p)}_{\substack{\text{tree fermion} \\ \text{propagator}}} + \underbrace{iS_F(p)ie_0^2\Sigma_{nth}(p)iS_F(p)}_{\substack{\text{last diagram of 1st row} \\ \text{plus 2nd row}}}}^{\text{proper diagrams}} \qquad (14\text{-}51)$$

$$+ \underbrace{iS_F(p)ie_0^2\Sigma_{nth}(p)iS_F(p)ie_0^2\Sigma_{nth}(p)iS_F(p)}_{\text{3rd row}} + \begin{pmatrix} \text{diagrams in higher} \\ \text{rows to nth order} \end{pmatrix}.$$

$iS_F^{nth}(p)$ of (14-51) therefore represents the fermion propagator corrected to nth order as shown in Fig. 14-7.

nth Order External Line Corrections

Fig. 14-8 shows a few higher order corrections to photon and fermion external lines.

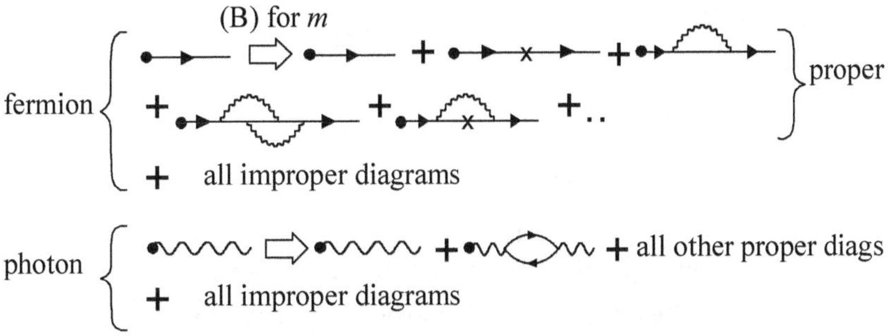

Figure 14-8. External Particles: Some of the Higher Order Divergent Corrections

nth order external line

Note that here we use the physical mass m, which we will deduce in Step (VIII) for the fermion propagator. As in the 2nd order case, we can get the same result using m_0 or m, but the latter is easier, so we choose it.

Following a parallel procedure to what was done for the photon and fermion propagators to nth order, we find

$$u_r(\mathbf{p}) \Rightarrow u_r^{nth}(\mathbf{p})$$
$$= u_r(\mathbf{p}) + iS_F(p)\left(ie_0^2\Sigma_{nth}(p) + i\delta m\right)u_r(\mathbf{p}) + \text{ all improper diagrams to } n\text{th order,} \tag{14-52}$$

with analogous results for $\bar{u}_r(\mathbf{p}), v_r(\mathbf{p})$, and $\bar{v}_r(\mathbf{p})$.

For the external photon, we have

$$\varepsilon_\mu(\mathbf{k}) \Rightarrow \varepsilon_\mu^{nth}(\mathbf{k})$$
$$= \varepsilon_\mu(\mathbf{k}) + iD_{F\alpha\beta}(k)ie_0^2\Pi_{nth}^{\alpha\beta}(k)\varepsilon_\mu(\mathbf{k}) + \text{ all improper diagrams to } n\text{th order.} \tag{14-53}$$

nth Order Vertex Corrections

In similar fashion, we can use Fig. 14-9 to find the nth order vertex correction.

nth order vertex

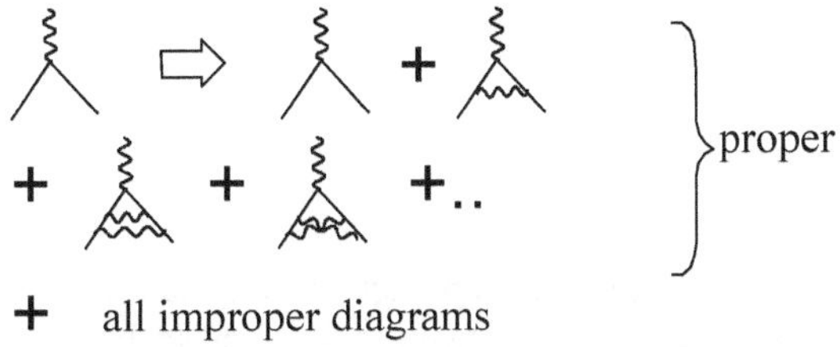

Figure 14-9. The Vertex: Some of the Higher Order Divergent Corrections

For the definition

$$\Lambda_{nth}^\mu(p,p') = \underbrace{\frac{-1}{(2\pi)^4}\int iD_{F\alpha\beta}(k)\gamma^\alpha\, iS_F(p'-k)\gamma^\mu iS_F(p-k)\gamma^\beta d^4k}_{\Lambda^\mu(p,p')} + \left(\begin{array}{c}\text{sum of Fig. 14-9 2nd}\\ \text{row contributions}\end{array}\right) \tag{14-54}$$

then

$$\gamma^\mu \Rightarrow \gamma_{nth}^\mu(p,p') = \gamma^\mu + e_0^2\Lambda_{nth}^\mu(p,p') + \left(\begin{array}{c}\text{diagrams in higher}\\ \text{rows to } n\text{th order}\end{array}\right). \tag{14-55}$$

(III) Loop Integrals

In this column of Wholeness Chart 14-5, we explicitly express the integrals $\Pi_{nth}^{\mu\nu}(k), \Sigma_{nth}(p), \Lambda_{nth}^\mu(p,p')$ of (14-45), (14-50), and (14-54), which represent the non-tree level proper diagrams to nth order.

Column (III) defines nth order proper diagrams integral symbol

(IV) Most General Form of the Loop Integrals

The most general possible form for $\Pi_{nth}^{\mu\nu}(k)$ is

$$\Pi_{nth}^{\mu\nu}(k) = -g^{\mu\nu}A_{nth}(k^2) + \underbrace{k^\mu k^\nu B_{nth}(k^2)}_{\text{drops out}} . \qquad (14\text{-}56)$$

In the 2nd order development we simply stated that the B term (analogous to B_{nth} above) didn't show up when we did the formal integration using regularization. In general, for higher orders, this is also true for the B_{nth} term. In the Appendix to this chapter we show why, and how, this occurs. The bottom line, as shown there, is that it is the result of current conservation.

Key Point

Note that the B_{nth} term dropping out enables the theory to be renormalizable. This drops out because of current conservation, and current conservation is a direct result of local gauge invariance (internal symmetry) of the Lagrangian. (See Chap. 11.) Renormalization depends on local symmetry. If we don't have a symmetric (gauge invariant) theory, we don't have a finite theory, i.e., we don't have a theory that works.

For fermions, similar to the 2nd order case, we can express the most general form for $\Sigma_{nth}(p)$ in terms of $(\not p - m_0)$ or $(\not p - m)$, but using the form with m makes the development easier. Thus, we have, similar to the 2nd order case, where again the A_{nth} and B_{nth} of (14-57) are different from those in (14-56),

$$\Sigma_{nth}(p) = A_{nth} + (\not p - m)B_{nth} + (\not p - m)\Sigma_{c\,nth}(\not p - m) . \qquad (14\text{-}57)$$

And again in parallel with the 2nd order case, the most general form for $\Lambda_{nth}^{\mu}(p,p')$ is

$$\Lambda_{nth}^{\mu}(p,p') = a_{nth}\gamma^\mu + b_{1\,nth}p^\mu + b_{2\,nth}p'^\mu . \qquad (14\text{-}58)$$

(V) Series Expansion of the Loop Integrals

The general forms of column (IV) are here expressed as Taylor series expansions. The photon relation is expanded in powers of k^2; the fermion one, in powers of $(\not p - m)$, just as we did in the 2nd order case. Like that case, these expansions contain divergent and convergent terms, the latter labeled with subscript "c".

For the photon, (14-56) becomes

$$\Pi_{nth}^{\mu\nu}(k) = -g^{\mu\nu}A_{nth}(k^2) = -g^{\mu\nu}\left(\underbrace{A_{nth}(0)}_{=0} + k^2 A'_{nth}(0) + k^2\Pi_{c\,nth}(k^2)\right) . \qquad (14\text{-}59)$$

In the 2nd order case we simply stated that the term $A(0)$ turned out to be zero when the integration was carried out via regularization. We can show this in general, and thus make it applicable to any order case, i.e., $A_{nth}(0) = 0$. But we need to wait until part (VI) below to do this.

The vertex form in column (V) is not exactly a series expansion but is the most general form in terms of L_{nth} and $\Lambda_{nth}^{\mu}(p,p')$, rather than the a_{nth}, b_{1nth}, and $b_{2\,nth}$ of the prior column. This form was found via the exact same steps used for the 2nd order vertex case in going from column (IV) to column (V) there, i.e., (14-58) becomes

$$\Lambda_{nth}^{\mu}(p,p') = L_{nth}\gamma^\mu + \Lambda_{c\,nth}^{\mu}(p,p') . \qquad (14\text{-}60)$$

(VI) Putting Expansions (V) into Propagator, Leg, Vertex Modifications (II)

Here we insert the expansions of the nth order integrals of column (V) into the 2nd order correction relations for the propagators and vertex of column (II).

nth Order Photon Propagator Expansion

For the photon, using (14-4), we take (where below our operators happen to be mere numbers)

$$\text{operator } \frac{1}{A} = \frac{-i}{k^2 + i\varepsilon} = \frac{1}{i(k^2 + i\varepsilon)} , \text{ and} \qquad (14\text{-}61)$$

operator $B = -ie_0^2 A_{nth}\left(k^2\right) = -ie_0^2\left(A_{nth}\left(0\right) + k^2 A'_{nth}\left(0\right) + k^2 \Pi_{c\,nth}\left(k^2\right)\right)$ (14-62)

Using (14-56) in the second line below, and (14-61) and (14-62) in the third, (14-46) then becomes

$$iD_{F\alpha\beta}^{nth}\left(k\right) = iD_{F\,\alpha\beta}\left(k\right) + iD_{F\,\alpha\mu}\left(k\right)ie_0^2\Pi_{nth}^{\mu\nu}\left(k\right)iD_{F\,\nu\beta}\left(k\right) + ...$$

Operator relation yields $iD_{F\alpha\beta}^{nth}\left(k\right)$ in terms of integral expansion quantities

$$= \frac{g_{\alpha\beta}}{i\left(k^2 + i\varepsilon\right)} + \frac{g_{\alpha\mu}}{i\left(k^2 + i\varepsilon\right)}ie_0^2\left(-g^{\mu\nu}\right)A_{nth}\left(k^2\right)\frac{g_{\nu\beta}}{i\left(k^2 + i\varepsilon\right)} + ...$$ (14-63)

$$= \frac{g_{\alpha\beta}}{i\left(k^2 + i\varepsilon\right)} - \frac{g_{\alpha\beta}}{i\left(k^2 + i\varepsilon\right)}ie_0^2 A_{nth}\left(k^2\right)\frac{1}{i\left(k^2 + i\varepsilon\right)} + ... = g_{\alpha\beta}\left(\frac{1}{A} + \frac{1}{A}B\frac{1}{A} + \frac{1}{A}B\frac{1}{A}B\frac{1}{A} + ...\right).$$

With (14-4) and A_{nth} defined as in (14-59) (or (14-62)), we have

$$iD_{F\alpha\beta}^{nth}\left(k\right) = \frac{1}{A - B}g_{\alpha\beta} = \frac{g_{\alpha\beta}}{i\left(k^2 + i\varepsilon\right) + i\left(e_0^2 A_{nth}\left(k^2 = 0\right) + e_0^2 k^2 A'_{nth}\left(k^2 = 0\right) + e_0^2 k^2 \Pi_{c\,nth}(k^2)\right)}.$$ (14-64)

Now consider the case where (14-64) approaches a representation of a real photon. That is, the virtual photon (propagator) it represents approaches being on-shell, i.e., $k^2 \to 0$. The real photon propagator has a pole at $k^2 = 0$. That is, it goes to infinity in this limit. (This is verified by experiment, i.e., by the predictions made for virtual photons approaching such limits.)

Taking $iD_{F\alpha\beta}^{nth}\left(k\right)$ in limit $k^2 \to 0$ proves quantity $A(0) = 0$.

All of the terms in the denominator of (14-64) go to zero as $k^2 \to 0$ except the term with $A(0)$ in it. If that term were non-zero, it would keep (14-64) finite in the real photon propagator limit, which does not happen. Therefore, $A(0)$ must equal zero, as we note in (14-59). Thus, (14-64) becomes

$$iD_{F\alpha\beta}^{nth}\left(k\right) = \frac{g_{\alpha\beta}}{i\left(k^2 + i\varepsilon\right) + i\left(e_0^2 k^2 A'_{nth} + e_0^2 k^2 \Pi_{c\,nth}(k^2)\right)} \approx \frac{-ig_{\alpha\beta}}{\left(k^2 + i\varepsilon\right)\left(1 + e_0^2 A'_{nth} + e_0^2 \Pi_{c\,nth}(k^2)\right)}$$

$$\approx \frac{1}{1 + e_0^2 A'_{nth}}\underbrace{\frac{-ig_{\alpha\beta}}{\left(k^2 + i\varepsilon\right)}}_{iD_{F\alpha\beta}(k)}\frac{1}{1 + e_0^2 \Pi_{c\,nth}(k^2)}.$$ (14-65)

Most useful form of $iD_{F\alpha\beta}^{nth}\left(k\right)$

nth Order Fermion Propagator Expansion

In (14-51) we take the relation (14-4) operators to be $1/A = iS_F(p) = i/\left(\not{p} - m_0 + i\varepsilon\right)$ and $B = ie_0^2\Sigma_{nth}\left(p\right)$. (See column (V)). This yields

Operator relation $\to iS_F^{nth}\left(p\right)$ in terms of integral expansion quantities

$$iS_F^{nth}\left(p\right) \approx \frac{1}{A - B} = \frac{i}{\underbrace{\not{p} - m_0 + e_0^2 A_{nth}}_{-m} + e_0^2\left(\not{p} - m\right)B_{nth} + e_0^2\left(\not{p} - m\right)\Sigma_{c\,nth} + i\varepsilon}$$

Renormalize mass in the process (see (VIII)

$$= \frac{i}{\not{p} - m + e_0^2\left(\not{p} - m\right)B_{nth} + e_0^2\left(\not{p} - m\right)\Sigma_{c\,nth} + i\varepsilon}$$ (14-66)

$$\approx \frac{i}{\left(\not{p} - m + i\varepsilon\right)\left(1 + e_0^2 B_{nth} + e_0^2\Sigma_{c\,nth}\right)} \approx \frac{1}{1 + e_0^2 B_{nth}}\underbrace{\frac{i}{\not{p} - m + i\varepsilon}}_{iS_F(p)}\frac{1}{1 + e_0^2\Sigma_{c\,nth}}.$$

Most useful form of $iS_F^{nth}\left(p\right)$

In the first line above, we renormalized mass in the same way we did it for the 2nd order case. More on this in (VIII).

nth Order External Particles in Terms of Expansion Quantities

Following logic exactly parallel to what we did for the 2nd order case, we find

$$u_r^{nth}\left(\mathbf{p}\right) = \frac{1}{\left(1 + e_0^2 B_{nth}\right)^{1/2}}u_r\left(\mathbf{p}\right) \quad \text{same for } \bar{u}_r^{nth}\left(\mathbf{p}\right), v_r^{nth}\left(\mathbf{p}\right), \bar{v}_r^{nth}\left(\mathbf{p}\right), \text{ and}$$ (14-67)

External line corrections to nth order

$$\varepsilon_\mu^{nth}\left(\mathbf{k}\right)=\frac{1}{\left(1+e_0^2 A'_{nth}\right)^{1/2}}\,\varepsilon_\mu\left(\mathbf{k}\right).\qquad(14\text{-}68)$$

nth Order Vertex Expansion

Again, exactly parallel to what we did for the 2nd order case, we find

$$\gamma_{nth}^\mu\left(p,p'\right)=\gamma^\mu\left(1+e_0^2 L_{nth}\right)+e_0^2\Lambda_{c\,nth}^\mu\left(p,p'\right)\approx\left(1+e_0^2 L_{nth}\right)\left(\gamma^\mu+e_0^2\Lambda_{c\,nth}^\mu\left(p,p'\right)\right).\qquad(14\text{-}69)$$

Most useful form of $\gamma_{nth}^\mu\left(p,p'\right)$

(VII) The Infinite Part of (VI) as $\Lambda\to\infty$

Similar to the 2nd order case, the unbounded integrals for nth order can be evaluated using regularization, but we won't get into that in this book. We do note that the quantities A'_{nth}, A_{nth}, B_{nth} and L_{nth} used in the expansions above are infinite. Quantities subscript "c" (convergent) are finite.

Column (VII) notes the infinite parts in the above relations and the equality $B_{nth}=L_{nth}$

We proved for the 2nd order case that $B=L$. It is also true that $B_{nth}=L_{nth}$, but we will not prove that here. As in the 2nd order case, we will find here that because these two quantities are equal, they drop out of our relations, and we will not need to actually determine their precise form.

(VIII) Fermion Mass m Renormalization

This column is only applicable to fermions, and as noted above, was already shown in the first line of (14-66).

Column (VIII) summarizes mass renormalization

Here, like the 2nd order case, we fold the unbounded value for A_{nth} in the denominator of $iS_F^{nth}\left(p\right)$ into the finite value for observed, physical mass m by assuming the unobserved, bare mass m_0 is also unbounded. This leaves us with $\not{p}-m$ in the denominator instead of $\not{p}-m_0+e_0^2 A_{nth}=\not{p}-m_0-\delta m$. Thus, the fermion mass is renormalized.

Again, like the 2nd order case, with this relationship, we can re-express our Lagrangian in terms of m. That is, in the original Lagrangian, we substitute $m_0=m-\delta m$. This results in an extra term in the Lagrangian with a factor of δm. This extra term leads to an extra Feynman diagram, in precisely the same way as shown in 2nd order column (I), Case (B). The term containing δm is our mass counterterm for nth order. We use the same symbol as for 2nd order though here we have A_{nth} in place of A.

The effect of using m, instead of m_0, in $\mathcal{L}$ summarized in 2nd order case, Column (I), Case B

This extra Feynman diagram, i.e., Case (B), was used as the starting point for external fermion line renormalization in column (I).

(IX) Definitions

In this column we define convenient shorthand symbols for quantities found in (14-65) to (14-69) that we will see repeatedly.

Column (IX) defines symbols that include the unbounded quantities

$$Z_\gamma^{nth}=\frac{1}{1+e_0^2 A'_{nth}}\qquad(14\text{-}70)$$

$$Z_f^{nth}=\frac{1}{1+e_0^2 B_{nth}}\qquad(14\text{-}71)$$

$$Z_V^{nth}=1+e_0^2 L_{nth}\qquad(14\text{-}72)$$

From the approximations on the RHS of (14-70) and (14-71), we see that our definitions here are consonant with the Z factors we found at second order ($n=2$). Note that because $B_{nth}=L_{nth}$, then

$$Z_V^{nth}=1/Z_f^{nth}.\qquad(14\text{-}73)$$

(X) Amplitude Determination

We will now examine assembling an nth order amplitude, given the foundation we have built for ourselves above. As in the 2nd order case, we start with the example of Compton scattering.

Column (X) summarizes using the above to assemble an amplitude (see below)

Compton Scattering Amplitude

We now want to express the tree level and divergent sub-amplitudes summation of (14-3) for Compton scattering in a convenient form.

Look at tree level plus divergent parts of Compton amplitude to nth order

$$\underbrace{\mathcal{M}_{C1}^{(2)} + \mathcal{M}_{C2}^{(2)}}_{\substack{\text{tree level, finite,}\\\text{irreducible}}} + \underbrace{\sum_{k=2}^{n}\left(\sum_{i}\mathcal{M}_{C1-i}^{(2k)} + \sum_{i}\mathcal{M}_{C2-i}^{(2k)}\right)}_{\substack{\text{contain divergent integrals,}\\\text{reducible}}} \qquad (14\text{-}74)$$

Let's begin by examining just the first way for Compton scattering, which we have denoted with subscripts $C1$. It may help to follow the diagrams in Fig. 14-5 as we progress.

Recall (see Wholeness Chart 14-1, pg. 350) that there are two routes we can take to do this. For one of these, we can add the tree level and all the 2nd to nth order contribution diagrams, as in

Two routes to finding nth order amplitude

$$\mathcal{M}_{C1}^{(2)} + \sum_{k=2}^{n}\sum_{i}\mathcal{M}_{C1-i}^{(2k)} = \mathcal{M}_{\substack{\text{Compton}\\\text{1st way, all}\\\text{reducibles to } e^{2n}}} \qquad \text{(first route).} \qquad (14\text{-}75)$$

Or we can simply substitute the *n*th order relations (which include tree level plus corrections) (14-65) to (14-69) for the propagators, vertices, and external lines into the tree level amplitude.

$$\mathcal{M}_{C1}^{(2)} \xrightarrow[\substack{\text{nth order propagators,}\\\text{external lines, and}\\\text{vertex relations}}]{\text{substitute}} \mathcal{M}_{\substack{\text{Compton}\\\text{1st way, all}\\\text{reducibles to } e^{2n}}} \qquad \text{(second route).} \qquad (14\text{-}76)$$

We use the simpler, second route

The second of these is far simpler. Using it, we find

$$\mathcal{M}_{C1}^{(2)} = -e_0^2 \bar{u}_{s'}(\mathbf{p}')\varepsilon_{\mu,r'}(\mathbf{k}')\gamma^\mu iS_F(p+k)\varepsilon_{\nu,r}(\mathbf{k})\gamma^\nu u_s(\mathbf{p}) \;\Rightarrow\; \mathcal{M}_{\substack{\text{Compton}\\\text{1st way, all}\\\text{reducibles to } e^{2n}}} \approx$$

$$\underbrace{\frac{1}{\left(1+e_0^2 B_{nth}\right)^{1/2}}}_{\left(Z_f^{nth}\right)^{1/2}} \underbrace{\frac{1}{\left(1+e_0^2 A'_{nth}\right)^{1/2}}}_{\left(Z_\gamma^{nth}\right)^{1/2}} \underbrace{\left(1+e_0^2 L_{nth}\right)}_{Z_V^{nth}} \underbrace{\frac{1}{1+e_0^2 B_{nth}}}_{Z_f^{nth}} \underbrace{\left(1+e_0^2 L_{nth}\right)}_{Z_V^{nth}} \underbrace{\frac{1}{\left(1+e_0^2 A'_{nth}\right)^{1/2}}}_{\left(Z_\gamma^{nth}\right)^{1/2}} \underbrace{\frac{1}{\left(1+e_0^2 B_{nth}\right)^{1/2}}}_{\left(Z_f^{nth}\right)^{1/2}} \quad (14\text{-}77)$$

$$\times \left(-e_0^2 \bar{u}_{s'}(\mathbf{p}')\varepsilon_{\mu,r'}(\mathbf{k}')\underbrace{\left(\gamma^\mu + e_0^2 \Lambda_{c\,nth}^\mu\right)}_{\substack{\gamma^\mu_{e_0\,Mod}\\nth}} iS_F(p+k)\underbrace{\frac{1}{\left(1+e_0^2\Sigma_{c\,nth}\right)}}_{\substack{iS_F^{nth}\\e_0\,Mod\\(p+k)}} \underbrace{\left(\gamma^\nu + e_0^2 \Lambda_{c\,nth}^\nu\right)}_{\substack{\gamma^\nu_{e_0\,Mod}\\nth}}\varepsilon_{\nu,r}(\mathbf{k})u_s(\mathbf{p}) \right),$$

where we define quantities in the last line of (14-77) that will be useful. To help keep notation straight, note that the "Mod" script designates only the convergent parts of (14-65) to (14-69).

Using these new symbols, we can express (14-77) more succinctly as

Re-expressing using Z symbols and "Mod" scripts

$$\mathcal{M}_{\substack{\text{Compton}\\\text{1st way, all}\\\text{reducibles to } e_0^{2n}}} \approx \left(Z_\gamma^{nth}\right)\left(Z_f^{nth}\right)^2\left(Z_V^{nth}\right)^2 \times$$

$$\underbrace{\left(-e_0^2 \bar{u}_{s'}(\mathbf{p}')\varepsilon_{\mu,r'}(\mathbf{k}')\gamma^\mu_{\substack{e_0\,Mod\\nth}} iS_F^{\substack{nth\\e_0\,Mod}}(p+k)\varepsilon_{\nu,r}(\mathbf{k})\gamma^\nu_{\substack{e_0\,Mod\\nth}} u_s(\mathbf{p}) \right)}_{\mathcal{M}_{\substack{C1\\e_0\,Mod,\,nth}}^{(2)}}, \quad (14\text{-}78)$$

where, in parallel with the 2nd order case, the last line is a modification to our tree level amplitude. That is, it is the same form as the first order in α ($=e_0^2$ order) expression but with modified expressions for the vertices and propagator. The RHS of the top row in (14-78) contains divergent quantities, the bottom row, all convergent quantities.

From (14-73), the fermion and vertex Z factors in (14-78) cancel out leaving

Fermion and vertex Z factors cancel leaving only Z_γ^{nth} in front

$$Z_\gamma^{nth}\,\mathcal{M}_{\substack{C1\\e_0\,Mod,\,nth}}^{(2)}. \qquad (14\text{-}79)$$

In directly parallel fashion, for the 2nd way for Compton scattering to occur (indicated with the subscript C2), one finds

$$\mathcal{M}_{\substack{\text{Compton}\\ \text{2nd way, all}\\ \text{reducibles to } e_0^{2n}}} = Z_\gamma^{nth} \mathcal{M}_{C2 \; \substack{e_0 \, Mod, \, nth}}^{(2)} \quad . \tag{14-80}$$

Fermion and vertex Z factors cancel in second way for Compton, too

That is, the fermion and vertex Z factors drop out there as well. Additionally, our vertex and propagator modifications to the tree level amplitude are the same as those in the last row of (14-77), though the tree level amplitude itself for the 2nd way for Compton scattering has somewhat different form.

Møller Scattering Example

By doing Prob. 7, you can show that the amplitude for the first way of Møller scattering to *nth* order in α (where we use the subscript B3 symbolism of Chap. 8) is

$$\mathcal{M}_{\substack{\text{Møller}\\ \text{1st way, all}\\ \text{reducibles to } e_0^{2n}}} = Z_\gamma^{nth} \left(Z_f^{nth} \right)^2 \left(Z_V^{nth} \right)^2 \mathcal{M}_{B3 \; \substack{e_0 \, Mod, \, nth}}^{(2)} = Z_\gamma^{nth} \mathcal{M}_{B3 \; \substack{e_0 \, Mod, \, nth}}^{(2)} \quad , \tag{14-81}$$

Fermion and vertex Z factors cancel in first way for Møller, just like in Compton

where the subscript on the last amplitude factor indicates the same substitutions as in the last row of (14-77) plus

$$iD_{F\mu\nu}^{\substack{e_0 \, mod\\ nth}}(k) = \frac{-ig_{\mu\nu}}{k^2 + i\varepsilon} \frac{1}{\left(1 + e_0^2 \Pi_{c\,nth}\right)} . \tag{14-82}$$

For the second way for Møller scattering to occur, in similar fashion, we get

$$\mathcal{M}_{\substack{\text{Møller}\\ \text{2nd way, all}\\ \text{reducibles to } e_0^{2n}}} = Z_\gamma^{nth} \left(Z_f^{nth} \right)^2 \left(Z_V^{nth} \right)^2 \mathcal{M}_{B4 \; \substack{e_0 \, Mod, \, nth}}^{(2)} = Z_\gamma^{nth} \mathcal{M}_{B4 \; \substack{e_0 \, Mod, \, nth}}^{(2)} \quad . \tag{14-83}$$

Same thing in second way for Møller scattering

Generalizing to Any Scattering Amplitude

As in the 2nd order case, it turns out that in every *nth* order amplitude calculation, the fermion and vertex factors in the divergent terms cancel and leave the photon factor Z_γ^{nth} as our only divergent quantity. Additionally, in every such amplitude we merely substitute the *nth* order modified relations of (14-65) to (14-69) for propagators and vertices in the tree level amplitude to get the modified amplitude.

Fermion and vertex Z factors cancel in every type amplitude

By doing Prob. 8, you can show that this same thing is true for Bhabha scattering.

Bottom line: In the divergent parts of our second order amplitude, for any type of QED scattering, the divergent fermion factor(s) will cancel with the divergent vertex factor(s), leaving only a photon divergent factor Z_γ^{nth}. The remainder of the amplitude can be found by substituting (finite) modifications to the propagator and vertex relations for those in the tree level amplitude.

(XI) Cancellation of *Z* Factors

In this column we simply note the first sentence of the Bottom line summary above.

Only Z factor remaining in every type of amplitude is the photon one

(XII) Charge Renormalization

For Compton scattering, first way, we have

$$\mathcal{M}_{\substack{\text{Compton}\\ \text{1st way, all}\\ \text{reducibles to } e_0^{2n}}} \approx Z_\gamma^{nth} \mathcal{M}_{C1 \; \substack{e_0 \, Mod, \, nth}}^{(2)}$$

$$= \frac{1}{1 + e_0^2 A'_{nth}} e_0^2 \left(-\bar{u}_{s'}(\mathbf{p}') \varepsilon_{\mu, r'}(\mathbf{k}') \gamma^\mu_{\substack{e_0 \, Mod\\ nth}} iS_F^{\substack{2nd\\ nth}}(p+k) \varepsilon_{\nu, r}(\mathbf{k}) \gamma^\nu_{\substack{e_0 \, Mod\\ nth}} u_s(\mathbf{p}) \right). \tag{14-84}$$

We renormalize charge, same for every case, by including Z_γ^{nth} in definition of physical charge

We now renormalize charge, as we saw in Chap. 12, and similar to what we did for the 2nd order case in this chapter.

$$e_0^2 \frac{1}{1 + e_0^2 \underbrace{A'_{nth}}_{\infty}} = e_0^2 Z_\gamma^{nth} = e^2 = \text{measured charge squared, or}$$

$$e_0 \frac{1}{1 + \frac{1}{2} e_0^2 \underbrace{A'_{nth}}_{\infty}} \approx e_0 \left(Z_f^{nth} \right)^{1/2} \approx e = \text{measured charge.} \tag{14-85}$$

Just use e^2 in place of e_0^2 at front of amplitude for every interaction from now on

The bare charge e_0 is zero, but e, due to the infinite quantity A'_{nth}, is finite and what we measure. As in the 2nd order case, e is a function of the energy level at which the charge measurement is made. (See Chap. 12.)

As an aside, you may be wondering that if A'_{nth} is infinite, then we shouldn't really have a problem with unboundedness in (14-84) since then $1 / 1 + e_0^2 A'_{nth} \to 0$ for non-zero e_0. But then, the amplitude (14-84), and every amplitude, would be zero, which does not reflect physical reality. In experiment, we find a dependence of physical charge e on energy level that can only be described by the assumption of (14-85). The approach of (14-85) simply works.

Note that given (14-85), with $e_0^2 A'_{nth} \ll 1$,

$$Z_\gamma^{nth} = \frac{1}{1 + e_0^2 A'_{nth}} \approx 1 - e_0^2 A'_{nth}, \tag{14-86}$$

For $n = 2$, nth order reduces to second order case

and our nth order case for $n = 2$ reduces to what we found before for second order.

Moreover, every reaction we encounter will have the factor $e_0^2 / \left(1 + e_0^2 A'_{nth} \right)$ in the nth order amplitude, as does (14-31), in light of (14-86).

<u>Bottom line:</u> So, from now on we can simply replace $e_0^2 / \left(1 + e_0^2 A'_{nth} \right)$ with e^2 in every QED amplitude calculation to nth order. We need to keep in mind that e depends on energy level (modestly via a log dependence).

(XIII) Substitution in Tree Amplitude to Get Amplitude to *n*th Order

One might expect we could then simply substitute the modified relations defined in the last row of (14-77) and (14-82). However, that leaves us with terms in e_0^2, which via our renormalization of charge should equal zero.

But propagator and vertex mods still have e_0^2 at this point

The resolution here is just like the resolution for the parallel situation at 2nd order. That is, at any given order giving us a factor of e_0^2, there are yet higher order diagrams that modify that factor. And they modify it in the same way we have seen e_0^2 modified above in (14-84) and (14-85). So, at any order, we can replace the e_0^2 term modified by higher orders with e^2, representing the physical charge squared (proportional to the QED coupling constant α).

Thus, for the photon propagator, where "*h.o.*" means "higher order" we have

When we take corrections beyond nth order into account, $e_0 \to e$ in these modifications

$$iD_{F\alpha\beta}^{\overset{e_0 \, mod}{nth}} (k) = \underbrace{\frac{-ig_{\alpha\beta}}{k^2 + i\varepsilon}}_{iD_{F\alpha\beta}(k)} \underbrace{\frac{1}{1 + e_0^2 \Pi_{c\,nth} + h.o.}}_{e^2 \Pi_{c\,nth}} \;\; \to \;\; iD_{F\alpha\beta}^{\overset{mod}{nth}} (k) = iD_{F\alpha\beta}(k) \frac{1}{1 + e^2 \Pi_{c\,nth}} . \tag{14-87}$$

Similarly,

So we redefine the modifications with e instead of e_0

$$iS_F^{\overset{mod}{nth}} (p) = \frac{i}{\not{p} - m + i\varepsilon} \frac{1}{\left(1 + e^2 \Sigma_{c\,nth} \right)}, \tag{14-88}$$

$$u_r(\mathbf{p}), \bar{u}_r(\mathbf{p}), v_r(\mathbf{p}), \bar{v}_r(\mathbf{p}), \text{ and } \varepsilon_\mu(\mathbf{k}) \text{ same as bare}, \tag{14-89}$$

$$\gamma_{\overset{mod}{nth}}^\mu (p, p') = \gamma^\mu + e^2 \Lambda_{c\,nth}^\mu . \tag{14-90}$$

We also define a <u>new symbol for the modified amplitude</u> $\mathcal{M}_{C1 \atop Mod,nth}^{(2)}$, now using physical charge throughout, including relations (14-87) to (14-90) instead of those in (14-77) and (14-82). Compare below with (14-31)

$$
\mathcal{M}_{\substack{\text{Compton} \\ \text{1st way, all} \\ \text{reducibles to } e_0^{2n}}} = Z_\gamma^{nth} \mathcal{M}_{\substack{C1 \\ e_0 \, Mod,nth}}^{(2)}
$$

$$
= \underbrace{\frac{1}{1+e_0^2 A'_{nth}}}_{e^2} e_0^2 \left(-\bar{u}_{s'}\left(\mathbf{p}'\right) \varepsilon_{\mu,r'}\left(\mathbf{k}'\right) \gamma_{\substack{Mod \\ nth}}^{\mu} i S_F^{\substack{Mod \\ nth}} \left(p+k\right) \varepsilon_{\nu,r}\left(\mathbf{k}\right) \gamma_{\substack{Mod \\ nth}}^{\nu} u_s\left(\mathbf{p}\right) \right) \quad (14\text{-}91)
$$

$$
= \mathcal{M}_{\substack{C1 \\ Mod,nth}}^{(2)} .
$$

Renormalizing the originally unbounded sub-amplitudes means $e_0 \to e$ & propagators and vertices $\to$ mod

Similar definitions hold for the second way of Compton scattering (with $C1 \to C2$ above) and any other type of scattering (e.g., above $C1 \to B1$, the first way of Bhabha scattering).

<u>Bottom line:</u> To re-normalize any (otherwise unbounded) interaction amplitude to nth order in α, in the tree level amplitude make the substitutions

- $e_0 \Rightarrow e(k)$

- $iS_F,\ \gamma^\mu,\ iD_{F\mu\nu} \Rightarrow$ (14-87) to (14-90) above.

14.7 The Total Amplitude to nth Order

From (14-3), the total Compton scattering amplitude to second order is then the finite value

The total amplitude for Compton scattering to nth order in α

$$
\mathcal{M}_{\substack{\text{Compton} \\ \text{all ways} \\ \text{total to } e_0^{2n}}} = \underbrace{\mathcal{M}_{C1}^{(2)} + \mathcal{M}_{C2}^{(2)}}_{\substack{\text{tree level, finite,} \\ \text{irreducible}}} + \underbrace{\sum_{k=2}^{n}\left(\sum_i \mathcal{M}_{C1-i}^{(2k)} + \sum_i \mathcal{M}_{C2-i}^{(2k)} \right)}_{\substack{\text{contain divergent integrals,} \\ \text{reducible}}} + \underbrace{\sum_{k=2}^{n}\left(\sum_i \mathcal{M}_{C1-\text{fin}\,i}^{(2k)} + \sum_i \mathcal{M}_{C2-\text{fin}\,i}^{(2k)} \right)}_{\substack{\text{finite, using } e_0^4 \to e_0^{2n} \\ \text{irreducible}}}
$$

$$
= \overbrace{\underbrace{\mathcal{M}_{\substack{C1 \\ Mod,nth}}^{(2)} + \mathcal{M}_{\substack{C2 \\ Mod,nth}}^{(2)}}_{\text{finite using } e^2}} + \overbrace{\sum_{k=2}^{n}\left(\sum_i \mathcal{M}_{C1-\text{fin}\,i}^{(2k)} + \sum_i \mathcal{M}_{C2-\text{fin}\,i}^{(2k)} \right)}^{\text{finite using } e^4 \to e^{2n}} .
$$

(14-92)

In the original sub-amplitudes that were finite, $e_0 \to e$ as well, due to higher order corrections

The irreducible (finite) amplitudes at the end of the first line of (14-92) originally had factors of e_0^4 to e_0^{2n}, but in similar manner as discussed for column (XIII) in the prior section, higher order corrections modify these to e^2 to e^{2n} (as in the second line of (14-92)). Each of these terms must be examined individually to determine whether it is negligible relative to other terms or not.

Thus, for any type of interaction in general, the renormalized amplitude to nth order is

$$
\mathcal{M}_{\substack{\text{Generic} \\ \text{all ways} \\ \text{total to } e_0^{2n}}} = \underbrace{\sum_{j}^{\#\text{ways}} \mathcal{M}_{\text{Gen}\,j}^{(2)}}_{\substack{\text{tree level, finite,} \\ \text{irreducible}}} + \underbrace{\sum_{k=2}^{n}\sum_{j,i} \mathcal{M}_{\text{Gen}\,j-i}^{(2k)}}_{\substack{\text{contain divergent integrals,} \\ \text{reducible}}} + \underbrace{\sum_{k=2}^{n}\sum_{j,i} \mathcal{M}_{\text{Gen}\,j-\text{fin}\,i}^{(2k)}}_{\substack{\text{finite, using } e_0^4 \to e_0^{2n}, \\ \text{irreducible}}}
$$

$$
= \overbrace{\sum_{j}^{\#\text{ways}} \mathcal{M}_{\substack{\text{Gen}\,j \\ Mod,nth}}^{(2)}}^{\text{finite using } e^2} + \overbrace{\sum_{k=2}^{n}\sum_{j,i} \mathcal{M}_{\text{Gen}\,j-\text{fin}\,i}^{(2k)}}^{\text{finite using } e^4 \to e^{2n}} .
$$

(14-93)

The total amplitude for any type scattering to nth order in α

Not only is (14-93) then finite, but its substitution in the transition amplitude S_{fi} equation (13-26), pg. 326, yields (to nth order accuracy) the actual probability $|S_{fi}|^2$ for the interaction to occur.

14.8 Renormalization to All Orders

The question remains about whether the amplitudes remain finite to *all* orders, i.e., when $n \to \infty$. In nature, all orders are included, so our theory needs to match that.

The resolution of this is fairly simple. We just take the limit of $n \to \infty$ in all of our relations for n^{th} order. Since, in ((IV) and subsequently, we used the most general form for including all the loop integrals that could arise, our results with A'_{nth}, A_{nth}, B_{nth}, and L_{nth} all being infinite are general results, good to all orders.

Thus, when the factors with B_{nth}, and L_{nth} cancel (the Z factors for fermions and vertices), they cancel at all orders. Similarly, our means for renormalizing mass, as in (14-66) is valid for A_{nth} at any order, and we can use renormalized mass m in all of the above relations at any, and all, orders. The same holds for renormalizing charge as in (14-32), good for any A'_{nth} at any, and all orders.

When we take $n \to \infty$, we use symbols such as $iD^{Mod}_{F\mu\nu}(k)$, $\gamma^{\mu}_{Mod}(p_1, p_2)$, and $iS^{mod}_F(p)$ without any nth sub or superscript. (As we did in Chap. 12.)

So taking $n \to \infty$ means our theory remains finite and our results are exact. Amplitudes use all finite terms (when expressed in terms of m and e) and correctly predict experiment. We have renormalized QED!

Taking $n \to \infty$ gives us the same renormalization results to all orders

$n \to \infty$ gives us the exact theory, which is renormalizable

14.9 Chapter Summary

We summarize the renormalization procedure for 2nd order on the next page, and for nth order on the page after that. Immediately below is a summary of how to solve problems once we have a renormalized theory.

14.9.1 Solving Scattering Problems to Order n

To find a transition amplitude for a given interaction, do the following.

1. Write down the tree level amplitude with bare quantities e_0 and m_0.

2. Replace e_0 and m_0 with renormalized values $e(p)$ and m, where p is the energy level of the interaction, and m is the measured particle (rest) mass.

3. Replace propagators with modified propagators and vertex relations with modified vertex relations (column (XIII) of Wholeness Charts 14-4 and 14-5).

4. Add amplitudes of above and non-negligible irreducible diagrams of four to $2n$ vertices (with $e_0, m_0 \to e, m$).

Summary of how to solve problems given our renormalization

14.9.2 An Important Result

- QED is only renormalizable because it has local gauge invariance (local symmetry).

  For $\Pi^{\mu\nu}_{nth}(k) = -g^{\mu\nu}A_{nth}(k^2) + k^{\mu}k^{\nu}B_{nth}(k^2)$, the last term drops out only because of current conservation (which is equivalent to local gauge invariance of $\mathcal{L}$). See Appendix. That term dropping out allows us to renormalize the theory.

  Additionally, as we saw in Chap. 13, the photon-photon scattering amplitude is finite because of the Ward identities, which are equivalent to current conservation and local gauge invariance.

  This principle re-appears at every level of QFT.

QED is a viable theory only because it is a symmetric theory

14.9.3 The Renormalization Procedure

Summary of how we renormalize

To 2nd Order

2nd order case

The process of renormalization to 2nd order comprises the following.

1. Draw all relevant Feynman diagrams with two and four vertices. Separate the four vertices (2nd order) diagrams into infinite amplitude (reducible) and finite amplitude (irreducible) diagrams.

2. For the infinite amplitude (reducible) four vertex contributions (as delineated in Wholeness Chart 14-4):

    i) Evaluate, via regularization, the corrected propagators, external lines, and vertex to 2nd order $[iD_{F\mu\nu}^{2nd}(k), iS_F^{2nd}(p), u_r^{2nd}(\mathbf{p}), \varepsilon_\mu^{2nd}(\mathbf{k})$, and $\gamma_{2nd}^\mu(p_1, p_2)$ of column (II)] in terms of a parameter Λ to find they have both infinite (as $\Lambda \to \infty$) parts and finite parts. (Columns (III) to (VII).)

    ii) Find the total amplitude to 2nd order by using the 2nd order correction (not the tree level contribution) results of i) to add all those 2nd order contribution amplitudes to the tree level amplitude. (Column (X))

    iii) Note that in the total amplitude, certain infinite contributions from the fermions and vertices cancel each other. (Column (XI).)

    iv) Eliminate the two other infinite contributions by renormalizing charge and mass, i.e., define physical e and m such that they are finite when $\Lambda \to \infty$. (e_0 and m_0 are re-defined in the process to make this work.) With this, e is dependent on interaction energy level p and is expressed as e (p). Use e and m in tree level amplitude in place of e_0 and m_0. (Columns (VIII) and (XII).)

    iv) Redefine propagators and vertex relations to include the finite parts from the regularization integration. Use these redefinitions (Column (XIII) $iD_{F\mu\nu}^{2nd}\overset{Mod}{}(k), iS_F^{2nd}\overset{Mod}{}(p), \gamma_{Mod}^\mu \atop 2nd (p_1, p_2)$) in tree level amplitude in place of original relations.

3. For the finite amplitude (irreducible diagrams) four vertex contributions:

    Recognize that higher order corrections will turn e_0 and m_0 into e and m, and making those changes in the amplitudes causes these terms to be small (higher order) relative to the terms of 2. above.

4. Thus, to 2nd order, we don't have to add Steps 2 and 3 above to get the total amplitude but can simply use Step 2. (See (14-42), repeated below.)

$$\mathcal{M}_{\substack{\text{Generic} \\ \text{all ways} \\ \text{total to } e_0^4}} = \underbrace{\sum_j^{\#\text{ways}} \mathcal{M}_{\substack{\text{Gen } j \\ Mod, 2nd}}^{(2)}}_{\substack{\text{Step 2 including tree} \\ \text{plus divergent} \\ \text{contributions}}} + \underbrace{\sum_{j,k} \mathcal{M}_{\substack{\text{Gen } j-k \\ Mod, 2nd}}^{(4)}}_{\substack{\text{Step 3 including convergent} \\ \text{(except tree) contributions,} \\ \text{negligible at 2nd order}}} \; .$$

Alternative route to 2 ii): Rather than adding relevant 1st and 2nd order diagrams, use the short cut of simply substituting full 2nd order (including tree level part) propagators, vertex, and external line relations of column (II) into the associated original 1st order relations in the tree level amplitude.

To nth Order

The process of renormalization to nth order comprises the following. *nth order case*

1. Draw all relevant infinite (reducible) and finite (irreducible) Feynman diagrams with $2n$ and fewer vertices. Separate reducible diagrams into non-composite (proper) and composite (improper) diagrams.

2. For the infinite amplitude (reducible) contributions (as delineated in Wholeness Chart 14-5):

   For non-composite (proper) diagrams alone:

   i) For each propagator, external line, and vertex add all non-composite (proper) diagram contributions except tree level (column (III).

   ii) Evaluate, via regularization, these in terms of a parameter Λ to find they have both infinite (as $\Lambda \to \infty$) parts and finite parts. (Columns (IV) to (VII).)

   For non-composite (proper) plus composite (improper) diagrams:

   iii) Use i) above to construct corrected propagators, external lines, and vertex to nth order $[iD_{F\mu\nu}^{nth}(k), iS_F^{nth}(p), u_r^{nth}(\mathbf{p}), \varepsilon_\mu^{nth}(\mathbf{k})$, and $\gamma_{nth}^\mu(p_1, p_2)$ of column (II)] including non-composite (proper) and composite (improper) diagram contributions.

   iv) Find the total amplitude to nth order by using the short cut route (see alternative route below). Substitute corrected propagators, external lines and vertices of iii) above in place of original associated relations in tree level amplitude. (Column (X).)

   v) Note that in the total amplitude, certain infinite contributions from the fermions and vertices cancel each other. (Column (XI).)

   vi) Eliminate the two other infinite contributions by renormalizing charge and mass, i.e., define physical e and m such that they are finite and non-zero when $\Lambda \to \infty$. (e_0 and m_0 are re-defined in the process to make this work.) With this, e is dependent on interaction energy level p and is expressed as $e(p)$. Use e and m in tree level amplitude in place of e_0 and m_0. (Columns (VIII) and (XII).)

   iv) Redefine propagators and vertex relations to include the finite parts from the regularization integration. Use these redefinitions (Column (XIII) $iD_{F\mu\nu}^{nth}\overset{Mod}{}(k), iS_F^{nth}\overset{Mod}{}(p), \gamma_{Mod,nth}^\mu(p_1, p_2)$) in tree level amplitude in place of original relations.

3. For the finite amplitude (irreducible diagrams) of four and higher vertices contributions:

   Recognize that higher order corrections will turn e_0 and m_0 into e and m, and make those changes in the amplitudes.

4. Add the results of Steps 2 and 3 above to get the total amplitude to nth order. (See (14-93), repeated below.) Note that we will have to examine terms in Step 3 individually to see if they are negligible relative to those of Step 2.

$$\mathcal{M}_{\substack{\text{Generic} \\ \text{all ways} \\ \text{total to } e^{2n}}} = \underbrace{\sum_{j}^{\#\text{ways}} \mathcal{M}_{\substack{\text{Gen } j \\ Mod,nth}}^{(2)}}_{\substack{\text{Step 2 including tree} \\ \text{plus divergent} \\ \text{contributions}}} + \underbrace{\sum_{k=2}^{n} \sum_{j,i} \mathcal{M}_{\text{Gen } j-\text{fin } i}^{(2k)}}_{\substack{\text{Step 3 including} \\ \text{convergent (except tree)} \\ \text{contributions}}} \quad .$$

Alternative to 2 iv): Rather than substituting the corrected propagator and vertex relations, one could add relevant diagrams. That, however, is far more difficult and cumbersome.

Wholeness Chart 14-4. Renormalization Steps to 2nd Order in α

	(I) Feynman Diagram Modification	(II) Propagator, Leg, Vertex Modification
Photon Self-Energy		$iD_{F\,\alpha\beta}(k) \;\Rightarrow\; iD^{2nd}_{F\,\alpha\beta}$ $= iD_{F\,\alpha\beta}(k) + iD_{F\,\alpha\mu}(k)ie_0^2\Pi^{\mu\nu}(k)iD_{F\,\nu\beta}(k)$
Fermion Self-Energy	(A) for m_0 OR (B) for m	(A) for m_0 $iS_F(p) \;\Rightarrow\; iS^{2nd}_F(p)$ $= iS_F(p) + iS_F(p)ie_0^2\Sigma(p)iS_F(p)$
External Line	(B) for m 	$u_r(\mathbf{p}) \;\Rightarrow\; u^{2nd}_r(\mathbf{p})$ $= u_r(\mathbf{p}) + iS_F(p)ie_0^2\Sigma(p)u_r(\mathbf{p}) + iS_F(p)(i\delta m)u_r(\mathbf{p})$ analogous for $\bar{u}_r(\mathbf{p}), v_r(\mathbf{p}), \bar{v}_r(\mathbf{p}), \varepsilon_\mu(\mathbf{k})$
Vertex Modification		$\gamma^\mu \Rightarrow \gamma^\mu_{2nd}(p,p') = \gamma^\mu + e_0^2\Lambda^\mu(p,p')$

Wholeness Chart 14-5. Renormalization Steps to nth Order

	(I) Feynman Diagram Modification	(II) Propagator, Leg, Vertex Modification
Photon Self-Energy	proper improper $+...$	$iD_{F\,\alpha\beta}(k) \;\Rightarrow\; iD^{nth}_{F\,\alpha\beta}(k)$ $= iD_{F\,\alpha\beta}(k) + iD_{F\,\alpha\mu}(k)ie_0^2\Pi^{\mu\nu}_{nth}(k)iD_{F\,\nu\beta}(k)$ $+ iD_{F\,\alpha\mu}(k)ie_0^2\Pi^{\mu\nu}_{nth}(k)iD_{F\,\nu\delta}(k)ie_0^2\Pi^{\delta\eta}_{nth}(k)iD_{F\,\eta\beta}(k)$ $+$ all other improper diagrams, 4th line on. $\Pi^{\mu\nu}_{nth}(k) =$ sum of proper diagrams except tree at left
Electron Self-Energy	(A) for m_0 proper improper $+...$	(A) for m_0 $iS_F(p) \;\Rightarrow\; iS^{nth}_F(p)$ $= iS_F(p) + iS_F(p)ie_0^2\Sigma_{nth}(p)iS_F(p)$ $+ iS_F(p)ie_0^2\Sigma_{nth}(p)iS_F(p)ie_0^2\Sigma_{nth}(p)iS_F(p)$ $+$ all other improper diagrams, 4th line on. $\Sigma_{nth}(p) =$ sum of proper diagrams except tree at left
External Lines	(B) for m proper $+$ all improper diagrams $+$ all other proper diags $+$ all improper diagrams	$u_r(\mathbf{p}) \;\Rightarrow\; u^{nth}_r(\mathbf{p})$ $= u_r(\mathbf{p}) + iS_F(p)\left(ie_0^2\Sigma_{nth}(p) + i\delta m\right)u_r(\mathbf{p})$ $+$ all improper diagrams, 3rd line on. $\Sigma_{nth}(p) =$ sum of proper diagrams except tree at left analogous for $\bar{u}_r(\mathbf{p}), v_r(\mathbf{p}), \bar{v}_r(\mathbf{p}), \varepsilon_\mu(\mathbf{k})$
Vertex Modification	proper $+$ all improper diagrams	$\gamma^\mu \Rightarrow \gamma^\mu_{nth}(p,p') = \gamma^\mu + e_0^2\Lambda^\mu_{nth}(p,p')$ $+$ all improper diagrams, 3rd line on. $\Lambda^\mu_{nth}(p,p') =$ from sum of diagrams in 2nd line at left

Wholeness Chart 14-4 (continued). Renormalization Steps to 2nd Order in α

(III) Loop Integral	(IV) Most Gen'l Form of $\leftarrow$	(V) Series Expansion of $\leftarrow$
$\Pi^{\mu\nu}(k) = -i(2\pi)^{-4}\,\mathrm{Tr}\int iS_F(p)\times$ $\gamma^{\mu} iS_F(p-k)\gamma^{\nu}d^4p$	$= -g^{\mu\nu}A\left(k^2\right) + \underbrace{k^{\mu}k^{\nu}B\left(k^2\right)}_{\substack{\text{not in this}\\\text{specific case}}}$	$\Pi^{\mu\nu}(k) = -g^{\mu\nu}\left(\underbrace{A(0)}_{=0} + k^2 A'(0) + k^2\Pi_c\left(k^2\right)\right)$
$\Sigma(p) = i(2\pi)^{-4}\int iD_{F\alpha\beta}(k)\times$ $\gamma^{\alpha} iS_F(p-k)\gamma^{\beta}d^4k$	$A,B\uparrow$ different from $A,B\downarrow$ $= A + \left(\not{p}-m\right)B$ $+\left(\not{p}-m\right)\Sigma_c\left(\not{p}-m\right)$	$\Sigma(p)$ as at left
Use above two	Use above two	Use above two
$\Lambda^{\mu}(p,p') = -(2\pi)^{-4}\int iD_{F\alpha\beta}(k)\gamma^{\alpha}\times$ $iS_F(p'-k)\gamma^{\mu}iS_F(p-k)\gamma^{\beta}d^4k$	$= a\gamma^{\mu} + b_1 p^{\mu} + b_2 p'^{\mu}$	$\Lambda^{\mu}(p,p') = L\gamma^{\mu} + \Lambda_c^{\mu}(p,p')$

Wholeness Chart 14-5 (continued). Renormalization Steps to nth Order

(III) Proper Diagram Integral	(IV) Most Gen'l Form of $\leftarrow$	(V) Series Expansion of $\leftarrow$
$\Pi^{\mu\nu}_{nth}(k) = -i(2\pi)^{-4}\,\mathrm{Tr}\int iS_F(p)\times$ $\gamma^{\mu} iS_F(p-k)\gamma^{\nu}d^4p$ $+e_0^2\left(\substack{\text{sum of 4 vertex}\\\text{diags, 2nd row}}\right)+\left(\substack{\text{3rd diag,}\\\text{2nd row}}\right)+$ $e_0^4\left(\substack{\text{sum of 6 vertex}\\\text{diags, 2nd row}}\right)+\left(\substack{\text{all higher 2nd}\\\text{row contribs}}\right)$	$\Pi^{\mu\nu}_{nth}(k) =$ $-g^{\mu\nu}A_{nth}\left(k^2\right) + \underbrace{k^{\mu}k^{\nu}B_{nth}\left(k^2\right)}_{\substack{\text{not in this}\\\text{specific case}}}$	$\Pi^{\mu\nu}_{nth}(k) =$ $-g^{\mu\nu}\left(\underbrace{A_{nth}(0)}_{=0} + k^2 A'_{nth}(0) + k^2\Pi_{c\,nth}\left(k^2\right)\right)$
$\Sigma_{nth}(p) = i(2\pi)^{-4}\int iD_{F\alpha\beta}(k)\times$ $\gamma^{\alpha} iS_F(p-k)\gamma^{\beta}d^4k$ $+e_0^2\left(\substack{\text{sum of 4 vertex}\\\text{diags, 2nd row}}\right)+$ $e_0^4\left(\substack{\text{sum of 6 vertex}\\\text{diags, 2nd row}}\right)+\left(\substack{\text{all higher 2nd}\\\text{row contribs}}\right)$	$A_{nth},B_{nth}\uparrow$ different from $A_{nth},B_{nth}\downarrow$ $\Sigma_{nth}(p) =$ $A_{nth} + \left(\not{p}-m\right)B_{nth}$ $+\left(\not{p}-m\right)\Sigma_{c\,nth}\left(\not{p}-m\right)$	$\Sigma_{nth}(p)$ as at left
Use above two	Use above two	Use above two
$\Lambda^{\mu}_{nth}(p,p') = -(2\pi)^{-4}\int iD_{F\alpha\beta}(k)\gamma^{\alpha}\times$ $iS_F(p'-k)\gamma^{\mu}iS_F(p-k)\gamma^{\beta}d^4k$ $+(\text{sum of higher order diags, 2nd row})$	$\Lambda^{\mu}_{nth}(p,p') =$ $a_{nth}\gamma^{\mu} + b_{1_{nth}} p^{\mu} + b_{2_{nth}} p'^{\mu}$	$\Lambda^{\mu}_{nth}(p,p') = L_{nth}\gamma^{\mu} + \Lambda_{c\,nth}^{\mu}(p,p')$

Wholeness Chart 14-4 (continued). Renormalization Steps to 2nd Order in α

(VI) Put Expansion (V) into (II)	(VII) ∞ Part as $\Lambda \to \infty$	(VIII) m Renorm	(IX) Define
$iD_{F\alpha\beta}^{2nd}(k) = \underbrace{\dfrac{-ig_{\alpha\beta}}{k^2+i\varepsilon}}_{iD_{F\alpha\beta}(k)}\left(1 - e_0^2 A' - e_0^2\Pi_c(k^2)\right)$	$A' = -2b_n \ln\dfrac{k}{\Lambda} + h.o.$	N/A	$Z_\gamma^{2nd} = 1 - e_0^2 A'$
use operator relation to get $\;\; iS_F^{2nd}(p) =$ $\dfrac{i}{\underbrace{\not{p}-m_0+e_0^2 A}_{-m}+e_0^2\left(\not{p}-m\right)B+e_0^2\left(\not{p}-m\right)\Sigma_c\left(\not{p}-m\right)+i\varepsilon}$	$A = -\dfrac{3m}{8\pi^2}\ln\dfrac{\Lambda}{m} + \mathcal{O}(1)$ $B \to \infty$	In (VI) $m_0 - e_0^2 A =$ $m_0 + \delta m = m$	$Z_f^{2nd} = 1 - e_0^2 B$ $\approx \dfrac{1}{1+e_0^2 B}$
$u_r^{2nd}(\mathbf{p}) = \left(1 - e_0^2 B\right)^{1/2} u_r(\mathbf{p})$, etc. $\varepsilon_\mu^{2nd}(\mathbf{k}) = \left(1 - e_0^2 A'\right)^{1/2}\varepsilon_\mu(\mathbf{k})$	As above	Above used for fermions N/A for γ	$\left(Z_f^{2nd}\right)^{1/2}, \left(Z_\gamma^{2nd}\right)^{1/2}$ used in (VI)
$\gamma_{2nd}^\mu(p,p') = \gamma^\mu\left(1+e_0^2 L\right) + e_0^2\Lambda_c^\mu(p,p')$ $\approx \left(1+e_0^2 L\right)\left(\gamma^\mu + e_0^2\Lambda_c^\mu(p,p')\right)$	$L \to \infty$ $B = L$	2nd block above used for fermion propagator part	$Z_V^{2nd} = 1 + e_0^2 L$ $Z_V^{2nd} = 1/Z_f^{2nd}$

Wholeness Chart 14-5 (continued). Renormalization Steps to nth Order

(VI) Put Expansion (V) into (II) (Use operator relation)	(VII) ∞ Part	(VIII) m Renorm	(IX) Define
$iD_{F\alpha\beta}^{nth}(k) == \dfrac{-ig_{\alpha\beta}}{\left(k^2+i\varepsilon\right)\left(1+e_0^2 A' + e_0^2\Pi_c(k^2)\right)}$ $\approx \dfrac{1}{\left(1+e_0^2 A'\right)}\underbrace{\dfrac{-ig_{\alpha\beta}}{\left(k^2+i\varepsilon\right)}}_{iD_{F\alpha\beta}(k)}\dfrac{1}{\left(1+e_0^2\Pi_c(k^2)\right)}$	$A'_{nth} \to \infty$	N/A	$Z_\gamma^{nth} = \dfrac{1}{1+e_0^2 A'_{nth}}$ $\approx 1 - e_0^2 A'_{nth}$
$iS_F^{nth}(p) =$ $\dfrac{i}{\underbrace{\not{p}-m_0+e_0^2 A_{nth}}_{-m}+e_0^2\left(\not{p}-m\right)B_{nth}+e_0^2\left(\not{p}-m\right)\Sigma_{c\,nth}\left(\not{p}-m\right)+i\varepsilon}$ $\approx \dfrac{1}{\left(1+e_0^2 B_{nth}\right)}\dfrac{i}{\left(\not{p}-m+i\varepsilon\right)}\dfrac{1}{\left(1+e_0^2\Sigma_{c\,nth}\left(\not{p}-m\right)\right)}$	$A_{nth} \to -\infty$ $B_{nth} \to \infty$	In (VI) $m_0 - e_0^2 A_{nth} =$ $m_0 + \delta m = m$	$Z_f^{nth} = \dfrac{1}{1+e_0^2 B_{nth}}$
$u_r^{nth}(\mathbf{p}) = \dfrac{1}{\left(1+e_0^2 B_{nth}\right)^{1/2}} u_r(\mathbf{p})$, etc. $\varepsilon_\mu^{nth}(\mathbf{k}) = \dfrac{1}{\left(1+e_0^2 A'_{nth}\right)^{1/2}}\varepsilon_\mu(\mathbf{k})$	As above	Above used for fermions and higher order corrections for γ	$\left(Z_f^{nth}\right)^{1/2}, \left(Z_\gamma^{nth}\right)^{1/2}$ used in (VI)
$\gamma_{nth}^\mu(p,p') = \gamma^\mu\left(1+e_0^2 L_{nth}\right) + e_0^2\Lambda_{c\,nth}^\mu(p,p')$ $\approx \left(1+e_0^2 L_{nth}\right)\left(\gamma^\mu + e_0^2\Lambda_{c\,nth}^\mu(p,p')\right)$	$L_{nth} \to \infty$ $B_{nth} = L_{nth}$	2nd block above used for fermion propagator part	$Z_V^{nth} = 1 + e_0^2 L_{nth}$ $Z_V^{nth} = 1/Z_f^{nth}$

Wholeness Chart 14-4 (continued). Renormalization Steps to 2nd Order in α

(X) Amplitude Determination	(XI) Cancellations	(XII) e Renormalization	(XIII) Substitute in Tree Amplits
Adding 1st & reducible 2nd order amplitudes [or alternatively, using (VI) in tree level amplitude], we get infinite term of form $$\mathcal{M}^{(2)} + \sum_{i=way} \mathcal{M}_i^{(4)} \approx$$ $$Z_\gamma^{2nd}\left(Z_f^{2nd}\right)^2\left(Z_V^{2nd}\right)^2 \mathcal{M}^{(2)}_{e_0 Mod, 2nd}$$ where the entities of (XIII) are used in $\mathcal{M}^{(2)}_{e_0 Mod, 2nd}$ (which contains terms in $e_0{}^2$).	From (X), all 2nd order amplitudes will have factors of $$e_0^2 Z_\gamma^{2nd}\left(Z_f^{2nd}\right)^2\left(Z_V^{2nd}\right)^2$$ which from last row of column (IX) equals $e_0^2 Z_\gamma^{2nd}$. Finding $B=L$ lets us cancel fermion and vertex infinite contributions.	Define e as finite physical charge. To 2nd order $$e^2 = e_0^2 Z_\gamma^{2nd}$$ $$= \underset{0}{e_0^2}\underbrace{\left(1 - e_0^2 A'\right)}_{\infty}$$ or $$e = e_0\left(Z_\gamma^{2nd}\right)^{1/2}$$ $$= e_0\left(1 - \tfrac{1}{2}e_0^2 A'\right)$$ Take $e_0 \to e$ in tree level amplitudes	$$iD^{\overset{mod}{2nd}}_{F\alpha\beta}(k) = \frac{-ig_{\alpha\beta}}{k^2+i\varepsilon}(1 - \underbrace{e_0^2\Pi_c}_{\to e^2\Pi_c} + h.o.)$$ $$iS^{\overset{mod}{2nd}}_F(p) = \frac{i}{\not{p}-m+i\varepsilon}(1-\underbrace{e_0^2\Sigma_c}_{\to e^2\Sigma_c}+h.o.)$$ $$u_r(\mathbf{p}), \bar{u}_r(\mathbf{p}), v_r(\mathbf{p}), \bar{v}_r(\mathbf{p})$$ $$\varepsilon_\mu(\mathbf{k})$$ Same as bare $$\gamma^\mu_{\underset{2nd}{mod}}(p,p') = \gamma^\mu + \underbrace{e_0^2\Lambda_c^\mu}_{\to e^2\Lambda_c^\mu} + h.o.$$

Wholeness Chart 14-5 (continued). Renormalization Steps to nth Order

(X) Amplitude Determination	(XI) Cancellations	(XII) e Renormalization	(XIII) Substitute in Tree Amplits
Adding 1st & reducible up to nth order amplitudes [or alternatively, using (VI) in tree level amplitude], we get infinite term of form $$\mathcal{M}^{(2)} + \sum_{i=way}\sum_{k=2}^{n} \mathcal{M}_i^{(2k)} \approx$$ $$Z_\gamma^{nth}\left(Z_f^{nth}\right)^2\left(Z_V^{nth}\right)^2 \mathcal{M}^{(2n)}_{e_0 Mod, nth}$$ where the entities of (XIII) are used in $\mathcal{M}^{(2n)}_{e_0 Mod, nth}$ (which contains terms in $e_0{}^{2n}$).	From (X), all amplitudes to nth order will have factors of $$e_0^2 Z_\gamma^{nth}\left(Z_f^{nth}\right)^2\left(Z_V^{nth}\right)^2$$ which from last row of column (IX) equals $e_0^2 Z_\gamma^{nth}$. Knowing $B_{nth}=L_{nth}$ lets us cancel fermion and vertex infinite contributions.	Define e as finite physical charge. To nth order $$e^2 = e_0^2 Z_\gamma^{nth}$$ $$= \underset{0}{e_0^2}\frac{1}{1 + e_0^2 \underbrace{A'_{nth}}_{\infty}}$$ or $$e = e_0\left(Z_\gamma^{nth}\right)^{1/2}$$ $$= e_0\frac{1}{\left(1 + e_0^2 A'_{nth}\right)^{1/2}}$$ Take $e_0 \to e$ in tree level amplitudes	$$iD^{\overset{Mod}{nth}}_{F\alpha\beta}(k) = \frac{-ig_{\alpha\beta}}{k^2+i\varepsilon}\frac{1}{1 + \underbrace{e_0^2\Pi_{c\,nth}}_{\to e^2\Pi_{c\,nth}} + h.o.}$$ $$iS^{\overset{Mod}{nth}}_F(p) = \frac{i}{\not{p}-m+i\varepsilon}\frac{1}{1 + \underbrace{e_0^2\Sigma_{c\,nth}}_{\to e^2\Sigma_{c\,nth}} + h.o.}$$ $$u_r(\mathbf{p}), \bar{u}_r(\mathbf{p}), v_r(\mathbf{p}), \bar{v}_r(\mathbf{p})$$ $$\varepsilon_\mu(\mathbf{k})$$ Same as bare $$\gamma'^\mu_{\underset{nth}{Mod}}(p,p') = \gamma^\mu + \underbrace{e_0^2\Lambda_{c\,nth}^\mu}_{\to e^2\Lambda_{c\,nth}^\mu} + h.o.$$

14.10 Appendix: Showing $k^\mu k^\nu B_{nth}$ Term Drops Out

To help show that the last term in (14-56) (repeated below),

$$\Pi_{nth}^{\mu\nu}(k) = -g^{\mu\nu} A_{nth}(k^2) + k^\mu k^\nu B_{nth}(k^2),$$ (14-94)

drops out, we start with the example of Møller's scattering, first way, as shown in the LHS of Fig. 14-3, pg. 347. We will work with the 2nd order case, as the nth order case is a simple extrapolation from that. The tree level amplitude is

$$\mathcal{M}_{B3}^{(2)} = e_0^2 \bar{u}_{r_1'}(\mathbf{p}_1')\gamma^\alpha u_{r_1}(\mathbf{p}_1) iD_{F\alpha\beta}(k)\bar{u}_{r_2'}(\mathbf{p}_2')\gamma^\beta u_{r_2}(\mathbf{p}_2).$$ (14-95)

To get the amplitude to 2nd order, we can substitute the 2nd order modifications of Wholeness Chart 14-4, column (II) into (14-95), yielding

$$\mathcal{M}_{B3}^{(2)} + \sum_i^{i=7} \mathcal{M}_{B3-i}^{(4)} = e_0^2 \bar{u}_{r_1'}(\mathbf{p}_1')\gamma_{2nd}^\alpha u_{r_1}(\mathbf{p}_1) \underbrace{iD_{F\alpha\beta}^{2nd}(k)}\bar{u}_{r_2'}(\mathbf{p}_2')\gamma_{2nd}^\beta u_{r_2}(\mathbf{p}_2)$$

$$= e_0^2 \bar{u}_{r_1'}(\mathbf{p}_1')\gamma_{2nd}^\alpha u_{r_1}(\mathbf{p}_1)\overbrace{\left(iD_{F\alpha\beta}(k) + i\underbrace{D_{F\alpha\mu}(k)}_{\frac{-g_{\alpha\mu}}{k^2+i\varepsilon}}ie_0^2\Pi^{\mu\nu}(k) i\underbrace{D_{F\nu\beta}(k)}_{\frac{-g_{\nu\beta}}{k^2+i\varepsilon}}\right)}\bar{u}_{r_2'}(\mathbf{p}_2')\gamma_{2nd}^\beta u_{r_2}(\mathbf{p}_2). \quad (14\text{-}96)$$

$$= e_0^2 \bar{u}_{r_1'}(\mathbf{p}_1')\gamma_{2nd}^\alpha u_{r_1}(\mathbf{p}_1)\left(iD_{F\alpha\beta}(k) - \frac{1}{(k^2+i\varepsilon)^2}ie_0^2\Pi_{\alpha\beta}(k)\right)\bar{u}_{r_2'}(\mathbf{p}_2')\gamma_{2nd}^\beta u_{r_2}(\mathbf{p}_2).$$

Using (14-94) in the last line of the above, we find terms of form

$$-e_0^2 \bar{u}_{r_1'}(\mathbf{p}_1')\gamma_{2nd}^\alpha u_{r_1}(\mathbf{p}_1)\frac{1}{(k^2+i\varepsilon)^2}ie_0^2 k_\alpha k_\beta B_{nth}\bar{u}_{r_2'}(\mathbf{p}_2')\gamma_{2nd}^\beta u_{r_2}(\mathbf{p}_2).$$ (14-97)

For the vertex with p_2 occurring after the vertex with p_1,

$$k_\beta = p_{2\beta}' - p_{2\beta}.$$ (14-98)

Now, every term in our amplitude corresponds to a term in the interaction Hamiltonian (or equivalently, the Lagrangian). But while (14-97) is expressed in 4-momentum space, the corresponding Hamiltonian term from which it is derived is expressed in 4D spacetime as the central portion of the amplitude (expressed in 4D spacetime),

$$\left\langle e_{r_1',\mathbf{p}_1'}, e_{r_2',\mathbf{p}_2'}\left|e_0^2\ \bar{\psi}(x_1)\gamma_{2nd}^\alpha\psi(x_1)\frac{1}{(k^2+i\varepsilon)^2}ie_0^2 k_\alpha k_\beta B_{nth}\underbrace{\bar{\psi}(x_2)\gamma_{2nd}^\beta\psi(x_2)}_{j_{2nd}^\beta(x_2)}\right|e_{r_1,\mathbf{p}_1}, e_{r_2,\mathbf{p}_2}\right\rangle,$$ (14-99)

where the current shown is the 2nd order equivalent of what we have seen before at 1st order. Given (14-98), we have a factor in (14-99) of

$$k_\beta j_{2nd}^\beta = (p_{2\beta}' - p_{2\beta}) j_{2nd}^\beta.$$ (14-100)

Now, current conservation holds at every order, so for the case being considered

$$\partial_\beta j_{2nd}^\beta = 0 = \partial_\beta\left(\bar{\psi}\ \gamma_{2nd}^\beta\psi\right) = \partial_\beta\left(c_r^\dagger(\mathbf{p}_2')\bar{u}_r(\mathbf{p}_2')e^{ip_2'x_2}\gamma_{2nd}^\beta c_r(\mathbf{p}_2)u_r(\mathbf{p}_2)e^{-ip_2 x_2}\right)$$

$$= i(p_{2\beta}' - p_{2\beta}) j_{2nd}^\beta = ik_\beta j_{2nd}^\beta = 0.$$ (14-101)

Thus, (14-100) is zero, and hence the last term in (14-94) (for $n = 2$) will net zero in our amplitude and can be ignored.

By continual trial and error for other cases, you can eventually see that every factor of (14-94) is contracted with a 4-current. You can also see that in every case, the conservation of current requirement leads to a result parallel to (14-101). That is, one of the factors in $k^\mu k^\nu$ (or equivalently, $k_\mu k_\nu$) is contracted with a current and that contraction equals zero. Hence, we can drop the term with B_{2nd} in it (and by extrapolation B_{nth} for higher order situations).

14.11 Problems

1. Pick several diagrams from the third row of Fig. 14-1 and show how they can be reduced to the irreducible diagram which is the second diagram of the first row.

2. Similar to what we did in (14-1) and (14-2) for the first diagram in the last row of Fig. 14-1, use Feynman rules to find the Feynman amplitude for the second diagram in the last row of the same figure, then show that amplitude is finite.

3. Show, for the second way for Compton scattering to occur, that to second order in α,

$$\mathcal{M}_{C2}^{(2)} + \sum_{i}^{i=7} \mathcal{M}_{C2-i}^{(4)} = Z_{\gamma}^{2nd} \left. \mathcal{M}_{C2}^{(2)} \right|_{e_0 \, Mod, 2nd} .$$

4. Show that for Bhabha scattering $\mathcal{M}_{B1}^{(2)} + \sum_{i}^{i=7} \mathcal{M}_{B1-i}^{(4)} \approx Z_{\gamma}^{2nd} \left. \mathcal{M}_{B1}^{(2)} \right|_{e_0 \, Mod, 2nd}$, by considering that

    we will get analogous results to (14-24) and Prob. 3 above. That is, don't go through all the steps of Prob. 3 but jump to a relation similar to (14-24) for Bhabha scattering. Use the relations of (14-24) and (14-28) for Z factors, propagators, and vertices in the tree level amplitude.

5. Sketch the remaining next higher order correction to the photon propagator not shown in Fig. 14-4.

6. Is a tree level graph (Feynman diagram) a skeleton graph? For each of Bhabha, Compton, and Møller scattering, draw a skeleton diagram that is not a tree level diagram. Are these diagrams reducible?

7. Show that the amplitude for the first way of Møller scattering to nth order is (14-81).

8. Show that in the Bhabha first way amplitude calculation to nth order, the fermion and vertex factors in the divergent terms cancel and leave the photon factor Z_{γ}^{nth} as the only divergent quantity. Show also that this amplitude has nth order modified relations of Wholeness Chart 14-5, column (XIII) for propagators and vertices in place of the propagator and vertex relations of the tree level amplitude.

Chapter 15

Regularization

*"Let Hercules himself do what he may,
the cat will mew and dog will have its day."*

Hamlet

15.0 Preliminaries

15.0.1 Background

The background for this chapter is the previous three chapters, where we renormalized QED by assuming we had integral tables that would tell us directly what our higher order correction loop integrals were in terms of a parameter Λ, which would ultimately go to infinity. To justify the assumption of those "integral tables", we now embark on regularization, the formal process of evaluating unbounded integrals in terms of the parameter Λ.

15.0.2 Ways to Regularization

There are a number of ways to regularize unbounded integrals. The four most common are

Four of the ways to regularize

- Cut off regularization

    Instead of integrating from $-\infty$ to $+\infty$, we integrate from $-\Lambda$ to $+\Lambda$. When we renormalize using the resulting relation, we take $\Lambda \to \infty$.

    Cut-off integrates to Λ, instead of ∞

- Pauli-Villars regularization

    We add in to QED an additional fictitious particle with mass Λ. So in the propagator for this particle, $m^2 = \Lambda^2$ appears as a term in the denominator (as mass does for all propagators). This adds an additional term to the Feynman amplitude (think of an extra Feynman diagram having this extra virtual particle). As it turns out, this causes the amplitude to converge over the $-\infty$ to $+\infty$ integration range. The result is in terms of Λ, but turns out to be divergent as $\Lambda \to \infty$. But we can use that result to renormalize. When we take $\Lambda \to \infty$ in the renormalization process, that means our fictitious particle has infinite mass, so it really never shows up anywhere in creation and drops out of the theory. Mathematically, the propagator denominator goes to infinity, so the term with that propagator makes zero contribution to the amplitude, at the end of the day.

    Pauli-Villars temporarily introduces fictitious particle of mass Λ

- Dimensional regularization

    Our unbounded loop integrals are over four dimensional spacetime. It turns out that for dimensions D other than $D = 4$ for spacetime, these integrals can be evaluated readily. So we take the same integrals over $D = 4 - \eta$, where $\eta \neq 0$. In the result, we get terms that are unbounded as $\eta \to 0$ (which corresponds to $\Lambda \to \infty$). We use these terms for renormalization in the same way as we do the results from any other method of regularization. Interestingly, η (and thus D) does not have to be an integer in this method. That is, we can have fractional dimensions. This seems weird, but mathematically it works.

    Dimensional method integrates over $D \neq 4$ dimensions

- Gauge lattice regularization (Wilson)

    This approach approximates continuous spacetime by breaking it into a lattice comprising a large number of small hyper-cubes (4D "cubes") of fixed grid size (cube

    Gauge lattice breaks spacetime into a 4D grid

edge). Thus, fields with wavelengths approaching, and smaller than, the cube grid length cannot be represented since the fields are approximated by their values at the boundaries of the cubes. This means particle 3-momenta and energies for shorter wavelengths are excluded. The result, for any given grid size, is a finite Feynman amplitude. After doing calculations on lattices with several different size grids, one can extrapolate to zero grid length, i.e., our natural universe with virtually unbounded possible particle energy and 3-momentum. The gauge lattice approach is an advanced topic and will not be covered in this book.

Ideally, every regularization method should give us the same result. More on this later.

15.0.3 Chapter Overview

In this chapter, we will cover

- some standard integrals for 4D spacetime,

- the Wick rotation transformation used to derive the above integrals from general math relations for D dimensional Euclidean spaces,

We'll look at all but gauge lattice approach

- Feynman parameterization, a trick used in evaluating loop integrals,

- finding the photon loop integral $\Pi^{\mu\nu}(k)$ via cut off regularization,

- comparing simple non-physical examples of Pauli-Villars and dimensional regularization,

- finding the photon loop integral $\Pi^{\mu\nu}(k)$ via dimensional regularization,

- finding the vertex loop correction integral $\Lambda^{\mu}(p',p)$ via dimensional regularization, and

- outlining the procedure to find the fermion loop integral $\Sigma(p)$ via dimensional regularization.

15.1 Relations We'll Need

15.1.1 Some Standard Integrals

The integrals shown in this section will be useful. It may be good to recall what the integral signs therein mean, i.e.,

$$\int f(p^{\mu})d^4p = \int_{-\Lambda}^{\Lambda}\int_{-\Lambda}^{\Lambda}\int_{-\Lambda}^{\Lambda}\int_{-\Lambda}^{\Lambda} f(p^{\mu})d^4p \qquad \Lambda \to \infty . \tag{15-1}$$

Recall also that for the Γ function used below, for the special case where n is an integer,

$$\Gamma(n) = (n-1)! \quad \text{from which it follows} \to \quad n\Gamma(n) = n(n-1)! = n! = \Gamma(n+1) \quad . \tag{15-2}$$

Gamma function for integers

The following integrals can simply be accepted, as we commonly do for any problem for which we use integral tables. However, in the following section, we derive one of these, i.e., (15-4), to illustrate a procedure called Wick rotation that plays a vital role in regularization. In the appendix we make some general comments related to the derivations of the other relations, for those who may be interested.

$$\int \frac{d^4p}{\left(p^2+s+i\varepsilon\right)} = i\pi^2\left(s\,ln(-s) - s\,ln\left(\Lambda^2-s\right) - \Lambda^2\right) \tag{15-3}$$

Useful integrals over spacetime

$$\int \frac{d^4p}{\left(p^2+s+i\varepsilon\right)^2} = -i\pi^2\left(ln(-s) - ln\left(\Lambda^2-s\right) + \frac{s}{\Lambda^2-s} + 1\right) \tag{15-4}$$

$$\int \frac{d^4p}{\left(p^2+s+i\varepsilon\right)^n} = i\pi^2\frac{\Gamma(n-2)}{\Gamma(n)}\frac{1}{s^{n-2}} \qquad n \geq 3 \tag{15-5}$$

$$\int \frac{p^{\mu}}{\left(p^2+s+i\varepsilon\right)^n}d^4p = 0 \qquad n \geq 3 \tag{15-6}$$

$$\int \frac{p^\mu p^\nu}{\left(p^2 + s + i\varepsilon\right)^n} d^4 p = i\pi^2 \frac{\Gamma(n-3)}{2\Gamma(n)} \frac{g^{\mu\nu}}{s^{n-3}} \qquad n \geq 4 \tag{15-7}$$

$$\int \frac{d^4 p}{\left(p^2 + 2pq + t + i\varepsilon\right)^n} = i\pi^2 \frac{\Gamma(n-2)}{\Gamma(n)} \frac{1}{\left(t - q^2\right)^{n-2}} \qquad n \geq 3 \tag{15-8}$$

$$\int \frac{p^\mu}{\left(p^2 + 2pq + t + i\varepsilon\right)^n} d^4 p = -i\pi^2 \frac{\Gamma(n-2)}{\Gamma(n)} \frac{q^\mu}{\left(t - q^2\right)^{n-2}} \qquad n \geq 3 \tag{15-9}$$

$$\int \frac{p^\mu p^\nu}{\left(p^2 + 2pq + t + i\varepsilon\right)^n} d^4 p = i\pi^2 \frac{\Gamma(n-3)}{2\Gamma(n)} \frac{\left(2(n-3)q^\mu q^\nu + \left(t - q^2\right)g^{\mu\nu}\right)}{\left(t - q^2\right)^{n-2}} \qquad n \geq 4 \tag{15-10}$$

$$\int p^\mu p^\nu d^4 p = \tfrac{1}{4} \int g^{\mu\nu} p^2 d^4 p \tag{15-11}$$

$$\int \frac{p^\mu p^\nu}{p^2 + s} d^4 p = \tfrac{1}{4} \int g^{\mu\nu} \frac{p^2}{p^2 + s} d^4 p \tag{15-12}$$

15.1.2 Deriving Spacetime Integrals Using Wick Rotation

The Issue

How above integrals derived

An integral in 4D Euclidean space is "relatively" easy to evaluate since we can convert from 4D Cartesian form of differential element $d^4 x = dwdxdydz$ to 4D spherical coordinates of differential element $dV_{4D} = d^4 r = 2\pi^2 r^3 dr$, where $2\pi^2 r^3$ is the 3D "surface" of a 4D hypersphere. (See Chap. 9, pg. 260.) Additionally, our radial distance is $r = \sqrt{w^2 + x^2 + y^2 + z^2}$ and that is simply the measured distance from the origin to a point.

Euclidean space of D dimension has simple interpretation of distance and integral

In 4D spacetime, however, things are not so simple because $r = \sqrt{t^2 - x^2 - y^2 - z^2}$ and there is no simple interpretation of that as a distance in 4D space. Further, defining a suitable differential element, due to the minus signs in the metric, is problematic.

Not so simple in spacetime because one dimension is time

To evaluate 4D spacetime integrals we use a trick called <u>Wick rotation</u>.

The Solution: Wick Rotation

Wick rotation converts spacetime to Euclidean space

Wick rotation is used to convert 4D spacetime to an associated 4D Euclidean space, in which a given integral is easier to evaluate. For it, we simply transform (or, equivalently, make a substitution of variables in the spacetime integral of) $E \rightarrow iE$.

That is, we multiply the $\mu = 0$ component (energy in this case) by i, leave the $\mu = 1,2,3$ components unchanged, and label the new coordinates with a subscript E (for Euclidean). Multiplication by i in the complex plane comprises a rotation by 90°. Hence the "rotation" part of the name for the transformation.

Wick rotation: $E \rightarrow iE$

The differential element $d^4 p$ has a factor of dE in it, so it transforms as shown in Wholeness Chart 15-1 below.

Note, for the last line in the chart if $p_E^2 = E^2 + \mathbf{p}^2 = E^2 + \left(p^1\right)^2 + \left(p^2\right)^2 + \left(p^3\right)^2$, then it is essentially the length squared in Euclidean energy-momentum space. And thus, the transformation is

Wick rotation, Euclidean space notation $p^\mu \rightarrow p_E^\mu$

$$p^2 = p^\mu p_\mu = E^2 - \sum\left(p^i\right)^2 = E^2 - \mathbf{p}^2 \xrightarrow[\text{Rotation}]{\text{Wick}} -E^2 - \mathbf{p}^2 = p_E^\mu p_{E\,\mu} = -p_E^2. \tag{15-13}$$

Wholeness Chart 15-1. Wick Rotation Summary

	Minkowski		**Euclidean**
Time component	E	Wick transform→	iE
Space component	p^i (or **p**)	Wick transform→	p^i (or **p**)
Differential element	d^4p	Wick transform→	id^4p_E
4D vector definition	$p^\mu = (E,\, \mathbf{p})$		$p_E{}^\mu = (E,\, \mathbf{p})$
4D vector transformation	$(E,\, \mathbf{p})$	Wick transform→	$(iE,\, \mathbf{p})$
Square of 4D vector definition	$p^2 = p^\mu p_\mu = E^2 - \mathbf{p}^2$		$p_E{}^2 = p_E{}^\mu p_{E\mu} = E^2 + \mathbf{p}^2$
Square of 4D vector, transformation	p^2	Wick transform→	$-p_E{}^2$

Getting the Integrals of Sect. 15.1.1

We use Wick rotation to determine the integrals (15-3) to (15-12). To illustrate, we will determine (15-4). We choose that, the second integral relation shown, instead of (15-3), because it is harder, and because I want to leave the easier of the two as a problem for you to do. (See Prob. 1.)

Use Wick rotation to convert D dimension Euclidean space integrals to D dimension spacetime

Integral (15-4)

First convert our integral, via the Wick transformation, to Euclidean coordinates (with imaginary quantities in places). We can ignore the $i\varepsilon$, or simply think of it as included temporarily in the constant s. In the last step on the RHS of (15-14) below, we convert from Cartesian 4D Euclidean coordinates to hyper-spherical 4D Euclidean coordinates, as discussed above.

An example

$$I = \int \frac{d^4p}{\left(p^2 + s\right)^2} = \int \frac{id^4p_E}{\left(-p_E^2 + s\right)^2} = i\int \frac{d^4p_E}{\left(p_E^2 - s\right)^2} = i2\pi^2 \int_0^\Lambda \frac{p_E^3\, dp_E}{\left(p_E^2 - s\right)^2}\ . \tag{15-14}$$

We then use the following relation from integral tables (or from manipulating the RHS integrands).

$$\int \frac{x^m dx}{\left(ax^n + c\right)^r} = \frac{1}{a}\int \frac{x^{m-n} dx}{\left(ax^n + c\right)^{r-1}} - \frac{c}{a}\int \frac{x^{m-n} dx}{\left(ax^n + c\right)^r}\,, \tag{15-15}$$

where $x = p_E$, $a = 1$, $c = -s$, $n = 2$, $m = 3$, and $r = 2$. With (15-15), (15-14) becomes

$$I = i2\pi^2 \underbrace{\int_0^\Lambda \frac{p_E\, dp_E}{\left(p_E^2 - s\right)}}_{\frac{1}{2} ln\left(p_E^2 - s\right)\big|_0^\Lambda} - \underbrace{(-s)i2\pi^2 \int_0^\Lambda \frac{p_E\, dp_E}{\left(p_E^2 - s\right)^2}}_{\text{Call this } I'}\ . \tag{15-16}$$

From integral tables, we find
$$\int x\left(ax^2 + c\right)^n dx = \frac{1}{2a}\frac{\left(ax^2 + c\right)^{n+1}}{n+1}\ . \tag{15-17}$$

Taking $x = p_E$, $a = 1$, $c = -s$, $n = -2$, the last part of (15-16) becomes

$$I' = i2\pi^2 \left(\frac{s}{2}\right)\frac{\left(p_E^2 - s\right)^{-1}}{-1}\Bigg|_0^\Lambda = -i\pi^2 s\, \frac{1}{\left(p_E^2 - s\right)}\Bigg|_0^\Lambda\ . \tag{15-18}$$

(15-16) is then equal to (15-4), i.e.,

$$I = i\pi^2\left[ln\left(p_E^2 - s\right) - s\frac{1}{\left(p_E^2 - s\right)}\right]_0^\Lambda = i\pi^2\left(ln\left(\Lambda^2 - s\right) - ln(-s) - \frac{s}{\Lambda^2 - s} - 1\right)$$

$$= -i\pi^2\left(ln(-s) - ln\left(\Lambda^2 - s\right) + \frac{s}{\Lambda^2 - s} + 1\right). \qquad \text{QED.} \tag{15-19}$$

For other integrals, see Prob. 1 and the appendix of this chapter.

15.1.3 Some Gamma Matrix Relations from Chapter 4

For convenience, we repeat some relations for gamma matrices as shown in Appendix A of Chap. 4.

$$\mathrm{Tr}\left(\gamma^{\alpha}\gamma^{\beta}\right) = 4g^{\alpha\beta} \tag{15-20}$$

$$\mathrm{Tr}\left(\gamma^{\sigma}\gamma^{\delta}\gamma^{\mu}...\right) = 0 \quad \text{for any odd number of gamma matrices} \tag{15-21}$$

$$\mathrm{Tr}\left(\gamma^{\alpha}\gamma^{\beta}\gamma^{\gamma}\gamma^{\rho}\right) = 4\left(g^{\alpha\beta}g^{\gamma\rho} - g^{\alpha\gamma}g^{\beta\rho} + g^{\alpha\rho}g^{\beta\gamma}\right) \tag{15-22}$$

or as we will see the indices later $\mathrm{Tr}\left(\gamma^{\sigma}\gamma^{\delta}\gamma^{\mu}\gamma^{\beta}\right) = 4\left(g^{\sigma\delta}g^{\mu\beta} - g^{\sigma\mu}g^{\delta\beta} + g^{\sigma\beta}g^{\delta\mu}\right)$.

$$
\begin{aligned}
\gamma_{\lambda}\gamma^{\lambda} &= 4, & \gamma_{\lambda}\gamma^{\alpha}\gamma^{\lambda} &= -2\gamma^{\alpha} \\
\gamma_{\lambda}\gamma^{\alpha}\gamma^{\beta}\gamma^{\lambda} &= 4g^{\alpha\beta} & \gamma_{\lambda}\gamma^{\alpha}\gamma^{\beta}\gamma^{\gamma}\gamma^{\lambda} &= -2\gamma^{\gamma}\gamma^{\beta}\gamma^{\alpha} \\
\gamma_{\lambda}\gamma^{\alpha}\gamma^{\beta}\gamma^{\gamma}\gamma^{\delta}\gamma^{\lambda} &= 2\left(\gamma^{\delta}\gamma^{\alpha}\gamma^{\beta}\gamma^{\gamma} + \gamma^{\gamma}\gamma^{\beta}\gamma^{\alpha}\gamma^{\delta}\right).
\end{aligned}
\tag{15-23}
$$

Restating some gamma matrices relations from Chap. 4 that we'll use in this chapter

15.1.4 Feynman Parameterization

Integrals (15-5) to (15-10) are actually not of the same form as the loop integrals and so, are not directly suitable to our purposes. But, thanks to a technique developed by Feynman, we can convert the loop integrals to forms for which we can use (15-5) to (15-10). The loop integrals typically have a product of several different polynomials multiplied in the denominator rather than a single such polynomial (raised to a power typically), as in the integrals (15-5) to (15-10).

A technique to convert loop integrals into format of standard integrals

The first of these useful relations is

$$\frac{1}{ab} = \frac{1}{b-a}\int_{a}^{b}\frac{dt}{t^2} \qquad = \frac{1}{b-a}\left(\frac{1}{a} - \frac{1}{b}\right) = \frac{1}{b-a}\left(\frac{b-a}{ab}\right) = \frac{1}{ab}. \tag{15-24}$$

Define the <u>Feynman parameter z</u> via

$$t = b + (a-b)z \quad \rightarrow \quad dt = (a-b)dz \tag{15-25}$$

where for the integration limits in (15-24), $t = a$ means $z = 1$, and $t = b$ means $z = 0$. The LHS of (15-24) then becomes (where the RHS of (15-26) follows from the symmetry of a and b on the LHS)

Feynman parameter helps us convert a product in a denominator to a form that will make our job easier

$$\frac{1}{ab} = -\int_{1}^{0}\frac{dz}{\left(b+(a-b)z\right)^2} = \int_{0}^{1}\frac{dz}{\left(b+(a-b)z\right)^2} = \int_{0}^{1}\frac{dz}{\left(a+(b-a)z\right)^2}. \tag{15-26}$$

You may be thinking that the RHS of (15-26) is an unwieldy way to express the LHS, and you would be right, but (15-26) will help us in what is to come.

Relation (15-26) readily extends to three factors (see Prob. 2), where x, y, and z are Feynman parameters like z above, i.e., they are dummy integration variables,

Similar relation for product of three factors in denominator

$$\frac{1}{abc} = 2\int_{0}^{1}dx\int_{0}^{x}dy\frac{1}{\left(a+(b-a)x+(c-b)y\right)^3} \qquad \text{(a)}$$

$$= 2\int_{0}^{1}dx\int_{0}^{1-x}dz\frac{1}{\left(a+(b-a)x+(c-a)z\right)^3} \qquad \text{(b)}.$$

$$\tag{15-27}$$

These results can be generalized, via induction, to

$$\frac{1}{a_0 a_1 a_2 ... a_n} = \Gamma(n+1)\int_{0}^{1}dz_1\int_{0}^{z_1}dz_2....\int_{0}^{z_{n-1}}dz_n\frac{1}{\left(a_0 + (a_1 - a_0)z_1 + ...(a_n - a_{n-1})z_n\right)^{n+1}}. \tag{15-28}$$

Extrapolating to any number of factors in denominator

15.1.5 Leading Log Approximations

Note that for a function of ε, $f(\varepsilon) = ln\,(\Lambda' + \varepsilon)$ where $\varepsilon << \Lambda'$,

$$ln(\Lambda' + \varepsilon) = f(\varepsilon) = f(0) + \varepsilon \left(\frac{\partial f(\varepsilon)}{\partial \varepsilon} \right)_{\varepsilon=0} + \frac{\varepsilon^2}{2} \left(\frac{\partial^2 f(\varepsilon)}{\partial \varepsilon^2} \right)_{\varepsilon=0} + ...$$

$$= ln(\Lambda' + \varepsilon)\big|_{\varepsilon=0} + \frac{\varepsilon}{(\Lambda' + \varepsilon)_{\varepsilon=0}} - \frac{\varepsilon^2}{2(\Lambda' + \varepsilon)^2_{\varepsilon=0}} + ... = ln\Lambda' + \frac{\varepsilon}{\Lambda'} - \frac{1}{2}\left(\frac{\varepsilon}{\Lambda'} \right)^2 + ... \quad (15\text{-}29)$$

$$\approx ln\Lambda' \quad \text{(leading log approximation)}.$$

Defining "leading log approximation"

15.2 Finding Photon Self Energy Factor Using the Cut-Off Method

Fig. 15-1 shows the tree level (first order in α) and the second order correction Feynman diagrams for the photon propagator.

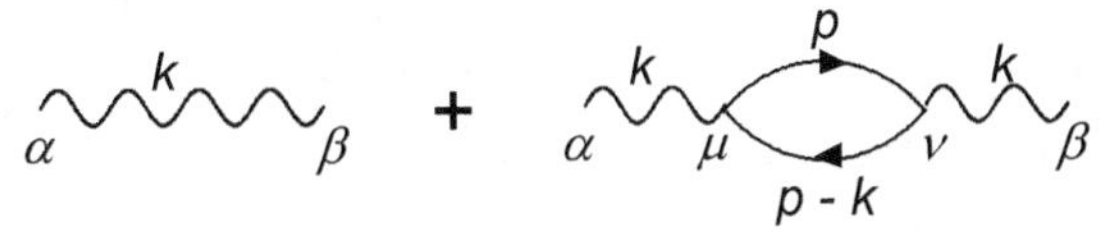

Finding 2nd order photon propagator via cut-off method

Figure 15-1. The Photon Propagator to Second Order

As we evaluate the higher order propagator, we will highlight, by providing an asterisk, the approximation steps whenever they occur.

* 1st Approximation: Use one loop (only evaluate the propagator to 2nd order in α).

The photon propagator is thus

2nd order means one loop added to 1st order propagator

$$iD^{2nd}_{F\alpha\beta}(k) = iD_{F\alpha\beta}(k) + iD_{F\alpha\mu}(k) \underbrace{\frac{(-1)}{(2\pi)^4} \left\{ \text{Tr} \int ie_0\gamma^\mu iS_F(p-k) ie_0 \gamma^\nu iS_F(p) d^4p \right\}}_{ie_0^2 \Pi^{\mu\nu}(k)} iD_{F\nu\beta}(k). \quad (15\text{-}30)$$

* 2nd Approximation: Ignore masses in propagators.

This approximation works because for the integration range where i) energy/momenta $\gg m$, mass is negligible; ii) energy/momenta $\ll m$, then mass in the fermion propagator in the denominator controls the integrand and makes the contribution small, so it can be ignored; iii) energy/momenta $\approx m$, the total range of integration for this is small compared to the entire integration range, so the contribution is negligible. These conclusions can be verified with a detailed analysis of the integrals, but we will not do that here.

We'll approximate and make calculation easier by ignoring masses

Ignoring masses, we can write (15-30) as

$$iD^{2nd}_{F\alpha\beta}(k) = iD_{F\alpha\beta}(k) + ie_0^2 \frac{-ig_{\alpha\mu}}{k^2 + i\varepsilon} \frac{-ig_{\nu\beta}}{k^2 + i\varepsilon} \underbrace{\left\{ \frac{(-1)}{(2\pi)^4} \text{Tr} \int \gamma^\mu i \frac{\not{p} - \not{k}}{(p-k)^2 + i\varepsilon} i\gamma^\nu i \frac{\not{p}}{p^2 + i\varepsilon} d^4p \right\}}_{\Pi^{\mu\nu}(k)}. \quad (15\text{-}31)$$

The 2nd order propagator showing what comprises $\Pi^{\mu\nu}(k)$

We need to evaluate $\Pi^{\mu\nu}(k)$. Strictly speaking, we should carry the $i\varepsilon$ terms through at every step and then take $\varepsilon \to 0$ at the very end, but for convenience, we will simply ignore such terms from here on. We do this because I know in advance that we will get the same answer either way, and hopefully, to save time, you can just take my word for it on this. Doing so, we have

$$\Pi^{\mu\nu}(k) = \frac{i}{(2\pi)^4} \left\{ \text{Tr} \int \frac{\gamma^\mu (\not{p} - \not{k})\gamma^\nu \not{p}}{(p-k)^2 p^2} d^4p \right\}. \quad (15\text{-}32)$$

Evaluating $\Pi^{\mu\nu}(k)$

The Denominator Re-arranged

We will re-express $(p-k)^2 p^2$ in the denominator of the integrand of (15-32) using (15-26) with $a = (p-k)^2$ and $b = p^2$. Thus,

Re-express the denominator using Feynman parameterization relations

$$\frac{1}{(p-k)^2 p^2} = \int_0^1 \frac{dz}{\left(p^2 + \left((p-k)^2 - p^2 \right)z \right)^2}. \quad (15\text{-}33)$$

We then re-express the squared part of the denominator of (15-33) as follows

$$p^2 + \left((p-k)^2 - p^2 \right) z = p^2 \underbrace{+zp^2}_{\text{cancels}} - 2zp^{\mu}k_{\mu} + zk^2 \underbrace{-p^2 z}_{\text{cancels}} = p^2 - 2zp^{\mu}k_{\mu} + zk^2 \tag{15-34}$$

$$= p^2 - 2zp^{\mu}k_{\mu} + z^2k^2 + zk^2 - z^2k^2 = \left(p - zk \right)^2 + z\left(1 - z \right)k^2 .$$

Then, with (15-34), (15-33) becomes

$$\frac{1}{\left(p-k \right)^2 p^2} = \int_0^1 \frac{dz}{\left(\left(p - zk \right)^2 + z\left(1 - z \right)k^2 \right)^2} . \tag{15-35}$$

So, (15-35) allows us to re-write (15-32) as

$$\Pi^{\mu\nu}\left(k \right) = \frac{i}{\left(2\pi \right)^4} \left\{ \text{Tr}\int \gamma^{\mu} \left(\not{p} - \not{k} \right) \gamma^{\nu} \not{p} \left(\int_0^1 \frac{dz}{\left(\left(p - zk \right)^2 + z\left(1 - z \right)k^2 \right)^2} \right) d^4 p \right\}$$

$$= \frac{i}{\left(2\pi \right)^4} \int_0^1 \left\{ \text{Tr}\int \frac{\gamma^{\mu} \left(\not{p} - \not{k} \right) \gamma^{\nu} \not{p}}{\left(\left(p - zk \right)^2 + z\left(1 - z \right)k^2 \right)^2} d^4 p \right\} dz . \tag{15-36}$$

Thus, we have re-expressed $\Pi^{\mu\nu}(k)$ so it is easier to evaluate

The Numerator and Denominator Re-arranged

We now use the substitution of variable technique[1] in (15-36) with $p \to p + zk$ (and thus $dp \to dp$) where we note the spacetime integral will still have limits of $\pm \infty$.

Use substitution of variable to re-express numerator and denominator

$$\Pi^{\mu\nu}\left(k \right) = \frac{i}{\left(2\pi \right)^4} \int_0^1 \left\{ \text{Tr}\int \frac{\gamma^{\mu} \left(\not{p} - \left(1 - z \right)\not{k} \right) \gamma^{\nu} \left(\not{p} + z\not{k} \right)}{\left(p^2 + z\left(1 - z \right)k^2 \right)^2} d^4 p \right\} dz . \tag{15-37}$$

The denominator of (15-37) is even in p, so all terms in the numerator that are odd in p will drop out. Thus, (15-37) reduces to

$$\Pi^{\mu\nu}\left(k \right) = \frac{i}{\left(2\pi \right)^4} \int_0^1 \left\{ \int \frac{\text{Tr}\left(\gamma^{\mu} \not{p} \gamma^{\nu} \not{p} \right) - z\left(1 - z \right)\text{Tr}\left(\gamma^{\mu} \not{k} \gamma^{\nu} \not{k} \right)}{\left(p^2 + z\left(1 - z \right)k^2 \right)^2} d^4 p \right\} dz . \tag{15-38}$$

Break into Two Integrals

Label the two integrals with respect to p in (15-38) as

Two terms in numerator → evaluate as two separate integrals

$$I_1^{\mu\nu} = \frac{i}{\left(2\pi \right)^4} \int_0^1 \left\{ \int \frac{\text{Tr}\left(\gamma^{\mu} \not{p} \gamma^{\nu} \not{p} \right)}{\left(p^2 + z\left(1 - z \right)k^2 \right)^2} d^4 p \right\} dz$$

$$I_2^{\mu\nu} = \frac{-i}{\left(2\pi \right)^4} \int_0^1 \left\{ \int \frac{z\left(1 - z \right)\text{Tr}\left(\gamma^{\mu} \not{k} \gamma^{\nu} \not{k} \right)}{\left(p^2 + z\left(1 - z \right)k^2 \right)^2} d^4 p \right\} dz . \tag{15-39}$$

So

$$\Pi^{\mu\nu}\left(k \right) = I_1^{\mu\nu} + I_2^{\mu\nu} \tag{15-40}$$

For $I_1^{\mu\nu}$

We use (15-11) in the first row below to get the second row.

First integral

---

[1] This is safe to do here, but in some theoretically more advanced situations it can destroy certain symmetries of the theory and lead to what is known as an "anomaly".

$$I_1{}^{\mu\nu} = \frac{i}{(2\pi)^4}\int_0^1 \left\{\int \frac{\text{Tr}\left(p_\delta p_\sigma \gamma^\mu \gamma^\delta \gamma^\nu \gamma^\sigma\right)}{\left(p^2 + z(1-z)k^2\right)^2} d^4 p\right\} dz$$

$$= \frac{i}{(2\pi)^4}\int_0^1 \left\{\int \frac{\text{Tr}\left(\frac{p^2}{4} g_{\delta\sigma} \gamma^\mu \gamma^\delta \gamma^\nu \gamma^\sigma\right)}{\left(p^2 + z(1-z)k^2\right)^2} d^4 p\right\} dz.$$

(15-41)

Use standard integrals and gamma matrices to re-express it

Then using (15-23) and (15-20), this becomes

$$I_1{}^{\mu\nu} = \frac{i}{(2\pi)^4}\int_0^1 \left\{\int \frac{p^2}{4} \frac{\text{Tr}(\gamma^\mu \overbrace{\gamma_\sigma \gamma^\nu \gamma^\sigma}^{-2\gamma^\nu})}{\left(p^2 + z(1-z)k^2\right)^2} d^4 p\right\} dz$$

(15-42)

$$= \frac{i}{(2\pi)^4}\int_0^1 \left\{\int \frac{p^2}{4} \frac{-2\,\text{Tr}\left(\overbrace{\gamma^\mu \gamma^\nu}^{4g^{\mu\nu}}\right)}{\left(p^2 + z(1-z)k^2\right)^2} d^4 p\right\} dz.$$

This can be re-expressed as

$$I_1{}^{\mu\nu} = -\frac{i}{(2\pi)^4}\int_0^1 \left\{\int 2g^{\mu\nu} \frac{p^2}{\left(p^2 + z(1-z)k^2\right)^2} d^4 p\right\} dz$$

$$= -2g^{\mu\nu}\frac{i}{(2\pi)^4}\int_0^1 \left\{\int \frac{p^2 + \overbrace{z(1-z)k^2 - z(1-z)k^2}^{0}}{\left(p^2 + z(1-z)k^2\right)^2} d^4 p\right\} dz$$

Re-expressing it yet again by adding terms that equal zero

$$= -2g^{\mu\nu}\frac{i}{(2\pi)^4}\int_0^1 \left\{\int \left\{\frac{1}{\left(p^2 + z(1-z)k^2\right)} - \frac{z(1-z)k^2}{\left(p^2 + z(1-z)k^2\right)^2}\right\} d^4 p\right\} dz.$$ (15-43)

Using (15-3) for the first integral above and (15-4) for the second, with $s = z(1-z)k^2$, we have

$$I_1{}^{\mu\nu} = -2g^{\mu\nu}\frac{i}{(2\pi)^4}\left(i\pi^2\right)\times$$

$$\int_0^1 \left\{z(1-z)k^2\left(\ln(-z) + \ln(1-z) + \ln k^2\right) - z(1-z)k^2 \ln\left(\Lambda^2 - z(1-z)k^2\right) - \Lambda^2\right.$$

$$+ z(1-z)k^2\left(\qquad '' \qquad\right) - z(1-z)k^2 \ln\left(\qquad '' \qquad\right)$$

$$\left. + z(1-z)k^2 + \frac{\left(z(1-z)k^2\right)^2}{\Lambda^2 - z(1-z)k^2}\right\} dz.$$

(15-44)

Using standard spacetime integrals

* 3rd Approximation: In terms with logarithms, only retain leading logs.

We will only retain the leading log terms (see Sect. 15.1.15, pg. 378), with $\Lambda^2 \gg k^2$. As shown in the appendix, the first two terms in the integral of (15-44) and the two directly below them are of order 1, so relative to Λ^2 (and k^2, since we are assuming $k^2 \gg m^2$ where $m^2 \gg 1$), they can be ignored. Doing this, we get (15-45).

Approximate log terms by dropping all but leading log parts

$$I_1{}^{\mu\nu} \approx -2g^{\mu\nu}i\frac{i\pi^2}{(2\pi)^4}\int_0^1 \left\{2k^2z(1-z)\ln k^2 - 2k^2z(1-z)\ln\Lambda^2 - \Lambda^2 + z(1-z)k^2\right\}dz$$

$$= g^{\mu\nu}\frac{1}{8\pi^2}\int_0^1 \left\{2k^2z(1-z)\ln\frac{k^2}{\Lambda^2} - \Lambda^2 + z(1-z)k^2\right\}dz$$

$$= \frac{g^{\mu\nu}}{8\pi^2}\left\{\underbrace{\int_0^1 z(1-z)dz}_{1/6}\left(2k^2\ln\frac{k^2}{\Lambda^2}\right) - \underbrace{\int_0^1\Lambda^2 dz}_{\Lambda^2} + k^2\underbrace{\int_0^1 z(1-z)dz}_{k^2/6}\right\}$$

$$= \frac{g^{\mu\nu}k^2}{8\pi^2}\left\{\frac{1}{3}\ln\frac{k^2}{\Lambda^2} - \frac{\Lambda^2}{k^2} + \underbrace{\frac{1}{6}}_{\text{negligible}}\right\} \approx \frac{g^{\mu\nu}k^2}{24\pi^2}\ln\frac{k^2}{\Lambda^2} - \frac{g^{\mu\nu}}{8\pi^2}\Lambda^2.$$

(15-45)

Gives us the first integral of our 2nd order relation

For $I_2{}^{\mu\nu}$

$I_2{}^{\mu\nu}$ of (15-38) and (15-39) can be re-arranged slightly as

Now, the second integral

$$I_2{}^{\mu\nu} = -\frac{i}{(2\pi)^4}k_\delta k_\sigma\left(\text{Tr}\left(\gamma^\mu\gamma^\delta\gamma^\nu\gamma^\sigma\right)\right)\int_0^1 z(1-z)\left\{\int\frac{1}{\left(p^2 + z(1-z)k^2\right)^2}d^4p\right\}dz. \quad (15\text{-}46)$$

Using (15-22), this becomes

Re-arranging and using a standard trace relation for gamma matrices

$$I_2{}^{\mu\nu} = -\frac{i}{(2\pi)^4}k_\delta k_\sigma 4\left(g^{\mu\delta}g^{\nu\sigma} - g^{\mu\nu}g^{\delta\sigma} + g^{\mu\sigma}g^{\delta\nu}\right)\times$$

$$\int_0^1 z(1-z)\left\{\int\frac{1}{\left(p^2 + z(1-z)k^2\right)^2}d^4p\right\}dz$$

$$= -\frac{i}{4\pi^4}\underbrace{\left(k^\mu k^\nu - g^{\mu\nu}k^2 + k^\mu k^\nu\right)}_{2k^\mu k^\nu - g^{\mu\nu}k^2}\int_0^1 z(1-z)\left\{\int\frac{1}{\left(p^2 + z(1-z)k^2\right)^2}d^4p\right\}dz. \quad (15\text{-}47)$$

We evaluate the integral with respect to p in (15-47) using (15-4) with $s = z(1-z)k^2$ to get

Use standard spacetime integral to evaluate

$$I_2{}^{\mu\nu} = -\frac{i}{4\pi^4}\left(2k^\mu k^\nu - k^2 g^{\mu\nu}\right)\int_0^1 z(1-z)\left(-i\pi^2\right)\times$$

$$\left\{\left(\ln(-z) + \ln(1-z) + \ln k^2\right) - \ln\left(\Lambda^2 - z(1-z)k^2\right) + \frac{z(1-z)k^2}{\Lambda^2 - z(1-z)k^2} + 1\right\}dz. \quad (15\text{-}48)$$

* 3rd Approximation again: Only retain leading logs

In the Appendix, where we elaborated on going from (15-44) to (15-45), we showed the integrals of the first two terms summed in (15-48) [see (15-141) and (15-139)] equal $\frac{i\pi}{6} - \frac{5}{18}$, which is negligible for large Λ.

For Λ large with respect to other quantities, keeping leading logs, we find (15-48) becomes

Retain only leading logs

$$I_2{}^{\mu\nu} = -\frac{1}{4\pi^2}\left(2k^\mu k^\nu - k^2 g^{\mu\nu}\right)\left\{\ln k^2 - \ln\Lambda^2 \underbrace{+1}_{\text{negligible}}\right\}\underbrace{\int_0^1 z(1-z)\,dz}_{1/6}$$

(15-49)

$$\approx -\frac{1}{24\pi^2}\left(2k^\mu k^\nu - k^2 g^{\mu\nu}\right)\ln\frac{k^2}{\Lambda^2} = -\frac{1}{24\pi^2}2k^\mu k^\nu\ln\frac{k^2}{\Lambda^2} + \frac{1}{24\pi^2}k^2 g^{\mu\nu}\ln\frac{k^2}{\Lambda^2}.$$

$\Pi^{\mu\nu} = I_1{}^{\mu\nu} + I_2{}^{\mu\nu}$

Adding (15-45) to (15-49) yields $\Pi^{\mu\nu}$ of (15-31), (15-32), and (15-40).

Add two integrals above to get $\Pi^{\mu\nu}(k)$

$$\Pi^{\mu\nu}(k) = I_1^{\mu\nu} + I_2^{\mu\nu} \approx \frac{g^{\mu\nu}}{24\pi^2}k^2 \ln\frac{k^2}{\Lambda^2} - \frac{g^{\mu\nu}}{8\pi^2}\Lambda^2 - \frac{1}{12\pi^2}k^\mu k^\nu \ln\frac{k^2}{\Lambda^2} + \frac{g^{\mu\nu}}{24\pi^2}k^2 \ln\frac{k^2}{\Lambda^2}$$

$$= \frac{g^{\mu\nu}}{12\pi^2}k^2\, 2\ln\frac{k}{\Lambda} - \frac{g^{\mu\nu}}{8\pi^2}\Lambda^2 - \frac{1}{12\pi^2}k^\mu k^\nu\, 2\ln\frac{k}{\Lambda} \qquad (15\text{-}50)$$

$$= \frac{g^{\mu\nu}}{6\pi^2}k^2 \ln\frac{k}{\Lambda} - \frac{g^{\mu\nu}}{8\pi^2}\Lambda^2 - \frac{1}{6\pi^2}k^\mu k^\nu \ln\frac{k}{\Lambda}.$$

where we take $k = +\sqrt{k^2}$ above. From (15-30) and (15-31), we have

Use $\Pi^{\mu\nu}(k)$ to determine $D_{F\alpha\beta}^{2nd}(k)$

$$iD_{F\alpha\beta}^{2nd}(k) = iD_{F\alpha\beta}(k) + iD_{F\alpha\mu}(k)ie_0^2\Pi^{\mu\nu}(k)iD_{F\nu\beta}(k)$$

$$= \frac{-ig_{\alpha\beta}}{k^2} + \frac{(-ig_{\alpha\mu})}{k^2}ie_0^2\left(\frac{g^{\mu\nu}}{6\pi^2}k^2 \ln\frac{k}{\Lambda} - \frac{g^{\mu\nu}}{8\pi^2}\Lambda^2 - \frac{1}{6\pi^2}k^\mu k^\nu \ln\frac{k}{\Lambda}\right)\frac{(-ig_{\nu\beta})}{k^2}$$

$$= \frac{-ig_{\alpha\beta}}{k^2} - ie_0^2\frac{g_{\alpha\beta}}{6\pi^2 k^2}\ln\frac{k}{\Lambda} + ie_0^2\frac{g_{\alpha\beta}}{8\pi^2 k^2}\frac{\Lambda^2}{k^2} + ie_0^2\frac{1}{6\pi^2 k^2}\frac{k_\alpha k_\beta}{k^2}\ln\frac{k}{\Lambda} \qquad (15\text{-}51)$$

Note $k_\alpha k_\beta$ term drops out

$$= \frac{-ig_{\alpha\beta}}{k^2}\left(1 + e_0^2\frac{1}{6\pi^2}\ln\frac{k}{\Lambda} - e_0^2\frac{1}{8\pi^2}\frac{\Lambda^2}{k^2}\right) \;+\; \underbrace{ie_0^2\frac{1}{6\pi^2 k^2}\frac{k_\alpha k_\beta}{k^2}\ln\frac{k}{\Lambda}}_{\text{Drops out. See Chap. 14 Appendix}}.$$

As we showed in the appendix of Chap. 14, any term in the above with a k_α factor drops out due to current conservation (gauge invariance). This leaves us with

$$iD_{F\alpha\beta}^{2nd}(k) = \underbrace{\frac{-ig_{\alpha\beta}}{k^2}}_{iD_{F\alpha\beta}(k)}\left(1 + e_0^2\frac{1}{6\pi^2}\ln\frac{k}{\Lambda} - e_0^2\frac{1}{8\pi^2}\frac{\Lambda^2}{k^2}\right). \qquad (15\text{-}52)$$

Thus (15-50), ignoring the last term, which drops out, becomes

from cut-off regularization $\quad \Pi^{\mu\nu}(k) = \underbrace{g^{\mu\nu}k^2\frac{1}{6\pi^2}\ln\frac{k}{\Lambda}}_{\text{Expected from Chap. 13}} - \underbrace{g^{\mu\nu}\frac{\Lambda^2}{8\pi^2}}_{?}. \qquad (15\text{-}53)$

This leaves $\Pi^{\mu\nu}(k)$ from cut-off method

Compare with What We Said Before We Would Find

Recall from Chap. 13 (see (13-7) on pg. 324) we stated without proof that $\Pi^{\mu\nu}$ could be written (with $b_{n=1} = 1/(12\pi^2)$ for only one electron/positron loop as we have here) as

Compare with $\Pi^{\mu\nu}(k)$ from Chap. 13 that agrees with experiment

from Chap. 13 $\quad \Pi^{\mu\nu}(k) = \underbrace{g^{\mu\nu}k^2\, 2b_n \ln\frac{k}{\Lambda}}_{-A'(k,\Lambda)} + \underbrace{g^{\mu\nu}k^2\frac{1}{2\pi^2}\int_0^1 z(1-z)\ln\left(z(1-z) - m^2/k^2\right)dz}_{-\Pi_c(k^2)}. \quad (15\text{-}54)$

Issues with (15-52)

1) There is no finite term of form $\Pi_c(k^2)$ in (15-53) as we indicated in earlier chapters there was.

2) The last term in (15-53) diverges quadratically as $\Lambda \to \infty$, but we stated from the beginning that the divergence was logarithmic. Experimental dependence of renormalized e with energy supports logarithmic dependence on energy level.

They are different: Λ^2 divergence and no finite term

Answers to Issues

1) Recall that at (15-31) we approximated by ignoring masses in our fermion propagators. But $\Pi_c(k^2)$ in (15-54) has mass m in it. In the more exact (but far more complicated) treatment, the missing $\Pi_c(k^2)$ would arise.

Ignoring mass lost finite term

2) The divergence with Λ^2 can be rectified by modifying the Lagrangian and essentially re-developing our entire theory of QFT with that new Lagrangian. We won't get into that here, but this would entail adding an infinite term to our Lagrangian with the appropriate sign and factors such that after re-doing the theory and arriving at a modified form of (15-53), the added term and the Λ^2 term therein would cancel one another.

Λ^2 divergence corrected by adding term to $\mathcal{L}$

Problems with Answer 2) Above

The problems with adding this term to the Lagrangian are that

- it is exceptionally complicated, and
- by modifying the Lagrangian, we lose gauge invariance

Recall all that we went through in Chap. 11 to find gauge transformations of ψ and A^μ that keep the Lagrangian symmetric (gauge invariant). We had to add precisely the correct interaction terms to the Lagrangian to make it work. Any extra terms added would destroy the symmetry. (We showed an example of this in Sect. 11.5.4, pg. 296 where a non-zero photon mass term, i.e., an extra term in $\mathcal{L}$, destroyed the gauge symmetry.)

No longer having gauge invariance means, from Noether's theorem, that we no longer have charge conservation. And we can't use things like the Ward identities, which help us renormalize, because they are simply charge conservation in different form.

This correction complicated and destroys gauge invariance

But gauge invariance a valued principle underlying much physics

Further Problem with Cut-Off Method

Another problem with the cut-off approach is that it is not Lorentz invariant. Λ, the upper limit of our 4-momentum integral, is an energy, and what one measures for energy varies from frame to frame. So, simply from this perspective, we should not expect the cut-off method to be viable.

Cut-off method also not Lorentz invariant

Bottom Line

With great complication, QFT might be reformulated in the manner discussed above, but we would lose gauge invariance. And the method, in general, is not Lorentz invariant.

There is a better way. In fact, there are at least three better ways: Pauli-Villars, dimensional, and gauge lattice regularization methods. These all honor both gauge and Lorentz invariance.

Other regularization methods preserve gauge and Lorentz invariance and are preferred

15.3 Pauli-Villars Regularization
15.3.1 The Concept

Note that in what follows, we assume the fermion mass m_0 has already been renormalized to m. In other words, we have already broken the m_0 term in the Lagrangian into two terms in m and δm, where the latter is the mass counterterm. (See Chap. 12, Sect. 12.6.1, pg. 312 and Chap. 13, Sect. 13.4, pg. 330.) So, our fermion propagators use m, not m_0.

Imagine there were an extremely heavy fermion, identical to the electron, muon, and tau in all qualities except mass. In our low energy (far below the mass of this fermion) experiments it could never play a role as a real particle, since all of them would have decayed into lighter particles at the beginning of the universe, and we would not have enough energy in the experiment to create another one. And it could never influence particle collision interactions (such as Compton scattering) as a tree level virtual particle since those interactions would never reach the mass-energy level of this fermion. We would never know it exists.

Pauli-Villars adds fictitious particle of heavy mass Λ to theory

However, in our higher order correction loop integrals we integrate to infinite energy levels and it could play a role there. If so, the effect would be like adding another Feynman diagram loop with that particular particle in addition to the one for the electron (and muon and tau) coupled with its antiparticle. That is, if the heavy fermion mass were Λ, then all our propagators in our theory would have to be modified as in (15-55). But, in order for this to work, we have to change the propagator of the so-called heavy particle a bit. We give it a negative sign and keep the "m" in the numerator without changing it to "Λ", like we do in the denominator. This makes the whole procedure work mathematically (which is our goal), even though it tarnishes the analogy to a heavy particle propagator a bit. The goal mathematically is first to perturb the propagator by adding a term to it containing Λ, then take Λ to infinity. For this, the form chosen in (15-55) works best.

Has effect of adding another Feynman diagram with this virtual particle

$$\frac{\not{p}+m}{p^2-m^2+i\varepsilon} \to \frac{\not{p}+m}{p^2-m^2+i\varepsilon} - \frac{\not{p}+m}{p^2-\Lambda^2+i\varepsilon} = \left(\not{p}+m\right)\frac{m^2-\Lambda^2}{\left(p^2-m^2+i\varepsilon\right)\left(p^2-\Lambda^2+i\varepsilon\right)} \quad (15\text{-}55)$$

Note that for large Λ compared to p, the RHS of (15-55) reduces to the LHS (in front of the arrow). For everyday purposes, the heavy fermion would not seem to exist. However, at high p levels, such as at higher levels of our loop integration range, the contribution from the heavy fermion propagator would kick in.

As mass $\Lambda \to \infty$, particle "disappears" from universe we know

The valuable part of all this, from the point of view of regularization, is that the bottom row of (15-55) falls off, at high p, with $1/p^3$, whereas the top row LHS falls off with $1/p$. This allows loop

Presence of new particle causes integral to converge faster for large p

integrals to converge as $p \to \infty$. They will diverge with Λ, however. But we take care of that with renormalization where, as $\Lambda \to \infty$ (and our heavy fermion cannot then exist in any sense), quantities like e and m take on physical, finite values.

15.3.2 A Simple (Unphysical) Example

As you have seen in Sect. 15.2, the mathematics of regularization gets unwieldy and cumbersome, and that is probably a gross understatement. To illustrate the basic concept of Pauli-Villars regularization, we will use a mathematical example that does not represent reality, but has the big advantage of being far simpler.

A simple math (not physically real) example

Suppose we had an integral of the form

$$\frac{1}{(2\pi)^4} \int \frac{1}{\left(p^2 - m^2 + i\varepsilon\right)^2} \, d^4 p \,. \tag{15-56}$$

A given integral over spacetime

We could use the Pauli-Villars methodology of (15-55) (applied just to the denominator of that expression) to turn this into (where we drop the small ε for convenience)

$$\frac{1}{(2\pi)^4} \int \frac{1}{\left(p^2 - m^2\right)^2} \, d^4 p \to \frac{1}{(2\pi)^4} \int \left(\frac{1}{\left(p^2 - m^2\right)^2} - \frac{1}{\left(p^2 - \Lambda^2\right)^2} \right) d^4 p \,. \tag{15-57}$$

Add in a Pauli-Villars particle

Using (15-4) in the above with the integration limit Λ in (15-4) now equal to infinity and $s = -m^2$ for the first term and $-\Lambda^2$ (now the heavy fermion mass squared) for the second, we get (using (15-29) under the last row below)

$$\frac{1}{(2\pi)^4} \int \frac{1}{\left(p^2 - m^2\right)^2} \, d^4 p \to$$

$$= \frac{1}{(2\pi)^4} \left(-i\pi^2 \left(ln\left(m^2\right) - ln\left(\infty^2 + m^2\right) + 1 \right) + i\pi^2 \left(ln\left(\Lambda^2\right) - ln\left(\infty^2 + \Lambda^2\right) + 1 \right) \right) \tag{15-58}$$

Use standard spacetime integral

$$= \frac{i}{(4\pi)^2} \left(\left(ln\frac{\Lambda^2}{m^2} \right) + \underbrace{ln\left(\infty^2 + m^2\right)}_{2ln\,\infty + \frac{m^2}{\infty^2} + ..} - \underbrace{ln\left(\infty^2 + \Lambda^2\right)}_{2ln\,\infty + \frac{\Lambda^2}{\infty^2} + ..} \right) \,.$$

This gives us

$$\frac{1}{(2\pi)^4} \int \frac{1}{\left(p^2 - m^2\right)^2} \, d^4 p \to \frac{i}{(4\pi)^2} \left(ln\frac{\Lambda^2}{m^2} \right) - \frac{i}{(4\pi)^2} \frac{\Lambda^2}{\infty^2} + ... = \frac{i2}{(4\pi)^2} \left(ln\Lambda - ln\,m - \frac{1}{2}\frac{\Lambda^2}{\infty^2} \right) + ... \tag{15-59}$$

For $\Lambda \to \infty$ this becomes

$$\frac{1}{(2\pi)^4} \int \frac{1}{\left(p^2 - m^2\right)^2} \, d^4 p \xrightarrow{\Lambda \to \infty} \frac{i2}{(4\pi)^2} ln\Lambda \,. \tag{15-60}$$

We find a $ln\Lambda$ dependence for large Λ

using Pauli-Villars regularization.

15.4 Dimensional Regularization

15.4.1 The Concept

In dimensional regularization, we take advantage of something mathematicians have discovered. That is, intractable integrals in one-dimension space often become tractable in a space of different dimensions. In fact, mathematically it can even be done in spaces of fractional dimensions such as 3.5 dimensions, or 4.1 dimensions. At first blush, this seems quite bizarre, but mathematically, it all works out, as hopefully, we will shortly come to realize.

Dimensional regularization takes spacetime $D = 4$ integrals to $D = 4 - \eta$ spacetime integrals

In our case, our unbounded loop integrals are over four dimensional spacetime. It turns out that for dimensions D other than $D = 4$ for spacetime, these integrals can be evaluated readily. So, we take the same integrals over $D = 4 - \eta$, where $\eta \neq 0$. In the result, we get terms that are unbounded as $\eta \to 0$ (which corresponds to $\Lambda \to \infty$). We then re-express the result in 4D and Λ for use in renormalization, just as we do for any other regularization technique.

Note that in $D \neq 4$ spacetime, one dimension is time and all the rest are spatial.

We will take $D \neq 4$ relations mathematicians give us without deriving them

15.4.2 Relations for Arbitrary Dimension Spacetime

The mathematics behind D dimension spaces is extensive and delving into it to any depth would consume considerable time and effort. Instead, we will simply cite certain results the mathematicians have provided to us physicists for use in renormalization. We start with integer values for D, and then will extrapolate, rather uncritically, to non-integer values.

First for integer D

Metrics in D Integer Dimensions

For D any integer, $g_{\mu\nu}$ is a DxD matrix. Parallel to $g_{\mu\nu}g^{\mu\nu} = 4$ for $D = 4$ spacetime, we have

$$g_{\mu\nu}g^{\mu\nu} = D . \tag{15-61}$$

Metric contraction for general D spacetime

Gamma Matrices in D Integer Dimensions

In D dimensions, where D is an integer, there are D gamma matrices labeled $\gamma^0, \gamma^1, ... \gamma^{D-1}$. These are $f(D)$x$f(D)$ matrices, where $f(D)$ is an integer that depends on D. For $D = 4$, $f(D) = 4$.

γ matrices for integer D satisfy anti-commutation relations similar to those seen before,

$$\gamma^\mu\gamma^\nu + \gamma^\nu\gamma^\mu = 2g^{\mu\nu} . \tag{15-62}$$

Gamma matrices relation for general D spacetime

From these, one can derive contraction and trace relations parallel to (15-20) to (15-23). That is

$$\gamma_\lambda\gamma^\lambda = D, \qquad\qquad \gamma_\lambda\gamma^\alpha\gamma^\lambda = -(D-2)\gamma^\alpha$$
$$\gamma_\lambda\gamma^\alpha\gamma^\beta\gamma^\lambda = -(D-4)\gamma^\alpha\gamma^\beta + 4g^{\alpha\beta} \qquad \text{etc.} \tag{15-63}$$

$$\text{Tr}\left(\gamma^\alpha\gamma^\beta\right) = f(D)g^{\alpha\beta}, \tag{15-64}$$

$$\text{Tr}\left(\gamma^\sigma\gamma^\delta\gamma^\mu ...\right) = 0 \quad \text{for any odd number of gamma matrices}, \tag{15-65}$$

$$\text{Tr}\left(\gamma^\alpha\gamma^\beta\gamma^\gamma\gamma^\rho\right) = f(D)\left(g^{\alpha\beta}g^{\gamma\rho} - g^{\alpha\gamma}g^{\beta\rho} + g^{\alpha\rho}g^{\beta\gamma}\right)$$
or as we will see the indices later $\text{Tr}\left(\gamma^\mu\gamma^\delta\gamma^\nu\gamma^\sigma\right) = f(D)\left(g^{\mu\delta}g^{\nu\sigma} - g^{\mu\nu}g^{\delta\sigma} + g^{\mu\sigma}g^{\delta\nu}\right).$ $\tag{15-66}$

Key Integrals in D Integer Dimensions

For a Euclidean space of arbitrary integer dimension D, the mathematicians have provided us with the integral

Demo: how we use Wick rotation to turn Euclidean D space integral into D spacetime integral

$$\int \frac{1}{(2\pi)^D} \frac{d^D p_E}{\left(p_E^2 - s\right)^2} = \frac{1}{(4\pi)^{D/2}} \frac{\Gamma\left(2 - \frac{D}{2}\right)}{\Gamma(2)} \frac{1}{s^{2-\frac{D}{2}}} . \tag{15-67}$$

We can use this to find its equivalent in D dimensional spacetime. First perform an inverse Wick rotation transformation (3rd step below) on the LHS of (15-67) to get

$$\int \frac{1}{(2\pi)^D} \frac{1}{\left(p_E^2 - s\right)^2} d^D p_E = \frac{1}{(2\pi)^D}\int \frac{1}{\left(-p_E^2 + s\right)^2} d^D p_E = \frac{-i}{(2\pi)^D}\int \frac{1}{\left(p^2 + s\right)^2} d^D p . \tag{15-68}$$

From (15-67), $\qquad \dfrac{1}{(2\pi)^D}\displaystyle\int \dfrac{1}{\left(p^2 + s\right)^2} d^D p = \dfrac{i}{(4\pi)^{D/2}} \dfrac{\Gamma\left(2 - \frac{D}{2}\right)}{\Gamma(2)} \dfrac{1}{s^{2-\frac{D}{2}}}$ $\tag{15-69}$

In similar fashion, one can deduce other relations for D dimensional spacetime parallel to (15-3) to (15-12). We list the most relevant of these (with the (2π) factors arranged differently from (15-69)) below where we use q instead of p to represent the general case. Note, for $n = 2$, (15-70) is (15-69). Note also that by doing Prob. 4, you can (fairly quickly) derive (15-73) from (15-72).

In similar fashion, other D spacetime integrals derived from Euclidean ones

$$\int \frac{1}{\left(q^2 + s\right)^n} d^D q = i\pi^{D/2} \frac{\Gamma\left(n - \frac{D}{2}\right)}{\Gamma(n)} \frac{1}{s^{n-\frac{D}{2}}} \tag{15-70}$$

$$\int \frac{q^\mu}{\left(q^2 + s\right)^n} d^D q = 0 \tag{15-71}$$

$$\int \frac{q^\mu q^\nu}{\left(q^2+s\right)^n}\,d^D q = i\pi^{D/2}\,\frac{\Gamma\left(n-1-\frac{D}{2}\right)}{2\Gamma(n)}\,\frac{g^{\mu\nu}}{s^{n-1-D/2}} \tag{15-72}$$

$$\int \frac{q^2}{\left(q^2+s\right)^n}\,d^D q = i\pi^{D/2}\,\frac{\Gamma\left(n-1-\frac{D}{2}\right)}{2\Gamma(n)}\,\frac{D}{s^{n-1-D/2}}\;. \tag{15-73}$$

Extrapolating to Non-Integer D Dimensions

Note that the gamma function Γ is also defined for non-integer D, so the RHS of integrals (15-69), (15-70), (15-72), and (15-73) remain valid in that case as well. So, we simply assume all of the relations (15-61) to (15-73) hold for both integer and non-integer D.

Now assume above relations work in non-integer D spaces

One may feel some unease with a metric $g_{\mu\nu}$ and γ matrices for dimension spaces where D is not an integer. However, when we use these entities in such cases, in our final result we always take $D \to 4$, so we can carry the symbols representing them along as we go, knowing that all will be OK in the end. Though this implies the seemingly weird process of integration over fractional dimension spaces, it does turn out to work, as we are about to see.

With this most general interpretation of the above relations, we can re-visit the simple example we regularized via the Pauli-Villars method, but this time, using dimensional regularization.

15.4.3 The Same Simple (Unphysical) Example Again

We can use (15-69) to deduce the integral (15-56) in D dimensional spacetime and then take $D \to 4$. Ignore the ε for convenience and use $s = -m^2$. Note that the LHS of (15-69) with $D = 4$ is the same integral (15-56) we evaluated using Pauli-Villars regularization.

Dimensional regularization applied to Pauli-Villars example

$$\frac{1}{(2\pi)^4}\int \frac{1}{\left(p^2-m^2\right)^2}\,d^4 p \longrightarrow \frac{1}{(2\pi)^D}\int \frac{1}{\left(p^2-m^2\right)^2}\,d^D p = \frac{i}{(4\pi)^{D/2}}\,\frac{\Gamma\left(2-\frac{D}{2}\right)}{\Gamma(2)}\left(\frac{1}{-m^2}\right)^{2-\frac{D}{2}} \tag{15-74}$$

Using standard integral for D spacetime

$\Gamma(z)$ has poles (goes to infinity) at $0, -1, -2, \ldots$, so (15-74) has poles at $D = 4, 6, 8, \ldots$ To examine the behavior around $D = 4$, define $\underline{\eta = 4 - D}$ and use the approximation (which hopefully you can accept like an integral from a table)

$$\Gamma\left(2-\frac{D}{2}\right) = \Gamma\left(\frac{\eta}{2}\right) \underset{\substack{\eta \to 0 \\ D \to 4}}{\to} \frac{2}{\eta} - \gamma + \mathcal{O}(\eta), \tag{15-75}$$

Gamma function in limit

where γ here is the Euler-Mascheroni constant ≈ 0.5772, which will always cancel, or be negligible, in observable quantities. We also use the standard relation

$$a^x = 1 + x\,ln\,a + \frac{(x\,ln\,a)^2}{2!} + \frac{(x\,ln\,a)^3}{3!} + \ldots \tag{15-76}$$

Expansion of a^x

with $a = 1/m^2$ and $x = \eta/2$ to obtain

$$\left(\frac{1}{-m^2}\right)^{2-\frac{D}{2}} = (-1)^{\frac{\eta}{2}}\left(\frac{1}{m^2}\right)^{\frac{\eta}{2}} \underset{\substack{\eta \to 0 \\ D \to 4}}{\to} = 1 + \frac{\eta}{2}\,ln\,\frac{1}{m^2} + \mathcal{O}\left(\eta^2\right) = 1 - \frac{\eta}{2}\,ln\,m^2 + \mathcal{O}\left(\eta^2\right). \tag{15-77}$$

Using limiting values as $D \to 4$

(15-74), with $\Gamma(2) = 1$, is then

$$\frac{1}{(2\pi)^D}\int \frac{1}{\left(p^2-m^2\right)^2}\,d^D p \underset{\substack{\eta \to 0 \\ D \to 4}}{\to} \frac{i}{(4\pi)^2}\left(\frac{2}{\eta} - \gamma + \mathcal{O}(\eta)\right)\left(1 - \frac{\eta}{2}\,ln\,m^2\right) = \frac{i2}{(4\pi)^2}\left(\frac{1}{\eta} - \frac{\gamma}{2} - ln\,m\right). \tag{15-78}$$

Of course, in the full limit $\eta \to 0$ and $D \to 4$, (15-74) then becomes

$$\frac{1}{(2\pi)^4}\int \frac{1}{\left(p^2-m^2\right)^2}\,d^4 p \underset{\substack{\eta \to 0 \\ D \to 4}}{\to} \frac{i2}{(4\pi)^2}\,\frac{1}{\eta}, \tag{15-79}$$

We get integral for $D = 4$

found via dimensional regularization.

15.4.4 Important Conclusion

We can compare our dimensional regularization result (15-78)-(15-79) to that for the same integral found via Pauli-Villars regularization (15-59)-(15-60), and if we assume they must give us the same result, we can conclude that,

Comparing result with Pauli-Villars to relate Λ to η

$$\text{for finite } \Lambda \text{ and } \eta \text{ small, } \quad ln\,\Lambda = \frac{1}{\eta} - \frac{\gamma}{2}. \qquad \text{For } \Lambda \rightarrow \infty \text{ and } \eta \rightarrow 0, \quad ln\,\Lambda = \frac{1}{\eta}. \qquad (15\text{-}80)$$

For this integral at least, $(1/\eta - \gamma/2)$ plays the role of $ln\,\Lambda$. Both go to infinity in the limiting condition, where the Euler-Mascheroni constant γ becomes negligible. This conclusion is true in general for regularization of any integral, though we won't prove that here. Hopefully, this one example will provide some justification for adopting (15-80) as an identity in what follows.

15.5 Comparing Various Regularization Approaches

15.5.1 Usefulness of the Different Approaches

Pauli-Villars regularization works well with QED, but no one has been able to make it work for weak or strong interactions. Dimensional regularization works for QED and weak interactions and is the favored approach for both, so we will devote the remainder of this chapter to it. In summary,

Usefulness of various approaches

- Cut-off method: Simple in concept but violates gauge and Lorentz invariance, and not very useful.

- Pauli-Villars: Works for QED, but not (to date) for weak or strong interactions.

- Dimensional regularization: Works for QED and weak, but not strong, interactions. (Because strong interaction theory is non-perturbative, so our renormalization scheme does not hold.)

- Gauge lattice: Works for QED, weak, and strong interactions, but is too advanced for this text.

15.5.2 Why Choose One Approach Over Others

The cut-off method violates gauge and Lorentz invariance, but the others do not. In addition, the others all give the same result straightforwardly. (This is a relative and restricted use of the term "straightforward", as the processes are lengthy.) Gauge invariance has become a cornerstone of physics at many levels, and the consensus is that we should stay with methodologies that preserve it.

Preserving gauge invariance key factor in choosing regularization method

So, other methods, such as the last three listed in the prior section, have become preferred over the cut-off method.

Cut-off method not preferred

15.5.3 Apology for Emphasizing Cut-Off Method Earlier

I may have led you to believe early on, in discussing regularization as if it were done readily by the cut-off method, that the cut-off method worked straightforwardly. As it turns out, that is not the case, and I misled you. I did it purposefully, however, as it was the simplest way to get the message across at that point in your developing understanding of renormalization. Had I engaged in a more detailed discussion of regularization then, it would in all likelihood have confused, rather than enlightened you, and done more harm than good. In any event, all of this should be a little clearer now.

15.6 Finding Photon Self Energy Factor Using Dimensional Regularization

With reference to Fig. 15-1 on pg. 379, we reproduce the second order photon propagator of (15-30) below.

Finding $\Pi^{\mu\nu}(k)$ via dimensional regularization

$$iD_{F\alpha\beta}^{2nd}\left(k\right) = iD_{F\alpha\beta}\left(k\right) + iD_{F\alpha\mu}\left(k\right)\underbrace{\frac{(-1)}{(2\pi)^4}\left\{\text{Tr}\int ie_0\gamma^\mu iS_F\left(p-k\right)ie_0\gamma^\nu iS_F\left(p\right)d^4p\right\}}_{ie_0^2\Pi^{\mu\nu}(k)}iD_{F\nu\beta}\left(k\right) \quad (15\text{-}81)$$

where

$$ie_0^2 \Pi^{\mu\nu}(k) = \frac{(-1)}{(2\pi)^4}\left\{ \mathrm{Tr}\int ie_0\gamma^\mu \, i\frac{\not{p}-\not{k}+m}{(p-k)^2-m^2+i\varepsilon} \, ie_0\gamma^\nu \, i\frac{\not{p}+m}{p^2-m^2+i\varepsilon}d^4 p \right\}$$

Expressing $\Pi^{\mu\nu}(k)$ in 4D spacetime

$$= \frac{-e_0^2}{(2\pi)^4}\int \frac{\overbrace{\mathrm{Tr}\left\{\gamma^\mu\left(\not{p}-\not{k}+m\right)\gamma^\nu\left(\not{p}+m\right)\right\}}^{N^{\mu\nu}(p,k)}}{\left((p-k)^2-m^2+i\varepsilon\right)\left(p^2-m^2+i\varepsilon\right)}d^4 p. \qquad (15\text{-}82)$$

Converting (15-82) to D dimensional space and dividing by ie_0^2, we have

Converting to general dimension D spacetime

$$\Pi^{\mu\nu}(k) = \frac{i}{(2\pi)^D}\int \frac{N^{\mu\nu}(p,k)}{\left((p-k)^2-m^2+i\varepsilon\right)\left(p^2-m^2+i\varepsilon\right)}d^D p. \qquad (15\text{-}83)$$

The Denominator Re-arranged

Now, using (15-26) in (15-83) with $a = (p-k)^2-m^2+i\varepsilon$ and $b = p^2-m^2+i\varepsilon$ and hence $a-b = k^2-2pk$ (where we note the short hand notation $pk = p_\rho k^\rho$),

Re-arranging the denominator

$$\Pi^{\mu\nu}(k) = \frac{i}{(2\pi)^D}\int \int_0^1 \frac{N^{\mu\nu}(p,k)}{\left(\underbrace{p^2-m^2+i\varepsilon}_{b} + \underbrace{\left(k^2-2pk\right)}_{a-b}z\right)^2}dz\,d^D p$$

Use Feynman parameterization relation

$$= \frac{i}{(2\pi)^D}\int_0^1 \int \frac{N^{\mu\nu}(p,k)}{\left(p^2-m^2+k^2z-2pkz+i\varepsilon\right)^2}d^D p\,dz. \qquad (15\text{-}84)$$

We introduce the new variable

Then, use substitution of variable q

$$q = p - kz \qquad dq = dp \;\rightarrow\; d^D q = d^D p \qquad p = q + kz, \qquad (15\text{-}85)$$

and substituting for p in (15-84) gives us

$$\Pi^{\mu\nu}(k) = \frac{i}{(2\pi)^D}\int_0^1 \int \frac{N^{\mu\nu}(q+kz,k)}{\left((q+kz)^2-m^2+k^2z-2(q+kz)kz+i\varepsilon\right)^2}d^D q\,dz$$

$$= \frac{i}{(2\pi)^D}\int_0^1 \int \frac{N^{\mu\nu}(q+kz,k)}{\left(q^2+2qkz+k^2z^2-m^2+k^2z-2qkz-2k^2z^2+i\varepsilon\right)^2}d^D q\,dz$$

$$= \frac{i}{(2\pi)^D}\int_0^1 \int \frac{N^{\mu\nu}(q+kz,k)}{\left(q^2-k^2z^2+k^2z-m^2+i\varepsilon\right)^2}d^D q\,dz$$

$$\qquad (15\text{-}86)$$

$$= \frac{i}{(2\pi)^D}\int_0^1 \int \frac{N^{\mu\nu}(q+kz,k)}{\left(q^2+k^2z(1-z)-m^2+i\varepsilon\right)^2}d^D q\,dz.$$

Useful expression for denominator

Note that the denominator is even in the integration variable q.

The Numerator Re-arranged

Using the trace relations of (15-64) to (15-66) in the numerator of (15-82) and (15-83) yields

Re-arranging the numerator

$$N^{\mu\nu}(p,k) = \text{Tr}\left\{\gamma^\mu\left(\not{p} - \not{k} + m\right)\gamma^\nu\left(\not{p} + m\right)\right\}$$

$$= \text{Tr}\left\{\gamma^\mu\left(\not{p} - \not{k}\right)\gamma^\nu\not{p} + \gamma^\mu\left(\not{p} - \not{k}\right)\gamma^\nu m + \gamma^\mu m \gamma^\nu\not{p} + \gamma^\mu\gamma^\nu m^2\right\}$$

$$= \Big\{(p_\delta - k_\delta)p_\sigma \underbrace{\text{Tr}\gamma^\mu\gamma^\delta\gamma^\nu\gamma^\sigma}_{\substack{f(D)(g^{\mu\delta}g^{\nu\sigma} \\ -g^{\mu\nu}g^{\delta\sigma}+g^{\mu\sigma}g^{\delta\nu})}} + (p_\delta - k_\delta)m\underbrace{\text{Tr}\gamma^\mu\gamma^\delta\gamma^\nu}_{=0}$$

Use trace relations in D space

(15-87)

$$+ p_\sigma m \underbrace{\text{Tr}\gamma^\mu\gamma^\nu\gamma^\sigma}_{=0} + m^2 \underbrace{\text{Tr}\gamma^\mu\gamma^\nu}_{f(D)g^{\mu\nu}}\Big\}.$$

This becomes,

$$N^{\mu\nu}(p,k) = f(D)\left\{\left(p^\mu - k^\mu\right)p^\nu - g^{\mu\nu}(p_\delta - k_\delta)p^\delta + \left(p^\nu - k^\nu\right)p^\mu + m^2 g^{\mu\nu}\right\}$$

$$= f(D)\left\{\left(p^\mu - k^\mu\right)p^\nu + \left(p^\nu - k^\nu\right)p^\mu + (m^2 - \underbrace{(p_\delta - k_\delta)p^\delta}_{(p-k)p})g^{\mu\nu}\right\}. \qquad (15\text{-}88)$$

The numerator of (15-86) in terms of q, found by substituting p of (15-85) into (15-88), is

Substitution of variable

$$N^{\mu\nu}(q+kz,k) = f(D)\left\{\left(q^\mu + k^\mu z - k^\mu\right)\left(q^\nu + k^\nu z\right) + \left(q^\nu + k^\nu z - k^\nu\right)\left(q^\mu + k^\mu z\right)\right.$$

$$\left. + \left(m^2 - (q+kz-k)(q+kz)\right)g^{\mu\nu}\right\}. \qquad (15\text{-}89)$$

All terms in numerator odd in q drop out

In expanding (15-89) we can drop all terms linear in q since they make the integrand in (15-86) odd and will therefore yield zero. We can then take $N^{\mu\nu}$ (where dots indicate terms that drop out) as

$$N^{\mu\nu}(q+kz,k) = f(D)\{q^\mu q^\nu + k^\mu k^\nu z^2 - k^\mu k^\nu z + q^\mu q^\nu + k^\mu k^\nu z^2 - k^\mu k^\nu z$$

$$+ m^2 g^{\mu\nu} - q^2 g^{\mu\nu} - k^2 z^2 g^{\mu\nu} + k^2 z g^{\mu\nu}\} + \dots \qquad (15\text{-}90)$$

$$= f(D)\{\underbrace{\left(2q^\mu q^\nu - q^2 g^{\mu\nu}\right)}_{N_1^{\mu\nu}} \underbrace{-2k^\mu k^\nu z(1-z)}_{N_2^{\mu\nu}} + \underbrace{\left(m^2 + k^2 z(1-z)\right)g^{\mu\nu}}_{N_3^{\mu\nu}}\} + \dots$$

Useful expression for numerator and shorthand notation for three numerator terms

where we label certain groups of terms in the numerator as shown.

<u>Returning to the Whole Integral and Breaking It into Three Parts</u>
(15-86) then becomes

Expressing whole integral in shorthand for the three terms

$$\Pi^{\mu\nu}(k) = \frac{i}{(2\pi)^D}f(D)\int_0^1\int\frac{N_1^{\mu\nu} + N_2^{\mu\nu} + N_3^{\mu\nu}}{\left(q^2 + k^2 z(1-z) - m^2 + i\varepsilon\right)^2}d^D q\, dz$$

(15-91)

$$= \frac{i}{(2\pi)^D}f(D)\int_0^1\left(I_1^{\mu\nu} + I_2^{\mu\nu} + I_3^{\mu\nu}\right)dz,$$

where each $I_i^{\mu\nu}$ represents an integral in D dimension spacetime over q.

Evaluate 1st integral using standard integral in general D dimensions

<u>For $I_1{}^{\mu\nu}$</u>

$$I_1^{\mu\nu} = \int\frac{2q^\mu q^\nu - q^2 g^{\mu\nu}}{\left(q^2 + k^2 z(1-z) - m^2 + i\varepsilon\right)^2}d^D q \qquad (15\text{-}92)$$

With (15-72) and (15-73), $s = k^2 z(1-z) - m^2 + i\varepsilon$, and $n = 2$, this becomes (dropping the tiny ε for convenience and using the RHS of (15-2) in the RHS of the second line below)

$$I_1^{\mu\nu} = i\pi^{D/2} \frac{\Gamma\left(1-\frac{D}{2}\right)}{2\underbrace{\Gamma(2)}_{1!}} \frac{2g^{\mu\nu}}{\left(k^2 z(1-z)-m^2\right)^{1-D/2}} - i\pi^{D/2} \frac{\Gamma\left(1-\frac{D}{2}\right)}{2\underbrace{\Gamma(2)}_{1!}} \frac{D\,g^{\mu\nu}}{\left(k^2 z(1-z)-m^2\right)^{1-D/2}}$$

$$= \frac{i\pi^{D/2}g^{\mu\nu}}{\left(k^2 z(1-z)-m^2\right)^{1-D/2}} \frac{\Gamma\left(1-\frac{D}{2}\right)}{2}(2-D) = \frac{i\pi^{D/2}g^{\mu\nu}}{\left(k^2 z(1-z)-m^2\right)^{1-D/2}} \underbrace{\left(1-\frac{D}{2}\right)\Gamma\left(1-\frac{D}{2}\right)}_{\Gamma\left(2-\frac{D}{2}\right)} \quad (15\text{-}93)$$

$$= \frac{i\pi^{D/2}\Gamma\left(2-\frac{D}{2}\right)}{\left(k^2 z(1-z)-m^2\right)^{1-D/2}} g^{\mu\nu} = \frac{i\pi^{D/2}\Gamma\left(2-\frac{D}{2}\right)}{\left(k^2 z(1-z)-m^2\right)^{2-D/2}} \left(k^2 z(1-z)-m^2\right) g^{\mu\nu}.$$

For $I_2{}^{\mu\nu}$

Evaluate 2nd integral using standard integral in general D dimensions

$$I_2^{\mu\nu} = \int \frac{-2k^\mu k^\nu z(1-z)}{\left(q^2 + k^2 z(1-z)-m^2+i\varepsilon\right)^2} d^D q . \tag{15-94}$$

From (15-70) with $n = 2$ and $s = k^2 z(1 - z) - m^2 + i\varepsilon$, this becomes

$$I_2^{\mu\nu} = -i\pi^{D/2} \frac{\Gamma\left(2-\frac{D}{2}\right)}{\Gamma(2)} \frac{2k^\mu k^\nu z(1-z)}{\left(k^2 z(1-z)-m^2\right)^{2-D/2}} = -\frac{i\pi^{D/2}\Gamma\left(2-\frac{D}{2}\right)}{\left(k^2 z(1-z)-m^2\right)^{2-D/2}} 2z(1-z)k^\mu k^\nu . \tag{15-95}$$

For $I_3{}^{\mu\nu}$

$$I_3^{\mu\nu} = \int \frac{\left(m^2 + k^2 z(1-z)\right)g^{\mu\nu}}{\left(q^2 + k^2 z(1-z)-m^2+i\varepsilon\right)^2} d^D q . \tag{15-96}$$

Using (15-70) again just as we did in (15-95) but with a different numerator (which is constant in both cases with respect to the q integration)

Evaluate 3rd integral using standard integral in general D dimensions

$$I_3^{\mu\nu} = \frac{i\pi^{D/2}\Gamma\left(2-\frac{D}{2}\right)}{\left(k^2 z(1-z)-m^2\right)^{2-D/2}} \left(m^2 + k^2 z(1-z)\right) g^{\mu\nu} , \tag{15-97}$$

$I_1{}^{\mu\nu} + I_2{}^{\mu\nu} + I_3{}^{\mu\nu}$

Adding the three integrals (15-93), (15-95), and (15-97) yields

$$I_1^{\mu\nu} + I_2^{\mu\nu} + I_3^{\mu\nu}$$

Adding the 3 integrals

$$= \frac{i\pi^{D/2}\Gamma\left(2-\frac{D}{2}\right)}{\left(k^2 z(1-z)-m^2\right)^{2-D/2}} \left(\left(k^2 z(1-z)-m^2\right)g^{\mu\nu} - 2z(1-z)k^\mu k^\nu + \left(m^2 + k^2 z(1-z)\right)g^{\mu\nu}\right) \tag{15-98}$$

$$= \frac{i\pi^{D/2}\Gamma\left(2-\frac{D}{2}\right)}{\left(k^2 z(1-z)-m^2\right)^{2-D/2}} 2z(1-z)\left(k^2 g^{\mu\nu} - k^\mu k^\nu\right).$$

$\Pi^{\mu\nu}(k)$

Using (15-98) in (15-91) we find

$$\Pi^{\mu\nu}(k) = \frac{i}{(2\pi)^D} f(D) \int_0^1 \left(\frac{i\pi^{D/2}\Gamma\left(2-\frac{D}{2}\right)}{\left(k^2 z(1-z)-m^2\right)^{2-D/2}} 2z(1-z)\left(k^2 g^{\mu\nu}-k^\mu k^\nu\right) \right) dz$$

Getting $\Pi^{\mu\nu}(k)$ from the addition of the 3 integrals

(15-99)

$$= \frac{-1}{2^{D-1}\pi^{D-D/2}} f(D)\Gamma\left(2-\frac{D}{2}\right)\left(k^2 g^{\mu\nu}-k^\mu k^\nu\right)\int_0^1 \frac{z(1-z)}{\left(k^2 z(1-z)-m^2\right)^{2-D/2}} dz.$$

Note that the integral over z is a function of k^2 and recall from the appendix of Chap. 14, we showed that for

$$\Pi^{\mu\nu}(k) = -g^{\mu\nu}A\left(k^2\right) + \underbrace{k^\mu k^\nu B\left(k^2\right)}_{\text{drops out of amplitude}}$$

(15-100)

The $k^\mu k^\nu$ term drops out due to gauge invariance

the term with $k^\mu k^\nu$ drops out due to current conservation (gauge invariance). So, we can ignore it from here on and consider that

$$\Pi^{\mu\nu}(k) = \frac{-1}{2^{D-1}\pi^{D/2}} f(D)\Gamma\left(2-\frac{D}{2}\right)k^2 g^{\mu\nu}\int_0^1 \frac{z(1-z)}{\left(k^2 z(1-z)-m^2\right)^{2-D/2}} dz .$$

(15-101)

Now we want to take the limit of (15-101) as $D \to 4$ (i.e., as $\eta \ (= 4 - D) \to 0$). From (15-76) with $a = (k^2 z(1-z) - m^2)$ we see that

Take the limit of the remainder for $D \to 4$

$$\lim_{\substack{D\to 4 \\ \eta\to 0}} \frac{1}{\left(k^2 z(1-z)-m^2\right)^{2-D/2}} = \lim_{\substack{D\to 4 \\ \eta\to 0}} \frac{1}{\left(k^2 z(1-z)-m^2\right)^{\eta/2}}$$

(15-102)

$$= \lim_{\substack{D\to 4 \\ \eta\to 0}} \underbrace{\left(k^2 z(1-z)-m^2\right)}_{a}^{\overset{x}{-\eta/2}} = 1 - \frac{\eta}{2}ln\left(k^2 z(1-z)-m^2\right) + ..$$

Recalling the behavior of the Γ function in that limit from (15-75), and now with $f(D) \to 4$, we have

$$\Pi^{\mu\nu}(k) = \frac{-1}{2^3 \pi^2} 4k^2 g^{\mu\nu}\left(\frac{2}{\eta} - \gamma \underbrace{+\mathcal{O}(\eta)}_{\text{negligible}}\right)\int_0^1 z(1-z)\left\{1-\frac{\eta}{2}ln\left(k^2 z(1-z)-m^2\right)\right\} dz$$

Use limit relations for gamma function

$$= \frac{-1}{2\pi^2} k^2 g^{\mu\nu}\left(\frac{2}{\eta} - \gamma\right)\int_0^1 z(1-z)\left\{1-\frac{\eta}{2}ln\left(k^2 z(1-z)-m^2\right)\right\} dz$$

(15-103)

$$= \frac{-1}{2\pi^2} k^2 g^{\mu\nu}\left(\left(\frac{2}{\eta} - \gamma\right)\int_0^1 z(1-z) dz - \int_0^1 z(1-z)ln\left(k^2 z(1-z)-m^2\right) dz\right).$$

where we again dropped terms of order η in the last line. Using (15-80), this becomes

$$\Pi^{\mu\nu}(k) = \frac{-1}{\pi^2} k^2 g^{\mu\nu} ln\Lambda \underbrace{\int_0^1 \left(z-z^2\right) dz}_{=1/6} + \frac{1}{2\pi^2} k^2 g^{\mu\nu}\int_0^1 z(1-z) \underbrace{ln\left(k^2 z(1-z)-m^2\right)}_{ln k^2\left(z(1-z)-m^2/k^2\right)} dz$$

$$= \frac{-1}{6\pi^2} k^2 g^{\mu\nu} ln\Lambda + \frac{1}{2\pi^2} k^2 g^{\mu\nu} \overbrace{ln k^2}^{2ln k} \underbrace{\int_0^1 z(1-z) dz}_{=1/6}$$

(15-104)

$$+ \frac{1}{2\pi^2} k^2 g^{\mu\nu}\int_0^1 z(1-z)ln\left(z(1-z)-m^2/k^2\right) dz$$

$\Pi^{\mu\nu}(k)$ from dimensional regularization

$$= \frac{g^{\mu\nu}}{6\pi^2} k^2 ln\frac{k}{\Lambda} + \frac{g^{\mu\nu}}{2\pi^2} k^2 \int_0^1 z(1-z)ln\left(z(1-z)-m^2/k^2\right) dz.$$

Compare with What We Said Before We Would Find

Compare this with our Chap. 13 expression (shown in this chapter as (15-54)) repeated again below as (15-105).

Compare with relation of Chap. 13 that agrees with experiment

$$\Pi^{\mu\nu}(k) = \underbrace{g^{\mu\nu}k^2\, 2b_n\, ln\frac{k}{\Lambda}}_{-A'(k,\Lambda)} + \underbrace{g^{\mu\nu}k^2\, \frac{1}{2\pi^2}\int_0^1 z(1-z) ln\Big(z(1-z) - m^2/k^2\Big)dz}_{-\Pi_c(k^2)}\ ,\qquad (15\text{-}105)$$

where $b_n = 1/(12\pi^2)$ for $n = 1$ (only electron-positron loop) as we have here. Dimensional regularization gives us what we find via experiment.

Same thing

15.7 Finding the Vertex Correction Factor Using Dimensional Regularization

Fig. 15-2 depicts what we have seen before, the 2nd order vertex correction diagrams.

Finding $\Lambda^{\mu}(p',p)$ via dimensional regularization

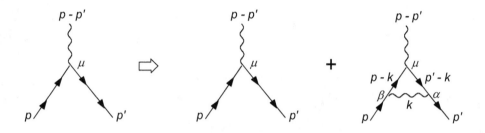

Figure 15-2. The Vertex Correction to Second Order

The vertex expression becomes (see Wholeness Chart 14-4 at the end of Chap. 14, columns (II) and (III) and Prob. 3 at the end of this chapter)

2nd order expression for vertex gamma function

$$\gamma^{\mu} \Rightarrow \gamma^{\mu}_{2nd}(p,p') = \gamma^{\mu} + e_0^2 \Lambda^{\mu}(p,p'),\qquad (15\text{-}106)$$

where from Feynman's rules, we have

$$\Lambda^{\mu}(p,p') = \frac{1}{(2\pi)^4}\int \frac{-ig_{\alpha\beta}}{k^2 + i\varepsilon} i\gamma^{\alpha} \frac{i}{(\not{p}' - \not{k}) - m + i\varepsilon}\gamma^{\mu}\frac{i}{(\not{p} - \not{k}) - m + i\varepsilon} i\gamma^{\beta} d^4k$$

Feynman's rules give us $\Lambda^{\mu}(p',p)$

$$(15\text{-}107)$$

$$= \frac{-i}{(2\pi)^4}\int \frac{\overbrace{g_{\alpha\beta}\gamma^{\alpha}\Big(\not{p}' - \not{k} + m\Big)\gamma^{\mu}\Big(\not{p} - \not{k} + m\Big)\gamma^{\beta}}^{N^{\mu}(p,p',k)}}{k^2\Big((p'-k)^2 - m^2\Big)\Big((p-k)^2 - m^2\Big)} d^4k\ .$$

where in the second line above we drop the $i\varepsilon$ terms for convenience.

Doing the contraction in the numerator of (15-107) yields

$$N^{\mu}(p,p',k) = \gamma_{\beta}\Big(\not{p}' - \not{k} + m\Big)\gamma^{\mu}\Big(\not{p} - \not{k} + m\Big)\gamma^{\beta}.\qquad (15\text{-}108)$$

Infrared Divergence

Note that if $k \to 0$ (low energy or infrared virtual photon in Fig. 15-2), then $p'^2 \to m^2$ and $p^2 \to m^2$. Thus, the denominator of the integrand in the last line of (15-107) approaches zero proportional to the sixth power, i.e., as 0^6, whereas the numerator (with d^4k taken as proportional to $k^3 dk$) as, at best, proportional to zero to the fifth power. So the integrand will diverge for k approaching zero.

Infrared divergence

We can correct for that with a trick. We temporarily assume the virtual photon has a mass λ. That keeps the photon propagator finite and prevents the integrand from diverging in the infrared regime. That is, we re-write (15-107) as

Temporarily using a fictitious photon propagator mass eliminates $k \to 0$ divergence problem

$$\Lambda^{\mu}(p,p') = \frac{-i}{(2\pi)^4}\int \frac{N^{\mu}(p,p',k)}{\Big(k^2 - \lambda^2\Big)\Big((p'-k)^2 - m^2\Big)\Big((p-k)^2 - m^2\Big)} d^4k\qquad (15\text{-}109)$$

and at the end of our evaluation of the integral, we take $\lambda = 0$. This works, though it may seem strange.

One might ask oneself, as I once did, why then does the integrand not diverge in the regime where $k^2 = \lambda^2$? It is a good question whose full mathematical answer would take us away from our task at hand. The short answer is that doing this can actually be justified, the integral is well defined, and the procedure results in the correct predictions of experiment.

The Denominator Re-arranged

Ignoring the small ε terms, jumping to D dimensions instead of 4, and using (15-27)(b), with

$$a = k^2 - \lambda^2 \qquad b = (p' - k)^2 - m^2 \qquad c = (p - k)^2 - m^2 \qquad (15\text{-}110)$$

in the denominator of (15-109), we can re-write (15-109) as

$$\Lambda^\mu(p, p') = \frac{-i}{(2\pi)^D} \iint_0^1 \int_0^{1-x} \frac{2N^\mu(p, p', k)}{\left(a + (b-a)x + (c-a)z\right)^3} \, dz \, dx \, d^D k$$

Aside: Express $\Lambda^\mu(p',p)$ in D dimensions and use Feynman parameterization relation to re-express denominator

$$= \frac{-2i}{(2\pi)^D} \iint_0^1 \int_0^{1-x} \frac{N^\mu(p, p', k)}{\underbrace{\left(\begin{array}{c} k^2 - \lambda^2 + \left((p'-k)^2 - m^2 - k^2 + \lambda^2\right)x \\ + \left((p-k)^2 - m^2 - k^2 + \lambda^2\right)z \end{array}\right)^3}_{\substack{k^2 - \lambda^2 + p'^2 x - 2p'kx + k^2 x - m^2 x + \lambda^2 x - k^2 x \\ + p^2 z - 2pkz + k^2 z - m^2 z - k^2 z + \lambda^2 z}}} \, dz \, dx \, d^D k$$

Introduce shorthand symbol r to make expressions less cumbersome

$$= \frac{-2i}{(2\pi)^D} \iint_0^1 \int_0^{1-x} \frac{N^\mu(p, p', k)}{\left(k^2 - 2k(p'x + pz) + \underbrace{x(p'^2 - m^2) + z(p^2 - m^2) - \lambda^2(1 - x - z)}_{-r}\right)^3} \, dz \, dx \, d^D k. \qquad (15\text{-}111)$$

where the symbol r is used as a shorthand symbol to keep notation streamlined. Recall that in the present case, photon $k = p - p' \to 0$, so the fermions in Fig. 15-2 approach on-shell, i.e., $(p')^2 \approx p^2 \approx m^2$.

If we re-express (15-111) in terms of a new variable t^μ (where a^μ is another shorthand symbol and is different from the a of (15-110)),

Substitution of variables t and a

$$t^\mu = k^\mu - \underbrace{(p'x + pz)^\mu}_{a^\mu} = k^\mu - a^\mu \qquad d^4 t = d^4 k \ \to \ d^D t = d^D k, \qquad (15\text{-}112)$$

so
$$t^2 = k^2 - 2k(p'x + pz) + (p'x + pz)^2 = k^2 - 2k(p'x + pz) + a^2, \qquad (15\text{-}113)$$

then (15-111) becomes

$$\Lambda^\mu(p, p') = \frac{-2i}{(2\pi)^D} \iint_0^1 \int_0^{1-x} \frac{N^\mu(p, p', t+a)}{(t^2 - r - a^2)^3} \, dz \, dx \, d^D t$$

$$= \frac{-2i}{(2\pi)^D} \int_0^1 \int_0^{1-x} \int \frac{N^\mu(p, p', t+a)}{(t^2 - r - a^2)^3} \, d^D t \, dz \, dx. \qquad (15\text{-}114)$$

$\Lambda^\mu(p',p)$ in D dimensions in terms of r, a, and t

The Numerator Re-arranged

With (15-112) solved for k^μ and substituted in (15-108), we have

$$N^\mu(p, p', t+a) = \gamma_\beta\left(\not{p}' - \not{t} - \not{a} + m\right)\gamma^\mu\left(\not{p} - \not{t} - \not{a} + m\right)\gamma^\beta$$

Re-expressing numerator in terms of r, a, and t

$$= \underbrace{\gamma_\beta\left(\not{p}' - \not{a} + m\right)\gamma^\mu\left(\not{p} - \not{a} + m\right)\gamma^\beta}_{N_0^\mu(p', p, a)} \qquad (15\text{-}115)$$

$$\underbrace{- \gamma_\beta\left(\not{t}\gamma^\mu\left(\not{p} - \not{a} + m\right) + \left(\not{p}' - \not{a} + m\right)\gamma^\mu \not{t}\right)\gamma^\beta}_{N_1^\mu(p', p, a, t)} + \underbrace{\gamma_\beta \not{t}\gamma^\mu \not{t}\gamma^\beta}_{N_2^\mu(p', p, a)}.$$

The subscripts i on the N_i^μ indicate the power to which t is raised in that particular term.

Back to the Whole Integral

(15-114) then becomes

$$\Lambda^\mu\left(p,p'\right) = \sum_{i=0}^{2} \Lambda_i^\mu\left(p,p'\right) = \sum_{i=0}^{2} \frac{-2i}{\left(2\pi\right)^D} \int_0^1 \int_0^{1-x} \int \frac{N_i^\mu}{\left(t^2 - r - a^2\right)^3}\, d^D t\, dz\, dx \ . \qquad (15\text{-}116)$$

The whole integral as a sum of 3 terms $\rightarrow \Lambda^\mu(p',p)$ as sum of 3 terms $\Lambda_i^\mu\left(p',p\right)$

For Λ_0^μ

From (15-115) and (15-116) we have

$$\Lambda_0^\mu\left(p,p'\right) = \frac{-2i}{\left(2\pi\right)^D} \int_0^1 \int_0^{1-x} \gamma_\beta\left(\not{p}' - \not{d} + m\right)\gamma^\mu\left(\not{p} - \not{d} + m\right)\gamma^\beta \left(\int \frac{1}{\left(t^2 - r - a^2\right)^3}\, d^D t\right) dz\, dx , \quad (15\text{-}117)$$

Finding $\Lambda_0^\mu(p',p)$

for which, since the numerator is not a function of t, we can use (15-70) with $q = t$ and $s = -r - a^2$. The result is

$$\Lambda_0^\mu\left(p,p'\right) = \frac{-2i}{\left(2\pi\right)^D} \int_0^1 \int_0^{1-x} \gamma_\beta\left(\not{p}' - \not{d} + m\right)\gamma^\mu\left(\not{p} - \not{d} + m\right)\gamma^\beta \times$$

Use standard integrals for D dimensions

$$\left(i\pi^{D/2} \frac{\Gamma\left(3 - \frac{D}{2}\right)}{\Gamma(3)} \frac{1}{\left(-r - a^2\right)^{3 - \frac{D}{2}}}\right) dz\, dx. \qquad (15\text{-}118)$$

For $D \rightarrow 4$, this becomes

$$\Lambda_0^\mu\left(p,p'\right) = \frac{-2i}{\left(2\pi\right)^4} \int_0^1 \int_0^{1-x} i\pi^2 \frac{\Gamma(1)}{\Gamma(3)} \frac{\gamma_\beta\left(\not{p}' - \not{d} + m\right)\gamma^\mu\left(\not{p} - \not{d} + m\right)\gamma^\beta}{\left(-r - a^2\right)}\, dz\, dx$$

Take limits for $D \rightarrow 4$

$$(15\text{-}119)$$

$$= -\frac{1}{2^4 \pi^2} \int_0^1 \int_0^{1-x} \frac{\gamma_\beta\left(\not{p}' - \not{d} + m\right)\gamma^\mu\left(\not{p} - \not{d} + m\right)\gamma^\beta}{\left(r + a^2\right)}\, dz\, dx,$$

where r and a are functions of x, z, p', and p. We see that (15-119) has a finite value and can be rewritten, using the gamma matrix relations of (15-23), as

$$\Lambda_0^\mu\left(p,p'\right) = -\frac{1}{16\pi^2} \int_0^1 \int_0^{1-x} \frac{1}{\left(r + a^2\right)} \left(\underbrace{\gamma_\beta \gamma^\sigma \gamma^\mu \gamma^\rho \gamma^\beta}_{-2\gamma^\rho \gamma^\mu \gamma^\sigma}\left(p'_\sigma - a_\sigma\right)\left(p_\rho - a_\rho\right) + \right.$$

Use gamma matrices relations

$$(15\text{-}120)$$

$$\left. \underbrace{\gamma_\beta \gamma^\sigma \gamma^\mu \gamma^\beta}_{4g^{\sigma\mu}}\left(p'_\sigma - a_\sigma\right)m + \underbrace{\gamma_\beta \gamma^\mu \gamma^\rho \gamma^\beta}_{4g^{\mu\rho}}\left(p_\rho - a_\rho\right)m + \underbrace{\gamma_\beta \gamma^\mu \gamma^\beta}_{-2\gamma^\mu} m^2 \right) dz\, dx.$$

Or, switching the dummy variable in the next to last term from ρ to σ,

$$\Lambda_0^\mu\left(p,p'\right) = \int_0^1 \int_0^{1-x} \frac{1}{\left(r + a^2\right)} \left(\frac{1}{8\pi^2} \gamma^\rho \gamma^\mu \gamma^\sigma \left(p'_\sigma - a_\sigma\right)\left(p_\rho - a_\rho\right) \right.$$

$$(15\text{-}121)$$

$$\left. - \frac{1}{4\pi^2} g^{\mu\sigma}\left(p'_\sigma + p_\sigma - 2a_\sigma\right)m + \frac{1}{8\pi^2} \gamma^\mu m^2 \right) dz\, dx.$$

Result for $\Lambda_0^\mu(p',p)$

For Λ_1^μ

From (15-115) and (15-116) we see that the numerator for $i = 1$ is odd in t, whereas the denominator is even. So this term will equal zero.

$\Lambda_i^\mu\left(p',p\right) = 0$

For $\Lambda_2{}^\mu$

For $i = 2$, we have

$$\Lambda_2^\mu(p,p') = \frac{-2i}{(2\pi)^D} \int_0^1 \int_0^{1-x} \int \frac{\overbrace{\gamma_\beta \gamma^\rho \gamma^\mu \gamma^\sigma \gamma^\beta t_\rho t_\sigma}^{}}{\gamma_\beta \slashed{t} \gamma^\mu \slashed{t} \gamma^\beta} \frac{}{(t^2 - r - a^2)^3} d^D t \, dz \, dx .$$

$$\Lambda_2^\mu(p,p') = \frac{-2i}{(2\pi)^D} \int_0^1 \int_0^{1-x} \int \frac{\gamma_\beta \slashed{t} \gamma^\mu \slashed{t} \gamma^\beta t_\rho t_\sigma}{(t^2 - r - a^2)^3} d^D t \, dz \, dx . \qquad (15\text{-}122)$$

Finding $\Lambda_2^\mu(p',p)$

Using (15-72) with $q = t$ makes this

$$\Lambda_2^\mu(p,p') = \frac{-2i}{(2\pi)^D} \gamma_\beta \gamma^\rho \gamma^\mu \gamma^\sigma \gamma^\beta \int_0^1 \int_0^{1-x} \int \frac{t_\rho t_\sigma}{(t^2 - r - a^2)^3} d^D t \, dz \, dx$$

$$= \frac{-2i}{(2\pi)^D} \gamma_\beta \gamma^\rho \gamma^\mu \gamma^\sigma \gamma^\beta \int_0^1 \int_0^{1-x} \left(i\pi^{D/2} \frac{\Gamma\left(\frac{\eta}{2} = 2 - \frac{D}{2}\right)}{2\Gamma(3)} \frac{g_{\rho\sigma}}{(-r - a^2)^{\frac{\eta}{2}}} \right) dz \, dx. \qquad (15\text{-}123)$$

Use standard integrals for D dimensions

Now take $D \to 4$ (i.e., $\eta \to 0$) and use (15-75) and (15-76) with a there $= -r - a^2$ above and $x = -\eta/2$. This, along with two gamma matrix relations of (15-23) yields

$$\Lambda_2^\mu(p,p') = \frac{2}{2^4 \pi^2} \underbrace{\gamma_\beta \gamma^\rho \gamma^\mu \gamma^\sigma \gamma^\beta}_{-2\gamma^\sigma \gamma^\mu \gamma^\rho} g_{\rho\sigma} \int_0^1 \int_0^{1-x} \lim_{\eta \to 0} \left(\frac{\Gamma\left(\frac{\eta}{2}\right)}{2\Gamma(3)} (-r - a^2)^{-\frac{\eta}{2}} \right) dz \, dx$$

Take limits for $D \to 4$

$$= \frac{-1}{4\pi^2 \cdot 2 \cdot 2} \underbrace{\gamma_\rho \gamma^\mu \gamma^\rho}_{-2\gamma^\mu} \int_0^1 \int_0^{1-x} \left(\frac{2}{\eta} - \gamma + \mathcal{O}(\eta) \right) \left(1 - \frac{\eta}{2} \ln(-r - a^2) + \mathcal{O}(\eta^2) \right) dz \, dx. \qquad (15\text{-}124)$$

Keeping only lowest order terms (i.e., dropping terms in η, η^2, etc.) gives us

$$\Lambda_2^\mu(p,p') = \frac{1}{8\pi^2} \gamma^\mu \int_0^1 \int_0^{1-x} \left(\frac{2}{\eta} - \gamma - \ln(-r - a^2) \right) dz \, dx . \qquad (15\text{-}125)$$

Keep lowest order terms

Using (15-80) yields

$$\Lambda_2^\mu(p,p') = \frac{1}{4\pi^2} \gamma^\mu \ln\Lambda \underbrace{\int_0^1 \int_0^{1-x} dz \, dx}_{\substack{1-x \\ 1/2}} - \frac{1}{8\pi^2} \gamma^\mu \int_0^1 \int_0^{1-x} \ln(-r - a^2) dz \, dx$$

$$(15\text{-}126)$$

$$= \frac{1}{8\pi^2} \gamma^\mu \ln\Lambda - \frac{1}{8\pi^2} \gamma^\mu \int_0^1 \int_0^{1-x} \ln(-r - a^2) dz \, dx.$$

Result for $\Lambda_2^\mu(p',p)$

$\Lambda_0{}^\mu + \Lambda_1{}^\mu + \Lambda_2{}^\mu$

Adding the $\Lambda_i^\mu(p,p')$ yields

Adding the $\Lambda_i^\mu(p',p)$ to get $\Lambda^\mu(p',p)$

$$\Lambda^\mu(p,p') = \Lambda_2^\mu(p,p') + \Lambda_1^\mu(p,p') + \Lambda_0^\mu(p,p') = \overbrace{\frac{1}{8\pi^2} \ln\Lambda \, \gamma^\mu}^{L(\Lambda)}$$

$$\left. - \frac{1}{8\pi^2} \gamma^\mu \int_0^1 \int_0^{1-x} \ln(-r - a^2) dz \, dx + \frac{1}{8\pi^2} \int_0^1 \int_0^{1-x} \frac{1}{(r + a^2)} \times \right\} = \Lambda_c^\mu(p,p') \qquad (15\text{-}127)$$

$$\left. \left(\gamma^\rho \gamma^\mu \gamma^\sigma (p'_\sigma - a_\sigma)(p_\rho - a_\rho) - 2(p'^\mu + p^\mu - 2a^\mu)m + \gamma^\mu m^2 \right) dz dx. \right.$$

Final result for $\Lambda^\mu(p',p)$

Compare with What We Said We Would Find

In Chap. 13, (13-9) on pg. 324, we expressed this as

Compare to Chap. 13 relation that agrees with experiment

$$\Lambda^{\mu}\left(p,p'\right) = \underbrace{\frac{1}{8\pi^2} \ln \Lambda \, \gamma^{\mu}}_{L(\Lambda)} + \underbrace{\Lambda_c^{\mu}\left(k^2\right)}_{\substack{\text{too complicated} \\ \text{to express}}} , \qquad (15\text{-}128)$$

They match

and perhaps now it is obvious why we didn't express $\Lambda_c^{\mu}(k^2)$ explicitly there.

15.8 Finding Fermion Self Energy Factor Using Dimensional Regularization

To save ourselves the tedium, similar to that of the photon and vertex correction, of going through pages of algebra to derive the fermion second order correction, we will simply summarize the steps one goes through and state the final result.

How to find $\Sigma(p)$

Figure 15-3. The Fermion Propagator to Second Order

From Fig. 15-3, one can use Feynman rules, pg. 236, (or simply refer to Wholeness Chart 14-4, pg. 368, columns (II) and (III)) to show

$$iS_F^{2nd}\left(p\right) = iS_F\left(p\right) + iS_F\left(p\right) ie_0^2 \Sigma\left(p\right) iS_F\left(p\right), \qquad (15\text{-}129)$$

where

$$\Sigma\left(p\right) = \frac{i}{\left(2\pi\right)^4} \int iD_{F\alpha\beta}\left(k\right) \gamma^{\alpha} iS_F\left(p-k\right) \gamma^{\beta} d^4k$$

Find $\Sigma(p)$ from Feynman rules

$$= \frac{i}{\left(2\pi\right)^4} \int \frac{-ig_{\alpha\beta}}{k^2 + i\varepsilon} \gamma^{\alpha} i \frac{\left(\not{p} - \not{k} + m_0\right)}{\left(p-k\right)^2 - m_0^2 + i\varepsilon} \gamma^{\beta} d^4k. \qquad (15\text{-}130)$$

Then, re-express (15-130) in D dimensions instead of 4 and evaluate the integral using the tricks we have used before, such as Feynman parameterization and D space integrals. Follow by taking the limit as $D \to 4$ in the result, and using the appropriate gamma matrix relations along with (15-80). The final result to leading log order, as we first stated in Chap. 13, (13-8) on pg. 324, is

Express in D dimensions, evaluate integral, take limit for $D \to 4$

$$\Sigma\left(p\right) = \underbrace{-\frac{3m}{8\pi^2} \ln \frac{\Lambda}{m}}_{A(\Lambda,m)} + \left(\not{p} - m\right) \underbrace{\frac{1}{8\pi^2} \ln \Lambda}_{B(\Lambda)} + \underbrace{\left(\not{p} - m\right) \Sigma_c \left(\not{p} - m\right)}_{\text{complicated}} , \qquad (15\text{-}131)$$

Will get Chap. 13 relation that agrees with experiment

15.9 Chapter Summary

There are a number of approaches to regularization, of which we have discussed four. These are summarized in Wholeness Chart 15-2.

Wholeness Chart 15-2. Comparison of Four Regularization Techniques

	Method	Result when renormalize	Usefulness
Cut-off	Integrate to Λ instead of ∞	Divergence with Λ^2 disagrees with experiment and other approaches	Violates gauge and Lorentz invariance. Not useful
Pauli-Villars	Add fictitious particle of mass Λ.	Agrees with experiment and other approaches except cut-off	Obeys gauge and Lorentz invariance. Useful in QED.
Dimensional	Integrate over $D = 4 - \eta$ dimensions	Agrees with experiment and other approaches except cut-off	Obeys gauge and Lorentz invariance. Useful in QED and weak interactions
Gauge lattice	Spacetime approximated by lattice of hyper-cubes	Agrees with experiment and other approaches except cut-off	Obeys gauge and Lorentz invariance. Useful in QED, weak, strong interactions

Summaries of the steps for the first three regularization methods above, which we examined in some depth in this chapter, are shown below.

Steps for Dimensional Regularization

1. Express the unbounded integral in four dimensional spacetime.

2. Convert the integral to arbitrary dimension D spacetime. $D = 4 - \eta$.

3. Re-arrange the denominator so that it has the form of the denominator in one of our standard D spacetime integrals ((15-70) to (15-73)). Do this using Feynman parameterization relations, such as

$$\frac{1}{ab} = \int_0^1 \frac{dz}{\left(b + (a-b)z\right)^2},$$

   and then typically doing substitution of variable, so the resulting expression comprises an integral over another variable such as q or t, rather than p. (Still in D dimensions.)

4. Re-arrange the numerator as convenient. Typically, by breaking it into separate terms and thus obtaining a separate integral for each term.

5. Re-express the gamma matrices in the numerator terms via the relations (15-61) to (15-66).

6. Evaluate each of the separate integrals (numerator and denominator of each together) using our standard integrals in D spacetime (15-70) to (15-73).

7. Take $D \to 4$ (i.e., $\eta \to 0$) in the result for each integral and retain only lowest order terms.

8. In each such result, take $1/\eta - \gamma/2 = \ln \Lambda$.

9. Add the results of these separate integrations to get the total integral result in terms of Λ.

(Note: Steps 3 to 5 can be done before or after step 2.)

Steps for Pauli-Villars Regularization

1. Express the unbounded integral in four dimensional spacetime.

2. Add term(s) to the integral to represent an additional (fictitious) particle of mass Λ in the theory. Specifically, an additional propagator having mass term Λ is included, as if we had additional Feynman diagrams, each a duplicate of an original diagram except the propagator has different mass Λ.

3. Re-arrange denominator and numerator as in steps 3 to 5 of dimensional regularization, but with $D = 4$.

4. Evaluate each of the separate integrals using our standard integrals in four dimensional spacetime (15-3) to (15-12). Note these are integrated over $-\infty$ to $+\infty$, not $-\Lambda$ to $+\Lambda$. (Λ in Pauli-Villars approach is mass, not integration limits.) The result converges for finite mass Λ.

5. Retain lowest order terms, assuming Λ is large.

6. Add the results of these separate integrations to get the total integral result in terms of Λ.

Steps for Cut-Off Regularization

1. Express the unbounded integral in four dimensional spacetime.

2. Optional approximation to make things simpler: ignore masses.

3. Re-arrange denominator and numerator as in steps 3 to 5 of dimensional regularization, but with $D = 4$.

4. Evaluate each of the separate integrals using our standard integrals in four dimensional spacetime (15-3) to (15-12). These are integrated over $-\Lambda$ to $+\Lambda$ instead of $-\infty$ to $+\infty$.

5. Retain lowest order terms, assuming Λ is large.

6. Add the results of these separate integrations to get the total integral result in terms of Λ.

Many more advanced texts delve into renormalization to greater depth. See, for example, Peskin, M. and Schroeder, D., *An Introduction to Quantum Field Theory* (Perseus, 1995), or Itzykson, C., and Zuber, J.B., *Quantum Field Theory* (McGraw Hill, 1985)

15.10 Appendix: Additional Notes on Integrals

For integral relations (15-5) to (15-10) note the following.

On the RHS of (15-5) to (15-10), as in (15-3) and (15-4), we take $\varepsilon = 0$, which is typically OK. In cases where it is not, we must, on the RHS, take $s \to s + i\varepsilon$ and $t \to t + i\varepsilon$.

(15-5) can be obtained for $n = 3$ by performing the k^0 integration as a contour integral and the **k** integration using spherical coordinates[1]. The $n = 4$ case is obtained by differentiating that with respect to s. Repeated such differentiations lead to the higher n expressions.

(15-6) equals zero because the denominator is even in k^μ and the numerator is odd. (15-8) and (15-9) follow from (15-5) and (15-6) by changing variables, respectively, as $k \to p = k - q$, and $s \to t = q^2 + s$.

Taking the derivative of (15-9) with respect to q_ν gives us (15-10). Taking $q = 0$ in (15-10) yields (15-7).

Integrals (15-11) and (15-12)

In relation (15-11),

$$I'' = \int p^\mu p^\nu d^4 p = \tfrac{1}{4} \int g^{\mu\nu} p^2 d^4 p \,, \qquad (15\text{-}132)$$

we can see that for $\mu \neq \nu$, the RHS is zero. For $\mu \neq \nu$ in the middle part, we have an odd factor of at least one 3-momentum component p^i, so the integral from $+\infty$ to $-\infty$ will be zero as well. So, we only have to worry about diagonal terms, i.e., terms with $\mu = \nu$. We can therefore express the middle part of (15-132), after Wick rotation, as

$$I'' = \int p^\mu p^\nu d^4 p = i \int \begin{bmatrix} -E^2 & & & \\ & \left(p^1\right)^2 & & \\ & & \left(p^2\right)^2 & \\ & & & \left(p^3\right)^2 \end{bmatrix} d^4 p_E \,. \qquad (15\text{-}133)$$

The integral of each of the four non-zero components in (15-133) is over $-\infty$ to $+\infty$. And they are all even. So, the absolute value of each integral is the same. That is,

$$\int (E)^2 i \, dE dp^1 dp^2 dp^3 = \int \left(p^1\right)^2 i \, dE dp^1 dp^2 dp^3 = \int \left(p^2\right)^2 i \, dE dp^1 dp^2 dp^3 = \int \left(p^3\right)^2 i \, dE dp^1 dp^2 dp^3 . (15\text{-}134)$$

With this, (15-133) becomes

$$I'' = i \int \begin{bmatrix} -E^2 & & & \\ & E^2 & & \\ & & E^2 & \\ & & & E^2 \end{bmatrix} d^4 p_E = -i \int g^{\mu\nu} E^2 d^4 p_E$$

$$= -\frac{i}{4} \int g^{\mu\nu} \left(E^2 + \left(p^1\right)^2 + \left(p^2\right)^2 + \left(p^3\right)^2 \right) d^4 p_E . \qquad .(15\text{-}135)$$

We now do a reverse Wick rotation on the last part of (15-135) to get back to 4D spacetime (Minkowski) coordinates. That is, take $E \to E/i$ and $d^4 p_E \to d^4 p/i$ to find

---

[1] See J.J. Sakurai, *Advanced Quantum Mechanics* (Addison-Wesley 1967), pg. 315.

$$I'' = -\frac{i}{4}\int g^{\mu\nu}\left(\left(\frac{E}{i}\right)^2 + \mathbf{p}^2\right)\frac{d^4p}{i} = \frac{1}{4}\int g^{\mu\nu}\left(E^2 - \mathbf{p}^2\right)d^4p = \frac{1}{4}\int g^{\mu\nu}p^2d^4p. \quad \text{QED. (15-136)}$$

Relation (15-12) follows in similar fashion.

Going from (15-44) to (15-45)

To start, note the following.

$$\int_0^1 z(1-z)dz = \int_0^1 (z - z^2)dz = \left[\frac{z^2}{2} - \frac{z^3}{3}\right]_0^1 = \frac{1}{2} - \frac{1}{3} = \frac{1}{6} \tag{15-137}$$

$$\int_0^1 z\ln z\,dz = \left[z^2\frac{\ln z}{2} - \frac{z^2}{4}\right]_0^1 = \frac{\ln 1}{2} - \frac{1}{4} = -\frac{1}{4} \qquad \int_0^1 z^2\ln z\,dz = \left[z^3\frac{\ln z}{3} - \frac{z^3}{9}\right]_0^1 = \frac{\ln 1}{3} - \frac{1}{9} = -\frac{1}{9} \tag{15-138}$$

$$\underbrace{\int_0^1 z(1-z)\ln(1-z)dz}_{\text{substitute } z=1-z'} = -\int_1^0 z'(1-z')\ln z'dz' = \underbrace{\int_0^1 z'\ln z'dz' - \int_0^1 z'^2\ln z'dz'}_{\text{see above}} = -\frac{1}{4} + \frac{1}{9} = -\frac{5}{36} \tag{15-139}$$

Also note the following for logs of negative numbers, where $z > 0$,

$$\ln(-1) = \ln(e^{i\pi}) = i\pi \quad \rightarrow \quad \ln(-z) = \ln(-1)z = \ln(-1) + \ln z = i\pi + \ln z. \tag{15-140}$$

In general, the log of a negative number is a complex number.

Now, look at the first two terms in the integrand of (15-44). First the first of those two terms, where after the first equal sign we use (15-140), and after the last equal sign we use (15-137) and (15-139),

$$\int_0^1 \{z(1-z)\ln(-z)dz = \int_0^1 \{z(1-z)(i\pi + \ln z)dz = i\pi\int_0^1 z(1-z)dz + \int_0^1 z(1-z)\ln z\,dz = \frac{i\pi}{6} - \frac{5}{36}. \tag{15-141}$$

For the second of the two terms, with (15-139)

$$\int_0^1 \{z(1-z)\ln(1-z)dz = -\frac{5}{36}. \tag{15-142}$$

Thus, adding (15-141) and (15-142), as we do in (15-44), we get

$$\int_0^1 \{z(1-z)(\ln(-z) + \ln(1-z))dz = \frac{i\pi}{6} - \frac{5}{36} - \frac{5}{36} = \frac{i\pi}{6} - \frac{5}{18}. \tag{15-143}$$

The first two terms in the second line of (15-44) are identical to the first two terms in the first line of (15-44), so they all sum to $i\pi/3 - 5/9$, and since this is of order 1, we can ignore all of these terms.

15.11 Problems

1. Derive (15-3) using Wick rotation and the following relations from standard integral tables.

    $$\int \frac{x^n dx}{ax^2 + c} = \frac{x^{n-1}}{a(n-1)} - \frac{c}{a}\int\frac{x^{n-2}dx}{ax^2+c} \quad n \neq 1 \qquad\qquad \int\frac{xdx}{ax^2+c} = \frac{1}{2a}\ln(ax^2+c)$$

2. Show (15-27)(a) by integrating with respect to y and using (15-26).

3. From Fig. 15-2 and Feynman's rules, determine (15-106) and (15-107).

4. (Problem added in revision of 2nd edition) From (15-72), derive (15-73). Hint: Multiply both sides of (15-72) by $g_{\mu\nu}$ and sum on repeated indices. Then use (15-61).

Part Four
Application to Experiment

"Knowledge is of no value unless you put it into practice."
Anton Chekhov

Chapter 16

Postdiction of Historical Experimental Results

"It is tough making predictions, especially about the future."

Robert Storm Petersen and others

16.0 Preliminaries

A <u>postdiction</u> (or <u>retrodiction</u>) in science occurs when already gathered data is accounted for by a later theoretical advance. Whereas a prediction is about something expected to be found in the future, a postdiction is about something that has already been found experimentally, but typically for which an adequate theory did not heretofore exist. An example of a postdiction is the perihelion shift of Mercury, which Newtonian mechanics/gravity was unable to account for, but Einstein's general relativity was.

Postdiction (after the experimental fact, not before) in this chapter

Other examples closer to home include the anomalous magnetic moment and the Lamb shift, for which QFT based calculations provided, for the first time, theoretical values matching the experimentally determined ones.

16.0.1 Background

The prior fifteen chapters have focused on the theory of quantum fields, with little discussion of experiment. We now turn our attention to how that theory was applied to determine real world empirical results that no prior theory had been able to explain. In the process we will also show how QFT, just as RQM did before it, postdicts the familiar Coulomb potential of classical physics.

16.0.2 Chapter Overview

In this chapter, from the perspective of the theory we have developed, we will examine

Experimental results postdicted by QFT that we'll examine

- the Coulomb potential in RQM and QFT,
- the anomalous magnetic moment,
- the Lamb shift, and
- concepts involving the hydrogen atom, and the accuracy difference between QED and RQM

16.1 Coulomb Potential in RQM

The Coulomb potential in relativistic theory, be it classical 4D electromagnetism or RQM, can be derived from Maxwell's free field equation in terms of the 4-potential,

Coulomb potential in RQM

$$\partial_\alpha \partial^\alpha A^\mu(x) = 0 . \tag{16-1}$$

For almost the entirety of this book, we have been dealing with plane waves, as depicted in Fig. 2-1, pg. 17. For photons, these are solutions to Maxwell's free field equation that have the specific dependence on time and space of form $e^{\pm ikx}$ in a Cartesian 3D space, i.e.,

Plane waves for most of book: Cartesian coordinates best

$$A^\mu(x) = \sum_{s,\mathbf{k}} \frac{1}{\sqrt{2V\omega_\mathbf{k}}} (\varepsilon_S^\mu(\mathbf{k}) a_S(\mathbf{k}) e^{-ikx} + \varepsilon_S^\mu(\mathbf{k}) a_S^\dagger(\mathbf{k}) e^{ikx}) , \tag{16-2}$$

where $A^\mu = (\Phi, \mathbf{A})$ with Φ being the electric field scalar potential and $\mathbf{A}$ being the magnetic field vector potential. For RQM, $a_s(\mathbf{k})$, $a_s^\dagger(\mathbf{k})$ are constants (usually represented by $A_s(\mathbf{k})$, $A_s^\dagger(\mathbf{k})$ in that theory), For QFT, they are operators (represented by lower case) that destroy and create photons. The plane wave solution form (16-2) is well suited to many problems and experiments.

16.1.1 Classical/RQM Derivation of the Coulomb Potential

But spherical coords best for Coulomb potential

Coulomb's potential, however, is different in that it describes a potential field extending radially outward from a source (charge), so using a spherical coordinate system would be far simpler. In addition, Maxwell's equation inside the charged region (see Chap. 7, relation (7-18), pg. 186) becomes

$$\partial^\alpha \partial_\alpha A^\mu = -e\bar\psi \gamma^\mu \psi \ . \tag{16-3}$$

For N (negatively charged) fermions occupying the charged region (such as electrons in a metallic sphere), we can use a modified form of (16-3),

$$\partial^\alpha \partial_\alpha A^\mu = -Ne\bar\psi \gamma^\mu \psi \ \left(= Ze\bar\psi \gamma^\mu \psi \text{ for } Z \text{ positively charged fermions}\right). \tag{16-4}$$

However, for the Coulomb potential this becomes simplified because that potential is measured in the region *outside* the charged region, where no charged fermion field ψ exists. That is, the fermion field carrying the charge extends throughout the source particle/object to its surface, but no further. Outside the surface, $\psi = 0$, and that is our region of interest.

Sourceless Maxwell's eq in spherical coords

So (16-1) governs in that region, and we prefer a spherical, rather than Cartesian, 3D coordinate system. In such coordinates, (16-1) can be expressed, with $\mu = 0, 1, 2, 3$ representing t, r, θ, ϕ, as

$$\partial^\alpha \partial_\alpha A^\mu = \frac{\partial^2}{\partial t^2} A^\mu - \frac{1}{r}\frac{\partial^2}{\partial r^2}\left(rA^\mu\right) - \frac{1}{r^2 \sin\theta}\frac{\partial}{\partial\theta}\left(\sin\theta \frac{\partial}{\partial\theta} A^\mu\right) - \frac{1}{r^2 \sin^2\theta}\frac{\partial^2}{\partial\phi^2} A^\mu = 0 \ . \tag{16-5}$$

But since the field is symmetric spherically about the origin, where the charge is located, A^μ can only be a function of r and t. The Coulomb potential is static (not a function of t), so (16-5) becomes

Reduces to simple form for A^μ a static function of r

$$\frac{\partial^2}{\partial r^2}\left(rA^\mu\right) = 0 \quad \text{for spherically symmetric, static source} \ . \tag{16-6}$$

The general solution to (16-6), readily shown by substitution, is $A^\mu = \varepsilon_s^\mu C/r + \varepsilon_s^\mu D$, where C and D are constants. Physically, the potential must vanish at infinity, so $D = 0$, and

Solution to that simple form

$$A^\mu \propto \frac{1}{r}\varepsilon_s^\mu \quad \text{or as column matrix,} \quad A^\mu = \begin{bmatrix} A^t \\ A^r \\ A^\theta \\ A^\phi \end{bmatrix} = \frac{1}{r}\begin{bmatrix} A_0^t \\ A_0^r \\ A_0^\theta \\ A_0^\phi \end{bmatrix} = \frac{1}{r}\begin{bmatrix} A_0^0 \\ A_0^1 \\ A_0^2 \\ A_0^3 \end{bmatrix} = \begin{bmatrix} \Phi_0 \\ \end{bmatrix} = \begin{bmatrix} \Phi(r) \\ \mathbf{A}(r) \end{bmatrix}. \tag{16-7}$$

Magnetic field $\mathbf{B}$ (curl of $\mathbf{A}$) in spherical coordinates

From the physical symmetry, the 3D vector potential can only have a radial direction, so it cannot have any component in the angular directions θ or ϕ, i.e., $A_0^\theta = A_0^\phi = 0$. Thus,

$$\mathbf{B} = \nabla \times \mathbf{A} = \varepsilon_s^r \frac{1}{r \sin\theta}\left(\frac{\partial}{\partial\theta}\left(A^\phi \sin\theta\right) - \frac{\partial A^\theta}{\partial\phi}\right) + \varepsilon_s^\theta\left(\frac{1}{r\sin\theta}\frac{\partial A^r}{\partial\phi} - \frac{1}{r}\frac{\partial}{\partial r}\left(rA^\phi\right)\right)$$

$\mathbf{B} = 0$ for spherically symmetric case

$$+ \varepsilon_s^\phi \frac{1}{r}\left(\frac{\partial}{\partial r}\left(rA^\theta\right) - \frac{\partial A^r}{\partial\theta}\right) = \mathbf{0}. \tag{16-8}$$

and no magnetic field is produced. To keep things simple, we can therefore just take $\mathbf{A} = 0$, (i.e., $A_0^r = 0$ also) without loss of generality (in this spherically symmetric case).

So, without loss of generality can assume $\mathbf{A} = 0$

From boundary conditions on the surface of the charged spherical source ($A^t = \Phi$ just on either side of the surface must be equal, though we won't go through the formal mathematics of it all), we obtain the constant A_0^t. We then end up with (16-7) having the well-known Coulomb potential (in Heaviside-Lorentz units) as the timelike component of the photon field A^μ,

$$A^\mu = \begin{bmatrix} \Phi \\ 0 \\ 0 \\ 0 \end{bmatrix} = \begin{bmatrix} -eN/(4\pi r) \\ 0 \\ 0 \\ 0 \end{bmatrix} \left(= \begin{bmatrix} Ze/(4\pi r) \\ 0 \\ 0 \\ 0 \end{bmatrix} \begin{array}{l} \text{for a nucleus of } Z \text{ positively} \\ \text{charged protons approximated} \\ \text{by Coulomb potential} \end{array} \right). \quad (16\text{-}9)$$

16.1.2 Using the Coulomb Potential in Relativistic Hydrogen Atom

Recall from Chap. 7 (Sects. 7.1.4, and 7.2.1 pgs. 184-186) that our governing interaction equations for coupled photon-fermion fields are (16-3) and the full (interacting) Dirac equation

$$\left(i\gamma^\mu \partial_\mu - m \right)\psi = -e\gamma^\mu \psi A_\mu. \quad (16\text{-}10)$$

To solve the H atom case exactly, we would need to solve coupled equations like (16-3)/(16-4) and (16-10), because the fermion field of the nucleus, the fermion field of the orbital electron, and the electromagnetic (photon) field all interact with one another throughout the entire region of the atom. Finding a closed form solution for this is essentially impossible.

To get a good approximation for the relativistic atom, however, we can assume the A^μ field, which really results from both the nucleus and the orbital electron (and for which the nuclear fermion field extends outside a clear spherical boundary of the nucleus), is just due to the nucleus and has Coulomb potential form as in (16-9). With that approximation substituted into (16-10), one then goes about solving the resulting equation. This is just the procedure we outlined in the first of the above referenced sections.

Doing that provides a more accurate solution (the relativistic solution) to the hydrogen atom, the orbital energy levels (eigenvalues), and thus the spectral line distribution seen in measurements. However, one might expect the resulting solution, due to the approximation (16-9) does not precisely match experiment. One would be right. One such discrepancy, a subtle but distinct one, is known as the Lamb shift, a slight shifting of the spectral lines in their actual measurement from that predicted by the above analysis approach. We discuss the Lamb shift and its successful postdiction via QFT later in this chapter.

RQM solution to H atom assumes Coulomb potential, though this is only an approximation

16.2 Coulomb Potential in QFT

One could simply assume in QFT that the form of Coulomb potential is same as that in RQM, since we found throughout our development of both theories that they paralleled one another in terms of the governing equations and solution forms, and differed only in the interpretation of the solution coefficients as constants or operators.

Doing so in the above described hydrogen atom analysis, for example, would have provided field eigen solutions of particular form, the same form as the state eigen solutions of RQM. The operators of those fields would create and destroy states mirroring those solutions, i.e., with the same eigen energies, spins, etc., and thus the same spectral line predictions.

However, for the sake of completeness, and to justify the parallel solutions argument for QFT, we present a derivation of the Coulomb potential from the perspective of QFT.

Coulomb potential in QFT

16.2.1 Repulsive Coulomb Scattering Equivalence to Møller Scattering

Repulsive Coulomb scattering can be represented by Møller scattering as shown in Fig. 16-1, where the source charge particle is spherical (has a radial distribution of its radiation.)

If the incoming particles in Fig. 16-1 are indistinguishable, such as two electrons, we need to include both diagrams to determine the amplitude. But, if they are distinguishable, such as an electron and a muon, then we only need to consider the LH diagram. (Because there is no indeterminancy in which original particle mutated into which final particle.) Further, the classical Coulomb potential is always between macro (distinguishable) objects. So, to make things simpler, we will assume the particles are distinguishable and examine the transition amplitude for only the LH diagram in Fig. 16-1.

We will also assume non-relativistic speeds of our incoming and outgoing particles, as that is typically the case for Coulomb scattering. (And it makes our calculations simpler.) One can think of the particle labeled 1 as the source, whose radiated virtual particle affects the particle labeled 2.

QFT repulsive Coulomb potential

Assume distinguishable particle to keep things simple

Assume non-relativistic speeds for electrons

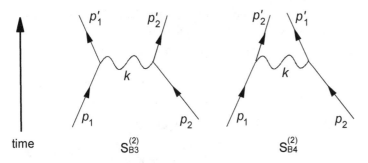

Figure 16-1. Møller Scattering

Analyze Møller scattering

There is one other thing to consider. Everything we have done so far in QFT has been with plane waves, not spherical waves, which one would expect to emanate from a point (or spherical) charge source, as the Coulomb potential does. A potential more suited to our theory, as developed so far, would be that due to an infinite (or effectively so for our purposes) plane on which a uniform charge is distributed. The potential between two flat plate conductors (like in many capacitors), at least one of which is charged, is a good example. Any related waves would then be plane waves.

Plane waves best suited for planar potentials

The potentials and the concomitant forces they produce, for certain geometric distributions of charge, which we should have seen in prior courses, are reviewed in the table below.

Table 16-1. Potentials and Resulting Force Fields for Various Geometries

Charge distribution	Potential V	Force $= -\partial V/\partial \mathbf{x}$
Planar	Kx	$-\partial V/\partial x = -K$
Cylindrical	$K' \ln r$	$-\partial V/\partial r = -K'/r$
Spherical	K''/r (Coulomb)	$-\partial V/\partial r = K''/r^2$

16.2.2 Relations We'll Need

Fourier Transform Pairs

Recall that the Fourier transforms of each other, $g(k)$ and $f(x)$, are defined as

$$g(k) = \int_{-\infty}^{+\infty} f(x) e^{-ikx} dx \quad \Leftrightarrow \quad f(x) = \frac{1}{2\pi} \int_{-\infty}^{+\infty} g(k) e^{ikx} dk . \tag{16-11}$$

From almost any table of <u>Fourier transform</u> pairs, we can get the following relations.

Fourier transform pairs we'll need

Fourier Transform Pairs		
$g(k)$		$f(x)$
$\dfrac{1}{k}$	$\Leftrightarrow$	$\dfrac{i}{2} sgn\, x \;\left(\dfrac{i}{2} \text{ for } x > 0 \quad -\dfrac{i}{2} \text{ for } x < 0 \right)$
$\dfrac{1}{k^2}$	$\Leftrightarrow$	$-\dfrac{1}{2}|x|$

A Result We'll Need from NRQM

In deriving the Coulomb potential using QFT under non-relativistic conditions, we will need to refer to a result from NRQM, namely the scattering of two charged particles from one another. In NRQM, actually, the charge on one particle is assumed to provide a potential $V(\mathbf{x})$ to which the other particle wave function responds. So, we consider the behavior of one wave function in a given potential.

We will be analyzing assuming non-relativistic conditions

Getting the transition amplitude in NRQM in terms of a potential field $V(\mathbf{x})$ is fairly extensive to develop[1], so we will have to satisfy ourselves by simply quoting the result here.

To get the NRQM transition amplitude, one usually employs the Born approximation, which assumes the incoming wave remains undistorted even within the scattering region. In fact, it is distorted, but modestly so, and the approximation gives results quite close to experiment. Thus, (with V the volume of the interaction, and $\tilde{V}(\mathbf{k})$ the Fourier transform of $V(\mathbf{x})$) the <u>Born approximation scattering amplitude in NRQM</u> in terms of a <u>single particle in a potential $V(\mathbf{x})$</u> is

So, can compare our results to the Born approx amplitude result from NRQM

$$S_{fi} = \frac{i}{V}\tilde{V}(\mathbf{k})2\pi\delta\left(E_f - E_i\right) \quad \text{where } \mathbf{k} = \mathbf{p}_i - \mathbf{p}_f \quad \left\{ \begin{array}{l} \text{NRQM, single particle in} \\ \text{external potential field } \tilde{V}(\mathbf{k}) \end{array} \right. \quad (16\text{-}12)(a)$$

This is true for a potential $V(\mathbf{x})$ of any form in $\mathbf{x}$. Note that the Fourier transform of that potential in momentum space is what appears in (16-12)(a).

From that comparison, we can pick out the potential energy part from our QFT amplitude

This result extends to the case where <u>two particles interact</u>, such as in Fig. 16-1, where $\tilde{V}(\mathbf{k})$ is the Fourier transform of $V(\mathbf{x})$, the potential field one particle feels due to the other.

$$S_{fi} = \frac{i}{V^2}\tilde{V}(\mathbf{k})(2\pi)^4\delta^{(4)}\left(p_f - p_i\right) \quad \text{where } \mathbf{k} = \mathbf{p}_i - \mathbf{p}_f \quad \left\{ \begin{array}{l} \text{two particles interacting, } \tilde{V}(\mathbf{k}) = \\ \text{potential field one feels from other} \end{array} \right. \quad (16\text{-}12)(b)$$

We will discuss the physics behind (16-12)(a) and (b) more in Chap. 17.

16.2.3 A Detour for the Planar Potential

As noted above, we have been dealing throughout this book with plane wave fields and particles. If such particles are charged, as electrons are, then potentials associated with them would be planar in form. So, it will be easier if we investigate plane wave potentials that arise from plane wave fields/particles, first. With that as training, in the next section, we will attack the Coulomb potential (which is spherical in form and more difficult to handle with plane waves). Thus, we now

i) derive the transition amplitude for plane waves (meaning a planar charge distribution) in Møller scattering using QFT at non-relativistic speeds, then

ii) compare the result with (16-12)(b) to deduce the potential in 3-momentum space, and finally,

iii) Fourier transform the result to obtain that potential in 3D physical space.

The transition amplitude for the LHS of Fig. 16-1 can be found using Feynman's rules, or by reference to Chap. 8, relation (8-69), pg. 229, to be (where the subscript "disting" means distinguishable particles)

$$S_{\substack{M\text{øller} \\ \text{disting}}} = \left(\prod_{\substack{\text{all external} \\ \text{fermions} \\ \mathbf{p}}} \sqrt{\frac{m}{VE_{\mathbf{p}}}} \right)(2\pi)^4\delta^{(4)}\left(p_1' + p_2' - p_1 - p_2\right)\mathcal{M}_{B3}^{(2)} \quad (16\text{-}13)$$

Our S matrix component and its Feynman amplitude

$$\mathcal{M}_{B3}^{(2)} = e^2\bar{u}_{r_1'}(\mathbf{p}_1')\gamma^\mu u_{r_1}(\mathbf{p}_1)iD_{F\mu\nu}\left(k = p_1 - p_1'\right)\bar{u}_{r_2'}(\mathbf{p}_2')\gamma^\nu u_{r_2}(\mathbf{p}_2).$$

where a subscript such as r_2 means the r spin state ($r_2 = 1, 2$) of particle #2. We will be considering non-relativistic speeds, where $E \approx m$ and $|\mathbf{p}| \ll m$ So our spinors will, to good approximation, become (with subscripts here referring to the spin state r)

$$u_{r=1}(\mathbf{p}) = \sqrt{\frac{E+m}{2m}}\begin{pmatrix} 1 \\ 0 \\ \dfrac{p^3}{E+m} \\ \dfrac{p^1+ip^2}{E+m} \end{pmatrix} \Rightarrow \begin{pmatrix} 1 \\ 0 \\ 0 \\ 0 \end{pmatrix} \qquad u_{r=2}(\mathbf{p}) = \sqrt{\frac{E+m}{2m}}\begin{pmatrix} 0 \\ 1 \\ \dfrac{p^1-ip^2}{E+m} \\ \dfrac{-p^3}{E+m} \end{pmatrix} \Rightarrow \begin{pmatrix} 0 \\ 1 \\ 0 \\ 0 \end{pmatrix}. \quad (16\text{-}14)$$

Spinors simplify for non-relativistic (NR) cases

For non-relativistic conditions, spinors become effectively independent of $\mathbf{p}$, so

---

[1] See Gasiorowicz, S., *Quantum Physics*, (Wiley 1974), pg. 398, relation (24-77) where he differs from us in that i) the delta function of his (24-76) is implicit in his (24-77) and (24-78), and ii) we have an extra factor of i due to differences in analysis method between NRQM and QFT (in both cases the probability of interaction, i.e., the complex conjugate of the amplitude times the amplitude, is the same).

$$u_{r_1=1}(\mathbf{p}_1) \approx u_{r_2=1}(\mathbf{p}_2) \qquad u_{r_1=2}(\mathbf{p}_1) \approx u_{r_2=2}(\mathbf{p}_2). \qquad (16\text{-}15)$$

Using γ^0 from

$$\gamma^0 = \begin{bmatrix} 1 & & & \\ & 1 & & \\ & & -1 & \\ & & & -1 \end{bmatrix} \gamma^1 = \begin{bmatrix} & & & 1 \\ & & 1 & \\ & -1 & & \\ -1 & & & \end{bmatrix} \gamma^2 = \begin{bmatrix} & & & -i \\ & & i & \\ & i & & \\ -i & & & \end{bmatrix} \gamma^3 = \begin{bmatrix} & & 1 & \\ & & & -1 \\ -1 & & & \\ & 1 & & \end{bmatrix}, \qquad (16\text{-}16)$$

Recalling gamma matrices

we find

$$\bar{u}_{r_1'=1}(\mathbf{p}_1') = u_{r_1'=1}^\dagger(\mathbf{p}_1')\gamma^0 \Rightarrow \begin{pmatrix} 1 & 0 & 0 & 0 \end{pmatrix} \qquad \bar{u}_{r_2'=1}(\mathbf{p}_2') = u_{r_2'=1}^\dagger(\mathbf{p}_2')\gamma^0 \Rightarrow \begin{pmatrix} 1 & 0 & 0 & 0 \end{pmatrix}$$

$$\bar{u}_{r_1'=2}(\mathbf{p}_1') = u_{r_1'=2}^\dagger(\mathbf{p}_1')\gamma^0 \Rightarrow \begin{pmatrix} 0 & 1 & 0 & 0 \end{pmatrix} \qquad \bar{u}_{r_2'=2}(\mathbf{p}_2') = u_{r_2'=2}^\dagger(\mathbf{p}_2')\gamma^0 \Rightarrow \begin{pmatrix} 0 & 1 & 0 & 0 \end{pmatrix} \;(16\text{-}17)$$

$$\bar{u}_{r_1'=1}(\mathbf{p}_1') \approx \bar{u}_{r_2'=1}(\mathbf{p}_2') \qquad \bar{u}_{r_1'=2}(\mathbf{p}_1') \approx \bar{u}_{r_2'=2}(\mathbf{p}_2').$$

Finding NR adjoint spinors

So all we have to worry about in (16-13) are the first two spinor components. Everything else is zero. Thus, we only have to consider the upper left 2X2 matrix part of the gamma matrices (16-16). These are zero for each γ^i (i = 1,2,3.) In other words, we only have to determine the part of (16-13) having $\gamma^\mu = \gamma^0$ and $\gamma^\nu = \gamma^0$, and the upper left 2 X 2 part of that is just the identity matrix.

All but upper two components contribute zero

So for non-relativistic conditions, (16-13) becomes, where we assume the case where $r_1 = 1$ and $r_2 = 1$,

NR amplitude

$$\mathcal{M}_{B3}^{(2)} = e^2 \underbrace{\bar{u}_{r_1'}(\mathbf{p}_1')}_{\substack{\text{must} = \\ (1\ 0)}} \begin{bmatrix} 1 & \\ & 1 \end{bmatrix}\begin{bmatrix} 1 \\ 0 \end{bmatrix} \frac{-ig_{00}}{k^2 + i\varepsilon} \underbrace{\bar{u}_{r_2'}(\mathbf{p}_2')}_{\substack{\text{must} = \\ (1\ 0)}} \begin{bmatrix} 1 & \\ & 1 \end{bmatrix}\begin{bmatrix} 1 \\ 0 \end{bmatrix} + \underbrace{\left(\text{terms with } \gamma^i \right)}_{\approx 0}. \qquad (16\text{-}18)$$

This is only non-zero if the adjoint spinors both equal (1 0). As an aside, this means the spins of the incoming and outgoing particles each remain unchanged.

Zero unless same spin out as in

Additionally, for elastic scattering in the center of mass (COM) system for two particles of equal mass, the two particles start with the same speed and end with the same speed, but they exchange velocities (directions). (We will look at this more closely in Chap. 17.) Thus, neither changes its kinetic energy, and thus, no energy is carried by the virtual particle from one real particle to the other. However, there is 3-momentum exchange because the direction of each particle's velocity changes. Hence $k^2 = -\mathbf{k}^2$ in the propagator. So (16-18) is

In COM frame, $k^2 = -\mathbf{k}^2$ in propagator

$$\mathcal{M}_{B3}^{(2)} = -ie^2 \frac{1}{-\mathbf{k}^2 + i\varepsilon}, \qquad (16\text{-}19)$$

Amplitude for this case

and the transition amplitude (S matrix element), ignoring the small ε, is

$$S_{\substack{M\o ller \\ disting}} = ie^2 \frac{1}{V^2}(2\pi)^4 \delta^{(4)}(p_1' + p_2' - p_1 - p_2)\frac{1}{\mathbf{k}^2} = \left\langle p_1', p_2' \left| S_{\substack{M\o ller \\ disting}}^{oper} \right| p_1, p_2 \right\rangle. \qquad (16\text{-}20)$$

Comparison with Born approx yields potential in $\mathbf{k}$ space

This, since it is for non-relativistic conditions, should equal (16-12)(b). Comparing the two, we conclude that the potential in 3-momentum space is

$$\tilde{V}(\mathbf{k}) = \frac{e^2}{\mathbf{k}^2}. \qquad (16\text{-}21)$$

To find the potential in physical space, consider $\mathbf{k}$ in the x^1 direction and use the second row of the table of Fourier pairs above to find

$$V(x^1) = -\tfrac{1}{2}e^2\left|x^1\right| \qquad = -\tfrac{1}{2}e^2 x^1 \quad \text{for } x^1 > 0. \qquad (16\text{-}22)$$

Fourier transform to position space gives us classical planar potential

(16-22) is of the form of a planar potential. This is not quite the same as that for a parallel plate capacitor. In the latter case, charged fields are confined to the parallel plates. In this example, the charged fields extend throughout the volume V, overlapping one another throughout the volume. However, the principle of plane waves carrying charge yielding a potential of form like that in the first line of Table 16-1 is seen to hold in QFT.

16.2.4 Finally, the Repulsive Coulomb Potential via QFT

Now, the QFT Coulomb potential

To get the Coulomb potential in QFT (without reference to RQM) we can take two routes.

1) Re-derive QFT in spherical coordinates.

2) Use the plane wave derivation, deduce the 3-momentum space potential $\tilde{V}(\mathbf{k})$, and in Fourier transforming that, change from Cartesian to spherical 3D coordinates.

Instead of re-doing QFT in spherical cords, we'll just go to spherical and Fourier transform to position space

The first of these would entail a great deal of work, not the least of which would be to express our four component spinors in spherical, not Cartesian, form. Ugh…, let's go with the second route.

For that, we can use the same low velocity Møller scattering amplitude (16-20) that we already derived. From that and (16-12)(b), we get the same momentum space potential (16-21).[1] The only issue remaining is to transform that to position space, expressed in spherical coordinates. Start with

*Start with plane wave potential in **k** space we already derived*

$$V(\mathbf{x}) = \frac{1}{(2\pi)^3}\int \tilde{V}(\mathbf{k})e^{i\mathbf{k}\cdot\mathbf{x}}d^3k = \frac{1}{(2\pi)^3}\int_{-\infty}^{\infty}\frac{e^2}{\mathbf{k}^2}e^{i\mathbf{k}\cdot\mathbf{x}}d^3k , \qquad (16\text{-}23)$$

and note that for spherical coordinates r, θ, ϕ, in $\mathbf{k}$ space we can align the k^3 axis (the $\theta = 0$ direction) with the direction of the position vector $\mathbf{x}$. With that alignment and with $\mathbf{x}$ representing any point in position space, $\mathbf{k}\cdot\mathbf{x} = |\mathbf{k}||\mathbf{x}|\cos\theta$, where we now use the symbol k as $= |\mathbf{k}|$ (usually it has been shorthand for k^μ) and $|\mathbf{x}| = r$, for the remainder of this section. Thus, $\mathbf{k}\cdot\mathbf{x} = kr\cos\theta$ and $\mathbf{k}^2 = k^2$ for the temporary notation.

The differential volume element d^3k in spherical coordinates in $\mathbf{k}$ space is $k^2\sin\theta d\theta d\phi dk$. Given this, (16-23) becomes

Fourier transform to spherical position coordinates

$$V(r) = \frac{1}{(2\pi)^3}\int_0^\infty\int_0^\pi\int_0^{2\pi}\frac{e^2}{k^2}e^{ikr\cos\theta}k^2\underbrace{\sin\theta d\theta}_{-d(\cos\theta)}d\phi dk = \frac{-e^2}{(2\pi)^3}2\pi\int_0^\infty\int_0^\pi e^{ikr\cos\theta}\underbrace{d(\cos\theta)}_{\text{take as }u}dk$$

$$= \frac{-e^2}{(2\pi)^2}\int_0^\infty\int_1^{-1}e^{ikru}du\,dk = \frac{-e^2}{(2\pi)^2}\int_0^\infty\left(\frac{e^{ikru}}{ikr}\right)_1^{-1}dk = \frac{-e^2}{(2\pi)^2}\int_0^\infty\left(\frac{e^{-ikr}-e^{ikr}}{ikr}\right)dk \qquad (16\text{-}24)$$

$$= \frac{e^2}{(2\pi)^2}\int_0^\infty\left(\frac{e^{ikr}-e^{-ikr}}{ikr}\right)dk .$$

The last part can then be re-expressed as an integral from $-\infty$ to $+\infty$, as follows (where in going to the last term in the first row below, we substitute variable $k \to -k$).

$$V(r) = \frac{e^2}{(2\pi)^2}\left(\int_0^\infty\left(\frac{e^{ikr}}{ikr}\right)dk - \int_0^\infty\left(\frac{e^{-ikr}}{ikr}\right)dk\right) = \frac{e^2}{(2\pi)^2}\left(\int_0^\infty\left(\frac{e^{ikr}}{ikr}\right)(dk) - \int_0^{-\infty}\left(\frac{e^{-i(-k)r}}{i(-k)r}\right)(-dk)\right)$$

$$= \frac{e^2}{(2\pi)^2}\left(\int_0^\infty\left(\frac{e^{ikr}}{ikr}\right)(dk) - \int_0^{-\infty}\left(\frac{e^{ikr}}{ikr}\right)dk\right) = \frac{e^2}{(2\pi)^2}\left(\int_0^\infty\left(\frac{e^{ikr}}{ikr}\right)(dk) + \int_{-\infty}^0\left(\frac{e^{ikr}}{ikr}\right)dk\right) \qquad (16\text{-}25)$$

$$= \frac{e^2}{(2\pi)^2}\int_{-\infty}^\infty\frac{e^{ikr}}{ikr}dk = \frac{e^2}{ir}\frac{1}{2\pi}\underbrace{\frac{1}{2\pi}\int_{-\infty}^\infty\frac{e^{ikr}}{k}dk}_{=\,i/2\text{ from table}} .$$

The last integral above is from the first row in our Table of Fourier Transform Pairs on pg. 405. Thus, we get

We get repulsive Coulomb potential

---

[1] In Coulomb scattering, however, we typically work with a small test particle acting under the influence of a very massive charged object, which due to its mass, has negligible change in velocity from the interaction. So, we effectively work in the frame of the large object producing the Coulomb potential. In that frame, the test particle final velocity changes direction from its initial velocity, but the magnitude stays effectively the same. Thus, there is effectively no change in kinetic energy of the test particle, so the virtual particle mediating the scattering carries no energy, only 3-momentum. Thus, again for this case, we take $k^2 = -\mathbf{k}^2$ as we did in (16-19). We review this in more detail in Chap. 17.

$$V(r) = \frac{e^2}{4\pi r}, \tag{16-26}$$

the Coulomb repulsive potential in Heaviside-Lorentz units (which in our case are also naturalized, although neither $c = 1$ nor $\hbar = 1$ appears in (16-26)). QED, for QED theory potential.

16.2.5 The Attractive Coulomb Potential via QFT

For oppositely charged particles, we have Bhabha scattering (without annihilation) instead of Møller, Fig. 16-2 instead of Fig. 16-1. In Chap. 8, we derived the transition amplitude for this, but our symbol use here is different because we want to match incoming and outgoing fermion symbols with those we used for Møller scattering in Fig. 16-1 and (16-13). So, best to deduce the amplitude with the symbols of Fig. 16-2 directly from Feynman rules (Box 8-3, pg. 236), rather than looking back at the (different symbol) transition amplitude we found in Chap. 8.

For attractive Coulomb potential, do same procedure but with Bhabha scattering

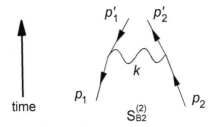

Figure 16-2. Bhabha Scattering with No Annihilation

Thus, the transition amplitude corresponding to Fig. 16-2 is

$$S^{(2)}_{\substack{Bhabha \\ no\ annih}} = S^{(2)}_{B2} = \left(\prod_{\mathbf{p}}^{\substack{all\ external \\ fermions}} \sqrt{\frac{m}{VE_{\mathbf{p}}}} \right) (2\pi)^4\, \delta^{(4)}\left(p'_1 + p'_2 - (p_1 + p_2)\right) \mathcal{M}^{(2)}_{B2} \tag{16-27}$$

$$\mathcal{M}^{(2)}_{B2} = -e^2 \bar{v}_{r_1}(\mathbf{p}_1) \gamma^\mu v_{r'_1}(\mathbf{p}'_1) i D_{F\mu\nu}\left(p_1 - p'_1\right) \bar{u}_{r'_2}(\mathbf{p}'_2) \gamma^\nu u_{r_2}(\mathbf{p}_2).$$

Associated Feynman amplitude

The sign of (16-27) is different from that of (16-13), because in (16-13) we would have had to move the destruction operator associated with $\mathbf{p}_1$ past one fermion operator (that associated with $\mathbf{p}'_2$) to have normal ordering of operators. In (16-27), to get the same ordering of 3-momenta as we had in (16-13), we would have to move the operator associated with $\mathbf{p}_1$ past two other fermion operators. Each switching of adjacent fermion operators multiplies the amplitude by -1.

So, (16-27) is similar to (16-13) except for the sign and except for the spinors on the left representing anti-fermions instead of fermions. At non-relativistic speeds, as we saw for (16-14), these type spinors reduce to

$$v_{r=1}(\mathbf{p}) = \sqrt{\frac{E+m}{2m}} \begin{pmatrix} \dfrac{p^3}{E+m} \\ \dfrac{p^1 + ip^2}{E+m} \\ 1 \\ 0 \end{pmatrix} \Rightarrow \begin{pmatrix} 0 \\ 0 \\ 1 \\ 0 \end{pmatrix} \qquad v_{r=2}(\mathbf{p}) = \sqrt{\frac{E+m}{2m}} \begin{pmatrix} \dfrac{p^1 - ip^2}{E+m} \\ \dfrac{-p^3}{E+m} \\ 0 \\ 1 \end{pmatrix} \Rightarrow \begin{pmatrix} 0 \\ 0 \\ 0 \\ 1 \end{pmatrix}. \tag{16-28}$$

By doing Prob. 1, you can show that, at low speeds, (16-27) reduces to

$$\mathcal{M}^{(2)}_{B2} = ie^2 \frac{1}{-\mathbf{k}^2 + i\varepsilon}, \tag{16-29}$$

*Associated NR amplitude in **k** space*

which is the same as (16-19) except for the sign. Following the steps we used to get to (16-21), we have

$$\tilde{V}(\mathbf{k}) = -\frac{e^2}{\mathbf{k}^2}, \tag{16-30}$$

Same as repulsive case except for sign

which only differs from (16-21) in sign. All of the steps (16-23) to (16-26) are exactly the same here, again except for sign, and so result in

$$V(r) = -\frac{e^2}{4\pi r}, \tag{16-31}$$

the Coulomb potential for attraction between unlike charges. QED.

16.2.6 Importance of the Sign of the Amplitude

Since the probability of interaction is the square of the absolute value of the transition amplitude, the sign of that amplitude is irrelevant for probability calculations. However, the sign, as we have seen above, does tell us whether the interaction is attractive or repulsive.

Note, from Feynman rule #9 (Chap. 8, Box 8-3, pg. 236), repeated below, that the sign of the amplitude depends on the number of fermion operator interchanges we need to normal order the S operator from which the S matrix component S_{fi} is determined.

Feynman rule #9: Multiply the expression by (–1) for each interchange of neighboring fermion operators (each associated with a particular spinor factor) which would be required to place the expression in appropriate normal order. "Appropriate", when we are adding (or comparing as in this chapter) sub amplitudes, means each sub amplitude must be in the same, not just any, normal order of destruction and creation operators.

In applying Feynman rules to get the Møller scattering amplitude (16-13), we needed to consider that the destruction operator $c_{r_1}(\mathbf{p}_1)$ associated with $u_{r_1}(\mathbf{p}_1)$ had to be interchanged with the creation operator $c_{r_2'}^{\dagger}(\mathbf{p}_2')$ associated with $\bar{u}_{r_2'}(\mathbf{p}_2')$ in order to get a normal ordered S operator.

This was the only exchange needed, and gave us an overall plus sign in front of the Feynman amplitude of (16-13).

Two such interchanges are needed for Bhabha scattering of (16-27), and this leaves us with an overall minus sign for the Feynman amplitude. Note the needed interchanges are those that place the 3-momenta in the same order in both sub-amplitudes. This is what is meant by "appropriate normal order" in Feynman rule #9.

The signs in each case are carried through to the final results (16-26) and (16-31), and importantly, we see that QFT gives us just what we have known about interacting charges since our earliest physics courses.

16.2.7 Coulomb Potential and Radiative Corrections

Recall from Chap. 12 that, due to higher order radiative corrections from renormalization, the measured value for e varies with energy level of the interaction. That is, $e = e(p)$, where p here symbolizes that energy level. In effect, at higher p, we have higher coupling e. We have a running coupling "constant".

But higher energy (and thus momentum) corresponds in measurement to shorter distances. So, e in (16-26) and (16-31) effectively gets greater for shorter distances. So, if we make measurements very close, for example, to the nucleus of the hydrogen atom, we would see a potential that varied from the classical Coulomb potential. The closer we got, the greater would be this variation.

At classical distance scales, e is effectively constant, and the pure Coulomb potential rules.

16.3 Other Potentials and Boson Types

Note that the scalar and photon propagators, $\Delta_F(k) = \dfrac{1}{k^2 - \mu^2 + i\varepsilon}$ and $D_{F\mu\nu}(k) = \dfrac{-g_{\mu\nu}}{k^2 + i\varepsilon}$,

have opposite signs. Assuming a particular scalar boson carried a force between fermions, the steps to find the associated potential would mirror those for finding the Coulomb potential above in virtually all regards except sign. There would be an associated scalar field charge, similar to the electromagnetic charge, but the final result, for particles with the same scalar field charge, would have a minus sign in the potential rather than a plus sign, as in (16-26). Thus, for forces carried by scalar bosons, like-charged real particles attract. (See Prob. 2 for details on a massive boson scalar force potential, known as the Yukawa potential.)

In more advanced QFT, one can show that a spin 2 boson (such as the graviton, hypothesized to mediate the gravitational force) also attracts particles that have like spin 2 field charges. For gravity, the spin 2 field charge is mass. So, positive mass particles attract one another.

In general, even number spin bosons mediate an attractive force between like charges. Odd number spin bosons mediate a repulsive force between like charges.

Wholeness Chart 16-1. Boson Spin and Like Charges

	Spin 0	Spin 1	Spin 2
Like charges	attract	repel	attract
Example	(pseudo) scalar mesons	photons	gravitons
Potential	Yukawa	Coulomb	gravitational
Charge type	Yukawa charge	electric	mass

16.4 Anomalous Magnetic Moment

16.4.1 Sophomore Physics Review

As a refresher, an elementary review of the electron magnetic moment is hereby provided.

Consider a circular loop of current I encompassing an area A, which acts like a magnetic dipole, i.e., acts just as if fictitious positive and negative "magnetic charges" were separated by a small distance. If the current loop is placed in an external magnetic field $\mathbf{B}^e$, the torque τ it experiences (which can be visualized as equal magnitude, opposite direction forces on the two fictitious "magnetic charges") is (where $\mathbf{A}$ has magnitude of area A and direction normal to the plane of the loop aligned with the thumb of the right hand when the fingers curve in the direction of the current)

$$\tau = I\mathbf{A} \times \mathbf{B}^e = \mu \times \mathbf{B}^e, \tag{16-32}$$

where $\mu = I\mathbf{A}$ is called the <u>magnetic moment of the current loop</u>. The energy of the loop/external field, with θ the angle between μ and $\mathbf{B}^e$, where we define the $\theta = \pi/2$ position as zero potential energy, is

$$E_{loop/field} = \int_{\pi/2}^{\theta} T d\theta' = \int_{\pi/2}^{\theta} \mu B^e \sin\theta' d\theta' = -\mu B^e \cos\theta = -\mu \cdot \mathbf{B}^e. \tag{16-33}$$

If the current is composed of a single particle of charge $-e$ (e.g., as one would have in the Bohr theory orbit of an electron in an atom), its speed is v, its time for one orbit is T_{orbit}, and the circular area A has radius r, then the magnetic moment due to the orbital angular momentum of the charge is

$$\mu = I\mathbf{A} = \frac{-e}{T_{orbit}} \pi r^2 \mathbf{e}_\perp = \frac{-ev}{2\pi r} \pi r^2 \mathbf{e}_\perp = -\frac{1}{2} evr \, \mathbf{e}_\perp \xrightarrow{\ L=mvr\ } \mu = -\frac{1}{2} \frac{e}{m} \mathbf{L}, \tag{16-34}$$

where $\mathbf{e}_\perp$ is a unit vector pointing in the direction of the right hand thumb above, and $\mathbf{L}$ is orbital angular momentum. In an atomic orbit, angular momentum magnitude is

$$L = \hbar m_l \tag{16-35}$$

with m_l an orbital quantum number. So with (16-34), we can define the <u>Bohr magneton</u> μ_B via

$$\text{(orbital)} \quad \mu = -\frac{1}{2} \frac{e}{m} \mathbf{L} = -\frac{1}{2} \frac{e\hbar}{m} m_l \mathbf{e}_\perp = -\mu_B m_l \mathbf{e}_\perp \qquad \mu_B = \frac{1}{2} \frac{e\hbar}{m}. \tag{16-36}$$

As to the electron itself, one can view it classically as a charge that is distributed internally, rather than being point-like, and that rotates, or "spins" around some internal axis. So, in effect, we would have a circulating current loop of sorts similar to that described above for an atom. In quantum theory, that spin of the electron is quantized, and the intrinsic (different from orbital contribution) angular momentum is spin angular momentum $\mathbf{S} = \hbar m_s \mathbf{e}_\perp = \pm \hbar/2 \, \mathbf{e}_\perp$ ($m_s = \pm \frac{1}{2}$ is spin quantum number). So, one might consider

$$\text{(spin)} \quad \mu = \pm\mu \mathbf{e}_\perp \xrightarrow[\text{loop?}]{\overset{?}{\text{current}}} = -\frac{1}{2} \frac{e}{m} \mathbf{S} = \pm \frac{1}{2} \frac{e\hbar}{m} \frac{1}{2} \mathbf{e}_\perp = \mu_B m_s \mathbf{e}_\perp = \pm \frac{\mu_B}{2} \mathbf{e}_\perp \qquad (\mu = |\mu|). \tag{16-37}$$

Anomalous magnetic moment

Review of basic physics

Torque on classical magnetic moment $\mu = I\mathbf{A}$

μ in terms of angular momentum $\mathbf{L}$

Magnitude of angular mom L of atomic orbit

Above used to define Bohr magneton μ_B

For electron, we know angular momentum (spin), but not internal charge distribution

However, the RHS of (16-37) is derived assuming charge is distributed as a neat current loop, as in (16-36), which is naïve. Given the unknown nature of this distribution, researchers introduced a constant g, called the gyromagnetic ratio[1] or the g-factor, which could be determined by experiment. So, the most general form for the magnetic moment μ of the electron and its magnitude μ would be

The unknown charge distribution contribution labeled as g, gyromagnetic ratio

$$\boldsymbol{\mu} = \pm \mu \mathbf{e}_\perp = \pm g \frac{\mu_B}{2} \mathbf{e}_\perp \qquad \mu = g \frac{\mu_B}{2} = g \frac{e\hbar}{4m} \left(= g \frac{e}{4m} \text{ in natural units} \right). \qquad (16\text{-}38)$$

If our naïve analysis (current distributed in a neat loop) were correct, then the gyromagnetic ratio g would be found equal to 1.

16.4.2 Result of Experiment

In fact, when Stern-Gerlach did an experiment to measure this, they got a value of approximately twice what naïve analysis predicted, i.e., $g \approx 2$. A simple factor of two was one thing, but perhaps more troubling was that it was not exactly two, but $g = 2.00232$ (for the electron).

NRQM can't predict g, but experiment measured it as 2.00232

16.4.3 Postdiction by RQM and QFT

Dirac's RQM came partially to the rescue. It gave a value of g = exactly 2. That was big step. The difference between 2 and 2.00232 was considered an anomaly, hence the famous nomenclature, the anomalous magnetic moment of the electron.

RQM predicts g = exactly 2

QFT put the final piece in the puzzle by yielding a theoretical value for the gyromagnetic ratio matching experiment to better than one part in a billion, making it the most accurately verified theoretical prediction in the history of physics.

QFT predicts g extremely accurately

The level of accuracy provided by QFT depends, as one might expect, on the perturbation order to which calculations are carried out. At lowest order, QFT, like RQM, predicts a value of precisely $g = 2$. At second and third orders, one obtains the values shown in Wholeness Chart 16-2 below.

**Wholeness Chart 16-2. Theoretical and Experimental Values
for Electron Gyromagnetic Ratio g**

Experiment	NRQM	RQM	QFT to order		
			α	α^2	α^3
2.002319304362	Cannot predict	2	2	2.00232	2.002319304

Thus, the correction from "2" can be thought of in QFT as due to the contributions of the higher order diagrams, and is often attributed to the presence of extra virtual particles associated with those diagrams.

We will only outline the steps, and not show the full RQM derivation, but will develop the QFT analysis completely for both first and for second order. The third order value is the result of 72 terms from many complicated Feynman diagrams, so we will leave that as an exercise to do in your leisure time. The 4th order involves 891 diagrams, and the 5th, over 12,000, so you probably won't have time to work those out.

For background, we note the experimentally determined values of the gyromagnetic ratio for the neutron (3.82608545), proton (5.585694713), and muon (2.0023318414). We will not carry out a theoretical determination of any of these herein. (We note in passing that theory and experiment are not precisely in agreement for the muon, leading some to surmise that yet more is to be learned.)

16.4.4 RQM and the Electron Magnetic Moment

We overview the steps taken in RQM to deduce the electron magnetic moment, without carrying out the actual, fairly extensive, analysis.

Overview of RQM solution for g = 2

---

[1] This term is used in the literature for two things, the g-factor described herein (which is dimensionless) and the ratio of magnetic dipole moment to angular momentum (which is often denoted by the symbol γ, and which has SI dimensions of radians per second per tesla). In this book, the term gyromagnetic ratio will be used as equivalent to g, as shown above.

Review

Recall from Chap. 3 that the Klein-Gordon equation was second order and contained the square of the Hamiltonian, $H^2\phi = -\dfrac{\partial^2}{\partial t^2}\phi$, where H^2 is taken over from classical relativity. In Chap. 4 we saw that Dirac's effort entailed expressing the wave equation in first order with H instead of H^2. So, the Dirac equation can be written as

$$i\frac{\partial}{\partial t}\psi_{state} = H\psi_{state} = \left(\boldsymbol{\alpha}\cdot\mathbf{p} + \beta m\right)\psi_{state} \quad \xleftrightarrow{\ \text{equivalent}\ } \quad \left(i\gamma^\mu\partial_\mu - m\right)\psi_{state} = 0 \ . \ (16\text{-}39)$$

Effectively, the Hamiltonian has 4X4 matrix form in spinor space.

Background

In classical relativistic theory, the Hamiltonian of a particle of charge $q = -e$ interacting with an applied field ($\mathbf{A}^e$ and Φ^e here are the external magnetic field vector potential and electric field potential, respectively) is, in the low speed limit, given by

NR limit of classical energy of electron in external e/m field

$$H = \underbrace{m}_{=mc^2} + \frac{1}{2m}\underbrace{\left(\mathbf{p} + e\mathbf{A}^e\right)^2}_{\substack{\text{3-momentum} \\ \text{including e/m} \\ \text{field}}} \underbrace{-\,e\Phi^e}_{\substack{\text{elect} \\ \text{field} \\ \text{energy}}} \underbrace{-\,\boldsymbol{\mu}\cdot\mathbf{B}^e}_{\substack{\text{mag dipole} \\ \text{energy in ext} \\ \text{mag field}}} \ . \quad (16\text{-}40)$$

Consonant with our quantization postulate, we can take the same Hamiltonian in RQM. In spinor space, as we saw in Chap. 4, the spin angular momentum of (16-37) entails use of the RQM spin operator Σ, where

Recalling spin operators

$$\Sigma_i = \frac{1}{2}\begin{bmatrix} \sigma_i & 0 \\ 0 & \sigma_i \end{bmatrix} \rightarrow \Sigma_1 = \frac{1}{2}\begin{bmatrix} & 1 & & \\ 1 & & & \\ & & & 1 \\ & & 1 & \end{bmatrix} \ \Sigma_2 = \frac{1}{2}\begin{bmatrix} & -i & & \\ i & & & \\ & & & -i \\ & & i & \end{bmatrix} \ \Sigma_3 = \frac{1}{2}\begin{bmatrix} 1 & & & \\ & -1 & & \\ & & 1 & \\ & & & -1 \end{bmatrix}, \quad (16\text{-}41)$$

such that, parallel with (16-38)

$$\mu = g\underbrace{\frac{1}{2}\frac{e}{m}}_{\mu_B}\underbrace{\Sigma}_{\text{spin}} = g\mu_B\Sigma\,. \quad (16\text{-}42)$$

μ and H in terms of spin operators

So we see that (16-40) is actually a 4X4 spinor space matrix for energy, as suggested earlier. (Other terms in (16-40) have 4X4 identity matrices associated with them.) Then, (16-40) becomes

$$H = m + \frac{1}{2m}\left(\mathbf{p} + e\mathbf{A}^e\right)^2 - e\Phi^e - g\mu_B\Sigma\cdot\mathbf{B}^e\,, \quad (16\text{-}43)$$

where we still don't know g.

The RQM Answer

However, if we start with the full interaction Dirac equation

Take interaction Dirac eq

$$\left(i\gamma^\mu\partial_\mu - m\right)\psi_{state} = \underbrace{-e\gamma^\mu A_\mu^e}_{\text{state}}\,\psi_{state}\,, \quad (16\text{-}44)$$

take the low velocity limit, as we did in Sect. 16.2.3 (so that things simplify as the spinors take forms like (1,0,0,0)), re-arrange a little, and take derivatives, we can solve this as a matrix eigenvector problem[1] (which is really what we are always doing when we solve the Dirac equation). We find the eigenvalue

Solve for NR limit; get g = 2

$$E = m + \frac{1}{2m}\left(\mathbf{p} + e\mathbf{A}^e\right)^2 - e\Phi^e - 2\mu_B\Sigma\cdot\mathbf{B}^e\,. \quad (16\text{-}45)$$

---

[1] For a little more detail, see Kaku, M., *Quantum Field Theory* (Oxford 1993), pgs 102-104 and Ryder, L., *Quantum Field Theory* (Cambridge 1996), pgs. 52-55.

Comparing (16-45) with (16-43), we see that in RQM, $g = 2$. The approximation, where g is not found exactly equal to the experimental value, is not due to the low velocity limit assumption, but to the fact that in RQM, we generally work only with first order relations. More on this later.

$g \neq 2.00232$
not due to NR
approximation

16.4.5 QFT's First Order Electron Magnetic Moment

We will derive the lowest order electron magnetic moment value in this section, and the second order value in the next.

1st order QFT
result for g

Note: A summary of this section can be found in the chapter summary on pg. 429.

Background

We need to know a few things before we can dive into determining the electron magnetic moment via QFT. In particular, we will be working a lot with the commutator $\left[\gamma^\mu, \gamma^\nu \right]$, and we will want to know what that equals for various indices μ and ν. It is common to define, and use, a new symbol related to this commutator,

$$\sigma^{\mu\nu} = \frac{i}{2}\left[\gamma^\mu, \gamma^\nu \right] = \frac{i}{2}\left(\gamma^\mu\gamma^\nu - \gamma^\nu\gamma^\mu \right). \tag{16-46}$$

Define $\sigma^{\mu\nu}$

For one example of (16-46) with $\mu = 1$ and $\nu = 2$, using (16-16), we can find

$$\sigma^{12} = \frac{i}{2}\gamma^1\gamma^2 - \frac{i}{2}\gamma^2\gamma^1$$

Finding $\sigma^{\mu\nu}$ in terms of spin operators

$$= \frac{i}{2}\begin{bmatrix} & & & 1 \\ & & 1 & \\ & -1 & & \\ -1 & & & \end{bmatrix}\begin{bmatrix} & & & -i \\ & & i & \\ & i & & \\ -i & & & \end{bmatrix} - \frac{i}{2}\begin{bmatrix} & & & -i \\ & & i & \\ & i & & \\ -i & & & \end{bmatrix}\begin{bmatrix} & & & 1 \\ & & 1 & \\ & -1 & & \\ -1 & & & \end{bmatrix} \tag{16-47}$$

$$= \frac{i}{2}\begin{bmatrix} -i & & & \\ & i & & \\ & & -i & \\ & & & i \end{bmatrix} - \frac{i}{2}\begin{bmatrix} i & & & \\ & -i & & \\ & & i & \\ & & & -i \end{bmatrix} = \begin{bmatrix} 1 & & & \\ & -1 & & \\ & & 1 & \\ & & & -1 \end{bmatrix} = \begin{bmatrix} \sigma_3 & \\ & \sigma_3 \end{bmatrix} = 2\Sigma_3,$$

where at the end, we have used the appropriate Pauli 2X2 matrix and shown this equals twice the spin operator Σ_3 of Chap. 4 (relation (4-39), pg. 93).

You can derive a couple more of (16-46) by doing Probs. 3, and 4. You can also note, from (16-46), that generally $\sigma^{\mu\nu} = -\sigma^{\nu\mu}$, and for $\mu = \nu$, $\sigma^{\mu\mu} = 0$ (no summation on μ). The result of all this is listed below,

$$\sigma^{00} = \sigma^{11} = \sigma^{22} = \sigma^{33} = 0$$

$$\sigma^{12} = -\sigma^{21} = \begin{bmatrix} 1 & & & \\ & -1 & & \\ & & 1 & \\ & & & -1 \end{bmatrix} = \begin{bmatrix} \sigma_3 & \\ & \sigma_3 \end{bmatrix} = 2\Sigma_3 \quad \sigma^{01} = -\sigma^{10} = \begin{bmatrix} & & & i \\ & & i & \\ & i & & \\ i & & & \end{bmatrix} = i\begin{bmatrix} & \sigma_1 \\ \sigma_1 & \end{bmatrix}$$

$$\sigma^{23} = -\sigma^{32} = \begin{bmatrix} & 1 & & \\ 1 & & & \\ & & & 1 \\ & & 1 & \end{bmatrix} = \begin{bmatrix} \sigma_1 & \\ & \sigma_1 \end{bmatrix} = 2\Sigma_1 \quad \sigma^{02} = -\sigma^{20} = \begin{bmatrix} & 1 & & \\ & & -1 & \\ 1 & & & \\ & & -1 & \end{bmatrix} = i\begin{bmatrix} & \sigma_2 \\ \sigma_2 & \end{bmatrix} \tag{16-48}$$

$$\sigma^{31} = -\sigma^{13} = \begin{bmatrix} & -i & & \\ i & & & \\ & & & -i \\ & & i & \end{bmatrix} = \begin{bmatrix} \sigma_2 & \\ & \sigma_2 \end{bmatrix} = 2\Sigma_2 \quad \sigma^{03} = -\sigma^{30} = \begin{bmatrix} & & i & \\ & & & -i \\ i & & & \\ & -i & & \end{bmatrix} = i\begin{bmatrix} & \sigma_3 \\ \sigma_3 & \end{bmatrix}$$

Easiest Feynman Diagram to Use

In all that we have done before, we have worked with Feynman diagrams representing complete interactions of real incoming and outgoing particles, such as Figs. 16-1 and 16-2. We could do that again to determine the gyromagnetic ratio, but it would be quite extensive and messy as it would include two vertices at which the magnetic moment of an electron and a magnetic field (from the photon virtual particle) interact.

Easiest Feynman diagram to work with is a truncated one, RH of Fig. 16-3

We can shorten the work by considering a single vertex Feynman diagram as in the RH of Fig. 16-3, which is a sort of truncated full Feynman diagram, and calculating the amplitude for that. There are some issues with doing this, which we discuss below. In the RH of Fig. 16-3, p and p' equal p_1 and p'_1 of the LHS, respectively. We drop the subscripts to streamline notation.

We will use the subscript "mm" to represent the S operator and amplitude associated with that particular diagram, as it will be used for our magnetic moment calculation.

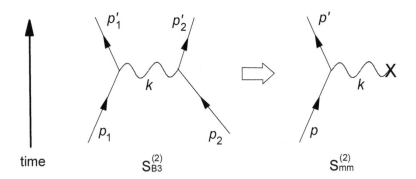

Figure 16-3. The Short Cut Feynman Diagram for Magnetic Moment Calculation

How We'll Go About It

Note that on the RHS of Fig. 16-3, we have an external electromagnetic field (the photon) interacting with an electron. The external field is, in essence, a potential field, as we have an electron scattered by that field. The electron has spin and an associated magnetic moment μ. The external electromagnetic field $A^e_\mu = (\Phi^e, \mathbf{A}^e)$ gives rise to a magnetic field via $\mathbf{B}^e = \nabla \times \mathbf{A}^e$. So, one term in the potential energy is $-\mu \cdot \mathbf{B}^e = -\mu \cdot (\nabla \times \mathbf{A}^e)$.

We will use the RH diagram of Fig. 16-3 to construct an amplitude and compare it with the Born approximation amplitude (16-12)(a), which is for a single particle interacting with a potential field, as we have in the RHS of Fig. 16-3. The $-\mu \cdot \mathbf{B}^e = -\mu \cdot (\nabla \times \mathbf{A}^e)$ part of the potential energy should show up in $\tilde{V}(\mathbf{k})$ of (16-12)(a) and its physical space counterpart $V(\mathbf{x})$.

We will find the QFT amplitude for NR case and compare with Born approx to deduce potential. From that we can see g value.

$\mathbf{B}^e$ here, and thus $\mathbf{A}^e$, are static fields. Since there is no time variation, the energy part of $e^{\pm ikx}$ in A^μ should play no role, and we can represent the field as $A^e_\mu(\mathbf{x})$ in position space and $A^e_\mu(\mathbf{k})$ in momentum space (as functions of space without time and 3-momentum without energy, respectively). Recall also, from Sect. 16.2.4, that in the frame of a large mass source of the field (the potential field is effectively immovable), the virtual photon carries zero energy.

We will still have energy conservation at the vertex, and this means the energies of the incoming and outgoing electrons are equal. Therefore, the magnitudes of their 3-momenta are also equal, i.e., $|\mathbf{p}| = |\mathbf{p}'|$. $\mathbf{k} = \mathbf{p}' - \mathbf{p}$ will be non-zero, however, since $\mathbf{p}'$ is not in the same direction as $\mathbf{p}$.

Changes to Feynman Rules for External Fields

In the RH diagram of Fig. 16-3, since the incoming and outgoing electrons are real, the photon must be off-shell and virtual. It may seem a strange animal, as it is not really an external photon, and not really a full propagator. So, we can't use the usual Feynman rules, as we cannot represent the photon in a Feynman amplitude for this diagram by either $\varepsilon_\mu(\mathbf{k})$ (external) or $D_{F\mu\nu}(k)$ (propagator). The photon here represents a <u>static external field</u> or <u>static potential field</u> (contrasted with an external particle, which is real, on-shell, and not static) and represented by the <u>symbol</u> A^e_μ.

A_μ^e is unique, not a full propagator and not a (const or destruct) operator field.

In the Appendix, we derive the transition amplitude for the RH diagram of Fig. 16-3, i.e., for scattering by a static external field, via the long Dyson-Wicks expansion method (rather than the short way of Feynman rules, since we don't yet have any Feynman rules for this.). From that derivation, we then deduce a <u>modification to Feynman's rules</u> applying to specific cases <u>involving static external fields</u>.

Need additional Feynman rules for A_μ^e, which we derive in Appendix

The result of all that is: Add the following Feynman rule #11 (Rule #10 appeared in Chap. 13, pg. 331 and was relevant to renormalization.)

New Feynman rule #11 for A_μ^e

Feynman rule #11. For each interaction of a charged particle with a static external photon field (a static potential field), write a factor

$$A_\mu^e(\mathbf{k}) = \int e^{-i\mathbf{k}\cdot\mathbf{x}} A_\mu^e(\mathbf{x}) d^3x, \qquad (16\text{-}49)(a)$$

and instead of $(2\pi)^4 \, \delta^{(4)}(\Sigma p_f - \Sigma p_i)$ in S_{fi}, use

$$2\pi\, \delta\left(\Sigma E_f - \Sigma E_i\right). \qquad (16\text{-}49)(b)$$

(16-49)(a) is much like a classical field. There are no operators and it has a field strength related to its amplitude. For a stronger external (potential) field, $A^e{}_\mu$ has greater value. In terms of virtual particles, we could think of (16-49)(a) as representing more photons for a stronger field, and fewer photons for a weaker one. This is the case physically. A stronger Coulomb field is due to more charges being present at its source, and thus more virtual photons being emitted from that source.

The Amplitude

We assume the renormalized value e (not the bare value e_0) is the relevant factor in our amplitude calculations, because that is what we measure physically. However, we assume there is no renormalization correction to the vertex factor γ^μ. That is, we analyze the vertex contribution of Fig. 16-3 as it is shown there, i.e., as first order in α. Thus, from the RHS of Fig. 16-3, we have

$$S_{mm} = \left(\prod_{\mathbf{p}''}^{\substack{\text{all external}\\\text{fermions}}} \sqrt{\frac{m}{VE_{\mathbf{p}''}}}\right) 2\pi\delta\left(E'-E\right)\mathcal{M}_{mm}^{(2)} = \underbrace{\frac{1}{V} 2\pi\delta\left(E'-E\right)\mathcal{M}_{mm}^{(2)}}_{E\approx m\text{ for non-relativistic}} \qquad (16\text{-}50)$$

Feynman amplitude for RH of Fig. 16-3

$$\mathcal{M}_{mm}^{(2)} = ie\bar{u}_{r'}(\mathbf{p}')\gamma^\mu u_r(\mathbf{p})A_\mu^e\left(\mathbf{k}=\mathbf{p}'-\mathbf{p}\right).$$

Use Gordon's identity, which we derived in Chap. 13,

$$\bar{u}_{r'}(\mathbf{p}')\gamma^\mu u_r(\mathbf{p}) = \frac{p^\mu+p'^\mu}{2m}\bar{u}_{r'}(\mathbf{p}')u_r(\mathbf{p}) + \frac{p_\nu-p'_\nu}{4m}\bar{u}_{r'}(\mathbf{p}')\left[\gamma^\mu,\gamma^\nu\right]u_r(\mathbf{p}), \qquad (16\text{-}51)$$

in (16-50) to obtain

Use Gordon identity to re-arrange

$$\mathcal{M}_{mm}^{(2)} = ie\bar{u}_{r'}(\mathbf{p}')\gamma^\mu u_r(\mathbf{p})A_\mu^e\left(\mathbf{k}=\mathbf{p}'-\mathbf{p}\right)$$

$$= \underbrace{ie\frac{p^\mu+p'^\mu}{2m}\bar{u}_{r'}(\mathbf{p}')u_r(\mathbf{p})A_\mu^e(\mathbf{k})}_{\substack{\text{will not contribute to spin}\\\text{related potential energy}}} + ie\frac{p_\nu-p'_\nu}{4m}\bar{u}_{r'}(\mathbf{p}')\underbrace{\left[\gamma^\mu,\gamma^\nu\right]}_{\substack{\text{contains spin}\\\text{operator}}}u_r(\mathbf{p})A_\mu^e(\mathbf{k}). \qquad (16\text{-}52)$$

Shows us which parts of amplitude correspond to spin (i.e., μ) contribution

Note that the first term in (16-52) has no spin operator Σ parts and therefore should contribute only to the non-spin related part of the potential energy. However, from what we saw in (16-46) and (16-48), we will find the commutator in the second term does have Σ parts to it. Given (16-42), i.e., $\mu = g\frac{1}{2}\frac{e}{m}\Sigma = g\mu_B\Sigma$, we conclude that this second term contains a magnetic moment factor. It also includes the external magnetic field $\mathbf{B}^e$ within the $\mathbf{A}^e$ of A_μ^e. So, the magnetic moment contribution to the potential energy in the presence of an external magnetic field must be in this term, and this term alone.

For incoming fermion with spin r = 1 (+z direction eigenstate)

Focusing on that term in (16-52), and restricting ourselves to the low velocity limit case with $r = 1$, which in that limit is a $+z$ direction spin eigenstate, we have

$$\mathcal{M}^{(2)}_{\substack{mm \\ \Sigma\,part \\ r=1}} = -ie\frac{k_\nu}{4m}\bar{u}_{r'}(\mathbf{p}')\underbrace{\frac{2}{i}\sigma^{\mu\nu}}_{[\gamma^\mu,\gamma^\nu]}u_{r=1}(\mathbf{p})A^e_\mu(\mathbf{k}) = -e\frac{k_\nu}{2m}\bar{u}_{r'}(\mathbf{p}')\sigma^{\mu\nu}\begin{bmatrix}1\\0\\0\\0\end{bmatrix}A^e_\mu(\mathbf{k}). \quad (16\text{-}53)$$

$\bar{u}_{r'}(\mathbf{p}')$ can only have forms [1 0 0 0] (for $r'=1$) or [0 1 0 0] (for $r'=2$). Note from (16-48) that for $r'=1$ with σ^{01},

$$\bar{u}_{r'=1}(\mathbf{p}')\sigma^{01}\begin{bmatrix}1\\0\\0\\0\end{bmatrix} = \begin{bmatrix}1 & 0 & 0 & 0\end{bmatrix}\begin{bmatrix}&&i&\\&&&i\\i&&&\\&i&&\end{bmatrix}\begin{bmatrix}1\\0\\0\\0\end{bmatrix} = \begin{bmatrix}1 & 0 & 0 & 0\end{bmatrix}\begin{bmatrix}0\\0\\i\\i\end{bmatrix} = 0. \quad (16\text{-}54)$$

All terms with $\sigma^{0\nu}$ yield zero

For $r'=2$ instead of 1 in (16-54), we get zero, as well. Thus, we conclude that the term with σ^{01} in the summation on μ and ν in (16-53) yields zero and we can ignore it. By doing Prob. 5, you can show that the same thing is true for any term where either μ or $\nu=0$ in $\sigma^{\mu\nu}$.

Additionally, from the first line of (16-48), we see that any term $\sigma^{\mu\mu}=0$. Hence, all we have to worry about are the six σ^{ij} for $i\neq j$. For σ^{12} (=$2\Sigma_3$), (16-53) becomes (for the case where $r'=1$)

All terms with $\sigma^{\mu\mu}$ yield zero

$$\mathcal{M}^{(2)}_{\substack{mm \\ r=1 \\ i=1,j=2 \\ (\Sigma_3\,term)}} = \frac{-e}{2m}k_2\begin{bmatrix}1 & 0 & 0 & 0\end{bmatrix}\underbrace{\begin{bmatrix}1&&&\\&-1&&\\&&1&\\&&&-1\end{bmatrix}}_{\sigma^{12}=2\Sigma_3}\begin{bmatrix}1\\0\\0\\0\end{bmatrix}A^e_1(\mathbf{k}) = \frac{-e}{2m}2\underbrace{\Sigma^3_{eig}}_{\frac{1}{2}}k_2A^e_1(\mathbf{k}) \quad (16\text{-}55)$$

Expressing one amplitude term in terms of spin operator and eigen value

where Σ^3_{eig} is the eigenvalue of the Σ_3 operator.

And from (16-48), reversing the indices reverses the sign, i.e.,

$$\mathcal{M}^{(2)}_{\substack{mm \\ r=1 \\ i=2,j=1 \\ (\Sigma_3\,term)}} = \frac{e}{2m}2\Sigma^3_{eig}k_1A^e_2(\mathbf{k}). \quad (16\text{-}56)$$

Continuing on to express all terms in terms of spin operator and eigen value

For $\sigma^{23}=2\Sigma_1$ and $r=1$ spin state, we get

$$\mathcal{M}^{(2)}_{\substack{mm \\ r=1 \\ i=2,j=3 \\ (\Sigma_1\,term)}} = \frac{-e}{2m}k_3\begin{bmatrix}1 & 0 & 0 & 0\end{bmatrix}\underbrace{\begin{bmatrix}&1&&\\1&&&\\&&&1\\&&1&\end{bmatrix}}_{\sigma^{23}=2\Sigma_1}\begin{bmatrix}1\\0\\0\\0\end{bmatrix}A^e_2(\mathbf{k}) = 0 \quad (16\text{-}57)$$

Similarly, the contributions from all other σ^{ij} are zero, as you can show by doing Prob. 6. So for the $r=1$ spin state, we only get contributions to the amplitude from $\sigma^{12}=2\Sigma_3$ and $\sigma^{21}=-2\Sigma_3$. Since Σ_3 is the z direction spin operator and $r=1$ is the z direction spin up state, this makes sense. We get zero for operators Σ_1 and Σ_2 (x and y direction spin operators) acting on a z direction spin state.

Thus, the total contribution for an $r=1$ spin state electron to the amplitude is (16-55) plus (16-56),

$$\mathcal{M}^{(2)}_{\substack{mm \\ r=1}} = \mathcal{M}^{(2)}_{\substack{mm \\ r=1 \\ i=1,j=2 \\ (\Sigma_3\,term)}} + \mathcal{M}^{(2)}_{\substack{mm \\ r=1 \\ i=2,j=1 \\ (\Sigma_3\,term)}} + 0+0+0+0 = \frac{e}{2m}2\underbrace{\Sigma^3_{eig}}_{\frac{1}{2}}\left(-k_2A^e_1(\mathbf{k})+k_1A^e_2(\mathbf{k})\right). \quad (16\text{-}58)$$

For incoming fermion with spin $r=2$ ($-z$ direction eigenstate)

By doing Prob. 7 you can find that for $u_{r=2}(\mathbf{p})$ (eigen spin in opposite direction, i.e., in negative z axis direction) and $i=1$, $j=2$, we get

$$\mathcal{M}^{(2)}_{\substack{mm \\ r=2 \\ i=1,j=2 \\ (\Sigma_3\ term)}} = \frac{-e}{2m} 2\ \underbrace{\Sigma^3_{eig}}_{-\frac{1}{2}} k_2 A^e_1\left(\mathbf{k}\right) \qquad\qquad \mathcal{M}^{(2)}_{\substack{mm \\ r=2 \\ i=2,j=1 \\ (\Sigma_3\ term)}} = \frac{e}{2m} 2\ \underbrace{\Sigma^3_{eig}}_{-\frac{1}{2}} k_1 A^e_2\left(\mathbf{k}\right). \qquad (16\text{-}59)$$

Note the spin eigenvalue has the opposite sign of the $r = 1$ state, as we would expect.

As before, the contributions from $\sigma^{23} = 2\,\Sigma_1$ and $\sigma^{31} = 2\,\Sigma_2$ and their siblings is zero. So the total contribution to the amplitude for an $r = 2$ spin state electron is

$$\mathcal{M}^{(2)}_{\substack{mm \\ r=2}} = \mathcal{M}^{(2)}_{\substack{mm \\ r=2 \\ i=2,j=1 \\ (\Sigma_3\ term)}} + \mathcal{M}^{(2)}_{\substack{mm \\ r=2 \\ i=1,j=2 \\ (\Sigma_3\ term)}} + 0+0+0+0 = \frac{e}{2m} 2\ \underbrace{\Sigma^3_{eig}}_{-\frac{1}{2}}\left(k_1 A^e_2\left(\mathbf{k}\right) - k_2 A^e_1\left(\mathbf{k}\right)\right). \quad (16\text{-}60)$$

For both r = 1 and r = 2 ($+z$ and $-z$ directions) incoming spin states in one expression

Since we would either have up spin ($r = 1$) or down spin ($r = 2$), we would like a single expression that would work for both cases. Comparing (16-60) to (16-58), we see the only difference is the sign of the eigenvalue. So, we can use the following expression, and it will work for both spin cases.

$$\mathcal{M}^{(2)}_{\substack{mm \\ either\ z\ direc}} = \frac{e}{2m} 2\Sigma^3_{eig}\left(k_1 A^e_2\left(\mathbf{k}\right) - k_2 A^e_1\left(\mathbf{k}\right)\right) \qquad (16\text{-}61)$$

For $+ x$ direction eigen spin state

By doing Prob. 8, you can show that for an eigen spin state in the positive x direction,

$$\mathcal{M}^{(2)}_{\substack{mm \\ +x\,direc \\ i=2,j=3 \\ (\Sigma_1\ term)}} = \frac{-e}{2m} 2\ \underbrace{\Sigma^1_{eig}}_{\frac{1}{2}} k_3 A^e_2\left(\mathbf{k}\right) \qquad\qquad \mathcal{M}^{(2)}_{\substack{mm \\ +x\,direc \\ i=3,j=2 \\ (\Sigma_1\ term)}} = \frac{e}{2m} 2\ \underbrace{\Sigma^1_{eig}}_{\frac{1}{2}} k_2 A^e_3\left(\mathbf{k}\right). \qquad (16\text{-}62)$$

All other amplitude contributions for other values of i and j yield zero.

For $- x$ direction eigen spin state

For an eigen spin state in the negative x direction, it follows (by analogy with the z direction case, or by cranking the algebra) that

$$\mathcal{M}^{(2)}_{\substack{mm \\ -x\,direc \\ i=2,j=3 \\ (\Sigma_1\ term)}} = \frac{-e}{2m} 2\ \underbrace{\Sigma^1_{eig}}_{-\frac{1}{2}} k_3 A^e_2\left(\mathbf{k}\right) \qquad\qquad \mathcal{M}^{(2)}_{\substack{mm \\ -x\,direc \\ i=3,j=2 \\ (\Sigma_1\ term)}} = \frac{e}{2m} 2\ \underbrace{\Sigma^1_{eig}}_{-\frac{1}{2}} k_2 A^e_3\left(\mathbf{k}\right). \qquad (16\text{-}63)$$

And as before, all other values for i and j yield zero.

For both $+ x$ and $- x$ direction incoming spin states in one expression

As with the z direction case, we can combine (16-62) and (16-63) into one expression that will work for either $+$ or $-x$ direction eigen spin state.

$$\mathcal{M}^{(2)}_{\substack{mm \\ either\ x\ direc}} = \mathcal{M}^{(2)}_{\substack{mm \\ either\ x\ direc \\ i=3,j=2 \\ (\Sigma_1\ term)}} + \mathcal{M}^{(2)}_{\substack{mm \\ either\ x\ direc \\ i=2,j=3 \\ (\Sigma_1\ term)}} = \frac{e}{2m} 2\Sigma^1_{eig}\left(k_2 A^e_3\left(\mathbf{k}\right) - k_3 A^e_2\left(\mathbf{k}\right)\right) \qquad (16\text{-}64)$$

For both $+ y$ and $- y$ direction incoming spin states in one expression

In similar fashion, for y direction eigen spin states, we get

$$\mathcal{M}^{(2)}_{\substack{mm \\ either\ y\ direc}} = \mathcal{M}^{(2)}_{\substack{mm \\ either\ y\ direc \\ i=1,j=3 \\ (\Sigma_2\ term)}} + \mathcal{M}^{(2)}_{\substack{mm \\ either\ y\ direc \\ i=3,j=1 \\ (\Sigma_2\ term)}} = \frac{e}{2m} 2\Sigma^2_{eig}\left(k_3 A^e_1\left(\mathbf{k}\right) - k_1 A^e_3\left(\mathbf{k}\right)\right) \qquad (16\text{-}65)$$

Adding All Components

Adding (16-61), (16-64), and (16-65) together we find (16-53) becomes

$$\mathcal{M}^{(2)}_{\substack{mm \\ \Sigma\,part}} = \frac{e}{2m} 2\Sigma^3_{eig}\left(k_1 A^e_2(\mathbf{k}) - k_2 A^e_1(\mathbf{k})\right)$$

$$+ \frac{e}{2m} 2\Sigma^1_{eig}\left(k_2 A^e_3(\mathbf{k}) - k_3 A^e_2(\mathbf{k})\right) + \frac{e}{2m} 2\Sigma^2_{eig}\left(k_3 A^e_1(\mathbf{k}) - k_1 A^e_3(\mathbf{k})\right).$$

(16-66)

The total spin part of the amplitude

Note, for example, that if our incoming particle were in a z direction spin eigen state $\Sigma^1_{eig} = \Sigma^2_{eig} = 0$, the only term we would have would be a term containing Σ^3_{eig}. Exactly parallel statements can be made for the x and y direction spin eigenstates.

Comparing with Born Approximation

In the Born approximation (16-12)(a) we have the momentum space potential energy

$$\tilde{V}(\mathbf{k}) = \tilde{V}_{\substack{non \\ spin}}(\mathbf{k}) + \tilde{V}_{spin}(\mathbf{k}),$$

(16-67)

Comparing to Born approx, we deduce the spin part of the potential

where the spin part is what we have been working on to get to (16-66). Each term of (16-67) has its corresponding relation in position space, and we will focus on the spin (magnetic moment) term here. So, we are interested in the spin part of (16-12)(a), i.e.,

$$S_{\substack{spin \\ part}} = \frac{i}{V}\tilde{V}_{spin}(\mathbf{k}) 2\pi\,\delta\left(E_f - E_i\right).$$

(16-68)

From (16-50), we have the equivalent of (16-68)

$$S_{\substack{mm \\ spin \\ part}} = \frac{1}{V} 2\pi\delta\left(E' - E\right)\mathcal{M}^{(2)}_{\substack{mm \\ \Sigma\,part}},$$

(16-69)

meaning

$$\tilde{V}_{spin}(\mathbf{k}) = \frac{1}{i}\mathcal{M}^{(2)}_{\substack{mm \\ \Sigma\,part}}.$$

(16-70)

Spin part of the potential in $\mathbf{k}$ space

With (16-66), this becomes

$$\tilde{V}_{spin}(\mathbf{k}) = \frac{1}{i}\mathcal{M}^{(2)}_{\substack{mm \\ \Sigma\,part}} = \frac{-ie}{2m} 2\Sigma^1_{eig}\left(k_2 A^e_3(\mathbf{k}) - k_3 A^e_2(\mathbf{k})\right)$$

$$- \frac{ie}{2m} 2\Sigma^2_{eig}\left(k_3 A^e_1(\mathbf{k}) - k_1 A^e_3(\mathbf{k})\right) - \frac{ie}{2m} 2\Sigma^3_{eig}\left(k_1 A^e_2(\mathbf{k}) - k_2 A^e_1(\mathbf{k})\right).$$

(16-71)

We should by now know that in converting a momentum space function to position space, momentum factors become spatial derivatives (with a factor of $-i$). But just for review, note for

$$F(\mathbf{x}) = \frac{1}{(2\pi)^3}\int_{-\infty}^{+\infty} e^{ik_i x^i}\tilde{F}(\mathbf{k})\,d\mathbf{k},$$

(16-72)

$$-i\frac{\partial}{\partial x^i}F(\mathbf{x}) = -\frac{1}{(2\pi)^3}\int_{-\infty}^{+\infty} i\frac{\partial}{\partial x^i} e^{ik_i x^i}\tilde{F}(\mathbf{k})\,d\mathbf{k} = \frac{1}{(2\pi)^3}\int_{-\infty}^{+\infty} e^{ik_i x^i} k_i\,\tilde{F}(\mathbf{k})\,d\mathbf{k}.$$

(16-73)

So the Fourier transform of the function $k_i\,\tilde{F}(\mathbf{k})$ is $-i\partial_i F(\mathbf{x})$. For us, $F(\mathbf{x}) = A^e_j(\mathbf{x})$.

Thus, each term with k_i in (16-71) in $\mathbf{k}$ space converts to $(-i)$ times a derivative with respect to x^i in position space. Thus, doing the Fourier transform of (16-71), we get

Fourier transform that to $\mathbf{x}$ space

$$V_{spin}(\mathbf{x}) = \frac{ie}{2m} 2\Sigma^1_{eig} \underbrace{\left(i\frac{\partial}{\partial x^2} A_3^e(\mathbf{x}) - i\frac{\partial}{\partial x^3} A_2^e(\mathbf{x}) \right)}_{i(\nabla\times\mathbf{A}^e)^1} + \frac{ie}{2m} 2\Sigma^2_{eig} \underbrace{\left(i\frac{\partial}{\partial x^3} A_1^e(\mathbf{x}) - i\frac{\partial}{\partial x^1} A_3^e(\mathbf{x}) \right)}_{i(\nabla\times\mathbf{A}^e)^2}$$

yields spin part of the potential in x space

$$+ \frac{ie}{2m} 2\Sigma^3_{eig} \underbrace{\left(i\frac{\partial}{\partial x^1} A_2^e(\mathbf{x}) - i\frac{\partial}{\partial x^2} A_1^e(\mathbf{x}) \right)}_{i(\nabla\times\mathbf{A}^e)^3}. \tag{16-74}$$

Recognizing the derivative portions represent a curl, and taking $\Sigma = \left(\Sigma^1_{eig}, \Sigma^2_{eig}, \Sigma^3_{eig} \right)$ to now represent the vector of spin eigenvalue components in the $x, y,$ and z directions (which in physical space is, for all intents and purposes, the same as $\Sigma = (\Sigma_1, \Sigma_2, \Sigma_3)$, the 3D vector of spin operators in spin space), we can re-write (16-74) (compare with (16-43)),as

Re-expressing that, we find $g = 2$

$$V_{spin}(\mathbf{x}) = -2\frac{e}{2m}\Sigma\cdot\left(\nabla\times\mathbf{A}^e\right) = -\underbrace{2}_{g}\mu_B\Sigma\cdot\mathbf{B}^e. \tag{16-75}$$

Our QED 1st order result

Result for Order α

Thus, we see that the electron gyromagnetic ratio g for the lowest order in QFT is 2.

16.4.6 QFT's Second Order Electron Magnetic Moment

Note: A summary of this section can be found in the chapter summary on pg. 429.

QED's 2nd order determination of g

As I'm sure you understand, we get the first refinement to the magnetic moment calculation by adding in the next higher order terms (associated with next higher order Feynman diagrams). When Julian Schwinger[1] first calculated this in 1948, and it matched experiment, the physics community was electrified. It was dramatic confirmation of QED.

Ways to Calculate the Second Order Correction to g

There are two ways we can calculate the gyromagnetic ratio g to second order, the first of which is probably what you would first assume we would do.

We will use a clever analysis that makes calculations much easier

1. Take all the amplitude terms from the various 2nd order Feynman diagrams (the last four diagrams in Fig. 16-4 below), find all the terms therein containing $\sigma^{\mu\nu}$ (using, in some cases, Gordon's identity to convert terms in γ^μ), add those terms, and use them in the way we did in the prior section to find g.

2. Use a quicker, simpler trick (to be described below.)

Method #1 above can be done, but it is extraordinarily long and messy. We will use method #2.

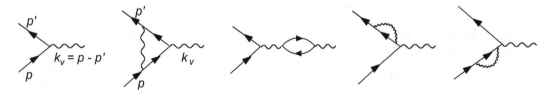

Figure 16-4. Contributions at Second Order to Single Vertex Interaction

The Method of Choice for Finding g to Second Order

The steps using method #2 above that we will follow are:

Step #1: Express the 2nd order amplitude in its most general form in terms of initially unknown functions $F_1(k_\nu^2)$ and $F_2(k_\nu^2)$. (We use k_ν for the 4-momentum of the external photon. We used k in the first order case, but in higher order cases, we typically use k for internal photon corrections such as the vertical photon in the 2nd diagram, Fig. 16-4.)

Steps in our analysis

[1] Schwinger, J., Quantum Electrodynamics III. The Electromagnetic Properties of the Electron – Radiative Corrections to Scattering, *Phys. Rev.* **76**(6), 790-817 (Sept 15, 1948).

Step #2: Examine the non-relativistic limit case where $k_v \rightarrow 0$ corresponding to experimentally determined gyromagnetic ratio g, and show $F_1(0) = 1$,

Step #3: Show $F_2(0)$ relates to g, so all we have to do is find $F_2(0)$ and the form of relevant F_2 terms.

Step #4: Because the term with $F_2(0)$ has no γ^μ factor, recognize that we don't need to evaluate any 2nd order Feynman diagrams that give us a term with a γ^μ factor.

Step #5: In taking $\gamma^\mu \rightarrow \gamma^\mu_{mod \atop 2nd}(p, p') = \gamma^\mu + e^2 \Lambda_c^\mu$ for 2nd order amplitude, we only need to consider terms in $\Lambda_c{}^\mu$ that have no γ^μ factor. Find the sum of those terms.

Step #6: Use the sum of terms in #5 in #3 to find g to 2nd order.

Step #1. Most General Form of the Amplitude

Consider Fig. 16-4 for time vertically upward and the electron emitting a photon. We will take it from the mathematicians that the most general expression for the total amplitude for electron scattering from an external field to any order (though we will focus on 2nd order as in Fig. 16-4) is

Step #1

$$\mathcal{M}_{mm}^{(nth)} = ie\bar{u}_{r'}(\mathbf{p}')\left(\gamma^\mu F_1\left(k_v^2\right) + i\frac{\sigma^{\mu\nu}k_v}{2m}F_2\left(k_v^2\right)\right)u_r(\mathbf{p})A_\mu^e(\mathbf{k}_v). \qquad (16\text{-}76)$$

Most general form of amplitude

with $k_v = p - p'$, and functions F_1 and F_2, about which we know little right now, are known as <u>form factors.</u> (If we treated photon absorption instead, then $k_v = p' - p$, and we would get the same result.)

Note that by using Gordon's identity

$$\bar{u}_{r'}(\mathbf{p}')\gamma^\mu u_r(\mathbf{p}) = \bar{u}_{r'}(\mathbf{p}')\left(\frac{p^\mu + p'^\mu}{2m} - \frac{i\sigma^{\mu\nu}k_v}{2m}\right)u_r(\mathbf{p}) \qquad (16\text{-}77)$$

to substitute for the γ^μ term in (16-76), we can express the amplitude of (16-76) solely as a function of a term that is a scalar in spinor space (having no γ^μ or $\sigma^{\mu\nu}$ factor) and a term having a $\sigma^{\mu\nu}$ factor. Alternatively, we could solve Gordon's identity for the $\sigma^{\mu\nu}$ factor term and substitute that in (16-76) to give us the amplitude solely as a function of a spinor space scalar term and a term in γ^μ.

With Gordon's identity, we can express the amplitude via any 2 of 3 possible types of terms

In other words, there are three possible types of terms we can have in the amplitude (16-76), but because of Gordon's identity, we can, as we find convenient, express (16-76) solely as a function of any two of them we like. We will use this to our advantage in what is coming.

Note also that we refer above to each part of (16-76) as a single term, but in actuality F_1 and F_2 are, in general, comprised of more than one term, so there are really more than two terms in (16-76) for all but the first order case.

We will from henceforth focus on the 2nd order case of (16-76). As an aside, we can see how the second order terms come in, if we temporarily use slightly different form factors F_A and F_B defined via

Temporarily re-express with slightly different form factors

$$F_1\left(k_v^2\right) = 1 + F_A\left(k_v^2\right) \qquad F_2\left(k_v^2\right) = F_B\left(k_v^2\right). \qquad (16\text{-}78)$$

So, (16-76) becomes

$$\mathcal{M}_{mm}^{(2+4)} = \mathcal{M}_{mm}^{(2)} + \mathcal{M}_{mm}^{(4)} = ie\bar{u}_{r'}(\mathbf{p}')(\gamma^\mu + \underbrace{\gamma^\mu F_A\left(k_v^2\right) + i\frac{\sigma^{\mu\nu}k_v}{2m}F_B\left(k_v^2\right)}_{\text{2nd order contibution}})u_r(\mathbf{p})A_\mu^e(\mathbf{k}_v), \quad (16\text{-}79)$$

With that re-expression, 1st and 2nd order parts appear separately

where the superscripts, as used in this book, refer to the powers of e involved. When we use the term "second order", we mean second order in α, i.e., 4th order in e. (16-79) expresses the amplitude to 2nd order in terms of the first order and second order contributions.

We now return to using the form factors F_1 and F_2, instead of F_A and F_B.

Return to form factors F_1 and F_2

NOTE: The following Step #2 and the equations therein have been changed since the May 2014 version of this book. The revised Step #2 below should be easier to understand than the prior one.

Step #2. Examine Non-relativistic Case Where $k_\nu \to 0$ and Show $F_1(0) = 1$

The experimental value for g cited herein is determined for non-relativistic speeds of the incoming and outgoing electron. Thus, $p^\mu \approx (m,0,0,0)$, $p'^\mu \approx (m,0,0,0)$, and

$$\text{non-relativistic } k_\nu = p_\nu - p'_\nu \approx \begin{bmatrix} m \\ 0 \\ 0 \\ 0 \end{bmatrix} - \begin{bmatrix} m \\ 0 \\ 0 \\ 0 \end{bmatrix} = \begin{bmatrix} 0 \\ 0 \\ 0 \\ 0 \end{bmatrix}. \tag{16-80}$$

So we will consider $k_\nu \to 0$, and then of course, $k_\nu^2 \to 0$. We use the symbols $F_1(0)$ and $F_2(0)$ to represent our non-relativistic form factors. With this, (16-76) becomes, where we keep in mind that the 3-momentum label $\mathbf{k}_\nu$ is non-zero, but in relativistic terms, vanishingly small.

$$\mathcal{M}^{(2+4)}_{\substack{mm \\ k_\nu \to 0}} = ie\bar{u}_{r'}(\mathbf{p}')\left(\gamma^\mu F_1(0) + i\frac{\sigma^{\mu\nu}k_\nu}{2m}F_2(0) \right)u_r(\mathbf{p})A^e_\mu(\mathbf{k}_\nu). \tag{16-81}$$

Using Gordon's identity (16-77) for the γ^μ term,

$$\mathcal{M}^{(2+4)}_{\substack{mm \\ k_\nu \to 0}} = ie\bar{u}_{r'}(\mathbf{p}')\left(\frac{p^\mu + p'^\mu}{2m}F_1(0) + i\frac{\sigma^{\mu\nu}k_\nu}{2m}\left(-F_1(0) + F_2(0)\right) \right)u_r(\mathbf{p})A^e_\mu(\mathbf{k}_\nu). \tag{16-82}$$

We now employ our non-relativistic assumption and use $p^\mu = p'^\mu = (m,0,0,0)$ to get

$$\mathcal{M}^{(2+4)}_{\substack{mm \\ k_\nu \to 0}} = ie\bar{u}_{r'}(\mathbf{p}')\left(A^e_0(\mathbf{k})F_1(0) + i\frac{\sigma^{\mu\nu}k_\nu}{2m}\left(-F_1(0) + F_2(0)\right)A^e_\mu(\mathbf{k}_\nu) \right)u_r(\mathbf{p}). \tag{16-83}$$

When we replace the zeroth component of the photon factor with Φ, representing the electric field potential, we have

$$\mathcal{M}^{(2+4)}_{\substack{mm \\ k_\nu \to 0}} = i\bar{u}_{r'}(\mathbf{p}')\left(\underbrace{e\Phi F_1(0)}_{\substack{\text{non-spin} \\ \text{contribution}}} + \underbrace{ie\frac{\sigma^{\mu\nu}k_\nu}{2m}\left(-F_1(0) + F_2(0)\right)A^e_\mu(\mathbf{k}_\nu)}_{\text{spin contribution, as it has } \sigma^{\mu\nu} \text{ in it}} \right)u_r(\mathbf{p}). \tag{16-84}$$

If we examine the non-spin contribution, all second order effects must be included in the renormalized value for e. We have no vertex or propagator renormalizations in that term to worry about, and the external line factors $u_r(\mathbf{p})$ and $\bar{u}_{r'}(\mathbf{p}')$ remain unchanged under renormalization. Thus, if e is the electron charge value measured in experiment (as it is), i.e., the renormalized value, then second order effects in the non-spin contribution of (16-84) are already included in the renormalization of e. Given that, the only relevant factors contributing to that part of the amplitude are e and Φ. Hence,

$$F_1(0) \text{ must } = 1. \tag{16-85}$$

Step #3. Form of Relevant Spin Terms

With $F_1(0) = 1$ and our results from the previous Sect. 16.4.5 (where we showed that terms having factors in $\sigma^{0\nu}$ drop out and those with factors of σ^{ij} can be re-expressed in terms of the spin operator Σ), relation (16-84), where the symbolism

$$\nabla \times \mathbf{A}^e(\mathbf{k}_\nu) \text{ means } \nabla \times \mathbf{A}^e(\mathbf{x}) \text{ Fourier transformed to momentum space,} \tag{16-86}$$

is

$$\mathcal{M}^{(2+4)}_{\substack{mm \\ k_\nu \to 0}} = i\bar{u}_{r'}(\mathbf{p}')\left(\underbrace{e\Phi}_{\substack{\text{non-spin} \\ \text{contribution}}} - \underbrace{\frac{e}{2m}i\sigma^{\mu\nu}k_\nu A^e_\mu(\mathbf{k}_\nu)}_{\substack{g=2 \text{ part of spin} \\ \text{contribution}}}\overset{\overset{\displaystyle\frac{g}{2}}{\overbrace{\frac{\mu_B}{e}-2\Sigma\cdot\nabla\times\mathbf{A}^e(\mathbf{k}_\nu)}}}{} + F_2(0)\underbrace{\frac{\mu_B}{e}\overset{-2\Sigma\cdot\nabla\times\mathbf{A}^e(\mathbf{k}_\nu)}{\overbrace{\qquad\qquad}}}{}\underbrace{\frac{e}{2m}i\sigma^{\mu\nu}k_\nu A^e_\mu(\mathbf{k}_\nu)}_{\substack{\text{higher order spin part} \\ \text{contribution to } g}} \right)u_r(\mathbf{p}), \tag{16-87}$$

or

$$\mathcal{M}_{\substack{mm \\ k_\nu \to 0}}^{(2+4)} = i\bar{u}_{r'}(\mathbf{p}')\left(e\Phi + (2\underbrace{-2F_2(0)}_{\substack{\text{higher order} \\ \text{correction} \\ \text{to 1st order } g}})\mu_B \Sigma \cdot \nabla \times \mathbf{A}^e(\mathbf{k}_\nu)\right)u_r(\mathbf{p}). \tag{16-88}$$

Note that 2nd order part of g in amplitude has factor of $F_2(0)$

We want to find $F_2(0)$. To help us do this, we need to note one more thing.

We employ Gordon's identity (16-77) again by solving it for the term with $\sigma^{\mu\nu}$ in it and substitute that into (16-81) to get

We only need to find $F_2(0)$

Use Gordon's identity to re-express amplitude

$$\mathcal{M}_{\substack{mm \\ k_\nu \to 0}}^{(2+4)} = ie\bar{u}_{r'}(\mathbf{p}')\left(\gamma^\mu - \gamma^\mu F_2(0) + \frac{p^\mu + p'^\mu}{2m}F_2(0)\right)u_r(\mathbf{p})A_\mu^e(\mathbf{k}_\nu). \tag{16-89}$$

We are only interested in $F_2(0)$, as it is all that contributes to the higher order spin correction to g at low k_ν. So, we can ignore all terms proportional to γ^μ and focus entirely on terms having form like that of the last term in (16-89). That is, we use Gordon's identity to re-arrange our amplitude into terms with a single gamma matrix γ^μ and terms that are spinor scalars. The factor $F_2(0)$ we seek is then the sum of every spinor scalar term divided by $(p^\mu + p'^\mu)/2m$. When we find $F_2(0)$, we can simply substitute it into (16-88) and determine the higher order contribution to the gyromagnetic ratio g.

To find $F_2(0)$ we can ignore terms in γ^μ and only evaluate those in $(p^\mu + p'^\mu)/2m$.

Step # 4. Contributing Diagrams

The diagrams to calculate the next higher order magnetic moment are shown in Fig. 16-4. The LH diagram is first order in α, which we used to find $g = 2$, and which we must augment with the remaining second order diagrams.

However, all but the second diagram in Fig. 16-4 are proportional to $\bar{u}_{r'}(\mathbf{p}')\gamma^\mu u_r(\mathbf{p})$, which we don't have to be concerned with. So we only need to consider the amplitude terms arising from that second diagram.

Step #4

All but 2nd diagram in Fig. 16-4 has γ^μ factor and can be ignored

Step #5. Contribution to F_2

Recall from Chap. 14 (see Wholeness Chart 14-4, Cols (V) and (XIII), pgs 369, 371), that the vertex correction in the amplitude that would include the first and second diagrams in Fig. 16-4 is

Step #5

$$\gamma_{\substack{mod \\ 2nd}}^\mu(p,p') = \underbrace{\gamma^\mu}_{\text{1st diag}} + \underbrace{e^2\Lambda_c^\mu}_{\text{2nd diag}}. \tag{16-90}$$

If we use (16-90) in place of γ^μ in the amplitude (16-50), and take the coupling constant as the physical charge e everywhere, we get the correction due to the second diagram in Fig. 16-4. That is, we get the first line of (16-91) below, where Λ_c^μ contains all terms with F_2 spin scalars (plus some other terms in γ^μ that we won't care about). The second line below is just (16-89) repeated, so we can see clearly what the scalar terms in $e^2\Lambda_c^\mu$ we seek look like.

Evaluating the contribution from that diagram to 2nd order by including Λ_c^μ from our renormalization work

$$\mathcal{M}_{\substack{mm \\ k_\nu \to 0}}^{(2+4)} = ie\bar{u}_{r'}(\mathbf{p}')\gamma_{\substack{mod \\ 2nd}}^\mu u_r(\mathbf{p})A_\mu^e(\mathbf{k}_\nu) = ie\bar{u}_{r'}(\mathbf{p}')(\gamma^\mu + \underbrace{e^2\Lambda_c^\mu}_{\substack{F_2 \text{ scalar} \\ + \gamma^\mu \text{ terms}}})u_r(\mathbf{p})A_\mu^e(\mathbf{k}_\nu)$$

$$= ie\bar{u}_{r'}(\mathbf{p}')(\gamma^\mu \underbrace{- \gamma^\mu F_2(0)}_{\substack{\gamma^\mu \text{ terms} \\ \text{of } e^2\Lambda_c^\mu}} + \underbrace{\frac{p^\mu + p'^\mu}{2m}F_2(0)}_{\text{scalar terms of } e^2\Lambda_c^\mu})u_r(\mathbf{p})A_\mu^e(\mathbf{k}_\nu). \tag{16-91}$$

So to find F_2, we need to evaluate the higher order part of the amplitude we will designate with a subscript Λ_c,

In evaluating that contribution ignore all parts with γ^μ factor

$$\mathcal{M}_{\substack{mm \\ \Lambda_c}}^{(4)} = ie\bar{u}_{r'}(\mathbf{p}')e^2\Lambda_c^\mu u_r(\mathbf{p})A_\mu^e(\mathbf{k}_\nu), \tag{16-92}$$

but only concern ourselves with the spin scalar terms therein.

From (15-127) in Chap. 15, pg. 396, we have

*Use expression
for $\Lambda_c{}^\mu$ found from
regularization in
Chap. 15*

$$\Lambda_c^\mu\left(p,p'\right) = \underbrace{-\frac{1}{8\pi^2}\gamma^\mu\int_0^1\int_0^{1-x}\ln\left(-r-a^2\right)dz\,dx}_{\gamma^\mu\text{ part}} \;+\; \frac{1}{8\pi^2}\int_0^1\int_0^{1-x}\frac{1}{\left(r+a^2\right)}\times$$

$$\underbrace{\left(\gamma^\rho\gamma^\mu\gamma^\sigma\left(p'_\sigma-a_\sigma\right)\left(p_\rho-a_\rho\right)\right)}_{\substack{F_2\text{ scalar}+\text{maybe }\gamma^\mu\text{ parts}\\\text{Call this whole term }G}}\quad \underbrace{-2\left(p'^\mu+p^\mu-2a^\mu\right)m}_{\substack{F_2\text{ scalar}+\text{maybe }\gamma^\mu\text{ parts}\\\text{Call this whole term }J}}\quad + \underbrace{\gamma^\mu m^2}_{\gamma^\mu\text{ part}}\Big)dz\,dx,$$

(16-93)

*Label certain
parts as G and J*

where

*Simplify with
symbols for
unwieldy relations*

$$r = -\underbrace{x\left(p'^2-m^2\right)}_{=0\text{ since on shell}} - \underbrace{z\left(p^2-m^2\right)}_{=0\text{ since on shell}} + \lambda^2\left(1-x-z\right)$$

$$a^\mu = p'^\mu x + p^\mu z \;\rightarrow\; a^2 = \underbrace{p'^2}_{m^2}x^2 + 2p'^\mu p_\mu xz + \underbrace{p^2}_{m^2}z^2$$

(16-94)

and λ is the fictitious photon mass we incorporated to avoid infra-red divergences, and where the incoming and outgoing particles are on shell, i.e., $p'^2 = p^2 = m^2$.

The first and last terms on the RH of the equal sign in (16-93) have a factor of γ^μ, and so we can ignore them. The task at hand, then, is to find the parts of the second and third terms in (16-93) that have the proper scalar form. We label those terms in (16-93) G and J, as shown there.

*Once again,
ignore all parts
with γ^μ factor*

Finding the J Part of the Amplitude

We will find J in (16-93) first because it is simpler. Substituting a^μ of (16-94) into that term in (16-93), we have

*Evaluating
J part*

$$J = -2\left(p'^\mu+p^\mu-2p'^\mu x-2p^\mu z\right)m = -2m\left(p'^\mu-p'^\mu x-p'^\mu x\right)-2m\left(p^\mu-p^\mu z-p^\mu z\right).\,(16\text{-}95)$$

Now, note from (16-93) and (16-94) that the denominator of the integral in (16-93) containing F_2 parts is symmetric in x and z. (Exchanging x and z leaves it unchanged.) By doing Prob. 10, you can show that

$$\int_0^1\int_0^{1-x}p^\mu x\,dz\,dx = \int_0^1\int_0^{1-x}p^\mu z\,dz\,dx \xrightarrow[\text{general}]{\text{more}} \int_0^1\int_0^{1-x}p^\mu f(x)\,dz\,dx = \int_0^1\int_0^{1-x}p^\mu f(z)\,dz\,dx,\,(16\text{-}96)$$

and thus, if we were to have a denominator $\mathcal{D}$ symmetric in x and z (as in (16-93)), then

$$\int_0^1\int_0^{1-x}\frac{p^\mu x}{\mathcal{D}}\,dz\,dx = \int_0^1\int_0^{1-x}\frac{p^\mu z}{\mathcal{D}}\,dz\,dx \xrightarrow[\text{general}]{\text{more}} \int_0^1\int_0^{1-x}\frac{p^\mu f(x)}{\mathcal{D}}\,dz\,dx = \int_0^1\int_0^{1-x}\frac{p^\mu f(z)}{\mathcal{D}}\,dz\,dx.\,(16\text{-}97)$$

From (16-96), (16-97), and (16-93), we can change (16-95) to

$$J = -2m\left(p'^\mu-p'^\mu x-p'^\mu z\right)-2m\left(p^\mu-p^\mu x-p^\mu z\right) = -2m\left(1-x-z\right)\left(p'^\mu+p^\mu\right).\,(16\text{-}98)$$

We can conclude that J consists of one spin space scalar term that will contribute to F_2 and none having a factor γ^μ. Note we still need to evaluate the integral over x and z in (16-93), but we will wait until the end to do that.

Finding the G Part of the Amplitude

*Evaluating
G part*

From (16-93) we have

$$G = \left(\not{p}-\not{a}\right)\gamma^\mu\left(\not{p}'-\not{a}\right) = \left(\not{p}-\not{p}'x-\not{p}z\right)\gamma^\mu\left(\not{p}'-\not{p}'x-\not{p}z\right)$$

$$= \left(\not{p}(1-z)-\not{p}'x\right)\gamma^\mu\left(\not{p}'(1-x)-\not{p}z\right).$$

(16-99)

If we remember that G is ultimately sandwiched between $\bar{u}_{r'}\left(\mathbf{p}'\right)$ and $u_r\left(\mathbf{p}\right)$ with (16-93) into (16-91), then we can take advantage of the relations we learned in Chap. 4 (see Appendix A there),

$$\left(\not{p}-m\right)u_r\left(\mathbf{p}\right) = 0 \qquad \bar{u}_r\left(\mathbf{p}\right)\left(\not{p}-m\right) = 0.$$

(16-100)

With (16-100) used for the $\not{p}'x$ and $\not{p}z$ parts of (16-99), we get

$$G = \left(\not{p}\left(1-z\right) - mx \right)\gamma^{\mu} \left(\not{p}'\left(1-x\right) - mz \right)$$

$$= \underbrace{\left(1-z\right)\left(1-x\right)\not{p}\,\gamma^{\mu}\not{p}'}_{G_a} \underbrace{-\,m\left(1-z\right)z\,\not{p}\,\gamma^{\mu} - m\left(1-x\right)x\,\gamma^{\mu}\not{p}'}_{G_b} + \underbrace{m^2 x z \gamma^{\mu}}_{\gamma^{\mu}\ \text{part, ignore}} \quad . \qquad (16\text{-}101)$$

Reduce that to G_a and G_b parts plus a part we can ignore

We can ignore the last term in (16-101) because it has a factor of γ^{μ}. We need to evaluate the terms labeled G_a and G_b. To do this we need to recall the anti-commutation relations for the gamma matrices below. The commutation relations are shown also (see (16-46)), for comparison, though we won't use them in what follows.

$$\left[\gamma^{\mu},\gamma^{\nu}\right]_{+} = \gamma^{\mu}\gamma^{\nu} + \gamma^{\nu}\gamma^{\mu} = 2g^{\mu\nu} \qquad \left[\gamma^{\mu},\gamma^{\nu}\right] = \gamma^{\mu}\gamma^{\nu} - \gamma^{\nu}\gamma^{\mu} = -2i\sigma^{\mu\nu} \qquad (16\text{-}102)$$

Finding G_a

From (16-101)

Find G_a using anti-commutation relations

$$G_a = \left(1-z\right)\left(1-x\right)p_{\nu}\gamma^{\nu}\gamma^{\mu}\gamma^{\sigma}p'_{\sigma} = \left(1-z\right)\left(1-x\right)p_{\nu}\gamma^{\nu}\left(-\gamma^{\sigma}\gamma^{\mu} + 2g^{\mu\sigma}\right)p'_{\sigma}$$

$$= \left(1-z\right)\left(1-x\right)\left(-p_{\nu}\gamma^{\nu}\,\gamma^{\sigma}\gamma^{\mu}p'_{\sigma} + 2p'^{\mu}\underbrace{p_{\nu}\gamma^{\nu}}_{\not{p}}\right). \qquad (16\text{-}103)$$

Using the RH side of (16-100) and LH of (16-102), we get

$$G_a = \left(1-z\right)\left(1-x\right)\left(-p_{\nu}\underbrace{\gamma^{\nu}\gamma^{\sigma}}_{\substack{\text{use anti} \\ \text{com rel}}}\gamma^{\mu}p'_{\sigma} + 2mp'^{\mu}\right)$$

$$= \left(1-z\right)\left(1-x\right)\left(p'_{\sigma}p_{\nu}\gamma^{\sigma}\gamma^{\nu}\gamma^{\mu} \underbrace{-2g^{\nu\sigma}p_{\nu}p'_{\sigma}\gamma^{\mu}}_{\gamma^{\mu}\ \text{part, ignore}} + 2mp'^{\mu}\right) \qquad (16\text{-}104)$$

$$= \left(1-z\right)\left(1-x\right)\left(\not{p}'\,\gamma^{\nu}\gamma^{\mu}p_{\nu} + 2mp'^{\mu}\right) = \left(1-z\right)\left(1-x\right)\left(m\gamma^{\nu}\gamma^{\mu}p_{\nu} + 2mp'^{\mu}\right),$$

where in the last step, we used (16-100) again. Using anti-commutation relations one more time in the last part of (16-104), we obtain

$$G_a = \left(1-z\right)\left(1-x\right)\left(m\left(-\gamma^{\mu}\gamma^{\nu} + 2g^{\nu\mu}\right)p_{\nu} + 2mp'^{\mu}\right)$$

$$= \left(1-z\right)\left(1-x\right)\left(\underbrace{-m\gamma^{\mu}\overset{m}{\not{p}}}_{\gamma^{\mu}\ \text{part, ignore}} + 2mp^{\mu} + 2mp'^{\mu}\right) = 2m\left(1-z\right)\left(1-x\right)\left(p'^{\mu} + p^{\mu}\right), \qquad (16\text{-}105)$$

which is of the form we seek for F_2.

Finding G_b

From (16-101), where we again use gamma matrix anti-commutators after the second equal sign below,

Find G_b using anti-commutation relations

$$G_b = -\,m\left(1-z\right)z\,p_{\nu}\gamma^{\nu}\gamma^{\mu} - m\left(1-x\right)x\gamma^{\mu}\gamma^{\nu}p'_{\nu}$$

$$= -\,m\left(1-z\right)z\,p_{\nu}\left(-\gamma^{\mu}\gamma^{\nu} + 2g^{\mu\nu}\right) - m\left(1-x\right)x\left(-\gamma^{\nu}\gamma^{\mu} + 2g^{\mu\nu}\right)p'_{\nu}. \qquad (16\text{-}106)$$

Using (16-100) after the second equal sign below, we have

$$G_b = m\left(1-z\right)z\left(\gamma^{\mu}\not{p} - 2p^{\mu}\right) + m\left(1-x\right)x\left(\not{p}'\,\gamma^{\mu} - 2p'^{\mu}\right)$$

$$= m\left(1-z\right)z\left(\underbrace{\gamma^{\mu}m}_{\substack{\gamma^{\mu}\ \text{part,} \\ \text{ignore}}} - 2p^{\mu}\right) + m\left(1-x\right)x\left(\underbrace{m\gamma^{\mu}}_{\substack{\gamma^{\mu}\ \text{part,} \\ \text{ignore}}} - 2p'^{\mu}\right) \qquad (16\text{-}107)$$

$$= -2m\left(1-z\right)z\,p^{\mu} - 2m\left(1-x\right)x\,p'^{\mu}.$$

Re-expressing, and using our symmetry in x and z, we find this becomes

$$G_b = -m(1-z)\,z\,p^\mu - m(1-x)\,x\,p'^\mu \quad \underbrace{-m(1-z)\,z\,p^\mu - m(1-x)\,x\,p'^\mu}_{x \leftrightarrow z \text{(valid in limit where } p^\mu \approx p'^\mu)} \qquad (16\text{-}108)$$

$$= -m(1-z)\,z\,p^\mu - m(1-x)\,x\,p'^\mu \overbrace{-m(1-x)\,x\,p^\mu - m(1-z)\,z\,p'^\mu}.$$

Or, finally,

$$G_b = -\big(m(1-z)\,z + m(1-x)\,x\big)\big(p'^\mu + p^\mu\big), \qquad (16\text{-}109)$$

which is the form we seek.

Adding $J + G_a + G_b$

Sum of all relevant terms

Note in all our calculations above, all terms were of forms having either a factor of γ^μ or being a spin space scalar, supporting our original conclusion from (16-89). Having thrown out the terms having γ^μ factors, we are left with

$$J + G = J + G_a + G_b = -2m(1-x-z)\big(p'^\mu + p^\mu\big) + 2m(1-z)(1-x)\big(p'^\mu + p^\mu\big)$$

$$-\big(m(1-z)\,z + m(1-x)\,x\big)\big(p'^\mu + p^\mu\big)$$

$$= m\big(p'^\mu + p^\mu\big)\big(-2(1-x-z) + 2(1-z)(1-x) - (1-z)\,z - (1-x)\,x\big)$$

$$= m\big(p'^\mu + p^\mu\big)\big(\underbrace{-2 + 2x + 2z + 2 - 2x - 2z}_{=0} + 2xz - z + z^2 - x + x^2\big) \qquad (16\text{-}110)$$

$$= m\big(p'^\mu + p^\mu\big)\big(-x - z + (x+z)^2\big) = m\big(p'^\mu + p^\mu\big)(-x-z)\big(1 - (x+z)\big)$$

$$= -m\big(p'^\mu + p^\mu\big)(x+z)(1-x-z).$$

Use this sum in non-ignorable part of Λ_c^μ

Using the Spin Scalar Parts of $J + G$ in Λ_c^μ

Put our result (16-110) into (16-93) to get the part of Λ_c^μ containing spin scalar type terms,

$$\Lambda_c^\mu_{\substack{spin\\scalar}}(p,p') = \frac{-m\big(p'^\mu + p^\mu\big)}{8\pi^2}\int_0^1\int_0^{1-x}\frac{1}{(r+a^2)}(x+z)(1-x-z)\,dz\,dx, \qquad (16\text{-}111)$$

which, with (16-94), becomes

Gives us an integral in dummy variables x and z

$$\Lambda_c^\mu_{\substack{spin\\scalar}}(p,p') = \frac{-m\big(p'^\mu + p^\mu\big)}{8\pi^2}\int_0^1\int_0^{1-x}\frac{(x+z)(1-x-z)}{\lambda^2(1-x-z) + m^2x^2 + 2p'^\mu p_\mu xz + m^2 z^2}\,dz\,dx. \qquad (16\text{-}112)$$

The integral in (16-112) converges for zero photon mass, i.e., $\lambda = 0$, so we can drop that term from the denominator. It turns out that some of the γ^μ type terms have integrals that do not converge, due to infra-red divergences, and the λ term would have to be retained during the integration, if one wished to evaluate those terms. Fortunately, we do not.

Recall that we are interested in finding the gyromagnetic ratio for the limiting case where $k_\nu \to 0$, i.e., where $p'^\mu \to p^\mu$, so $p'^\mu p_\mu \approx m^2$. Additionally, for non-relativistic speeds $p^\mu \approx p'^\mu \approx (m,0,0,0)$, so again, $p'^\mu p_\mu \approx m^2$. Using that in (16-112), we find

Invoking our limit $k_\nu \to 0$. simplifies the integral

$$\Lambda_c^\mu_{\substack{spin\\scalar}}(p,p') = \frac{-m\big(p'^\mu + p^\mu\big)}{8\pi^2}\int_0^1\int_0^{1-x}\frac{(x+z)(1-x-z)}{\underbrace{m^2x^2 + 2m^2xz + m^2z^2}_{m^2(x+z)^2}}\,dz\,dx$$

The final relevant part of Λ_c^μ in terms of the integral

$$= \frac{-\big(p'^\mu + p^\mu\big)}{8\pi^2 m}\int_0^1\int_0^{1-x}\frac{(1-x-z)}{(x+z)}\,dz\,dx. \qquad (16\text{-}113)$$

Evaluating the Integral in x and z

Finding the integral in (16-113) is work for a freshman calculus course, but to save you tedium and time, I do it here. The quickest way to prove the first part of the 2nd line, without searching through integral tables, is to simply take its derivative with respect to z and show it equals the integrand on the RHS of line 1.

$$\int_0^1 \int_0^{1-x} \frac{(1-x-z)}{(x+z)} \, dz \, dx = \int_0^1 \left(\int_0^{1-x} \frac{1}{(x+z)} - \frac{x}{(x+z)} - \frac{z}{(x+z)} \, dz \right) dx$$

$$= \int_0^1 \left(\left(ln(x+z) - x\, ln(x+z) - \left\{ z - x\, ln(x+z) \right\} \right)_0^{1-x} \right) dx = \int_0^1 \left(ln(x+z) - z \right)_0^{1-x} dx \qquad (16\text{-}114)$$

$$= \int_0^1 \left(\underbrace{ln1}_{0} - 1 + x - ln\, x \right) dx = \left(-x + \frac{x^2}{2} - x\, ln\, x + x \right)_0^1 = -1 + \frac{1}{2} - 0 + 1 - 0 - 0 - 0 - 0 = \frac{1}{2}.$$

Putting the Pieces Together

Use (16-114) in (16-113) to get

$$e^2 \Lambda_{c \atop \substack{spin \\ scalar}}^\mu (p, p') = -\frac{e^2}{8\pi^2} \frac{\left(p'^\mu + p^\mu \right)}{2m}. \qquad (16\text{-}115)$$

Compare this with (16-91) to see $\qquad F_2(0) = -\frac{e^2}{8\pi^2} = -\frac{\alpha}{2\pi}. \qquad (16\text{-}116)$

The Final Result at 2nd Order

Then use this in (16-88) to find that, to second order

$$g = 2 + \frac{\alpha}{\pi} = 2.00232. \qquad (16\text{-}117)$$

Result of evaluation of the integral

That result gives us Λ_c^μ

Comparing that result to earlier amplitude expression gives us $F_2(0)$

And from that, the 2nd order value for g

16.5 The Lamb Shift

The Lamb shift is a subtle shifting of orbital electron energy levels (thus spectral lines) in atoms, not predicted by RQM, but found experimentally by Willis Lamb and his graduate student Robert Retherford in 1947. Later in the same year, Hans Bethe carried out a reasonably good estimate, though non-relativistic, of the Lamb shift in hydrogen. Analysis in subsequent years by others, using the full machinery of QED to incorporate higher order radiative corrections, yielded excellent agreement with experiment and was another dramatically convincing corroboration of the theory.

Lamb received the Nobel prize in 1955 for the discovery. (See how well graduate student contributions are valued?)

Determining the Lamb shift theoretically is a significant endeavor, entailing considerably more work and complexity than what we have been through for the anomalous magnetic moment. So, though some advantage might be gained in studying it in detail, far more will probably be gained by spending the time learning other things. Hence, we will not be doing the Lamb shift analysis here. (Do I hear sighs of relief?) Those interested in pursuing it further can see the references below[1].

Lamb shift postdiction by QED was dramatic confirmation of theory

Lamb shift QED calculation very long and will not be done here

16.6 A Note on QED Successes Over RQM

As noted in Chap. 1, QFT holds significant advantages over RQM. For example, in its electroweak theory aspect (not treated in this text), it handles particle decay and other transmutations of particles from one type to another. Even in QED, it does this, e.g., in the interaction $e^- e^+ \to \mu^- \mu^+$, with a virtual photon mediating the transmutation. RQM cannot do this.

It is also often said that QED is more accurate than RQM because its higher order corrections give us the correct values for the anomalous magnetic moment, the Lamb shift, the dependence of e (or equivalently, α) on energy level, and even modifications to the Coulomb potential. We tend to unconsciously assume this means QED has some fundamental advantage in accuracy over RQM, inherent in the base theory. But is this really the case?

Reasons for QED accuracy edge over RQM

---

[1] Itzykson, C. and Zuber, J.B., *Quantum Field Theory* (McGraw-Hill, 1985), Sect. 7.3.2, pgs 358-365. Mandl, F. and Shaw, G., *Quantum Field Theory*, 2nd ed (Wiley, 2010), 187-191.

Consider that the governing equations, coupling the fermion and photon fields, are the same in both theories, i.e.,

$$\partial^{\alpha}\partial_{\alpha}A^{\mu} = -e\bar{\psi}\gamma^{\mu}\psi \tag{16-118}$$

$$\left(i\gamma^{\mu}\partial_{\mu} - m\right)\psi = -e\gamma^{\mu}\psi A_{\mu}, \tag{16-119}$$

where ψ and A_{μ} are states in RQM and (operator) fields in QFT/QED. The starting point is the same, save for our interpretation of the dependent functions in each case. Then, why shouldn't they have equal predictive power?

The reason is that, for full accuracy, we need to solve the two coupled equations simultaneously. The value of ψ is affected by A_{μ}, and the value of A_{μ} is, in turn, affected by ψ. A closed solution for all practical problems is simply not possible. So, we must turn to perturbation theory.

RQM was not readily amenable to perturbation, so for it, researchers generally assumed a value for A_{μ} that was unaffected by ψ. This is not true, but it is often a good approximation. It was a good enough approximation to yield the relativistic hydrogen atom spectrum (to better accuracy than NRQM, though still not perfect, as it did not, for one, include the Lamb shift.)

In the RQM hydrogen atom analysis, the assumption was $A_{\mu} = (\Phi,0,0,0)$, where Φ is the Coulomb potential. (See (16-9) and Sect. 16.1.2. pg. 404.) Due to the influence of the electron (i.e. the function ψ), A_{μ} is actually modified from this value. But by ignoring that modification, we can still get pretty accurate results. That is, we assume the above is the solution to the A_{μ} field in (16-118) and (16-119), and then simply solve (16-119) for ψ. That gives us the RQM solution to the hydrogen atom.

However, in QFT/QED we use a perturbation approach (each Feynman diagram represents another term in the perturbative expansion.) We use the same two equations (16-118) and (16-119) as in RQM, but can solve them simultaneously (in a perturbative sense), so we are not restricted by assuming a fixed A_{μ}.

The point is this. Had we been able to solve (16-118) and (16-119) perturbatively in RQM, we could have gotten the same result. The advantage of QED is that it allows us to use perturbation theory. There is no deeper advantage in the theory itself. (Again, we are limiting discussion to electrodynamics, not including weak or strong interactions.)

QED has this advantage (that QED is amenable to perturbation) over RQM because the solutions to the field equations (16-118) and (16-119) are operators that create and destroy states. This allows us to use the machinery of transition amplitudes and all the rest, which led us to Feynman diagrams and rules. They enable us to add higher order perturbative corrections to our lower order estimates.

QED's accuracy edge over RQM is because QED amenable to perturbation solutions

<u>Bottom line:</u> RQM and QED share the same fundamental field equations and in theory, should yield the same results, if we could obtain closed solutions to those equations (which we cannot). However, because we can use perturbation theory in QED, we get higher accuracy than in RQM.

<u>Caveat:</u> As I have done several places in this text, I have presented above my personal interpretation, which I have not seen elsewhere in the literature, but believe to be true. As I have also done before, I warn you to think this issue through and form your own conclusion.

16.7 Chapter Summary

Coulomb Potential in RQM

Found from solving the uncoupled field equation $\partial^{\alpha}\partial_{\alpha}A^{\mu} = 0$ for the state A_{μ} in spherical coordinates.

Coulomb Potential in QFT

Found from finding the amplitude for electrons scattered by a virtual photon (Møller scattering) for non-relativistic speeds. This amplitude is evaluated in momentum space and compared with the Born approximation for the same scattering found in NRQM, which contains the potential in momentum space. From the comparison, the form of the Coulomb potential in momentum space $\tilde{V}(\mathbf{k})$ is gleaned. This is then Fourier transformed to spherical coordinates in position space to obtain the Coulomb potential in position space $V(r) = \pm e^2/r$.

For repulsive (like charges) force, the sign above is +.

For attractive (opposite charges) force, the sign above is –.

The sign difference arises from the number of adjacent fermion field interchanges needed in each case to bring the associated expression for the amplitude into normal order (as required by Wick's theorem.)

Anomalous Magnetic Moment

Wholeness Chart 16-2, reproduced below, summarizes the magnetic moment affair.

Theoretical and Experimental Values for Electron Gyromagnetic Ratio g

Experiment	NRQM	RQM	QFT to order		
			α	α^2	α^3
2.002319304362	Cannot predict	2	2	2.00232	2.002319304

RQM value:

The RQM value is determined by finding, from solving the full Dirac interaction equation, the terms contributing to the total energy. One of these terms, that from magnetic moment interacting with an external magnetic field, has form $g\mu_B \Sigma \cdot \mathbf{B}^e$ with $g = 2$.

QED value to order α:

The first order QED result is found by using a truncated Feynman diagram version of Møller scattering with only a single incoming and outgoing electron and an external virtual photon field A_μ^e. A new Feynman rule is used for the external field, which is neither an external particle nor a full propagator. Non-relativistic electron speeds are assumed.

The single γ^μ term in the amplitude is converted via Gordon's identity to a term with no gamma matrices and a term in $\sigma^{\mu\nu}$. The latter term turns out to be expressible as factors of Σ (the vector spin operator), the Bohr magneton μ_B, and the external field A_μ^e.

This amplitude term (in momentum space) is compared to the Born approximation amplitude to extract the form of the potential energy in momentum space due to the electromagnetic moment and the external field. This expression is then Fourier transformed to position space where the potential energy is found to be $g\mu_B \Sigma \cdot \mathbf{B}^e$ with $g = 2$.

QED value to order α^2:

Steps #1 to 6 beginning on pg. 420 are a good summary of this.

Rather than proceed, as we did for the α case, by converting all γ^μ terms using Gordon's identity, we use a trick that greatly simplifies the calculations. To do this, we need to restrict calculation to the lowest energy levels (non-relativistic) for the external field, which means $k_\nu = p - p' \to 0$. Like e, the magnetic interaction contribution will depend on k_ν^2, but we can simplify things greatly by taking the limiting case (for which the experimental value is relevant).

We start with the most general form for the magnetic moment part of the amplitude in terms of $\gamma^\mu F_1$ and $\sigma^{\mu\nu} k_\nu F_2/2m$ with initially unknown form factors $F_1(k_\nu^2)$ and $F_2(k_\nu^2)$. In the low speed limiting case, using Gordon's identity, we see that $F_1(0) = 1$. So, all we need to do is find F_2.

To do that, we then use Gordon's identity again, only this time in reverse, to convert the terms in F_2 to two terms in $\gamma^\mu F_2$ and $((p^\mu + p'^\mu)/2m) F_2$. Thus, to find F_2, we can simply evaluate all terms in the amplitude having a $(p^\mu + p'^\mu)$ factor and no γ^μ factor. That is, we can ignore all terms with γ^μ in them for this determination. All other terms (the ones we care about) will have no gamma matrices.

We then focus on the second order vertex correction diagram amplitude contribution Λ_c^μ to find F_2. We use the expression for Λ_c^μ from Chap. 15 where we evaluated it using regularization. After a great deal of mathematical manipulation, we find $F_2 = -e^2/8\pi^2$.

With this value for F_2 inserted into the general form of our amplitude, we find, in addition to the first order magnetic moment interaction term, an additional term of similar form (except for the factor in front), $\dfrac{\alpha}{\pi}\mu_B\Sigma\bullet\mathbf{B}^e$. When we add this to the first order term, we find $g = 2 + \alpha/\pi = 2.00232$.

Lamb Shift

The Lamb shift comprises a shift in atomic energy levels from their RQM values that was found experimentally before it was evaluated theoretically. QED, using higher order radiative corrections, postdicts this shift.

Addendum

We should keep in mind the distinction between the commutation and anti-commutation relations for the gamma matrices (see (16-102), repeated below), each of which we use at different points in our development of QFT.

$$\left[\gamma^\mu,\gamma^\nu\right]_+ = \gamma^\mu\gamma^\nu + \gamma^\nu\gamma^\mu = 2g^{\mu\nu} \qquad \left[\gamma^\mu,\gamma^\nu\right] = \gamma^\mu\gamma^\nu - \gamma^\nu\gamma^\mu = -2i\sigma^{\mu\nu}. \qquad (16\text{-}120)$$

16.8 Appendix: Deriving Feynman Rules for Static, External (Potential) Field

16.8.1 Comparing Four-Momenta in Different Types of Interactions

Before deriving the transition amplitude for the RHS of Fig. 16-3 (repeated below as Diagram (F) of Fig. 16-5), we point out its salient (four-momentum) characteristics by comparing it to other types of interactions. In Fig. 16-5 we show purely elastic types of classical interactions in the upper part, and each corresponding quantum elastic interaction below its classical counterpart.

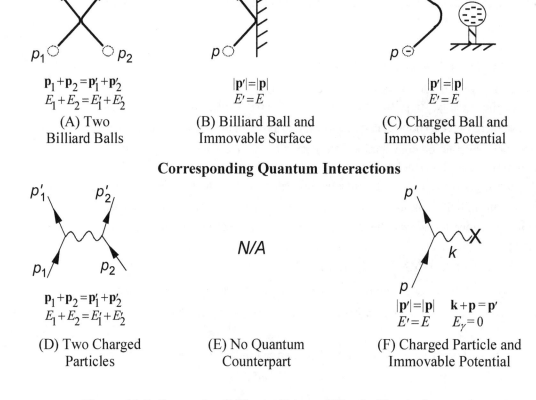

Classical Interactions

$\mathbf{p}_1+\mathbf{p}_2=\mathbf{p}_1'+\mathbf{p}_2'$
$E_1+E_2=E_1'+E_2'$

(A) Two
Billiard Balls

$|\mathbf{p}'|=|\mathbf{p}|$
$E'=E$

(B) Billiard Ball and
Immovable Surface

$|\mathbf{p}'|=|\mathbf{p}|$
$E'=E$

(C) Charged Ball and
Immovable Potential

Corresponding Quantum Interactions

$\mathbf{p}_1+\mathbf{p}_2=\mathbf{p}_1'+\mathbf{p}_2'$
$E_1+E_2=E_1'+E_2'$

(D) Two Charged
Particles

N/A

(E) No Quantum
Counterpart

$|\mathbf{p}'|=|\mathbf{p}| \quad \mathbf{k}+\mathbf{p}=\mathbf{p}'$
$E'=E \qquad E_\gamma=0$

(F) Charged Particle and
Immovable Potential

**Figure 16-5. Comparing Different Types of Elastic Classical
Interactions with Their Quantum Counterparts**

Diagrams (A) and (D)

The energy and 3-momentum balances in Diagrams (A) and (D) should not need explanation. No external forces (potentials) act, so total initial 4-momentum of the particles equals total final 4-momentum.

Diagrams (B) and (E)

In the classical case of an object (billiard ball of Diagram (B) in Fig. 16-5) rebounding (fully elastically) off of a solid (immovable) surface, the surface picks up no energy. It doesn't begin to move as a result of the collision, so it has no kinetic energy. There is no stored compressive energy because the interaction is elastic. Hence, the surface neither gains nor loses energy. Thus, the outgoing particle must have the same kinetic energy it came in with. $E' = E$. The object changed direction so there was a change in velocity direction (and thus 3-momentum direction). Since $E = (\mathbf{p})^2 / 2m$ and $E' = (\mathbf{p}')^2 / 2m$, we must have $|\mathbf{p}| = |\mathbf{p}'|$, i.e., the magnitude of the 3-momentum is unchanged. No energy was transferred to the object. The surface transmitted 3-momentum, but not energy. It exerted a force on the object for a time (leading to 3-momentum change) but no force acting over a distance (which would lead to kinetic energy change) since it moved zero distance.

There are no solid surfaces in the quantum realm, so there is no quantum counterpart to Diagram (B).

Diagrams (C) and (F)

The classical interaction (scattering) of a charged object (ball in Diagram (C)) by an immovable, static potential (arising from the collection of charges on the stationary ball in Diagram (C)) parallels the case of Diagram (B). That is, the stationary source of the field does not move, so it picks up no kinetic energy. No energy is absorbed by the field from the interaction with the object. So neither the stationary source nor its field gain or lose energy. Hence there is no energy transfer in the process.

Jumping to the quantum realm (Diagram (F)), the same principles remain. The source of the potential (static, external) field does not move and so neither it nor its field gains kinetic energy (or any other kind of energy). But the potential field can exert a force on a charged particle and thus change the particle's 3-momentum. Since the particle energy is purely kinetic (ignoring rest mass-energy, which is always the same for a given particle), and that energy does not increase or decrease, then no energy is transmitted from the field to the particle. That is, the energy of the virtual photon E_γ equals zero. But the virtual photon must carry 3-momentum since it results in the change in direction (but not magnitude) of the charged particle 3-momentum. $E' = E$, $E_\gamma = 0$, $|\mathbf{p}'| = |\mathbf{p}|$, $\mathbf{k} = \mathbf{p}' - \mathbf{p}$.

Bottom line: In our QFT transition amplitude, we would expect the virtual photon of Diagram (F), which is off shell, to have $\mathbf{k} = \mathbf{p}' - \mathbf{p}$ and $E_\gamma = 0$.

16.8.2 The Transition Amplitude for Diagram (F)

We need to start from scratch to find S_{fi}, for Diagram (F), as we have no ready-made Feynman rules for this case. This will parallel the development of (8-12) to ((8-18) on pgs. 217-218 (see Fig. 8-1 there), except that we cannot represent the photon here by the free field solution $A_\mu(x)$, which we have grown familiar with, and which we used there. The photon there was a free field, but in Diagram (F) above it is not free, but related to the external force the potential source supplies. In the present case, A_μ is virtual (it must be off shell, for the initial and final particles to be on shell) and *not* a free field. Additionally, it is static, so has no dependence on time, and thus contains no factor of $e^{\pm i\omega_k t}$. That is, a static, external, potential e/m field must have generic form $A_\mu(\mathbf{x})$, where $\mathbf{x}$ is the 3D position vector. Further, we cannot specify the precise dependence of A_μ on $\mathbf{x}$, as that will vary with the characteristics of the potential source (shape, number of charges, etc.)

Using (8-12), pg. 217, with our initial multiparticle state containing a real electron and a virtual photon emanating from the source of the potential along with a final real electron, we have a transition amplitude

$$S_{fi} = \left\langle f \left| S^{(1)} \right| i \right\rangle = \left\langle e^-_{\mathbf{p}',r'} \left| (-i) \int d^4 x_1 N \left\{ -e\,\overline{\psi}(x) \gamma^\mu A^e_\mu(x) \psi(x) \right\}_{x_1} \right| e^-_{\mathbf{p},r}, \gamma_{\mathbf{k},s} \right\rangle \quad (\gamma_{\mathbf{k},s}\ \text{off shell}), \quad (16\text{-}121)$$

where the fermion fields are the usual free fields we have been working with since Chap. 4, and

$$A^e_\mu(\mathbf{x}) = \sum_{s'',\mathbf{k}''} \left(a^e_{s''}(\mathbf{k}'') A^e_{\mathbf{k}''\mu}(\mathbf{x}) + a^{e\dagger}_{s''}(\mathbf{k}'') A^{\dagger e}_{\mathbf{k}''\mu}(\mathbf{x}) \right), \quad (16\text{-}122)$$

Note the numerical coefficients $1/\sqrt{2V\omega_\mathbf{k}}$ and the $e^{\pm ikx}$ factors we usually see for terms in $A_\mu(x)$ are replaced by the $A^e_{\mathbf{k}''\mu}(\mathbf{x})$ factors, whose precise form is not specified and unknown at this point. The construction and destruction operators in (16-122) create and destroy virtual photons emitted from the external field source. If we were to express these photon states mathematically, they would have the same dependence on $\mathbf{x}$ as $A^e_{\mathbf{k}''\mu}(\mathbf{x})$.

Using (16-122) in (16-121), we get (by a process that hopefully is getting somewhat familiar to you, i.e., the ket particles are destroyed and a sum of final particle kets are created, but only the final particle ket that matches the bra final particle gives a non-zero result)

$$S_{fi} = S_{\substack{static \\ ext\ pot}} = ie \left\langle e^-_{\mathbf{p}',r'} \left| \sum_{r'',\mathbf{p}''} \int \left(\sqrt{\frac{m}{VE_{\mathbf{p}''}}} \sqrt{\frac{m}{VE_{\mathbf{p}}}} A^e_{\mathbf{k}\mu}(\mathbf{x}) \overline{u}_{r''}(\mathbf{p}'') \gamma^\mu u_r(\mathbf{p}) e^{ip''x} e^{-ipx} \right) d^4x \right| e^-_{\mathbf{p}'',r''} \right\rangle$$

$$= 0 + ... + ie \int \left(\sqrt{\frac{m}{VE_{\mathbf{p}'}}} \sqrt{\frac{m}{VE_{\mathbf{p}}}} A^e_{\mathbf{k}\mu}(\mathbf{x}) \overline{u}_{r'}(\mathbf{p}') \gamma^\mu u_r(\mathbf{p}) e^{ip'x} e^{-ipx} \right) d^4x \underbrace{\left\langle e^-_{\mathbf{p}',r'} \middle| e^-_{\mathbf{p}',r'} \right\rangle}_{=1} + 0 + ... \quad (16\text{-}123)$$

Re-arranging, we have

$$S_{\substack{static \\ ext\ pot}} = ie \sqrt{\frac{m}{VE_{\mathbf{p}'}}} \sqrt{\frac{m}{VE_{\mathbf{p}}}} \, \overline{u}_{r'}(\mathbf{p}') \gamma^\mu u_r(\mathbf{p}) \int A^e_{\mathbf{k}\mu}(\mathbf{x}) e^{-i\mathbf{p}'\cdot\mathbf{x}} e^{i\mathbf{p}\cdot\mathbf{x}} \underbrace{\left(\int e^{iE't} e^{-iEt} dt \right)}_{2\pi\delta(E'-E)} d^3x$$

$$= 2\pi\delta(E'-E) ie \sqrt{\frac{m}{VE_{\mathbf{p}'}}} \sqrt{\frac{m}{VE_{\mathbf{p}}}} \, \overline{u}_{r'}(\mathbf{p}') \gamma^\mu u_r(\mathbf{p}) \underbrace{\int A^e_{\mathbf{k}\mu}(\mathbf{x}) e^{-i\mathbf{p}'\cdot\mathbf{x}} e^{i\mathbf{p}\cdot\mathbf{x}} d^3x}_{\substack{\int A^e_{\mathbf{k}\mu}(\mathbf{x}) e^{-i(\mathbf{p}'-\mathbf{p})\cdot\mathbf{x}} d^3x \\ \text{Fourier transform with } \mathbf{k}=\mathbf{p}'-\mathbf{p}}}. \quad (16\text{-}124)$$

Using the Fourier transform relation (16-11), we end up with the transition amplitude for Diagram (F),

$$S_{\substack{static \\ ext\ pot}} = 2\pi\delta(E'-E) ie \sqrt{\frac{m}{VE_{\mathbf{p}'}}} \sqrt{\frac{m}{VE_{\mathbf{p}}}} \, \overline{u}_{r'}(\mathbf{p}') \gamma^\mu u_r(\mathbf{p}) A^e_\mu(\mathbf{k}). \quad (16\text{-}125)$$

16.8.3 Feynman Rules for Static, External (Potential) Field

We can see that (16-125) can be arrived at by using the usual Feynman rules except that we substitute rule #11 ((16-49)(a) and (b)) for the delta function and photon parts. One can then generalize ((16-49)(a) and (b)) to apply to any Feynman diagram with a static, external (potential) field.

Note that had we considered the photon to be outgoing rather than incoming, we would have obtained the same result.

16.9 Problems

1. Use (16-28) to show that at low speeds, the first adjoint anti-fermion factor in (16-27) has the following possible forms.

$$\overline{v}_{r_1=1}(\mathbf{p}_1) = v^\dagger_{r_1=1}(\mathbf{p}_1)\gamma^0 \Rightarrow \begin{pmatrix} 0 & 0 & -1 & 0 \end{pmatrix} \qquad \overline{v}_{r_1=2}(\mathbf{p}_1) = v^\dagger_{r_1=2}(\mathbf{p}_1)\gamma^0 \Rightarrow \begin{pmatrix} 0 & 0 & 0 & -1 \end{pmatrix}$$

Then assuming $r_1 = 1$ and $r_2 = 1$, use (16-14), (16-16), (16-17), and (16-27) to get (16-29).

2.   Consider the massive scalar boson propagator, $\dfrac{1}{k^2 - \mu^2 + i\varepsilon}$, where μ is boson mass, and that there is a scalar boson to fermion coupling constant g similar to e for photon to fermion coupling. Assume all else is similar to QED amplitude determination except that $\bar{u}_A(\mathbf{p}_A)\gamma^\mu u_B(\mathbf{p}_B)$ goes to $\bar{u}_A(\mathbf{p}_A)u_B(\mathbf{p}_B)$ for scalars, since the scalar propagator has no

g$\mu\nu$. Follow steps similar to those of (16-18) to (16-21) to show that $\tilde{V}_{Yukawa}(\mathbf{k}) = \dfrac{-g^2}{\mathbf{k}^2 + \mu^2}$.

Then, follow steps similar to (16-23) to (16-26), and use the Fourier transform pair relation $\dfrac{k}{k^2 + \mu^2} \Leftrightarrow \dfrac{i}{2}e^{-\mu r}$ (for $r > 0$) to show that the potential in position space for a massive scalar

force carrier boson, known as the Yukawa potential, is $V_{Yukawa}(r) = -g^2 \dfrac{e^{-\mu r}}{4\pi r}$.

Note that due to the minus sign, this is attractive. Similar particles (two fermions, with no anti-fermion, interacting) attract. If there were no positive and negative Yukawa "charges" (all particles had the same Yukawa charge g and no negative charge, i.e., no $- g$, exists), then the force mediated by the massive scalar boson would always be attractive. And this is what occurs inside nuclei, according to the Yukawa model. Protons and neutrons exchange massive scalar virtual mesons, and those exchanges hold the nucleus together.

Note also that due to the exponential dependence on r in the Yukawa potential, it falls off far more rapidly than the Coulomb potential. The Yukawa force is considered a "short range" force. The limited range is a result of the virtual boson mediating the force having a mass μ. For $\mu = 0$, the force would be long range, like the electromagnetic force (with zero mass virtual photons mediating.)

3.   Find σ^{31} of (16-48).

4.   Find σ^{01} of (16-48).

5.   Show that for σ^{02} the value of (16-53) is zero. Can you see rather quickly, without doing the actual matrix multiplication, that for σ^{03}, the value is zero, as well?

6.   Show that for an $r = 1$ spin state electron, with $i = 3$, $j = 1$, the contribution to the amplitude from spin terms is zero. Since we know that switching index values for i and j merely changes sign in the amplitude term, can we not say that for $r = 1$, all index values other than 1 and 2 yield zero amplitude terms?

7.   Find (16-59). That is, for the low velocity $u_{r=2}(\mathbf{p})$ (eigen spin in opposite direction, i.e in negative x^3 axis direction), find the spin contribution to the Feynman amplitude of the RHS of Fig. 16-3.

8.   Find the LHS of (16-62). That is, find the spin contribution to the Feynman amplitude for a $+x$ direction low velocity eigen spin state electron in the RHS of Fig. 16-3. Hint: This eigenstate is $\dfrac{1}{\sqrt{2}}\begin{bmatrix}1 & 1 & 0 & 0\end{bmatrix}^T = \dfrac{1}{\sqrt{2}}u_{r=1}(\mathbf{p}) + \dfrac{1}{\sqrt{2}}u_{r=2}(\mathbf{p})$. Check that it is so by operating on it with Σ_1 of (16-41) (or (16-48).

9.   This problem was deleted after Sect. 16.4.6 of the May 2014 book version was revised.

10. a) Show the LHS of (16-96). b) In the x-z plane, sketch the area over which the integration is carried out. (Hint: It is a right isosceles triangle. c) Sketch the 3D plot of the function $f(x,z) = Kx$ over the x-z plane, where K is a constant. Do the same for $f(x,z) = Kz$. Is it obvious that the volume under $f(x,z)$ is the same in each case? d) Is volume under the function $f(x,z) = Kx/\mathcal{D}$, where $\mathcal{D} = x^2 + z^2$ the same as the volume under the function $f(x,z) = Kz/\mathcal{D}$? e) Is it apparent that the volume under either function in d) is the same whenever $\mathcal{D}$ is symmetric in x and z. e) Can we readily extrapolate our results to the RHS of (16-96), i.e., for any case where the 1st numerator is the same function of x as the 2nd is of z ? (Hint: Consider integration volumes.)

Chapter 17

Scattering

"A law of nature is expected to hold true without exceptions; it is given up as soon as one is convinced that one of its conclusions is incompatible with a single experiment."

Albert Einstein

17.0 Preliminaries

17.0.1 Background

Every interaction between two or more particles is a scattering event, so all of the previous chapter actually entailed analyzing <u>scattering</u>. However, in another context, the word pertains specifically to laboratory scattering experiments, such as found in particle accelerators at Fermi Lab's Tevatron and the Large Hadron Collider (LHC) at CERN. It is this latter sense to which the title of this chapter refers, and with which the material in it is concerned.

Scattering defs:
#1: any interaction between particles
#2: scattering experiment (sense in this chapter)

17.0.2 Chapter Overview

In this chapter we will

- define and develop the concept of scattering cross-section
- overview scattering in various physical theories
    - classical mechanical (billiard ball type) collisions (Review)
    - classical scattering by a potential field, such as Coulomb field (Review)
    - NRQM scattering (Review)
    - QFT scattering (Preview),
- apply QFT to laboratory scattering experiments, and
- discuss how cross-section contributions from infrared divergences cancel out.

We'll review scattering in physical theories before QFT

Then examine it in QFT

17.1 The Cross-section

The simplest way to understand the <u>cross-section</u>, symbolized by the Greek letter σ is with reference to a collision between two classical particles. From there, we can extend its application to systems containing many particles, to scattering by potential fields, and to quantum scattering.

17.1.1 Classical Collision Cross-section

<u>Cross-section as an Area</u>

Consider a single billiard ball hanging by a thread being shot at with a BB gun (which shoots very small spherical metallic objects called BBs), as shown in Fig. 17-1. The constraints we now impose on the gun and the target may seem unrealistic for a shooting range, but they reflect the situation in laboratory scattering. The constraints are these: 1) the BB must travel through an <u>area</u> of size A_t ("t" for target) that is perpendicular to the path of travel of the BB, and 2) other than this restriction the BB gun can be considered to be shot by a blindfolded person. That is, the BB can randomly pass anywhere through the area A_t.

The target particle, a billiard ball in Fig. 17-1, has <u>cross-sectional area</u> (through the center of the ball and perpendicular to the flight path of the BB) of $\underline{\sigma}$. Thus, the

Cross-section the simple way: as area of target in classical collision with point-like particle

probability of beam particle (BB) hitting target particle (billiard ball) $= \dfrac{\sigma}{A_t}$.   (17-1)

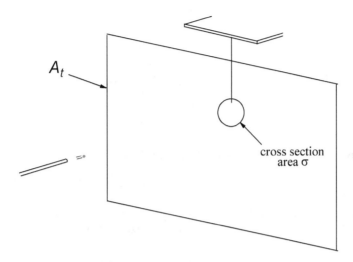

Figure 17-1. Small BB Shot at Billiard Ball

In its simplest, classical, interpretation, the cross-section σ is just the cross-sectional area of a target particle when the beam particle is point-like. A larger cross-section means a greater probability of a collision (interaction) occurring.

Cross-section as a Probability

Now consider that if A_t is of unit area size, i.e., $A_t = 1$ (cm^2 in cgs system, for example), then the

For target particle inside unit area, cross-section σ = probability of interaction

$$\boxed{\text{probability of single beam particle hitting single target particle in unit area } = \sigma}$$.   (17-2)

In this sense, the cross-section σ represents a probability, being greater for larger targets, and hence, the phrase "larger cross-section" is synonymous with "higher probability of interaction".

Larger cross-section $\rightarrow$ greater probability of interaction

Note that in (17-1), σ has units of area, whereas probability, as in (17-2), is unitless. The units of σ tell us what units are used for the unit area. For $\sigma = 0.1$ cm^2 there would be a probability of 0.1 of one beam particle interacting with one target with $A_t = 1$ cm^2. The same cross-section could be expressed as 0.00001 m^2, meaning a 0.00001 probability for the same two particles for $A_t = 1$ m^2.

Units of σ indicate units of A_t for probability interpretation

Effective Cross-section

Consider the case where the beam particle is not a tiny, virtually point-like, BB, but another billiard ball. Since the beam particle itself has greater size, the probability for collision is greater than that for a point-like beam particle. We can then define an effective cross-section σ as shown in Fig. 17-2. The effective cross-section is simply the area through which the center point of a spherical beam particle must pass in order for a collision to occur. It is the area that would correspond to what would be needed for deflection of a point-like beam particle.

In most cases, classically or quantum mechanically, we deal with non point-like beam particles, and so, we deal with effective cross-sections. In practice, the two are rarely distinguished, and so the term "cross-section" and the symbol σ are used for both effective and actual cross-sections.

For non point-like beam particle, cross-section is effective target area if beam were point-like

Common Usage of Term "Cross-section"

In either the actual or effective case, we can think of the cross-section as defined in (17-2). That is its most useful form and the one you should really lock into memory. Even when you hear "cross-section" expressed as an area, it is really the (17-2) interpretation people are thinking of and the area units then simply mean that much (effective) target cross-sectional area per unit area perpendicular to the beam path.

Probability definition of σ true for actual or effective case

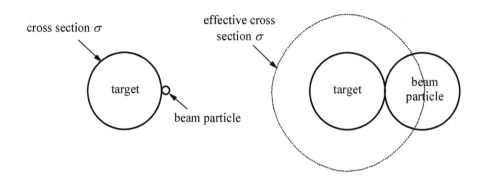

Figure 17-2. Effective Cross-section for Non Point-like Beam Particle

Cross-section Related to Type of Particles

Even though we might have the same cross-sectional area for a target particle, we would have different (effective) cross-sections, depending on the type (actually the size in this case) of beam particle.

So, when we specify a cross-section, it must be associated with particular beam and target particles. A BB and billiard ball interaction would have cross-section σ_1, but a golf ball and billiard ball interaction would have a different cross-section σ_2.

So, cross-section (as a probability) depends on the type (size here) of both beam and target particles

Size of Typical Cross-sections

In atomic and subatomic scattering experiments, we are generally considering scales on the order of nucleons, which have diameters of about one Fermi, abbreviated by "F" (also called a femtometer, abbreviated by "fm"), equal to 10^{-13} cm (10^{-15} m). The associated cross-sectional area is then on the order of 10^{-26} cm^2.

When some of the first scattering experiments were carried out, and the associated cross-sections found, one was so much bigger than expected that it led one physicist to exclaim "it's as big as a barn!". The term stuck, so now cross-sections are commonly measured in barns. A barn = 10^{-24} cm^2, so a nucleon cross-section is about 0.01 barns. Values in typical atomic and subatomic experiments range between 0.001 and 1,000 barns.

In subatomic scattering, cross-sections incredibly small. Unit = barn.

More Than One Target Particle

In practice, we usually have a collection of target particles, like nucleons, in a volume $V_t = A_t d$, where d is the depth perpendicular to A_t. Also, in practice, the target particles are usually very, very small, like nucleons, and the spaces between them enormous by comparison. Thus, it would be highly unlikely that the cross-sectional area of any target particle in a volume would shield any other particle behind it from the beam particle(s).

So, if the number of particles in the target volume is N_t, and the density of the particles therein is n_t target particles/unit volume, then the

$$\text{probability of single beam particle hitting any target particle} = \frac{\sigma}{A_t} N_t = \frac{\sigma}{A_t} n_t \underbrace{V_t}_{A_t d} . \qquad (17\text{-}3)$$

Probability of interaction with more than one target particle

More Than One Beam Particle

Also, in practice, we typically have a shower of beam particles, as in Fig. 17-3. If $N_{b\,tot}$ is the total number of beam particles passing through the volume V_t, then the expected number of beam particles hitting target particles equals

$$\text{expected total collisions} = N_{b\,tot} \frac{\sigma}{A_t} N_t = N_{b\,tot} \frac{\sigma}{A_t} n_t V_t . \qquad (17\text{-}4)$$

And with more than one beam particle

Note that if the beam volume extends outside the target volume in a direction perpendicular to the beam direction, the volume where interactions take place is the target volume, and N_t is the number of target particles in that volume. $N_{b\,tot}$ is the total number of beam particles that enter the target volume, not the total number of beam particles.

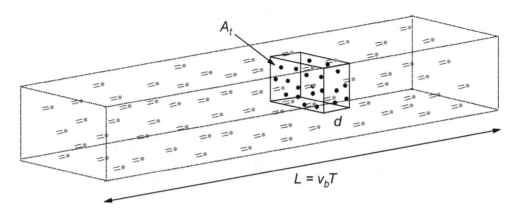

Figure 17-3. Many Beam Particles and Many Target Particles

For a stream of particles, we generally want to consider the collisions per second for such a stream, rather than the total number of collisions. In doing that, it helps to work with the beam particle number density, n_b (= N_b/V_b with N_b = number of beam particles in beam volume V_b).

The total number of beam particles (see Fig. 17-3) passing through V_t, with v_b the velocity of a beam particle (assuming all particles have the same velocity) and T the time duration that the beam is turned on, is

$$N_{b\,tot} = n_b A_t L = n_b A_t v_b T .$$ (17-5)

Total number of beam particles passing through V_t

So the collisions per second can be found by using (17-5) in (17-4) and dividing by the total time T the beam is on.

$$\frac{\text{expected total collisions}}{\text{second}} = \frac{N_{b\,tot}}{T}\frac{\sigma}{A_t}N_t = \frac{n_b A_t v_b T}{T}\frac{\sigma}{A_t}n_t V_t = n_b v_b \sigma n_t V_t$$

$$\boxed{= f_b \sigma n_t V_t = f_b \sigma N_t = \text{transition rate (for a given experiment)}} \; ,$$ (17-6)

Expected total interactions/sec (= transition rate) for a given experiment

where $n_b v_b = f_b$ is the incident beam flux (particles/sec passing through unit area). From henceforth, we will use the term "collisions" interchangeably with "expected collisions", and "transition rate", with "expected transition rate", because our experiments will have so many individual transitions that the measured, actual rates will be, to high accuracy, equal to the expected rates.

Note that though we have portrayed the target as a rectangular solid, any shape volume works for V_t. Any A_t works, because it cancels out. And the key number above with regard to the target is N_t, the total number of particles in the target. It equals $n_t V_t$ for any shape of target volume V_t.

Particular Experimental Configuration vs Cross-section

Note that for any given type of interaction, such as BBs and billiard balls classically, or alpha particles and gold nuclei quantum mechanically, we have a particular cross-section σ that characterizes the way a single particle of one particular type interacts with another single particle of a second particular type. This is what is expressed in (17-2). If the two interact readily (higher probability of interaction), the cross-section of (17-2) is higher than if those two particle types don't interact so readily (lower probability of interaction).

In different experimental configurations, on the other hand, two given particle types will produce a different number of collisions per second (transition rate), depending on how many target particles N_t there are and what the beam particle flux f_b is. More target particles mean more collisions per second. More particles per unit volume of the beam (yielding higher beam flux) means more collisions per second. And classically at least, higher particle velocity (yielding higher beam flux) in the beam means more collisions per second. This is what is expressed in (17-6).

So, in any given experiment, we know the total number of target particles ($N_t = n_t V_t = n_t A_t d$) and the beam flux f_b. We measure the total collisions per second. Then, from (17-6), we can calculate the cross-section σ. In another experimental setup involving the same two particle types,

Same two particle types will generate different transition rates in different experiments

σ is independent of experimental setup, so is a good way to compare intrinsic tendency to interact of two particle types

we would typically have a different beam flux, a different number of target particles, and a different measured transition rate. However, the cross-section σ for the two given particle types would be the same for any experimental arrangement, and it is this number that allows us to draw conclusions about how strongly two particular particle types interact intrinsically. See Prob. 1.

17.1.2 Classical Collision Differential Cross-section

Beam particles colliding inside V_t will scatter at different angles. Classically, this angle would depend on where on their surfaces a spherical beam particle and a spherical target particle make contact, the masses of the particles, and their relative velocities. For given particles, relative velocity, and contact locations, one could calculate the direction at which the beam particle would be deflected (as well as the direction the target particle would be nudged in). This direction could, of course, be specified in a spherical coordinate system by two angles, θ and ϕ. See Fig. 17-4. It is convenient to align the z direction axis with the beam path, as in Fig. 17-5.

It is useful to describe angles at which beam particle is scattered

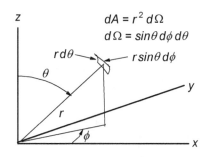

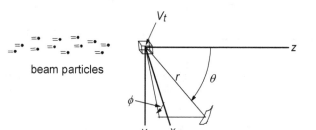

Figure 17-4. Spherical Coordinates Showing Solid Angle $d\Omega$

Figure 17-5. Aligning Coordinates So Beam Particles Scatter into θ and ϕ Angles

Use spherical coordinates

For a large group of particles, we could calculate the expected distribution of beam particles scattered per unit solid angle $d\Omega$ (units of steradians, symbol sr) as a function of θ and ϕ. Imagine a spherical surface with the target volume in the center and scattered particles passing through that surface in different numbers per second at different places on the spherical surface. (See Fig. 17-5.) Measuring those numbers, i.e., the distribution with respect to angular location on the spherical surface, would prove or disprove the theory we used to calculate what those numbers should be.

Want to consider transitions per unit solid angle

The bottom line is that we will find it useful to be able to calculate (and in experiments, measure) a probability density with respect to angular direction of scattered particles.

Recall from (17-2) that the cross-section σ represents the probability of a single beam particle hitting a single target particle within a unit area. So, σ represents the *total* probability to scatter at all, i.e., it represents beam particle scattering in all possible directions. If we consider the probability density for σ per unit solid angle to be $\rho(\theta,\phi)$, then

$$\sigma = \int_0^{2\pi}\int_0^\pi \underbrace{\rho(\theta,\phi)}_{\frac{d\sigma}{d\Omega}}\underbrace{sin\theta\, d\theta\, d\phi}_{d\Omega} = \int_0^{2\pi}\int_0^\pi \underbrace{\frac{d\sigma}{d\Omega}}_{\substack{\text{generally a}\\\text{function}\\\text{of }\theta,\phi}} d\Omega = \int_0^{2\pi}\int_0^\pi d\sigma \; , \qquad (17\text{-}7)$$

where we define the <u>differential cross-section</u> as

Differential cross-section definition

probability of detecting a scattered particle per unit solid angle $\left.\vphantom{\begin{array}{c}a\\b\end{array}}\right\}$ = $\dfrac{d\sigma}{d\Omega}$. (17-8)
(for a single target particle and a single beam particle in unit area)

In this case, "per unit solid angle" is more convenient and appropriate than "per unit volume", which we are more accustomed to for probability densities. Differential cross-section, for any two beam and target particle types, is, in general, a particular function of θ and ϕ.

Units

Note from (17-2) that the units for the differential cross-section, if we consider σ as probability, are probability per steradian (for one beam particle hitting one target particle where both particles are confined to unit area and the unit area is normal to the path of the beam particle.)

Alternatively, as we have noted, if we express σ in terms of area units, as in (17-1), then the differential cross-section has units of area per steradian.

In either case, a higher differential cross-section at a given angular direction means a greater probability of finding a scattered particle in that direction.

Typical Dependence Only on θ

For classical collisions between spherically shaped beam and target particles, the probability of scattering into any given ϕ direction is the same as that for any other ϕ direction, if both have the same θ value. (Think of spherical objects colliding as aligned in Fig. 17-5 where the collision contact point varies over the target sphere surface.) Newtonian (or relativistic) theory, for given incoming particle velocity, predicts purely random variation of the ϕ direction at given θ. Thus, there is no dependence on ϕ for the differential cross-section. The only dependence is on θ.

Due to symmetry, transition rate, σ, and $d\sigma/d\Omega$ usually depend only on θ and not ϕ.

As we will see, this is a general rule that extends into the quantum realm. So, in almost all cases, and certainly all we will deal with,

$$\frac{d\sigma}{d\Omega} = \frac{d\sigma}{d\Omega}(\theta) \qquad \text{(in almost all cases only a function of } \theta\text{)}. \qquad (17\text{-}9)$$

In Any Given Experimental Setup

If (17-6) is for total cross-section σ and total transition rate, we can readily deduce that

$$\boxed{\frac{\text{transition rate}}{\text{unit solid angle}} \text{ (at } \theta \text{ for a given experiment)} = f_b \frac{d\sigma}{d\Omega} n_t V_t = f_b \frac{d\sigma}{d\Omega} N_t} \qquad (17\text{-}10)$$

Transition rate per unit solid angle for a given experiment as a function of $d\sigma/d\Omega$

As with σ, $\dfrac{d\sigma}{d\Omega}$ is independent of the experimental configuration. Different experiments will have different f_b, $N_t = n_t V_t$, and transition rates per steradian, but for the same types of beam and target particles, they will have the same differential cross-section (with the same functional dependence on θ). We can perform the experiment and solve (17-10) for $\dfrac{d\sigma}{d\Omega}(\theta)$.

Useful Plots

In any given scattering experiment, or any theoretical analysis of such an experiment, it can be useful to plot differential cross-section vs. θ. If we analyze a single beam particle hitting a single target particle inside unit area, we can (as we will see) determine that plot theoretically and compare that result with different experiments performed at different labs at different times.

Good evaluations of theory can be made with $d\sigma/d\Omega$ vs. θ plots of data

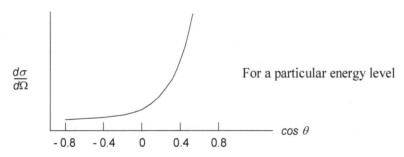

Figure 17-6. Typical Differential Cross-section θ Dependence for $e^- e^+$ Scattering

An example of this, not for classical collisions as we have considered so far, but for Bhabha scattering, is shown in Fig. 17-6. The dependence there is more conveniently expressed in terms of $\cos\theta$ than θ. Note that as $\theta \to 0$, the differential cross-section gets huge (relatively). This means most beam particles pass through the target volume with little interaction, a result that should not be too surprising.

17.1.3 Cross-section for Classical Potentials

The concepts of (total) cross-section and differential cross-section are readily extrapolated from collision (solid objects colliding) theory to particles interacting with classical potentials, such as a charged macroscopic object in a Coulomb field.

In the latter case, the Coulomb field would not be a second solid object from which, along with the charged object, we could calculate the (effective) cross-section, as in Fig. 17-2. However, the

Extend σ and $d\sigma/d\Omega$ concepts to scattering off a potential

charged object would be deflected by the Coulomb potential in a manner similar to a classical collision. See Fig. 17-7. The final path of the charged object is similar in all regards to the final path from a collision between hard objects.

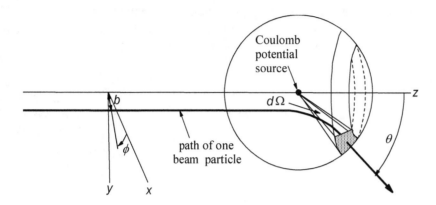

Figure 17-7. Classical Scattering of a Particle by a Coulomb Field

Fig. 17-7 shows only a single particle in the beam and how it is scattered into a solid angle section $d\Omega$ at θ and ϕ. The "impact parameter" b is the distance between the z axis (which passes through the center of the Coulomb potential source) and the beam particle path. The impact parameter, the beam particle velocity, the beam particle mass, and the strength of the Coulomb potential all have an effect on the final direction of the particle after scattering. We will not derive the relationship between these quantities (which can be found in almost any classical mechanics text), but simply draw some conclusions about the entire process.

Imagine, rather than a single beam particle, a shower of particles coming in from the left that are scattered by the Coulomb field. The beam particles are distributed randomly, so from symmetry for a large number of beam particles, we would expect the scattered distribution to be independent of ϕ. At any given θ, the particles per unit solid angle would be the same for any ϕ. This is just as it was for hard objects collision theory (see (17-9)).

Further, we can apply (17-10) to this case, where we take $N_t = 1$, since there is only one target potential (playing the role of one target particle). Thus, for a given beam flux f_b, where the impact parameter is randomly distributed among beam particles,

$$\frac{\text{transition rate}}{\text{unit solid angle}} \left(\begin{array}{l}\text{at } \theta \text{ for a given classical} \\ \text{potential experiment}\end{array}\right) = \frac{\text{scattered particles/sec}}{\text{unit solid angle}} = f_b \frac{d\sigma}{d\Omega} \underbrace{N_t}_{\substack{=1 \\ \text{here}}} . \quad (17\text{-}11)$$

For beam of particles, relation between transition rate, σ, and dσ/dΩ same as before

We can perform an experiment to measure the LHS (same as the center part) of (17-11) and solve for $\frac{d\sigma}{d\Omega}$, the differential cross-section as a function of θ. We can then integrate that over all θ and ϕ to get σ.

We could, of course, also determine theoretically what the LHS of (17-11) should be and thus solve for a theoretical value for $\frac{d\sigma}{d\Omega}$, but we won't do that here.

<u>Bottom line:</u> We can use the same mathematical tools (differential cross-section and cross-section) to describe the interaction of a classical particle and a classical potential.

17.1.4 Cross-sections for Relativistic Classical Scattering

All of the general arguments and conclusions of Sects. 17.1.1, 17.1.2, and 17.1.3 for classical scattering such as (17-6), (17-10), and (17-11) are valid for both non-relativistic and relativistic cases. The actual theoretical analyses determining cross-sections and differential cross-sections will differ, though, depending on whether the underlying principles applied are Newtonian or Einsteinian. So will the transformations of the results to other coordinate frames.

Same concepts and relations extend to classical relativistic scattering

17.1.5 Cross-section for Non-Relativistic Quantum Mechanics

In NRQM, we typically describe the incoming particle as a plane wave that interacts with the potential, typically a single Coulomb field, and is scattered. The scattered wave is then most easily described as a spherical wave expanding away from the center of the potential, as shown in Fig. 17-8. (Remember that θ is the angle away from the z axis. It looks like it has different directions in Figs. 17-8 and 17-7 because the values for ϕ in the two figures are different.)

NRQM scattering off a potential

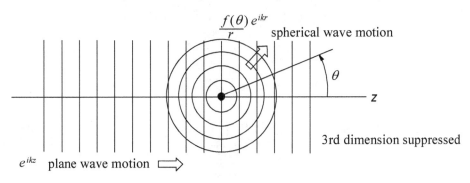

Figure 17-8. Scattering of a NRQM Plane Wave by Spherically Symmetric Potential

Quantum mechanics is different from the classical theory of the previous two sections in that a given wave function collapses to a point (or effectively a point) when a measurement of position is performed. The probability density (= complex conjugate of wave function times wave function) inherent in the scattered wave can be expressed as a function of θ and ϕ and with units of probability per unit solid angle. Note from (17-7) and (17-8) that this is precisely how the differential cross-section $\dfrac{d\sigma}{d\Omega}$ for hard objects collision was defined.

So if we have a large number of beam particles with flux f_b scattering off a single potential, we can again measure the scattered particles per second per unit solid angle at any θ and ϕ. Thus, we can again use (17-10) (or (17-11)) to find the NRQM $\dfrac{d\sigma}{d\Omega}$, i.e.,

Expected transition rate, σ, and $d\sigma/d\Omega$ relations same as before

$$\frac{\text{transition rate}}{\text{unit solid angle}}\begin{pmatrix}\text{at }\theta\text{ for a given NRQM}\\ \text{scattering experiment}\end{pmatrix}=\frac{\text{scattered particles/sec}}{\text{unit solid angle}}=f_b\frac{d\sigma}{d\Omega}\underbrace{N_t}_{\substack{=1\\ \text{here}}} \qquad (17\text{-}12)$$

<u>Theoretical NRQM Scattering</u>

Fig. 17-8 displays, with the third dimension suppressed, an incoming NR quantum plane wave being scattered off of a spherically symmetric potential. Due to symmetry, we have no variation in ϕ in the scattered wave. The scattered wave is expressed as a spherical wave and we adopt the Born approximation, which assumes the incoming wave is unaffected in the process. In effect, the change in the incoming wave is so small, we can ignore it. Thus, the wave function for the interaction, which solves the Schrödinger equation (actually the first term on the RHS below solves the Schrödinger equation in Cartesian coordinates and the second, the Schrödinger equation expressed in spherical coordinates), is

But now, need NRQM theory to determine $d\sigma/d\Omega$

Total wave function for beam and scattered wave using Born approximation

$$\psi\left(r,\theta\right)=\psi_{incom}+\psi_{scat}=C\left[e^{ikz}+f\left(\theta\right)\frac{e^{ikr}}{r}\right] \quad \text{for large } r. \qquad (17\text{-}13)$$

$f\left(\theta\right)$ is, of course, quite different from beam flux f_b. Note that $f\left(\theta\right) \ll 1$ is almost invariably the case, and we assume that here. Consider the incoming wave to be of finite length L ($= v_b\tau$, where τ is the time for the wave to pass a point in space and v_b is wave speed) and to extend over area A in the x-y plane. For these assumptions, (17-13) is normalized if we take $C=1/\sqrt{AL}$. (See Prob. 2.) Thus, the probability of measuring the scattered wave per unit volume is

Probability density for scattered wave

$$\text{probability density} = \frac{\text{probability}}{\text{unit volume}}=\psi_{scat}^{*}\psi_{scat}=\frac{1}{AL}f^{*}\left(\theta\right)\frac{e^{-ikr}}{r}f\left(\theta\right)\frac{e^{ikr}}{r}=\frac{1}{AL}\left|f\left(\theta\right)\right|^{2}\frac{1}{r^{2}}. \qquad (17\text{-}14)$$

Consider that the scattered wave has the same wave speed as the incoming wave. Thus, the scattered wave will have finite extension in the r direction, over a radial length $\Delta r = L = r_2 - r_1$. The probability of detecting the wave inside this region[1] is

$$\left.\begin{array}{c}\text{probability of measuring scattered particle} \\ \text{inside its volume (between } r_1 \text{ and } r_2)\end{array}\right\} = \int_{r_1 \, to \, r_2} \psi^*_{scat}\psi_{scat}dV$$

$$= \int_{r_1}^{r_2}\int_0^{2\pi}\int_0^\pi \frac{1}{AL}\left|f(\theta)\right|^2 \frac{1}{r^2}r^2\sin\theta\,d\theta\,d\phi\,dr = \overbrace{(r_2-r_1)}^{L}\int_0^{2\pi}\int_0^\pi \frac{1}{AL}\left|f(\theta)\right|^2 \overbrace{\sin\theta\,d\theta\,d\phi}^{d\Omega} \quad (17\text{-}15)$$

$$= \frac{1}{A}\int_0^{2\pi}\int_0^\pi \left|f(\theta)\right|^2 d\Omega.$$

If we consider the incoming wave area A to be of unit area size, so $A = 1$, then (17-15) becomes our definition of the cross-section σ of (17-2), i.e.,

$$\left.\begin{array}{c}\text{probability of measuring a particle scattered} \\ \text{off one target for one beam particle in unit area}\end{array}\right\} = \underbrace{\int_0^{2\pi}\int_0^\pi \left|f(\theta)\right|^2 d\Omega}_{\frac{d\sigma}{d\Omega}} = \sigma. \quad (17\text{-}16)$$

Cross-section
for NRQM

Hence, we see that the differential cross-section is

$$\frac{d\sigma}{d\Omega} = \left|f(\theta)\right|^2. \quad (17\text{-}17)$$

Differential
cross-section
for NRQM

<u>Bottom line</u>: We can use the same mathematical tools (differential cross-section and cross-section) to describe the interaction of a non-relativistic quantum wave off of a potential.

<u>Note</u>: To find the functional form of $f(\theta)$, one uses (17-13) in the Schrödinger equation with the particular potential $V(r)$, assumes the Born approximation, and solves the resulting equation.

17.1.6 Cross-section for Relativistic Quantum Mechanics

RQM scattering is similar to NRQM scattering but a bit more complicated because, of course, it is relativistic. We will not do anything further with RQM scattering except to note the following conclusion, which we would reach if we were to explore it in depth.

Similar for RQM,
but need
relativistic theory
to predict dσ/dΩ

<u>Bottom line</u>: We can use the same mathematical tools (differential cross-section and cross-section) to describe the interaction of a relativistic quantum wave off a potential.

17.1.7 Cross-section for Quantum Field Theory

<u>Experimental QFT Scattering</u>

Like all our other physical theories, scattering described by QFT is measured in experiment using the same concepts of cross-section (17-2) and differential cross-section (17-8), and the same relations between them and total scattering rate (17-6) and scattering rate per unit solid angle (17-10). However, each physical theory (solid objects, object and a potential, quantum wave and a potential, where each has a non-relativistic and a relativistic branch) has its own approach to theoretical analysis of scattering. QFT is no exception, and we take our first look at how it handles scattering below. Prior to that, however, we note how we would determine differential cross-section and total cross-section experimentally.

QFT: Same
concepts and
relations between
transition rate,
σ, and dσ/dΩ
as before

<u>Determining Cross-sections in QFT Experimentally</u>

In a scattering experiment we wish to analyze via QFT, we know the beam flux f_b, the target particle density n_t, and the target volume V_t. With many scatterings, and many detections at different θ, one can determine the transition rate (scatterings per second) per unit solid angle for different θ. So, one can build a histogram with dependence on θ. From (17-10) (repeated below), we can then calculate the differential cross-section $\frac{d\sigma}{d\Omega}(\theta)$ at specific values of θ.

Experimental
considerations as
in other theories

---

[1] Here we assume the plane wave of Fig. 17-8 is finite in the x and y directions and does not extend into the region between r_1 and r_2 we are interested in.

$$\frac{\text{transition rate}}{\text{unit solid angle}} \text{ (at } \theta \text{ for a given experiment)} = f_b \frac{d\sigma}{d\Omega} n_t V_t = f_b \frac{d\sigma}{d\Omega} N_t, \qquad \text{same as } (17\text{-}10).$$

Expected transition rate relation same as in other theories

We can integrate the differential cross-section (add discrete approximations of $d\sigma/d\Omega$ from each piece of our histogram) over a complete spherical shell to get total cross-section σ.

Theoretical QFT Scattering

Key Point in Using QFT to Determine Cross-sections

Note that the cross-section and differential cross-section contained in relations (17-6) and (17-10) are the same for any experiment having the same particles (and, as we will see, for QFT scattering, at the same energy level). So different hypothetical scattering experiments between particle type A and particle type B, which we wish to analyze via QFT, will yield the same σ and $d\sigma/d\Omega$. This is true even though volume V and time T would be different, and even though target particle number N_t and beam flux $f_b = (N_b/V)v_b$ would be different.

So, in using theory to determine (differential) cross-section between particle types A and B, we can, and should, devise our hypothetical experiment that we will analyze theoretically in a way that will make our analysis simpler.

In our earlier work, we derived Feynman amplitudes for single beam and single target particles. So, if we want to use all of that earlier analytic work, we need to apply it to a hypothetical experiment that has only one beam and one target particle, i.e., $N_b = N_t = 1$. We also assumed $V, T \to \infty$ in order to get mathematically tractable analytical results. (Integrals over infinity tend to reduce to simpler form, e.g., delta functions.) In addition, we assumed that both particles (waves) occupy the same volume V during all of T.

In fact, only with these assumptions can we make use of the transition amplitude S_{fi} we worked so hard to determine in earlier chapters, because we used those assumptions to derive S_{fi} via the Dyson-Wicks expansion.

QFT analysis considers scattering like in Feynman diagrams

<u>Key Point Bottom line</u>: We will analyze a hypothetical experiment where test parameters are the same as those we used to develop quantum field theory. σ and $d\sigma/d\Omega$ determined from such analysis will be found to agree with those measured in any real-world scattering experiment between the same particles. The values we will use are

$$V_b = V_t = V \text{ very large}; \quad T \text{ very long}; \quad N_b = N_t = 1; \quad \text{both particles in } V \text{ for all } T \qquad (17\text{-}18)$$

We made these assumptions in Chaps. 7 and 8 because 1) they simplified the math, and 2) because the theoretical results, though perhaps unrealistic for many real-world conditions (with relatively small V for example), could be used to predict real world experimental results. We promised then that we would justify the usefulness of the $V, T \to \infty$ assumption, and here we have delivered on that promise.

Brief Digression: Related Comment on Chapter 16 Analyses

Our postdiction analyses of Chap. 16 can be justified in the same way. The magnetic moment and the Coulomb potential, for example, would be the same for effectively infinite volume/time as they would be for volume/time in any real-world experiment. So we can use QFT developed for $V, T \to \infty$, $N_b = N_t = 1$, and both particles always in V to make meaningful theoretical post/predictions.

Hypothetical Experimental Setup Mimicking Feynman Diagrams

In QFT interactions as described by Feynman diagrams and their associated transition amplitudes, every field (and the particles created/destroyed by the fields) is considered to be spread through the same volume V (which is taken very large and approaching infinity in our Dyson-Wicks expansion). In our cases, the fields have been represented by plane waves inside that volume. The time duration T the fields are in the same V is taken as long (approaching infinity in our Dyson-Wicks expansion.)

But, one could then ask, if the two particles and their associated fields have relative velocity with respect to one another, they must move away from one another and so, could not continually share the same volume. However, if we have a beam of particles, one following the other sequentially, and each occupying a volume $V_b = V_t = V$ that moves with the particle, we get a situation like that of Fig. 17-9.

One beam and one target particle in volume $V_b = V_t = V$

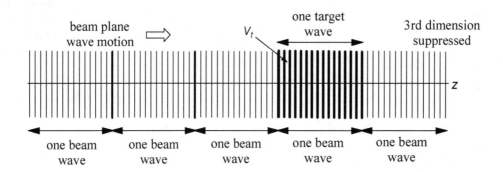

*How to keep
effectively
one beam and
one target
particle in V*

Figure 17-9. Arrangement Where Effectively One Beam Wave Always Inside Target Volume

At any given instant we have, effectively, one beam particle/field inside the volume, with velocity v_b, and that is true continuously in time. And hence we can use the machinery of QFT, which assumes each particle (beam and target) occupy the same volume indefinitely, to analyze that case, even though the target and beam particles have different velocities.

Same Feynman Diagram Represents Various Final States

The next thing we need to be clear on is that one Feynman diagram (and associated transition amplitude) represents many possible particle velocity alignments, because we have different 3-momenta associated with each particle. In particular, the final scattered particles can have different $\mathbf{p}'_1$ and $\mathbf{p}'_2$. Different directions for these three-vectors means different angular locations at which our detectors would pick them up. See Fig. 17-10 (where for simplicity, we represent states as particle-like, rather than wave-like.) So, one Feynman diagram represents many different final states, comprised of the same particle types, but different momenta.

*Same Feynman
diagram
represents
many different
final states*

For the cases we have been discussing, the target particle (think particle #2 in Fig. 17-10) has been at rest, so its three-momentum would be zero. Fig. 17-10 represents a more general case where both beam and target particles have velocity.

The point of Fig. 17-10 is that the transition amplitude S_{fi} we would calculate with the aid of the Feynman diagram would have the same form algebraically for both cases on the RH of Fig. 17-10. But the numerical value would be different because the values for $\mathbf{p}'_1$ and $\mathbf{p}'_2$ would be different.

Since direction is inherent in $\mathbf{p}'_1$ and $\mathbf{p}'_2$, they can be represented in terms of $|\mathbf{p}'_1|$ and $|\mathbf{p}'_2|$ along with their angular directions expressed in terms of θ and ϕ of our spherical coordinate system. If particle #1 with $\mathbf{p}'_1$ is our beam particle, then its θ and ϕ directions (at which our detectors would pick it up) represent the direction it was scattered into.

*Different final
states have
different **p** at
different θ*

For a given experiment, we want to be able to predict the probability of measuring a value of $\mathbf{p}'_1$ (which includes the probability of measuring it at a given θ and ϕ in addition to the probability of it having a certain magnitude). As scattering experiment symmetries typically lead to probabilities that do not vary with ϕ, we are essentially only concerned with the θ value inherent in $\mathbf{p}'_1$.

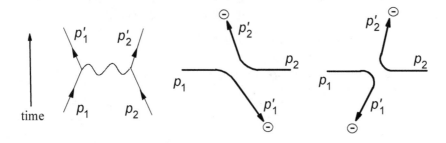

Feynman diagram   Physical space, different final states
                                 represented by same Feynman diagram

**Figure 17-10. The Same Feynman Diagram Represents Particles
with Different Values for 3-Momenta**

The overall point of this subsection is that a given transition amplitude S_{fi} (where $|S_{fi}|^2$ is the probability of the transition occurring) can be expressed in terms of θ. Determining that relational dependence is a primary goal of QFT scattering analysis, because that can lead to determination of the differential cross-section (which is a probability density that depends on angular direction.)

So, can express S_{fi} in terms of θ

Plane, Not Spherical, Scattered Waves in QFT

Scattering problems in NRQM are usually expressed in terms of spherical scattered waves, as in Fig. 17-8, when the potential is spherical in form. In QFT theory, however, we have been dealing almost exclusively with plane waves, which are considered to interact with one another (rather than with a potential field fixed in space), so scattered waves are evaluated as plane waves. (Fig. 17-11.)

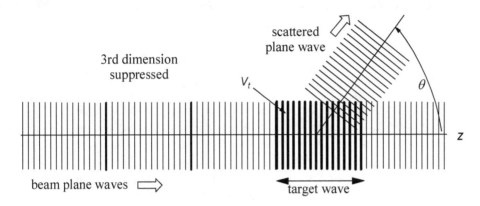

In QFT, unlike NRQM, scattered wave treated as a plane wave

Figure 17-11. In QFT, Scattered Waves are Treated as Plane Waves

Finding Differential Cross-section Theoretically in QFT

To analyze scattering of incoming beam particles by a stationary target and with QFT, we assume a configuration like that of Figs. 17-9 and 17-11. That is, the target volume contains a single particle (wave) and there is effectively always a single beam particle (wave) in the same volume having the beam particle velocity v_b. So, $N_b = N_t = 1$, and $V_b = V_t = V$. This corresponds to our usual Feynman diagrams, as in the LH of Fig. 17-10, and thus to our analysis of transition amplitude S_{fi}. (Note that f here is different for each of the cases on the RH side of Fig. 17-10, though S_{fi} has the same algebraic form in terms of $\mathbf{p}'_1$ and $\mathbf{p}'_2$. Different values for $\mathbf{p}'_1$ and $\mathbf{p}'_2$ correspond to different values for the subscript f.)

We will use the subscript F to denote the final scattered particle states (i.e., f states) inside $d\Omega$ in a given angular direction, and whose 3-momentum lies in the range between $\mathbf{p}_f$ and $\mathbf{p}_f + d\mathbf{p}_f$. We designate the probability of measuring *any* scattered state inside $d\Omega$ by $d|S_{Fi}|^2$, so that $d|S_{Fi}|^2/T$ is the probability per unit time, where T is the total time of interaction (in our case, as we assumed in Chaps. 7 and 8, approaching infinity). So, (17-10) becomes

$$\frac{\text{transition rate}}{\text{unit solid angle}} \ (\text{at } \theta) = \frac{d|S_{Fi}|^2/T}{d\Omega}$$

$$= f_b \frac{d\sigma}{d\Omega} N_t = n_b v_b \frac{d\sigma}{d\Omega} N_t = \frac{\overset{1}{\overbrace{\frac{N_b}{V_b}}} v_b \frac{d\sigma}{d\Omega} \overset{1}{\overbrace{N_t}} = \frac{1}{V} v_b \frac{d\sigma}{d\Omega}.$$

(17-19)

Transition rate in terms of transition amplitude S_{Fi} (for case of one beam and one target particle in V)

Thus,

$$\frac{d|S_{Fi}|^2}{T} = \frac{1}{V} v_b \, d\sigma .$$

(17-20)

For dN_f = the number of states for the scattered particle between $\mathbf{p}_f$ and $\mathbf{p}_f + d\mathbf{p}_f$, where $\mathbf{p}_f = \mathbf{p}'_1$ in our prior examples, (see Fig. 17-12)

$$d|S_{Fi}|^2 = |S_{fi}|^2 \, dN_f .$$

(17-21)

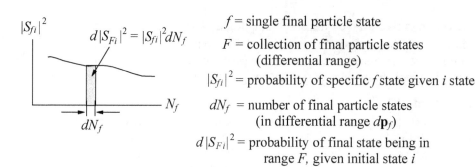

f = single final particle state

F = collection of final particle states
(differential range)

$|S_{fi}|^2$ = probability of specific f state given i state

dN_f = number of final particle states
(in differential range $d\mathbf{p}_f$)

$d|S_{Fi}|^2$ = probability of final state being in
range F, given initial state i

Expressing (differential) probability as probability for an individual final state f times number of final states in differential range

Figure 17-12. Total Probability for dN_f Number of Scattered Particle States

Thus, (17-20) becomes[1]

$$d\sigma = \frac{V}{v_bT}|S_{fi}|^2 \, dN_f = \begin{cases} \text{cross section for scattered particle states in} \\ \text{differential range of 3-momentum inside } d\Omega \end{cases}. \qquad (17\text{-}22)$$

Expressing QFT $d\sigma$ in terms of differential probability, beam particle velocity, volume V, and time T

It would be appropriate to call "$d\sigma$" the differential cross-section, except that particular name has been reserved, through custom, for $d\sigma/d\Omega$. The latter would better be deemed the "cross-section density" (cross-section per solid angle), but best to stick to terminology that is most common in the field.

We can evaluate (17-22) because we have learned how to evaluate the transition amplitude S_{fi} in terms of the initial and final 3-momenta of the particles. $d\sigma$ is the cross-section for all scattered states between $\mathbf{p}_f$ and $\mathbf{p}_f + d\mathbf{p}_f$ in a given θ, ϕ direction. By expressing the 3-momentum $\mathbf{p}_f$ of the scattered particle in terms of θ, we can determine the value of $d\sigma$ for a given $d\theta$, and thus, a given $d\Omega$. Hence, we can find $\dfrac{d\sigma}{d\Omega}$. We can also integrate (17-22) over all possible final states to get total cross-section σ.

The evaluation of (17-22) is fairly extensive, so we save that for Sect. 17.4.

17.1.8 Summary of Scattering for Different Theories

Wholeness Chart 17-1 summarizes the definitions and interpretations of cross-section and differential cross-section, which are true in any physical theory. Wholeness Chart 17-2 summarizes all of the prior material on scattering off a stationary target for different theories.

Wholeness Chart 17-1. Summary of Definitions and Interpretations of σ and $d\sigma/d\Omega$

Definitions	
Cross-section σ	#1: Effective area of target particle point-like beam particle would hit. (Units of area.) #2: Probability of single beam particle hitting single target particle in unit area. (Unitless.)
Differential Cross-section $d\sigma/d\Omega$	Usually taken like #2 above but per unit solid angle through which scattered particle passes. Typically a function of θ in a spherical coordinate system.
Interpretation	Higher cross-section means greater probability of scattering to occur for given experimental setup. Cross-section is independent of experimental setup, for given particle types.

---

[1] Some authors, such as Kaku (*Quantum Field Theory*, Oxford 1993), define $d\sigma$ as what we would call $d\sigma/V$. Mandl and Shaw (*Quantum Field Theory*, Wiley 1984) use our definition.

Wholeness Chart 17-2. Scattering Off Stationary Target for Different Physical Theories

	Experiment	**Theory**
Non-Relativistic		
Hard Objects **(Figs. 17-1, 2, 3)**	total transition rate $= f_b \sigma N_t$ $\quad$ (A) $\dfrac{\text{transition rate}}{\text{unit solid angle}} = f_b \dfrac{d\sigma}{d\Omega} N_t$ $\quad$ (B)	Using coordinates of point of contact on surface of each object, and their relative velocity, Newtonian particle mechanics can predict angle θ of scattered object.
Object & Potential **(Fig. 17-7)**	Same as above	From impact parameter, beam particle velocity, and form of potential (e.g. Coulomb), Newtonian field mechanics can predict θ of scattered object.
Quantum Wave & **Potential** **(Fig. 17-8)**	Same as above	Spherical scattered wave. For $\psi = C\left[e^{ikz} + f(\theta) e^{ikr}/r \right]$, $\dfrac{d\sigma}{d\Omega} = \left| f(\theta) \right|^2$
Relativistic of Above	Same as above	Parallels above, but with relativity theory replacing non-relativistic theory.
QFT **(Figs. 17-9, 10, 11, 12)**	Same as above	Plane scattered wave. Use (B) at left top box for Feynman amplitude with $N_b = N_t = 1$; $V_b = V_t = V$; $V, T \to \infty$ $d\sigma = \dfrac{V}{v_b T} \left| S_{fi} \right|^2 dN_f$. Express quantities in terms of θ and $d\Omega$ to find $\dfrac{d\sigma}{d\Omega}$. Integrate to find σ.

17.2 Review of Interaction Conservation Laws

Energy and 3-momentum conservation review

We now do a brief review of energy and momentum conservation in classical and quantum collisions/interactions/scattering. (All three words mean essentially the same thing.) This is summarized in Wholeness Charts 17-3 and 17-4.

17.2.1 Classical Scattering

Classical hard object collisions, no external potential, two particles

In any classical interaction with no external potential (no external forces), total energy conservation and total 3-momentum conservation hold. That is, with our usual convention where unprimed quantities indicate "prior to scattering" and primed indicate "post scattering", for two particles in and two particles out,

All such cases

$$
\begin{array}{lll}
\text{Elastic and inelastic} & \left\{ \begin{array}{ll} p_1^i + p_2^i = p_1'^{\,i} + p_2'^{\,i} & (a) \\ E_1 + E_2 = E_1' + E_2' & (b) \end{array} \right. & \begin{array}{l} \text{Total 3-momentum conservation} \\ \text{Total energy conservation.} \end{array}
\end{array} \quad (17\text{-}23)
$$

If no kinetic energy is converted to, or from, potential energy, heat, mass (via $E = mc^2$) or other forms of energy, then the interaction is <u>elastic</u>, and kinetic energy is conserved. For that case,

Elastic case

$$
\begin{array}{lll}
\text{Elastic cases} & \left\{ \begin{array}{ll} p_1^i + p_2^i = p_1'^{\,i} + p_2'^{\,i} & (a) \\ KE_1 + KE_2 = KE_1' + KE_2' & (b) \end{array} \right. & \begin{array}{l} \text{Total 3-momentum conservation} \\ \text{Total kinetic energy conservation.} \end{array}
\end{array} \quad (17\text{-}24)
$$

We will only look at elastic collisions here. In Wholeness Charts 17-3 and 17-4, we summarize results for 1D, 2D, and 3D cases, but here we will skip directly to the 3D case, the most relevant for QFT. Note that there is significant information in Wholeness Charts 17-3 and 17-4 that we do not review in the text. That information is all review, generally at a freshman physics level, but is included to make those charts a more complete and holistic overview.

<u>Non-relativistic Classical Elastic Scattering in 3D</u>

Non-relativistic, elastic

From henceforth, we will use our 4-vector symbol p^0 for energy E, but remember that in elastic cases $p^0 = E = KE$, i.e., all energy is kinetic.

Equations (17-24) comprise four scalar equations, one for each component of 3-momentum and one for kinetic energy. But assuming we know the momenta and energy of the incoming particles, there are eight scalar unknowns, (3 of) $p'_1{}^i$, (3 of) $p'_2{}^i$, $p'_1{}^0$, and $p'_2{}^0$.

However, if we know the 3-momentum of a particle, we can find its energy via

$$KE' = p'^0 = \frac{(\mathbf{p}')^2}{2m} = \frac{p'^i p'^i}{2m} \quad \text{and} \quad KE = p^0 = \frac{\mathbf{p}^2}{2m} = \frac{p^i p^i}{2m}. \tag{17-25}$$

Substituting the LHS of (17-25) into (17-24) for primed particles 1 and 2, we can eliminate 2 unknowns, and end up with four equations in six unknowns, $p'_1{}^i$ and $p'_2{}^i$. With the RHS also,

Conservation laws in terms of unknown 3-momenta of final particles

Elastic cases
$$\begin{cases} p_1^i + p_2^i = p'_1{}^i + p'_2{}^i & (a) \quad \text{Total 3-momentum conservation} \\ \dfrac{p_1^i p_1^i}{2m_1} + \dfrac{p_2^i p_2^i}{2m_2} = \dfrac{p'_1{}^i p'_1{}^i}{2m_1} + \dfrac{p'_2{}^i p'_2{}^i}{2m_2} & (b) \quad \text{Total kinetic energy conservation.} \end{cases} \tag{17-26}$$

From conservation laws alone, since we have four equations in six unknowns, we cannot predict the result of the scattering in three dimensions.

More Theory

However, if we had other information, such as knowing, for hard objects, where on their surfaces they make contact; or, for an object scattered by a potential, where exactly relative to the centerline of Fig. 17-7 the object's velocity vector lies, we would then be able to determine exact values for momenta and energy of the final particles. In effect, geometry would give us enough extra independent equations to make that determination.

If we knew surface point of contact, Newton theory could predict final $\mathbf{p}'_1$, $\mathbf{p}'_2$

For a shower of hard beam objects hitting random locations on hard target object(s), adding statistical analysis to Newtonian mechanics analysis would give us expected transition rates per unit solid angle at given θ values. Similarly, for a shower of beam objects in a potential field where the beam particles have random variation in impact parameter, we could incorporate statistical analysis to predict the same kinds of transition rates.

For random collisions, can calculate probability of given $\mathbf{p}'_1$, $\mathbf{p}'_2$

Experiment

For two hard object interactions, if we have no way of knowing where on their surfaces they make contact, we could measure any two values of $p'_1{}^i$ and from them and our four independent conservation equations, calculate the missing $p'_1{}^i$ component, as well as the three $p'_2{}^i$ components. That is, from measurement of the momentum (actually only two components of momentum) of either final object, we can calculate that of the other.

If measure any two components of $\mathbf{p}'_1$ and $\mathbf{p}'_2$, can calculate other four

Switching Unknowns

Instead of representing the 3-momentum of a scattered object as components in Cartesian coordinates, $p'_1{}^i$, we can express that same momentum in terms of spherical coordinates.

$\mathbf{p}$ represented in terms of p^1, p^2, p^3 or in terms of p, θ, ϕ with $p = |\mathbf{p}|$

$$p^1 = p \sin\theta \cos\phi \qquad p^2 = p \sin\theta \sin\phi \qquad p^3 = p \cos\theta. \tag{17-27}$$

Can express $\mathbf{p}'_1$ and $\mathbf{p}'_2$, in spherical coords

Substituting the bottom row of (17-27) for primed particles into (17-26) yields four equations in six unknowns (p'_1, θ'_1, ϕ'_1) and (p'_2, θ'_2, ϕ'_2). So, unless we have some additional theory beyond conservation equations to help us, we cannot predict the scattered objects' properties. With such theory (as described above in "More Theory" subsection), we can make such determinations.

Then if measure any two scattering angles, can calculate magnitudes and other angles of $\mathbf{p}'_1$ and $\mathbf{p}'_2$

In experiment, by measuring the two coordinate angles at which a scattered object leaves the scattering region, we can, via our four conservation equations (17-26) expressed in spherical coordinates, determine the momentum magnitude of both scattered objects and the angular direction of the second one. That is, by measuring θ'_1 and ϕ'_1, we can calculate p'_1, p'_2, θ'_2, and ϕ'_2.

Relativistic Classical Elastic Scattering in 3D

Analysis of relativistic elastic scattering for macroscopic objects is identical to that of non-relativistic scattering above except for the relationship (17-25) between kinetic energy and three momentum. That is, instead of (17-25), we have, in natural units,

Relativistic, elastic

$$\text{from } (p'^0)^2 = m^2 + \mathbf{p}'^2 \rightarrow KE_{rel} = \underbrace{p'^0}_{\substack{\text{total} \\ \text{energy}}} - \underbrace{m}_{\substack{\text{rest} \\ \text{energy}}} = (m^2 + \mathbf{p}'^2)^{1/2} - m = (m^2 + p'^i p'^i)^{1/2} - m, \tag{17-28}$$

so we can once again express the four equations (17-24) solely in terms of the six $p'_1{}^i, p'_2{}^i$.

We can also express those equations in terms of $(p'_1, \theta'_1, \phi'_1)$ and $(p'_2, \theta'_2, \phi'_2)$, and all conclusions above about measuring two angles of the direction of 3-momentum of a scattered particle carry over to relativistic scattering.

Same general conclusions as for non-relativistic, elastic case

In relativistic cases, the math is easier to work by expressing (17-23) and (17-24) in terms of particle velocities rather than momenta. But the same conclusions of the above two paragraphs hold, since if we know velocity (and mass), we know momentum, be it relativistic or not.

But need to use relativistic energy and 3-momentum relations

In terms of velocity, the relativistic kinetic energy can be found as part of

$$p'^0 = \frac{m'c^2}{\sqrt{1-v'^2/c^2}} = m'c^2\left(1+\frac{1}{2}\frac{v'^2}{c^2}+\frac{3}{8}\frac{v'^4}{c^4}+...\right) = \underbrace{m'c^2}_{\text{rest energy}} + \underbrace{\frac{1}{2}m'v'^2+\frac{3}{8}m'\frac{v'^4}{c^2}+...}_{\text{relativistic kinetic energy}}, \quad (17\text{-}29)$$

so

$$KE_{rel} = p'^0 - m'c^2 = \frac{m'c^2}{\sqrt{1-v'^2/c^2}} - m'c^2 \qquad c=1 \text{ in natural units}. \quad (17\text{-}30)$$

Center of Mass vs Target Stationary Frames

We can examine elastic scattering in any frame we like and conservation laws (17-24) will hold in that frame. (We can also examine inelastic scattering in any frame we like and (17-23) will hold.) Typical frames that will be of value in our work are i) the center of mass (COM) frame for two incoming particles that scatter off of one another, ii) the target stationary frame, and iii) the lab frame. The lab frame, depending on the particular experiment, can be the COM, the target stationary frame, or neither. Note that many authors call the target stationary frame the lab frame.

Two useful frames for analysis: COM and target stationary frames

Analysis may be easier, for example, in the COM frame than the lab frame. But since we measure in the lab frame, we would then want to convert the answer we found in the COM frame to the lab frame coordinate system to compare experiment to theory.

In the COM frame of a system of particles, the sum total of all particles' momenta equals zero. (See Prob. 12-1 of Chap. 12.) For our case of two incoming particles,

$$\mathbf{p}_1 + \mathbf{p}_2 = 0 \qquad \text{COM frame}. \quad (17\text{-}31)$$

In COM frame, total 3-momentum = 0

Sometimes the COM frame is the same as the lab frame, as in CERN's LEP (Large Electron Positron) Collider where electrons with 3-momentum $\mathbf{p}_1$ collide with positrons of 3-momentum $\mathbf{p}_2 = -\mathbf{p}_1$, or in CERN's LHC (Large Hadron Collider), which typically collides protons with momentum $\mathbf{p}$ with other protons of momentum $-\mathbf{p}$.

Sometimes COM frame = lab frame

Sometimes, such as bombarding gold nuclei with alpha particles, the target particles (gold nuclei) are originally stationary in the lab frame, and the lab frame is the target stationary frame. In this example, $\mathbf{p}_1 \neq -\mathbf{p}_2$ (in the lab/target stationary frame). This should be obvious since, in the target stationary frame, gold nucleus momentum $\mathbf{p}_2 = 0$, but alpha particle momentum $\mathbf{p}_1 \neq 0$ so (17-31) doesn't hold.

Sometimes target stationary frame = lab frame

In Center of Mass Frame, Elastic Case, Particles Keep Same Speed (Same KE)
Non-Relativistic Case

In the center of mass frame, (17-24) become, for low velocities,

Closer look at COM frame

Non-relativistic

$$\mathbf{p}_1 + \mathbf{p}_2 = \mathbf{p}'_1 + \mathbf{p}'_2 = 0 \qquad (a)$$
$$\frac{|\mathbf{p}_1|^2}{2m_1} + \frac{|\mathbf{p}_2|^2}{2m_2} = \frac{|\mathbf{p}'_1|^2}{2m_1} + \frac{|\mathbf{p}'_2|^2}{2m_2} \qquad (b)$$

COM frame, non relativistic, elastic. (17-32)

From the first row above

$$\mathbf{p}_1 = -\mathbf{p}_2 \qquad \mathbf{p}'_1 = -\mathbf{p}'_2, \quad (17\text{-}33)$$

so, the second row becomes

$$\frac{|\mathbf{p}_1|^2}{2m_1} + \frac{|\mathbf{p}_1|^2}{2m_2} = \frac{|\mathbf{p}_1'|^2}{2m_1} + \frac{|\mathbf{p}_1'|^2}{2m_2} \;\;\rightarrow\;\; \frac{1}{2}\left(\frac{1}{m_1} + \frac{1}{m_2}\right)|\mathbf{p}_1|^2 = \frac{1}{2}\left(\frac{1}{m_1} + \frac{1}{m_2}\right)|\mathbf{p}_1'|^2$$

$$\rightarrow \;\; |\mathbf{p}_1|^2 = |\mathbf{p}_1'|^2 \;\;\rightarrow\;\; \frac{|\mathbf{p}_1|^2}{2m_1} = \frac{|\mathbf{p}_1'|^2}{2m_1} \;\;\rightarrow\;\; KE_1 = KE_1'.$$

(17-34)

In COM, elastic case, each particle final KE = initial KE (i.e., final |$\mathbf{p}$| = initial |$\mathbf{p}$| for each)

The kinetic energy of incoming particle #1 is unchanged by the interaction. Since total kinetic energy is conserved, that must mean the kinetic energy of particle #2 is unchanged, as well. Each particle must have the same magnitude for its velocity afterwards as it did before. No energy is exchanged between the two particles.

However, the direction of the velocity of each particle can change. But it must do so in a way that conserves total 3-momentum. Since the total 3-momentum in the COM frame is zero (see (17-32)(a)) before interaction, it must be zero after interaction. Thus, the **p** vectors for the two particles must be equal and opposite before scattering and after scattering. Typical possible scattering scenarios, as viewed from the COM frame are shown in Fig. 17-13.

But each particle final $\mathbf{p}$ direction can change

Relativistic Case

Although it is somewhat more complicated to derive, the same conclusions hold for the relativistic case. There, (17-32)(a) holds and we have (17-33). (17-32)(b) is different, however, because relativistic KE is expressed as in (17-28) or (17-30). Nevertheless, we would end up with relations parallel in concept, though different in form, from (17-34). Do Prob. 3, if you want to work this out for yourself.

Same conclusions for relativistic elastic scattering in COM frame

Bottom Line: For two particle elastic collisions viewed in the COM frame, each particle has the same speed after interaction as it did before, and the final 3-momenta of the particles must be equal and opposite. Zero energy is exchanged between the two particles.

| $\mathbf{p}_1$|= |$\mathbf{p}_2$| and |$\mathbf{p}_1'$| = |$\mathbf{p}_2'$| always true in COM frame

| $\mathbf{p}_1$|= |$\mathbf{p}_1'$| = |$\mathbf{p}_2$| = |$\mathbf{p}_2'$| in elastic interaction, COM frame, $\longrightarrow$ $KE_1 = KE_1'$, $KE_2 = KE_2'$

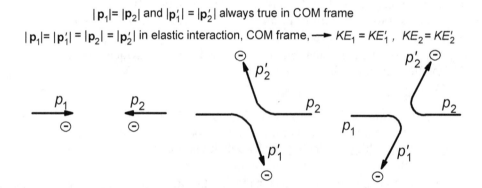

Initial particle states, equal   Different possible final states, all with
and opposite 3-momenta   equal and opposite 3-momenta

Figure 17-13. Elastic Scattering in COM Frame: No Energy Exchange

17.2.2 Quantum Elastic Scattering

Scattering in NRQM and RQM

In NRQM scattering, we typically examined a single particle/wave scattered by a potential, such as the Coulomb potential. Scattering in NRQM (and RQM) courses is usually restricted to such cases. For a potential field fixed in space, a force is exerted on the wave/particle, so 3-momentum is not conserved and we cannot use the usual conservation laws analysis method.

In NRQM and RQM scattering, generally analyze one wave/particle in a potential field

However, if we were to analyze NRQM (and RQM) scattering as initial and final particles that scatter off of one another, then those conservation laws would hold. This is not a topic that is discussed in most physics circles, and we won't do so here either. Such scattering of particles off of one another is best handled by QFT, and we look at that in some depth in the rest of this chapter.

In QFT scattering, analyze one wave/particle scattering off another

Scattering in QFT

In QFT, we saw that for both elastic (particles maintain their original identities) and inelastic (particles transmute from one type to another, so rest masses are different and KE changes to rest energy, or vice versa) scattering, our transition amplitudes had the form

$$S_{fi} = \delta_{fi} + \left((2\pi)^4 \delta^{(4)} \left(P_f - P_i \right) \left(\prod^{\substack{\text{all ext} \\ \text{bosons}}} \sqrt{\frac{1}{2V\omega}} \right) \left(\prod^{\substack{\text{all ext} \\ \text{fermions}}} \sqrt{\frac{m}{VE}} \right) \right) \mathcal{M} \qquad \begin{aligned} P_f &= p_1'^{\mu} + p_2'^{\mu}, \\ \underline{P_i} &= \underline{p_1^{\mu} + p_2^{\mu}} \\ & \text{for 2 particles in; 2 out} \end{aligned} \quad .(17\text{-}35)$$

From transition amplitude, total energy and total 3-momentum must be conserved

This is zero unless total scattered particles 4-momentum P_f equals total incoming 4-momentum P_i. But four-momentum is simply one component of energy and three components of 3-momentum. Thus, both energy and 3-momentum conservation of (17-23) and the pictorial expression of that conservation in Fig. 17-13 hold quantum mechanically.

Hence, we can draw the same bottom line for QFT elastic scattering as we did above for classical elastic scattering. In the COM frame, zero energy is exchanged between particles in elastic interaction between two particles. In QFT language, that means that for a virtual particle mediating an elastic interaction, that virtual particle carries no energy. But momentum direction can change, so the virtual particle can carry 3-momentum, but no energy.

COM frame, elastic scattering: Initial KE = final KE for each particle

Recall that in Chap. 16, when we analyzed planar and Coulomb potentials in the COM frame, we took the energy in the virtual photon propagator to be zero. (See Sect. 16.2.3, pg 407, just above (16-19)) We noted there that we would look more closely at this in the present chapter, and that is what we have done above.

Virtual energy = 0 (even though virtual 3-momentum $\neq 0$)

Bottom Line: In QFT COM frame elastic scattering of two particles via a virtual particle, that virtual particle carries 3-momentum but no energy.

17.3 Another Look at Macroscopic Charged Particles Interacting

Same conclusions for QFT in COM, elastic scattering

Recall, however, from Chap. 8, Sect. 8.10.2, pgs. 244-246, that for the case of two macroscopic charged objects approaching one another, we interpreted classical potential energy of the electric field between them as the sum of the energies of the virtual photons mediating the electric interaction between the two objects. Fig. 8-12 in the referenced section depicts these two objects in the COM frame of the two. But, if so, from the logic of the section before this one, one might conclude that the virtual photons should have zero energy, contrary to what we claimed in Chap. 8.

But, in Chap. 8, we said macro charged objects in COM frame exchange virtuals with non-zero energy

Note that if no energy could be exchanged via the virtual photons in Fig. 8-12, then neither macro object could slow down (for repulsion case) as they approached one another. Slowing down means decreased kinetic energy. But our conservation relations seem, via the reasoning of Sect. 17.2.2, to prohibit this very well-known macroscopic effect.

This issue can be resolved with the aid of Fig. 17-14. Note that if we have two particles scattering, and two, rather than one, virtual photons are exchanged, each virtual photon can carry the same amount of energy, so the net energy exchange between the two real particles after scattering is zero. This would generally be a higher order interaction.

If more than one virtual is exchanged in COM, elastic case, they can carry energy

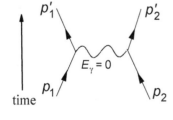 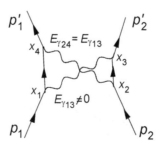

Feynman diagram for   Feynman diagram for
single virtual photon   two virtual photons

But sum of energy exchanges must total zero after scattering

Figure 17-14. COM Elastic Scattering in QFT (Total Energy Exchange = 0)

Wholeness Chart 17-3. Two Particle Elastic Collisions (Non-relativistic)

	NR Classical		NRQM	
Laws	momentum conserved (always true)	kinetic energy conserved (only true if elastic, i.e., no friction, deformation)	momentum conserved (always true)	total energy conserved (K.E. also conserved as no particle transmutation)
One Spatial Dimension				
(primed = final)	$m_1 v_1 + m_2 v_2$ $= m_1 v_1' + m_2 v_2'$	$\frac{1}{2}m_1 v_1^2 + \frac{1}{2}m_2 v_2^2$ $= \frac{1}{2}m_1 v_1'^2 + \frac{1}{2}m_2 v_2'^2$	$p_1^i + p_2^i = p_1'^{\,i} + p_2'^{\,i}$ (1 equation; $i = 1$)	$p_1^0 + p_2^0 = p_1'^{\,0} + p_2'^{\,0}$ (1 equation)
	2 eqs in 2 unknowns, v_1' and v_2'. Can determine result via theory alone.		1D NRQM (and RQM) scattering typically comprises a quantum wave scattering off a potential, where a force acts on the wave. In that case, neither kinetic energy nor momentum is conserved.	
Reduce to	↑ (above relation)	$v_1 - v_2 = -\left(v_1' - v_2'\right)$		
Special cases	Use above two relations to derive			
$m_1 = m_2$	$v_2' = v_1$ $v_1' = v_2$ (particles exchange velocities)		For one quantum wave scattering off of another (with no external potential), however, total energy (which would be all kinetic) and momentum of the two waves/particles would be conserved. Analysis would parallel that at left.	
↑ + $v_2 = 0$	$v_2' = v_1$ $v_1' = 0$			
$m_1 \ne m_2$ + $v_2 = 0$	$v_2' = v_1\left(\dfrac{2m_1}{m_1 + m_2}\right)$	$v_1' = v_1\left(\dfrac{m_1 - m_2}{m_1 + m_2}\right)$		
↑ + $m_1 \ll m_2$	$v_2' \approx 0$ $v_1' = -v_1$		Material at left supplied for reference only.	
COM system	$m_1 v_1 + m_2 v_2$ $= m_1 v_1' + m_2 v_2' = 0$	Using K.E. conservation $v_1 - v_2 = -\left(v_1' - v_2'\right)$	No need to review it in depth at this time.	
↑ + $m_1 = m_2$	$v_1 = -v_2$ $v_2' = -v_1' = v_1$			
Two or Three Spatial Dimensions				
Laws	One momentum conservation eq for each x,y,z direction	One K.E. conservation eq similar to top above	$p_1^i + p_2^i = p_1'^{\,i} + p_2'^{\,i}$ (2 or 3 equations)	$p_1^0 + p_2^0 = p_1'^{\,0} + p_2'^{\,0}$ (1 equation)
In 2D	3 eqs, 4 unknowns $v_{1x}' = v_1' \cos\theta_1'$ $v_{1y}' = v_1' \sin\theta_1'$ $v_{2x}' = v_2' \cos\theta_2'$ $v_{2y}' = v_2' \sin\theta_2'$ If measure θ_1', can calculate v_1', v_2', and θ_2'.		2D NRQM (and RQM) scattering parallel to 1D discussion above. For two waves/particles, NRQM would be same as that at left.	
In 3D	4 eqs, 6 unknowns $v_1' = \dfrac{\lvert\mathbf{p}_1'\rvert}{m_1}, v_2' = \dfrac{\lvert\mathbf{p}_2'\rvert}{m_2}, \theta_1', \theta_2', \phi_1',$ and ϕ_2'. If measure θ_1', ϕ_1', can calculate others.		3D NRQM (and RQM) scattering parallel to 1D and 2D discussions above. For two waves/particles, NRQM would be same as that at left.	
	Cannot determine result via theory alone.		For two waves/particles, would be as at left	
Special case	Sometimes simple rels between final $\mathbf{p}_j'$, i.e., $\mathbf{v}_j'$		For two waves/particles, would be as at left	
COM system	$\mathbf{p}_1 + \mathbf{p}_2 = \mathbf{p}_1' + \mathbf{p}_2' = 0 \;\rightarrow\; \mathbf{p}_1 = -\mathbf{p}_2$ $\mathbf{p}_1' = -\mathbf{p}_2'$ KE conservation $\rightarrow \lvert\mathbf{p}_1'\rvert = \lvert\mathbf{p}_1\rvert = \lvert\mathbf{p}_2'\rvert = \lvert\mathbf{p}_2\rvert$		For two waves/particles, would be as at left	

Wholeness Chart 17-4. Two Particle Elastic Collisions (Relativistic)

	Relativistic Classical		QFT	
Laws	momentum conserved (always true)	kinetic energy conserved (only true if elastic, i.e., no friction, deformation)	momentum conserved (always true)	total energy conserved (K.E. also conserved if no particle transmutation)
One Spatial Dimension				
(primed = final)	$p_1^1 + p_2^1 = p_1'^1 + p_2'^1$ (1 equation)	$p_1^0 + p_2^0 = p_1'^0 + p_2'^0$ (1 equation)	Same as at left	
	2 eqs in 2 unknowns, $p_1'^1$ and $p_2'^1$ (Since $(p'^0)^2 = m^2 + (p'^1)^2$.) Can determine result via theory alone.		Same as at left	
COM sys	$p_1^1 + p_2^1 = p_1'^1 + p_2'^1 = 0 \rightarrow p_1^1 = -p_2^1 \quad p_1'^1 = -p_2'^1$			
$\uparrow$ + $m_1 = m_2$	$v_1 = -v_2 \quad v_2' = -v_1' = v_1$			
Two or Three Spatial Dimensions				
Laws	$p_1^i + p_2^i = p_1'^i + p_2'^i$ (2 or 3 equations)	$p_1^0 + p_2^0 = p_1'^0 + p_2'^0$ (1 equation)	Same as at left	
In 2D	3 eqs, 4 unknowns $p_1'^i, p_2'^i$ (since $(p'^0)^2 = m^2 + \mathbf{p}'^2$), which can be expressed as $|\mathbf{p}_1'|, |\mathbf{p}_2'|, \theta_1'$, and θ_2' If measure θ_1', can calculate $|\mathbf{p}_1'|, |\mathbf{p}_2'|$, and θ_2'.		Same as at left	
In 3D	4 eqs, 6 unknowns $p_1'^i, p_2'^i$ (since $(p'^0)^2 = m^2 + \mathbf{p}'^2$), which can be expressed as $|\mathbf{p}_1'|, |\mathbf{p}_2'|, \theta_1', \theta_2', \phi_1'$, and ϕ_2'. If measure θ_1', ϕ_1', can calculate others.		Same as at left	
	Cannot determine result via theory alone.		Same as at left	
Special case	But, sometimes simple relations between final $\mathbf{p}_j'$		Same as at left	
COM sys	$\mathbf{p}_1 + \mathbf{p}_2 = \mathbf{p}_1' + \mathbf{p}_2' = 0 \rightarrow \mathbf{p}_1 = -\mathbf{p}_2 \quad \mathbf{p}_1' = -\mathbf{p}_2'$ K.E. energy conservation $\rightarrow |\mathbf{p}_1'| = |\mathbf{p}_1| = |\mathbf{p}_2'| = |\mathbf{p}_2|$		Same as at left	
$\uparrow$ + $m_1 = m_2$	$\mathbf{v}_1 = -\mathbf{v}_2 \quad \mathbf{v}_2' = -\mathbf{v}_1' \quad |\mathbf{v}_1'| = |\mathbf{v}_1| = |\mathbf{v}_2'| = |\mathbf{v}_2|$		Same as at left	

Note, however, the semi-classical heuristic (as the afore referenced section of Chap. 8 admittedly was) picture of Fig. 17-15. In it, we have many real particles in each macro object, and in each object, the particles are bound together. (Binding represented by rod-like structures in Fig. 17-15.) At a given time, any of these mutually bound particles in one object may have emitted virtual particles, but that object may have yet to receive the virtual particles emitted from the other.

After all virtual particles have been exchanged, i.e., when the macro objects are again far apart, the net energy exchange between macro objects would be zero. However, during the time the macro objects are influencing one another, many virtual particles exist between the two, and these can carry energy. Because the virtual particles carry energy, the macro objects lose (or gain) kinetic energy and slow down (or speed up) during the exchange process.

For repulsion, the virtual particle energies would be positive; for attraction, negative, as discussed in Chap. 8.

virtual photons

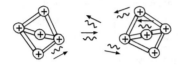

Macro object #1   Macro object #2

For interacting macro objects, many virtual photons exchanged, so they can carry energy

Figure 17-15. Macro Charged Object Acts Like Single Particle in RHS of Fig. 17-14
(Exchanges multiple virtual photons with other charged macro object)

<u>Bottom line</u>: For elastic scattering of two elementary particles in the COM frame with a single virtual particle exchanged, the virtual particle carries no energy, so neither particle loses speed (though it can, and usually does, change direction.) However, for scattering of macro objects in the COM frame, during the interaction, many virtual particles are exchanged, and each can carry energy. However, the sum total of energy exchanges after all have occurred (after the macro object scattering is finished) must equal zero. The KE (i.e., speed) of each macro object before interaction must equal that of the same object after interaction. During interaction, they can slow down (or speed up) depending on how much and what type (positive or negative) energy is carried by the many virtual particles.

17.4 Scattering in QFT: An In Depth Look

QFT Scattering

Now we will finally, after hundreds of pages of preparation, deduce the QFT relations for differential cross-section and cross-section. It should help if, as we do this, you follow along with the Wholeness Chart 17-6 summary/overview at the end of the chapter, pg. 483.

Follow Wholeness Chart 17-6 as we study this

We will start with (17-22), derived earlier, but first, we need just a little more background.

17.4.1 Things We'll Need to Know

<u>Using Renormalized *e* Alone in Tree Level Feynman Amplitude Yields High Accuracy</u>

Use renormalized e and m

In doing scattering calculations, one can usually get very good accuracy by assuming the renormalized value for the electromagnetic coupling "constant" e (p) at the energy of the test, symbolized by p here. For our purposes, we typically do not have to incorporate renormalized values for γ^{μ}, iS_F, or $iD_{F\mu\nu}$. This may be obvious if we recall that the incorporation of γ^{μ}_{Mod} for γ^{μ} in the magnetic moment calculations gave us a correction at second order of about 0.1%.

High accuracy without renormalized forms of vertex and propagator

So, to keep things simple, without losing much accuracy, we will use tree level Feynman diagrams and associated transition amplitude relations with renormalized value for e (and m).

<u>Renormalization Made S_{fi} Unitary Again</u>

Recall (Chap. 7, pgs. 196-197) that, for a final general state $|F\rangle$ composed of basis states $|f\rangle$, $|F\rangle = \sum_f S_{fi}|f\rangle$ where $\langle f|f\rangle = 1$. Now if we insist on a normalized final general state, then $\langle F|F\rangle = 1$, and this ensures that our interpretation of $|S_{fi}|^2$ as a probability is consistent, as it results in conservation of probability via

After renormalization, S_{fi} is unitary

$$\sum_f \left|S_{fi}\right|^2 = 1 \quad \text{conservation of probability} . \qquad (17\text{-}36)$$

Any operator such as S_{fi} that conserves probability is a unitary operator.

However, in our naïve, pre-renormalization derivation of the transition amplitude, we saw that S_{fi} went to infinity when higher order corrections were included, so (17-36) obviously could not hold. But after renormalization of our theory, we had eliminated the infinities and restored the unitarity of S_{fi}. Thus, (17-36) holds in the renormalized theory.

Natural vs cgs Units for σ

We have been using natural units throughout the book, yet in this chapter have been discussing cross-sections in units of cm^2. We will do our analysis in the following sections assuming natural units. After we determine σ in those units, we can convert to cm^2, and thus to barns, which are the standard ways in which cross-sections are expressed.

QFT analysis in nat units, convert σ to cm^2 after

Use Relative Velocity When Both Particles Moving

When we deduced the relations for cross-section and differential cross-section, (17-6) and (17-10) (also in Wholeness Chart 17-2, pg. 445, Experiment column), we assumed the target particle was stationary. And we used that assumption in deriving the QFT theoretical expression for $d\sigma$, (17-22) (also in same chart, bottom right block).

$v_b \to v_{rel}$ in our transition rate formula when target not stationary

However, in the COM frame, and an infinite number of other frames, the target particle(s) is (are) not stationary. Note that for the target stationary frame (typically, the lab), each of the afore noted relations (17-6), (17-10), and (17-22) contains a factor of the beam particle velocity v_b. (In the first two of these relations, we have $f_b = n_b v_b$.) This velocity is actually the velocity *relative* to the target, since the target is stationary.

It is therefore a simple extension of the theory to scattering cases where both particles are moving, such as would be seen in the COM frame. We simply take v_b to be the relative velocity between a beam and a target particle in the frame of observation. So from here on, we take

$$\mathbf{v}_b \to \mathbf{v}_{rel} = \mathbf{v}_b - \mathbf{v}_t \quad |v_{rel}| = |\mathbf{v}_b - \mathbf{v}_t| \quad \text{most general case}$$
$$v_b \to v_{rel} = v_b - v_t \quad \text{for co-linear velocities case.} \tag{17-37}$$

The relative velocity is *not* found via relativistic velocity addition/subtraction, because it is the difference between the velocities of two particles as observed in the same frame (which we will label with *obs* here). That is, in the co-linear case, for b symbolizing the frame of the first particle, t that of the second, we have $v_{rel} = v_{b/obs} - v_{t/obs}$, which is (17-37). For relativistic velocity addition, we would be looking at the velocity of b relative to *obs*, found from the velocity of b relative to t and the velocity of t relative to *obs*, i.e., $v_{b/obs} = (v_{b/t} - v_{t/obs})/(1 - (v_{b/t}v_{t/obs})/c^2)$.

Also, when both particles are moving, and we wish to use Feynman diagram/transition amplitude analysis that assume both occupy the same volume V indefinitely, we can assume both particles move like the beam particle in Fig. 17-9 on pg. 442. That is, there is a stream of particles of each type configured so that effectively one particle wave of each type is inside the volume V at any one time.

Expressing v_{rel} in Terms of 3 Momenta and Energy

It will help later if we can express v_{rel} in terms of $\mathbf{p}_1$ and $\mathbf{p}_2$. To that end, we re-express the general case of (17-37), where we now define our particle labeling so that v_{rel} is positive and use natural units with $c = 1$, as

Useful expressions for v_{rel}

General expression for v_{rel}

$$v_{rel} = |\mathbf{v}_b - \mathbf{v}_t| = |\mathbf{v}_1 - \mathbf{v}_2| = \left| \frac{m_1\mathbf{v}_1 / \sqrt{1 - v_1^2}}{m_1 / \sqrt{1 - v_1^2}} - \frac{m_2\mathbf{v}_2 / \sqrt{1 - v_2^2}}{m_2 / \sqrt{1 - v_2^2}} \right| = \left| \frac{\mathbf{p}_1}{E_1} - \frac{\mathbf{p}_2}{E_2} \right| \tag{17-38}$$

COM Frame

In the center of mass frame, $\mathbf{p}_1 = -\mathbf{p}_2$. For that case, we can then express (17-38) as

v_{rel} in COM frame

$$v_{rel} = \left| \frac{\mathbf{p}_1}{E_1} + \frac{\mathbf{p}_1}{E_2} \right| = \frac{|\mathbf{p}_1|}{E_1} + \frac{|\mathbf{p}_1|}{E_2} = |\mathbf{p}_1|\left(\frac{1}{E_1} + \frac{1}{E_2} \right) = |\mathbf{p}_1|\left(\frac{E_1 + E_2}{E_1 E_2} \right) \quad \begin{cases} \text{COM frame,} \\ \text{2 initial particles.} \end{cases} \tag{17-39}$$

In tests, particle velocities are effectively co-linear ($\mathbf{p}_1 = -\mathbf{p}_2$ merely implies parallel trajectories, which could be displaced significantly from one another) in order for scattering to occur.

Target at Rest Frame (Often the Lab)

In the frame where the target particle (which we have been labeling with "2") is stationary, $\mathbf{p}_2 = \mathbf{v}_2 = 0$). In this frame, (17-38) becomes

v_{rel} in target stationary frame

$$v_{rel} = \frac{|\mathbf{p}_1|}{E_1} \quad \begin{cases} \text{target stationary frame} \\ \text{2 initial particles.} \end{cases} \tag{17-40}$$

<u>Relating Delta Functions to V and T</u>

Recall that the delta function in our transition amplitude (17-35) was obtained from a factor in that transition amplitude (see Chap. 8) of the LHS below.

Useful relations between V, T and delta functions

$$\int_{V=\infty}\int_{T=\infty} e^{-i(P_f - P_i)x}\,dVdT = (2\pi)^4\,\delta^{(4)}\left(P_f - P_i\right). \tag{17-41}$$

If we assume our volume V of interaction and the time T of interaction are very large, but not infinite, then (17-41) becomes an approximate relation

$$(2\pi)^4\,\delta^{(4)}\left(P_f - P_i\right) \approx \int_{V\,\text{large}}\int_{T\,\text{large}} e^{-i(P_f - P_i)x}\,dVdT \quad \text{where} \; \approx\to= \; \text{as } V\to\infty,\; T\to\infty.\,(17\text{-}42)$$

So the correct factor in our transition amplitude is actually the RHS of (17-42), but for large, finite volume and time duration, we approximate it with the LHS.

Note that if we had a situation where $P_f = P_i$, then our integral in (17-42) would become

$$\int_{V\,\text{large}}\int_{T\,\text{large}} e^{-i(P_f - P_i)x}\,dVdT = VT \qquad \text{for } P_f = P_i. \tag{17-43}$$

That is, for cases where $P_f = P_i$, for finite volume and time, from (17-42) and (17-43), we can make the following substitution in our expression for the transition amplitude, if we like.

$$(2\pi)^4\,\delta^{(4)}\left(P_f - P_i\right) \to VT \qquad \text{for finite } V \text{ and } T \text{ when } P_f = P_i. \tag{17-44}$$

In our analyses, we have assumed volume size and time duration are large enough so that (17-42) holds. If, in addition $P_f = P_i$,, then we can use (17-44).

<u>Average Over Initial Spins</u>

Scattering experiments typically do not measure the spins of particles. So what we do is consider the beam of incoming particles to be in an "average" spin state. What we mean by average spin state is that half of the incoming particles are taken to be in the $s = 1$ spin state and half in the s=2 spin state. This will reflect our lack of knowledge of the individual spin state of each incoming particle.

For experiments where spin not measured, need to use average initial spin state

<u>Sum Over Final Spins</u>

For the final scattered states, however, we need to sum probabilities over all possible final states (assuming we do not measure the final spins). This is because for any given final **p** and s (spin) state, say s =1 for example, we will get a certain probability (a certain differential cross-section). But for that same final **p** state we will get a separate probability for another spin state, in our example, for s = 2. The sum total of the two will give us our probability for measuring a final state of **p** of any possible spin.

Those experiments will measure sum of final spin states, so need to calculate that sum

In determining differential cross-sections, we will, however, first assume we know the initial and final spin states. At the end, we will then average over initial spins and sum over final spins.

<u>Averaging and Summing Over Polarization States</u>

If one or more of the real particles is a photon, as in Compton scattering, the same logic used above for fermion spin states applies to photon polarization states. We average over initial polarization states, but sum over final polarization states.

For experiments where photon polarization not measured, treat like spin above

<u>Expressing Number of Final States dN_f in Terms of V and $d^3\mathbf{p}$</u>

In Appendix A of Chap. 10, we noted that for a single particle, the wave number of the N_{f1} state (we used the symbol n instead of N_{f1} there, and N_{f1} = 1,2,3, …) in the x^1 direction (where l_1 is the length of the volume V in that direction and $p_{N_{f1}}$ is 3-momentum for that state) is

Finding a useful expression for dN_f

$$k_{N_{f1}} = \frac{2\pi}{\lambda_{N_{f1}}} = \frac{2\pi N_{f1}}{l_1} = p_{N_{f1}} \quad = \text{3-momentum in } x^1 \text{ direction, state } N_f, \text{ natural units}. \,(17\text{-}45)$$

The differential of (17-45) is

$$\frac{2\pi dN_{f1}}{l_1} = dp_{N_{f1}}, \tag{17-46}$$

where dN_{f1} is the number of states between $p_{N_{f1}}$ and $p_{N_{f1}} + dp_{N_{f1}}$.

In three dimensions, the total number of states would be $dN_f = dN_{f1}\,dN_{f2}\,dN_{f3}$. For example, if for each of 3 states in the x^1 direction, the same particle had 2 in the x^2 direction, and for each of those combinations, there were 4 states in the x^3 direction, then there would be a total of 3 X 2 X 4 = 24 distinct possible states for the particle in the volume. So, in 3 dimensions, (17-46) becomes

$$\frac{(2\pi)^3\,dN_{f1}dN_{f2}dN_{f3}}{l_1 l_2 l_3} = \underbrace{dp_{N_{f1}}dp_{N_{f2}}dp_{N_{f3}}}_{\text{label this } d^3\mathbf{p}_f} \;\rightarrow\; \frac{(2\pi)^3\,dN_f}{V} = d^3\mathbf{p}_f\,, \qquad (17\text{-}47)$$

where dN_f is the number of states between $\mathbf{p}_f$ and $\mathbf{p}_f + d^3\mathbf{p}_f$. We re-arrange (17-47) to get

$$dN_f = \frac{V\,d^3\mathbf{p}_f}{(2\pi)^3} \quad \text{for a single final scattered particle}\,. \qquad (17\text{-}48)$$

For multiple final particles, using similar logic to that in the paragraph before (17-47), we would have (with subscript f' signifying a single final particle, subscript f signifying a generally multi-particle final state, and (17-48) is a special case of (17-49) for a single f')

$$dN_f = \prod_{f'}^{M}\frac{V\,d^3\mathbf{p}_{f'}}{(2\pi)^3} \quad \text{for } M \text{ final particles}\,. \qquad (17\text{-}49)$$

dN_f in terms of V and $d^3\mathbf{p}_{f'}$ of each particle

17.4.2 Finding $d\sigma$ in Terms of Feynman Amplitude and Final 3-Momenta

dσ in terms of $\mathcal{M}$ and $\mathbf{p}_{f'}$

<u>The Most General Relation for $d\sigma$ for Two Initial, M Final Particle Collisions</u>

We begin our derivation of the differential cross-section in QFT with (17-22), restated in the first row of (17-50) below with $v_b \to v_{rel}$ for reasons discussed above. We insert (17-35) and (17-49) into that, where $\delta_{fi} = 0$, since for us the initial and final (scattered) states are different.

Finding most general relations for dσ for two initial particle, M final particles scattering

$$d\sigma = \frac{V}{v_{rel}T}\big|S_{fi}\big|^2 dN_f$$

$$= \frac{V}{v_{rel}T}\left|\left((2\pi)^4\,\delta^{(4)}\big(P_f - P_i\big)\left(\overset{\substack{\text{all ext}\\\text{bosons}}}{\prod}\sqrt{\frac{1}{2V\omega}}\right)\left(\overset{\substack{\text{all ext}\\\text{ferms}}}{\prod}\sqrt{\frac{m}{VE}}\right)\right)\mathcal{M}\right|^2 \prod_{f'}^{M}\frac{V\,d^3\mathbf{p}_{f'}}{(2\pi)^3} \qquad (17\text{-}50)$$

$$= \frac{V}{v_{rel}T}(2\pi)^4\,\delta^{(4)}\big(P_f - P_i\big)(2\pi)^4\,\delta^{(4)}\big(P_f - P_i\big)\left(\overset{\substack{\text{all ext}\\\text{bosons}}}{\prod}\frac{1}{2V\omega}\right)\left(\overset{\substack{\text{all ext}\\\text{ferms}}}{\prod}\frac{m}{VE}\right)|\mathcal{M}|^2 \prod_{f'}^{M}\frac{V\,d^3\mathbf{p}_{f'}}{(2\pi)^3}\,.$$

We have, all along, assumed large V and T in order for the delta functions in (17-50) to hold. One of them alone ensures $P_f = P_i$. Thus, we can use (17-44) for the other delta function to get

$$d\sigma = \frac{V}{v_{rel}T}VT(2\pi)^4\,\delta^{(4)}\big(P_f - P_i\big)\left(\overset{\substack{\text{all ext}\\\text{bosons}}}{\prod}\frac{1}{2V\omega}\right)\left(\overset{\substack{\text{all ext}\\\text{ferms}}}{\prod}\frac{m}{VE}\right)|\mathcal{M}|^2 \prod_{f'}^{M}\frac{V\,d^3\mathbf{p}_{f'}}{(2\pi)^3}\,. \qquad (17\text{-}51)$$

Relation (17-51) is good for two particle collisions, elastic or inelastic, with any number of final products, although in our work we will only be considering two final particles.

We have two incoming external particles and M outgoing external ones, so in (17-51), we have a) two V factors in the numerator; b) one V factor in the denominator for each incoming particle (2 in all); c) one V factor in the numerator for each outgoing particle (M in all); and d) one V factor in the denominator for each such outgoing particle (M in all). Thus, cancelling the V and T factors in (17-51) and multiplying by 2 in the numerator and denominator because it will help us later, yields

$$d\sigma = \frac{1}{v_{rel}}(2\pi)^4\,\delta^{(4)}\big(P_f - P_i\big)\left(\overset{\substack{\text{all ext}\\\text{bosons}}}{\prod}\frac{1}{2\omega}\right)\left(\overset{\substack{\text{all ext}\\\text{ferms}}}{\prod}\frac{2m}{2E}\right)\prod_{f'}^{M}\frac{d^3\mathbf{p}_{f'}}{(2\pi)^3}|\mathcal{M}|^2\,. \qquad (17\text{-}52)$$

If we now take our incoming particles to have energy E_1 and E_2, whether they be fermions or bosons, (17-52) becomes (where we begin using more formal, and more correct than our earlier "loose", notation for the product of external fermion masses)

$$d\sigma = \left(2\pi\right)^4 \delta^{(4)}\left(P_f - P_i\right)\frac{1}{4E_1E_2v_{rel}}\left(\prod_l^{\text{all ext ferms}} 2m_l\right)\left(\prod_{f'}^{M}\frac{d^3\mathbf{p}'_{f'}}{\left(2\pi\right)^3 2E'_{f'}}\right)\left|\mathcal{M}\right|^2 \left\{\begin{matrix}\text{2 initial,}\\ \text{M final}\\ \text{particles,}\\ \text{elastic or}\\ \text{inelastic}\end{matrix}\right. \quad (17\text{-}53)$$

Most general $d\sigma$ for two initial, M final particles scattering

(17-53) is good for two incoming particles (either boson or fermion) and any number of outgoing particles (either boson or fermion), for elastic or inelastic cases. Note that total energy is conserved via the delta function in (17-53), so even if rest mass-energy is interchanged with kinetic energy (as in inelastic scattering), conservation of total energy still holds.

Relation for Two Initial and Two Final Particles Scattering

We now restrict ourselves in (17-53) to the common QED two particle scattering experiment case of two final particles. For this case, (17-53) becomes (where l runs over all external fermions)

$d\sigma$ for two initial and two final particles scattering

$$d\sigma = \left(2\pi\right)^4 \delta^{(4)}\left(P_f - P_i\right)\frac{1}{4E_1E_2v_{rel}}\left(\prod_l^{\text{all ext ferms}} 2m_l\right)\frac{d^3\mathbf{p}'_1}{\left(2\pi\right)^3 2E'_1}\frac{d^3\mathbf{p}'_2}{\left(2\pi\right)^3 2E'_2}\left|\mathcal{M}\right|^2$$

$$= \delta^{(4)}\left(p'_1 + p'_2 - p_1 - p_2\right)\frac{1}{64\pi^2 v_{rel}E_1E_2E'_1E'_2}\left(\prod_l^{\text{all ext ferms}} 2m_l\right)d^3\mathbf{p}'_1 d^3\mathbf{p}'_2\left|\mathcal{M}\right|^2 \left\{\begin{matrix}\text{2 initial and 2}\\ \text{final particles,}\\ \text{elastic or inelastic}\end{matrix}\right.$$

(17-54)

Relation in Terms of Only One Final Particle 3-Momentum

Note that, as we saw earlier and summarized for the elastic case in Wholeness Charts 17-3 and 17-4 (see row titled "In 3D") energy and 3-momentum conservation mean that the $\mathbf{p}'_1$ and $\mathbf{p}'_2$ of (17-54) are not independent. For the elastic case, in terms of their components, we have 6 unknowns and 4 conservation equations linking them. So, for the elastic case, there are really only two unknowns, since if we determine any two, the other four can be found via the 4 conservation equations.

Conservation equations mean all components of $\mathbf{p}'_1$ and $\mathbf{p}'_2$ not independent unknowns

The inelastic case is a bit more complicated, but the same general rule holds, i.e., six final components and four conservation equations means two unknowns.

So, in either the elastic or inelastic case, we can re-express (17-54) in terms of 3-momentum components of a single final particle, say $\mathbf{p}'_1$. Only two of those components will, of course, be independent, but this means we can simplify by focusing on only one of the scattered particles. We can, at the end, then calculate components of the other final particle from the conservation laws.

So, use conservation equations to express $\mathbf{p}'_2$ in terms of $\mathbf{p}'_1$

And if we can express $\mathbf{p}'_1$ in terms of magnitude $|\mathbf{p}'_1|$ and spherical coordinate angles θ and ϕ, we can get $d\sigma$ as a function of θ, which will lead to $d\sigma/d\Omega$ as a function of θ.

To eliminate the $d^3\mathbf{p}'_2$ in (17-54), we can integrate (17-54) over $\mathbf{p}'_2$. In doing this, we note

$$\delta^{(4)}\left(p'_1 + p'_2 - p_1 - p_2\right) = \delta\left(E'_1 + E'_2 - E_1 - E_2\right)\delta^{(3)}\left(\mathbf{p}'_1 + \mathbf{p}'_2 - \mathbf{p}_1 - \mathbf{p}_2\right) \quad (17\text{-}55)$$

so for any integral of a function of $\mathbf{p}'_2$, $g(\mathbf{p}'_2)$,

$$\int_{\mathbf{p}'_2} \delta^{(4)}\left(p'_1 + p'_2 - p_1 - p_2\right)g\left(\mathbf{p}'_2\right)d^3\mathbf{p}'_2$$

$$= \delta\left(E'_1 + E'_2 - E_1 - E_2\right)\int_{\mathbf{p}'_2}\delta^{(3)}\left(\mathbf{p}'_1 + \mathbf{p}'_2 - \mathbf{p}_1 - \mathbf{p}_2\right)g\left(\mathbf{p}'_2\right)d^3\mathbf{p}'_2 \quad (17\text{-}56)$$

$$= \delta\left(E'_1 + E'_2 - E_1 - E_2\right)g\left(\mathbf{p}_1 + \mathbf{p}_2 - \mathbf{p}'_1\right).$$

Thus, where $g(\mathbf{p}'_2)$ is all of the last row of (17-54) except for the delta function and $d^3\mathbf{p}'_1$, our integral of (17-54) becomes (for which, everywhere we had $\mathbf{p}'_2$ before we now have $\mathbf{p}_1 + \mathbf{p}_2 - \mathbf{p}'_1$)

$d\sigma$ in terms of $\mathbf{p}'_1$, two initial, two final particles

$$d\sigma = \delta\left(E'_1 + E'_2 - E_1 - E_2\right)\frac{1}{64\pi^2 v_{rel}E_1E_2E'_1E'_2}\left(\prod_l^{\text{all ext ferms}} 2m_l\right)d^3\mathbf{p}'_1\left|\mathcal{M}\right|^2 \text{ where } \mathbf{p}'_2 = \mathbf{p}_1 + \mathbf{p}_2 - \mathbf{p}'_1. (17\text{-}57)$$

We now have $d\sigma$ in terms of the single unknown $\mathbf{p}'_1$.

One may wonder why we didn't integrate $d\sigma$, the LHS of (17-54) in the same way we did the RHS. In effect the integration is simply the imposition of the constraint equation $\mathbf{p}'_2 = \mathbf{p}_1 + \mathbf{p}_2 - \mathbf{p}'_1$

in the RHS of (17-54). In essence, our integration on the RHS used the conservation law to replace $\mathbf{p}'_2$ with its equivalent in terms of $\mathbf{p}'_1$, $\mathbf{p}_1$, and $\mathbf{p}_2$. The same thing occurred with the LHS, so now $d\sigma$ is the same quantity it was before, except now we think of it as expressed solely in terms of the one unknown $\mathbf{p}'_1$ (and the known values of $\mathbf{p}_1$ and $\mathbf{p}_2$) rather than the two unknowns $\mathbf{p}'_1$ and $\mathbf{p}'_2$.

17.4.3 Finding dσ/dΩ in Terms of Feynman Amplitude and Final 3-Momentum

<u>dσ Relation in Terms of Solid Angle</u>

We ultimately want to find $d\sigma/d\Omega$, so we want (17-57) in terms of $d\Omega$ rather than $d^3\mathbf{p}'_1$. To do this we will want to express $d^3\mathbf{p}'_1$ in spherical coordinates (see Fig. 17-4, pg. 436, where $dV = d^3\mathbf{r} = |\mathbf{r}|^2 \sin\theta\, d\theta\, d\phi\, d|\mathbf{r}| = r^2 d\Omega\, dr$)

Getting dσ in terms dΩ'₁, to obtain dσ/dΩ'₁

$$d^3\mathbf{p}'_1 = |\mathbf{p}'_1|^2 \sin\theta'_1\, d\theta'_1\, d\phi'_1\, d|\mathbf{p}'_1| = |\mathbf{p}'_1|^2 d\Omega'_1\, d|\mathbf{p}'_1|\,. \tag{17-58}$$

With (17-58), (17-57) becomes

$$d\sigma = \delta\left(E'_1 + E'_2 - E_1 - E_2\right)\frac{1}{64\pi^2 v_{rel} E_1 E_2 E'_1 E'_2}\left(\overset{\overset{\text{extern}}{\text{ferms}}}{\underset{l}{\prod}} 2m_l\right)|\mathbf{p}'_1|^2 d\Omega'_1\, d|\mathbf{p}'_1|\,|\mathcal{M}|^2 \tag{17-59}$$

where in $|\mathcal{M}|^2$, $\mathbf{p}'_2 = \mathbf{p}_1 + \mathbf{p}_2 - \mathbf{p}'_1$.

We want (17-59) in terms of the single differential $d\Omega'_1$ without the additional differential factor $d|\mathbf{p}'_1|$. We can do this by integrating (17-59) over $|\mathbf{p}'_1|$. To do this, we need to employ the general math relation

$$\int \underbrace{f(x,y)}_{\substack{\text{all else in}\\ d\sigma \text{ relation}}} \underbrace{\delta\left[g(x,y)\right]}_{\delta\left(E'_1+E'_2-E_1-E_2\right)} \underbrace{dx}_{d|\mathbf{p}'_1|} = \int f(x,y)\delta\left[g(x,y)\right]\left(\frac{\partial x}{\partial g}\right)_y dg$$

$$= f(x,y)\left(\frac{\partial x}{\partial g}\right)_{g=0} = d\sigma \text{ (for us)}. \tag{17-60}$$

For us, in (17-60), $x = |\mathbf{p}'_1|$, $y =$ any other dependences on $\mathbf{p}_1$, $\mathbf{p}_2$, or whatever, and $g = E'_1 + E'_2 - E_1 - E_2$. Note that from $E' = \sqrt{m^2 + \mathbf{p}'^2}$, E'_1 can be expressed in terms of $|\mathbf{p}'_1|$; and E'_2, in terms of $|\mathbf{p}'_2|$. However, from the last line of (17-59), we know we can express $|\mathbf{p}'_2|$ in terms of $|\mathbf{p}'_1|$. Therefore, for us the argument of the δ function in (17-60), i.e., g, is a function of $|\mathbf{p}'_1|$ ($= x$ in (17-60)).

With (17-60), (17-59) becomes

$$d\sigma = \frac{1}{64\pi^2 v_{rel} E_1 E_2 E'_1 E'_2}\left(\overset{\overset{\text{ext}}{\text{ferms}}}{\underset{l}{\prod}} 2m_l\right)|\mathcal{M}|^2 |\mathbf{p}'_1|^2 d\Omega'_1\left(\frac{\partial|\mathbf{p}'_1|}{\partial\left(E'_1 + E'_2 - E_1 - E_2\right)}\right)_{E'_1+E'_2-E_1-E_2=0}$$

$$= \frac{1}{64\pi^2 v_{rel} E_1 E_2 E'_1 E'_2}\left(\overset{\overset{\text{ext}}{\text{ferms}}}{\underset{l}{\prod}} 2m_l\right)|\mathcal{M}|^2 |\mathbf{p}'_1|^2 d\Omega'_1\left(\frac{\partial\left(E'_1 + E'_2 - E_1 - E_2\right)}{\partial|\mathbf{p}'_1|}\right)^{-1}_{E'_1+E'_2-E_1-E_2=0}. \tag{17-61}$$

Or, since neither E_1 nor E_2 is a function of $|\mathbf{p}'_1|$, with both sides divided by $d\Omega'_1$, we obtain the differential cross-section

$$\frac{d\sigma}{d\Omega'_1} = \frac{1}{64\pi^2 v_{rel} E_1 E_2 E'_1 E'_2}\left(\overset{\overset{\text{ext}}{\text{ferms}}}{\underset{l}{\prod}} 2m_l\right)|\mathcal{M}|^2 |\mathbf{p}'_1|^2\left(\frac{\partial\left(E'_1 + E'_2\right)}{\partial|\mathbf{p}'_1|}\right)^{-1}\begin{cases}\text{any frame,}\\ \text{2 initial and 2}\\ \text{final particles,}\\ \text{elastic or inelastic}\end{cases}, \tag{17-62}$$

General relation for dσ/dΩ'₁ for 2 initial and 2 final particles in terms of p'₁

where $p'_2 = p_1 + p_2 - p'_1$, so all dependencies on E'_2 and $\mathbf{p}'_2$ can be expressed in terms of E'_1 and $\mathbf{p}'_1$, where E'_1 is a function of $\mathbf{p}'_1$. The partial derivative is evaluated with the polar angles θ'_1 and ϕ'_1 of the vector $\mathbf{p}'_1$ constant.

COM Frame, Two Initial and Two Final Particles

Getting $d\sigma/d\Omega'_1$
above for
COM frame

In the COM frame, $\mathbf{p}'_1 = -\mathbf{p}'_2$, and from

$$\left(E'_1\right)^2 = \left(m_1\right)^2 + \left|\mathbf{p}'_1\right|^2, \tag{17-63}$$

we have

$$\frac{\partial\left(E'_1\right)^2}{\partial\left|\mathbf{p}'_1\right|} = 2E'_1\frac{\partial E'_1}{\partial\left|\mathbf{p}'_1\right|} \quad \text{also} \quad \frac{\partial\left(E'_1\right)^2}{\partial\left|\mathbf{p}'_1\right|} = \frac{\partial\left(\left(m_1\right)^2 + \left|\mathbf{p}'_1\right|^2\right)}{\partial\left|\mathbf{p}'_1\right|} = 2\left|\mathbf{p}'_1\right|\frac{\partial\left|\mathbf{p}'_1\right|}{\partial\left|\mathbf{p}'_1\right|} = 2\left|\mathbf{p}'_1\right|$$

$$\rightarrow \quad 2E'_1\frac{\partial E'_1}{\partial\left|\mathbf{p}'_1\right|} = 2\left|\mathbf{p}'_1\right| \quad \rightarrow \quad \frac{\partial E'_1}{\partial\left|\mathbf{p}'_1\right|} = \frac{\left|\mathbf{p}'_1\right|}{E'_1}. \tag{17-64}$$

Similarly,

$$\frac{\partial E'_2}{\partial\left|\mathbf{p}'_2\right|} = \frac{\left|\mathbf{p}'_2\right|}{E'_2} \quad \text{and with } \left|\mathbf{p}'_1\right| = \left|\mathbf{p}'_2\right| \text{ in COM} \quad \rightarrow \frac{\partial E'_2}{\partial\left|\mathbf{p}'_1\right|} = \frac{\left|\mathbf{p}'_1\right|}{E'_2}. \tag{17-65}$$

Thus, in (17-62), we have

$$\left(\frac{d\sigma}{d\Omega'_1}\right)_{COM} = \frac{1}{64\pi^2 v_{rel} E_1 E_2 E'_1 E'_2}\left(\prod_l^{\substack{\text{extern} \\ \text{ferms}}} 2m_l\right)\left|\mathcal{M}\right|^2\left|\mathbf{p}'_1\right|^2\underbrace{\left(\frac{\left|\mathbf{p}'_1\right|}{E'_1} + \frac{\left|\mathbf{p}'_1\right|}{E'_2}\right)^{-1}}_{\left(\left|\mathbf{p}'_1\right|\frac{E'_1+E'_2}{E'_1 E'_2}\right)^{-1}}. \tag{17-66}$$

We use v_{rel} of (17-39) along with $E'_1 + E'_2 = E_1 + E_2$ in (17-66) to obtain

$$\left(\frac{d\sigma}{d\Omega'_1}\right)_{COM} = \frac{1}{64\pi^2 \underbrace{\left|\mathbf{p}_1\right|\left(\frac{E_1+E_2}{E_1 E_2}\right)}_{v_{rel}} E_1 E_2 E'_1 E'_2}\left(\prod_l^{\substack{\text{extern} \\ \text{ferms}}} 2m_l\right)\left|\mathcal{M}\right|^2\left|\mathbf{p}'_1\right|^2\left(\frac{1}{\left|\mathbf{p}'_1\right|}\frac{E'_1 E'_2}{E_1 + E_2}\right) \tag{17-67}$$

Cancelling factors, we end up with a representation of (17-62) in the COM frame,

$d\sigma/d\Omega'_1$ of
(17-62) for
COM frame

$$\boxed{\left(\frac{d\sigma}{d\Omega'_1}\right)_{COM} = \frac{1}{64\pi^2\left(E_1 + E_2\right)^2}\frac{\left|\mathbf{p}'_1\right|}{\left|\mathbf{p}_1\right|}\left(\prod_l^{\substack{\text{extern} \\ \text{ferms}}} 2m_l\right)\left|\mathcal{M}\right|^2} \quad \left\{\begin{array}{l}\text{COM, 2 initial \&} \\ \text{2 final particles,} \\ \text{elastic or inelastic.}\end{array}\right. \tag{17-68}$$

The above expression is enclosed in a box because it is a key relation, but unlike certain other boxed in expressions in this text, it is not critical to memorize it.

Target Stationary Frame, Two Initial and Two Final Particles

Getting $d\sigma/d\Omega'_1$
of (17-62) for
target stationary
frame

For the frame where the target is stationary (often called the lab frame in other texts, though as we have seen, the COM can be the lab frame), we label the target as the #2 particle, which, since it is stationary, cannot be a photon and must have mass. Thus, $E_2 = m_2$. Additionally, we need to use (17-40) instead of (17-39) for v_{rel}.

With these relations, (17-62) becomes

$$\boxed{\left(\frac{d\sigma}{d\Omega'_1}\right)_{\substack{\#2 \\ stat}} = \frac{1}{64\pi^2\left|\mathbf{p}_1\right| m_2 E'_1 E'_2}\left(\prod_l^{\substack{\text{extern} \\ \text{ferms}}} 2m_l\right)\left|\mathcal{M}\right|^2\left|\mathbf{p}'_1\right|^2\left(\frac{\partial E'_1}{\partial\left|\mathbf{p}'_1\right|} + \frac{\partial E'_2}{\partial\left|\mathbf{p}'_1\right|}\right)^{-1}} \left\{\begin{array}{l}\text{#2 stationary,} \\ \text{2 initial \& 2} \\ \text{final particles,} \\ \text{elastic or inelastic,}\end{array}\right. \tag{17-69}$$

As with (17-68), you do not need to memorize (17-69).

If either of the final particles is a photon then this relation simplifies since, for example, $\left|\mathbf{p}'_1\right| = \left|\mathbf{k}'_1\right| = E'_1 = \omega_1$. Alternatively, if either is a fermion, it also simplifies, since $\left|\mathbf{p}'_2\right| / E'_2 = \left|\mathbf{v}'_2\right|$.

17.4.4 Spin and Polarization Averages and Sums

As noted earlier (pg. 436), many experiments do not measure particle spins or photon polarizations. So, for reasons delineated in that earlier section, for such experiments, we need to work with the average of all possible incoming fermion spin (and/or photon polarization) states and sum over all possible outgoing fermion spin (photon polarization) states. Note that we do both averaging and summing at once for a given amplitude calculation and that process of doing both is typically called, for short, spin sums (polarization sums).

Many experiments don't measure spin or polarization

Note that in some experiments, spin (polarization) of external particles is actually measured, so for those we would not do spin (polarization) sums.

Spin sums = average over initial & sum over final spins

Spin Sums

Note that only the Feynman amplitude part $\mathcal{M}$ of the transition amplitude S_{fi} contains factors related to spin, so for our spin sums analysis, we need only examine $\mathcal{M}$.

For a typical external fermion example, consider Compton scattering $e^-\gamma \rightarrow e^-\gamma$. There are two contributing Feynman amplitudes (see pg. 225). In order to generalize later, and to keep notation simple, we represent all of the Feynman amplitude except the external fermions by the symbol Γ.

An example of spin sums: Compton scattering

$$\mathcal{M}_{C1}^{(2)} = \bar{u}_{s'}(\mathbf{p}')\underbrace{(-e^2)\varepsilon_{\mu,r'}(\mathbf{k}')\gamma^\mu iS_F(q=p+k)\varepsilon_{\nu,r}(\mathbf{k})\gamma^\nu}_{\Gamma_1}u_s(\mathbf{p}) = \bar{u}_{s'}(\mathbf{p}')\Gamma_1 u_s(\mathbf{p})$$

$$\mathcal{M}_{C2}^{(2)} = \bar{u}_{s'}(\mathbf{p}')\underbrace{(-e^2)\varepsilon_{\mu,r}(\mathbf{k})\gamma^\mu iS_F(q=p-k')\varepsilon_{\nu,r'}(\mathbf{k}')\gamma^\nu}_{\Gamma_2}u_s(\mathbf{p}) = \bar{u}_{s'}(\mathbf{p}')\Gamma_2 u_s(\mathbf{p}).$$
(17-70)

$$\mathcal{M} = \mathcal{M}_{C1}^{(2)} + \mathcal{M}_{C2}^{(2)} = \bar{u}_{s'}(\mathbf{p}')\underbrace{(\Gamma_1 + \Gamma_2)}_{\Gamma}u_s(\mathbf{p}) = \bar{u}_{s'}(\mathbf{p}')\Gamma u_s(\mathbf{p}).$$
(17-71)

The index s (= 1 or 2) indicates the spin state of the incoming electron. The index s' (= 1 or 2) indicates the spin state of the outgoing electron. Again, to keep things simple, we ignore the external photon polarization sums for now. (And as noted above, some experiments might be performed where photon polarizations are measured, but fermion spins are not.)

In our probability determination for specific spin states, we take the square of the absolute value of the amplitude (17-71). For indeterminate incoming particle spin state, we need to use the average of all possible incoming s states, since we will have many incoming particles of varying spin in any given experiment. For a given average incoming state, we will have general outgoing particle spin states, comprised of two basis states (s' =1,2), each having a particular probability. The total probability of measuring either one in a given measurement will be the sum of the two probabilities.

Since there are two distinct incoming states, we add their probabilities and divide by two to get the average, i.e., where we define X as shown,

probability of measuring final particle $\mathbf{p}'_1$ state of any spin $\propto \underbrace{\sum_{s'=1}^{2}}_{\substack{\text{sum over} \\ \text{outgoing}}} \underbrace{\left(\frac{1}{2}\sum_{s=1}^{2}|\mathcal{M}|^2\right)}_{\substack{\text{average over} \\ \text{incoming}}}.$ (17-72)

Using our newly defined Γ symbol from (17-71) in (17-72), we have

$$\frac{1}{2}\sum_{s'}\sum_{s}|\mathcal{M}|^2 = \sum_{s'=1}^{2}\left(\frac{1}{2}\sum_{s=1}^{2}\mathcal{M}\mathcal{M}^\dagger\right) = \frac{1}{2}\sum_{s'=1}^{2}\sum_{s=1}^{2}\left(\bar{u}_{s'}(\mathbf{p}')\Gamma u_s(\mathbf{p})\right)(\underbrace{\bar{u}_{s'}(\mathbf{p}')}_{u^\dagger_{s'}(\mathbf{p}')\gamma^0}\Gamma u_s(\mathbf{p}))^\dagger$$

$$= \frac{1}{2}\sum_{s'=1}^{2}\sum_{s=1}^{2}\left(\bar{u}_{s'}(\mathbf{p}')\Gamma u_s(\mathbf{p})\right)(u^\dagger_s(\mathbf{p})\underbrace{\gamma^0\gamma^0}_{=1}\Gamma^\dagger\underbrace{\gamma^{0\dagger}}_{=\gamma^0}u_{s'}(\mathbf{p}'))$$
(17-73)

$$= \frac{1}{2}\sum_{s'=1}^{2}\sum_{s=1}^{2}\left(\bar{u}_{s'}(\mathbf{p}')\Gamma u_s(\mathbf{p})\right)(\bar{u}_s(\mathbf{p})\underbrace{\gamma^0\Gamma^\dagger\gamma^0}_{\tilde{\Gamma}}u_{s'}(\mathbf{p}')),$$

where we define the symbol $\tilde{\Gamma}$ in the last row above. Writing out spinor indices in the last row of (17-73), we get the 1st row of (17-74). Taking the matrix elements as their equivalent (commutable) scalar elements, we can re-arrange the relation to serve our purposes, as in the 2nd row of (17-74).

$$\frac{1}{2}\sum_{s'}\sum_{s}|\mathcal{M}|^2 = \frac{1}{2}\sum_{s'=1}^{2}\sum_{s=1}^{2}\left(\overline{u}_{s',\alpha}\left(\mathbf{p}'\right)\Gamma_{\alpha\beta}u_{s,\beta}\left(\mathbf{p}\right)\right)\left(\overline{u}_{s,\gamma}\left(\mathbf{p}\right)\tilde{\Gamma}_{\gamma\delta}u_{s',\delta}\left(\mathbf{p}'\right)\right)$$

(17-74)

$$= \frac{1}{2}\sum_{s'=1}^{2}\sum_{s=1}^{2}\left(u_{s',\delta}\left(\mathbf{p}'\right)\overline{u}_{s',\alpha}\left(\mathbf{p}'\right)\right)\Gamma_{\alpha\beta}\left(u_{s,\beta}\left(\mathbf{p}\right)\overline{u}_{s,\gamma}\left(\mathbf{p}\right)\right)\tilde{\Gamma}_{\gamma\delta}.$$

From Appendix A of Chap. 4, under the heading "Spinor outer product relations", we have

$$u_s(\mathbf{p})\overline{u}_s(\mathbf{p}) = \frac{\not{p}+m}{2m} = \frac{\gamma^\mu p_\mu + m}{2m} = \frac{\gamma^\mu_{\alpha\beta}p_\mu + mI_{\alpha\beta}}{2m} = u_{s,\alpha}(\mathbf{p})\overline{u}_{s,\beta}(\mathbf{p}) \quad \text{sum on } s \quad (17\text{-}75)$$

$$v_s(\mathbf{p})\overline{v}_s(\mathbf{p}) = \frac{\not{p}-m}{2m} = \frac{\gamma^\mu p_\mu - m}{2m} = \frac{\gamma^\mu_{\alpha\beta}p_\mu - mI_{\alpha\beta}}{2m} = v_{s,\alpha}(\mathbf{p})\overline{v}_{s,\beta}(\mathbf{p}) \quad \text{sum on } s . \quad (17\text{-}76)$$

Using (17-75) in the bottom row of (17-74), we get

$$\frac{1}{2}\sum_{s'}\sum_{s}|\mathcal{M}|^2 = \frac{1}{2}\sum_{s'=1}^{2}\underbrace{\left(u_{s',\delta}\left(\mathbf{p}'\right)\overline{u}_{s',\alpha}\left(\mathbf{p}'\right)\right)}_{\left(\frac{\not{p}'+m}{2m}\right)_{\delta\alpha}}\Gamma_{\alpha\beta}\sum_{s=1}^{2}\underbrace{\left(u_{s,\beta}\left(\mathbf{p}\right)\overline{u}_{s,\gamma}\left(\mathbf{p}\right)\right)}_{\left(\frac{\not{p}+m}{2m}\right)_{\beta\gamma}}\tilde{\Gamma}_{\gamma\delta}$$

Spins sums for Compton: one initial and one final electron

.(17-77)

$$= \frac{1}{2}\underbrace{\left(\frac{\not{p}'+m}{2m}\right)_{\delta\alpha}\Gamma_{\alpha\beta}\left(\frac{\not{p}+m}{2m}\right)_{\beta\gamma}\tilde{\Gamma}_{\gamma\delta}}_{\text{sum on }\delta} = \frac{1}{2}\underbrace{\text{Tr}\left(\left(\frac{\not{p}'+m}{2m}\right)\Gamma\left(\frac{\not{p}+m}{2m}\right)\tilde{\Gamma}\right)}_{\text{trace in spinor space}}.$$

In similar fashion, for other types of interactions with external electrons and positrons, we would get the relations shown below in Wholeness Chart 17-5. Doing Prob. 4 can provide insight justifying these. In Sect. 17.5.1 we will show where the last row comes from.

Note that in general,

probability for spin indeterminate fermions, $\Big\}$ any number initial and final

$$\propto \underbrace{\sum_{s'_1=1}^{2}\sum_{s'_2=1}^{2}\cdots}_{\substack{\text{sum over out}\\\text{going fermions}}}\underbrace{\left(\frac{1}{2}\right)^n\sum_{s_1=1}^{2}\sum_{s_2=1}^{2}\cdots\sum_{s_n=1}^{2}|\mathcal{M}|^2}_{\substack{\text{average over } n\\\text{incoming fermions}}}. \quad (17\text{-}78)$$

General relations for spin sums for any number of initial and final fermions

Wholeness Chart 17-5. Fermion Spin Sum Relations

Interaction	$\mathcal{M}$	$\left(\frac{1}{2}\right)^n\sum_{s'}\cdots\sum_{s}\cdots|\mathcal{M}|^2$ where $\tilde{\Gamma}=\gamma^0\Gamma^\dagger\gamma^0$	General Relations n = number of incoming fermions
$e^-\gamma \to e^-\gamma$	$\overline{u}_{s'}(\mathbf{p}')\Gamma u_s(\mathbf{p})$	$\frac{1}{2}\text{Tr}\left(\frac{\not{p}'+m}{2m}\right)\Gamma\left(\frac{\not{p}+m}{2m}\right)\tilde{\Gamma}$	In $\mathcal{M} \to$ In $\left(\frac{1}{2}\right)^n\sum_{s'}\sum_{s}\cdots|\mathcal{M}|^2$
$e^+\gamma \to e^+\gamma$	$\overline{v}_s(\mathbf{p})\Gamma v_{s'}(\mathbf{p}')$	$\frac{1}{2}\text{Tr}\left(\frac{\not{p}-m}{2m}\right)\Gamma\left(\frac{\not{p}'-m}{2m}\right)\tilde{\Gamma}$	$\downarrow \qquad\qquad \downarrow$
$\gamma\gamma \to e^-e^+$	$\overline{u}_{s'_1}(\mathbf{p}'_1)\Gamma v_{s'_2}(\mathbf{p}'_2)$	$\text{Tr}\left(\frac{\not{p}_1'+m}{2m}\right)\Gamma\left(\frac{\not{p}_2'-m}{2m}\right)\tilde{\Gamma}$	$u_s(\mathbf{p}) \to \left(\frac{\not{p}+m}{2m}\right)$
$e^-e^+ \to \gamma\gamma$	$\overline{v}_{s_2}(\mathbf{p}_2)\Gamma u_{s_1}(\mathbf{p}_1)$	$\frac{1}{4}\text{Tr}\left(\frac{\not{p}_2-m}{2m}\right)\Gamma\left(\frac{\not{p}_1+m}{2m}\right)\tilde{\Gamma}$	$v_s(\mathbf{p}) \to \left(\frac{\not{p}-m}{2m}\right)$
$e^-e^+ \to e^-e^+$ annihilation part only $\to$	$\overline{u}_{s'_2}(\mathbf{p}'_2)\gamma_\alpha v_{s'_1}(\mathbf{p}'_1)\Gamma\times$ $\overline{v}_{s_1}(\mathbf{p}_1)\gamma^\alpha u_{s_2}(\mathbf{p}_2)$	$\frac{1}{4}\text{Tr}\left(\frac{\not{p}_2'+m}{2m}\gamma_\alpha\frac{\not{p}_1'-m}{2m}\gamma_\beta\right)\Gamma\times$ $\text{Tr}\left(\frac{\not{p}_1-m}{2m}\gamma^\alpha\frac{\not{p}_2+m}{2m}\gamma^\beta\right)\Gamma^*$	$\overline{u}_s(\mathbf{p}) \to \left(\frac{\not{p}+m}{2m}\right)$ $\overline{v}_s(\mathbf{p}) \to \left(\frac{\not{p}-m}{2m}\right)$

Polarization Sums

For photons, when polarization is not fixed and measured, we must carry out a similar procedure as we did for fermions, except we are now concerned with polarization instead of spin. In similar fashion, we will consider separating the Feynman amplitude into parts, the external photon factors like $\varepsilon_{\mu, r}(\mathbf{k})$ and $\varepsilon_{\nu, r'}(\mathbf{k'})$, and all the rest of the Feynman amplitude, for which, in Chap. 13, we used the symbol $\mathcal{M}^{\mu, \nu, \ldots}$, where the number of Greek superscripts equals the number of external photon factors in a particular amplitude. Consider, for example, Compton scattering of (17-70), where we switch dummy variables in the second line

Example of polarization sums: Compton scattering again

$$\mathcal{M}_{C1}^{(2)} = \varepsilon_{\mu, r'}(\mathbf{k'}) \varepsilon_{\nu, r}(\mathbf{k}) \underbrace{\left(-e^2\right) \overline{u}_{s'}(\mathbf{p'}) \gamma^{\mu} iS_F(p+k) \gamma^{\nu} u_s(\mathbf{p})}_{\mathcal{M}_1^{\mu\nu}} = \varepsilon_{\mu, r'}(\mathbf{k'}) \varepsilon_{\nu, r}(\mathbf{k}) \mathcal{M}_1^{\mu\nu}$$

$$\mathcal{M}_{C2}^{(2)} = \varepsilon_{\mu, r'}(\mathbf{k'}) \varepsilon_{\nu, r}(\mathbf{k}) \underbrace{\left(-e^2\right) \overline{u}_{s'}(\mathbf{p'}) \gamma^{\nu} iS_F(p-k') \gamma^{\mu} u_s(\mathbf{p})}_{\mathcal{M}_2^{\nu\mu}} = \varepsilon_{\mu, r'}(\mathbf{k'}) \varepsilon_{\nu, r}(\mathbf{k}) \mathcal{M}_2^{\nu\mu} \tag{17-79}$$

$$\mathcal{M} = \mathcal{M}_{C1}^{(2)} + \mathcal{M}_{C2}^{(2)} = \varepsilon_{\mu, r'}(\mathbf{k'}) \varepsilon_{\nu, r}(\mathbf{k}) \left(\mathcal{M}_1^{\mu\nu} + \mathcal{M}_2^{\nu\mu} \right) = \varepsilon_{\mu, r'}(\mathbf{k'}) \varepsilon_{\nu, r}(\mathbf{k}) \mathcal{M}^{\mu\nu}. \tag{17-80}$$

It can be easier keeping track of signs in what we are about to do, if we use the contravariant form for the polarization vectors, i.e., $\varepsilon^{\mu}_{r}(\mathbf{k})$. The Feynman amplitude is Lorentz invariant, so we can raise and lower indices of (17-80) to get

$$\mathcal{M} = \varepsilon^{\mu}_{r'}(\mathbf{k'}) \varepsilon^{\nu}_{r}(\mathbf{k}) \mathcal{M}_{\mu\nu}. \tag{17-81}$$

Parallel with (17-72), we have,

probability for two photons, one initial, one final, with unmeasured polarization $\left. \right\} \propto \underbrace{\sum_{r'=1}^{2}}_{\substack{\text{sum over} \\ \text{outgoing } \gamma}} \left(\underbrace{\left(\frac{1}{2}\right) \sum_{r=1}^{2} |\mathcal{M}|^2}_{\substack{\text{average over} \\ \text{incoming } \gamma}} \right). \tag{17-82}$

Average over initial photon polarizations, sum over final

More generally, we have

probability for photons with unmeasured polarization, any number initial and final $\left. \right\} \propto \underbrace{\sum_{r'_1=1}^{2} \sum_{r'_2=1}^{2} \cdots}_{\substack{\text{sum over} \\ \text{outgoing } \gamma}} \left(\underbrace{\left(\frac{1}{2}\right)^m \sum_{r_1=1}^{2} \sum_{r_2=1}^{2} \cdots \sum_{r_m=1}^{2} |\mathcal{M}|^2}_{\substack{\text{average over} \\ m \text{ incoming } \gamma\text{s}}} \right). \tag{17-83}$

General relations for polarization sums for any number of initial and final photons

We will need to recall some things. First, from Chap. 13, pg. 328, the Ward identities, which tell us that if we replace a polarization vector factor in the amplitude with its associated 4-momentum, the result equals zero.

Ward identities $\qquad \varepsilon^{\mu}_{r}\left(\mathbf{k}_j\right) \mathcal{M}_{\mu}\left(\mathbf{k}_1, \ldots, \mathbf{k}_j, \ldots\right) \rightarrow k^{\mu}_j \mathcal{M}_{\mu}\left(\mathbf{k}_1, \ldots, \mathbf{k}_j, \ldots\right) = 0. \tag{17-84}$

Second, from Chap. 5, pgs. 142-143, we will express the photon polarization vectors in the special photon aligned (more formally, the "photon-polarization vectors-axes aligned") coordinate system of Fig. 5-1(d). Since our amplitude is Lorentz invariant, our final result will be good for any coordinate system, but we want to do our analyses in the easiest one. For the chosen coordinates,

Express polarization vectors in easiest to analyze frame

$$\varepsilon^{\mu}_0 = (1,0,0,0) \qquad \underbrace{\varepsilon^{\mu}_1 = (0,1,0,0) \qquad \varepsilon^{\mu}_2 = (0,0,1,0)}_{\substack{\text{transverse components} \\ \text{of real (external) } \gamma\text{s}}} \qquad \varepsilon^{\mu}_3 = (0,0,0,1) \qquad k^{\mu} = |\mathbf{k}|(1,0,0,1) \tag{17-85}$$

Keep things simple at first by considering only one external photon

Now, to keep things simple, consider only one photon factor forming an inner product with the remainder of the amplitude,

$$\mathcal{M} = \varepsilon^{\mu}_{r}(\mathbf{k}) \mathcal{M}_{\mu}. \tag{17-86}$$

Now, put (17-86) into the r summation part of (17-82) i.e.,

$$\sum_{r=1}^{2} |\mathcal{M}|^2 = \mathcal{M}_{\mu} \mathcal{M}_{\nu}^* \sum_{r=1}^{2} \varepsilon^{\mu}_{r}(\mathbf{k}) \varepsilon^{\nu}_{r}(\mathbf{k}), \tag{17-87}$$

$\Sigma |\mathcal{M}|^2$ *in terms of sums on polarization vectors*

where from (17-85), we can express the polarization vectors part of (17-87) as (where we carry the matrix form along to make things easier to understand).

$$\sum_{r=1}^{2}\varepsilon_r^{\mu}(\mathbf{k})\,\varepsilon_r^{\nu}(\mathbf{k})=\begin{bmatrix}0\\1\\0\\0\end{bmatrix}\begin{bmatrix}0&1&0&0\end{bmatrix}+\begin{bmatrix}0\\0\\1\\0\end{bmatrix}\begin{bmatrix}0&0&1&0\end{bmatrix}=\begin{bmatrix}0&0&0&0\\0&1&0&0\\0&0&1&0\\0&0&0&0\end{bmatrix}$$

Evaluating those polarization vector sums

$$=\underbrace{\begin{bmatrix}-1&0&0&0\\0&1&0&0\\0&0&1&0\\0&0&0&1\end{bmatrix}}_{-g^{\mu\nu}}+\underbrace{\begin{bmatrix}-1&0&0&-1\\0&0&0&0\\0&0&0&0\\-1&0&0&-1\end{bmatrix}}_{A}+\underbrace{\begin{bmatrix}1&0&0&0\\0&0&0&0\\0&0&0&0\\1&0&0&0\end{bmatrix}}_{B}+\underbrace{\begin{bmatrix}1&0&0&1\\0&0&0&0\\0&0&0&0\\0&0&0&0\end{bmatrix}}_{C}. \qquad (17\text{-}88)$$

We can re-express A as

$$A=-\begin{bmatrix}1&0&0&1\\0&0&0&0\\0&0&0&0\\1&0&0&1\end{bmatrix}=-\begin{bmatrix}1\\0\\0\\1\end{bmatrix}\begin{bmatrix}1&0&0&1\end{bmatrix}=-\frac{k^{\mu}k^{\nu}}{|\mathbf{k}|^{2}}, \qquad (17\text{-}89)$$

and B and C as

$$B=\begin{bmatrix}1\\0\\0\\1\end{bmatrix}\begin{bmatrix}1&0&0&0\end{bmatrix}=\frac{k^{\mu}\varepsilon_0^{\nu}}{|\mathbf{k}|}\qquad C=\begin{bmatrix}1\\0\\0\\0\end{bmatrix}\begin{bmatrix}1&0&0&1\end{bmatrix}=\frac{\varepsilon_0^{\mu}k^{\nu}}{|\mathbf{k}|}, \qquad (17\text{-}90)$$

which makes (17-88)

$$\sum_{r=1}^{2}\varepsilon_r^{\mu}(\mathbf{k})\,\varepsilon_r^{\nu}(\mathbf{k})=-g^{\mu\nu}-\frac{k^{\mu}k^{\nu}}{|\mathbf{k}|^{2}}+\frac{k^{\mu}\varepsilon_0^{\nu}}{|\mathbf{k}|}+\frac{\varepsilon_0^{\mu}k^{\nu}}{|\mathbf{k}|}. \qquad (17\text{-}91)$$

Using (17-91) in (17-87), we get

$$\sum_{r=1}^{2}|\mathcal{M}|^{2}=\mathcal{M}_{\mu}\mathcal{M}_{\nu}^{*}\left(-g^{\mu\nu}-\frac{k^{\mu}k^{\nu}}{|\mathbf{k}|^{2}}+\frac{k^{\mu}\varepsilon_0^{\nu}}{|\mathbf{k}|}+\frac{\varepsilon_0^{\mu}k^{\nu}}{|\mathbf{k}|}\right). \qquad (17\text{-}92)$$

Ward identities simplify $\Sigma|\mathcal{M}|^2$

But from the Ward identities (17-84), the last three terms in (17-92) must equal zero. Therefore,

$$\sum_{r=1}^{2}|\mathcal{M}|^{2}=-\mathcal{M}_{\mu}\mathcal{M}_{\nu}^{*}g^{\mu\nu}=-\mathcal{M}^{\nu}\mathcal{M}_{\nu}^{*}. \qquad (17\text{-}93)$$

This result, for a single external photon, is readily extrapolated to multiple external photons as

Generalize single external photon $\Sigma|\mathcal{M}|^2$ results to any number

$$\sum_{r_1'=1}^{2}\sum_{r_2'=1}^{2}\cdots\sum_{r_1=1}^{2}\sum_{r_2=1}^{2}\cdots|\mathcal{M}|^{2}=(-1)^{n_{\gamma}}\,\mathcal{M}^{\mu\nu\rho\ldots}\mathcal{M}_{\mu\nu\rho\ldots}^{*}\quad n_{\gamma}=\text{total number of external photons}, \quad (17\text{-}94)$$

and (17-82) becomes

probability for two photons, one initial, one final, $\left.\right\}$ $\propto\dfrac{1}{2}\displaystyle\sum_{r'=1}^{2}\sum_{r=1}^{2}|\mathcal{M}|^{2}=\dfrac{1}{2}\mathcal{M}^{\mu\nu}\mathcal{M}_{\mu\nu}^{*}.$ (17-95)
with unmeasured polarization

Polarization sums for one initial and one final external photon

The more general relation (17-83) becomes

probability for n_{γ} external photons, unmeasured polarization $\left.\right\}$ $\propto\underbrace{\displaystyle\sum_{r_1'=1}^{2}\sum_{r_2'=1}^{2}\cdots}_{\substack{\text{sum over}\\\text{outgoing }\gamma\text{s}}}\underbrace{\left(\frac{1}{2}\right)^{m}\displaystyle\sum_{r_1=1}^{2}\sum_{r_2=1}^{2}\cdots\sum_{r_m=1}^{2}|\mathcal{M}|^{2}}_{\substack{\text{average over}\\m\text{ incoming }\gamma\text{s}}}=(-1)^{n_{\gamma}}\left(\frac{1}{2}\right)^{m}\mathcal{M}^{\mu\nu\ldots}\mathcal{M}_{\mu\nu\ldots}^{*}.$ (17-96)

Generalize polarization sums to any number of external photons

17.5 Scattering in QFT: Some Examples

17.5.1 Electron-Positron Interactions Producing Muons and Taus

One might think we would want to look at Bhabha scattering $e^+e^- \to e^+e^-$, which is elastic, first, but it turns out that the interactions $e^+e^- \to \mu^+\mu^-$ and $e^+e^- \to \tau^+\tau^-$, which are inelastic, are actually easier to evaluate. That is because there are two possible ways for Bhabha scattering to occur, one where the electron and positron annihilate to produce a virtual photon, and one where they exchange a virtual photon, but do not annihilate one another. So, we would need to add two separate Feynman amplitudes. On the other hand, the muon and tau production interactions can each only occur with a single (2nd order) Feynman diagram/amplitude, in which the electron and positron annihilate to produce a virtual photon, which in turn produces a heavier lepton and its antiparticle.

Spin sums applied to $e^+e^- \to l^+ l^-$ scattering, where $e \ne l$

$$e^+\left(\mathbf{p}_1, s_1\right) + e^-\left(\mathbf{p}_2, s_2\right) \to l^+\left(\mathbf{p}_1', s_1'\right) + l^-\left(\mathbf{p}_2', s_2'\right) \tag{17-97}$$

At this point, you should be able to readily draw the relevant Feynman diagram for this interaction, and with Feynman rules determine the Feynman amplitude, where the label $l = \mu$ or τ.

$$\mathcal{M} = \mathcal{M}_{e^-e^+\to l^-l^+} = -e^2 \left(\bar{u}_{s_2'}(\mathbf{p}_2')\gamma^\alpha v_{s_1'}(\mathbf{p}_1')\right)_{(l)} iD_{F\alpha\beta}\left(k = p_1 + p_2\right)\left(\bar{v}_{s_1}(\mathbf{p}_1)\gamma^\beta u_{s_2}(\mathbf{p}_2)\right)_{(e)}$$

$$= \Big(\underbrace{\bar{u}_{s_2'}(\mathbf{p}_2')}_{u^\dagger_{s_2'}(\mathbf{p}_2')\gamma^0}\gamma_\alpha v_{s_1'}(\mathbf{p}_1')\Big)_{(l)}\underbrace{\frac{ie^2}{(p_1+p_2)^2}}_{\Gamma}\Big(\underbrace{\bar{v}_{s_1}(\mathbf{p}_1)}_{v^\dagger_{s_1}(\mathbf{p}_1)\gamma^0}\gamma^\alpha u_{s_2}(\mathbf{p}_2)\Big)_{(e)}. \tag{17-98}$$

The relevant amplitude $\mathcal{M}$

We have dropped the $i\varepsilon$ in the denominator of the propagator because this term is only significant when $(p_1 + p_2)^2 = 0$, but in the present case, we will always have $(p_1 + p_2)^2 \ge 4m_e^2$

Spin Sums

We will now derive a more general case than the last row of Wholeness Chart 17-5, where the final particles can be any lepton pair, electron/positron, muon/anti-muon, or tau/anti-tau. So when we are all done, the label l can stand for electron/positron pair as well as the others.

Carrying out spin sums for this case

For the unpolarized (this word is used in a general context to include both fermion spin and photon polarization) cross-section, we require (17-78) for two incoming and two outgoing fermions,

$$\frac{1}{4}\sum_{s_1'=1}^{2}\sum_{s_2'=1}^{2}\sum_{s_1=1}^{2}\sum_{s_2=1}^{2}|\mathcal{M}|^2 . \tag{17-99}$$

Using $\gamma^{0\dagger} = \gamma^0$, the hermiticity conditions from Chap. 4, Appendix A, $\gamma^\alpha = \gamma^0\gamma^{\alpha\dagger}\gamma^0$, their covariant equivalent $\gamma_\alpha = \gamma^0\gamma_\alpha^\dagger\gamma^0$, and (17-98) with dummy index $\alpha \to \beta$, we obtain

$$\mathcal{M}^* = (v^\dagger_{s_1'}(\mathbf{p}_1')\underbrace{\gamma^0\gamma^0}_{=1}\overbrace{\gamma^0\gamma_\beta^\dagger\gamma^{0\dagger}}^{\gamma^0\gamma_\beta}u_{s_2'}(\mathbf{p}_2'))_{(l)}\frac{-ie^2}{(p_1+p_2)^2}(u^\dagger_{s_2}(\mathbf{p}_2)\overbrace{\gamma^0\gamma^0\gamma^{\beta\dagger}\gamma^{0\dagger}}^{\gamma^0\gamma^\beta}v_{s_1}(\mathbf{p}_1))_{(e)}$$

Find $\mathcal{M}^*$

$$\tag{17-100}$$

$$= (\bar{v}_{s_1'}(\mathbf{p}_1')\gamma_\beta u_{s_2'}(\mathbf{p}_2'))_{(l)}\underbrace{\frac{-ie^2}{(p_1+p_2)^2}}_{\Gamma^*}(\bar{u}_{s_2}(\mathbf{p}_2)\gamma^\beta v_{s_1}(\mathbf{p}_1))_{(e)}.$$

We recognize Γ and Γ^* are scalars, and so are the quantities inside parentheses with subscripts (e) and (l), so they can be placed anywhere in the expression. Thus, from (17-98) and (17-100), (17-99) becomes,

Carry out the algebra for $(1/4)\Sigma|\mathcal{M}|^2$

$$\frac{1}{4}\sum_{s_1'=1}^{2}\sum_{s_2'=1}^{2}\sum_{s_1=1}^{2}\sum_{s_2=1}^{2}\mathcal{M}\mathcal{M}^* = \frac{1}{4}|\Gamma|^2\underbrace{\sum_{s_1'=1}^{2}\sum_{s_2'=1}^{2}\left(\bar{u}_{s_2'}(\mathbf{p}_2')\gamma_\alpha v_{s_1'}(\mathbf{p}_1')\right)_{(l)}\left(\bar{v}_{s_1'}(\mathbf{p}_1')\gamma_\beta u_{s_2'}(\mathbf{p}_2')\right)_{(l)}}_{A_{(l)\alpha\beta}} \times$$

$$\underbrace{\sum_{s_1=1}^{2}\sum_{s_2=1}^{2}\left(\bar{v}_{s_1}(\mathbf{p}_1)\gamma^\alpha u_{s_2}(\mathbf{p}_2)\right)_{(e)}\left(\bar{u}_{s_2}(\mathbf{p}_2)\gamma^\beta v_{s_1}(\mathbf{p}_1)\right)_{(e)}}_{B_{(e)}^{\alpha\beta}}. \tag{17-101}$$

Re-expressing the spinor space factor on the first line above with spinor indices, and using (17-75) and (17-76), we have

$$A_{(l)\alpha\beta} = \sum_{s_1'=1}^{2}\sum_{s_2'=1}^{2}\left(\bar{u}_{s_2'\,\delta}(\mathbf{p}_2')(\gamma_\alpha)_{\delta\eta}v_{s_1'\,\eta}(\mathbf{p}_1')\right)_{(l)}\left(\bar{v}_{s_1'\,\rho}(\mathbf{p}_1')(\gamma_\beta)_{\rho\sigma}u_{s_2'\,\sigma}(\mathbf{p}_2')\right)_{(l)}$$

$$= \underbrace{\left(\sum_{s_2'=1}^{2}\left(u_{s_2'\,\sigma}(\mathbf{p}_2')\bar{u}_{s_2'\,\delta}(\mathbf{p}_2')\right)_{(l)}\right)}_{\left((\not{p}_2'+m_l)/2m_l\right)_{\sigma\delta}}(\gamma_\alpha)_{\delta\eta}\underbrace{\left(\sum_{s_1'=1}^{2}\left(v_{s_1'\,\eta}(\mathbf{p}_1')\bar{v}_{s_1'\,\rho}(\mathbf{p}_1')\right)_{(l)}\right)}_{\left((\not{p}_1'-m_l)/2m_l\right)_{\eta\rho}}(\gamma_\beta)_{\rho\sigma}, \quad (17\text{-}102)$$

i.e.,

$$A_{(l)\alpha\beta} = \text{Tr}\left\{\frac{\not{p}_2'+m_l}{2m_l}\gamma_\alpha\frac{\not{p}_1'-m_l}{2m_l}\gamma_\beta\right\}. \quad (17\text{-}103)$$

By doing Prob. 5, you can show that, in similar fashion,

$$B_{(e)}^{\alpha\beta} = \text{Tr}\left\{\frac{\not{p}_1-m_e}{2m_e}\gamma^\alpha\frac{\not{p}_2+m_e}{2m_e}\gamma^\beta\right\}. \quad (17\text{-}104)$$

Again referring to Chap. 4, Appendix A, we can evaluate the gamma matrix traces in (17-103) and (17-104). First, we note that traces of an odd number of gamma matrices equal zero so all factors of 4-momentum with mass drop out. This gives us

$$A_{(l)\alpha\beta} = \frac{1}{4m_l^2}\left(\text{Tr}\{\not{p}_2'\,\gamma_\alpha\not{p}_1'\,\gamma_\beta\} - m_l^2\text{Tr}\{\gamma_\alpha\gamma_\beta\}\right)$$

Using gamma matrix algebra to re-express parts of $(1/4)\,\Sigma|\mathcal{M}|^2$

$$= \frac{1}{4m_l^2}(p_2'^{\,\eta}p_1'^{\,\rho}\underbrace{\text{Tr}\{\gamma_\eta\gamma_\alpha\gamma_\rho\gamma_\beta\}}_{4\left(g_{\eta\alpha}g_{\rho\beta}-g_{\eta\rho}g_{\alpha\beta}+g_{\eta\beta}g_{\alpha\rho}\right)} - m_l^2\underbrace{\text{Tr}\{\gamma_\alpha\gamma_\beta\}}_{4g_{\alpha\beta}})$$

$$= \frac{1}{m_l^2}(p_2'^{\,}_\alpha p_1'^{\,}_\beta - \underbrace{p_2'^{\,}_\rho p_1'^{\,\rho}}_{p_2'p_1'}g_{\alpha\beta} + p_2'^{\,}_\beta p_1'^{\,}_\alpha - m_l^2 g_{\alpha\beta}) \quad (17\text{-}105)$$

$$= \frac{1}{m_l^2}\left(p_1'^{\,}_\alpha p_2'^{\,}_\beta + p_2'^{\,}_\alpha p_1'^{\,}_\beta - \left(m_l^2 + p_2'p_1'\right)g_{\alpha\beta}\right).$$

Similarly, by doing Prob. 6, you can show (17-104) reduces to

$$B_{(e)}^{\alpha\beta} = \frac{1}{m_e^2}\left(p_1^\alpha p_2^\beta + p_2^\alpha p_1^\beta - \left(m_e^2 + p_1p_2\right)g^{\alpha\beta}\right). \quad (17\text{-}106)$$

Putting (17-105) and (17-106) into (17-101) gives us

$$\frac{1}{4}\sum_{s_1'=1}^{2}\sum_{s_2'=1}^{2}\sum_{s_1=1}^{2}\sum_{s_2=1}^{2}|\mathcal{M}|^2 = \frac{1}{4}|\Gamma|^2\frac{1}{m_l^2}\left(p_1'^{\,}_\alpha p_2'^{\,}_\beta + p_2'^{\,}_\alpha p_1'^{\,}_\beta - \left(m_l^2 + p_1'p_2'\right)g_{\alpha\beta}\right)\times$$

More algebra

$$\frac{1}{m_e^2}\left(p_1^\alpha p_2^\beta + p_2^\alpha p_1^\beta - \left(m_e^2 + p_1p_2\right)g^{\alpha\beta}\right)$$

$$= \frac{1}{4}|\Gamma|^2\frac{1}{m_l^2 m_e^2}\big(\underbrace{p_1'^{\,}_\alpha p_2'^{\,}_\beta p_1^\alpha p_2^\beta}_{(p_1p_1')(p_2p_2')} + \underbrace{p_1'^{\,}_\alpha p_2'^{\,}_\beta p_2^\alpha p_1^\beta}_{(p_1p_2')(p_2p_1')} - \underbrace{p_1'^{\,}_\alpha p_2'^{\,}_\beta g^{\alpha\beta}}_{p_1'p_2'}\left(m_e^2 + p_1p_2\right)$$

$$+ \underbrace{p_2'^{\,}_\alpha p_1'^{\,}_\beta p_1^\alpha p_2^\beta}_{(p_1p_2')(p_2p_1')} + \underbrace{p_2'^{\,}_\alpha p_1'^{\,}_\beta p_2^\alpha p_1^\beta}_{(p_1p_1')(p_2p_2')} - \underbrace{p_2'^{\,}_\alpha p_1'^{\,}_\beta g^{\alpha\beta}}_{p_1'p_2'}\left(m_e^2 + p_1p_2\right) - \left(m_l^2 + p_1'p_2'\right)\underbrace{g_{\alpha\beta}p_1^\alpha p_2^\beta}_{p_1p_2} \quad (17\text{-}107)$$

$$- \left(m_l^2 + p_1'p_2'\right)\underbrace{g_{\alpha\beta}p_2^\alpha p_1^\beta}_{p_1p_2} + \left(m_l^2 + p_1'p_2'\right)\underbrace{g_{\alpha\beta}g^{\alpha\beta}}_{4}\left(m_e^2 + p_1p_2\right)\big).$$

Combining terms gives us

$$= \frac{1}{4}|\Gamma|^2\frac{1}{m_l^2 m_e^2}\big(2(p_1p_1')(p_2p_2') + 2(p_1p_2')(p_2p_1') - 2p_1'p_2'm_e^2 - 2p_1'p_2'p_1p_2$$

$$- 2p_1p_2m_l^2 - 2p_1p_2p_1'p_2' + 4m_l^2m_e^2 + 4p_1p_2m_l^2 + 4p_1'p_2'm_e^2 + 4p_1'p_2'p_1p_2\big). \quad (17\text{-}108)$$

Eliminating all the terms that cancel, we end up with our final spin relation

$$\frac{1}{4}\sum_{s'_1=1}^{2}\sum_{s'_2=1}^{2}\sum_{s_1=1}^{2}\sum_{s_2=1}^{2}|\mathcal{M}|^2$$

$$=|\Gamma|^2\frac{1}{2m_l^2 m_e^2}\Big((p_1 p'_1)(p_2 p'_2)+(p_1 p'_2)(p_2 p'_1)+(p'_1 p'_2)m_e^2+(p_1 p_2)m_l^2+2m_l^2 m_e^2\Big).$$

(17-109)

Our final spin sums relation for any frame

COM Frame

So far our results have been good for any frame. Since $e^+ e^-$ scattering experiments are almost invariably done in the COM frame, we will now evaluate the differential cross-section in that frame.

Fig. 17-16 represents a summary of the various parameters involved. Note that in the COM frame the incoming particle 3-momenta are of equal magnitude $|\mathbf{p}|$, and opposite direction. Since they are antiparticles of one another, they have the same mass, and via $E=\sqrt{m_e^2+|\mathbf{p}|^2}$, they therefore have the same energy E. Since total 3-momentum is conserved, outgoing particles 3-momenta must sum to zero, as the incoming 3-momenta summed to zero. Thus, the outgoing leptons 3-momenta are of equal magnitude and opposite direction. Since energy is conserved, and there was $2E$ total incoming energy, there must be $2E$ outgoing energy. Because both outgoing particles have the same mass and same $|\mathbf{p}'|$, they must each have the same energy, i.e., half of $2E$. Note that $|\mathbf{p}|\neq|\mathbf{p}'|$, since $m_e\neq m_l$. Note also that $|\mathbf{p}'|$ is a known quantity, since $|\mathbf{p}'|=\sqrt{E^2-m_l^2}$. And thus, $|\mathbf{p}'|$ is independent of θ.

Now apply above results in COM frame

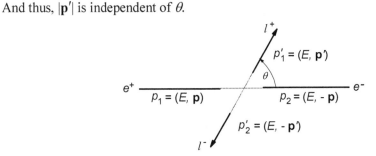

Figure 17-16. Kinematics and Dynamics for the Process $e^+ e^- \rightarrow l^+ l^-$ **in COM Frame**

Using Fig. 17-16, we can find the expressions for the quantities in (17-109).

$$p_1 p'_1 = p_2 p'_2 = E^2-\mathbf{p}\cdot\mathbf{p}' = E^2-|\mathbf{p}||\mathbf{p}'|\cos\theta \qquad p_1 p'_2 = p_2 p'_1 = E^2+|\mathbf{p}||\mathbf{p}'|\cos\theta$$

$$p_1 p_2 = E^2+|\mathbf{p}|^2 \qquad p'_1 p'_2 = E^2+|\mathbf{p}'|^2 \qquad (p_1+p_2)^2 = 4E^2$$

(17-110)

Key 4-momenta relations in COM frame

In addition, since $E\geq m_\mu\approx 207 m_e$, we can approximate $|\mathbf{p}|\approx E$ to high accuracy. For the same reason, we can ignore terms proportional to m_e^2 inside the brackets of (17-109), i.e., the 3rd and 5th terms. With these approximations, we substitute (17-110) and Γ of (17-98) into (17-109).

$E\geq m_\mu$ allows us to simplify since $|\mathbf{p}|\approx E$

$$\frac{1}{4}\sum_{s'_1=1}^{2}\sum_{s'_2=1}^{2}\sum_{s_1=1}^{2}\sum_{s_2=1}^{2}|\mathcal{M}|^2\approx\overbrace{\frac{e^4}{\big((p_1+p_2)^2\big)^2}}^{|\Gamma|^2}\frac{1}{2m_l^2 m_e^2}\Big((p_1 p'_1)(p_2 p'_2)+(p_1 p'_2)(p_2 p'_1)+(p_1 p_2)m_l^2\Big)$$

$$=\frac{e^4}{2m_l^2 m_e^2}\frac{1}{\big(4E^2\big)^2}\left(\Big(E^2-\overbrace{|\mathbf{p}|}^{E}|\mathbf{p}'|\cos\theta\Big)^2+\Big(E^2+\overbrace{|\mathbf{p}|}^{E}|\mathbf{p}'|\cos\theta\Big)^2+\Big(E^2+\overbrace{|\mathbf{p}|^2}^{E^2}\Big)m_l^2\right)$$

(17-111)

Using 4-momenta relations and $|\mathbf{p}|\approx E$ in $(1/4)\Sigma|\mathcal{M}|^2$ in COM

$$=\frac{e^4}{32m_l^2 m_e^2 E^2}\Big(\big(E-|\mathbf{p}'|\cos\theta\big)^2+\big(E+|\mathbf{p}'|\cos\theta\big)^2+2m_l^2\Big)$$

$$=\frac{e^4}{32m_l^2 m_e^2 E^2}\Big(2E^2+2|\mathbf{p}'|^2\cos^2\theta+2m_l^2\Big)=\frac{e^4}{16m_l^2 m_e^2 E^2}\Big(E^2+m_l^2+|\mathbf{p}'|^2\cos^2\theta\Big).$$

We now need to use (17-111) in (17-68), where we use $|\mathbf{p}_1| = |\mathbf{p}| \approx E$, $E_1 = E_2 = E$, and $\mathbf{p}'_1 = \mathbf{p}'$.

$$\left(\frac{d\sigma}{d\Omega'_1}\right)_{\substack{COM \\ e^-e^+ \to l^-l^+}} = \frac{1}{64\pi^2 (E_1+E_2)^2} \frac{|\mathbf{p}'|}{|\mathbf{p}|} \left(\prod_l^{\text{ext ferms}} 2m_l\right)\left(\frac{1}{4}\sum_{s'_1=1}^2\sum_{s'_2=1}^2\sum_{s_1=1}^2\sum_{s_2=1}^2 |\mathcal{M}|^2\right)$$

$$= \frac{e^4}{64\pi^2 4E^2}\frac{|\mathbf{p}'|}{E} 4m_l^2 4m_e^2 \frac{1}{16m_l^2 m_e^2 E^2}\left(E^2+m_l^2+|\mathbf{p}'|^2 \cos^2\theta\right). \qquad (17\text{-}112)$$

$$= \left(\frac{e^2}{4\pi}\right)^2 \frac{|\mathbf{p}'|}{E}\frac{1}{16E^4}\left(E^2+m_l^2+|\mathbf{p}'|^2 \cos^2\theta\right).$$

Or finally, the differential cross-section,

$$\left(\frac{d\sigma}{d\Omega'_1}\right)_{\substack{COM \\ e^-e^+ \to l^-l^+}} = \frac{\alpha^2}{16E^4}\frac{|\mathbf{p}'|}{E}\left(E^2+m_l^2+|\mathbf{p}'|^2 \cos^2\theta\right) \begin{cases} \text{unknown spins, } l\neq e, \\ |\mathbf{p}'|=\sqrt{E^2-m_l^2} \end{cases}. \qquad (17\text{-}113)$$

Differential cross-section for unmeasured spins

The total cross-section, from (17-7), is

$$\sigma_{\substack{COM \\ e^-e^+ \to l^-l^+}} = \int_0^{2\pi}\int_0^\pi \left(\frac{d\sigma}{d\Omega'_1}\right)_{\substack{COM \\ e^-e^+ \to l^-l^+}} d\Omega$$

$$= \int_0^{2\pi}\int_0^\pi \left(\frac{\alpha^2}{16E^4}\frac{|\mathbf{p}'|}{E}\left(E^2+m_l^2+|\mathbf{p}'|^2 \cos^2\theta\right)\right)\underbrace{\sin\theta\, d\theta}_{-d(\cos\theta)=-du}\, d\phi \qquad (17\text{-}114)$$

$$= -\int_0^{2\pi}\int_1^{-1}\left(\frac{\alpha^2}{16E^4}\frac{|\mathbf{p}'|}{E}\left(E^2+m_l^2+|\mathbf{p}'|^2 u^2\right)\right) du\, d\phi.$$

Or finally,

$$\sigma_{\substack{COM \\ e^-e^+ \to l^-l^+}} = \frac{\pi\alpha^2}{4E^4}\frac{|\mathbf{p}'|}{E}\left(E^2+m_l^2+\tfrac{1}{3}|\mathbf{p}'|^2\right) \begin{cases} \text{unknown spins, } l\neq e, \\ |\mathbf{p}'|=\sqrt{E^2-m_l^2} \end{cases}. \qquad (17\text{-}115)$$

Total cross-section for unmeasured spins

In the highly relativistic limit, $E \gg m_l$, so we can ignore terms with m_l and take $|\mathbf{p}'| \approx E$, and get

$$\left(\frac{d\sigma}{d\Omega'_1}\right)_{\substack{COM \\ e^-e^+ \to l^-l^+}} = \frac{\alpha^2}{16E^2}\left(1+\cos^2\theta\right) \qquad \sigma_{\substack{COM \\ e^-e^+ \to l^-l^+}} = \frac{\pi\alpha^2}{3E^2} \begin{cases} \text{unknown spins} \\ E \gg m_l,\ l\neq e \end{cases}. \qquad (17\text{-}116)$$

High speed limit differential cross-section, unknown spins

These expressions are often written in terms of the total energy $E_{tot} = 2E$ of the interaction. So, for instance, (17-116) would be

$$\left(\frac{d\sigma}{d\Omega'_1}\right)_{\substack{COM \\ e^-e^+ \to l^-l^+}} = \frac{\alpha^2}{4E_{tot}^2}\left(1+\cos^2\theta\right) \qquad \sigma_{\substack{COM \\ e^-e^+ \to l^-l^+}} = \frac{4\pi\alpha^2}{3E_{tot}^2} \begin{cases} \text{unknown spins} \\ E_{tot} \gg 2m_l,\ l\neq e \end{cases}. \qquad (17\text{-}117)$$

High speed limit total cross-section, unknown spins

Note that the differential cross-section and cross-sections for this interaction would be zero for $E_{tot} < 2m_l$, because there would not be enough energy to supply the mass needed to produce the leptons. The above relations are thus approximations (we made approximations all along the way) that are good for energies somewhat above the masses of the outgoing leptons.

Cross-section = 0 for $E_{tot} < 2m_l$

We have reached a major milestone in your evolution as a field theorist. We have now, for the first time, calculated the cross-section for a real-world scattering experiment.

Comparison with Experiment

Many experiments have been done to verify the above relations at many energy levels. The results for relativistic energy track the theoretical curves for (17-117) shown below in Fig. 17-17. Note from the LHS that scattering is symmetric in the forward ($\theta < 90°$) and backward ($\theta > 90°$) directions.

Agrees well with experiment

Note also from both the LHS and the RHS, that interpreting cross-section as the probability of two particles inside unit area scattering, two such higher energy initial particles are less likely to interact than those with lower energy. One can interpret this as follows: higher velocity particles spend less time near one another as they pass, and so are less likely to interact. The longer they are in each other's vicinity, the greater the probability they would exchange a virtual photon. This differs from hard object collisions where only the cross-sectional areas of the particles matter, and the scattering cross-section is the same for any particle speeds.

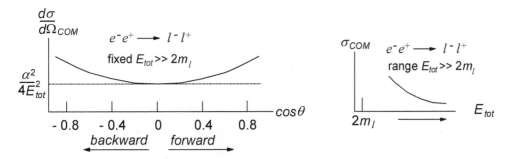

Fig. 17-17. Plots of Relations (17-117), $e \ne l$

17.5.2 Bhabha Scattering

We will now determine the (differential) cross-section for $e^+ e^- \to e^+ e^-$. This interaction is more complicated than those of the prior section, since it encompasses two sub amplitudes, represented by the Feynman diagrams of Fig. 17-18. That is, there is the annihilation diagram similar to that for $e^+ e^- \to l^+ l^-$ ($e \ne l$), and also the non-annihilation diagram as well.

Another example of spin sums: Bhabha scattering

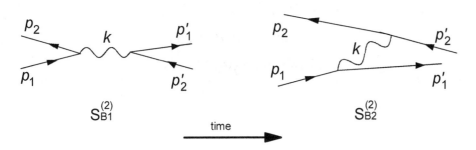

More complicated because of extra sub amplitude (RHS in figure)

Figure 17-18. Two Feynman Diagrams for Bhabha Scattering

From Feynman's rules, pg. 236, (or by looking back at Bhabha scattering analysis in Chap. 8), the two sub-amplitudes for Bhabha scattering, where we assume that e and m_e are renormalized, and that gamma matrix and propagator modifications from higher order terms are negligible, are

$$\mathcal{M}_{B1} = +ie^2 \bar{u}_{s_1'}(\mathbf{p}_1')\gamma_\mu v_{s_2'}(\mathbf{p}_2')\frac{1}{(p_1 + p_2)^2}\bar{v}_{s_2}(\mathbf{p}_2)\gamma^\mu u_{s_1}(\mathbf{p}_1)$$

$$\mathcal{M}_{B2} = -ie^2 \bar{u}_{s_1'}(\mathbf{p}_1')\gamma_\nu u_{s_1}(\mathbf{p}_1)\frac{1}{(p_2 - p_2')^2}\bar{v}_{s_2}(\mathbf{p}_2)\gamma^\nu v_{s_2'}(\mathbf{p}_2').$$

(17-118)

The two sub amplitudes

The sign of each line in (17-118) is not important, as we will be taking the square of the absolute value of each. However, the relative difference in sign between the two is important, as we will see below. That is, the signs of $\mathcal{M}_{B1}$ and $\mathcal{M}_{B2}$ must be opposite, and this is assured by Feynman rule #9.

Bhabha scattering is elastic, so much simplifies

To evaluate our cross-section in the COM frame, we start with (17-68), where the collision is elastic, so we can use $\mathbf{p}_1 = -\mathbf{p}_2$, $\mathbf{p}_2' = -\mathbf{p}_1'$, $|\mathbf{p}_1'| = |\mathbf{p}_1| = |\mathbf{p}_2'| = |\mathbf{p}_2|$. (See Wholeness Chart 17-4, pg. 451 and Fig. 17-13, pg. 448.) Since, $m_1 = m_2 = m$, we also have $E_1 = E_2 = E_1' = E_2' = E$.

Examine COM

$$\left(\frac{d\sigma}{d\Omega_1'}\right)_{\substack{COM \\ e^-e^+\to e^-e^+}} = \frac{1}{64\pi^2\left(E_1+E_2\right)^2}\frac{|\mathbf{p}_1'|}{|\mathbf{p}_1|}\left(\prod_l^{\substack{extern \\ ferms}} 2m_l\right)|\mathcal{M}|^2 = \frac{m^4}{16\pi^2 E^2}|\mathcal{M}|^2 \quad \binom{measured}{spins}$$

$$= \frac{m^4}{16\pi^2 E^2}\left|\mathcal{M}_{B1}+\mathcal{M}_{B2}\right|^2 = \frac{m^4}{16\pi^2 E^2}\left(\left|\mathcal{M}_{B1}\right|^2+\left|\mathcal{M}_{B2}\right|^2+\mathcal{M}_{B1}\mathcal{M}_{B2}^*+\mathcal{M}_{B2}\mathcal{M}_{B1}^*\right).$$

Differential cross-section for measured spins (17-119)

If spins are not measured (which is typical), we need to do our spins sums on (17-119), to get

$$\left(\frac{d\sigma}{d\Omega_1'}\right)_{\substack{COM \\ e^-e^+\to e^-e^+}} = \frac{m^4}{16\pi^2 E^2}\left(\begin{array}{c}\frac{1}{4}\sum_{spins}\left|\mathcal{M}_{B1}\right|^2+\frac{1}{4}\sum_{spins}\left|\mathcal{M}_{B2}\right|^2 \\ +\frac{1}{4}\sum_{spins}\mathcal{M}_{B1}\mathcal{M}_{B2}^*+\frac{1}{4}\sum_{spins}\mathcal{M}_{B2}\mathcal{M}_{B1}^*\end{array}\right)\binom{unmeasured}{spins}.(17\text{-}120)$$

Differential cross-section for more typical case of unmeasured spins

The first term in (17-120) is simply (17-111) with $m_l=m$. If, in addition, for simplicity, we restrict ourselves to the important case of relativistic high energies with $E \gg m$, (17-111) becomes

Further simplify by considering only very high speeds

$$\frac{1}{4}\sum_{spins}\left|\mathcal{M}_{B1}\right|^2 = \frac{1}{4}\sum_{s_1'=1}^{2}\sum_{s_2'=1}^{2}\sum_{s_1=1}^{2}\sum_{s_2=1}^{2}\left|\mathcal{M}_{B1}\right|^2 \approx \frac{e^4}{8m^4}\left(\frac{1+\cos^2\theta}{2}+\mathcal{O}\left(\frac{m^2}{E^2}\right)\right) \quad (E\gg m).(17\text{-}121)$$

First term in (17-120) taken from prior section

The determination of the second term in (17-120) follows similar steps to what we used for (17-111). Doing Prob. 7 (which is very long, so you might want to skip it for now[1]) shows that to be

$$\frac{1}{4}\sum_{spins}\left|\mathcal{M}_{B2}\right|^2 = \frac{1}{4}\sum_{s_1'=1}^{2}\sum_{s_2'=1}^{2}\sum_{s_1=1}^{2}\sum_{s_2=1}^{2}\left|\mathcal{M}_{B2}\right|^2 \approx \frac{e^4}{8m^4\sin^4\left(\theta/2\right)}\left(1+\cos^4\frac{\theta}{2}+\mathcal{O}\left(\frac{m^2}{E^2}\right)\right) \quad (E\gg m).(17\text{-}122)$$

Second term in (17-120) computed in similar manner and just given here

The third term in (17-120) is (where the last line uses (4-148), pg. 122, $\gamma^{\mu\dagger}\gamma^0=\gamma^0\gamma^\mu$)

Third term in (17-120) more complicated

$$\frac{1}{4}\sum_{spins}\mathcal{M}_{B1}\mathcal{M}_{B2}^* = \frac{-e^4}{\underbrace{\left(p_2-p_2'\right)^2\left(p_1+p_2\right)^2}_{\left(\Gamma_1\Gamma_2^*\right)}}\frac{1}{4}\sum_{spins}\left\{\begin{array}{c}\left(\overline{u}_{s_1'}(\mathbf{p}_1')\gamma_\mu v_{s_2'}(\mathbf{p}_2')\overline{v}_{s_2}(\mathbf{p}_2)\gamma^\mu u_{s_1}(\mathbf{p}_1)\right)\times \\ \left(\overline{u}_{s_1'}(\mathbf{p}_1')\gamma_\nu u_{s_1}(\mathbf{p}_1)\overline{v}_{s_2}(\mathbf{p}_2)\gamma^\nu v_{s_2'}(\mathbf{p}_2')\right)^* \\ \underbrace{\left(\overline{u}_{s_1'}(\mathbf{p}_1')\gamma_\nu u_{s_1}(\mathbf{p}_1)\right)^\dagger\left(\overline{v}_{s_2}(\mathbf{p}_2)\gamma^\nu v_{s_2'}(\mathbf{p}_2')\right)^\dagger}\end{array}\right\}$$

(17-123)

$$= \left(\Gamma_1\Gamma_2^*\right)\frac{1}{4}\sum_{spins}\left\{\begin{array}{c}\left(\overline{u}_{s_1'\,\beta}(\mathbf{p}_1')\left(\gamma_\mu\right)_{\beta\delta}v_{s_2'\,\delta}(\mathbf{p}_2')\overline{v}_{s_2\,\varepsilon}(\mathbf{p}_2)\left(\gamma^\mu\right)_{\varepsilon\eta}u_{s_1\eta}(\mathbf{p}_1)\right)\times \\ \left(\overline{u}_{s_1\,\kappa}(\mathbf{p}_1)\left(\gamma_\nu\right)_{\kappa\lambda}u_{s_1'\,\lambda}(\mathbf{p}_1')\overline{v}_{s_2'\,\rho}(\mathbf{p}_2')\left(\gamma^\nu\right)_{\rho\xi}v_{s_2\,\xi}(\mathbf{p}_2)\right)\end{array}\right\}.$$

Rearranged with (17-75) and (17-76) incorporated, this becomes

$$\frac{1}{4}\sum_{spins}\mathcal{M}_{B1}\mathcal{M}_{B2}^* = \frac{1}{4}\sum_{s_1'=1}^{2}\sum_{s_2'=1}^{2}\sum_{s_1=1}^{2}\sum_{s_2=1}^{2}\mathcal{M}_{B1}\mathcal{M}_{B2}^*$$

$$= \frac{\left(\Gamma_1\Gamma_2^*\right)}{4}\left\{\underbrace{\sum_{s_1'=1}^{2}u_{s_1'\,\lambda}(\mathbf{p}_1')\overline{u}_{s_1'\,\beta}(\mathbf{p}_1')}_{\left(\frac{\not{p}_1'+m}{2m}\right)_{\lambda\beta}}\left(\gamma_\mu\right)_{\beta\delta}\underbrace{\left(\sum_{s_2'=1}^{2}v_{s_2'\,\delta}(\mathbf{p}_2')\overline{v}_{s_2'\,\rho}(\mathbf{p}_2')\right)}_{\left(\frac{\not{p}_2'-m}{2m}\right)_{\delta\rho}}\left(\gamma^\nu\right)_{\rho\xi}\times \right.$$

(17-124)

$$\left.\underbrace{\left(\sum_{s_2=1}^{2}v_{s_2\,\xi}(\mathbf{p}_2)\overline{v}_{s_2\,\varepsilon}(\mathbf{p}_2)\right)}_{\left(\frac{\not{p}_2-m}{2m}\right)_{\xi\varepsilon}}\left(\gamma^\mu\right)_{\varepsilon\eta}\underbrace{\left(\sum_{s_1=1}^{2}u_{s_1\eta}(\mathbf{p}_1)\overline{u}_{s_1\,\kappa}(\mathbf{p}_1)\right)}_{\left(\frac{\not{p}_1+m}{2m}\right)_{\eta\kappa}}\left(\gamma_\nu\right)_{\kappa\lambda}\right\}.$$

In short hand notation, without the spinor indices, where we note that in (17-124), we can move the last two factors to the front and we still have a trace on the first and last dummy spinor indices,

----

[1] The solution to this problem is posted on the web site for this book. URL on pg. xvi, opposite pg. 1.

$$\frac{1}{4}\sum_{spins} \mathcal{M}_{B1}\mathcal{M}_{B2}^* = \frac{\left(\Gamma_1\Gamma_2^*\right)}{4} \text{Tr}\left\{\left(\frac{\not{p}_1+m}{2m}\right)\gamma_\nu\left(\frac{\not{p}_1'+m}{2m}\right)\gamma_\mu\left(\frac{\not{p}_2'-m}{2m}\right)\gamma^\nu\left(\frac{\not{p}_2-m}{2m}\right)\gamma^\mu\right\}. \quad (17\text{-}125)$$

Third term as function of 4-momenta and mass relations

Traces of an odd number of gamma matrices equal zero, and the spinor space identity matrix associated with the m factors in the number effectively don't count as gamma matrices. So, we only have to worry about terms with four, two, or zero factors of 4-momentum.

The term with four such factors will have no m factors. The term with two 4-momentum factors will have two factors of m (i.e., one factor of m^2). But since we are considering the highly relativistic case where $E \approx |\mathbf{p}| \gg m$, that term will be negligible compared to the term with four factors of 4-momentum. That is, one term will be of order E^4, and the other, of order E^2m^2. So, we can ignore the second of these. The term with zero factors of 4-momentum will be of order m^4, and thus even more negligible than the term of order E^2m^2. With these considerations, (17-125) becomes

Since relativistic energies assumed, we can drop many terms

$$\frac{1}{4}\sum_{spins} \mathcal{M}_{B1}\mathcal{M}_{B2}^* = \frac{\left(\Gamma_1\Gamma_2^*\right)}{64m^4} \text{Tr}\left\{\not{p}_1\gamma_\nu\not{p}_1'\,\gamma_\mu\not{p}_2'\,\gamma^\nu\not{p}_2\gamma^\mu + \text{negligible terms if } E \gg m\right\}. \quad (17\text{-}126)$$

This has eight gamma matrices in it. Using the contraction and trace identities in Appendix A of Chap. 4, we have

Use gamma matrix algebra to simplify

$$\frac{1}{4}\sum_{spins} \mathcal{M}_{B1}\mathcal{M}_{B2}^* \approx \frac{\left(\Gamma_1\Gamma_2^*\right)}{64m^4} \text{Tr}\left\{p_1^\alpha\,\gamma_\alpha\gamma_\nu\,p_1'^\beta\,\gamma_\beta\gamma_\mu\,p_2'^\delta\,\gamma_\delta\gamma^\nu\,p_2^\eta\,\gamma_\eta\gamma^\mu\right\}$$

$$= \frac{\left(\Gamma_1\Gamma_2^*\right)}{64m^4} \text{Tr}\left\{p_1^\alpha\,p_1'^\beta\,p_2'^\delta\,p_2^\eta\,\gamma_\alpha\underbrace{\gamma_\nu\,\gamma_\beta\gamma_\mu\,\gamma_\delta\gamma^\nu}_{-2\gamma_\delta\gamma_\mu\gamma_\beta}\gamma_\eta\,\gamma^\mu\right\}$$

$$= -\frac{\left(\Gamma_1\Gamma_2^*\right)}{32m^4} \text{Tr}\left\{p_1^\alpha\,p_1'^\beta\,p_2'^\delta\,p_2^\eta\,\gamma_\alpha\,\gamma_\delta\underbrace{\gamma_\mu\gamma_\beta\gamma_\eta\,\gamma^\mu}_{4g_{\beta\eta}}\right\} \qquad\qquad (17\text{-}127)$$

$$= -\frac{\left(\Gamma_1\Gamma_2^*\right)}{8m^4} \text{Tr}\left\{p_1^\alpha\,p_2'^\delta\,\underbrace{p_1'^\beta\,p_2^\eta\,g_{\beta\eta}}_{\substack{p_1'^\beta p_{2\beta}\\=p_1'\,p_2}}\gamma_\alpha\,\gamma_\delta\right\} = -\frac{\left(\Gamma_1\Gamma_2^*\right)}{8m^4}\left(p_1'p_2\right)\underbrace{\text{Tr}\left\{\not{p}_1\not{p}_2'\right\}}_{\substack{4p_1^\alpha p_{2\alpha}'\\=4p_1p_2'}} = -\frac{\left(\Gamma_1\Gamma_2^*\right)}{2m^4}\left(p_1'p_2\right)\left(p_1p_2'\right).$$

From (17-110), this becomes, with the help of a trig identity and $|\mathbf{p}| \approx E$, for relativistic energies,

$$\frac{1}{4}\sum_{spins} \mathcal{M}_{B1}\mathcal{M}_{B2}^* \approx -\frac{\left(\Gamma_1\Gamma_2^*\right)}{2m^4}\underbrace{\left(E^2 + |\mathbf{p}||\mathbf{p}'|\cos\theta\right)^2}_{E^2} = -\frac{\left(\Gamma_1\Gamma_2^*\right)}{2m^4}E^4\frac{\left(1+\cos\theta\right)^2}{\left(2\cos^2(\theta/2)\right)^2}$$

Third term spins sums evaluated

$$(17\text{-}128)$$

$$= -\frac{\left(\Gamma_1\Gamma_2^*\right)}{m^4}2E^4\cos^4\left(\frac{\theta}{2}\right) \qquad (E \gg m).$$

Thus, (17-128), the third term in (17-120), is real since $\Gamma_1\Gamma_2^*$ is real by inspection of (17-123). The fourth term in (17-120) is the complex conjugate of the third, so therefore it is also real and equals the third term, i.e., (17-128).

Fourth term same as third

Before putting all our pieces together, we need to evaluate $\Gamma_1\Gamma_2^*$. To do this, we'll need to calculate $\left(p_2 - p_2'\right)^2$, which we digress to do below. From Fig. 17-16, pg. 465, we get

Evaluating $\Gamma_1\Gamma_2^*$ part of third term

$$\left(p_2 - p_2'\right)^2 = (E-E)^2 - \left(\mathbf{p}_2 - \mathbf{p}_2'\right)^2 = 0 - \left(\begin{pmatrix}-|\mathbf{p}|\\0\\0\end{pmatrix} - \begin{pmatrix}-|\mathbf{p}|\cos\theta\\-|\mathbf{p}|\sin\theta\\0\end{pmatrix}\right)^2 = -\begin{pmatrix}-|\mathbf{p}|+|\mathbf{p}|\cos\theta\\|\mathbf{p}|\sin\theta\\0\end{pmatrix}^2$$

$$= -|\mathbf{p}|^2\left((1-\cos\theta)^2 + \sin^2\theta\right) = -|\mathbf{p}|^2\left(1 - 2\cos\theta + \cos^2\theta + \sin^2\theta\right) \qquad (17\text{-}129)$$

$$= -2|\mathbf{p}|^2\left(1-\cos\theta\right) = -4|\mathbf{p}|^2\sin^2\left(\theta/2\right) \approx -4E^2\sin^2\left(\theta/2\right).$$

With (17-129), we find

$$\Gamma_1 \Gamma_2^* = \frac{-e^4}{\underbrace{(p_2 - p_2')^2}_{\approx -4E^2 sin^2(\theta/2)} \underbrace{(p_1 + p_2)^2}_{\approx (2E)^2}} \approx \frac{e^4}{16E^4 sin^2(\theta/2)}. \qquad (17\text{-}130)$$

So, (17-128) becomes

$$\frac{1}{4}\sum_{spins} \mathcal{M}_{B1}\mathcal{M}_{B2}^* \approx -\frac{e^4}{16E^4 sin^2(\theta/2)}\frac{2E^4}{m^4}cos^4\left(\frac{\theta}{2}\right) = -\frac{e^4}{8m^4}\frac{cos^4(\theta/2)}{sin^2(\theta/2)}. \qquad (17\text{-}131)$$

Result for third term

Putting (17-121), (17-122), and (17-131) (twice) into (17-120), we get

$$\left(\frac{d\sigma}{d\Omega_1'}\right)_{\substack{COM \\ e^-e^+ \to e^-e^+}} = \frac{m^4}{16\pi^2 E^2}\left(\frac{e^4}{8m^4}\frac{1+cos^2\theta}{2} + \frac{e^4}{8m^4 sin^4(\theta/2)}\left(1 + cos^4\frac{\theta}{2}\right)\right)$$

$$- 2\frac{e^4}{8m^4}\frac{cos^4(\theta/2)}{sin^2(\theta/2)}. \qquad (17\text{-}132)$$

Adding all four terms

Or finally, since $\alpha = e^2/4\pi$,

$$\left(\frac{d\sigma}{d\Omega_1'}\right)_{\substack{COM \\ e^-e^+ \to e^-e^+}} = \underbrace{\frac{\alpha^2}{8E^2}}_{\frac{\alpha^2}{2E_{tot}^2}}\left(\underbrace{\frac{1+cos^2\theta}{2}}_{\substack{\text{annihilation} \\ \text{contribution}}} + \underbrace{\frac{1+cos^4(\theta/2)}{sin^4(\theta/2)}}_{\substack{\text{non-annihilation} \\ \text{(exchange)} \\ \text{contribution}}} - \underbrace{\frac{2cos^4(\theta/2)}{sin^2(\theta/2)}}_{\substack{\text{interference} \\ \text{contribution}}}\right) \begin{cases}\text{unmeasured} \\ \text{spins,} \\ E_{tot} = 2E \gg 2m\end{cases}, (17\text{-}133)$$

Final result: differential cross-section for Bhabha scattering

a result first found by Homi J. Bhabha in 1935, the reason the interaction bears his name.

Note that the first term in (17-133), the annihilation contribution represented by the LHS of Fig. 17-18 is the same as what we found for the $e^+e^- \to \mu^+\mu^-$ and $e^+e^- \to \tau^+\tau^-$ interactions, (17-117), where there was only an annihilation amplitude.

Note also that the annihilation contribution must have a positive sign, regardless of the sign of $\mathcal{M}_{B1}$ in (17-118) because it results from $\mathcal{M}_{B1}\mathcal{M}^*_{B1}$. For cross-section determination, the sign of $\mathcal{M}_{B1}$ is arbitrary, since though it must result from normal ordering of the operators associated with each spinor, we can for example exchange the order of two adjacent destruction operators and still have a normal ordered result, but with opposite sign. However, we must normal order the operators/spinors in the same order for $\mathcal{M}_{B2}$ as we did for $\mathcal{M}_{B1}$. This gives us a relative sign difference between $\mathcal{M}_{B1}$ and $\mathcal{M}_{B2}$. This relative sign difference doesn't matter in finding the non-annihilation contribution $\mathcal{M}_{B2}\mathcal{M}^*_{B2}$, as that quantity will always be positive. But it does give us a minus sign in $\mathcal{M}_{B1}\mathcal{M}^*_{B2}$, and that leads to a minus sign in the last term in (17-133), which is very important. That last term arises from the wave interference between the annihilation and non-annihilation interactions, and so is termed the "interference" contribution.

Relative sign between $\mathcal{M}_{B1}$ and $\mathcal{M}_{B2}$ is important

(17-133) is plotted in Fig. 17-19 along with data taken in one experiment that measured Bhabha scattering. Note that by plotting $(d\sigma/d\Omega) \times (E_{tot})^2$ instead of just $(d\sigma/d\Omega)$ on the vertical axis, we get a curve that is independent of the energy level, so we can compare results from tests at different energies readily.

Multiplying y axis $(d\sigma/d\Omega)$ by $(E_{tot})^2$ gives same plot for any experiment at any E_{tot}

The non-annihilation term, often called the <u>exchange</u> term (because it involves a simple exchange of a virtual photon rather than production of a virtual photon from annihilation), along with the interference term result in a front-back asymmetry, unlike the muon/tau producing interactions. (17-133) is not symmetric in $cos\theta$, whereas (17-117) is.

"Exchange" term for virtual photon exchange part

At small angles, the non-annihilation (exchange) term dominates. The probability for little or no change in the incoming e^+e^- is higher. This corresponds to the virtual (exchanged) photon carrying little 4-momentum, so its $k^2 = (p_2 - p'_2)^2 \to 0$. (See the bottom row of (17-118).) This makes $\mathcal{M}_{B2}$ large, since the photon propagator gets large. At large angles, the annihilation and exchange terms are of comparable importance.

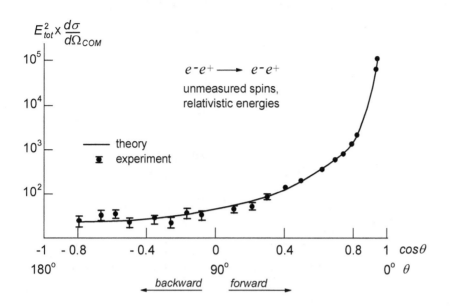

Analytical result compares well with experiment

Figure 17-19. Differential Cross-section for Bhabha Scattering
Adapted from H. J. Behrend *et al, Phys. Lett.* **103B** (1981), 148, which had
experimental E_{tot} = 34 GeV. Units on vertical axis, GeV^2-nanobarns/steradian.

17.5.3 Compton Scattering

Compton scattering, $e^+ \gamma \rightarrow e^+ \gamma$, has two sub-amplitude contributions to the total amplitude (see (17-70)), represented by the two Feynman diagrams in Fig. 17-20,

Compton scattering cross-section determination

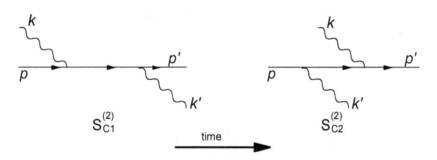

Figure 17-20. Feynman Diagrams for Compton Scattering

We will consider the common case where the electron is initially stationary in the lab and take (θ, ϕ) as the polar angles of the scattered photon, i.e., as indicating the direction of $\mathbf{k}'$. The initial photon 3-momentum $\mathbf{k}$ will be aligned with the polar coordinate axis (z axis). See Fig. 17-21.

Use frame where initial e^- stationary

Final photon $\theta_1' = \theta$

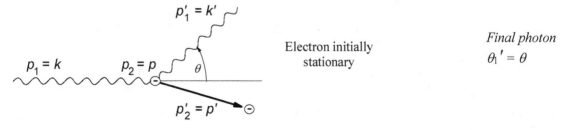

Figure 17-21. Compton Scattering When Initial Electron is Stationary in the Lab

For this frame, we have, where we note that because photons are massless (and $k^\mu k_\mu = 0$), $|\mathbf{p}_1| = |\mathbf{k}| = \omega$, and $|\mathbf{p'}_1| = |\mathbf{k'}| = \omega'$,

$$p_1 = k = \begin{pmatrix} \omega \\ \mathbf{k} \end{pmatrix} = \begin{pmatrix} \omega \\ 0 \\ 0 \\ |\mathbf{k}| \end{pmatrix} = \begin{pmatrix} \omega \\ 0 \\ 0 \\ \omega \end{pmatrix} \qquad\qquad p_2 = p = \begin{pmatrix} m \\ \mathbf{p} \end{pmatrix} = \begin{pmatrix} m \\ 0 \\ 0 \\ 0 \end{pmatrix}$$

4-momenta in terms of initial ω, $\mathbf{k}$, m, and final ω', $\mathbf{k'}$, θ

(17-134)

$$p_1' = \begin{pmatrix} E_1' \\ \mathbf{p}_1' \end{pmatrix} = k' = \begin{pmatrix} \omega' \\ \mathbf{k'} \end{pmatrix} = \begin{pmatrix} \omega' \\ |\mathbf{k'}|\sin\theta \\ 0 \\ |\mathbf{k'}|\cos\theta \end{pmatrix} \qquad p_2' = \begin{pmatrix} E_2' \\ \mathbf{p}_2' \end{pmatrix} = p' = \begin{pmatrix} E' \\ \mathbf{p'} \end{pmatrix} = \begin{pmatrix} \omega + m - \omega' \\ -|\mathbf{k'}|\sin\theta \\ 0 \\ |\mathbf{k}| - |\mathbf{k'}|\cos\theta \end{pmatrix} \left.\begin{matrix} \\ \\ \\ \end{matrix}\right\} \begin{matrix} p' \\ \text{from} \\ \text{conserv} \\ \text{laws.} \end{matrix}$$

From (17-134), and/or Fig. 17-21, we see

$$\mathbf{k}\cdot\mathbf{k'} = |\mathbf{k}||\mathbf{k'}|\cos\theta \;\; (=\omega\omega'\cos\theta). \tag{17-135}$$

We will look at obtaining the Compton cross-section for three cases in the common test frame case where the initial electron is stationary in the lab.

1. Spins and polarizations specific and measured
2. Spins and polarizations random (incoming) and unmeasured (outgoing) [called <u>unpolarized</u>]
3. Polarizations specific and measured, but spins random (incoming) and unmeasured (outgoing)

3 cases: depending on spin and polarization measurements

The first of these is difficult to do experimentally, but its analysis provides a good starting point. The second and third of these are common in experiments. In the third case, photon polarizations are measured, which is considerably easier to do than measuring fermion spins.

<u>1. Spins and Polarizations Specific and Measured</u>

(17-69), when one of the final particles is a photon, repeated below as the first line of (17-136) is

Case 1: all spins and polarizations measured

$$\left(\frac{d\sigma}{d\Omega_1'}\right)_{\substack{\#2 \\ stat}} = \frac{1}{64\pi^2 m_2 E_1' E_2'} \left(\prod_l^{\substack{\text{extern} \\ \text{ferms}}} 2m_l\right) |\mathcal{M}|^2 \frac{|\mathbf{p}_1'|^2}{|\mathbf{p}_1|} \left(\frac{\partial E_1'}{\partial|\mathbf{p}_1'|} + \frac{\partial E_2'}{\partial|\mathbf{p}_1'|}\right)^{-1} \left\{\begin{matrix} \text{\#2 stationary,} \\ \text{2 initial and 2} \\ \text{final particles,} \\ \text{elastic/inelastic} \end{matrix}\right.$$

Start with basic differential cross-section expression with #2 particle stationary

$$\rightarrow \left(\frac{d\sigma}{d\Omega_1'}\right)_{\substack{e^-\gamma\to e^-\gamma \\ 1st\, e^-\, stat}} = \frac{1}{64\pi^2 m\,\omega' E'} 4m^2 |\mathcal{M}|^2 \frac{|\mathbf{k'}|^2}{|\mathbf{k}|} \left(\frac{\partial\omega'}{\partial|\mathbf{k'}|} + \frac{\partial E'}{\partial|\mathbf{k'}|}\right)^{-1} \tag{17-136}$$

$$= \frac{m}{16\pi^2 \omega' E'} \frac{\omega'^2}{\omega} \left(\frac{\partial\omega'}{\partial\omega'} + \frac{\partial E'}{\partial|\mathbf{k'}|}\right)^{-1} |\mathcal{M}|^2 = \frac{m}{16\pi^2 E'} \frac{\omega'}{\omega} \left(1 + \frac{\partial E'}{\partial|\mathbf{k'}|}\right)^{-1} |\mathcal{M}|^2.$$

To evaluate $\partial E'/\partial|\mathbf{k'}|$, we need E' as a function of $|\mathbf{k'}|$. With $\mathbf{p}_2 = 0$ in $\mathbf{p}_1 + \mathbf{p}_2 = \mathbf{p'}_1 + \mathbf{p'}_2$,

Evaluate energy expression in this frame

$$\left(E_2'\right)^2 = \left((m_2)^2 + |\mathbf{p}_2'|^2\right) = \left((m_2)^2 + |\mathbf{p}_1 - \mathbf{p}_1'|^2\right) \rightarrow (E')^2 = \left(m^2 + |\mathbf{p'}|^2\right) = \left(m^2 + |\mathbf{k} - \mathbf{k'}|^2\right) \tag{17-137}$$

$$\rightarrow E' = \left(m^2 + |\mathbf{k}|^2 + |\mathbf{k'}|^2 - 2\mathbf{k}\cdot\mathbf{k'}\right)^{1/2} = \left(m^2 + |\mathbf{k}|^2 + |\mathbf{k'}|^2 - 2|\mathbf{k}||\mathbf{k'}|\cos\theta\right)^{1/2}.$$

From (17-137), we find (using the last part of (17-134) in the second line below)

$$\frac{\partial E'}{\partial|\mathbf{k'}|} = \frac{1}{2}\underbrace{\left(m^2 + |\mathbf{k}|^2 + |\mathbf{k'}|^2 - 2|\mathbf{k}||\mathbf{k'}|\cos\theta\right)^{-1/2}}_{E'^{-1}}\left(2|\mathbf{k'}| - 2|\mathbf{k}|\cos\theta\right) = \frac{\left(|\mathbf{k'}| - |\mathbf{k}|\cos\theta\right)}{E'}$$

(17-138)

$$\rightarrow 1 + \frac{\partial E'}{\partial|\mathbf{k'}|} = \frac{E' + \left(|\mathbf{k'}| - |\mathbf{k}|\cos\theta\right)}{E'} = \frac{\omega + m - \omega' + \omega' - \omega\cos\theta}{E'} = \frac{m + \omega(1 - \cos\theta)}{E'}.$$

Then, we want to find a simpler form for the numerator of the last part of (17-138). To do this, we use our conservation relation

$$p + k = p' + k', \tag{17-139}$$

along with (which you can prove by doing Prob. 8)

$$pk = p'k' \quad \text{i.e.,} \quad p^\mu k_\mu = p'^\mu k'_\mu, \tag{17-140}$$

A useful relation

to show
$$pk = p'k' = p'^\mu k'_\mu = \left(p^\mu + k^\mu - k'^\mu\right)k'_\mu = p^\mu k'_\mu + k^\mu k'_\mu - \underbrace{k'^\mu k'_\mu}_{0} = pk' + kk'. \tag{17-141}$$

(17-141) is true in any frame, but in our frame where p and k are as found in (17-134), we have

$$pk = m\omega - \underbrace{|\mathbf{p}|\cdot|\mathbf{k}|}_{0} = pk' + kk' = p^\mu k'_\mu + k^\mu k'_\mu$$

With the useful relation, we can simplify things

$$\rightarrow \quad m\omega = m\omega' - 0 + \omega\omega' - \mathbf{k}\cdot\mathbf{k}' = m\omega' + \omega\omega' - \omega\omega' \cos\theta \tag{17-142}$$

$$\rightarrow \quad \frac{m\omega}{\omega'} = m + \omega(1 - \cos\theta).$$

We now use the last row of (17-142) in the numerator of the last part of (17-138) to get

$$\left(1 + \frac{\partial E'}{\partial|\mathbf{k}'|}\right)_{\theta\phi} = \frac{m\omega}{E'\omega'}. \tag{17-143}$$

With (17-143), (17-136) becomes

$$\left(\frac{d\sigma}{d\Omega'}\right)_{\substack{e^-\gamma\rightarrow e^-\gamma \\ \text{1st } e^- \text{ stat}}} = \frac{m}{16\pi^2 E'}\frac{\omega'}{\omega}\frac{E'\omega'}{m\omega}|\mathcal{M}|^2 = \frac{1}{(4\pi)^2}\left(\frac{\omega'}{\omega}\right)^2 |\mathcal{M}|^2 \left\{\begin{array}{l}\text{fully polarized, i.e.,}\\ \text{measured spins}\\ \text{and polarizations}\end{array}\right. .\tag{17-144}$$

Result for Case 1

All we need to do further is evaluate $|\mathcal{M}|^2$ assuming spins and polarizations are known. We will not do that here but proceed to the more typical experimental condition where neither spins nor polarizations are measured.

2. Spins and Polarizations Not Measured ("Unpolarized" Case)

We need to take both spin sums and polarization sums into account. Using (17-95), we find

Case 2: no spins nor polarizations measured

$$\underbrace{\frac{1}{2}\sum_{s'=1}^{2}\sum_{s=1}^{2}}_{\text{spins}}\underbrace{\frac{1}{2}\sum_{r'=1}^{2}\sum_{r=1}^{2}}_{\text{polariz}}|\mathcal{M}|^2 = \frac{1}{4}\underbrace{\sum_{s'=1}^{2}\sum_{s=1}^{2}}_{\text{spins}}\mathcal{M}^{\mu\nu}\mathcal{M}^{*}_{\mu\nu}, \tag{17-145}$$

Need to do spin and polarization sums

where for Compton scattering, $\mathcal{M}^{\mu\nu}$ is defined in (17-79) and (17-80), as

$$\mathcal{M}^{\mu\nu} = \underbrace{-e^2\bar{u}_{s'}(\mathbf{p}')\gamma^\mu iS_F(p+k)\gamma^\nu u_s(\mathbf{p})}_{\mathcal{M}_1^{\mu\nu}} \underbrace{- e^2\bar{u}_{s'}(\mathbf{p}')\gamma^\nu iS_F(p-k')\gamma^\mu u_s(\mathbf{p})}_{\mathcal{M}_2^{\nu\mu}}$$

Compton scattering amplitude

$$= \bar{u}_{s'}(\mathbf{p}')\left(\underbrace{-e^2\gamma^\mu iS_F(p+k)\gamma^\nu}_{\Gamma_1=\Gamma_1^{\mu\nu}}\underbrace{-e^2\gamma^\nu iS_F(p-k')\gamma^\mu}_{\Gamma_2=\Gamma_2^{\nu\mu}}\right)u_s(\mathbf{p}). \tag{17-146}$$

$$\underbrace{\Gamma=\Gamma^{\mu\nu}=\Gamma_1^{\mu\nu}+\Gamma_2^{\nu\mu}}$$

From the 1st row of Wholeness Chart 17-5, pg. 460, with Γ there $= \Gamma^{\mu\nu}$ here, (17-145) becomes

$$\frac{1}{4}\sum_{s'=1}^{2}\sum_{s=1}^{2}\mathcal{M}^{\mu\nu}\mathcal{M}^{*}_{\mu\nu} = \frac{1}{4}\text{Tr}\left\{\left(\frac{\not{p}'+m}{2m}\right)\Gamma^{\mu\nu}\left(\frac{\not{p}+m}{2m}\right)\tilde{\Gamma}_{\mu\nu}\right\}$$

$$= \frac{1}{4}\text{Tr}\left\{\left(\frac{\not{p}'+m}{2m}\right)\left(\Gamma_1^{\mu\nu}+\Gamma_2^{\nu\mu}\right)\left(\frac{\not{p}+m}{2m}\right)\left(\tilde{\Gamma}_{1\mu\nu}+\tilde{\Gamma}_{2\nu\mu}\right)\right\}$$

Spin and polarization sums in terms of traces $\rightarrow$ 4 terms

$$= \underbrace{\frac{1}{4}\text{Tr}\left\{\left(\frac{\not{p}'+m}{2m}\right)\Gamma_1^{\mu\nu}\left(\frac{\not{p}+m}{2m}\right)\tilde{\Gamma}_{1\mu\nu}\right\}}_{e^4 X_{11}} + \underbrace{\frac{1}{4}\text{Tr}\left\{\left(\frac{\not{p}'+m}{2m}\right)\Gamma_2^{\nu\mu}\left(\frac{\not{p}+m}{2m}\right)\tilde{\Gamma}_{2\nu\mu}\right\}}_{e^4 X_{22}} \tag{17-147}$$

Label the 4 traces

$$+ \underbrace{\frac{1}{4}\text{Tr}\left\{\left(\frac{\not{p}'+m}{2m}\right)\Gamma_1^{\mu\nu}\left(\frac{\not{p}+m}{2m}\right)\tilde{\Gamma}_{2\nu\mu}\right\}}_{e^4 X_{12}} + \underbrace{\frac{1}{4}\text{Tr}\left\{\left(\frac{\not{p}'+m}{2m}\right)\Gamma_2^{\nu\mu}\left(\frac{\not{p}+m}{2m}\right)\tilde{\Gamma}_{1\mu\nu}\right\}}_{e^4 X_{21}} .$$

We want to re-write (17-147) using the definition from the underbrackets in (17-146), the expressions for the propagators, and $\tilde{\Gamma}^{\mu\nu} = \gamma^0 \Gamma^{\dagger\mu\nu} \gamma^0$ from Wholeness Chart 17-5, top row. Thus,

$$\frac{1}{4}\sum_{s'=1}^{2}\sum_{s=1}^{2}\mathcal{M}^{\mu\nu}\mathcal{M}_{\mu\nu}^{*} = \frac{e^4}{4}\left(X_{11} + X_{22} + X_{12} + X_{21}\right) \tag{17-148}$$

where we can switch upper and lower locations of the indices μ and ν since it is an inner product,

Evaluating first trace

$$X_{11} = \text{Tr}\left\{\left(\frac{\not{p}'+m}{2m}\right)\gamma_\mu\,(-i)\left(\frac{\not{p}+\not{k}+m}{(p+k)^2-m^2+i\varepsilon}\right)\gamma_\nu\left(\frac{\not{p}+m}{2m}\right)\underbrace{\gamma^0\gamma^{\dagger\nu}}_{\gamma^\nu\gamma^0}\left(i\underbrace{\frac{\not{p}^\dagger+\not{k}^\dagger+m}{(p+k)^2-m^2-i\varepsilon}}_{(p_\alpha+k_\alpha)\gamma^{\dagger\alpha}}\right)\underbrace{\gamma^{\dagger\mu}\gamma^0}_{\gamma^0\gamma^\mu}\right\},$$

$$\underbrace{i\gamma^\nu\frac{\left((p_\alpha+k_\alpha)\gamma^0\gamma^{\dagger\alpha}\gamma^0+m\gamma^0\gamma^0\right)}{(p+k)^2-m^2-i\varepsilon}\gamma^\mu}$$

or, with $\gamma^0 \gamma^{\dagger\alpha} \gamma^0 = \gamma^\alpha$ and $\gamma^0 \gamma^0 = 1$,

$$X_{11} = \text{Tr}\left\{\left(\frac{\not{p}'+m}{2m}\right)\gamma_\mu\left(\frac{\not{p}+\not{k}+m}{(p+k)^2-m^2+i\varepsilon}\right)\gamma_\nu\left(\frac{\not{p}+m}{2m}\right)\gamma^\nu\left(\frac{\not{p}+\not{k}+m}{(p+k)^2-m^2-i\varepsilon}\right)\gamma^\mu\right\}.$$

Since the trace means we are summing on the first and last spinor indices of the quantity inside the bracket, we can move the first sub bracket factor above to the end without changing the trace summation. Thus, where we switch upper and lower locations of the indices again,

$$X_{11} = \text{Tr}\left\{\gamma^\mu\left(\frac{\not{p}+\not{k}+m}{(p+k)^2-m^2+i\varepsilon}\right)\gamma^\nu\left(\frac{\not{p}+m}{2m}\right)\gamma_\nu\left(\frac{\not{p}+\not{k}+m}{(p+k)^2-m^2-i\varepsilon}\right)\gamma_\mu\left(\frac{\not{p}'+m}{2m}\right)\right\}. \tag{17-149}$$

Note also that we can simplify the denominator of (17-149) with the first line of (17-150) below.

$$(p+k)^2 - m^2 = p^2 + 2p^\eta k_\eta + k^2 - m^2 = m^2 + 2p^\eta k_\eta + 0 - m^2 = 2p^\eta k_\eta = 2pk$$

$$(p-k')^2 - m^2 = -2pk'. \tag{17-150}$$

With (17-150), (17-149) becomes, where we ignore the tiny ε,

Result for first trace expressed in terms of trace of gamma matrices

$$X_{11} = \frac{1}{16m^2(pk)^2}\underbrace{\text{Tr}\left\{\gamma^\mu\left(\not{p}+\not{k}+m\right)\gamma^\nu\left(\not{p}+m\right)\gamma_\nu\left(\not{p}+\not{k}+m\right)\gamma_\mu\left(\not{p}'+m\right)\right\}}_{\text{call this quantity } Y \text{ for future reference}} \tag{17-151}(a)$$

Similarly, as you can show by doing Probs. 9 and 10,

Second to fourth traces expressed in terms of trace of gamma matrices

$$X_{22} = \frac{1}{16m^2(pk')^2}\text{Tr}\left\{\gamma^\mu\left(\not{p}-\not{k}'+m\right)\gamma^\nu\left(\not{p}+m\right)\gamma_\nu\left(\not{p}-\not{k}'+m\right)\gamma_\mu\left(\not{p}'+m\right)\right\} \tag{17-151}(b)$$

$$X_{12} = -\frac{1}{16m^2(pk)(pk')}\text{Tr}\left\{\gamma^\mu\left(\not{p}+\not{k}+m\right)\gamma^\nu\left(\not{p}+m\right)\gamma_\mu\left(\not{p}-\not{k}'+m\right)\gamma_\nu\left(\not{p}'+m\right)\right\} \tag{17-151}(c)$$

$$X_{21} = -\frac{1}{16m^2(pk)(pk')}\text{Tr}\left\{\gamma^\mu\left(\not{p}-\not{k}'+m\right)\gamma^\nu\left(\not{p}+m\right)\gamma_\mu\left(\not{p}+\not{k}+m\right)\gamma_\nu\left(\not{p}'+m\right)\right\}. \tag{17-151}(d)$$

Note that if we exchange $k \leftrightarrow -k'$ in (17-151), then $X_{11} \leftrightarrow X_{22}$ and $X_{12} \leftrightarrow X_{21}$. So, we only need to find X_{11} and X_{12} and then exchange $k \leftrightarrow -k'$ in our answers to get X_{22} and X_{21}.

Symmetries → can find X_{22} if know X_{11} and X_{21} if know X_{12}

Finding X_{11}

We need to reduce the eight gamma matrices of the quantity we designate as Y in (17-151)(a). Recalling again the gamma matrices relations from Appendix A of Chap. 4, with which you should be gaining some comfort, we have

Evaluating first trace

$$Y = \gamma^\mu \left(\not p + \not k + m\right) \underbrace{\gamma^\nu \left(p_\alpha \gamma^\alpha + m\right)\gamma_\nu}_{-p_\alpha 2\gamma^\alpha + 4m} \left(\not p + \not k + m\right)\gamma_\mu \left(\not p' + m\right)$$

Cranking the algebra

$$= \left\{\gamma^\mu \left(\left(\not p + \not k\right) + m\right)\left(-2\not p + 4m\right)\left(\left(\not p + \not k\right) + m\right)\gamma_\mu\right\}\left(\not p' + m\right)$$

$$= \left\{
\begin{aligned}
&-2\underbrace{\gamma^\mu \left(\not p + \not k\right)\not p \left(\not p + \not k\right)\gamma_\mu}_{-2\left(\not p + \not k\right)\not p\left(\not p + \not k\right)} - 2\underbrace{\gamma^\mu \left(\not p + \not k\right)\not p\, m\, \gamma_\mu}_{4m\left(p+k\right)p} \\
&+\underbrace{\gamma^\mu \left(\not p + \not k\right)4m\left(\not p + \not k\right)\gamma_\mu}_{16m\left(p+k\right)\left(p+k\right)} + 4\underbrace{\gamma^\mu \left(\not p + \not k\right)m^2 \gamma_\mu}_{-2m^2\left(\not p + \not k\right)} - 2\underbrace{\gamma^\mu m\not p\left(\not p + \not k\right)\gamma_\mu}_{4mp\left(p+k\right)} \\
&-2\underbrace{\gamma^\mu \not p\, m^2 \gamma_\mu}_{-2m^2 \not p} + 4\underbrace{\gamma^\mu m^2\left(\not p + \not k\right)\gamma_\mu}_{-2m^2\left(\not p + \not k\right)} + 4\underbrace{\gamma^\mu m^3 \gamma_\mu}_{4m^3}
\end{aligned}
\right\}\left(\not p' + m\right). \quad (17\text{-}152)$$

Or

$$Y = \left\{4\left(\not p + \not k\right)\not p\left(\not p + \not k\right) - 8m\left(p+k\right)p + 16m\left(p+k\right)\left(p+k\right) - 8m^2\left(\not p + \not k\right)\right.$$
$$\left. -8mp\left(p+k\right) + 4m^2 \not p - 8m^2\left(\not p + \not k\right) + 16m^3\right\}\left(\not p' + m\right)$$

$$= \left\{4\left(\not p + \not k\right)\not p\left(\not p + \not k\right) - 16m\left(p+k\right)p + 16m\left(p+k\right)^2\right.$$
$$\left. -16m^2\left(\not p + \not k\right) + 4m^2 \not p + 16m^3\right\}\left(\not p' + m\right)$$

$$= \left\{
\begin{aligned}
&4\left(\not p + \not k\right)\not p\left(\not p + \not k\right)\not p' + 4\left(\not p + \not k\right)\not p\left(\not p + \not k\right)m - 16m\left(p+k\right)p\not p' \\
&-16m\left(p+k\right)pm + 16m\left(p+k\right)^2 \not p' + 16m\left(p+k\right)^2 m - 16m^2\left(\not p + \not k\right)\not p' \\
&-16m^2\left(\not p + \not k\right)m + 4m^2 \not p\not p' + 4m^2 \not p m + 16m^3 \not p' + 16m^4
\end{aligned}
\right\}. \quad (17\text{-}153)$$

With (17-153) in (17-151)(a), we note that all traces of an odd number of gamma matrix factors yield zero, and

$$X_{11} = \frac{1}{16m^2 \left(pk\right)^2}\operatorname{Tr}\left\{Y\right\} = \frac{1}{16m^2\left(pk\right)^2}\operatorname{Tr}\left\{
\begin{aligned}
&4\left(\not p + \not k\right)\not p\left(\not p + \not k\right)\not p' - 16m\left(p+k\right)pm \\
&+16m\left(p+k\right)^2 m - 16m^2\left(\not p + \not k\right)\not p' + 4m^2 \not p\not p' + 16m^4
\end{aligned}
\right\}$$

$$= \frac{1}{4m^2\left(pk\right)^2}\left\{
\begin{aligned}
&\underbrace{\operatorname{Tr}\left\{\left(\not p + \not k\right)\not p\left(\not p + \not k\right)\not p'\right\}}_{4\left\{
\begin{aligned}
&\left(p^\alpha + k^\alpha\right)p_\alpha\left(p^\beta + k^\beta\right)p'_\beta \\
&-\left(p^\alpha + k^\alpha\right)\left(p_\alpha + k_\alpha\right)p^\beta p'_\beta \\
&+\left(p^\alpha + k^\alpha\right)p'_\alpha\left(p^\beta + k^\beta\right)p_\beta
\end{aligned}
\right\}} - 4m^2 \underbrace{\operatorname{Tr}\left\{\left(p+k\right)p\right\}}_{4\left(p^\alpha + k^\alpha\right)p_\alpha} + 4m^2 \underbrace{\operatorname{Tr}\left\{\left(p+k\right)^2\right\}}_{4\left(p+k\right)^2} \\
&-4m^2 \underbrace{\operatorname{Tr}\left\{\left(\not p + \not k\right)\not p'\right\}}_{4\left(p^\alpha + k^\alpha\right)p'_\alpha} + m^2 \underbrace{\operatorname{Tr}\left\{\not p\not p'\right\}}_{4p^\alpha p'_\alpha} + 4m^4 \underbrace{\operatorname{Tr}\left\{I\right\}}_{4}
\end{aligned}
\right\}$$

$$= X_{11} = \frac{1}{m^2\left(pk\right)^2}\left\{
\begin{aligned}
&\left(p^\alpha + k^\alpha\right)p_\alpha\left(p^\beta + k^\beta\right)p'_\beta - \left(p^\alpha + k^\alpha\right)\left(p_\alpha + k_\alpha\right)p^\beta p'_\beta \\
&+\left(p^\alpha + k^\alpha\right)p'_\alpha\left(p^\beta + k^\beta\right)p_\beta - 4m^2\left(p^\alpha + k^\alpha\right)p_\alpha + 4m^2\left(p+k\right)^2 \\
&-4m^2\left(p^\alpha + k^\alpha\right)p'_\alpha + m^2 p^\alpha p'_\alpha + 4m^4
\end{aligned}
\right\}. (17\text{-}154)$$

If we express all quantities in terms of three linearly independent scalars (where the second relation below is (17-140) and you should have proven both it and the third relation in Prob. 8),

$$p^2 = p'^2 = m^2 \qquad pk = p'k' \qquad pk' = p'k , \qquad (17\text{-}155)$$

Useful relations

then (17-154) becomes

$$X_{11} = \frac{1}{m^2 (pk)^2} \left\{ \begin{array}{l} (m^2 + kp)(pp' + kp') - (m^2 + 2pk + 0)(pp') + (pp' + kp')(m^2 + kp) \\ -4m^2(m^2 + kp) + 4m^2(m^2 + 2pk + 0) - 4m^2(pp' + kp') + m^2 pp' + 4m^4 \end{array} \right\} . (17\text{-}156)$$

This reduces to

$$X_{11} = \frac{1}{m^2 (pk)^2} \left\{ \begin{array}{l} m^2 pp' + m^2 kp' + (kp)(pp') + (kp)(kp') - m^2 pp' - 2(pk)(pp') \\ + pp'm^2 + (pp')(kp) + kp'm^2 + (kp')(kp) - 4m^4 - 4m^2 kp \\ + 4m^4 + 8m^2 pk - 4m^2 pp' - 4m^2 kp' + m^2 pp' + 4m^4 \end{array} \right\}$$

$$= \frac{1}{m^2 (pk)^2} \left\{ \begin{array}{l} m^2 pp' - m^2 pp' + pp'm^2 - 4m^2 pp' + m^2 pp' \\ + m^2 kp' + kp'm^2 - 4m^2 kp' \\ + (kp)(pp') - 2(pk)(pp') + (pp')(kp) \\ + (kp)(kp') + (kp')(kp) \\ -4m^4 + 4m^4 \\ -4m^2 kp + 8m^2 pk \\ +4m^4 \end{array} \right\} = \frac{1}{m^2 (pk)^2} \left\{ \begin{array}{l} -2m^2 pp' \\ -2m^2 kp' \\ +0 \\ +2(kp)(kp') \\ +0 \\ +4m^2 pk \\ +4m^4 \end{array} \right\} . \quad (17\text{-}157)$$

And by substituting further, this reduces further to

$$X_{11} = \frac{1}{m^2 (pk)^2} \left\{ -2m^2 \underbrace{(p+k)}_{p'+k'} p' + 2(kp)(kp') + 4m^2 pk + 4m^4 \right\}$$

$$= \frac{1}{m^2 (pk)^2} \left\{ -2m^2 \underbrace{p'p'}_{m^2} - 2m^2 \underbrace{k'p'}_{pk} + 2(kp)\underbrace{(kp')}_{pk'} + 4m^2 pk + 4m^4 \right\}$$

$$= \frac{1}{m^2 (pk)^2} \left\{ 2m^4 + 2m^2 pk + 2(pk)(pk') \right\}$$

Or finally

$$X_{11} = \frac{2}{m^2 (pk)^2} \left\{ m^4 + m^2 pk + (pk)(pk') \right\} . \qquad (17\text{-}158)(a)$$

Finding X_{22}

As we saw in (17-151)(a) and (b), by exchanging k with $-k'$ in X_{11} we get X_{22}. Thus,

$$X_{22} = \frac{2}{m^2 (pk')^2} \left\{ m^4 - m^2 pk' + (pk)(pk') \right\} . \qquad (17\text{-}158)(b)$$

Finding X_{12}

With another marathon of algebra, similar to what we did to find (17-158)(a), you would find

$$X_{12} = -\frac{1}{(pk)(pk')} \left\{ 2m^2 + pk - pk' \right\} . \qquad (17\text{-}158)(c)$$

If you need to prove (17-158)(c) to yourself, please do so. I suggest you just accept it (as I have from others), as the time could be better spent learning more physics, rather than slogging through extensive algebraic tedium.

Finding X_{21}

Exchanging k with $-k'$ in X_{12} we get X_{21}.

$$X_{21} = -\frac{1}{(pk)(pk')} \left\{ 2m^2 + pk - pk' \right\} , \qquad (17\text{-}158)(d)$$

and we note that $X_{12} = X_{21}$.

The Cross-section

Substituting (17-158)(a)-(d) into (17-148), we obtain

$$\frac{1}{4}\sum_{s'=1}^{2}\sum_{s=1}^{2}\mathcal{M}^{\mu\nu}\mathcal{M}^{*}_{\mu\nu}$$

$$=\frac{e^4}{4}\left(\begin{array}{c}\dfrac{2}{m^2(pk)^2}\{m^4+m^2pk+(pk)(pk')\}+\dfrac{2}{m^2(pk')^2}\{m^4-m^2pk'+(pk)(pk')\}\\[12pt]-\dfrac{2}{(pk)(pk')}\{2m^2+pk-pk'\}\end{array}\right)$$

$$=\frac{e^4}{2}\left(\frac{m^2}{(pk)^2}+\frac{1}{pk}+\frac{pk'}{m^2pk}+\frac{m^2}{(pk')^2}-\frac{1}{pk'}+\frac{pk}{m^2pk'}-\frac{2m^2}{(pk)(pk')}-\frac{1}{pk'}+\frac{1}{pk}\right)$$

$$=\frac{e^4}{2m^2}\left(\left(\frac{pk}{pk'}+\frac{pk'}{pk}\right)+2m^2\left(\frac{1}{pk}-\frac{1}{pk'}\right)+m^4\left(\frac{1}{pk}-\frac{1}{pk'}\right)^2\right). \tag{17-159}$$

In the frame where the electron is initially stationary, from (17-134), we have $pk = m\omega$ and $pk' = m\omega'$, so

$$\frac{1}{4}\sum_{s'=1}^{2}\sum_{s=1}^{2}\mathcal{M}^{\mu\nu}\mathcal{M}^{*}_{\mu\nu}=\frac{e^4}{2m^2}\left(\left(\frac{\omega}{\omega'}+\frac{\omega'}{\omega}\right)+2m\left(\frac{1}{\omega}-\frac{1}{\omega'}\right)+m^2\left(\frac{1}{\omega}-\frac{1}{\omega'}\right)^2\right) \tag{17-160}$$

From the last row of (17-142)

$$\frac{1}{\omega}-\frac{1}{\omega'}=\frac{1}{m}(\cos\theta-1), \tag{17-161}$$

so (17-160) becomes

$$\frac{1}{4}\sum_{s'=1}^{2}\sum_{s=1}^{2}\mathcal{M}^{\mu\nu}\mathcal{M}^{*}_{\mu\nu}=\frac{e^4}{2m^2}\left(\left(\frac{\omega}{\omega'}+\frac{\omega'}{\omega}\right)+2(\cos\theta-1)+\underbrace{\frac{(\cos\theta-1)^2}{\underbrace{\cos^2\theta-2\cos\theta+1}_{\cos^2\theta-1}}}\right) \tag{17-162}$$

$$=\frac{e^4}{2m^2}\left(\frac{\omega}{\omega'}+\frac{\omega'}{\omega}-\sin^2\theta\right).$$

Using (17-162) in (17-144) in place of $|\mathcal{M}|^2$ yields the Compton cross-section for unmeasured spins and polarizations,

$$\left(\frac{d\sigma}{d\Omega'_1}\right)_{\substack{e^-\gamma\to e^-\gamma\\ \text{1st }e^-\text{ stat}}}=\underbrace{\frac{e^4}{(4\pi)^2}}_{\alpha^2}\frac{1}{2m^2}\left(\frac{\omega'}{\omega}\right)^2\left(\frac{\omega}{\omega'}+\frac{\omega'}{\omega}-\sin^2\theta\right)\quad\left\{\begin{array}{c}\text{fully unmeasured spins}\\\text{and polarizations}\end{array}\right. \tag{17-163}$$

By solving (17-142) for ω' and inserting that into (17-163), we can get the differential cross-section solely in terms of θ (and known quantities such as ω and m.) That relationship is cumbersome, however, and does not lend itself to any particular insight, so we will not deduce it here. A computer analysis using it can result in a plot of differential cross-section vs θ. Such plots match experiment to high degree.

Thomson Scattering

In the low energy case, i.e., $\omega << m$, we have $\omega' \approx \omega$. That is, the kinetic energy of the electron after recoiling is negligible, and (17-163) reduces to the well-known Thomson scattering relation

$$\left(\frac{d\sigma}{d\Omega'_1}\right)_{\substack{e^-\gamma\to e^-\gamma\\ \text{Thomson}}}=\frac{\alpha^2}{2m^2}(1+1-\sin^2\theta)=\frac{\alpha^2}{2m^2}(1+\cos^2\theta)\quad\left\{\begin{array}{c}\text{fully unmeasured}\\\text{spins/polarizations,}\\\omega\ll m\end{array}\right. \tag{17-164}$$

3. Polarizations Specific and Measured, Spins Unmeasured

It is relatively easy to polarize photons for experiments and to measure photon polarization, whereas it is not easy to force fermions into specific spin states or measure fermion spins. So, it is common to have scattering experiments where one knows the photon polarization states, but not the fermion spin states.

In such cases, one sums and averages over fermion spins only, while calculating the cross-section for definite photon polarizations. We will not go through this analysis here, but merely state the final result. You now have the tools to determine this yourself, if you choose to, though you may be better served timewise if you simply accept what others before you have already done. Mandl and Shaw[1] provide an abbreviated proof for those who may be interested.

Just state final result without doing all the algebra

The result is known as the <u>Klein-Nishina formula</u>, where ε^μ represents the initial photon polarization vector and ε'_μ the final photon polarization vector,

$$\left(\frac{d\sigma}{d\Omega'_1}\right)_{\substack{e^-\gamma \to e^-\gamma \\ \text{1st } e^- \text{ stat}}} = \frac{\alpha^2}{4m^2}\left(\frac{\omega'}{\omega}\right)^2\left(\frac{\omega}{\omega'}+\frac{\omega'}{\omega}+4\varepsilon^\mu\varepsilon'_\mu - 2\right) \quad \begin{cases} \text{measured polarizations,} \\ \text{unmeasured spins} \end{cases} \quad (17\text{-}165)$$

17.5.4 Other Kinds of Scattering

We could, of course, go on to deduce differential cross-sections for Compton positron, Møller, $e^+e^- \to \gamma\gamma$, $\gamma\gamma \to e^+e^-$, and other types of scattering. However, as you have seen for other cases, the effort and tedium in deducing each one is considerable, so we will forego those endeavors with the aim of maximizing efficiency in learning. The time such exercises in algebra would consume can probably be better spent elsewhere. If you have followed closely the differential cross-section calculations done earlier in this chapter, you should be fairly well grounded in the general methodology, and that, of course, is the purpose of the chapter.

Other kinds of particle-particle scattering evaluated in similar fashion

We do, however, overview certain well-known scattering cases, like Rutherford scattering for example, in sections to follow.

17.5.5 Electrons Scattered by an External Field

The QFT scattering we have been considering so far in this chapter comprises two particles interacting with one another as depicted in Fig. 17-11, pg. 443. We will now consider the case of electron scattering by an external field (rather than by another particle), such as Rutherford scattering of an electron by an atomic nucleus (its field, actually). This is depicted in Fig. 17-7, pg. 438, for classical scattering, and Fig. 17-8, pg. 439, for NRQM scattering. However, the scattered wave in QFT will be analyzed here as a plane wave (see Fig. 17-11 where the target particle is replaced by a potential field) rather than a spherical wave as in NRQM. Note that in the spirit of Fourier, a spherical wave can be represented as a summation of plane waves, and vice versa, so the two methods are ultimately consonant. We generally want to choose the method that is easiest to use within a given theoretical structure.

Electron scattering by potential field (rather than another particle)

Here, the external field is represented in the transition amplitude as we did in Chap. 16, pgs. 415-416, in particular Feynman rule #11 for such a field. Also, see the RHS of Fig. 16-3, pg. 415.

Use Feynman rule #11 (for external potential field)

One then goes through a rather lengthy analysis similar to what we did from (17-19) to (17-22) and from (17-41) to (17-62) for one initial particle, one external field, and one final particle (instead of two initial and two final particles), in the frame of the external field potential, to derive

Analysis similar to particle-particle scattering

$$\left(\frac{d\sigma}{d\Omega'}\right)_{\substack{e^- \& \\ \text{extern} \\ \text{field}}} = \left(\frac{m}{2\pi}\right)^2|\mathcal{M}|^2 = \left(\frac{me}{2\pi}\right)^2\left|\bar{u}_{s'}(\mathbf{p}')\mathcal{A}_e(\mathbf{k})u_s(\mathbf{p})\right|^2. \quad (17\text{-}166)$$

Result of that analysis for arbitrary external field

Treating the nucleus as heavy (no recoil, effectively) and as a point charge located at the origin, one takes the Coulomb potential, in physical and momentum space, as

$$A_e^\alpha(x) = \left(\frac{Ze}{4\pi|\mathbf{x}|} \quad 0 \quad 0 \quad 0\right) \to A_e^\alpha(\mathbf{k}) = \left(\frac{Ze}{|\mathbf{k}|^2} \quad 0 \quad 0 \quad 0\right). \quad (17\text{-}167)$$

Examine Coulomb field case

where Z is the number of protons in the nucleus. (See Sect. 16.2.4, pg. 408.)

---

[1] Mandl, F. and Shaw, G. (*Quantum Field Theory*, Wiley 1984) pgs 157-159.

Substituting the RHS of (17-167) in (17-166) and summing/averaging over electron spins, as Prob. 12 asks you to do, yields the unpolarized differential cross-section

$$\left(\frac{d\sigma}{d\Omega'}\right)_{\substack{e^- \,\&\\ \text{extern}\\ \text{field}}} = \left(\frac{2mZ}{|\mathbf{k}|^2}\right)^2 \underbrace{\frac{e^4}{16\pi^2}}_{\alpha^2} \frac{1}{2}\sum_{s'}\sum_{s}\left|\bar{u}_{s'}(\mathbf{p}')\gamma^0 u_s(\mathbf{p})\right|^2$$

Spin sum for Coulomb field case

(17-168)

$$= 2\left(\frac{m\alpha Z}{|\mathbf{k}|^2}\right)^2 \text{Tr}\left\{\left(\frac{\not{p}'+m}{2m}\right)\gamma^0\left(\frac{\not{p}+m}{2m}\right)\gamma^0\right\}.$$

After evaluating the trace (by doing Prob. 13), this becomes

$$\left(\frac{d\sigma}{d\Omega'}\right)_{\substack{e^- \,\&\\ \text{extern}\\ \text{field}}} = 2\left(\frac{\alpha Z}{|\mathbf{k}|^2}\right)^2\left(EE' + \mathbf{p}\cdot\mathbf{p}' + m^2\right).$$

(17-169)

Using kinematics and conservation laws to simplify in terms of θ, final e^- scatter angle

The scattering angle θ is the angle of the final electron (the initial electron approaches along the z axis). Using $|\mathbf{p}| = |\mathbf{p}'|$, because the nucleus is assumed stationary, we see that

$$\mathbf{p}\cdot\mathbf{p}' = |\mathbf{p}|^2 \cos\theta = |\mathbf{p}|^2\left(1 - 2\sin^2(\theta/2)\right)$$

$$|\mathbf{k}|^2 = |\mathbf{p}' - \mathbf{p}|^2 = \mathbf{p}'\cdot\mathbf{p}' - 2\mathbf{p}'\cdot\mathbf{p} + \mathbf{p}\cdot\mathbf{p} = 2|\mathbf{p}|^2 - 2|\mathbf{p}|^2\cos\theta = 4|\mathbf{p}|^2\sin^2(\theta/2),$$

(17-170)

and thus (17-169) (where $E = E'$) becomes

$$\left(\frac{d\sigma}{d\Omega'}\right)_{\substack{e^- \,\&\\ \text{extern}\\ \text{field}}} = 2\left(\frac{\alpha Z}{4E^2 v^2 \sin^2(\theta/2)}\right)^2\left(E^2 + |\mathbf{p}|^2\left(1 - 2\sin^2(\theta/2)\right) + m^2\right).$$

(17-171)

General cross-section for e^- in Coulomb field

Mott Scattering (any energy level)

Since $|\mathbf{p}|^2 + m^2 = E^2$ and $|\mathbf{p}| = Ev$, (17-171) becomes the <u>Mott scattering</u> relation, good for relativistic or non-relativistic cases,

$$\left(\frac{d\sigma}{d\Omega'}\right)_{\substack{e^- \,\&\\ \text{extern}\\ \text{field}}} = \frac{(\alpha Z)^2}{4E^2 v^4 \sin^4(\theta/2)}\left(1 - v^2\sin^2(\theta/2)\right) \quad \left\{\begin{array}{l}\text{relativistic or not, field}\\ \text{of heavy pointlike nucleus,}\\ \text{unmeasured spins}\end{array}\right.$$

(17-172)

Relativistic or non-relativistic scattering

Non-Relativistic (Rutherford) Scattering

For non-relativistic scattering $E \approx m \gg |\mathbf{p}|$ and $1 \gg v^2$ (natural units), so (17-171) simplifies into the famous <u>Rutherford scattering</u> relation

$$\left(\frac{d\sigma}{d\Omega'}\right)_{\substack{e^- \,\&\\ \text{extern}\\ \text{field}}} = \frac{(\alpha Z)^2}{4m^2 v^4 \sin^4(\theta/2)} \quad \left\{\begin{array}{l}\text{non relativistic, field of}\\ \text{heavy pointlike nucleus,}\\ \text{unmeasured spins}\end{array}\right. .$$

(17-173)

Special case: non-relativistic scattering

17.6 Bremsstrahlung and Infrared Divergences

17.6.1 Bremsstrahlung

When a charged particle slows down, it loses energy and momentum. This energy lost by the particle is carried away by radiation it emits, i.e., by photon(s). This process, and the radiation emitted, goes by the name <u>bremsstrahlung</u>, which in German literally means "braking radiation". In fact, any change in momentum of a charged particle, such as an electron scattered by a nucleus or anything else, entails a deflection, i.e., a change in momentum, for the particle. Any deceleration entails radiation emission and the term bremsstrahlung is generalized to include any such behavior.

Bremsstrahlung overview

Figure 17-22 shows the two Feynman diagrams associated with the two sub-amplitudes for the case of bremsstrahlung comprising both an external field (like that of the nuclear Coulomb potential field), represented by the photon with an X at its origin, and an emitted quantized photon field. Note that the process is inelastic because a new particle (the emitted photon) is created.

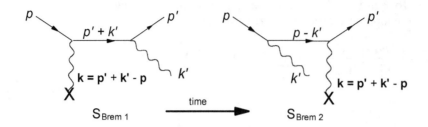

Figure 17-22. Two Feynman Diagrams for Bremsstrahlung

The transition amplitude for this process, found via Feynman's rules, where $\mathbf{k} = \mathbf{p}' + \mathbf{k}' - \mathbf{p}$, is

$$S_{Brems} = 2\pi\,\delta\left(E' - \omega - E\,\right))\left(\frac{m}{VE}\right)^{1/2}\left(\frac{m}{VE'}\right)^{1/2}\left(\frac{1}{2V\omega}\right)^{1/2}\left(\mathcal{M}_{Brem1} + \mathcal{M}_{Brem2}\right)$$

$$\mathcal{M} = \mathcal{M}_{Brem1} + \mathcal{M}_{Brem2} \tag{17-174}$$

$$= -e^2 \bar{u}_{s'}(\mathbf{p}')\left\{ \displaystyle{\not}\epsilon(\mathbf{k}')\,iS_F\left(p' + k'\right)\displaystyle{\not}A_e(\mathbf{k}) + \displaystyle{\not}A_e(\mathbf{k})\,iS_F\left(p - k'\right)\displaystyle{\not}\epsilon(\mathbf{k}')\right\} u_s(\mathbf{p}).$$

The derivation of the cross-section relation is similar to what we have done before, where, in doing so, we sum and average over electron spins. However, in the course of evaluating the amplitude of (17-174) in order to find that, we see the fermion propagators have form

$$S_F\left(p' + k'\right) = \frac{\displaystyle{\not}p' + \displaystyle{\not}k' + m}{\left(p' + k'\right)^2 - m^2} = \frac{\displaystyle{\not}p' + \displaystyle{\not}k' + m}{m^2 + 2p'k' + (k')^2 - m^2} = \frac{\displaystyle{\not}p' + \displaystyle{\not}k' + m}{2p'k'} = \frac{\displaystyle{\not}p' + \displaystyle{\not}k' + m}{2\left(E'\omega' - \mathbf{p}'\boldsymbol{\cdot}\mathbf{k}'\right)}$$

$$S_F\left(p - k'\right) = \frac{\displaystyle{\not}p - \displaystyle{\not}k' + m}{-2pk'} = \frac{\displaystyle{\not}p - \displaystyle{\not}k' + m}{-2\left(E\omega' - \mathbf{p}\boldsymbol{\cdot}\mathbf{k}'\right)}. \tag{17-175}$$

For the limiting case of the emitted <u>soft photon</u> ($\omega' = |\mathbf{k}'| \approx 0$), $\mathcal{M}$ diverges, and hence, so does the cross-section. We have an <u>infrared divergence</u>, which we alluded to in earlier chapters, but now address.

The first step in seeing how this infrared-caused infinity issue is resolved entails temporarily assuming the photon has a non-zero, small mass μ. In this case, the lowest value ω' can have is μ, so (17-175) will not diverge.

*No divergence if
assume photon
has mass μ*

When we do this and plow through all the algebra, we find the bremsstrahlung cross-section, where θ' and $d\Omega'$ refer to the final electron, to be

$$\left(\frac{d\sigma}{d\Omega'}\right)_{Brems} = \left(\frac{d\sigma}{d\Omega'}\right)_{\substack{e^- \,\& \\ \text{extern} \\ \text{field}}} \frac{\alpha}{\pi} ln\frac{-k^2}{m^2} ln\frac{E^2}{\mu^2}. \tag{17-176}$$

where the differential cross-section just after the equal sign is (17-166). We have not restricted (17-176) to a Coulomb potential (which would make the cross-section after the equal sign (17-171)), so it is good for soft bremsstrahlung in an arbitrary static external field.

For the second step, we need to look anew at the calculation we made to find the magnetic moment, involving the vertex correction of Fig. 16-4, repeated below as Fig. 17-23.

17.6.2 The Vertex Correction and Its Infrared Divergence

In Chap. 16 (see comment below (16-112). pg. 426), when we calculated the magnetic moment, we were able to use Gordon's identity in a clever way to avoid having to calculate some terms in the amplitude associated with Fig. 17-23 that were infrared divergent.

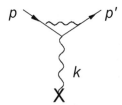

Figure 17-23. Vertex Correction Feynman Diagram for Scattering by External Field

*Lowest order
radiative
correction
Feynman
diagram*

However, if we want to calculate the cross-section for electron scattering by an external field, then we need to include these terms. The infrared divergence in the amplitude makes the cross-section infrared divergent.

To hold those infinities at bay, we temporarily, as we have done before, assume the photon has a non-zero mass μ. When we do this, and carry out the rather lengthy algebra involved, one finds that including the lowest order radiative correction leads to

$$\left(\frac{d\sigma}{d\Omega'}\right)_{\substack{e^- \ \& \\ \text{extern} \\ \text{field, lowest} \\ \text{correction}}} = \left(\frac{d\sigma}{d\Omega'}\right)_{\substack{e^- \ \& \\ \text{extern} \\ \text{field}}} \left(1 - \frac{\alpha}{\pi} ln\frac{-k^2}{m^2} ln\frac{-k^2}{\mu^2}\right). \tag{17-177}$$

Recall that the differential cross-section factor just after the equal sign is the result (17-171) (i.e., (17-166), which assumed no vertex correction in Fig. 17-23. And we note that (17-177) represents an elastic interaction. No new particles are created.

17.6.3 Cancelling of Infrared Divergences

Note that for vanishingly small energy for the bremsstrahlung photon, that photon is undetectable physically. So in that limit, the two processes (bremsstrahlung as in Fig. 17-22 and elastic charged particle scattering from a potential field as in Fig. 17-23) cannot be discerned experimentally, and thus we must add their cross-sections to get the measured cross-section.

<u>Lowest Order Radiative Correction</u>

Hence, the total cross-section for an electron scattered by an external field, including the infrared limit for very soft photons, is the sum of (17-176) and (17-177), i.e.,

$$\left(\frac{d\sigma}{d\Omega'}\right)_{\substack{e^- \ \& \\ \text{extern} \\ \text{field,} \\ \text{experim}}} = \overbrace{\left(\frac{d\sigma}{d\Omega'}\right)_{Brems}}^{\text{inelastic}} + \overbrace{\left(\frac{d\sigma}{d\Omega'}\right)_{\substack{e^- \ \& \\ \text{extern} \\ \text{field, lowest} \\ \text{correction}}}}^{\text{elastic}}$$

$$= \left(\frac{d\sigma}{d\Omega'}\right)_{\substack{e^- \ \& \\ \text{extern} \\ \text{field}}} \left(1 + \frac{\alpha}{\pi} ln\frac{-k^2}{m^2}\left(ln E^2 - ln(-k^2) + ln\frac{1}{\mu^2} - ln\frac{1}{\mu^2}\right)\right). \tag{17-178}$$

The last two terms cancel, so taking $\mu \to 0$, (17-178) remains finite for massless photons.

The infinite contribution from radiative correction of order α in elastic scattering cancels the infinite contribution from the inelastic scattering of very low energy bremsstrahlung radiation. The so-called <u>infrared catastrophe</u> arises from considering soft bremsstrahlung and elastic scattering as separate processes in perturbation theory. The separation is artificial. One is always present with the other, and the total cross-section from both combined is finite.

<u>All Orders of Radiative Corrections</u>

It is well beyond the scope of this book to deduce the equivalent of (17-178) to all orders of radiative corrections. It has, however, been proven, as stated in the famous <u>Bloch-Nordsieck theorem</u>. This asserts that the infrared divergences cancel exactly at all orders of perturbation theory, resulting in a finite correction to the cross-section of any order in α.

From some perspectives, this cancellation seems miraculous. Regardless, it does save QFT.

*Another infrared
divergence arises
from the
correction*

*No divergence if
assume photon
has mass μ*

*For this, all the
algebra yields
cross-section
including lowest
order correction*

*Cannot
distinguish very
low energy
bremsstrahlung
from pure elastic
e^- scattering*

*So must add their
cross-section
contributions*

*The terms with
photon mass μ
cancel, so in
limit $\mu \to 0$,
cross-section
is finite*

*Infrared
catastrophe arises
from considering
the two processes
separately.
Disappears when
they are combined.*

*Others have shown
the cancellation
works to all orders
of radiative
corrections*

17.7 Closure

But for the chapter summary below, this brings to an end our study of the canonical quantization approach to QFT and its application to QED. We summarize this approach in Chap. 19, and we offer a very brief introduction to the path integral approach in Chap. 18.

If you have gotten this far, congratulations on a fine effort!

This ends this book's treatment of canonical quantization QFT

17.8 Chapter Summary

The <u>cross-section σ</u> is the probability of two initial particles interacting with both confined to a unit sized area, and where that area is perpendicular to the relative velocity (as seen in the lab) between the two particles. The units (cm^2, barns, whatever) for the unit sized area are those of σ.

<u>Differential cross-section $d\sigma/d\Omega$</u> (a function of θ) is, under the above conditions, the probability per unit solid angle of measuring a particular final particle whose velocity has an angle θ with respect to the line of action of the relative velocity between the two initial particles.

Both σ and $d\sigma/d\Omega$ are dependent only on the particle types and their initial energies. For given particle types and energies, they are the same in different experiments and so provide a good measure of the inherent strength of interaction between said particle types. Higher cross-section means two particle types are more likely to interact.

The σ is the integral of $d\sigma/d\Omega$ over the surface of a sphere,

$$\sigma = \int_{\theta,\phi} \frac{d\sigma}{d\Omega}(\theta)\, d\Omega . \qquad (17\text{-}179)$$

In experiment, the <u>transition rate</u> is (for large numbers of transitions) effectively equal to the probability of a transition per unit time for two only initial particles times the number of first initial particles times the number of second initial particles. This is expressed, where t represents target particle and b beam particle, V is the volume the particle waves occupy, N represents the number of particles, and v_b is beam particle speed, as

$$\text{total transition rate} = f_b \sigma N_t \qquad \frac{\text{transition rate}}{\text{unit solid angle}} = f_b \frac{d\sigma}{d\Omega} N_t \qquad f_b = \frac{N_b}{V_b} v_b . \qquad (17\text{-}180)$$

When both particles are moving with collinear velocities, $v_b \rightarrow$ magnitude of relative velocity between them.

With 3-momenta of two initial particles known, the energy-momentum conservation laws (4 of them) mean we can calculate four final unknown momentum values if we measure (or determine from theory) the rest of the final unknowns. For two final particles in 3D, there are 6 unknown 3-momentum components. Thus, by measuring any two, we can determine the other four. If we express the six momentum components in terms of 3-momentum magnitudes and polar angles θ, ϕ, then we have six different unknowns $|\mathbf{p}'_1|$, θ_1', ϕ_1', $|\mathbf{p}'_2|$, θ_2', and ϕ_2'. By measuring θ_1' and ϕ_1', we can determine all the others via conservations laws. Since scattering problems are set up to be symmetric in ϕ (i.e., differential cross-sections are independent of ϕ), we seek differential cross-sections as functions of θ_1'.

Wholeness Chart 17-2, pg. 445, summarizes non-relativistic and relativistic scattering analyses for classical (hard objects, objects in potential fields) and quantum (NRQM, RQM, QFT) cases.

For QFT, we can use our prior development of the theory as a special case of (17-180). That is, we analyze a particular hypothetical experimental setup that matches the conditions we assumed in developing QFT in Parts 1 to 3 of this book. See Step 2 in Wholeness Chart 17-6 below. Since σ and $d\sigma/d\Omega$ for the same types of interacting particles are unaffected by any particular experimental setup, we choose a setup (the QFT case setup) that makes analysis tractable.

Wholeness Chart 17-6 (Parts 1 and 2) summarizes the steps to deduce differential cross-sections in QFT as presented in this chapter.

<u>Infrared divergences</u> that arise in <u>bremsstrahlung</u> scattering cross-sections are cancelled to all orders by divergent radiative vertex correction terms that, almost miraculously, have equal magnitude but opposite sign from the bremsstrahlung terms.

Wholeness Chart 17-6. Differential Cross-section Determination in QFT

Part 1: Two Initial Particles

Step	Procedure	In Text	Example(s)
1	$$\frac{\text{transition rate}}{\text{unit solid angle}} = f_b \frac{d\sigma}{d\Omega} N_t , \quad f_b = \frac{N_b}{V_b} v_b$$	(17-10)	Every two initial particles case (any type particles; classical, NRQM, RQM, QFT; theory & experiment)
2	To apply QFT as developed, take $$N_b = N_t = 1; V_b = V_t = V; V,T \to \infty;$$ $$\frac{\text{transition rate}}{\text{unit solid angle}} \text{ (at } \theta) = \frac{d\left(|S_{Fi}|^2 / T\right)}{d\Omega}$$	Pg. 441 (17-19)	Every one of our QFT scattering cases
	Use step 2 in step 1 with $d|S_{Fi}|^2 = |S_{fi}|^2 dN_f$ to get ✓		
3	$$d\sigma = \frac{V}{v_b T} |S_{fi}|^2 dN_f \quad \text{(for } dN_f \text{ final states)}$$	(17-22)	As above.
4	Chose frame. If both initial particles moving, $v_b \to v_{rel}$	(17-39)	COM frame $v_{rel} = |\mathbf{p}_1|\left(\frac{E_1 + E_2}{E_1 E_2}\right)$
	Use relation between N_f state and its 3-momentum to get ✓		
5	$$dN_f = \prod_{f'}^{M} \frac{V d^3 \mathbf{p}_{f'}}{(2\pi)^3} \quad \text{for } M \text{ final particles}$$	(17-49)	$M = 2$ in all cases in this book
	Use step 4, step 5 with $M = 2$, S_{fi} from QFT, $(2\pi)^4 \delta^{(4)}\left(P_f - P_i\right) \to VT$, in step 3 to get step 6 ✓		
6	$$d\sigma = \delta^{(4)}\left(p_1' + p_2' - p_1 - p_2\right) \frac{1}{64\pi^2 v_{rel} E_1 E_2 E_1' E_2'}$$ $$\times \left(\prod_{l}^{\substack{\text{extern} \\ \text{ferms}}} 2m_l\right) d^3\mathbf{p}_1' d^3\mathbf{p}_2' |\mathcal{M}|^2 \left\{\begin{array}{l}\text{2 initial and 2} \\ \text{final particles,} \\ \text{elastic or inelast}\end{array}\right.$$	(17-54)	$e^+ e^- \to e^+ e^-$; $e^+ e^- \to l^+ l^-$; $e^- \gamma \to e^- \gamma$; $e^- e^- \to e^- e^-$; $e^+ e^- \to \gamma\gamma$; $\gamma\gamma \to e^+ e^-$
	Integrate step 6 over $\mathbf{p}_2'$ to get ↓ in terms of single unknown $\mathbf{p}_1'$		
7	$$d\sigma = \delta\left(E_1' + E_2' - E_1 - E_2\right)$$ $$\times \frac{1}{64\pi^2 v_{rel} E_1 E_2 E_1' E_2'}\left(\prod_{l}^{\text{ext ferms}} 2m_l\right) d^3\mathbf{p}_1' |\mathcal{M}|^2$$	(17-57)	As above.
	Use $d^3\mathbf{p}_1' = |\mathbf{p}_1'|^2 d\Omega_1' d|\mathbf{p}_1'|$ to get step 7 in terms of $|\mathbf{p}_1'|$ and $d\Omega_1'$, use a math trick (17-60) with the delta function, then divide by $d\Omega_1'$ to get ↓		

8	$$\frac{d\sigma}{d\Omega_1'} = \frac{1}{64\pi^2 v_{rel} E_1 E_2 E_1' E_2'} \left(\overset{\text{extern}}{\underset{l}{\overset{\text{ferms}}{\prod}}} 2m_l \right) |\mathcal{M}|^2 \times$$ $$|\mathbf{p}_1'|^2 \left(\frac{\partial(E_1' + E_2')}{\partial|\mathbf{p}_1'|} \right)^{-1} \begin{cases} \text{any frame,} \\ \text{2 initial and 2} \\ \text{final particles,} \\ \text{elastic or inelastic} \end{cases}$$	(17-62)	As above
	Choose a frame to express energies and evaluate $(\ )^{-1}$ bracket above, to get step 9 ✔		
9	$$\left(\frac{d\sigma}{d\Omega_1'} \right)_{\substack{chosen \\ frame}}$$	(17-68) (17-69)	$$\left(\frac{d\sigma}{d\Omega_1'} \right)_{COM} = \frac{1}{64\pi^2 (E_1 + E_2)^2}$$ $$\times \frac{|\mathbf{p}_1'|}{|\mathbf{p}_1|} \left(\overset{\text{extern}}{\underset{l}{\overset{\text{ferms}}{\prod}}} 2m_l \right) |\mathcal{M}|^2$$
10	Find $\mathcal{M}$. (Use e not e_0. Assume negligible propagator & vertex radiative corrections.)	(17-118)	Bhabha scattering example from here on. $\mathcal{M} = \mathcal{M}_{B1} + \mathcal{M}_{B2}$ $(E_1 = E_2 = E)$
11a.	If spins and polarizations measured, find $|\mathcal{M}|^2$	(17-119)	$$\left(\frac{d\sigma}{d\Omega_1'} \right)_{\substack{COM \\ e^- e^+ \to e^- e^+}} = \frac{m^4}{16\pi^2 E^2} \left| \mathcal{M}_{B1} + \mathcal{M}_{B2} \right|^2$$
11b	If not, do spin/polarization sums and averages. fermions: $$\underbrace{\sum_{s_1'=1}^{2} \cdots \sum_{s_q'=1}^{2}}_{q \text{ final}} \left(\left(\frac{1}{2}\right)^n \underbrace{\sum_{s_1=1}^{2} \cdots \sum_{s_n=1}^{2} |\mathcal{M}|^2}_{n \text{ initial fermions}} \right)$$ photons: m initial, p tot $(-1)^p \left(\frac{1}{2}\right)^m \mathcal{M}^{\mu\nu\cdots} \mathcal{M}^*_{\mu\nu\cdots}$	(17-78) (17-96)	$$\left(\frac{d\sigma}{d\Omega_1'} \right)_{\substack{COM \\ e^- e^+ \\ \to e^- e^+}} = \frac{m^4}{16\pi^2 E^2} \left(\frac{1}{4} \sum_{spins} \left(\left| \mathcal{M}_{B1} \right|^2 \right. \right.$$ $$\left. \left. + \left| \mathcal{M}_{B2} \right|^2 + \mathcal{M}_{B1}\mathcal{M}_{B2}^* + \mathcal{M}_{B2}\mathcal{M}_{B1}^* \right) \right)$$ $\uparrow$ (17-120)
12	For step 11b, evaluate traces involved using gamma matrix relations		
	In step 11a or step 12, express $\mathbf{p}$ and $\mathbf{p}_1'$ in terms of $|\mathbf{p}|$, $|\mathbf{p}_1'|$, and θ $(= \theta_1')$, then that result in terms of incoming particle energies and θ, to get $\downarrow$		
13	$$\left(\frac{d\sigma}{d\Omega_1'} \right)_{\substack{chosen \\ frame}}$$ as a function of θ	(17-133)	$$\left(\frac{d\sigma}{d\Omega_1'} \right)_{\substack{COM \\ e^- e^+ \to e^- e^+}} = \frac{\alpha^2}{2E_{tot}^2} \left(\frac{1 + cos^2\theta}{2} + \right.$$ $$\left. \frac{1 + cos^4(\theta/2)}{sin^4(\theta/2)} - \frac{2cos^4(\theta/2)}{sin^2(\theta/2)} \right) \begin{cases} E_{tot} \gg 2m \\ \text{spin} \\ \text{unmeas} \end{cases}$$

Part 2: Initial Fermion and External Potential Field

Step	Procedure	In Text	Example
1 to 10	Similar to steps 1 to 10 in Part 1 above, but 1 initial particle and 1 external potential field $A_e(\mathbf{k})$. Use Feynman rule #11 to get $$\left(\frac{d\sigma}{d\Omega'_1}\right)_{\substack{extern\ field\\ frame}}$$	(17-166)	Example in all boxes below: only 1 final particle, and it is a fermion. $$\left(\frac{d\sigma}{d\Omega'}\right)_{\substack{ext\ field\\ frame,\ e^-\\ \&\ ext\ field}} = \left(\frac{me}{2\pi}\right)^2 \left| \bar{u}_{s'}(\mathbf{p}')A_e(\mathbf{k})u_s(\mathbf{p})\right|^2$$
11	Assume particular form for potential	(17-167)	Coulomb potential $A_e^\alpha(\mathbf{k}) = \left(\dfrac{Ze}{|\mathbf{k}|^2}\ \ 0\ \ 0\ \ 0\right)$
11a	If spins measured, find $\mathcal{M}|^2$		
11b	If not, do spin sums	(17-168)	$$\left(\frac{d\sigma}{d\Omega'}\right)_{\substack{ext\ field\\ frame,\ e^-\\ \&\ ext\ field}} = 2\left(\frac{m\alpha Z}{|\mathbf{k}|^2}\right)^2 \mathrm{Tr}\left\{\left(\frac{\not{p}'+m}{2m}\right)\gamma^0\left(\frac{\not{p}+m}{2m}\right)\gamma^0\right\}$$
12	For step 11b, evaluate traces involved using gamma matrix relations		
	In step 11a or step 12, express $\mathbf{k}$, $\mathbf{p}$, and $\mathbf{p}'$ in terms of $|\mathbf{p}|$, $|\mathbf{p}'|$, and θ ($=\theta'$), then that result in terms of incoming particle energy E and θ, to get ↓		
13	$$\left(\frac{d\sigma}{d\Omega'_1}\right)_{\substack{extern\ field\\ frame}}$$ as a function of θ	(17-172) (17-173)	$$\left(\frac{d\sigma}{d\Omega'}\right)_{\substack{Rutherford\\ scattering}} = \frac{(\alpha Z)^2}{4m^2v^4\sin^4(\theta/2)} \begin{cases} E\approx m,\\ spins\\ unmeas \end{cases}$$

17.9 Problems

1. For a given experiment where beam particles of type A are scattered off of resting target particles of type B, the measured transitions per second equals 2×10^2. Beam flux $f_b = 10^7$ particles per second per cm^2, and total target particles inside the beam flux $N_t = 10^{21}$. Find the cross-section σ.

    For a second experiment between the same two particle types, measured transition rate = 120 detections/second, $f_b = 3\times10^5$, and $N_t = 2\times10^{22}$. Find σ.

    What significance is there in the values you found for σ in both cases?

2. Show that the NRQM total wave function for the incident and scattered waves of Fig. 17-8 on pg. 439, i.e., (17-13), has normalization factor $C = 1/\sqrt{AL}$.

3. Show that for relativistic, elastic scattering of two particles in the COM frame, that each particle has the same kinetic energy after scattering as before. That is, each particle can have different direction for its velocity after scattering, but it must have the same magnitude for its velocity. Hint: Use $E_1^2 - \mathbf{p}_1^2 = m_1^2 = E_1'^2 - \mathbf{p}_1'^2$, a similar relation for particle #2, and the 3-momentum balance equation [where total 3-momentum is zero in the COM] to find $E_1^2 - E_1'^2 = E_2^2 - E_2'^2$. Divide this by the energy balance equation. Then add and subtract the energy balance equation from your result.

4. Derive the spin sum for the second row of Wholeness Chart 17-5, i.e., for positron photon scattering.

5.   Derive (17-104).

6.   Derive (17-106).

7.   Derive (17-122).

8.   Show that, for Compton scattering, $p^\mu k_\mu = p'^\mu k'_\mu$. Hint: Square the conservation relation (17-139). That is, find $\left(p^\mu + k^\mu\right)\left(p_\mu + k_\mu\right) = \left(p'^\mu + k'^\mu\right)\left(p'_\mu + k'_\mu\right)$. Also, show that $p'^\mu k_\mu = p^\mu k'_\mu$. Hint: Start by re-arranging the conservation relation.

9.   Show (17-151)(c). i.e., Show

$$X_{12} = -\frac{1}{16m^2 (pk)(pk')} \text{Tr}\left\{ \gamma^\mu \left(\not{p} + \not{k} + m \right) \gamma^\nu \left(\not{p} + m \right) \gamma_\mu \left(\not{p} - \not{k'} + m \right) \gamma_\nu \left(\not{p}' + m \right) \right\}$$

10. By means of analogy, deduce X_{22} and X_{21} of (17-151)(b) and (d) in one step from X_{11} and X_{12} of (17-151)(a) and (c), without going through the algebraic steps to derive X_{11} and X_{12} of the text and Prob. 9.

11. (Problem added to revision of 2nd edition.) Show that, in any Lorentz frame where two particles have collinear 3-momenta, that the magnitude of the relative velocity between them (where subscripts 1 and 2 refer to the first and second particles) is

$$v_{rel} = \frac{\sqrt{(p_1 p_2)^2 - m_1^2 m_2^2}}{E_1 E_2} .$$

Hint: Start with (17-38) $v_{rel} = \left| \mathbf{v}_1 - \mathbf{v}_2 \right| = \left| \dfrac{\mathbf{p}_1}{E_1} - \dfrac{\mathbf{p}_2}{E_2} \right|$, solve for

$E_1 E_2 v_{rel} = \left| E_2 \mathbf{p}_1 - E_1 \mathbf{p}_2 \right| = \sqrt{\left(E_2 \mathbf{p}_1 - E_1 \mathbf{p}_2 \right) \cdot \left(E_2 \mathbf{p}_1 - E_1 \mathbf{p}_2 \right)}$, and carry out the inner product in the

latter expression. Then use $\left(p_1 p_2 \right)^2 = \left(E_1 E_2 - \mathbf{p}_1 \cdot \mathbf{p}_2 \right)^2 = E_1^2 E_2^2 + \left| \mathbf{p}_1 \right|^2 \left| \mathbf{p}_2 \right|^2 - 2 E_1 E_2 \mathbf{p}_1 \cdot \mathbf{p}_2$ (where the

last part is valid when 3-momenta are co-linear) and crank the algebra.

Then, show that the special cases for v_{rel} for the COM frame and particle #2 stationary are (17-39) and (17-40), respectively.

12. (Problem added to revision of 2nd edition.) Find (17-168). Hint: (4-153), pg. 122 can be useful.

13. (Problem added to revision of 2nd edition.) Find (17-169). Hint: Trace results of pg. 124 can be useful.

Addenda

*"My father says that almost everyone is asleep. Those few who
are awake go around being amazed all the time."*

Meg Ryan's character in
movie *Joe Versus the Volcano*

*Chapter 18 Path Integrals in Quantum Theories:
A Pedagogic First Step*

*Chapter 19 Looking Backward and Looking Forward:
Book Summary and What's Next*

Chapter 18

Path Integrals in Quantum Theories:
A Pedagogic First Step

The universe in each dimension
is vast beyond all comprehension.
A myriad of mysteries,
a multitude of histories ...

From *Divine Intentions*
by R. Klauber

18.0 Preliminaries

As I mentioned on the first page of the book, I strongly believe it is far easier, and more meaningful, for students to learn quantum field theory (QFT) first by the canonical quantization method, and once that has been digested, move on to what is termed the path integral (functional integral, many paths, or sum over histories) approach (<u>functional quantization</u>). The rest of the book is devoted to the first of these; the present chapter, to a brief introduction to the second.

Two approaches to (ways to quantize) QFT:
1) canonical
2) path integral

18.0.1 Chapter Overview

This chapter was composed so it can be read independently of (without reading) the rest of the book. So, some things may be defined/discussed again herein that are covered elsewhere in the text.

In this chapter, we will define

- the functional and
- the functional integral,

then, with regard to non-relativistic quantum mechanics (NRQM),

- transition amplitudes for position eigenstates,
- the role of the Lagrangian and the wave function peak,
- the central idea in Feynman's path integral approach,
- expressing that idea mathematically, including Feynman's postulates,
- comparing the path integral approach in NRQM to Schrödinger and Heisenberg's,
- determining the transition amplitude from the functional integral, and
- applying the theory to an example.

Then, with regard to QFT, we will investigate

- comparing particle theory (NRQM) to field theory (QFT)
- "derivation" of the many paths approach to QFT, and
- deducing the form of the transition amplitude for QFT

#1 simpler, rest of book;
#2 introduced in this chapter

We'll examine path integrals:
- math behind
- NRQM
- QFT

18.1 Background Math

18.1.1 Integrating Functions of a Function

Functionals form the mathematical roots of Feynman's many paths approach to quantum theories. To help in understanding the concept, consider first a function of another function, such as the Lagrangian of a particle, which is typically a function of particle position x and its time derivative $\dot{x}$. Position x, in turn, is a function of time t, i.e., $x(t)$, and finding that functional dependence on time comprises typical problems to be solved.

There are several ways we can integrate such a function of another function, two being shown in Wholeness Chart 18-1 (Part A) below. The figures and comments in that chart should be self-explanatory. Mathematically, L can be any function of a function, but for our purposes, it will generally be the Lagrangian.

Integrating a function of a function

Wholeness Chart 18-1. From a Function of a Function to the Functional Integral – Part A

Case	Procedure	Graphically	Math	Comment
1.	Integration above the path in $x(t)$ vs t space = area shown		$\int_{s_a}^{s_b} L ds$ where s is spacetime distance along the path	L is a function of the function x (and $\dot{x}$), and the functional dependence of x on t is typically the problem to be solved. Integration shown is not relevant for us.
2.	Integration over t = projection of the area in case #1 onto the L-t plane		$F = \int_{t_a}^{t_b} L dt$ F is a <u>functional</u>	If L is the Lagrangian, then this integral $F = S$, the action. Classically, S = minimum (or stationary) for classical physical paths

18.1.2 Defining "Functional"

In the path integral approach to quantum physics, we use a narrower definition of a functional than the general mathematical definition[1]. We define the integration of case #2 above, the integral of the function (L) of a function ($x(t)$) with respect to the independent variable (t) between fixed limits t_a and t_b as a <u>functional</u>, and designate it as F. It is a number that depends on the form of the function $x(t)$, on t_a, and on t_b. It is different for different paths.

Our definition of a functional F tailored to quantum physics

$$F = \int_{t_a}^{t_b} L dt \quad \text{(for a particular path)} \tag{18-1}$$

That definition

Functionals are often symbolized by enclosing their arguments in square brackets.

$$\underline{\text{Symbolism}:} \qquad F\big[x(t)\big] \quad \text{or} \quad F\big[x\big], \tag{18-2}$$

though you may see functionals written with normal, rather than square, brackets.

If L is the Lagrangian, then the functional $F = S$, the action.

If L = Lagrangian, F = S, the action

18.2 Defining Functional Integral

A functional (our definition) is a definite integral, i.e., a number obtained by integrating between the end points of a certain path. Yet, because we get a different such number for each different path in x-t space, we can integrate those numbers over all possible paths. In other words, the functional, an integral for us, can itself be integrated. Such integrations are not simple, nor is their purpose at all obvious at this point. They are visualized in cases #4 and #7 below and are called <u>functional integrals</u>. We devote much of this chapter to explaining their origin, value, and means to evaluate. For now, just let the general concept sink in, without straining to analyze it too much.

The functional integral is an integral (over all paths) of the functional F (itself an integral)

---

[1] Mathematically, a functional is a function of a vector space to a scalar field, i.e., a functional maps a vector to a scalar. Spatial functions of time, i.e., paths, form a vector space by themselves, so our narrower definition is in line with the general definition. In our case, the mapping involves an integration.

Wholeness Chart 18-1 (continued). From a Function of a Function to the Functional Integral – Part B

Case	Procedure	Graphically	Math	Comment
3.	Sum F values as in case #2 above for a number of discrete paths between **a** and **b**.	$L(x(t))$ 4 of an infinite number of paths t **p** **a** $x(t)$	$\sum_{n=1}^{4} F_n =$ $\sum_{n=1}^{4} \int_{t_a}^{t_b} L_n dt$	Not relevant for us.
4.	Integrate F over all possible (continuous range of) paths between **a** and **b**.	Hard to show visually.	$\int_{x_a}^{x_b} F \mathcal{D}x(t)$	Not relevant for us. $\mathcal{D}x(t)$ implies all paths.
5.	Another function of F (i.e., where F is the argument), e.g. exponentiation of F.	Not graphic. Raise e to i times value F for a given path.	$e^{iF[x(t)]} = e^{i\int_{t_a}^{t_b} L dt}$	Relevant for us.
6.	Sum e^{iF} values for a number of discrete paths, like in case #3 above.	Same paths as in #3.	$\sum_{n=1}^{4} e^{iF_n} = \sum_{n=1}^{4} e^{i\int_{t_a}^{t_b} L_n dt}$	Relevant for us.
7.	Integration like case #4 above over all possible paths in $x(t)$ vs t space.	Hard to show visually. Same paths as in case #4.	$\int_{x_a}^{x_b} e^{iF} \mathcal{D}x(t)$	Feynman QM path integral approach. All paths, not just classical.

The chart above should be relatively self-explanatory. In summary, we can add the values F_n for a discrete number N (= 4 in case #3) of paths. In the limit of adding all paths, we pass to an integral (don't worry how for now), where we use the <u>symbol $\mathcal{D}x(t)$</u> to represent that functional integration.

$$\sum_{n=1}^{N} F_n = \sum_{n=1}^{N} \int_{t_a}^{t_b} L_n dt \xrightarrow[\text{total paths } N \to \infty]{\text{limit as}} \int_{x_a}^{x_b} F \mathcal{D}x(t) \qquad \text{(not relevant for us)} . \qquad (18\text{-}3)$$

Adding F for all paths → integration over all paths = functional integration

Alternatively, we can do the same thing for a function of F, such as e^{iF} (as in cases #6 and #7 above). Note that e^{iF} can itself be considered a functional, as it comprises a mapping from $x(t)$ to a (complex) scalar.

$$\sum_{n=1}^{N} e^{iF_n} = \sum_{n=1}^{N} e^{i\int_{t_a}^{t_b} L_n dt} \xrightarrow[\text{total paths } N \to \infty]{\text{limit as}} \int_{x_a}^{x_b} e^{iF} \mathcal{D}x(t) \qquad \text{(will be relevant for us)} . \quad (18\text{-}4)$$

Instead of F, do functional integration of e^{iF} over all paths. This is important in quantum physics

We will evaluate (18-4) for a free quantum particle later in this chapter.

<u>Alternative nomenclature:</u> Because <u>functional integration</u> involves integration over paths (in *x-t* space), Feynman's approach is often also referred to as the <u>path integral</u> approach.

18.3 The Transition Amplitude

18.3.1 General Wave Functions (States)

Review of states, amplitudes, & probability

Recall from QM wave mechanics, that for a general normalized wave function ψ equal to a superposition of energy eigenfunction waves (which are each also normalized),

$$\psi = A_1 \psi_1 + A_2 \psi_2 + A_3 \psi_3 , \qquad (18\text{-}5)$$

$\psi \to \psi_1$ transition amplitude = A_1

A_1 is the (complex) amplitude of ψ_1, so the probability of finding ψ_1 upon measuring is

$$A_1^* A_1 = |A_1|^2 . \qquad (18\text{-}6)$$

If we were to start with ψ initially, and measure ψ_1 later, the wave function would have collapsed, i.e., underwent a transition to a new state. (18-6) would be the transition probability.

Probability of transition = $|A_1|^2$

<u>Definition</u>: The <u>transition amplitude</u> is that complex number, the square of the absolute magnitude of which is the probability of measuring a transition from a given initial state to a specific final state. (As discussed in Chaps. 1, 7, 8, etc.)

<u>Symbolism</u>: The transition amplitude for a time of interaction approaching infinity, as in the canonical quantization approach, is typically written as S_{fi} (see chapters cited above). However, in the path integral approach, where elapsed time T between measurements of the initial state ψ_i and final state ψ_f is commonly finite, it is more typical to write

Square of absolute value of transition amplitude = probability of transition

$$U\left(\psi_i, \psi_f; T\right) \qquad \left(\text{for } T \to \infty,\; U = S_{fi} \text{ of canonical quantization}\right). \qquad (18\text{-}7)$$

Symbol U for path integrals with T finite; for $T \to \infty$, $U = S_{fi}$ of canonical case

This terminology carries over to inelastic cases (where particles change types). (Most of QFT, as seen in the rest of this book, is devoted to determining the transition amplitudes for the different possible interactions between particles.)

<u>Schrödinger Approach – Transition Amplitudes</u>

The Schrödinger approach to QM leads to an expression of the transition amplitude of form (note the parallel with (7-62), pg. 198)

U for NRQM Schrödinger wave mechanics approach

$$U\left(\psi_i, \psi_f; T\right) = \underbrace{\langle\psi_f|}_{\substack{\text{final state} \\ \text{measured} \\ \text{at } T + t_a}} e^{-iHT/\hbar} \underbrace{|\psi_i\rangle}_{\substack{\text{inital state at } t_a}} , \qquad (18\text{-}8)$$

$$\underbrace{\qquad\qquad\qquad\qquad}_{\text{evolved state at } T + t_a}$$

where H is the Hamiltonian operator, and we retain the symbol $\hbar$ even though $\hbar = 1$ in natural units.

<u>Alternative nomenclature</u>: The transition amplitude U is sometimes called the <u>propagator</u> (though *not* the QFT Feynman propagator). It projects the wave function at $T + t_a$ that evolved from the initial state $|\psi_i\rangle$ at t_a onto the final state $|\psi_f\rangle$ at time $T + t_a$. It "propagates" the particle from i to f.

18.3.2 Position Eigenstates

When the particle has a definite position, e.g., x_i, the ket is an eigenstate of position, written $|x_i\rangle$. The transition amplitude for measuring a particle initially at x_i, and finally at x_f, would take the form

$|x_i\rangle$, eigenstate of position, in x space rep, is a delta function; which can be effectively represented by a steep, narrow wave packet

$$U\left(x_i, x_f; T\right) = \langle x_f| \underbrace{e^{-iHT/\hbar}|x_i\rangle}_{\substack{\text{evolved state,} \\ \text{in } x \text{ space} = \psi}} . \qquad (18\text{-}9)$$

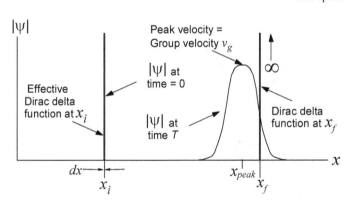

$|\psi|$

Peak velocity = Group velocity v_g

$|\psi|$ at time = 0

Effective Dirac delta function at x_i

$|\psi|$ at time T

∞

Dirac delta function at x_f

dx

x_i

x_{peak}

x_f

x

Figure 18-1. Propagation of an Effectively Initial Position Eigenstate Quantum Wave

A position eigenstate such as $|x_f\rangle$ is, in x space, a delta function of form[1] $\delta(x - x_f)$, schematically represented on the right in Fig. 18-1. As the initial state at x_i evolves into ψ, however, it, like wave packets generally do, spreads, and its peak diminishes (wave function envelope at right in Fig. 18-1.) The amplitude for measuring the particle at time T at x_f, i.e., for measuring $|x_f\rangle$ that collapsed from ψ, is (18-9).

We can re-write (18-9), in wave mechanics notation, as

It spreads as it evolves

When measured at x_f, wave packet collapses to $|x_f\rangle$, eigenstate of position, i.e., a delta function

So, U for position eigenstate at $x_f \to |U|^2 =$ probability density at x_f

$$U(x_i, x_f; T) = \int_{-\infty}^{+\infty} \delta\left(x - x_f\right)\psi\left(x, T\right)dx = \psi(x_f, T) \qquad (18\text{-}10)$$

---

[1] There are different ways to normalize position eigenstates. Here we use what is easiest to understand for our purposes. Also, in practice, a position measurement is always over finite Δx, not dx, so our initial delta function actually corresponds to a very narrow, very high real-world wave packet (with the standard normalization, such that the square of its absolute value is probability density).

Thus, $\left|U\left(x_i,x_f;T\right)\right|^2 = \left|\psi\left(x_f,T\right)\right|^2 = \psi^*(x_f,T)\psi(x_f,T) = \begin{cases}\text{probability density of measuring} \\ \text{particle at } x_f \text{ at time } T.\end{cases}$ (18-11)

<u>Modification to definition:</u> Hence, from (18-10), the square of the absolute value of the <u>transition</u> <u>amplitude for eigenstates of position</u> (with the chosen normalization and considering the initial state a very high, very narrow wave packet), is *probability density* (probability per unit *x*), *not probability*, as was the case for energy eigenstate wave functions of form (18-5).[1]

As we will see, the value found using the RHS of (18-9), i.e., that of the Schrödinger approach, is the same as the value found using Feynman's many paths approach.

18.4 Expressing the Wave Function Peak in Terms of the Lagrangian

18.4.1 Background

Path integral approach expresses wave function peak in terms of Lagrangian

One of Feynman's assumptions for his path integral approach to NRQM, RQM (relativistic quantum mechanics), and QFT was to express the wave function value at the peak of a wave packet (see Fig. 18-1) in terms of the Lagrangian (exact relation shown at the end of this section 18.4). I have never seen much justification for this in the literature, other than it is simply an assumption that works (so learn to live with it and move on!)

In the present section I take a different tack, by providing rationale for why we could expect Feynman's expression for the value of the wave function peak to work. The logic herein may well parallel what went on in Feynman's mind as he was developing his path integral approach.

18.4.2 Deducing Feynman's Phase Peak Relationship

<u>The Simplified, Heuristic Argument</u>

Heuristic way to deduce ψ_{peak}
$= Ae^{i\int\frac{L}{\hbar}dt}$

In NRQM, the plane wave function solution to the Schrödinger equation,

$$\psi = Ae^{-i\left(Et-\mathbf{p}\cdot\mathbf{x}\right)/\hbar} \, ,$$ (18-12)

means the phase angle, at any given $\mathbf{x}$ and t, is

$$\phi = -\left(Et-\mathbf{p}\cdot\mathbf{x}\right)/\hbar \, .$$ (18-13)

If we have a particle wave packet, it is an aggregate of many such waves, so it is not in an energy or momentum eigenstate. However, it does have energy and momentum expectation values that correspond to the classical values for the particle. The wave packet peak travels at the wave packet group velocity, which corresponds to the classical particle velocity.

Now, imagine that we approximate the wave packet with a (spatially short) wave function such as ψ, where E and $\mathbf{p}$ take on the values of the wave packet expectation values for energy and momentum, respectively. If $\mathbf{x}$ represents the position of the wave packet "peak" (the middle of our approximated wave function ψ), the time rate of change of phase at $\mathbf{x}$ is then

$$\frac{d\phi}{dt} = \frac{-\left(E-\mathbf{p}\cdot\mathbf{v}\right)}{\hbar} = \frac{-T-V+\mathbf{p}\cdot\mathbf{v}}{\hbar} \, ,$$ (18-14)

where $\mathbf{v}$ is the velocity of the wave peak, T is kinetic energy, and V is potential energy. Non-relativistically,

$$T = \tfrac{1}{2}mv^2 \qquad \mathbf{p} = m\mathbf{v} \quad \rightarrow \quad \mathbf{p}\cdot\mathbf{v} = 2T \, ,$$ (18-15)

so, in terms of the classical Lagrangian L, (18-14) becomes

$$\frac{d\phi}{dt} = \frac{T-V}{\hbar} = \frac{L}{\hbar} \, .$$ (18-16)

More formally, using the Legendre transformation

$$H = p_i\dot{q}_i - L \quad \left(E = \mathbf{p}\cdot\mathbf{v} - L \text{ here}\right),$$ (18-17)

directly in (18-14), after the first equality, we get (18-16).

Thus, from (18-16), the phase difference between two events the particle traverses is

---

[1] This definition of U differs from that of other authors. We address this in Sect. 18.9.1 and the appendix.

$$\phi = \int \frac{L}{\hbar} dt = \frac{S}{\hbar} \ , \tag{18-18}$$

where S is the classical action of Hamilton. The classical path between two events is that for which the Hamiltonian action is least. Note that (18-18) is an integral of type 2 in Wholeness Chart 18-1.

Hence, the wave function at the peak could be written in terms of the Lagrangian as

$$\psi_{peak} = Ae^{i\int \frac{L}{\hbar} dt} = Ae^{i\frac{S}{\hbar}} \tag{18-19}$$

This is the typical starting point assumption when teaching the Feynman path integral approach (still to be developed beginning in Section 18.5).

In RQM and QFT, we get a solution form similar to (18-12) (differing only in the normalization factor A), and thus (18-14) is also true relativistically. Further, since (18-17) is true relativistically, as well, then so are (18-16), (18-18), and (18-19).

More Precise Argument

The precise expression for a QM particle wave packet[1], where overbars designate expectation (classical) values; v_g, the group (peak, classical) velocity; and $g(p)$, the momentum space distribution, is

More formal way to deduce ψ_{peak} $= Ae^{i\int \frac{L}{\hbar} dt}$

$$\psi(x,t) = e^{-\frac{i}{\hbar}(\bar{E}t - \bar{p}x)} \underbrace{\frac{1}{2\pi\hbar} \int_{-\infty}^{+\infty} \underbrace{g(p)}_{\substack{real}} \underbrace{e^{-\frac{i}{\hbar}(v_g t - x)(p-\bar{p})}}_{\substack{=1 \text{ for } x= \text{peak,} \\ \text{i.e., for } x=v_g t}} \underbrace{e^{-\frac{i}{2\hbar}\frac{t}{m}(p-\bar{p})^2}}_{\substack{\text{time depend} \\ \& \text{ complex}}} dp}_{\text{call this } A(t) \text{ for } x=x_{peak}} \ . \tag{18-20}$$

We are interested in the value of (18-20) at the peak, $\psi(x_{peak},t)$, where $x_{peak} = v_g t$. To begin, note that with $x = x_{peak}$ inside the integral, the exponent of the second factor in the integrand equals zero, and so that factor equals one. The function $g(p)$ is typically a real, Gaussian distribution in $p - \bar{p}$, and independent of time. The third factor in the integrand is complex and time dependent.

Thus, with $x=x_{peak}$, the integral in (18-20) is a function (generally complex) only of time, which, along with the factor in front, we will designate as $A(t)$. Thus, for the entire history of the wave packet, the wave function value at the peak is

$$\psi(x_{peak},t) = A(t)e^{-\frac{i}{\hbar}(\bar{E}t - \bar{p}x_{peak})} \ . \tag{18-21}$$

Except for the time dependence in $A(t)$, this is equivalent to (18-12), as the expectation values for E and p equal the classical values for the particle. So, with regard to the exponent factor in (18-21), all of the logic from (18-13) through (18-19) applies here as well. The final result is so important, we repeat it below, with L being the classical particle Lagrangian, T representing the time when the peak is detected, and phase at $t = 0$ taken as zero. The RHS comes from (18-10).

$$\boxed{\psi(x_{peak},T) = A(T)e^{i\int_0^T \frac{L}{\hbar} dt} = A(T)e^{i\frac{S}{\hbar}} = U(x_i, x_{peak}, T)} \tag{18-22}$$

$\psi_{peak}(T) = U(x_i, x_{peak}, T)$

Definition: Borrowing a term from electrical engineering, we will herein refer to $e^{i\phi}$ as a phasor.

18.5 Feynman's Path Integral Approach: The Central Idea

Feynman's remarkable idea takes a little getting used to. He reasoned that a particle/wave (such as an electron) traveling a path (world line in spacetime) between two events could actually be considered to be traveling along all possible paths (infinite in number) between those events.

Feynman's idea: particle travels all paths in spacetime simultaneously

---

[1] Merzbacher, E., *Quantum Mechanics.* 2nd ed. John Wiley & Sons (1970). See Chap 2, Sec 3.

Difficult as it may be, initially, to believe, we will see below that adding the phasors from all of these individual paths gives us the same result as using the standard QM theory of Schrödinger with a single wave. The two different approaches are therefore equivalent.

<u>Definition:</u> Feynman's method is called the "<u>path integral</u>", "<u>functional integral</u>", "<u>many paths</u>", or "<u>sum over histories</u>" approach to QM (and QFT).

Superimposing all $e^{i\int \frac{L}{\hbar} dt}$ from all those paths, gives same $|U|$ as other methods

Note that the paths do not have to satisfy physical laws like conservation of energy, least action, etc. Moreover, each path is considered equally probable.

We will lead into the formal mathematics of the many paths approach by first examining simple situations with a finite number of paths between two events.

18.6 Superimposing a Finite Number of Paths[1]

18.6.1 The Rotating Phasor

The phasor of (18-22) can be expressed in the complex plane as a unit length vector with angle ϕ relative to the real positive (horizontal) axis. As time evolves this vector rotates at the rate $L/\hbar$, i.e.,

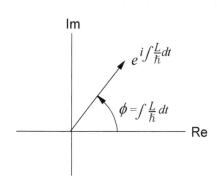

Figure 18-2. Rotating Phasor

the total phase $\phi = \int \frac{L}{\hbar} dt$. So, we can picture the phasor as a unit length vector rotating like a hand on a clock in a 2D complex plane (though here it is a counterclockwise rotation).

Visualizing Feynman's rotating phasor

For the purposes of Feynman's approach, we can consider the particle as a wave packet with phase at the peak determined by (18-22), and our final measurement a position eigenstate at the packet peak. We then imagine a different wave packet following each one of the infinitely many paths between two specific events. We visualize the phasor at the wave packet peak for each of these paths as a vector rotating in the complex plane as time passes (i.e., as the wave packet peak moves along the path), eventually having a particular value at the final event, the arrival place and time. Each path will have a different final phase.

18.6.2 Several Paths Graphically

Fig. 24 in Feynman's book *QED: The Strange Theory of Light and Matter*[2], is an insightful, somewhat heuristic, illustration of the many paths concept for light. Since we wish to focus, at this point, on non-relativistic quanta, we employ a similar, and at least equally heuristic, illustration in Fig. 18-3 for an electron rather than a photon. In Fig. 18-3 an electron is emitted at event a, reflected, like light from a mirror, off of a scattering surface, and detected at event b. The scattering surface might be difficult to construct in practice, but one can imagine a surface densely packed with tightly bound negative charge.

Graphical visualization of one example of many paths superposition

We look at a representative 15 different paths for the electron, out of the infinite number in the many paths approach, and label them with letters A to O. Each path takes the same time T. Note that path H is the classical path, having equal angles of incidence and reflection. Since it is the shortest, particle speed for that path is lowest.

The Lagrangian here is simply the kinetic energy, and this is constant, though different, for each path. Since speed is least for the classical path H, it has the smallest Lagrangian, and thus the least action. The other paths do not obey the usual classical laws, such as least action, equal angles of incidence and reflection, etc. But according to Feynman's approach, we have to include all of them.

---

[1] Much of the material in this section parallels "Action on Stage: Historical Introduction", Ogborn, J., Hanc, J. and Taylor, E.F., and "A First Introduction to Quantum Behavior", Ogborn, J., both from The Girep Conference 2006, Modeling Physics and Physics Education, Universiteit van Amsterdam.

[2] Feynman, R., *QED: The strange theory of light and matter.* Penguin Books, London (1985). Feynman's Fig. 24 is actually for a slightly different form of the least action principle than we investigate here, but the underlying concept is the same

Many Paths Electron Reflection

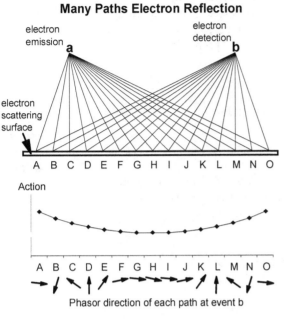

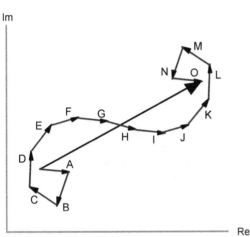

Phasor direction of each path at event b

Phase Addition of All Paths at Event b

Figure 18-3. Graphical Justification for Many Paths Approach

From (18-22) and Fig. 18-2, we can determine the phasor $e^{iS/\hbar}$ of (18-22) for the particle/wave arriving at event b, for each path, where $S = LT = \frac{1}{2}\, mv^2 T$. The phasor direction in complex space, for each path at the detection event b, is depicted in the middle of Fig. 18-3.

The bottom part of Fig. 18-3 shows the addition of the final event phasors for all 15 paths. Note that the paths further from the classical path H tend to cancel each other out, because they are out of phase. Conversely, H and the paths close to H are close to being in phase, and thus, reinforce each other via constructive interference. So, the primary contributions to the phasor sum are from those paths close to the classical path.

If we were to increase the number of paths, the jaggedness of the curve formed by the 15 phasors would smooth out, but its basic overall shape would remain essentially the same. If we were to increase the Lagrangian, while keeping speed the same for each path (i.e., increasing mass of the particle), phasors now near the middle of the curve would shift towards the ends, and thus, be cancelled out via interference. In other words, increasing mass brings us closer to the classical case, and the paths closer to classical then make greater contributions to the final sum. A similar effect would occur if the value for Planck's constant were smaller. As $\hbar \to 0$, all paths but H would tend to cancel out.

Clarification

I once thought that increasing mass, and thus getting closer to the classical situation, would bring the phase angle of the sum-of-all-paths phasor in directional alignment with H, the classical path phasor (or at least with U of (18-22).) However, this is not the case. The important thing in Feynman's approach is not the phase of the sum-of-all-paths phasor, but its *length*, which is proportional to $|U|$. And this length gets greater contribution from paths nearer H than from those further away.

Note that in order to get a graphically significant Fig. 18-3, I had to use a value for $\hbar$ almost eight orders of magnitude greater than the actual value. Otherwise the phase angles between adjacent paths, for the relatively large spacing between paths of the figure, would have resulted in a seemingly random hodgepodge of phasors, and obscured, rather than illumined, the real physics involved.

If you would like to experiment with changing values for mass, $\hbar$, and number of paths yourself, download the Excel spreadsheet from the website for this book[1].

End of Clarification

For probability: Magnitude of final sum over all paths (total phasor length) is important thing. Total phasor angle not relevant.

---

[1] See pg. xvi, opposite pg. 1. Click on link under Auxiliary Material.

Feynman intuited that the amplitude of the final phasor sum was extremely meaningful. That is, the square of its absolute value (i.e., the square of its length in complex space) was proportional (approximately, for a finite number of paths; exactly, for an infinite number) to the probability density for measuring the photon/particle at event b. What we mean by "proportional" should become clearer after the following three sections.

18.6.3 Many Paths Mathematically

Math expression of our example of many paths superposition

Consider particle paths similar to those of Fig. 18-3, where the wave function peak for path number 1, with $A_1(T)$ as in (18-22), is

$$\psi_1^{peak} = A_1(T)e^{iS_1/\hbar}. \tag{18-23}$$

In the spirit of the prior section, one considers the phasor of (18-23) *without* $A_1(T)$ as representing the particle, AND that particle is considered to simultaneously travel many paths between events a and b. Then, the summation of the final phasors for each path is expressed mathematically as

$$e^{iS_1/\hbar} + e^{iS_2/\hbar} + e^{iS_3/\hbar} + ... = A_b e^{i\phi_{sum}} \tag{18-24}$$

where A_b is the amplitude of the sum. As the number of paths approaches infinity, $|A_b|^2$ becomes proportional to the probability density of measuring the particular final state at event b. That is,

$$\lim_{N \to \infty} \sum_{j=1}^{N} e^{iS_j/\hbar} = A_b e^{i\phi_{sum}} \propto U(x_i, x_f, T) \quad \left|A_b e^{i\phi_{sum}}\right|^2 = |A_b|^2 \propto |U|^2 \text{ (probability density)}. \tag{18-25}$$

We will learn how to evaluate the limit in (18-25).

18.6.4 Another Example

Another example: double slit experiment

Consider a double slit experiment with a classical Huygens wave analysis showing alternating fringes of light and dark, which via the classical interpretation are caused by constructive and destructive interference of light/electron waves.

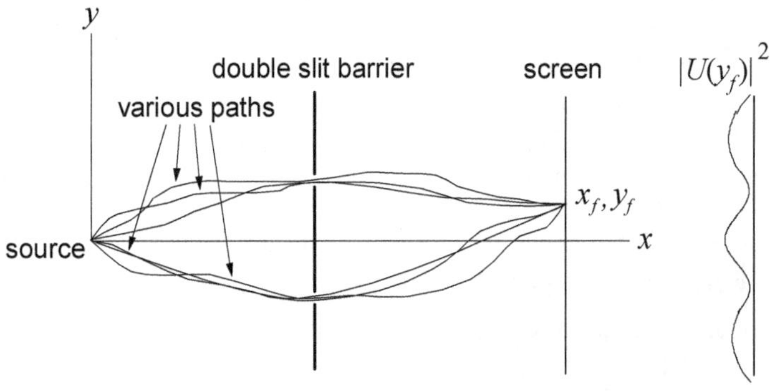

Figure 18-4. Double Slit Experiment in Many Paths Approach

By the Schrödinger (probability) wave approach, a single quantum wave travels through both slits, interferes with itself, either constructively or destructively, to result in a wave amplitude that varies with location along the receiving screen. The probability density (square of the absolute value of the amplitude) of finding a photon/electron also varies with that screen location. So, as the quantum waves collapse, one at a time, on the screen, they tend to collapse more often in the high probability (high magnitude amplitude) regions. These correspond to the bright fringe regions, which, with enough individual quanta collapsing on the screen, are seen by the human eye.

In the many paths approach, for any particular spot on the screen, we would add the phasors of every "possible" path from the emission point, through one slit, to that spot (x_f, y_f), plus all paths through the other slit to the spot. See Fig. 18-4. The result would be proportional to the amplitude at the spot found in the Schrödinger approach. That is, the sum of all phasors at (x_f, y_f) (see (18-25)) yields

$$C \lim_{N \to \infty} \sum_{j=1}^{N} e^{iS_j/\hbar} = U(x_i, y_i; x_f, y_f; T), \tag{18-26}$$

where C is some constant.

We would then repeat that procedure for every other point on the screen. For a fixed source at (x_i, y_i), and a fixed x_f for the screen, the amplitude would be spatially only a function of y_f, and we could express it simply as $U(y_f)$.

18.6.5 Finding the Proportionality Constant: By Example

Feynman result of summation is only proportional to $|U|^2$. Need to find proportionality constant another way.

The square of the absolute value of the amplitude U is the probability density. So, we can normalize U over the length of the screen, i.e.,

$$\int_{y_f=-\infty}^{y_f=+\infty} \left| C \lim_{N \to \infty} \sum_{j=1}^{N} e^{iS_j/\hbar} \right|^2 dy_f = \int_{y_f=-\infty}^{y_f=+\infty} \left| U(y_f) \right|^2 dy_f = 1, \tag{18-27}$$

and thus, once the value of the limit is determined, readily find the proportionality constant C.

18.7 Summary of Approaches

18.7.1 Feynman's Postulates

Richard Feynman was probably well aware of much of the foregoing when he speculated on the viability of the following four postulates for his many paths approach. Subsequent extensive analysis by Feynman and many others has validated his initial speculation.

Path integral four starting postulates

The postulates of the many paths approach to quantum theories are:

1. A particle is assumed classical in the sense that it can be considered a point-like object, with both its position and its 3-momentum well defined along each individual path, so those values determine the Lagrangian at any point and time along any given path. However, the particle is assumed quantum mechanical in that, like a wave function, it has a phase (at the point).

2. The phasor value at any final event is equal to $e^{iS/\hbar}$ where the action S is calculated along a particular path beginning with a particular initial event.

3. The probability density for the final event is given by the square of the magnitude of a typically complex amplitude.

4. That amplitude is found by adding together the phasor values at that final event from all paths between the initial and final events, including classically impossible paths. The amplitude of the resultant summation must then be normalized relative to all other possible final events, and it is this normalized form of the amplitude referred to in 3.

Note two things.

First, there is no weighting of the various path phasors. The nearly classical paths are not weighted more heavily than the paths that are far from classical. That is, the different individual paths in the summation do not have different amplitudes (see (18-24) and Fig. 18-3). The correlation with the classical result comes from destructive interference among the paths far from classical, and constructive interference among the paths close to classical.

Phasors are not weighted when summing them

Second, time on all paths (all histories) must move forward. This is implicit in the exponent phase value of (18-19), where the integral of L is over time, with time moving forward. Our paths do not include particles zig-zagging backward and forward through time[1].

18.7.2 Comparison of Approaches to QM

Wholeness Chart 18-2 summarizes the major similarities and differences between alternative approaches to NRQM.

Comparing 3 equivalent approaches to NRQM

---

[1] Caveat: A famous quote by Freeman Dyson states that Feynman, while speculating on this approach, told him that one particle travels all paths, including those going backward in time. But the usual development of the theory (see Section 18.6) only includes paths forward in time. Perhaps all paths backward in time sum to zero and so are simply ignored. In such case, Dyson's quote would be accurate. But I have not personally investigated this and do not know for sure.

Wholeness Chart 18-2. Equivalent Approaches to Non-relativistic Quantum Mechanics

	Schrödinger Wave Mechanics	Heisenberg Matrix Mechanics	Feynman Many Paths
Probability Density of Position Eigenstates	$\lvert \text{amplitude}\rvert^2$		$\lvert \text{amplitude}\rvert^2$
Transition Amplitude	$U\left(x_i,x_f;T\right)=\left\langle x_f\left\lvert e^{-iHT/\hbar}\right\rvert x_i\right\rangle$	Same results as other two approaches.	$U\left(x_i,x_f;T\right)=C\;\underset{N\to\infty}{lim}\sum_{j=1}^{N}e^{iS_j/\hbar}$ $=C\int_{x_i}^{x_f}e^{i\int_0^T\frac{L}{\hbar}dt}\mathcal{D}x(t)$
Comments	Above interpretation assumes $\lvert x_i\rangle$ is high narrow wave packet and $\lvert x_f\rangle$ is a pure delta function in position space		Need to determine C. Some others include C in definition of $\mathcal{D}x(t)$. We haven't done the integral part yet.

18.8 Finite Sums to Functional Integrals

18.8.1 Time Slicing: The Concept

After all of the foregoing groundwork, it is time to extend the phasor sum of a finite number of paths, such as we saw in Fig. 18-3 and (18-24), over into an infinite sum, or in other words, an integral. To do this, we first consider finite "slices" of time, for a finite number of paths in one spatial dimension, as shown in Fig. 18-5 where, for convenience, we plot time vertically and space horizontally. As opposed to our spatially 2D example in Fig. 18-3, different paths in Fig. 18-5 actually refer to the particle traveling along the x axis only between i and f, though at varying (both positive and negative) velocities. The paths between each slice are straight lines, but there is no loss in generality, as one can take the time between slices $\Delta t \to dt$, and thus, any possible shape path can be included.

Slicing time into "pieces" for discrete time analysis

As noted earlier, for any <u>single path</u>, the

$$\text{phasor at }\mathbf{f} = \underbrace{e^{i\int_{t_i}^{t_f}\frac{L}{\hbar}dt}}_{\text{one path}}=e^{iS/\hbar}\;,\qquad(18\text{-}28)$$

A simple example

The amplitude U for the transition from $\mathbf{i}$ to $\mathbf{f}$ is proportional to the sum of (18-28) for <u>all paths</u>,

$$\text{sum of }\infty\text{ phasors at }\mathbf{f} = \underset{N\to\infty}{lim}\sum_{j=1}^{N}e^{iS_j/\hbar}\;.\qquad(18\text{-}29)$$

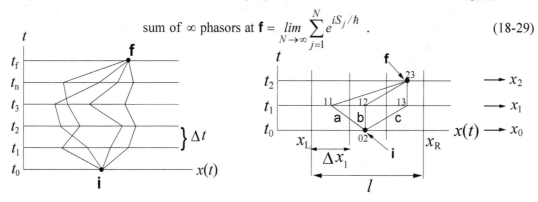

Figure 18-5. Time Slicing for Finite Number of Paths   **Figure 18-6. Space Slicing for Three Discrete Paths**

18.8.2 Space Slicing: Simple Paths with Discrete Approximation

To evaluate (18-29), we next also discretize ("slice") space, and consider a small number (three) of paths over a small number of discrete events in spacetime, as in Fig. 18-6. We label the paths a, b, and c, and the events with two numbers, such that the first number represents the time slice, and the second the space slice. The continuous range of x values at time t_1 will be designated x_1; at t_2, x_2; etc. We limit the spatial range for paths considered to $x_R - x_L = l$, where the number of paths N $= 3 = l/\Delta x_1$. Each path passes through the center of one Δx_1 segment. Here there is only one time slice (at t_1), with two time intervals of Δt, since time is fixed at t_0, t_2; and a single Δx value of Δx_1.

Slicing space into "pieces" for discrete space analysis

We then assume the phase ϕ_{02} at **i** is zero, and find the phasors at **f** for each of the three paths by subsequently adding the phase difference between discrete events along a given path, as in the second column of Wholeness Chart 18-3 below.

Note that in the last line of column two in Wholeness Chart 18-3, the Lagrangian L without subscript is assumed to be the L for the particular sub-path being integrated (common notation).

In column three, we approximate the integrals of L over t, e.g., for path a over an interval Δt,

$$S_a \approx L_a^{apprx}\Delta t \tag{18-30}$$

where, for the first sub-path,

Approximate (discretize) L like we approximate (discretize) space and time

$$L_a = \tfrac{1}{2}m\dot{x}^2 - V(x) \approx \tfrac{1}{2}m\left(\frac{x_{11}-x_{02}}{\Delta t}\right)^2 - V\left(\frac{x_{11}+x_{02}}{2}\right) = L_a^{apprx}(x_{11},x_{02}) \tag{18-31}$$

Similar relations hold for the other sub-paths, and are shown in Wholeness Chart 18-3.

Note that (18-31) is solely a function of x_{11} and x_{02}. The summation of all three paths in the last row of column three in Wholeness Chart 18-3 is solely a function of x_{02}, x_{23}, and the three intermediate event x values x_{11}, x_{12}, and x_{13}. Since x_{02} and x_{23} are the positions of the initial and final events, which are fixed and the same for all paths, the final summation approximation in Wholeness Chart 18-3 is really only a function of the three x_{1j}. It will, however, serve a future purpose if we keep x_{02} and x_{23} in the relationship for the time being.

Summing approximations of phasors

Wholeness Chart 18-3. Adding Phasors at the Final Event for Three Discrete Paths

Path	Phasor at **f**	Phasor at **f** in Terms of Approx L
a	$e^{i\phi_{a23}} = e^{i(\phi_{02\to11}+\phi_{11\to23})} = e^{i\int_{02}^{11}\frac{L_a}{\hbar}dt}\,e^{i\int_{11}^{23}\frac{L_a}{\hbar}dt}$	$\approx e^{\frac{i}{\hbar}\left\{\frac{1}{2}m\left(\frac{x_{11}-x_{02}}{\Delta t}\right)^2 -V\left(\frac{x_{11}+x_{02}}{2}\right)\right\}\Delta t}\,e^{\frac{i}{\hbar}\left\{\frac{1}{2}m\left(\frac{x_{23}-x_{11}}{\Delta t}\right)^2 -V\left(\frac{x_{23}+x_{11}}{2}\right)\right\}\Delta t}$ $= e^{\frac{i}{\hbar}f(x_{02},x_{11})}\,e^{\frac{i}{\hbar}f(x_{11},x_{23})}$
b	$e^{i\phi_{b23}} = e^{i(\phi_{02\to12}+\phi_{12\to23})} = e^{i\int_{02}^{12}\frac{L_b}{\hbar}dt}\,e^{i\int_{12}^{23}\frac{L_b}{\hbar}dt}$	$\approx e^{\frac{i}{\hbar}\left\{\frac{1}{2}m\left(\frac{x_{12}-x_{02}}{\Delta t}\right)^2 -V\left(\frac{x_{12}+x_{02}}{2}\right)\right\}\Delta t}\,e^{\frac{i}{\hbar}\left\{\frac{1}{2}m\left(\frac{x_{23}-x_{12}}{\Delta t}\right)^2 -V\left(\frac{x_{23}+x_{12}}{2}\right)\right\}\Delta t}$ $= e^{\frac{i}{\hbar}f(x_{02},x_{12})}\,e^{\frac{i}{\hbar}f(x_{12},x_{23})}$
c	$e^{i\phi_{c23}} = e^{i(\phi_{02\to13}+\phi_{13\to23})} = e^{i\int_{02}^{13}\frac{L_c}{\hbar}dt}\,e^{i\int_{13}^{23}\frac{L_c}{\hbar}dt}$	$\approx e^{\frac{i}{\hbar}\left\{\frac{1}{2}m\left(\frac{x_{13}-x_{02}}{\Delta t}\right)^2 -V\left(\frac{x_{13}+x_{02}}{2}\right)\right\}\Delta t}\,e^{\frac{i}{\hbar}\left\{\frac{1}{2}m\left(\frac{x_{23}-x_{13}}{\Delta t}\right)^2 -V\left(\frac{x_{23}+x_{13}}{2}\right)\right\}\Delta t}$ $= e^{\frac{i}{\hbar}f(x_{02},x_{13})}\,e^{\frac{i}{\hbar}f(x_{13},x_{23})}$
Sum of a, b, c	$= e^{i\int_{02}^{11}\frac{L_a}{\hbar}dt}\,e^{i\int_{11}^{23}\frac{L_a}{\hbar}dt} + e^{i\int_{02}^{12}\frac{L_b}{\hbar}dt}\,e^{i\int_{12}^{23}\frac{L_b}{\hbar}dt}$ $+ e^{i\int_{02}^{13}\frac{L_c}{\hbar}dt}\,e^{i\int_{13}^{23}\frac{L_c}{\hbar}dt}$ $= \sum_{j=1}^{N=3} e^{i\int_{02}^{1j}\frac{L}{\hbar}dt}\,e^{i\int_{1j}^{23}\frac{L}{\hbar}dt}$	$= e^{\frac{i}{\hbar}f(x_{02},x_{11})}\,e^{\frac{i}{\hbar}f(x_{11},x_{23})} + e^{\frac{i}{\hbar}f(x_{02},x_{12})}\,e^{\frac{i}{\hbar}f(x_{12},x_{23})}$ $+ e^{\frac{i}{\hbar}f(x_{02},x_{13})}\,e^{\frac{i}{\hbar}f(x_{13},x_{23})}$ $= \sum_{j=1}^{N=3} e^{\frac{i}{\hbar}f(x_{02},x_{1j})}\,e^{\frac{i}{\hbar}f(x_{1j},x_{23})}$

The final relationship in Wholeness Chart 18-3 is approximately proportional to the transition amplitude, i.e.,

$$U\left(i,f;T=t_f-t_i\right) \approx C \sum_{j=1}^{N=3} e^{\frac{i}{\hbar}f\left(x_{02},x_{1j}\right)} e^{\frac{i}{\hbar}f\left(x_{1j},x_{23}\right)} = C \sum_{j=1}^{N=3} e^{\frac{i}{\hbar}S^{apprx}\left(x_{02},x_{1j}\right)} e^{\frac{i}{\hbar}S^{apprx}\left(x_{1j},x_{23}\right)} , \quad (18\text{-}32)$$

Final form of approximate U for our example

where C is some constant, and what we designated as a function f in Wholeness Chart 18-3, in order to emphasize its independent variables, is actually an approximation to the action S.

Since U is *proportional* to the sum of the phasors, we can multiply the RHS of (18-32) by any constant we like and the proportionality still holds. To aid us in taking limits to get an integral, we multiply (18-32) by Δx_1, and get

$$U\left(i,f;T\right) \approx C' \sum_{j=1}^{N=3} e^{\frac{i}{\hbar}S^{apprx}\left(x_{02},x_{1j}\right)} e^{\frac{i}{\hbar}S^{apprx}\left(x_{1j},x_{23}\right)} \Delta x_1 , \quad (18\text{-}33)$$

Can multiply by constant Δx_1 since our sum proportional to U

where C' is a new constant. Taking the limit where $\Delta x_1 \to dx_1$ means taking the number of paths $N \to \infty$, where we assume one path point at the center of each dx_1. And thus,

$$U\left(i,f;T\right) \approx C' \lim_{N\to\infty} \sum_{j=1}^{N} e^{\frac{i}{\hbar}S^{apprx}\left(x_{02},x_{1j}\right)} e^{\frac{i}{\hbar}S^{apprx}\left(x_{1j},x_{23}\right)} \Delta x_1$$

$$= C' \int_{x_1=x_L}^{x_1=x_R} e^{\frac{i}{\hbar}S^{apprx}\left(x_{02},x_1\right)} e^{\frac{i}{\hbar}S^{apprx}\left(x_1,x_{23}\right)} dx_1 \approx C' \int_{x_1=x_L}^{x_1=x_R} e^{i\int_{t_{02}}^{t_{23}}\frac{L}{\hbar}dt} dx_1 .$$

$(18\text{-}34)$

Take number of paths $\to \infty$ (but still have discretized time)

where our discrete values x_{1j} have become a continuum x_1, and it is implicit that the L of the last part of (18-34) is that over the appropriate path corresponding to each increment of dx_1. (18-34) is still only approximately proportional to the amplitude because time is still discretized in Δt intervals and we limit the integration range to $x_L > x_1 > x_R$. Before extending those limits, however, we must consider a slightly more complicated set of paths.

18.8.3 From Simple Discrete Paths to the General Case

In Fig. 18-7 we introduce one more time interval between the initial and final events, resulting in nine discrete paths.

En route to making time continuous, examine our example with another time interval

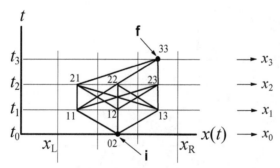

Figure 18-7. Nine Discrete Paths between Two Events
(Two time slices at t_1 and t_2 with three time intervals Δt and thus, with Δx_1 and Δx_2)

Repeating the logic from the previous section (use Wholeness Chart 18-3 as an aide), the phasor of the first path ($02 \to 11 \to 21 \to 33$) is simply

$$\underbrace{e^{i\phi_{33}}}_{\substack{\text{1st}\\\text{path}\\\text{only}}} = e^{i\left(\phi_{02\to11}+\phi_{11\to21}++\phi_{21\to33}\right)}$$

$(18\text{-}35)$

$$= e^{i\int_{02}^{11}\frac{L}{\hbar}dt} e^{i\int_{11}^{21}\frac{L}{\hbar}dt} e^{i\int_{21}^{33}\frac{L}{\hbar}dt} .$$

We repeat this for the other eight paths, approximate L along sub-paths as before, and take k below to indicate the kth Δx_2 segment. This results in a phasor summation from all paths at event $\mathbf{f}\,(=33)$ [compare with last row, last column of Wholeness Chart 18-3 and (18-32)] proportional to the amplitude, i.e.,

$$U(i,f;T) \approx C \sum_{j=1}^{N=3} \sum_{k=1}^{N=3} e^{\frac{i}{\hbar}S^{apprx}(x_{02},x_{1j})} e^{\frac{i}{\hbar}S^{apprx}(x_{1j},x_{2k})} e^{\frac{i}{\hbar}S^{apprx}(x_{2k},x_{33})}. \tag{18-36}$$

Note that (18-36) depends on the discrete values of both x_{1j} and x_{2k}. So, as we did with (18-33), we can multiply (18-36) by one or more constants without changing the proportionality. We choose to multiply by Δx_1 and Δx_2. We follow by taking limits $\Delta x_1 \to dx_1$ and $\Delta x_2 \to dx_2$ (i.e., $N \to \infty$ for each dx_1 and dx_2), [compare with (18-34)] which results in

$$U(i,f;T) \approx C' \lim_{N \to \infty} \sum_{j=1}^{N} \sum_{k=1}^{N} e^{\frac{i}{\hbar}S^{apprx}(x_{02},x_{1j})} e^{\frac{i}{\hbar}S^{apprx}(x_{1j},x_{2k})} e^{\frac{i}{\hbar}S^{apprx}(x_{2k},x_{33})} \Delta x_1 \Delta x_2$$

$$= C' \int_{x_2=x_L}^{x_2=x_R} \int_{x_1=x_L}^{x_1=x_R} e^{\frac{i}{\hbar}S^{apprx}(x_{02},x_1)} e^{\frac{i}{\hbar}S^{apprx}(x_1,x_2)} e^{\frac{i}{\hbar}S^{apprx}(x_2,x_{33})} dx_1 dx_2 \tag{18-37}$$

$$\approx C' \int_{x_2=x_L}^{x_2=x_R} \int_{x_1=x_L}^{x_1=x_R} e^{i\int_{t_{02}}^{t_{33}}\frac{L}{\hbar}dt} dx_1 dx_2 .$$

Result for this example of 2 time slices

We can readily generalize (18-37) to any number n of time slices as

$$\boxed{U(i,f;T=t_f-t_i) \approx C \int_{x_n=x_L}^{x_n=x_R} \cdots \int_{x_2=x_L}^{x_2=x_R} \int_{x_1=x_L}^{x_1=x_R} e^{i\int_{t_i}^{t_f}\frac{L}{\hbar}dt} dx_1 dx_2 ... dx_n}$$
$$\text{Approximation for Transition Amplitude} \tag{18-38}$$

Result for this example if had any number of time slices

where, as before, it is implicit that L in the integral is for each particular segment in a given path from $\mathbf{i}$ to x_1 to x_2, etc. through each respective time slice to $\mathbf{f}$.

18.8.4 From Approximate to Exact

To get an exact, not approximate, relation for the RHS of (18-38) we have to do two things.

1. Take the x range from l to infinity, i.e., $x_L \to -\infty$ and $x_R \to \infty$, and

2. Take $\Delta t \to dt$ for the same T (time between events) so the number of time slices $\to \infty$.

Extending to exact case: length $\to \infty$ and time continuous

Doing this, (18-38) becomes

$$\boxed{U(i,f;T=t_f-t_i) = C \int_{x=x_i}^{x=x_f} e^{i\int_{t_i}^{t_f}\frac{L}{\hbar}dt} \mathcal{D}x}$$
$$\underbrace{}$$
$$\text{integ limits along with } \mathcal{D} \text{ symbol imply } all \text{ paths between i and f}$$
$$\text{Exact Expression for Transition Amplitude} \tag{18-39}$$

Yields exact relation for transition amplitude via path integral approach (but with undetermined constant C)

The symbol $\mathcal{D}$, as noted earlier, represents integration over all paths. With this, the integration limits designate the initial and final x values and do not imply a constraint on the x dimension during the integration (as was the case with (18-38).) In (18-39) we have, at long last, obtained the relation of integration type #7 in Wholeness Chart 18-1, pg. 490, where

$$e^{iF} = e^{i\int_{t_i}^{t_f}\frac{L}{\hbar}dt} . \tag{18-40}$$

18.8.5 Practicality and Calculations

Practically, for the first approximation addressed in Section 18.8.4, we really don't have to take l to infinity, as we know that paths outside of a reasonably large range from the initial and final spatial locations will sum to very close to zero. So, we can live with significant, but not infinite, l.

Practically can take l large enough to give accurate answer

For the second approximation, we only need small enough Δt such that taking a smaller value does not change our answer much.

If we use (18-38), with judicious choices for Δt and l, we can, in many cases, obtain valid closed form solutions for the amplitude. We can also obtain numerical solutions with a digital computer by using approximations for L between time slices, as we did previously. That is, we can approximate the RHS of (18-38) in the manner we did for the first line of (18-37), but extending the approximation of (18-37) from 3 to n time slices.

And can take Δt small as needed as well

18.9 An Example: Free Particle

We will first determine the amplitude (and thus detection probability density) of a free particle via the Schrödinger approach and then compare it to that for Feynman's many paths approach.

Free particle example in NRQM

18.9.1 Schrödinger Transition Amplitude

Recall, from Section 18.3.2, that, in the Schrödinger approach, a position eigenstate is effectively a delta function, and as it evolves, the wave function envelope spreads and the peak diminishes. $|U|^2$ for such functions is the probability density at the final point x_f, after time T. We should then expect $|U|^2$ to decrease as T increases, and to effectively equal infinity at x_{peak} when $T = 0$.

First, via Schrödinger wave mechanics approach

We start with the Schrödinger transition amplitude relation (18-9),

$$U\left(x_i, x_f; T\right) = \left\langle x_f \left| e^{-iHT/\hbar} \right| x_i \right\rangle \tag{18-41}$$

where we take the bra to be a pure delta function and the ket, a normalized wave packet approximation to a delta function. It is simpler mathematically to use a pure delta function to represent $|x_i\rangle$, but then we have to normalize it in a manner similar to a wave packet. There is a lot behind this that we summarize in the appendix (pg. 509), but here, we simply use A to represent the normalization factor in the ket of (18-41).

$$U\left(x_i, x_f; T\right) = \int_{-\infty}^{\infty} \left(\delta\left(x - x_f\right) e^{-iHT/\hbar} A\,\delta\left(x - x_i\right)\right) dx, \tag{18-42}$$

with the well-known relations

$$\delta\left(x - x_i\right) = \frac{1}{2\pi} \int_{-\infty}^{+\infty} e^{ik(x-x_i)} dk = \frac{1}{2\pi\hbar} \int_{-\infty}^{+\infty} e^{i\frac{p}{\hbar}(x-x_i)} dp. \tag{18-43}$$

(Box 2-3, pg. 27, explains the use of operators in exponents. In essence, one can express the exponential quantity as a Taylor series expanded about $T = 0$, i.e., $f(T) = e^{-iTH/\hbar} = 1 - iTH/\hbar - \frac{1}{2} T^2 H^2/\hbar^2 + \dots$ Then, operate on the ket/state term by term [getting terms in $iET/\hbar$ to various powers], and finally re-express the resulting Taylor series as an exponential in $iET/\hbar$. We have taken the ket with time $t_i = 0$ to make things simpler, but even if you think of the Hamiltonian operator as a time derivative, when it acts on that ket, it functions as an energy operator and still yields the energy.)

For the exponential with the H operator acting on the initial state, and $E = p^2/2m$, (18-42) is

$$U\left(x_i, x_f; T\right) = \int_{-\infty}^{\infty} \left(\left(\frac{1}{2\pi\hbar} \int_{-\infty}^{+\infty} e^{i\frac{p'}{\hbar}(x-x_f)} dp'\right)\left(A\frac{1}{2\pi\hbar} \int_{-\infty}^{+\infty} e^{-\frac{i}{\hbar}TH} e^{i\frac{p}{\hbar}(x-x_i)} dp\right)\right) dx$$

$$= \int_{-\infty}^{\infty} \left(\left(\frac{1}{2\pi\hbar} \int_{-\infty}^{+\infty} e^{i\frac{p'}{\hbar}(x_f-x)} dp'\right)\left(A\frac{1}{2\pi\hbar} \int_{-\infty}^{+\infty} e^{-iTp^2/2m\hbar} e^{i\frac{p}{\hbar}(x-x_i)} dp\right)\right) dx, \tag{18-44}$$

where we took $(x - x_f) \rightarrow (x_f - x)$ in the 2nd line to help us later. (It makes no difference in the integral from $-\infty$ to $+\infty$.) We then re-arrange (18-44) to get

$$U\left(x_i,x_f;T\right) = A\frac{1}{2\pi\hbar}\iint e^{-iTp^2/2m\hbar}\underbrace{\left(\frac{1}{2\pi\hbar}\int e^{i\frac{x}{\hbar}(p-p')}dx\right)}_{\delta(p-p')}e^{\frac{i}{\hbar}p'x_f}e^{-\frac{i}{\hbar}px_i}dp'dp$$

$$= A\frac{1}{2\pi\hbar}\int e^{-iTp^2/2m\hbar}e^{\frac{i}{\hbar}p\left(x_f-x_i\right)}dp. \tag{18-45}$$

Using the integral formula

$$\int_{-\infty}^{+\infty}e^{-ax^2+bx}dx = \sqrt{\frac{\pi}{a}}e^{b^2/4a} \qquad Re(a)>0,, \tag{18-46}$$

we find
$$U\left(x_i,x_f;T\right) = A\sqrt{\frac{m}{i2\pi\hbar T}}e^{\frac{i}{\hbar}\frac{m}{2T}\left(x_f-x_i\right)^2}. \tag{18-47}$$

The astute reader may question whether (18-46), with complex a and b, converges. It does because the integrand oscillation rate increases with larger $|p|$ in such a way as to make successive cycles shorter. As $|p|$ gets very large, the cycles become so short that the contribution from each cycle (think area under a sine curve) tends to zero, and it does so in a manner that allows the integral to converge. Said another way, the smaller and smaller contributions as $|p|$ gets large alternate between positive and negative values (for both real and complex portions), and thus convergence is assured.

From (18-47), the probability density at event **f** is

$$\left|U\left(x_i,x_f;T\right)\right|^2 = A^2\frac{m}{2\pi\hbar T}, \tag{18-48}$$

which, as we said it must, decreases with increasing T, and equals infinity for $T=0$[1].

Probability density at final event for free particle via Schrödinger approach

18.9.2 Many Paths Transition Amplitude

We now seek to derive (18-47) using the many paths approach. (There is an alternative derivation, which you may find easier, on this book's website. See URL on pg. xvi, opposite pg. 1.)

A free, non-relativistic particle has Lagrangian (all values are wave packet expectation values, e.g., $x_f = \bar{x} = x_{peak}$, $v = v_g$)

Now, for free particle via path integral approach

$$L = \tfrac{1}{2}mv^2 \approx \tfrac{1}{2}m\left(\frac{x(t+\Delta t)-x(t)}{\Delta t}\right)^2, \tag{18-49}$$

where the RHS is an approximation between adjacent time slices. Taking $t_i=0$, and $l\to\infty$ (see Fig. 18-6, pg. 498), (18-38) becomes

$$U\left(i,f;T\right) \approx C\int_{x_n=-\infty}^{x_n=\infty}\cdots\int_{x_2=-\infty}^{x_2=\infty}\int_{x_1=-\infty}^{x_1=\infty}e^{i\int_{t_n}^{t_f=T}\frac{L}{\hbar}dt}e^{i\int_{t_{n-1}}^{t_n}\frac{L}{\hbar}dt}\ldots e^{i\int_{t_1}^{t_2}\frac{L}{\hbar}dt}e^{i\int_0^{t_1}\frac{L}{\hbar}dt}dx_1dx_2\ldots dx_n$$

$$\approx C\int_{x_n=-\infty}^{x_n=\infty}\cdots\int_{x_2=-\infty}^{x_2=\infty}\int_{x_1=-\infty}^{x_1=\infty}e^{\frac{i}{\hbar}\left\{\frac{1}{2}m\left(\frac{x_f-x_n}{\Delta t}\right)^2\right\}\Delta t}\ldots e^{\frac{i}{\hbar}\left\{\frac{1}{2}m\left(\frac{x_2-x_1}{\Delta t}\right)^2\right\}\Delta t}e^{\frac{i}{\hbar}\left\{\frac{1}{2}m\left(\frac{x_1-x_i}{\Delta t}\right)^2\right\}\Delta t}dx_1dx_2\ldots dx_n. \tag{18-50}$$

$$= C\int_{x_n=-\infty}^{x_n=\infty}\underbrace{e^{\frac{im}{2\hbar(\Delta t)}\left(x_f-x_n\right)^2}}_{f_\zeta}\cdots\int_{x_2=-\infty}^{x_2=\infty}\underbrace{e^{\frac{im}{2\hbar(\Delta t)}\left(x_3-x_2\right)^2}}_{f_\gamma}\underbrace{\int_{x_1=-\infty}^{x_1=\infty}\underbrace{e^{\frac{im}{2\hbar(\Delta t)}\left(x_2-x_1\right)^2}}_{f_\beta}\underbrace{e^{\frac{im}{2\hbar(\Delta t)}\left(x_1-x_i\right)^2}}_{f_\alpha}}_{f(x_2)}dx_1\,dx_2\ldots dx_n,$$

---

[1] Note that probability density for a wave function that is an exact delta function at time $T=0$, is a straight line any time $T>0$. This may seem confusing, but that is what (18-48) (with no x_f dependence) tells us. For a wave packet approximation to a delta function (instead of an exact delta function), for $T>0$, we have the behavior as in Fig. 18-1, pg. 491.

where the under-bracket notation will help us in subsequent sections.

Background Math

Look, for the moment, at the last two factors (functions f_α and f_β) in the integral. They must be integrated over x_1, and that result is a function of x_2. When one of the two functions in such a procedure is a function of $x_2 - x_1$, as it is here, the integral is called a *convolution integral*. (See http://www-structmed.cimr.cam.ac.uk/Course/Convolution/convolution.html.)

In mathematical circles (search "Borel's Theorem"), it is well known that the Fourier (and also, the Laplace) transform of such an integral equals the product of the Fourier (or Laplace) transforms of the two functions. That is, for $\mathcal{F}$ representing Fourier transform,

$$\mathcal{F}\left\{\int f_\beta (x_2 - x_1) f_\alpha (x_1) dx_1\right\} = \mathcal{F}\{f_\beta\}\mathcal{F}\{f_\alpha\}.$$
(18-51)

Note that although f_α is a function of $x_1 - x_i$, we can write $f_\alpha(x_1)$ because x_i is fixed.

Each factor in the last row of (18-50), as one moves leftward, plays the part of f_β in the theorem above for the next convolution integral, where the prior convolution integral plays the role of f_α. We get, in essence, a series of nested convolution integrals. Using (18-51), you should be able to prove to yourself that the transform of (18-50) equals the product of the transforms of the exponential factors in (18-50). If you can't, or don't want to bother, proving it, then just accept that a corollary to (18-51) is

$$\mathcal{F}\left\{\int....\int \int f_\zeta (x_f - x_n)....f_\gamma (x_3 - x_2) f_\beta (x_2 - x_1) f_\alpha (x_1) dx_1 dx_2...dx_n\right\}$$
$$= \mathcal{F}\{f_\zeta\}...\mathcal{F}\{f_\gamma\}\mathcal{F}\{f_\beta\}\mathcal{F}\{f_\alpha\}.$$
(18-52)

Evaluating the Amplitude

So, to evaluate (18-50), using (18-52), we i) transform each exponential factor f_μ, ii) multiply those transforms together, and iii) take the inverse transform of the result to get U (actually U/C of (18-50)). This is made simpler, because each f_μ has the same form, so each transform is the same, i.e.,

$$\mathcal{F}\{f_\alpha\} = \mathcal{F}\{f_\beta\} = ... = \mathcal{F}\{f_\zeta\} .$$
(18-53)

The Fourier transform of a function f_α is

$$\mathcal{F}\{f_\alpha (x_1)\} = \tilde{f}_\alpha (p) = \int_{-\infty}^{\infty} f_\alpha (x_1) e^{-\frac{i}{\hbar}px_1} dx_1 .$$
(18-54)

For the f_α of (18-50), and for convenience, taking the coordinate $x_i = 0$, this is

$$\tilde{f}_\alpha (p) = \int_{-\infty}^{\infty} e^{\frac{i}{\hbar}\frac{m}{2(\Delta t)}x_1^2} e^{-\frac{i}{\hbar}px_1} dx_1 ,$$
(18-55)

where here and throughout this section, p acts as merely a dummy variable allowing us to carry out the math. Using (18-46), we find (18-55) becomes

$$\tilde{f}_\alpha (p) = \sqrt{\frac{i2\pi\hbar(\Delta t)}{m}} e^{-\frac{i}{\hbar}\frac{\Delta t}{2m}p^2} ,$$
(18-56)

and thus, from (18-50), (18-52), and (18-53), where we define a new $N = n + 1 =$ number of functions $f_{Greek\ index}$.

$$\tilde{U}(p) \approx C \tilde{f}_\zeta (p)..... \tilde{f}_\beta (p) \tilde{f}_\alpha (p) = C\left(\frac{i2\pi\hbar(\Delta t)}{m}\right)^{N/2} e^{-\frac{i}{\hbar}\frac{T}{2m}p^2} .$$
(18-57)

The inverse Fourier transform of (18-57), is

$$U\left(x_i,x_f;T\right)=\frac{1}{2\pi\hbar}\int_{-\infty}^{\infty}\tilde{U}\left(p\right)e^{\frac{i}{\hbar}p\left(x_f-x_i\right)}dp$$

$$\approx\frac{1}{2\pi\hbar}\left(\frac{i2\pi\hbar(\Delta t)}{m}\right)^{N/2}C\int_{-\infty}^{\infty}e^{-\frac{i}{\hbar}\frac{T}{2m}p^2}e^{\frac{i}{\hbar}p\left(x_f-x_i\right)}dp. \tag{18-58}$$

In (18-58), we could have simply used x_f in the exponent, as we have been taking $x_i = 0$, and our result would have been in terms of x_f. In that case, x_f would have been the distance between x_i and x_f, i.e., $x_f - x_i$. In order to frame our final result in the most general terms, we re-introduced x_i as having any coordinate value in (18-58).

With (18-46) again, (18-58) becomes

$$U\left(x_i,x_f;T\right)\approx C\left(\frac{i2\pi\hbar(\Delta t)}{m}\right)^{N/2}\sqrt{\frac{m}{i2\pi\hbar T}}e^{\frac{i}{\hbar}\frac{m}{2T}\left(x_f-x_i\right)^2}. \tag{18-59}$$

By comparison with (18-47), we see the phase and dependence on T is the same as in the wave mechanics approach. Using that comparison, we can see that the constant of proportionality is

$$C = A\left(\frac{m}{i2\pi\hbar(\Delta t)}\right)^{N/2}. \tag{18-60}$$

And thus, the probability density at the final event **f** is the same as (18-48), i.e.,

$$\left|U\left(x_i,x_f;T\right)\right|^2 = A^2\frac{m}{2\pi\hbar T}, \tag{18-61}$$

where the equal sign is appropriate for $N \to \infty$.

We can find the normalization factor A by integrating $|U|^2$ over all space and setting the result to one, as is usual in NRQM. (See the appendix, pg. 509, for more on this.)

Note that for $v = (x_f - x_i)/T$, the amplitude (18-59) can be expressed in terms of the classical action as

Path integral approach = Schrödinger approach

$$U\left(x_i,x_f;T\right)=A\sqrt{\frac{m}{i2\pi\hbar T}}e^{\frac{i}{\hbar}\frac{mv^2}{2}T}=A\sqrt{\frac{m}{i2\pi\hbar T}}e^{\frac{i}{\hbar}LT}=A\sqrt{\frac{m}{i2\pi\hbar T}}e^{\frac{i}{\hbar}S}\qquad x_{peak}=\overline{x}=x_f. \tag{18-62}$$

18.9.3 The Message

It has probably not escaped the reader that the evaluation of a free particle using Feynman's many paths approach is considerably more complicated and lengthy than the Schrödinger approach. This is true for most, if not all, problems in NRQM and RQM.

Pluses and minuses of path integral method

The <u>disadvantages</u> of the many paths approach in NRQM and RQM are these.

1. It is generally more mathematically cumbersome and time consuming than the wave mechanics approach.

2. The quantity calculated is only proportional to the amplitude, and further analysis is required to determine the precise amplitude.

More limited and generally harder

3. The approach is suitable primarily for position eigenstates and is not readily amenable to more general states, so it is generally not as encompassing in nature.

The <u>advantages</u> of the many paths approach are these.

1. The approach also applies to QFT. In a number of instances therein, development of the theory is more direct, and calculation of amplitudes is easier, than with the alternative approach (canonical quantization).

Has some advantages for QFT

2. Philosophically, we see that there is more than one way to skin a cat. We learn anew that the physical world can be modeled in different, equivalent ways. We learn caution with regard to interpreting a given model as an actual picture of reality.

18.10 Quantum Field Theory via Path Integrals

So far, we have dealt primarily with NRQM, but the many paths approach is also applicable to RQM, and as noted above, to QFT. (For any readers who have not read Chap. 1, similarities and differences between the two are delineated in Wholeness Chart 1-2, pg. 7-8. Further similarities and differences are illustrated in Wholeness Chart 18-4, below.)

Path integral approach applied to QFT

We will not go deeply into QFT, and only outline, in a broad overview, how the theory presented herein is applicable therein. This should help those students who continue on to the standard texts for the many paths approach keep the forest in view while examining the trees.

18.10.1 Particle Theory (NRQM/RQM) vs Field Theory (QFT)

For the many paths approach, we want to make the jump from NRQM/RQM, which are quantized versions of particle theory, to QFT, which is a quantized version of field theory. Wholeness Chart 18-4 below can help us do that. In it, the 2nd and 3rd columns compare particle theory entities/concepts to corresponding field theory entities/concepts. The first part of the chart (this page), as indicated, summarizes classical theory (non-quantum, and implicitly including special relativity). The latter part (next page) summarizes quantum theory *via approaches other than many paths*. The chart should be relatively self-explanatory, so we will not comment much on it.

Comparing analogous quantities in NRQM and QFT (without reference to path integrals)

We compare the quantum approaches of Wholeness Chart 18-4 to the many paths approach in the next section.

Wholeness Chart 18-4. Comparing Particle Theory to Field Theory: Classical and Quantum

	Particle Theory		Field Theory
	Classical Theory		
Independent variables	<u>1D</u> t	<u>3D</u> t	<u>3D</u> x,y,z,t
Dependent variables	$x(t)$ position	$x(t), y(t), z(t)$	$\phi(x,y,z,t)$ field
Dynamical variables	Particle total value: $\mathbf{p}, E, L$ functions of $x,\dot{x},t$ (or $\mathbf{r},\dot{\mathbf{r}},t$)		Density values (per unit vol): $\mathbf{p}/V, \mathcal{E}, \mathcal{L}$ functions of $\phi,\phi_{,\mu},x,y,z,t$ $E = \int \mathcal{E}d^3x$, etc.
Equations of motion	$\mathbf{F} = m a$ or equivalently, Euler-Lagrange formulation, $\dfrac{d}{dt}\left(\dfrac{\partial L}{\partial \dot{x}}\right) - \dfrac{\partial L}{\partial x} = 0$		$f = \rho\mathbf{a}$ (force/vol) for media; Maxwell's eqs for e/m, or equivalently, for $\mathcal{L}$ of either, $\dfrac{\partial}{\partial x^{\mu}}\left(\dfrac{\partial \mathcal{L}}{\partial \phi_{,\mu}}\right) - \dfrac{\partial \mathcal{L}}{\partial \phi} = 0$
Variable correspondences particle ↔ field	$t \leftrightarrow x,y,z,t$ $x \leftrightarrow \phi$ total values ↔ density values		

	Particle Theory	**Field Theory**
	Quantum Theories	
	NRQM and RQM via Wave Mechanics	QFT via Wave Mechanics = Canonical Quantization
Quantum character change	x and all dynamical variables $\rightarrow$ operators	ϕ and all dynamical variables $\rightarrow$ operators
New quantum entity	state $\left\vert\psi\right\rangle =$ wave function ψ	state $\left\vert\phi\right\rangle$ different from (operator) field ϕ
Note		Fields create & destroy states. States can be multi-particle $\left(\left\vert\phi_1,\phi_2,...\right\rangle\right)$
Operators	functions of $x,\dot{x},t$	functions of $\phi,\phi_{,\mu},t$
Expectation values of operators	$\overline{E}=\left\langle\psi\right\vert H\left\vert\psi\right\rangle$ etc. for other operators	$\overline{E}=\left\langle\phi\right\vert H\left\vert\phi\right\rangle$ or for multi-particle state $\overline{E}=\left\langle\phi_1,\phi_2...\right\vert H\left\vert\phi_1,\phi_2...\right\rangle$
Equations of motion	For wave function ψ QM: Schrödinger eq RQM: Klein-Gordon, Dirac, Maxwell, Proça eqs or equivalently, Euler-Lagrange formulations	For quantum field ϕ QFT: Klein-Gordon, Dirac, Maxwell, Proça eqs or equivalently, Euler-Lagrange formulations
Macro equations of motion	Deduced from above and expectation values of force, acceleration	Deduced from above and expectation values of relevant quantities
Transition amplitude U (finite T)	$U\left(x_i,x_f;T\right)=\left\langle x_f\right\vert e^{-iHT}\left\vert x_i\right\rangle$ i & f are eigenstates of position	$U\left(\phi_i,\phi_f;T\right)=\left\langle\phi_f\right\vert e^{-iHT}\left\vert\phi_i\right\rangle$ i & f states can be multi-particle
$\lvert U\rvert^2 =$	probability density (for normalizations chosen herein)	probability

18.10.2 "Derivation" of Many Paths Approach for QFT

Extend the same analogies to path integrals

From the next to last row of Wholeness Chart 18-4, we see that the transition amplitude for the QFT canonical approach, which is essentially a wave mechanics approach for relativistic fields, is similar in form to that of the NRQM/RQM wave mechanics approach, given the correspondence $x \rightarrow \phi$ between NRQM/RQM and QFT. An additional fundamental difference between NRQM and QFT is the form of the Hamiltonian H. In NRQM, H is a non-relativistic function of $x, \dot{x},$ and (rarely) t. In QFT, it is a relativistic function of $\phi, \phi_{,\mu},$ and (rarely) x, y, z, t.

Since the canonical (wave mechanics) QFT approach mirrors the wave mechanics NRQM/RQM approach, one could postulate (and Feynman probably did) that the many paths approach in QFT would mirror the many paths approach in NRQM/RQM. (See Wholeness Chart 18-2 in Section 18.7.2 for the corresponding NRQM transition amplitudes using each approach.) Simply using the same correspondences $x \rightarrow \phi$ and $H_{nonrel} \rightarrow H_{rel}$ (and thus, $L_{nonrel} \rightarrow L_{rel}$) for the many paths approach yields Wholeness Chart 18-5.

Wholeness Chart 18-5. Comparing NRQM/RQM to QFT for the Many Paths Approach

	Quantum Theories	
	NRQM and RQM via Many Paths	QFT via Many Paths
Transition amplitude	$U\left(x_i,x_f;T\right)=C\lim_{N\to\infty}\sum_{j=1}^{N}e^{iS_j/\hbar}$ $=C\int_{x_i}^{x_f}e^{i\int_0^T\frac{L}{\hbar}dt}\mathcal{D}x(t)$	$U\left(\phi_i,\phi_f;T\right)=C\lim_{N\to\infty}\sum_{j=1}^{N}e^{iS_j/\hbar}$ $=C\int_{\phi_i}^{\phi_f}e^{i\int_0^T\frac{\mathcal{L}}{\hbar}d^4x}\mathcal{D}\phi\left(x^\mu\right)$
Note	Above is from Wholeness Chart 18-2 in Section 18.7.2	Above is a simplified example for a single scalar field.

In the RH column above, all paths, comprising all configurations of the entire field ϕ over all space between its initial and final configurations, are added (integrated). S here is the action for the entire field. $\mathcal{L}$ is the (relativistic) Lagrangian *density* for the field, which, integrated as it is above over all space d^3x, yields L.

That extension gives us a viable path integral formulation of QFT

Of course, the many paths transition amplitude for QFT of Wholeness Chart 18-5 is, at this point, only a guess. However, decades of research, first by Feynman and then by many others, have proven that it is completely valid.

To summarize, briefly

Wholeness Chart 18-6. Super Simple Summary

Correspondences	$x\to\phi$ $H_{nonrel}\to H_{rel}$	
Wave mechanics amplitude	NRQM $\to$ QFT	canonical quantization QFT
Many paths amplitude	NRQM $\to$ QFT	functional quantization QFT

18.10.3 Time Slicing in QFT

Using the same correspondences as in Wholeness Chart 18-6, and the time slicing approximation for QM of (18-38), we find, for QFT,

We get a QFT U analogous to NRQM U, for approximation case

$$U\left(i,f;T=t_f-t_i\right)\approx C\int\ ....\int\ \int e^{i\int_{t_i}^{t_f}\frac{\mathcal{L}}{\hbar}d^4x}\ d\phi_1 d\phi_2...d\phi_n \qquad , \qquad (18\text{-}63)$$

QFT Approximation for Transition Amplitude for Single Field

where the subscripts refer to different time slices, not to different fields. This example is for only a single field.

The exact form of the transition amplitude, obtained from (18-39), is given in Wholeness Chart 18-5, and is repeated here,

And in the limit, we get a QFT U analogous to NRQM U that is exact

$$U\left(i,f;T\right)=C\int_{\phi_i}^{\phi_f}e^{i\int_0^T\frac{\mathcal{L}}{\hbar}d^4x}\mathcal{D}\phi\left(x^\mu\right) \qquad . \qquad (18\text{-}64)$$

QFT Exact Expression for Transition Amplitude for Single Field

One can extrapolate (18-64) to more than one field, but we will not do this here.

18.10.4 More Ahead in Path Integral QFT

Note that we have only scratched the surface of the many paths approach to QFT. There is a great deal more, including some fairly fundamental concepts. However, hopefully, all of the above will provide a solid foundation for that, by explaining more simply, more completely, and in smaller steps of development what traditional introductions to the subject often treat more concisely.

There is more to learn about QFT path integrals. This was an intro.

18.11 Chapter Summary

It is time to try your hand at creating a wholeness chart summary by doing Prob. 1.

18.12 Appendix

There are issues with normalization of position eigenstates that complicate the interpretation of the resulting transition amplitude for an initial position eigenstate transitioning to a final position eigenstate. To investigate this, we first consider A to be a normalization constant to be determined in

We will compare normalization constants for different methods of normalization for position eigenstates

$$\left| x_j \right\rangle \xrightarrow[\text{space}]{\text{in position}} = A\delta\left(x - x_j\right), \tag{18-65}$$

and then determine A for different ways to normalize.

18.12.1 Standard NRQM/RQM Normalization

In standard NRQM, eigenstates are generally orthonormal. For (18-65), this means

Standard NRQM normalization

$$\left\langle x_j \middle| x_k \right\rangle = \delta_{jk} \xrightarrow[\text{space for } j=k]{\text{in position}} \underbrace{\int A^* \delta\left(x - x_j\right) \overbrace{A\delta\left(x - x_j\right)}^{f(x)} dx = 1}_{\text{probability density}}, \tag{18-66}$$

where the square of the absolute value of the wave function equals probability density and the total probability of measuring the position eigenstate anywhere in space is one, as it should be. If we consider one of the delta functions to be just like any function of x, $f(x)$, then (18-66) leads to

Standard NRQM normalization constant

$$|A|^2 f\left(x_j\right) = |A|^2 \delta\left(x_j - x_j\right) = |A|^2 \delta(0) = 1 \quad \rightarrow \quad A = \frac{1}{\sqrt{\delta(0)}}. \tag{18-67}$$

While at first blush it may seem strange to have a factor with the square root of infinity in the denominator, it is not much different from having a wave function like $Ae^{-i(\omega t - kx)}$ that extends from $-\infty$ to $+\infty$ along the x axis. In that case, $A = 1/\sqrt{\infty}$ as well. So, if we can live with this hypothetically pure position eigenstate, then for NRQM, as usually done, (18-65) becomes

Standard NRQM normalized position eigenstate

$$\left| x_j \right\rangle \xrightarrow[\text{in position space}]{\text{NRQM normalization}} = \frac{1}{\sqrt{\delta(0)}} \delta\left(x - x_j\right), \tag{18-68}$$

and probability density is

$$\rho_{NRQM}\left(x\right) = \frac{\left(\delta\left(x - x_j\right)\right)^2}{\delta(0)} = \delta\left(x - x_j\right). \tag{18-69}$$

Standard NRQM total probability is one

(18-69) is infinite at $x = x_j$ and zero elsewhere. Total probability, its integral over all space, is one.

For this normalization of both bra and ket, the transition amplitude U and $|U|^2$ are

Standard NRQM $|U|^2$ is total transition probability

$$U_{NRQM}\left(x_i, x_f; T\right) = \left\langle x_f \middle| e^{-iHT/\hbar} \middle| x_i \right\rangle \quad \rightarrow \quad |U_{NRQM}|^2 = \text{total probability of transition}, \tag{18-70}$$

which is what we have come to expect $|U|^2$ to represent.

18.12.2 Normalization Found in Other QFT Texts

Other QFT texts, when discussing the path integral approach, use a different normalization[1],

---

[1] See Peskin. M. & Schroeder, D., *An Introduction to Quantum Field Theory* (Perseus 1995), pg. 277, (9-3) LHS and the first sentence in the paragraph beginning after (9-7) on pg. 279. See Zee, A., *Quantum Field Theory in a Nutshell* (Princeton 2010), pg. 10, 3rd line down under heading "Dirac's formulation".

$$\langle x_j | x_k \rangle = \delta(x_j - x_k) \xrightarrow[\text{space for } j=k]{\text{in position}} \int A^* \delta(x - x_j) A \delta(x - x_j) dx = \delta(0). \qquad (18\text{-}71)$$

Other texts normalization

Taking one of the delta functions on the RHS of (18-71) as $f(x)$ as we did above, we find

$$|A|^2 \delta(x_j - x_j) = |A|^2 \delta(0) = \delta(0) \quad \rightarrow \quad A = 1, \qquad (18\text{-}72)$$

Other texts normalization constant

so (18-65) becomes

$$|x_j\rangle \xrightarrow[\text{in position space}]{\text{other texts normalization}} = \delta(x - x_j). \qquad (18\text{-}73)$$

Other texts normalized position eigenstate

Note that what we generally consider probability density is

$$\rho_{\substack{other \\ texts}}(x) = \big(\delta(x - x_j)\big)^2, \qquad (18\text{-}74)$$

Other texts total probability is infinity

and the integral of (18-74) over all space, what we usually interpret as total probability, is infinite (and thus cannot represent total probability).

For this normalization of both bra and ket, the transition amplitude U and $|U|^2$ are

$$U_{\substack{other \\ texts}}(x_i, x_f; T) = \langle x_f | e^{-iHT/\hbar} | x_i \rangle \xrightarrow[\text{for } T=0, \text{ i.e., } i=f]{\text{in position space}} \left| U_{\substack{other \\ texts, i=f}} \right|^2 = \infty^2 \left(\begin{array}{c} \text{no physical} \\ \text{interpretation} \end{array} \right). (18\text{-}75)$$

Other texts $|U|^2$ has no physical meaning

If there is no physical interpretation when $T = 0$, it follows that there is none when $T \neq 0$.

18.12.3 Hybrid Normalization Found in This Text

This text assumes ket is normalized as in standard NRQM. i.e., as if it were a wave packet

In Sect. 18.3.2, pg. 491, we considered a surrogate for the initial position eigenstate ket to be a high, narrow wave packet approximating a delta function, but, importantly, normalized as is usual in NRQM (see (18-66)), i.e.,

$$\langle x_i | x_i \rangle = 1 \quad |x_i\rangle \text{ a high, narrow wave packet approx to position eigenstate}. \qquad (18\text{-}76)$$

That, along with considering the bra to be a pure delta function, as is (18-73), let us interpret $|U|^2$ of (18-11) as probability density. Had we taken both the ket and the bra as pure delta functions, such as in (18-73), we would have no readily comprehensible physical meaning for $|U|^2$. (See (18-75).) Since the path integral approach yields a quantity that is proportional to the probability density, I, the author, felt it best to present the background NRQM material in a manner amenable to correlating it with that approach.

But this text assumes bra is normalized as in other texts, i.e., as if it were a pure delta function

However, in Sect. 18.9.1, pg. 502, the math is greatly simplified by using an actual delta function for the initial ket, rather than a limiting case wave packet. But, we then need to normalize our initial delta function ket to satisfy (18-76). That is, we need the initial ket of delta function form to have $A = 1/\sqrt{\delta(0)}$, i.e., to be of form (18-68).

This gives us a $|U|^2$ that is probability density and can thus be readily related to path integral result

So, the normalization constant A in Sects. 18.9.1 and 18.9.2 equals $1/\sqrt{\delta(0)}$, but I felt that introducing the square root of infinity at that point would be inordinately confusing and take us away from the main purpose of the section.

These three approaches to normalization of position eigenstates, and the ramifications of each, are summarized in Wholeness Chart 18-7 on the next page.

18.12.4 Bottom Line

The important thing: what we find in path integral approach is proportional to probability density

All of this appendix is focused on NRQM, and is not so relevant to QFT, except as part of an introduction to the path integral methodology. It is just background for the most important concept in the chapter, stated below.

Bottom line: The path integral approach result is proportional to probability density. We only discuss position eigenstates as an aid to developing that concept and later, to extrapolating it to QFT.

Wholeness Chart 18-7. Comparing Normalization Methods for Position Eigenstates

	Standard	**Some Other Texts**	**This Text**
Position Eigenstate	$\left\lvert x_j \right\rangle$	as at left	as at left
In position space	$\dfrac{1}{\sqrt{\delta_L(0)}}\,\delta(x-x_j)$	$\delta(x-x_j)$	ket: $\psi_{x_k}(x)$ = high, narrow normalized wave packet bra: $\delta(x-x_j)$
Normalization	$\left\langle x_j \middle\vert x_k \right\rangle = \delta_{jk}$	$\left\langle x_j \middle\vert x_k \right\rangle = \delta(x_j - x_k)$	$\left\langle x_j \middle\vert \psi_{x_k}(x) \right\rangle = \psi_{x_k}(x_j)$ = wave function value at x_j
Total probab of measuring (take $j = k$)	$\left\langle x_k \middle\vert x_k \right\rangle$ = unity	$\left\langle x_k \middle\vert x_k \right\rangle$ = infinity	$\left\langle x_k \middle\vert \psi_{x_k}(x) \right\rangle$ not total probability, but wave function value at x_k
Transition Amplitude U	$U(x_i, x_f, T) = \left\langle x_f \middle\vert e^{-iHT} \middle\vert x_i \right\rangle$ in each case		
$\lvert U \rvert^2$ represents	Total probability of measuring x_f state at time T.	Not total probability, nor probability density of measuring x_f at time T.	Probability density of evolved state.
Pro	1) Usual NRQM analysis. 2) Easy to visualize.	What most authors use.	1) Easy to visualize. 2) Easier to accept probability density than ∞ as substitute for total probability. 3) Easier to relate to path integral result, which is proportional to probability density.
Con	1) Not easy to see how total probability related to probability density, which is what we want to get from path integral approach. 2) Trying to do this with ∞ in denominator can be confusing.	1) ∞ probability, contradicts NRQM. 2) Impossible to visualize meaning of $\lvert U \rvert^2$. Students have no idea what U means.	1) Probability density not total probability, contradicts NRQM. 2) Not what most (any other?) authors use.
Bottom line	1) Path integral approach result is proportional to probability density. Don't need to use any of above approaches if set integral of that result over all space equal to one and solve for proportionality constant C. (But then no real proof that path integral approach yields what is claimed.) 2) Can compare any of above three wave mechanics interpretations to path integral approach to determine proportionality constant C. Just need to interpret final result as equal to that of the particular wave mechanics approach used, i.e., the meaning in "$\lvert U \rvert^2$ represents" row above. (Provides a proof that wave mechanics and path integral approaches are consonant.)		

18.13 Problem

1.   Create a wholeness chart summarizing this chapter.

Chapter 19

Looking Backward and Looking Forward:
Book Summary and What's Next

"Our notions of mind and matter must go through many a phase as yet unimagined."

F. W. H. Myers (1903)

19.0 Preliminaries

Not only was Myers right, in spades, but his statement still holds true today.

19.0.1 Background

An old adage in education is this. "First, tell them what you are going to tell them. Then, tell them. Finally, tell them what it is you just told them." Or, as I like to say it, "Start with wholeness and end with wholeness. In between, do the parts."

For proper understanding we need to know the "big picture", the master overview of what a particular field of study means in its essence and entirety. We also need to have traversed the details, the step-by-(sometimes grueling)-step mathematics in which the big picture is structured, and by which it is glued together.

Chap. 1 presents a broad-brush schematic of QFT, a simplified framework, *sans* detail, which, hopefully, can quickly provide newcomers with some level of comfort in the theory, its workings, and its purpose, before they begin the arduous trek through extensive mathematical minutiae. The present chapter provides a similar overview, except that it is post trek and references many of the finer points of the theory covered in the intervening chapters.

There are at least three ways one can review and summarize the contents of this book.

  1) Peruse the wholeness charts, one by one. (A Table of Wholeness Charts follows the Table of Contents at the beginning of the book.)
  2) Scrutinize the chapter summaries, one by one.
  3) Read this chapter (which references particular wholeness charts and chapter summaries).

Regardless of which way(s) you choose, I recommend first going back and re-reading Chap. 1. Given the ground you have covered, it should not take long, and may help in solidifying your understanding of how the various QFT pieces fit together.

19.0.2 Chapter Overview

In this chapter, we will first present a summary of the book, broken into

- the four main parts of QFT, with
- reference to key summary wholeness charts, and
- emphasis on QED.

We will then present a brief overview of where a student of field theory should be heading from here, i.e.,

- what's next.

19.1 Book Summary

19.1.1 Pre QFT

Chaps. 1 and 2 i) reviewed and organized pre-requisite fields of study (classical physics, NRQM), ii) compared them to RQM and QFT, and iii) introduced concepts needed to begin study of QFT.

These are summarized in

Quantization Summary, pg. 4	First quantization (particles) and second quantization (fields)
	1) Hamiltonian/Lagrangian/governing equation same for classical and quantum theories (density values for fields, total values for particles)
	2) Poisson brackets $\rightarrow$ commutators
Wholeness Chart 1-1, pg. 5	The Overall Structure of Physics (Deducing quantum theories from classical via quantization)
Wholeness Chart 1-2, pgs. 7-8	Comparison of NRQM, RQM, and QFT
Wholeness Chart 2-1 and Sect. 2.1.7, pg. 14	Natural Units
Wholeness Chart 2-2, pgs. 20-21	Summary of Classical (Variational) Mechanics
Wholeness Chart 2-5, pgs. 30-31	Summary of Quantum (Variational) Theories
Wholeness Chart 2-4, pg. 28	Schrödinger vs. Heisenberg Pictures

With the above as background, one proceeds as follows to develop QFT.

Steps to our goal

2nd quantization postulates $\rightarrow$ QFT theory $\rightarrow$ transition amplitude calculation $\rightarrow$ probability
$\rightarrow$ scattering, decay, other experimental results $\rightarrow$ confirmation of QFT

Note that decay is, for the most part, a weak interaction phenomenon, and since this book is devoted primarily to QED, is not treated herein.

19.1.2 QFT in a Kernel (Canonical Quantization Approach)

The remainder of the book focused in part on RQM, but mostly on QFT. The latter can be broken down into four major branches,

1) Free Fields (Chaps. 3 to 6, pgs. 39-180)
2) Interacting Fields (Chaps. 7 to 11, pgs. 181-301)
3) Renormalization (Chaps. 12 to 15, pgs. 303-400)
4) Application to Experiment (Chaps. 16 and 17, pgs. 401-486).

We summarize each of these in the following pages.

Part 1: Free Fields

Steps to QFT for Free Bosonic Fields (Heisenberg Picture)

(Only scalars and particle [not antiparticle] coefficients shown for brevity)

1st postulate $\rightarrow$ Classical $\mathcal{H}_0$ = Quantum $\mathcal{H}_0$ $\rightarrow$ Classical $\mathcal{L}_0$ = Quantum $\mathcal{L}_0$ $\rightarrow$

QFT free field equation $\rightarrow$ solutions $\phi^r(\mathbf{x},t)$ $\rightarrow$ $\mathcal{H}_0$ in terms of $\phi^r(\mathbf{x},t)$, $\pi_s(\mathbf{y},t)$

2nd postulate $\rightarrow$ $[\phi^r(\mathbf{x},t), \pi_s(\mathbf{y},t)] = i\delta^r{}_s\,\delta(\mathbf{x}-\mathbf{y})$ $\rightarrow$

$[a(\mathbf{k}), a^\dagger(\mathbf{k'})] = \delta_{\mathbf{kk'}}$ $\rightarrow$ $H_0 = \int \mathcal{H}_0\, d^3x$ in terms of $a^\dagger(\mathbf{k})\, a(\mathbf{k})$ $\rightarrow$ vacuum energy plus

$N_a(\mathbf{k}) = a^\dagger(\mathbf{k})a(\mathbf{k})$ as number operator $\rightarrow$ $a^\dagger(\mathbf{k})$, $a(\mathbf{k})$ as creation/destruction operators

$\downarrow$ $\swarrow$ $\searrow$

form of observable operators   the propagator   interaction theory (Part 2)
(for real particles)   (for virtual particles)   (for real & virtual particles)

Steps to QFT for Free Fermionic Fields (Heisenberg Picture)

Fermions cannot occupy the same state, so cannot form macroscopically observed fields. Thus, they do not have classical field theory counterparts from which to quantize a quantum theory. Dirac deduced the spin ½ field equation from other considerations. Further, the Pauli exclusion principle and other fermionic properties only arise by assuming anti-commutation for the coefficients of solutions to the Dirac equation (whereas commutation is needed for correct boson properties).

(Only particle [not antiparticle] coefficients shown below for brevity)

1st postulate = QFT Dirac field equation $\rightarrow$

Dirac QFT $\mathcal{L}_0{}^{1/2}$ and solutions $\psi(\mathbf{x},t)$ $\rightarrow$ $\mathcal{H}_0{}^{1/2}$ in terms of $\psi(\mathbf{x},t)$, $\pi^{1/2}(\mathbf{y},t)$

2nd postulate = $[c(\mathbf{p}), c^\dagger(\mathbf{p'})]_+ = \delta_{\mathbf{pp'}}$ $\rightarrow$

$H_0{}^{1/2} = \int \mathcal{H}_0{}^{1/2}\, d^3x$ in terms of $c^\dagger(\mathbf{p})\, c(\mathbf{p})$ $\rightarrow$ vacuum energy plus

$N_r(\mathbf{p}) = c^\dagger(\mathbf{p})c(\mathbf{p})$ as number operator $\rightarrow$ $c^\dagger(\mathbf{p})$, $c(\mathbf{p})$ as creation/destruction operators

$\downarrow$ $\swarrow$ $\searrow$

form of observable operators   the propagator   interaction theory (Part 2)
(for real particles)   (for virtual particles)   (for real & virtual particles)

Wholeness Chart 5-4, pgs. 156-160, has an extensive summary of free field theory.

Part 2: Interacting Fields

Steps of QFT for Interacting Fields (Interaction Picture)

Using the I.P., we have (superscript I = interaction picture; subscript I = interaction term in H)

Time evolution of <u>states</u> governed by H_I^I: $\quad i\dfrac{d}{dt}|\Psi\rangle = H_I^I|\Psi\rangle$ $\qquad\qquad$ (19-1)

Time evolution of <u>operators</u> governed by $H_0^I = H_0$: i.e, by free field equations

For interactions, if employ I.P., can use

1. Part 1 free field operator solutions ϕ, ψ, $A\mu$; free field operator creation and destruction properties; free field number operators; free field observables operators; free field Feynman propagators, and

2. state equations of motion in H_I^I, (19-1), to determine change in state in time (due to interactions).

The S Matrix and the S Operator

$$S_{fi} = \langle f|S_{oper}|i\rangle = \langle f|\Psi(t_f)\rangle \text{ where general final state } |\Psi(t_f)\rangle = \sum_{f'}|f'\rangle S_{f'i} \quad (19\text{-}2)$$

I.P. state eq of motion (19-1) for $|\Psi(t)\rangle = S_{oper}|i\rangle = S_{oper}|\Psi(t_i)\rangle$ yields

$$i\frac{dS_{oper}}{dt} = H_I^I S_{oper} \quad \text{with solution } S_{oper} = e^{-i\int_{t_i}^{t_f} H_I^I dt} = e^{-i\int_{t_i}^{t_f}\int_V \mathcal{H}_I^I d^4x} \quad (19\text{-}3)$$

Define $S = S_{oper}$ with $V \to \infty$, $t_i \to -\infty$, $t_f \to \infty$.

$\qquad$ (This seems unrealistic, but makes math tractable and will allow calculation
$\qquad\qquad\qquad$ of experimental results, as shown in Part 4.)

Then,

$\qquad\qquad\qquad\qquad$ Dyson expansion $\rightarrow$

S in terms of time ordering operator T and H_I^I (which is composed of field
construction/destruction operator solutions of free field equation) $\rightarrow$

$\qquad\qquad\qquad\qquad$ Wick's theorem $\rightarrow$

Easier form of S to evaluate in terms of normal ordering operator N, field
operators, and contractions $\rightarrow$

Evaluate S_{fi} of (19-2), using Wick's form of S for any given $\langle f|$ and $|i\rangle$ $\rightarrow$

$\qquad\qquad$ Probability of interaction occurring = $|S_{fi}|^2$

QED

For quantum electrodynamics, in S_{fi} above, use

$$\mathcal{H}_I^I = -e\bar{\psi}A_\mu\gamma^\mu\psi = -e\left(\bar{\psi}^+ + \bar{\psi}^-\right)\left(A^+ + A^-\right)\left(\psi^+ + \psi^-\right). \quad (19\text{-}4)$$

Feynman's Rules

The lengthy procedure to calculate the form of any given S_{fi} is reduced enormously by using the short cut of Feynman's Rules (see next page).

Note that Feynman diagrams depict particles as point-like objects traveling through spacetime. However, they are treated mathematically in QFT (and manifest physically, for the most part) as waves. In QFT as developed herein, all such waves in a given interaction occupy the same volume V, for the same time T.

Wholeness Chart 8-4, pgs. 248-251, has an extensive summary of interacting fields theory.

Feynman's Rules for QED

For Particles Interacting with Particles

A. The S matrix element (the transition amplitude) for a given interaction is

$$S_{fi} = \delta_{fi} + \left((2\pi)^4 \, \delta^{(4)} \left(\sum p_f - \sum p_i \right) \left(\overbrace{\prod}^{\substack{\text{all external}\\\text{bosons}}} \sqrt{\frac{1}{2V\omega_k}} \right) \left(\overbrace{\prod}^{\substack{\text{all external}\\\text{fermions}}} \sqrt{\frac{m}{VE_p}} \right) \right) \mathcal{M} \qquad \mathcal{M} = \sum_{n=1}^{\infty} \mathcal{M}^{(n)} \quad (19\text{-}5)$$

where P_f is the total 4-momentum of all final particles, P_i is the total 4-momentum of all initial particles, and the contribution $\mathcal{M}^{(n)}$ comes from the nth order perturbation term of the S operator, $S^{(n)}$.

B. The Feynman amplitude $\mathcal{M}^{(n)}$ is obtained from all of the topologically distinct, connected (i.e., all lines connected to one another in a given diagram) Feynman diagrams which contain n vertices. The contribution to each $\mathcal{M}^{(n)}$ is obtained by the following.

1. For each vertex, include a factor $ie\gamma^\mu$.

2. For each internal photon line, labeled by 4-momentum k, include a factor $iD_{F\mu\nu}(k) = i\dfrac{-g_{\mu\nu}}{k^2 + i\varepsilon}$

3. For each internal fermion line, labeled by 4-momentum p, write a factor $iS_F(p) = i\dfrac{\not{p} + m}{p^2 - m^2 + i\varepsilon}$

4. For each external line, write one of the following spinor factors, where $\mathbf{p}$ and $\mathbf{k}$ indicate basis states of corresponding 3-momenta, r represents spin state for fermions and polarization state for photons, [see (4-93) and (4-94), pg. 111]

        a) for each initial electron: $u_r(\mathbf{p})$ b) for each final electron: $\bar{u}_r(\mathbf{p})$

        c) for each initial positron: $\bar{v}_r(\mathbf{p})$ d) for each final positron: $v_r(\mathbf{p})$

        e) for each initial photon: $\varepsilon_{r,\mu}(\mathbf{k})$ f) for each final photon: $\varepsilon_{r,\mu}(\mathbf{k})$

5. The spinor factors (γ matrices, S_F functions, spinors) for each fermion line are ordered so that, reading from right to left, they occur in the same sequence as following the fermion line in the direction of its arrows through the vertex. (Order is important as it conveys spinor matrix multiplication order when we do not show spinor indices.)

6. The four-momenta at each vertex are conserved (same total after as before).

6a. For each anti-particle fermion propagator label the Feynman diagram 4-momentum with opposite sign of what it has physically and use this negative of the physical value in the propagator.

7. For each closed loop of internal fermions only (without photons inside the loop itself, like what we call a "photon loop" which internally has an electron and a positron), take the trace (in spinor space) of the resulting matrix and multiply by a factor of (-1).

8. For each 4-momentum q which is not fixed by 4-momentum conservation, carry out the integration $\left(1/(2\pi)^4\right)\int d^4q$ One such integration for each closed loop (fermion/fermion or fermion/photon loop).

9. Multiply the expression by (-1) for each interchange of neighboring fermion operators (each associated with a particular spinor factor) which would be required to place the expression in appropriate normal order. "Appropriate", when we are adding sub amplitudes, means each sub amplitude must be in the same, not just any, normal order of destruction and creation operators.

10. (Only needed when doing renormalization. Can ignore otherwise.)
Add a mass counter term to any fermion line in a Feynman diagram $\rightarrow$ ———▶—✗—▶———
For each mass counter term diagram, add a term to the amplitude with a factor equal to $i\delta m$.

For Particles Interacting with an External Potential Field

11. For each interaction of a charged particle with a static external photon field (a static potential field), a) write a factor $A_\mu^e(\mathbf{k}) = \int e^{-i\mathbf{k}\cdot\mathbf{x}} A_\mu^e(\mathbf{x}) d^3x$, and

       b) instead of $(2\pi)^4 \, \delta^{(4)}(\Sigma p_f - \Sigma p_i)$ in (19-5), use $2\pi \, \delta(\Sigma E_f - \Sigma E_i)$.

Part 3: Renormalization in QED

The Problem

Three integrals we can run into in evaluating S_{fi} in QED are divergent. These integrals arise from loops in the photon propagator (photon self-energy), the Dirac propagator (fermion self-energy), and the vertex correction (vertex loop). Renormalization tames these infinities by re-defining e and m to absorb infinite quantities in those new definitions, and thereby leaves a finite S_{fi}, with a probability $|S_{fi}|^2$ between 0 and 1 for any interaction. In the process, we end up with (fairly complicated) modified versions of our propagator and vertex relations, which are finite in the ultraviolet limit.

The Renormalization Procedure

The process of renormalization to a given order n in α (i.e., in e^2) comprises the following.

1. Draw all relevant Feynman diagrams with two to $2n$ vertices.

2. Add all the related Feynman amplitudes.

3. Evaluate all the propagator integrals in the amplitudes via regularization, yielding quantities e and m dependent on e_0 and m_0 respectively, and a parameter Λ, wherein e and m become infinite when $\Lambda \to \infty$. Other quantities will be obtained that are finite and not dependent on Λ. These other quantities modify propagators and vertex relations.

4. Redefine e_0 and m_0 such that, as $\Lambda \to \infty$, e and m equal the finite, physical values for the QED charge constant and fermion particle (rest) mass. e is dependent on interaction energy level p and is expressed as $e(p)$.

A more extensive, but still simplified summary of the renormalization procedure can be found on pgs. 366-367.

An even more extensive summary, with relevant equations, can be found in Wholeness Charts 14-4 and 14-5, pgs. 368-371.

Solving Interaction Problems to Order n

To find a transition amplitude for a given interaction, do the following.

1. Write down the tree level amplitude with bare quantities e_0 and m_0.
2. Replace e_0 and m_0 with renormalized values $e(p)$ and m, where p is the energy level of the interaction. m is the measured particle (rest) mass. $e(p)$ is measured charge at energy level p.
3. Replace propagators with modified propagators.
4. Replace the vertex relations with modified vertex relations.

Regularization

Regularization is the process of evaluating divergent integrals in terms of a parameter, typically designated by Λ, which is actually infinite, but, if taken as finite, leaves a particular divergent integral finite. When $\Lambda \to \infty$, the integral also goes to infinity.

In QED renormalization, with the troublesome integrals temporarily expressed as finite in terms of a finite Λ, we redefine e and m in such a way that the resulting integral becomes finite as $\Lambda \to \infty$, its true value. The infinities become hidden in our new values for e and m, which we take to be the measured values one finds in experiment.

As a byproduct, e becomes dependent on energy level and is typically written as $e(p)$ where p represents energy level. (Other symbols are used, as well.)

There are many types of regularization. We discussed four, the first three of which we examined closely. These are

1. Dimensional regularization (the most useful; employs $D \neq 4$ dimensions).
2. Pauli-Villars regularization (adding in a fictitious particle of mass Λ)
3. Cutoff regularization (the simplest, but it destroys gauge invariance and Lorentz invariance, both underpinnings of QFT)
4. Gauge lattice regularization (very useful in strong interaction theory)

The steps to do each of 1), 2), and 3) are summarized on pg. 398.

Part 4: Application to Experiment
Postdiction of Historical Experimental Results (Chap. 16)

Coulomb Potential in QFT

Amplitude $S_{M\o ller}$ (electrons scattered by a virtual photon) →

non-relativistic speeds → simplifies spinor math → evaluate in momentum space →

compare with Born approximation for same scattering found in NRQM (which contains the

potential in momentum space) → form of the Coulomb potential in momentum space $\tilde{V}(\mathbf{k})$ →

Fourier transform to spherical coordinates in position space → $V(r) = \pm e^2/r$

For repulsive (like charges) force, the sign above is +; attractive (opposite charges), sign is −.
The sign difference arises from the number of adjacent fermion field interchanges needed in each
case to bring the associated expression for the amplitude into the same normal order (as required
by Wick's theorem and Feynman's rules.)

Anomalous Magnetic Moment

Wholeness Chart 16-2, reproduced below, summarizes the experimental and theoretical
magnetic moment values

Theoretical and Experimental Values for Electron Gyromagnetic Ratio g

Experiment	NRQM	RQM	QFT to order		
			α	α^2	α^3
2.002319304362	Cannot predict	2	2	2.00232	2.002319304

QED value to order α:

$S_{mm}^{(2)}$ for initial electron scattering off external virtual photon field A_μ^e →

Feynman rule #11 → non-relativistic electron speeds → simplifies spinor math →

single γ^μ term in amplitude → Gordon's identity →

a term with no gamma matrices and a term in $\sigma^{\mu\nu}$ →

express as factors of Σ (spin operator), Bohr magneton μ_B, and A_μ^e →

compare to Born approximation of NRQM → spin part of potential energy (momentum space) →

Fourier transform to position space → spin potential energy $= - \underbrace{2}_{g} \mu_B \Sigma \cdot \mathbf{B}^e$

QED value to order α^2:

We don't convert γ^μ terms to $\sigma^{\mu\nu}$ terms using Gordon's identity, as we did above, but use a trick
that greatly simplifies the calculations, as summarized in steps below.

Step #1: Express $S_{mm}^{(2+4)}$ in its most general form in terms of initially unknown functions $F_1(k_v^2)$
and $F_2(k_v^2)$, where k_v is the 4-momentum of the external photon.

Step #2: Examine low energy case where $k_v \to 0$ corresponding to experimentally determined
gyromagnetic ratio g and show that $F_1(0) = 1$.

Step #3: Show $F_2(0)$ relates to g, so all we need do is find $F_2(0)$ and form of relevant F_2 terms.

Step #4: Because the term with $F_2(0)$ has no γ^μ factor, recognize that we don't need to evaluate
any 2nd order Feynman diagrams that give us a term with a γ^μ factor.

Step #5: In taking $\gamma^\mu \to \gamma_{mod}^\mu (p,p') = \gamma^\mu + e^2 \Lambda_c^\mu$ for 2nd order amplitude, only need to
${}_{2nd}$

consider terms in Λ_c^μ that have no γ^μ factor. Find the sum of those terms.

Step #6: Use the sum of terms from #5 in #3 to find, to 2nd order, $g = 2 + \alpha/\pi = 2.00232$.

Lamb Shift

We did not show it, but QED, using higher order radiative corrections, postdicts the subtle shift of orbital electron energy levels found experimentally by Lamb and Retherford.

Scattering Experiments: Determining Cross Sections (Chap. 17)

True in Every Scattering Case

For any theory (classical objects, classical object and a potential field, NRQM, RQM, QFT), relativistic or not, and any experiment,

$$\text{total transition rate} = f_b \sigma N_t \qquad \frac{\text{transition rate}}{\text{unit solid angle}} = f_b \frac{d\sigma}{d\Omega} N_t, \qquad (19\text{-}6)$$

where

$$f_b = \frac{N_b}{V_b} v_b \qquad \sigma = \int_{\theta,\phi} \frac{d\sigma}{d\Omega}(\theta) d\Omega. \qquad (19\text{-}7)$$

When both particles are moving $v_b \to v_{rel}$, magnitude of relative velocity between them.

For same particle types interacting, σ (and $d\sigma/d\Omega$) is the same in any experiment.

QFT Case, Particles Interacting with Each Other

To use QFT as developed $\to$

analyze hypothetical experiment with $N_b = N_t = 1$; $V_b = V_t = V$; $V,T \approx \infty$ $\to$

take $|S_{fi}|^2/T$ as total transition rate (probability per unit time of transition) in LHS of (19-6) $\to$

$d(|S_{fi}|^2/T)/d\Omega$ = transition rate/unit solid angle in RHS relation of (19-6) $\to$

solve for $d\sigma/d\Omega$ $\to$ cancel $d\Omega$ on both sides to get $d\sigma$ $\to$

express $d\sigma$ in terms of final particles' energies and 3-momenta $\to$

use 4-momentum conservation to express in terms of one final particle 3-momentum $\mathbf{p}'_1$ and

while doing that take $d^3\mathbf{p}'_1 = |\mathbf{p}'_1|^2 d\Omega'_1 d|\mathbf{p}'_1|$ $\to$ divide both sides by $d\Omega'_1$ to get $d\sigma/d\Omega'_1$

choose frame to evaluate final particle energies in terms of $\mathbf{p}'_1$ $\to$

if not measuring initial spins/polarizations, average initial spins/polarizations $\to$

if not measuring final spins/polarizations, do final spin/polarization sums $\to$

express $\mathbf{p}'_1$ in terms of $|\mathbf{p}'_1|$, and θ'_1 $\to$ $d\sigma/d\Omega'_1$ as function of θ'_1

QFT Case, Particles Interacting with External (Potential) Field

Similar to case above for particles interacting with each other, except in formulating S_{fi} use Feynman rule #11 for external field scattering of real particle(s). Take relevant potential form for $A_e^\mu(\mathbf{k})$, such as the Coulomb potential.

Wholeness Chart 17-6 (Parts 1 and 2), pgs. 483-485 provides an in depth summary, with equations and an example, of how to deduce differential cross sections in QFT.

Infrared Divergences

Infrared divergences that arise in bremsstrahlung scattering cross sections are cancelled to all orders by divergent radiative vertex correction terms that, almost miraculously, have equal magnitude but opposite sign from the bremsstrahlung terms.

19.1.3 Symmetry (Chaps. 6 and 11)

For symmetry and transformations in general

- Symmetry is the propensity for non-change with superficial change
- Mathematically, symmetry is invariance under transformation
- Wholeness Chart 6-1, pg. 166 compares symmetric with non-symmetric transformations

    Scalar value at a point is always invariant. Scalar function form invariant only under symmetry transformation

    Vector components at a point vary covariantly.

Transformations in QFT (see Wholeness Chart 6-3, pg. 172)

- Scalar and vector quantum fields transform like classical ones; spinors do not exist classically, but have their own form for QFT transformations
- Symmetry of QFT Lagrangian density $\mathcal{L}$ under Lorentz transformation means field equation (law of nature) is invariant in form for different observers
- Noether's theorem: If $\mathcal{L}$ is symmetric under a change of a parameter, then there is an associated quantity that is conserved
- There are three ways to determine if a quantity is conserved (see Wholeness Chart 6-4, pg. 177), though Noether's theorem method is the most useful.
- A gauge theory is a field theory for which the Lagrangian (and thus all measurables) remains invariant under a transformation of the underlying unmeasurable gauge field
- A gauge symmetry is an internal symmetry; a Lorentz symmetry is an external symmetry

Noether's theorem mathematically

If $\mathcal{L}\left(\phi^r, \phi^r{}_{,\mu}\right)$ is symmetric in form with respect to a transformation in ϕ^r which is a function of a parameter α, i.e., $\phi^r(x^\mu) \to \phi^r(x^\mu, \alpha)$, then the 4-current (using $\phi^r(x^\mu, \alpha)$)

$$j^\mu\left(\phi^r, \phi^r{}_{,\nu}\right) = \frac{\partial \mathcal{L}}{\partial \phi^r{}_{,\mu}} \frac{\partial \phi^r}{\partial \alpha} \quad \text{(sum on } r) \tag{19-8}$$

has zero four-divergence, $\partial_\mu j^\mu = 0$, and thus, $\int j^0 d^3x$ is conserved.

Symmetry for Interacting Fields

An alternative form for the free Lagrangian $\mathcal{L}_0$ that is more suitable for QED interactions yields the full Lagrangian (see (11-6), pg. 288 for the definition of $F^{\mu\nu}$)

$$\mathcal{L} = \underbrace{-\tfrac{1}{4} F^{\mu\nu} F_{\mu\nu} + \bar{\psi}\left(i\gamma^\mu \partial_\mu - m\right)\psi}_{\mathcal{L}_0} + \underbrace{e\bar{\psi}\gamma^\mu \psi A_\mu}_{\mathcal{L}_I} \tag{19-9}$$

This results in the following.

- External (Lorentz) symmetry of the full Lagrangian (including interaction term)
- Local gauge invariance (local symmetry), which is essential for renormalizing our theory.
- Charge conservation in interactions as a result of Noether's theorem
- Two perspectives, which only work for $\mathcal{L}_0$ having form shown in (19-9)

    1st: Local symmetry for the full Lagrangian $\mathcal{L}$ means we must add a term ($\mathcal{L}_I$) to the free Lagrangian ($\mathcal{L}_0$) that can only have one form, which turns out to be the correct form for the interaction as found in experiment. Local symmetry dictates the form of the interaction.

    2nd perspective: After deriving the correct interaction theory from other principles unrelated to symmetry, we find the theory is locally symmetric. Nature seems to love symmetry.

- Requiring the QED Lagrangian to be gauge invariant restricts the photon to being massless.
- Substituting the gauge covariant derivative $\partial_\nu \to D_\nu = \partial_\nu - ieA_\nu$ into $\mathcal{L}_0$ of (19-9) is called minimal substitution and results in the correct form of the full Lagrangian $\mathcal{L}$.

Global and local symmetries for internal and external transformations are summarized in Wholeness Chart 11-1, pg. 293.

Local symmetry applied to interaction theory is summarized in Wholeness Chart 11-2, pg. 295

We repeat Wholeness Chart 11-3 below.

Wholeness Chart 11-3. Summary of Symmetry Effects for Interactions

	External Symmetry in Full $\mathcal{L}$		Internal Symmetry in Full $\mathcal{L}$	
Type	Global	Local	Global	Local
Result	Laws of nature (field equations) invariant (same for all observers)	Not treated herein.	Conserved charge exists	i) Must add correct interaction term to $\mathcal{L}_0$ to make $\mathcal{L}$ symmetric $\rightarrow$ correct theory arises ii) Conserved charge exists

19.1.4 The Vacuum (Chap. 10)

Wholeness Chart 10-1, pg. 278, summarizes and compares the theory behind the four scenarios referred to in the literature as vacuum fluctuations, the last three of which arise in the formal development of QFT.

1. Virtual particle vacuum pair popping
2. Zero-point-energy quanta
3. Virtual particle vacuum bubbles with three particles at a vertex
4. Radiative corrections

None of #2 to #4 above correspond to the popular lay literature scenario of #1. No irrefutable empirical evidence has been found for true vacuum fluctuations (which #4 is not), though a number of experiments are often claimed to do so. Advanced theories may lead to particles popping in and out of the vacuum that influence the physical world, but it appears traditional QFT does not.

An update on experiments and theoretical articles on the subject was added as Appendix F in the 2018 version of this text. Given that additional material, I have reached the same conclusion I had in the original version of the text (summarized in the prior paragraph), though I again caution that each reader must form her/his own opinion on the subject, as it remains controversial.

19.1.5 That's All Folks

This concludes this book's summary of the canonical quantization approach to QFT and QED.

19.2 What's Next

We have laid the groundwork in this text for a comprehensive study of QFT. However, we have limited ourselves herein to the simplest branch of QFT, QED, which deals with the electromagnetic interaction.

19.2.1 The Standard Model

The full, present scope of QFT extends to include weak and strong interactions as well. The field of theoretical physics encompassing all three interactions is known as the standard model (SM). Understanding that should be your next goal.

Standard Model

↙   ↓   ↘

QED   Weak   Strong

Note that weak interaction theory actually is closely meshed with electromagnetic interaction theory, but full discussion of that is beyond the scope of this volume. As you are no doubt aware to some degree, weak interactions are responsible for particle decays/radioactivity, whereas strong interactions hold the quarks inside hadrons (baryons such as protons and neutrons, and mesons such as the pi meson) together. As a residual effect, strong forces associated with baryons hold nuclei together.

As you have probably also heard, the virtual particles mediating weak interactions are called intermediate vector bosons. They are 4D vectors like the photon A^μ and come in three varieties, $W^{\mu+}$, $W^{\mu-}$, and Z^μ, which carry a different kind of charge from that in electromagnetism called, appropriately, weak charge. The W's also carry electric charge as indicated by their superscripts, whereas the Z boson has zero electric charge (which is easily remembered by the symbol "Z", for zero).

The strong force virtual particles are called gluons, and they carry a strong force charge designated color charge. Color charge comes in three kinds, symbolized by the colors red, green, and blue.

The gravitational interaction is not part of the standard model, at least as of the publication date of this version of this text. You are almost certainly aware of the challenge incorporating it into our quantum theories has been presenting to the physics community for quite some time now.

For those moving on to study the SM, perhaps the most student friendly presentation of the weak and strong interactions is in the latter part of the text by Mandl and Shaw[1]. And of course, if I have the time and energy to write a subsequent volume to this one on weak and strong interactions, then it should be no surprise that I would recommend that one.

Note that some topics in the SM beyond QED are treated under advanced topics on the web site of this book (see URL on pg. xvi, opposite pg. 1).

19.2.2 More on Symmetry

Physics Pervaded by Symmetry

I also recommend the superb, highly pedagogic book on the all-pervasive nature of symmetry in physics by Schwichtenberg[2]. In addition to showing how many diverse aspects of physics fit under one umbrella of symmetry, its extensive treatment of group theory is an excellent prep for the SM.

Supersymmetry

One branch of QFT with possible broad implications for quantum theories of gravity called supersymmetry (SUSY) posits a symmetry between fermions and bosons (a transformation changing bosons $\leftrightarrow$ fermions leaving the Lagrangian unchanged). However, as of this text version date, SUSY has failed to find support in test results at CERN and may be falling out of favor.

You may wish to consider Aitchison's text[3] as an introduction to SUSY.

19.2.3 Gravity and QFT

Gravitational Effects in Classical Spacetime

Certain gravitational effects can be analyzed effectively using QFT in a classical spacetime. This is not quantum gravity, but a model using quantum fields interacting on a non-Minkowski (curved) spacetime background. Mukhanov and Winitzki[4] provide an exceptional introduction to this topic, and I have written pedagogic notes, posted on the website of this book, as an additional aid.

Theories of Quantum Gravity

There are a number of different theories, and branches to each thereof, advanced in efforts to unite gravity with quantum theory. None, as of this writing, have succeeded. The two major ones are superstring theory (now more appropriately called M theory) and loop quantum gravity.

After gaining some mastery over the standard model, you, if you wish to be a field theorist, will, in high likelihood, join the effort of trying to bring gravity under the QM tent. At that point, you will have to choose to focus on one of the above two major research fields, or perhaps one of presently lesser renown.

Whatever your path, good luck!

---

[1] Mandl, F. and Shaw, G., *Quantum Field Theory* (2nd ed., John Wiley, 2010)

[2] Schwichtenberg, J. *Physics from Symmetry* (2nd ed., Springer, 2018)

[3] Aitchison, I., *Supersymmetry in Particle Physics: An Elementary Introduction* (Cambridge, 2007).

[4] Mukhanov, V. F., and Winitzki, S., *Introduction to Quantum Effects in Gravity* (Cambridge, 2007)

Index

CPSIA information can be obtained
at www.ICGtesting.com
Printed in the USA
LVHW100240250320
651139LV00015B/87